Tips for Customizing Your RSMeans Cost Data By Location

The cost data found throughout the book reflects national averages.

Customize your cost data for over 930 locations throughout the U.S. and Canada.

- ☑ To adjust costs to your region, use the Location Factors found in the back of the book.
- ☑ To compare costs from region-to-region, use the City Cost Indexes found in the back of the book.
- ☑ For additional guidance, refer to the How RSMeans Data Works section towards the front of the book.

RSMeans
Reed Construction Data

Metric Conversion Formulas

Length

cm = 0.3937 in.	in. = 2.5400 cm.
meter = 3.2808 ft.	ft. = 0.3048 m.
meter = 1.0936 yd.	yd. = 0.9144 m.
Km. = 0.6214 mile	mile = 1.6093 km.

Area

sq. cm. = 0.1550 sq. in.	sq.in. = 6.4516 sq. cm.
sq. m. = 10.7639 sq. ft.	sq. ft. = 0.0929 sq. m.
sq. m. = 1.1960 sq. yd.	sq. yd. = 0.8361 sq. m.
hectare = 2.4710 acres	acre = 0.4047 hectare
sq. km. = 0.3861 sq. mile	sq. mile = 2.5900 sq. km.

Weight

gram = 15.4324 grains	grain = 0.0648 g.
gram = 0.0353 oz.	oz. = 28.3495 g.
kg. = 2.2046 lbs.	lb. = 0.4536 kg.

Weight, cont.

kg. = 0.0011 ton (short)	ton (short) = 907.1848 kg.
ton (met.) = 1.1023 ton (short)	ton (short) = 0.9072 ton (met.)
ton (met.) = 0.9842 ton (large)	ton (large) = 1.0160 ton (met.)

Common Units of Measure

Abbreviation	Measurement	Formula
SF	Square feet	Length (in feet) x Width (in feet)
SY	Square yards	Square feet / 9
CF	Cubic feet	Length (in feet) x Width (in feet) x Depth (in feet)
CY	Cubic yards	Cubic feet / 27
BF	Board foot	Length (in inches) x width (in inches) x thickness (in inches)/144
MBF	Thousand board foot	Board foot / 1,000
LF	Linear feet	12 inches of the item
SFCA	Square foot of contact area	A square foot of concrete formwork
SQ	Square	100 square feet
Ton	Ton	2,000 pounds

RSMeans®
Electrical Cost Data
36TH ANNUAL EDITION

188.95 per copy (in United States)
'ice is subject to change without prior notice.

033

2013

SMeans®
division of Reed Construction Data, LLC
onstruction Publishers & Consultants
00 Longwater Drive
orwell, MA 02061
SA
-800-334-3509
ww.rsmeans.com

inted in the United States of America
SN 1540-6318
BN 978-1-936335-58-9

Senior Editor
Adrian C. Charest, PE

Contributing Editors
Christopher Babbitt
Ted Baker
Gary W. Christensen
David G. Drain, PE
Cheryl Elsmore
Robert J. Kuchta
Robert C. McNichols
Genevieve Medeiros
Melville J. Mossman, PE
Jeannene D. Murphy
Marilyn Phelan, AIA
Stephen C. Plotner
Eugene R. Spencer
Phillip R. Waier, PE

Cover Design
Wendy McBay
Paul Robertson
Wayne Engebretson

President & CEO
Iain Melville

Chief Operating Officer
Richard Remington

Chief Customer Officer
Lisa Fiondella

Vice President, Marketing & Business Development
Steven Ritchie

Vice President, Finance & Operations
Renée Pinczes

Engineering Director
Bob Mewis, CCC

Director, Product Management
Jennifer Johnson

Product Manager
Andrea Sillah

Product Marketing Manager
Ryan Cummings

Senior Director, Project & Process Management
Rodney L. Sharples

Production Manager
Michael Kokernak

Editing Supervisor
Jill Goodman

Production Supervisor
Paula Reale-Camelio

Production
Jonathan Forgit
Mary Lou Geary
Debbie Panarelli
Caroline Sullivan

Technical Support
Christopher Anderson
Gary L. Hoitt
Kathryn S. Rodriguez

This book is printed using soy-based printing ink.
This book is recyclable.

PRINTED WITH SOY INK

Related RSMeans Products and Services

The engineers at RSMeans suggest the following products and services as companion information resources to *RSMeans Electrical Cost Data:*

Construction Cost Data Books

Electrical Change Order Cost Data 2013
Building Construction Cost Data 2013
Mechanical Cost Data 2013

Reference Books

Electrical Estimating Methods
Estimating Building Costs
RSMeans Estimating Handbook
Green Building: Project Planning & Estimating
How to Estimate with RSMeans Data
Plan Reading & Material Takeoff
Project Scheduling & Management for Construction

Seminars and In-House Training

Mechanical & Electrical Estimating
Unit Price Estimating
Means CostWorks® Training
RSMeans Data for Job Order Contracting (JOC)
Practical Project Management for Construction Professionals
Scheduling with MSProject for Construction Professionals

RSMeans on the Internet

Visit RSMeans at **www.rsmeans.com** for a one-stop portal for the most reliable and current resources available on the market. Here you can learn about our more than 20 annual cost data books, also available on CD and online versions, as well as our libary of reference books and RSMeans seminars.

RSMeans Electronic Data

Get the information found in RSMeans cost books electronically at **MeansCostWorks.com**.

RSMeans Business Solutions

Cost engineers and analysts conduct facility life cycle and benchmark research studies and predictive cost modeling, as well as offer consultation services for conceptual estimating, real property management, and job order contracting. Research studies are designed with the application of proprietary cost and project data from Reed Construction Data and RSMeans' extensive North American databases. Analysts offer building product manufacturers qualitative and quantitative market research, as well as time/motion studies for new products, and Web-based dashboards for market opportunity analysis. Clients are from both the public and private sectors, including federal, state, and municipal agencies; corporations; institutions; construction management firms; hospitals; and associations.

RSMeans for Job Order Contracting (JOC)

Best practice job order contracting (JOC) services support the owner in the development of their program. RSMeans develops all required contracting documents, works with stakeholders, and develops qualified and experienced contractor lists. RSMeans JOCWorks® software is the best of class in the functional tools that meet JOC specific requirements. RSMeans data is organized to meet JOC cost estimating requirements.

Construction Costs for Software Applications

More than 25 unit price and assemblies cost databases are available through a number of leading estimating and facilities management software providers (listed below). For more information see the "Other RSMeans Products" pages at the back of this publication.

RSMeansData™ is also available to federal, state, and local government agencies as multi-year, multi-seat licenses.

- 4Clicks-Solutions, LLC
- ArenaSoft Estimating
- Assetworks
- Beck Technology
- BSD – Building Systems Design, Inc.
- CMS – Construction Management Software
- Corecon Technologies, Inc.
- CorVet Systems
- Hard Dollar Corporation
- Maxwell Systems
- MC^2
- Parsons Corporation
- ProEst
- Sage Timberline Office
- UDA Technologies, Inc.
- US Cost, Inc.
- VFA – Vanderweil Facility Advisers
- WinEstimator, Inc.
- R & K Solutions
- Tririga

Table of Contents

Foreword

Our Mission

Since 1942, RSMeans has been actively engaged in construction cost publishing and consulting throughout North America.

Today, more than 70 years after RSMeans began, our primary objective remains the same: to provide you, the construction and facilities professional, with the most current and comprehensive construction cost data possible.

Whether you are a contractor, owner, architect, engineer, facilities manager, or anyone else who needs a reliable construction cost estimate, you'll find this publication to be a highly useful and necessary tool.

With the constant flow of new construction methods and materials today, it's difficult to find the time to look at and evaluate all the different construction cost possibilities. In addition, because labor and material costs keep changing, last year's cost information is not a reliable basis for today's estimate or budget.

That's why so many construction professionals turn to RSMeans. We keep track of the costs for you, along with a wide range of other key information, from city cost indexes . . . to productivity rates . . . to crew composition . . . to contractor's overhead and profit rates.

RSMeans performs these functions by collecting data from all facets of the industry and organizing it in a format that is instantly accessible to you. From the preliminary budget to the detailed unit price estimate, you'll find the data in this book useful for all phases of construction cost determination.

The Staff, the Organization, and Our Services

When you purchase one of RSMeans' publications, you are, in effect, hiring the services of a full-time staff of construction and engineering professionals.

Our thoroughly experienced and highly qualified staff works daily at collecting, analyzing, and disseminating comprehensive cost information for your needs. These staff members have years of practical construction experience and engineering training prior to joining the firm. As a result, you can count on them not only for accurate cost figures, but also for additional background reference information that will help you create a realistic estimate.

The RSMeans organization is always prepared to help you solve construction problems through its variety of data solutions, including online, CD, and print book formats, as well as cost estimating expertise available via our business solutions, training, and seminars.

Besides a full array of construction cost estimating books, RSMeans also publishes a number of other reference works for the construction industry. Subjects include construction estimating and project and business management, special topics such as green building and job order contracting, and a library of facility management references.

In addition, you can access all of our construction cost data electronically in convenient CD format or on the web. Visit **RSMeansonline.com** for more information on our 24/7 online cost data.

What's more, you can increase your knowledge and improve your construction estimating and management performance with an RSMeans construction seminar or in-house training program. These two-day seminar programs offer unparalleled opportunities for everyone in your organization to become updated on a wide variety of construction-related issues.

RSMeans is also a worldwide provider of construction cost management and analysis services for commercial and government owners.

In short, RSMeans can provide you with the tools and expertise for constructing accurate and dependable construction estimates and budgets in a variety of ways.

Robert Snow Means Established a Tradition of Quality That Continues Today

Robert Snow Means spent years building RSMeans, making certain he always delivered a quality product.

Today, at RSMeans, we do more than talk about the quality of our data and the usefulness of our books. We stand behind all of our data, from historical cost indexes to construction materials and techniques to current costs.

If you have any questions about our products or services, please call us toll-free at 1-800-334-3509. Our customer service representatives will be happy to assist you. You can also visit our website at **www.rsmeans.com**

How the Book Is Built: An Overview

The Construction Specifications Institute (CSI) and Construction Specifications Canada (CSC) have produced the 2010 edition of MasterFormat™, a system of titles and numbers used extensively to organize construction information.

All unit price data in the RSMeans cost data books is now arranged in the 50-division MasterFormat™ 2010 system.

A Powerful Construction Tool

You have in your hands one of the most powerful construction tools available today. A successful project is built on the foundation of an accurate and dependable estimate. This book will enable you to construct just such an estimate.

For the casual user the book is designed to be:

- quickly and easily understood so you can get right to your estimate.
- filled with valuable information so you can understand the necessary factors that go into the cost estimate.

For the regular user, the book is designed to be:

- a handy desk reference that can be quickly referred to for key costs.
- a comprehensive, fully reliable source of current construction costs and productivity rates so you'll be prepared to estimate any project.
- a source book for preliminary project cost, product selections, and alternate materials and methods.

To meet all of these requirements, we have organized the book into the following clearly defined sections.

Quick Start

See our "Quick Start" instructions on the following page to get started right away.

Estimating with RSMeans Unit Price Cost Data

Please refer to these steps for guidance on completing an estimate using RSMeans unit price cost data.

How to Use the Book: The Details

This section contains an in-depth explanation of how the book is arranged . . . and how you can use it to determine a reliable construction cost estimate. It includes information about how we develop our cost figures and how to completely prepare your estimate.

Unit Price Section

All cost data has been divided into the 50 divisions according to the MasterFormat system of classification and numbering.
For a listing of these divisions and an outline of their subdivisions, see the Unit Price Section Table of Contents.

Estimating tips are included at the beginning of each division.

Assemblies Section

The cost data in this section has been organized in an "Assemblies" format. These assemblies are the functional elements of a building and are arranged according to the 7 elements of the UNIFORMAT II classification system. For a complete explanation of a typical "Assemblies" page, see "How to Use the Assemblies Cost Tables."

Reference Section

This section includes information on Equipment Rental Costs, Crew Listings, Historical Cost Indexes, City Cost Indexes, Location Factors, Reference Tables, Change Orders, Square Foot Costs, and a listing of Abbreviations.

Equipment Rental Costs: This section contains the average costs to rent and operate hundreds of pieces of construction equipment.

Crew Listings: This section lists all of the crews referenced in the book. For the purposes of this book, a crew is composed of more than one trade classification and/or the addition of power equipment to any trade classification. Power equipment is included in the cost of the crew. Costs are shown both with bare labor rates and with the installing contractor's overhead and profit added. For each, the total crew cost per eight-hour day and the composite cost per labor-hour are listed.

Historical Cost Indexes: These indexes provide you with data to adjust construction costs over time.

City Cost Indexes: All costs in this book are U.S. national averages. Costs vary because of the regional economy. You can adjust costs by CSI Division to over 700 locations throughout the U.S. and Canada by using the data in this section.

Location Factors: You can adjust total project costs to over 900 locations throughout the U.S. and Canada by using the data in this section.

Reference Tables: At the beginning of selected major classifications in the Unit Price section are reference numbers shown in a shaded box. These numbers refer you to related information in the Reference Section. In this section, you'll find reference tables, explanations, estimating information that support how we develop the unit price data, technical data, and estimating procedures.

Change Orders: This section includes information on the factors that influence the pricing of change orders.

Square Foot Costs: This section contains costs for 59 different building types that allow you to make a rough estimate for the overall cost of a project or its major components.

Abbreviations: A listing of abbreviations used throughout this book, along with the terms they represent, is included in this section.

Index

A comprehensive listing of all terms and subjects in this book will help you quickly find what you need when you are not sure where it occurs in MasterFormat.

The Scope of This Book

This book is designed to be as comprehensive and as easy to use as possible. To that end we have made certain assumptions and limited its scope in two key ways:

1. We have established material prices based on a national average.
2. We have computed labor costs based on a 30-city national average of union wage rates.

For a more detailed explanation of how the cost data is developed, see "How to Use the Book: The Details."

Project Size/Type

The material prices in Means data cost books are "contractor's prices." They are the prices that contractors can expect to pay at the lumberyards, suppliers/distributers warehouses, etc. Small orders of speciality items would be higher than the costs shown, while very large orders, such as truckload lots, would be less. The variation would depend on the size, timing, and negotiating power of the contractor. The labor costs are primarily for new construction or major renovation rather than repairs or minor alterations.

With reasonable exercise of judgment, the figures can be used for any building work.

Absolute Essentials for a Quick Start

If you feel you are ready to use this book and don't think you will need the detailed instructions that begin on the following page, this Absolute Essentials for a Quick Start page is for you. These steps will allow you to get started estimating in a matter of minutes.

1 Scope

Think through the project that you will be estimating, and identify the many individual work tasks that will need to be covered in your estimate.

2 Quantify

Determine the number of units that will be required for each work task that you identified.

3 Pricing

Locate individual Unit Price line items that match the work tasks you identified. The Unit Price Section Table of Contents that begins on page 1 and the Index in the back of the book will help you find these line items.

4 Multiply

Multiply the Total Incl O&P cost for a Unit Price line item in the book by your quantity for that item. The price you calculate will be an estimate for a completed item of work performed by a subcontractor. Keep adding line items in this manner to build your estimate.

5 Project Overhead

Include project overhead items in your estimate. These items are needed to make the job run and are typically, but not always, provided by the General Contractor. They can be found in Division 1. An alternate method of estimating project overhead costs is to apply a percentage of the total project cost.

Include rented tools not included in crews, waste, rubbish handling, and cleanup.

6 Estimate Summary

Include General Contractor's markup on subcontractors, General Contractor's office overhead and profit, and sales tax on materials and equipment.

Adjust your estimate to the project's location by using the City Cost Indexes or Location Factors found in the Reference Section.

Editors' Note: We urge you to spend time reading and understanding the supporting material in the front of this book. An accurate estimate requires experience, knowledge, and careful calculation. The more you know about how we at RSMeans developed the data, the more accurate your estimate will be. In addition, it is important to take into consideration the reference material in the back of the book such as Equipment Listings, Crew Listings, City Cost Indexes, Location Factors, and Reference Tables.

Estimating with RSMeans Unit Price Cost Data

Following these steps will allow you to complete an accurate estimate using RSMeans Unit Price cost data.

1 Scope Out the Project

- Identify the individual work tasks that will need to be covered in your estimate.
- The Unit Price data in this book has been divided into 50 Divisions according to the CSI MasterFormat 2010—their titles are listed on the back cover of your book.
- Think through the project and identify those CSI Divisions needed in your estimate.
- The Unit Price Section Table of Contents on page 1 may also be helpful when scoping out your project.
- Experienced estimators find it helpful to begin with Division 2 and continue through completion. Division 1 can be estimated after the full project scope is known.

2 Quantify

- Determine the number of units required for each work task that you identified.
- Experienced estimators include an allowance for waste in their quantities. (Waste is not included in RSMeans Unit Price line items unless so stated.)

3 Price the Quantities

- Use the Unit Price Table of Contents, and the Index, to locate individual Unit Price line items for your estimate.
- Reference Numbers indicated within a Unit Price section refer to additional information that you may find useful.
- The crew indicates who is performing the work for that task. Crew codes are expanded in the Crew Listings in the Reference Section to include all trades and equipment that comprise the crew.
- The Daily Output is the amount of work the crew is expected to do in one day.
- The Labor-Hours value is the amount of time it will take for the crew to install one unit of work.
- The abbreviated Unit designation indicates the unit of measure upon which the crew, productivity, and prices are based.
- Bare Costs are shown for materials, labor, and equipment needed to complete the Unit Price line item. Bare costs do not include waste, project overhead, payroll insurance, payroll taxes, main office overhead, or profit.
- The Total Incl O&P cost is the billing rate or invoice amount of the installing contractor or subcontractor who performs the work for the Unit Price line item.

4 Multiply

- Multiply the total number of units needed for your project by the Total Incl O&P cost for each Unit Price line item.
- Be careful that your take off unit of measure matches the unit of measure in the Unit column.
- The price you calculate is an estimate for a completed item of work.
- Keep scoping individual tasks, determining the number of units required for those tasks, matching each task with individual Unit Price line items in the book, and multiply quantities by Total Incl O&P costs.
- An estimate completed in this manner is priced as if a subcontractor, or set of subcontractors, is performing the work. The estimate does not yet include Project Overhead or Estimate Summary components such as General Contractor markups on subcontracted work, General Contractor office overhead and profit, contingency, and location factor.

5 Project Overhead

- Include project overhead items from Division 1 – General Requirements.
- These items are needed to make the job run. They are typically, but not always, provided by the General Contractor. Items include, but are not limited to, field personnel, insurance, performance bond, permits, testing, temporary utilities, field office and storage facilities, temporary scaffolding and platforms, equipment mobilization and demobilization, temporary roads and sidewalks, winter protection, temporary barricades and fencing, temporary security, temporary signs, field engineering and layout, final cleaning and commissioning.
- Each item should be quantified, and matched to individual Unit Price line items in Division 1, and then priced and added to your estimate.
- An alternate method of estimating project overhead costs is to apply a percentage of the total project cost, usually 5% to 15% with an average of 10% (see General Conditions, page ix).
- Include other project related expenses in your estimate such as:
  - Rented equipment not itemized in the Crew Listings
  - Rubbish handling throughout the project (see 02 41 19.19)

6 Estimate Summary

- Include sales tax as required by laws of your state or county.
- Include the General Contractor's markup on self-performed work, usually 5% to 15% with an average of 10%.
- Include the General Contractor's markup on subcontracted work, usually 5% to 15% with an average of 10%.
- Include General Contractor's main office overhead and profit:
  - RSMeans gives general guidelines on the General Contractor's main office overhead (see section 01 31 13.50 and Reference Number R013113-50).
  - RSMeans gives no guidance on the General Contractor's profit.
  - Markups will depend on the size of the General Contractor's operations, his projected annual revenue, the level of risk he is taking on, and on the level of competition in the local area and for this project in particular.
- Include a contingency, usually 3% to 5%, if appropriate.
- Adjust your estimate to the project's location by using the City Cost Indexes or the Location Factors in the Reference Section:
  - Look at the rules on the pages for How to Use the City Cost Indexes to see how to apply the Indexes for your location.
  - When the proper Index or Factor has been identified for the project's location, convert it to a multiplier by dividing it by 100, and then multiply that multiplier by your estimated total cost. The original estimated total cost will now be adjusted up or down from the national average to a total that is appropriate for your location.

Editors' Notes:

1) *We urge you to spend time reading and understanding the supporting material in the front of this book. An accurate estimate requires experience, knowledge, and careful calculation. The more you know about how we at RSMeans developed the data, the more accurate your estimate will be. In addition, it is important to take into consideration the reference material in the back of the book such as Equipment Listings, Crew Listings, City Cost Indexes, Location Factors, and Reference Tables.*
2) *Contractors who are bidding or are involved in JOC, DOC, SABER, or IDIQ type contracts are cautioned that workers' compensation insurance, federal and state payroll taxes, waste, project supervision, project overhead, main office overhead, and profit are not included in bare costs. The coefficient or multiplier must cover these costs.*

How to Use the Book: The Details

What's Behind the Numbers? The Development of Cost Data

The staff at RSMeans continually monitors developments in the construction industry in order to ensure reliable, thorough, and up-to-date cost information. While overall construction costs may vary relative to general economic conditions, price fluctuations within the industry are dependent upon many factors. Individual price variations may, in fact, be opposite to overall economic trends. Therefore, costs are constantly tracked and complete updates are published yearly. Also, new items are frequently added in response to changes in materials and methods.

Costs—$ (U.S.)

All costs represent U.S. national averages and are given in U.S. dollars. The RSMeans City Cost Indexes can be used to adjust costs to a particular location. The City Cost Indexes for Canada can be used to adjust U.S. national averages to local costs in Canadian dollars. No exchange rate conversion is necessary.

G The processes or products identified by the green symbol in our publications have been determined to be environmentally responsible and/or resource-efficient solely by the RSMeans engineering staff. The inclusion of the green symbol does not represent compliance with any specific industry association or standard.

Material Costs

The RSMeans staff contacts manufacturers, dealers, distributors, and contractors all across the U.S. and Canada to determine national average material costs. If you have access to current material costs for your specific location, you may wish to make adjustments to reflect differences from the national average. Included within material costs are fasteners for a normal installation. RSMeans engineers use manufacturers' recommendations, written specifications, and/or standard construction practice for size and spacing of fasteners. Adjustments to material costs may be required for your specific application or location. Material costs do not include sales tax.

Labor Costs

Labor costs are based on the average of wage rates from 30 major U.S. cities. Rates are determined from labor union agreements or prevailing wages for construction trades for the current year. Rates, along with overhead and profit markups, are listed on the inside back cover of this book.

- If wage rates in your area vary from those used in this book, or if rate increases are expected within a given year, labor costs should be adjusted accordingly.

Labor costs reflect productivity based on actual working conditions. In addition to actual installation, these figures include time spent during a normal weekday on tasks such as, material receiving and handling, mobilization at site, site movement, breaks, and cleanup. Productivity data is developed over an extended period so as not to be influenced by abnormal variations and reflects a typical average.

Equipment Costs

Equipment costs include not only rental, but also operating costs for equipment under normal use. The operating costs include parts and labor for routine servicing such as repair and replacement of pumps, filters, and worn lines. Normal operating expendables, such as fuel, lubricants, tires, and electricity (where applicable), are also included. Extraordinary operating expendables with highly variable wear patterns, such as diamond bits and blades, are excluded. These costs are included under materials. Equipment rental rates are obtained from industry sources throughout North America—contractors, suppliers, dealers, manufacturers, and distributors.

Rental rates can also be treated as reimbursement costs for contractor-owned equipment. Owned equipment costs include depreciation, loan payments, interest, taxes, insurance, storage and major repairs.

Equipment costs do not include operators' wages; nor do they include the cost to move equipment to a job site (mobilization) or from a job site (demobilization).

Equipment Cost/Day——The cost of power equipment required for each crew is included in the Crew Listings in the Reference Section (small tools that are considered as essential everyday tools are not listed out separately). The Crew Listings itemize specialized tools and heavy equipment along with labor trades. The daily cost of itemized equipment included in a crew is based on dividing the weekly bare rental rate by 5 (number of working days per week) and then adding the hourly operating cost times 8 (the number of hours per day). This Equipment Cost/Day is shown in the last column of the Equipment Rental Cost pages in the Reference Section.

Mobilization/Demobilization——The cost to move construction equipment from an equipment yard or rental company to the job site and back again is not included in equipment costs. Mobilization (to the site) and demobilization (from the site) costs can be found in the Unit Price Section. If a piece of equipment is already at the job site, it is not appropriate to utilize mob./demob. costs again in an estimate.

Overhead and Profit

Total Cost including O&P for the *Installing Contractor* is shown in the last column on the Unit Price and/or the Assemblies pages of this book. This figure is the sum of the bare material cost plus 10% for profit, the bare labor cost plus total overhead and profit, and the bare equipment cost plus 10% for profit. Details for the calculation of Overhead and Profit on labor are shown on the inside back cover and in the Reference Section of this book. (See "How RSMeans Data Works" for an example of this calculation.)

General Conditions

Cost data in this book is presented in two ways: Bare Costs and Total Cost including O&P (Overhead and Profit). General Conditions, or General Requirements, of the contract should also be added to the Total Cost including O&P when applicable. Costs for General Conditions are listed in Division 1 of the Unit Price Section and the Reference

Section of this book. General Conditions for the *Installing Contractor* may range from 0% to 10% of the Total Cost including O&P. For the *General* or *Prime Contractor*, costs for General Conditions may range from 5% to 15% of the Total Cost including O&P, with a figure of 10% as the most typical allowance. If applicable, the Assemblies and Models sections of this book use costs that include the installing contractor's overhead and profit (O&P).

Factors Affecting Costs

Costs can vary depending upon a number of variables. Here's how we have handled the main factors affecting costs.

Quality—The prices for materials and the workmanship upon which productivity is based represent sound construction work. They are also in line with U.S. government specifications.

Overtime—We have made no allowance for overtime. If you anticipate premium time or work beyond normal working hours, be sure to make an appropriate adjustment to your labor costs.

Productivity—The productivity, daily output, and labor-hour figures for each line item are based on working an eight-hour day in daylight hours in moderate temperatures. For work that extends beyond normal work hours or is performed under adverse conditions, productivity may decrease. (See "How RSMeans Data Works" for more on productivity.)

Size of Project—The size, scope of work, and type of construction project will have a significant impact on cost. Economies of scale can reduce costs for large projects. Unit costs can often run higher for small projects.

Location—Material prices in this book are for metropolitan areas. However, in dense urban areas, traffic and site storage limitations may increase costs. Beyond a 20-mile radius of large cities, extra trucking or transportation charges may also increase the material costs slightly. On the other hand, lower wage rates may be in effect. Be sure to consider both of these factors when preparing an estimate, particularly if the job site is located in a central city or remote rural location. In addition, highly specialized subcontract items may require travel and per-diem expenses for mechanics.

Other Factors—

- season of year
- contractor management
- weather conditions
- local union restrictions
- building code requirements
- availability of:
  - adequate energy
  - skilled labor
  - building materials
- owner's special requirements/restrictions
- safety requirements
- environmental considerations

Unpredictable Factors—General business conditions influence "in-place" costs of all items. Substitute materials and construction methods may have to be employed. These may affect the installed cost and/or life cycle costs. Such factors may be difficult to evaluate and cannot necessarily be predicted on the basis of the job's location in a particular section of the country. Thus, where these factors apply, you may find significant but unavoidable cost variations for which you will have to apply a measure of judgment to your estimate.

Rounding of Costs

In general, all unit prices in excess of $5.00 have been rounded to make them easier to use and still maintain adequate precision of the results. The rounding rules we have chosen are in the following table.

Prices from . . .	Rounded to the nearest . . .
$.01 to $5.00	$.01
$5.01 to $20.00	$.05
$20.01 to $100.00	$.50
$100.01 to $300.00	$1.00
$300.01 to $1,000.00	$5.00
$1,000.01 to $10,000.00	$25.00
$10,000.01 to $50,000.00	$100.00
$50,000.01 and above	$500.00

How Subcontracted Items Affect Costs

A considerable portion of all large construction jobs is usually subcontracted. In fact, the percentage done by subcontractors is constantly increasing and may run over 90%. Since the workers employed by these companies do nothing else but install their particular product, they soon become expert in that line. The result is, installation by these firms is accomplished so efficiently that the total in-place cost, even adding the general contractor's overhead and profit, is no more, and often less, than if the principal contractor had handled the installation himself/herself. Companies that deal with construction specialties are anxious to have their product perform well and, consequently, the installation will be the best possible.

Contingencies

The allowance for contingencies generally provides for unforeseen construction difficulties. On alterations or repair jobs, 20% is not too much. If drawings are final and only field contingencies are being considered, 2% or 3% is probably sufficient, and often nothing need be added. Contractually, changes in plans will be covered by extras. The contractor should consider inflationary price trends and possible material shortages during the course of the job. These escalation factors are dependent upon both economic conditions and the anticipated time between the estimate and actual construction. If drawings are not complete or approved, or a budget cost is wanted, it is wise to add 5% to 10%. Contingencies, then, are a matter of judgment. Additional allowances for contingencies are shown in Division 1.

Important Estimating Considerations

The "productivity," or daily output, of each craftsman includes mobilization and cleanup time, break time and plan layout time, as well as an allowance to carry stock from the storage trailer or location on the job site up to 200′ into the building and to the first or second floor. If material has to be transported over greater distances or to higher floors, an additional allowance should be considered by the estimator. An allowance has also been included in the piping and fittings installation time for leak checking and minor tightening. Equipment installation time includes the following applicable items: positioning, leveling and securing the unit in place, connecting all associated piping, ducts, vents, etc., which shall have been estimated separately, connecting to an adjacent power source, filling/bleeding, startup, adjusting the controls up and down to ensure proper response, setting the integral controls/valves/regulators/thermostats for proper operation (does not include external building type control systems, DDC systems, etc.), explaining/training owner's operator, and warranty. A reasonable breakdown of the labor costs is as follows:

1.	Movement into building, installation/setting of equipment	35%
2.	Connection to piping/duct/power, etc.	25%
3.	Filling/flushing/cleaning/touchup, etc.	15%
4.	Startup/running adjustments	5%
5.	Training owner's representative	5%
6.	Warranty/call back/service	15%

Note that cranes or other lifting equipment are not included on any lines in the Mechanical divisions. For example, if a crane is required to lift a heavy piece of pipe into place high above a gym floor, or to put a rooftop unit on the roof of a four-story building, etc., it must be added. Due to the potential for extreme variation—from nothing additional required to a major crane or helicopter—we feel that including a nominal amount for "lifting contingency" would be useless and detract from the accuracy of the estimate. When using equipment rental from RSMeans, remember to include the cost of the operator(s).

Estimating Labor-Hours

The labor-hours expressed in this publication are based on Average Installation time, using an efficiency level of approximately 60%-65% (see item 7), which has been found reasonable and acceptable by many contractors. The book uses this national efficiency average to establish a consistent benchmark. For bid situations, adjustments to this efficiency level should be the responsibility of the contractor bidding the project. The unit labor-hour is divided in the following manner. A typical day for a journeyman might be:

1.	Study Plans	3%	14.4 min.
2.	Material Procurement	3%	14.4 min.
3.	Receiving and Storing	3%	14.4 min
4.	Mobilization	5%	24.0 min.
5.	Site Movement	5%	24.0 min.
6.	Layout and Marking	8%	38.4 min.
7.	Actual Installation	64%	307.2 min.
8.	Cleanup	3%	14.4 min.
9.	Breaks, Nonproductive	6%	28.8 min.
		100%	480.0 min.

If any of the percentages expressed in this breakdown do not apply to the particular work or project situation, then that percentage or a portion of it may be deducted from or added to labor-hours.

Final Checklist

Estimating can be a straightforward process provided you remember the basics. Here's a checklist of some of the steps you should remember to complete before finalizing your estimate. Did you remember to . . .

- factor in the City Cost Index for your locale?
- take into consideration which items have been marked up and by how much?
- mark up the entire estimate sufficiently for your purposes?
- read the background information on techniques and technical matters that could impact your project time span and cost?
- include all components of your project in the final estimate?
- double check your figures for accuracy?
- call RSMeans if you have any questions about your estimate or the data you've found in our publications? Remember, RSMeans stands behind its publications. If you have any questions about your estimate . . . about the costs you've used from our books . . . or even about the technical aspects of the job that may affect your estimate, feel free to call the RSMeans editors at 1-800-334-3509.

Unit Price Section

Table of Contents

Table of Contents (cont.)

How RSMeans Data Works

All RSMeans unit price data is organized in the same way.

It is important to understand the structure, so that you can find information easily and use it correctly.

RSMeans **Line Numbers** consist of 12 characters, which identify a unique location in the database for each task. The first 6 or 8 digits conform to the Construction Specifications Institute MasterFormat 2010. The remainder of the digits are a further breakdown by RSMeans in order to arrange items in understandable groups of similar tasks. Line numbers are consistent across all RSMeans publications, so a line number in any RSMeans product will always refer to the same unit of work.

RSMeans engineers have created **reference** information to assist you in your estimate. If there is information that applies to a section, it will be indicated at the start of the section. In this case, R033105-10 provides information on the proportionate quantities of formwork, reinforcing, and concrete used in cast-in-place concrete items such as footings, slabs, beams, and columns. The Reference Tables is located in the back of the book on the pages with a gray edge.

RSMeans **Descriptions** are shown in a hierarchical structure to make them readable. In order to read a complete description, read up through the indents to the top of the section. Include everything that is above and to the left that is not contradicted by information below. For instance, the complete description for line 03 30 53.40 3550 is "Concrete in place, including forms (4 uses), reinforcing steel, concrete, placement, and finishing unless otherwise indicated; Equipment pad (3000 psi), 4' x 4' x 6" thick."

When using **RSMeans data**, it is important to read through an entire section to ensure that you use the data that most closely matches your work. Note that sometimes there is additional information shown in the section that may improve your price. There are frequently lines that further describe, add to, or adjust data for specific situations.

03 30 Cast-In-Place Concrete

03 30 53 – Miscellaneous Cast-In-Place Concrete

03 30 53.40 Concrete In Place

0010	**CONCRETE IN PLACE**	R033105-10
0020	Including forms (4 uses), Grade 60 rebar, concrete (Portland cement	R033105-20
0050	Type I), placement and finishing unless otherwise indicated	R033105-50
0300	Beams (3500 psi), 5 kip per L.F., 10' span	R033105-65
0350	25' span	R033105-70
0500	Chimney foundations (5000 psi), over 5 C.Y.	R033105-85
0510	(3500 psi), under 5 C.Y.	
0700	Columns, square (4000 psi), 12" x 12", less than 2% reinforcing	
3450	Over 10,000 S.F.	
3500	Add per floor for 3 to 6 stories high	
3520	For 7 to 20 stories high	
3540	Equipment pad (3000 psi), 3' x 3' x 6" thick	
3550	4' x 4' x 6" thick	
3560	5' x 5' x 8" thick	
3570	6' x 6' x 8" thick	

The data published in RSMeans print books represents a "national average" cost. This data should be modified to the project location using the **City Cost Indexes** or **Location Factors** tables found in the reference section (see pages 490–538). Use the location factors to adjust estimate totals if the project covers multiple trades. Use the city cost indexes (CCI) for single trade projects or projects where a more detailed analysis is required. All figures in the two tables are derived from the same research. The last row of data in the CCI, the weighted average, is the same as the numbers reported for each location in the location factor table.

Crews include labor or labor and equipment necessary to accomplish each task. In this case, Crew C-14H is used. RSMeans selects a crew to represent the workers and equipment that are typically used for that task. In this case, Crew C-14H consists of one carpenter foreman (outside), two carpenters, one rodman, one laborer, one cement finisher, and one gas engine vibrator. Details of all crews can be found in the reference section.

Crews

Crew No.	Bare Costs		Incl. Subs O&P		Cost Per Labor-Hour	
Crew C-14H	Hr.	Daily	Hr.	Daily	Bare Costs	Incl. O&P
1 Carpenter Foreman (outside)	$46.90	$375.20	$72.25	$578.00	$44.17	$67.88
2 Carpenters	44.90	718.40	69.15	1106.40		
1 Rodman (reinf.)	49.80	398.40	78.70	629.60		
1 Laborer	35.45	283.60	54.60	436.80		
1 Cement Finisher	43.05	344.40	63.40	507.20		
1 Gas Engine Vibrator		33.00		36.30	.69	.76
48 L.H., Daily Totals		$2153.00		$3294.30	$44.85	$68.63

The **Daily Output** is the amount of work that the crew can do in a normal 8-hour workday, including mobilization, layout, movement of materials, and cleanup. In this case, crew C-14H can install thirty 4' x 4' x 6" thick concrete pads in a day. Daily output is variable, based on many factors, including the size of the job, location, and environmental conditions. RSMeans data represents work done in daylight (or adequate lighting) and temperate conditions.

Bare Costs are the costs of materials, labor, and equipment that the installing contractor pays. They represent the cost, in U.S. dollars, for one unit of work. They do not include any markups for profit or labor burden.

Crew	Daily Output	Labor-Hours	Unit	2013 Bare Costs Material	Labor	Equipment	Total	Total Incl O&P
C-14A	15.62	12.804	C.Y.	315	575	48	938	1,300
"	18.55	10.782		330	485	40.50	855.50	1,150
C-14C	32.22	3.476		144	149	1.01	294.01	390
"	23.71	4.724		170	203	1.37	374.37	505
C-14A	11.96	16.722		365	755	62.50	1,182.50	1,625
	2200	.025		.75	1.01	.33	2.09	2.71
	31800	.002			.07	.02	.09	.13
	21200	.003			.10	.03	.13	.20
C-14H	45	1.067	Ea.	44.50	47	.74	92.24	122
	30	1.600		66	70.50	1.10	137.60	183
	18	2.667		117	118	1.84	236.84	310
	14	3.429		157	151	2.37	310.37	410

The **Total Incl O&P column** is the total cost, including overhead and profit, that the installing contractor will charge the customer. This represents the cost of materials plus 10% profit, the cost of labor plus labor burden and 10% profit, and the cost of equipment plus 10% profit. It does not include the general contractor's overhead and profit. Note: See the inside back cover for details of how RSMeans calculates labor burden.

The **Total column** represents the total bare cost for the installing contractor, in U.S. dollars. In this case, the sum of $66 for material + $70.50 for labor + $1.10 for equipment is $137.60.

The figure in the **Labor Hours** column is the amount of labor required to perform one unit of work—in this case the amount of labor required to construct one 4' x 4' equipment pad. This figure is calculated by dividing the number of hours of labor in the crew by the daily output (48 labor hours divided by 30 pads = 1.6 hours of labor per pad). Multiply 1.600 times 60 to see the value in minutes: 60 x 1.6 = 96 minutes. Note: the labor hour figure is not dependent on the crew size. A change in crew size will result in a corresponding change in daily output, but the labor hours per unit of work will not change.

All RSMeans unit cost data includes the typical **Unit of Measure** used for estimating that item. For concrete-in-place the typical unit is cubic yards (C.Y.) or each (Ea.). For installing broadloom carpet it is square yard, and for gypsum board it is square foot. The estimator needs to take special care that the unit in the data matches the unit in the take-off. Unit conversions may be found in the Reference Section.

How RSMeans Data Works

Sample Estimate

This sample demonstrates the elements of an estimate, including a tally of the RSMeans data lines, and a summary of the markups on a contractor's work to arrive at a total cost to the owner. The RSMeans Location Factor is added at the bottom of the estimate to adjust the cost of the work to a specific location.

Work Performed: The body of the estimate shows the RSMeans data selected, including line number, a brief description of each item, its take-off unit and quantity, and the bare costs of materials, labor, and equipment. This estimate also includes a column titled "SubContract." This data is taken from the RSMeans column "Total Incl O&P," and represents the total that a subcontractor would charge a general contractor for the work, including the sub's markup for overhead and profit.

Division 1, General Requirements: This is the first division numerically, but the last division estimated. Division 1 includes project-wide needs provided by the general contractor. These requirements vary by project, but may include temporary facilities and utilities, security, testing, project cleanup, etc. For small projects a percentage can be used, typically between 5% and 15% of project cost. For large projects the costs may be itemized and priced individually.

Bonds: Bond costs should be added to the estimate. The figures here represent a typical performance bond, ensuring the owner that if the general contractor does not complete the obligations in the construction contract the bonding company will pay the cost for completion of the work.

Location Adjustment: RSMeans published data is based on national average costs. If necessary, adjust the total cost of the project using a location factor from the "Location Factor" table or the "City Cost Index" table. Use location factors if the work is general, covering multiple trades. If the work is by a single trade (e.g., masonry) use the more specific data found in the "City Cost Indexes."

Project Name: Pre-Engineered Steel Building		Architect: As Shown		
Location:	**Anywhere, USA**			
Line Number	**Description**	**Qty**	**Unit**	**Material**
03 30 53.40 3940	Strip footing, 12" x 24", reinforced	34	C.Y.	$4,488.00
03 30 53.40 3950	Strip footing, 12" x 36", reinforced	15	C.Y.	$1,905.00
03 11 13.65 3000	Concrete slab edge forms	500	L.F.	$155.00
03 22 05.50 0200	Welded wire fabric reinforcing	150	C.S.F.	$2,602.50
03 31 05.35 0300	Ready mix concrete, 4000 psi for slab on grade	186	C.Y.	$18,972.00
03 31 05.70 4300	Place, strike off & consolidate concrete slab	186	C.Y.	$0.00
03 35 29.30 0250	Machine float & trowel concrete slab	15,000	S.F.	$0.00
03 35 29.35 0160	Cut control joints in concrete slab	950	L.F.	$66.50
03 39 23.13 0300	Sprayed concrete curing membrane	150	C.S.F.	$1,207.50
Division 03	**Subtotal**			**$29,396.50**
08 36 13.10 2650	Manual 10' x 10' steel sectional overhead door	8	Ea.	$8,800.00
08 36 13.10 2860	Insulation and steel back panel for OH door	800	S.F.	$3,680.00
Division 08	**Subtotal**			**$12,480.00**
13 34 19.50 1100	Pre-Engineered Steel Building, 100' x 150' x 24'	15,000	SF Flr.	$0.00
13 34 19.50 6050	Framing for PESB door opening, 3' x 7'	4	Opng.	$0.00
13 34 19.50 6100	Framing for PESB door opening, 10' x 10'	8	Opng.	$0.00
13 34 19.50 6200	Framing for PESB window opening, 4' x 3'	6	Opng.	$0.00
13 34 19.50 5750	PESB door, 3' x 7', single leaf	4	Opng.	$2,320.00
13 34 19.50 7750	PESB sliding window, 4' x 3' with screen	6	Opng.	$2,220.00
13 34 19.50 6550	PESB gutter, eave type, 26 ga., painted	300	L.F.	$1,950.00
13 34 19.50 8650	PESB roof vent, 12" wide x 10' long	15	Ea.	$517.50
13 34 19.50 6900	PESB insulation, vinyl faced, 4" thick	27,400	S.F.	$10,686.00
Division 13	**Subtotal**			**$17,693.50**
		Subtotal		**$59,570.00**
Division 01	**General Requirements @ 7%**			**4,169.90**
		Estimate Subtotal		**$63,739.90**
		Sales Tax @ 5%		**3,187.00**
		Subtotal		**66,926.90**
		GC O & P		**6,692.69**
		Subtotal		**73,619.58**
		Contingency @ 5%		
		Subtotal		
		Bond @ $12/1000 +10% O&P		
		Subtotal		
		Location Adjustment Factor		
		Grand Total		

This estimate is based on an interactive spreadsheet. A copy of this spreadsheet is located on the RSMeans website at **http://www.reedconstructiondata.com/rsmeans/extras/546011**. You are free to download it and adjust it to your methodology.

This example shows the cost to construct a pre-engineered steel building. The foundation, doors, windows, and insulation will be installed by the general contractor. A subcontractor will install the structural steel, roofing, and siding.

		01/01/13	STD	
Labor	Equipment	SubContract	Estimate Total	
$3,400.00	$23.12	$0.00		
$1,200.00	$8.10	$0.00		
$1,135.00	$0.00	$0.00		
$3,825.00	$0.00	$0.00		
$0.00	$0.00	$0.00		
$3,003.90	$111.60	$0.00		
$8,550.00	$450.00	$0.00		
$408.50	$104.50	$0.00		
$892.50	$0.00	$0.00		
$22,414.90	**$697.32**	**$0.00**	**$52,508.72**	**Division 03**
$3,200.00	$0.00	$0.00		
$0.00	$0.00	$0.00		
$3,200.00	**$0.00**	**$0.00**	**$15,680.00**	**Division 08**
$0.00	$0.00	$337,500.00		
$0.00	$0.00	$2,220.00		
$0.00	$0.00	$9,000.00		
$0.00	$0.00	$3,270.00		
$640.00	$0.00	$0.00		
$546.00	$66.60	$0.00		
$750.00	$0.00	$0.00		
$3,000.00	$0.00	$0.00		
$8,494.00	$0.00	$0.00		
$13,430.00	**$66.60**	**$351,990.00**	**$383,180.10**	**Division 13**
$39,044.90	**$763.92**	**$351,990.00**	**$451,368.82**	**Subtotal**
2,733.14	**53.47**	**24,639.30**		**Gen. Requirements**
$41,778.04	**$817.39**	**$376,629.30**	**$451,368.82**	**Estimate Subtotal**
	40.87	**9,415.73**		**Sales tax**
41,778.04	**858.26**	**386,045.03**		**Subtotal**
22,810.81	**85.83**	**38,604.50**		**GC O&P**
64,588.85	**944.09**	**424,649.54**	**$563,802.07**	**Subtotal**
			28,190.10	**Contingency**
			$591,992.17	**Subtotal**
			7,814.30	**Bond**
			$599,806.47	**Subtotal**
102.30			**13,795.55**	**Location Adjustment**
			$613,602.01	**Grand Total**

Sales Tax: If the work is subject to state or local sales taxes, the amount must be added to the estimate. Sales tax may be added to material costs, equipment costs, and subcontracted work. In this case, sales tax was added in all three categories. It was assumed that approximately half the subcontracted work would be material cost, so the tax was applied to 50% of the subcontract total.

GC O&P: This entry represents the general contractor's markup on material, labor, equipment, and subcontractor costs. RSMeans' standard markup on materials, equipment, and subcontracted work is 10%. In this estimate, the markup on the labor performed by the GC's workers uses "Skilled Workers Average" shown in Column F on the table "Installing Contractor's Overhead & Profit," which can be found on the inside-back cover of the book.

Contingency: A factor for contingency may be added to any estimate to represent the cost of unknowns that may occur between the time that the estimate is performed and the time the project is constructed. The amount of the allowance will depend on the stage of design at which the estimate is done, and the contractor's assessment of the risk involved. Refer to section 01 21 16.50 for contigency allowances.

Estimating Tips

01 20 00 Price and Payment Procedures

- Allowances that should be added to estimates to cover contingencies and job conditions that are not included in the national average material and labor costs are shown in section 01 21.
- When estimating historic preservation projects (depending on the condition of the existing structure and the owner's requirements), a 15%–20% contingency or allowance is recommended, regardless of the stage of the drawings.

01 30 00 Administrative Requirements

- Before determining a final cost estimate, it is a good practice to review all the items listed in Subdivisions 01 31 and 01 32 to make final adjustments for items that may need customizing to specific job conditions.
- Requirements for initial and periodic submittals can represent a significant cost to the General Requirements of a job. Thoroughly check the submittal specifications when estimating a project to determine any costs that should be included.

01 40 00 Quality Requirements

- All projects will require some degree of Quality Control. This cost is not included in the unit cost of construction listed in each division. Depending upon the terms of the contract, the various costs of inspection and testing can be the responsibility of either the owner or the contractor. Be sure to include the required costs in your estimate.

01 50 00 Temporary Facilities and Controls

- Barricades, access roads, safety nets, scaffolding, security, and many more requirements for the execution of a safe project are elements of direct cost. These costs can easily be overlooked when preparing an estimate. When looking through the major classifications of this subdivision, determine which items apply to each division in your estimate.
- Construction Equipment Rental Costs can be found in the Reference Section in section 01 54 33. Operators' wages are not included in equipment rental costs.
- Equipment mobilization and demobilization costs are not included in equipment rental costs and must be considered separately in section 01 54 36.50.
- The cost of small tools provided by the installing contractor for his workers is covered in the "Overhead" column on the "Installing Contractor's Overhead and Profit" table that lists labor trades, base rates and markups and, therefore, is included in the "Total Incl. O&P" cost of any Unit Price line item. For those users who are constrained by contract terms to use only bare costs, there are two line items in section 01 54 39.70 for small tools as a percentage of bare labor cost. If some of those users are further constrained by contract terms to refrain from using line items from Division 1, you are advised to cover the cost of small tools within your coefficient or multiplier.

01 70 00 Execution and Closeout Requirements

- When preparing an estimate, thoroughly read the specifications to determine the requirements for Contract Closeout. Final cleaning, record documentation, operation and maintenance data, warranties and bonds, and spare parts and maintenance materials can all be elements of cost for the completion of a contract. Do not overlook these in your estimate.

Reference Numbers

Reference numbers are shown in shaded boxes at the beginning of some major classifications. These numbers refer to related items in the Reference Section. The reference information may be an estimating procedure, an alternate pricing method, or technical information.

Note: Not all subdivisions listed here necessarily appear in this publication.

01 11 Summary of Work

01 11 31 – Professional Consultants

01 11 31.20 Construction Management Fees		Crew	Daily Output	Labor-Hours	Unit	2013 Bare Costs Material	Labor	Equipment	Total	Total Incl O&P
0010	**CONSTRUCTION MANAGEMENT FEES**									
0020	$1,000,000 job, minimum				Project				4.50%	4.50%
0050	Maximum								7.50%	7.50%
0060	For work to $100,000								10%	10%
0300	$50,000,000 job, minimum								2.50%	2.50%
0350	Maximum				▼				4%	4%

01 11 31.30 Engineering Fees

		Crew	Daily Output	Labor-Hours	Unit	Material	Labor	Equipment	Total	Total Incl O&P
0010	**ENGINEERING FEES** R011110-30									
0020	Educational planning consultant, minimum				Project				.50%	.50%
0100	Maximum				"				2.50%	2.50%
0200	Electrical, minimum				Contrct				4.10%	4.10%
0300	Maximum								10.10%	10.10%
0400	Elevator & conveying systems, minimum								2.50%	2.50%
0500	Maximum								5%	5%
0600	Food service & kitchen equipment, minimum								8%	8%
0700	Maximum								12%	12%
1000	Mechanical (plumbing & HVAC), minimum								4.10%	4.10%
1100	Maximum				▼				10.10%	10.10%

01 21 Allowances

01 21 16 – Contingency Allowances

01 21 16.50 Contingencies

		Crew	Daily Output	Labor-Hours	Unit	Material	Labor	Equipment	Total	Total Incl O&P
0010	**CONTINGENCIES**, Add to estimate									
0020	Conceptual stage				Project				20%	20%
0050	Schematic stage								15%	15%
0100	Preliminary working drawing stage (Design Dev.)								10%	10%
0150	Final working drawing stage				▼				3%	3%

01 21 53 – Factors Allowance

01 21 53.50 Factors

		Crew	Daily Output	Labor-Hours	Unit	Material	Labor	Equipment	Total	Total Incl O&P
0010	**FACTORS** Cost adjustments									
0100	Add to construction costs for particular job requirements									
0500	Cut & patch to match existing construction, add, minimum				Costs	2%	3%			
0550	Maximum					5%	9%			
0800	Dust protection, add, minimum					1%	2%			
0850	Maximum					4%	11%			
1100	Equipment usage curtailment, add, minimum					1%	1%			
1150	Maximum					3%	10%			
1400	Material handling & storage limitation, add, minimum					1%	1%			
1450	Maximum					6%	7%			
1700	Protection of existing work, add, minimum					2%	2%			
1750	Maximum					5%	7%			
2000	Shift work requirements, add, minimum						5%			
2050	Maximum						30%			
2300	Temporary shoring and bracing, add, minimum					2%	5%			
2350	Maximum					5%	12%			
2400	Work inside prisons and high security areas, add, minimum						30%			
2450	Maximum				▼		50%			

01 21 Allowances

01 21 55 – Job Conditions Allowance

	01 21 53.50 Factors	Crew	Daily Output	Labor-Hours	Unit	Material (2013 Bare Costs)	Labor (2013 Bare Costs)	Equipment (2013 Bare Costs)	Total (2013 Bare Costs)	Total Incl O&P
0010	**JOB CONDITIONS** Modifications to applicable									
0020	cost summaries									
0100	Economic conditions, favorable, deduct				Project				2%	2%
0200	Unfavorable, add								5%	5%
0300	Hoisting conditions, favorable, deduct								2%	2%
0400	Unfavorable, add								5%	5%
0700	Labor availability, surplus, deduct								1%	1%
0800	Shortage, add								10%	10%
0900	Material storage area, available, deduct								1%	1%
1000	Not available, add								2%	2%
1100	Subcontractor availability, surplus, deduct								5%	5%
1200	Shortage, add								12%	12%
1300	Work space, available, deduct								2%	2%
1400	Not available, add								5%	5%

01 21 57 – Overtime Allowance

	01 21 57.50 Overtime	Crew	Daily Output	Labor-Hours	Unit	Material	Labor	Equipment	Total	Total Incl O&P
0010	**OVERTIME** for early completion of projects or where R012909-90									
0020	labor shortages exist, add to usual labor, up to				Costs		100%			

01 21 61 – Cost Indexes

	01 21 61.10 Construction Cost Index	Crew	Daily Output	Labor-Hours	Unit	Material	Labor	Equipment	Total	Total Incl O&P
0010	**CONSTRUCTION COST INDEX** (Reference) over 930 zip code locations in									
0020	the U.S. and Canada, total bldg. cost, min. (Guymon, OK)				%					67.40%
0050	Average									100%
0100	Maximum (New York, NY)									131.90%

	01 21 61.20 Historical Cost Indexes	Crew	Daily Output	Labor-Hours	Unit	Material	Labor	Equipment	Total	Total Incl O&P
0010	**HISTORICAL COST INDEXES** (See Reference Section)									

	01 21 61.30 Labor Index	Crew	Daily Output	Labor-Hours	Unit	Material	Labor	Equipment	Total	Total Incl O&P
0010	**LABOR INDEX** (Reference) For over 930 zip code locations in									
0020	the U.S. and Canada, minimum (Guymon, OK)				%		30.70%			
0050	Average						100%			
0100	Maximum (New York, NY)						167%			

	01 21 61.50 Material Index	Crew	Daily Output	Labor-Hours	Unit	Material	Labor	Equipment	Total	Total Incl O&P
0011	**MATERIAL INDEX** For over 930 zip code locations in									
0020	the U.S. and Canada, minimum (Vale, OR)				%	91.80%				
0040	Average					100%				
0060	Maximum (Ketchikan, AK)					132.80%				

01 21 63 – Taxes

	01 21 63.10 Taxes	Crew	Daily Output	Labor-Hours	Unit	Material	Labor	Equipment	Total	Total Incl O&P
0010	**TAXES** R012909-80									
0020	Sales tax, State, average				%	5.03%				
0050	Maximum R012909-85					7.25%				
0300	Unemployment, combined Federal and State, minimum						.80%			
0350	Average						7.80%			
0400	Maximum						13.50%			

01 21 Allowances

01 21 63 – Taxes

01 21 63.10 Taxes		Crew	Daily Output	Labor-Hours	Unit	Material	2013 Bare Costs Labor	Equipment	Total	Total Incl O&P
0010	**FIELD PERSONNEL**									
0020	Clerk, average				Week		500		500	775
0100	Field engineer, minimum						1,025		1,025	1,575
0120	Average						1,325		1,325	2,050
0140	Maximum						1,500		1,500	2,325
0160	General purpose laborer, average						1,425		1,425	2,175
0180	Project manager, minimum						1,875		1,875	2,900
0200	Average						2,150		2,150	3,350
0220	Maximum						2,475		2,475	3,825
0240	Superintendent, minimum						1,825		1,825	2,825
0260	Average						2,000		2,000	3,100
0280	Maximum						2,300		2,300	3,550
0290	Timekeeper, average						1,175		1,175	1,800

01 31 13.30 Insurance

		Crew	Daily Output	Labor-Hours	Unit	Material	Labor	Equipment	Total	Total Incl O&P
0010	**INSURANCE** R013113-40									
0020	Builders risk, standard, minimum				Job				.24%	.24%
0050	Maximum R013113-60								.64%	.64%
0200	All-risk type, minimum								.25%	.25%
0250	Maximum								.62%	.62%
0400	Contractor's equipment floater, minimum				Value				.50%	.50%
0450	Maximum				"				1.50%	1.50%
0600	Public liability, average				Job				2.02%	2.02%
0800	Workers' compensation & employer's liability, average									
0850	by trade, carpentry, general				Payroll		15.13%			
1000	Electrical						5.68%			
1150	Insulation						11.44%			
1450	Plumbing						6.66%			
1550	Sheet metal work (HVAC)						8.56%			

01 31 13.50 General Contractor's Mark-Up

		Crew	Daily Output	Labor-Hours	Unit	Material	Labor	Equipment	Total	Total Incl O&P
0010	**GENERAL CONTRACTOR'S MARK-UP** on Change Orders									
0200	Extra work, by subcontractors, add				%				10%	10%
0250	By General Contractor, add								15%	15%
0400	Omitted work, by subcontractors, deduct all but								5%	5%
0450	By General Contractor, deduct all but								7.50%	7.50%
0600	Overtime work, by subcontractors, add								15%	15%
0650	By General Contractor, add								10%	10%

01 31 13.60 Installing Contractor's Main Office Overhead

		Crew	Daily Output	Labor-Hours	Unit	Material	Labor	Equipment	Total	Total Incl O&P
0010	**INSTALLING CONTRACTOR'S MAIN OFFICE OVERHEAD**									
0020	As percent of direct costs, minimum				%				5%	
0050	Average								13%	
0100	Maximum								30%	

01 31 13.80 Overhead and Profit

		Crew	Daily Output	Labor-Hours	Unit	Material	Labor	Equipment	Total	Total Incl O&P
0010	**OVERHEAD & PROFIT** Allowance to add to items in this									
0020	book that do not include Subs O&P, average				%				25%	
0100	Allowance to add to items in this book that									
0110	do include Subs O&P, minimum				%				5%	5%
0150	Average								10%	10%
0200	Maximum								15%	15%
0300	Typical, by size of project, under $100,000								30%	
0350	$500,000 project								25%	
0400	$2,000,000 project								20%	
0450	Over $10,000,000 project								15%	

01 31 Project Management and Coordination

01 31 13 – Project Coordination

01 31 13.90 Performance Bond		Crew	Daily Output	Labor-Hours	Unit	2013 Bare Costs Material	Labor	Equipment	Total	Total Incl O&P
0010	**PERFORMANCE BOND** R013113-80									
0020	For buildings, minimum				Job				.60%	.60%
0100	Maximum				"				2.50%	2.50%

01 32 Construction Progress Documentation

01 32 33 – Photographic Documentation

01 32 33.50 Photographs		Crew	Daily Output	Labor-Hours	Unit	Material	Labor	Equipment	Total	Total Incl O&P
0010	**PHOTOGRAPHS**									
0020	8" x 10", 4 shots, 2 prints ea., std. mounting				Set	475			475	520
0100	Hinged linen mounts					530			530	580
0200	8" x 10", 4 shots, 2 prints each, in color					415			415	460
0300	For I.D. slugs, add to all above				↓	5.30			5.30	5.85
1500	Time lapse equipment, camera and projector, buy				Ea.	2,650			2,650	2,925
1550	Rent per month				"	1,575			1,575	1,725
1700	Cameraman and film, including processing, B.&W.				Day	1,375			1,375	1,525
1720	Color				"	1,375			1,375	1,525

01 41 Regulatory Requirements

01 41 26 – Permit Requirements

01 41 26.50 Permits		Crew	Daily Output	Labor-Hours	Unit	Material	Labor	Equipment	Total	Total Incl O&P
0010	**PERMITS**									
0020	Rule of thumb, most cities, minimum				Job				.50%	.50%
0100	Maximum				"				2%	2%

01 51 Temporary Utilities

01 51 13 – Temporary Electricity

01 51 13.50 Temporary Power Equip (Pro-Rated Per Job)		Crew	Daily Output	Labor-Hours	Unit	Material	Labor	Equipment	Total	Total Incl O&P
0010	**TEMPORARY POWER EQUIP (PRO-RATED PER JOB)** R015113-65									
0020	Service, overhead feed, 3 use									
0030	100 Amp	1 Elec	1.25	6.400	Ea.	755	335		1,090	1,325
0040	200 Amp		1	8		990	420		1,410	1,725
0050	400 Amp		.75	10.667		1,850	560		2,410	2,850
0060	600 Amp	↓	.50	16	↓	2,675	840		3,515	4,200
0100	Underground feed, 3 use									
0110	100 Amp	1 Elec	2	4	Ea.	710	210		920	1,100
0120	200 Amp		1.15	6.957		970	365		1,335	1,625
0130	400 Amp		1	8		1,800	420		2,220	2,625
0140	600 Amp		.75	10.667		2,250	560		2,810	3,325
0150	800 Amp		.50	16		3,450	840		4,290	5,050
0160	1000 Amp		.35	22.857		3,850	1,200		5,050	6,025
0170	1200 Amp		.25	32		4,300	1,675		5,975	7,250
0180	2000 Amp	↓	.20	40	↓	5,225	2,100		7,325	8,875
0200	Transformers, 3 use									
0210	30 kVA	1 Elec	1	8	Ea.	1,650	420		2,070	2,450
0220	45 kVA		.75	10.667		1,975	560		2,535	3,000
0230	75 kVA		.50	16		3,025	840		3,865	4,575
0240	112.5 kVA	↓	.40	20	↓	3,275	1,050		4,325	5,200

01 51 Temporary Utilities

01 51 13 – Temporary Electricity

01 51 13.50 Temporary Power Equip (Pro-Rated Per Job)		Crew	Daily Output	Labor-Hours	Unit	2013 Bare Costs Material	Labor	Equipment	Total	Total Incl O&P
0250	Feeder, PVC, CU wire in trench									
0260	60 Amp	1 Elec	96	.083	L.F.	4.52	4.37		8.89	11.50
0270	100 Amp		85	.094		8.80	4.93		13.73	17.10
0280	200 Amp		59	.136		21.50	7.10		28.60	34
0290	400 Amp		42	.190		47.50	10		57.50	67.50
0300	Feeder, PVC, aluminum wire in trench									
0310	60 Amp	1 Elec	96	.083	L.F.	4.67	4.37		9.04	11.70
0320	100 Amp		85	.094		5.20	4.93		10.13	13.10
0330	200 Amp		59	.136		11.20	7.10		18.30	23
0340	400 Amp		42	.190		23.50	10		33.50	40.50
0350	Feeder, EMT, CU wire									
0360	60 Amp	1 Elec	90	.089	L.F.	3.57	4.66		8.23	10.90
0370	100 Amp		80	.100		8.35	5.25		13.60	17.05
0380	200 Amp		60	.133		17.55	7		24.55	30
0390	400 Amp		35	.229		51.50	12		63.50	74.50
0400	Feeder, EMT, alum. wire									
0410	60 Amp	1 Elec	90	.089	L.F.	4.72	4.66		9.38	12.15
0420	100 Amp		80	.100		5.95	5.25		11.20	14.40
0430	200 Amp		60	.133		16.40	7		23.40	28.50
0440	400 Amp		35	.229		28	12		40	49
0500	Equipment, 3 use									
0510	Spider box 50 Amp	1 Elec	8	1	Ea.	670	52.50		722.50	815
0520	Lighting cord 100′		8	1		90	52.50		142.50	177
0530	Light stanchion		8	1		68.50	52.50		121	154
0540	Temporary cords, 100′, 3 use									
0550	Feeder cord, 50 Amp	1 Elec	16	.500	Ea.	385	26		411	465
0560	Feeder cord, 100 Amp		12	.667		535	35		570	645
0570	Tap cord, 50 Amp		12	.667		221	35		256	296
0580	Tap cord, 100 Amp		6	1.333		535	70		605	695
0590	Temporary cords, 50′, 3 use									
0600	Feeder cord, 50 Amp	1 Elec	16	.500	Ea.	179	26		205	236
0610	Feeder cord, 100 Amp		12	.667		370	35		405	460
0620	Tap cord, 50 Amp		12	.667		109	35		144	173
0630	Tap cord, 100 Amp		6	1.333		370	70		440	510
0700	Connections									
0710	Compressor or pump									
0720	30 Amp	1 Elec	7	1.143	Ea.	27	60		87	120
0730	60 Amp		5.30	1.509		51	79		130	174
0740	100 Amp		4	2		112	105		217	280
0750	Tower crane									
0760	60 Amp	1 Elec	4.50	1.778	Ea.	51	93		144	195
0770	100 Amp	"	3	2.667	"	112	140		252	330
0780	Manlift									
0790	Single	1 Elec	3	2.667	Ea.	49	140		189	263
0800	Double	"	2	4	"	88.50	210		298.50	410
0810	Welder with disconnect									
0820	50 Amp	1 Elec	5	1.600	Ea.	310	84		394	465
0830	100 Amp		3.80	2.105		515	110		625	730
0840	200 Amp		2.50	3.200		925	168		1,093	1,275
0850	400 Amp		1	8		2,225	420		2,645	3,075
0860	Office trailer									
0870	60 Amp	1 Elec	4.50	1.778	Ea.	118	93		211	269
0880	100 Amp		3	2.667		229	140		369	460

01 51 Temporary Utilities

01 51 13 – Temporary Electricity

01 51 13.50 Temporary Power Equip (Pro-Rated Per Job)		Crew	Daily Output	Labor-Hours	Unit	2013 Bare Costs Material	Labor	Equipment	Total	Total Incl O&P
0890	200 Amp	1 Elec	2	4	Ea.	770	210		980	1,150
0900	Lamping, add per floor				Total				625	690
0910	Maintenance, total temp. power cost				Job				5%	5%

01 51 13.80 Temporary Utilities		Crew	Daily Output	Labor-Hours	Unit	Material	Labor	Equipment	Total	Total Incl O&P
0010	**TEMPORARY UTILITIES**									
0100	Heat, incl. fuel and operation, per week, 12 hrs. per day	1 Skwk	100	.080	CSF Flr	27.50	3.70		31.20	36
0200	24 hrs. per day	"	60	.133		53	6.15		59.15	67.50
0350	Lighting, incl. service lamps, wiring & outlets, minimum	1 Elec	34	.235		2.80	12.35		15.15	21.50
0360	Maximum	"	17	.471		5.85	24.50		30.35	43.50
0400	Power for temp lighting only, 6.6 KWH, per month								.92	1.01
0430	11.8 KWH, per month								1.65	1.82
0450	23.6 KWH, per month								3.30	3.63
0600	Power for job duration incl. elevator, etc., minimum								47	51.50
0650	Maximum								110	121
1000	Toilet, portable, see Equip. Rental 01 54 33 in Reference Section									

01 52 Construction Facilities

01 52 13 – Field Offices and Sheds

01 52 13.20 Office and Storage Space		Crew	Daily Output	Labor-Hours	Unit	Material	Labor	Equipment	Total	Total Incl O&P
0010	**OFFICE AND STORAGE SPACE**									
0020	Office trailer, furnished, no hookups, 20′ x 8′, buy	2 Skwk	1	16	Ea.	8,550	740		9,290	10,600
0250	Rent per month					153			153	169
0300	32′ x 8′, buy	2 Skwk	.70	22.857		13,600	1,050		14,650	16,600
0350	Rent per month					190			190	209
0400	50′ x 10′, buy	2 Skwk	.60	26.667		22,300	1,225		23,525	26,500
0450	Rent per month					291			291	320
0500	50′ x 12′, buy	2 Skwk	.50	32		27,000	1,475		28,475	32,000
0550	Rent per month					340			340	375
0700	For air conditioning, rent per month, add					46			46	50.50
0800	For delivery, add per mile				Mile	10.60			10.60	11.65
0890	Delivery each way				Ea.	200			200	220
0900	Bunk house trailer, 8′ x 40′ duplex dorm with kitchen, no hookups, buy	2 Carp	1	16		37,300	720		38,020	42,100
0910	9 man with kitchen and bath, no hookups, buy		1	16		38,000	720		38,720	42,900
0920	18 man sleeper with bath, no hookups, buy		1	16		49,000	720		49,720	55,000
1000	Portable buildings, prefab, on skids, economy, 8′ x 8′		265	.060	S.F.	29.50	2.71		32.21	36.50
1100	Deluxe, 8′ x 12′		150	.107	"	24	4.79		28.79	34
1200	Storage boxes, 20′ x 8′, buy	2 Skwk	1.80	8.889	Ea.	3,025	410		3,435	3,950
1250	Rent per month					74			74	81.50
1300	40′ x 8′, buy	2 Skwk	1.40	11.429		3,975	530		4,505	5,200
1350	Rent per month					102			102	112

01 52 13.40 Field Office Expense		Crew	Daily Output	Labor-Hours	Unit	Material	Labor	Equipment	Total	Total Incl O&P
0010	**FIELD OFFICE EXPENSE**									
0100	Office equipment rental average				Month	200			200	220
0120	Office supplies, average				"	75			75	82.50
0125	Office trailer rental, see Section 01 52 13.20									
0140	Telephone bill; avg. bill/month incl. long dist.				Month	81			81	89
0160	Lights & HVAC				"	152			152	167

01 52 Construction Facilities

01 52 13 – Field Offices and Sheds

01 52 13.40 Field Office Expense		Crew	Daily Output	Labor-Hours	Unit	2013 Bare Costs Material	Labor	Equipment	Total	Total Incl O&P
0010	**WEEKLY FORKLIFT CREW**									
0100	All-terrain forklift, 45′ lift, 35′ reach, 9000 lb. capacity	A-3P	.20	40	Week		1,825	2,550	4,375	5,600

01 54 19 – Temporary Cranes

01 54 19.50 Daily Crane Crews

		Crew	Daily Output	Labor-Hours	Unit	Material	Labor	Equipment	Total	Total Incl O&P
0010	**DAILY CRANE CREWS** for small jobs, portal to portal									
0100	12-ton truck-mounted hydraulic crane	A-3H	1	8	Day		390	855	1,245	1,525
0200	25-ton	A-3I	1	8			390	980	1,370	1,675
0300	40-ton	A-3J	1	8			390	1,225	1,615	1,925
0400	55-ton	A-3K	1	16			730	1,650	2,380	2,900
0500	80-ton	A-3L	1	16			730	2,350	3,080	3,675
0600	100-ton	A-3M	1	16			730	2,350	3,080	3,675
0900	If crane is needed on a Saturday, Sunday or Holiday									
0910	At time-and-a-half, add				Day		50%			
0920	At double time, add				"		100%			

01 54 19.60 Monthly Tower Crane Crew

		Crew	Daily Output	Labor-Hours	Unit	Material	Labor	Equipment	Total	Total Incl O&P
0010	**MONTHLY TOWER CRANE CREW**, excludes concrete footing									
0100	Static tower crane, 130′ high, 106′ jib, 6200 lb. capacity	A-3N	.05	176	Month		8,600	23,600	32,200	38,900

01 54 23 – Temporary Scaffolding and Platforms

01 54 23.70 Scaffolding

		Crew	Daily Output	Labor-Hours	Unit	Material	Labor	Equipment	Total	Total Incl O&P
0010	**SCAFFOLDING** R015423-10									
0015	Steel tube, regular, no plank, labor only to erect & dismantle									
0090	Building exterior, wall face, 1 to 5 stories, 6′-4″ x 5′ frames	3 Carp	8	3	C.S.F.		135		135	207
0200	6 to 12 stories	4 Carp	8	4			180		180	277
0301	13 to 20 stories	5 Clab	8	5			177		177	273
0460	Building interior, wall face area, up to 16′ high	3 Carp	12	2			90		90	138
0560	16′ to 40′ high		10	2.400			108		108	166
0800	Building interior floor area, up to 30′ high		150	.160	C.C.F.		7.20		7.20	11.05
0900	Over 30′ high	4 Carp	160	.200	"		9		9	13.85
0906	Complete system for face of walls, no plank, material only rent/mo				C.S.F.	36			36	39.50
0908	Interior spaces, no plank, material only rent/mo				C.C.F.	3.42			3.42	3.76
0910	Steel tubular, heavy duty shoring, buy									
0920	Frames 5′ high 2′ wide				Ea.	81.50			81.50	90
0925	5′ high 4′ wide					93			93	102
0930	6′ high 2′ wide					93.50			93.50	103
0935	6′ high 4′ wide					109			109	120
0940	Accessories									
0945	Cross braces				Ea.	15.85			15.85	17.40
0950	U-head, 8″ x 8″					19.10			19.10	21
0955	J-head, 4″ x 8″					13.90			13.90	15.30
0960	Base plate, 8″ x 8″					15.50			15.50	17.05
0965	Leveling jack					33.50			33.50	36.50
1000	Steel tubular, regular, buy									
1100	Frames 3′ high 5′ wide				Ea.	58.50			58.50	64.50
1150	5′ high 5′ wide					66.50			66.50	73
1200	6′-4″ high 5′ wide					76			76	83.50
1350	7′-6″ high 6′ wide					158			158	173
1500	Accessories cross braces					15.50			15.50	17.05
1550	Guardrail post					16.40			16.40	18.05
1600	Guardrail 7′ section					6.30			6.30	6.95
1650	Screw jacks & plates					21.50			21.50	24
1700	Sidearm brackets					25.50			25.50	28.50

01 54 23 – Temporary Scaffolding and Platforms

01 54 23.70 Scaffolding		Crew	Daily Output	Labor-Hours	Unit	2013 Bare Costs Material	Labor	Equipment	Total	Total Incl O&P
1750	8" casters				Ea.	31			31	34
1800	Plank 2" x 10" x 16'-0"					51.50			51.50	56.50
1900	Stairway section					286			286	315
1910	Stairway starter bar					32			32	35
1920	Stairway inside handrail					53			53	58
1930	Stairway outside handrail					81.50			81.50	90
1940	Walk-thru frame guardrail				↓	41.50			41.50	46
2000	Steel tubular, regular, rent/mo.									
2100	Frames 3' high 5' wide				Ea.	5			5	5.50
2150	5' high 5' wide					5			5	5.50
2200	6'-4" high 5' wide					5.15			5.15	5.65
2250	7'-6" high 6' wide					7			7	7.70
2500	Accessories, cross braces					1			1	1.10
2550	Guardrail post					1			1	1.10
2600	Guardrail 7' section					1			1	1.10
2650	Screw jacks & plates					2			2	2.20
2700	Sidearm brackets					2			2	2.20
2750	8" casters					8			8	8.80
2800	Outrigger for rolling tower					3			3	3.30
2850	Plank 2" x 10" x 16'-0"					6			6	6.60
2900	Stairway section					40			40	44
2940	Walk-thru frame guardrail				↓	2.50			2.50	2.75
3000	Steel tubular, heavy duty shoring, rent/mo.									
3250	5' high 2' & 4' wide				Ea.	5			5	5.50
3300	6' high 2' & 4' wide					5			5	5.50
3500	Accessories, cross braces					1			1	1.10
3600	U-head, 8" x 8"					1			1	1.10
3650	J-head, 4" x 8"					1			1	1.10
3700	Base plate, 8" x 8"					1			1	1.10
3750	Leveling jack					2			2	2.20
5700	Planks, 2" x 10" x 16'-0", labor only to erect & remove to 50' H	3 Carp	72	.333			14.95		14.95	23
5800	Over 50' high	4 Carp	80	.400	↓		17.95		17.95	27.50

01 54 23.75 Scaffolding Specialties		Crew	Daily Output	Labor-Hours	Unit	Material	Labor	Equipment	Total	Total Incl O&P
0010	**SCAFFOLDING SPECIALTIES**									
1500	Sidewalk bridge using tubular steel scaffold frames including									
1510	planking (material cost is rent/month)	3 Carp	45	.533	L.F.	5.60	24		29.60	43
1600	For 2 uses per month, deduct from all above					50%				
1700	For 1 use every 2 months, add to all above					100%				
1900	Catwalks, 20" wide, no guardrails, 7' span, buy				Ea.	133			133	146
2000	10' span, buy					186			186	205
3720	Putlog, standard, 8' span, with hangers, buy					66.50			66.50	73
3750	12' span, buy					99			99	109
3760	Trussed type, 16' span, buy					228			228	251
3790	22' span, buy					274			274	300
3795	Rent per month					30			30	33
3800	Rolling ladders with handrails, 30" wide, buy, 2 step					204			204	225
4000	7 step					635			635	700
4050	10 step					880			880	965
4100	Rolling towers, buy, 5' wide, 7' long, 10' high					1,075			1,075	1,200
4200	For 5' high added sections, to buy, add				↓	164			164	180
4300	Complete incl. wheels, railings, outriggers,									
4350	21' high, to buy				Ea.	1,825			1,825	2,000

01 54 Construction Aids

01 54 23 – Temporary Scaffolding and Platforms

01 54 23.75 Scaffolding Specialties		Crew	Daily Output	Labor-Hours	Unit	2013 Bare Costs Material	Labor	Equipment	Total	Total Incl O&P
4400	Rent/month = 5% of purchase cost				Ea.	113			113	124

01 54 26 – Temporary Swing Staging

01 54 26.50 Swing Staging

		Crew	Daily Output	Labor-Hours	Unit	Material	Labor	Equipment	Total	Total Incl O&P
0010	**SWING STAGING**, 500 lb. cap., 2′ wide to 24′ long, hand operated									
0020	steel cable type, with 60′ cables, buy				Ea.	4,900			4,900	5,375
0030	Rent per month				"	490			490	540
2200	Move swing staging (setup and remove)	E-4	2	16	Move		810	72	882	1,525

01 54 36 – Equipment Mobilization

01 54 36.50 Mobilization

		Crew	Daily Output	Labor-Hours	Unit	Material	Labor	Equipment	Total	Total Incl O&P
0010	**MOBILIZATION** (Use line item again for demobilization) R015433-10									
0015	Up to 25 mi. haul dist (50 mi. RT for mob/demob crew)									
0020	Dozer, loader, backhoe, excav., grader, paver, roller, 70 to 150 H.P.	B-34N	4	2	Ea.		73	142	215	269
0100	Above 150 HP	B-34K	3	2.667			97.50	320	417.50	505
0900	Shovel or dragline, 3/4 C.Y.	"	3.60	2.222			81.50	267	348.50	420
1100	Small equipment, placed in rear of, or towed by pickup truck	A-3A	8	1			35.50	20.50	56	77.50
1150	Equip up to 70 HP, on flatbed trailer behind pickup truck	A-3D	4	2			71.50	69.50	141	186
2000	Crane, truck-mounted, up to 75 ton, (driver only, one-way)	1 Eqhv	7.20	1.111			54		54	82
2100	Crane, truck-mounted, over 75 ton	A-3E	2.50	6.400			273	66	339	490
2200	Crawler-mounted, up to 75 ton	A-3F	2	8			340	500	840	1,075
2300	Over 75 ton	A-3G	1.50	10.667	↓		455	750	1,205	1,525
2500	For each additional 5 miles haul distance, add						10%	10%		
3000	For large pieces of equipment, allow for assembly/knockdown									
3100	For mob/demob of micro-tunneling equip, see Section 33 05 23.19									

01 54 39 – Construction Equipment

01 54 39.70 Small Tools

		Crew	Daily Output	Labor-Hours	Unit	Material	Labor	Equipment	Total	Total Incl O&P
0010	**SMALL TOOLS**									
0020	As % of contractor's bare labor cost for project, minimum				Total		.50%			
0100	Maximum				"		2%			

01 55 Vehicular Access and Parking

01 55 23 – Temporary Roads

01 55 23.50 Roads and Sidewalks

		Crew	Daily Output	Labor-Hours	Unit	Material	Labor	Equipment	Total	Total Incl O&P
0010	**ROADS AND SIDEWALKS** Temporary									
0050	Roads, gravel fill, no surfacing, 4″ gravel depth	B-14	715	.067	S.Y.	4.35	2.52	.51	7.38	9.20
0100	8″ gravel depth	"	615	.078	"	8.70	2.93	.60	12.23	14.75
1000	Ramp, 3/4″ plywood on 2″ x 6″ joists, 16″ O.C.	2 Carp	300	.053	S.F.	1.33	2.39		3.72	5.15
1100	On 2″ x 10″ joists, 16″ O.C.	"	275	.058	"	1.85	2.61		4.46	6.05

01 55 Vehicular Access and Parking

01 55 23 – Temporary Roads

01 55 23.50 Roads and Sidewalks		Crew	Daily Output	Labor-Hours	Unit	Material	Labor	Equipment	Total	Total Incl O&P
						2013 Bare Costs				
0010	**WINTER PROTECTION**									
0100	Framing to close openings	2 Clab	500	.032	S.F.	.40	1.13		1.53	2.19
0200	Tarpaulins hung over scaffolding, 8 uses, not incl. scaffolding		1500	.011		.25	.38		.63	.86
0300	Prefab fiberglass panels, steel frame, 8 uses		1200	.013		2.20	.47		2.67	3.15

01 56 16 – Temporary Dust Barriers

01 56 16.10 Dust Barriers, Temporary

		Crew	Daily Output	Labor-Hours	Unit	Material	Labor	Equipment	Total	Total Incl O&P
0010	**DUST BARRIERS, TEMPORARY**									
0020	Spring loaded telescoping pole & head, to 12′, erect and dismantle	1 Clab	240	.033	Ea.		1.18		1.18	1.82
0025	Cost per day (based upon 250 days)				Day	.24			.24	.26
0030	To 21′, erect and dismantle	1 Clab	240	.033	Ea.		1.18		1.18	1.82
0035	Cost per day (based upon 250 days)				Day	.39			.39	.42
0040	Accessories, caution tape reel, erect and dismantle	1 Clab	480	.017	Ea.		.59		.59	.91
0045	Cost per day (based upon 250 days)				Day	.05			.05	.05
0060	Foam rail and connector, erect and dismantle	1 Clab	240	.033	Ea.		1.18		1.18	1.82
0065	Cost per day (based upon 250 days)				Day	.10			.10	.11
0070	Caution tape	1 Clab	384	.021	C.L.F.	2.70	.74		3.44	4.11
0080	Zipper, standard duty		60	.133	Ea.	8	4.73		12.73	16.10
0090	Heavy duty		48	.167	″	9.50	5.90		15.40	19.55
0100	Polyethylene sheet, 4 mil		37	.216	Sq.	2.92	7.65		10.57	15
0110	6 mil		37	.216	″	4.03	7.65		11.68	16.25
1000	Dust partition, 6 mil polyethylene, 1″ x 3″ frame	2 Carp	2000	.008	S.F.	.27	.36		.63	.84
1080	2″ x 4″ frame	″	2000	.008	″	.31	.36		.67	.89

01 56 23 – Temporary Barricades

01 56 23.10 Barricades

		Crew	Daily Output	Labor-Hours	Unit	Material	Labor	Equipment	Total	Total Incl O&P
0010	**BARRICADES**									
0020	5′ high, 3 rail @ 2″ x 8″, fixed	2 Carp	20	.800	L.F.	5.35	36		41.35	61.50
0150	Movable		30	.533		4.47	24		28.47	42
1000	Guardrail, wooden, 3′ high, 1″ x 6″, on 2″ x 4″ posts		200	.080		1.30	3.59		4.89	7
1100	2″ x 6″, on 4″ x 4″ posts		165	.097		2.24	4.35		6.59	9.15
1200	Portable metal with base pads, buy					15.55			15.55	17.10
1250	Typical installation, assume 10 reuses	2 Carp	600	.027		2.55	1.20		3.75	4.65
1300	Barricade tape, polyethylene, 7 mil, 3″ wide x 500′ long roll				Ea.	25			25	27.50
3000	Detour signs, set up and remove									
3010	Reflective aluminum, MUTCD, 24″ x 24″, post mounted	1 Clab	20	.400	Ea.	2.24	14.20		16.44	24.50

01 56 26 – Temporary Fencing

01 56 26.50 Temporary Fencing

		Crew	Daily Output	Labor-Hours	Unit	Material	Labor	Equipment	Total	Total Incl O&P
0010	**TEMPORARY FENCING**									
0020	Chain link, 11 ga., 4′ high	2 Clab	400	.040	L.F.	2.59	1.42		4.01	5.05
0100	6′ high		300	.053		2.59	1.89		4.48	5.75
0200	Rented chain link, 6′ high, to 1000′ (up to 12 mo.)		400	.040		4.39	1.42		5.81	7
0250	Over 1000′ (up to 12 mo.)		300	.053		4.29	1.89		6.18	7.65
0350	Plywood, painted, 2″ x 4″ frame, 4′ high	A-4	135	.178		6.20	7.60		13.80	18.50
0400	4″ x 4″ frame, 8′ high	″	110	.218		10.95	9.35		20.30	26.50
0500	Wire mesh on 4″ x 4″ posts, 4′ high	2 Carp	100	.160		9.30	7.20		16.50	21.50
0550	8′ high	″	80	.200		14.10	9		23.10	29.50

01 56 29 – Temporary Protective Walkways

01 56 29.50 Protection

		Crew	Daily Output	Labor-Hours	Unit	Material	Labor	Equipment	Total	Total Incl O&P
0010	**PROTECTION**									
0020	Stair tread, 2″ x 12″ planks, 1 use	1 Carp	75	.107	Tread	3.97	4.79		8.76	11.75
0100	Exterior plywood, 1/2″ thick, 1 use		65	.123		1.74	5.55		7.29	10.40

01 56 Temporary Barriers and Enclosures

01 56 29 – Temporary Protective Walkways

01 56 29.50 Protection		Crew	Daily Output	Labor-Hours	Unit	Material	Labor	Equipment	Total	Total Incl O&P
						2013 Bare Costs				
0200	3/4" thick, 1 use	1 Carp	60	.133	Tread	2.32	6		8.32	11.75
2200	Sidewalks, 2" x 12" planks, 2 uses		350	.023	S.F.	.66	1.03		1.69	2.31
2300	Exterior plywood, 2 uses, 1/2" thick		750	.011		.29	.48		.77	1.06
2400	5/8" thick		650	.012		.36	.55		.91	1.25
2500	3/4" thick		600	.013		.39	.60		.99	1.35

01 58 Project Identification

01 58 13 – Temporary Project Signage

01 58 13.50 Signs

		Crew	Daily Output	Labor-Hours	Unit	Material	Labor	Equipment	Total	Total Incl O&P
0010	**SIGNS**									
0020	High intensity reflectorized, no posts, buy				S.F.	34			34	37.50

01 74 Cleaning and Waste Management

01 74 13 – Progress Cleaning

01 74 13.20 Cleaning Up

		Crew	Daily Output	Labor-Hours	Unit	Material	Labor	Equipment	Total	Total Incl O&P
0010	**CLEANING UP**									
0020	After job completion, allow, minimum				Job				.30%	.30%
0040	Maximum				"				1%	1%
0050	Cleanup of floor area, continuous, per day, during const.	A-5	24	.750	M.S.F.	1.77	26.50	2.77	31.04	46
0100	Final by GC at end of job	"	11.50	1.565	"	1.87	55.50	5.80	63.17	94

01 91 Commissioning

01 91 13 – General Commissioning Requirements

01 91 13.50 Building Commissioning

		Crew	Daily Output	Labor-Hours	Unit	Material	Labor	Equipment	Total	Total Incl O&P
0010	**BUILDING COMMISSIONING**									
0100	Basic building commissioning, minimum				%				.25%	.25%
0150	Maximum								.50%	.50%
0200	Enhanced building commissioning, minimum								.50%	.50%
0250	Maximum								1%	1%

01 93 Facility Maintenance

01 93 13 – Facility Maintenance Procedures

01 93 13.16 Electrical Facilities Maintenance

		Crew	Daily Output	Labor-Hours	Unit	Material	Labor	Equipment	Total	Total Incl O&P
0010	**ELECTRICAL FACILITIES MAINTENANCE**									
0700	Cathodic protection systems									
0720	Check and adjust reading on rectifier	1 Elec	20	.400	Ea.		21		21	31.50
0730	Check pipe to soil potential		20	.400			21		21	31.50
0740	Replace lead connection		4	2			105		105	157
0800	Control device, install		5.70	1.404			73.50		73.50	110
0810	Disassemble, clean and reinstall		7	1.143			60		60	89.50
0820	Replace		10.70	.748			39		39	58.50
0830	Trouble shoot		10	.800			42		42	62.50
0900	Demolition, for electrical demolition see Section 26 05 05.10									
1000	Distribution systems and equipment install or repair a breaker									
1010	In power panels up to 200 amps	1 Elec	7	1.143	Ea.		60		60	89.50
1020	Over 200 amps		2	4			210		210	315

01 93 13 – Facility Maintenance Procedures

01 93 13.16 Electrical Facilities Maintenance		Crew	Daily Output	Labor-Hours	Unit	Material	2013 Bare Costs Labor	Equipment	Total	Total Incl O&P
1030	Reset breaker or replace fuse	1 Elec	20	.400	Ea.		21		21	31.50
1100	Megger test MCC (each stack)		4	2			105		105	157
1110	MCC vacuum and clean (each stack)		5.30	1.509			79		79	118
2500	Remove and replace or maint., road fixture & lamp		3	2.667		905	140		1,045	1,200
2510	Fluorescent fixture		7	1.143		62	60		122	158
2515	Relamp (fluor.) facility area each tube		60	.133		5.55	7		12.55	16.60
2518	Fluorescent fixture, clean (area)		44	.182		.09	9.55		9.64	14.35
2520	Incandescent fixture		11	.727		60.50	38		98.50	124
2530	Lamp (incandescent or fluorescent)		60	.133		4.69	7		11.69	15.60
2535	Replace cord in socket lamp		13	.615		2.45	32.50		34.95	51
2540	Ballast electronic type for two tubes		8	1		36.50	52.50		89	119
2541	Starter		30	.267		.93	13.95		14.88	22
2545	Replace other lighting parts		11	.727		15.80	38		53.80	74.50
2550	Switch		11	.727		8.70	38		46.70	66.50
2555	Receptacle		11	.727		9.60	38		47.60	67.50
2560	Floodlight		4	2		267	105		372	450
2570	Christmas lighting, indoor, per string		16	.500			26		26	39
2580	Outdoor		13	.615			32.50		32.50	48.50
2590	Test battery operated emergency lights		40	.200			10.50		10.50	15.70
2600	Repair and replace component in communication system		6	1.333		53.50	70		123.50	164
2700	Repair misc. appliances (incl. clocks, vent fan, blower, etc.)		6	1.333			70		70	105
2710	Reset clocks & timers		50	.160			8.40		8.40	12.55
2720	Adjust time delay relays		16	.500			26		26	39
2730	Test specific gravity of lead-acid batteries	↓	80	.100	↓		5.25		5.25	7.85
3000	Motors and generators									
3020	Disassemble, clean and reinstall motor, up to 1/4 HP	1 Elec	4	2	Ea.		105		105	157
3030	Up to 3/4 HP		3	2.667			140		140	209
3040	Up to 10 HP		2	4			210		210	315
3050	Replace part, up to 1/4 HP		6	1.333			70		70	105
3060	Up to 3/4 HP		4	2			105		105	157
3070	Up to 10 HP		3	2.667			140		140	209
3080	Megger test motor windings		5.33	1.501			78.50		78.50	118
3082	Motor vibration check		16	.500			26		26	39
3084	Oil motor bearings		25	.320			16.75		16.75	25
3086	Run test emergency generator for 30 minutes		11	.727			38		38	57
3090	Rewind motor, up to 1/4 HP		3	2.667			140		140	209
3100	Up to 3/4 HP		2	4			210		210	315
3110	Up to 10 HP		1.50	5.333			279		279	420
3150	Generator, repair or replace part		4	2			105		105	157
3160	Repair DC generator		2	4			210		210	315
4000	Stub pole, install or remove		3	2.667			140		140	209
4500	Transformer maintenance up to 15 kVA	↓	2.70	2.963	↓		155		155	232

Division Notes

	CREW	DAILY OUTPUT	LABOR-HOURS	UNIT	2012 BARE COSTS				TOTAL INCL O&P
					MAT.	LABOR	EQUIP.	TOTAL	

Estimating Tips

02 30 00 Subsurface Investigation

In preparing estimates on structures involving earthwork or foundations, all information concerning soil characteristics should be obtained. Look particularly for hazardous waste, evidence of prior dumping of debris, and previous stream beds.

02 40 00 Demolition and Structure Moving

The costs shown for selective demolition do not include rubbish handling or disposal. These items should be estimated separately using RSMeans data or other sources.

- Historic preservation often requires that the contractor remove materials from the existing structure, rehab them, and replace them. The estimator must be aware of any related measures and precautions that must be taken when doing selective demolition and cutting and patching. Requirements may include special handling and storage, as well as security.
- In addition to Subdivision 02 41 00, you can find selective demolition items in each division. Example: Roofing demolition is in Division 7.

02 40 00 Building Deconstruction

This section provides costs for the careful dismantling and recycling of most of low-rise building materials.

02 50 00 Containment of Hazardous Waste

This section addresses on-site hazardous waste disposal costs.

02 80 00 Hazardous Material Disposal/ Remediation

This subdivision includes information on hazardous waste handling, asbestos remediation, lead remediation, and mold remediation. See reference R028213-20 and R028319-60 for further guidance in using these unit price lines.

02 90 00 Monitoring Chemical Sampling, Testing Analysis

This section provides costs for on-site sampling and testing hazardous waste.

Reference Numbers

Reference numbers are shown in shaded boxes at the beginning of some major classifications. These numbers refer to related items in the Reference Section. The reference information may be an estimating procedure, an alternate pricing method, or technical information.

Note: Not all subdivisions listed here necessarily appear in this publication.

02 41 Demolition

02 41 13 – Selective Site Demolition

02 41 13.17 Demolish, Remove Pavement and Curb

	02 41 13.17 Demolish, Remove Pavement and Curb	Crew	Daily Output	Labor-Hours	Unit	Material	2013 Bare Costs Labor	Equipment	Total	Total Incl O&P
0010	**DEMOLISH, REMOVE PAVEMENT AND CURB** R024119-10									
5010	Pavement removal, bituminous roads, 3″ thick	B-38	690	.058	S.Y.		2.34	1.86	4.20	5.60
5050	4″ to 6″ thick		420	.095			3.84	3.06	6.90	9.20
5100	Bituminous driveways		640	.063			2.52	2.01	4.53	6.05
5200	Concrete to 6″ thick, hydraulic hammer, mesh reinforced		255	.157			6.35	5.05	11.40	15.20
5300	Rod reinforced		200	.200			8.05	6.45	14.50	19.40
5400	Concrete, 7″ to 24″ thick, plain		33	1.212	C.Y.		49	39	88	118
5500	Reinforced		24	1.667	″		67	53.50	120.50	162
5600	With hand held air equipment, bituminous, to 6″ thick	B-39	1900	.025	S.F.		.95	.12	1.07	1.58
5700	Concrete to 6″ thick, no reinforcing		1600	.030			1.13	.14	1.27	1.89
5800	Mesh reinforced		1400	.034			1.29	.16	1.45	2.15
5900	Rod reinforced		765	.063			2.35	.30	2.65	3.94

02 41 13.33 Minor Site Demolition

	02 41 13.33 Minor Site Demolition	Crew	Daily Output	Labor-Hours	Unit	Material	Labor	Equipment	Total	Total Incl O&P
0010	**MINOR SITE DEMOLITION** R024119-10									
0015	No hauling, abandon catch basin or manhole	B-6	7	3.429	Ea.		133	52.50	185.50	262
0020	Remove existing catch basin or manhole, masonry		4	6			233	92	325	455
0030	Catch basin or manhole frames and covers, stored		13	1.846			72	28.50	100.50	141
0040	Remove and reset		7	3.429			133	52.50	185.50	262
4000	Sidewalk removal, bituminous, 2″ thick		350	.069	S.Y.		2.67	1.05	3.72	5.25
4010	2-1/2″ thick		325	.074			2.87	1.13	4	5.65
4050	Brick, set in mortar		185	.130			5.05	1.99	7.04	9.90
4100	Concrete, plain, 4″		160	.150			5.85	2.30	8.15	11.45
4110	Plain, 5″		140	.171			6.65	2.62	9.27	13.10
4120	Plain, 6″		120	.200			7.80	3.06	10.86	15.25
4200	Mesh reinforced, concrete, 4 ″		150	.160			6.20	2.45	8.65	12.20
4210	5 ″ thick		131	.183			7.15	2.80	9.95	14
4220	6 ″ thick		112	.214			8.35	3.28	11.63	16.35

02 41 19 – Selective Demolition

02 41 19.13 Selective Building Demolition

	02 41 19.13 Selective Building Demolition	Crew	Daily Output	Labor-Hours	Unit	Material	Labor	Equipment	Total	Total Incl O&P
0010	**SELECTIVE BUILDING DEMOLITION**									
0020	Costs related to selective demolition of specific building components									
0025	are included under Common Work Results (XX 05)									
0030	in the component's appropriate division.									

02 41 19.16 Selective Demolition, Cutout

	02 41 19.16 Selective Demolition, Cutout	Crew	Daily Output	Labor-Hours	Unit	Material	Labor	Equipment	Total	Total Incl O&P
0010	**SELECTIVE DEMOLITION, CUTOUT** R024119-10									
0020	Concrete, elev. slab, light reinforcement, under 6 C.F.	B-9	65	.615	C.F.		22	3.50	25.50	38
0050	Light reinforcing, over 6 C.F.	″	75	.533	″		19.10	3.03	22.13	33
6000	Walls, interior, not including re-framing,									
6010	openings to 5 S.F.									
6100	Drywall to 5/8″ thick	1 Clab	24	.333	Ea.		11.80		11.80	18.20
6200	Paneling to 3/4″ thick		20	.400			14.20		14.20	22
6300	Plaster, on gypsum lath		20	.400			14.20		14.20	22
6340	On wire lath		14	.571			20.50		20.50	31

02 41 19.19 Selective Facility Services Demolition

	02 41 19.19 Selective Facility Services Demolition	Crew	Daily Output	Labor-Hours	Unit	Material	Labor	Equipment	Total	Total Incl O&P
0010	**SELECTIVE FACILITY SERVICES DEMOLITION**, Rubbish Handling R024119-10									
0020	The following are to be added to the demolition prices									
0600	Dumpster, weekly rental, 1 dump/week, 6 C.Y. capacity (2 Tons)				Week	460			460	505
0700	10 C.Y. capacity (3 Tons)					535			535	590
0725	20 C.Y. capacity (5 Tons) R024119-20					630			630	695
0800	30 C.Y. capacity (7 Tons)					810			810	890
0840	40 C.Y. capacity (10 Tons)					860			860	945

02 41 Demolition

02 41 19 – Selective Demolition

02 41 19.19 Selective Facility Services Demolition		Crew	Daily Output	Labor-Hours	Unit	Material	2013 Bare Costs Labor	Equipment	Total	Total Incl O&P
2000	Load, haul, dump and return, 50′ haul, hand carried	2 Clab	24	.667	C.Y.		23.50		23.50	36.50
2005	Wheeled		37	.432			15.35		15.35	23.50
2040	51′ to 100′ haul, hand carried		16.50	.970			34.50		34.50	53
2045	Wheeled		25	.640			22.50		22.50	35
2080	Over 100′ haul, add per 100 L.F., hand carried		35.50	.451			16		16	24.50
2085	Wheeled		54	.296			10.50		10.50	16.20
2120	In elevators, per 10 floors, add		140	.114			4.05		4.05	6.25
2130	Load, haul, dump and return, up to 50′ haul, incl. up to 5 rsr stair, hand		23	.696			24.50		24.50	38
2135	Wheeled		35	.457			16.20		16.20	25
2140	6 – 10 riser stairs, hand carried		22	.727			26		26	39.50
2145	Wheeled		34	.471			16.70		16.70	25.50
2150	11 – 20 riser stairs, hand carried		20	.800			28.50		28.50	43.50
2155	Wheeled		31	.516			18.30		18.30	28
2160	21 – 40 riser stairs, hand carried		16	1			35.50		35.50	54.50
2165	Wheeled		24	.667			23.50		23.50	36.50
2170	100′ haul, incl. 5 riser stair, hand carried		15	1.067			38		38	58
2175	Wheeled		23	.696			24.50		24.50	38
2180	6 – 10 riser stair, hand carried		14	1.143			40.50		40.50	62.50
2185	Wheeled		21	.762			27		27	41.50
2190	11 – 20 riser stair, hand carried		12	1.333			47.50		47.50	73
2195	Wheeled		18	.889			31.50		31.50	48.50
2200	21 – 40 riser stair, hand carried		8	2			71		71	109
2205	Wheeled		12	1.333			47.50		47.50	73
2210	Over 100′ haul, add per 100 L.F., hand carried		35.50	.451			16		16	24.50
2215	Wheeled		54	.296			10.50		10.50	16.20
2220	For each additional flight of stairs, up to 5 risers, add		550	.029	Flight		1.03		1.03	1.59
2225	6 – 10 risers, add		275	.058			2.06		2.06	3.18
2230	11 – 20 risers, add		138	.116			4.11		4.11	6.35
2235	21 – 40 risers, add		69	.232			8.20		8.20	12.65
3000	Loading & trucking, including 2 mile haul, chute loaded	B-16	45	.711	C.Y.		26	15.40	41.40	56.50
3040	Hand loading truck, 50′ haul	"	48	.667			24	14.45	38.45	53
3080	Machine loading truck	B-17	120	.267			10.20	6.55	16.75	23
5000	Haul, per mile, up to 8 C.Y. truck	B-34B	1165	.007			.25	.59	.84	1.04
5100	Over 8 C.Y. truck	"	1550	.005			.19	.45	.64	.78

02 41 19.20 Selective Demolition, Dump Charges		Crew	Daily Output	Labor-Hours	Unit	Material	Labor	Equipment	Total	Total Incl O&P
0010	**SELECTIVE DEMOLITION, DUMP CHARGES** R024119-10									
0020	Dump charges, typical urban city, tipping fees only									
0100	Building construction materials				Ton	82			82	90
0200	Trees, brush, lumber					70			70	77
0300	Rubbish only					70			70	77
0500	Reclamation station, usual charge					82			82	90

02 41 19.27 Selective Demolition, Torch Cutting		Crew	Daily Output	Labor-Hours	Unit	Material	Labor	Equipment	Total	Total Incl O&P
0010	**SELECTIVE DEMOLITION, TORCH CUTTING** R024119-10									
0020	Steel, 1″ thick plate	1 Clab	360	.022	L.F.	.38	.79		1.17	1.63
0040	1″ diameter bar	"	210	.038	Ea.		1.35		1.35	2.08
1000	Oxygen lance cutting, reinforced concrete walls									
1040	12″ to 16″ thick walls	1 Clab	10	.800	L.F.		28.50		28.50	43.50
1080	24″ thick walls	"	6	1.333	"		47.50		47.50	73

02 82 Asbestos Remediation

02 82 13 – Asbestos Abatement

02 82 13.47 Asbestos Waste Pkg., Handling, and Disp.		Crew	Daily Output	Labor-Hours	Unit	2013 Bare Costs Material	Labor	Equipment	Total	Total Incl O&P
0010	**ASBESTOS WASTE PACKAGING, HANDLING, AND DISPOSAL**									
0100	Collect and bag bulk material, 3 C.F. bags, by hand	A-9	400	.160	Ea.	1.60	7.95		9.55	14.10
0200	Large production vacuum loader	A-12	880	.073		.81	3.62	.88	5.31	7.45
1000	Double bag and decontaminate	A-9	960	.067		1.60	3.32		4.92	6.90
2000	Containerize bagged material in drums, per 3 C.F. drum	"	800	.080		17.80	3.98		21.78	26
3000	Cart bags 50′ to dumpster	2 Asbe	400	.040			1.99		1.99	3.09
5000	Disposal charges, not including haul, minimum				C.Y.				60	66
5020	Maximum				"				350	385
9000	For type B (supplied air) respirator equipment, add				%				10%	10%

Estimating Tips

General

- Carefully check all the plans and specifications. Concrete often appears on drawings other than structural drawings, including mechanical and electrical drawings for equipment pads. The cost of cutting and patching is often difficult to estimate. See Subdivision 03 81 for Concrete Cutting, Subdivision 02 41 19.16 for Cutout Demolition, Subdivision 03 05 05.10 for Concrete Demolition, and Subdivision 02 41 19.19 for Rubbish Handling (handling, loading and hauling of debris).
- Always obtain concrete prices from suppliers near the job site. A volume discount can often be negotiated, depending upon competition in the area. Remember to add for waste, particularly for slabs and footings on grade.

03 10 00 Concrete Forming and Accessories

- A primary cost for concrete construction is forming. Most jobs today are constructed with prefabricated forms. The selection of the forms best suited for the job and the total square feet of forms required for efficient concrete forming and placing are key elements in estimating concrete construction. Enough forms must be available for erection to make efficient use of the concrete placing equipment and crew.
- Concrete accessories for forming and placing depend upon the systems used. Study the plans and specifications to ensure that all special accessory requirements have been included in the cost estimate, such as anchor bolts, inserts, and hangers.
- Included within costs for forms-in-place are all necessary bracing and shoring.

03 20 00 Concrete Reinforcing

- Ascertain that the reinforcing steel supplier has included all accessories, cutting, bending, and an allowance for lapping, splicing, and waste. A good rule of thumb is 10% for lapping, splicing, and waste. Also, 10% waste should be allowed for welded wire fabric.
- The unit price items in the subdivisions for Reinforcing In Place, Glass Fiber Reinforcing, and Welded Wire Fabric include the labor to install accessories such as beam and slab bolsters, high chairs, and bar ties and tie wire. The material cost for these accessories is not included; they may be obtained from the Accessories Division.

03 30 00 Cast-In-Place Concrete

- When estimating structural concrete, pay particular attention to requirements for concrete additives, curing methods, and surface treatments. Special consideration for climate, hot or cold, must be included in your estimate. Be sure to include requirements for concrete placing equipment, and concrete finishing.
- For accurate concrete estimating, the estimator must consider each of the following major components individually: forms, reinforcing steel, ready-mix concrete, placement of the concrete, and finishing of the top surface. For faster estimating, Subdivision 03 30 53.40 for Concrete-In-Place can be used; here, various items of concrete work are presented that include the costs of all five major components (unless specifically stated otherwise).

03 40 00 Precast Concrete

03 50 00 Cast Decks and Underlayment

- The cost of hauling precast concrete structural members is often an important factor. For this reason, it is important to get a quote from the nearest supplier. It may become economically feasible to set up precasting beds on the site if the hauling costs are prohibitive.

Reference Numbers

Reference numbers are shown in shaded boxes at the beginning of some major classifications. These numbers refer to related items in the Reference Section. The reference information may be an estimating procedure, an alternate pricing method, or technical information.

Note: Not all subdivisions listed here necessarily appear in this publication.

03 01 Maintenance of Concrete

03 01 30 – Maintenance of Cast-In-Place Concrete

03 01 30.62 Concrete Patching		Crew	Daily Output	Labor-Hours	Unit	Material	Labor (2013 Bare Costs)	Equipment	Total	Total Incl O&P
0010	**CONCRETE PATCHING**									
0100	Floors, 1/4" thick, small areas, regular grout	1 Cefi	170	.047	S.F.	1.29	2.03		3.32	4.40
0150	Epoxy grout	"	100	.080	"	7.80	3.44		11.24	13.60
2000	Walls, including chipping, cleaning and epoxy grout									
2100	1/4" deep	1 Cefi	65	.123	S.F.	7.20	5.30		12.50	15.70
2150	1/2" deep		50	.160		14.40	6.90		21.30	26
2200	3/4" deep		40	.200		21.50	8.60		30.10	36.50

03 11 Concrete Forming

03 11 13 – Structural Cast-In-Place Concrete Forming

03 11 13.40 Forms In Place, Equipment Foundations

		Crew	Daily Output	Labor-Hours	Unit	Material	Labor	Equipment	Total	Total Incl O&P
0010	**FORMS IN PLACE, EQUIPMENT FOUNDATIONS**									
0020	1 use	C-2	160	.300	SFCA	3.08	13.10		16.18	23.50
0050	2 use		190	.253		1.70	11.05		12.75	18.85
0100	3 use		200	.240		1.23	10.50		11.73	17.50
0150	4 use		205	.234		1.01	10.20		11.21	16.85

03 11 13.45 Forms In Place, Footings

		Crew	Daily Output	Labor-Hours	Unit	Material	Labor	Equipment	Total	Total Incl O&P
0010	**FORMS IN PLACE, FOOTINGS**									
0020	Continuous wall, plywood, 1 use	C-1	375	.085	SFCA	6.20	3.63		9.83	12.40
0050	2 use		440	.073		3.40	3.09		6.49	8.50
0100	3 use		470	.068		2.47	2.90		5.37	7.20
0150	4 use		485	.066		2.01	2.81		4.82	6.55
5000	Spread footings, job-built lumber, 1 use		305	.105		2	4.46		6.46	9.05
5050	2 use		371	.086		1.11	3.67		4.78	6.85
5100	3 use		401	.080		.80	3.39		4.19	6.15
5150	4 use		414	.077		.65	3.29		3.94	5.75

03 11 13.65 Forms In Place, Slab On Grade

		Crew	Daily Output	Labor-Hours	Unit	Material	Labor	Equipment	Total	Total Incl O&P
0010	**FORMS IN PLACE, SLAB ON GRADE**									
3000	Edge forms, wood, 4 use, on grade, to 6" high	C-1	600	.053	L.F.	.31	2.27		2.58	3.83
6000	Trench forms in floor, wood, 1 use		160	.200	SFCA	1.66	8.50		10.16	14.95
6050	2 use		175	.183		.92	7.80		8.72	13
6100	3 use		180	.178		.67	7.55		8.22	12.40
6150	4 use		185	.173		.54	7.35		7.89	11.95

03 15 Concrete Accessories

03 15 05 – Concrete Forming Accessories

03 15 05.75 Sleeves and Chases

		Crew	Daily Output	Labor-Hours	Unit	Material	Labor	Equipment	Total	Total Incl O&P
0010	**SLEEVES AND CHASES**									
0100	Plastic, 1 use, 12" long, 2" diameter	1 Carp	100	.080	Ea.	2.09	3.59		5.68	7.85
0150	4" diameter		90	.089		6	3.99		9.99	12.75
0200	6" diameter		75	.107		12.20	4.79		16.99	21
0250	12" diameter		60	.133		40	6		46	53
5000	Sheet metal, 2" diameter G		100	.080		1.57	3.59		5.16	7.25
5100	4" diameter G		90	.089		1.96	3.99		5.95	8.30
5150	6" diameter G		75	.107		1.96	4.79		6.75	9.55
5200	12" diameter G		60	.133		3.90	6		9.90	13.50
6000	Steel pipe, 2" diameter G		100	.080		5.10	3.59		8.69	11.15
6100	4" diameter G		90	.089		18	3.99		21.99	26
6150	6" diameter G		75	.107		30.50	4.79		35.29	41

03 15 Concrete Accessories

03 15 05 – Concrete Forming Accessories

03 15 05.75 Sleeves and Chases			Crew	Daily Output	Labor-Hours	Unit	Material (2013 Bare Costs)	Labor (2013 Bare Costs)	Equipment (2013 Bare Costs)	Total (2013 Bare Costs)	Total Incl O&P
6200	12″ diameter	G	1 Carp	60	.133	Ea.	60.50	6		66.50	75.50

03 15 19 – Cast-In Concrete Anchors

03 15 19.10 Anchor Bolts

Line	Description		Crew	Daily Output	Labor-Hours	Unit	Material	Labor	Equipment	Total	Total Incl O&P
0010	**ANCHOR BOLTS**										
0015	Made from recycled materials										
0025	Single bolts installed in fresh concrete, no templates										
0030	Hooked w/nut and washer, 1/2″ diameter, 8″ long	G	1 Carp	132	.061	Ea.	1.43	2.72		4.15	5.75
0040	12″ long	G		131	.061		1.59	2.74		4.33	5.95
0050	5/8″ diameter, 8″ long	G		129	.062		3.11	2.78		5.89	7.70
0060	12″ long	G		127	.063		3.82	2.83		6.65	8.55
0070	3/4″ diameter, 8″ long	G		127	.063		3.82	2.83		6.65	8.55
0080	12″ long	G		125	.064		4.78	2.87		7.65	9.70

03 15 19.45 Machinery Anchors

Line	Description		Crew	Daily Output	Labor-Hours	Unit	Material	Labor	Equipment	Total	Total Incl O&P
0010	**MACHINERY ANCHORS**, heavy duty, incl. sleeve, floating base nut,										
0020	lower stud & coupling nut, fiber plug, connecting stud, washer & nut.										
0030	For flush mounted embedment in poured concrete heavy equip. pads.										
0200	Stud & bolt, 1/2″ diameter	G	E-16	40	.400	Ea.	71	20.50	3.60	95.10	119
0300	5/8″ diameter	G		35	.457		79	23.50	4.11	106.61	133
0500	3/4″ diameter	G		30	.533		91	27	4.80	122.80	154
0600	7/8″ diameter	G		25	.640		99.50	32.50	5.75	137.75	174
0800	1″ diameter	G		20	.800		104	41	7.20	152.20	196
0900	1-1/4″ diameter	G		15	1.067		139	54.50	9.60	203.10	261

03 21 Reinforcement Bars

03 21 10 – Uncoated Reinforcing Steel

03 21 10.60 Reinforcing In Place

Line	Description		Crew	Daily Output	Labor-Hours	Unit	Material	Labor	Equipment	Total	Total Incl O&P
0015	**REINFORCING IN PLACE**, 50-60 ton lots, A615 Grade 60										
0020	Includes labor, but not material cost, to install accessories										
0030	Made from recycled materials	G									
0502	Footings, #4 to #7	G	4 Rodm	4200	.008	Lb.	.50	.38		.88	1.15
0552	#8 to #18	G		7200	.004		.50	.22		.72	.90
0602	Slab on grade, #3 to #7	G		4200	.008		.50	.38		.88	1.15
0900	For other than 50 – 60 ton lots										
1000	Under 10 ton job, #3 to #7, add						25%	10%			
1010	#8 to #18, add						20%	10%			
1050	10 – 50 ton job, #3 to #7, add						10%				
1060	#8 to #18, add						5%				
1100	60 – 100 ton job, #3 to #7, deduct						5%				
1110	#8 to #18, deduct						10%				
1150	Over 100 ton job, #3 to #7, deduct						10%				
1160	#8 to #18, deduct						15%				

03 22 Fabric and Grid Reinforcing

03 22 05 – Uncoated Welded Wire Fabric

03 22 05.50 Welded Wire Fabric		Crew	Daily Output	Labor-Hours	Unit	2013 Bare Costs Material	Labor	Equipment	Total	Total Incl O&P
0010	**WELDED WIRE FABRIC** ASTM A185									
0020	Includes labor, but not material cost, to install accessories									
0050	Sheets									
0100	6 x 6 - W1.4 x W1.4 (10 x 10) 21 lb. per C.S.F. [G]	2 Rodm	35	.457	C.S.F.	14.60	23		37.60	52

03 30 Cast-In-Place Concrete

03 30 53 – Miscellaneous Cast-In-Place Concrete

03 30 53.40 Concrete In Place

03 30 53.40 Concrete In Place		Crew	Daily Output	Labor-Hours	Unit	Material	Labor	Equipment	Total	Total Incl O&P
0010	**CONCRETE IN PLACE**									
0020	Including forms (4 uses), Grade 60 rebar, concrete (Portland cement									
0050	Type I), placement and finishing unless otherwise indicated									
3540	Equipment pad (3000 psi), 3′ x 3′ x 6″ thick	C-14H	45	1.067	Ea.	44.50	47	.74	92.24	122
3550	4′ x 4′ x 6″ thick		30	1.600		66	70.50	1.10	137.60	183
3560	5′ x 5′ x 8″ thick		18	2.667		117	118	1.84	236.84	310
3570	6′ x 6′ x 8″ thick		14	3.429		157	151	2.37	310.37	410
3580	8′ x 8′ x 10″ thick		8	6		330	265	4.14	599.14	775
3590	10′ x 10′ x 12″ thick		5	9.600		560	425	6.60	991.60	1,275
3800	Footings (3000 psi), spread under 1 C.Y.	C-14C	28	4	C.Y.	162	172	1.16	335.16	445
3825	1 C.Y. to 5 C.Y.		43	2.605		197	112	.76	309.76	390
3850	Over 5 C.Y.		75	1.493		179	64	.43	243.43	296
3900	Footings, strip (3000 psi), 18″ x 9″, unreinforced	C-14L	40	2.400		117	100	.82	217.82	284
3920	18″ x 9″, reinforced	C-14C	35	3.200		141	137	.93	278.93	370
3925	20″ x 10″, unreinforced	C-14L	45	2.133		114	89	.73	203.73	263
3930	20″ x 10″, reinforced	C-14C	40	2.800		134	120	.81	254.81	335
3935	24″ x 12″, unreinforced	C-14L	55	1.745		112	73	.59	185.59	236
3940	24″ x 12″, reinforced	C-14C	48	2.333		132	100	.68	232.68	300
3945	36″ x 12″, unreinforced	C-14L	70	1.371		109	57.50	.47	166.97	209
3950	36″ x 12″, reinforced	C-14C	60	1.867		127	80	.54	207.54	264
4000	Foundation mat (3000 psi), under 10 C.Y.		38.67	2.896		202	124	.84	326.84	415
4050	Over 20 C.Y.		56.40	1.986		176	85	.58	261.58	325
4650	Slab on grade (3500 psi), not including finish, 4″ thick	C-14E	60.75	1.449		119	64	.55	183.55	231
4700	6″ thick	″	92	.957		114	42	.36	156.36	191
4701	Thickened slab edge (3500 psi), for slab on grade poured									
4702	monolithically with slab; depth is in addition to slab thickness;									
4703	formed vertical outside edge, earthen bottom and inside slope									
4705	8″ deep x 8″ wide bottom, unreinforced	C-14L	2190	.044	L.F.	3.27	1.83	.01	5.11	6.45
4710	8″ x 8″, reinforced	C-14C	1670	.067		5.60	2.88	.02	8.50	10.65
4715	12″ deep x 12″ wide bottom, unreinforced	C-14L	1800	.053		6.65	2.23	.02	8.90	10.75
4720	12″ x 12″, reinforced	C-14C	1310	.086		10.95	3.67	.02	14.64	17.75
4725	16″ deep x 16″ wide bottom, unreinforced	C-14L	1440	.067		11.20	2.78	.02	14	16.65
4730	16″ x 16″, reinforced	C-14C	1120	.100		16.35	4.29	.03	20.67	24.50
4735	20″ deep x 20″ wide bottom, unreinforced	C-14L	1150	.083		17	3.49	.03	20.52	24
4740	20″ x 20″, reinforced	C-14C	920	.122		23.50	5.20	.04	28.74	34
4745	24″ deep x 24″ wide bottom, unreinforced	C-14L	930	.103		24	4.31	.04	28.35	33
4750	24″ x 24″, reinforced	C-14C	740	.151		32.50	6.50	.04	39.04	45.50

03 31 Structural Concrete

03 31 05 – Normal Weight Structural Concrete

03 31 05.35 Normal Weight Concrete, Ready Mix		Crew	Daily Output	Labor-Hours	Unit	2013 Bare Costs Material	Labor	Equipment	Total	Total Incl O&P
0010	**NORMAL WEIGHT CONCRETE, READY MIX**, delivered									
0012	Includes local aggregate, sand, Portland cement (Type I) and water									
0015	Excludes all additives and treatments									
0020	2000 psi				C.Y.	91			91	100
0100	2500 psi					93.50			93.50	103
0150	3000 psi					97			97	107
0200	3500 psi					99			99	109
0300	4000 psi					102			102	112
1000	For high early strength (Portland cement Type III), add					10%				
1300	For winter concrete (hot water), add					4.50			4.50	4.95
1400	For hot weather concrete (ice), add					9.35			9.35	10.25
1410	For mid-range water reducer, add					3.79			3.79	4.17
1420	For high-range water reducer/superplasticizer, add					6.10			6.10	6.70
1430	For retarder, add					2.67			2.67	2.94
1440	For non-Chloride accelerator, add					4.90			4.90	5.40
1450	For Chloride accelerator, per 1%, add					3.31			3.31	3.64
1460	For fiber reinforcing, synthetic (1 lb./C.Y.), add					6.70			6.70	7.40
1500	For Saturday delivery, add					8.85			8.85	9.70
1510	For truck holding/waiting time past 1st hour per load, add				Hr.	82			82	90.50
1520	For short load (less than 4 C.Y.), add per load				Ea.	112			112	123
2000	For all lightweight aggregate, add				C.Y.	45%				

03 31 05.70 Placing Concrete		Crew	Daily Output	Labor-Hours	Unit	Material	Labor	Equipment	Total	Total Incl O&P
0010	**PLACING CONCRETE**									
0020	Includes labor and equipment to place, level (strike off) and consolidate									
1900	Footings, continuous, shallow, direct chute	C-6	120	.400	C.Y.		14.80	.55	15.35	23
1950	Pumped	C-20	150	.427			16.30	5.20	21.50	31
2000	With crane and bucket	C-7	90	.800			31	13.40	44.40	61.50
2100	Footings, continuous, deep, direct chute	C-6	140	.343			12.70	.47	13.17	19.90
2150	Pumped	C-20	160	.400			15.25	4.89	20.14	29
2200	With crane and bucket	C-7	110	.655			25.50	10.95	36.45	50.50
2400	Footings, spread, under 1 C.Y., direct chute	C-6	55	.873			32.50	1.20	33.70	51
2450	Pumped	C-20	65	.985			37.50	12.05	49.55	71
2500	With crane and bucket	C-7	45	1.600			62	27	89	124
2600	Over 5 C.Y., direct chute	C-6	120	.400			14.80	.55	15.35	23
2650	Pumped	C-20	150	.427			16.30	5.20	21.50	31
2700	With crane and bucket	C-7	100	.720			28	12.05	40.05	56
2900	Foundation mats, over 20 C.Y., direct chute	C-6	350	.137			5.10	.19	5.29	7.95
2950	Pumped	C-20	400	.160			6.10	1.96	8.06	11.45
3000	With crane and bucket	C-7	300	.240			9.25	4.02	13.27	18.50

03 35 Concrete Finishing

03 35 29 – Tooled Concrete Finishing

03 35 29.30 Finishing Floors		Crew	Daily Output	Labor-Hours	Unit	Material	Labor	Equipment	Total	Total Incl O&P
0010	**FINISHING FLOORS**									
0012	Finishing requires that concrete first be placed, struck off & consolidated									
0015	Basic finishing for various unspecified flatwork									
0100	Bull float only	C-10	4000	.006	S.F.		.24		.24	.36
0125	Bull float & manual float		2000	.012			.49		.49	.73
0150	Bull float, manual float, & broom finish, w/edging & joints		1850	.013			.53		.53	.78
0200	Bull float, manual float & manual steel trowel		1265	.019			.77		.77	1.15
0210	For specified Random Access Floors in ACI Classes 1, 2, 3 and 4 to achieve									

03 35 Concrete Finishing

03 35 29 – Tooled Concrete Finishing

03 35 29.30 Finishing Floors		Crew	Daily Output	Labor-Hours	Unit	Material	2013 Bare Costs Labor	Equipment	Total	Total Incl O&P
0215	Composite Overall Floor Flatness and Levelness values up to F35/F25									
0250	Bull float, machine float & machine trowel (walk-behind)	C-10C	1715	.014	S.F.		.57	.03	.60	.88
0300	Power screed, bull float, machine float & trowel (walk-behind)	C-10D	2400	.010			.41	.05	.46	.65
0350	Power screed, bull float, machine float & trowel (ride-on)	C-10E	4000	.006			.24	.07	.31	.43
0352	For specified Random Access Floors in ACI Classes 5, 6, 7 and 8 to achieve									
0354	Composite Overall Floor Flatness and Levelness values up to F50/F50									
0356	Add for two-dimensional restraightening after power float	C-10	6000	.004	S.F.		.16		.16	.24
0358	For specified Random or Defined Access Floors in ACI Class 9 to achieve									
0360	Composite Overall Floor Flatness and Levelness values up to F100/F100									
0362	Add for two-dimensional restraightening after bull float & power float	C-10	3000	.008	S.F.		.32		.32	.48
0364	For specified Superflat Defined Access Floors in ACI Class 9 to achieve									
0366	Minimum Floor Flatness and Levelness values of F100/F100									
0368	Add for 2-dim'l restraightening after bull float, power float, power trowel	C-10	2000	.012	S.F.		.49		.49	.73

03 35 29.35 Control Joints, Saw Cut		Crew	Daily Output	Labor-Hours	Unit	Material	Labor	Equipment	Total	Total Incl O&P
0010	**CONTROL JOINTS, SAW CUT**									
0100	Sawcut control joints in green concrete									
0120	1" depth	C-27	2000	.008	L.F.	.03	.34	.09	.46	.65
0140	1-1/2" depth		1800	.009		.05	.38	.10	.53	.72
0160	2" depth		1600	.010		.07	.43	.11	.61	.82
0180	Sawcut joint reservoir in cured concrete									
0182	3/8" wide x 3/4" deep, with single saw blade	C-27	1000	.016	L.F.	.05	.69	.18	.92	1.25
0184	1/2" wide x 1" deep, with double saw blades		900	.018		.10	.77	.19	1.06	1.45
0186	3/4" wide x 1-1/2" deep, with double saw blades		800	.020		.20	.86	.22	1.28	1.73
0190	Water blast joint to wash away laitance, 2 passes	C-29	2500	.003			.11	.03	.14	.20
0200	Air blast joint to blow out debris and air dry, 2 passes	C-28	2000	.004			.17	.01	.18	.26
0300	For backer rod, see Section 07 91 23.10									
0342	For joint sealant, see Section 07 92 13.20									

03 54 Cast Underlayment

03 54 16 – Hydraulic Cement Underlayment

03 54 16.50 Cement Underlayment		Crew	Daily Output	Labor-Hours	Unit	Material	Labor	Equipment	Total	Total Incl O&P
0010	**CEMENT UNDERLAYMENT**									
2510	Underlayment, P.C based, self-leveling, 4100 psi, pumped, 1/4" thick	C-8	20000	.003	S.F.	1.46	.11	.04	1.61	1.81
2520	1/2" thick		19000	.003		2.91	.12	.04	3.07	3.42
2530	3/4" thick		18000	.003		4.37	.12	.04	4.53	5.05
2540	1" thick		17000	.003		5.85	.13	.04	6.02	6.65
2550	1-1/2" thick		15000	.004		8.75	.15	.05	8.95	9.85
2560	Hand placed, 1/2" thick	C-18	450	.020		2.91	.71	.14	3.76	4.45
2610	Topping, P.C based, self-leveling, 6100 psi, pumped, 1/4" thick	C-8	20000	.003		2.13	.11	.04	2.28	2.56
2620	1/2" thick		19000	.003		4.27	.12	.04	4.43	4.91
2630	3/4" thick		18000	.003		6.40	.12	.04	6.56	7.30
2660	1" thick		17000	.003		8.55	.13	.04	8.72	9.65
2670	1-1/2" thick		15000	.004		12.80	.15	.05	13	14.35
2680	Hand placed, 1/2" thick	C-18	450	.020		4.27	.71	.14	5.12	5.95

03 63 Epoxy Grouting

03 63 05 – Grouting of Dowels and Fasteners

03 63 05.10 Epoxy Only		Crew	Daily Output	Labor-Hours	Unit	Material	2013 Bare Costs Labor	Equipment	Total	Total Incl O&P
0010	**EPOXY ONLY**									
1500	Chemical anchoring, epoxy cartridge, excludes layout, drilling, fastener									
1530	For fastener 3/4" diam. x 6" embedment	2 Skwk	72	.222	Ea.	4.61	10.25		14.86	21
1535	1" diam. x 8" embedment		66	.242		6.90	11.20		18.10	25
1540	1-1/4" diam. x 10" embedment		60	.267		13.85	12.30		26.15	34.50
1545	1-3/4" diam. x 12" embedment		54	.296		23	13.70		36.70	46.50
1550	14" embedment		48	.333		27.50	15.40		42.90	54.50
1555	2" diam. x 12" embedment		42	.381		37	17.60		54.60	67.50
1560	18" embedment		32	.500		46	23		69	86

03 81 Concrete Cutting

03 81 13 – Flat Concrete Sawing

03 81 13.50 Concrete Floor/Slab Cutting		Crew	Daily Output	Labor-Hours	Unit	Material	Labor	Equipment	Total	Total Incl O&P
0010	**CONCRETE FLOOR/SLAB CUTTING**									
0300	Saw cut concrete slabs, plain, up to 3" deep	B-89	1060	.015	L.F.	.12	.61	.49	1.22	1.60
0320	Each additional inch of depth		3180	.005		.04	.20	.16	.40	.53
0400	Mesh reinforced, up to 3" deep		980	.016		.13	.67	.53	1.33	1.75
0420	Each additional inch of depth		2940	.005		.04	.22	.18	.44	.59
0500	Rod reinforced, up to 3" deep		800	.020		.17	.81	.65	1.63	2.14
0520	Each additional inch of depth		2400	.007		.06	.27	.22	.55	.71

03 82 Concrete Boring

03 82 13 – Concrete Core Drilling

03 82 13.10 Core Drilling		Crew	Daily Output	Labor-Hours	Unit	Material	Labor	Equipment	Total	Total Incl O&P
0010	**CORE DRILLING**									
0020	Reinf. conc slab, up to 6" thick, incl. bit, layout & set up									
0100	1" diameter core	B-89A	17	.941	Ea.	.18	38.50	6.60	45.28	67
0150	Each added inch thick in same hole, add		1440	.011		.03	.45	.08	.56	.82
0200	2" diameter core		16.50	.970		.29	39.50	6.80	46.59	69
0250	Each added inch thick in same hole, add		1080	.015		.05	.60	.10	.75	1.09
0300	3" diameter core		16	1		.41	41	7.05	48.46	71
0350	Each added inch thick in same hole, add		720	.022		.07	.91	.16	1.14	1.65
0500	4" diameter core		15	1.067		.50	43.50	7.50	51.50	76
0550	Each added inch thick in same hole, add		480	.033		.08	1.36	.23	1.67	2.45
0700	6" diameter core		14	1.143		.83	46.50	8.05	55.38	82
0750	Each added inch thick in same hole, add		360	.044		.14	1.81	.31	2.26	3.29
0900	8" diameter core		13	1.231		1.08	50.50	8.65	60.23	88
0950	Each added inch thick in same hole, add		288	.056		.18	2.27	.39	2.84	4.13
1100	10" diameter core		12	1.333		1.58	54.50	9.35	65.43	96
1150	Each added inch thick in same hole, add		240	.067		.26	2.72	.47	3.45	5
1300	12" diameter core		11	1.455		1.73	59.50	10.25	71.48	105
1350	Each added inch thick in same hole, add		206	.078		.29	3.17	.55	4.01	5.80
1500	14" diameter core		10	1.600		2.19	65.50	11.25	78.94	116
1550	Each added inch thick in same hole, add		180	.089		.37	3.63	.62	4.62	6.70
1700	18" diameter core		9	1.778		2.93	72.50	12.50	87.93	129
1750	Each added inch thick in same hole, add		144	.111		.49	4.54	.78	5.81	8.40
1754	24" diameter core		8	2		4.13	81.50	14.05	99.68	146
1756	Each added inch thick in same hole, add		120	.133		.69	5.45	.94	7.08	10.20
1760	For horizontal holes, add to above						20%	20%		

03 82 Concrete Boring

03 82 13 – Concrete Core Drilling

03 82 13.10 Core Drilling		Crew	Daily Output	Labor-Hours	Unit	2013 Bare Costs Material	Labor	Equipment	Total	Total Incl O&P
1770	Prestressed hollow core plank, 8″ thick									
1780	1″ diameter core	B-89A	17.50	.914	Ea.	.25	37.50	6.45	44.20	65
1790	Each added inch thick in same hole, add		3840	.004		.03	.17	.03	.23	.32
1794	2″ diameter core		17.25	.928		.39	38	6.50	44.89	66
1796	Each added inch thick in same hole, add		2880	.006		.05	.23	.04	.32	.44
1800	3″ diameter core		17	.941		.55	38.50	6.60	45.65	67.50
1810	Each added inch thick in same hole, add		1920	.008		.07	.34	.06	.47	.67
1820	4″ diameter core		16.50	.970		.66	39.50	6.80	46.96	69
1830	Each added inch thick in same hole, add		1280	.013		.08	.51	.09	.68	.98
1840	6″ diameter core		15.50	1.032		1.10	42	7.25	50.35	74
1850	Each added inch thick in same hole, add		960	.017		.14	.68	.12	.94	1.33
1860	8″ diameter core		15	1.067		1.43	43.50	7.50	52.43	77
1870	Each added inch thick in same hole, add		768	.021		.18	.85	.15	1.18	1.67
1880	10″ diameter core		14	1.143		2.11	46.50	8.05	56.66	83
1890	Each added inch thick in same hole, add		640	.025		.26	1.02	.18	1.46	2.06
1900	12″ diameter core		13.50	1.185		2.30	48.50	8.35	59.15	86
1910	Each added inch thick in same hole, add	↓	548	.029	↓	.29	1.19	.21	1.69	2.39

03 82 16 – Concrete Drilling

03 82 16.10 Concrete Impact Drilling

		Crew	Daily Output	Labor-Hours	Unit	Material	Labor	Equipment	Total	Total Incl O&P
0010	**CONCRETE IMPACT DRILLING**									
0050	Up to 4″ deep in conc/brick floor/wall, incl. bit & layout, no anchor									
0100	Holes, 1/4″ diameter	1 Carp	75	.107	Ea.	.06	4.79		4.85	7.45
0150	For each additional inch of depth, add		430	.019		.01	.84		.85	1.31
0200	3/8″ diameter		63	.127		.05	5.70		5.75	8.85
0250	For each additional inch of depth, add		340	.024		.01	1.06		1.07	1.64
0300	1/2″ diameter		50	.160		.05	7.20		7.25	11.10
0350	For each additional inch of depth, add		250	.032		.01	1.44		1.45	2.22
0400	5/8″ diameter		48	.167		.08	7.50		7.58	11.65
0450	For each additional inch of depth, add		240	.033		.02	1.50		1.52	2.32
0500	3/4″ diameter		45	.178		.11	8		8.11	12.40
0550	For each additional inch of depth, add		220	.036		.03	1.63		1.66	2.54
0600	7/8″ diameter		43	.186		.14	8.35		8.49	13
0650	For each additional inch of depth, add		210	.038		.04	1.71		1.75	2.67
0700	1″ diameter		40	.200		.15	9		9.15	14
0750	For each additional inch of depth, add		190	.042		.04	1.89		1.93	2.95
0800	1-1/4″ diameter		38	.211		.23	9.45		9.68	14.80
0850	For each additional inch of depth, add		180	.044		.06	2		2.06	3.13
0900	1-1/2″ diameter		35	.229		.35	10.25		10.60	16.20
0950	For each additional inch of depth, add	↓	165	.048	↓	.09	2.18		2.27	3.45
1000	For ceiling installations, add						40%			

Estimating Tips

05 05 00 Common Work Results for Metals

- Nuts, bolts, washers, connection angles, and plates can add a significant amount to both the tonnage of a structural steel job and the estimated cost. As a rule of thumb, add 10% to the total weight to account for these accessories.
- Type 2 steel construction, commonly referred to as "simple construction," consists generally of field-bolted connections with lateral bracing supplied by other elements of the building, such as masonry walls or x-bracing. The estimator should be aware, however, that shop connections may be accomplished by welding or bolting. The method may be particular to the fabrication shop and may have an impact on the estimated cost.

05 10 00 Structural Steel

- Steel items can be obtained from two sources: a fabrication shop or a metals service center. Fabrication shops can fabricate items under more controlled conditions than can crews in the field. They are also more efficient and can produce items more economically. Metal service centers serve as a source of long mill shapes to both fabrication shops and contractors.
- Most line items in this structural steel subdivision, and most items in 05 50 00 Metal Fabrications, are indicated as being shop fabricated. The bare material cost for these shop fabricated items is the "Invoice Cost" from the shop and includes the mill base price of steel plus mill extras, transportation to the shop, shop drawings and detailing where warranted, shop fabrication and handling, sandblasting and a shop coat of primer paint, all necessary structural bolts, and delivery to the job site. The bare labor cost and bare equipment cost for these shop fabricated items is for field installation or erection.
- Line items in Subdivision 05 12 23.40 Lightweight Framing, and other items scattered in Division 5, are indicated as being field fabricated. The bare material cost for these field fabricated items is the "Invoice Cost" from the metals service center and includes the mill base price of steel plus mill extras, transportation to the metals service center, material handling, and delivery of long lengths of mill shapes to the job site. Material costs for structural bolts and welding rods should be added to the estimate. The bare labor cost and bare equipment cost for these items is for both field fabrication and field installation or erection, and include time for cutting, welding and drilling in the fabricated metal items. Drilling into concrete and fasteners to fasten field fabricated items to other work are not included and should be added to the estimate.

05 20 00 Steel Joist Framing

- In any given project the total weight of open web steel joists is determined by the loads to be supported and the design. However, economies can be realized in minimizing the amount of labor used to place the joists. This is done by maximizing the joist spacing, and therefore minimizing the number of joists required to be installed on the job. Certain spacings and locations may be required by the design, but in other cases maximizing the spacing and keeping it as uniform as possible will keep the costs down.

05 30 00 Steel Decking

- The takeoff and estimating of metal deck involves more than simply the area of the floor or roof and the type of deck specified or shown on the drawings. Many different sizes and types of openings may exist. Small openings for individual pipes or conduits may be drilled after the floor/roof is installed, but larger openings may require special deck lengths as well as reinforcing or structural support. The estimator should determine who will be supplying this reinforcing. Additionally, some deck terminations are part of the deck package, such as screed angles and pour stops, and others will be part of the steel contract, such as angles attached to structural members and cast-in-place angles and plates. The estimator must ensure that all pieces are accounted for in the complete estimate.

05 50 00 Metal Fabrications

- The most economical steel stairs are those that use common materials, standard details, and most importantly, a uniform and relatively simple method of field assembly. Commonly available A36 channels and plates are very good choices for the main stringers of the stairs, as are angles and tees for the carrier members. Risers and treads are usually made by specialty shops, and it is most economical to use a typical detail in as many places as possible. The stairs should be pre-assembled and shipped directly to the site. The field connections should be simple and straightforward to be accomplished efficiently, and with minimum equipment and labor.

Reference Numbers

Reference numbers are shown in shaded boxes at the beginning of some major classifications. These numbers refer to related items in the Reference Section. The reference information may be an estimating procedure, an alternate pricing method, or technical information. *Note:* Not all subdivisions listed here necessarily appear in this publication.

05 05 19 – Post-Installed Concrete Anchors

05 05 19.10 Chemical Anchors		Crew	Daily Output	Labor-Hours	Unit	2013 Bare Costs Material	Labor	Equipment	Total	Total Incl O&P
0010	**CHEMICAL ANCHORS**									
0020	Includes layout & drilling									
1430	Chemical anchor, w/rod & epoxy cartridge, 3/4" diam. x 9-1/2" long	B-89A	27	.593	Ea.	8.90	24	4.17	37.07	52
1435	1" diameter x 11-3/4" long		24	.667		16.75	27	4.69	48.44	65.50
1440	1-1/4" diameter x 14" long		21	.762		35.50	31	5.35	71.85	93
1445	1-3/4" diameter x 15" long		20	.800		58	32.50	5.60	96.10	120
1450	18" long		17	.941		69.50	38.50	6.60	114.60	143
1455	2" diameter x 18" long		16	1		87.50	41	7.05	135.55	167
1460	24" long		15	1.067		113	43.50	7.50	164	200

05 05 19.20 Expansion Anchors		Crew	Daily Output	Labor-Hours	Unit	Material	Labor	Equipment	Total	Total Incl O&P
0010	**EXPANSION ANCHORS**									
0100	Anchors for concrete, brick or stone, no layout and drilling									
0200	Expansion shields, zinc, 1/4" diameter, 1-5/16" long, single G	1 Carp	90	.089	Ea.	.45	3.99		4.44	6.65
0300	1-3/8" long, double G		85	.094		.55	4.23		4.78	7.10
0400	3/8" diameter, 1-1/2" long, single G		85	.094		.77	4.23		5	7.35
0500	2" long, double G		80	.100		1.24	4.49		5.73	8.25
0600	1/2" diameter, 2-1/16" long, single G		80	.100		1.47	4.49		5.96	8.50
0700	2-1/2" long, double G		75	.107		1.98	4.79		6.77	9.60
0800	5/8" diameter, 2-5/8" long, single G		75	.107		2.67	4.79		7.46	10.35
0900	2-3/4" long, double G		70	.114		2.82	5.15		7.97	11
1000	3/4" diameter, 2-3/4" long, single G		70	.114		3.07	5.15		8.22	11.30
1100	3-15/16" long, double G		65	.123		5.40	5.55		10.95	14.45
2100	Hollow wall anchors for gypsum wall board, plaster or tile									
2300	1/8" diameter, short G	1 Carp	160	.050	Ea.	.22	2.25		2.47	3.70
2400	Long G		150	.053		.21	2.39		2.60	3.92
2500	3/16" diameter, short G		150	.053		.37	2.39		2.76	4.10
2600	Long G		140	.057		.55	2.57		3.12	4.56
2700	1/4" diameter, short G		140	.057		.58	2.57		3.15	4.59
2800	Long G		130	.062		.57	2.76		3.33	4.89
3000	Toggle bolts, bright steel, 1/8" diameter, 2" long G		85	.094		.20	4.23		4.43	6.70
3100	4" long G		80	.100		.25	4.49		4.74	7.20
3400	1/4" diameter, 3" long G		75	.107		.32	4.79		5.11	7.75
3500	6" long G		70	.114		.48	5.15		5.63	8.45
3600	3/8" diameter, 3" long G		70	.114		.78	5.15		5.93	8.75
3700	6" long G		60	.133		1.13	6		7.13	10.45
3800	1/2" diameter, 4" long G		60	.133		1.66	6		7.66	11.05
3900	6" long G		50	.160		2.16	7.20		9.36	13.45
4000	Nailing anchors									
4100	Nylon nailing anchor, 1/4" diameter, 1" long	1 Carp	3.20	2.500	C	13.05	112		125.05	187
4200	1-1/2" long		2.80	2.857		17.60	128		145.60	217
4300	2" long		2.40	3.333		22.50	150		172.50	256
4400	Metal nailing anchor, 1/4" diameter, 1" long G		3.20	2.500		16.40	112		128.40	191
4500	1-1/2" long G		2.80	2.857		21	128		149	221
4600	2" long G		2.40	3.333		26	150		176	260
5000	Screw anchors for concrete, masonry,									
5100	stone & tile, no layout or drilling included									
5700	Lag screw shields, 1/4" diameter, short G	1 Carp	90	.089	Ea.	.38	3.99		4.37	6.55
5900	3/8" diameter, short G		85	.094		.69	4.23		4.92	7.25
6100	1/2" diameter, short G		80	.100		1.12	4.49		5.61	8.15
6300	5/8" diameter, short G		70	.114		1.84	5.15		6.99	9.90
6600	Lead, #6 & #8, 3/4" long G		260	.031		.17	1.38		1.55	2.32
6700	#10 - #14, 1-1/2" long G		200	.040		.27	1.80		2.07	3.07

05 05 Common Work Results for Metals

05 05 19 – Post-Installed Concrete Anchors

05 05 19.20 Expansion Anchors			Crew	Daily Output	Labor-Hours	Unit	Material	Labor (2013 Bare Costs)	Equipment	Total	Total Incl O&P
6800	#16 & #18, 1-1/2" long	G	1 Carp	160	.050	Ea.	.37	2.25		2.62	3.87
6900	Plastic, #6 & #8, 3/4" long			260	.031		.04	1.38		1.42	2.17
7100	#10 & #12, 1" long			220	.036		.04	1.63		1.67	2.55
8000	Wedge anchors, not including layout or drilling										
8050	Carbon steel, 1/4" diameter, 1-3/4" long	G	1 Carp	150	.053	Ea.	.39	2.39		2.78	4.12
8100	3-1/4" long	G		140	.057		.52	2.57		3.09	4.52
8150	3/8" diameter, 2-1/4" long	G		145	.055		.51	2.48		2.99	4.38
8200	5" long	G		140	.057		.89	2.57		3.46	4.93
8250	1/2" diameter, 2-3/4" long	G		140	.057		1.01	2.57		3.58	5.05
8300	7" long	G		125	.064		1.72	2.87		4.59	6.30
8350	5/8" diameter, 3-1/2" long	G		130	.062		1.54	2.76		4.30	5.95
8400	8-1/2" long	G		115	.070		3.28	3.12		6.40	8.40
8450	3/4" diameter, 4-1/4" long	G		115	.070		2.64	3.12		5.76	7.70
8500	10" long	G		95	.084		6	3.78		9.78	12.40
8550	1" diameter, 6" long	G		100	.080		9.05	3.59		12.64	15.50
8575	9" long	G		85	.094		11.75	4.23		15.98	19.45
8600	12" long	G		75	.107		12.70	4.79		17.49	21.50
8650	1-1/4" diameter, 9" long	G		70	.114		26.50	5.15		31.65	37
8700	12" long	G		60	.133		34	6		40	46.50
8750	For type 303 stainless steel, add						350%				
8800	For type 316 stainless steel, add						450%				
8950	Self-drilling concrete screw, hex washer head, 3/16" diam. x 1-3/4" long	G	1 Carp	300	.027	Ea.	.20	1.20		1.40	2.06
8960	2-1/4" long	G		250	.032		.21	1.44		1.65	2.44
8970	Phillips flat head, 3/16" diam. x 1-3/4" long	G		300	.027		.19	1.20		1.39	2.05
8980	2-1/4" long	G		250	.032		.21	1.44		1.65	2.44

05 05 21 – Fastening Methods for Metal

05 05 21.15 Drilling Steel		Crew	Daily Output	Labor-Hours	Unit	Material	Labor	Equipment	Total	Total Incl O&P
0010	**DRILLING STEEL**									
1910	Drilling & layout for steel, up to 1/4" deep, no anchor									
1920	Holes, 1/4" diameter	1 Sswk	112	.071	Ea.	.08	3.58		3.66	6.50
1925	For each additional 1/4" depth, add		336	.024		.08	1.19		1.27	2.22
1930	3/8" diameter		104	.077		.08	3.85		3.93	6.95
1935	For each additional 1/4" depth, add		312	.026		.08	1.28		1.36	2.38
1940	1/2" diameter		96	.083		.09	4.17		4.26	7.55
1945	For each additional 1/4" depth, add		288	.028		.09	1.39		1.48	2.58
1950	5/8" diameter		88	.091		.13	4.55		4.68	8.25
1955	For each additional 1/4" depth, add		264	.030		.13	1.52		1.65	2.85
1960	3/4" diameter		80	.100		.16	5		5.16	9.15
1965	For each additional 1/4" depth, add		240	.033		.16	1.67		1.83	3.16
1970	7/8" diameter		72	.111		.21	5.55		5.76	10.15
1975	For each additional 1/4" depth, add		216	.037		.21	1.85		2.06	3.54
1980	1" diameter		64	.125		.21	6.25		6.46	11.40
1985	For each additional 1/4" depth, add		192	.042		.21	2.09		2.30	3.96
1990	For drilling up, add						40%			

05 05 23 – Metal Fastenings

05 05 23.10 Bolts and Hex Nuts			Crew	Daily Output	Labor-Hours	Unit	Material	Labor	Equipment	Total	Total Incl O&P
0010	**BOLTS & HEX NUTS**, Steel, A307										
0100	1/4" diameter, 1/2" long	G	1 Sswk	140	.057	Ea.	.06	2.86		2.92	5.15
0200	1" long	G		140	.057		.07	2.86		2.93	5.20
0300	2" long	G		130	.062		.10	3.08		3.18	5.60
0400	3" long	G		130	.062		.15	3.08		3.23	5.65
0500	4" long	G		120	.067		.17	3.34		3.51	6.15

05 05 Common Work Results for Metals

05 05 23 – Metal Fastenings

05 05 23.10 Bolts and Hex Nuts			Crew	Daily Output	Labor-Hours	Unit	Material	Labor	Equipment	Total	Total Incl O&P
							2013 Bare Costs				
0600	3/8″ diameter, 1″ long	G	1 Sswk	130	.062	Ea.	.14	3.08		3.22	5.65
0700	2″ long	G		130	.062		.18	3.08		3.26	5.70
0800	3″ long	G		120	.067		.24	3.34		3.58	6.20
0900	4″ long	G		120	.067		.30	3.34		3.64	6.30
1000	5″ long	G		115	.070		.38	3.48		3.86	6.60
1100	1/2″ diameter, 1-1/2″ long	G		120	.067		.40	3.34		3.74	6.40
1200	2″ long	G		120	.067		.46	3.34		3.80	6.45
1300	4″ long	G		115	.070		.75	3.48		4.23	7.05
1400	6″ long	G		110	.073		1.05	3.64		4.69	7.65
1500	8″ long	G		105	.076		1.38	3.81		5.19	8.30
1600	5/8″ diameter, 1-1/2″ long	G		120	.067		1.01	3.34		4.35	7.05
1700	2″ long	G		120	.067		1.12	3.34		4.46	7.20
1800	4″ long	G		115	.070		1.62	3.48		5.10	8
1900	6″ long	G		110	.073		2.08	3.64		5.72	8.80
2000	8″ long	G		105	.076		3.09	3.81		6.90	10.20
2100	10″ long	G		100	.080		3.90	4		7.90	11.45
2200	3/4″ diameter, 2″ long	G		120	.067		1.12	3.34		4.46	7.20
2300	4″ long	G		110	.073		1.62	3.64		5.26	8.30
2400	6″ long	G		105	.076		2.09	3.81		5.90	9.10
2500	8″ long	G		95	.084		3.17	4.21		7.38	11
2600	10″ long	G		85	.094		4.17	4.71		8.88	13
2700	12″ long	G		80	.100		4.89	5		9.89	14.35
2800	1″ diameter, 3″ long	G		105	.076		2.69	3.81		6.50	9.75
2900	6″ long	G		90	.089		3.94	4.45		8.39	12.30
3000	12″ long	G		75	.107		7.10	5.35		12.45	17.35
3100	For galvanized, add						75%				
3200	For stainless, add						350%				

05 05 23.30 Lag Screws

			Crew	Daily Output	Labor-Hours	Unit	Material	Labor	Equipment	Total	Total Incl O&P
0010	**LAG SCREWS**										
0020	Steel, 1/4″ diameter, 2″ long	G	1 Carp	200	.040	Ea.	.08	1.80		1.88	2.86
0100	3/8″ diameter, 3″ long	G		150	.053		.28	2.39		2.67	4
0200	1/2″ diameter, 3″ long	G		130	.062		.58	2.76		3.34	4.90
0300	5/8″ diameter, 3″ long	G		120	.067		1.06	2.99		4.05	5.80

05 05 23.35 Machine Screws

			Crew	Daily Output	Labor-Hours	Unit	Material	Labor	Equipment	Total	Total Incl O&P
0010	**MACHINE SCREWS**										
0020	Steel, round head, #8 x 1″ long	G	1 Carp	4.80	1.667	C	4.02	75		79.02	119
0110	#8 x 2″ long	G		2.40	3.333		6.30	150		156.30	238
0200	#10 x 1″ long	G		4	2		6.20	90		96.20	145
0300	#10 x 2″ long	G		2	4		8.65	180		188.65	287

05 05 23.50 Powder Actuated Tools and Fasteners

			Crew	Daily Output	Labor-Hours	Unit	Material	Labor	Equipment	Total	Total Incl O&P
0010	**POWDER ACTUATED TOOLS & FASTENERS**										
0020	Stud driver, .22 caliber, single shot					Ea.	145			145	160
0100	.27 caliber, semi automatic, strip					″	425			425	470
0300	Powder load, single shot, .22 cal, power level 2, brown					C	5.05			5.05	5.55
0400	Strip, .27 cal, power level 4, red						7.25			7.25	8
0600	Drive pin, .300 x 3/4″ long	G	1 Carp	4.80	1.667		4.63	75		79.63	120
0700	.300 x 3″ long with washer	G	″	4	2		11.95	90		101.95	151

05 05 23.55 Rivets

			Crew	Daily Output	Labor-Hours	Unit	Material	Labor	Equipment	Total	Total Incl O&P
0010	**RIVETS**										
0100	Aluminum rivet & mandrel, 1/2″ grip length x 1/8″ diameter	G	1 Carp	4.80	1.667	C	6.60	75		81.60	122

05 05 Common Work Results for Metals

05 05 23 – Metal Fastenings

05 05 23.70 Structural Blind Bolts		Crew	Daily Output	Labor-Hours	Unit	Material	2013 Bare Costs Labor	Equipment	Total	Total Incl O&P
0010	**STRUCTURAL BLIND BOLTS**									
0100	1/4" diameter x 1/4" grip G	1 Sswk	240	.033	Ea.	1.24	1.67		2.91	4.34
0150	1/2" grip G		216	.037		1.33	1.85		3.18	4.77
0200	3/8" diameter x 1/2" grip G		232	.034		1.75	1.73		3.48	5
0250	3/4" grip G		208	.038		1.84	1.92		3.76	5.45
0300	1/2" diameter x 1/2" grip G		224	.036		3.99	1.79		5.78	7.60
0350	3/4" grip G		200	.040		5.60	2		7.60	9.70
0400	5/8" diameter x 3/4" grip G		216	.037		8.25	1.85		10.10	12.40
0450	1" grip G		192	.042		9.50	2.09		11.59	14.15

05 12 Structural Steel Framing

05 12 23 – Structural Steel for Buildings

05 12 23.60 Pipe Support Framing		Crew	Daily Output	Labor-Hours	Unit	Material	Labor	Equipment	Total	Total Incl O&P
0010	**PIPE SUPPORT FRAMING**									
0020	Under 10#/L.F., shop fabricated G	E-4	3900	.008	Lb.	1.74	.42	.04	2.20	2.70
0200	10.1 to 15#/L.F. G		4300	.007		1.72	.38	.03	2.13	2.60
0400	15.1 to 20#/L.F. G		4800	.007		1.69	.34	.03	2.06	2.49
0600	Over 20#/L.F. G		5400	.006		1.66	.30	.03	1.99	2.39

05 35 Raceway Decking Assemblies

05 35 13 – Steel Cellular Decking

05 35 13.50 Cellular Decking		Crew	Daily Output	Labor-Hours	Unit	Material	Labor	Equipment	Total	Total Incl O&P
0010	**CELLULAR DECKING**									
0015	Made from recycled materials G									
0200	Cellular units, galv, 1-1/2" deep, Type BC, 20-20 ga., over 15 squares G	E-4	1460	.022	S.F.	7.60	1.11	.10	8.81	10.50
0250	18-20 ga. G		1420	.023		8.65	1.14	.10	9.89	11.65
0300	18-18 ga. G		1390	.023		8.90	1.16	.10	10.16	11.95
0320	16-18 ga. G		1360	.024		10.60	1.19	.11	11.90	13.90
0340	16-16 ga. G		1330	.024		11.80	1.22	.11	13.13	15.25
0400	3" deep, Type NC, galvanized, 20-20 ga. G		1375	.023		8.40	1.18	.10	9.68	11.40
0500	18-20 ga. G		1350	.024		10.10	1.20	.11	11.41	13.40
0600	18-18 ga. G		1290	.025		10.10	1.25	.11	11.46	13.45
0700	16-18 ga. G		1230	.026		11.40	1.32	.12	12.84	15
0800	16-16 ga. G		1150	.028		12.40	1.41	.13	13.94	16.30
1000	4-1/2" deep, Type JC, galvanized, 18-20 ga. G		1100	.029		11.70	1.47	.13	13.30	15.60
1100	18-18 ga. G		1040	.031		11.60	1.56	.14	13.30	15.70
1200	16-18 ga. G		980	.033		13.10	1.65	.15	14.90	17.50
1300	16-16 ga. G		935	.034		14.25	1.73	.15	16.13	18.95
1900	For multi-story or congested site, add						50%			

Division Notes

	CREW	DAILY OUTPUT	LABOR-HOURS	UNIT	2012 BARE COSTS MAT.	LABOR	EQUIP.	TOTAL	TOTAL INCL O&P

Estimating Tips

06 05 00 Common Work Results for Wood, Plastics, and Composites

- Common to any wood-framed structure are the accessory connector items such as screws, nails, adhesives, hangers, connector plates, straps, angles, and hold-downs. For typical wood-framed buildings, such as residential projects, the aggregate total for these items can be significant, especially in areas where seismic loading is a concern. For floor and wall framing, the material cost is based on 10 to 25 lbs. per MBF. Hold-downs, hangers, and other connectors should be taken off by the piece.

  Included with material costs are fasteners for a normal installation. RSMeans engineers use manufacturer's recommendations, written specifications, and/or standard construction practice for size and spacing of fasteners. Prices for various fasteners are shown for informational purposes only. Adjustments should be made if unusual fastening conditions exist.

06 10 00 Carpentry

- Lumber is a traded commodity and therefore sensitive to supply and demand in the marketplace. Even in "budgetary" estimating of wood-framed projects, it is advisable to call local suppliers for the latest market pricing.
- Common quantity units for wood-framed projects are "thousand board feet" (MBF). A board foot is a volume of wood, 1" x 1' x 1', or 144 cubic inches. Board-foot quantities are generally calculated using nominal material dimensions—dressed sizes are ignored. Board foot per lineal foot of any stick of lumber can be calculated by dividing the nominal cross-sectional area by 12. As an example, 2,000 lineal feet of 2 x 12 equates to 4 MBF by dividing the nominal area, 2 x 12, by 12, which equals 2, and multiplying by 2,000 to give 4,000 board feet. This simple rule applies to all nominal dimensioned lumber.
- Waste is an issue of concern at the quantity takeoff for any area of construction. Framing lumber is sold in even foot lengths, i.e., 10', 12', 14', 16', and depending on spans, wall heights, and the grade of lumber, waste is inevitable. A rule of thumb for lumber waste is 5%–10% depending on material quality and the complexity of the framing.
- Wood in various forms and shapes is used in many projects, even where the main structural framing is steel, concrete, or masonry. Plywood as a back-up partition material and 2x boards used as blocking and cant strips around roof edges are two common examples. The estimator should ensure that the costs of all wood materials are included in the final estimate.

06 20 00 Finish Carpentry

- It is necessary to consider the grade of workmanship when estimating labor costs for erecting millwork and interior finish. In practice, there are three grades: premium, custom, and economy. The RSMeans daily output for base and case moldings is in the range of 200 to 250 L.F. per carpenter per day. This is appropriate for most average custom-grade projects. For premium projects, an adjustment to productivity of 25%–50% should be made, depending on the complexity of the job.

Reference Numbers

Reference numbers are shown in shaded boxes at the beginning of some major classifications. These numbers refer to related items in the Reference Section. The reference information may be an estimating procedure, an alternate pricing method, or technical information.

Note: Not all subdivisions listed here necessarily appear in this publication.

06 05 Common Work Results for Wood, Plastics, and Composites

06 05 23 – Wood, Plastic, and Composite Fastenings

06 05 23.10 Nails		Crew	Daily Output	Labor-Hours	Unit	2013 Bare Costs Material	Labor	Equipment	Total	Total Incl O&P
0010	**NAILS**, material only, based upon 50# box purchase									
0020	Copper nails, plain				Lb.	10.10			10.10	11.15
0400	Stainless steel, plain					7.90			7.90	8.65
0500	Box, 3d to 20d, bright					1.26			1.26	1.39
0520	Galvanized					1.79			1.79	1.97
0600	Common, 3d to 60d, plain					1.31			1.31	1.44
0700	Galvanized					1.58			1.58	1.74
0800	Aluminum					5.85			5.85	6.45
1000	Annular or spiral thread, 4d to 60d, plain					1.82			1.82	2
1200	Galvanized					3.03			3.03	3.33
1400	Drywall nails, plain					3.63			3.63	3.99
1600	Galvanized					2.21			2.21	2.43
1800	Finish nails, 4d to 10d, plain					1.37			1.37	1.51
2000	Galvanized					1.81			1.81	1.99
2100	Aluminum					5.85			5.85	6.45
2300	Flooring nails, hardened steel, 2d to 10d, plain					3.13			3.13	3.44
2400	Galvanized					4.04			4.04	4.44
2500	Gypsum lath nails, 1-1/8", 13 ga. flathead, blued					3.63			3.63	3.99
2600	Masonry nails, hardened steel, 3/4" to 3" long, plain					2.22			2.22	2.44
2700	Galvanized					3.79			3.79	4.17
3600	Siding nails, plain shank, galvanized					2.02			2.02	2.22
3800	Aluminum					5.85			5.85	6.45
5000	Add to prices above for cement coating					.11			.11	.12
5200	Zinc or tin plating					.14			.14	.15
5500	Vinyl coated sinkers, 8d to 16d				↓	.66			.66	.73

06 05 23.40 Sheet Metal Screws		Crew	Daily Output	Labor-Hours	Unit	Material	Labor	Equipment	Total	Total Incl O&P
0010	**SHEET METAL SCREWS**									
0020	Steel, standard, #8 x 3/4", plain				C	2.95			2.95	3.25
0100	Galvanized					2.95			2.95	3.25
0300	#10 x 1", plain					4.07			4.07	4.48
0400	Galvanized					4.07			4.07	4.48
0600	With washers, #14 x 1", plain					12.45			12.45	13.70
0700	Galvanized					12.45			12.45	13.70
0900	#14 x 2", plain					17.05			17.05	18.80
1000	Galvanized					17.05			17.05	18.80
1500	Self-drilling, with washers, (pinch point) #8 x 3/4", plain					7.85			7.85	8.65
1600	Galvanized					7.85			7.85	8.65
1800	#10 x 3/4", plain					9			9	9.90
1900	Galvanized					9			9	9.90
3000	Stainless steel w/aluminum or neoprene washers, #14 x 1", plain					30.50			30.50	33.50
3100	#14 x 2", plain				↓	40.50			40.50	44.50

06 05 23.50 Wood Screws		Crew	Daily Output	Labor-Hours	Unit	Material	Labor	Equipment	Total	Total Incl O&P
0010	**WOOD SCREWS**									
0020	#8 x 1" long, steel				C	3.44			3.44	3.78
0100	Brass					10.65			10.65	11.70
0200	#8, 2" long, steel					5.30			5.30	5.85
0300	Brass					20			20	22
0400	#10, 1" long, steel					3.69			3.69	4.06
0500	Brass					14.30			14.30	15.75
0600	#10, 2" long, steel					5.95			5.95	6.50
0700	Brass					24.50			24.50	27
0800	#10, 3" long, steel					9.90			9.90	10.90

06 05 Common Work Results for Wood, Plastics, and Composites

06 05 23 – Wood, Plastic, and Composite Fastenings

06 05 23.50 Wood Screws		Crew	Daily Output	Labor-Hours	Unit	2013 Bare Costs Material	Labor	Equipment	Total	Total Incl O&P
1000	#12, 2" long, steel				C	8.10			8.10	8.95
1100	Brass					29.50			29.50	32.50
1500	#12, 3" long, steel					11.75			11.75	12.95
2000	#12, 4" long, steel				▼	21.50			21.50	24

06 16 Sheathing

06 16 36 – Wood Panel Product Sheathing

06 16 36.10 Sheathing		Crew	Daily Output	Labor-Hours	Unit	Material	Labor	Equipment	Total	Total Incl O&P
0010	**SHEATHING**									
0012	Plywood on roofs, CDX									
0030	5/16" thick	2 Carp	1600	.010	S.F.	.55	.45		1	1.29
0035	Pneumatic nailed		1952	.008		.55	.37		.92	1.17
0050	3/8" thick		1525	.010		.51	.47		.98	1.29
0055	Pneumatic nailed		1860	.009		.51	.39		.90	1.15
0100	1/2" thick		1400	.011		.58	.51		1.09	1.43
0105	Pneumatic nailed		1708	.009		.58	.42		1	1.29
0200	5/8" thick		1300	.012		.73	.55		1.28	1.65
0205	Pneumatic nailed		1586	.010		.73	.45		1.18	1.50
0300	3/4" thick		1200	.013		.77	.60		1.37	1.77
0305	Pneumatic nailed		1464	.011		.77	.49		1.26	1.61
0500	Plywood on walls with exterior CDX, 3/8" thick		1200	.013		.51	.60		1.11	1.48
0505	Pneumatic nailed		1488	.011		.51	.48		.99	1.30
0600	1/2" thick		1125	.014		.58	.64		1.22	1.62
0605	Pneumatic nailed		1395	.011		.58	.52		1.10	1.43
0700	5/8" thick		1050	.015		.73	.68		1.41	1.85
0705	Pneumatic nailed		1302	.012		.73	.55		1.28	1.65
0800	3/4" thick		975	.016		.77	.74		1.51	1.98
0805	Pneumatic nailed	▼	1209	.013	▼	.77	.59		1.36	1.76

Division Notes

	CREW	DAILY OUTPUT	LABOR-HOURS	UNIT	2012 BARE COSTS MAT.	LABOR	EQUIP.	TOTAL	TOTAL INCL O&P

Estimating Tips

07 10 00 Dampproofing and Waterproofing

- Be sure of the job specifications before pricing this subdivision. The difference in cost between waterproofing and dampproofing can be great. Waterproofing will hold back standing water. Dampproofing prevents the transmission of water vapor. Also included in this section are vapor retarding membranes.

07 20 00 Thermal Protection

- Insulation and fireproofing products are measured by area, thickness, volume or R-value. Specifications may give only what the specific R-value should be in a certain situation. The estimator may need to choose the type of insulation to meet that R-value.

07 30 00 Steep Slope Roofing

07 40 00 Roofing and Siding Panels

- Many roofing and siding products are bought and sold by the square. One square is equal to an area that measures 100 square feet.

This simple change in unit of measure could create a large error if the estimator is not observant. Accessories necessary for a complete installation must be figured into any calculations for both material and labor.

07 50 00 Membrane Roofing

07 60 00 Flashing and Sheet Metal

07 70 00 Roofing and Wall Specialties and Accessories

- The items in these subdivisions compose a roofing system. No one component completes the installation, and all must be estimated. Built-up or single-ply membrane roofing systems are made up of many products and installation trades. Wood blocking at roof perimeters or penetrations, parapet coverings, reglets, roof drains, gutters, downspouts, sheet metal flashing, skylights, smoke vents, and roof hatches all need to be considered along with the roofing material. Several different installation trades will need to work together on the roofing system. Inherent difficulties in the scheduling and coordination of various trades must be accounted for when estimating labor costs.

07 90 00 Joint Protection

- To complete the weather-tight shell, the sealants and caulkings must be estimated. Where different materials meet—at expansion joints, at flashing penetrations, and at hundreds of other locations throughout a construction project—they provide another line of defense against water penetration. Often, an entire system is based on the proper location and placement of caulking or sealants. The detailed drawings that are included as part of a set of architectural plans show typical locations for these materials. When caulking or sealants are shown at typical locations, this means the estimator must include them for all the locations where this detail is applicable. Be careful to keep different types of sealants separate, and remember to consider backer rods and primers if necessary.

Reference Numbers

Reference numbers are shown in shaded boxes at the beginning of some major classifications. These numbers refer to related items in the Reference Section. The reference information may be an estimating procedure, an alternate pricing method, or technical information.

Note: Not all subdivisions listed here necessarily appear in this publication.

07 84 Firestopping

07 84 13 – Penetration Firestopping

07 84 13.10 Firestopping		Crew	Daily Output	Labor-Hours	Unit	2013 Bare Costs Material	Labor	Equipment	Total	Total Incl O&P
0010	**FIRESTOPPING** R078413-30									
0100	Metallic piping, non insulated									
0110	Through walls, 2″ diameter	1 Carp	16	.500	Ea.	15.60	22.50		38.10	51.50
0120	4″ diameter		14	.571		24	25.50		49.50	65.50
0130	6″ diameter		12	.667		32	30		62	81.50
0140	12″ diameter		10	.800		56.50	36		92.50	118
0150	Through floors, 2″ diameter		32	.250		9.45	11.25		20.70	27.50
0160	4″ diameter		28	.286		13.60	12.85		26.45	34.50
0170	6″ diameter		24	.333		17.85	14.95		32.80	42.50
0180	12″ diameter		20	.400		30	17.95		47.95	60.50
0190	Metallic piping, insulated									
0200	Through walls, 2″ diameter	1 Carp	16	.500	Ea.	22	22.50		44.50	59
0210	4″ diameter		14	.571		30	25.50		55.50	72.50
0220	6″ diameter		12	.667		38.50	30		68.50	88
0230	12″ diameter		10	.800		63	36		99	125
0240	Through floors, 2″ diameter		32	.250		16	11.25		27.25	35
0250	4″ diameter		28	.286		20	12.85		32.85	42
0260	6″ diameter		24	.333		24.50	14.95		39.45	50
0270	12″ diameter		20	.400		30	17.95		47.95	60.50
0280	Non metallic piping, non insulated									
0290	Through walls, 2″ diameter	1 Carp	12	.667	Ea.	64	30		94	117
0300	4″ diameter		10	.800		80.50	36		116.50	144
0310	6″ diameter		8	1		112	45		157	192
0330	Through floors, 2″ diameter		16	.500		50.50	22.50		73	90
0340	4″ diameter		6	1.333		62.50	60		122.50	161
0350	6″ diameter		6	1.333		75	60		135	175
0370	Ductwork, insulated & non insulated, round									
0380	Through walls, 6″ diameter	1 Carp	12	.667	Ea.	32.50	30		62.50	82
0390	12″ diameter		10	.800		65	36		101	127
0400	18″ diameter		8	1		106	45		151	185
0410	Through floors, 6″ diameter		16	.500		17.85	22.50		40.35	54
0420	12″ diameter		14	.571		32.50	25.50		58	75.50
0430	18″ diameter		12	.667		56.50	30		86.50	109
0440	Ductwork, insulated & non insulated, rectangular									
0450	With stiffener/closure angle, through walls, 6″ x 12″	1 Carp	8	1	Ea.	27	45		72	98.50
0460	12″ x 24″		6	1.333		36	60		96	132
0470	24″ x 48″		4	2		103	90		193	251
0480	With stiffener/closure angle, through floors, 6″ x 12″		10	.800		14.65	36		50.65	71.50
0490	12″ x 24″		8	1		26.50	45		71.50	98
0500	24″ x 48″		6	1.333		51.50	60		111.50	149
0510	Multi trade openings									
0520	Through walls, 6″ x 12″	1 Carp	2	4	Ea.	56.50	180		236.50	340
0530	12″ x 24″	″	1	8		229	360		589	805
0540	24″ x 48″	2 Carp	1	16		920	720		1,640	2,100
0550	48″ x 96″	″	.75	21.333		3,675	960		4,635	5,525
0560	Through floors, 6″ x 12″	1 Carp	2	4		56.50	180		236.50	340
0570	12″ x 24″	″	1	8		229	360		589	805
0580	24″ x 48″	2 Carp	.75	21.333		920	960		1,880	2,475
0590	48″ x 96″	″	.50	32		3,675	1,425		5,100	6,275
0600	Structural penetrations, through walls									
0610	Steel beams, W8 x 10	1 Carp	8	1	Ea.	36	45		81	109
0620	W12 x 14		6	1.333		56.50	60		116.50	155
0630	W21 x 44		5	1.600		114	72		186	236

07 84 Firestopping

07 84 13 – Penetration Firestopping

07 84 13.10 Firestopping		Crew	Daily Output	Labor-Hours	Unit	Material	2013 Bare Costs Labor	Equipment	Total	Total Incl O&P
0640	W36 x 135	1 Carp	3	2.667	Ea.	277	120		397	490
0650	Bar joists, 18" deep		6	1.333		52	60		112	150
0660	24" deep		6	1.333		65	60		125	164
0670	36" deep		5	1.600		97.50	72		169.50	218
0680	48" deep	↓	4	2	↓	114	90		204	263
0690	Construction joints, floor slab at exterior wall									
0700	Precast, brick, block or drywall exterior									
0710	2" wide joint	1 Carp	125	.064	L.F.	8.10	2.87		10.97	13.40
0720	4" wide joint	"	75	.107	"	16.25	4.79		21.04	25.50
0730	Metal panel, glass or curtain wall exterior									
0740	2" wide joint	1 Carp	40	.200	L.F.	19.25	9		28.25	35
0750	4" wide joint	"	25	.320	"	26.50	14.35		40.85	51
0760	Floor slab to drywall partition									
0770	Flat joint	1 Carp	100	.080	L.F.	8	3.59		11.59	14.35
0780	Fluted joint		50	.160		16.25	7.20		23.45	29
0790	Etched fluted joint	↓	75	.107	↓	10.55	4.79		15.34	19
0800	Floor slab to concrete/masonry partition									
0810	Flat joint	1 Carp	75	.107	L.F.	17.85	4.79		22.64	27
0820	Fluted joint	"	50	.160	"	21	7.20		28.20	34
0830	Concrete/CMU wall joints									
0840	1" wide	1 Carp	100	.080	L.F.	9.75	3.59		13.34	16.25
0850	2" wide		75	.107		17.85	4.79		22.64	27
0860	4" wide	↓	50	.160	↓	34	7.20		41.20	48.50
0870	Concrete/CMU floor joints									
0880	1" wide	1 Carp	200	.040	L.F.	4.89	1.80		6.69	8.15
0890	2" wide		150	.053		8.90	2.39		11.29	13.50
0900	4" wide	↓	100	.080	↓	17.05	3.59		20.64	24.50

07 91 Preformed Joint Seals

07 91 13 – Compression Seals

07 91 13.10 Compression Seals		Crew	Daily Output	Labor-Hours	Unit	Material	Labor	Equipment	Total	Total Incl O&P
0010	**COMPRESSION SEALS**									
4900	O-ring type cord, 1/4"	1 Bric	472	.017	L.F.	.48	.76		1.24	1.68
4910	1/2"		440	.018		1.79	.81		2.60	3.20
4920	3/4"		424	.019		3.62	.85		4.47	5.25
4930	1"		408	.020		6.95	.88		7.83	9
4940	1-1/4"		384	.021		13.75	.93		14.68	16.50
4950	1-1/2"		368	.022		17.40	.97		18.37	20.50
4960	1-3/4"		352	.023		36.50	1.02		37.52	41.50
4970	2"	↓	344	.023	↓	64.50	1.04		65.54	72

07 91 23 – Backer Rods

07 91 23.10 Backer Rods		Crew	Daily Output	Labor-Hours	Unit	Material	Labor	Equipment	Total	Total Incl O&P
0010	**BACKER RODS**									
0030	Backer rod, polyethylene, 1/4" diameter	1 Bric	4.60	1.739	C.L.F.	2.75	78		80.75	121
0050	1/2" diameter		4.60	1.739		5.35	78		83.35	124
0070	3/4" diameter		4.60	1.739		5.20	78		83.20	124
0090	1" diameter	↓	4.60	1.739	↓	10.75	78		88.75	130

07 91 Preformed Joint Seals

07 91 26 – Joint Fillers

07 91 26.10 Joint Fillers		Crew	Daily Output	Labor-Hours	Unit	Material	2013 Bare Costs Labor	Equipment	Total	Total Incl O&P
0010	JOINT FILLERS									
4360	Butyl rubber filler, 1/4" x 1/4"	1 Bric	290	.028	L.F.	.15	1.24		1.39	2.03
4365	1/2" x 1/2"		250	.032		.59	1.43		2.02	2.82
4370	1/2" x 3/4"		210	.038		.89	1.71		2.60	3.57
4375	3/4" x 3/4"		230	.035		1.33	1.56		2.89	3.83
4380	1" x 1"		180	.044		1.78	1.99		3.77	4.98
4390	For coloring, add					12%				
4980	Polyethylene joint backing, 1/4" x 2"	1 Bric	2.08	3.846	C.L.F.	12	172		184	274
4990	1/4" x 6"		1.28	6.250	"	28	280		308	455
5600	Silicone, room temp vulcanizing foam seal, 1/4" x 1/2"		1312	.006	L.F.	.33	.27		.60	.78
5610	1/2" x 1/2"		656	.012		.67	.55		1.22	1.57
5620	1/2" x 3/4"		442	.018		1	.81		1.81	2.33
5630	3/4" x 3/4"		328	.024		1.50	1.09		2.59	3.30
5640	1/8" x 1"		1312	.006		.33	.27		.60	.78
5650	1/8" x 3"		442	.018		1	.81		1.81	2.33
5670	1/4" x 3"		295	.027		2.01	1.22		3.23	4.05
5680	1/4" x 6"		148	.054		4.01	2.42		6.43	8.10
5690	1/2" x 6"		82	.098		8	4.37		12.37	15.40
5700	1/2" x 9"		52.50	.152		12.05	6.85		18.90	23.50
5710	1/2" x 12"		33	.242		16.05	10.85		26.90	34

07 92 Joint Sealants

07 92 13 – Elastomeric Joint Sealants

07 92 13.20 Caulking and Sealant Options

Code	Description	Crew	Daily Output	Labor-Hours	Unit	Material	Labor	Equipment	Total	Total Incl O&P
0010	CAULKING AND SEALANT OPTIONS									
0050	Latex acrylic based, bulk				Gal.	29.50			29.50	32.50
0055	Bulk in place 1/4" x 1/4" bead	1 Bric	300	.027	L.F.	.09	1.19		1.28	1.91
0060	1/4" x 3/8"		294	.027		.16	1.22		1.38	2.02
0065	1/4" x 1/2"		288	.028		.21	1.24		1.45	2.11
0075	3/8" x 3/8"		284	.028		.23	1.26		1.49	2.16
0080	3/8" x 1/2"		280	.029		.31	1.28		1.59	2.28
0085	3/8" x 5/8"		276	.029		.39	1.30		1.69	2.40
0095	3/8" x 3/4"		272	.029		.46	1.32		1.78	2.51
0100	1/2" x 1/2"		275	.029		.41	1.30		1.71	2.43
0105	1/2" x 5/8"		269	.030		.52	1.33		1.85	2.59
0110	1/2" x 3/4"		263	.030		.62	1.36		1.98	2.74
0115	1/2" x 7/8"		256	.031		.72	1.40		2.12	2.92
0120	1/2" x 1"		250	.032		.83	1.43		2.26	3.08
0125	3/4" x 3/4"		244	.033		.93	1.47		2.40	3.25
0130	3/4" x 1"		225	.036		1.24	1.59		2.83	3.78
0135	1" x 1"		200	.040		1.66	1.79		3.45	4.53
3200	Polyurethane, 1 or 2 component				Gal.	51			51	56
3300	Cartridges				"	70			70	77
3500	Bulk, in place, 1/4" x 1/4"	1 Bric	300	.027	L.F.	.17	1.19		1.36	1.99
3655	1/2" x 1/4"		288	.028		.33	1.24		1.57	2.24
3800	3/4" x 3/8"		272	.029		.75	1.32		2.07	2.83
3900	1" x 1/2"		250	.032		1.33	1.43		2.76	3.63

07 92 Joint Sealants

07 92 19 – Acoustical Joint Sealants

07 92 19.10 Acoustical Sealant		Crew	Daily Output	Labor-Hours	Unit	Material	2013 Bare Costs Labor	Equipment	Total	Total Incl O&P
0010	**ACOUSTICAL SEALANT**									
0020	Acoustical sealant, elastomeric, cartridges				Ea.	6.25			6.25	6.90
0025	In place, 1/4" x 1/4"	1 Bric	300	.027	L.F.	.26	1.19		1.45	2.09
0030	1/4" x 1/2"		288	.028		.51	1.24		1.75	2.44
0035	1/2" x 1/2"		275	.029		1.02	1.30		2.32	3.09
0040	1/2" x 3/4"		263	.030		1.53	1.36		2.89	3.75
0045	3/4" x 3/4"		244	.033		2.30	1.47		3.77	4.75
0050	1" x 1"		200	.040		4.09	1.79		5.88	7.20

Division Notes

		CREW	DAILY OUTPUT	LABOR-HOURS	UNIT	2012 BARE COSTS MAT.	LABOR	EQUIP.	TOTAL	TOTAL INCL O&P

Estimating Tips

08 10 00 Doors and Frames

All exterior doors should be addressed for their energy conservation (insulation and seals).

- Most metal doors and frames look alike, but there may be significant differences among them. When estimating these items, be sure to choose the line item that most closely compares to the specification or door schedule requirements regarding:
  - — type of metal
  - — metal gauge
  - — door core material
  - — fire rating
  - — finish
- Wood and plastic doors vary considerably in price. The primary determinant is the veneer material. Lauan, birch, and oak are the most common veneers. Other variables include the following:
  - — hollow or solid core
  - — fire rating
  - — flush or raised panel
  - — finish
- Door pricing includes bore for cylindrical lockset and mortise for hinges.

08 30 00 Specialty Doors and Frames

- There are many varieties of special doors, and they are usually priced per each. Add frames, hardware, or operators required for a complete installation.

08 40 00 Entrances, Storefronts, and Curtain Walls

- Glazed curtain walls consist of the metal tube framing and the glazing material. The cost data in this subdivision is presented for the metal tube framing alone or the composite wall. If your estimate requires a detailed takeoff of the framing, be sure to add the glazing cost.

08 50 00 Windows

- Most metal windows are delivered preglazed. However, some metal windows are priced without glass. Refer to 08 80 00 Glazing for glass pricing. The grade C indicates commercial grade windows, usually ASTM C-35.
- All wood windows and vinyl are priced preglazed. The glazing is insulating glass. Some wood windows may have single pane float glass. Add the cost of screens and grills if required, and not already included.

08 70 00 Hardware

- Hardware costs add considerably to the cost of a door. The most efficient method to determine the hardware requirements for a project is to review the door schedule.
- Door hinges are priced by the pair, with most doors requiring 1-1/2 pairs per door. The hinge prices do not include installation labor because it is included in door installation. Hinges are classified according to the frequency of use.

08 80 00 Glazing

- Different openings require different types of glass. The most common types are:
  - — float
  - — tempered
  - — insulating
  - — impact-resistant
  - — ballistic-resistant
- Most exterior windows are glazed with insulating glass. Entrance doors and window walls, where the glass is less than 18″ from the floor, are generally glazed with tempered glass. Interior windows and some residential windows are glazed with float glass.
- Coastal communities require the use of impact-resistant glass, dependant on wind speed.
- The insulation or 'u' value is a strong consideration, along with solar heat gain, to determine total energy efficiency.

Reference Numbers

Reference numbers are shown in shaded boxes at the beginning of some major classifications. These numbers refer to related items in the Reference Section. The reference information may be an estimating procedure, an alternate pricing method, or technical information.

Note: Not all subdivisions listed here necessarily appear in this publication.

08 31 Access Doors and Panels

08 31 13 – Access Doors and Frames

08 31 13.10 Types of Framed Access Doors		Crew	Daily Output	Labor-Hours	Unit	2013 Bare Costs Material	Labor	Equipment	Total	Total Incl O&P
0010	**TYPES OF FRAMED ACCESS DOORS**									
4000	Recessed door for drywall									
4100	Metal 12″ x 12″	1 Carp	6	1.333	Ea.	80	60		140	180
4150	12″ x 24″		5.50	1.455		114	65.50		179.50	226
4200	24″ x 36″		5	1.600		182	72		254	310

08 74 Access Control Hardware

08 74 13 – Card Key Access Control Hardware

08 74 13.50 Card Key Access

		Crew	Daily Output	Labor-Hours	Unit	Material	Labor	Equipment	Total	Total Incl O&P
0010	**CARD KEY ACCESS**, does not include door hardware									
0020	Card type, 1 time zone, minimum				Ea.	685			685	750
0040	Maximum					1,775			1,775	1,950
0060	3 time zones, minimum					2,225			2,225	2,425
0080	Maximum					2,475			2,475	2,725
0100	System with printer, and control console, 3 zones				Total	9,275			9,275	10,200
0120	6 zones				″	12,100			12,100	13,300
0140	For each door, minimum, add				Ea.	1,525			1,525	1,675
0160	Maximum, add					2,125			2,125	2,350
0320	Card reader, int/ext, basic, excl striker/power/wiring	1 Elec	3	2.667		300	140		440	540
0330	Tag reader, int/ext, basic, excl striker/power/wiring	″	3	2.667		200	140		340	430

08 74 13.60 Entrance Card Systems

		Crew	Daily Output	Labor-Hours	Unit	Material	Labor	Equipment	Total	Total Incl O&P
0010	**ENTRANCE CARD SYSTEMS**									
0100	Entrance card, barium ferrite				Ea.	9			9	9.90
0120	Credential					8			8	8.80
0140	Proximity					8.50			8.50	9.35
0160	Weigand					8.75			8.75	9.65
0500	Entrance card reader, barium ferrite	R-19	4	5		290	263		553	715
0520	Credential		4	5		360	263		623	790
0540	Proximity		4	5		310	263		573	735
0560	Weigand		4	5		330	263		593	760
0600	Local processor for card system		4	5		1,900	263		2,163	2,500
0650	Scanner, eye retina		4	5		6,200	263		6,463	7,225
0700	Gate opener, cantilever	R-18	3	8.667		2,800	355		3,155	3,625
0710	Average		3	8.667		1,400	355		1,755	2,100
0712	Custom		3	8.667		1,900	355		2,255	2,650
0750	Switch, tamper	R-19	6	3.333		22	175		197	286
0900	Accessories, electric door strike/bolt		5	4		77.50	210		287.50	400
0920	Electromagnetic lock		5	4		159	210		369	490
0940	Keypad for card reader		4	5		640	263		903	1,100

08 74 16 – Keypad Access Control Hardware

08 74 16.50 Keypad Access

		Crew	Daily Output	Labor-Hours	Unit	Material	Labor	Equipment	Total	Total Incl O&P
0010	**KEYPAD ACCESS**									
0340	Digital keypad, int/ext, basic, excl. striker/power/wiring	1 Elec	3	2.667	Ea.	63	140		203	279
0350	Lockset, mechanical push-button type, complete, incl hardware	1 Carp	4	2	″	330	90		420	505

08 74 19 – Biometric Identity Access Control Hardware

08 74 19.50 Biometric Identity Access

		Crew	Daily Output	Labor-Hours	Unit	Material	Labor	Equipment	Total	Total Incl O&P
0010	**BIOMETRIC IDENTITY ACCESS**									
0200	Fingerprint scanner unit, excl striker/power supply	1 Elec	4	2	Ea.	1,400	105		1,505	1,700
0210	Fingerprint scanner unit, for computer keyboard access		8	1		1,100	52.50		1,152.50	1,275
0220	Hand geometry scanner, mem of 512 users, excl striker/power		3	2.667		2,100	140		2,240	2,500

08 74 Access Control Hardware

08 74 19 – Biometric Identity Access Control Hardware

08 74 19.50 Biometric Identity Access		Crew	Daily Output	Labor-Hours	Unit	Material	2013 Bare Costs Labor	Equipment	Total	Total Incl O&P
0230	Memory upgrade for, adds 9,700 user profiles	1 Elec	8	1	Ea.	300	52.50		352.50	410
0240	Adds 32,500 user profiles		8	1		600	52.50		652.50	740
0250	Prison type, memory of 256 users, excl striker, power		3	2.667		2,600	140		2,740	3,050
0260	Memory upgrade for, adds 3,300 user profiles		8	1		250	52.50		302.50	355
0270	Adds 9,700 user profiles		8	1		460	52.50		512.50	585
0280	Adds 27,900 user profiles		8	1		610	52.50		662.50	750
0290	All weather, mem of 512 users, excl striker/power		3	2.667		3,900	140		4,040	4,500
0300	Facial & fingerprint scanner, combination unit, excl striker/power		3	2.667		4,200	140		4,340	4,825
0310	Access for, for initial setup, excl striker/power		3	2.667		1,100	140		1,240	1,400

08 74 23 – Access Control Accessories

08 74 23.50 Security Access Control Accessories		Crew	Daily Output	Labor-Hours	Unit	Material	Labor	Equipment	Total	Total Incl O&P
0010	**SECURITY ACCESS CONTROL ACCESSORIES**									
0360	Scanner/reader access, power supply/transf, 110 V to 12/24 V	1 Elec	4	2	Ea.	300	105		405	485
0370	Elec/mag strikers, 12/24V		4	2		80	105		185	245
0380	1 hour battery backup power supply		4	2		45	105		150	207
0390	Deadbolt, digital, batt-operated, indoor/outdoor, complete, incl hardware	1 Carp	4	2		148	90		238	300

Division Notes

	CREW	DAILY OUTPUT	LABOR-HOURS	UNIT	2012 BARE COSTS MAT.	LABOR	EQUIP.	TOTAL	TOTAL INCL O&P

Estimating Tips

General

- Room Finish Schedule: A complete set of plans should contain a room finish schedule. If one is not available, it would be well worth the time and effort to obtain one.

09 20 00 Plaster and Gypsum Board

- Lath is estimated by the square yard plus a 5% allowance for waste. Furring, channels, and accessories are measured by the linear foot. An extra foot should be allowed for each accessory miter or stop.
- Plaster is also estimated by the square yard. Deductions for openings vary by preference, from zero deduction to 50% of all openings over 2 feet in width. The estimator should allow one extra square foot for each linear foot of horizontal interior or exterior angle located below the ceiling level. Also, double the areas of small radius work.
- Drywall accessories, studs, track, and acoustical caulking are all measured by the linear foot. Drywall taping is figured by the square foot. Gypsum wallboard is estimated by the square foot. No material deductions should be made for door or window openings under 32 S.F.

09 60 00 Flooring

- Tile and terrazzo areas are taken off on a square foot basis. Trim and base materials are measured by the linear foot. Accent tiles are listed per each. Two basic methods of installation are used. Mud set is approximately 30% more expensive than thin set. In terrazzo work, be sure to include the linear footage of embedded decorative strips, grounds, machine rubbing, and power cleanup.
- Wood flooring is available in strip, parquet, or block configuration. The latter two types are set in adhesives with quantities estimated by the square foot. The laying pattern will influence labor costs and material waste. In addition to the material and labor for laying wood floors, the estimator must make allowances for sanding and finishing these areas unless the flooring is prefinished.
- Sheet flooring is measured by the square yard. Roll widths vary, so consideration should be given to use the most economical width, as waste must be figured into the total quantity. Consider also the installation methods available, direct glue down or stretched.

09 70 00 Wall Finishes

- Wall coverings are estimated by the square foot. The area to be covered is measured, length by height of wall above baseboards, to calculate the square footage of each wall. This figure is divided by the number of square feet in the single roll which is being used. Deduct, in full, the areas of openings such as doors and windows. Where a pattern match is required allow 25%–30% waste.

09 80 00 Acoustic Treatment

- Acoustical systems fall into several categories. The takeoff of these materials should be by the square foot of area with a 5% allowance for waste. Do not forget about scaffolding, if applicable, when estimating these systems.

09 90 00 Painting and Coating

- A major portion of the work in painting involves surface preparation. Be sure to include cleaning, sanding, filling, and masking costs in the estimate.
- Protection of adjacent surfaces is not included in painting costs. When considering the method of paint application, an important factor is the amount of protection and masking required. These must be estimated separately and may be the determining factor in choosing the method of application.

Reference Numbers

Reference numbers are shown in shaded boxes at the beginning of some major classifications. These numbers refer to related items in the Reference Section. The reference information may be an estimating procedure, an alternate pricing method, or technical information.

Note: Not all subdivisions listed here necessarily appear in this publication.

09 22 Supports for Plaster and Gypsum Board

09 22 03 – Fastening Methods for Finishes

09 22 03.20 Drilling Plaster/Drywall		Crew	Daily Output	Labor-Hours	Unit	2013 Bare Costs Material	Labor	Equipment	Total	Total Incl O&P
0010	DRILLING PLASTER/DRYWALL									
1100	Drilling & layout for drywall/plaster walls, up to 1" deep, no anchor									
1200	Holes, 1/4" diameter	1 Carp	150	.053	Ea.	.01	2.39		2.40	3.70
1300	3/8" diameter		140	.057		.01	2.57		2.58	3.96
1400	1/2" diameter		130	.062		.01	2.76		2.77	4.27
1500	3/4" diameter		120	.067		.01	2.99		3	4.62
1600	1" diameter		110	.073		.02	3.27		3.29	5.05
1700	1-1/4" diameter		100	.080		.03	3.59		3.62	5.60
1800	1-1/2" diameter	↓	90	.089		.04	3.99		4.03	6.20
1900	For ceiling installations, add				↓		40%			

09 69 Access Flooring

09 69 13 – Rigid-Grid Access Flooring

09 69 13.10 Access Floors		Crew	Daily Output	Labor-Hours	Unit	Material	Labor	Equipment	Total	Total Incl O&P
0010	ACCESS FLOORS									
0015	Access floor package including panel, pedestal, stringers & laminate cover									
0100	Computer room, greater than 6,000 S.F.	4 Carp	750	.043	S.F.	7.50	1.92		9.42	11.20
0110	Less than 6,000 S.F.	2 Carp	375	.043		8.15	1.92		10.07	11.95
0120	Office, greater than 6,000 S.F.	4 Carp	1050	.030		5.35	1.37		6.72	8
0250	Panels, particle board or steel, 1250# load, no covering, under 6,000 S.F.	2 Carp	600	.027		3.75	1.20		4.95	5.95
0300	Over 6,000 S.F.		640	.025		3.21	1.12		4.33	5.25
0400	Aluminum, 24" panels	↓	500	.032		32.50	1.44		33.94	37.50
0600	For carpet covering, add					8.70			8.70	9.55
0700	For vinyl floor covering, add					8.95			8.95	9.85
0900	For high pressure laminate covering, add					7.45			7.45	8.20
0910	For snap on stringer system, add	2 Carp	1000	.016	↓	1.54	.72		2.26	2.80
0950	Office applications, steel or concrete panels,									
0960	no covering, over 6,000 S.F.	2 Carp	960	.017	S.F.	10.40	.75		11.15	12.60
1000	Machine cutouts after initial installation	1 Carp	50	.160	Ea.	20	7.20		27.20	33
1050	Pedestals, 6" to 12"	2 Carp	85	.188		8.20	8.45		16.65	22
1100	Air conditioning grilles, 4" x 12"	1 Carp	17	.471		66.50	21		87.50	106
1150	4" x 18"	"	14	.571	↓	91	25.50		116.50	140
1200	Approach ramps, minimum	2 Carp	60	.267	S.F.	24.50	11.95		36.45	45.50
1300	Maximum	"	40	.400	"	33.50	17.95		51.45	64
1500	Handrail, 2 rail, aluminum	1 Carp	15	.533	L.F.	98.50	24		122.50	146

Estimating Tips

General

- The items in this division are usually priced per square foot or each.
- Many items in Division 10 require some type of support system or special anchors that are not usually furnished with the item. The required anchors must be added to the estimate in the appropriate division.
- Some items in Division 10, such as lockers, may require assembly before installation. Verify the amount of assembly required. Assembly can often exceed installation time.

10 20 00 Interior Specialties

- Support angles and blocking are not included in the installation of toilet compartments, shower/dressing compartments, or cubicles. Appropriate line items from Divisions 5 or 6 may need to be added to support the installations.
- Toilet partitions are priced by the stall. A stall consists of a side wall, pilaster, and door with hardware. Toilet tissue holders and grab bars are extra.
- The required acoustical rating of a folding partition can have a significant impact on costs. Verify the sound transmission coefficient rating of the panel priced to the specification requirements.
- Grab bar installation does not include supplemental blocking or backing to support the required load. When grab bars are installed at an existing facility, provisions must be made to attach the grab bars to solid structure.

Reference Numbers

Reference numbers are shown in shaded boxes at the beginning of some major classifications. These numbers refer to related items in the Reference Section. The reference information may be an estimating procedure, an alternate pricing method, or technical information.

Note: Not all subdivisions listed here necessarily appear in this publication.

10 28 16 – Bath Accessories

10 28 16.20 Medicine Cabinets		Crew	Daily Output	Labor-Hours	Unit	Material	Labor	Equipment	Total	Total Incl O&P
						2013 Bare Costs				
0010	MEDICINE CABINETS									
0020	With mirror, sst frame, 16" x 22", unlighted	1 Carp	14	.571	Ea.	81.50	25.50		107	130
0100	Wood frame		14	.571		136	25.50		161.50	190
0300	Sliding mirror doors, 20" x 16" x 4-3/4", unlighted		7	1.143		125	51.50		176.50	216
0400	24" x 19" x 8-1/2", lighted		5	1.600		179	72		251	310
0600	Triple door, 30" x 32", unlighted, plywood body		7	1.143		320	51.50		371.50	430
0700	Steel body		7	1.143		350	51.50		401.50	465
0900	Oak door, wood body, beveled mirror, single door		7	1.143		195	51.50		246.50	293
1000	Double door		6	1.333		390	60		450	520
1200	Hotel cabinets, stainless, with lower shelf, unlighted		10	.800		200	36		236	275
1300	Lighted	▼	5	1.600	▼	305	72		377	445

Division 11 - Equipment

Estimating Tips

General

- The items in this division are usually priced per square foot or each. Many of these items are purchased by the owner for installation by the contractor. Check the specifications for responsibilities and include time for receiving, storage, installation, and mechanical and electrical hookups in the appropriate divisions.
- Many items in Division 11 require some type of support system that is not usually furnished with the item. Examples of these systems include blocking for the attachment of casework and support angles for ceiling-hung projection screens. The required blocking or supports must be added to the estimate in the appropriate division.
- Some items in Division 11 may require assembly or electrical hookups. Verify the amount of assembly required or the need for a hard electrical connection and add the appropriate costs.

Reference Numbers

Reference numbers are shown in shaded boxes at the beginning of some major classifications. These numbers refer to related items in the Reference Section. The reference information may be an estimating procedure, an alternate pricing method, or technical information.

Note: Not all subdivisions listed here necessarily appear in this publication.

11 12 Parking Control Equipment

11 12 13 – Parking Key and Card Control Units

11 12 13.10 Parking Control Units		Crew	Daily Output	Labor-Hours	Unit	Material	Labor	Equipment	Total	Total Incl O&P
						2013 Bare Costs				
0010	**PARKING CONTROL UNITS**									
5100	Card reader	1 Elec	2	4	Ea.	1,925	210		2,135	2,450
5120	Proximity with customer display	2 Elec	1	16		5,500	840		6,340	7,300
6000	Parking control software, minimum	1 Elec	.50	16		23,600	840		24,440	27,200
6020	Maximum	"	.20	40		103,000	2,100		105,100	116,500

11 12 16 – Parking Ticket Dispensers

11 12 16.10 Ticket Dispensers

		Crew	Daily Output	Labor-Hours	Unit	Material	Labor	Equipment	Total	Total Incl O&P
0010	**TICKET DISPENSERS**									
5900	Ticket spitter with time/date stamp, standard	2 Elec	2	8	Ea.	6,400	420		6,820	7,650
5920	Mag stripe encoding	"	2	8	"	18,600	420		19,020	21,000

11 12 26 – Parking Fee Collection Equipment

11 12 26.13 Parking Fee Coin Collection Equipment

		Crew	Daily Output	Labor-Hours	Unit	Material	Labor	Equipment	Total	Total Incl O&P
0010	**PARKING FEE COIN COLLECTION EQUIPMENT**									
5200	Cashier booth, average	B-22	1	30	Ea.	10,200	1,250	206	11,656	13,300
5300	Collector station, pay on foot	2 Elec	.20	80		109,500	4,200		113,700	127,000
5320	Credit card only	"	.50	32		20,200	1,675		21,875	24,800

11 12 26.23 Fee Equipment

		Crew	Daily Output	Labor-Hours	Unit	Material	Labor	Equipment	Total	Total Incl O&P
0010	**FEE EQUIPMENT**									
5600	Fee computer	1 Elec	1.50	5.333	Ea.	14,000	279		14,279	15,700

11 12 33 – Parking Gates

11 12 33.13 Lift Arm Parking Gates

		Crew	Daily Output	Labor-Hours	Unit	Material	Labor	Equipment	Total	Total Incl O&P
0010	**LIFT ARM PARKING GATES**									
5000	Barrier gate with programmable controller	2 Elec	3	5.333	Ea.	3,325	279		3,604	4,075
5020	Industrial		3	5.333		4,950	279		5,229	5,850
5500	Exit verifier		1	16		17,400	840		18,240	20,500
5700	Full sign, 4" letters	1 Elec	2	4		1,225	210		1,435	1,650
5800	Inductive loop	2 Elec	4	4		169	210		379	500
5950	Vehicle detector, microprocessor based	1 Elec	3	2.667		420	140		560	675

11 23 Commercial Laundry and Dry Cleaning Equipment

11 23 13 – Dry Cleaning Equipment

11 23 13.13 Dry Cleaners

		Crew	Daily Output	Labor-Hours	Unit	Material	Labor	Equipment	Total	Total Incl O&P
0010	**DRY CLEANERS** Not incl. rough-in									
2000	Dry cleaners, electric, 20 lb. capacity	L-1	.20	80	Ea.	33,300	4,325		37,625	43,100
2050	25 lb. capacity		.17	94.118		47,200	5,100		52,300	59,500
2100	30 lb. capacity		.15	106		49,800	5,775		55,575	63,500
2150	60 lb. capacity		.09	177		78,000	9,625		87,625	100,500

11 23 19 – Finishing Equipment

11 23 19.13 Folders and Spreaders

		Crew	Daily Output	Labor-Hours	Unit	Material	Labor	Equipment	Total	Total Incl O&P
0010	**FOLDERS AND SPREADERS**									
3500	Folders, blankets & sheets, minimum	1 Elec	.17	47.059	Ea.	32,200	2,475		34,675	39,100
3700	King size with automatic stacker		.10	80		58,000	4,200		62,200	70,500
3800	For conveyor delivery, add		.45	17.778		13,900	930		14,830	16,700

11 23 23 – Commercial Ironing Equipment

11 23 23.13 Irons and Pressers

		Crew	Daily Output	Labor-Hours	Unit	Material	Labor	Equipment	Total	Total Incl O&P
0010	**IRONS AND PRESSERS**									
4500	Ironers, institutional, 110", single roll	1 Elec	.20	40	Ea.	31,200	2,100		33,300	37,400

11 23 Commercial Laundry and Dry Cleaning Equipment

11 23 26 – Commercial Washers and Extractors

11 23 26.13 Washers and Extractors		Crew	Daily Output	Labor-Hours	Unit	Material	Labor	Equipment	Total	Total Incl O&P
						2013 Bare Costs				
0010	**WASHERS AND EXTRACTORS**, not including rough-in									
6000	Combination washer/extractor, 20 lb. capacity	L-6	1.50	8	Ea.	5,775	435		6,210	7,025
6100	30 lb. capacity		.80	15		9,050	820		9,870	11,200
6200	50 lb. capacity		.68	17.647		10,600	965		11,565	13,200
6300	75 lb. capacity		.30	40		20,000	2,175		22,175	25,300
6350	125 lb. capacity	↓	.16	75	↓	27,300	4,100		31,400	36,300

11 23 33 – Coin-Operated Laundry Equipment

11 23 33.13 Coin Operated Washers and Dryers		Crew	Daily Output	Labor-Hours	Unit	Material	Labor	Equipment	Total	Total Incl O&P
0010	**COIN OPERATED WASHERS AND DRYERS**									
5290	Clothes washer									
5300	Commercial, coin operated, average	1 Plum	3	2.667	Ea.	1,250	149		1,399	1,600

11 26 Unit Kitchens

11 26 13 – Metal Unit Kitchens

11 26 13.10 Commercial Unit Kitchens		Crew	Daily Output	Labor-Hours	Unit	Material	Labor	Equipment	Total	Total Incl O&P
0010	**COMMERCIAL UNIT KITCHENS**									
1500	Combination range, refrigerator and sink, 30" wide, minimum	L-1	2	8	Ea.	1,125	435		1,560	1,875
1550	Maximum		1	16		3,875	865		4,740	5,550
1570	60" wide, average		1.40	11.429		2,800	620		3,420	4,000
1590	72" wide, average		1.20	13.333		4,400	720		5,120	5,925
1600	Office model, 48" wide		2	8		2,725	435		3,160	3,650
1620	Refrigerator and sink only	↓	2.40	6.667	↓	2,850	360		3,210	3,675
1640	Combination range, refrigerator, sink, microwave									
1660	Oven and ice maker	L-1	.80	20	Ea.	4,675	1,075		5,750	6,750

11 31 Residential Appliances

11 31 13 – Residential Kitchen Appliances

11 31 13.13 Cooking Equipment		Crew	Daily Output	Labor-Hours	Unit	Material	Labor	Equipment	Total	Total Incl O&P
0010	**COOKING EQUIPMENT**									
0020	Cooking range, 30" free standing, 1 oven, minimum	2 Clab	10	1.600	Ea.	435	56.50		491.50	570
0050	Maximum		4	4		1,700	142		1,842	2,100
0700	Free-standing, 1 oven, 21" wide range, minimum		10	1.600		385	56.50		441.50	510
0750	21" wide, maximum	↓	4	4		420	142		562	680
0900	Countertop cooktops, 4 burner, standard, minimum	1 Elec	6	1.333		275	70		345	410
0950	Maximum		3	2.667		1,175	140		1,315	1,475
1050	As above, but with grill and griddle attachment, minimum		6	1.333		1,100	70		1,170	1,325
1100	Maximum		3	2.667		3,575	140		3,715	4,125
1200	Induction cooktop, 30" wide		3	2.667		1,175	140		1,315	1,500
1250	Microwave oven, minimum		4	2		110	105		215	278
1300	Maximum	↓	2	4	↓	435	210		645	795

11 31 Residential Appliances

11 31 13 – Residential Kitchen Appliances

	11 31 13.33 Kitchen Cleaning Equipment	Crew	Daily Output	Labor-Hours	Unit	2013 Bare Costs Material	Labor	Equipment	Total	Total Incl O&P
0010	**KITCHEN CLEANING EQUIPMENT**									
2750	Dishwasher, built-in, 2 cycles, minimum	L-1	4	4	Ea.	227	216		443	575
2800	Maximum		2	8		430	435		865	1,125
2950	4 or more cycles, minimum		4	4		340	216		556	700
2960	Average		4	4		455	216		671	825
3000	Maximum		2	8		1,075	435		1,510	1,825
	11 31 13.43 Waste Disposal Equipment									
0010	**WASTE DISPOSAL EQUIPMENT**									
3300	Garbage disposal, sink type, minimum	L-1	10	1.600	Ea.	86	86.50		172.50	225
3350	Maximum	"	10	1.600	"	217	86.50		303.50	370
	11 31 13.53 Kitchen Ventilation Equipment									
0010	**KITCHEN VENTILATION EQUIPMENT**									
4150	Hood for range, 2 speed, vented, 30" wide, minimum	L-3	5	3.200	Ea.	63	156		219	310
4200	Maximum		3	5.333		760	261		1,021	1,225
4300	42" wide, minimum		5	3.200		163	156		319	420
4330	Custom		5	3.200		1,725	156		1,881	2,150
4350	Maximum		3	5.333		2,100	261		2,361	2,725
4500	For ventless hood, 2 speed, add					18.45			18.45	20.50
4650	For vented 1 speed, deduct from maximum					48.50			48.50	53.50

11 31 23 – Residential Laundry Appliances

	11 31 23.13 Washers	Crew	Daily Output	Labor-Hours	Unit	Material	Labor	Equipment	Total	Total Incl O&P
0010	**WASHERS**									
5000	Residential, 4 cycle, average	1 Plum	3	2.667	Ea.	855	149		1,004	1,175

11 31 33 – Miscellaneous Residential Appliances

	11 31 33.23 Water Heaters	Crew	Daily Output	Labor-Hours	Unit	Material	Labor	Equipment	Total	Total Incl O&P
0010	**WATER HEATERS**									
6900	Electric, glass lined, 30 gallon, minimum	L-1	5	3.200	Ea.	625	173		798	950
6950	Maximum		3	5.333		870	289		1,159	1,400
7100	80 gallon, minimum		2	8		1,125	435		1,560	1,900
7150	Maximum		1	16		1,575	865		2,440	3,025
	11 31 33.43 Air Quality									
0010	**AIR QUALITY**									
2450	Dehumidifier, portable, automatic, 15 pint	1 Elec	4	2	Ea.	150	105		255	320
2550	40 pint					229			229	252
3550	Heater, electric, built-in, 1250 watt, ceiling type, minimum	1 Elec	4	2		106	105		211	273
3600	Maximum		3	2.667		172	140		312	400
3700	Wall type, minimum		4	2		170	105		275	345
3750	Maximum		3	2.667		183	140		323	410
3900	1500 watt wall type, with blower		4	2		170	105		275	345
3950	3000 watt		3	2.667		350	140		490	595
4850	Humidifier, portable, 8 gallons per day					173			173	191
5000	15 gallons per day					208			208	229

11 41 Foodservice Storage Equipment

11 41 13 – Refrigerated Food Storage Cases

11 41 13.20 Refrigerated Food Storage Equipment		Crew	Daily Output	Labor-Hours	Unit	Material	Labor	Equipment	Total	Total Incl O&P
						2013 Bare Costs				
0010	**REFRIGERATED FOOD STORAGE EQUIPMENT**									
2350	Cooler, reach-in, beverage, 6' long	Q-1	6	2.667	Ea.	3,475	134		3,609	4,000

11 42 Food Preparation Equipment

11 42 10 – Commercial Food Preparation Equipment

11 42 10.10 Choppers, Mixers and Misc. Equipment		Crew	Daily Output	Labor-Hours	Unit	Material	Labor	Equipment	Total	Total Incl O&P
0010	**CHOPPERS, MIXERS AND MISC. EQUIPMENT**									
1850	Coffee urn, twin 6 gallon urns	1 Plum	2	4	Ea.	2,100	223		2,323	2,625
3800	Food mixers, bench type, 20 quarts	L-7	7	4		2,700	173		2,873	3,250
3900	60 quarts	"	5	5.600		11,000	242		11,242	12,500

11 44 Food Cooking Equipment

11 44 13 – Commercial Ranges

11 44 13.10 Cooking Equipment		Crew	Daily Output	Labor-Hours	Unit	Material	Labor	Equipment	Total	Total Incl O&P
0010	**COOKING EQUIPMENT**									
0020	Bake oven, gas, one section	Q-1	8	2	Ea.	5,050	100		5,150	5,700
1300	Broiler, without oven, standard	"	8	2		3,675	100		3,775	4,175
1550	Infrared	L-7	4	7		9,450	305		9,755	10,900
6350	Kettle, w/steam jacket, tilting, w/positive lock, SS, 20 gallons		7	4		7,550	173		7,723	8,575
6600	60 gallons		6	4.667		10,900	202		11,102	12,300
8850	Steamer, electric 27 KW		7	4		8,500	173		8,673	9,625
9100	Electric, 10 KW or gas 100,000 BTU		5	5.600		6,250	242		6,492	7,250
9150	Toaster, conveyor type, 16-22 slices per minute					935			935	1,025
9200	For deluxe models of above equipment, add					75%				
9400	Rule of thumb: Equipment cost based									
9410	on kitchen work area									
9420	Office buildings, minimum	L-7	77	.364	S.F.	89.50	15.75		105.25	123
9450	Maximum		58	.483		151	21		172	198
9550	Public eating facilities, minimum		77	.364		117	15.75		132.75	153
9600	Maximum		46	.609		191	26.50		217.50	251
9750	Hospitals, minimum		58	.483		121	21		142	165
9800	Maximum		39	.718		222	31		253	292

11 46 Food Dispensing Equipment

11 46 16 – Service Line Equipment

11 46 16.10 Commercial Food Dispensing Equipment		Crew	Daily Output	Labor-Hours	Unit	Material	Labor	Equipment	Total	Total Incl O&P
0010	**COMMERCIAL FOOD DISPENSING EQUIPMENT**									
3300	Food warmer, counter, 1.2 KW				Ea.	510			510	560
3550	1.6 KW					2,275			2,275	2,500
5750	Ice dispenser 567 pound	Q-1	6	2.667		5,150	134		5,284	5,875

11 48 Cleaning and Disposal Equipment

11 48 13 – Commercial Dishwashers

11 48 13.10 Dishwashers		Crew	Daily Output	Labor-Hours	Unit	2013 Bare Costs Material	Labor	Equipment	Total	Total Incl O&P
0010	**DISHWASHERS**									
2700	Dishwasher, commercial, rack type									
2720	10 to 12 racks per hour	Q-1	3.20	5	Ea.	4,475	251		4,726	5,300
2730	Energy star rated, 35 to 40 racks/hour G		1.30	12.308		4,900	620		5,520	6,325
2740	50 to 60 racks/hour G		1.30	12.308		10,700	620		11,320	12,600

11 52 Audio-Visual Equipment

11 52 16 – Projectors

11 52 16.10 Movie Equipment

Line	Description	Crew	Daily Output	Labor-Hours	Unit	Material	Labor	Equipment	Total	Total Incl O&P
0010	**MOVIE EQUIPMENT**									
0020	Changeover, minimum				Ea.	465			465	510
0100	Maximum					905			905	995
0800	Lamphouses, incl. rectifiers, xenon, 1,000 watt	1 Elec	2	4		6,650	210		6,860	7,625
0900	1,600 watt		2	4		7,100	210		7,310	8,125
1000	2,000 watt		1.50	5.333		7,625	279		7,904	8,800
1100	4,000 watt		1.50	5.333		9,425	279		9,704	10,800
3700	Sound systems, incl. amplifier, mono, minimum		.90	8.889		3,300	465		3,765	4,325
3800	Dolby/Super Sound, maximum		.40	20		18,100	1,050		19,150	21,500
4100	Dual system, 2 channel, front surround, minimum		.70	11.429		4,625	600		5,225	6,000
4200	Dolby/Super Sound, 4 channel, maximum		.40	20		16,500	1,050		17,550	19,800
5300	Speakers, recessed behind screen, minimum		2	4		1,050	210		1,260	1,500
5400	Maximum		1	8		3,100	420		3,520	4,025
7000	For automation, varying sophistication, minimum		1	8	System	2,375	420		2,795	3,250
7100	Maximum	2 Elec	.30	53.333	"	5,575	2,800		8,375	10,300

11 52 16.20 Movie Equipment- Digital

Line	Description	Crew	Daily Output	Labor-Hours	Unit	Material	Labor	Equipment	Total	Total Incl O&P
0010	**MOVIE EQUIPMENT- DIGITAL**									
1000	Digital 2K projection system, 98" DMD	1 Elec	2	4	Ea.	44,000	210		44,210	48,700
1100	OEM lens		2	4		5,375	210		5,585	6,225
2000	Pedestal with power distribution		2	4		2,050	210		2,260	2,575
3000	Software		2	4		1,725	210		1,935	2,225

11 53 Laboratory Equipment

11 53 19 – Laboratory Sterilizers

11 53 19.13 Sterilizers

Line	Description	Crew	Daily Output	Labor-Hours	Unit	Material	Labor	Equipment	Total	Total Incl O&P
0010	**STERILIZERS**									
0700	Glassware washer, undercounter, minimum	L-1	1.80	8.889	Ea.	6,175	480		6,655	7,525
0710	Maximum	"	1	16	"	12,700	865		13,565	15,300

11 53 33 – Emergency Safety Appliances

11 53 33.13 Emergency Equipment

Line	Description	Crew	Daily Output	Labor-Hours	Unit	Material	Labor	Equipment	Total	Total Incl O&P
0010	**EMERGENCY EQUIPMENT**									
1400	Safety equipment, eye wash, hand held				Ea.	405			405	445
1450	Deluge shower				"	760			760	840

11 53 43 – Service Fittings and Accessories

11 53 43.13 Fittings

Line	Description	Crew	Daily Output	Labor-Hours	Unit	Material	Labor	Equipment	Total	Total Incl O&P
0010	**FITTINGS**									
8000	Alternate pricing method: as percent of lab furniture									
8050	Installation, not incl. plumbing & duct work				% Furn.				22%	22%
8100	Plumbing, final connections, simple system								10%	10%

11 53 Laboratory Equipment

11 53 43 – Service Fittings and Accessories

11 53 43.13 Fittings		Crew	Daily Output	Labor-Hours	Unit	Material	2013 Bare Costs Labor	Equipment	Total	Total Incl O&P
8110	Moderately complex system				% Furn.				15%	15%
8120	Complex system								20%	20%
8150	Electrical, simple system								10%	10%
8160	Moderately complex system								20%	20%
8170	Complex system								35%	35%

11 61 Broadcast, Theater, and Stage Equipment

11 61 33 – Rigging Systems and Controls

11 61 33.10 Controls

		Crew	Daily Output	Labor-Hours	Unit	Material	Labor	Equipment	Total	Total Incl O&P
0010	**CONTROLS**									
0050	Control boards with dimmers and breakers, minimum	1 Elec	1	8	Ea.	12,500	420		12,920	14,300
0100	Average		.50	16		39,200	840		40,040	44,400
0150	Maximum		.20	40		127,500	2,100		129,600	143,500

11 66 Athletic Equipment

11 66 43 – Interior Scoreboards

11 66 43.10 Scoreboards

		Crew	Daily Output	Labor-Hours	Unit	Material	Labor	Equipment	Total	Total Incl O&P
0010	**SCOREBOARDS**									
7000	Baseball, minimum	R-3	1.30	15.385	Ea.	4,075	800	106	4,981	5,800
7200	Maximum		.05	400		17,900	20,800	2,750	41,450	54,000
7300	Football, minimum		.86	23.256		5,275	1,200	160	6,635	7,775
7400	Maximum		.20	100		15,700	5,200	685	21,585	25,800
7500	Basketball (one side), minimum		2.07	9.662		2,375	500	66.50	2,941.50	3,450
7600	Maximum		.30	66.667		3,475	3,450	455	7,380	9,500
7700	Hockey-basketball (four sides), minimum		.25	80		5,600	4,150	550	10,300	13,000
7800	Maximum		.15	133		5,675	6,925	915	13,515	17,700

11 71 Medical Sterilizing Equipment

11 71 10 – Medical Sterilizers & Distillers

11 71 10.10 Sterilizers and Distillers

		Crew	Daily Output	Labor-Hours	Unit	Material	Labor	Equipment	Total	Total Incl O&P
0010	**STERILIZERS AND DISTILLERS**									
6200	Steam generators, electric 10 kW to 180 kW, freestanding									
6250	Minimum	1 Elec	3	2.667	Ea.	8,675	140		8,815	9,750
6300	Maximum	"	.70	11.429	"	28,800	600		29,400	32,600

11 72 Examination and Treatment Equipment

11 72 53 – Treatment Equipment

11 72 53.13 Medical Treatment Equipment

		Crew	Daily Output	Labor-Hours	Unit	Material	Labor	Equipment	Total	Total Incl O&P
0010	**MEDICAL TREATMENT EQUIPMENT**									
6700	Surgical lights, doctor's office, single arm	2 Elec	2	8	Ea.	2,325	420		2,745	3,175
6750	Dual arm	"	1	16	"	4,150	840		4,990	5,825

11 76 Operating Room Equipment

11 76 10 – Operating Room Equipment

11 76 10.10 Surgical Equipment		Crew	Daily Output	Labor-Hours	Unit	2013 Bare Costs Material	Labor	Equipment	Total	Total Incl O&P
0010	**SURGICAL EQUIPMENT**									
6800	Surgical lights, major operating room, dual head, minimum	2 Elec	1	16	Ea.	4,450	840		5,290	6,125
6850	Maximum	"	1	16	"	29,700	840		30,540	34,000

11 78 Mortuary Equipment

11 78 13 – Mortuary Refrigerators

11 78 13.10 Mortuary and Autopsy Equipment		Crew	Daily Output	Labor-Hours	Unit	2013 Bare Costs Material	Labor	Equipment	Total	Total Incl O&P
0010	**MORTUARY AND AUTOPSY EQUIPMENT**									
0015	Autopsy table, standard	1 Plum	1	8	Ea.	9,175	445		9,620	10,800

11 78 16 – Crematorium Equipment

11 78 16.10 Crematory		Crew	Daily Output	Labor-Hours	Unit	2013 Bare Costs Material	Labor	Equipment	Total	Total Incl O&P
0010	**CREMATORY**									
1500	Crematory, not including building, 1 place	Q-3	.20	160	Ea.	71,500	8,500		80,000	92,000
1750	2 place	"	.10	320	"	102,500	17,000		119,500	138,000

11 82 Solid Waste Handling Equipment

11 82 26 – Facility Waste Compactors

11 82 26.10 Compactors		Crew	Daily Output	Labor-Hours	Unit	2013 Bare Costs Material	Labor	Equipment	Total	Total Incl O&P
0010	**COMPACTORS**									
0020	Compactors, 115 volt, 250#/hr., chute fed	L-4	1	24	Ea.	11,900	1,000		12,900	14,700
1000	Heavy duty industrial compactor, 0.5 C.Y. capacity		1	24		10,000	1,000		11,000	12,600
1050	1.0 C.Y. capacity		1	24		15,000	1,000		16,000	18,100
1100	3.0 C.Y. capacity		.50	48		25,500	2,025		27,525	31,200
1150	5.0 C.Y. capacity		.50	48		32,000	2,025		34,025	38,300
1200	Combination shredder/compactor (5,000 lb./hr.)	↓	.50	48		62,500	2,025		64,525	72,000
1400	For handling hazardous waste materials, 55 gallon drum packer, std.					19,500			19,500	21,500
1410	55 gallon drum packer w/HEPA filter					24,300			24,300	26,700
1420	55 gallon drum packer w/charcoal & HEPA filter					32,400			32,400	35,600
1430	All of the above made explosion proof, add					1,425			1,425	1,575
5800	Shredder, industrial, minimum					23,700			23,700	26,100
5850	Maximum					127,000			127,000	139,500
5900	Baler, industrial, minimum					9,500			9,500	10,500
5950	Maximum				↓	554,000			554,000	609,500

Estimating Tips

General

- The items in this division are usually priced per square foot or each. Most of these items are purchased by the owner and installed by the contractor. Do not assume the items in Division 12 will be purchased and installed by the contractor. Check the specifications for responsibilities and include receiving, storage, installation, and mechanical and electrical hookups in the appropriate divisions.
- Some items in this division require some type of support system that is not usually furnished with the item. Examples of these systems include blocking for the attachment of casework and heavy drapery rods. The required blocking must be added to the estimate in the appropriate division.

Reference Numbers

Reference numbers are shown in shaded boxes at the beginning of some major classifications. These numbers refer to related items in the Reference Section. The reference information may be an estimating procedure, an alternate pricing method, or technical information.

Note: Not all subdivisions listed here necessarily appear in this publication.

12 46 Furnishing Accessories

12 46 19 – Clocks

12 46 19.50 Wall Clocks		Crew	Daily Output	Labor-Hours	Unit	2013 Bare Costs Material	Labor	Equipment	Total	Total Incl O&P
0010	**WALL CLOCKS**									
0080	12" diameter, single face	1 Elec	8	1	Ea.	127	52.50		179.50	218
0100	Double face	"	6.20	1.290	"	294	67.50		361.50	425

Estimating Tips

General

- The items and systems in this division are usually estimated, purchased, supplied, and installed as a unit by one or more subcontractors. The estimator must ensure that all parties are operating from the same set of specifications and assumptions, and that all necessary items are estimated and will be provided. Many times the complex items and systems are covered, but the more common ones, such as excavation or a crane, are overlooked for the very reason that everyone assumes nobody could miss them. The estimator should be the central focus and be able to ensure that all systems are complete.
- Another area where problems can develop in this division is at the interface between systems. The estimator must ensure, for instance, that anchor bolts, nuts, and washers are estimated and included for the air-supported structures and pre-engineered buildings to be bolted to their foundations. Utility supply is a common area where essential items or pieces of equipment can be missed or overlooked due to the fact that each subcontractor may feel it is another's responsibility. The estimator should also be aware of certain items which may be supplied as part of a package but installed by others, and ensure that the installing contractor's estimate includes the cost of installation. Conversely, the estimator must also ensure that items are not costed by two different subcontractors, resulting in an inflated overall estimate.

13 30 00 Special Structures

- The foundations and floor slab, as well as rough mechanical and electrical, should be estimated, as this work is required for the assembly and erection of the structure. Generally, as noted in the book, the pre-engineered building comes as a shell. Pricing is based on the size and structural design parameters stated in the reference section. Additional features, such as windows and doors with their related structural framing, must also be included by the estimator. Here again, the estimator must have a clear understanding of the scope of each portion of the work and all the necessary interfaces.

Reference Numbers

Reference numbers are shown in shaded boxes at the beginning of some major classifications. These numbers refer to related items in the Reference Section. The reference information may be an estimating procedure, an alternate pricing method, or technical information.

Note: Not all subdivisions listed here necessarily appear in this publication.

13 11 Swimming Pools

13 11 13 – Below-Grade Swimming Pools

13 11 13.50 Swimming Pools		Crew	Daily Output	Labor-Hours	Unit	2013 Bare Costs Material	Labor	Equipment	Total	Total Incl O&P
0010	**SWIMMING POOLS** Residential in-ground, vinyl lined, concrete									
0020	Sides including equipment, sand bottom	B-52	300	.187	SF Surf	19.30	7.75	1.98	29.03	35.50
0100	Metal or polystyrene sides	B-14	410	.117		16.15	4.39	.90	21.44	25.50
0200	Add for vermiculite bottom					1.23			1.23	1.35
0500	Gunite bottom and sides, white plaster finish									
0600	12' x 30' pool	B-52	145	.386	SF Surf	36	16	4.09	56.09	68.50
0720	16' x 32' pool		155	.361		32.50	14.95	3.83	51.28	62.50
0750	20' x 40' pool		250	.224		29	9.30	2.37	40.67	49
0810	Concrete bottom and sides, tile finish									
0820	12' x 30' pool	B-52	80	.700	SF Surf	36.50	29	7.40	72.90	92.50
0830	16' x 32' pool		95	.589		30	24.50	6.25	60.75	77.50
0840	20' x 40' pool		130	.431		24	17.85	4.57	46.42	59
1100	Motel, gunite with plaster finish, incl. medium									
1150	capacity filtration & chlorination	B-52	115	.487	SF Surf	44.50	20	5.15	69.65	85
1200	Municipal, gunite with plaster finish, incl. high									
1250	capacity filtration & chlorination	B-52	100	.560	SF Surf	57	23	5.95	85.95	105
1350	Add for formed gutters				L.F.	84			84	92.50
1360	Add for stainless steel gutters				"	249			249	274
1700	Filtration and deck equipment only, as % of total				Total				20%	20%
1800	Deck equipment, rule of thumb, 20' x 40' pool				SF Pool				1.18	1.30
1900	5000 S.F. pool				"				1.73	1.90

13 11 46 – Swimming Pool Accessories

13 11 46.50 Swimming Pool Equipment		Crew	Daily Output	Labor-Hours	Unit	Material	Labor	Equipment	Total	Total Incl O&P
0010	**SWIMMING POOL EQUIPMENT**									
0020	Diving stand, stainless steel, 3 meter	2 Carp	.40	40	Ea.	11,700	1,800		13,500	15,600
2100	Lights, underwater, 12 volt, with transformer, 300 watt	1 Elec	1	8		290	420		710	945
2200	110 volt, 500 watt, standard		1	8		242	420		662	890
2400	Low water cutoff type		1	8		281	420		701	935
2800	Heaters, see Section 23 52 28.10									

13 21 Controlled Environment Rooms

13 21 13 – Clean Rooms

13 21 13.50 Clean Room Components		Crew	Daily Output	Labor-Hours	Unit	Material	Labor	Equipment	Total	Total Incl O&P
0010	**CLEAN ROOM COMPONENTS**									
1100	Clean room, soft wall, 12' x 12', Class 100	1 Carp	.18	44.444	Ea.	17,200	2,000		19,200	22,100
1110	Class 1,000		.18	44.444		14,000	2,000		16,000	18,500
1120	Class 10,000		.21	38.095		12,400	1,700		14,100	16,200
1130	Class 100,000		.21	38.095		11,600	1,700		13,300	15,400
2800	Ceiling grid support, slotted channel struts 4'-0" O.C., ea. way				S.F.				5.90	6.50
3000	Ceiling panel, vinyl coated foil on mineral substrate									
3020	Sealed, non-perforated				S.F.				1.27	1.40
4000	Ceiling panel seal, silicone sealant, 150 L.F./gal.	1 Carp	150	.053	L.F.	.32	2.39		2.71	4.04
4100	Two sided adhesive tape	"	240	.033	"	.13	1.50		1.63	2.44
4200	Clips, one per panel				Ea.	1.04			1.04	1.14
6000	HEPA filter, 2' x 4', 99.97% eff., 3" dp beveled frame (silicone seal)					480			480	530
6040	6" deep skirted frame (channel seal)					420			420	460
6100	99.99% efficient, 3" deep beveled frame (silicone seal)					555			555	610
6140	6" deep skirted frame (channel seal)					435			435	475
6200	99.999% efficient, 3" deep beveled frame (silicone seal)					625			625	690
6240	6" deep skirted frame (channel seal)					450			450	495

13 21 Controlled Environment Rooms

13 21 13 – Clean Rooms

13 21 13.50 Clean Room Components		Crew	Daily Output	Labor-Hours	Unit	Material	Labor	Equipment	Total	Total Incl O&P
						2013 Bare Costs				
7000	Wall panel systems, including channel strut framing									
7020	Polyester coated aluminum, particle board				S.F.				18.20	20
7100	Porcelain coated aluminum, particle board								32	35
7400	Wall panel support, slotted channel struts, to 12′ high				▼				16.35	18

13 24 Special Activity Rooms

13 24 16 – Saunas

13 24 16.50 Saunas and Heaters

		Crew	Daily Output	Labor-Hours	Unit	Material	Labor	Equipment	Total	Total Incl O&P
0010	**SAUNAS AND HEATERS**									
0020	Prefabricated, incl. heater & controls, 7′ high, 6′ x 4′, C/C	L-7	2.20	12.727	Ea.	4,900	550		5,450	6,225
0050	6′ x 4′, C/P		2	14		4,450	605		5,055	5,825
0400	6′ x 5′, C/C		2	14		5,400	605		6,005	6,850
0450	6′ x 5′, C/P		2	14		4,925	605		5,530	6,350
0600	6′ x 6′, C/C		1.80	15.556		5,725	675		6,400	7,325
0650	6′ x 6′, C/P		1.80	15.556		5,225	675		5,900	6,775
0800	6′ x 9′, C/C		1.60	17.500		7,125	755		7,880	8,975
0850	6′ x 9′, C/P		1.60	17.500		6,450	755		7,205	8,250
1000	8′ x 12′, C/C		1.10	25.455		11,400	1,100		12,500	14,200
1050	8′ x 12′, C/P		1.10	25.455		10,300	1,100		11,400	13,100
1200	8′ x 8′, C/C		1.40	20		8,425	865		9,290	10,600
1250	8′ x 8′, C/P		1.40	20		7,800	865		8,665	9,900
1400	8′ x 10′, C/C		1.20	23.333		9,350	1,000		10,350	11,900
1450	8′ x 10′, C/P		1.20	23.333		8,550	1,000		9,550	11,000
1600	10′ x 12′, C/C		1	28		12,000	1,200		13,200	15,100
1650	10′ x 12′, C/P	▼	1	28		10,800	1,200		12,000	13,700
2500	Heaters only (incl. above), wall mounted, to 200 C.F.					630			630	690
2750	To 300 C.F.					830			830	915
3000	Floor standing, to 720 C.F., 10,000 watts, w/controls	1 Elec	3	2.667		2,100	140		2,240	2,500
3250	To 1,000 C.F., 16,000 watts	"	3	2.667	▼	3,275	140		3,415	3,825

13 24 26 – Steam Baths

13 24 26.50 Steam Baths and Components

		Crew	Daily Output	Labor-Hours	Unit	Material	Labor	Equipment	Total	Total Incl O&P
0010	**STEAM BATHS AND COMPONENTS**									
0020	Heater, timer & head, single, to 140 C.F.	1 Plum	1.20	6.667	Ea.	2,200	370		2,570	2,975
0500	To 300 C.F.		1.10	7.273		2,500	405		2,905	3,350
1000	Commercial size, with blow-down assembly, to 800 C.F.		.90	8.889		9,000	495		9,495	10,600
1500	To 2500 C.F.	▼	.80	10		11,400	560		11,960	13,300
2000	Multiple, motels, apts., 2 baths, w/blow-down assm., 500 C.F.	Q-1	1.30	12.308		7,750	620		8,370	9,450
2500	4 baths	"	.70	22.857	▼	9,050	1,150		10,200	11,700

13 49 Radiation Protection

13 49 13 – Integrated X-Ray Shielding Assemblies

13 49 13.50 Lead Sheets		Crew	Daily Output	Labor-Hours	Unit	2013 Bare Costs Material	Labor	Equipment	Total	Total Incl O&P
0010	**LEAD SHEETS**									
0300	Lead sheets, 1/16" thick	2 Lath	135	.119	S.F.	10.05	4.71		14.76	17.95
0400	1/8" thick		120	.133		20	5.30		25.30	30
0500	Lead shielding, 1/4" thick		135	.119		49	4.71		53.71	61
0550	1/2" thick		120	.133		83.50	5.30		88.80	99.50
0950	Lead headed nails (average 1 lb. per sheet)				Lb.	7.60			7.60	8.35
1000	Butt joints in 1/8" lead or thicker, 2" batten strip x 7' long	2 Lath	240	.067	Ea.	23.50	2.65		26.15	30
1200	X-ray protection, average radiography or fluoroscopy									
1210	room, up to 300 S.F. floor, 1/16" lead, minimum	2 Lath	.25	64	Total	9,200	2,550		11,750	13,800
1500	Maximum, 7'-0" walls	"	.15	106	"	11,100	4,225		15,325	18,400
1600	Deep therapy X-ray room, 250 kV capacity,									
1800	up to 300 S.F. floor, 1/4" lead, minimum	2 Lath	.08	200	Total	25,700	7,950		33,650	40,000
1900	Maximum, 7'-0" walls	"	.06	266	"	31,700	10,600		42,300	50,500

13 49 19 – Lead-Lined Materials

13 49 19.50 Shielding Lead		Crew	Daily Output	Labor-Hours	Unit	Material	Labor	Equipment	Total	Total Incl O&P
0010	**SHIELDING LEAD**									
0100	Laminated lead in wood doors, 1/16" thick, no hardware				S.F.	65.50			65.50	72
0200	Lead lined door frame, not incl. hardware,									
0210	1/16" thick lead, butt prepared for hardware	1 Lath	2.40	3.333	Ea.	770	132		902	1,050
0850	Window frame with 1/16" lead and voice passage, 36" x 60"	2 Glaz	2	8		1,550	345		1,895	2,225
0870	24" x 36" frame		4	4		1,275	173		1,448	1,650
0900	Lead gypsum board, 5/8" thick with 1/16" lead		160	.100	S.F.	9.30	4.33		13.63	16.75
0910	1/8" lead		140	.114		22	4.95		26.95	31.50
0930	1/32" lead	2 Lath	200	.080		6.20	3.18		9.38	11.50

13 49 21 – Lead Glazing

13 49 21.50 Lead Glazing		Crew	Daily Output	Labor-Hours	Unit	Material	Labor	Equipment	Total	Total Incl O&P
0010	**LEAD GLAZING**									
0600	Lead glass, 1/4" thick, 2.0 mm LE, 12" x 16"	2 Glaz	13	1.231	Ea.	305	53.50		358.50	415
0700	24" x 36"		8	2		1,100	86.50		1,186.50	1,350
0800	36" x 60"		2	8		3,375	345		3,720	4,225
2000	X-ray viewing panels, clear lead plastic									
2010	7 mm thick, 0.3 mm LE, 2.3 lb./S.F.	H-3	139	.115	S.F.	224	4.44		228.44	253
2020	12 mm thick, 0.5 mm LE, 3.9 lb./S.F.		82	.195		305	7.50		312.50	345
2030	18 mm thick, 0.8 mm LE, 5.9 lb./S.F.		54	.296		330	11.40		341.40	375
2040	22 mm thick, 1.0 mm LE, 7.2 lb./S.F.		44	.364		335	14		349	390
2050	35 mm thick, 1.5 mm LE, 11.5 lb./S.F.		28	.571		375	22		397	450
2060	46 mm thick, 2.0 mm LE, 15.0 lb./S.F.		21	.762		495	29.50		524.50	590
2090	For panels 12 S.F. to 48 S.F., add crating charge				Ea.					50

13 49 23 – Integrated RFI/EMI Shielding Assemblies

13 49 23.50 Modular Shielding Partitions		Crew	Daily Output	Labor-Hours	Unit	Material	Labor	Equipment	Total	Total Incl O&P
0010	**MODULAR SHIELDING PARTITIONS**									
4000	X-ray barriers, modular, panels mounted within framework for									
4002	attaching to floor, wall or ceiling, upper portion is clear lead									
4005	plastic window panels 48"H, lower portion is opaque leaded									
4008	steel panels 36"H, structural supports not incl.									
4010	1-section barrier, 36"W x 84"H overall									
4020	0.5 mm LE panels	H-3	6.40	2.500	Ea.	7,075	96.50		7,171.50	7,925
4030	0.8 mm LE panels		6.40	2.500		7,525	96.50		7,621.50	8,450
4040	1.0 mm LE panels		5.33	3.002		7,675	116		7,791	8,625
4050	1.5 mm LE panels		5.33	3.002		8,175	116		8,291	9,175
4060	2-section barrier, 72"W x 84"H overall									

13 49 Radiation Protection

13 49 23 – Integrated RFI/EMI Shielding Assemblies

13 49 23.50 Modular Shielding Partitions		Crew	Daily Output	Labor-Hours	Unit	Material	2013 Bare Costs Labor	Equipment	Total	Total Incl O&P
4070	0.5 mm LE panels	H-3	4	4	Ea.	14,400	154		14,554	16,100
4080	0.8 mm LE panels		4	4		15,300	154		15,454	17,000
4090	1.0 mm LE panels		3.56	4.494		15,600	173		15,773	17,500
5000	1.5 mm LE panels	↓	3.20	5	↓	17,700	193		17,893	19,800
5010	3-section barrier, 108"W x 84"H overall									
5020	0.5 mm LE panels	H-3	3.20	5	Ea.	21,700	193		21,893	24,200
5030	0.8 mm LE panels		3.20	5		22,900	193		23,093	25,500
5040	1.0 mm LE panels		2.67	5.993		23,300	231		23,531	26,000
5050	1.5 mm LE panels	↓	2.46	6.504	↓	25,300	251		25,551	28,200
7000	X-ray barriers, mobile, mounted within framework w/casters on									
7005	bottom, clear lead plastic window panels on upper portion,									
7010	opaque on lower, 30"W x 75"H overall, incl. framework									
7020	24"H upper w/0.5 mm LE, 48"H lower w/0.8 mm LE	1 Carp	16	.500	Ea.	3,775	22.50		3,797.50	4,175
7030	48"W x 75"H overall, incl. framework									
7040	36"H upper w/0.5 mm LE, 36"H lower w/0.8 mm LE	1 Carp	16	.500	Ea.	6,900	22.50		6,922.50	7,625
7050	36"H upper w/1.0 mm LE, 36"H lower w/1.5 mm LE	"	16	.500	"	8,200	22.50		8,222.50	9,050
7060	72"W x 75"H overall, incl. framework									
7070	36"H upper w/0.5 mm LE, 36"H lower w/0.8 mm LE	1 Carp	16	.500	Ea.	8,175	22.50		8,197.50	9,025
7080	36"H upper w/1.0 mm LE, 36"H lower w/1.5 mm LE	"	16	.500	"	10,200	22.50		10,222.50	11,300

13 49 33 – Radio Frequency Shielding

13 49 33.50 Shielding, Radio Frequency		Crew	Daily Output	Labor-Hours	Unit	Material	Labor	Equipment	Total	Total Incl O&P
0010	**SHIELDING, RADIO FREQUENCY**									
0020	Prefabricated galvanized steel	2 Carp	375	.043	SF Surf	4.46	1.92		6.38	7.85
0050	5 oz., copper wall/ceiling panel		155	.103		3.68	4.64		8.32	11.20
0100	12 oz., copper floor panel		470	.034	↓	7.85	1.53		9.38	11
0150	Door, copper/wood laminate, 4' x 7'	↓	1.50	10.667	Ea.	5,775	480		6,255	7,100

Division Notes

		CREW	DAILY OUTPUT	LABOR-HOURS	UNIT	2012 BARE COSTS MAT.	LABOR	EQUIP.	TOTAL	TOTAL INCL O&P

Estimating Tips

Pipe for fire protection and all uses is located in Subdivisions 21 11 13 and 22 11 13.

The labor adjustment factors listed in Subdivision 22 01 02.20 also apply to Division 21.

Many, but not all, areas in the U.S. require backflow protection in the fire system. It is advisable to check local building codes for specific requirements.

For your reference, the following is a list of the most applicable Fire Codes and Standards which may be purchased from the NFPA, 1 Batterymarch Park, Quincy, MA 02169-7471.

NFPA 1: Uniform Fire Code
NFPA 10: Portable Fire Extinguishers
NFPA 11: Low-, Medium-, and High-Expansion Foam
NFPA 12: Carbon Dioxide Extinguishing Systems (Also companion 12A)
NFPA 13: Installation of Sprinkler Systems (Also companion 13D, 13E, and 13R)
NFPA 14: Installation of Standpipe and Hose Systems
NFPA 15: Water Spray Fixed Systems for Fire Protection
NFPA 16: Installation of Foam-Water Sprinkler and Foam-Water Spray Systems
NFPA 17: Dry Chemical Extinguishing Systems (Also companion 17A)
NFPA 18: Wetting Agents
NFPA 20: Installation of Stationary Pumps for Fire Protection
NFPA 22: Water Tanks for Private Fire Protection
NFPA 24: Installation of Private Fire Service Mains and their Appurtenances
NFPA 25: Inspection, Testing and Maintenance of Water-Based Fire Protection

Reference Numbers

Reference numbers are shown in shaded boxes at the beginning of some major classifications. These numbers refer to related items in the Reference Section. The reference information may be an estimating procedure, an alternate pricing method, or technical information.

Note: Not all subdivisions listed here necessarily appear in this publication.

Note: **Trade Service,** *in part, has been used as a reference source for some of the material prices used in Division 21.*

21 21 Carbon-Dioxide Fire-Extinguishing Systems

21 21 16 – Carbon-Dioxide Fire-Extinguishing Equipment

21 21 16.50 CO2 Fire Extinguishing System		Crew	Daily Output	Labor-Hours	Unit	2013 Bare Costs Material	Labor	Equipment	Total	Total Incl O&P
0010	**CO_2 FIRE EXTINGUISHING SYSTEM**									
0100	Control panel, single zone with batteries (2 zones det., 1 suppr.)	1 Elec	1	8	Ea.	1,600	420		2,020	2,375
0150	Multizone (4) with batteries (8 zones det., 4 suppr.)	"	.50	16		3,025	840		3,865	4,600
1000	Dispersion nozzle, CO_2, 3" x 5"	1 Plum	18	.444		62	25		87	106
2000	Extinguisher, CO_2 system, high pressure, 75 lb. cylinder	Q-1	6	2.667		1,175	134		1,309	1,500
2100	100 lb. cylinder	"	5	3.200		1,200	161		1,361	1,575
3000	Electro/mechanical release	L-1	4	4		155	216		371	495
3400	Manual pull station	1 Plum	6	1.333		55.50	74.50		130	174
4000	Pneumatic damper release	"	8	1	↓	206	56		262	310

21 22 Clean-Agent Fire-Extinguishing Systems

21 22 16 – Clean-Agent Fire-Extinguishing Equipment

21 22 16.50 FM200 Fire Extinguishing System		Crew	Daily Output	Labor-Hours	Unit	Material	Labor	Equipment	Total	Total Incl O&P
0010	**FM200 FIRE EXTINGUISHING SYSTEM**									
1100	Dispersion nozzle FM200, 1-1/2"	1 Plum	14	.571	Ea.	62	32		94	116
2400	Extinguisher, FM200 system, filled, with mounting bracket									
2460	26 lb. container	Q-1	8	2	Ea.	2,125	100		2,225	2,475
2480	44 lb. container		7	2.286		2,825	115		2,940	3,300
2500	63 lb. container		6	2.667		3,300	134		3,434	3,825
2520	101 lb. container		5	3.200		4,425	161		4,586	5,125
2540	196 lb. container	↓	4	4	↓	7,175	201		7,376	8,200
6000	Average FM200 system, minimum				C.F.	1.63			1.63	1.79
6020	Maximum				"	3.24			3.24	3.56

Estimating Tips

22 10 00 Plumbing Piping and Pumps

This subdivision is primarily basic pipe and related materials. The pipe may be used by any of the mechanical disciplines, i.e., plumbing, fire protection, heating, and air conditioning.

Note: CPVC plastic piping approved for fire protection is located in 21 11 13.

- The labor adjustment factors listed in Subdivision 22 01 02.20 apply throughout Divisions 21, 22, and 23. CAUTION: the correct percentage may vary for the same items. For example, the percentage add for the basic pipe installation should be based on the maximum height that the craftsman must install for that particular section. If the pipe is to be located 14′ above the floor but it is suspended on threaded rod from beams, the bottom flange of which is 18′ high (4′ rods), then the height is actually 18′ and the add is 20%. The pipe coverer, however, does not have to go above the 14′, and so his or her add should be 10%.
- Most pipe is priced first as straight pipe with a joint (coupling, weld, etc.) every 10′ and a hanger usually every 10′. There are exceptions with hanger spacing such as for cast iron pipe (5′) and plastic pipe (3 per 10′). Following each type of pipe there are several lines listing sizes and the amount to be subtracted to delete couplings and hangers. This is for pipe that is to be buried or supported together on trapeze hangers. The reason that the couplings are deleted is that these runs are usually long, and frequently longer lengths of pipe are used. By deleting the couplings, the estimator is expected to look up and add back the correct reduced number of couplings.
- When preparing an estimate, it may be necessary to approximate the fittings. Fittings usually run between 25% and 50% of the cost of the pipe. The lower percentage is for simpler runs, and the higher number is for complex areas, such as mechanical rooms.
- For historic restoration projects, the systems must be as invisible as possible, and pathways must be sought for pipes, conduit, and ductwork. While installations in accessible spaces (such as basements and attics) are relatively straightforward to estimate, labor costs may be more difficult to determine when delivery systems must be concealed.

22 40 00 Plumbing Fixtures

- Plumbing fixture costs usually require two lines: the fixture itself and its "rough-in, supply, and waste."
- In the Assemblies Section (Plumbing D2010) for the desired fixture, the System Components Group at the center of the page shows the fixture on the first line. The rest of the list (fittings, pipe, tubing, etc.) will total up to what we refer to in the Unit Price section as "Rough-in, supply, waste, and vent." Note that for most fixtures we allow a nominal 5′ of tubing to reach from the fixture to a main or riser.
- Remember that gas- and oil-fired units need venting.

Reference Numbers

Reference numbers are shown in shaded boxes at the beginning of some major classifications. These numbers refer to related items in the Reference Section. The reference information may be an estimating procedure, an alternate pricing method, or technical information.

Note: Not all subdivisions listed here necessarily appear in this publication.

22 01 02 – Labor Adjustments

22 01 02.10 Boilers, General		Crew	Daily Output	Labor-Hours	Unit	2013 Bare Costs Material	Labor	Equipment	Total	Total Incl O&P
0010	**BOILERS, GENERAL**, Prices do not include flue piping, elec. wiring,									
0020	gas or oil piping, boiler base, pad, or tankless unless noted									
0100	Boiler H.P.: 10 KW = 34 lb./steam/hr. = 33,475 BTU/hr.									
0150	To convert SFR to BTU rating: Hot water, 150 x SFR;									
0160	Forced hot water, 180 x SFR; steam, 240 x SFR									

22 01 02.20 Labor Adjustment Factors		Crew	Daily Output	Labor-Hours	Unit	2013 Bare Costs Material	Labor	Equipment	Total	Total Incl O&P
0010	**LABOR ADJUSTMENT FACTORS**, (For Div. 21, 22 and 23) R220102-20									
0100	Labor factors, The below are reasonable suggestions, however									
0110	each project must be evaluated for its own peculiarities, and									
0120	the adjustments be increased or decreased depending on the									
0130	severity of the special conditions.									
1000	Add to labor for elevated installation (Above floor level)									
1080	10′ to 14.5′ high						10%			
1100	15′ to 19.5′ high						20%			
1120	20′ to 24.5′ high						25%			
1140	25′ to 29.5′ high						35%			
1160	30′ to 34.5′ high						40%			
1180	35′ to 39.5′ high						50%			
1200	40′ and higher						55%			
2000	Add to labor for crawl space									
2100	3′ high						40%			
2140	4′ high						30%			
3000	Add to labor for multi-story building									
3100	Add per floor for floors 3 thru 19						2%			
3140	Add per floor for floors 20 and up						4%			
4000	Add to labor for working in existing occupied buildings									
4100	Hospital						35%			
4140	Office building						25%			
4180	School						20%			
4220	Factory or warehouse						15%			
4260	Multi dwelling						15%			
5000	Add to labor, miscellaneous									
5100	Cramped shaft						35%			
5140	Congested area						15%			
5180	Excessive heat or cold						30%			
9010	each project should be evaluated for its own peculiarities.									
9100	Other factors to be considered are:									
9140	Movement of material and equipment through finished areas									
9180	Equipment room									
9220	Attic space									
9260	No service road									
9300	Poor unloading/storage area									
9340	Congested site area/heavy traffic									

	22 05 29.10 Hangers & Supp. for Plumb'g/HVAC Pipe/Equip.	Crew	Daily Output	Labor-Hours	Unit	Material	2013 Bare Costs Labor	Equipment	Total	Total Incl O&P
0010	**HANGERS AND SUPPORTS FOR PLUMB'G/HVAC PIPE/EQUIP.**									
0011	TYPE numbers per MSS-SP58									
0050	Brackets									
0060	Beam side or wall, malleable iron, TYPE 34									
0070	3/8" threaded rod size	1 Plum	48	.167	Ea.	3.62	9.30		12.92	18
0080	1/2" threaded rod size		48	.167		6.40	9.30		15.70	21
0090	5/8" threaded rod size		48	.167		9.80	9.30		19.10	25
0100	3/4" threaded rod size		48	.167		11.85	9.30		21.15	27
0110	7/8" threaded rod size		48	.167		13.95	9.30		23.25	29.50
0120	For concrete installation, add						30%			
0150	Wall, welded steel, medium, TYPE 32									
0160	0 size, 12" wide, 18" deep	1 Plum	34	.235	Ea.	241	13.15		254.15	285
0170	1 size, 18" wide, 24" deep		34	.235		286	13.15		299.15	335
0180	2 size, 24" wide, 30" deep		34	.235		380	13.15		393.15	435
0300	Clamps									
0310	C-clamp, for mounting on steel beam flange, w/locknut, TYPE 23									
0320	3/8" threaded rod size	1 Plum	160	.050	Ea.	3.51	2.79		6.30	8.05
0330	1/2" threaded rod size		160	.050		4.37	2.79		7.16	9
0340	5/8" threaded rod size		160	.050		6.70	2.79		9.49	11.55
0350	3/4" threaded rod size		160	.050		9	2.79		11.79	14.10
0750	Riser or extension pipe, carbon steel, TYPE 8									
0760	3/4" pipe size	1 Plum	48	.167	Ea.	5.05	9.30		14.35	19.55
0770	1" pipe size		47	.170		5.25	9.50		14.75	20
0780	1-1/4" pipe size		46	.174		6.45	9.70		16.15	21.50
0790	1-1/2" pipe size		45	.178		6.95	9.90		16.85	22.50
0800	2" pipe size		43	.186		7.20	10.40		17.60	23.50
0810	2-1/2" pipe size		41	.195		7.85	10.90		18.75	25
0820	3" pipe size		40	.200		8.70	11.15		19.85	26.50
0830	3-1/2" pipe size		39	.205		10.60	11.45		22.05	29
0840	4" pipe size		38	.211		10.85	11.75		22.60	29.50
0850	5" pipe size		37	.216		14.70	12.05		26.75	34.50
0860	6" pipe size		36	.222		18.20	12.40		30.60	38.50
1150	Insert, concrete									
1160	Wedge type, carbon steel body, malleable iron nut, galvanized									
1170	1/4" threaded rod size	1 Plum	96	.083	Ea.	2.25	4.65		6.90	9.50
1180	3/8" threaded rod size		96	.083		2.26	4.65		6.91	9.50
1190	1/2" threaded rod size		96	.083		2.54	4.65		7.19	9.80
1200	5/8" threaded rod size		96	.083		3.02	4.65		7.67	10.30
1210	3/4" threaded rod size		96	.083		6.05	4.65		10.70	13.65
1220	7/8" threaded rod size		96	.083		7.85	4.65		12.50	15.65
2650	Rods, carbon steel									
2660	Continuous thread									
2670	1/4" thread size	1 Plum	144	.056	L.F.	1.53	3.10		4.63	6.35
2680	3/8" thread size		144	.056		1.63	3.10		4.73	6.45
2690	1/2" thread size		144	.056		2.57	3.10		5.67	7.50
2700	5/8" thread size		144	.056		3.65	3.10		6.75	8.70
2710	3/4" thread size		144	.056		6.40	3.10		9.50	11.70
2720	7/8" thread size		144	.056		8.05	3.10		11.15	13.50
2725	1/4" thread size, bright finish		144	.056		1.90	3.10		5	6.75
2726	1/2" thread size, bright finish		144	.056		3.42	3.10		6.52	8.45
2730	For galvanized, add					40%				
2750	Both ends machine threaded 18" length									
2760	3/8" thread size	1 Plum	240	.033	Ea.	6.25	1.86		8.11	9.70

22 05 29 – Hangers and Supports for Plumbing Piping and Equipment

22 05 29.10 Hangers & Supp. for Plumb'g/HVAC Pipe/Equip.		Crew	Daily Output	Labor-Hours	Unit	2013 Bare Costs Material	Labor	Equipment	Total	Total Incl O&P
2770	1/2" thread size	1 Plum	240	.033	Ea.	10.05	1.86		11.91	13.85
2780	5/8" thread size		240	.033		14.40	1.86		16.26	18.65
2790	3/4" thread size		240	.033		23	1.86		24.86	28
2800	7/8" thread size		240	.033		31	1.86		32.86	37.50
2810	1" thread size	↓	240	.033	↓	37	1.86		38.86	43.50
4400	U-bolt, carbon steel									
4410	Standard, with nuts, TYPE 42									
4420	1/2" pipe size	1 Plum	160	.050	Ea.	2.09	2.79		4.88	6.50
4430	3/4" pipe size		158	.051		2.09	2.83		4.92	6.55
4450	1" pipe size		152	.053		2.15	2.94		5.09	6.80
4460	1-1/4" pipe size		148	.054		2.63	3.02		5.65	7.45
4470	1-1/2" pipe size		143	.056		2.75	3.12		5.87	7.75
4480	2" pipe size		139	.058		3	3.21		6.21	8.15
4490	2-1/2" pipe size		134	.060		4.97	3.33		8.30	10.45
4500	3" pipe size		128	.063		5.20	3.49		8.69	10.95
4510	3-1/2" pipe size		122	.066		5.50	3.66		9.16	11.55
4520	4" pipe size		117	.068		5.60	3.82		9.42	11.90
4530	5" pipe size		114	.070		5.65	3.92		9.57	12.10
4540	6" pipe size	↓	111	.072	↓	11.55	4.02		15.57	18.75
4580	For plastic coating on 1/2" thru 6" size, add					150%				
8800	Wire cable support system									
8810	Cable with hook terminal and locking device									
8830	2 mm, (.079") dia cable, (100 lb. cap.)									
8840	1 m, (3.3') length, with hook	1 Shee	96	.083	Ea.	3.03	4.44		7.47	10.10
8850	2 m, (6.6') length, with hook		84	.095		3.50	5.10		8.60	11.60
8860	3 m, (9.9') length, with hook	↓	72	.111		3.97	5.90		9.87	13.40
8870	5 m, (16.4') length, with hook	Q-9	60	.267		4.97	12.80		17.77	25
8880	10 m, (32.8') length, with hook	"	30	.533	↓	7.20	25.50		32.70	47
8900	3mm, (.118") dia cable, (200 lb. cap.)									
8910	1 m, (3.3') length, with hook	1 Shee	96	.083	Ea.	3.88	4.44		8.32	11
8920	2 m, (6.6') length, with hook		84	.095		4.36	5.10		9.46	12.55
8930	3 m, (9.9') length, with hook	↓	72	.111		4.81	5.90		10.71	14.35
8940	5 m, (16.4') length, with hook	Q-9	60	.267		5.95	12.80		18.75	26
8950	10 m, (32.8') length, with hook	"	30	.533	↓	8.55	25.50		34.05	48.50
9000	Cable system accessories									
9010	Anchor bolt, 3/8", with nut	1 Shee	140	.057	Ea.	1.16	3.05		4.21	5.95
9020	Air duct corner protector		160	.050		.52	2.67		3.19	4.64
9030	Air duct support attachment	↓	140	.057	↓	.99	3.05		4.04	5.75
9040	Flange clip, hammer-on style									
9044	For flange thickness 3/32" - 9/64", 160 lb. cap.	1 Shee	180	.044	Ea.	.29	2.37		2.66	3.93
9048	For flange thickness 1/8" - 1/4", 200 lb. cap.		160	.050		.29	2.67		2.96	4.39
9052	For flange thickness 5/16" - 1/2", 200 lb. cap.		150	.053		.56	2.84		3.40	4.96
9056	For flange thickness 9/16" - 3/4", 200 lb. cap.		140	.057	↓	.75	3.05		3.80	5.50
9060	Wire insulation protection tube	↓	180	.044	L.F.	.34	2.37		2.71	3.98
9070	Wire cutter				Ea.	41.50			41.50	45.50

22 33 Electric Domestic Water Heaters

22 33 13 – Instantaneous Electric Domestic Water Heaters

22 33 13.10 Hot Water Dispensers		Crew	Daily Output	Labor-Hours	Unit	2013 Bare Costs Material	Labor	Equipment	Total	Total Incl O&P
0010	**HOT WATER DISPENSERS**									
0160	Commercial, 100 cup, 11.3 amp	1 Plum	14	.571	Ea.	510	32		542	615
3180	Household, 60 cup	"	14	.571	"	269	32		301	345

22 33 30 – Residential, Electric Domestic Water Heaters

22 33 30.13 Residential, Small-Capacity Elec. Water Heaters

		Crew	Daily Output	Labor-Hours	Unit	Material	Labor	Equipment	Total	Total Incl O&P
0010	**RESIDENTIAL, SMALL-CAPACITY ELECTRIC DOMESTIC WATER HEATERS**									
1000	Residential, electric, glass lined tank, 5 yr., 10 gal., single element R224000-10	1 Plum	2.30	3.478	Ea.	305	194		499	625
1040	20 gallon, single element R224000-20		2.20	3.636		380	203		583	720
1060	30 gallon, double element		2.20	3.636		695	203		898	1,075
1080	40 gallon, double element		2	4		740	223		963	1,150
1100	52 gallon, double element		2	4		830	223		1,053	1,250
1120	66 gallon, double element		1.80	4.444		1,125	248		1,373	1,600
1140	80 gallon, double element		1.60	5		1,250	279		1,529	1,800
1180	120 gallon, double element		1.40	5.714		1,750	320		2,070	2,400

22 33 33 – Light-Commercial Electric Domestic Water Heaters

22 33 33.10 Commercial Electric Water Heaters

		Crew	Daily Output	Labor-Hours	Unit	Material	Labor	Equipment	Total	Total Incl O&P
0010	**COMMERCIAL ELECTRIC WATER HEATERS**									
4000	Commercial, 100° rise. NOTE: for each size tank, a range of									
4010	heaters between the ones shown are available									
4020	Electric									
4100	5 gal., 3 kW, 12 GPH, 208 volt	1 Plum	2	4	Ea.	2,625	223		2,848	3,200
4120	10 gal., 6 kW, 25 GPH, 208 volt		2	4		2,900	223		3,123	3,525
4140	50 gal., 9 kW, 37 GPH, 208 volt		1.80	4.444		3,975	248		4,223	4,750
4160	50 gal., 36 kW, 148 GPH, 208 volt		1.80	4.444		6,075	248		6,323	7,075
4180	80 gal., 12 kW, 49 GPH, 208 volt		1.50	5.333		4,900	298		5,198	5,850
4200	80 gal., 36 kW, 148 GPH, 208 volt		1.50	5.333		6,775	298		7,073	7,925
4220	100 gal., 36 kW, 148 GPH, 208 volt		1.20	6.667		7,075	370		7,445	8,325
4240	120 gal., 36 kW, 148 GPH, 208 volt		1.20	6.667		7,375	370		7,745	8,650
4260	150 gal., 15 kW , 61 GPH, 480 volt		1	8		17,300	445		17,745	19,700
4280	150 gal., 120 kW, 490 GPH, 480 volt		1	8		24,300	445		24,745	27,500
4300	200 gal., 15 kW, 61 GPH, 480 volt	Q-1	1.70	9.412		18,800	475		19,275	21,300
4320	200 gal., 120 kW , 490 GPH, 480 volt		1.70	9.412		25,600	475		26,075	28,900
4340	250 gal., 15 kW, 61 GPH, 480 volt		1.50	10.667		19,300	535		19,835	22,100
4360	250 gal., 150 kW, 615 GPH, 480 volt		1.50	10.667		28,200	535		28,735	31,800
4380	300 gal., 30 kW, 123 GPH, 480 volt		1.30	12.308		21,300	620		21,920	24,300
4400	300 gal., 180 kW, 738 GPH, 480 volt		1.30	12.308		36,800	620		37,420	41,300
4420	350 gal., 30 kW, 123 GPH, 480 volt		1.10	14.545		22,500	730		23,230	25,900
4440	350 gal., 180 kW, 738 GPH, 480 volt		1.10	14.545		31,700	730		32,430	36,000
4460	400 gal., 30 kW, 123 GPH, 480 volt		1	16		25,500	805		26,305	29,300
4480	400 gal., 210 kW, 860 GPH, 480 volt		1	16		36,600	805		37,405	41,500
4500	500 gal., 30 kW, 123 GPH, 480 volt		.80	20		29,800	1,000		30,800	34,200
4520	500 gal., 240 kW, 984 GPH, 480 volt		.80	20		52,500	1,000		53,500	59,000
4540	600 gal., 30 kW, 123 GPH, 480 volt	Q-2	1.20	20		26,000	1,050		27,050	30,200
4560	600 gal., 300 kW, 1230 GPH, 480 volt		1.20	20		34,400	1,050		35,450	39,400
4580	700 gal., 30 kW, 123 GPH, 480 volt		1	24		23,800	1,250		25,050	28,100
4600	700 gal., 300 kW, 1230 GPH, 480 volt		1	24		35,800	1,250		37,050	41,300
4620	800 gal., 60 kW, 245 GPH, 480 volt		.90	26.667		25,800	1,400		27,200	30,500
4640	800 gal., 300 kW, 1230 GPH, 480 volt		.90	26.667		36,600	1,400		38,000	42,400
4660	1000 gal., 60 kW, 245 GPH, 480 volt		.70	34.286		28,200	1,775		29,975	33,700
4680	1000 gal., 480 kW, 1970 GPH, 480 volt		.70	34.286		46,800	1,775		48,575	54,000
4700	1200 gal., 60 kW, 245 GPH, 480 volt		.60	40		48,000	2,075		50,075	56,000
4720	1200 gal., 480 kW, 1970 GPH, 480 volt		.60	40		73,000	2,075		75,075	83,000

22 33 Electric Domestic Water Heaters

22 33 33 – Light-Commercial Electric Domestic Water Heaters

22 33 33.10 Commercial Electric Water Heaters		Crew	Daily Output	Labor-Hours	Unit	Material	2013 Bare Costs Labor	Equipment	Total	Total Incl O&P
4740	1500 gal., 60 kW, 245 GPH, 480 volt	Q-2	.50	48	Ea.	63,000	2,500		65,500	73,500
4760	1500 gal., 480 kW, 1970 GPH, 480 volt	↓	.50	48		87,500	2,500		90,000	100,500
5400	Modulating step control, 2-5 steps	1 Elec	5.30	1.509		900	79		979	1,100
5440	6-10 steps		3.20	2.500		1,150	131		1,281	1,475
5460	11-15 steps		2.70	2.963		1,625	155		1,780	2,025
5480	16-20 steps	↓	1.60	5	↓	2,400	262		2,662	3,050

22 34 Fuel-Fired Domestic Water Heaters

22 34 30 – Residential Gas Domestic Water Heaters

22 34 30.13 Residential, Atmos, Gas Domestic Wtr Heaters		Crew	Daily Output	Labor-Hours	Unit	Material	Labor	Equipment	Total	Total Incl O&P
0010	**RESIDENTIAL, ATMOSPHERIC, GAS DOMESTIC WATER HEATERS**									
2000	Gas fired, foam lined tank, 10 yr., vent not incl.									
2040	30 gallon	1 Plum	2	4	Ea.	895	223		1,118	1,325
2060	40 gallon		1.90	4.211		895	235		1,130	1,350
2080	50 gallon		1.80	4.444		935	248		1,183	1,400
2090	60 gallon		1.70	4.706		1,275	263		1,538	1,800
2100	75 gallon		1.50	5.333		1,350	298		1,648	1,950
2120	100 gallon		1.30	6.154		1,525	345		1,870	2,200
2900	Water heater, safety-drain pan, 26" round	↓	20	.400	↓	37	22.50		59.50	74.50

22 34 46 – Oil-Fired Domestic Water Heaters

22 34 46.10 Residential Oil-Fired Water Heaters		Crew	Daily Output	Labor-Hours	Unit	Material	Labor	Equipment	Total	Total Incl O&P
0010	**RESIDENTIAL OIL-FIRED WATER HEATERS**									
3000	Oil fired, glass lined tank, 5 yr., vent not included, 30 gallon	1 Plum	2	4	Ea.	1,050	223		1,273	1,475
3040	50 gallon		1.80	4.444		1,250	248		1,498	1,750
3060	70 gallon	↓	1.50	5.333	↓	1,775	298		2,073	2,425

Estimating Tips

The labor adjustment factors listed in Subdivision 22 01 02.20 also apply to Division 23.

23 10 00 Facility Fuel Systems

- The prices in this subdivision for above- and below-ground storage tanks do not include foundations or hold-down slabs, unless noted. The estimator should refer to Divisions 3 and 31 for foundation system pricing. In addition to the foundations, required tank accessories, such as tank gauges, leak detection devices, and additional manholes and piping, must be added to the tank prices.

23 50 00 Central Heating Equipment

- When estimating the cost of an HVAC system, check to see who is responsible for providing and installing the temperature control system. It is possible to overlook controls, assuming that they would be included in the electrical estimate.
- When looking up a boiler, be careful on specified capacity. Some manufacturers rate their products on output while others use input.
- Include HVAC insulation for pipe, boiler, and duct (wrap and liner).
- Be careful when looking up mechanical items to get the correct pressure rating and connection type (thread, weld, flange).

23 70 00 Central HVAC Equipment

- Combination heating and cooling units are sized by the air conditioning requirements. (See Reference No. R236000-20 for preliminary sizing guide.)
- A ton of air conditioning is nominally 400 CFM.
- Rectangular duct is taken off by the linear foot for each size, but its cost is usually estimated by the pound. Remember that SMACNA standards now base duct on internal pressure.
- Prefabricated duct is estimated and purchased like pipe: straight sections and fittings.
- Note that cranes or other lifting equipment are not included on any lines in Division 23. For example, if a crane is required to lift a heavy piece of pipe into place high above a gym floor, or to put a rooftop unit on the roof of a four-story building, etc., it must be added. Due to the potential for extreme variation—from nothing additional required to a major crane or helicopter—we feel that including a nominal amount for "lifting contingency" would be useless and detract from the accuracy of the estimate. When using equipment rental cost data from RSMeans, do not forget to include the cost of the operator(s).

Reference Numbers

Reference numbers are shown in shaded boxes at the beginning of some major classifications. These numbers refer to related items in the Reference Section. The reference information may be an estimating procedure, an alternate pricing method, or technical information.

Note: Not all subdivisions listed here necessarily appear in this publication.

Note: **Trade Service,** *in part, has been used as a reference source for some of the material prices used in Division 23.*

23 05 Common Work Results for HVAC

23 05 02 – HVAC General

23 05 02.10 Air Conditioning, General		Crew	Daily Output	Labor-Hours	Unit	2013 Bare Costs Material	Labor	Equipment	Total	Total Incl O&P
0010	**AIR CONDITIONING, GENERAL** Prices are for standard efficiencies (SEER 13)									
0020	for upgrade to SEER 14 add					10%				

23 09 Instrumentation and Control for HVAC

23 09 23 – Direct-Digital Control System for HVAC

23 09 23.10 Control Components/DDC Systems		Crew	Daily Output	Labor-Hours	Unit	Material	Labor	Equipment	Total	Total Incl O&P
0010	**CONTROL COMPONENTS/DDC SYSTEMS** (Sub's quote incl. M & L)									
0100	Analog inputs									
0110	Sensors (avg. 50′ run in 1/2″ EMT)									
0120	Duct temperature				Ea.				360	395
0130	Space temperature								575	635
0140	Duct humidity, +/- 3%								605	665
0150	Space humidity, +/- 2%								925	1,025
0160	Duct static pressure								490	540
0170	CFM/Transducer								660	730
0180	KW/Transducer								1,175	1,300
0182	KWH totalization (not incl. elec. meter pulse xmtr.)								545	600
0190	Space static pressure								925	1,025
1000	Analog outputs (avg. 50′ run in 1/2″ EMT)									
1010	P/I Transducer				Ea.				550	605
1020	Analog output, matl. in MUX								265	291
1030	Pneumatic (not incl. control device)								560	615
1040	Electric (not incl control device)								330	365
2000	Status (Alarms)									
2100	Digital inputs (avg. 50′ run in 1/2″ EMT)									
2110	Freeze				Ea.				380	415
2120	Fire								340	375
2130	Differential pressure, (air)								520	570
2140	Differential pressure, (water)								850	935
2150	Current sensor								380	415
2160	Duct high temperature thermostat								495	545
2170	Duct smoke detector								615	675
2200	Digital output (avg. 50′ run in 1/2″ EMT)									
2210	Start/stop				Ea.				297	325
2220	On/off (maintained contact)				"				510	560
3000	Controller MUX panel, incl. function boards									
3100	48 point				Ea.				4,575	5,050
3110	128 point				"				6,300	6,925
3200	DDC controller (avg. 50′ run in conduit)									
3210	Mechanical room									
3214	16 point controller (incl. 120 volt/1 phase power supply)				Ea.				1,925	3,125
3229	32 point controller (incl. 120 volt/1 phase power supply)				"				4,725	5,200
3230	Includes software programming and checkout									
3260	Space									
3266	VAV terminal box (incl. space temp. sensor)				Ea.				735	805
3280	Host computer (avg. 50′ run in conduit)									
3281	Package complete with PC, keyboard,									
3282	printer, color CRT, modem, basic software				Ea.				8,500	9,375
4000	Front end costs									
4100	Computer (P.C.)/software program				Ea.				5,675	6,250
4200	Color graphics software								3,400	3,750

23 09 23 – Direct-Digital Control System for HVAC

23 09 23.10 Control Components/DDC Systems		Crew	Daily Output	Labor-Hours	Unit	2013 Bare Costs Material	Labor	Equipment	Total	Total Incl O&P
4300	Color graphics slides				Ea.				425	470
4350	Additional dot matrix printer				↓				850	935
4400	Communications trunk cable				L.F.				3.31	3.64
4500	Engineering labor, (not incl. dftg.)				Point				82	90
4600	Calibration labor								105	115
4700	Start-up, checkout labor				↓				105	115
4800	Programming labor, as req'd									
5000	Communications bus (data transmission cable)									
5010	#18 twisted shielded pair in 1/2" EMT conduit				C.L.F.				330	365
8000	Applications software									
8050	Basic maintenance manager software (not incl. data base entry)				Ea.				1,700	1,875
8100	Time program				Point				5.95	6.55
8120	Duty cycle								11.85	13.05
8140	Optimum start/stop								36	39.50
8160	Demand limiting								17.80	19.55
8180	Enthalpy program				↓				36	39.50
8200	Boiler optimization				Ea.				1,075	1,175
8220	Chiller optimization				"				1,425	1,550
8240	Custom applications									
8260	Cost varies with complexity									

23 09 43 – Pneumatic Control System for HVAC

23 09 43.10 Pneumatic Control Systems		Crew	Daily Output	Labor-Hours	Unit	Material	Labor	Equipment	Total	Total Incl O&P
0010	**PNEUMATIC CONTROL SYSTEMS**									
0011	Including a nominal 50 ft. of tubing. Add control panelboard if req'd.									
0100	Heating and ventilating, split system									
0200	Mixed air control, economizer cycle, panel readout, tubing									
0220	Up to 10 tons G	Q-19	.68	35.294	Ea.	3,925	1,825		5,750	7,050
0240	For 10 to 20 tons G		.63	37.915		4,200	1,950		6,150	7,550
0260	For over 20 tons G		.58	41.096		4,550	2,125		6,675	8,175
0270	Enthalpy cycle, up to 10 tons		.50	48.387		4,325	2,500		6,825	8,525
0280	For 10 to 20 tons		.46	52.174		4,675	2,675		7,350	9,175
0290	For over 20 tons	↓	.42	56.604	↓	5,075	2,900		7,975	9,950
0300	Heating coil, hot water, 3 way valve,									
0320	Freezestat, limit control on discharge, readout	Q-5	.69	23.088	Ea.	2,900	1,175		4,075	4,975
0500	Cooling coil, chilled water, room									
0520	Thermostat, 3 way valve	Q-5	2	8	Ea.	1,300	410		1,710	2,050
0600	Cooling tower, fan cycle, damper control,									
0620	Control system including water readout in/out at panel	Q-19	.67	35.821	Ea.	5,150	1,850		7,000	8,450
1000	Unit ventilator, day/night operation,									
1100	freezestat, ASHRAE, cycle 2	Q-19	.91	26.374	Ea.	2,850	1,350		4,200	5,200
2000	Compensated hot water from boiler, valve control,									
2100	readout and reset at panel, up to 60 GPM	Q-19	.55	43.956	Ea.	5,350	2,250		7,600	9,275
2120	For 120 GPM		.51	47.059		5,725	2,425		8,150	9,925
2140	For 240 GPM		.49	49.180		5,975	2,525		8,500	10,400
3000	Boiler room combustion air, damper to 5 S.F., controls		1.37	17.582		2,575	905		3,480	4,175
3500	Fan coil, heating and cooling valves, 4 pipe control system		3	8		1,150	410		1,560	1,900
3600	Heat exchanger system controls	↓	.86	27.907	↓	2,500	1,425		3,925	4,900
3900	Multizone control (one per zone), includes thermostat, damper									
3910	motor and reset of discharge temperature	Q-5	.51	31.373	Ea.	2,575	1,600		4,175	5,225
4000	Pneumatic thermostat, including controlling room radiator valve	"	2.43	6.593		775	335		1,110	1,350
4040	Program energy saving optimizer G	Q-19	1.21	19.786		6,450	1,025		7,475	8,625
4060	Pump control system	"	3	8		1,200	410		1,610	1,925

23 09 Instrumentation and Control for HVAC

23 09 43 – Pneumatic Control System for HVAC

23 09 43.10 Pneumatic Control Systems		Crew	Daily Output	Labor-Hours	Unit	Material	2013 Bare Costs Labor	Equipment	Total	Total Incl O&P
4080	Reheat coil control system, not incl coil	Q-5	2.43	6.593	Ea.	1,000	335		1,335	1,600
4500	Air supply for pneumatic control system									
4600	Tank mounted duplex compressor, starter, alternator,									
4620	piping, dryer, PRV station and filter									
4630	1/2 HP	Q-19	.68	35.139	Ea.	9,575	1,800		11,375	13,200
4640	3/4 HP		.64	37.383		10,000	1,925		11,925	13,900
4650	1 HP		.61	39.539		11,000	2,025		13,025	15,200
4660	1-1/2 HP		.58	41.739		11,700	2,150		13,850	16,100
4680	3 HP		.55	43.956		15,900	2,250		18,150	20,900
4690	5 HP	↓	.42	57.143	↓	27,800	2,950		30,750	35,000
4800	Main air supply, includes 3/8" copper main and labor	Q-5	1.82	8.791	C.L.F.	320	450		770	1,025
4810	If poly tubing used, deduct								30%	30%
7000	Static pressure control for air handling unit, includes pressure									
7010	sensor, receiver controller, readout and damper motors	Q-19	.64	37.383	Ea.	7,625	1,925		9,550	11,300
7020	If return air fan requires control, add								70%	70%
8600	VAV boxes, incl. thermostat, damper motor, reheat coil & tubing	Q-5	1.46	10.989		1,200	560		1,760	2,175
8610	If no reheat coil, deduct				↓					204

23 09 53 – Pneumatic and Electric Control System for HVAC

23 09 53.10 Control Components		Crew	Daily Output	Labor-Hours	Unit	Material	2013 Bare Costs Labor	Equipment	Total	Total Incl O&P
0010	**CONTROL COMPONENTS**									
0700	Controller, receiver									
0850	Electric, single snap switch	1 Elec	4	2	Ea.	455	105		560	655
0860	Dual snap switches	"	3	2.667	"	610	140		750	885
3590	Sensor, electric operated									
3620	Humidity	1 Elec	8	1	Ea.	121	52.50		173.50	212
3650	Pressure		8	1		1,700	52.50		1,752.50	1,950
3680	Temperature	↓	10	.800	↓	107	42		149	180
5000	Thermostats									
5200	24 hour, automatic, clock G	1 Shee	8	1	Ea.	153	53.50		206.50	250
5220	Electric, low voltage, 2 wire	1 Elec	13	.615		36.50	32.50		69	88.50
5230	3 wire	"	10	.800	↓	33.50	42		75.50	99.50
7090	Valves, motor controlled, including actuator									
7100	Electric motor actuated									
7200	Brass, two way, screwed									
7210	1/2" pipe size	L-6	36	.333	Ea.	270	18.20		288.20	325
7220	3/4" pipe size		30	.400		475	22		497	555
7230	1" pipe size		28	.429		570	23.50		593.50	660
7240	1-1/2" pipe size		19	.632		620	34.50		654.50	735
7250	2" pipe size	↓	16	.750	↓	900	41		941	1,050
7350	Brass, three way, screwed									
7360	1/2" pipe size	L-6	33	.364	Ea.	280	19.90		299.90	335
7370	3/4" pipe size		27	.444		325	24.50		349.50	395
7380	1" pipe size		25.50	.471		425	25.50		450.50	505
7384	1-1/4" pipe size		21	.571		540	31		571	635
7390	1-1/2" pipe size		17	.706		615	38.50		653.50	735
7400	2" pipe size	↓	14	.857	↓	820	47		867	970
7550	Iron body, two way, flanged									
7560	2-1/2" pipe size	L-6	4	3	Ea.	1,225	164		1,389	1,600
7570	3" pipe size		3	4		1,275	219		1,494	1,750
7580	4" pipe size	↓	2	6	↓	2,225	330		2,555	2,950
7850	Iron body, three way, flanged									
7860	2-1/2" pipe size	L-6	3	4	Ea.	1,225	219		1,444	1,675

23 09 Instrumentation and Control for HVAC

23 09 53 – Pneumatic and Electric Control System for HVAC

23 09 53.10 Control Components		Crew	Daily Output	Labor-Hours	Unit	Material	Labor	Equipment	Total	Total Incl O&P
						2013 Bare Costs				
7870	3" pipe size	L-6	2.50	4.800	Ea.	1,425	262		1,687	1,975
7880	4" pipe size	↓	2	6	↓	1,775	330		2,105	2,450

23 34 HVAC Fans

23 34 13 – Axial HVAC Fans

23 34 13.10 Axial Flow HVAC Fans

		Crew	Daily Output	Labor-Hours	Unit	Material	Labor	Equipment	Total	Total Incl O&P
0010	**AXIAL FLOW HVAC FANS**									
0020	Air conditioning and process air handling									
0030	Axial flow, compact, low sound, 2.5" S.P.									
0050	3,800 CFM, 5 HP	Q-20	3.40	5.882	Ea.	4,900	287		5,187	5,800
0080	6,400 CFM, 5 HP		2.80	7.143		5,475	350		5,825	6,525
0100	10,500 CFM, 7-1/2 HP		2.40	8.333		6,800	405		7,205	8,100
0120	15,600 CFM, 10 HP		1.60	12.500		8,575	610		9,185	10,400
0140	23,000 CFM, 15 HP		.70	28.571		13,200	1,400		14,600	16,600
0160	28,000 CFM, 20 HP		.40	50		14,700	2,450		17,150	19,800
1500	Vaneaxial, low pressure, 2000 CFM, 1/2 HP		3.60	5.556		2,100	271		2,371	2,700
1520	4,000 CFM, 1 HP		3.20	6.250		2,450	305		2,755	3,150
1540	8,000 CFM, 2 HP		2.80	7.143		3,100	350		3,450	3,925
1560	16,000 CFM, 5 HP	↓	2.40	8.333	↓	4,350	405		4,755	5,400

23 34 14 – Blower HVAC Fans

23 34 14.10 Blower Type HVAC Fans

		Crew	Daily Output	Labor-Hours	Unit	Material	Labor	Equipment	Total	Total Incl O&P
0010	**BLOWER TYPE HVAC FANS**									
2000	Blowers, direct drive with motor, complete									
2020	1045 CFM @ .5" S.P., 1/5 HP	Q-20	18	1.111	Ea.	294	54.50		348.50	410
2040	1385 CFM @ .5" S.P., 1/4 HP		18	1.111		296	54.50		350.50	410
2060	1640 CFM @ .5" S.P., 1/3 HP		18	1.111		264	54.50		318.50	375
2080	1760 CFM @ .5" S.P., 1/2 HP	↓	18	1.111	↓	275	54.50		329.50	385
2090	4 speed									
2100	1164 to 1739 CFM @ .5" S.P., 1/3 HP	Q-20	16	1.250	Ea.	283	61		344	405
2120	1467 to 2218 CFM @ 1.0" S.P., 3/4 HP	"	14	1.429	"	330	70		400	465
2500	Ceiling fan, right angle, extra quiet, 0.10" S.P.									
2520	95 CFM	Q-20	20	1	Ea.	273	49		322	375
2540	210 CFM		19	1.053		325	51.50		376.50	435
2560	385 CFM		18	1.111		410	54.50		464.50	535
2580	885 CFM		16	1.250		810	61		871	985
2600	1,650 CFM		13	1.538		1,125	75		1,200	1,350
2620	2,960 CFM	↓	11	1.818		1,500	89		1,589	1,775
2680	For speed control switch, add	1 Elec	16	.500	↓	149	26		175	203
7500	Utility set, steel construction, pedestal, 1/4" S.P.									
7520	Direct drive, 150 CFM, 1/8 HP	Q-20	6.40	3.125	Ea.	820	153		973	1,125
7540	485 CFM, 1/6 HP		5.80	3.448		1,025	168		1,193	1,375
7560	1950 CFM, 1/2 HP		4.80	4.167		1,200	204		1,404	1,625
7580	2410 CFM, 3/4 HP		4.40	4.545		2,225	222		2,447	2,775
7600	3328 CFM, 1-1/2 HP	↓	3	6.667	↓	2,500	325		2,825	3,250
7680	V-belt drive, drive cover, 3 phase									
7700	800 CFM, 1/4 HP	Q-20	6	3.333	Ea.	925	163		1,088	1,275
7720	1,300 CFM, 1/3 HP		5	4		970	195		1,165	1,375
7740	2,000 CFM, 1 HP		4.60	4.348		1,150	212		1,362	1,600
7760	2,900 CFM, 3/4 HP		4.20	4.762		1,550	233		1,783	2,050
7780	3,600 CFM, 3/4 HP		4	5		1,900	244		2,144	2,475

23 34 HVAC Fans

23 34 14 – Blower HVAC Fans

23 34 14.10 Blower Type HVAC Fans		Crew	Daily Output	Labor-Hours	Unit	Material	2013 Bare Costs Labor	Equipment	Total	Total Incl O&P
7800	4,800 CFM, 1 HP	Q-20	3.50	5.714	Ea.	2,250	279		2,529	2,900
7820	6,700 CFM, 1-1/2 HP		3	6.667		2,775	325		3,100	3,550
7830	7,500 CFM, 2 HP		2.50	8		3,775	390		4,165	4,750
7840	11,000 CFM, 3 HP		2	10		5,025	490		5,515	6,300
7860	13,000 CFM, 3 HP		1.60	12.500		5,125	610		5,735	6,550
7880	15,000 CFM, 5 HP		1	20		5,300	975		6,275	7,300
7900	17,000 CFM, 7-1/2 HP		.80	25		5,675	1,225		6,900	8,075
7920	20,000 CFM, 7-1/2 HP	↓	.80	25	↓	6,750	1,225		7,975	9,275

23 34 16 – Centrifugal HVAC Fans

23 34 16.10 Centrifugal Type HVAC Fans		Crew	Daily Output	Labor-Hours	Unit	Material	Labor	Equipment	Total	Total Incl O&P
0010	**CENTRIFUGAL TYPE HVAC FANS**									
0200	In-line centrifugal, supply/exhaust booster									
0220	aluminum wheel/hub, disconnect switch, 1/4" S.P.									
0240	500 CFM, 10" diameter connection	Q-20	3	6.667	Ea.	1,200	325		1,525	1,825
0260	1,380 CFM, 12" diameter connection		2	10		1,275	490		1,765	2,175
0280	1,520 CFM, 16" diameter connection		2	10		1,400	490		1,890	2,300
0300	2,560 CFM, 18" diameter connection		1	20		1,525	975		2,500	3,150
0320	3,480 CFM, 20" diameter connection		.80	25		1,800	1,225		3,025	3,825
0326	5,080 CFM, 20" diameter connection	↓	.75	26.667	↓	1,975	1,300		3,275	4,125
3500	Centrifugal, airfoil, motor and drive, complete									
3520	1000 CFM, 1/2 HP	Q-20	2.50	8	Ea.	1,800	390		2,190	2,575
3540	2,000 CFM, 1 HP		2	10		2,025	490		2,515	2,975
3560	4,000 CFM, 3 HP		1.80	11.111		2,575	545		3,120	3,650
3580	8,000 CFM, 7-1/2 HP		1.40	14.286		3,850	700		4,550	5,300
3600	12,000 CFM, 10 HP	↓	1	20	↓	5,150	975		6,125	7,125
4500	Corrosive fume resistant, plastic									
4600	roof ventilators, centrifugal, V belt drive, motor									
4620	1/4" S.P., 250 CFM, 1/4 HP	Q-20	6	3.333	Ea.	4,300	163		4,463	5,000
4640	895 CFM, 1/3 HP		5	4		4,675	195		4,870	5,450
4660	1630 CFM, 1/2 HP		4	5		5,525	244		5,769	6,450
4680	2240 CFM, 1 HP		3	6.667		5,775	325		6,100	6,850
4700	3810 CFM, 2 HP		2	10		6,400	490		6,890	7,800
4710	5000 CFM, 2 HP		1.80	11.111		13,500	545		14,045	15,600
4715	8000 CFM, 5 HP		1.40	14.286		15,000	700		15,700	17,600
4720	11760 CFM, 5 HP		1	20		15,200	975		16,175	18,300
4740	18810 CFM, 10 HP	↓	.70	28.571	↓	15,800	1,400		17,200	19,500
4800	For intermediate capacity, motors may be varied									
4810	For explosion proof motor, add				Ea.	15%				
5000	Utility set, centrifugal, V belt drive, motor									
5020	1/4" S.P., 1200 CFM, 1/4 HP	Q-20	6	3.333	Ea.	1,550	163		1,713	1,950
5040	1520 CFM, 1/3 HP		5	4		1,925	195		2,120	2,425
5060	1850 CFM, 1/2 HP		4	5		1,925	244		2,169	2,500
5080	2180 CFM, 3/4 HP		3	6.667		2,300	325		2,625	3,050
5100	1/2" S.P., 3600 CFM, 1 HP		2	10		2,625	490		3,115	3,625
5120	4250 CFM, 1-1/2 HP		1.60	12.500		2,900	610		3,510	4,125
5140	4800 CFM, 2 HP		1.40	14.286		3,375	700		4,075	4,750
5160	6920 CFM, 5 HP		1.30	15.385		4,125	750		4,875	5,675
5180	7700 CFM, 7-1/2 HP	↓	1.20	16.667	↓	4,425	815		5,240	6,075
5200	For explosion proof motor, add					15%				
5500	Fans, industrial exhauster, for air which may contain granular matl.									
5520	1000 CFM, 1-1/2 HP	Q-20	2.50	8	Ea.	2,725	390		3,115	3,575
5540	2000 CFM, 3 HP		2	10		3,300	490		3,790	4,375

23 34 16 – Centrifugal HVAC Fans

23 34 16.10 Centrifugal Type HVAC Fans		Crew	Daily Output	Labor-Hours	Unit	2013 Bare Costs Material	Labor	Equipment	Total	Total Incl O&P
5560	4000 CFM, 7-1/2 HP	Q-20	1.80	11.111	Ea.	4,600	545		5,145	5,875
5580	8000 CFM, 15 HP		1.40	14.286		5,925	700		6,625	7,575
5600	12,000 CFM, 30 HP		1	20		9,625	975		10,600	12,100
7000	Roof exhauster, centrifugal, aluminum housing, 12" galvanized									
7020	curb, bird screen, back draft damper, 1/4" S.P.									
7100	Direct drive, 320 CFM, 11" sq. damper	Q-20	7	2.857	Ea.	650	140		790	925
7120	600 CFM, 11" sq. damper		6	3.333		810	163		973	1,150
7140	815 CFM, 13" sq. damper		5	4		810	195		1,005	1,200
7160	1450 CFM, 13" sq. damper		4.20	4.762		1,350	233		1,583	1,825
7180	2050 CFM, 16" sq. damper		4	5		1,600	244		1,844	2,125
7200	V-belt drive, 1650 CFM, 12" sq. damper		6	3.333		1,250	163		1,413	1,625
7220	2750 CFM, 21" sq. damper		5	4		1,475	195		1,670	1,925
7230	3500 CFM, 21" sq. damper		4.50	4.444		1,675	217		1,892	2,175
7240	4910 CFM, 23" sq. damper		4	5		2,050	244		2,294	2,625
7260	8525 CFM, 28" sq. damper		3	6.667		2,700	325		3,025	3,475
7280	13,760 CFM, 35" sq. damper		2	10		3,775	490		4,265	4,900
7300	20,558 CFM, 43" sq. damper		1	20		7,600	975		8,575	9,825
7320	For 2 speed winding, add					15%				
7340	For explosionproof motor, add					545			545	600
7360	For belt driven, top discharge, add					15%				
8500	Wall exhausters, centrifugal, auto damper, 1/8" S.P.									
8520	Direct drive, 610 CFM, 1/20 HP	Q-20	14	1.429	Ea.	385	70		455	525
8540	796 CFM, 1/12 HP		13	1.538		825	75		900	1,025
8560	822 CFM, 1/6 HP		12	1.667		975	81.50		1,056.50	1,200
8580	1,320 CFM, 1/4 HP		12	1.667		1,200	81.50		1,281.50	1,450
8600	1756 CFM, 1/4 HP		11	1.818		1,150	89		1,239	1,375
8620	1983 CFM, 1/4 HP		10	2		1,175	97.50		1,272.50	1,450
8640	2900 CFM, 1/2 HP		9	2.222		1,250	109		1,359	1,550
8660	3307 CFM, 3/4 HP		8	2.500		1,350	122		1,472	1,650
9500	V-belt drive, 3 phase									
9520	2,800 CFM, 1/4 HP	Q-20	9	2.222	Ea.	1,750	109		1,859	2,100
9540	3,740 CFM, 1/2 HP		8	2.500		1,800	122		1,922	2,175
9560	4400 CFM, 3/4 HP		7	2.857		1,825	140		1,965	2,225
9580	5700 CFM, 1-1/2 HP		6	3.333		1,900	163		2,063	2,350

23 34 23 – HVAC Power Ventilators

23 34 23.10 HVAC Power Circulators and Ventilators

Line	Description		Crew	Daily Output	Labor-Hours	Unit	Material	Labor	Equipment	Total	Total Incl O&P
0010	**HVAC POWER CIRCULATORS AND VENTILATORS**										
3000	Paddle blade air circulator, 3 speed switch										
3020	42", 5,000 CFM high, 3000 CFM low	G	1 Elec	2.40	3.333	Ea.	163	175		338	440
3040	52", 6,500 CFM high, 4000 CFM low	G	"	2.20	3.636	"	170	191		361	470
3100	For antique white motor, same cost										
3200	For brass plated motor, same cost										
3300	For light adaptor kit, add	G				Ea.	41			41	45
6000	Propeller exhaust, wall shutter										
6020	Direct drive, one speed, .075" S.P.										
6100	653 CFM, 1/30 HP		Q-20	10	2	Ea.	174	97.50		271.50	340
6120	1033 CFM, 1/20 HP			9	2.222		260	109		369	450
6140	1323 CFM, 1/15 HP			8	2.500		290	122		412	505
6160	2444 CFM, 1/4 HP			7	2.857		365	140		505	610
6300	V-belt drive, 3 phase										
6320	6175 CFM, 3/4 HP		Q-20	5	4	Ea.	1,050	195		1,245	1,450
6340	7500 CFM, 3/4 HP			5	4		1,100	195		1,295	1,525

23 34 HVAC Fans

23 34 23 – HVAC Power Ventilators

	23 34 23.10 HVAC Power Circulators and Ventilators	Crew	Daily Output	Labor-Hours	Unit	2013 Bare Costs Material	Labor	Equipment	Total	Total Incl O&P
6360	10,100 CFM, 1 HP	Q-20	4.50	4.444	Ea.	1,250	217		1,467	1,700
6380	14,300 CFM, 1-1/2 HP		4	5		1,500	244		1,744	2,025
6400	19,800 CFM, 2 HP		3	6.667		1,800	325		2,125	2,475
6420	26,250 CFM, 3 HP		2.60	7.692		1,950	375		2,325	2,725
6440	38,500 CFM, 5 HP		2.20	9.091		2,300	445		2,745	3,225
6460	46,000 CFM, 7-1/2 HP		2	10		2,475	490		2,965	3,475
6480	51,500 CFM, 10 HP		1.80	11.111		2,575	545		3,120	3,650
6650	Residential, bath exhaust, grille, back draft damper									
6660	50 CFM	Q-20	24	.833	Ea.	57	40.50		97.50	125
6670	110 CFM		22	.909		90.50	44.50		135	168
6680	Light combination, squirrel cage, 100 watt, 70 CFM		24	.833		100	40.50		140.50	172
6700	Light/heater combination, ceiling mounted									
6710	70 CFM, 1450 watt	Q-20	24	.833	Ea.	149	40.50		189.50	226
6800	Heater combination, recessed, 70 CFM		24	.833		60.50	40.50		101	129
6820	With 2 infrared bulbs		23	.870		94.50	42.50		137	169
6900	Kitchen exhaust, grille, complete, 160 CFM		22	.909		100	44.50		144.50	178
6910	180 CFM		20	1		83.50	49		132.50	166
6920	270 CFM		18	1.111		153	54.50		207.50	251
6930	350 CFM		16	1.250		119	61		180	224
6940	Residential roof jacks and wall caps									
6944	Wall cap with back draft damper									
6946	3" & 4" diam. round duct	1 Shee	11	.727	Ea.	24	39		63	85
6948	6" diam. round duct	"	11	.727	"	59	39		98	124
6958	Roof jack with bird screen and back draft damper									
6960	3" & 4" diam. round duct	1 Shee	11	.727	Ea.	24	39		63	85.50
6962	3-1/4" x 10" rectangular duct	"	10	.800	"	44.50	42.50		87	114
6980	Transition									
6982	3-1/4" x 10" to 6" diam. round	1 Shee	20	.400	Ea.	29	21.50		50.50	64
8020	Attic, roof type									
8030	Aluminum dome, damper & curb									
8040	6" diameter, 300 CFM	1 Elec	16	.500	Ea.	475	26		501	565
8050	7" diameter, 450 CFM		15	.533		520	28		548	610
8060	9" diameter, 900 CFM		14	.571		570	30		600	670
8080	12" diameter, 1000 CFM (gravity)		10	.800		520	42		562	635
8090	16" diameter, 1500 CFM (gravity)		9	.889		625	46.50		671.50	760
8100	20" diameter, 2500 CFM (gravity)		8	1		770	52.50		822.50	925
8110	26" diameter, 4000 CFM (gravity)		7	1.143		930	60		990	1,125
8120	32" diameter, 6500 CFM (gravity)		6	1.333		1,275	70		1,345	1,500
8130	38" diameter, 8000 CFM (gravity)		5	1.600		1,900	84		1,984	2,200
8140	50" diameter, 13,000 CFM (gravity)		4	2		2,750	105		2,855	3,175
8160	Plastic, ABS dome									
8180	1050 CFM	1 Elec	14	.571	Ea.	153	30		183	213
8200	1600 CFM	"	12	.667	"	229	35		264	305
8240	Attic, wall type, with shutter, one speed									
8250	12" diameter, 1000 CFM	1 Elec	14	.571	Ea.	375	30		405	460
8260	14" diameter, 1500 CFM		12	.667		405	35		440	500
8270	16" diameter, 2000 CFM		9	.889		460	46.50		506.50	575
8290	Whole house, wall type, with shutter, one speed									
8300	30" diameter, 4800 CFM	1 Elec	7	1.143	Ea.	985	60		1,045	1,175
8310	36" diameter, 7000 CFM		6	1.333		1,075	70		1,145	1,275
8320	42" diameter, 10,000 CFM		5	1.600		1,200	84		1,284	1,450
8330	48" diameter, 16,000 CFM		4	2		1,500	105		1,605	1,800
8340	For two speed, add					89.50			89.50	98.50

23 34 HVAC Fans

23 34 23 – HVAC Power Ventilators

23 34 23.10	HVAC Power Circulators and Ventilators	Crew	Daily Output	Labor-Hours	Unit	2013 Bare Costs Material	Labor	Equipment	Total	Total Incl O&P
8350	Whole house, lay-down type, with shutter, one speed									
8360	30" diameter, 4500 CFM	1 Elec	8	1	Ea.	1,050	52.50		1,102.50	1,225
8370	36" diameter, 6500 CFM		7	1.143		1,125	60		1,185	1,350
8380	42" diameter, 9000 CFM		6	1.333		1,250	70		1,320	1,450
8390	48" diameter, 12,000 CFM		5	1.600		1,400	84		1,484	1,675
8440	For two speed, add					67.50			67.50	74
8450	For 12 hour timer switch, add	1 Elec	32	.250		67.50	13.10		80.60	93.50

23 52 Heating Boilers

23 52 13 – Electric Boilers

23 52 13.10 Electric Boilers, ASME

23 52 13.10	Electric Boilers, ASME	Crew	Daily Output	Labor-Hours	Unit	Material	Labor	Equipment	Total	Total Incl O&P
0010	**ELECTRIC BOILERS, ASME**, Standard controls and trim									
1000	Steam, 6 KW, 20.5 MBH	Q-19	1.20	20	Ea.	3,800	1,025		4,825	5,725
1040	9 KW, 30.7 MBH		1.20	20		3,900	1,025		4,925	5,850
1060	18 KW, 61.4 MBH		1.20	20		4,025	1,025		5,050	5,975
1080	24 KW, 81.8 MBH		1.10	21.818		4,675	1,125		5,800	6,800
1120	36 KW, 123 MBH		1.10	21.818		5,150	1,125		6,275	7,350
1160	60 KW, 205 MBH		1	24		6,450	1,225		7,675	8,950
1220	112 KW, 382 MBH		.75	32		8,850	1,650		10,500	12,200
1240	148 KW, 505 MBH		.65	36.923		9,325	1,900		11,225	13,100
1260	168 KW, 573 MBH		.60	40		19,800	2,050		21,850	24,900
1280	222 KW, 758 MBH		.55	43.636		23,100	2,250		25,350	28,800
1300	296 KW, 1010 MBH		.45	53.333		24,300	2,750		27,050	30,900
1320	300 KW, 1023 MBH		.40	60		25,600	3,075		28,675	32,900
1340	370 KW, 1263 MBH		.35	68.571		28,300	3,525		31,825	36,500
1360	444 KW, 1515 MBH		.30	80		30,400	4,125		34,525	39,700
1380	518 KW, 1768 MBH	Q-21	.36	88.889		31,700	4,700		36,400	41,900
1400	592 KW, 2020 MBH		.34	94.118		34,300	4,975		39,275	45,200
1420	666 KW, 2273 MBH		.32	100		35,700	5,275		40,975	47,200
1460	740 KW, 2526 MBH		.28	114		37,200	6,025		43,225	50,000
1480	814 KW, 2778 MBH		.25	128		39,400	6,750		46,150	53,500
1500	962 KW, 3283 MBH		.22	145		43,600	7,675		51,275	59,500
1520	1036 KW, 3536 MBH		.20	160		45,700	8,450		54,150	63,000
1540	1110 KW, 3788 MBH		.19	168		48,100	8,875		56,975	66,500
1560	2070 KW, 7063 MBH		.18	177		66,000	9,375		75,375	87,000
1580	2250 KW, 7677 MBH		.17	188		77,500	9,925		87,425	100,500
1600	2,340 KW, 7984 MBH		.16	200		84,000	10,600		94,600	108,500
2000	Hot water, 7.5 KW, 25.6 MBH	Q-19	1.30	18.462		4,800	950		5,750	6,700
2020	15 KW, 51.2 MBH		1.30	18.462		4,825	950		5,775	6,750
2040	30 KW, 102 MBH		1.20	20		5,150	1,025		6,175	7,225
2060	45 KW, 164 MBH		1.20	20		5,250	1,025		6,275	7,325
2070	60 KW, 205 MBH		1.20	20		5,375	1,025		6,400	7,450
2080	75 KW, 256 MBH		1.10	21.818		5,725	1,125		6,850	7,975
2100	90 KW, 307 MBH		1.10	21.818		5,825	1,125		6,950	8,100
2120	105 KW, 358 MBH		1	24		6,075	1,225		7,300	8,525
2140	120 KW, 410 MBH		.90	26.667		6,200	1,375		7,575	8,850
2160	135 KW, 461 MBH		.75	32		7,200	1,650		8,850	10,400
2180	150 KW, 512 MBH		.65	36.923		7,275	1,900		9,175	10,900
2200	165 KW, 563 MBH		.60	40		7,375	2,050		9,425	11,200
2220	296 KW, 1010 MBH		.55	43.636		14,900	2,250		17,150	19,800
2280	370 KW, 1263 MBH		.40	60		16,500	3,075		19,575	22,900

23 52 Heating Boilers

23 52 13 – Electric Boilers

23 52 13.10 Electric Boilers, ASME		Crew	Daily Output	Labor-Hours	Unit	Material	Labor	Equipment	Total	Total Incl O&P
						2013 Bare Costs				
2300	444 KW, 1515 MBH	Q-19	.35	68.571	Ea.	19,300	3,525		22,825	26,500
2340	518 KW, 1768 MBH	Q-21	.44	72.727		21,300	3,825		25,125	29,300
2360	592 KW, 2020 MBH		.43	74.419		24,200	3,925		28,125	32,500
2400	666 KW, 2273 MBH		.40	80		25,200	4,225		29,425	34,200
2420	740 KW, 2526 MBH		.39	82.051		26,400	4,325		30,725	35,500
2440	814 KW, 2778 MBH		.38	84.211		28,200	4,450		32,650	37,700
2460	888 KW, 3031 MBH		.37	86.486		29,400	4,550		33,950	39,300
2480	962 KW, 3283 MBH		.36	88.889		31,400	4,700		36,100	41,700
2500	1036 KW, 3536 MBH		.34	94.118		33,500	4,975		38,475	44,400
2520	1110 KW, 3788 MBH		.33	96.970		36,600	5,125		41,725	48,000
2540	1440 KW, 4915 MBH		.32	100		41,700	5,275		46,975	54,000
2560	1560 KW, 5323 MBH		.31	103		44,600	5,450		50,050	57,000
2580	1680 KW, 5733 MBH		.30	106		47,600	5,625		53,225	61,000
2600	1800 KW, 6143 MBH		.29	110		51,000	5,825		56,825	65,000
2620	1980 KW, 6757 MBH		.28	114		55,500	6,025		61,525	70,000
2640	2100 KW, 7167 MBH		.27	118		56,500	6,250		62,750	72,000
2660	2220 KW, 7576 MBH		.26	123		62,000	6,500		68,500	78,000
2680	2400 KW, 8191 MBH		.25	128		64,500	6,750		71,250	81,000
2700	2610 KW, 8905 MBH		.24	133		67,500	7,025		74,525	84,500
2720	2790 KW, 9519 MBH		.23	139		72,000	7,350		79,350	90,500
2740	2970 KW, 10133 MBH		.21	152		73,500	8,050		81,550	93,000
2760	3150 KW, 10748 MBH		.19	168		78,500	8,875		87,375	100,000
2780	3240 KW, 11055 MBH		.18	177		80,500	9,375		89,875	102,500
2800	3420 KW, 11669 MBH		.17	188		86,000	9,925		95,925	109,500
2820	3600 KW, 12,283 MBH		.16	200		89,500	10,600		100,100	114,500

23 52 28 – Swimming Pool Boilers

23 52 28.10 Swimming Pool Heaters		Crew	Daily Output	Labor-Hours	Unit	Material	Labor	Equipment	Total	Total Incl O&P
0010	**SWIMMING POOL HEATERS**, Not including wiring, external									
0020	piping, base or pad,									
2000	Electric, 12 KW, 4,800 gallon pool	Q-19	3	8	Ea.	2,075	410		2,485	2,900
2020	15 KW, 7,200 gallon pool		2.80	8.571		2,100	440		2,540	3,000
2040	24 KW, 9,600 gallon pool		2.40	10		2,425	515		2,940	3,450
2060	30 KW, 12,000 gallon pool		2	12		2,475	615		3,090	3,650
2080	36 KW, 14,400 gallon pool		1.60	15		2,850	770		3,620	4,275
2100	57 KW, 24,000 gallon pool		1.20	20		3,575	1,025		4,600	5,475
9000	To select pool heater: 12 BTUH x S.F. pool area									
9010	X temperature differential = required output									
9050	For electric, KW = gallons x 2.5 divided by 1000									
9100	For family home type pool, double the									
9110	Rated gallon capacity = 1/2°F rise per hour									

23 54 Furnaces

23 54 13 – Electric-Resistance Furnaces

23 54 13.10 Electric Furnaces		Crew	Daily Output	Labor-Hours	Unit	Material	2013 Bare Costs Labor	Equipment	Total	Total Incl O&P
0010	**ELECTRIC FURNACES**, Hot air, blowers, std. controls									
1000	Electric, UL listed									
1100	34.1 MBH	Q-20	4.40	4.545	Ea.	560	222		782	950
1120	51.6 MBH		4.20	4.762		630	233		863	1,050
1140	68.3 MBH		4	5		650	244		894	1,075
1160	85.3 MBH	↓	3.80	5.263	↓	660	257		917	1,125

23 82 Convection Heating and Cooling Units

23 82 16 – Air Coils

23 82 16.20 Duct Heaters		Crew	Daily Output	Labor-Hours	Unit	Material	2013 Bare Costs Labor	Equipment	Total	Total Incl O&P
0010	**DUCT HEATERS**, Electric, 480 V, 3 Ph.									
0020	Finned tubular insert, 500°F									
0100	8″ wide x 6″ high, 4.0 kW	Q-20	16	1.250	Ea.	770	61		831	945
0120	12″ high, 8.0 kW		15	1.333		1,275	65		1,340	1,500
0140	18″ high, 12.0 kW		14	1.429		1,800	70		1,870	2,075
0160	24″ high, 16.0 kW		13	1.538		2,300	75		2,375	2,650
0180	30″ high, 20.0 kW		12	1.667		2,825	81.50		2,906.50	3,225
0300	12″ wide x 6″ high, 6.7 kW		15	1.333		820	65		885	1,000
0320	12″ high, 13.3 kW		14	1.429		1,325	70		1,395	1,550
0340	18″ high, 20.0 kW		13	1.538		1,850	75		1,925	2,175
0360	24″ high, 26.7 kW		12	1.667		2,375	81.50		2,456.50	2,750
0380	30″ high, 33.3 kW		11	1.818		2,900	89		2,989	3,325
0500	18″ wide x 6″ high, 13.3 kW		14	1.429		880	70		950	1,075
0520	12″ high, 26.7 kW		13	1.538		1,525	75		1,600	1,800
0540	18″ high, 40.0 kW		12	1.667		2,000	81.50		2,081.50	2,350
0560	24″ high, 53.3 kW		11	1.818		2,675	89		2,764	3,075
0580	30″ high, 66.7 kW		10	2		3,350	97.50		3,447.50	3,825
0700	24″ wide x 6″ high, 17.8 kW		13	1.538		970	75		1,045	1,200
0720	12″ high, 35.6 kW		12	1.667		1,650	81.50		1,731.50	1,925
0740	18″ high, 53.3 kW		11	1.818		2,300	89		2,389	2,650
0760	24″ high, 71.1 kW		10	2		2,975	97.50		3,072.50	3,425
0780	30″ high, 88.9 kW		9	2.222		3,650	109		3,759	4,175
0900	30″ wide x 6″ high, 22.2 kW		12	1.667		1,025	81.50		1,106.50	1,250
0920	12″ high, 44.4 kW		11	1.818		1,725	89		1,814	2,025
0940	18″ high, 66.7 kW		10	2		2,450	97.50		2,547.50	2,850
0960	24″ high, 88.9 kW		9	2.222		3,000	109		3,109	3,475
0980	30″ high, 111.0 kW	↓	8	2.500	↓	3,850	122		3,972	4,425
1400	Note decreased kW available for									
1410	each duct size at same cost									
1420	See line 5000 for modifications and accessories									
2000	Finned tubular flange with insulated									
2020	terminal box, 500°F									
2100	12″ wide x 36″ high, 54 kW	Q-20	10	2	Ea.	3,800	97.50		3,897.50	4,325
2120	40″ high, 60 kW		9	2.222		4,375	109		4,484	4,975
2200	24″ wide x 36″ high, 118.8 kW		9	2.222		4,850	109		4,959	5,525
2220	40″ high, 132 kW		8	2.500		5,100	122		5,222	5,800
2400	36″ wide x 8″ high, 40 kW		11	1.818		1,925	89		2,014	2,250
2420	16″ high, 80 kW		10	2		2,575	97.50		2,672.50	2,975
2440	24″ high, 120 kW		9	2.222		3,375	109		3,484	3,875
2460	32″ high, 160 kW		8	2.500		4,325	122		4,447	4,925
2480	36″ high, 180 kW		7	2.857		5,325	140		5,465	6,050

23 82 Convection Heating and Cooling Units

23 82 16 – Air Coils

23 82 16.20 Duct Heaters		Crew	Daily Output	Labor-Hours	Unit	Material	Labor	Equipment	Total	Total Incl O&P
						2013 Bare Costs				
2500	40" high, 200 kW	Q-20	6	3.333	Ea.	5,900	163		6,063	6,750
2600	40" wide x 8" high, 45 kW		11	1.818		2,025	89		2,114	2,350
2620	16" high, 90 kW		10	2		2,775	97.50		2,872.50	3,200
2640	24" high, 135 kW		9	2.222		3,525	109		3,634	4,050
2660	32" high, 180 kW		8	2.500		4,700	122		4,822	5,350
2680	36" high, 202.5 kW		7	2.857		5,400	140		5,540	6,150
2700	40" high, 225 kW		6	3.333		6,150	163		6,313	7,025
2800	48" wide x 8" high, 54.8 kW		10	2		2,125	97.50		2,222.50	2,475
2820	16" high, 109.8 kW		9	2.222		2,925	109		3,034	3,400
2840	24" high, 164.4 kW		8	2.500		3,725	122		3,847	4,275
2860	32" high, 219.2 kW		7	2.857		4,975	140		5,115	5,675
2880	36" high, 246.6 kW		6	3.333		5,725	163		5,888	6,550
2900	40" high, 274 kW		5	4		6,475	195		6,670	7,425
3000	56" wide x 8" high, 64 kW		9	2.222		2,400	109		2,509	2,825
3020	16" high, 128 kW		8	2.500		3,350	122		3,472	3,850
3040	24" high, 192 kW		7	2.857		4,050	140		4,190	4,650
3060	32" high, 256 kW		6	3.333		5,650	163		5,813	6,475
3080	36" high, 288kW		5	4		6,475	195		6,670	7,425
3100	40" high, 320kW		4	5		7,175	244		7,419	8,275
3200	64" wide x 8" high, 74kW		8	2.500		2,475	122		2,597	2,900
3220	16" high, 148kW		7	2.857		3,425	140		3,565	3,975
3240	24" high, 222kW		6	3.333		4,375	163		4,538	5,050
3260	32" high, 296kW		5	4		5,850	195		6,045	6,750
3280	36" high, 333kW		4	5		6,975	244		7,219	8,050
3300	40" high, 370kW	↓	3	6.667	↓	7,700	325		8,025	8,975
3800	Note decreased kW available for									
3820	each duct size at same cost									
5000	Duct heater modifications and accessories									
5120	T.C.O. limit auto or manual reset	Q-20	42	.476	Ea.	126	23.50		149.50	175
5140	Thermostat		28	.714		540	35		575	650
5160	Overheat thermocouple (removable)		7	2.857		765	140		905	1,050
5180	Fan interlock relay		18	1.111		184	54.50		238.50	286
5200	Air flow switch		20	1		159	49		208	249
5220	Split terminal box cover	↓	100	.200	↓	52.50	9.75		62.25	73
8000	To obtain BTU multiply kW by 3413									

23 82 27 – Infrared Units

23 82 27.10 Infrared Type Heating Units		Crew	Daily Output	Labor-Hours	Unit	Material	Labor	Equipment	Total	Total Incl O&P
0010	**INFRARED TYPE HEATING UNITS**									
2000	Electric, single or three phase									
2050	6 kW, 20,478 BTU	1 Elec	2.30	3.478	Ea.	575	182		757	910
2100	13.5 KW, 40,956 BTU		2.20	3.636		830	191		1,021	1,200
2150	24 KW, 81,912 BTU	↓	2	4	↓	1,675	210		1,885	2,175

23 83 33.10 Electric Heating		Crew	Daily Output	Labor-Hours	Unit	Material	2013 Bare Costs Labor	Equipment	Total	Total Incl O&P
0010	**ELECTRIC HEATING**, not incl. conduit or feed wiring									
1100	Rule of thumb: Baseboard units, including control	1 Elec	4.40	1.818	kW	101	95.50		196.50	254
1300	Baseboard heaters, 2' long, 350 watt		8	1	Ea.	30	52.50		82.50	112
1400	3' long, 750 watt		8	1		46	52.50		98.50	129
1600	4' long, 1000 watt		6.70	1.194		41	62.50		103.50	139
1800	5' long, 935 watt		5.70	1.404		51	73.50		124.50	166
2000	6' long, 1500 watt		5	1.600		56	84		140	187
2200	7' long, 1310 watt		4.40	1.818		84	95.50		179.50	236
2400	8' long, 2000 watt		4	2		69.50	105		174.50	233
2600	9' long, 1680 watt		3.60	2.222		104	116		220	288
2800	10' long, 1875 watt		3.30	2.424		193	127		320	400
2950	Wall heaters with fan, 120 to 277 volt									
3160	Recessed, residential, 750 watt	1 Elec	6	1.333	Ea.	134	70		204	252
3170	1000 watt		6	1.333		134	70		204	252
3180	1250 watt		5	1.600		134	84		218	272
3190	1500 watt		4	2		134	105		239	305
3210	2000 watt		4	2		134	105		239	305
3230	2500 watt		3.50	2.286		335	120		455	550
3240	3000 watt		3	2.667		495	140		635	750
3250	4000 watt		2.70	2.963		495	155		650	775
3260	Commercial, 750 watt		6	1.333		181	70		251	305
3270	1000 watt		6	1.333		181	70		251	305
3280	1250 watt		5	1.600		181	84		265	325
3290	1500 watt		4	2		181	105		286	355
3300	2000 watt		4	2		181	105		286	355
3310	2500 watt		3.50	2.286		181	120		301	380
3320	3000 watt		3	2.667		330	140		470	570
3330	4000 watt		2.70	2.963		330	155		485	590
3600	Thermostats, integral		16	.500		30	26		56	72
3800	Line voltage, 1 pole		8	1		28.50	52.50		81	110
3810	2 pole		8	1		29.50	52.50		82	111
4000	Heat trace system, 400 degree R238313-10									
4020	115 V, 2.5 watts per L.F.	1 Elec	530	.015	L.F.	7.70	.79		8.49	9.70
4030	5 watts per L.F. R238313-20		530	.015		7.70	.79		8.49	9.70
4050	10 watts per L.F.		530	.015		7.70	.79		8.49	9.70
4060	208 V, 5 watts per L.F.		530	.015		7.70	.79		8.49	9.70
4080	480 V, 8 watts per L.F.		530	.015		7.70	.79		8.49	9.70
4200	Heater raceway									
4260	Heat transfer cement									
4280	1 gallon				Ea.	57			57	62.50
4300	5 gallon				"	220			220	242
4320	Snap band, clamp									
4340	3/4" pipe size	1 Elec	470	.017	Ea.		.89		.89	1.33
4360	1" pipe size		444	.018			.94		.94	1.41
4380	1-1/4" pipe size		400	.020			1.05		1.05	1.57
4400	1-1/2" pipe size		355	.023			1.18		1.18	1.77
4420	2" pipe size		320	.025			1.31		1.31	1.96
4440	3" pipe size		160	.050			2.62		2.62	3.92
4460	4" pipe size		100	.080			4.19		4.19	6.25
4480	Thermostat NEMA 3R, 22 amp, 0-150 Deg, 10' cap.		8	1		241	52.50		293.50	345
4500	Thermostat NEMA 4X, 25 amp, 40 Deg, 5-1/2' cap.		7	1.143		241	60		301	355
4520	Thermostat NEMA 4X, 22 amp, 25-325 Deg, 10' cap.		7	1.143		650	60		710	805
4540	Thermostat NEMA 4X, 22 amp, 15-140 Deg,		6	1.333		545	70		615	705

23 83 33.10 Electric Heating		Crew	Daily Output	Labor-Hours	Unit	2013 Bare Costs Material	Labor	Equipment	Total	Total Incl O&P
4580	Thermostat NEMA 4,7,9, 22 amp, 25-325 Deg, 10′ cap.	1 Elec	3.60	2.222	Ea.	800	116		916	1,050
4600	Thermostat NEMA 4,7,9, 22 amp, 15-140 Deg,		3	2.667		795	140		935	1,075
4720	Fiberglass application tape, 36 yard roll		11	.727		77.50	38		115.50	143
5000	Radiant heating ceiling panels, 2′ x 4′, 500 watt		16	.500		279	26		305	345
5050	750 watt		16	.500		300	26		326	370
5200	For recessed plaster frame, add		32	.250		106	13.10		119.10	136
5300	Infrared quartz heaters, 120 volts, 1000 watts		6.70	1.194		230	62.50		292.50	345
5350	1500 watt		5	1.600		230	84		314	380
5400	240 volts, 1500 watt		5	1.600		230	84		314	380
5450	2000 watt		4	2		230	105		335	410
5500	3000 watt		3	2.667		255	140		395	490
5550	4000 watt		2.60	3.077		255	161		416	520
5570	Modulating control		.80	10		275	525		800	1,100
5600	Unit heaters, heavy duty, with fan & mounting bracket									
5650	Single phase, 208-240-277 volt, 3 kW	1 Elec	3.20	2.500	Ea.	420	131		551	655
5750	5 kW		2.40	3.333		440	175		615	745
5800	7 kW		1.90	4.211		680	221		901	1,075
5850	10 kW		1.30	6.154		775	320		1,095	1,325
5950	15 kW		.90	8.889		1,250	465		1,715	2,075
6000	480 volt, 3 kW		3.30	2.424		465	127		592	700
6020	4 kW		3	2.667		470	140		610	730
6040	5 kW		2.60	3.077		490	161		651	775
6060	7 kW		2	4		730	210		940	1,125
6080	10 kW		1.40	5.714		815	299		1,114	1,350
6100	13 kW		1.10	7.273		1,250	380		1,630	1,950
6120	15 kW		1	8		1,250	420		1,670	2,000
6140	20 kW		.90	8.889		1,625	465		2,090	2,500
6300	3 phase, 208-240 volt, 5 kW		2.40	3.333		415	175		590	715
6320	7 kW		1.90	4.211		645	221		866	1,025
6340	10 kW		1.30	6.154		720	320		1,040	1,275
6360	15 kW		.90	8.889		1,325	465		1,790	2,150
6380	20 kW		.70	11.429		1,650	600		2,250	2,725
6400	25 kW		.50	16		1,950	840		2,790	3,400
6500	480 volt, 5 kW		2.60	3.077		585	161		746	880
6520	7 kW		2	4		735	210		945	1,125
6540	10 kW		1.40	5.714		775	299		1,074	1,300
6560	13 kW		1.10	7.273		1,425	380		1,805	2,150
6580	15 kW		1	8		1,425	420		1,845	2,200
6600	20 kW		.90	8.889		1,850	465		2,315	2,725
6620	25 kW		.60	13.333		1,950	700		2,650	3,200
6630	30 kW		.70	11.429		2,250	600		2,850	3,400
6640	40 kW		.60	13.333		2,900	700		3,600	4,225
6650	50 kW		.50	16		3,475	840		4,315	5,075
6800	Vertical discharge heaters, with fan									
6820	Single phase, 208-240-277 volt, 10 kW	1 Elec	1.30	6.154	Ea.	720	320		1,040	1,275
6840	15 kW		.90	8.889		1,200	465		1,665	2,025
6900	3 phase, 208-240 volt, 10 kW		1.30	6.154		720	320		1,040	1,275
6920	15 kW		.90	8.889		1,200	465		1,665	2,025
6940	20 kW		.70	11.429		1,725	600		2,325	2,800
6960	25 kW		.50	16		1,950	840		2,790	3,400
6980	30 kW		.40	20		2,350	1,050		3,400	4,150
7000	40 kW		.36	22.222		2,900	1,175		4,075	4,925
7020	50 kW		.32	25		3,600	1,300		4,900	5,925

23 83 33 – Electric Radiant Heaters

23 83 33.10 Electric Heating		Crew	Daily Output	Labor-Hours	Unit	Material	Labor	Equipment	Total	Total Incl O&P
7100	480 volt, 10 kW	1 Elec	1.40	5.714	Ea.	775	299		1,074	1,300
7120	15 kW		1	8		1,250	420		1,670	2,000
7140	20 kW		.90	8.889		1,625	465		2,090	2,500
7160	25 kW		.60	13.333		1,950	700		2,650	3,200
7180	30 kW		.50	16		2,250	840		3,090	3,750
7200	40 kW		.40	20		2,900	1,050		3,950	4,750
7220	50 kW		.35	22.857		3,475	1,200		4,675	5,625
7410	Sill height convector heaters, 5" high x 2' long, 500 watt		6.70	1.194		275	62.50		337.50	400
7420	3' long, 750 watt		6.50	1.231		325	64.50		389.50	455
7430	4' long, 1000 watt		6.20	1.290		375	67.50		442.50	515
7440	5' long, 1250 watt		5.50	1.455		425	76		501	585
7450	6' long, 1500 watt		4.80	1.667		485	87.50		572.50	660
7460	8' long, 2000 watt		3.60	2.222		655	116		771	900
7470	10' long, 2500 watt	↓	3	2.667	↓	815	140		955	1,100
7900	Cabinet convector heaters, 240 volt									
7920	3' long, 2000 watt	1 Elec	5.30	1.509	Ea.	1,900	79		1,979	2,225
7940	3000 watt		5.30	1.509		2,000	79		2,079	2,300
7960	4000 watt		5.30	1.509		2,050	79		2,129	2,375
7980	6000 watt		4.60	1.739		2,125	91		2,216	2,450
8000	8000 watt		4.60	1.739		2,200	91		2,291	2,550
8020	4' long, 4000 watt		4.60	1.739		2,075	91		2,166	2,400
8040	6000 watt		4	2		2,150	105		2,255	2,525
8060	8000 watt		4	2		2,225	105		2,330	2,600
8080	10,000 watt	↓	4	2	↓	2,250	105		2,355	2,625
8100	Available also in 208 or 277 volt									
8200	Cabinet unit heaters, 120 to 277 volt, 1 pole,									
8220	wall mounted, 2 kW	1 Elec	4.60	1.739	Ea.	1,850	91		1,941	2,150
8230	3 kW		4.60	1.739		1,875	91		1,966	2,175
8240	4 kW		4.40	1.818		1,925	95.50		2,020.50	2,275
8250	5 kW		4.40	1.818		2,000	95.50		2,095.50	2,350
8260	6 kW		4.20	1.905		2,000	100		2,100	2,350
8270	8 kW		4	2		2,075	105		2,180	2,450
8280	10 kW		3.80	2.105		2,125	110		2,235	2,525
8290	12 kW		3.50	2.286		2,300	120		2,420	2,725
8300	13.5 kW		2.90	2.759		2,350	145		2,495	2,800
8310	16 kW		2.70	2.963		2,400	155		2,555	2,875
8320	20 kW		2.30	3.478		3,475	182		3,657	4,100
8330	24 kW		1.90	4.211		3,525	221		3,746	4,200
8350	Recessed, 2 kW		4.40	1.818		1,800	95.50		1,895.50	2,125
8370	3 kW		4.40	1.818		1,875	95.50		1,970.50	2,200
8380	4 kW		4.20	1.905		1,875	100		1,975	2,200
8390	5 kW		4.20	1.905		2,100	100		2,200	2,450
8400	6 kW		4	2		2,200	105		2,305	2,575
8410	8 kW		3.80	2.105		2,200	110		2,310	2,600
8420	10 kW		3.50	2.286		2,675	120		2,795	3,125
8430	12 kW		2.90	2.759		2,700	145		2,845	3,200
8440	13.5 kW		2.70	2.963		2,850	155		3,005	3,375
8450	16 kW		2.30	3.478		2,875	182		3,057	3,450
8460	20 kW		1.90	4.211		2,900	221		3,121	3,525
8470	24 kW		1.60	5		3,450	262		3,712	4,200
8490	Ceiling mounted, 2 kW		3.20	2.500		1,875	131		2,006	2,250
8510	3 kW		3.20	2.500		1,950	131		2,081	2,350
8520	4 kW		3	2.667		1,975	140		2,115	2,375

Labor, Equipment columns under "2013 Bare Costs".

23 83 Radiant Heating Units

23 83 33 – Electric Radiant Heaters

23 83 33.10 Electric Heating		Crew	Daily Output	Labor-Hours	Unit	2013 Bare Costs Material	Labor	Equipment	Total	Total Incl O&P
8530	5 kW	1 Elec	3	2.667	Ea.	2,075	140		2,215	2,475
8540	6 kW		2.80	2.857		2,075	150		2,225	2,525
8550	8 kW		2.40	3.333		2,150	175		2,325	2,625
8560	10 kW		2.20	3.636		2,125	191		2,316	2,625
8570	12 kW		2	4		2,250	210		2,460	2,800
8580	13.5 kW		1.50	5.333		2,275	279		2,554	2,925
8590	16 kW		1.30	6.154		2,400	320		2,720	3,125
8600	20 kW		.90	8.889		3,375	465		3,840	4,400
8610	24 kW		.60	13.333		3,450	700		4,150	4,850
8630	208 to 480 V, 3 pole									
8650	Wall mounted, 2 kW	1 Elec	4.60	1.739	Ea.	2,000	91		2,091	2,325
8670	3 kW		4.60	1.739		2,075	91		2,166	2,425
8680	4 kW		4.40	1.818		2,125	95.50		2,220.50	2,500
8690	5 kW		4.40	1.818		2,200	95.50		2,295.50	2,575
8700	6 kW		4.20	1.905		2,225	100		2,325	2,600
8710	8 kW		4	2		2,300	105		2,405	2,675
8720	10 kW		3.80	2.105		2,325	110		2,435	2,750
8730	12 kW		3.50	2.286		2,375	120		2,495	2,800
8740	13.5 kW		2.90	2.759		2,425	145		2,570	2,875
8750	16 kW		2.70	2.963		2,425	155		2,580	2,875
8760	20 kW		2.30	3.478		3,075	182		3,257	3,650
8770	24 kW		1.90	4.211		3,600	221		3,821	4,300
8790	Recessed, 2 kW		4.40	1.818		2,000	95.50		2,095.50	2,350
8810	3 kW		4.40	1.818		2,075	95.50		2,170.50	2,450
8820	4 kW		4.20	1.905		1,975	100		2,075	2,325
8830	5 kW		4.20	1.905		2,200	100		2,300	2,575
8840	6 kW		4	2		2,225	105		2,330	2,600
8850	8 kW		3.80	2.105		2,300	110		2,410	2,700
8860	10 kW		3.50	2.286		2,325	120		2,445	2,750
8870	12 kW		2.90	2.759		2,375	145		2,520	2,850
8880	13.5 kW		2.70	2.963		2,425	155		2,580	2,875
8890	16 kW		2.30	3.478		3,500	182		3,682	4,125
8900	20 kW		1.90	4.211		3,600	221		3,821	4,300
8920	24 kW		1.60	5		3,650	262		3,912	4,400
8940	Ceiling mount, 2 kW		3.20	2.500		2,000	131		2,131	2,400
8950	3 kW		3.20	2.500		2,075	131		2,206	2,500
8960	4 kW		3	2.667		2,125	140		2,265	2,550
8970	5 kW		3	2.667		2,200	140		2,340	2,625
8980	6 kW		2.80	2.857		2,225	150		2,375	2,675
8990	8 kW		2.40	3.333		2,300	175		2,475	2,775
9000	10 kW		2.20	3.636		2,325	191		2,516	2,850
9020	13.5 kW		1.50	5.333		2,425	279		2,704	3,075
9030	16 kW		1.30	6.154		3,550	320		3,870	4,400
9040	20 kW		.90	8.889		3,550	465		4,015	4,600
9060	24 kW		.60	13.333		3,425	700		4,125	4,825

Estimating Tips

26 05 00 Common Work Results for Electrical

- Conduit should be taken off in three main categories—power distribution, branch power, and branch lighting—so the estimator can concentrate on systems and components, therefore making it easier to ensure all items have been accounted for.
- For cost modifications for elevated conduit installation, add the percentages to labor according to the height of installation, and only to the quantities exceeding the different height levels, not to the total conduit quantities.
- Remember that aluminum wiring of equal ampacity is larger in diameter than copper and may require larger conduit.
- If more than three wires at a time are being pulled, deduct percentages from the labor hours of that grouping of wires.
- When taking off grounding systems, identify separately the type and size of wire, and list each unique type of ground connection.
- The estimator should take the weights of materials into consideration when completing a takeoff. Topics to consider include: How will the materials be supported? What methods of support are available? How high will the support structure have to reach? Will the final support structure be able to withstand the total burden? Is the support material included or separate from the fixture, equipment, and material specified?
- Do not overlook the costs for equipment used in the installation. If scaffolding or highlifts are available in the field, contractors may use them in lieu of the proposed ladders and rolling staging.

26 20 00 Low-Voltage Electrical Transmission

- Supports and concrete pads may be shown on drawings for the larger equipment, or the support system may be only a piece of plywood for the back of a panelboard. In either case, it must be included in the costs.

26 40 00 Electrical and Cathodic Protection

- When taking off cathodic protections systems, identify the type and size of cable, and list each unique type of anode connection.

26 50 00 Lighting

- Fixtures should be taken off room by room, using the fixture schedule, specifications, and the ceiling plan. For large concentrations of lighting fixtures in the same area, deduct the percentages from labor hours.

Reference Numbers

Reference numbers are shown in shaded boxes at the beginning of some major classifications. These numbers refer to related items in the Reference Section. The reference information may be an estimating procedure, an alternate pricing method, or technical information.

Note: Not all subdivisions listed here necessarily appear in this publication.

Note: **Trade Service,** *in part, has been used as a reference source for some of the material prices used in Division 26.*

26 01 Operation and Maintenance of Electrical Systems

26 01 40 – Operation and Maintenance of Electrical and Cathodic Protection Systems

26 01 40.51 Electrical Systems Repair and Replacement		Crew	Daily Output	Labor-Hours	Unit	Material	2013 Bare Costs Labor	Equipment	Total	Total Incl O&P
0010	**ELECTRICAL SYSTEMS REPAIR AND REPLACEMENT**									
3000	Remove and replace (reinstall), switch cover	1 Elec	60	.133	Ea.		7		7	10.45
3020	Outlet cover	"	60	.133	"		7		7	10.45

26 01 50 – Operation and Maintenance of Lighting

26 01 50.81 Luminaire Replacement

		Crew	Daily Output	Labor-Hours	Unit	Material	Labor	Equipment	Total	Total Incl O&P
0010	**LUMINAIRE REPLACEMENT**									
0962	Remove and install new replacement lens, 2′ x 2′ troffer	1 Elec	16	.500	Ea.	13.95	26		39.95	54.50
0964	2′ x 4′ troffer		16	.500		13.65	26		39.65	54
3200	Remove and replace (reinstall), lighting fixture		4	2			105		105	157

26 05 Common Work Results for Electrical

26 05 05 – Selective Electrical Demolition

26 05 05.10 Electrical Demolition

		Crew	Daily Output	Labor-Hours	Unit	Material	Labor	Equipment	Total	Total Incl O&P
0010	**ELECTRICAL DEMOLITION**									
0020	Conduit to 15′ high, including fittings & hangers R260105-30									
0100	Rigid galvanized steel, 1/2″ to 1″ diameter	1 Elec	242	.033	L.F.		1.73		1.73	2.59
0120	1-1/4″ to 2″	"	200	.040			2.10		2.10	3.14
0140	2-1/2″ to 3-1/2″	2 Elec	302	.053			2.78		2.78	4.15
0160	4″ to 6″	"	160	.100			5.25		5.25	7.85
0170	PVC #40, 1/2″ to 1″	1 Elec	410	.020			1.02		1.02	1.53
0172	1-1/4″ to 2″		350	.023			1.20		1.20	1.79
0174	2-1/2″		250	.032			1.68		1.68	2.51
0176	3″ to 3-1/2″	2 Elec	340	.047			2.47		2.47	3.69
0178	4″ to 6″	"	230	.070			3.65		3.65	5.45
0200	Electric metallic tubing (EMT), 1/2″ to 1″	1 Elec	394	.020			1.06		1.06	1.59
0220	1-1/4″ to 1-1/2″		326	.025			1.29		1.29	1.92
0240	2″ to 3″		236	.034			1.78		1.78	2.66
0260	3-1/2″ to 4″	2 Elec	310	.052			2.70		2.70	4.05
0270	Armored cable, (BX) avg. 50′ runs									
0280	#14, 2 wire	1 Elec	690	.012	L.F.		.61		.61	.91
0290	#14, 3 wire		571	.014			.73		.73	1.10
0300	#12, 2 wire		605	.013			.69		.69	1.04
0310	#12, 3 wire		514	.016			.82		.82	1.22
0320	#10, 2 wire		514	.016			.82		.82	1.22
0330	#10, 3 wire		425	.019			.99		.99	1.48
0340	#8, 3 wire		342	.023			1.23		1.23	1.83
0350	Non metallic sheathed cable (Romex)									
0360	#14, 2 wire	1 Elec	720	.011	L.F.		.58		.58	.87
0370	#14, 3 wire		657	.012			.64		.64	.95
0380	#12, 2 wire		629	.013			.67		.67	1
0390	#10, 3 wire		450	.018			.93		.93	1.39
0400	Wiremold raceway, including fittings & hangers									
0420	No. 3000	1 Elec	250	.032	L.F.		1.68		1.68	2.51
0440	No. 4000		217	.037			1.93		1.93	2.89
0460	No. 6000		166	.048			2.53		2.53	3.78
0462	Plugmold with receptacle		114	.070			3.68		3.68	5.50
0465	Telephone/power pole		12	.667	Ea.		35		35	52.50
0470	Non-metallic, straight section		480	.017	L.F.		.87		.87	1.31
0500	Channels, steel, including fittings & hangers									
0520	3/4″ x 1-1/2″	1 Elec	308	.026	L.F.		1.36		1.36	2.04

26 05 Common Work Results for Electrical

26 05 05 – Selective Electrical Demolition

26 05 05.10 Electrical Demolition		Crew	Daily Output	Labor-Hours	Unit	Material	2013 Bare Costs Labor	Equipment	Total	Total Incl O&P
0540	1-1/2" x 1-1/2"	1 Elec	269	.030	L.F.		1.56		1.56	2.33
0560	1-1/2" x 1-7/8"		229	.035			1.83		1.83	2.74
0600	Copper bus duct, indoor, 3 phase									
0610	Including hangers & supports									
0620	225 amp	2 Elec	135	.119	L.F.		6.20		6.20	9.30
0640	400 amp		106	.151			7.90		7.90	11.85
0660	600 amp		86	.186			9.75		9.75	14.60
0680	1000 amp		60	.267			13.95		13.95	21
0700	1600 amp		40	.400			21		21	31.50
0720	3000 amp		10	1.600			84		84	125
0800	Plug-in switches, 600V 3 ph., incl. disconnecting									
0820	wire, conduit terminations, 30 amp	1 Elec	15.50	.516	Ea.		27		27	40.50
0840	60 amp		13.90	.576			30		30	45
0850	100 amp		10.40	.769			40.50		40.50	60.50
0860	200 amp		6.20	1.290			67.50		67.50	101
0880	400 amp	2 Elec	5.40	2.963			155		155	232
0900	600 amp		3.40	4.706			247		247	370
0920	800 amp		2.60	6.154			320		320	480
0940	1200 amp		2	8			420		420	625
0960	1600 amp		1.70	9.412			495		495	740
1010	Safety switches, 250 or 600V, incl. disconnection									
1050	of wire & conduit terminations									
1100	30 amp	1 Elec	12.30	.650	Ea.		34		34	51
1120	60 amp		8.80	.909			47.50		47.50	71.50
1140	100 amp		7.30	1.096			57.50		57.50	86
1160	200 amp		5	1.600			84		84	125
1180	400 amp	2 Elec	6.80	2.353			123		123	184
1200	600 amp	"	4.60	3.478			182		182	273
1202	Explosion proof, 60 amp	1 Elec	8.80	.909			47.50		47.50	71.50
1204	100 amp		5	1.600			84		84	125
1206	200 amp		3.33	2.402			126		126	188
1210	Panel boards, incl. removal of all breakers,									
1220	conduit terminations & wire connections									
1230	3 wire, 120/240 V, 100A, to 20 circuits	1 Elec	2.60	3.077	Ea.		161		161	241
1240	200 amps, to 42 circuits	2 Elec	2.60	6.154			320		320	480
1250	400 amps, to 42 circuits	"	2.20	7.273			380		380	570
1260	4 wire, 120/208 V, 125A, to 20 circuits	1 Elec	2.40	3.333			175		175	261
1270	200 amps, to 42 circuits	2 Elec	2.40	6.667			350		350	525
1280	400 amps, to 42 circuits		1.92	8.333			435		435	655
1285	600 amps, to 42 circuits		1.60	10			525		525	785
1300	Transformer, dry type, 1 phase, incl. removal of									
1320	supports, wire & conduit terminations									
1340	1 kVA	1 Elec	7.70	1.039	Ea.		54.50		54.50	81.50
1360	5 kVA		4.70	1.702			89		89	133
1380	10 kVA		3.60	2.222			116		116	174
1400	37.5 kVA	2 Elec	3	5.333			279		279	420
1420	75 kVA	"	2.50	6.400			335		335	500
1440	3 phase to 600V, primary									
1460	3 kVA	1 Elec	3.85	2.078	Ea.		109		109	163
1480	15 kVA	2 Elec	4.20	3.810			200		200	299
1500	30 kVA		3.50	4.571			240		240	360
1510	45 kVA		3.10	5.161			270		270	405
1520	75 kVA		2.70	5.926			310		310	465

26 05 Common Work Results for Electrical

26 05 05 – Selective Electrical Demolition

26 05 05.10 Electrical Demolition		Crew	Daily Output	Labor-Hours	Unit	Material	2013 Bare Costs Labor	Equipment	Total	Total Incl O&P
1530	112.5 kVA	R-3	2.90	6.897	Ea.		360	47.50	407.50	585
1540	150 kVA		2.70	7.407			385	51	436	630
1550	300 kVA		1.80	11.111			575	76	651	950
1560	500 kVA		1.40	14.286			740	98	838	1,200
1570	750 kVA		1.10	18.182			945	125	1,070	1,550
1600	Pull boxes & cabinets, sheet metal, incl. removal									
1620	of supports and conduit terminations									
1640	6″ x 6″ x 4″	1 Elec	31.10	.257	Ea.		13.50		13.50	20
1660	12″ x 12″ x 4″		23.30	.343			18		18	27
1680	24″ x 24″ x 6″		12.30	.650			34		34	51
1700	36″ x 36″ x 8″		7.70	1.039			54.50		54.50	81.50
1720	Junction boxes, 4″ sq. & oct.		80	.100			5.25		5.25	7.85
1740	Handy box		107	.075			3.92		3.92	5.85
1760	Switch box		107	.075			3.92		3.92	5.85
1780	Receptacle & switch plates		257	.031			1.63		1.63	2.44
1790	Receptacles & switches, 15 to 30 amp		135	.059			3.11		3.11	4.65
1800	Wire, THW-THWN-THHN, removed from									
1810	in place conduit, to 15′ high									
1830	#14	1 Elec	65	.123	C.L.F.		6.45		6.45	9.65
1840	#12		55	.145			7.60		7.60	11.40
1850	#10		45.50	.176			9.20		9.20	13.80
1860	#8		40.40	.198			10.40		10.40	15.50
1870	#6		32.60	.245			12.85		12.85	19.25
1880	#4	2 Elec	53	.302			15.80		15.80	23.50
1890	#3		50	.320			16.75		16.75	25
1900	#2		44.60	.359			18.80		18.80	28
1910	1/0		33.20	.482			25.50		25.50	38
1920	2/0		29.20	.548			28.50		28.50	43
1930	3/0		25	.640			33.50		33.50	50
1940	4/0		22	.727			38		38	57
1950	250 kcmil		20	.800			42		42	62.50
1960	300 kcmil		19	.842			44		44	66
1970	350 kcmil		18	.889			46.50		46.50	69.50
1980	400 kcmil		17	.941			49.50		49.50	74
1990	500 kcmil		16.20	.988			52		52	77.50
2000	Interior fluorescent fixtures, incl. supports									
2010	& whips, to 15′ high									
2100	Recessed drop-in 2′ x 2′, 2 lamp	2 Elec	35	.457	Ea.		24		24	36
2120	2′ x 4′, 2 lamp		33	.485			25.50		25.50	38
2140	2′ x 4′, 4 lamp		30	.533			28		28	42
2160	4′ x 4′, 4 lamp		20	.800			42		42	62.50
2180	Surface mount, acrylic lens & hinged frame									
2200	1′ x 4′, 2 lamp	2 Elec	44	.364	Ea.		19.05		19.05	28.50
2220	2′ x 2′, 2 lamp		44	.364			19.05		19.05	28.50
2260	2′ x 4′, 4 lamp		33	.485			25.50		25.50	38
2280	4′ x 4′, 4 lamp		23	.696			36.50		36.50	54.50
2300	Strip fixtures, surface mount									
2320	4′ long, 1 lamp	2 Elec	53	.302	Ea.		15.80		15.80	23.50
2340	4′ long, 2 lamp		50	.320			16.75		16.75	25
2360	8′ long, 1 lamp		42	.381			19.95		19.95	30
2380	8′ long, 2 lamp		40	.400			21		21	31.50
2400	Pendant mount, industrial, incl. removal									
2410	of chain or rod hangers, to 15′ high									

26 05 05 – Selective Electrical Demolition

26 05 05.10 Electrical Demolition		Crew	Daily Output	Labor-Hours	Unit	Material	2013 Bare Costs Labor	Equipment	Total	Total Incl O&P
2420	4' long, 2 lamp	2 Elec	35	.457	Ea.		24		24	36
2440	8' long, 2 lamp	"	27	.593	"		31		31	46.50
2460	Interior incandescent, surface, ceiling									
2470	or wall mount, to 12' high									
2480	Metal cylinder type, 75 Watt	2 Elec	62	.258	Ea.		13.50		13.50	20
2500	150 Watt		62	.258			13.50		13.50	20
2502	300 Watt		53	.302			15.80		15.80	23.50
2520	Metal halide, high bay									
2540	400 Watt	2 Elec	15	1.067	Ea.		56		56	83.50
2560	1000 Watt		12	1.333			70		70	105
2580	150 Watt, low bay		20	.800			42		42	62.50
2585	Globe type, ceiling mount		62	.258			13.50		13.50	20
2586	Can type, recessed mount		62	.258			13.50		13.50	20
2600	Exterior fixtures, incandescent, wall mount									
2620	100 Watt	2 Elec	50	.320	Ea.		16.75		16.75	25
2640	Quartz, 500 Watt		33	.485			25.50		25.50	38
2660	1500 Watt		27	.593			31		31	46.50
2680	Wall pack, mercury vapor									
2700	175 Watt	2 Elec	25	.640	Ea.		33.50		33.50	50
2720	250 Watt	"	25	.640			33.50		33.50	50
7000	Weatherhead/mast, 2"	1 Elec	16	.500			26		26	39
7002	3"		10	.800			42		42	62.50
7004	3-1/2"		8.50	.941			49.50		49.50	74
7006	4"		8	1			52.50		52.50	78.50
7100	Service entry cable, #6, +#6 neutral		420	.019	L.F.		1		1	1.49
7102	#4, +#4 neutral		360	.022			1.16		1.16	1.74
7104	#2, +#4 neutral		345	.023			1.22		1.22	1.82
7106	#2, +#2 neutral		330	.024			1.27		1.27	1.90
9900	Add to labor for higher elevated installation									
9910	15' to 20' high, add						10%			
9920	20' to 25' high, add						20%			
9930	25' to 30' high, add						25%			
9940	30' to 35' high, add						30%			
9950	35' to 40' high, add						35%			
9960	Over 40' high, add						40%			

26 05 05.15 Electrical Demolition, Grounding

		Crew	Daily Output	Labor-Hours	Unit	Material	Labor	Equipment	Total	Total Incl O&P
0010	**ELECTRICAL DEMOLITION, GROUNDING** Addition									
0100	Ground clamp, bronze	1 Elec	64	.125	Ea.		6.55		6.55	9.80
0140	Water pipe ground clamp, bronze, heavy duty		24	.333			17.45		17.45	26
0150	Ground rod, 8' to 10'		13	.615			32.50		32.50	48.50
0200	Ground wire, bare armored	2 Elec	900	.018	L.F.		.93		.93	1.39
0240	Bare copper or aluminum	"	2800	.006	"		.30		.30	.45

26 05 05.20 Electrical Demolition, Wiring Methods

		Crew	Daily Output	Labor-Hours	Unit	Material	Labor	Equipment	Total	Total Incl O&P
0010	**ELECTRICAL DEMOLITION, WIRING METHODS** Addition									
0100	Armored cable, w/PVC jacket, in cable tray									
0110	#6 R024119-10	1 Elec	9.30	.860	C.L.F.		45		45	67.50
0120	#4	2 Elec	16.20	.988			52		52	77.50
0130	#2		13.80	1.159			61		61	91
0140	#1		12	1.333			70		70	105
0150	1/0		10.80	1.481			77.50		77.50	116
0160	2/0		10.20	1.569			82		82	123
0170	3/0		9.60	1.667			87.50		87.50	131

26 05 05 – Selective Electrical Demolition

26 05 05.20 Electrical Demolition, Wiring Methods		Crew	Daily Output	Labor-Hours	Unit	2013 Bare Costs Material	Labor	Equipment	Total	Total Incl O&P
0180	4/0	2 Elec	9	1.778	C.L.F.		93		93	139
0190	250 kcmil	3 Elec	10.80	2.222			116		116	174
0210	350 kcmil		9.90	2.424			127		127	190
0230	500 kcmil		9	2.667			140		140	209
0240	750 kcmil		8.10	2.963			155		155	232
1100	Control cable, 600 V or less									
1110	3 wires	1 Elec	24	.333	C.L.F.		17.45		17.45	26
1120	5 wires		20	.400			21		21	31.50
1130	7 wires		16.50	.485			25.50		25.50	38
1140	9 wires		15	.533			28		28	42
1150	12 wires	2 Elec	26	.615			32.50		32.50	48.50
1160	15 wires		22	.727			38		38	57
1170	19 wires		19	.842			44		44	66
1180	25 wires		16	1			52.50		52.50	78.50
1500	Mineral insulated (MI) cable, 600 V									
1510	#10	1 Elec	4.80	1.667	C.L.F.		87.50		87.50	131
1520	#8		4.50	1.778			93		93	139
1530	#6		4.20	1.905			100		100	149
1540	#4	2 Elec	7.20	2.222			116		116	174
1550	#2		6.60	2.424			127		127	190
1560	#1		6.30	2.540			133		133	199
1570	1/0		6	2.667			140		140	209
1580	2/0		5.70	2.807			147		147	220
1590	3/0		5.40	2.963			155		155	232
1600	4/0		4.80	3.333			175		175	261
1610	250 kcmil	3 Elec	7.20	3.333			175		175	261
1620	500 kcmil	″	5.90	4.068			213		213	320
2000	Shielded cable, XLP shielding, to 35 kV									
2010	#4	2 Elec	13.20	1.212	C.L.F.		63.50		63.50	95
2020	#1		12	1.333			70		70	105
2030	1/0		11.40	1.404			73.50		73.50	110
2040	2/0		10.80	1.481			77.50		77.50	116
2050	4/0		9.60	1.667			87.50		87.50	131
2060	250 kcmil	3 Elec	13.50	1.778			93		93	139
2070	350 kcmil		11.70	2.051			107		107	161
2080	500 kcmil		11.20	2.143			112		112	168
2090	750 kcmil		10.80	2.222			116		116	174
3000	Modular flexible wiring									
3010	Cable set	1 Elec	120	.067	Ea.		3.49		3.49	5.25
3020	Conversion module		48	.167			8.75		8.75	13.05
3030	Switching assembly		96	.083			4.37		4.37	6.55
3200	Undercarpet									
3210	Power or telephone, flat cable	1 Elec	1320	.006	L.F.		.32		.32	.48
3220	Transition block assemblies w/fitting		75	.107	Ea.		5.60		5.60	8.35
3230	Floor box with fitting		60	.133			7		7	10.45
3300	Data system, cable with connection		50	.160			8.40		8.40	12.55
4000	Cable tray, including fitting & support									
4010	Galvanized steel, 6″ wide	2 Elec	310	.052	L.F.		2.70		2.70	4.05
4020	9″ wide		295	.054			2.84		2.84	4.25
4030	12″ wide		285	.056			2.94		2.94	4.40
4040	18″ wide		270	.059			3.11		3.11	4.65
4050	24″ wide		260	.062			3.22		3.22	4.82
4060	30″ wide		240	.067			3.49		3.49	5.25

26 05 Common Work Results for Electrical

26 05 05 – Selective Electrical Demolition

26 05 05.20	Electrical Demolition, Wiring Methods	Crew	Daily Output	Labor-Hours	Unit	Material	2013 Bare Costs Labor	Equipment	Total	Total Incl O&P
4070	36" wide	2 Elec	220	.073	L.F.		3.81		3.81	5.70
4110	Aluminum, 6" wide		420	.038			2		2	2.99
4120	9" wide		415	.039			2.02		2.02	3.02
4130	12" wide		390	.041			2.15		2.15	3.22
4140	18" wide		370	.043			2.27		2.27	3.39
4150	24" wide		350	.046			2.40		2.40	3.58
4160	30" wide		325	.049			2.58		2.58	3.86
4170	36" wide		300	.053			2.79		2.79	4.18
4310	Cable channel, aluminum, 4" wide straight	1 Elec	240	.033			1.75		1.75	2.61
4320	Cable tray fittings, 6" to 9" wide	2 Elec	27	.593	Ea.		31		31	46.50
4330	12" to 18" wide	"	19	.842	"		44		44	66
5000	Conduit nipples, with locknuts and bushings									
5020	1/2"	1 Elec	108	.074	Ea.		3.88		3.88	5.80
5040	3/4"		96	.083			4.37		4.37	6.55
5060	1"		81	.099			5.20		5.20	7.75
5080	1-1/4"		69	.116			6.10		6.10	9.10
5100	1-1/2"		60	.133			7		7	10.45
5120	2"		54	.148			7.75		7.75	11.60
5140	2-1/2"		45	.178			9.30		9.30	13.95
5160	3"		36	.222			11.65		11.65	17.40
5180	3-1/2"		33	.242			12.70		12.70	19
5200	4"		27	.296			15.55		15.55	23
5220	5"		21	.381			19.95		19.95	30
5240	6"		18	.444			23.50		23.50	35
5500	Electric nonmetallic tubing (ENT), flexible, 1/2" to 1" diameter		690	.012	L.F.		.61		.61	.91
5510	1-1/4" to 2" diameter		300	.027			1.40		1.40	2.09
5600	Flexible metallic tubing, steel, 3/8" to 3/4" diameter		600	.013			.70		.70	1.05
5610	1" to 1-1/4" diameter		260	.031			1.61		1.61	2.41
5620	1-1/2" to 2" diameter		140	.057			2.99		2.99	4.48
5630	2-1/2" diameter		100	.080			4.19		4.19	6.25
5640	3" to 3-1/2" diameter	2 Elec	150	.107			5.60		5.60	8.35
5650	4" diameter	"	100	.160			8.40		8.40	12.55
5700	Sealtite flexible conduit, 3/8" to 3/4" diameter	1 Elec	420	.019			1		1	1.49
5710	1" to 1-1/4" diameter		180	.044			2.33		2.33	3.48
5720	1-1/2" to 2" diameter		120	.067			3.49		3.49	5.25
5730	2-1/2" diameter		80	.100			5.25		5.25	7.85
5740	3" diameter	2 Elec	150	.107			5.60		5.60	8.35
5750	4" diameter	"	100	.160			8.40		8.40	12.55
5800	Wiring duct, plastic, 1-1/2" to 2-1/2" wide	1 Elec	360	.022			1.16		1.16	1.74
5810	3" wide		330	.024			1.27		1.27	1.90
5820	4" wide		300	.027			1.40		1.40	2.09
6000	Floor box and carpet flange		9	.889	Ea.		46.50		46.50	69.50
6300	Wireway, with fittings and supports, to 15' high									
6310	2-1/2" x 2-1/2"	1 Elec	180	.044	L.F.		2.33		2.33	3.48
6320	4" x 4"	"	160	.050			2.62		2.62	3.92
6330	6" x 6"	2 Elec	240	.067			3.49		3.49	5.25
6340	8" x 8"		160	.100			5.25		5.25	7.85
6350	10" x 10"		120	.133			7		7	10.45
6360	12" x 12"		80	.200			10.50		10.50	15.70
6400	Cable reel with receptacle, 120 V/208 V		13	1.231	Ea.		64.50		64.50	96.50
6500	Equipment connection, to 10 HP	1 Elec	23	.348			18.25		18.25	27.50
6510	To 15 to 30 HP		18	.444			23.50		23.50	35
6520	To 40 to 60 HP		15	.533			28		28	42

26 05 05 – Selective Electrical Demolition

26 05 05.20 Electrical Demolition, Wiring Methods		Crew	Daily Output	Labor-Hours	Unit	Material	2013 Bare Costs Labor	Equipment	Total	Total Incl O&P
6530	To 75 to 100 HP	1 Elec	10	.800	Ea.		42		42	62.50
7008	Receptacle, explosionproof, 120 V, to 30 A		16	.500			26		26	39
7020	Dimmer switch, 2000 W or less		62	.129			6.75		6.75	10.10

26 05 05.25 Electrical Demolition, Electrical Power		Crew	Daily Output	Labor-Hours	Unit	Material	Labor	Equipment	Total	Total Incl O&P
0010	ELECTRICAL DEMOLITION, ELECTRICAL POWER R024119-10									
0100	Meter centers and sockets									
0120	Meter socket, 4 terminal R260105-30	1 Elec	8.40	.952	Ea.		50		50	74.50
0140	Trans-socket, 13 terminal, 400 A	"	3.60	2.222			116		116	174
0160	800 A	2 Elec	4.40	3.636			191		191	285
0200	Meter center, 400 A		5.80	2.759			145		145	216
0210	600 A		4	4			210		210	315
0220	800 A		3.30	4.848			254		254	380
0230	1200 A		2.80	5.714			299		299	450
0240	1600 A		2.50	6.400			335		335	500
0300	Base meter devices, 3 meter		3.60	4.444			233		233	350
0310	4 meter		3.30	4.848			254		254	380
0320	5 meter		2.90	5.517			289		289	435
0330	6 meter		2.20	7.273			380		380	570
0340	7 meter		2	8			420		420	625
0350	8 meter		1.90	8.421			440		440	660
0400	Branch meter devices									
0410	Socket w/circuit breaker 200 A, 2 meter	2 Elec	3.30	4.848	Ea.		254		254	380
0420	3 meter		2.90	5.517			289		289	435
0430	4 meter		2.60	6.154			320		320	480
0450	Main circuit breaker, 400 A		5.80	2.759			145		145	216
0460	600 A		4	4			210		210	315
0470	800 A		3.30	4.848			254		254	380
0480	1200 A		2.80	5.714			299		299	450
0490	1600 A		2.50	6.400			335		335	500
0500	Main lug terminal box, 800 A		3.40	4.706			247		247	370
0510	1200 A		2.60	6.154			320		320	480
1000	Motors, 230/460 V, 60 Hz, 3/4 HP	1 Elec	10.70	.748			39		39	58.50
1010	5 HP		9	.889			46.50		46.50	69.50
1020	10 HP		8	1			52.50		52.50	78.50
1030	15 HP		6.40	1.250			65.50		65.50	98
1040	20 HP	2 Elec	10.40	1.538			80.50		80.50	121
1050	50 HP		9.60	1.667			87.50		87.50	131
1060	75 HP		5.60	2.857			150		150	224
1070	100 HP	3 Elec	5.40	4.444			233		233	350
1080	150 HP		3.60	6.667			350		350	525
1090	200 HP		3	8			420		420	625
1200	Variable frequency drive, 460 V, for 5 HP motor size	1 Elec	3.20	2.500			131		131	196
1210	10 HP motor size	"	2.70	2.963			155		155	232
1220	20 HP motor size	2 Elec	3.60	4.444			233		233	350
1230	50 HP motor size	"	2.10	7.619			400		400	595
1240	75 HP motor size	R-3	2.20	9.091			470	62.50	532.50	775
1250	100 HP motor size		2	10			520	68.50	588.50	855
1260	150 HP motor size		2	10			520	68.50	588.50	855
1270	200 HP motor size		1.70	11.765			610	80.50	690.50	1,000
2000	Generator set w/accessories, 3 phase 4 wire, 277/480 V									
2010	7.5 kW	3 Elec	2	12	Ea.		630		630	940
2020	20 kW		1.70	14.118			740		740	1,100

26 05 Common Work Results for Electrical

26 05 05 – Selective Electrical Demolition

26 05 05.25	Electrical Demolition, Electrical Power	Crew	Daily Output	Labor-Hours	Unit	Material	2013 Bare Costs Labor	Equipment	Total	Total Incl O&P
2040	30 kW	3 Elec	1.33	18.045	Ea.		945		945	1,425
2060	50 kW		1	24			1,250		1,250	1,875
2080	100 kW		.75	32			1,675		1,675	2,500
2100	150 kW		.63	38.095			2,000		2,000	2,975
2120	250 kW		.59	40.678			2,125		2,125	3,200
2140	400 kW		.50	48			2,525		2,525	3,775
2160	500 kW		.44	54.545			2,850		2,850	4,275
2180	750 kW	4 Elec	.56	57.143			3,000		3,000	4,475
2200	1000 kW	″	.46	69.565			3,650		3,650	5,450
3000	Uninterruptible power supply system (UPS)									
3010	Single phase, 120 V, 1 kVA	1 Elec	3.20	2.500	Ea.		131		131	196
3020	2 kVA	2 Elec	3.60	4.444			233		233	350
3030	5 kVA	3 Elec	2.50	9.600			505		505	755
3040	10 kVA		2.30	10.435			545		545	820
3050	15 kVA		1.80	13.333			700		700	1,050
4000	Transformer, incl support, wire & conduit termination									
4010	Buck-boost, single phase, 120/240 V, 0.1 kVA	1 Elec	25	.320	Ea.		16.75		16.75	25
4020	0.5 kVA		12.50	.640			33.50		33.50	50
4030	1 kVA		6.30	1.270			66.50		66.50	99.50
4040	5 kVA		3.80	2.105			110		110	165
4800	5 kV or 15 kV primary, 277/480 V second, 112.5 kVA	R-3	3.50	5.714			296	39	335	490
4810	150 kVA		2.70	7.407			385	51	436	630
4820	225 kVA		2.30	8.696			450	59.50	509.50	740
4830	500 kVA		1.45	13.793			715	94.50	809.50	1,175
4840	750 kVA		1.33	15.038			780	103	883	1,300
4850	1000 kVA		1.25	16			830	110	940	1,375
4860	2000 kVA		1.05	19.048			990	131	1,121	1,625
4870	3000 kVA		.75	26.667			1,375	183	1,558	2,275
6010	Isolation panel, 3 kVA	1 Elec	2.30	3.478			182		182	273
6020	5 kVA		2.20	3.636			191		191	285
6030	7.5 kVA		2.10	3.810			200		200	299
6040	10 kVA		1.80	4.444			233		233	350
6050	15 kVA		1.40	5.714			299		299	450
7000	Power filters & conditioners									
7100	Automatic voltage regulator	2 Elec	6	2.667	Ea.		140		140	209
7200	Capacitor, 1 kVAR	1 Elec	8.60	.930			48.50		48.50	73
7210	5 kVAR		5.80	1.379			72.50		72.50	108
7220	10 kVAR		4.80	1.667			87.50		87.50	131
7230	15 kVAR		4.20	1.905			100		100	149
7240	20 kVAR		3.50	2.286			120		120	179
7250	30 kVAR		3.40	2.353			123		123	184
7260	50 kVAR		3.20	2.500			131		131	196
7400	Computer isolator transformer									
7410	Single phase, 120/240 V, 0.5 kVAR	1 Elec	12.80	.625	Ea.		33		33	49
7420	1 kVAR		8.50	.941			49.50		49.50	74
7430	2.5 kVAR		6.40	1.250			65.50		65.50	98
7440	5 kVAR		3.70	2.162			113		113	170
7500	Computer regulator transformer									
7510	Single phase, 240 V, 0.5 kVAR	1 Elec	8.50	.941	Ea.		49.50		49.50	74
7520	1 kVAR		6.40	1.250			65.50		65.50	98
7530	2 kVAR		3.20	2.500			131		131	196
7540	Single phase, plug-in unit 120 V, 0.5 kVAR		26	.308			16.10		16.10	24
7550	1 kVAR		17	.471			24.50		24.50	37

26 05 05 – Selective Electrical Demolition

26 05 05.25 Electrical Demolition, Electrical Power		Crew	Daily Output	Labor-Hours	Unit	2013 Bare Costs Material	Labor	Equipment	Total	Total Incl O&P
7600	Power conditioner transformer									
7610	Single phase 115 V - 240 V, 3 kVA	2 Elec	5.10	3.137	Ea.		164		164	246
7620	5 kVA	″	3.70	4.324			227		227	340
7630	7.5 kVA	3 Elec	4.80	5			262		262	390
7640	10 kVA	″	4.30	5.581			292		292	440
7700	Transient suppressor/voltage regulator									
7710	Single phase 115 or 220 V, 1 kVA	1 Elec	8.50	.941	Ea.		49.50		49.50	74
7720	2 kVA		7.30	1.096			57.50		57.50	86
7730	4 kVA		6.80	1.176			61.50		61.50	92
7800	Transient voltage suppressor transformer									
7810	Single phase, 115 or 220 V, 3.6 kVA	1 Elec	12.80	.625	Ea.		33		33	49
7820	7.2 kVA		11.50	.696			36.50		36.50	54.50
7830	14.4 kVA		10.20	.784			41		41	61.50
7840	Single phase, plug-in, 120 V, 1.8 kVA		26	.308			16.10		16.10	24
8000	Power measurement & control									
8010	Switchboard instruments, 3 phase 4 wire, indicating unit	1 Elec	25	.320	Ea.		16.75		16.75	25
8020	Recording unit		12	.667			35		35	52.50
8100	3 current transformers, 3 phase 4 wire, 5 to 800 A		6.40	1.250			65.50		65.50	98
8110	1000 to 1500 A		4.20	1.905			100		100	149
8120	2000 to 4000 A		3.20	2.500			131		131	196
26 05 05.30 Electrical Demolition, Transmission and Distribution										
0010	**ELECTRICAL DEMOLITION, TRANSMISSION & DISTRIBUTION**									
0100	Load interrupter switch, 600 A, NEMA 1, 4.8 kV	R-3	1.33	15.038	Ea.		780	103	883	1,300
0120	13.8 kV R024119-10	″	1.27	15.748			815	108	923	1,350
0200	Lightning arrester, 4.8 kV	1 Elec	9	.889			46.50		46.50	69.50
0220	13.8 kV R260105-30		6.67	1.199			63		63	94
0300	Alarm or option items		3.33	2.402			126		126	188
26 05 05.35 Electrical Demolition, L.V. Distribution										
0010	**ELECTRICAL DEMOLITION, L.V. DISTRIBUTION** Addition R024119-10									
0100	Circuit breakers in enclosure									
0120	Enclosed (NEMA 1), 600 V, 3 pole, 30 A R260105-30	1 Elec	12.30	.650	Ea.		34		34	51
0140	60 A		8.80	.909			47.50		47.50	71.50
0160	100 A		7.30	1.096			57.50		57.50	86
0180	225 A		5	1.600			84		84	125
0200	400 A	2 Elec	6.80	2.353			123		123	184
0220	600 A		4.60	3.478			182		182	273
0240	800 A		3.60	4.444			233		233	350
0260	1200 A		3.10	5.161			270		270	405
0280	1600 A		2.80	5.714			299		299	450
0300	2000 A		2.50	6.400			335		335	500
0400	Enclosed (NEMA 7), 600 V, 3 pole, 50 A	1 Elec	8.80	.909			47.50		47.50	71.50
0410	100 A		5.80	1.379			72.50		72.50	108
0420	150 A		3.80	2.105			110		110	165
0430	250 A	2 Elec	6.20	2.581			135		135	202
0440	400 A	″	4.60	3.478			182		182	273
0460	Manual motor starter, NEMA 1	1 Elec	24	.333			17.45		17.45	26
0480	NEMA 4 or NEMA 7		15	.533			28		28	42
0500	Time switches, single pole single throw		15.40	.519			27		27	40.50
0520	Photo cell		30	.267			13.95		13.95	21
0540	Load management device, 4 loads		7.70	1.039			54.50		54.50	81.50
0550	8 loads		3.80	2.105			110		110	165
0560	Master light control panel	2 Elec	2	8			420		420	625

26 05 Common Work Results for Electrical

26 05 05 – Selective Electrical Demolition

26 05 05.35 Electrical Demolition, L.V. Distribution		Crew	Daily Output	Labor-Hours	Unit	2013 Bare Costs Material	Labor	Equipment	Total	Total Incl O&P
0600	Transfer switches, enclosed, 30 A	1 Elec	12	.667	Ea.		35		35	52.50
0610	60 A		9.50	.842			44		44	66
0620	100 A		6.50	1.231			64.50		64.50	96.50
0630	150 A	2 Elec	12	1.333			70		70	105
0640	260 A		10	1.600			84		84	125
0660	400 A		8	2			105		105	157
0670	600 A		5	3.200			168		168	251
0680	800 A		4	4			210		210	315
0690	1200 A		3.50	4.571			240		240	360
0700	1600 A		3	5.333			279		279	420
0710	2000 A		2.50	6.400			335		335	500
0800	Air terminal with base and connector	1 Elec	24	.333			17.45		17.45	26
1000	Enclosed controller, NEMA 1, 30 A		13.80	.580			30.50		30.50	45.50
1020	60 A		11.50	.696			36.50		36.50	54.50
1040	100 A		9.60	.833			43.50		43.50	65.50
1060	150 A		7.70	1.039			54.50		54.50	81.50
1080	200 A		5.40	1.481			77.50		77.50	116
1100	400 A	2 Elec	6.90	2.319			122		122	182
1120	600 A		4.60	3.478			182		182	273
1140	800 A		3.80	4.211			221		221	330
1160	1200 A		3.10	5.161			270		270	405
1200	Control station, NEMA 1	1 Elec	30	.267			13.95		13.95	21
1220	NEMA 7		23	.348			18.25		18.25	27.50
1230	Control switches, push button		69	.116			6.10		6.10	9.10
1240	Indicating light unit		120	.067			3.49		3.49	5.25
1250	Relay		15	.533			28		28	42
2000	Motor control center components									
2020	Starter, NEMA 1, size 1	1 Elec	9	.889	Ea.		46.50		46.50	69.50
2040	Size 2	2 Elec	13.30	1.203			63		63	94.50
2060	Size 3		6.70	2.388			125		125	187
2080	Size 4		5.30	3.019			158		158	237
2100	Size 5		3.30	4.848			254		254	380
2120	NEMA 7, size 1	1 Elec	8.70	.920			48		48	72
2140	Size 2	2 Elec	12.70	1.260			66		66	99
2160	Size 3		6.30	2.540			133		133	199
2180	Size 4		5	3.200			168		168	251
2200	Size 5		3.20	5			262		262	390
2300	Fuse, light contactor, NEMA 1, 30 A	1 Elec	9	.889			46.50		46.50	69.50
2310	60 A		6.70	1.194			62.50		62.50	93.50
2320	100 A		3.30	2.424			127		127	190
2330	200 A		2.70	2.963			155		155	232
2350	Motor control center, incoming section	2 Elec	4	4			210		210	315
2352	Structure per section	″	4.20	3.810			200		200	299
2400	Starter & structure, 10 HP	1 Elec	9	.889			46.50		46.50	69.50
2410	25 HP	2 Elec	13.30	1.203			63		63	94.50
2420	50 HP		6.70	2.388			125		125	187
2430	75 HP		5.30	3.019			158		158	237
2440	100 HP		4.70	3.404			178		178	267
2450	200 HP		3.30	4.848			254		254	380
2460	400 HP		2.70	5.926			310		310	465
2500	Motor starter & control									
2510	Motor starter, NEMA 1, 5 HP	1 Elec	8.90	.899	Ea.		47		47	70.50
2520	10 HP	″	6.20	1.290			67.50		67.50	101

26 05 05 – Selective Electrical Demolition

26 05 05.35 Electrical Demolition, L.V. Distribution		Crew	Daily Output	Labor-Hours	Unit	Material	2013 Bare Costs Labor	Equipment	Total	Total Incl O&P
2530	25 HP	2 Elec	8.40	1.905	Ea.		100		100	149
2540	50 HP		6.90	2.319			122		122	182
2550	100 HP		4.60	3.478			182		182	273
2560	200 HP		3.50	4.571			240		240	360
2570	400 HP		3.10	5.161			270		270	405
2610	Motor starter, NEMA 7, 5 HP	1 Elec	6.20	1.290			67.50		67.50	101
2620	10 HP	"	4.20	1.905			100		100	149
2630	25 HP	2 Elec	6.90	2.319			122		122	182
2640	50 HP		4.60	3.478			182		182	273
2650	100 HP		3.50	4.571			240		240	360
2660	200 HP		1.90	8.421			440		440	660
2710	Combination control unit, NEMA 1, 5 HP	1 Elec	6.90	1.159			61		61	91
2720	10 HP	"	5	1.600			84		84	125
2730	25 HP	2 Elec	7.70	2.078			109		109	163
2740	50 HP		5.10	3.137			164		164	246
2750	100 HP		3.10	5.161			270		270	405
2810	NEMA 7, 5 HP	1 Elec	5	1.600			84		84	125
2820	10 HP	"	3.90	2.051			107		107	161
2830	25 HP	2 Elec	5.10	3.137			164		164	246
2840	50 HP		3.10	5.161			270		270	405
2850	100 HP		2.30	6.957			365		365	545
2860	200 HP		1.50	10.667			560		560	835
3000	Panelboard or load center circuit breaker									
3010	Bolt-on or plug in, 15 A to 50 A	1 Elec	20	.400	Ea.		21		21	31.50
3020	60 A to 70 A		16	.500			26		26	39
3030	Bolt-on, 80 A to 100 A		14	.571			30		30	45
3040	Up to 250 A		7.30	1.096			57.50		57.50	86
3050	Motor operated, 30 A		13	.615			32.50		32.50	48.50
3060	60 A		10	.800			42		42	62.50
3070	100 A		8	1			52.50		52.50	78.50
3200	Switchboard circuit breaker									
3210	15 A - 60 A	1 Elec	19	.421	Ea.		22		22	33
3220	70 A - 100 A		14	.571			30		30	45
3230	125 A - 400 A		10	.800			42		42	62.50
3240	450 A - 600 A		5.30	1.509			79		79	118
3250	700 A - 800 A		4.30	1.860			97.50		97.50	146
3260	1000 A		3.30	2.424			127		127	190
3270	1200 A		2.70	2.963			155		155	232
3500	Switchboard, incoming section, 400 A	2 Elec	3.70	4.324			227		227	340
3510	600 A		3.30	4.848			254		254	380
3520	800 A		2.90	5.517			289		289	435
3530	1200 A		2.40	6.667			350		350	525
3540	1600 A		2.20	7.273			380		380	570
3550	2000 A		2.10	7.619			400		400	595
3560	3000 A		1.90	8.421			440		440	660
3570	4000 A		1.70	9.412			495		495	740
3610	Distribution section, 600 A		4	4			210		210	315
3620	800 A		3.60	4.444			233		233	350
3630	1200 A		3.10	5.161			270		270	405
3640	1600 A		2.90	5.517			289		289	435
3650	2000 A		2.70	5.926			310		310	465
3710	Transition section, 600 A		3.80	4.211			221		221	330
3720	800 A		3.30	4.848			254		254	380

26 05 Common Work Results for Electrical

26 05 05 – Selective Electrical Demolition

26 05 05.35 Electrical Demolition, L.V. Distribution		Crew	Daily Output	Labor-Hours	Unit	Material	2013 Bare Costs Labor	Equipment	Total	Total Incl O&P
3730	1200 A	2 Elec	2.70	5.926	Ea.		310		310	465
3740	1600 A		2.40	6.667			350		350	525
3750	2000 A		2.20	7.273			380		380	570
3760	2500 A		2.10	7.619			400		400	595
3770	3000 A		1.90	8.421			440		440	660
4000	Bus duct, aluminum or copper, 30 A	1 Elec	200	.040	L.F.		2.10		2.10	3.14
4020	60 A		160	.050			2.62		2.62	3.92
4040	100 A		140	.057			2.99		2.99	4.48
5000	Feedrail, trolley busway, up to 60 A		160	.050			2.62		2.62	3.92
5020	100 A		120	.067			3.49		3.49	5.25
5040	Busway, 50 A		200	.040			2.10		2.10	3.14
6000	Fuse, 30 A		133	.060	Ea.		3.15		3.15	4.72
6010	60 A		133	.060			3.15		3.15	4.72
6020	100 A		105	.076			3.99		3.99	5.95
6030	200 A		95	.084			4.41		4.41	6.60
6040	400 A		80	.100			5.25		5.25	7.85
6050	600 A		53	.151			7.90		7.90	11.85
6060	601 A - 1200 A		42	.190			10		10	14.95
6070	1500 A - 1600 A		34	.235			12.35		12.35	18.45
6080	1800 A - 2500 A		26	.308			16.10		16.10	24
6090	4000 A		21	.381			19.95		19.95	30
6100	4500 A - 5000 A		18	.444			23.50		23.50	35
6120	6000 A		15	.533			28		28	42
6200	Fuse plug or fustat		100	.080			4.19		4.19	6.25

26 05 05.50 Electrical Demolition, Lighting		Crew	Daily Output	Labor-Hours	Unit	Material	Labor	Equipment	Total	Total Incl O&P
0010	**ELECTRICAL DEMOLITION, LIGHTING** Addition R024119-10									
0100	Fixture hanger, flexible, 1/2" diameter	1 Elec	36	.222	Ea.		11.65		11.65	17.40
0120	3/4" diameter R260105-30	"	30	.267	"		13.95		13.95	21
3000	Light pole, anchor base, excl concrete bases									
3010	Metal light pole, 10'	2 Elec	24	.667	Ea.		35		35	52.50
3020	16'	"	18	.889			46.50		46.50	69.50
3030	20'	R-3	8.70	2.299			119	15.75	134.75	196
3040	40'	"	6	3.333			173	23	196	284
3100	Wood light pole, 10'	2 Elec	36	.444			23.50		23.50	35
3120	20'	"	24	.667			35		35	52.50
3140	Bollard light, 42"	1 Elec	9	.889			46.50		46.50	69.50
3160	Walkway luminaire	"	8.10	.988			52		52	77.50
4000	Explosionproof									
4010	Metal halide, 175 W	1 Elec	8.70	.920	Ea.		48		48	72
4020	250 W	"	8.10	.988			52		52	77.50
4030	400 W	2 Elec	14.40	1.111			58		58	87
4050	High pressure sodium, 70 W	1 Elec	9	.889			46.50		46.50	69.50
4060	100 W		9	.889			46.50		46.50	69.50
4070	150 W		8.10	.988			52		52	77.50
4100	Incandescent		8.70	.920			48		48	72
4200	Flurorescent		8.10	.988			52		52	77.50
5000	Ballast, fluorescent fixture		24	.333			17.45		17.45	26
5040	High intensity discharge fixture		24	.333			17.45		17.45	26
5300	Exit and emergency lighting									
5310	Exit light	1 Elec	24	.333	Ea.		17.45		17.45	26
5320	Emergency battery pack lighting unit		12	.667			35		35	52.50
5330	Remote lamp only		80	.100			5.25		5.25	7.85

26 05 Common Work Results for Electrical

26 05 05 – Selective Electrical Demolition

26 05 05.50 Electrical Demolition, Lighting		Crew	Daily Output	Labor-Hours	Unit	Material	2013 Bare Costs Labor	Equipment	Total	Total Incl O&P
5340	Self-contained fluorescent lamp pack	1 Elec	30	.267	Ea.		13.95		13.95	21
5500	Track lighting, 8′ section	2 Elec	40	.400			21		21	31.50
5510	Track lighting fxiture	1 Elec	64	.125			6.55		6.55	9.80
5800	Energy saving devices									
5810	Occupancy sensor	1 Elec	21	.381	Ea.		19.95		19.95	30
5820	Automatic wall switch		72	.111			5.80		5.80	8.70
5830	Remote power pack		30	.267			13.95		13.95	21
5840	Photoelectric control		24	.333			17.45		17.45	26
5850	Fixture whip		100	.080			4.19		4.19	6.25
6000	Lamps									
6010	Fluorescent	1 Elec	200	.040	Ea.		2.10		2.10	3.14
6030	High intensity discharge lamp, up to 400 W		68	.118			6.15		6.15	9.20
6040	Up to 1000 W		45	.178			9.30		9.30	13.95
6050	Quartz		90	.089			4.66		4.66	6.95
6070	Incandescent		360	.022			1.16		1.16	1.74
6080	Exterior, PAR		290	.028			1.45		1.45	2.16
6090	Guards for fluorescent lamp		24	.333			17.45		17.45	26

26 05 13 – Medium-Voltage Cables

26 05 13.10 Cable Terminations		Crew	Daily Output	Labor-Hours	Unit	Material	Labor	Equipment	Total	Total Incl O&P
0010	**CABLE TERMINATIONS**, 5 kV to 35 kV									
0100	Indoor, insulation diameter range .64″ to 1.08″									
0300	Padmount, 5 kV	1 Elec	8	1	Ea.	85	52.50		137.50	172
0400	10 kV		6.40	1.250		97	65.50		162.50	205
0500	15 kV		6	1.333		126	70		196	243
0600	25 kV		5.60	1.429		155	75		230	282
0700	insulation diameter range 1.05″ to 1.8″									
0800	Padmount, 5 kV	1 Elec	8	1	Ea.	98	52.50		150.50	187
0900	10 kV		6	1.333		129	70		199	247
1000	15 kV		5.60	1.429		149	75		224	276
1100	25 kV		5.30	1.509		191	79		270	330
1200	insulation diameter range 1.53″ to 2.32″									
1300	Padmount, 5 KV	1 Elec	7.40	1.081	Ea.	116	56.50		172.50	213
1400	10 kV		5.60	1.429		163	75		238	292
1500	15 kV		5.30	1.509		191	79		270	330
1600	25 kV		5	1.600		228	84		312	375
1700	Outdoor systems, #4 stranded to 1/0 stranded									
1800	5 kV	1 Elec	7.40	1.081	Ea.	106	56.50		162.50	202
1900	15 kV		5.30	1.509		132	79		211	263
2000	25 kV		5	1.600		177	84		261	320
2100	35 kV		4.80	1.667		238	87.50		325.50	395
2200	#1 solid to 4/0 stranded, 5 kV		6.90	1.159		141	61		202	246
2300	15 kV		5	1.600		169	84		253	310
2400	25 kV		4.80	1.667		227	87.50		314.50	380
2500	35 kV		4.60	1.739		251	91		342	415
2600	2/0 solid to 350 kcmil stranded, 5 kV		6.40	1.250		190	65.50		255.50	305
2700	15 kV		4.80	1.667		190	87.50		277.50	340
2800	25 kV		4.60	1.739		294	91		385	460
2900	35 kV		4.40	1.818		310	95.50		405.50	485
3000	400 kcmil compact to 750 kcmil stranded, 5 kV		6	1.333		193	70		263	315
3100	15 kV		4.60	1.739		229	91		320	390
3200	25 kV		4.40	1.818		335	95.50		430.50	515
3300	35 kV		4.20	1.905		345	100		445	525

26 05 13.10 Cable Terminations

	26 05 13.10 Cable Terminations	Crew	Daily Output	Labor-Hours	Unit	2013 Bare Costs Material	Labor	Equipment	Total	Total Incl O&P
3400	1000 kcmil, 5 kV	1 Elec	5.60	1.429	Ea.	310	75		385	450
3500	15 kV		4.40	1.818		310	95.50		405.50	485
3600	25 kV		4.20	1.905		345	100		445	525
3700	35 kV		4	2		345	105		450	530

26 05 13.16 Medium-Voltage, Single Cable

	26 05 13.16 Medium-Voltage, Single Cable	Crew	Daily Output	Labor-Hours	Unit	Material	Labor	Equipment	Total	Total Incl O&P
0010	**MEDIUM-VOLTAGE, SINGLE CABLE** Splicing & terminations not included									
0040	Copper, XLP shielding, 5 kV, #6	2 Elec	4.40	3.636	C.L.F.	162	191		353	465
0050	#4		4.40	3.636		210	191		401	515
0100	#2		4	4		248	210		458	590
0200	#1		4	4		285	210		495	630
0400	1/0		3.80	4.211		315	221		536	675
0600	2/0		3.60	4.444		395	233		628	785
0800	4/0		3.20	5		515	262		777	955
1000	250 kcmil	3 Elec	4.50	5.333		635	279		914	1,125
1200	350 kcmil		3.90	6.154		820	320		1,140	1,375
1400	500 kcmil		3.60	6.667		1,000	350		1,350	1,625
1600	15 kV, ungrounded neutral, #1	2 Elec	4	4		345	210		555	695
1800	1/0		3.80	4.211		415	221		636	785
2000	2/0		3.60	4.444		470	233		703	870
2200	4/0		3.20	5		625	262		887	1,075
2400	250 kcmil	3 Elec	4.50	5.333		695	279		974	1,175
2600	350 kcmil		3.90	6.154		875	320		1,195	1,450
2800	500 kcmil		3.60	6.667		1,075	350		1,425	1,700
3000	25 kV, grounded neutral, #1/0	2 Elec	3.60	4.444		570	233		803	980
3200	2/0		3.40	4.706		630	247		877	1,075
3400	4/0		3	5.333		790	279		1,069	1,300
3600	250 kcmil	3 Elec	4.20	5.714		980	299		1,279	1,525
3800	350 kcmil		3.60	6.667		1,150	350		1,500	1,800
3900	500 kcmil		3.30	7.273		1,350	380		1,730	2,050
4000	35 kV, grounded neutral, #1/0	2 Elec	3.40	4.706		605	247		852	1,025
4200	2/0		3.20	5		710	262		972	1,175
4400	4/0		2.80	5.714		895	299		1,194	1,425
4600	250 kcmil	3 Elec	3.90	6.154		1,050	320		1,370	1,625
4800	350 kcmil		3.30	7.273		1,250	380		1,630	1,950
5000	500 kcmil		3	8		1,475	420		1,895	2,250
5050	Aluminum, XLP shielding, 5 kV, #2	2 Elec	5	3.200		175	168		343	445
5070	#1		4.40	3.636		180	191		371	485
5090	1/0		4	4		209	210		419	545
5100	2/0		3.80	4.211		234	221		455	590
5150	4/0		3.60	4.444		277	233		510	655
5200	250 kcmil	3 Elec	4.80	5		335	262		597	760
5220	350 kcmil		4.50	5.333		395	279		674	855
5240	500 kcmil		3.90	6.154		505	320		825	1,050
5260	750 kcmil		3.60	6.667		670	350		1,020	1,275
5300	15 kV aluminum, XLP, #1	2 Elec	4.40	3.636		224	191		415	530
5320	1/0		4	4		231	210		441	570
5340	2/0		3.80	4.211		276	221		497	635
5360	4/0		3.60	4.444		305	233		538	685
5380	250 kcmil	3 Elec	4.80	5		365	262		627	790
5400	350 kcmil		4.50	5.333		410	279		689	870
5420	500 kcmil		3.90	6.154		565	320		885	1,100
5440	750 kcmil		3.60	6.667		755	350		1,105	1,350

26 05 19 – Low-Voltage Electrical Power Conductors and Cables

26 05 19.13 Undercarpet Electrical Power Cables		Crew	Daily Output	Labor-Hours	Unit	Material	Labor	Equipment	Total	Total Incl O&P
						2013 Bare Costs				
0010	**UNDERCARPET ELECTRICAL POWER CABLES** R260519-80									
0020	Power System									
0100	Cable flat, 3 conductor, #12, w/attached bottom shield	1 Elec	982	.008	L.F.	5.05	.43		5.48	6.20
0200	Shield, top, steel		1768	.005	"	5.40	.24		5.64	6.30
0250	Splice, 3 conductor		48	.167	Ea.	16.40	8.75		25.15	31
0300	Top shield		96	.083		1.47	4.37		5.84	8.15
0350	Tap		40	.200		21	10.50		31.50	38.50
0400	Insulating patch, splice, tap, & end		48	.167		52	8.75		60.75	70
0450	Fold		230	.035			1.82		1.82	2.73
0500	Top shield, tap & fold		96	.083		1.47	4.37		5.84	8.15
0700	Transition, block assembly		77	.104		75.50	5.45		80.95	91
0750	Receptacle frame & base		32	.250		41.50	13.10		54.60	65
0800	Cover receptacle		120	.067		3.54	3.49		7.03	9.15
0850	Cover blank		160	.050		4.16	2.62		6.78	8.50
0860	Receptacle, direct connected, single		25	.320		89	16.75		105.75	123
0870	Dual		16	.500		146	26		172	199
0880	Combination high & low, tension		21	.381		107	19.95		126.95	148
0900	Box, floor with cover		20	.400		90	21		111	131
0920	Floor service w/barrier		4	2		254	105		359	435
1000	Wall, surface, with cover		20	.400		59.50	21		80.50	96.50
1100	Wall, flush, with cover		20	.400		41.50	21		62.50	77
1450	Cable flat, 5 conductor #12, w/attached bottom shield		800	.010	L.F.	8.25	.52		8.77	9.90
1550	Shield, top, steel		1768	.005	"	8.20	.24		8.44	9.40
1600	Splice, 5 conductor		48	.167	Ea.	26.50	8.75		35.25	42
1650	Top shield		96	.083		1.47	4.37		5.84	8.15
1700	Tap		48	.167		34.50	8.75		43.25	51
1750	Insulating patch, splice tap, & end		83	.096		52	5.05		57.05	64.50
1800	Transition, block assembly		77	.104		54.50	5.45		59.95	68
1850	Box, wall, flush with cover		20	.400		55	21		76	92.50
1900	Cable flat, 4 conductor, #12		933	.009	L.F.	6.70	.45		7.15	8
1950	3 conductor #10		982	.008		5.80	.43		6.23	7
1960	4 conductor #10		933	.009		7.55	.45		8	8.95
1970	5 conductor #10		884	.009		9.25	.47		9.72	10.90
2500	Telephone System									
2510	Transition fitting wall box, surface	1 Elec	24	.333	Ea.	44.50	17.45		61.95	75
2520	Flush		24	.333		44.50	17.45		61.95	75
2530	Flush, for PC board		24	.333		44.50	17.45		61.95	75
2540	Floor service box		4	2		231	105		336	410
2550	Cover, surface					14.20			14.20	15.60
2560	Flush					14.20			14.20	15.60
2570	Flush for PC board					14.20			14.20	15.60
2700	Floor fitting w/duplex jack & cover	1 Elec	21	.381		43.50	19.95		63.45	78
2720	Low profile		53	.151		15.25	7.90		23.15	28.50
2740	Miniature w/duplex jack		53	.151		23.50	7.90		31.40	38
2760	25 pair kit		21	.381		46	19.95		65.95	80.50
2780	Low profile		53	.151		15.60	7.90		23.50	29
2800	Call director kit for 5 cable		19	.421		73.50	22		95.50	114
2820	4 pair kit		19	.421		87	22		109	129
2840	3 pair kit		19	.421		92.50	22		114.50	135
2860	Comb. 25 pair & 3 conductor power		21	.381		73.50	19.95		93.45	111
2880	5 conductor power		21	.381		83.50	19.95		103.45	122
2900	PC board, 8 per 3 pair		161	.050		63.50	2.60		66.10	74

26 05 19 – Low-Voltage Electrical Power Conductors and Cables

26 05 19.13 Undercarpet Electrical Power Cables		Crew	Daily Output	Labor-Hours	Unit	Material	Labor	Equipment	Total	Total Incl O&P
						2013 Bare Costs				
2920	6 per 4 pair	1 Elec	161	.050	Ea.	63.50	2.60		66.10	74
2940	3 pair adapter		161	.050		57.50	2.60		60.10	67.50
2950	Plug		77	.104		2.70	5.45		8.15	11.10
2960	Couplers		321	.025		7.55	1.31		8.86	10.30
3000	Bottom shield for 25 pair cable		4420	.002	L.F.	.76	.09		.85	.98
3020	4 pair		4420	.002		.37	.09		.46	.55
3040	Top shield for 25 pair cable		4420	.002		.76	.09		.85	.98
3100	Cable assembly, double-end, 50′, 25 pair		11.80	.678	Ea.	223	35.50		258.50	298
3110	3 pair		23.60	.339		66	17.75		83.75	99
3120	4 pair		23.60	.339		73.50	17.75		91.25	107
3140	Bulk 3 pair		1473	.005	L.F.	1.13	.28		1.41	1.67
3160	4 pair		1473	.005	″	1.41	.28		1.69	1.98
3500	Data System									
3520	Cable 25 conductor w/connection 40′, 75 ohm	1 Elec	14.50	.552	Ea.	66	29		95	116
3530	Single lead		22	.364		180	19.05		199.05	227
3540	Dual lead		22	.364		224	19.05		243.05	275
3560	Shields same for 25 conductor as 25 pair telephone									
3570	Single & dual, none required									
3590	BNC coax connectors, Plug	1 Elec	40	.200	Ea.	9.85	10.50		20.35	26.50
3600	TNC coax connectors, Plug	″	40	.200	″	12.65	10.50		23.15	29.50
3700	Cable-bulk									
3710	Single lead	1 Elec	1473	.005	L.F.	2.41	.28		2.69	3.08
3720	Dual lead	″	1473	.005	″	3.48	.28		3.76	4.26
3730	Hand tool crimp				Ea.	500			500	550
3740	Hand tool notch				″	17.20			17.20	18.90
3750	Boxes & floor fitting same as telephone									
3790	Data cable notching, 90°	1 Elec	97	.082	Ea.		4.32		4.32	6.45
3800	180°		60	.133			7		7	10.45
8100	Drill floor		160	.050		1.92	2.62		4.54	6.05
8200	Marking floor		1600	.005	L.F.		.26		.26	.39
8300	Tape, hold down		6400	.001	″	.16	.07		.23	.28
8350	Tape primer, 500 ft. per can		96	.083	Ea.	29	4.37		33.37	38.50
8400	Tool, splicing				″	232			232	255

26 05 19.20 Armored Cable		Crew	Daily Output	Labor-Hours	Unit	Material	Labor	Equipment	Total	Total Incl O&P
0010	**ARMORED CABLE** R260519-20									
0050	600 volt, copper (BX), #14, 2 conductor, solid	1 Elec	2.40	3.333	C.L.F.	61	175		236	330
0100	3 conductor, solid		2.20	3.636		94	191		285	390
0120	4 conductor, solid		2	4		130	210		340	460
0150	#12, 2 conductor, solid		2.30	3.478		61.50	182		243.50	340
0200	3 conductor, solid		2	4		100	210		310	425
0220	4 conductor, solid		1.80	4.444		137	233		370	500
0250	#10, 2 conductor, solid		2	4		113	210		323	440
0300	3 conductor, solid		1.60	5		156	262		418	560
0320	4 conductor, solid		1.40	5.714		242	299		541	715
0340	#8, 2 conductor, stranded		1.50	5.333		295	279		574	745
0350	3 conductor, stranded		1.30	6.154		295	320		615	805
0370	4 conductor, stranded		1.10	7.273		420	380		800	1,025
0380	#6, 2 conductor, stranded		1.30	6.154		310	320		630	820
0400	3 conductor with PVC jacket, in cable tray, #6		3.10	2.581		390	135		525	630
0450	#4	2 Elec	5.40	2.963		500	155		655	780
0500	#2		4.60	3.478		665	182		847	1,000
0550	#1		4	4		905	210		1,115	1,300

26 05 19 – Low-Voltage Electrical Power Conductors and Cables

26 05 19.20 Armored Cable		Crew	Daily Output	Labor-Hours	Unit	Material	Labor	Equipment	Total	Total Incl O&P
						2013 Bare Costs				
0600	1/0	2 Elec	3.60	4.444	C.L.F.	1,075	233		1,308	1,525
0650	2/0		3.40	4.706		1,300	247		1,547	1,800
0700	3/0		3.20	5		1,500	262		1,762	2,050
0750	4/0		3	5.333		1,725	279		2,004	2,325
0800	250 kcmil	3 Elec	3.60	6.667		1,950	350		2,300	2,650
0850	350 kcmil		3.30	7.273		2,600	380		2,980	3,425
0900	500 kcmil		3	8		3,375	420		3,795	4,350
0910	4 conductor with PVC jacket, in cable tray, #6	1 Elec	2.70	2.963		495	155		650	775
0920	#4	2 Elec	4.60	3.478		635	182		817	975
0930	#2		4	4		790	210		1,000	1,175
0940	#1		3.60	4.444		1,150	233		1,383	1,625
0950	1/0		3.40	4.706		1,300	247		1,547	1,825
0960	2/0		3.20	5		1,500	262		1,762	2,050
0970	3/0		3	5.333		1,775	279		2,054	2,375
0980	4/0		2.40	6.667		2,275	350		2,625	3,025
0990	250 kcmil	3 Elec	3.30	7.273		2,700	380		3,080	3,550
1000	350 kcmil		3	8		3,200	420		3,620	4,150
1010	500 kcmil		2.70	8.889		4,400	465		4,865	5,525
1050	5 kV, copper, 3 conductor with PVC jacket,									
1060	non-shielded, in cable tray, #4	2 Elec	380	.042	L.F.	7.65	2.21		9.86	11.75
1100	#2		360	.044		10	2.33		12.33	14.50
1200	#1		300	.053		12.70	2.79		15.49	18.15
1400	1/0		290	.055		14.65	2.89		17.54	20.50
1600	2/0		260	.062		16.95	3.22		20.17	23.50
2000	4/0		240	.067		22.50	3.49		25.99	30.50
2100	250 kcmil	3 Elec	330	.073		31	3.81		34.81	39.50
2150	350 kcmil		315	.076		38	3.99		41.99	48
2200	500 kcmil		270	.089		50.50	4.66		55.16	62.50
2400	15 kV, copper, 3 conductor with PVC jacket galv., steel armored									
2500	grounded neutral, in cable tray, #2	2 Elec	300	.053	L.F.	16.10	2.79		18.89	22
2600	#1		280	.057		17.15	2.99		20.14	23.50
2800	1/0		260	.062		19.75	3.22		22.97	26.50
2900	2/0		220	.073		26	3.81		29.81	34
3000	4/0		190	.084		29.50	4.41		33.91	39
3100	250 kcmil	3 Elec	270	.089		33	4.66		37.66	43.50
3150	350 kcmil		240	.100		39	5.25		44.25	50.50
3200	500 kcmil		210	.114		52	6		58	66.50
3400	15 kV, copper, 3 conductor with PVC jacket,									
3450	ungrounded neutral, in cable tray, #2	2 Elec	260	.062	L.F.	17.35	3.22		20.57	24
3500	#1		230	.070		19.25	3.65		22.90	26.50
3600	1/0		200	.080		22	4.19		26.19	30.50
3700	2/0		190	.084		27	4.41		31.41	36.50
3800	4/0		160	.100		32.50	5.25		37.75	44
4000	250 kcmil	3 Elec	210	.114		38	6		44	51
4050	350 kcmil		195	.123		50	6.45		56.45	64.50
4100	500 kcmil		180	.133		61	7		68	77.50
4200	600 volt, aluminum, 3 conductor in cable tray with PVC jacket									
4300	#2	2 Elec	540	.030	L.F.	4.06	1.55		5.61	6.80
4400	#1		460	.035		4.50	1.82		6.32	7.70
4500	#1/0		400	.040		5.60	2.10		7.70	9.30
4600	#2/0		360	.044		5.70	2.33		8.03	9.75
4700	#3/0		340	.047		6.65	2.47		9.12	11
4800	#4/0		320	.050		8.05	2.62		10.67	12.75

26 05 19 – Low-Voltage Electrical Power Conductors and Cables

26 05 19.20 Armored Cable		Crew	Daily Output	Labor-Hours	Unit	2013 Bare Costs Material	Labor	Equipment	Total	Total Incl O&P
4900	250 kcmil	3 Elec	450	.053	L.F.	9.65	2.79		12.44	14.85
5000	350 kcmil		360	.067		11.50	3.49		14.99	17.90
5200	500 kcmil		330	.073		14.35	3.81		18.16	21.50
5300	750 kcmil		285	.084		18.45	4.41		22.86	27
5400	600 volt, aluminum, 4 conductor in cable tray with PVC jacket									
5410	#2	2 Elec	520	.031	L.F.	4.66	1.61		6.27	7.55
5430	#1		440	.036		5.70	1.91		7.61	9.10
5450	1/0		380	.042		6.65	2.21		8.86	10.60
5470	2/0		340	.047		6.80	2.47		9.27	11.15
5480	3/0		320	.050		8	2.62		10.62	12.70
5500	4/0		300	.053		9.35	2.79		12.14	14.50
5520	250 kcmil	3 Elec	420	.057		10.05	2.99		13.04	15.55
5540	350 kcmil		330	.073		13	3.81		16.81	20
5560	500 kcmil		300	.080		16.25	4.19		20.44	24
5580	750 kcmil		270	.089		22.50	4.66		27.16	31.50
5600	5 kV, aluminum, unshielded in cable tray, #2 with PVC jacket	2 Elec	380	.042		5.70	2.21		7.91	9.55
5700	#1 with PVC jacket		360	.044		6.35	2.33		8.68	10.50
5800	1/0 with PVC jacket		300	.053		6.50	2.79		9.29	11.35
6000	2/0 with PVC jacket		290	.055		6.65	2.89		9.54	11.65
6200	3/0 with PVC jacket		260	.062		8	3.22		11.22	13.60
6300	4/0 with PVC jacket		240	.067		9.35	3.49		12.84	15.55
6400	250 kcmil with PVC jacket	3 Elec	330	.073		10.25	3.81		14.06	17
6500	350 kcmil with PVC jacket		315	.076		12.05	3.99		16.04	19.20
6600	500 kcmil with PVC jacket		300	.080		14.25	4.19		18.44	22
6800	750 kcmil with PVC jacket		270	.089		17.60	4.66		22.26	26.50
6900	15 kV, aluminum, shielded-grounded, #2 with PVC jacket	2 Elec	320	.050		12.90	2.62		15.52	18.10
7000	#1 with PVC jacket		300	.053		13.30	2.79		16.09	18.80
7200	1/0 with PVC jacket		280	.057		14.35	2.99		17.34	20
7300	2/0 with PVC jacket		260	.062		14.55	3.22		17.77	21
7400	3/0 with PVC jacket		240	.067		16.30	3.49		19.79	23
7500	4/0 with PVC jacket		220	.073		16.75	3.81		20.56	24
7600	250 kcmil with PVC jacket	3 Elec	300	.080		18.40	4.19		22.59	26.50
7700	350 kcmil with PVC jacket		270	.089		21.50	4.66		26.16	30.50
7800	500 kcmil with PVC jacket		240	.100		26	5.25		31.25	36.50
8000	750 kcmil with PVC jacket		204	.118		30.50	6.15		36.65	43
8200	15 kV, aluminum, shielded-ungrounded, #1 with PVC jacket	2 Elec	250	.064		16.15	3.35		19.50	23
8300	1/0 with PVC jacket		230	.070		16.75	3.65		20.40	24
8400	2/0 with PVC jacket		210	.076		18.40	3.99		22.39	26
8500	3/0 with PVC jacket		200	.080		18.70	4.19		22.89	27
8600	4/0 with PVC jacket		190	.084		20.50	4.41		24.91	29
8700	250 kcmil with PVC jacket	3 Elec	270	.089		22	4.66		26.66	31
8800	350 kcmil with PVC jacket		240	.100		25	5.25		30.25	35.50
8900	500 kcmil with PVC jacket		210	.114		30	6		36	42
8950	750 kcmil with PVC jacket		174	.138		37.50	7.25		44.75	52
9010	600 volt, copper (MC) steel clad, #14, 2 wire	1 Elec	2.40	3.333	C.L.F.	62	175		237	330
9020	3 wire		2.20	3.636		98.50	191		289.50	395
9030	4 wire		2	4		137	210		347	465
9040	#12, 2 wire		2.30	3.478		62.50	182		244.50	340
9050	3 wire		2	4		109	210		319	435
9060	4 wire		1.80	4.444		147	233		380	510
9070	#10, 2 wire		2	4		135	210		345	465
9080	3 wire		1.60	5		188	262		450	595
9090	4 wire		1.40	5.714		295	299		594	775

26 05 19 – Low-Voltage Electrical Power Conductors and Cables

26 05 19.20 Armored Cable		Crew	Daily Output	Labor-Hours	Unit	2013 Bare Costs Material	Labor	Equipment	Total	Total Incl O&P
9091	For health care facilities cable, add, minimum					5%				
9092	Maximum					20%				
9100	#8, 2 wire, stranded	1 Elec	1.80	4.444	C.L.F.	252	233		485	625
9110	3 wire, stranded		1.30	6.154		355	320		675	870
9120	4 wire, stranded		1.10	7.273		480	380		860	1,100
9130	#6, 2 wire, stranded		1.30	6.154		370	320		690	890
9200	600 volt, copper (MC) aluminum clad, #14, 2 wire		2.65	3.019		61.50	158		219.50	305
9210	3 wire		2.45	3.265		98.50	171		269.50	365
9220	4 wire		2.20	3.636		137	191		328	435
9230	#12, 2 wire		2.55	3.137		61	164		225	315
9240	3 wire		2.20	3.636		109	191		300	405
9250	4 wire		2	4		147	210		357	475
9260	#10, 2 wire		2.20	3.636		135	191		326	435
9270	3 wire		1.80	4.444		188	233		421	555
9280	4 wire		1.55	5.161		295	270		565	730
9600	Alum (MC) aluminum clad, #6, 3 conductor w/#6 grnd		1.67	4.790		335	251		586	745
9610	4 conductor w/#6 grnd		1.64	4.878		365	256		621	780
9620	#4, 3 conductor w/#6 grnd	2 Elec	2.86	5.594		350	293		643	825
9630	4 conductor w/#6 grnd		2.82	5.674		375	297		672	860
9640	#2, 3 conductor w/#4 grnd		2.50	6.400		400	335		735	940
9650	4 conductor w/#4 grnd		2.47	6.478		450	340		790	1,000
9660	#1, 3 conductor w/#4 grnd		2	8		515	420		935	1,200
9670	4 conductor w/#4 grnd		1.98	8.081		615	425		1,040	1,300
9680	1/0, 3 conductor w/#4 grnd		1.82	8.791		590	460		1,050	1,350
9690	4 conductor w/#4 grnd		1.79	8.939		715	470		1,185	1,500
9700	2/0, 3 conductor w/#4 grnd		1.75	9.143		620	480		1,100	1,400
9710	4 conductor w/#4 grnd		1.71	9.357		800	490		1,290	1,625
9720	3/0, 3 conductor w/#4 grnd		1.71	9.357		795	490		1,285	1,600
9730	4 conductor w/#4 grnd		1.67	9.581		995	500		1,495	1,850
9740	4/0, 3 conductor w/#2 grnd		1.67	9.581		895	500		1,395	1,725
9750	4 conductor w/#2 grnd		1.61	9.938		1,125	520		1,645	2,000
9760	250 kcmil, 3 conductor w/#1 grnd	3 Elec	2.42	9.917		1,150	520		1,670	2,025
9770	4 conductor w/#1 grnd		2.36	10.169		1,550	535		2,085	2,500
9775	300 kcmil, 4 conductor w/#1 grnd		2.22	10.811		1,650	565		2,215	2,675
9780	350 kcmil, 3 conductor w/1/0 grnd		2.19	10.959		1,300	575		1,875	2,275
9790	4 conductor w/1/0 grnd		2.10	11.429		1,900	600		2,500	2,975
9800	500 kcmil, 3 conductor w/#1 grnd		2.10	11.429		1,875	600		2,475	2,975
9810	4 conductor w/2/0 grnd		2	12		2,300	630		2,930	3,475
9840	750 kcmil, 3 conductor w/1/0 grnd		1.95	12.308		2,725	645		3,370	3,975
9850	4 conductor w/3/0 grnd		1.89	12.698		3,225	665		3,890	4,525

26 05 19.25 Cable Connectors

26 05 19.25 Cable Connectors		Crew	Daily Output	Labor-Hours	Unit	Material	Labor	Equipment	Total	Total Incl O&P
0010	**CABLE CONNECTORS**									
0100	600 volt, nonmetallic, #14-2 wire	1 Elec	160	.050	Ea.	1.41	2.62		4.03	5.45
0200	#14-3 wire to #12-2 wire		133	.060		1.41	3.15		4.56	6.25
0300	#12-3 wire to #10-2 wire		114	.070		1.41	3.68		5.09	7.05
0400	#10-3 wire to #14-4 and #12-4 wire		100	.080		1.41	4.19		5.60	7.80
0500	#8-3 wire to #10-4 wire		80	.100		2.66	5.25		7.91	10.80
0600	#6-3 wire		40	.200		3.57	10.50		14.07	19.65
0800	SER, 3 #8 insulated + 1 #8 ground		32	.250		3.22	13.10		16.32	23
0900	3 #6 + 1 #6 ground		24	.333		4.03	17.45		21.48	30.50
1000	3 #4 + 1 #6 ground		22	.364		5	19.05		24.05	34
1100	3 #2 + 1 #4 ground		20	.400		9.15	21		30.15	41.50

26 05 19 – Low-Voltage Electrical Power Conductors and Cables

26 05 19.25 Cable Connectors		Crew	Daily Output	Labor-Hours	Unit	Material	2013 Bare Costs Labor	Equipment	Total	Total Incl O&P
1200	3 1/0 + 1 #2 ground	1 Elec	18	.444	Ea.	23.50	23.50		47	61
1400	3 2/0 + 1 #1 ground		16	.500		28.50	26		54.50	70.50
1600	3 4/0 + 1 #2/0 ground		14	.571		31.50	30		61.50	79.50
1800	600 volt, armored, #14-2 wire		80	.100		.77	5.25		6.02	8.70
2200	#14-4, #12-3 and #10-2 wire		40	.200		.78	10.50		11.28	16.55
2400	#12-4, #10-3 and #8-2 wire		32	.250		1.81	13.10		14.91	21.50
2600	#8-3 and #10-4 wire		26	.308		3.11	16.10		19.21	27.50
2650	#8-4 wire		22	.364		4.76	19.05		23.81	34
2700	PVC jacket connector, #6-3 wire, #6-4 wire		16	.500		11.05	26		37.05	51
2800	#4-3 wire, #4-4 wire		16	.500		11.05	26		37.05	51
2900	#2-3 wire		12	.667		11.05	35		46.05	64.50
3000	#1-3 wire, #2-4 wire		12	.667		17	35		52	71
3200	1/0-3 wire		11	.727		17	38		55	75.50
3400	2/0-3 wire, 1/0-4 wire		10	.800		17	42		59	81
3500	3/0-3 wire, 2/0-4 wire		9	.889		28	46.50		74.50	101
3600	4/0-3 wire, 3/0-4 wire		7	1.143		28	60		88	121
3800	250 kcmil-3 wire, 4/0-4 wire		6	1.333		52	70		122	162
4000	350 kcmil-3 wire, 250 kcmil-4 wire		5	1.600		52	84		136	182
4100	350 kcmil-4 wire		4	2		176	105		281	350
4200	500 kcmil-3 wire		4	2		176	105		281	350
4250	500 kcmil-4 wire, 750 kcmil-3 wire		3.50	2.286		240	120		360	440
4300	750 kcmil-4 wire		3	2.667		240	140		380	470
4400	5 kV, armored, #4		8	1		65.50	52.50		118	151
4600	#2		8	1		65.50	52.50		118	151
4800	#1		8	1		81.50	52.50		134	168
5000	1/0		6.40	1.250		101	65.50		166.50	209
5200	2/0		5.30	1.509		101	79		180	229
5500	4/0		4	2		135	105		240	305
5600	250 kcmil		3.60	2.222		135	116		251	320
5650	350 kcmil		3.20	2.500		162	131		293	375
5700	500 kcmil		2.50	3.200		201	168		369	475
5720	750 kcmil		2.20	3.636		248	191		439	560
5750	1000 kcmil		2	4		305	210		515	650
5800	15 kV, armored, #1		4	2		110	105		215	278
5900	1/0		4	2		137	105		242	310
6000	3/0		3.60	2.222		162	116		278	350
6100	4/0		3.40	2.353		162	123		285	365
6200	250 kcmil		3.20	2.500		191	131		322	405
6300	350 kcmil		2.70	2.963		246	155		401	505
6400	500 kcmil		2	4		244	210		454	585

26 05 19.30 Cable Splicing

		Crew	Daily Output	Labor-Hours	Unit	Material	Labor	Equipment	Total	Total Incl O&P
0010	**CABLE SPLICING** URD or similar, ideal conditions									
0100	#6 stranded to #1 stranded, 5 kV	1 Elec	4	2	Ea.	139	105		244	310
0120	15 kV		3.60	2.222		139	116		255	325
0140	25 kV		3.20	2.500		139	131		270	350
0200	#1 stranded to 4/0 stranded, 5 kV		3.60	2.222		154	116		270	345
0210	15 kV		3.20	2.500		154	131		285	365
0220	25 kV		2.80	2.857		139	150		289	375
0300	4/0 stranded to 500 kcmil stranded, 5 kV		3.30	2.424		154	127		281	360
0310	15 kV		2.90	2.759		251	145		396	495
0320	25 kV		2.50	3.200		251	168		419	530
0400	500 kcmil, 5 kV		3.20	2.500		251	131		382	475

26 05 19.30 Cable Splicing		Crew	Daily Output	Labor-Hours	Unit	2013 Bare Costs Material	Labor	Equipment	Total	Total Incl O&P
0410	15 kV	1 Elec	2.80	2.857	Ea.	251	150		401	500
0420	25 kV		2.30	3.478		251	182		433	550
0500	600 kcmil, 5 kV		2.90	2.759		251	145		396	495
0510	15 kV		2.40	3.333		251	175		426	540
0520	25 kV		2	4		251	210		461	590
0600	750 kcmil, 5 kV		2.60	3.077		251	161		412	520
0610	15 kV		2.20	3.636		251	191		442	560
0620	25 kV		1.90	4.211		251	221		472	605
0700	1000 kcmil, 5 kV		2.30	3.478		251	182		433	550
0710	15 kV		1.90	4.211		273	221		494	630
0720	25 kV		1.60	5		360	262		622	785

26 05 19.35 Cable Terminations		Crew	Daily Output	Labor-Hours	Unit	Material	Labor	Equipment	Total	Total Incl O&P
0010	**CABLE TERMINATIONS**									
0015	Wire connectors, screw type, #22 to #14	1 Elec	260	.031	Ea.	.06	1.61		1.67	2.48
0020	#18 to #12		240	.033		.09	1.75		1.84	2.71
0025	#18 to #10		240	.033		.13	1.75		1.88	2.75
0030	Screw-on connectors, insulated, #18 to #12		240	.033		.20	1.75		1.95	2.83
0035	#16 to #10		230	.035		.27	1.82		2.09	3.03
0040	#14 to #8		210	.038		.33	2		2.33	3.35
0045	#12 to #6		180	.044		.62	2.33		2.95	4.16
0050	Terminal lugs, solderless, #16 to #10		50	.160		.40	8.40		8.80	13
0100	#8 to #4		30	.267		.80	13.95		14.75	22
0150	#2 to #1		22	.364		1.03	19.05		20.08	29.50
0200	1/0 to 2/0		16	.500		1.90	26		27.90	41
0250	3/0		12	.667		2.50	35		37.50	55.50
0300	4/0		11	.727		3.06	38		41.06	60.50
0350	250 kcmil		9	.889		3.67	46.50		50.17	73.50
0400	350 kcmil		7	1.143		4.44	60		64.44	94.50
0450	500 kcmil		6	1.333		9.15	70		79.15	115
0500	600 kcmil		5.80	1.379		10.25	72.50		82.75	119
0550	750 kcmil		5.20	1.538		10.95	80.50		91.45	133
0600	Split bolt connectors, tapped, #6		16	.500		4.35	26		30.35	44
0650	#4		14	.571		5.20	30		35.20	50.50
0700	#2		12	.667		7.90	35		42.90	61
0750	#1		11	.727		9.90	38		47.90	68
0800	1/0		10	.800		9.90	42		51.90	73.50
0850	2/0		9	.889		16.20	46.50		62.70	87.50
0900	3/0		7.20	1.111		22.50	58		80.50	112
1000	4/0		6.40	1.250		27.50	65.50		93	128
1100	250 kcmil		5.70	1.404		27.50	73.50		101	140
1200	300 kcmil		5.30	1.509		50	79		129	173
1400	350 kcmil		4.60	1.739		50	91		141	191
1500	500 kcmil		4	2		62.50	105		167.50	226
1600	Crimp 1 hole lugs, copper or aluminum, 600 volt									
1620	#14	1 Elec	60	.133	Ea.	.67	7		7.67	11.20
1630	#12		50	.160		1.08	8.40		9.48	13.75
1640	#10		45	.178		1.08	9.30		10.38	15.15
1780	#8		36	.222		2.24	11.65		13.89	19.85
1800	#6		30	.267		2.93	13.95		16.88	24
2000	#4		27	.296		3.98	15.55		19.53	27.50
2200	#2		24	.333		6.30	17.45		23.75	33
2400	#1		20	.400		6.45	21		27.45	38.50

26 05 19.35 Cable Terminations		Crew	Daily Output	Labor-Hours	Unit	2013 Bare Costs Material	Labor	Equipment	Total	Total Incl O&P
2500	1/0	1 Elec	17.50	.457	Ea.	6.85	24		30.85	43.50
2600	2/0		15	.533		8	28		36	51
2800	3/0		12	.667		9.45	35		44.45	63
3000	4/0		11	.727		10.50	38		48.50	68.50
3200	250 kcmil		9	.889		12.35	46.50		58.85	83
3400	300 kcmil		8	1		14.15	52.50		66.65	94
3500	350 kcmil		7	1.143		15.95	60		75.95	107
3600	400 kcmil		6.50	1.231		17.80	64.50		82.30	116
3800	500 kcmil		6	1.333		21	70		91	128
4000	600 kcmil		5.80	1.379		37	72.50		109.50	149
4200	700 kcmil		5.50	1.455		44	76		120	163
4400	750 kcmil		5.20	1.538		44	80.50		124.50	170
4500	Crimp 2-way connectors, copper or alum., 600 volt,									
4510	#14	1 Elec	60	.133	Ea.	2.53	7		9.53	13.25
4520	#12		50	.160		2.69	8.40		11.09	15.50
4530	#10		45	.178		2.84	9.30		12.14	17.05
4540	#8		27	.296		2.87	15.55		18.42	26
4600	#6		25	.320		4.95	16.75		21.70	30.50
4800	#4		23	.348		5.65	18.25		23.90	33.50
5000	#2		20	.400		7.20	21		28.20	39.50
5200	#1		16	.500		8.15	26		34.15	48
5400	1/0		13	.615		8.95	32.50		41.45	58.50
5420	2/0		12	.667		9.40	35		44.40	63
5440	3/0		11	.727		11.20	38		49.20	69.50
5460	4/0		10	.800		11.95	42		53.95	75.50
5480	250 kcmil		9	.889		15.95	46.50		62.45	87
5500	300 kcmil		8.50	.941		17.55	49.50		67.05	93.50
5520	350 kcmil		8	1		18.55	52.50		71.05	99
5540	400 kcmil		7.30	1.096		26.50	57.50		84	115
5560	500 kcmil		6.20	1.290		27	67.50		94.50	131
5580	600 kcmil		5.50	1.455		39	76		115	157
5600	700 kcmil		4.50	1.778		51	93		144	195
5620	750 kcmil		4	2		51	105		156	213
7000	Compression equipment adapter, aluminum wire, #6		30	.267		10.70	13.95		24.65	33
7020	#4		27	.296		11.10	15.55		26.65	35.50
7040	#2		24	.333		11.90	17.45		29.35	39
7060	#1		20	.400		13.50	21		34.50	46.50
7080	1/0		18	.444		14.10	23.50		37.60	50.50
7100	2/0		15	.533		27	28		55	71.50
7140	4/0		11	.727		26.50	38		64.50	86.50
7160	250 kcmil		9	.889		26.50	46.50		73	99
7180	300 kcmil		8	1		32	52.50		84.50	114
7200	350 kcmil		7	1.143		32	60		92	125
7220	400 kcmil		6.50	1.231		41.50	64.50		106	142
7240	500 kcmil		6	1.333		41	70		111	151
7260	600 kcmil		5.80	1.379		56.50	72.50		129	171
7280	750 kcmil		5.20	1.538		56.50	80.50		137	183
8000	Compression tool, hand					1,175			1,175	1,300
8100	Hydraulic					1,525			1,525	1,675
8500	Hydraulic dies					320			320	355

26 05 19 – Low-Voltage Electrical Power Conductors and Cables

26 05 19.50 Mineral Insulated Cable		Crew	Daily Output	Labor-Hours	Unit	2013 Bare Costs Material	Labor	Equipment	Total	Total Incl O&P
0010	**MINERAL INSULATED CABLE** 600 volt									
0100	1 conductor, #12	1 Elec	1.60	5	C.L.F.	385	262		647	810
0200	#10		1.60	5		495	262		757	935
0400	#8		1.50	5.333		550	279		829	1,025
0500	#6		1.40	5.714		655	299		954	1,175
0600	#4	2 Elec	2.40	6.667		880	350		1,230	1,500
0800	#2		2.20	7.273		1,250	380		1,630	1,950
0900	#1		2.10	7.619		1,425	400		1,825	2,175
1000	1/0		2	8		1,675	420		2,095	2,475
1100	2/0		1.90	8.421		2,000	440		2,440	2,850
1200	3/0		1.80	8.889		2,400	465		2,865	3,350
1400	4/0		1.60	10		2,775	525		3,300	3,825
1410	250 kcmil	3 Elec	2.40	10		3,125	525		3,650	4,225
1420	350 kcmil		1.95	12.308		3,575	645		4,220	4,900
1430	500 kcmil		1.95	12.308		4,600	645		5,245	6,025
1500	2 conductor, #12	1 Elec	1.40	5.714		800	299		1,099	1,325
1600	#10		1.20	6.667		985	350		1,335	1,600
1800	#8		1.10	7.273		1,125	380		1,505	1,825
2000	#6		1.05	7.619		1,550	400		1,950	2,300
2100	#4	2 Elec	2	8		2,100	420		2,520	2,925
2200	3 conductor, #12	1 Elec	1.20	6.667		1,000	350		1,350	1,625
2400	#10		1.10	7.273		1,125	380		1,505	1,825
2600	#8		1.05	7.619		1,425	400		1,825	2,150
2800	#6		1	8		1,875	420		2,295	2,675
3000	#4	2 Elec	1.80	8.889		2,325	465		2,790	3,275
3100	4 conductor, #12	1 Elec	1.20	6.667		1,050	350		1,400	1,675
3200	#10		1.10	7.273		1,275	380		1,655	1,975
3400	#8		1	8		1,650	420		2,070	2,450
3600	#6		.90	8.889		2,150	465		2,615	3,075
3620	7 conductor, #12		1.10	7.273		1,350	380		1,730	2,050
3640	#10		1	8		1,775	420		2,195	2,575
3800	Terminations, 600 volt, 1 conductor, #12		8	1	Ea.	16.90	52.50		69.40	97
4000	#10		7.60	1.053		16.90	55		71.90	101
4100	#8		7.30	1.096		16.90	57.50		74.40	105
4200	#6		6.70	1.194		16.90	62.50		79.40	112
4400	#4		6.20	1.290		16.90	67.50		84.40	120
4600	#2		5.70	1.404		25.50	73.50		99	138
4800	#1		5.30	1.509		25.50	79		104.50	146
5000	1/0		5	1.600		25.50	84		109.50	153
5100	2/0		4.70	1.702		25.50	89		114.50	161
5200	3/0		4.30	1.860		25.50	97.50		123	174
5400	4/0		4	2		56.50	105		161.50	219
5410	250 kcmil		4	2		56.50	105		161.50	219
5420	350 kcmil		4	2		89	105		194	255
5430	500 kcmil		4	2		89	105		194	255
5500	2 conductor, #12		6.70	1.194		16.90	62.50		79.40	112
5600	#10		6.40	1.250		25.50	65.50		91	126
5800	#8		6.20	1.290		25.50	67.50		93	129
6000	#6		5.70	1.404		25.50	73.50		99	138
6200	#4		5.30	1.509		56.50	79		135.50	180
6400	3 conductor, #12		5.70	1.404		25.50	73.50		99	138
6500	#10		5.50	1.455		25.50	76		101.50	142

26 05 19.50 Mineral Insulated Cable		Crew	Daily Output	Labor-Hours	Unit	Material	2013 Bare Costs Labor	Equipment	Total	Total Incl O&P
6600	#8	1 Elec	5.20	1.538	Ea.	25.50	80.50		106	149
6800	#6		4.80	1.667		25.50	87.50		113	159
7200	#4		4.60	1.739		56.50	91		147.50	198
7400	4 conductor, #12		4.60	1.739		28.50	91		119.50	167
7500	#10		4.40	1.818		28.50	95.50		124	174
7600	#8		4.20	1.905		28.50	100		128.50	180
8400	#6		4	2		59.50	105		164.50	223
8500	7 conductor, #12		3.50	2.286		28.50	120		148.50	210
8600	#10	↓	3	2.667		61.50	140		201.50	277
8800	Crimping tool, plier type					58.50			58.50	64
9000	Stripping tool					220			220	242
9200	Hand vise				↓	65.50			65.50	72

26 05 19.55 Non-Metallic Sheathed Cable		Crew	Daily Output	Labor-Hours	Unit	Material	Labor	Equipment	Total	Total Incl O&P
0010	**NON-METALLIC SHEATHED CABLE** 600 volt									
0100	Copper with ground wire, (Romex)									
0150	#14, 2 conductor	1 Elec	2.70	2.963	C.L.F.	25	155		180	260
0200	3 conductor		2.40	3.333		35	175		210	300
0220	4 conductor		2.20	3.636		51	191		242	340
0250	#12, 2 conductor		2.50	3.200		38	168		206	293
0300	3 conductor		2.20	3.636		53.50	191		244.50	345
0320	4 conductor		2	4		79	210		289	400
0350	#10, 2 conductor		2.20	3.636		60.50	191		251.50	350
0400	3 conductor		1.80	4.444		85.50	233		318.50	445
0420	4 conductor		1.60	5		122	262		384	525
0430	#8, 2 conductor		1.60	5		92.50	262		354.50	490
0450	3 conductor		1.50	5.333		136	279		415	570
0500	#6, 3 conductor	↓	1.40	5.714		220	299		519	695
0520	#4, 3 conductor	2 Elec	2.40	6.667		495	350		845	1,075
0540	#2, 3 conductor	"	2.20	7.273	↓	745	380		1,125	1,400
0550	SE type SER aluminum cable, 3 RHW and									
0600	1 bare neutral, 3 #8 & 1 #8	1 Elec	1.60	5	C.L.F.	162	262		424	570
0650	3 #6 & 1 #6	"	1.40	5.714		183	299		482	650
0700	3 #4 & 1 #6	2 Elec	2.40	6.667		169	350		519	710
0750	3 #2 & 1 #4		2.20	7.273		300	380		680	905
0800	3 #1/0 & 1 #2		2	8		460	420		880	1,125
0850	3 #2/0 & 1 #1		1.80	8.889		540	465		1,005	1,300
0900	3 #4/0 & 1 #2/0		1.60	10		770	525		1,295	1,625
1000	URD - triplex underground distribution cable, alum. 2 #4 + #4 neutral		2.80	5.714		87	299		386	545
1010	2 #2 + #4 neutral		2.65	6.038		107	315		422	590
1020	2 #2 + #2 neutral		2.55	6.275		109	330		439	610
1030	2 1/0 + #2 neutral		2.40	6.667		135	350		485	675
1040	2 1/0 + 1/0 neutral		2.30	6.957		148	365		513	710
1050	2 2/0 + #1 neutral		2.20	7.273		159	380		539	745
1060	2 2/0 + 2/0 neutral		2.10	7.619		172	400		572	785
1070	2 3/0 + 1/0 neutral		1.95	8.205		189	430		619	855
1080	2 3/0 + 3/0 neutral		1.95	8.205		219	430		649	885
1090	2 4/0 + 2/0 neutral		1.85	8.649		219	455		674	920
1100	2 4/0 + 4/0 neutral	↓	1.85	8.649		243	455		698	945
1450	UF underground feeder cable, copper with ground, #14, 2 conductor	1 Elec	4	2		29	105		134	189
1500	#12, 2 conductor		3.50	2.286		44	120		164	228
1550	#10, 2 conductor		3	2.667		70	140		210	286
1600	#14, 3 conductor		3.50	2.286		41.50	120		161.50	225

26 05 Common Work Results for Electrical

26 05 19 – Low-Voltage Electrical Power Conductors and Cables

26 05 19.55 Non-Metallic Sheathed Cable		Crew	Daily Output	Labor-Hours	Unit	Material	2013 Bare Costs Labor	Equipment	Total	Total Incl O&P
1650	#12, 3 conductor	1 Elec	3	2.667	C.L.F.	62.50	140		202.50	278
1700	#10, 3 conductor		2.50	3.200		99	168		267	360
1710	#8, 3 conductor		2	4		160	210		370	490
1720	#6, 3 conductor		1.80	4.444		260	233		493	635
2400	SEU service entrance cable, copper 2 conductors, #8 + #8 neutral		1.50	5.333		119	279		398	550
2600	#6 + #8 neutral		1.30	6.154		177	320		497	675
2800	#6 + #6 neutral		1.30	6.154		201	320		521	700
3000	#4 + #6 neutral	2 Elec	2.20	7.273		279	380		659	875
3200	#4 + #4 neutral		2.20	7.273		310	380		690	915
3400	#3 + #5 neutral		2.10	7.619		365	400		765	995
3600	#3 + #3 neutral		2.10	7.619		395	400		795	1,025
3800	#2 + #4 neutral		2	8		430	420		850	1,100
4000	#1 + #1 neutral		1.90	8.421		490	440		930	1,200
4200	1/0 + 1/0 neutral		1.80	8.889		780	465		1,245	1,550
4400	2/0 + 2/0 neutral		1.70	9.412		975	495		1,470	1,825
4600	3/0 + 3/0 neutral		1.60	10		1,225	525		1,750	2,125
4620	4/0 + 4/0 neutral		1.45	11.034		1,375	580		1,955	2,400
4800	Aluminum 2 conductors, #8 + #8 neutral	1 Elec	1.60	5		134	262		396	535
5000	#6 + #6 neutral	"	1.40	5.714		135	299		434	600
5100	#4 + #6 neutral	2 Elec	2.50	6.400		158	335		493	675
5200	#4 + #4 neutral		2.40	6.667		173	350		523	715
5300	#2 + #4 neutral		2.30	6.957		212	365		577	780
5400	#2 + #2 neutral		2.20	7.273		231	380		611	825
5450	1/0 + #2 neutral		2.10	7.619		335	400		735	965
5500	1/0 + 1/0 neutral		2	8		350	420		770	1,000
5550	2/0 + #1 neutral		1.90	8.421		375	440		815	1,075
5600	2/0 + 2/0 neutral		1.80	8.889		405	465		870	1,150
5800	3/0 + 1/0 neutral		1.70	9.412		445	495		940	1,225
6000	3/0 + 3/0 neutral		1.70	9.412		480	495		975	1,275
6200	4/0 + 2/0 neutral		1.60	10		520	525		1,045	1,350
6400	4/0 + 4/0 neutral		1.60	10		565	525		1,090	1,400
6500	Service entrance cap for copper SEU									
6600	100 amp	1 Elec	12	.667	Ea.	11.50	35		46.50	65
6700	150 amp		10	.800		21	42		63	85.50
6800	200 amp		8	1		32	52.50		84.50	114

26 05 19.70 Portable Cord		Crew	Daily Output	Labor-Hours	Unit	Material	Labor	Equipment	Total	Total Incl O&P
0010	**PORTABLE CORD** 600 volt									
0100	Type SO, #18, 2 conductor	1 Elec	980	.008	L.F.	.42	.43		.85	1.10
0110	3 conductor		980	.008		.51	.43		.94	1.20
0120	#16, 2 conductor		840	.010		.50	.50		1	1.30
0130	3 conductor		840	.010		.70	.50		1.20	1.52
0140	4 conductor		840	.010		.99	.50		1.49	1.84
0240	#14, 2 conductor		840	.010		.81	.50		1.31	1.64
0250	3 conductor		840	.010		1.02	.50		1.52	1.87
0260	4 conductor		840	.010		1.31	.50		1.81	2.19
0280	#12, 2 conductor		840	.010		1.21	.50		1.71	2.08
0290	3 conductor		840	.010		1.47	.50		1.97	2.37
0300	4 conductor		840	.010		1.76	.50		2.26	2.69
0320	#10, 2 conductor		765	.010		1.66	.55		2.21	2.65
0330	3 conductor		765	.010		2	.55		2.55	3.02
0340	4 conductor		765	.010		2.26	.55		2.81	3.31
0360	#8, 2 conductor		555	.014		3.45	.76		4.21	4.93

26 05 19 – Low-Voltage Electrical Power Conductors and Cables

26 05 19.70 Portable Cord		Crew	Daily Output	Labor-Hours	Unit	2013 Bare Costs Material	Labor	Equipment	Total	Total Incl O&P
0370	3 conductor	1 Elec	540	.015	L.F.	3.36	.78		4.14	4.86
0380	4 conductor		525	.015		4.62	.80		5.42	6.30
0400	#6, 2 conductor		525	.015		3.53	.80		4.33	5.05
0410	3 conductor		490	.016		4.72	.86		5.58	6.50
0420	4 conductor		415	.019		6.65	1.01		7.66	8.80
0440	#4, 2 conductor	2 Elec	830	.019		4.64	1.01		5.65	6.60
0450	3 conductor		700	.023		8	1.20		9.20	10.60
0460	4 conductor		660	.024		9.10	1.27		10.37	11.90
0480	#2, 2 conductor		450	.036		7.70	1.86		9.56	11.25
0490	3 conductor		350	.046		9.40	2.40		11.80	13.90
0500	4 conductor		280	.057		12.25	2.99		15.24	17.95

26 05 19.75 Modular Flexible Wiring System

Line	Description	Crew	Daily Output	Labor-Hours	Unit	Material	Labor	Equipment	Total	Total Incl O&P
0010	**MODULAR FLEXIBLE WIRING SYSTEM**									
0020	Commercial system grid ceiling									
0100	Conversion Module	1 Elec	32	.250	Ea.	9.55	13.10		22.65	30
0120	Fixture cable, for fixture to fixture, 3 conductor, 15′ long		40	.200		38	10.50		48.50	57.50
0150	Extender cable, 3 conductor, 15′ long		48	.167		37	8.75		45.75	54
0200	Switch drop, 1 level, 9′ long		32	.250		33	13.10		46.10	55.50
0220	2 level, 9′ long		32	.250		32.50	13.10		45.60	55.50
0250	Power tee, 9′ long		32	.250		42	13.10		55.10	65.50
1020	Industrial system open ceiling									
1100	Converter, interface between hardwiring and modular wiring	1 Elec	32	.250	Ea.	15.65	13.10		28.75	37
1120	Fixture cable, for fixture to fixture, 3 conductor, 21′ long		32	.250		94	13.10		107.10	124
1125	3 conductor, 25′ long		24	.333		108	17.45		125.45	144
1130	3 conductor, 31′ long		19	.421		103	22		125	146
1150	Fixture cord drop, 10′ long		40	.200		27.50	10.50		38	46

26 05 19.90 Wire

Line	Description	Crew	Daily Output	Labor-Hours	Unit	Material	Labor	Equipment	Total	Total Incl O&P
0010	**WIRE** R260519-90									
0020	600 volt, copper type THW, solid, #14	1 Elec	13	.615	C.L.F.	8.25	32.50		40.75	57.50
0030	#12		11	.727		12.45	38		50.45	70.50
0040	#10		10	.800		19.70	42		61.70	84
0050	Stranded, #14 R260519-91		13	.615		9.30	32.50		41.80	58.50
0100	#12		11	.727		14.25	38		52.25	72.50
0120	#10 R260519-92		10	.800		22.50	42		64.50	87
0140	#8		8	1		37	52.50		89.50	120
0160	#6 R260519-93		6.50	1.231		63	64.50		127.50	166
0180	#4	2 Elec	10.60	1.509		99	79		178	227
0200	#3 R260519-94		10	1.600		125	84		209	262
0220	#2		9	1.778		157	93		250	310
0240	#1 R260533-22		8	2		199	105		304	375
0260	1/0		6.60	2.424		248	127		375	465
0280	2/0		5.80	2.759		310	145		455	555
0300	3/0		5	3.200		390	168		558	680
0350	4/0		4.40	3.636		495	191		686	830
0400	250 kcmil	3 Elec	6	4		580	210		790	955
0420	300 kcmil		5.70	4.211		695	221		916	1,100
0450	350 kcmil		5.40	4.444		850	233		1,083	1,275
0480	400 kcmil		5.10	4.706		975	247		1,222	1,450
0490	500 kcmil		4.80	5		1,150	262		1,412	1,675
0500	600 kcmil		3.90	6.154		1,375	320		1,695	2,000
0510	750 kcmil		3.30	7.273		1,950	380		2,330	2,725
0520	1000 kcmil		2.70	8.889		2,800	465		3,265	3,775

26 05 19.90 Wire		Crew	Daily Output	Labor-Hours	Unit	2013 Bare Costs Material	Labor	Equipment	Total	Total Incl O&P
0540	600 volt, aluminum type THHN, stranded, #6	1 Elec	8	1	C.L.F.	32.50	52.50		85	114
0560	#4	2 Elec	13	1.231		40	64.50		104.50	141
0580	#2		10.60	1.509		54.50	79		133.50	178
0600	#1		9	1.778		80	93		173	227
0620	1/0		8	2		95.50	105		200.50	262
0640	2/0		7.20	2.222		113	116		229	298
0680	3/0		6.60	2.424		140	127		267	345
0700	4/0		6.20	2.581		156	135		291	375
0720	250 kcmil	3 Elec	8.70	2.759		191	145		336	425
0740	300 kcmil		8.10	2.963		263	155		418	520
0760	350 kcmil		7.50	3.200		268	168		436	545
0780	400 kcmil		6.90	3.478		315	182		497	620
0800	500 kcmil		6	4		345	210		555	695
0850	600 kcmil		5.70	4.211		435	221		656	810
0880	700 kcmil		5.10	4.706		505	247		752	925
0900	750 kcmil		4.80	5		585	262		847	1,025
0910	1000 kcmil		3.78	6.349		870	335		1,205	1,450
0920	600 volt, copper type THWN-THHN, solid, #14	1 Elec	13	.615		8.25	32.50		40.75	57.50
0940	#12		11	.727		12.45	38		50.45	70.50
0960	#10		10	.800		19.70	42		61.70	84
1000	Stranded, #14		13	.615		10.10	32.50		42.60	59.50
1200	#12		11	.727		15.50	38		53.50	74
1250	#10		10	.800		24	42		66	89
1300	#8		8	1		37	52.50		89.50	119
1350	#6		6.50	1.231		63	64.50		127.50	166
1400	#4	2 Elec	10.60	1.509		101	79		180	229
1450	#3		10	1.600		125	84		209	263
1500	#2		9	1.778		157	93		250	310
1550	#1		8	2		206	105		311	385
1600	1/0		6.60	2.424		250	127		377	465
1650	2/0		5.80	2.759		315	145		460	560
1700	3/0		5	3.200		395	168		563	685
2000	4/0		4.40	3.636		495	191		686	830
2200	250 kcmil	3 Elec	6	4		595	210		805	970
2400	300 kcmil		5.70	4.211		710	221		931	1,100
2600	350 kcmil		5.40	4.444		830	233		1,063	1,275
2700	400 kcmil		5.10	4.706		945	247		1,192	1,425
2800	500 kcmil		4.80	5		1,150	262		1,412	1,650
2802	600 kcmil		3.90	6.154		1,375	320		1,695	2,000
2804	750 kcmil		3.30	7.273		2,325	380		2,705	3,125
2805	1000 kcmil		1.94	12.371		3,075	650		3,725	4,350
2900	600 volt, copper type XHHW, solid, #14	1 Elec	13	.615		14.10	32.50		46.60	64
2920	#12		11	.727		22	38		60	81.50
2940	#10		10	.800		34.50	42		76.50	101
3000	Stranded, #14		13	.615		14.70	32.50		47.20	64.50
3020	#12		11	.727		22	38		60	81
3040	#10		10	.800		33.50	42		75.50	99
3060	#8		8	1		51	52.50		103.50	135
3080	#6		6.50	1.231		84.50	64.50		149	190
3100	#4	2 Elec	10.60	1.509		131	79		210	262
3120	#2		9	1.778		204	93		297	365
3140	#1		8	2		279	105		384	460
3160	1/0		6.60	2.424		345	127		472	570

26 05 19 – Low-Voltage Electrical Power Conductors and Cables

26 05 19.90 Wire		Crew	Daily Output	Labor-Hours	Unit	Material	2013 Bare Costs Labor	Equipment	Total	Total Incl O&P
3180	2/0	2 Elec	5.80	2.759	C.L.F.	435	145		580	690
3200	3/0		5	3.200		540	168		708	845
3220	4/0		4.40	3.636		680	191		871	1,025
3240	250 kcmil	3 Elec	6	4		710	210		920	1,100
3260	300 kcmil		5.70	4.211		800	221		1,021	1,200
3280	350 kcmil		5.40	4.444		890	233		1,123	1,325
3300	400 kcmil		5.10	4.706		1,025	247		1,272	1,500
3320	500 kcmil		4.80	5		1,200	262		1,462	1,700
3340	600 kcmil		3.90	6.154		1,400	320		1,720	2,025
3360	750 kcmil		3.30	7.273		2,325	380		2,705	3,150
3380	1000 kcmil		2.40	10		3,425	525		3,950	4,550
5020	600 volt, aluminum type XHHW, stranded, #6	1 Elec	8	1		36	52.50		88.50	118
5040	#4	2 Elec	13	1.231		44.50	64.50		109	146
5060	#2		10.60	1.509		60.50	79		139.50	185
5080	#1		9	1.778		88	93		181	236
5100	1/0		8	2		106	105		211	273
5120	2/0		7.20	2.222		125	116		241	310
5140	3/0		6.60	2.424		155	127		282	360
5160	4/0		6.20	2.581		173	135		308	390
5180	250 kcmil	3 Elec	8.70	2.759		211	145		356	450
5200	300 kcmil		8.10	2.963		291	155		446	550
5220	350 kcmil		7.50	3.200		320	168		488	600
5240	400 kcmil		6.90	3.478		345	182		527	655
5260	500 kcmil		6	4		380	210		590	735
5280	600 kcmil		5.70	4.211		485	221		706	860
5300	700 kcmil		5.40	4.444		560	233		793	965
5320	750 kcmil		5.10	4.706		565	247		812	990
5340	1000 kcmil		3.60	6.667		840	350		1,190	1,450
5390	600 volt, copper type XLPE-USE (RHW), solid, #14	1 Elec	12	.667		16.25	35		51.25	70.50
5400	#12		11	.727		24	38		62	83.50
5420	#10		10	.800		35.50	42		77.50	102
5440	Stranded, #14		13	.615		24	32.50		56.50	74.50
5460	#12		11	.727		28.50	38		66.50	88.50
5480	#10		10	.800		42.50	42		84.50	110
5500	#8		8	1		58.50	52.50		111	143
5520	#6		6.50	1.231		98	64.50		162.50	205
5540	#4	2 Elec	10.60	1.509		148	79		227	281
5560	#2		9	1.778		241	93		334	405
5580	#1		8	2		325	105		430	510
5600	1/0		6.60	2.424		405	127		532	635
5620	2/0		5.80	2.759		510	145		655	775
5640	3/0		5	3.200		640	168		808	950
5660	4/0		4.40	3.636		730	191		921	1,075
5680	250 kcmil	3 Elec	6	4		765	210		975	1,150
5700	300 kcmil		5.70	4.211		915	221		1,136	1,325
5720	350 kcmil		5.40	4.444		1,075	233		1,308	1,525
5740	400 kcmil		5.10	4.706		1,225	247		1,472	1,725
5760	500 kcmil		4.80	5		1,425	262		1,687	1,975
5780	600 kcmil		3.90	6.154		1,725	320		2,045	2,350
5800	750 kcmil		3.30	7.273		3,050	380		3,430	3,950
5820	1000 kcmil		2.70	8.889		4,050	465		4,515	5,150
5840	600 volt, aluminum type XLPE-USE (RHW), stranded, #6	1 Elec	8	1		44	52.50		96.50	127
5860	#4	2 Elec	13	1.231		50.50	64.50		115	152

26 05 19 – Low-Voltage Electrical Power Conductors and Cables

26 05 19.90 Wire

	26 05 19.90 Wire	Crew	Daily Output	Labor-Hours	Unit	2013 Bare Costs Material	Labor	Equipment	Total	Total Incl O&P
5880	#2	2 Elec	10.60	1.509	C.L.F.	69.50	79		148.50	195
5900	#1		9	1.778		96.50	93		189.50	245
5920	1/0		8	2		119	105		224	288
5940	2/0		7.20	2.222		140	116		256	330
5960	3/0		6.60	2.424		165	127		292	370
5980	4/0	↓	6.20	2.581		183	135		318	405
6000	250 kcmil	3 Elec	8.70	2.759		248	145		393	490
6020	300 kcmil		8.10	2.963		325	155		480	585
6040	350 kcmil		7.50	3.200		330	168		498	615
6060	400 kcmil		6.90	3.478		370	182		552	685
6080	500 kcmil		6	4		445	210		655	805
6100	600 kcmil		5.70	4.211		570	221		791	955
6110	700 kcmil		5.40	4.444		655	233		888	1,075
6120	750 kcmil	↓	5.10	4.706	↓	660	247		907	1,100

26 05 23 – Control-Voltage Electrical Power Cables

26 05 23.10 Control Cable

	26 05 23.10 Control Cable	Crew	Daily Output	Labor-Hours	Unit	Material	Labor	Equipment	Total	Total Incl O&P
0010	**CONTROL CABLE**									
0020	600 volt, copper, #14 THWN wire with PVC jacket, 2 wires	1 Elec	9	.889	C.L.F.	23	46.50		69.50	95
0030	3 wires		8	1		32	52.50		84.50	114
0100	4 wires		7	1.143		39.50	60		99.50	133
0150	5 wires		6.50	1.231		49	64.50		113.50	151
0200	6 wires		6	1.333		74.50	70		144.50	187
0300	8 wires		5.30	1.509		91	79		170	218
0400	10 wires		4.80	1.667		107	87.50		194.50	249
0500	12 wires		4.30	1.860		125	97.50		222.50	284
0600	14 wires		3.80	2.105		149	110		259	330
0700	16 wires		3.50	2.286		153	120		273	350
0800	18 wires		3.30	2.424		169	127		296	375
0810	19 wires		3.10	2.581		189	135		324	410
0900	20 wires		3	2.667		206	140		346	435
1000	22 wires	↓	2.80	2.857	↓	212	150		362	455

26 05 23.20 Special Wires and Fittings

	26 05 23.20 Special Wires and Fittings	Crew	Daily Output	Labor-Hours	Unit	Material	Labor	Equipment	Total	Total Incl O&P
0010	**SPECIAL WIRES & FITTINGS**									
0100	Fixture TFFN 600 volt 90°C stranded, #18	1 Elec	13	.615	C.L.F.	9.50	32.50		42	59
0150	#16		13	.615		13	32.50		45.50	63
0500	Thermostat, jacket non-plenum, twisted, #18-2 conductor		8	1		14.75	52.50		67.25	94.50
0550	#18-3 conductor		7	1.143		22	60		82	114
0600	#18-4 conductor		6.50	1.231		30.50	64.50		95	130
0650	#18-5 conductor		6	1.333		33.50	70		103.50	142
0700	#18-6 conductor		5.50	1.455		39.50	76		115.50	158
0750	#18-7 conductor		5	1.600		46	84		130	176
0800	#18-8 conductor	↓	4.80	1.667	↓	52	87.50		139.50	188
1500	Fire alarm FEP teflon 150 volt to 200°C									
1550	#22, 1 pair	1 Elec	10	.800	C.L.F.	61.50	42		103.50	130
1600	2 pair		8	1		102	52.50		154.50	191
1650	4 pair		7	1.143		157	60		217	262
1700	6 pair		6	1.333		204	70		274	330
1750	8 pair		5.50	1.455		256	76		332	395
1800	10 pair		5	1.600		310	84		394	465
1850	#18, 1 pair		8	1		70.50	52.50		123	156
1900	2 pair		6.50	1.231		135	64.50		199.50	245
1950	4 pair		4.80	1.667		204	87.50		291.50	355

26 05 23.20	Special Wires and Fittings	Crew	Daily Output	Labor-Hours	Unit	2013 Bare Costs Material	Labor	Equipment	Total	Total Incl O&P
2000	6 pair	1 Elec	4	2	C.L.F.	268	105		373	450
2050	8 pair		3.50	2.286		350	120		470	565
2100	10 pair		3	2.667		400	140		540	650
2460	Tray cable, type TC, copper, #16-2 conductor		9.40	.851		18.80	44.50		63.30	87
2464	#16-3 conductor		8.40	.952		24	50		74	101
2468	#16-4 conductor		7.30	1.096		29	57.50		86.50	118
2472	#16-5 conductor		6.70	1.194		36.50	62.50		99	134
2476	#16-7 conductor		5.50	1.455		50	76		126	169
2480	#16-9 conductor		5	1.600		62	84		146	193
2484	#16-12 conductor	2 Elec	8.80	1.818		81	95.50		176.50	232
2488	#16-15 conductor		7.20	2.222		109	116		225	294
2492	#16-19 conductor		6.60	2.424		133	127		260	335
2496	#16-25 conductor		6.40	2.500		190	131		321	405
2500	#14-2 conductor	1 Elec	9	.889		23	46.50		69.50	95
2520	#14-3 conductor		8	1		32	52.50		84.50	114
2540	#14-4 conductor		7	1.143		39.50	60		99.50	133
2560	#14-5 conductor		6.50	1.231		49	64.50		113.50	151
2564	#14-7 conductor		5.30	1.509		74.50	79		153.50	200
2568	#14-9 conductor	2 Elec	9.60	1.667		91	87.50		178.50	231
2572	#14-12 conductor		8.60	1.860		125	97.50		222.50	284
2576	#14-15 conductor		7.20	2.222		149	116		265	340
2578	#14-19 conductor		6.20	2.581		189	135		324	410
2582	#14-25 conductor		5.30	3.019		246	158		404	510
2590	#12-2 conductor	1 Elec	8.40	.952		33	50		83	111
2592	#12-3 conductor		7.60	1.053		48	55		103	135
2594	#12-4 conductor		6.60	1.212		60	63.50		123.50	161
2596	#12-5 conductor		6.20	1.290		74.50	67.50		142	183
2598	#12-7 conductor	2 Elec	10.40	1.538		109	80.50		189.50	241
2602	#12-9 conductor		8.80	1.818		141	95.50		236.50	298
2604	#12-12 conductor		7.80	2.051		183	107		290	365
2606	#12-15 conductor		6.80	2.353		225	123		348	430
2608	#12-19 conductor		6	2.667		275	140		415	515
2610	#12-25 conductor	3 Elec	7.70	3.117		375	163		538	655
2618	#10-2 conductor	1 Elec	8	1		49	52.50		101.50	133
2622	#10-3 conductor	"	7.30	1.096		72.50	57.50		130	166
2624	#10-4 conductor	2 Elec	12.80	1.250		94	65.50		159.50	201
2626	#10-5 conductor		11.80	1.356		117	71		188	234
2628	#10-7 conductor		9.40	1.702		169	89		258	320
2630	#10-9 conductor		8.40	1.905		221	100		321	390
2632	#10-12 conductor		7.20	2.222		277	116		393	480
2640	300 V, copper braided shield, PVC jacket									
2650	2 conductor #18 stranded	1 Elec	7	1.143	C.L.F.	55	60		115	150
2660	3 conductor #18 stranded	"	6	1.333	"	77.50	70		147.50	191
3000	Strain relief grip for cable									
3050	Cord, top, #12-3	1 Elec	40	.200	Ea.	14.65	10.50		25.15	32
3060	#12-4		40	.200		14.65	10.50		25.15	32
3070	#12-5		39	.205		15.55	10.75		26.30	33
3100	#10-3		39	.205		15.55	10.75		26.30	33
3110	#10-4		38	.211		15.55	11.05		26.60	33.50
3120	#10-5		38	.211		16.90	11.05		27.95	35
3200	Bottom, #12-3		40	.200		36.50	10.50		47	55.50
3210	#12-4		40	.200		36.50	10.50		47	55.50
3220	#12-5		39	.205		37	10.75		47.75	56.50

26 05 23 – Control-Voltage Electrical Power Cables

	26 05 23.20 Special Wires and Fittings	Crew	Daily Output	Labor-Hours	Unit	Material	2013 Bare Costs Labor	Equipment	Total	Total Incl O&P
3230	#10-3	1 Elec	39	.205	Ea.	37	10.75		47.75	56.50
3300	#10-4		38	.211		37	11.05		48.05	57
3310	#10-5		38	.211		46	11.05		57.05	67.50
3400	Cable ties, standard, 4" length		190	.042		.15	2.21		2.36	3.47
3410	7" length		160	.050		.19	2.62		2.81	4.13
3420	14.5" length		90	.089		.46	4.66		5.12	7.45
3430	Heavy, 14.5" length		80	.100		.64	5.25		5.89	8.55

26 05 26 – Grounding and Bonding for Electrical Systems

26 05 26.80 Grounding

	26 05 26.80 Grounding	Crew	Daily Output	Labor-Hours	Unit	Material	Labor	Equipment	Total	Total Incl O&P
0010	**GROUNDING** R260526-80									
0030	Rod, copper clad, 8' long, 1/2" diameter	1 Elec	5.50	1.455	Ea.	20	76		96	136
0040	5/8" diameter		5.50	1.455		19.20	76		95.20	135
0050	3/4" diameter		5.30	1.509		33.50	79		112.50	155
0080	10' long, 1/2" diameter		4.80	1.667		22	87.50		109.50	156
0090	5/8" diameter		4.60	1.739		24	91		115	163
0100	3/4" diameter		4.40	1.818		38.50	95.50		134	186
0130	15' long, 3/4" diameter		4	2		70.50	105		175.50	235
0150	Coupling, bronze, 1/2" diameter					4.54			4.54	4.99
0160	5/8" diameter					5.05			5.05	5.55
0170	3/4" diameter					14.55			14.55	16
0190	Drive studs, 1/2" diameter					10.50			10.50	11.55
0210	5/8" diameter					13.70			13.70	15.10
0220	3/4" diameter					15.65			15.65	17.25
0230	Clamp, bronze, 1/2" diameter	1 Elec	32	.250		5.20	13.10		18.30	25.50
0240	5/8" diameter		32	.250		6.30	13.10		19.40	26.50
0250	3/4" diameter		32	.250		7.40	13.10		20.50	28
0260	Wire ground bare armored, #8-1 conductor		2	4	C.L.F.	106	210		316	430
0270	#6-1 conductor		1.80	4.444		127	233		360	490
0280	#4-1 conductor		1.60	5		172	262		434	580
0320	Bare copper wire, #14 solid		14	.571		7.80	30		37.80	53.50
0330	#12		13	.615		12.10	32.50		44.60	62
0340	#10		12	.667		19.15	35		54.15	73.50
0350	#8		11	.727		31	38		69	91
0360	#6		10	.800		54.50	42		96.50	123
0370	#4		8	1		94	52.50		146.50	182
0380	#2		5	1.600		149	84		233	289
0390	Bare copper wire, stranded, #8		11	.727		32	38		70	92.50
0400	#6		10	.800		56	42		98	125
0450	#4	2 Elec	16	1		94.50	52.50		147	183
0600	#2		10	1.600		141	84		225	280
0650	#1		9	1.778		194	93		287	355
0700	1/0		8	2		213	105		318	390
0750	2/0		7.20	2.222		267	116		383	470
0800	3/0		6.60	2.424		335	127		462	560
1000	4/0		5.70	2.807		425	147		572	685
1200	250 kcmil	3 Elec	7.20	3.333		505	175		680	815
1210	300 kcmil		6.60	3.636		570	191		761	910
1220	350 kcmil		6	4		700	210		910	1,075
1230	400 kcmil		5.70	4.211		840	221		1,061	1,250
1240	500 kcmil		5.10	4.706		1,000	247		1,247	1,475
1260	750 kcmil		3.60	6.667		1,700	350		2,050	2,400
1270	1000 kcmil		3	8		2,275	420		2,695	3,125

26 05 26.80 Grounding		Crew	Daily Output	Labor-Hours	Unit	2013 Bare Costs Material	Labor	Equipment	Total	Total Incl O&P
1360	Bare aluminum, stranded, #6	1 Elec	9	.889	C.L.F.	14.60	46.50		61.10	85.50
1370	#4	2 Elec	16	1		22	52.50		74.50	103
1380	#2		13	1.231		29.50	64.50		94	129
1390	#1		10.60	1.509		39	79		118	161
1400	1/0		9	1.778		44	93		137	188
1410	2/0		8	2		54.50	105		159.50	217
1420	3/0		7.20	2.222		67.50	116		183.50	248
1430	4/0		6.60	2.424		84	127		211	283
1440	250 kcmil	3 Elec	9.30	2.581		116	135		251	330
1450	300 kcmil		8.70	2.759		134	145		279	365
1460	400 kcmil		7.50	3.200		158	168		326	425
1470	500 kcmil		6.90	3.478		187	182		369	480
1480	600 kcmil		6	4		211	210		421	545
1490	700 kcmil		5.70	4.211		231	221		452	585
1500	750 kcmil		5.10	4.706		240	247		487	635
1510	1000 kcmil		4.80	5		305	262		567	725
1800	Water pipe ground clamps, heavy duty									
2000	Bronze, 1/2" to 1" diameter	1 Elec	8	1	Ea.	25	52.50		77.50	106
2100	1-1/4" to 2" diameter		8	1		33	52.50		85.50	115
2200	2-1/2" to 3" diameter		6	1.333		57.50	70		127.50	169
2730	Exothermic weld, 4/0 wire to 1" ground rod		7	1.143		11.45	60		71.45	102
2740	4/0 wire to building steel		7	1.143		9.15	60		69.15	99.50
2750	4/0 wire to motor frame		7	1.143		9.15	60		69.15	99.50
2760	4/0 wire to 4/0 wire		7	1.143		7.95	60		67.95	98.50
2770	4/0 wire to #4 wire		7	1.143		7.95	60		67.95	98.50
2780	4/0 wire to #8 wire		7	1.143		7.95	60		67.95	98.50
2790	Mold, reusable, for above					137			137	151
2800	Brazed connections, #6 wire	1 Elec	12	.667		15.85	35		50.85	70
3000	#2 wire		10	.800		21	42		63	86
3100	3/0 wire		8	1		32	52.50		84.50	114
3200	4/0 wire		7	1.143		36.50	60		96.50	130
3400	250 kcmil wire		5	1.600		42.50	84		126.50	172
3600	500 kcmil wire		4	2		52.50	105		157.50	215
3700	Insulated ground wire, copper #14		13	.615	C.L.F.	10.10	32.50		42.60	59.50
3710	#12		11	.727		15.50	38		53.50	74
3720	#10		10	.800		24	42		66	89
3730	#8		8	1		37	52.50		89.50	119
3740	#6		6.50	1.231		63	64.50		127.50	166
3750	#4	2 Elec	10.60	1.509		101	79		180	229
3770	#2		9	1.778		157	93		250	310
3780	#1		8	2		206	105		311	385
3790	1/0		6.60	2.424		250	127		377	465
3800	2/0		5.80	2.759		315	145		460	560
3810	3/0		5	3.200		395	168		563	685
3820	4/0		4.40	3.636		495	191		686	830
3830	250 kcmil	3 Elec	6	4		595	210		805	970
3840	300 kcmil		5.70	4.211		710	221		931	1,100
3850	350 kcmil		5.40	4.444		830	233		1,063	1,275
3860	400 kcmil		5.10	4.706		945	247		1,192	1,425
3870	500 kcmil		4.80	5		1,150	262		1,412	1,650
3880	600 kcmil		3.90	6.154		1,375	320		1,695	2,000
3890	750 kcmil		3.30	7.273		1,950	380		2,330	2,725
3900	1000 kcmil		2.70	8.889		2,800	465		3,265	3,775

26 05 26 – Grounding and Bonding for Electrical Systems

26 05 26.80 Grounding

	26 05 26.80 Grounding	Crew	Daily Output	Labor-Hours	Unit	Material	Labor (2013 Bare Costs)	Equipment	Total	Total Incl O&P
3960	Insulated ground wire, aluminum, #6	1 Elec	8	1	C.L.F.	32.50	52.50		85	114
3970	#4	2 Elec	13	1.231		40	64.50		104.50	141
3980	#2		10.60	1.509		54.50	79		133.50	178
3990	#1		9	1.778		80	93		173	227
4000	1/0		8	2		95.50	105		200.50	262
4010	2/0		7.20	2.222		113	116		229	298
4020	3/0		6.60	2.424		140	127		267	345
4030	4/0		6.20	2.581		156	135		291	375
4040	250 kcmil	3 Elec	8.70	2.759		191	145		336	425
4050	300 kcmil		8.10	2.963		263	155		418	520
4060	350 kcmil		7.50	3.200		268	168		436	545
4070	400 kcmil		6.90	3.478		315	182		497	620
4080	500 kcmil		6	4		345	210		555	695
4090	600 kcmil		5.70	4.211		435	221		656	810
4100	700 kcmil		5.10	4.706		505	247		752	925
4110	750 kcmil		4.80	5		585	262		847	1,025
5000	Copper Electrolytic ground rod system									
5010	Includes augering hole, mixing bentonite clay,									
5020	Installing rod, and terminating ground wire									
5100	Straight Vertical type, 2" diam.									
5120	8.5' long, clamp connection	1 Elec	2.67	2.996	Ea.	755	157		912	1,075
5130	With exothermic weld connection		1.95	4.103		755	215		970	1,150
5140	10' long		2.35	3.404		880	178		1,058	1,225
5150	With exothermic weld connection		1.78	4.494		880	236		1,116	1,325
5160	12' long		2.16	3.704		995	194		1,189	1,400
5170	With exothermic weld connection		1.67	4.790		995	251		1,246	1,475
5180	20' long		1.74	4.598		1,425	241		1,666	1,925
5190	With exothermic weld connection		1.40	5.714		1,725	299		2,024	2,350
5195	40' long with exothermic weld connection	2 Elec	2	8		3,000	420		3,420	3,925
5200	L-Shaped, 2" diam.									
5220	4' Vert. x 10' Horz., clamp connection	1 Elec	5.33	1.501	Ea.	1,250	78.50		1,328.50	1,525
5230	With exothermic weld connection	"	3.08	2.597	"	1,250	136		1,386	1,600
5300	Protective Box at grade level, with breather slots									
5320	Round 12" long, fiberlyte	1 Elec	32	.250	Ea.	64	13.10		77.10	89.50
5330	Concrete	"	16	.500		98.50	26		124.50	147
5400	Bentonite Clay, 50# bag, 1 per 10' of rod					40.50			40.50	44.50
5500	Equipotential earthing bar	1 Elec	2	4		173	210		383	505

26 05 29 – Hangers and Supports for Electrical Systems

26 05 29.20 Hangers

	26 05 29.20 Hangers	Crew	Daily Output	Labor-Hours	Unit	Material	Labor (2013 Bare Costs)	Equipment	Total	Total Incl O&P
0010	**HANGERS** R260533-20									
0015	See section 22 05 29.10 for additional items									
0030	Conduit supports									
0050	Strap w/2 holes, rigid steel conduit									
0100	1/2" diameter	1 Elec	470	.017	Ea.	.17	.89		1.06	1.52
0150	3/4" diameter		440	.018		.20	.95		1.15	1.65
0200	1" diameter		400	.020		.31	1.05		1.36	1.91
0300	1-1/4" diameter		355	.023		.48	1.18		1.66	2.30
0350	1-1/2" diameter		320	.025		.55	1.31		1.86	2.57
0400	2" diameter		266	.030		.59	1.58		2.17	3.01
0500	2-1/2" diameter		160	.050		1.15	2.62		3.77	5.20
0550	3" diameter		133	.060		1.50	3.15		4.65	6.35
0600	3-1/2" diameter		100	.080		2.40	4.19		6.59	8.90

26 05 Common Work Results for Electrical

26 05 29 – Hangers and Supports for Electrical Systems

26 05 29.20 Hangers		Crew	Daily Output	Labor-Hours	Unit	2013 Bare Costs Material	Labor	Equipment	Total	Total Incl O&P
0650	4″ diameter	1 Elec	80	.100	Ea.	2.03	5.25		7.28	10.10
0700	EMT, 1/2″ diameter		470	.017		.19	.89		1.08	1.54
0800	3/4″ diameter		440	.018		.25	.95		1.20	1.71
0850	1″ diameter		400	.020		.44	1.05		1.49	2.05
0900	1-1/4″ diameter		355	.023		.71	1.18		1.89	2.55
0950	1-1/2″ diameter		320	.025		.75	1.31		2.06	2.79
1000	2″ diameter		266	.030		1.07	1.58		2.65	3.54
1100	2-1/2″ diameter		160	.050		1.84	2.62		4.46	5.95
1150	3″ diameter		133	.060		2.46	3.15		5.61	7.45
1200	3-1/2″ diameter		100	.080		2.77	4.19		6.96	9.30
1250	4″ diameter		80	.100		3.13	5.25		8.38	11.30
1400	Hanger, with bolt, 1/2″ diameter		200	.040		.73	2.10		2.83	3.94
1450	3/4″ diameter		190	.042		.76	2.21		2.97	4.14
1500	1″ diameter		176	.045		1.24	2.38		3.62	4.92
1550	1-1/4″ diameter		160	.050		1.75	2.62		4.37	5.85
1600	1-1/2″ diameter		140	.057		2.17	2.99		5.16	6.85
1650	2″ diameter		130	.062		2.55	3.22		5.77	7.65
1700	2-1/2″ diameter		100	.080		2.94	4.19		7.13	9.50
1750	3″ diameter		64	.125		3.44	6.55		9.99	13.60
1800	3-1/2″ diameter		50	.160		5.10	8.40		13.50	18.15
1850	4″ diameter		40	.200		11.55	10.50		22.05	28.50
1900	Riser clamps, conduit, 1/2″ diameter		40	.200		10.95	10.50		21.45	28
1950	3/4″ diameter		36	.222		11.90	11.65		23.55	30.50
2000	1″ diameter		30	.267		12	13.95		25.95	34
2100	1-1/4″ diameter		27	.296		14.65	15.55		30.20	39
2150	1-1/2″ diameter		27	.296		15.35	15.55		30.90	40
2200	2″ diameter		20	.400		16.10	21		37.10	49
2250	2-1/2″ diameter		20	.400		17	21		38	50
2300	3″ diameter		18	.444		18.50	23.50		42	55.50
2350	3-1/2″ diameter		18	.444		21	23.50		44.50	58
2400	4″ diameter		14	.571		25	30		55	72.50
2500	Threaded rod, painted, 1/4″ diameter		260	.031	L.F.	2.46	1.61		4.07	5.10
2600	3/8″ diameter		200	.040		4	2.10		6.10	7.55
2700	1/2″ diameter		140	.057		5.20	2.99		8.19	10.25
2800	5/8″ diameter		100	.080		7.05	4.19		11.24	14
2900	3/4″ diameter		60	.133		10.50	7		17.50	22
2940	Couplings painted, 1/4″ diameter				C	425			425	470
2960	3/8″ diameter					515			515	570
2970	1/2″ diameter					960			960	1,050
2980	5/8″ diameter					1,500			1,500	1,650
2990	3/4″ diameter					1,925			1,925	2,125
3000	Nuts, galvanized, 1/4″ diameter					15.10			15.10	16.60
3050	3/8″ diameter					20.50			20.50	22.50
3100	1/2″ diameter					50.50			50.50	55.50
3150	5/8″ diameter					95.50			95.50	105
3200	3/4″ diameter					142			142	156
3250	Washers, galvanized, 1/4″ diameter					15.10			15.10	16.60
3300	3/8″ diameter					21.50			21.50	23.50
3350	1/2″ diameter					34			34	37.50
3400	5/8″ diameter					77.50			77.50	85.50
3450	3/4″ diameter					113			113	125
3500	Lock washers, galvanized, 1/4″ diameter					13.60			13.60	14.95
3550	3/8″ diameter					18			18	19.80

26 05 29 – Hangers and Supports for Electrical Systems

26 05 29.20 Hangers		Crew	Daily Output	Labor-Hours	Unit	Material	Labor	Equipment	Total	Total Incl O&P
						2013 Bare Costs				
3600	1/2" diameter				C	21			21	23
3650	5/8" diameter					38			38	41.50
3700	3/4" diameter				↓	72			72	79.50
3800	Channels, steel, 3/4" x 1-1/2"	1 Elec	80	.100	L.F.	5.70	5.25		10.95	14.10
3900	1-1/2" x 1-1/2"		70	.114		7.50	6		13.50	17.20
4000	1-7/8" x 1-1/2"		60	.133		25.50	7		32.50	38.50
4100	3" x 1-1/2"		50	.160	↓	36.50	8.40		44.90	52.50
4200	Spring nuts, long, 1/4"		120	.067	Ea.	1.62	3.49		5.11	7.05
4250	3/8"		100	.080		1.70	4.19		5.89	8.10
4300	1/2"		80	.100		1.84	5.25		7.09	9.85
4350	Spring nuts, short, 1/4"		120	.067		1.60	3.49		5.09	7
4400	3/8"		100	.080		1.65	4.19		5.84	8.05
4450	1/2"		80	.100	↓	1.70	5.25		6.95	9.70
4500	Closure strip		200	.040	L.F.	4.80	2.10		6.90	8.45
4550	End cap		60	.133	Ea.	1.57	7		8.57	12.20
4600	End connector 3/4" conduit		40	.200		5.60	10.50		16.10	22
4650	Junction box, 1 channel		16	.500		45	26		71	88
4700	2 channel		14	.571		54	30		84	105
4750	3 channel		12	.667		63	35		98	122
4800	4 channel		10	.800		68.50	42		110.50	138
4850	Splice plate		40	.200		12.60	10.50		23.10	29.50
4900	Continuous concrete insert, 1-1/2" deep, 1' long		16	.500		17.85	26		43.85	58.50
4950	2' long		14	.571		22	30		52	69
5000	3' long		12	.667		24.50	35		59.50	79.50
5050	4' long		10	.800		30.50	42		72.50	96
5100	6' long		8	1		46	52.50		98.50	129
5150	3/4" deep, 1' long		16	.500		14	26		40	54.50
5200	2' long		14	.571		17.45	30		47.45	64
5250	3' long		12	.667		18.15	35		53.15	72.50
5300	4' long		10	.800		24	42		66	89
5350	6' long		8	1		36	52.50		88.50	118
5400	90° angle fitting 2-1/8" x 2-1/8"		60	.133		3.36	7		10.36	14.15
5450	Supports, suspension rod type, small		60	.133		18.40	7		25.40	30.50
5500	Large		40	.200		20.50	10.50		31	38
5550	Beam clamp, small		60	.133		7.80	7		14.80	19.05
5600	Large		40	.200		8.75	10.50		19.25	25.50
5650	U-support, small		60	.133		4.62	7		11.62	15.55
5700	Large		40	.200		11.05	10.50		21.55	28
5750	Concrete insert, cast, for up to 1/2" threaded rod		16	.500		4.48	26		30.48	44
5800	Beam clamp, 1/4" clamp, for 1/4" threaded drop rod		32	.250		2.56	13.10		15.66	22.50
5900	3/8" clamp, for 3/8" threaded drop rod		32	.250		8.05	13.10		21.15	28.50
6000	Strap, rigid conduit, 1/2" diameter		540	.015		1.56	.78		2.34	2.88
6050	3/4" diameter		440	.018		1.74	.95		2.69	3.34
6100	1" diameter		420	.019		1.92	1		2.92	3.60
6150	1-1/4" diameter		400	.020		2.16	1.05		3.21	3.95
6200	1-1/2" diameter		400	.020		2.61	1.05		3.66	4.44
6250	2" diameter		267	.030		2.82	1.57		4.39	5.45
6300	2-1/2" diameter		267	.030		3.26	1.57		4.83	5.95
6350	3" diameter		160	.050		3.59	2.62		6.21	7.85
6400	3-1/2" diameter		133	.060		4.77	3.15		7.92	9.95
6450	4" diameter		100	.080		5	4.19		9.19	11.75
6500	5" diameter		80	.100		10.40	5.25		15.65	19.30
6550	6" diameter	↓	60	.133	↓	15.85	7		22.85	28

26 05 29 – Hangers and Supports for Electrical Systems

26 05 29.20 Hangers		Crew	Daily Output	Labor-Hours	Unit	Material	Labor	Equipment	Total	Total Incl O&P
						2013 Bare Costs				
6600	EMT, 1/2" diameter	1 Elec	540	.015	Ea.	1.56	.78		2.34	2.88
6650	3/4" diameter		440	.018		1.67	.95		2.62	3.27
6700	1" diameter		420	.019		1.86	1		2.86	3.54
6750	1-1/4" diameter		400	.020		2.09	1.05		3.14	3.87
6800	1-1/2" diameter		400	.020		2.57	1.05		3.62	4.40
6850	2" diameter		267	.030		2.91	1.57		4.48	5.55
6900	2-1/2" diameter		267	.030		3.10	1.57		4.67	5.75
6950	3" diameter		160	.050		3.36	2.62		5.98	7.60
6970	3-1/2" diameter		133	.060		3.76	3.15		6.91	8.85
6990	4" diameter		100	.080		4.32	4.19		8.51	11
7000	Clip, 1 hole for rigid conduit, 1/2" diameter		500	.016		.66	.84		1.50	1.98
7050	3/4" diameter		470	.017		.96	.89		1.85	2.39
7100	1" diameter		440	.018		1.20	.95		2.15	2.75
7150	1-1/4" diameter		400	.020		2.58	1.05		3.63	4.41
7200	1-1/2" diameter		355	.023		2.97	1.18		4.15	5.05
7250	2" diameter		320	.025		5.80	1.31		7.11	8.30
7300	2-1/2" diameter		266	.030		12.90	1.58		14.48	16.55
7350	3" diameter		160	.050		18.45	2.62		21.07	24.50
7400	3-1/2" diameter		133	.060		26.50	3.15		29.65	33.50
7450	4" diameter		100	.080		59	4.19		63.19	71.50
7500	5" diameter		80	.100		184	5.25		189.25	211
7550	6" diameter		60	.133		198	7		205	228
7820	Conduit hangers, with bolt & 12" rod, 1/2" diameter		150	.053		4.73	2.79		7.52	9.40
7830	3/4" diameter		145	.055		4.76	2.89		7.65	9.60
7840	1" diameter		135	.059		5.25	3.11		8.36	10.40
7850	1-1/4" diameter		120	.067		5.75	3.49		9.24	11.60
7860	1-1/2" diameter		110	.073		6.15	3.81		9.96	12.50
7870	2" diameter		100	.080		7.75	4.19		11.94	14.80
7880	2-1/2" diameter		80	.100		8.15	5.25		13.40	16.85
7890	3" diameter		60	.133		8.65	7		15.65	20
7900	3-1/2" diameter		45	.178		10.30	9.30		19.60	25.50
7910	4" diameter		35	.229		16.80	12		28.80	36.50
7920	5" diameter		30	.267		18.65	13.95		32.60	41.50
7930	6" diameter		25	.320		36	16.75		52.75	64.50
7950	Jay clamp, 1/2" diameter		32	.250		4.73	13.10		17.83	25
7960	3/4" diameter		32	.250		4.76	13.10		17.86	25
7970	1" diameter		32	.250		6.40	13.10		19.50	26.50
7980	1-1/4" diameter		30	.267		6.95	13.95		20.90	28.50
7990	1-1/2" diameter		30	.267		8.65	13.95		22.60	30.50
8000	2" diameter		30	.267		11.60	13.95		25.55	34
8010	2-1/2" diameter		28	.286		16.55	14.95		31.50	41
8020	3" diameter		28	.286		16.55	14.95		31.50	41
8030	3-1/2" diameter		25	.320		22	16.75		38.75	49.50
8040	4" diameter		25	.320		22	16.75		38.75	49.50
8050	5" diameter		20	.400		94.50	21		115.50	136
8060	6" diameter		16	.500		178	26		204	235
8070	Channels, 3/4" x 1-1/2" w/12" rods for 1/2" to 1" conduit		30	.267		12.55	13.95		26.50	35
8080	1-1/2" x 1-1/2" w/12" rods for 1-1/4" to 2" conduit		28	.286		13.35	14.95		28.30	37
8090	1-1/2" x 1-1/2" w/12" rods for 2-1/2" to 4" conduit		26	.308		16.40	16.10		32.50	42
8100	1-1/2" x 1-7/8" w/12" rods for 5" to 6" conduit		24	.333		61.50	17.45		78.95	93.50
8110	Beam clamp, conduit, plastic coated steel, 1/2" diam.		30	.267		19.75	13.95		33.70	42.50
8120	3/4" diameter		30	.267		21	13.95		34.95	44
8130	1" diameter		30	.267		23.50	13.95		37.45	47

26 05 29.20 Hangers		Crew	Daily Output	Labor-Hours	Unit	2013 Bare Costs Material	Labor	Equipment	Total	Total Incl O&P
8140	1-1/4" diameter	1 Elec	28	.286	Ea.	28.50	14.95		43.45	54
8150	1-1/2" diameter		28	.286		35	14.95		49.95	61
8160	2" diameter		28	.286		49	14.95		63.95	76.50
8170	2-1/2" diameter		26	.308		50	16.10		66.10	79
8180	3" diameter		26	.308		54	16.10		70.10	83.50
8190	3-1/2" diameter		23	.348		58	18.25		76.25	91
8200	4" diameter		23	.348		63	18.25		81.25	97
8210	5" diameter		18	.444		204	23.50		227.50	260
8220	Channels, plastic coated									
8250	3/4" x 1-1/2", w/12" rods for 1/2" to 1" conduit	1 Elec	28	.286	Ea.	34	14.95		48.95	60
8260	1-1/2" x 1-1/2", w/12" rods for 1-1/4" to 2" conduit		26	.308		38	16.10		54.10	66
8270	1-1/2" x 1-1/2", w/12" rods for 2-1/2" to 3-1/2" conduit		24	.333		41.50	17.45		58.95	72
8280	1-1/2" x 1-7/8", w/12" rods for 4" to 5" conduit		22	.364		69.50	19.05		88.55	105
8290	1-1/2" x 1-7/8", w/12" rods for 6" conduit		20	.400		76	21		97	115
8320	Conduit hangers, plastic coated steel, with bolt & 12" rod, 1/2" diam.		140	.057		19.60	2.99		22.59	26
8330	3/4" diameter		135	.059		20	3.11		23.11	26.50
8340	1" diameter		125	.064		20.50	3.35		23.85	27.50
8350	1-1/4" diameter		110	.073		22	3.81		25.81	29.50
8360	1-1/2" diameter		100	.080		24.50	4.19		28.69	33.50
8370	2" diameter		90	.089		27	4.66		31.66	36.50
8380	2-1/2" diameter		70	.114		32	6		38	44
8390	3" diameter		50	.160		39	8.40		47.40	55
8400	3-1/2" diameter		35	.229		39.50	12		51.50	61
8410	4" diameter		25	.320		57.50	16.75		74.25	88.50
8420	5" diameter		20	.400		63	21		84	101
9000	Parallel type, conduit beam clamp, 1/2"		32	.250		4.24	13.10		17.34	24.50
9010	3/4"		32	.250		4.67	13.10		17.77	25
9020	1"		32	.250		4.99	13.10		18.09	25
9030	1-1/4"		30	.267		6.80	13.95		20.75	28.50
9040	1-1/2"		30	.267		7.75	13.95		21.70	29.50
9050	2"		30	.267		10.10	13.95		24.05	32
9060	2-1/2"		28	.286		13.40	14.95		28.35	37.50
9070	3"		28	.286		17.35	14.95		32.30	41.50
9090	4"		25	.320		21.50	16.75		38.25	49
9110	Right angle, conduit beam clamp, 1/2"		32	.250		2.79	13.10		15.89	22.50
9120	3/4"		32	.250		2.88	13.10		15.98	23
9130	1"		32	.250		3.20	13.10		16.30	23
9140	1-1/4"		30	.267		3.95	13.95		17.90	25.50
9150	1-1/2"		30	.267		4.46	13.95		18.41	26
9160	2"		30	.267		6.45	13.95		20.40	28
9170	2-1/2"		28	.286		8	14.95		22.95	31.50
9180	3"		28	.286		9.15	14.95		24.10	32.50
9190	3-1/2"		25	.320		11.55	16.75		28.30	37.50
9200	4"		25	.320		12.45	16.75		29.20	38.50
9230	Adjustable, conduit hanger, 1/2"		32	.250		6.25	13.10		19.35	26.50
9240	3/4"		32	.250		6.40	13.10		19.50	26.50
9250	1"		32	.250		6.50	13.10		19.60	27
9260	1-1/4"		30	.267		6.60	13.95		20.55	28.50
9270	1-1/2"		30	.267		7.15	13.95		21.10	29
9280	2"		30	.267		6.50	13.95		20.45	28
9290	2-1/2"		28	.286		13.50	14.95		28.45	37.50
9300	3"		28	.286		14.20	14.95		29.15	38
9310	3-1/2"		25	.320		15.20	16.75		31.95	41.50

26 05 29 – Hangers and Supports for Electrical Systems

26 05 29.20 Hangers		Crew	Daily Output	Labor-Hours	Unit	Material	2013 Bare Costs Labor	Equipment	Total	Total Incl O&P
9320	4"	1 Elec	25	.320	Ea.	24	16.75		40.75	51
9330	5"		20	.400		27.50	21		48.50	61.50
9340	6"		16	.500		40.50	26		66.50	83.50
9350	Combination conduit hanger, 3/8"		32	.250		11	13.10		24.10	31.50
9360	Adjustable flange 3/8"	↓	32	.250	↓	12.95	13.10		26.05	34

26 05 33 – Raceway and Boxes for Electrical Systems

26 05 33.13 Conduit

26 05 33.13 Conduit		Crew	Daily Output	Labor-Hours	Unit	Material	2013 Bare Costs Labor	Equipment	Total	Total Incl O&P
0010	**CONDUIT** To 15' high, includes 2 terminations, 2 elbows, R260533-20									
0020	11 beam clamps, and 11 couplings per 100 L.F.									
0300	Aluminum, 1/2" diameter	1 Elec	100	.080	L.F.	1.97	4.19		6.16	8.40
0500	3/4" diameter R260533-21		90	.089		2.64	4.66		7.30	9.85
0700	1" diameter		80	.100		3.66	5.25		8.91	11.85
1000	1-1/4" diameter		70	.114		4.68	6		10.68	14.10
1030	1-1/2" diameter		65	.123		5.80	6.45		12.25	16.05
1050	2" diameter		60	.133		8	7		15	19.25
1070	2-1/2" diameter	↓	50	.160		12.45	8.40		20.85	26.50
1100	3" diameter	2 Elec	90	.178		17.25	9.30		26.55	33
1130	3-1/2" diameter		80	.200		22.50	10.50		33	40.50
1140	4" diameter		70	.229		27.50	12		39.50	48
1150	5" diameter		50	.320		63.50	16.75		80.25	94.50
1160	6" diameter	↓	40	.400	↓	91	21		112	132
1161	Field bends, 45° to 90°, 1/2" diameter	1 Elec	53	.151	Ea.		7.90		7.90	11.85
1162	3/4" diameter		47	.170			8.90		8.90	13.35
1163	1" diameter		44	.182			9.55		9.55	14.25
1164	1-1/4" diameter		23	.348			18.25		18.25	27.50
1165	1-1/2" diameter		21	.381			19.95		19.95	30
1166	2" diameter		16	.500			26		26	39
1170	Elbows, 1/2" diameter		40	.200		10.35	10.50		20.85	27
1200	3/4" diameter		32	.250		14.10	13.10		27.20	35
1230	1" diameter		28	.286		19.65	14.95		34.60	44
1250	1-1/4" diameter		24	.333		31	17.45		48.45	60
1270	1-1/2" diameter		20	.400		41.50	21		62.50	77
1300	2" diameter		16	.500		61.50	26		87.50	107
1330	2-1/2" diameter		12	.667		103	35		138	167
1350	3" diameter		8	1		158	52.50		210.50	253
1370	3-1/2" diameter		6	1.333		248	70		318	380
1400	4" diameter		5	1.600		294	84		378	450
1410	5" diameter		4	2		870	105		975	1,125
1420	6" diameter		2.50	3.200		1,200	168		1,368	1,575
1430	Couplings, 1/2" diameter		320	.025		3.40	1.31		4.71	5.70
1450	3/4" diameter		192	.042		5.15	2.18		7.33	8.95
1470	1" diameter		160	.050		6.80	2.62		9.42	11.35
1500	1-1/4" diameter		120	.067		8.35	3.49		11.84	14.45
1530	1-1/2" diameter		96	.083		9.60	4.37		13.97	17.10
1550	2" diameter		80	.100		13.60	5.25		18.85	23
1570	2-1/2" diameter		69	.116		30.50	6.10		36.60	42.50
1600	3" diameter		64	.125		40	6.55		46.55	54
1630	3-1/2" diameter		56	.143		55	7.50		62.50	71.50
1650	4" diameter		53	.151		66.50	7.90		74.40	85
1670	5" diameter		48	.167		224	8.75		232.75	260
1690	6" diameter	↓	46	.174	↓	350	9.10		359.10	400
1691	See note on line 26 05 33.13 9995									

26 05 33.13 Conduit		Crew	Daily Output	Labor-Hours	Unit	2013 Bare Costs Material	Labor	Equipment	Total	Total Incl O&P
1750	Rigid galvanized steel, 1/2" diameter	1 Elec	90	.089	L.F.	2.56	4.66		7.22	9.75
1770	3/4" diameter		80	.100		2.69	5.25		7.94	10.80
1800	1" diameter		65	.123		3.94	6.45		10.39	14
1830	1-1/4" diameter		60	.133		5.35	7		12.35	16.30
1850	1-1/2" diameter		55	.145		6.35	7.60		13.95	18.40
1870	2" diameter		45	.178		7.90	9.30		17.20	22.50
1900	2-1/2" diameter		35	.229		14.95	12		26.95	34.50
1930	3" diameter	2 Elec	50	.320		18.25	16.75		35	45
1950	3-1/2" diameter		44	.364		23	19.05		42.05	53.50
1970	4" diameter		40	.400		27	21		48	61
1980	5" diameter		30	.533		52	28		80	99.50
1990	6" diameter		20	.800		80	42		122	151
1991	Field bends, 45° to 90°, 1/2" diameter	1 Elec	44	.182	Ea.		9.55		9.55	14.25
1992	3/4" diameter		40	.200			10.50		10.50	15.70
1993	1" diameter		36	.222			11.65		11.65	17.40
1994	1-1/4" diameter		19	.421			22		22	33
1995	1-1/2" diameter		18	.444			23.50		23.50	35
1996	2" diameter		13	.615			32.50		32.50	48.50
2000	Elbows, 1/2" diameter		32	.250		9.25	13.10		22.35	30
2030	3/4" diameter		28	.286		9.65	14.95		24.60	33
2050	1" diameter		24	.333		14.85	17.45		32.30	42.50
2070	1-1/4" diameter		18	.444		20.50	23.50		44	57.50
2100	1-1/2" diameter		16	.500		25	26		51	66.50
2130	2" diameter		12	.667		36.50	35		71.50	92.50
2150	2-1/2" diameter		8	1		67.50	52.50		120	153
2170	3" diameter		6	1.333		94	70		164	209
2200	3-1/2" diameter		4.20	1.905		149	100		249	315
2220	4" diameter		4	2		169	105		274	340
2230	5" diameter		3.50	2.286		465	120		585	695
2240	6" diameter		2	4		700	210		910	1,100
2250	Couplings, 1/2" diameter		267	.030		2.28	1.57		3.85	4.86
2270	3/4" diameter		160	.050		2.79	2.62		5.41	7
2300	1" diameter		133	.060		4.12	3.15		7.27	9.25
2330	1-1/4" diameter		100	.080		5.15	4.19		9.34	11.95
2350	1-1/2" diameter		80	.100		6.50	5.25		11.75	15
2370	2" diameter		67	.119		8.65	6.25		14.90	18.85
2400	2-1/2" diameter		57	.140		21.50	7.35		28.85	34.50
2430	3" diameter		53	.151		27.50	7.90		35.40	42.50
2450	3-1/2" diameter		47	.170		37	8.90		45.90	54
2470	4" diameter		44	.182		37	9.55		46.55	55.50
2480	5" diameter		40	.200		86	10.50		96.50	110
2490	6" diameter		38	.211		120	11.05		131.05	149
2491	See note on line 26 05 33.13 9995									
2500	Steel, intermediate conduit (IMC), 1/2" diameter	1 Elec	100	.080	L.F.	1.82	4.19		6.01	8.25
2530	3/4" diameter		90	.089		2.17	4.66		6.83	9.35
2550	1" diameter		70	.114		3.18	6		9.18	12.45
2570	1-1/4" diameter		65	.123		4.07	6.45		10.52	14.15
2600	1-1/2" diameter		60	.133		5.10	7		12.10	16.10
2630	2" diameter		50	.160		6.15	8.40		14.55	19.35
2650	2-1/2" diameter		40	.200		11.65	10.50		22.15	28.50
2670	3" diameter	2 Elec	60	.267		15.65	13.95		29.60	38.50
2700	3-1/2" diameter		54	.296		21	15.55		36.55	46
2730	4" diameter		50	.320		22.50	16.75		39.25	50

26 05 33.13 Conduit		Crew	Daily Output	Labor-Hours	Unit	2013 Bare Costs Material	Labor	Equipment	Total	Total Incl O&P
2731	Field bends, 45° to 90°, 1/2" diameter	1 Elec	44	.182	Ea.		9.55		9.55	14.25
2732	3/4" diameter		40	.200			10.50		10.50	15.70
2733	1" diameter		36	.222			11.65		11.65	17.40
2734	1-1/4" diameter		19	.421			22		22	33
2735	1-1/2" diameter		18	.444			23.50		23.50	35
2736	2" diameter		13	.615			32.50		32.50	48.50
2750	Elbows, 1/2" diameter		32	.250		12.45	13.10		25.55	33.50
2770	3/4" diameter		28	.286		13.25	14.95		28.20	37
2800	1" diameter		24	.333		19.35	17.45		36.80	47.50
2830	1-1/4" diameter		18	.444		29.50	23.50		53	67.50
2850	1-1/2" diameter		16	.500		36	26		62	79
2870	2" diameter		12	.667		39	35		74	95.50
2900	2-1/2" diameter		8	1		70.50	52.50		123	156
2930	3" diameter		6	1.333		102	70		172	217
2950	3-1/2" diameter		4.20	1.905		216	100		316	385
2970	4" diameter		4	2		208	105		313	385
3000	Couplings, 1/2" diameter		293	.027		2.28	1.43		3.71	4.65
3030	3/4" diameter		176	.045		2.79	2.38		5.17	6.65
3050	1" diameter		147	.054		4.12	2.85		6.97	8.80
3070	1-1/4" diameter		110	.073		5.15	3.81		8.96	11.40
3100	1-1/2" diameter		88	.091		6.50	4.76		11.26	14.30
3130	2" diameter		73	.110		8.65	5.75		14.40	18.10
3150	2-1/2" diameter		63	.127		21.50	6.65		28.15	33.50
3170	3" diameter		59	.136		27.50	7.10		34.60	41
3200	3-1/2" diameter		52	.154		37	8.05		45.05	52.50
3230	4" diameter		49	.163		37	8.55		45.55	54
3231	See note on line 26 05 33.13 9995									
4100	Rigid steel, plastic coated, 40 mil thick									
4130	1/2" diameter	1 Elec	80	.100	L.F.	7.25	5.25		12.50	15.80
4150	3/4" diameter		70	.114		8	6		14	17.75
4170	1" diameter		55	.145		9.95	7.60		17.55	22.50
4200	1-1/4" diameter		50	.160		12.50	8.40		20.90	26.50
4230	1-1/2" diameter		45	.178		15.15	9.30		24.45	30.50
4250	2" diameter		35	.229		20.50	12		32.50	40.50
4270	2-1/2" diameter		25	.320		30	16.75		46.75	58
4300	3" diameter	2 Elec	44	.364		37.50	19.05		56.55	69.50
4330	3-1/2" diameter		40	.400		45.50	21		66.50	81.50
4350	4" diameter		36	.444		52.50	23.50		76	93
4370	5" diameter		30	.533		118	28		146	172
4400	Elbows, 1/2" diameter	1 Elec	28	.286	Ea.	17.20	14.95		32.15	41.50
4430	3/4" diameter		24	.333		17.65	17.45		35.10	45.50
4450	1" diameter		18	.444		20.50	23.50		44	57.50
4470	1-1/4" diameter		16	.500		25	26		51	66.50
4500	1-1/2" diameter		12	.667		30.50	35		65.50	86
4530	2" diameter		8	1		42.50	52.50		95	126
4550	2-1/2" diameter		6	1.333		81	70		151	194
4570	3" diameter		4.20	1.905		129	100		229	291
4600	3-1/2" diameter		4	2		169	105		274	345
4630	4" diameter		3.80	2.105		183	110		293	365
4650	5" diameter		3.50	2.286		445	120		565	665
4680	Couplings, 1/2" diameter		213	.038		5.05	1.97		7.02	8.50
4700	3/4" diameter		128	.063		5.10	3.28		8.38	10.50
4730	1" diameter		107	.075		6.65	3.92		10.57	13.15

26 05 33.13 Conduit		Crew	Daily Output	Labor-Hours	Unit	Material	2013 Bare Costs Labor	Equipment	Total	Total Incl O&P
4750	1-1/4" diameter	1 Elec	80	.100	Ea.	7.85	5.25		13.10	16.45
4770	1-1/2" diameter		64	.125		9.20	6.55		15.75	19.90
4800	2" diameter		53	.151		13.35	7.90		21.25	26.50
4830	2-1/2" diameter		46	.174		33	9.10		42.10	50
4850	3" diameter		43	.186		40.50	9.75		50.25	59
4870	3-1/2" diameter		38	.211		51.50	11.05		62.55	73
4900	4" diameter		36	.222		60.50	11.65		72.15	84
4950	5" diameter		32	.250		196	13.10		209.10	236
4951	See note on line 26 05 33.13 9995									
5000	Electric metallic tubing (EMT), 1/2" diameter	1 Elec	170	.047	L.F.	.62	2.47		3.09	4.38
5020	3/4" diameter		130	.062		.96	3.22		4.18	5.90
5040	1" diameter		115	.070		1.69	3.65		5.34	7.30
5060	1-1/4" diameter		100	.080		2.82	4.19		7.01	9.35
5080	1-1/2" diameter		90	.089		3.68	4.66		8.34	11
5100	2" diameter		80	.100		4.69	5.25		9.94	13
5120	2-1/2" diameter		60	.133		11.15	7		18.15	22.50
5140	3" diameter	2 Elec	100	.160		12.95	8.40		21.35	27
5160	3-1/2" diameter		90	.178		16.40	9.30		25.70	32
5180	4" diameter		80	.200		18	10.50		28.50	35.50
5200	Field bends, 45° to 90°, 1/2" diameter	1 Elec	89	.090	Ea.		4.71		4.71	7.05
5220	3/4" diameter		80	.100			5.25		5.25	7.85
5240	1" diameter		73	.110			5.75		5.75	8.60
5260	1-1/4" diameter		38	.211			11.05		11.05	16.50
5280	1-1/2" diameter		36	.222			11.65		11.65	17.40
5300	2" diameter		26	.308			16.10		16.10	24
5320	Offsets, 1/2" diameter		65	.123			6.45		6.45	9.65
5340	3/4" diameter		62	.129			6.75		6.75	10.10
5360	1" diameter		53	.151			7.90		7.90	11.85
5380	1-1/4" diameter		30	.267			13.95		13.95	21
5400	1-1/2" diameter		28	.286			14.95		14.95	22.50
5420	2" diameter		20	.400			21		21	31.50
5700	Elbows, 1" diameter		40	.200		7.45	10.50		17.95	24
5720	1-1/4" diameter		32	.250		9.20	13.10		22.30	30
5740	1-1/2" diameter		24	.333		10.75	17.45		28.20	38
5760	2" diameter		20	.400		15.75	21		36.75	49
5780	2-1/2" diameter		12	.667		42.50	35		77.50	99.50
5800	3" diameter		9	.889		57	46.50		103.50	133
5820	3-1/2" diameter		7	1.143		76.50	60		136.50	174
5840	4" diameter		6	1.333		90	70		160	204
6200	Couplings, set screw, steel, 1/2" diameter		235	.034		1.79	1.78		3.57	4.64
6220	3/4" diameter		188	.043		2.70	2.23		4.93	6.30
6240	1" diameter		157	.051		4.38	2.67		7.05	8.80
6260	1-1/4" diameter		118	.068		9.35	3.55		12.90	15.60
6280	1-1/2" diameter		94	.085		13.30	4.46		17.76	21.50
6300	2" diameter		78	.103		17.75	5.35		23.10	27.50
6320	2-1/2" diameter		67	.119		51	6.25		57.25	66
6340	3" diameter		59	.136		56	7.10		63.10	72
6360	3-1/2" diameter		47	.170		63	8.90		71.90	83
6380	4" diameter		43	.186		69.50	9.75		79.25	91
6381	See note on line 26 05 33.13 9995									
6500	Box connectors, set screw, steel, 1/2" diameter	1 Elec	120	.067	Ea.	1.37	3.49		4.86	6.75
6520	3/4" diameter		110	.073		2.23	3.81		6.04	8.15
6540	1" diameter		90	.089		4.22	4.66		8.88	11.60

26 05 33.13 Conduit		Crew	Daily Output	Labor-Hours	Unit	Material	Labor	Equipment	Total	Total Incl O&P
						2013 Bare Costs				
6560	1-1/4" diameter	1 Elec	70	.114	Ea.	8.20	6		14.20	18
6580	1-1/2" diameter		60	.133		11.85	7		18.85	23.50
6600	2" diameter		50	.160		16.40	8.40		24.80	30.50
6620	2-1/2" diameter		36	.222		57.50	11.65		69.15	80.50
6640	3" diameter		27	.296		68.50	15.55		84.05	98.50
6680	3-1/2" diameter		21	.381		96	19.95		115.95	135
6700	4" diameter		16	.500		107	26		133	156
6740	Insulated box connectors, set screw, steel, 1/2" diameter		120	.067		1.88	3.49		5.37	7.30
6760	3/4" diameter		110	.073		2.97	3.81		6.78	8.95
6780	1" diameter		90	.089		5.45	4.66		10.11	12.95
6800	1-1/4" diameter		70	.114		9.95	6		15.95	19.85
6820	1-1/2" diameter		60	.133		14.05	7		21.05	26
6840	2" diameter		50	.160		20.50	8.40		28.90	35
6860	2-1/2" diameter		36	.222		96.50	11.65		108.15	123
6880	3" diameter		27	.296		114	15.55		129.55	149
6900	3-1/2" diameter		21	.381		159	19.95		178.95	205
6920	4" diameter		16	.500		174	26		200	230
7000	EMT to conduit adapters, 1/2" diameter (compression)		70	.114		4.38	6		10.38	13.75
7020	3/4" diameter		60	.133		6.20	7		13.20	17.30
7040	1" diameter		50	.160		9.50	8.40		17.90	23
7060	1-1/4" diameter		40	.200		16.50	10.50		27	34
7080	1-1/2" diameter		30	.267		20.50	13.95		34.45	43.50
7100	2" diameter		25	.320		29.50	16.75		46.25	57.50
7200	EMT to Greenfield adapters, 1/2" to 3/8" diameter (compression)		90	.089		2.98	4.66		7.64	10.25
7220	1/2" diameter		90	.089		5.60	4.66		10.26	13.10
7240	3/4" diameter		80	.100		7.20	5.25		12.45	15.75
7260	1" diameter		70	.114		19	6		25	30
7270	1-1/4" diameter		60	.133		23.50	7		30.50	36.50
7280	1-1/2" diameter		50	.160		26.50	8.40		34.90	41.50
7290	2" diameter		40	.200		40	10.50		50.50	59.50
7400	EMT,LB, LR or LL fittings with covers, 1/2" dia, set screw		24	.333		7.50	17.45		24.95	34.50
7420	3/4" diameter		20	.400		9.20	21		30.20	41.50
7440	1" diameter		16	.500		13.75	26		39.75	54
7450	1-1/4" diameter		13	.615		21	32.50		53.50	71.50
7460	1-1/2" diameter		11	.727		27	38		65	87
7470	2" diameter		9	.889		45	46.50		91.50	119
7600	EMT, "T" fittings with covers, 1/2" diameter, set screw		16	.500		10.90	26		36.90	51
7620	3/4" diameter		15	.533		13.95	28		41.95	57.50
7640	1" diameter		12	.667		19.35	35		54.35	74
7650	1-1/4" diameter		11	.727		26.50	38		64.50	86
7660	1-1/2" diameter		10	.800		33	42		75	99
7670	2" diameter		8	1		43.50	52.50		96	127
8000	EMT, expansion fittings, no jumper, 1/2" diameter		24	.333		58.50	17.45		75.95	90.50
8020	3/4" diameter		20	.400		69	21		90	108
8040	1" diameter		16	.500		81.50	26		107.50	129
8060	1-1/4" diameter		13	.615		110	32.50		142.50	170
8080	1-1/2" diameter		11	.727		150	38		188	222
8100	2" diameter		9	.889		222	46.50		268.50	315
8110	2-1/2" diameter		7	1.143		355	60		415	480
8120	3" diameter		6	1.333		435	70		505	585
8140	4" diameter		5	1.600		760	84		844	960
8200	Split adapter, 1/2" diameter		110	.073		3.14	3.81		6.95	9.15
8210	3/4" diameter		90	.089		2.79	4.66		7.45	10

26 05 33.13 Conduit		Crew	Daily Output	Labor-Hours	Unit	2013 Bare Costs Material	Labor	Equipment	Total	Total Incl O&P
8220	1" diameter	1 Elec	70	.114	Ea.	3.85	6		9.85	13.20
8230	1-1/4" diameter		60	.133		6.20	7		13.20	17.25
8240	1-1/2" diameter		50	.160		9.55	8.40		17.95	23
8250	2" diameter		36	.222		28	11.65		39.65	48.50
8300	1 hole clips, 1/2" diameter		500	.016		.54	.84		1.38	1.84
8320	3/4" diameter		470	.017		.71	.89		1.60	2.11
8340	1" diameter		444	.018		1.30	.94		2.24	2.84
8360	1-1/4" diameter		400	.020		1.73	1.05		2.78	3.47
8380	1-1/2" diameter		355	.023		2.65	1.18		3.83	4.69
8400	2" diameter		320	.025		4.16	1.31		5.47	6.55
8420	2-1/2" diameter		266	.030		7.75	1.58		9.33	10.85
8440	3" diameter		160	.050		9.70	2.62		12.32	14.60
8460	3-1/2" diameter		133	.060		15.35	3.15		18.50	21.50
8480	4" diameter		100	.080		19.80	4.19		23.99	28.50
8500	Clamp back spacers, 1/2" diameter		500	.016		1.61	.84		2.45	3.02
8510	3/4" diameter		470	.017		1.93	.89		2.82	3.45
8520	1" diameter		444	.018		3.38	.94		4.32	5.15
8530	1-1/4" diameter		400	.020		6.05	1.05		7.10	8.20
8540	1-1/2" diameter		355	.023		6.75	1.18		7.93	9.20
8550	2" diameter		320	.025		11.40	1.31		12.71	14.50
8560	2-1/2" diameter		266	.030		22	1.58		23.58	26.50
8570	3" diameter		160	.050		31.50	2.62		34.12	38.50
8580	3-1/2" diameter		133	.060		42.50	3.15		45.65	51
8590	4" diameter		100	.080		92	4.19		96.19	107
8600	Offset connectors, 1/2" diameter		40	.200		5.95	10.50		16.45	22.50
8610	3/4" diameter		32	.250		8.60	13.10		21.70	29
8620	1" diameter		24	.333		9.45	17.45		26.90	36.50
8650	90° pulling elbows, female, 1/2" diameter, with gasket		24	.333		6.60	17.45		24.05	33.50
8660	3/4" diameter		20	.400		11.30	21		32.30	44
8700	Couplings, compression, 1/2" diameter, steel		78	.103		4.29	5.35		9.64	12.75
8710	3/4" diameter		67	.119		6	6.25		12.25	15.95
8720	1" diameter		59	.136		10.30	7.10		17.40	22
8730	1-1/4" diameter		47	.170		19.50	8.90		28.40	35
8740	1-1/2" diameter		38	.211		28.50	11.05		39.55	47.50
8750	2" diameter		31	.258		38.50	13.50		52	62.50
8760	2-1/2" diameter		24	.333		173	17.45		190.45	216
8770	3" diameter		21	.381		216	19.95		235.95	268
8780	3-1/2" diameter		19	.421		350	22		372	420
8790	4" diameter		16	.500		360	26		386	435
8791	See note on line 26 05 33.13 9995									
8800	Box connectors, compression, 1/2" diam., steel	1 Elec	120	.067	Ea.	3.55	3.49		7.04	9.15
8810	3/4" diameter		110	.073		4.99	3.81		8.80	11.20
8820	1" diameter		90	.089		7.40	4.66		12.06	15.10
8830	1-1/4" diameter		70	.114		18.85	6		24.85	30
8840	1-1/2" diameter		60	.133		22	7		29	35
8850	2" diameter		50	.160		32	8.40		40.40	47.50
8860	2-1/2" diameter		36	.222		79	11.65		90.65	104
8870	3" diameter		27	.296		109	15.55		124.55	143
8880	3-1/2" diameter		21	.381		164	19.95		183.95	211
8890	4" diameter		16	.500		168	26		194	224
8900	Box connectors, insulated compression, 1/2" diam., steel		120	.067		3.43	3.49		6.92	9
8910	3/4" diameter		110	.073		4.77	3.81		8.58	10.95
8920	1" diameter		90	.089		8	4.66		12.66	15.75

26 05 33.13 Conduit		Crew	Daily Output	Labor-Hours	Unit	Material	Labor	Equipment	Total	Total Incl O&P
						2013 Bare Costs				
8930	1-1/4" diameter	1 Elec	70	.114	Ea.	16.15	6		22.15	26.50
8940	1-1/2" diameter		60	.133		24.50	7		31.50	37
8950	2" diameter		50	.160		35	8.40		43.40	51
8960	2-1/2" diameter		36	.222		91.50	11.65		103.15	118
8970	3" diameter		27	.296		120	15.55		135.55	155
8980	3-1/2" diameter		21	.381		176	19.95		195.95	223
8990	4" diameter		16	.500		179	26		205	236
9100	PVC, schedule 40, 1/2" diameter		190	.042	L.F.	1.04	2.21		3.25	4.45
9110	3/4" diameter		145	.055		1.15	2.89		4.04	5.60
9120	1" diameter		125	.064		2.14	3.35		5.49	7.35
9130	1-1/4" diameter		110	.073		2.89	3.81		6.70	8.90
9140	1-1/2" diameter		100	.080		3.28	4.19		7.47	9.85
9150	2" diameter		90	.089		4.15	4.66		8.81	11.50
9160	2-1/2" diameter		65	.123		7	6.45		13.45	17.35
9170	3" diameter	2 Elec	110	.145		7.90	7.60		15.50	20
9180	3-1/2" diameter		100	.160		10.55	8.40		18.95	24
9190	4" diameter		90	.178		11.70	9.30		21	27
9200	5" diameter		70	.229		15.85	12		27.85	35.50
9210	6" diameter		60	.267		21.50	13.95		35.45	44.50
9220	Elbows, 1/2" diameter	1 Elec	50	.160	Ea.	1.24	8.40		9.64	13.90
9225	3/4" diameter		42	.190		1.22	10		11.22	16.30
9230	1" diameter		35	.229		1.85	12		13.85	19.95
9235	1-1/4" diameter		28	.286		2.84	14.95		17.79	25.50
9240	1-1/2" diameter		20	.400		3.67	21		24.67	35.50
9245	2" diameter		16	.500		4.97	26		30.97	44.50
9250	2-1/2" diameter		11	.727		8.95	38		46.95	67
9255	3" diameter		9	.889		14.95	46.50		61.45	86
9260	3-1/2" diameter		7	1.143		19.35	60		79.35	111
9265	4" diameter		6	1.333		24	70		94	131
9270	5" diameter		4	2		39	105		144	200
9275	6" diameter		3	2.667		89.50	140		229.50	310
9312	Couplings, 1/2" diameter		50	.160		.22	8.40		8.62	12.80
9314	3/4" diameter		42	.190		.30	10		10.30	15.30
9316	1" diameter		35	.229		.47	12		12.47	18.40
9318	1-1/4" diameter		28	.286		.65	14.95		15.60	23
9320	1-1/2" diameter		20	.400		.65	21		21.65	32
9322	2" diameter		16	.500		1.37	26		27.37	40.50
9324	2-1/2" diameter		11	.727		1.93	38		39.93	59
9326	3" diameter		9	.889		2.58	46.50		49.08	72.50
9328	3-1/2" diameter		7	1.143		3.26	60		63.26	93
9330	4" diameter		6	1.333		4.01	70		74.01	109
9332	5" diameter		4	2		7.55	105		112.55	165
9334	6" diameter		3	2.667		10.20	140		150.20	220
9335	See note on line 26 05 33.13 9995									
9340	Field bends, 45° & 90°, 1/2" diameter	1 Elec	45	.178	Ea.		9.30		9.30	13.95
9350	3/4" diameter		40	.200			10.50		10.50	15.70
9360	1" diameter		35	.229			12		12	17.90
9370	1-1/4" diameter		32	.250			13.10		13.10	19.60
9380	1-1/2" diameter		27	.296			15.55		15.55	23
9390	2" diameter		20	.400			21		21	31.50
9400	2-1/2" diameter		16	.500			26		26	39
9410	3" diameter		13	.615			32.50		32.50	48.50
9420	3-1/2" diameter		12	.667			35		35	52.50

26 05 33.13 Conduit		Crew	Daily Output	Labor-Hours	Unit	Material	2013 Bare Costs Labor	Equipment	Total	Total Incl O&P
9430	4″ diameter	1 Elec	10	.800	Ea.		42		42	62.50
9440	5″ diameter		9	.889			46.50		46.50	69.50
9450	6″ diameter		8	1			52.50		52.50	78.50
9460	PVC adapters, 1/2″ diameter		50	.160		.35	8.40		8.75	12.95
9470	3/4″ diameter		42	.190		.58	10		10.58	15.60
9480	1″ diameter		38	.211		.77	11.05		11.82	17.35
9490	1-1/4″ diameter		35	.229		1.03	12		13.03	19.05
9500	1-1/2″ diameter		32	.250		1.34	13.10		14.44	21
9510	2″ diameter		27	.296		1.69	15.55		17.24	25
9520	2-1/2″ diameter		23	.348		2.96	18.25		21.21	31
9530	3″ diameter		18	.444		4.11	23.50		27.61	39.50
9540	3-1/2″ diameter		13	.615		6.35	32.50		38.85	55.50
9550	4″ diameter		11	.727		7.25	38		45.25	65
9560	5″ diameter		8	1		15.55	52.50		68.05	95.50
9570	6″ diameter	↓	6	1.333	↓	26	70		96	134
9580	PVC-LB, LR or LL fittings & covers									
9590	1/2″ diameter	1 Elec	20	.400	Ea.	4.22	21		25.22	36
9600	3/4″ diameter		16	.500		6.40	26		32.40	46
9610	1″ diameter		12	.667		6.95	35		41.95	60
9620	1-1/4″ diameter		9	.889		10.80	46.50		57.30	81.50
9630	1-1/2″ diameter		7	1.143		12.85	60		72.85	104
9640	2″ diameter		6	1.333		20.50	70		90.50	128
9650	2-1/2″ diameter		6	1.333		51	70		121	161
9660	3″ diameter		5	1.600		56	84		140	187
9670	3-1/2″ diameter		4	2		73.50	105		178.50	238
9680	4″ diameter	↓	3	2.667	↓	91	140		231	310
9690	PVC-tee fitting & cover									
9700	1/2″	1 Elec	14	.571	Ea.	7.45	30		37.45	53
9710	3/4″		13	.615		8.05	32.50		40.55	57.50
9720	1″		10	.800		9.10	42		51.10	72.50
9730	1-1/4″		9	.889		15.85	46.50		62.35	87
9740	1-1/2″		8	1		18.65	52.50		71.15	99
9750	2″	↓	7	1.143		28.50	60		88.50	121
9760	PVC-reducers, 3/4″ x 1/2″ diameter					1.96			1.96	2.16
9770	1″ x 1/2″ diameter					4.28			4.28	4.71
9780	1″ x 3/4″ diameter					4.62			4.62	5.10
9790	1-1/4″ x 3/4″ diameter					5.70			5.70	6.25
9800	1-1/4″ x 1″ diameter					6			6	6.60
9810	1-1/2″ x 1-1/4″ diameter					6.25			6.25	6.90
9820	2″ x 1-1/4″ diameter					7.55			7.55	8.30
9830	2-1/2″ x 2″ diameter					23.50			23.50	26
9840	3″ x 2″ diameter					24			24	26.50
9850	4″ x 3″ diameter					27.50			27.50	30.50
9860	Cement, quart					23			23	25.50
9870	Gallon					102			102	112
9880	Heat bender, to 6″ diameter				↓	1,575			1,575	1,725
9900	Add to labor for higher elevated installation									
9910	15′ to 20′ high, add						10%			
9920	20′ to 25′ high, add						20%			
9930	25′ to 30′ high, add						25%			
9940	30′ to 35′ high, add						30%			
9950	35′ to 40′ high, add						35%			
9960	Over 40′ high, add						40%			

26 05 33.13 Conduit		Crew	Daily Output	Labor-Hours	Unit	2013 Bare Costs Material	Labor	Equipment	Total	Total Incl O&P
9980	Allowance for conduit fittings 5% min.-20% max.									
9995	Do not include labor when adding couplings to a fitting installation									

26 05 33.14 Conduit

		Crew	Daily Output	Labor-Hours	Unit	Material	Labor	Equipment	Total	Total Incl O&P
0010	**CONDUIT** To 15′ high, includes 11 CLP per 100′ R260533-23									
0200	Electric metallic tubing, 1/2″diameter	1 Elec	435	.018	L.F.	.43	.96		1.39	1.91
0220	3/4″ diameter		253	.032		.77	1.66		2.43	3.32
0240	1″ diameter		207	.039		1.29	2.03		3.32	4.45
0260	1-1/4″ diameter		173	.046		2.30	2.42		4.72	6.15
0280	1-1/2″ diameter		153	.052		3.04	2.74		5.78	7.45
0300	2″ diameter		130	.062		3.84	3.22		7.06	9.05
0320	2-1/2″ diameter		92	.087		8.70	4.56		13.26	16.40
0340	3″ diameter	2 Elec	148	.108		9.95	5.65		15.60	19.45
0360	3-1/2″ diameter		134	.119		12.40	6.25		18.65	23
0380	4″ diameter		114	.140		13.15	7.35		20.50	25.50
0500	Steel rigid galvanized, 1/2″ diameter R260533-24	1 Elec	146	.055		1.84	2.87		4.71	6.30
0520	3/4″ diameter		125	.064		1.95	3.35		5.30	7.15
0540	1″ diameter		93	.086		2.90	4.51		7.41	9.95
0560	1-1/4″ diameter		88	.091		4.12	4.76		8.88	11.70
0580	1-1/2″ diameter		80	.100		4.83	5.25		10.08	13.15
0600	2″ diameter		65	.123		5.75	6.45		12.20	16
0620	2-1/2″ diameter		48	.167		11.45	8.75		20.20	25.50
0640	3″ diameter	2 Elec	64	.250		14.10	13.10		27.20	35
0660	3-1/2″ diameter		60	.267		17.90	13.95		31.85	40.50
0680	4″ diameter		52	.308		19.85	16.10		35.95	46
0700	5″ diameter		50	.320		38	16.75		54.75	67
0720	6″ diameter		48	.333		55.50	17.45		72.95	87
1000	Steel intermediate conduit (IMC), 1/2 diameter R260533-25	1 Elec	155	.052		1.13	2.70		3.83	5.30
1010	3/4″ diameter		130	.062		1.40	3.22		4.62	6.35
1020	1″ diameter		100	.080		2.10	4.19		6.29	8.55
1030	1-1/4″ diameter		93	.086		2.72	4.51		7.23	9.75
1040	1-1/2″ diameter		85	.094		3.46	4.93		8.39	11.20
1050	2″ diameter		70	.114		4.39	6		10.39	13.80
1060	2-1/2″ diameter		53	.151		9.30	7.90		17.20	22
1070	3″ diameter	2 Elec	80	.200		11.95	10.50		22.45	29
1080	3-1/2″ diameter		70	.229		14.60	12		26.60	34
1090	4″ diameter		60	.267		15.65	13.95		29.60	38

26 05 33.15 Conduit Nipples

		Crew	Daily Output	Labor-Hours	Unit	Material	Labor	Equipment	Total	Total Incl O&P
0010	**CONDUIT NIPPLES** With locknuts and bushings									
0100	Aluminum, 1/2″ diameter, close	1 Elec	36	.222	Ea.	7.30	11.65		18.95	25.50
0120	1-1/2″ long		36	.222		7.35	11.65		19	25.50
0140	2″ long		36	.222		7.70	11.65		19.35	26
0160	2-1/2″ long		36	.222		8.35	11.65		20	26.50
0180	3″ long		36	.222		8.60	11.65		20.25	27
0200	3-1/2″ long		36	.222		9	11.65		20.65	27.50
0220	4″ long		36	.222		9.45	11.65		21.10	28
0240	5″ long		36	.222		10.60	11.65		22.25	29
0260	6″ long		36	.222		10.80	11.65		22.45	29.50
0280	8″ long		36	.222		12.95	11.65		24.60	31.50
0300	10″ long		36	.222		14.85	11.65		26.50	34
0320	12″ long		36	.222		16.65	11.65		28.30	35.50
0340	3/4″ diameter, close		32	.250		10	13.10		23.10	30.50
0360	1-1/2″ long		32	.250		10.20	13.10		23.30	31

26 05 33.15 Conduit Nipples		Crew	Daily Output	Labor-Hours	Unit	Material	Labor	Equipment	Total	Total Incl O&P
						2013 Bare Costs				
0380	2" long	1 Elec	32	.250	Ea.	10.40	13.10		23.50	31
0400	2-1/2" long		32	.250		10.75	13.10		23.85	31.50
0420	3" long		32	.250		11.15	13.10		24.25	32
0440	3-1/2" long		32	.250		11.55	13.10		24.65	32.50
0460	4" long		32	.250		11.70	13.10		24.80	32.50
0480	5" long		32	.250		13.10	13.10		26.20	34
0500	6" long		32	.250		14.05	13.10		27.15	35
0520	8" long		32	.250		16.80	13.10		29.90	38
0540	10" long		32	.250		18.75	13.10		31.85	40
0560	12" long		32	.250		26	13.10		39.10	48
0580	1" diameter, close		27	.296		15.20	15.55		30.75	40
0600	2" long		27	.296		15.80	15.55		31.35	40.50
0620	2-1/2" long		27	.296		16.40	15.55		31.95	41
0640	3" long		27	.296		18.75	15.55		34.30	43.50
0660	3-1/2" long		27	.296		18.10	15.55		33.65	43
0680	4" long		27	.296		20.50	15.55		36.05	46
0700	5" long		27	.296		20.50	15.55		36.05	45.50
0720	6" long		27	.296		22.50	15.55		38.05	47.50
0740	8" long		27	.296		25.50	15.55		41.05	51
0760	10" long		27	.296		30	15.55		45.55	56
0780	12" long		27	.296		33.50	15.55		49.05	60
0800	1-1/4" diameter, close		23	.348		21	18.25		39.25	50.50
0820	2" long		23	.348		21.50	18.25		39.75	51
0840	2-1/2" long		23	.348		22	18.25		40.25	52
0860	3" long		23	.348		23	18.25		41.25	53
0880	3-1/2" long		23	.348		25	18.25		43.25	55
0900	4" long		23	.348		26	18.25		44.25	56
0920	5" long		23	.348		28	18.25		46.25	58.50
0940	6" long		23	.348		30.50	18.25		48.75	61
0960	8" long		23	.348		38.50	18.25		56.75	69.50
0980	10" long		23	.348		39.50	18.25		57.75	71
1000	12" long		23	.348		53.50	18.25		71.75	86.50
1020	1-1/2" diameter, close		20	.400		29.50	21		50.50	64
1040	2" long		20	.400		30	21		51	64.50
1060	2-1/2" long		20	.400		31	21		52	65.50
1080	3" long		20	.400		32	21		53	67
1100	3-1/2" long		20	.400		35	21		56	70.50
1120	4" long		20	.400		35.50	21		56.50	70.50
1140	5" long		20	.400		38.50	21		59.50	73.50
1160	6" long		20	.400		40	21		61	75.50
1180	8" long		20	.400		46	21		67	82
1200	10" long		20	.400		55.50	21		76.50	92.50
1220	12" long		20	.400		67	21		88	106
1240	2" diameter, close		18	.444		42	23.50		65.50	81.50
1260	2-1/2" long		18	.444		43.50	23.50		67	83
1280	3" long		18	.444		45	23.50		68.50	84.50
1300	3-1/2" long		18	.444		48	23.50		71.50	87.50
1320	4" long		18	.444		48	23.50		71.50	88
1340	5" long		18	.444		52	23.50		75.50	92.50
1360	6" long		18	.444		55	23.50		78.50	95.50
1380	8" long		18	.444		69.50	23.50		93	112
1400	10" long		18	.444		69	23.50		92.50	111
1420	12" long		18	.444		78	23.50		101.50	121

26 05 33.15 Conduit Nipples		Crew	Daily Output	Labor-Hours	Unit	2013 Bare Costs Material	Labor	Equipment	Total	Total Incl O&P
1440	2-1/2" diameter, close	1 Elec	15	.533	Ea.	91.50	28		119.50	142
1460	3" long		15	.533		92	28		120	143
1480	3-1/2" long		15	.533		95	28		123	146
1500	4" long		15	.533		97	28		125	149
1520	5" long		15	.533		101	28		129	153
1540	6" long		15	.533		104	28		132	157
1560	8" long		15	.533		115	28		143	168
1580	10" long		15	.533		125	28		153	180
1600	12" long		15	.533		135	28		163	190
1620	3" diameter, close		12	.667		108	35		143	171
1640	3" long		12	.667		116	35		151	180
1660	3-1/2" long		12	.667		116	35		151	181
1680	4" long		12	.667		123	35		158	189
1700	5" long		12	.667		123	35		158	188
1720	6" long		12	.667		130	35		165	196
1740	8" long		12	.667		158	35		193	226
1760	10" long		12	.667		157	35		192	225
1780	12" long		12	.667		178	35		213	249
1800	3-1/2" diameter, close		11	.727		200	38		238	277
1820	4" long		11	.727		211	38		249	290
1840	5" long		11	.727		218	38		256	297
1860	6" long		11	.727		226	38		264	305
1880	8" long		11	.727		241	38		279	320
1900	10" long		11	.727		259	38		297	340
1920	12" long		11	.727		293	38		331	375
1940	4" diameter, close		9	.889		246	46.50		292.50	340
1960	4" long		9	.889		268	46.50		314.50	365
1980	5" long		9	.889		276	46.50		322.50	375
2000	6" long		9	.889		275	46.50		321.50	375
2020	8" long		9	.889		310	46.50		356.50	415
2040	10" long		9	.889		325	46.50		371.50	425
2060	12" long		9	.889		335	46.50		381.50	440
2080	5" diameter, close		7	1.143		475	60		535	615
2100	5" long		7	1.143		495	60		555	630
2120	6" long		7	1.143		500	60		560	640
2140	8" long		7	1.143		535	60		595	675
2160	10" long		7	1.143		565	60		625	710
2180	12" long		7	1.143		595	60		655	745
2200	6" diameter, close		6	1.333		745	70		815	925
2220	5" long		6	1.333		780	70		850	965
2240	6" long		6	1.333		790	70		860	975
2260	8" long		6	1.333		840	70		910	1,025
2280	10" long		6	1.333		880	70		950	1,075
2300	12" long		6	1.333		905	70		975	1,100
2320	Rigid galvanized steel, 1/2" diameter, close		32	.250		3.18	13.10		16.28	23
2340	1-1/2" long		32	.250		3.56	13.10		16.66	23.50
2360	2" long		32	.250		3.81	13.10		16.91	24
2380	2-1/2" long		32	.250		4	13.10		17.10	24
2400	3" long		32	.250		4.25	13.10		17.35	24.50
2420	3-1/2" long		32	.250		4.50	13.10		17.60	24.50
2440	4" long		32	.250		4.76	13.10		17.86	25
2460	5" long		32	.250		5.15	13.10		18.25	25.50
2480	6" long		32	.250		5.90	13.10		19	26

26 05 33 – Raceway and Boxes for Electrical Systems

26 05 33.15 Conduit Nipples		Crew	Daily Output	Labor-Hours	Unit	Material	2013 Bare Costs Labor	Equipment	Total	Total Incl O&P
2500	8″ long	1 Elec	32	.250	Ea.	9.15	13.10		22.25	29.50
2520	10″ long		32	.250		10.35	13.10		23.45	31
2540	12″ long		32	.250		11.65	13.10		24.75	32.50
2560	3/4″ diameter, close		27	.296		4.21	15.55		19.76	27.50
2580	2″ long		27	.296		4.65	15.55		20.20	28
2600	2-1/2″ long		27	.296		4.96	15.55		20.51	28.50
2620	3″ long		27	.296		5.20	15.55		20.75	29
2640	3-1/2″ long		27	.296		5.35	15.55		20.90	29
2660	4″ long		27	.296		5.85	15.55		21.40	29.50
2680	5″ long		27	.296		6.40	15.55		21.95	30
2700	6″ long		27	.296		7.15	15.55		22.70	31
2720	8″ long		27	.296		10.55	15.55		26.10	34.50
2740	10″ long		27	.296		12.25	15.55		27.80	36.50
2760	12″ long		27	.296		13.60	15.55		29.15	38
2780	1″ diameter, close		23	.348		6.80	18.25		25.05	35
2800	2″ long		23	.348		7.10	18.25		25.35	35.50
2820	2-1/2″ long		23	.348		7.40	18.25		25.65	35.50
2840	3″ long		23	.348		7.95	18.25		26.20	36
2860	3-1/2″ long		23	.348		8.55	18.25		26.80	37
2880	4″ long		23	.348		8.95	18.25		27.20	37.50
2900	5″ long		23	.348		9.65	18.25		27.90	38
2920	6″ long		23	.348		10.25	18.25		28.50	39
2940	8″ long		23	.348		14.20	18.25		32.45	43
2960	10″ long		23	.348		17.45	18.25		35.70	46.50
2980	12″ long		23	.348		19.35	18.25		37.60	49
3000	1-1/4″ diameter, close		20	.400		9.40	21		30.40	42
3020	2″ long		20	.400		9.75	21		30.75	42.50
3040	3″ long		20	.400		10.65	21		31.65	43
3060	3-1/2″ long		20	.400		11.40	21		32.40	44
3080	4″ long		20	.400		11.80	21		32.80	44.50
3100	5″ long		20	.400		12.85	21		33.85	45.50
3120	6″ long		20	.400		13.85	21		34.85	47
3140	8″ long		20	.400		19.50	21		40.50	53
3160	10″ long		20	.400		23.50	21		44.50	57
3180	12″ long		20	.400		26.50	21		47.50	60.50
3200	1-1/2″ diameter, close		18	.444		13.05	23.50		36.55	49.50
3220	2″ long		18	.444		13.35	23.50		36.85	49.50
3240	2-1/2″ long		18	.444		14.05	23.50		37.55	50.50
3260	3″ long		18	.444		14.55	23.50		38.05	51
3280	3-1/2″ long		18	.444		15.55	23.50		39.05	52
3300	4″ long		18	.444		16.30	23.50		39.80	53
3320	5″ long		18	.444		17.30	23.50		40.80	54
3340	6″ long		18	.444		19.60	23.50		43.10	56.50
3360	8″ long		18	.444		26	23.50		49.50	63.50
3380	10″ long		18	.444		29.50	23.50		53	67.50
3400	12″ long		18	.444		31.50	23.50		55	69.50
3420	2″ diameter, close		16	.500		17.15	26		43.15	58
3440	2-1/2″ long		16	.500		18.30	26		44.30	59
3460	3″ long		16	.500		19.55	26		45.55	60.50
3480	3-1/2″ long		16	.500		21	26		47	62
3500	4″ long		16	.500		22	26		48	63
3520	5″ long		16	.500		24	26		50	65
3540	6″ long		16	.500		25.50	26		51.50	67

26 05 33.15 Conduit Nipples		Crew	Daily Output	Labor-Hours	Unit	Material	Labor	Equipment	Total	Total Incl O&P
						2013 Bare Costs				
3560	8" long	1 Elec	16	.500	Ea.	32.50	26		58.50	74.50
3580	10" long		16	.500		36.50	26		62.50	79
3600	12" long		16	.500		40	26		66	83
3620	2-1/2" diameter, close		13	.615		47.50	32.50		80	101
3640	3" long		13	.615		47	32.50		79.50	101
3660	3-1/2" long		13	.615		51	32.50		83.50	105
3680	4" long		13	.615		52	32.50		84.50	106
3700	5" long		13	.615		56.50	32.50		89	111
3720	6" long		13	.615		60	32.50		92.50	115
3740	8" long		13	.615		70	32.50		102.50	125
3760	10" long		13	.615		76.50	32.50		109	133
3780	12" long		13	.615		84	32.50		116.50	141
3800	3" diameter, close		12	.667		58	35		93	117
3820	3" long		12	.667		47.50	35		82.50	105
3900	3-1/2" long		12	.667		62.50	35		97.50	121
3920	4" long		12	.667		64.50	35		99.50	124
3940	5" long		12	.667		69	35		104	129
3960	6" long		12	.667		74	35		109	134
3980	8" long		12	.667		85	35		120	146
4000	10" long		12	.667		93.50	35		128.50	156
4020	12" long		12	.667		106	35		141	170
4040	3-1/2" diameter, close		10	.800		96	42		138	169
4060	4" long		10	.800		103	42		145	177
4080	5" long		10	.800		108	42		150	182
4100	6" long		10	.800		114	42		156	188
4120	8" long		10	.800		125	42		167	200
4140	10" long		10	.800		136	42		178	213
4160	12" long		10	.800		147	42		189	225
4180	4" diameter, close		8	1		109	52.50		161.50	199
4200	4" long		8	1		116	52.50		168.50	207
4220	5" long		8	1		123	52.50		175.50	214
4240	6" long		8	1		128	52.50		180.50	220
4260	8" long		8	1		140	52.50		192.50	233
4280	10" long		8	1		155	52.50		207.50	249
4300	12" long		8	1		170	52.50		222.50	266
4320	5" diameter, close		6	1.333		219	70		289	345
4340	5" long		6	1.333		246	70		316	375
4360	6" long		6	1.333		253	70		323	385
4380	8" long		6	1.333		271	70		341	405
4400	10" long		6	1.333		290	70		360	425
4420	12" long		6	1.333		320	70		390	455
4440	6" diameter, close		5	1.600		425	84		509	590
4460	5" long		5	1.600		455	84		539	625
4480	6" long		5	1.600		465	84		549	635
4500	8" long		5	1.600		485	84		569	655
4520	10" long		5	1.600		520	84		604	695
4540	12" long		5	1.600		535	84		619	715
4560	Plastic coated, 40 mil thick, 1/2" diameter, 2" long		32	.250		16.40	13.10		29.50	37.50
4580	2-1/2" long		32	.250		19	13.10		32.10	40.50
4600	3" long		32	.250		19.25	13.10		32.35	40.50
4680	3-1/2" long		32	.250		22	13.10		35.10	43.50
4700	4" long		32	.250		21.50	13.10		34.60	43
4720	5" long		32	.250		21.50	13.10		34.60	43.50

26 05 33 – Raceway and Boxes for Electrical Systems

26 05 33.15 Conduit Nipples		Crew	Daily Output	Labor-Hours	Unit	Material	Labor	Equipment	Total	Total Incl O&P
						2013 Bare Costs				
4740	6″ long	1 Elec	32	.250	Ea.	23	13.10		36.10	44.50
4760	8″ long		32	.250		22	13.10		35.10	43.50
4780	10″ long		32	.250		22.50	13.10		35.60	44
4800	12″ long		32	.250		23.50	13.10		36.60	45
4820	3/4″ diameter, 2″ long		26	.308		17.50	16.10		33.60	43.50
4840	2-1/2″ long		26	.308		19.90	16.10		36	46
4860	3″ long		26	.308		19.35	16.10		35.45	45.50
4880	3-1/2″ long		26	.308		22.50	16.10		38.60	49
4900	4″ long		26	.308		22.50	16.10		38.60	48.50
4920	5″ long		26	.308		23.50	16.10		39.60	50
4940	6″ long		26	.308		22.50	16.10		38.60	48.50
4960	8″ long		26	.308		23	16.10		39.10	49.50
4980	10″ long		26	.308		24	16.10		40.10	50.50
5000	12″ long		26	.308		27	16.10		43.10	54
5020	1″ diameter, 2″ long		22	.364		19.35	19.05		38.40	50
5040	2-1/2″ long		22	.364		23	19.05		42.05	54
5060	3″ long		22	.364		22	19.05		41.05	53
5080	3-1/2″ long		22	.364		24	19.05		43.05	54.50
5100	4″ long		22	.364		26	19.05		45.05	57
5120	5″ long		22	.364		26.50	19.05		45.55	57.50
5140	6″ long		22	.364		25	19.05		44.05	56
5160	8″ long		22	.364		26	19.05		45.05	57
5180	10″ long		22	.364		27.50	19.05		46.55	58.50
5200	12″ long		22	.364		32.50	19.05		51.55	64
5220	1-1/4″ diameter, 2″ long		18	.444		24.50	23.50		48	62
5240	2-1/2″ long		18	.444		27	23.50		50.50	64.50
5260	3″ long		18	.444		28	23.50		51.50	66
5280	3-1/2″ long		18	.444		28	23.50		51.50	65.50
5300	4″ long		18	.444		29.50	23.50		53	67.50
5320	5″ long		18	.444		30	23.50		53.50	68
5340	6″ long		18	.444		33.50	23.50		57	71.50
5360	8″ long		18	.444		31.50	23.50		55	69.50
5380	10″ long		18	.444		35	23.50		58.50	73.50
5400	12″ long		18	.444		44.50	23.50		68	84
5420	1-1/2″ diameter, 2″ long		16	.500		28	26		54	70
5440	2-1/2″ long		16	.500		30.50	26		56.50	72.50
5460	3″ long		16	.500		30.50	26		56.50	72.50
5480	3-1/2″ long		16	.500		32	26		58	74
5500	4″ long		16	.500		35	26		61	77.50
5520	5″ long		16	.500		36.50	26		62.50	79
5540	6″ long		16	.500		38	26		64	80.50
5560	8″ long		16	.500		39	26		65	82
5580	10″ long		16	.500		46.50	26		72.50	90
5600	12″ long		16	.500		58	26		84	103
5620	2″ diameter, 2-1/2″ long		14	.571		38	30		68	86.50
5640	3″ long		14	.571		37.50	30		67.50	86.50
5660	3-1/2″ long		14	.571		39.50	30		69.50	88.50
5680	4″ long		14	.571		42	30		72	91.50
5700	5″ long		14	.571		45	30		75	94.50
5720	6″ long		14	.571		47	30		77	96.50
5740	8″ long		14	.571		50.50	30		80.50	101
5760	10″ long		14	.571		61	30		91	112
5780	12″ long		14	.571		72	30		102	124

26 05 33 – Raceway and Boxes for Electrical Systems

26 05 33.15 Conduit Nipples		Crew	Daily Output	Labor-Hours	Unit	Material	Labor	Equipment	Total	Total Incl O&P
						2013 Bare Costs				
5800	2-1/2" diameter, 3-1/2" long	1 Elec	12	.667	Ea.	74.50	35		109.50	135
5820	4" long		12	.667		76.50	35		111.50	137
5840	5" long		12	.667		88	35		123	150
5860	6" long		12	.667		95	35		130	157
5880	8" long		12	.667		101	35		136	164
5900	10" long		12	.667		110	35		145	174
5920	12" long		12	.667		124	35		159	190
5940	3" diameter, 3-1/2" long		11	.727		91	38		129	157
5960	4" long		11	.727		93	38		131	159
5980	5" long		11	.727		99.50	38		137.50	166
6000	6" long		11	.727		108	38		146	176
6020	8" long		11	.727		120	38		158	189
6040	10" long		11	.727		137	38		175	208
6060	12" long		11	.727		157	38		195	230
6080	3-1/2" diameter, 4" long		9	.889		140	46.50		186.50	224
6100	5" long		9	.889		141	46.50		187.50	225
6120	6" long		9	.889		156	46.50		202.50	241
6140	8" long		9	.889		168	46.50		214.50	255
6160	10" long		9	.889		185	46.50		231.50	273
6180	12" long		9	.889		201	46.50		247.50	291
6200	4" diameter, 4" long		7.50	1.067		153	56		209	252
6220	5" long		7.50	1.067		157	56		213	256
6240	6" long		7.50	1.067		170	56		226	271
6260	8" long		7.50	1.067		190	56		246	292
6280	10" long		7.50	1.067		224	56		280	330
6300	12" long		7.50	1.067		234	56		290	340
6320	5" diameter, 5" long		5.50	1.455		281	76		357	425
6340	6" long		5.50	1.455		300	76		376	445
6360	8" long		5.50	1.455		315	76		391	460
6380	10" long		5.50	1.455		340	76		416	490
6400	12" long		5.50	1.455		345	76		421	495
6420	6" diameter, 5" long		4.50	1.778		500	93		593	690
6440	6" long		4.50	1.778		520	93		613	710
6460	8" long		4.50	1.778		540	93		633	730
6480	10" long		4.50	1.778		565	93		658	760
6500	12" long		4.50	1.778		595	93		688	795

26 05 33.16 Outlet Boxes		Crew	Daily Output	Labor-Hours	Unit	Material	Labor	Equipment	Total	Total Incl O&P
0010	**OUTLET BOXES** R260533-65									
0020	Pressed steel, octagon, 4"	1 Elec	20	.400	Ea.	3.61	21		24.61	35.50
0040	For Romex or BX		20	.400		5.40	21		26.40	37.50
0050	For Romex or BX, with bracket		20	.400		8.55	21		29.55	41
0060	Covers, blank		64	.125		1.51	6.55		8.06	11.45
0100	Extension rings		40	.200		6	10.50		16.50	22.50
0150	Square, 4"		20	.400		2.46	21		23.46	34
0160	For Romex or BX		20	.400		9.65	21		30.65	42
0170	For Romex or BX, with bracket		20	.400		10.50	21		31.50	43
0200	Extension rings		40	.200		6.05	10.50		16.55	22.50
0220	2-1/8" deep, 1" KO		20	.400		8.05	21		29.05	40.50
0250	Covers, blank		64	.125		1.64	6.55		8.19	11.60
0260	Raised device		64	.125		3.08	6.55		9.63	13.20
0300	Plaster rings		64	.125		3.32	6.55		9.87	13.45
0350	Square, 4-11/16"		20	.400		8.45	21		29.45	41

26 05 33.16 Outlet Boxes		Crew	Daily Output	Labor-Hours	Unit	2013 Bare Costs Material	Labor	Equipment	Total	Total Incl O&P
0370	2-1/8" deep, 3/4" to 1-1/4" KO	1 Elec	20	.400	Ea.	10.20	21		31.20	42.50
0400	Extension rings		40	.200		13.60	10.50		24.10	30.50
0450	Covers, blank		53	.151		3.06	7.90		10.96	15.20
0460	Raised device		53	.151		9.15	7.90		17.05	22
0500	Plaster rings		53	.151		9.05	7.90		16.95	22
0550	Handy box		27	.296		3.99	15.55		19.54	27.50
0560	Covers, device		64	.125		1.40	6.55		7.95	11.35
0600	Extension rings		54	.148		4.76	7.75		12.51	16.85
0650	Switchbox		27	.296		5.80	15.55		21.35	29.50
0660	Romex or BX		27	.296		8.30	15.55		23.85	32
0670	with bracket		27	.296		9.55	15.55		25.10	33.50
0680	Partition, metal		27	.296		3.62	15.55		19.17	27
0700	Masonry, 1 gang, 2-1/2" deep		27	.296		11.05	15.55		26.60	35
0710	3-1/2" deep		27	.296		11.05	15.55		26.60	35
0750	2 gang, 2-1/2" deep		20	.400		20.50	21		41.50	54.50
0760	3-1/2" deep		20	.400		16.65	21		37.65	50
0800	3 gang, 2-1/2" deep		13	.615		25	32.50		57.50	76
0850	4 gang, 2-1/2" deep		10	.800		28.50	42		70.50	94
0860	5 gang, 2-1/2" deep		9	.889		30.50	46.50		77	103
0870	6 gang, 2-1/2" deep		8	1		62.50	52.50		115	147
0880	Masonry thru-the-wall, 1 gang, 4" block		16	.500		41	26		67	84
0890	6" block		16	.500		61.50	26		87.50	107
0900	8" block		16	.500		69.50	26		95.50	116
0920	2 gang, 6" block		16	.500		92.50	26		118.50	141
0940	Bar hanger with 3/8" stud, for wood and masonry boxes		53	.151		7.75	7.90		15.65	20.50
0950	Concrete, set flush, 4" deep		20	.400		17	21		38	50
1000	Plate with 3/8" stud		80	.100		11.20	5.25		16.45	20
1100	Concrete, floor, 1 gang		5.30	1.509		105	79		184	233
1150	2 gang		4	2		159	105		264	330
1200	3 gang		2.70	2.963		225	155		380	480
1250	For duplex receptacle, pedestal mounted, add		24	.333		110	17.45		127.45	147
1270	Flush mounted, add		27	.296		28	15.55		43.55	54
1300	For telephone, pedestal mounted, add		30	.267		147	13.95		160.95	182
1350	Carpet flange, 1 gang		53	.151		49.50	7.90		57.40	66.50
1400	Cast, 1 gang, FS (2" deep), 1/2" hub		12	.667		18	35		53	72.50
1410	3/4" hub		12	.667		18.85	35		53.85	73
1420	FD (2-11/16" deep), 1/2" hub		12	.667		18	35		53	72.50
1430	3/4" hub		12	.667		19.55	35		54.55	74
1450	2 gang, FS, 1/2" hub		10	.800		32	42		74	98
1460	3/4" hub		10	.800		34.50	42		76.50	100
1470	FD, 1/2" hub		10	.800		40	42		82	107
1480	3/4" hub		10	.800		38	42		80	105
1500	3 gang, FS, 3/4" hub		9	.889		56	46.50		102.50	131
1510	Switch cover, 1 gang, FS		64	.125		4.72	6.55		11.27	15
1520	2 gang		53	.151		7.20	7.90		15.10	19.75
1530	Duplex receptacle cover, 1 gang, FS		64	.125		4.81	6.55		11.36	15.10
1540	2 gang, FS		53	.151		7.30	7.90		15.20	19.90
1542	Weatherproof blank cover, 1 gang		64	.125		1.93	6.55		8.48	11.90
1544	2 gang		53	.151		4.11	7.90		12.01	16.35
1550	Weatherproof switch cover, 1 gang		64	.125		8.15	6.55		14.70	18.80
1554	2 gang		53	.151		17.30	7.90		25.20	31
1600	Weatherproof receptacle cover, 1 gang		64	.125		6.65	6.55		13.20	17.15
1604	2 gang		53	.151		17.25	7.90		25.15	31

26 05 33 – Raceway and Boxes for Electrical Systems

26 05 33.16 Outlet Boxes

	26 05 33.16 Outlet Boxes	Crew	Daily Output	Labor-Hours	Unit	Material	Labor	Equipment	Total	Total Incl O&P
						2013 Bare Costs				
1620	Weatherproof receptacle cover, tamper resistant, 1 gang	1 Elec	58	.138	Ea.	13.50	7.25		20.75	25.50
1624	2 gang		48	.167		27	8.75		35.75	42.50
1750	FSC, 1 gang, 1/2" hub		11	.727		19.90	38		57.90	79
1760	3/4" hub		11	.727		22	38		60	81
1770	2 gang, 1/2" hub		9	.889		35	46.50		81.50	108
1780	3/4" hub		9	.889		38	46.50		84.50	111
1790	FDC, 1 gang, 1/2" hub		11	.727		22.50	38		60.50	82
1800	3/4" hub		11	.727		25	38		63	84.50
1810	2 gang, 1/2" hub		9	.889		48.50	46.50		95	123
1820	3/4" hub		9	.889		52.50	46.50		99	127
1850	Weatherproof in-use cover, 1 gang		64	.125		19.60	6.55		26.15	31.50
1870	2 gang		53	.151		36.50	7.90		44.40	52
2000	Poke-thru fitting, fire rated, for 3-3/4" floor		6.80	1.176		121	61.50		182.50	225
2040	For 7" floor		6.80	1.176		124	61.50		185.50	229
2100	Pedestal, 15 amp, duplex receptacle & blank plate		5.25	1.524		135	80		215	267
2120	Duplex receptacle and telephone plate		5.25	1.524		135	80		215	267
2140	Pedestal, 20 amp, duplex recept. & phone plate		5	1.600		136	84		220	274
2160	Telephone plate, both sides		5.25	1.524		128	80		208	260
2200	Abandonment plate	↓	32	.250	↓	36	13.10		49.10	59

26 05 33.17 Outlet Boxes, Plastic

	26 05 33.17 Outlet Boxes, Plastic	Crew	Daily Output	Labor-Hours	Unit	Material	Labor	Equipment	Total	Total Incl O&P
0010	**OUTLET BOXES, PLASTIC**									
0050	4" diameter, round with 2 mounting nails	1 Elec	25	.320	Ea.	3.04	16.75		19.79	28.50
0100	Bar hanger mounted		25	.320		6.25	16.75		23	32
0200	4", square with 2 mounting nails		25	.320		5.15	16.75		21.90	30.50
0300	Plaster ring		64	.125		2.37	6.55		8.92	12.40
0400	Switch box with 2 mounting nails, 1 gang		30	.267		4.31	13.95		18.26	25.50
0500	2 gang		25	.320		4.34	16.75		21.09	30
0600	3 gang		20	.400		6.80	21		27.80	39
0700	Old work box		30	.267		7.90	13.95		21.85	29.50
1400	PVC, FSS, 1 gang, 1/2" hub		14	.571		22.50	30		52.50	70
1410	3/4" hub		14	.571		18.45	30		48.45	65.50
1420	FD, 1 gang for variable terminations		14	.571		15.20	30		45.20	61.50
1450	FS, 2 gang for variable terminations		12	.667		17.95	35		52.95	72.50
1480	Weatherproof blank cover, FS, 1 gang		64	.125		6.20	6.55		12.75	16.60
1500	2 gang		53	.151		6.80	7.90		14.70	19.30
1510	Weatherproof switch cover, FS, 1 gang		64	.125		11.40	6.55		17.95	22.50
1520	2 gang		53	.151		20.50	7.90		28.40	34.50
1530	Weatherproof duplex receptacle cover, FS, 1 gang		64	.125		17.10	6.55		23.65	28.50
1540	2 gang		53	.151		19.35	7.90		27.25	33.50
1750	FSC, 1 gang, 1/2" hub		13	.615		15	32.50		47.50	65
1760	3/4" hub		13	.615		16.15	32.50		48.65	66.50
1770	FSC, 2 gang, 1/2" hub		11	.727		23	38		61	82.50
1780	3/4" hub		11	.727		23	38		61	82
1790	FDC, 1 gang, 1/2" hub		13	.615		15.45	32.50		47.95	65.50
1800	3/4" hub		13	.615		16.90	32.50		49.40	67
1810	Weatherproof, T box w/3 holes		14	.571		13.30	30		43.30	59.50
1820	4" diameter round w/5 holes		14	.571		13.15	30		43.15	59.50
1850	In-use cover, 1 gang		64	.125		8.35	6.55		14.90	19
1870	2 gang	↓	53	.151	↓	12.25	7.90		20.15	25.50

26 05 33.18 Pull Boxes

	26 05 33.18 Pull Boxes	Crew	Daily Output	Labor-Hours	Unit	Material	Labor	Equipment	Total	Total Incl O&P
0010	**PULL BOXES** R260533-70									
0100	Sheet metal, pull box, NEMA 1, type SC, 6" W x 6" H x 4" D	1 Elec	8	1	Ea.	12.70	52.50		65.20	92.50

26 05 33 – Raceway and Boxes for Electrical Systems

26 05 33.18 Pull Boxes		Crew	Daily Output	Labor-Hours	Unit	2013 Bare Costs Material	Labor	Equipment	Total	Total Incl O&P
0180	8" W x 6" H x 4" D	1 Elec	8	1	Ea.	17.85	52.50		70.35	98
0200	8" W x 8" H x 4" D		8	1		17.95	52.50		70.45	98.50
0210	10" W x 10" H x 4" D		7	1.143		23.50	60		83.50	116
0220	12" W x 12" H x 4" D		6.50	1.231		30.50	64.50		95	130
0230	15" W x 15" H x 4" D		5.20	1.538		42.50	80.50		123	168
0240	18" W x 18" H x 4" D		4.40	1.818		38.50	95.50		134	186
0250	6" W x 6" H x 6" D		8	1		10.90	52.50		63.40	90.50
0260	8" W x 8" H x 6" D		7.50	1.067		14.65	56		70.65	99.50
0270	10" W x 10" H x 6" D		5.50	1.455		19.15	76		95.15	135
0300	10" W x 12" H x 6" D		5.30	1.509		29	79		108	150
0310	12" W x 12" H x 6" D		5.20	1.538		24	80.50		104.50	148
0320	15" W x 15" H x 6" D		4.60	1.739		42.50	91		133.50	183
0330	18" W x 18" H x 6" D		4.20	1.905		43	100		143	197
0340	24" W x 24" H x 6" D		3.20	2.500		86.50	131		217.50	291
0350	12" W x 12" H x 8" D		5	1.600		27.50	84		111.50	155
0360	15" W x 15" H x 8" D		4.50	1.778		49.50	93		142.50	194
0370	18" W x 18" H x 8" D		4	2		71.50	105		176.50	236
0380	24" W x 18" H x 6" D		3.70	2.162		79.50	113		192.50	258
0400	16" W x 20" H x 8" D		4	2		102	105		207	269
0500	20" W x 24" H x 8" D		3.20	2.500		106	131		237	310
0510	24" W x 24" H x 8" D		3	2.667		132	140		272	355
0600	24" W x 36" H x 8" D		2.70	2.963		156	155		311	405
0610	30" W x 30" H x 8" D		2.70	2.963		179	155		334	430
0620	36" W x 36" H x 8" D		2	4		203	210		413	540
0630	24" W x 24" H x 10" D		2.50	3.200		139	168		307	405
0650	Pull box, hinged , NEMA 1, 6" W x 6" H x 4" D		8	1		10.55	52.50		63.05	90
0660	8" W x 8" H x 4" D		8	1		14.90	52.50		67.40	95
0670	10" W x 10" H x 4" D		7	1.143		20.50	60		80.50	112
0680	12" W x 12" H x 4" D		6	1.333		25	70		95	133
0690	15" W x 15" H x 4" D		5.20	1.538		35.50	80.50		116	160
0700	18" W x 18" H x 4" D		4.40	1.818		40.50	95.50		136	188
0710	6" W x 6" H x 6" D		8	1		13.70	52.50		66.20	93.50
0720	8" W x 8" H x 6" D		7.50	1.067		18.20	56		74.20	104
0730	10" W x 10" H x 6" D		5.50	1.455		24.50	76		100.50	141
0740	12" W x 12" H x 6" D		5.20	1.538		29	80.50		109.50	153
0800	12" W x 16" H x 6" D		4.70	1.702		37.50	89		126.50	174
0810	15" W x 15" H x 6" D		4.60	1.739		37.50	91		128.50	178
0820	18" W x 18" H x 6" D		4.20	1.905		53	100		153	207
1000	20" W x 20" H x 6" D		3.60	2.222		65	116		181	246
1010	24" W x 24" H x 6" D		3.20	2.500		104	131		235	310
1020	12" W x 12" H x 8" D		5	1.600		61	84		145	193
1030	15" W x 15" H x 8" D		4.50	1.778		79	93		172	226
1040	18" W x 18" H x 8" D		4	2		111	105		216	279
1200	20" W x 20" H x 8" D		3.20	2.500		130	131		261	340
1210	24" W x 24" H x 8" D		3	2.667		148	140		288	370
1220	30" W x 30" H x 8" D		2.70	2.963		211	155		366	465
1400	24" W x 36" H x 8" D		2.70	2.963		208	155		363	460
1600	24" W x 42" H x 8" D		2	4		315	210		525	660
1610	36" W x 36" H x 8" D		2	4		310	210		520	660
2100	Pull box, NEMA 3R, type SC, raintight & weatherproof									
2150	6" L x 6" W x 6" D	1 Elec	10	.800	Ea.	22	42		64	87
2200	8" L x 6" W x 6" D		8	1		32.50	52.50		85	115
2250	10" L x 6" W x 6" D		7	1.143		43.50	60		103.50	137

26 05 33 – Raceway and Boxes for Electrical Systems

26 05 33.18 Pull Boxes		Crew	Daily Output	Labor-Hours	Unit	2013 Bare Costs Material	Labor	Equipment	Total	Total Incl O&P
2300	12" L x 12" W x 6" D	1 Elec	5	1.600	Ea.	52	84		136	183
2350	16" L x 16" W x 6" D		4.50	1.778		122	93		215	273
2400	20" L x 20" W x 6" D		4	2		167	105		272	340
2450	24" L x 18" W x 8" D		3	2.667		147	140		287	370
2500	24" L x 24" W x 10" D		2.50	3.200		189	168		357	460
2550	30" L x 24" W x 12" D		2	4		445	210		655	805
2600	36" L x 36" W x 12" D		1.50	5.333		500	279		779	970
2800	Cast iron, pull boxes for surface mounting R260533-75									
3000	NEMA 4, watertight & dust tight									
3050	6" L x 6" W x 6" D	1 Elec	4	2	Ea.	239	105		344	420
3100	8" L x 6" W x 6" D		3.20	2.500		315	131		446	540
3150	10" L x 6" W x 6" D		2.50	3.200		375	168		543	660
3200	12" L x 12" W x 6" D		2.30	3.478		680	182		862	1,025
3250	16" L x 16" W x 6" D		1.30	6.154		1,325	320		1,645	1,950
3300	20" L x 20" W x 6" D		.80	10		2,475	525		3,000	3,500
3350	24" L x 18" W x 8" D		.70	11.429		3,050	600		3,650	4,275
3400	24" L x 24" W x 10" D		.50	16		4,700	840		5,540	6,400
3450	30" L x 24" W x 12" D		.40	20		7,850	1,050		8,900	10,200
3500	36" L x 36" W x 12" D		.20	40		8,275	2,100		10,375	12,200
3510	NEMA 4 clamp cover, 6" L x 6" W x 4" D		4	2		232	105		337	410
3520	8" L x 6" W x 4" D		4	2		270	105		375	455
4000	NEMA 7, explosionproof									
4050	6" L x 6" W x 6" D	1 Elec	2	4	Ea.	730	210		940	1,125
4100	8" L x 6" W x 6" D		1.80	4.444		880	233		1,113	1,325
4150	10" L x 6" W x 6" D		1.60	5		1,200	262		1,462	1,725
4200	12" L x 12" W x 6" D		1	8		2,050	420		2,470	2,875
4250	16" L x 14" W x 6" D		.60	13.333		2,875	700		3,575	4,200
4300	18" L x 18" W x 8" D		.50	16		5,325	840		6,165	7,125
4350	24" L x 18" W x 8" D		.40	20		6,775	1,050		7,825	9,050
4400	24" L x 24" W x 10" D		.30	26.667		9,250	1,400		10,650	12,300
4450	30" L x 24" W x 12" D		.20	40		13,400	2,100		15,500	17,900
5000	NEMA 9, dust tight 6" L x 6" W x 6" D		3.20	2.500		310	131		441	540
5050	8" L x 6" W x 6" D		2.70	2.963		375	155		530	640
5100	10" L x 6" W x 6" D		2	4		500	210		710	865
5150	12" L x 12" W x 6" D		1.60	5		985	262		1,247	1,475
5200	16" L x 16" W x 6" D		1	8		1,700	420		2,120	2,500
5250	18" L x 18" W x 8" D		.70	11.429		2,675	600		3,275	3,850
5300	24" L x 18" W x 8" D		.60	13.333		3,775	700		4,475	5,200
5350	24" L x 24" W x 10" D		.40	20		4,850	1,050		5,900	6,925
5400	30" L x 24" W x 12" D		.30	26.667		7,650	1,400		9,050	10,500
6000	J.I.C. wiring boxes, NEMA 12, dust tight & drip tight									
6050	6" L x 8" W x 4" D	1 Elec	10	.800	Ea.	55.50	42		97.50	124
6100	8" L x 10" W x 4" D		8	1		69	52.50		121.50	155
6150	12" L x 14" W x 6" D		5.30	1.509		117	79		196	247
6200	14" L x 16" W x 6" D		4.70	1.702		137	89		226	284
6250	16" L x 20" W x 6" D		4.40	1.818		264	95.50		359.50	435
6300	24" L x 30" W x 6" D		3.20	2.500		390	131		521	620
6350	24" L x 30" W x 8" D		2.90	2.759		415	145		560	670
6400	24" L x 36" W x 8" D		2.70	2.963		460	155		615	735
6450	24" L x 42" W x 8" D		2.30	3.478		505	182		687	830
6500	24" L x 48" W x 8" D		2	4		550	210		760	920

26 05 33 – Raceway and Boxes for Electrical Systems

26 05 33.23 Wireway		Crew	Daily Output	Labor-Hours	Unit	2013 Bare Costs Material	Labor	Equipment	Total	Total Incl O&P
0010	**WIREWAY** to 15′ high R260533-60									
0020	For higher elevations, see Section 26 05 36.40									
0100	NEMA 1, Screw cover w/fittings and supports, 2-1/2″ x 2-1/2″	1 Elec	45	.178	L.F.	10.40	9.30		19.70	25.50
0200	4″ x 4″	″	40	.200		11.50	10.50		22	28.50
0400	6″ x 6″	2 Elec	60	.267		19.60	13.95		33.55	42.50
0600	8″ x 8″		40	.400		33	21		54	68
0620	10″ x 10″		30	.533		48	28		76	94.50
0640	12″ x 12″		20	.800		61	42		103	130
0800	Elbows, 90°, 2-1/2″	1 Elec	24	.333	Ea.	32.50	17.45		49.95	61.50
1000	4″		20	.400		36.50	21		57.50	72
1200	6″		18	.444		41.50	23.50		65	80.50
1400	8″		16	.500		66.50	26		92.50	113
1420	10″		12	.667		94	35		129	156
1440	12″		10	.800		121	42		163	196
1500	Elbows, 45°, 2-1/2″		24	.333		32.50	17.45		49.95	61.50
1510	4″		20	.400		39.50	21		60.50	75
1520	6″		18	.444		41.50	23.50		65	80.50
1530	8″		16	.500		66.50	26		92.50	113
1540	10″		12	.667		95	35		130	157
1550	12″		10	.800		169	42		211	248
1600	″T″ box, 2-1/2″		18	.444		37	23.50		60.50	76
1800	4″		16	.500		45	26		71	88.50
2000	6″		14	.571		50.50	30		80.50	101
2200	8″		12	.667		93	35		128	155
2220	10″		10	.800		124	42		166	199
2240	12″		8	1		175	52.50		227.50	272
2300	Cross, 2-1/2″		16	.500		41.50	26		67.50	84.50
2310	4″		14	.571		50.50	30		80.50	101
2320	6″		12	.667		62.50	35		97.50	121
2400	Panel adapter, 2-1/2″		24	.333		12.95	17.45		30.40	40
2600	4″		20	.400		15.95	21		36.95	49
2800	6″		18	.444		19	23.50		42.50	56
3000	8″		16	.500		27	26		53	69
3020	10″		14	.571		43.50	30		73.50	92.50
3040	12″		12	.667		55.50	35		90.50	114
3200	Reducer, 4″ to 2-1/2″		24	.333		20	17.45		37.45	48
3400	6″ to 4″		20	.400		38.50	21		59.50	74
3600	8″ to 6″		18	.444		45	23.50		68.50	84.50
3620	10″ to 8″		16	.500		51.50	26		77.50	95.50
3640	12″ to 10″		14	.571		64.50	30		94.50	116
3780	End cap, 2-1/2″		24	.333		4.62	17.45		22.07	31
3800	4″		20	.400		5.70	21		26.70	38
4000	6″		18	.444		6.85	23.50		30.35	42.50
4200	8″		16	.500		9.30	26		35.30	49.50
4220	10″		14	.571		15.10	30		45.10	61.50
4240	12″		12	.667		21	35		56	75.50
4300	U-connector, 2-1/2″		200	.040		4.62	2.10		6.72	8.25
4320	4″		200	.040		5.70	2.10		7.80	9.40
4340	6″		180	.044		6.85	2.33		9.18	11
4360	8″		170	.047		13.85	2.47		16.32	18.95
4380	10″		150	.053		18.75	2.79		21.54	24.50
4400	12″		130	.062		28	3.22		31.22	35.50

26 05 33.23 Wireway		Crew	Daily Output	Labor-Hours	Unit	2013 Bare Costs Material	Labor	Equipment	Total	Total Incl O&P
4420	Hanger, 2-1/2″	1 Elec	100	.080	Ea.	10.20	4.19		14.39	17.50
4430	4″		100	.080		13.05	4.19		17.24	20.50
4440	6″		80	.100		18.85	5.25		24.10	28.50
4450	8″		65	.123		24.50	6.45		30.95	36.50
4460	10″		50	.160		44.50	8.40		52.90	61.50
4470	12″		40	.200		65	10.50		75.50	87
4475	NEMA 3R, Screw cover w/fittings and supports, 4″ x 4″		36	.222	L.F.	23	11.65		34.65	42.50
4480	6″ x 6″	2 Elec	55	.291		29	15.25		44.25	54.50
4485	8″ x 8″		36	.444		47	23.50		70.50	87
4490	12″ x 12″		18	.889		65.50	46.50		112	142
4500	Hinged cover, with fittings and supports, 2-1/2″ x 2-1/2″	1 Elec	60	.133		11.45	7		18.45	23
4520	4″ x 4″	″	45	.178		14.50	9.30		23.80	30
4540	6″ x 6″	2 Elec	80	.200		23.50	10.50		34	41
4560	8″ x 8″		60	.267		37.50	13.95		51.45	62.50
4580	10″ x 10″		50	.320		48	16.75		64.75	78
4600	12″ x 12″		24	.667		69	35		104	129
4700	Elbows 90°, 2-1/2″ x 2-1/2″	1 Elec	32	.250	Ea.	40.50	13.10		53.60	64
4720	4″		27	.296		46.50	15.55		62.05	74.50
4730	6″		23	.348		54.50	18.25		72.75	87.50
4740	8″		18	.444		85	23.50		108.50	129
4750	10″		14	.571		95	30		125	149
4760	12″		12	.667		151	35		186	219
4800	Tee box, hinged cover, 2-1/2″ x 2-1/2″		23	.348		47	18.25		65.25	79
4810	4″		20	.400		56.50	21		77.50	93.50
4820	6″		18	.444		63.50	23.50		87	105
4830	8″		16	.500		118	26		144	169
4840	10″		12	.667		135	35		170	201
4860	12″		10	.800		226	42		268	310
4880	Cross box, hinged cover, 2-1/2″ x 2-1/2″		18	.444		51	23.50		74.50	91.50
4900	4″		16	.500		64	26		90	109
4920	6″		13	.615		79.50	32.50		112	136
4940	8″		11	.727		116	38		154	185
4960	10″		10	.800		200	42		242	283
4980	12″		9	.889		218	46.50		264.50	310
5000	NEMA 12, Hinged cover, 2-1/2″ x 2-1/2″		40	.200	L.F.	38.50	10.50		49	58
5020	4″ x 4″		35	.229		45	12		57	67.50
5040	6″ x 6″	2 Elec	60	.267		64.50	13.95		78.45	92
5060	8″ x 8″	″	50	.320		88	16.75		104.75	122
5120	Elbows 90°, flanged, 2-1/2″ x 2-1/2″	1 Elec	23	.348	Ea.	75.50	18.25		93.75	111
5140	4″		20	.400		95	21		116	136
5160	6″		18	.444		121	23.50		144.50	168
5180	8″		15	.533		176	28		204	235
5240	Tee box, flanged, 2-1/2″ x 2-1/2″		18	.444		103	23.50		126.50	148
5260	4″		16	.500		121	26		147	172
5280	6″		15	.533		163	28		191	222
5300	8″		13	.615		238	32.50		270.50	310
5360	Cross box, flanged, 2-1/2″ x 2-1/2″		15	.533		126	28		154	181
5380	4″		13	.615		164	32.50		196.50	230
5400	6″		12	.667		213	35		248	287
5420	8″		10	.800		251	42		293	340
5480	Flange gasket, 2-1/2″		160	.050		3.74	2.62		6.36	8.05
5500	4″		80	.100		5.20	5.25		10.45	13.55
5520	6″		53	.151		7.20	7.90		15.10	19.75

26 05 33.23 Wireway		Crew	Daily Output	Labor-Hours	Unit	Material	2013 Bare Costs Labor	Equipment	Total	Total Incl O&P
5530	8"	1 Elec	40	.200	Ea.	9.30	10.50		19.80	26

26 05 33.25 Conduit Fittings for Rigid Galvanized Steel		Crew	Daily Output	Labor-Hours	Unit	Material	2013 Bare Costs Labor	Equipment	Total	Total Incl O&P
0010	**CONDUIT FITTINGS FOR RIGID GALVANIZED STEEL**									
0050	Standard, locknuts, 1/2" diameter				Ea.	.22			.22	.24
0100	3/4" diameter					.38			.38	.42
0300	1" diameter					.61			.61	.67
0500	1-1/4" diameter					.84			.84	.92
0700	1-1/2" diameter					1.39			1.39	1.53
1000	2" diameter					2.02			2.02	2.22
1030	2-1/2" diameter					5.65			5.65	6.20
1050	3" diameter					7.20			7.20	7.95
1070	3-1/2" diameter					12.15			12.15	13.40
1100	4" diameter					15.20			15.20	16.75
1110	5" diameter					32.50			32.50	36
1120	6" diameter					63			63	69.50
1130	Bushings, plastic, 1/2" diameter	1 Elec	40	.200		.22	10.50		10.72	15.95
1150	3/4" diameter		32	.250		.37	13.10		13.47	20
1170	1" diameter		28	.286		.60	14.95		15.55	23
1200	1-1/4" diameter		24	.333		.90	17.45		18.35	27
1230	1-1/2" diameter		18	.444		1.22	23.50		24.72	36.50
1250	2" diameter		15	.533		2.27	28		30.27	44.50
1270	2-1/2" diameter		13	.615		5.25	32.50		37.75	54.50
1300	3" diameter		12	.667		5.45	35		40.45	58.50
1330	3-1/2" diameter		11	.727		7.15	38		45.15	65
1350	4" diameter		9	.889		8.80	46.50		55.30	79
1360	5" diameter		7	1.143		19.65	60		79.65	111
1370	6" diameter		5	1.600		42.50	84		126.50	172
1390	Steel, 1/2" diameter		40	.200		.49	10.50		10.99	16.25
1400	3/4" diameter		32	.250		.59	13.10		13.69	20.50
1430	1" diameter		28	.286		1.06	14.95		16.01	23.50
1450	Steel insulated, 1-1/4" diameter		24	.333		6.35	17.45		23.80	33
1470	1-1/2" diameter		18	.444		8.05	23.50		31.55	44
1500	2" diameter		15	.533		11.45	28		39.45	54.50
1530	2-1/2" diameter		13	.615		21.50	32.50		54	72
1550	3" diameter		12	.667		29	35		64	84
1570	3-1/2" diameter		11	.727		37	38		75	97.50
1600	4" diameter		9	.889		46.50	46.50		93	121
1610	5" diameter		7	1.143		102	60		162	202
1620	6" diameter		5	1.600		159	84		243	300
1630	Sealing locknuts, 1/2" diameter		40	.200		1.29	10.50		11.79	17.10
1650	3/4" diameter		32	.250		1.54	13.10		14.64	21.50
1670	1" diameter		28	.286		2.21	14.95		17.16	25
1700	1-1/4" diameter		24	.333		3.81	17.45		21.26	30
1730	1-1/2" diameter		18	.444		4.79	23.50		28.29	40.50
1750	2" diameter		15	.533		6.20	28		34.20	49
1760	Grounding bushing, insulated, 1/2" diameter		32	.250		5.70	13.10		18.80	26
1770	3/4" diameter		28	.286		6.55	14.95		21.50	29.50
1780	1" diameter		20	.400		8.15	21		29.15	40.50
1800	1-1/4" diameter		18	.444		10.35	23.50		33.85	46.50
1830	1-1/2" diameter		16	.500		11.20	26		37.20	51.50
1850	2" diameter		13	.615		16.75	32.50		49.25	67
1870	2-1/2" diameter		12	.667		25.50	35		60.50	80.50

26 05 33 – Raceway and Boxes for Electrical Systems

26 05 33.25 Conduit Fittings for Rigid Galvanized Steel		Crew	Daily Output	Labor-Hours	Unit	Material	Labor	Equipment	Total	Total Incl O&P
						2013 Bare Costs				
1900	3" diameter	1 Elec	11	.727	Ea.	28.50	38		66.50	88.50
1930	3-1/2" diameter		9	.889		32.50	46.50		79	106
1950	4" diameter		8	1		39.50	52.50		92	122
1960	5" diameter		6	1.333		104	70		174	219
1970	6" diameter		4	2		159	105		264	330
1990	Coupling, with set screw, 1/2" diameter		50	.160		4.85	8.40		13.25	17.90
2000	3/4" diameter		40	.200		6.15	10.50		16.65	22.50
2030	1" diameter		35	.229		10.15	12		22.15	29
2050	1-1/4" diameter		28	.286		17.35	14.95		32.30	41.50
2070	1-1/2" diameter		23	.348		22.50	18.25		40.75	52
2090	2" diameter		20	.400		49.50	21		70.50	86
2100	2-1/2" diameter		18	.444		107	23.50		130.50	153
2110	3" diameter		15	.533		127	28		155	181
2120	3-1/2" diameter		12	.667		181	35		216	252
2130	4" diameter		10	.800		235	42		277	320
2140	5" diameter		9	.889		380	46.50		426.50	490
2150	6" diameter		8	1		505	52.50		557.50	635
2160	Box connector with set screw, plain, 1/2" diameter		70	.114		3.18	6		9.18	12.45
2170	3/4" diameter		60	.133		4.53	7		11.53	15.45
2180	1" diameter		50	.160		7.20	8.40		15.60	20.50
2190	Insulated, 1-1/4" diameter		40	.200		14.85	10.50		25.35	32
2200	1-1/2" diameter		30	.267		24	13.95		37.95	47
2210	2" diameter		20	.400		42.50	21		63.50	78
2220	2-1/2" diameter		18	.444		103	23.50		126.50	148
2230	3" diameter		15	.533		140	28		168	196
2240	3-1/2" diameter		12	.667		196	35		231	269
2250	4" diameter		10	.800		253	42		295	340
2260	5" diameter		9	.889		305	46.50		351.50	405
2270	6" diameter		8	1		390	52.50		442.50	510
2280	LB, LR or LL fittings & covers, 1/2" diameter		16	.500		9.25	26		35.25	49
2290	3/4" diameter		13	.615		11.10	32.50		43.60	60.50
2300	1" diameter		11	.727		16.65	38		54.65	75.50
2330	1-1/4" diameter		8	1		26.50	52.50		79	108
2350	1-1/2" diameter		6	1.333		34.50	70		104.50	143
2370	2" diameter		5	1.600		56.50	84		140.50	188
2380	2-1/2" diameter		4	2		113	105		218	281
2390	3" diameter		3.50	2.286		146	120		266	340
2400	3-1/2" diameter		3	2.667		236	140		376	470
2410	4" diameter		2.50	3.200		263	168		431	540
2420	T fittings, with cover, 1/2" diameter		12	.667		11.75	35		46.75	65.50
2430	3/4" diameter		11	.727		13.95	38		51.95	72.50
2440	1" diameter		9	.889		23	46.50		69.50	95
2450	1-1/4" diameter		6	1.333		30.50	70		100.50	139
2470	1-1/2" diameter		5	1.600		41	84		125	170
2500	2" diameter		4	2		61	105		166	224
2510	2-1/2" diameter		3.50	2.286		129	120		249	320
2520	3" diameter		3	2.667		165	140		305	390
2530	3-1/2" diameter		2.50	3.200		264	168		432	540
2540	4" diameter		2	4		335	210		545	685
2550	Nipples chase, plain, 1/2" diameter		40	.200		.96	10.50		11.46	16.75
2560	3/4" diameter		32	.250		1.20	13.10		14.30	21
2570	1" diameter		28	.286		2.44	14.95		17.39	25
2600	Insulated, 1-1/4" diameter		24	.333		9.05	17.45		26.50	36

26 05 33 – Raceway and Boxes for Electrical Systems

26 05 33.25 Conduit Fittings for Rigid Galvanized Steel		Crew	Daily Output	Labor-Hours	Unit	Material	2013 Bare Costs Labor	Equipment	Total	Total Incl O&P
2630	1-1/2" diameter	1 Elec	18	.444	Ea.	12	23.50		35.50	48
2650	2" diameter		15	.533		18	28		46	62
2660	2-1/2" diameter		12	.667		43	35		78	99.50
2670	3" diameter		10	.800		46.50	42		88.50	114
2680	3-1/2" diameter		9	.889		68	46.50		114.50	145
2690	4" diameter		8	1		99.50	52.50		152	189
2700	5" diameter		7	1.143		293	60		353	415
2710	6" diameter		6	1.333		450	70		520	600
2720	Nipples offset, plain, 1/2" diameter		40	.200		6.10	10.50		16.60	22.50
2730	3/4" diameter		32	.250		6.35	13.10		19.45	26.50
2740	1" diameter		24	.333		8.35	17.45		25.80	35
2750	Insulated, 1-1/4" diameter		20	.400		34	21		55	69
2760	1-1/2" diameter		18	.444		42.50	23.50		66	82
2770	2" diameter		16	.500		66	26		92	112
2780	3" diameter		14	.571		135	30		165	193
2850	Coupling, expansion, 1/2" diameter		12	.667		43.50	35		78.50	101
2880	3/4" diameter		10	.800		51	42		93	119
2900	1" diameter		8	1		60	52.50		112.50	145
2920	1-1/4" diameter		6.40	1.250		81	65.50		146.50	188
2940	1-1/2" diameter		5.30	1.509		111	79		190	240
2960	2" diameter		4.60	1.739		165	91		256	320
2980	2-1/2" diameter		3.60	2.222		262	116		378	460
3000	3" diameter		3	2.667		320	140		460	565
3020	3-1/2" diameter		2.80	2.857		475	150		625	745
3040	4" diameter		2.40	3.333		565	175		740	880
3060	5" diameter		2	4		1,175	210		1,385	1,625
3080	6" diameter		1.80	4.444		1,900	233		2,133	2,450
3100	Expansion deflection, 1/2" diameter		12	.667		232	35		267	310
3120	3/4" diameter		12	.667		247	35		282	325
3140	1" diameter		10	.800		275	42		317	370
3160	1-1/4" diameter		6.40	1.250		330	65.50		395.50	460
3180	1-1/2" diameter		5.30	1.509		385	79		464	540
3200	2" diameter		4.60	1.739		475	91		566	660
3220	2-1/2" diameter		3.60	2.222		645	116		761	885
3240	3" diameter		3	2.667		795	140		935	1,075
3260	3-1/2" diameter		2.80	2.857		1,025	150		1,175	1,350
3280	4" diameter		2.40	3.333		1,150	175		1,325	1,500
3300	5" diameter		2	4		1,775	210		1,985	2,275
3320	6" diameter		1.80	4.444		3,100	233		3,333	3,750
3340	Ericson, 1/2" diameter		16	.500		6.85	26		32.85	46.50
3360	3/4" diameter		14	.571		8.75	30		38.75	54.50
3380	1" diameter		11	.727		16.15	38		54.15	75
3400	1-1/4" diameter		8	1		34.50	52.50		87	117
3420	1-1/2" diameter		7	1.143		41.50	60		101.50	135
3440	2" diameter		5	1.600		83	84		167	216
3460	2-1/2" diameter		4	2		201	105		306	380
3480	3" diameter		3.50	2.286		305	120		425	515
3500	3-1/2" diameter		3	2.667		510	140		650	770
3520	4" diameter		2.70	2.963		600	155		755	890
3540	5" diameter		2.50	3.200		1,100	168		1,268	1,475
3560	6" diameter		2.30	3.478		1,475	182		1,657	1,900
3580	Split, 1/2" diameter		32	.250		4.57	13.10		17.67	24.50
3600	3/4" diameter		27	.296		5.70	15.55		21.25	29.50

26 05 33.25	Conduit Fittings for Rigid Galvanized Steel	Crew	Daily Output	Labor-Hours	Unit	2013 Bare Costs Material	Labor	Equipment	Total	Total Incl O&P
3620	1" diameter	1 Elec	20	.400	Ea.	10.50	21		31.50	43
3640	1-1/4" diameter		16	.500		15.25	26		41.25	56
3660	1-1/2" diameter		14	.571		19	30		49	66
3680	2" diameter		12	.667		38	35		73	94.50
3700	2-1/2" diameter		10	.800		88.50	42		130.50	160
3720	3" diameter		9	.889		133	46.50		179.50	216
3740	3-1/2" diameter		8	1		214	52.50		266.50	315
3760	4" diameter		7	1.143		252	60		312	365
3780	5" diameter		6	1.333		435	70		505	585
3800	6" diameter		5	1.600		585	84		669	770
4600	Reducing bushings, 3/4" to 1/2" diameter		54	.148		2.01	7.75		9.76	13.80
4620	1" to 3/4" diameter		46	.174		3.09	9.10		12.19	17.05
4640	1-1/4" to 1" diameter		40	.200		7.20	10.50		17.70	23.50
4660	1-1/2" to 1-1/4" diameter		36	.222		10.30	11.65		21.95	29
4680	2" to 1-1/2" diameter		32	.250		18.70	13.10		31.80	40
4740	2-1/2" to 2" diameter		30	.267		21	13.95		34.95	44
4760	3" to 2-1/2" diameter		28	.286		23	14.95		37.95	47.50
4800	Through-wall seal, 1/2" diameter		8	1		186	52.50		238.50	284
4820	3/4" diameter		7.50	1.067		209	56		265	315
4840	1" diameter		6.50	1.231		209	64.50		273.50	325
4860	1-1/4" diameter		5.50	1.455		246	76		322	385
4880	1-1/2" diameter		5	1.600		320	84		404	480
4900	2" diameter		4.20	1.905		291	100		391	470
4920	2-1/2" diameter		3.50	2.286		395	120		515	610
4940	3" diameter		3	2.667		420	140		560	670
4960	3-1/2" diameter		2.50	3.200		635	168		803	945
4980	4" diameter		2	4		645	210		855	1,025
5000	5" diameter		1.50	5.333		830	279		1,109	1,325
5020	6" diameter		1	8		830	420		1,250	1,525
5100	Cable supports, 2 or more wires									
5120	1-1/2" diameter	1 Elec	8	1	Ea.	115	52.50		167.50	206
5140	2" diameter		6	1.333		143	70		213	263
5160	2-1/2" diameter		4	2		157	105		262	330
5180	3" diameter		3.50	2.286		200	120		320	400
5200	3-1/2" diameter		2.60	3.077		272	161		433	540
5220	4" diameter		2	4		335	210		545	685
5240	5" diameter		1.50	5.333		545	279		824	1,025
5260	6" diameter		1	8		885	420		1,305	1,600
5280	Service entrance cap, 1/2" diameter		16	.500		7.80	26		33.80	47.50
5300	3/4" diameter		13	.615		9	32.50		41.50	58.50
5320	1" diameter		10	.800		10.90	42		52.90	74.50
5340	1-1/4" diameter		8	1		14.60	52.50		67.10	94.50
5360	1-1/2" diameter		6.50	1.231		43.50	64.50		108	145
5380	2" diameter		5.50	1.455		43	76		119	162
5400	2-1/2" diameter		4	2		117	105		222	285
5420	3" diameter		3.40	2.353		190	123		313	395
5440	3-1/2" diameter		3	2.667		272	140		412	510
5460	4" diameter		2.70	2.963		355	155		510	620
5600	Fire stop fittings, to 3/4" diameter		24	.333		137	17.45		154.45	177
5610	1" diameter		22	.364		166	19.05		185.05	212
5620	1-1/2" diameter		20	.400		208	21		229	261
5640	2" diameter		16	.500		325	26		351	395
5660	3" diameter		12	.667		405	35		440	500

26 05 33.25 Conduit Fittings for Rigid Galvanized Steel		Crew	Daily Output	Labor-Hours	Unit	Material	Labor	Equipment	Total	Total Incl O&P
						2013 Bare Costs				
5680	4″ diameter	1 Elec	10	.800	Ea.	540	42		582	655
5700	6″ diameter		8	1		930	52.50		982.50	1,100
5750	90° pull elbows steel, female, 1/2″ diameter		16	.500		9.05	26		35.05	49
5760	3/4″ diameter		13	.615		10.60	32.50		43.10	60
5780	1″ diameter		11	.727		17.95	38		55.95	77
5800	1-1/4″ diameter		8	1		26.50	52.50		79	108
5820	1-1/2″ diameter		6	1.333		38	70		108	147
5840	2″ diameter		5	1.600		61	84		145	193
6000	Explosion proof, flexible coupling									
6010	1/2″ diameter, 4″ long	1 Elec	12	.667	Ea.	111	35		146	175
6020	6″ long		12	.667		119	35		154	184
6050	12″ long		12	.667		152	35		187	220
6070	18″ long		12	.667		187	35		222	259
6090	24″ long		12	.667		222	35		257	297
6110	30″ long		12	.667		294	35		329	380
6130	36″ long		12	.667		288	35		323	370
6140	3/4″ diameter, 4″ long		10	.800		136	42		178	213
6150	6″ long		10	.800		146	42		188	223
6180	12″ long		10	.800		185	42		227	267
6200	18″ long		10	.800		240	42		282	325
6220	24″ long		10	.800		288	42		330	380
6240	30″ long		10	.800		385	42		427	485
6260	36″ long		10	.800		380	42		422	485
6270	1″ diameter, 6″ long		8	1		315	52.50		367.50	425
6300	12″ long		8	1		330	52.50		382.50	440
6320	18″ long		8	1		395	52.50		447.50	515
6340	24″ long		8	1		485	52.50		537.50	615
6360	30″ long		8	1		650	52.50		702.50	795
6380	36″ long		8	1		715	52.50		767.50	865
6390	1-1/4″ diameter, 12″ long		6.40	1.250		550	65.50		615.50	705
6410	18″ long		6.40	1.250		685	65.50		750.50	850
6430	24″ long		6.40	1.250		815	65.50		880.50	995
6450	30″ long		6.40	1.250		915	65.50		980.50	1,100
6470	36″ long		6.40	1.250		1,225	65.50		1,290.50	1,450
6480	1-1/2″ diameter, 12″ long		5.30	1.509		665	79		744	850
6500	18″ long		5.30	1.509		780	79		859	975
6520	24″ long		5.30	1.509		1,000	79		1,079	1,225
6540	30″ long		5.30	1.509		1,150	79		1,229	1,375
6560	36″ long		5.30	1.509		1,350	79		1,429	1,625
6570	2″ diameter, 12″ long		4.60	1.739		1,025	91		1,116	1,250
6590	18″ long		4.60	1.739		1,275	91		1,366	1,525
6610	24″ long		4.60	1.739		1,375	91		1,466	1,650
6630	30″ long		4.60	1.739		1,500	91		1,591	1,775
6650	36″ long		4.60	1.739		1,825	91		1,916	2,150
7000	Close up plug, 1/2″ diameter, explosion proof		40	.200		2.34	10.50		12.84	18.25
7010	3/4″ diameter		32	.250		2.71	13.10		15.81	22.50
7020	1″ diameter		28	.286		3.16	14.95		18.11	26
7030	1-1/4″ diameter		24	.333		3.41	17.45		20.86	30
7040	1-1/2″ diameter		18	.444		4.69	23.50		28.19	40
7050	2″ diameter		15	.533		8.10	28		36.10	51
7060	2-1/2″ diameter		13	.615		14.05	32.50		46.55	64
7070	3″ diameter		12	.667		20.50	35		55.50	75.50
7080	3-1/2″ diameter		11	.727		25.50	38		63.50	85

26 05 33.25 Conduit Fittings for Rigid Galvanized Steel		Crew	Daily Output	Labor-Hours	Unit	Material	2013 Bare Costs Labor	Equipment	Total	Total Incl O&P
7090	4" diameter	1 Elec	9	.889	Ea.	34	46.50		80.50	107
7091	Elbow female, 45°, 1/2"		16	.500		10.85	26		36.85	51
7092	3/4"		13	.615		13.50	32.50		46	63.50
7093	1"		11	.727		16.10	38		54.10	75
7094	1-1/4"		8	1		24.50	52.50		77	106
7095	1-1/2"		6	1.333		26.50	70		96.50	134
7096	2"		5	1.600		30.50	84		114.50	159
7097	2-1/2"		4.50	1.778		89.50	93		182.50	238
7098	3"		4.20	1.905		90.50	100		190.50	249
7099	3-1/2"		4	2		140	105		245	310
7100	4"		3.80	2.105		155	110		265	335
7101	90°, 1/2"		16	.500		10.95	26		36.95	51
7102	3/4"		13	.615		11.85	32.50		44.35	61.50
7103	1"		11	.727		16.15	38		54.15	75
7104	1-1/4"		8	1		26	52.50		78.50	107
7105	1-1/2"		6	1.333		36	70		106	145
7106	2"		5	1.600		56.50	84		140.50	187
7107	2-1/2"		4.50	1.778		132	93		225	284
7110	Elbows 90°, long male & female, 1/2" diameter, explosion proof		16	.500		16	26		42	56.50
7120	3/4" diameter		13	.615		18.50	32.50		51	69
7130	1" diameter		11	.727		24	38		62	83.50
7140	1-1/4" diameter		8	1		28.50	52.50		81	110
7150	1-1/2" diameter		6	1.333		35.50	70		105.50	144
7160	2" diameter		5	1.600		53.50	84		137.50	184
7170	Capped elbow, 1/2" diameter, explosion proof		11	.727		17.55	38		55.55	76.50
7180	3/4" diameter		8	1		21.50	52.50		74	102
7190	1" diameter		6	1.333		23.50	70		93.50	131
7200	1-1/4" diameter		5	1.600		47	84		131	177
7210	Pulling elbow, 1/2" diameter, explosion proof		11	.727		69.50	38		107.50	134
7220	3/4" diameter		8	1		127	52.50		179.50	219
7230	1" diameter		6	1.333		185	70		255	310
7240	1-1/4" diameter		5	1.600		237	84		321	385
7250	1-1/2" diameter		5	1.600		289	84		373	445
7260	2" diameter		4	2		291	105		396	475
7270	2-1/2" diameter		3.50	2.286		725	120		845	975
7280	3" diameter		3	2.667		975	140		1,115	1,275
7290	3-1/2" diameter		2.50	3.200		1,225	168		1,393	1,600
7300	4" diameter		2.20	3.636		1,300	191		1,491	1,700
7310	LB conduit body, 1/2" diameter		11	.727		41.50	38		79.50	103
7320	3/4" diameter		8	1		50.50	52.50		103	135
7330	T conduit body, 1/2" diameter		9	.889		45.50	46.50		92	120
7340	3/4" diameter		6	1.333		53.50	70		123.50	164
7350	Explosion proof, round box w/cover, 3 threaded hubs, 1/2" diameter		8	1		43.50	52.50		96	127
7351	3/4" diameter		8	1		58.50	52.50		111	143
7352	1" diameter		7.50	1.067		67	56		123	157
7353	1-1/4" diameter		7	1.143		91	60		151	190
7354	1-1/2" diameter		7	1.143		198	60		258	310
7355	2" diameter		6	1.333		205	70		275	330
7356	Round box w/cover & mtng flange, 3 threaded hubs, 1/2" diameter		8	1		57.50	52.50		110	142
7357	3/4" diameter		8	1		60.50	52.50		113	145
7358	4 threaded hubs, 1" diameter		7	1.143		59	60		119	155
7400	Unions, 1/2" diameter		20	.400		10.90	21		31.90	43.50
7410	3/4" - 1/2" diameter		16	.500		15.75	26		41.75	56.50

26 05 33 – Raceway and Boxes for Electrical Systems

26 05 33.25 Conduit Fittings for Rigid Galvanized Steel		Crew	Daily Output	Labor-Hours	Unit	2013 Bare Costs Material	Labor	Equipment	Total	Total Incl O&P
7420	3/4" diameter	1 Elec	16	.500	Ea.	14.95	26		40.95	55.50
7430	1" diameter		14	.571		27	30		57	75
7440	1-1/4" diameter		12	.667		40	35		75	96.50
7450	1-1/2" diameter		10	.800		51	42		93	119
7460	2" diameter		8.50	.941		65.50	49.50		115	146
7480	2-1/2" diameter		8	1		103	52.50		155.50	192
7490	3" diameter		7	1.143		136	60		196	240
7500	3-1/2" diameter		6	1.333		248	70		318	380
7510	4" diameter		5	1.600		250	84		334	400
7680	Reducer, 3/4" to 1/2"		54	.148		2.28	7.75		10.03	14.10
7690	1" to 1/2"		46	.174		4.10	9.10		13.20	18.15
7700	1" to 3/4"		46	.174		4.16	9.10		13.26	18.25
7710	1-1/4" to 3/4"		40	.200		5.15	10.50		15.65	21.50
7720	1-1/4" to 1"		40	.200		5.95	10.50		16.45	22.50
7730	1-1/2" to 1"		36	.222		9.65	11.65		21.30	28
7740	1-1/2" to 1-1/4"		36	.222		9.75	11.65		21.40	28
7750	2" to 3/4"		32	.250		11.60	13.10		24.70	32.50
7760	2" to 1-1/4"		32	.250		11.35	13.10		24.45	32
7770	2" to 1-1/2"		32	.250		12.05	13.10		25.15	33
7780	2-1/2" to 1-1/2"		30	.267		20.50	13.95		34.45	43.50
7790	3" to 2"		30	.267		20.50	13.95		34.45	43.50
7800	3-1/2" to 2-1/2"		28	.286		42	14.95		56.95	68.50
7810	4" to 3"		28	.286		50.50	14.95		65.45	78
7820	Sealing fitting, vertical/horizontal, 1/2" diameter		14.50	.552		14.80	29		43.80	60
7830	3/4" diameter		13.30	.602		17.45	31.50		48.95	66
7840	1" diameter		11.40	.702		22.50	37		59.50	79.50
7850	1-1/4" diameter		10	.800		27	42		69	92
7860	1-1/2" diameter		8.80	.909		41.50	47.50		89	117
7870	2" diameter		8	1		53.50	52.50		106	137
7880	2-1/2" diameter		6.70	1.194		75	62.50		137.50	176
7890	3" diameter		5.70	1.404		93	73.50		166.50	212
7900	3-1/2" diameter		4.70	1.702		246	89		335	405
7910	4" diameter		4	2		400	105		505	595
7920	Sealing hubs, 1" by 1-1/2"		12	.667		35.50	35		70.50	91.50
7930	1-1/4" by 2"		10	.800		46	42		88	114
7940	1-1/2" by 2"		9	.889		67.50	46.50		114	144
7950	2" by 2-1/2"		8	1		81	52.50		133.50	168
7960	3" by 4"		7	1.143		121	60		181	223
7970	4" by 5"		6	1.333		271	70		341	405
7980	Drain, 1/2"		32	.250		80.50	13.10		93.60	109
7990	Breather, 1/2"		32	.250		80.50	13.10		93.60	109
8000	Plastic coated 40 mil thick									
8010	LB, LR or LL conduit body w/cover, 1/2" diameter	1 Elec	13	.615	Ea.	47.50	32.50		80	101
8020	3/4" diameter		11	.727		52	38		90	115
8030	1" diameter		8	1		70	52.50		122.50	156
8040	1-1/4" diameter		6	1.333		102	70		172	218
8050	1-1/2" diameter		5	1.600		125	84		209	262
8060	2" diameter		4.50	1.778		183	93		276	340
8070	2-1/2" diameter		4	2		335	105		440	525
8080	3" diameter		3.50	2.286		425	120		545	650
8090	3-1/2" diameter		3	2.667		625	140		765	900
8100	4" diameter		2.50	3.200		700	168		868	1,025
8150	T conduit body with cover, 1/2" diameter		11	.727		57.50	38		95.50	121

26 05 33.25 Conduit Fittings for Rigid Galvanized Steel		Crew	Daily Output	Labor-Hours	Unit	Material	Labor	Equipment	Total	Total Incl O&P
						2013 Bare Costs				
8160	3/4" diameter	1 Elec	9	.889	Ea.	68	46.50		114.50	145
8170	1" diameter		6	1.333		81.50	70		151.50	195
8180	1-1/4" diameter		5	1.600		116	84		200	252
8190	1-1/2" diameter		4.50	1.778		144	93		237	297
8200	2" diameter		4	2		209	105		314	385
8210	2-1/2" diameter		3.50	2.286		370	120		490	590
8220	3" diameter		3	2.667		485	140		625	740
8230	3-1/2" diameter		2.50	3.200		760	168		928	1,100
8240	4" diameter		2	4		825	210		1,035	1,225
8300	FS conduit body, 1 gang, 3/4" diameter		11	.727		56	38		94	119
8310	1" diameter		10	.800		59	42		101	127
8350	2 gang, 3/4" diameter		9	.889		101	46.50		147.50	181
8360	1" diameter		8	1		116	52.50		168.50	206
8400	Duplex receptacle cover		64	.125		40.50	6.55		47.05	54.50
8410	Switch cover		64	.125		59	6.55		65.55	75
8420	Switch, vaportight cover		53	.151		168	7.90		175.90	196
8430	Blank, cover		64	.125		39.50	6.55		46.05	53
8520	FSC conduit body, 1 gang, 3/4" diameter		10	.800		63.50	42		105.50	132
8530	1" diameter		9	.889		71.50	46.50		118	149
8550	2 gang, 3/4" diameter		8	1		114	52.50		166.50	205
8560	1" diameter		7	1.143		118	60		178	220
8590	Conduit hubs, 1/2" diameter		18	.444		35.50	23.50		59	74
8600	3/4" diameter		16	.500		40.50	26		66.50	83.50
8610	1" diameter		14	.571		50.50	30		80.50	101
8620	1-1/4" diameter		12	.667		59.50	35		94.50	118
8630	1-1/2" diameter		10	.800		66.50	42		108.50	136
8640	2" diameter		8.80	.909		96	47.50		143.50	177
8650	2-1/2" diameter		8.50	.941		150	49.50		199.50	239
8660	3" diameter		8	1		206	52.50		258.50	305
8670	3-1/2" diameter		7.50	1.067		271	56		327	385
8680	4" diameter		7	1.143		325	60		385	450
8690	5" diameter	↓	6	1.333	↓	465	70		535	620
8700	Plastic coated 40 mil thick									
8710	Pipe strap, stamped 1 hole, 1/2" diameter	1 Elec	470	.017	Ea.	10.85	.89		11.74	13.30
8720	3/4" diameter		440	.018		10.05	.95		11	12.50
8730	1" diameter		400	.020		10.35	1.05		11.40	12.95
8740	1-1/4" diameter		355	.023		17.05	1.18		18.23	20.50
8750	1-1/2" diameter		320	.025		18	1.31		19.31	22
8760	2" diameter		266	.030		25	1.58		26.58	30.50
8770	2-1/2" diameter		200	.040		34.50	2.10		36.60	41
8780	3" diameter		133	.060		45.50	3.15		48.65	54.50
8790	3-1/2" diameter		110	.073		75	3.81		78.81	88
8800	4" diameter		90	.089		80	4.66		84.66	95.50
8810	5" diameter		70	.114		355	6		361	400
8840	Clamp back spacers, 3/4" diameter		440	.018		16.45	.95		17.40	19.55
8850	1" diameter		400	.020		19.80	1.05		20.85	23.50
8860	1-1/4" diameter		355	.023		27	1.18		28.18	31.50
8870	1-1/2" diameter		320	.025		37	1.31		38.31	42.50
8880	2" diameter		266	.030		56.50	1.58		58.08	64.50
8900	3" diameter		133	.060		131	3.15		134.15	149
8920	4" diameter	↓	90	.089		143	4.66		147.66	164
8950	Touch-up plastic coating, spray, 12 oz.					49			49	54
8960	Sealing fittings, 1/2" diameter	1 Elec	11	.727		58.50	38		96.50	121

26 05 33.25 Conduit Fittings for Rigid Galvanized Steel

	26 05 33.25 Conduit Fittings for Rigid Galvanized Steel	Crew	Daily Output	Labor-Hours	Unit	2013 Bare Costs Material	Labor	Equipment	Total	Total Incl O&P
8970	3/4" diameter	1 Elec	9	.889	Ea.	60.50	46.50		107	136
8980	1" diameter		7.50	1.067		72.50	56		128.50	163
8990	1-1/4" diameter		6.50	1.231		91.50	64.50		156	197
9000	1-1/2" diameter		5.50	1.455		119	76		195	245
9010	2" diameter		4.80	1.667		144	87.50		231.50	289
9020	2-1/2" diameter		4	2		246	105		351	430
9030	3" diameter		3.50	2.286		263	120		383	470
9040	3-1/2" diameter		3	2.667		710	140		850	990
9050	4" diameter		2.50	3.200		1,075	168		1,243	1,425
9060	5" diameter		1.70	4.706		1,900	247		2,147	2,450
9070	Unions, 1/2" diameter		18	.444		51.50	23.50		75	92
9080	3/4" diameter		15	.533		50	28		78	97
9090	1" diameter		13	.615		75	32.50		107.50	131
9100	1-1/4" diameter		11	.727		117	38		155	186
9110	1-1/2" diameter		9.50	.842		141	44		185	222
9120	2" diameter		8	1		173	52.50		225.50	270
9130	2-1/2" diameter		7.50	1.067		279	56		335	390
9140	3" diameter		6.80	1.176		365	61.50		426.50	490
9150	3-1/2" diameter		5.80	1.379		480	72.50		552.50	640
9160	4" diameter		4.80	1.667		580	87.50		667.50	765
9170	5" diameter		4	2		810	105		915	1,050

26 05 33.30 Electrical Nonmetallic Tubing (ENT)

	Description	Crew	Daily Output	Labor-Hours	Unit	Material	Labor	Equipment	Total	Total Incl O&P
0010	**ELECTRICAL NONMETALLIC TUBING (ENT)**									
0050	Flexible, 1/2" diameter	1 Elec	270	.030	L.F.	.73	1.55		2.28	3.12
0100	3/4" diameter		230	.035		1.18	1.82		3	4.03
0200	1" diameter		145	.055		1.99	2.89		4.88	6.50
0210	1-1/4" diameter		125	.064		2.15	3.35		5.50	7.35
0220	1-1/2" diameter		100	.080		2.91	4.19		7.10	9.45
0230	2" diameter		75	.107		3.81	5.60		9.41	12.55
0300	Connectors, to outlet box, 1/2" diameter		230	.035	Ea.	1.41	1.82		3.23	4.28
0310	3/4" diameter		210	.038		2.49	2		4.49	5.75
0320	1" diameter		200	.040		3.85	2.10		5.95	7.40
0400	Couplings, to conduit, 1/2" diameter		145	.055		1.49	2.89		4.38	5.95
0410	3/4" diameter		130	.062		1.76	3.22		4.98	6.75
0420	1" diameter		125	.064		3.68	3.35		7.03	9.05

26 05 33.35 Flexible Metallic Conduit

	Description	Crew	Daily Output	Labor-Hours	Unit	Material	Labor	Equipment	Total	Total Incl O&P
0010	**FLEXIBLE METALLIC CONDUIT**									
0050	Steel, 3/8" diameter	1 Elec	200	.040	L.F.	.60	2.10		2.70	3.80
0100	1/2" diameter		200	.040		.67	2.10		2.77	3.88
0200	3/4" diameter		160	.050		.92	2.62		3.54	4.93
0250	1" diameter		100	.080		1.67	4.19		5.86	8.10
0300	1-1/4" diameter		70	.114		2.16	6		8.16	11.35
0350	1-1/2" diameter		50	.160		3.53	8.40		11.93	16.45
0370	2" diameter		40	.200		4.31	10.50		14.81	20.50
0380	2-1/2" diameter		30	.267		5.25	13.95		19.20	27
0390	3" diameter	2 Elec	50	.320		9.05	16.75		25.80	35
0400	3-1/2" diameter		40	.400		10.25	21		31.25	43
0410	4" diameter		30	.533		11.70	28		39.70	55
0420	Connectors, plain, 3/8" diameter	1 Elec	100	.080	Ea.	2.72	4.19		6.91	9.25
0430	1/2" diameter		80	.100		3.14	5.25		8.39	11.30
0440	3/4" diameter		70	.114		3.61	6		9.61	12.90
0450	1" diameter		50	.160		7.65	8.40		16.05	21

26 05 33.35 Flexible Metallic Conduit		Crew	Daily Output	Labor-Hours	Unit	Material	Labor	Equipment	Total	Total Incl O&P
						2013 Bare Costs				
0452	1-1/4" diameter	1 Elec	45	.178	Ea.	9.40	9.30		18.70	24.50
0454	1-1/2" diameter		40	.200		13.85	10.50		24.35	31
0456	2" diameter		28	.286		21	14.95		35.95	45.50
0458	2-1/2" diameter		25	.320		35.50	16.75		52.25	64
0460	3" diameter		20	.400		51	21		72	87.50
0462	3-1/2" diameter		16	.500		148	26		174	202
0464	4" diameter		13	.615		189	32.50		221.50	256
0490	Insulated, 1" diameter		40	.200		8.35	10.50		18.85	25
0500	1-1/4" diameter		40	.200		14.95	10.50		25.45	32
0550	1-1/2" diameter		32	.250		21.50	13.10		34.60	43
0600	2" diameter		23	.348		38.50	18.25		56.75	70
0610	2-1/2" diameter		20	.400		101	21		122	143
0620	3" diameter		17	.471		133	24.50		157.50	184
0630	3-1/2" diameter		13	.615		227	32.50		259.50	298
0640	4" diameter		10	.800		298	42		340	395
0650	Connectors 90°, plain, 3/8" diameter		80	.100		2.78	5.25		8.03	10.90
0660	1/2" diameter		60	.133		4.68	7		11.68	15.60
0700	3/4" diameter		50	.160		7.75	8.40		16.15	21
0750	1" diameter		40	.200		13.55	10.50		24.05	30.50
0790	Insulated, 1" diameter		40	.200		17.25	10.50		27.75	34.50
0800	1-1/4" diameter		30	.267		40.50	13.95		54.45	65.50
0850	1-1/2" diameter		23	.348		75	18.25		93.25	110
0900	2" diameter		18	.444		92	23.50		115.50	136
0910	2-1/2" diameter		16	.500		176	26		202	232
0920	3" diameter		14	.571		226	30		256	294
0930	3-1/2" diameter		11	.727		665	38		703	785
0940	4" diameter		8	1		905	52.50		957.50	1,075
0960	Couplings, to flexible conduit, 1/2" diameter		50	.160		2.05	8.40		10.45	14.80
0970	3/4" diameter		40	.200		3.50	10.50		14	19.55
0980	1" diameter		35	.229		6.10	12		18.10	24.50
0990	1-1/4" diameter		28	.286		13.45	14.95		28.40	37.50
1000	1-1/2" diameter		23	.348		17.35	18.25		35.60	46.50
1010	2" diameter		20	.400		35	21		56	70
1020	2-1/2" diameter		18	.444		41	23.50		64.50	80
1030	3" diameter		15	.533		103	28		131	155
1032	Aluminum, 3/8" diameter		210	.038	L.F.	.57	2		2.57	3.62
1034	1/2" diameter		210	.038		.65	2		2.65	3.71
1036	3/4" diameter		165	.048		.90	2.54		3.44	4.79
1038	1" diameter		105	.076		1.72	3.99		5.71	7.85
1040	1-1/4" diameter		75	.107		2.39	5.60		7.99	11
1042	1-1/2" diameter		53	.151		3.39	7.90		11.29	15.60
1044	2" diameter		42	.190		4.33	10		14.33	19.70
1046	2-1/2" diameter		32	.250		5.55	13.10		18.65	26
1048	3" diameter	2 Elec	53	.302		9.10	15.80		24.90	33.50
1050	3-1/2" diameter		42	.381		10.50	19.95		30.45	41.50
1052	4" diameter		32	.500		11.70	26		37.70	52
1070	Sealtite, 3/8" diameter	1 Elec	140	.057		1.46	2.99		4.45	6.10
1080	1/2" diameter		140	.057		1.58	2.99		4.57	6.20
1090	3/4" diameter		100	.080		2.21	4.19		6.40	8.70
1100	1" diameter		70	.114		3.34	6		9.34	12.60
1200	1-1/4" diameter		50	.160		4.57	8.40		12.97	17.60
1300	1-1/2" diameter		40	.200		5.30	10.50		15.80	21.50
1400	2" diameter		30	.267		6.65	13.95		20.60	28.50

26 05 33 – Raceway and Boxes for Electrical Systems

26 05 33.35 Flexible Metallic Conduit		Crew	Daily Output	Labor-Hours	Unit	2013 Bare Costs Material	Labor	Equipment	Total	Total Incl O&P
1410	2-1/2″ diameter	1 Elec	27	.296	L.F.	11.35	15.55		26.90	35.50
1420	3″ diameter	2 Elec	50	.320		15.70	16.75		32.45	42.50
1440	4″ diameter	″	30	.533		23.50	28		51.50	68
1490	Connectors, plain, 3/8″ diameter	1 Elec	70	.114	Ea.	3.30	6		9.30	12.60
1500	1/2″ diameter		70	.114		5.20	6		11.20	14.65
1700	3/4″ diameter		50	.160		7.70	8.40		16.10	21
1900	1″ diameter		40	.200		13.55	10.50		24.05	30.50
1910	Insulated, 1″ diameter		40	.200		17.10	10.50		27.60	34.50
2000	1-1/4″ diameter		32	.250		27	13.10		40.10	49
2100	1-1/2″ diameter		27	.296		35	15.55		50.55	61.50
2200	2″ diameter		20	.400		62.50	21		83.50	100
2210	2-1/2″ diameter		15	.533		370	28		398	445
2220	3″ diameter		12	.667		410	35		445	510
2240	4″ diameter		8	1		510	52.50		562.50	640
2290	Connectors, 90°, 3/8″ diameter		70	.114		3.60	6		9.60	12.90
2300	1/2″ diameter		70	.114		5.75	6		11.75	15.25
2400	3/4″ diameter		50	.160		8.90	8.40		17.30	22.50
2600	1″ diameter		40	.200		18.10	10.50		28.60	35.50
2790	Insulated, 1″ diameter		40	.200		30.50	10.50		41	49
2800	1-1/4″ diameter		32	.250		49.50	13.10		62.60	74
3000	1-1/2″ diameter		27	.296		61	15.55		76.55	90
3100	2″ diameter		20	.400		85	21		106	125
3110	2-1/2″ diameter		14	.571		400	30		430	485
3120	3″ diameter		11	.727		390	38		428	485
3140	4″ diameter		7	1.143		595	60		655	745
4300	Coupling sealtite to rigid, 1/2″ diameter		20	.400		4.96	21		25.96	37
4500	3/4″ diameter		18	.444		7	23.50		30.50	42.50
4800	1″ diameter		14	.571		9.60	30		39.60	55.50
4900	1-1/4″ diameter		12	.667		16	35		51	70
5000	1-1/2″ diameter		11	.727		32	38		70	92.50
5100	2″ diameter		10	.800		49.50	42		91.50	117
5110	2-1/2″ diameter		9.50	.842		191	44		235	277
5120	3″ diameter		9	.889		204	46.50		250.50	295
5130	3-1/2″ diameter		9	.889		228	46.50		274.50	320
5140	4″ diameter		8.50	.941		251	49.50		300.50	350

26 05 33.95 Cutting and Drilling		Crew	Daily Output	Labor-Hours	Unit	Material	Labor	Equipment	Total	Total Incl O&P
0010	**CUTTING AND DRILLING**									
0100	Hole drilling to 10′ high, concrete wall									
0110	8″ thick, 1/2″ pipe size	R-31	12	.667	Ea.	.25	35	4.01	39.26	57
0120	3/4″ pipe size		12	.667		.25	35	4.01	39.26	57
0130	1″ pipe size		9.50	.842		.39	44	5.05	49.44	72
0140	1-1/4″ pipe size		9.50	.842		.39	44	5.05	49.44	72
0150	1-1/2″ pipe size		9.50	.842		.39	44	5.05	49.44	72
0160	2″ pipe size		4.40	1.818		.55	95.50	10.95	107	156
0170	2-1/2″ pipe size		4.40	1.818		.55	95.50	10.95	107	156
0180	3″ pipe size		4.40	1.818		.55	95.50	10.95	107	156
0190	3-1/2″ pipe size		3.30	2.424		.66	127	14.55	142.21	207
0200	4″ pipe size		3.30	2.424		.66	127	14.55	142.21	207
0500	12″ thick, 1/2″ pipe size		9.40	.851		.37	44.50	5.10	49.97	72.50
0520	3/4″ pipe size		9.40	.851		.37	44.50	5.10	49.97	72.50
0540	1″ pipe size		7.30	1.096		.59	57.50	6.60	64.69	94
0560	1-1/4″ pipe size		7.30	1.096		.59	57.50	6.60	64.69	94

26 05 33.95 Cutting and Drilling		Crew	Daily Output	Labor-Hours	Unit	Material	2013 Bare Costs Labor	Equipment	Total	Total Incl O&P
0570	1-1/2" pipe size	R-31	7.30	1.096	Ea.	.59	57.50	6.60	64.69	94
0580	2" pipe size		3.60	2.222		.82	116	13.35	130.17	190
0590	2-1/2" pipe size		3.60	2.222		.82	116	13.35	130.17	190
0600	3" pipe size		3.60	2.222		.82	116	13.35	130.17	190
0610	3-1/2" pipe size		2.80	2.857		.99	150	17.15	168.14	244
0630	4" pipe size		2.50	3.200		.99	168	19.25	188.24	273
0650	16" thick, 1/2" pipe size		7.60	1.053		.49	55	6.35	61.84	90
0670	3/4" pipe size		7	1.143		.49	60	6.85	67.34	97.50
0690	1" pipe size		6	1.333		.78	70	8	78.78	115
0710	1-1/4" pipe size		5.50	1.455		.78	76	8.75	85.53	124
0730	1-1/2" pipe size		5.50	1.455		.78	76	8.75	85.53	124
0750	2" pipe size		3	2.667		1.09	140	16.05	157.14	228
0770	2-1/2" pipe size		2.70	2.963		1.09	155	17.80	173.89	253
0790	3" pipe size		2.50	3.200		1.09	168	19.25	188.34	273
0810	3-1/2" pipe size		2.30	3.478		1.32	182	21	204.32	297
0830	4" pipe size		2	4		1.32	210	24	235.32	345
0850	20" thick, 1/2" pipe size		6.40	1.250		.62	65.50	7.50	73.62	107
0870	3/4" pipe size		6	1.333		.62	70	8	78.62	114
0890	1" pipe size		5	1.600		.98	84	9.60	94.58	137
0910	1-1/4" pipe size		4.80	1.667		.98	87.50	10	98.48	143
0930	1-1/2" pipe size		4.60	1.739		.98	91	10.45	102.43	149
0950	2" pipe size		2.70	2.963		1.37	155	17.80	174.17	253
0970	2-1/2" pipe size		2.40	3.333		1.37	175	20	196.37	285
0990	3" pipe size		2.20	3.636		1.37	191	22	214.37	310
1010	3-1/2" pipe size		2	4		1.65	210	24	235.65	345
1030	4" pipe size		1.70	4.706		1.65	247	28.50	277.15	405
1050	24" thick, 1/2" pipe size		5.50	1.455		.74	76	8.75	85.49	124
1070	3/4" pipe size		5.10	1.569		.74	82	9.45	92.19	134
1090	1" pipe size		4.30	1.860		1.17	97.50	11.20	109.87	160
1110	1-1/4" pipe size		4	2		1.17	105	12	118.17	171
1130	1-1/2" pipe size		4	2		1.17	105	12	118.17	171
1150	2" pipe size		2.40	3.333		1.64	175	20	196.64	285
1170	2-1/2" pipe size		2.20	3.636		1.64	191	22	214.64	310
1190	3" pipe size		2	4		1.64	210	24	235.64	345
1210	3-1/2" pipe size		1.80	4.444		1.99	233	26.50	261.49	380
1230	4" pipe size		1.50	5.333		1.99	279	32	312.99	460
1500	Brick wall, 8" thick, 1/2" pipe size		18	.444		.25	23.50	2.67	26.42	38
1520	3/4" pipe size		18	.444		.25	23.50	2.67	26.42	38
1540	1" pipe size		13.30	.602		.39	31.50	3.62	35.51	51.50
1560	1-1/4" pipe size		13.30	.602		.39	31.50	3.62	35.51	51.50
1580	1-1/2" pipe size		13.30	.602		.39	31.50	3.62	35.51	51.50
1600	2" pipe size		5.70	1.404		.55	73.50	8.45	82.50	120
1620	2-1/2" pipe size		5.70	1.404		.55	73.50	8.45	82.50	120
1640	3" pipe size		5.70	1.404		.55	73.50	8.45	82.50	120
1660	3-1/2" pipe size		4.40	1.818		.66	95.50	10.95	107.11	156
1680	4" pipe size		4	2		.66	105	12	117.66	171
1700	12" thick, 1/2" pipe size		14.50	.552		.37	29	3.32	32.69	47.50
1720	3/4" pipe size		14.50	.552		.37	29	3.32	32.69	47.50
1740	1" pipe size		11	.727		.59	38	4.37	42.96	62.50
1760	1-1/4" pipe size		11	.727		.59	38	4.37	42.96	62.50
1780	1-1/2" pipe size		11	.727		.59	38	4.37	42.96	62.50
1800	2" pipe size		5	1.600		.82	84	9.60	94.42	137
1820	2-1/2" pipe size		5	1.600		.82	84	9.60	94.42	137

26 05 33.95 Cutting and Drilling		Crew	Daily Output	Labor-Hours	Unit	2013 Bare Costs Material	Labor	Equipment	Total	Total Incl O&P
1840	3" pipe size	R-31	5	1.600	Ea.	.82	84	9.60	94.42	137
1860	3-1/2" pipe size		3.80	2.105		.99	110	12.65	123.64	180
1880	4" pipe size		3.30	2.424		.99	127	14.55	142.54	207
1900	16" thick, 1/2" pipe size		12.30	.650		.49	34	3.91	38.40	56
1920	3/4" pipe size		12.30	.650		.49	34	3.91	38.40	56
1940	1" pipe size		9.30	.860		.78	45	5.15	50.93	74
1960	1-1/4" pipe size		9.30	.860		.78	45	5.15	50.93	74
1980	1-1/2" pipe size		9.30	.860		.78	45	5.15	50.93	74
2000	2" pipe size		4.40	1.818		1.09	95.50	10.95	107.54	156
2010	2-1/2" pipe size		4.40	1.818		1.09	95.50	10.95	107.54	156
2030	3" pipe size		4.40	1.818		1.09	95.50	10.95	107.54	156
2050	3-1/2" pipe size		3.30	2.424		1.32	127	14.55	142.87	207
2070	4" pipe size		3	2.667		1.32	140	16.05	157.37	228
2090	20" thick, 1/2" pipe size		10.70	.748		.62	39	4.49	44.11	64
2110	3/4" pipe size		10.70	.748		.62	39	4.49	44.11	64
2130	1" pipe size		8	1		.98	52.50	6	59.48	86
2150	1-1/4" pipe size		8	1		.98	52.50	6	59.48	86
2170	1-1/2" pipe size		8	1		.98	52.50	6	59.48	86
2190	2" pipe size		4	2		1.37	105	12	118.37	172
2210	2-1/2" pipe size		4	2		1.37	105	12	118.37	172
2230	3" pipe size		4	2		1.37	105	12	118.37	172
2250	3-1/2" pipe size		3	2.667		1.65	140	16.05	157.70	228
2270	4" pipe size		2.70	2.963		1.65	155	17.80	174.45	253
2290	24" thick, 1/2" pipe size		9.40	.851		.74	44.50	5.10	50.34	73
2310	3/4" pipe size		9.40	.851		.74	44.50	5.10	50.34	73
2330	1" pipe size		7.10	1.127		1.17	59	6.75	66.92	97
2350	1-1/4" pipe size		7.10	1.127		1.17	59	6.75	66.92	97
2370	1-1/2" pipe size		7.10	1.127		1.17	59	6.75	66.92	97
2390	2" pipe size		3.60	2.222		1.64	116	13.35	130.99	191
2410	2-1/2" pipe size		3.60	2.222		1.64	116	13.35	130.99	191
2430	3" pipe size		3.60	2.222		1.64	116	13.35	130.99	191
2450	3-1/2" pipe size		2.80	2.857		1.99	150	17.15	169.14	245
2470	4" pipe size	↓	2.50	3.200	↓	1.99	168	19.25	189.24	274
3000	Knockouts to 8' high, metal boxes & enclosures									
3020	With hole saw, 1/2" pipe size	1 Elec	53	.151	Ea.		7.90		7.90	11.85
3040	3/4" pipe size		47	.170			8.90		8.90	13.35
3050	1" pipe size		40	.200			10.50		10.50	15.70
3060	1-1/4" pipe size		36	.222			11.65		11.65	17.40
3070	1-1/2" pipe size		32	.250			13.10		13.10	19.60
3080	2" pipe size		27	.296			15.55		15.55	23
3090	2-1/2" pipe size		20	.400			21		21	31.50
4010	3" pipe size		16	.500			26		26	39
4030	3-1/2" pipe size		13	.615			32.50		32.50	48.50
4050	4" pipe size		11	.727			38		38	57
4070	With hand punch set, 1/2" pipe size		40	.200			10.50		10.50	15.70
4090	3/4" pipe size		32	.250			13.10		13.10	19.60
4110	1" pipe size		30	.267			13.95		13.95	21
4130	1-1/4" pipe size		28	.286			14.95		14.95	22.50
4150	1-1/2" pipe size		26	.308			16.10		16.10	24
4170	2" pipe size		20	.400			21		21	31.50
4190	2-1/2" pipe size		17	.471			24.50		24.50	37
4200	3" pipe size		15	.533			28		28	42
4220	3-1/2" pipe size		12	.667			35		35	52.50

26 05 33 – Raceway and Boxes for Electrical Systems

26 05 33.95 Cutting and Drilling		Crew	Daily Output	Labor-Hours	Unit	Material	Labor	Equipment	Total	Total Incl O&P
						2013 Bare Costs				
4240	4″ pipe size	1 Elec	10	.800	Ea.		42		42	62.50
4260	With hydraulic punch, 1/2″ pipe size		44	.182			9.55		9.55	14.25
4280	3/4″ pipe size		38	.211			11.05		11.05	16.50
4300	1″ pipe size		38	.211			11.05		11.05	16.50
4320	1-1/4″ pipe size		38	.211			11.05		11.05	16.50
4340	1-1/2″ pipe size		38	.211			11.05		11.05	16.50
4360	2″ pipe size		32	.250			13.10		13.10	19.60
4380	2-1/2″ pipe size		27	.296			15.55		15.55	23
4400	3″ pipe size		23	.348			18.25		18.25	27.50
4420	3-1/2″ pipe size		20	.400			21		21	31.50
4440	4″ pipe size		18	.444			23.50		23.50	35

26 05 36 – Cable Trays for Electrical Systems

26 05 36.10 Cable Tray Ladder Type		Crew	Daily Output	Labor-Hours	Unit	Material	Labor	Equipment	Total	Total Incl O&P
0010	**CABLE TRAY LADDER TYPE** w/ftngs. & supports, 4″ dp., to 15′ elev.									
0100	For higher elevations, see Section 26 05 36.40									
0160	Galvanized steel tray R260536-10									
0170	4″ rung spacing, 6″ wide	2 Elec	98	.163	L.F.	17.80	8.55		26.35	32.50
0180	9″ wide R260536-11		92	.174		19.40	9.10		28.50	35
0200	12″ wide		86	.186		21.50	9.75		31.25	38
0400	18″ wide		82	.195		25	10.20		35.20	43
0600	24″ wide		78	.205		28.50	10.75		39.25	47.50
0650	30″ wide		68	.235		36	12.35		48.35	58.50
0700	36″ wide		60	.267		40	13.95		53.95	65
0800	6″ rung spacing, 6″ wide		100	.160		16.20	8.40		24.60	30.50
0850	9″ wide		94	.170		17.80	8.90		26.70	33
0860	12″ wide		88	.182		18.65	9.55		28.20	35
0870	18″ wide		84	.190		21.50	10		31.50	38.50
0880	24″ wide		80	.200		24	10.50		34.50	41.50
0890	30″ wide		70	.229		28.50	12		40.50	49
0900	36″ wide		64	.250		31	13.10		44.10	53.50
0910	9″ rung spacing, 6″ wide		102	.157		15.20	8.20		23.40	29
0920	9″ wide		98	.163		15.60	8.55		24.15	30
0930	12″ wide		94	.170		17.90	8.90		26.80	33
0940	18″ wide		90	.178		20	9.30		29.30	36
0950	24″ wide		86	.186		23	9.75		32.75	40
0960	30″ wide		80	.200		25	10.50		35.50	43
0970	36″ wide		74	.216		27	11.35		38.35	46.50
0980	12″ rung spacing, 6″ wide		106	.151		15	7.90		22.90	28.50
0990	9″ wide		104	.154		15.35	8.05		23.40	29
1000	12″ wide		100	.160		16.20	8.40		24.60	30.50
1010	18″ wide		96	.167		17.40	8.75		26.15	32
1020	24″ wide		94	.170		18.65	8.90		27.55	34
1030	30″ wide		88	.182		21	9.55		30.55	38
1040	36″ wide		84	.190		22.50	10		32.50	39.50
1041	18″ rung spacing, 6″ wide		108	.148		14.85	7.75		22.60	28
1042	9″ wide		106	.151		15.10	7.90		23	28.50
1043	12″ wide		102	.157		16.10	8.20		24.30	30
1044	18″ wide		98	.163		16.90	8.55		25.45	31.50
1045	24″ wide		96	.167		17.40	8.75		26.15	32
1046	30″ wide		90	.178		18.35	9.30		27.65	34
1047	36″ wide		86	.186		18.85	9.75		28.60	35.50
1050	Elbows horiz. 9″ rung spacing, 90°, 12″ radius, 6″ wide		9.60	1.667	Ea.	69	87.50		156.50	207

26 05 36 – Cable Trays for Electrical Systems

26 05 36.10 Cable Tray Ladder Type		Crew	Daily Output	Labor-Hours	Unit	2013 Bare Costs Material	Labor	Equipment	Total	Total Incl O&P
1060	9" wide	2 Elec	8.40	1.905	Ea.	73.50	100		173.50	230
1070	12" wide		7.60	2.105		79.50	110		189.50	253
1080	18" wide		6.20	2.581		105	135		240	315
1090	24" wide		5.40	2.963		117	155		272	360
1100	30" wide		4.80	3.333		155	175		330	430
1110	36" wide		4.20	3.810		180	200		380	495
1120	90° 24" radius, 6" wide		9.20	1.739		143	91		234	293
1130	9" wide		8	2		147	105		252	320
1140	12" wide		7.20	2.222		151	116		267	340
1150	18" wide		5.80	2.759		164	145		309	395
1160	24" wide		5	3.200		176	168		344	445
1170	30" wide		4.40	3.636		191	191		382	495
1180	36" wide		3.80	4.211		227	221		448	580
1190	90° 36" radius, 6" wide		8.80	1.818		157	95.50		252.50	315
1200	9" wide		7.60	2.105		166	110		276	345
1210	12" wide		6.80	2.353		174	123		297	375
1220	18" wide		5.40	2.963		199	155		354	450
1230	24" wide		4.60	3.478		218	182		400	515
1240	30" wide		4	4		262	210		472	605
1250	36" wide		3.40	4.706		285	247		532	685
1260	45° 12" radius, 6" wide		13.20	1.212		53.50	63.50		117	154
1270	9" wide		11	1.455		55.50	76		131.50	175
1280	12" wide		9.60	1.667		58	87.50		145.50	195
1290	18" wide		7.60	2.105		61	110		171	232
1300	24" wide		6.20	2.581		78.50	135		213.50	289
1310	30" wide		5.40	2.963		100	155		255	340
1320	36" wide		4.60	3.478		105	182		287	390
1330	45° 24" radius, 6" wide		12.80	1.250		67.50	65.50		133	173
1340	9" wide		10.60	1.509		72	79		151	197
1350	12" wide		9.20	1.739		81	91		172	225
1360	18" wide		7.20	2.222		83	116		199	265
1370	24" wide		5.80	2.759		100	145		245	325
1380	30" wide		5	3.200		120	168		288	385
1390	36" wide		4.20	3.810		137	200		337	450
1400	45° 36" radius, 6" wide		12.40	1.290		97	67.50		164.50	208
1410	9" wide		10.20	1.569		103	82		185	236
1420	12" wide		8.80	1.818		105	95.50		200.50	258
1430	18" wide		6.80	2.353		113	123		236	310
1440	24" wide		5.40	2.963		120	155		275	365
1450	30" wide		4.60	3.478		155	182		337	445
1460	36" wide		3.80	4.211		172	221		393	520
1470	Elbows horizontal, 4" rung spacing, use 9" rung x 1.50									
1480	6" rung spacing use 9" rung x 1.20									
1490	12" rung spacing use 9" rung x .93									
1500	Elbows vert. 9" rung spacing, 90°, 12" radius, 6" wide	2 Elec	9.60	1.667	Ea.	95	87.50		182.50	235
1510	9" wide		8.40	1.905		97	100		197	256
1520	12" wide		7.60	2.105		100	110		210	275
1530	18" wide		6.20	2.581		105	135		240	315
1540	24" wide		5.40	2.963		109	155		264	350
1550	30" wide		4.80	3.333		117	175		292	390
1560	36" wide		4.20	3.810		122	200		322	435
1570	24" radius, 6" wide		9.20	1.739		134	91		225	284
1580	9" wide		8	2		138	105		243	310

26 05 36 – Cable Trays for Electrical Systems

26 05 36.10 Cable Tray Ladder Type		Crew	Daily Output	Labor-Hours	Unit	Material	2013 Bare Costs Labor	Equipment	Total	Total Incl O&P
1590	12" wide	2 Elec	7.20	2.222	Ea.	143	116		259	330
1600	18" wide		5.80	2.759		151	145		296	380
1610	24" wide		5	3.200		159	168		327	425
1620	30" wide		4.40	3.636		166	191		357	465
1630	36" wide		3.80	4.211		187	221		408	535
1640	36" radius, 6" wide		8.80	1.818		180	95.50		275.50	340
1650	9" wide		7.60	2.105		187	110		297	370
1660	12" wide		6.80	2.353		189	123		312	390
1670	18" wide		5.40	2.963		206	155		361	460
1680	24" wide		4.60	3.478		208	182		390	500
1690	30" wide		4	4		235	210		445	575
1700	36" wide		3.40	4.706		252	247		499	645
1710	Elbows vertical, 4" rung spacing, use 9" rung x 1.25									
1720	6" rung spacing, use 9" rung x 1.15									
1730	12" rung spacing, use 9" rung x .90									
1740	Tee horizontal, 9" rung spacing, 12" radius, 6" wide	2 Elec	5	3.200	Ea.	147	168		315	415
1750	9" wide		4.60	3.478		157	182		339	445
1760	12" wide		4.40	3.636		166	191		357	465
1770	18" wide		4	4		189	210		399	525
1780	24" wide		3.60	4.444		222	233		455	595
1790	30" wide		3.40	4.706		266	247		513	665
1800	36" wide		3	5.333		300	279		579	750
1810	24" radius, 6" wide		4.60	3.478		229	182		411	525
1820	9" wide		4.20	3.810		239	200		439	560
1830	12" wide		4	4		248	210		458	585
1840	18" wide		3.60	4.444		269	233		502	645
1850	24" wide		3.20	5		292	262		554	710
1860	30" wide		3	5.333		315	279		594	770
1870	36" wide		2.60	6.154		405	320		725	930
1880	36" radius, 6" wide		4.20	3.810		340	200		540	675
1890	9" wide		3.80	4.211		355	221		576	720
1900	12" wide		3.60	4.444		380	233		613	770
1910	18" wide		3.20	5		405	262		667	840
1920	24" wide		2.80	5.714		460	299		759	955
1930	30" wide		2.60	6.154		500	320		820	1,025
1940	36" wide		2.20	7.273		540	380		920	1,175
1980	Tee vertical, 9" rung spacing, 12" radius, 6" wide		5.40	2.963		250	155		405	505
1990	9" wide		5.20	3.077		252	161		413	520
2000	12" wide		5	3.200		254	168		422	530
2010	18" wide		4.60	3.478		256	182		438	555
2020	24" wide		4.40	3.636		262	191		453	575
2030	30" wide		4	4		279	210		489	620
2040	36" wide		3.60	4.444		290	233		523	670
2050	24" radius, 6" wide		5	3.200		450	168		618	745
2060	9" wide		4.80	3.333		455	175		630	760
2070	12" wide		4.60	3.478		460	182		642	785
2080	18" wide		4.20	3.810		475	200		675	825
2090	24" wide		4	4		490	210		700	855
2100	30" wide		3.60	4.444		500	233		733	900
2110	36" wide		3.20	5		525	262		787	965
2120	36" radius, 6" wide		4.60	3.478		840	182		1,022	1,200
2130	9" wide		4.40	3.636		850	191		1,041	1,225
2140	12" wide		4.20	3.810		860	200		1,060	1,250

26 05 36 – Cable Trays for Electrical Systems

26 05 36.10 Cable Tray Ladder Type		Crew	Daily Output	Labor-Hours	Unit	2013 Bare Costs Material	Labor	Equipment	Total	Total Incl O&P
2150	18" wide	2 Elec	3.80	4.211	Ea.	875	221		1,096	1,300
2160	24" wide		3.60	4.444		885	233		1,118	1,325
2170	30" wide		3.20	5		915	262		1,177	1,400
2180	36" wide	↓	2.80	5.714	↓	925	299		1,224	1,475
2190	Tee, 4" rung spacing, use 9" rung x 1.30									
2200	6" rung spacing, use 9" rung x 1.20									
2210	12" rung spacing, use 9" rung x .90									
2220	Cross horizontal, 9" rung spacing, 12" radius, 6" wide	2 Elec	4	4	Ea.	181	210		391	515
2230	9" wide		3.80	4.211		190	221		411	540
2240	12" wide		3.60	4.444		208	233		441	580
2250	18" wide		3.40	4.706		227	247		474	620
2260	24" wide		3	5.333		252	279		531	695
2270	30" wide		2.80	5.714		305	299		604	785
2280	36" wide		2.60	6.154		340	320		660	850
2290	24" radius, 6" wide		3.60	4.444		355	233		588	740
2300	9" wide		3.40	4.706		365	247		612	770
2310	12" wide		3.20	5		375	262		637	800
2320	18" wide		3	5.333		400	279		679	860
2330	24" wide		2.60	6.154		425	320		745	945
2340	30" wide		2.40	6.667		450	350		800	1,025
2350	36" wide		2.20	7.273		525	380		905	1,150
2360	36" radius, 6" wide		3.20	5		490	262		752	930
2370	9" wide		3	5.333		510	279		789	985
2380	12" wide		2.80	5.714		535	299		834	1,050
2390	18" wide		2.60	6.154		590	320		910	1,125
2400	24" wide		2.20	7.273		655	380		1,035	1,300
2410	30" wide		2	8		710	420		1,130	1,400
2420	36" wide	↓	1.80	8.889	↓	765	465		1,230	1,525
2430	Cross horizontal, 4" rung spacing, use 9" rung x 1.30									
2440	6" rung spacing, use 9" rung x 1.20									
2450	12" rung spacing, use 9" rung x .90									
2460	Reducer, 9" to 6" wide tray	2 Elec	13	1.231	Ea.	101	64.50		165.50	208
2470	12" to 9" wide tray		12	1.333		103	70		173	218
2480	18" to 12" wide tray		10.40	1.538		103	80.50		183.50	234
2490	24" to 18" wide tray		9	1.778		105	93		198	254
2500	30" to 24" wide tray		8	2		107	105		212	275
2510	36" to 30" wide tray		7	2.286		120	120		240	310
2511	Reducer, 18" to 6" wide tray		10.40	1.538		107	80.50		187.50	239
2512	24" to 12" wide tray		9	1.778		109	93		202	259
2513	30" to 18" wide tray		8	2		113	105		218	282
2514	30" to 12" wide tray		8	2		118	105		223	286
2515	36" to 24" wide tray		7	2.286		122	120		242	315
2516	36" to 18" wide tray		7	2.286		122	120		242	315
2517	36" to 12" wide tray		7	2.286		122	120		242	315
2520	Dropout or end plate, 6" wide		32	.500		11.20	26		37.20	51.50
2530	9" wide		28	.571		13.20	30		43.20	59.50
2540	12" wide		26	.615		14.50	32.50		47	64.50
2550	18" wide		22	.727		17	38		55	75.50
2560	24" wide		20	.800		20.50	42		62.50	85
2570	30" wide		18	.889		22.50	46.50		69	94
2580	36" wide		16	1		26	52.50		78.50	107
2590	Tray connector		48	.333	↓	19.95	17.45		37.40	48
3200	Aluminum tray, 4" deep, 6" rung spacing, 6" wide		134	.119	L.F.	17.65	6.25		23.90	29

26 05 36.10 Cable Tray Ladder Type		Crew	Daily Output	Labor-Hours	Unit	Material	Labor	Equipment	Total	Total Incl O&P
						2013 Bare Costs				
3210	9" wide	2 Elec	128	.125	L.F.	18.70	6.55		25.25	30.50
3220	12" wide		124	.129		19.75	6.75		26.50	32
3230	18" wide		114	.140		22	7.35		29.35	35
3240	24" wide		106	.151		25.50	7.90		33.40	40
3250	30" wide		100	.160		28	8.40		36.40	43.50
3260	36" wide		94	.170		30.50	8.90		39.40	47
3270	9" rung spacing, 6" wide		140	.114		14.35	6		20.35	25
3280	9" wide		134	.119		15	6.25		21.25	26
3290	12" wide		130	.123		15.30	6.45		21.75	26.50
3300	18" wide		122	.131		17.85	6.85		24.70	30
3310	24" wide		116	.138		20.50	7.25		27.75	34
3320	30" wide		108	.148		23	7.75		30.75	37
3330	36" wide		100	.160		24.50	8.40		32.90	39
3340	12" rung spacing, 6" wide		146	.110		15.20	5.75		20.95	25.50
3350	9" wide		140	.114		15.40	6		21.40	26
3360	12" wide		134	.119		15.90	6.25		22.15	27
3370	18" wide		128	.125		16.80	6.55		23.35	28.50
3380	24" wide		124	.129		18.70	6.75		25.45	30.50
3390	30" wide		114	.140		20	7.35		27.35	33
3400	36" wide		106	.151		21	7.90		28.90	35
3401	18" rung spacing, 6" wide		150	.107		15.10	5.60		20.70	25
3402	9" wide tray		144	.111		15.30	5.80		21.10	25.50
3403	12" wide tray		140	.114		15.90	6		21.90	26.50
3404	18" wide tray		134	.119		16.70	6.25		22.95	28
3405	24" wide tray		130	.123		18	6.45		24.45	29.50
3406	30" wide tray		120	.133		18.65	7		25.65	31
3407	36" wide tray		110	.145		19.50	7.60		27.10	33
3410	Elbows horiz. 9" rung spacing, 90° 12" radius, 6" wide		9.60	1.667	Ea.	68	87.50		155.50	206
3420	9" wide		8.40	1.905		70	100		170	226
3430	12" wide		7.60	2.105		82.50	110		192.50	256
3440	18" wide		6.20	2.581		93	135		228	305
3450	24" wide		5.40	2.963		115	155		270	360
3460	30" wide		4.80	3.333		122	175		297	395
3470	36" wide		4.20	3.810		147	200		347	460
3480	24" radius, 6" wide		9.20	1.739		121	91		212	269
3490	9" wide		8	2		126	105		231	295
3500	12" wide		7.20	2.222		132	116		248	320
3510	18" wide		5.80	2.759		143	145		288	375
3520	24" wide		5	3.200		157	168		325	425
3530	30" wide		4.40	3.636		171	191		362	475
3540	36" wide		3.80	4.211		194	221		415	545
3550	90°, 36" radius, 6" wide		8.80	1.818		133	95.50		228.50	289
3560	9" wide		7.60	2.105		136	110		246	315
3570	12" wide		6.80	2.353		152	123		275	350
3580	18" wide		5.40	2.963		171	155		326	420
3590	24" wide		4.60	3.478		192	182		374	485
3600	30" wide		4	4		212	210		422	550
3610	36" wide		3.40	4.706		244	247		491	640
3620	45° 12" radius, 6" wide		13.20	1.212		44.50	63.50		108	144
3630	9" wide		11	1.455		46.50	76		122.50	165
3640	12" wide		9.60	1.667		48	87.50		135.50	184
3650	18" wide		7.60	2.105		57	110		167	228
3660	24" wide		6.20	2.581		70.50	135		205.50	280

26 05 36 – Cable Trays for Electrical Systems

26 05 36.10 Cable Tray Ladder Type		Crew	Daily Output	Labor-Hours	Unit	2013 Bare Costs Material	Labor	Equipment	Total	Total Incl O&P
3670	30" wide	2 Elec	5.40	2.963	Ea.	81	155		236	320
3680	36" wide		4.60	3.478		84.50	182		266.50	365
3690	45° 24" radius, 6" wide		12.80	1.250		59.50	65.50		125	164
3700	9" wide		10.60	1.509		68	79		147	193
3710	12" wide		9.20	1.739		69.50	91		160.50	213
3720	18" wide		7.20	2.222		74.50	116		190.50	256
3730	24" wide		5.80	2.759		86	145		231	310
3740	30" wide		5	3.200		103	168		271	365
3750	36" wide		4.20	3.810		109	200		309	420
3760	45° 36" radius, 6" wide		12.40	1.290		81	67.50		148.50	191
3770	9" wide		10.20	1.569		83	82		165	214
3780	12" wide		8.80	1.818		84.50	95.50		180	236
3790	18" wide		6.80	2.353		93	123		216	286
3800	24" wide		5.40	2.963		114	155		269	360
3810	30" wide		4.60	3.478		121	182		303	405
3820	36" wide		3.80	4.211		133	221		354	475
3830	Elbows horizontal, 4" rung spacing, use 9" rung x 1.50									
3840	6" rung spacing, use 9" rung x 1.20									
3850	12" rung spacing, use 9" rung x .93									
3860	Elbows vertical 9" rung spacing, 90° 12" radius, 6" wide	2 Elec	9.60	1.667	Ea.	94.50	87.50		182	235
3870	9" wide		8.40	1.905		99.50	100		199.50	258
3880	12" wide		7.60	2.105		101	110		211	276
3890	18" wide		6.20	2.581		104	135		239	315
3900	24" wide		5.40	2.963		111	155		266	355
3910	30" wide		4.80	3.333		116	175		291	390
3920	36" wide		4.20	3.810		119	200		319	430
3930	24" radius, 6" wide		9.20	1.739		103	91		194	249
3940	9" wide		8	2		106	105		211	274
3950	12" wide		7.20	2.222		109	116		225	294
3960	18" wide		5.80	2.759		116	145		261	345
3970	24" wide		5	3.200		123	168		291	385
3980	30" wide		4.40	3.636		129	191		320	425
3990	36" wide		3.80	4.211		138	221		359	480
4000	36" radius, 6" wide		8.80	1.818		159	95.50		254.50	320
4010	9" wide		7.60	2.105		162	110		272	345
4020	12" wide		6.80	2.353		172	123		295	375
4030	18" wide		5.40	2.963		179	155		334	430
4040	24" wide		4.60	3.478		186	182		368	475
4050	30" wide		4	4		200	210		410	535
4060	36" wide		3.40	4.706		209	247		456	600
4070	Elbows vertical, 4" rung spacing, use 9" rung x 1.25									
4080	6" rung spacing, use 9" rung x 1.15									
4090	12" rung spacing, use 9" rung x .90									
4100	Tee horizontal 9" rung spacing, 12" radius, 6" wide	2 Elec	5	3.200	Ea.	105	168		273	365
4110	9" wide		4.60	3.478		113	182		295	395
4120	12" wide		4.40	3.636		125	191		316	420
4130	18" wide		4.20	3.810		136	200		336	450
4140	24" wide		4	4		149	210		359	480
4150	30" wide		3.60	4.444		174	233		407	540
4160	36" wide		3.40	4.706		209	247		456	600
4170	24" radius, 6" wide		4.60	3.478		179	182		361	470
4180	9" wide		4.20	3.810		189	200		389	505
4190	12" wide		4	4		194	210		404	530

26 05 36.10 Cable Tray Ladder Type		Crew	Daily Output	Labor-Hours	Unit	2013 Bare Costs Material	Labor	Equipment	Total	Total Incl O&P
4200	18" wide	2 Elec	3.80	4.211	Ea.	212	221		433	565
4210	24" wide		3.60	4.444		232	233		465	605
4220	30" wide		3.20	5		252	262		514	665
4230	36" wide		3	5.333		390	279		669	845
4240	36" radius, 6" wide		4.20	3.810		253	200		453	580
4250	9" wide		3.80	4.211		260	221		481	615
4260	12" wide		3.60	4.444		285	233		518	665
4270	18" wide		3.40	4.706		300	247		547	700
4280	24" wide		3.20	5		330	262		592	755
4290	30" wide		2.80	5.714		420	299		719	910
4300	36" wide		2.60	6.154		465	320		785	990
4310	Tee vertical 9" rung spacing, 12" radius, 6" wide		5.40	2.963		235	155		390	490
4320	9" wide		5.20	3.077		244	161		405	510
4330	12" wide		5	3.200		247	168		415	525
4340	18" wide		4.60	3.478		250	182		432	550
4350	24" wide		4.40	3.636		270	191		461	580
4360	30" wide		4.20	3.810		275	200		475	605
4370	36" wide		4	4		285	210		495	630
4380	24" radius, 6" wide		5	3.200		400	168		568	690
4390	9" wide		4.80	3.333		405	175		580	705
4400	12" wide		4.60	3.478		410	182		592	725
4410	18" wide		4.20	3.810		420	200		620	765
4420	24" wide		4	4		430	210		640	790
4430	30" wide		3.80	4.211		445	221		666	820
4440	36" wide		3.60	4.444		480	233		713	880
4450	36" radius, 6" wide		4.60	3.478		880	182		1,062	1,250
4460	9" wide		4.40	3.636		895	191		1,086	1,275
4470	12" wide		4.20	3.810		930	200		1,130	1,325
4480	18" wide		3.80	4.211		940	221		1,161	1,350
4490	24" wide		3.60	4.444		965	233		1,198	1,400
4500	30" wide		3.40	4.706		975	247		1,222	1,450
4510	36" wide		3.20	5		990	262		1,252	1,500
4520	Tees, 4" rung spacing, use 9" rung x 1.30									
4530	6" rung spacing, use 9" rung x 1.20									
4540	12" rung spacing, use 9" rung x .90									
4550	Cross horizontal 9" rung spacing, 12" radius, 6" wide	2 Elec	4.40	3.636	Ea.	146	191		337	445
4560	9" wide		4.20	3.810		162	200		362	480
4570	12" wide		4	4		180	210		390	515
4580	18" wide		3.60	4.444		199	233		432	570
4590	24" wide		3.40	4.706		205	247		452	595
4600	30" wide		3	5.333		242	279		521	685
4610	36" wide		2.80	5.714		300	299		599	780
4620	24" radius, 6" wide		4	4		280	210		490	625
4630	9" wide		3.80	4.211		290	221		511	650
4640	12" wide		3.60	4.444		298	233		531	680
4650	18" wide		3.20	5		315	262		577	735
4660	24" wide		3	5.333		340	279		619	795
4670	30" wide		2.60	6.154		365	320		685	880
4680	36" wide		2.40	6.667		415	350		765	980
4690	36" radius, 6" wide		3.60	4.444		330	233		563	715
4700	9" wide		3.40	4.706		345	247		592	750
4710	12" wide		3.20	5		365	262		627	790
4720	18" wide		2.80	5.714		390	299		689	880

26 05 36 – Cable Trays for Electrical Systems

26 05 36.10 Cable Tray Ladder Type		Crew	Daily Output	Labor-Hours	Unit	Material	Labor	Equipment	Total	Total Incl O&P
						2013 Bare Costs				
4730	24" wide	2 Elec	2.60	6.154	Ea.	480	320		800	1,000
4740	30" wide		2.20	7.273		535	380		915	1,150
4750	36" wide		2	8		595	420		1,015	1,275
4760	Cross horizontal, 4" rung spacing, use 9" rung x 1.30									
4770	6" rung spacing, use 9" rung x 1.20									
4780	12" rung spacing, use 9" rung x .90									
4790	Reducer, 9" to 6" wide tray	2 Elec	16	1	Ea.	78	52.50		130.50	164
4800	12" to 9" wide tray		14	1.143		81	60		141	179
4810	18" to 12" wide tray		12.40	1.290		81	67.50		148.50	191
4820	24" to 18" wide tray		10.60	1.509		83	79		162	209
4830	30" to 24" wide tray		9.20	1.739		84.50	91		175.50	229
4840	36" to 30" wide tray		8	2		89.50	105		194.50	256
4841	Reducer, 18" to 6" wide tray		12.40	1.290		81	67.50		148.50	191
4842	24" to 12" wide tray		10.60	1.509		83	79		162	209
4843	30" to 18" wide tray		9.20	1.739		84.50	91		175.50	229
4844	30" to 12" wide tray		9.20	1.739		84.50	91		175.50	229
4845	36" to 24" wide tray		8	2		89.50	105		194.50	256
4846	36" to 18" wide tray		8	2		89.50	105		194.50	256
4847	36" to 12" wide tray		8	2		89.50	105		194.50	256
4850	Dropout or end plate, 6" wide		32	.500		10.30	26		36.30	50.50
4860	9" wide tray		28	.571		10.95	30		40.95	57
4870	12" wide tray		26	.615		11.95	32.50		44.45	61.50
4880	18" wide tray		22	.727		15.90	38		53.90	74.50
4890	24" wide tray		20	.800		18.70	42		60.70	83
4900	30" wide tray		18	.889		22	46.50		68.50	94
4910	36" wide tray		16	1		24.50	52.50		77	106
4920	Tray connector		48	.333		14.90	17.45		32.35	42.50
8000	Elbow 36" radius horiz., 60°, 6" wide tray		10.60	1.509		113	79		192	242
8010	9" wide tray		9	1.778		123	93		216	274
8020	12" wide tray		7.80	2.051		126	107		233	300
8030	18" wide tray		6.20	2.581		134	135		269	350
8040	24" wide tray		5	3.200		157	168		325	425
8050	30" wide tray		4.40	3.636		161	191		352	460
8060	30°, 6" wide tray		14	1.143		86	60		146	185
8070	9" wide tray		11.40	1.404		91	73.50		164.50	210
8080	12" wide tray		9.80	1.633		93	85.50		178.50	230
8090	18" wide tray		7.40	2.162		97.50	113		210.50	278
8100	24" wide tray		5.80	2.759		119	145		264	345
8110	30" wide tray		4.80	3.333		123	175		298	395
8120	Adjustable, 6" wide tray		12.40	1.290		114	67.50		181.50	227
8130	9" wide tray		10.20	1.569		121	82		203	256
8140	12" wide tray		8.80	1.818		123	95.50		218.50	278
8150	18" wide tray		6.80	2.353		134	123		257	330
8160	24" wide tray		5.40	2.963		142	155		297	390
8170	30" wide tray		4.60	3.478		159	182		341	450
8180	Wye 36" radius horiz., 45°, 6" wide tray		4.60	3.478		124	182		306	410
8190	9" wide tray		4.40	3.636		131	191		322	430
8200	12" wide tray		4.20	3.810		136	200		336	450
8210	18" wide tray		3.80	4.211		154	221		375	500
8220	24" wide tray		3.60	4.444		172	233		405	540
8230	30" wide tray		3.40	4.706		192	247		439	580
8240	Elbow 36" radius vert. in/outside, 60°, 6" wide tray		10.60	1.509		124	79		203	255
8250	9" wide tray		9	1.778		129	93		222	281

26 05 36 – Cable Trays for Electrical Systems

26 05 36.10 Cable Tray Ladder Type		Crew	Daily Output	Labor-Hours	Unit	Material	2013 Bare Costs Labor	Equipment	Total	Total Incl O&P
8260	12" wide tray	2 Elec	7.80	2.051	Ea.	131	107		238	305
8270	18" wide tray		6.20	2.581		136	135		271	350
8280	24" wide tray		5	3.200		144	168		312	410
8290	30" wide tray		4.40	3.636		154	191		345	455
8300	45°, 6" wide tray		12.40	1.290		112	67.50		179.50	224
8310	9" wide tray		10.20	1.569		119	82		201	254
8320	12" wide tray		8.80	1.818		121	95.50		216.50	276
8330	18" wide tray		6.80	2.353		124	123		247	320
8340	24" wide tray		5.40	2.963		131	155		286	375
8350	30" wide tray		4.60	3.478		134	182		316	420
8360	30°, 6" wide tray		14	1.143		98	60		158	198
8370	9" wide tray		11.40	1.404		100	73.50		173.50	220
8380	12" wide tray		9.80	1.633		107	85.50		192.50	246
8390	18" wide tray		7.40	2.162		109	113		222	289
8400	24" wide tray		5.80	2.759		111	145		256	340
8410	30" wide tray		4.80	3.333		117	175		292	390
8660	Adjustable, 6" wide tray		12.40	1.290		112	67.50		179.50	224
8670	9" wide tray		10.20	1.569		119	82		201	254
8680	12" wide tray		8.80	1.818		121	95.50		216.50	276
8690	18" wide tray		6.80	2.353		134	123		257	330
8700	24" wide tray		5.40	2.963		142	155		297	390
8710	30" wide tray		4.60	3.478		157	182		339	445
8720	Cross, vertical, 24" radius, 6" wide tray		3.60	4.444		765	233		998	1,200
8730	9" wide tray		3.40	4.706		780	247		1,027	1,225
8740	12" wide tray		3.20	5		800	262		1,062	1,275
8750	18" wide tray		2.80	5.714		840	299		1,139	1,375
8760	24" wide tray		2.60	6.154		865	320		1,185	1,425
8770	30" wide tray		2.20	7.273		920	380		1,300	1,575
9200	Splice plate	1 Elec	48	.167	Pr.	15.55	8.75		24.30	30
9210	Expansion joint		48	.167		18.70	8.75		27.45	33.50
9220	Horizontal hinged		48	.167		15.55	8.75		24.30	30
9230	Vertical hinged		48	.167		18.70	8.75		27.45	33.50
9240	Ladder hanger, vertical		28	.286	Ea.	4.63	14.95		19.58	27.50
9250	Ladder to channel connector		24	.333		54.50	17.45		71.95	86
9260	Ladder to box connector 30" wide		19	.421		54.50	22		76.50	93
9270	24" wide		20	.400		54.50	21		75.50	91.50
9280	18" wide		21	.381		52.50	19.95		72.45	88
9290	12" wide		22	.364		49	19.05		68.05	82.50
9300	9" wide		23	.348		44	18.25		62.25	76
9310	6" wide		24	.333		39	17.45		56.45	69
9320	Ladder floor flange		24	.333		35.50	17.45		52.95	65
9330	Cable roller for tray 30" wide		10	.800		289	42		331	385
9340	24" wide		11	.727		246	38		284	325
9350	18" wide		12	.667		234	35		269	310
9360	12" wide		13	.615		184	32.50		216.50	252
9370	9" wide		14	.571		163	30		193	225
9380	6" wide		15	.533		133	28		161	188
9390	Pulley, single wheel		12	.667		300	35		335	390
9400	Triple wheel		10	.800		605	42		647	730
9440	Nylon cable tie, 14" long		80	.100		.46	5.25		5.71	8.35
9450	Ladder, hold down clamp		60	.133		9.95	7		16.95	21.50
9460	Cable clamp		60	.133		10.10	7		17.10	21.50
9470	Wall bracket, 30" wide tray		19	.421		53.50	22		75.50	91.50

26 05 Common Work Results for Electrical

26 05 36 – Cable Trays for Electrical Systems

26 05 36.10 Cable Tray Ladder Type		Crew	Daily Output	Labor-Hours	Unit	2013 Bare Costs Material	Labor	Equipment	Total	Total Incl O&P
9480	24" wide tray	1 Elec	20	.400	Ea.	41.50	21		62.50	77
9490	18" wide tray		21	.381		37	19.95		56.95	71
9500	12" wide tray		22	.364		35	19.05		54.05	67
9510	9" wide tray		23	.348		30.50	18.25		48.75	61
9520	6" wide tray	↓	24	.333	↓	29.50	17.45		46.95	58.50

26 05 36.20 Cable Tray Solid Bottom		Crew	Daily Output	Labor-Hours	Unit	Material	Labor	Equipment	Total	Total Incl O&P
0010	**CABLE TRAY SOLID BOTTOM** w/ftngs. & supports, 3" dp, to 15' high									
0200	For higher elevations, see Section 26 05 36.40									
0220	Galvanized steel, tray, 6" wide R260536-10	2 Elec	120	.133	L.F.	14.35	7		21.35	26.50
0240	12" wide		100	.160		18.40	8.40		26.80	33
0260	18" wide		70	.229		22.50	12		34.50	42.50
0280	24" wide R260536-11		60	.267		26.50	13.95		40.45	50
0300	30" wide		50	.320		31.50	16.75		48.25	59.50
0320	36" wide		44	.364	↓	44.50	19.05		63.55	77.50
0340	Elbow horizontal 90°, 12" radius, 6" wide		9.60	1.667	Ea.	109	87.50		196.50	251
0360	12" wide		6.80	2.353		122	123		245	320
0370	18" wide		5.40	2.963		140	155		295	385
0380	24" wide		4.40	3.636		170	191		361	470
0390	30" wide		3.80	4.211		204	221		425	555
0400	36" wide		3.40	4.706		231	247		478	625
0420	24" radius, 6" wide		9.20	1.739		155	91		246	305
0440	12" wide		6.40	2.500		180	131		311	395
0450	18" wide		5	3.200		210	168		378	480
0460	24" wide		4	4		243	210		453	585
0470	30" wide		3.40	4.706		285	247		532	685
0480	36" wide		3	5.333		320	279		599	775
0500	36" radius, 6" wide		8.80	1.818		227	95.50		322.50	390
0520	12" wide		6	2.667		256	140		396	490
0530	18" wide		4.60	3.478		315	182		497	620
0540	24" wide		3.60	4.444		340	233		573	720
0550	30" wide		3	5.333		410	279		689	870
0560	36" wide		2.60	6.154		460	320		780	985
0580	Elbow vertical 90°, 12" radius, 6" wide		9.60	1.667		132	87.50		219.50	276
0600	12" wide		6.80	2.353		143	123		266	340
0610	18" wide		5.40	2.963		155	155		310	400
0620	24" wide		4.40	3.636		166	191		357	465
0630	30" wide		3.80	4.211		176	221		397	525
0640	36" wide		3.40	4.706		183	247		430	570
0670	24" radius, 6" wide		9.20	1.739		187	91		278	340
0690	12" wide		6.40	2.500		206	131		337	420
0700	18" wide		5	3.200		222	168		390	495
0710	24" wide		4	4		241	210		451	580
0720	30" wide		3.40	4.706		256	247		503	650
0730	36" wide		3	5.333		279	279		558	725
0750	36" radius, 6" wide		8.80	1.818		256	95.50		351.50	425
0770	12" wide		6.60	2.424		285	127		412	505
0780	18" wide		4.60	3.478		315	182		497	620
0790	24" wide		3.60	4.444		340	233		573	725
0800	30" wide		3	5.333		375	279		654	830
0810	36" wide		2.60	6.154		405	320		725	925
0840	Tee horizontal, 12" radius, 6" wide		5	3.200		153	168		321	420
0860	12" wide	↓	4	4	↓	166	210		376	495

26 05 36 – Cable Trays for Electrical Systems

26 05 36.20 Cable Tray Solid Bottom		Crew	Daily Output	Labor-Hours	Unit	2013 Bare Costs Material	Labor	Equipment	Total	Total Incl O&P
0870	18" wide	2 Elec	3.40	4.706	Ea.	197	247		444	585
0880	24" wide		2.80	5.714		220	299		519	690
0890	30" wide		2.60	6.154		252	320		572	755
0900	36" wide		2.20	7.273		287	380		667	885
0940	24" radius, 6" wide		4.60	3.478		237	182		419	535
0960	12" wide		3.60	4.444		277	233		510	655
0970	18" wide		3	5.333		310	279		589	760
0980	24" wide		2.40	6.667		420	350		770	985
0990	30" wide		2.20	7.273		455	380		835	1,075
1000	36" wide		1.80	8.889		490	465		955	1,225
1020	36" radius, 6" wide		4.20	3.810		375	200		575	710
1040	12" wide		3.20	5		420	262		682	850
1050	18" wide		2.60	6.154		455	320		775	980
1060	24" wide		2.20	7.273		585	380		965	1,200
1070	30" wide		2	8		635	420		1,055	1,325
1080	36" wide		1.60	10		655	525		1,180	1,500
1100	Tee vertical, 12" radius, 6" wide		5	3.200		252	168		420	530
1120	12" wide		4	4		262	210		472	605
1130	18" wide		3.60	4.444		266	233		499	645
1140	24" wide		3.40	4.706		292	247		539	690
1150	30" wide		3	5.333		315	279		594	765
1160	36" wide		2.60	6.154		325	320		645	835
1180	24" radius, 6" wide		4.60	3.478		370	182		552	685
1200	12" wide		3.60	4.444		390	233		623	775
1210	18" wide		3.20	5		405	262		667	840
1220	24" wide		3	5.333		430	279		709	890
1230	30" wide		2.60	6.154		470	320		790	995
1240	36" wide		2.20	7.273		490	380		870	1,100
1260	36" radius, 6" wide		4.20	3.810		580	200		780	935
1280	12" wide		3.20	5		590	262		852	1,050
1290	18" wide		2.80	5.714		635	299		934	1,150
1300	24" wide		2.60	6.154		655	320		975	1,200
1310	30" wide		2.20	7.273		765	380		1,145	1,400
1320	36" wide		2	8		810	420		1,230	1,525
1340	Cross horizontal, 12" radius, 6" wide		4	4		183	210		393	515
1360	12" wide		3.40	4.706		199	247		446	590
1370	18" wide		2.80	5.714		233	299		532	705
1380	24" wide		2.40	6.667		262	350		612	815
1390	30" wide		2	8		292	420		712	945
1400	36" wide		1.80	8.889		320	465		785	1,050
1420	24" radius, 6" wide		3.60	4.444		335	233		568	715
1440	12" wide		3	5.333		375	279		654	830
1450	18" wide		2.40	6.667		420	350		770	985
1460	24" wide		2	8		535	420		955	1,225
1470	30" wide		1.80	8.889		580	465		1,045	1,325
1480	36" wide		1.60	10		610	525		1,135	1,450
1500	36" radius, 6" wide		3.20	5		545	262		807	990
1520	12" wide		2.60	6.154		600	320		920	1,150
1530	18" wide		2	8		655	420		1,075	1,350
1540	24" wide		1.80	8.889		775	465		1,240	1,550
1550	30" wide		1.60	10		840	525		1,365	1,700
1560	36" wide		1.40	11.429		895	600		1,495	1,875
1580	Drop out or end plate, 6" wide		32	.500		22	26		48	63.50

26 05 36.20 Cable Tray Solid Bottom		Crew	Daily Output	Labor-Hours	Unit	Material	Labor	Equipment	Total	Total Incl O&P
						2013 Bare Costs				
1600	12" wide	2 Elec	26	.615	Ea.	27.50	32.50		60	79
1610	18" wide		22	.727		29.50	38		67.50	89.50
1620	24" wide		20	.800		35	42		77	101
1630	30" wide		18	.889		39	46.50		85.50	113
1640	36" wide		16	1		42	52.50		94.50	125
1660	Reducer, 12" to 6" wide		12	1.333		107	70		177	223
1680	18" to 12" wide		10.60	1.509		109	79		188	238
1700	18" to 6" wide		10.60	1.509		109	79		188	238
1720	24" to 18" wide		9.20	1.739		113	91		204	261
1740	24" to 12" wide		9.20	1.739		113	91		204	261
1760	30" to 24" wide		8	2		122	105		227	291
1780	30" to 18" wide		8	2		122	105		227	291
1800	30" to 12" wide		8	2		122	105		227	291
1820	36" to 30" wide		7.20	2.222		126	116		242	310
1840	36" to 24" wide		7.20	2.222		126	116		242	310
1860	36" to 18" wide		7.20	2.222		126	116		242	310
1880	36" to 12" wide		7.20	2.222		126	116		242	310
2000	Aluminum tray, 6" wide		150	.107	L.F.	13.75	5.60		19.35	23.50
2020	12" wide		130	.123		18	6.45		24.45	29.50
2030	18" wide		100	.160		22.50	8.40		30.90	37.50
2040	24" wide		90	.178		28.50	9.30		37.80	45.50
2050	30" wide		70	.229		33.50	12		45.50	55
2060	36" wide		64	.250		42	13.10		55.10	66
2080	Elbow horizontal 90°, 12" radius, 6" wide		9.60	1.667	Ea.	110	87.50		197.50	252
2100	12" wide		7.60	2.105		124	110		234	300
2110	18" wide		6.80	2.353		152	123		275	350
2120	24" wide		5.80	2.759		174	145		319	405
2130	30" wide		5	3.200		212	168		380	485
2140	36" wide		4.40	3.636		239	191		430	545
2160	24" radius, 6" wide		9.20	1.739		157	91		248	310
2180	12" wide		7.20	2.222		186	116		302	380
2190	18" wide		6.40	2.500		212	131		343	430
2200	24" wide		5.40	2.963		262	155		417	520
2210	30" wide		4.60	3.478		300	182		482	605
2220	36" wide		4	4		345	210		555	695
2240	36" radius, 6" wide		8.80	1.818		232	95.50		327.50	400
2260	12" wide		6.80	2.353		278	123		401	490
2270	18" wide		6	2.667		330	140		470	575
2280	24" wide		5	3.200		355	168		523	640
2290	30" wide		4.20	3.810		400	200		600	740
2300	36" wide		3.60	4.444		470	233		703	870
2320	Elbow vertical 90°, 12" radius, 6" wide		9.60	1.667		131	87.50		218.50	275
2340	12" wide		7.60	2.105		136	110		246	315
2350	18" wide		6.80	2.353		154	123		277	355
2360	24" wide		5.80	2.759		162	145		307	395
2370	30" wide		5	3.200		172	168		340	440
2380	36" wide		4.40	3.636		177	191		368	480
2400	24" radius, 6" wide		9.20	1.739		186	91		277	340
2420	12" wide		7.20	2.222		202	116		318	395
2430	18" wide		6.40	2.500		217	131		348	435
2440	24" wide		5.40	2.963		235	155		390	490
2450	30" wide		4.60	3.478		250	182		432	550
2460	36" wide		4	4		275	210		485	620

26 05 36.20 Cable Tray Solid Bottom		Crew	Daily Output	Labor-Hours	Unit	Material	2013 Bare Costs Labor	Equipment	Total	Total Incl O&P
2480	36" radius, 6" wide	2 Elec	8.80	1.818	Ea.	245	95.50		340.50	415
2500	12" wide		6.80	2.353		275	123		398	490
2510	18" wide		6	2.667		300	140		440	540
2520	24" wide		5	3.200		310	168		478	595
2530	30" wide		4.20	3.810		345	200		545	680
2540	36" wide		3.60	4.444		360	233		593	745
2560	Tee horizontal, 12" radius, 6" wide		5	3.200		171	168		339	440
2580	12" wide		4.40	3.636		207	191		398	515
2590	18" wide		4	4		232	210		442	570
2600	24" wide		3.60	4.444		260	233		493	635
2610	30" wide		3	5.333		300	279		579	750
2620	36" wide		2.40	6.667		345	350		695	905
2640	24" radius, 6" wide		4.60	3.478		282	182		464	585
2660	12" wide		4	4		315	210		525	665
2670	18" wide		3.60	4.444		360	233		593	745
2680	24" wide		3	5.333		440	279		719	905
2690	30" wide		2.40	6.667		500	350		850	1,075
2700	36" wide		2.20	7.273		560	380		940	1,175
2720	36" radius, 6" wide		4.20	3.810		455	200		655	805
2740	12" wide		3.60	4.444		500	233		733	900
2750	18" wide		3.20	5		575	262		837	1,025
2760	24" wide		2.60	6.154		675	320		995	1,225
2770	30" wide		2	8		740	420		1,160	1,450
2780	36" wide		1.80	8.889		825	465		1,290	1,600
2800	Tee vertical, 12" radius, 6" wide		5	3.200		244	168		412	520
2820	12" wide		4.40	3.636		249	191		440	560
2830	18" wide		4.20	3.810		258	200		458	585
2840	24" wide		4	4		275	210		485	620
2850	30" wide		3.60	4.444		300	233		533	680
2860	36" wide		3	5.333		315	279		594	770
2880	24" radius, 6" wide		4.60	3.478		360	182		542	670
2900	12" wide		4	4		380	210		590	730
2910	18" wide		3.80	4.211		405	221		626	775
2920	24" wide		3.60	4.444		440	233		673	835
2930	30" wide		3.20	5		480	262		742	920
2940	36" wide		2.60	6.154		510	320		830	1,050
2960	36" radius, 6" wide		4.20	3.810		560	200		760	915
2980	12" wide		3.40	4.706		585	247		832	1,025
2990	18" wide		3.40	4.706		620	247		867	1,050
3000	24" wide		3.20	5		640	262		902	1,100
3010	30" wide		2.80	5.714		690	299		989	1,200
3020	36" wide		2.20	7.273		760	380		1,140	1,400
3040	Cross horizontal, 12" radius, 6" wide		4.40	3.636		217	191		408	525
3060	12" wide		4	4		245	210		455	585
3070	18" wide		3.40	4.706		278	247		525	675
3080	24" wide		2.80	5.714		315	299		614	800
3090	30" wide		2.60	6.154		360	320		680	875
3100	36" wide		2.20	7.273		395	380		775	1,000
3120	24" radius, 6" wide		4	4		380	210		590	730
3140	12" wide		3.60	4.444		440	233		673	835
3150	18" wide		3	5.333		475	279		754	945
3160	24" wide		2.40	6.667		575	350		925	1,150
3170	30" wide		2.20	7.273		640	380		1,020	1,275

26 05 Common Work Results for Electrical

26 05 36 – Cable Trays for Electrical Systems

26 05 36.20 Cable Tray Solid Bottom		Crew	Daily Output	Labor-Hours	Unit	2013 Bare Costs Material	Labor	Equipment	Total	Total Incl O&P
3180	36" wide	2 Elec	1.80	8.889	Ea.	665	465		1,130	1,425
3200	36" radius, 6" wide		3.60	4.444		665	233		898	1,075
3220	12" wide		3.20	5		695	262		957	1,150
3230	18" wide		2.60	6.154		775	320		1,095	1,325
3240	24" wide		2	8		845	420		1,265	1,550
3250	30" wide		1.80	8.889		915	465		1,380	1,700
3260	36" wide		1.60	10		1,075	525		1,600	1,950
3280	Dropout, or end plate, 6" wide		32	.500		22	26		48	63
3300	12" wide		26	.615		24.50	32.50		57	75.50
3310	18" wide		22	.727		29.50	38		67.50	89.50
3320	24" wide		20	.800		32.50	42		74.50	98
3330	30" wide		18	.889		38	46.50		84.50	111
3340	36" wide		16	1		44	52.50		96.50	127
3380	Reducer, 12" to 6" wide		14	1.143		107	60		167	208
3400	18" to 12" wide		12	1.333		110	70		180	226
3420	18" to 6" wide		12	1.333		110	70		180	226
3440	24" to 18" wide		10.60	1.509		117	79		196	247
3460	24" to 12" wide		10.60	1.509		117	79		196	247
3480	30" to 24" wide		9.20	1.739		128	91		219	276
3500	30" to 18" wide		9.20	1.739		128	91		219	276
3520	30" to 12" wide		9.20	1.739		128	91		219	276
3540	36" to 30" wide		8	2		134	105		239	305
3560	36" to 24" wide		8	2		136	105		241	305
3580	36" to 18" wide		8	2		136	105		241	305
3600	36" to 12" wide		8	2		139	105		244	310

26 05 36.30 Cable Tray Trough

		Crew	Daily Output	Labor-Hours	Unit	Material	Labor	Equipment	Total	Total Incl O&P
0010	**CABLE TRAY TROUGH** vented, w/ftngs & supports, 6" dp, to 15' high									
0020	For higher elevations, see Section 26 05 36.40									
0200	Galvanized steel, tray, 6" wide R260536-10	2 Elec	90	.178	L.F.	18.90	9.30		28.20	35
0240	12" wide		80	.200		23	10.50		33.50	41
0260	18" wide		70	.229		28	12		40	48.50
0280	24" wide R260536-11		60	.267		32	13.95		45.95	56
0300	30" wide		50	.320		45.50	16.75		62.25	75
0320	36" wide		40	.400		51	21		72	87.50
0340	Elbow horizontal 90°, 12" radius, 6" wide		7.60	2.105	Ea.	117	110		227	293
0360	12" wide		5.60	2.857		138	150		288	375
0370	18" wide		4.40	3.636		159	191		350	460
0380	24" wide		3.60	4.444		197	233		430	565
0390	30" wide		3.20	5		222	262		484	635
0400	36" wide		2.80	5.714		258	299		557	735
0420	24" radius, 6" wide		7.20	2.222		172	116		288	365
0440	12" wide		5.20	3.077		199	161		360	460
0450	18" wide		4	4		239	210		449	580
0460	24" wide		3.20	5		271	262		533	690
0470	30" wide		2.80	5.714		315	299		614	795
0480	36" wide		2.40	6.667		365	350		715	925
0500	36" radius, 6" wide		6.80	2.353		258	123		381	470
0520	12" wide		4.80	3.333		294	175		469	585
0530	18" wide		3.60	4.444		355	233		588	740
0540	24" wide		2.80	5.714		375	299		674	860
0550	30" wide		2.40	6.667		420	350		770	985
0560	36" wide		2	8		470	420		890	1,150

26 05 36.30 Cable Tray Trough		Crew	Daily Output	Labor-Hours	Unit	Material	2013 Bare Costs Labor	Equipment	Total	Total Incl O&P
0580	Elbow vertical 90°, 12″ radius, 6″ wide	2 Elec	7.60	2.105	Ea.	151	110		261	330
0600	12″ wide		5.60	2.857		168	150		318	410
0610	18″ wide		4.40	3.636		170	191		361	470
0620	24″ wide		3.60	4.444		195	233		428	565
0630	30″ wide		3.20	5		199	262		461	610
0640	36″ wide		2.80	5.714		210	299		509	680
0660	24″ radius, 6″ wide		7.20	2.222		216	116		332	410
0680	12″ wide		5.20	3.077		229	161		390	495
0690	18″ wide		4	4		254	210		464	595
0700	24″ wide		3.20	5		262	262		524	680
0710	30″ wide		2.80	5.714		294	299		593	775
0720	36″ wide		2.40	6.667		310	350		660	865
0740	36″ radius, 6″ wide		6.80	2.353		292	123		415	505
0760	12″ wide		4.80	3.333		310	175		485	600
0770	18″ wide		3.60	4.444		345	233		578	730
0780	24″ wide		2.80	5.714		370	299		669	855
0790	30″ wide		2.40	6.667		380	350		730	940
0800	36″ wide		2	8		420	420		840	1,075
0820	Tee horizontal, 12″ radius, 6″ wide		4	4		176	210		386	510
0840	12″ wide		3.20	5		197	262		459	605
0850	18″ wide		2.80	5.714		222	299		521	695
0860	24″ wide		2.40	6.667		258	350		608	810
0870	30″ wide		2.20	7.273		294	380		674	895
0880	36″ wide		2	8		325	420		745	980
0900	24″ radius, 6″ wide		3.60	4.444		292	233		525	670
0920	12″ wide		2.80	5.714		320	299		619	800
0930	18″ wide		2.40	6.667		355	350		705	915
0940	24″ wide		2	8		460	420		880	1,125
0950	30″ wide		1.80	8.889		500	465		965	1,250
0960	36″ wide		1.60	10		555	525		1,080	1,400
0980	36″ radius, 6″ wide		3.20	5		420	262		682	850
1000	12″ wide		2.40	6.667		480	350		830	1,050
1010	18″ wide		2	8		520	420		940	1,200
1020	24″ wide		1.60	10		630	525		1,155	1,475
1030	30″ wide		1.40	11.429		665	600		1,265	1,625
1040	36″ wide		1.20	13.333		785	700		1,485	1,925
1060	Tee vertical, 12″ radius, 6″ wide		4	4		294	210		504	640
1080	12″ wide		3.20	5		298	262		560	720
1090	18″ wide		3	5.333		305	279		584	755
1100	24″ wide		2.80	5.714		325	299		624	805
1110	30″ wide		2.60	6.154		340	320		660	855
1120	36″ wide		2.20	7.273		350	380		730	955
1140	24″ radius, 6″ wide		3.60	4.444		395	233		628	785
1160	12″ wide		2.80	5.714		415	299		714	905
1170	18″ wide		2.60	6.154		435	320		755	960
1180	24″ wide		2.40	6.667		470	350		820	1,050
1190	30″ wide		2.20	7.273		490	380		870	1,100
1200	36″ wide		1.80	8.889		535	465		1,000	1,275
1220	36″ radius, 6″ wide		3.20	5		630	262		892	1,075
1240	12″ wide		2.40	6.667		645	350		995	1,225
1250	18″ wide		2.20	7.273		665	380		1,045	1,300
1260	24″ wide		2	8		705	420		1,125	1,400
1270	30″ wide		1.80	8.889		820	465		1,285	1,600

26 05 36 – Cable Trays for Electrical Systems

26 05 36.30 Cable Tray Trough		Crew	Daily Output	Labor-Hours	Unit	Material	2013 Bare Costs Labor	Equipment	Total	Total Incl O&P
1280	36" wide	2 Elec	1.40	11.429	Ea.	830	600		1,430	1,800
1300	Cross horizontal, 12" radius, 6" wide		3.20	5		227	262		489	640
1320	12" wide		2.80	5.714		229	299		528	700
1330	18" wide		2.40	6.667		256	350		606	805
1340	24" wide		2	8		269	420		689	920
1350	30" wide		1.80	8.889		300	465		765	1,025
1360	36" wide		1.60	10		340	525		865	1,150
1380	24" radius, 6" wide		2.80	5.714		355	299		654	840
1400	12" wide		2.40	6.667		375	350		725	935
1410	18" wide		2	8		410	420		830	1,075
1420	24" wide		1.60	10		520	525		1,045	1,350
1430	30" wide		1.40	11.429		575	600		1,175	1,525
1440	36" wide		1.20	13.333		610	700		1,310	1,725
1460	36" radius, 6" wide		2.40	6.667		610	350		960	1,200
1480	12" wide		2	8		630	420		1,050	1,325
1490	18" wide		1.60	10		645	525		1,170	1,500
1500	24" wide		1.20	13.333		775	700		1,475	1,900
1510	30" wide		1	16		850	840		1,690	2,175
1520	36" wide		.80	20		915	1,050		1,965	2,575
1540	Dropout or end plate, 6" wide		26	.615		26	32.50		58.50	77.50
1560	12" wide		22	.727		32	38		70	92.50
1580	18" wide		20	.800		35.50	42		77.50	102
1600	24" wide		18	.889		40.50	46.50		87	114
1620	30" wide		16	1		43.50	52.50		96	127
1640	36" wide		13.40	1.194		48	62.50		110.50	147
1660	Reducer, 12" to 6" wide		9.40	1.702		110	89		199	254
1680	18" to 12" wide		8.40	1.905		115	100		215	275
1700	18" to 6" wide		8.40	1.905		115	100		215	275
1720	24" to 18" wide		7.20	2.222		122	116		238	310
1740	24" to 12" wide		7.20	2.222		122	116		238	310
1760	30" to 24" wide		6.40	2.500		127	131		258	335
1780	30" to 18" wide		6.40	2.500		127	131		258	335
1800	30" to 12" wide		6.40	2.500		127	131		258	335
1820	36" to 30" wide		5.80	2.759		135	145		280	365
1840	36" to 24" wide		5.80	2.759		135	145		280	365
1860	36" to 18" wide		5.80	2.759		137	145		282	365
1880	36" to 12" wide		5.80	2.759		137	145		282	365
2000	Aluminum, tray, vented, 6" wide		120	.133	L.F.	18	7		25	30.50
2010	9" wide		110	.145		20.50	7.60		28.10	34
2020	12" wide		100	.160		22.50	8.40		30.90	37.50
2030	18" wide		90	.178		27	9.30		36.30	44
2040	24" wide		80	.200		32	10.50		42.50	51
2050	30" wide		70	.229		42.50	12		54.50	65
2060	36" wide		60	.267		47	13.95		60.95	72.50
2080	Elbow horiz. 90°, 12" radius, 6" wide		7.60	2.105	Ea.	119	110		229	296
2090	9" wide		7	2.286		129	120		249	320
2100	12" wide		6.20	2.581		138	135		273	355
2110	18" wide		5.60	2.857		161	150		311	400
2120	24" wide		4.60	3.478		186	182		368	475
2130	30" wide		4	4		229	210		439	565
2140	36" wide		3.60	4.444		250	233		483	625
2160	24" radius, 6" wide		7.20	2.222		172	116		288	365
2180	12" wide		5.80	2.759		202	145		347	440

26 05 36.30 Cable Tray Trough		Crew	Daily Output	Labor-Hours	Unit	2013 Bare Costs Material	Labor	Equipment	Total	Total Incl O&P
2190	18" wide	2 Elec	5.20	3.077	Ea.	232	161		393	495
2200	24" wide		4.20	3.810		265	200		465	590
2210	30" wide		3.60	4.444		305	233		538	685
2220	36" wide		3.20	5		330	262		592	750
2240	36" radius, 6" wide		6.80	2.353		244	123		367	450
2260	12" wide		5.40	2.963		277	155		432	535
2270	18" wide		4.80	3.333		315	175		490	605
2280	24" wide		3.80	4.211		345	221		566	710
2290	30" wide		3.40	4.706		405	247		652	815
2300	36" wide		2.80	5.714		440	299		739	935
2320	Elbow vertical 90°, 12" radius, 6" wide		7.60	2.105		144	110		254	325
2330	9" wide		7	2.286		154	120		274	350
2340	12" wide		6.20	2.581		156	135		291	375
2350	18" wide		5.60	2.857		161	150		311	400
2360	24" wide		4.60	3.478		177	182		359	470
2370	30" wide		4	4		186	210		396	520
2380	36" wide		3.60	4.444		187	233		420	555
2400	24" radius, 6" wide		7.20	2.222		195	116		311	390
2420	12" wide		5.80	2.759		212	145		357	450
2430	18" wide		5.20	3.077		230	161		391	495
2440	24" wide		4.20	3.810		234	200		434	555
2450	30" wide		3.60	4.444		247	233		480	620
2460	36" wide		3.20	5		258	262		520	675
2480	36" radius, 6" wide		6.80	2.353		242	123		365	450
2500	12" wide		5.40	2.963		258	155		413	515
2510	18" wide		4.80	3.333		290	175		465	580
2520	24" wide		3.80	4.211		315	221		536	675
2530	30" wide		3.40	4.706		345	247		592	750
2540	36" wide		2.80	5.714		355	299		654	840
2560	Tee horizontal, 12" radius, 6" wide		4	4		181	210		391	515
2570	9" wide		3.80	4.211		186	221		407	535
2580	12" wide		3.60	4.444		202	233		435	570
2590	18" wide		3.20	5		234	262		496	645
2600	24" wide		2.80	5.714		268	299		567	745
2610	30" wide		2.40	6.667		288	350		638	840
2620	36" wide		2.20	7.273		330	380		710	930
2640	24" radius, 6" wide		3.60	4.444		275	233		508	655
2660	12" wide		3.20	5		315	262		577	735
2670	18" wide		2.80	5.714		345	299		644	830
2680	24" wide		2.40	6.667		430	350		780	1,000
2690	30" wide		2	8		465	420		885	1,125
2700	36" wide		1.80	8.889		545	465		1,010	1,300
2720	36" radius, 6" wide		3.20	5		440	262		702	875
2740	12" wide		2.80	5.714		510	299		809	1,000
2750	18" wide		2.40	6.667		575	350		925	1,150
2760	24" wide		2	8		675	420		1,095	1,375
2770	30" wide		1.60	10		715	525		1,240	1,575
2780	36" wide		1.40	11.429		825	600		1,425	1,800
2800	Tee vertical, 12" radius, 6" wide		4	4		249	210		459	590
2810	9" wide		3.80	4.211		250	221		471	605
2820	12" wide		3.60	4.444		268	233		501	645
2830	18" wide		3.40	4.706		270	247		517	665
2840	24" wide		3.20	5		277	262		539	695

26 05 36.30 Cable Tray Trough		Crew	Daily Output	Labor-Hours	Unit	Material	Labor	Equipment	Total	Total Incl O&P
						2013 Bare Costs				
2850	30" wide	2 Elec	3	5.333	Ea.	290	279		569	740
2860	36" wide		2.60	6.154		300	320		620	810
2880	24" radius, 6" wide		3.60	4.444		350	233		583	735
2900	12" wide		3.20	5		380	262		642	805
2910	18" wide		3	5.333		405	279		684	865
2920	24" wide		2.80	5.714		430	299		729	925
2930	30" wide		2.60	6.154		465	320		785	990
2940	36" wide		2.20	7.273		475	380		855	1,100
2960	36" radius, 6" wide		3.20	5		545	262		807	990
2980	12" wide		2.80	5.714		575	299		874	1,075
2990	18" wide		2.60	6.154		595	320		915	1,125
3000	24" wide		2.40	6.667		640	350		990	1,225
3010	30" wide		2.20	7.273		690	380		1,070	1,325
3020	36" wide		1.80	8.889		725	465		1,190	1,500
3040	Cross horizontal, 12" radius, 6" wide		3.60	4.444		225	233		458	600
3050	9" wide		3.40	4.706		242	247		489	635
3060	12" wide		3.20	5		249	262		511	665
3070	18" wide		2.80	5.714		260	299		559	735
3080	24" wide		2.40	6.667		290	350		640	845
3090	30" wide		2.20	7.273		350	380		730	955
3100	36" wide		1.80	8.889		425	465		890	1,150
3120	24" radius, 6" wide		3.20	5		405	262		667	835
3140	12" wide		2.80	5.714		440	299		739	935
3150	18" wide		2.40	6.667		475	350		825	1,050
3160	24" wide		2	8		550	420		970	1,225
3170	30" wide		1.80	8.889		595	465		1,060	1,350
3180	36" wide		1.40	11.429		675	600		1,275	1,625
3200	36" radius, 6" wide		2.80	5.714		675	299		974	1,200
3220	12" wide		2.40	6.667		705	350		1,055	1,300
3230	18" wide		2	8		765	420		1,185	1,475
3240	24" wide		1.60	10		890	525		1,415	1,750
3250	30" wide		1.40	11.429		965	600		1,565	1,950
3260	36" wide		1.20	13.333		1,125	700		1,825	2,275
3280	Dropout, or end plate, 6" wide		26	.615		22.50	32.50		55	73.50
3300	12" wide		22	.727		26.50	38		64.50	86.50
3310	18" wide		20	.800		32	42		74	98
3320	24" wide		18	.889		39	46.50		85.50	112
3330	30" wide		16	1		41.50	52.50		94	124
3340	36" wide		14	1.143		47.50	60		107.50	142
3370	Reducer, 9" to 6" wide		12	1.333		110	70		180	226
3380	12" to 6" wide		11.40	1.404		116	73.50		189.50	238
3390	12" to 9" wide		11.40	1.404		116	73.50		189.50	238
3400	18" to 12" wide		9.60	1.667		124	87.50		211.50	268
3420	18" to 6" wide		9.60	1.667		124	87.50		211.50	268
3430	18" to 9" wide		9.60	1.667		124	87.50		211.50	268
3440	24" to 18" wide		8.40	1.905		133	100		233	295
3460	24" to 12" wide		8.40	1.905		133	100		233	295
3470	24" to 9" wide		8.40	1.905		134	100		234	297
3475	24" to 6" wide		8.40	1.905		134	100		234	297
3480	30" to 24" wide		7.20	2.222		136	116		252	325
3500	30" to 18" wide		7.20	2.222		136	116		252	325
3520	30" to 12" wide		7.20	2.222		139	116		255	325
3540	36" to 30" wide		6.40	2.500		141	131		272	350

26 05 36.30 Cable Tray Trough		Crew	Daily Output	Labor-Hours	Unit	2013 Bare Costs Material	Labor	Equipment	Total	Total Incl O&P
3560	36″ to 24″ wide	2 Elec	6.40	2.500	Ea.	141	131		272	350
3580	36″ to 18″ wide		6.40	2.500		141	131		272	350
3600	36″ to 12″ wide		6.40	2.500		141	131		272	350
3610	Elbow horizontal 60°, 12″ radius, 6″ wide		7.80	2.051		96	107		203	267
3620	9″ wide		7.20	2.222		105	116		221	290
3630	12″ wide		6.40	2.500		116	131		247	325
3640	18″ wide		5.80	2.759		124	145		269	355
3650	24″ wide		4.80	3.333		152	175		327	430
3680	Elbow horizontal 45°, 12″ radius, 6″ wide		8	2		83	105		188	248
3690	9″ wide		7.40	2.162		88	113		201	267
3700	12″ wide		6.60	2.424		93	127		220	292
3710	18″ wide		6	2.667		104	140		244	325
3720	24″ wide		5	3.200		121	168		289	385
3750	Elbow horizontal, 30° 12″ radius, 6″ wide		8.20	1.951		72.50	102		174.50	233
3760	9″ wide		7.60	2.105		76	110		186	249
3770	12″ wide		6.80	2.353		81	123		204	274
3780	18″ wide		6.20	2.581		88	135		223	299
3790	24″ wide		5.20	3.077		96	161		257	345
3820	Elbow vertical 60° in/outside, 12″ radius, 6″ wide		7.80	2.051		119	107		226	292
3830	9″ wide		7.20	2.222		121	116		237	305
3840	12″ wide		6.40	2.500		123	131		254	330
3850	18″ wide		5.80	2.759		128	145		273	355
3860	24″ wide		4.80	3.333		133	175		308	405
3890	Elbow vertical 45° in/outside, 12″ radius, 6″ wide		8	2		96	105		201	263
3900	9″ wide		7.40	2.162		102	113		215	282
3910	12″ wide		6.60	2.424		105	127		232	305
3920	18″ wide		6	2.667		109	140		249	330
3930	24″ wide		5	3.200		119	168		287	380
3960	Elbow vertical 30° in/outside, 12″ radius, 6″ wide		8.20	1.951		82.50	102		184.50	244
3970	9″ wide		7.60	2.105		88	110		198	262
3980	12″ wide		6.80	2.353		89.50	123		212.50	283
3990	18″ wide		6.20	2.581		93	135		228	305
4000	24″ wide		5.20	3.077		94.50	161		255.50	345
4250	Reducer, left or right hand, 24″ to 18″ wide		8.40	1.905		126	100		226	288
4260	24″ to 12″ wide		8.40	1.905		126	100		226	288
4270	24″ to 9″ wide		8.40	1.905		126	100		226	288
4280	24″ to 6″ wide		8.40	1.905		128	100		228	289
4290	18″ to 12″ wide		9.60	1.667		118	87.50		205.50	260
4300	18″ to 9″ wide		9.60	1.667		118	87.50		205.50	260
4310	18″ to 6″ wide		9.60	1.667		118	87.50		205.50	260
4320	12″ to 9″ wide		11.40	1.404		113	73.50		186.50	234
4330	12″ to 6″ wide		11.40	1.404		113	73.50		186.50	234
4340	9″ to 6″ wide		12	1.333		109	70		179	225
4350	Splice plate	1 Elec	48	.167		10.45	8.75		19.20	24.50
4360	Splice plate, expansion joint		48	.167		10.70	8.75		19.45	25
4370	Splice plate, hinged, horizontal		48	.167		8.60	8.75		17.35	22.50
4380	Vertical		48	.167		12.25	8.75		21	26.50
4390	Trough, hanger, vertical		28	.286		34.50	14.95		49.45	60.50
4400	Box connector, 24″ wide		20	.400		46.50	21		67.50	82.50
4410	18″ wide		21	.381		39	19.95		58.95	72.50
4420	12″ wide		22	.364		37	19.05		56.05	69.50
4430	9″ wide		23	.348		35.50	18.25		53.75	66.50
4440	6″ wide		24	.333		33.50	17.45		50.95	63

26 05 Common Work Results for Electrical

26 05 36 – Cable Trays for Electrical Systems

26 05 36.30 Cable Tray Trough		Crew	Daily Output	Labor-Hours	Unit	Material	Labor	Equipment	Total	Total Incl O&P
						2013 Bare Costs				
4450	Floor flange	1 Elec	24	.333	Ea.	35.50	17.45		52.95	65
4460	Hold down clamp		60	.133		4.07	7		11.07	14.95
4520	Wall bracket, 24" wide tray		20	.400		35.50	21		56.50	70.50
4530	18" wide tray		21	.381		35	19.95		54.95	68.50
4540	12" wide tray		22	.364		18.55	19.05		37.60	49
4550	9" wide tray		23	.348		16.55	18.25		34.80	46
4560	6" wide tray		24	.333		15.05	17.45		32.50	42.50
5000	Cable channel aluminum, vented, 1-1/4" deep, 4" wide, straight		80	.100	L.F.	13.10	5.25		18.35	22.50
5010	Elbow horizontal, 36" radius, 90°		5	1.600	Ea.	236	84		320	385
5020	60°		5.50	1.455		181	76		257	315
5030	45°		6	1.333		147	70		217	267
5040	30°		6.50	1.231		127	64.50		191.50	236
5050	Adjustable		6	1.333		121	70		191	238
5060	Elbow vertical, 36" radius, 90°		5	1.600		254	84		338	405
5070	60°		5.50	1.455		198	76		274	330
5080	45°		6	1.333		163	70		233	285
5090	30°		6.50	1.231		144	64.50		208.50	256
5100	Adjustable		6	1.333		121	70		191	238
5110	Splice plate, hinged, horizontal		48	.167		9.05	8.75		17.80	23
5120	Splice plate, hinged, vertical		48	.167		12.80	8.75		21.55	27
5130	Hanger, vertical		28	.286		14.15	14.95		29.10	38
5140	Single		28	.286		21.50	14.95		36.45	46
5150	Double		20	.400		22	21		43	55.50
5160	Channel to box connector		24	.333		29	17.45		46.45	57.50
5170	Hold down clip		80	.100		4.07	5.25		9.32	12.35
5180	Wall bracket, single		28	.286		11.75	14.95		26.70	35.50
5190	Double		20	.400		15.05	21		36.05	48
5200	Cable roller		16	.500		184	26		210	242
5210	Splice plate		48	.167		5.65	8.75		14.40	19.25

26 05 36.40 Cable Tray, Covers and Dividers		Crew	Daily Output	Labor-Hours	Unit	Material	Labor	Equipment	Total	Total Incl O&P
0010	**CABLE TRAY, COVERS AND DIVIDERS** To 15' high R260536-10									
0011	For higher elevations, see lines 9900 – 9960									
0100	Covers, ventilated galv. steel, straight, 6" wide tray size	2 Elec	520	.031	L.F.	6.90	1.61		8.51	10
0200	9" wide tray size		460	.035		8.50	1.82		10.32	12.10
0300	12" wide tray size R260536-11		400	.040		10.20	2.10		12.30	14.35
0400	18" wide tray size		300	.053		13.85	2.79		16.64	19.40
0500	24" wide tray size		220	.073		17	3.81		20.81	24.50
0600	30" wide tray size		180	.089		20.50	4.66		25.16	29.50
0700	36" wide tray size		160	.100		23.50	5.25		28.75	34
1000	Elbow horizontal 90°, 12" radius, 6" wide tray size		150	.107	Ea.	51	5.60		56.60	65
1020	9" wide tray size		128	.125		57	6.55		63.55	72.50
1040	12" wide tray size		108	.148		60	7.75		67.75	77.50
1060	18" wide tray size		84	.190		83	10		93	106
1080	24" wide tray size		66	.242		100	12.70		112.70	129
1100	30" wide tray size		60	.267		124	13.95		137.95	158
1120	36" wide tray size		50	.320		151	16.75		167.75	191
1160	24" radius, 6" wide tray size		136	.118		85	6.15		91.15	103
1180	9" wide tray size		116	.138		87.50	7.25		94.75	107
1200	12" wide tray size		96	.167		97	8.75		105.75	120
1220	18" wide tray size		76	.211		120	11.05		131.05	149
1240	24" wide tray size		60	.267		146	13.95		159.95	182
1260	30" wide tray size		52	.308		192	16.10		208.10	235

26 05 36.40 Cable Tray, Covers and Dividers		Crew	Daily Output	Labor-Hours	Unit	2013 Bare Costs Material	Labor	Equipment	Total	Total Incl O&P
1280	36" wide tray size	2 Elec	44	.364	Ea.	222	19.05		241.05	274
1320	36" radius, 6" wide tray size		120	.133		124	7		131	147
1340	9" wide tray size		104	.154		137	8.05		145.05	163
1360	12" wide tray size		84	.190		147	10		157	177
1380	18" wide tray size		72	.222		187	11.65		198.65	222
1400	24" wide tray size		52	.308		222	16.10		238.10	269
1420	30" wide tray size		46	.348		264	18.25		282.25	320
1440	36" wide tray size		40	.400		310	21		331	370
1480	Elbow horizontal 45°, 12" radius, 6" wide tray size		150	.107		36.50	5.60		42.10	48.50
1500	9" wide tray size		128	.125		43.50	6.55		50.05	58
1520	12" wide tray size		108	.148		47.50	7.75		55.25	63.50
1540	18" wide tray size		88	.182		58	9.55		67.55	78
1560	24" wide tray size		76	.211		67.50	11.05		78.55	91
1580	30" wide tray size		66	.242		79.50	12.70		92.20	107
1600	36" wide tray size		60	.267		89.50	13.95		103.45	120
1640	24" radius, 6" wide tray size		136	.118		53.50	6.15		59.65	68
1660	9" wide tray size		116	.138		60	7.25		67.25	77
1680	12" wide tray size		96	.167		67.50	8.75		76.25	87.50
1700	18" wide tray size		80	.200		77.50	10.50		88	101
1720	24" wide tray size		70	.229		89.50	12		101.50	116
1740	30" wide tray size		60	.267		114	13.95		127.95	146
1760	36" wide tray size		52	.308		124	16.10		140.10	161
1800	36" radius, 6" wide tray size		120	.133		77.50	7		84.50	95.50
1820	9" wide tray size		104	.154		87.50	8.05		95.55	108
1840	12" wide tray size		84	.190		96	10		106	121
1860	18" wide tray size		76	.211		110	11.05		121.05	138
1880	24" wide tray size		62	.258		134	13.50		147.50	168
1900	30" wide tray size		52	.308		146	16.10		162.10	185
1920	36" wide tray size		48	.333		176	17.45		193.45	220
1960	Elbow vertical 90°, 12" radius, 6" wide tray size		150	.107		43	5.60		48.60	55.50
1980	9" wide tray size		128	.125		43.50	6.55		50.05	58
2000	12" wide tray size		108	.148		47	7.75		54.75	63
2020	18" wide tray size		88	.182		53.50	9.55		63.05	73.50
2040	24" wide tray size		68	.235		54.50	12.35		66.85	78.50
2060	30" wide tray size		60	.267		60	13.95		73.95	87
2080	36" wide tray size		50	.320		77.50	16.75		94.25	110
2120	24" radius, 6" wide tray size		136	.118		52.50	6.15		58.65	66.50
2140	9" wide tray size		116	.138		58	7.25		65.25	74.50
2160	12" wide tray size		96	.167		60	8.75		68.75	79
2180	18" wide tray size		80	.200		79.50	10.50		90	103
2200	24" wide tray size		62	.258		89.50	13.50		103	119
2220	30" wide tray size		52	.308		102	16.10		118.10	136
2240	36" wide tray size		44	.364		116	19.05		135.05	156
2280	36" radius, 6" wide tray size		120	.133		61	7		68	77.50
2300	9" wide tray size		104	.154		77.50	8.05		85.55	97
2320	12" wide tray size		84	.190		87.50	10		97.50	111
2340	18" wide tray size		76	.211		110	11.05		121.05	138
2350	24" wide tray size		54	.296		122	15.55		137.55	157
2360	30" wide tray size		46	.348		146	18.25		164.25	189
2370	36" wide tray size		40	.400		172	21		193	222
2400	Tee horizontal, 12" radius, 6" wide tray size		92	.174		77.50	9.10		86.60	99
2410	9" wide tray size		80	.200		79.50	10.50		90	103
2420	12" wide tray size		68	.235		90	12.35		102.35	117

26 05 36 – Cable Trays for Electrical Systems

26 05 36.40 Cable Tray, Covers and Dividers		Crew	Daily Output	Labor-Hours	Unit	Material	2013 Bare Costs Labor	Equipment	Total	Total Incl O&P
2430	18" wide tray size	2 Elec	60	.267	Ea.	109	13.95		122.95	141
2440	24" wide tray size		52	.308		137	16.10		153.10	175
2460	30" wide tray size		36	.444		162	23.50		185.50	213
2470	36" wide tray size		30	.533		195	28		223	257
2500	24" radius, 6" wide tray size		88	.182		122	9.55		131.55	148
2510	9" wide tray size		76	.211		140	11.05		151.05	171
2520	12" wide tray size		64	.250		144	13.10		157.10	178
2530	18" wide tray size		56	.286		185	14.95		199.95	226
2540	24" wide tray size		48	.333		283	17.45		300.45	335
2560	30" wide tray size		32	.500		325	26		351	395
2570	36" wide tray size		26	.615		355	32.50		387.50	440
2600	36" radius, 6" wide tray size		84	.190		220	10		230	257
2610	9" wide tray size		72	.222		225	11.65		236.65	264
2620	12" wide tray size		60	.267		250	13.95		263.95	296
2630	18" wide tray size		52	.308		292	16.10		308.10	345
2640	24" wide tray size		44	.364		385	19.05		404.05	450
2660	30" wide tray size		28	.571		420	30		450	505
2670	36" wide tray size		22	.727		475	38		513	575
2700	Cross horizontal, 12" radius, 6" wide tray size		68	.235		116	12.35		128.35	145
2710	9" wide tray size		64	.250		122	13.10		135.10	154
2720	12" wide tray size		60	.267		136	13.95		149.95	171
2730	18" wide tray size		52	.308		162	16.10		178.10	202
2740	24" wide tray size		36	.444		197	23.50		220.50	252
2760	30" wide tray size		30	.533		225	28		253	289
2770	36" wide tray size		28	.571		262	30		292	335
2800	24" radius, 6" wide tray size		64	.250		220	13.10		233.10	262
2810	9" wide tray size		60	.267		239	13.95		252.95	284
2820	12" wide tray size		56	.286		258	14.95		272.95	305
2830	18" wide tray size		48	.333		310	17.45		327.45	365
2840	24" wide tray size		32	.500		375	26		401	455
2860	30" wide tray size		26	.615		425	32.50		457.50	515
2870	36" wide tray size		24	.667		475	35		510	580
2900	36" radius, 6" wide tray size		60	.267		370	13.95		383.95	425
2910	9" wide tray size		56	.286		380	14.95		394.95	445
2920	12" wide tray size		52	.308		400	16.10		416.10	465
2930	18" wide tray size		44	.364		460	19.05		479.05	540
2940	24" wide tray size		28	.571		590	30		620	695
2960	30" wide tray size		22	.727		635	38		673	750
2970	36" wide tray size		20	.800		675	42		717	810
3000	Reducer, 9" to 6" wide tray size		128	.125		49	6.55		55.55	64
3010	12" to 6" wide tray size		108	.148		51	7.75		58.75	68
3020	12" to 9" wide tray size		108	.148		51	7.75		58.75	68
3030	18" to 12" wide tray size		88	.182		54.50	9.55		64.05	74.50
3050	18" to 6" wide tray size		88	.182		54.50	9.55		64.05	74.50
3060	24" to 18" wide tray size		80	.200		76.50	10.50		87	99.50
3070	24" to 12" wide tray size		80	.200		70	10.50		80.50	92.50
3090	30" to 24" wide tray size		70	.229		81	12		93	107
3100	30" to 18" wide tray size		70	.229		81	12		93	107
3110	30" to 12" wide tray size		70	.229		70	12		82	95
3140	36" to 30" wide tray size		64	.250		88	13.10		101.10	117
3150	36" to 24" wide tray size		64	.250		88	13.10		101.10	117
3160	36" to 18" wide tray size		64	.250		88	13.10		101.10	117
3170	36" to 12" wide tray size		64	.250		88	13.10		101.10	117

26 05 36.40 Cable Tray, Covers and Dividers		Crew	Daily Output	Labor-Hours	Unit	Material	2013 Bare Costs Labor	Equipment	Total	Total Incl O&P
3250	Covers, aluminum, straight, 6" wide tray size	2 Elec	520	.031	L.F.	5.30	1.61		6.91	8.25
3270	9" wide tray size		460	.035		6.35	1.82		8.17	9.75
3290	12" wide tray size		400	.040		7.60	2.10		9.70	11.55
3310	18" wide tray size		320	.050		10.10	2.62		12.72	15
3330	24" wide tray size		260	.062		12.60	3.22		15.82	18.65
3350	30" wide tray size		200	.080		13.90	4.19		18.09	21.50
3370	36" wide tray size		180	.089		14.75	4.66		19.41	23
3400	Elbow horizontal 90°, 12" radius, 6" wide tray size		150	.107	Ea.	39.50	5.60		45.10	52
3410	9" wide tray size		128	.125		41.50	6.55		48.05	55.50
3420	12" wide tray size		108	.148		44.50	7.75		52.25	60.50
3430	18" wide tray size		88	.182		58.50	9.55		68.05	79
3440	24" wide tray size		70	.229		74	12		86	99.50
3460	30" wide tray size		64	.250		89	13.10		102.10	117
3470	36" wide tray size		54	.296		109	15.55		124.55	142
3500	24" radius, 6" wide tray size		136	.118		54.50	6.15		60.65	69
3510	9" wide tray size		116	.138		67.50	7.25		74.75	85
3520	12" wide tray size		96	.167		74	8.75		82.75	94.50
3530	18" wide tray size		80	.200		88	10.50		98.50	112
3540	24" wide tray size		64	.250		109	13.10		122.10	139
3560	30" wide tray size		56	.286		133	14.95		147.95	169
3570	36" wide tray size		48	.333		164	17.45		181.45	206
3600	36" radius, 6" wide tray size		120	.133		93	7		100	112
3610	9" wide tray size		104	.154		102	8.05		110.05	124
3620	12" wide tray size		84	.190		115	10		125	142
3630	18" wide tray size		76	.211		138	11.05		149.05	168
3640	24" wide tray size		56	.286		167	14.95		181.95	207
3660	30" wide tray size		50	.320		192	16.75		208.75	236
3670	36" wide tray size		44	.364		225	19.05		244.05	277
3700	Elbow horizontal 45°, 12" radius, 6" wide tray size		150	.107		29.50	5.60		35.10	41
3710	9" wide tray size		128	.125		30	6.55		36.55	43
3720	12" wide tray size		108	.148		34	7.75		41.75	48.50
3730	18" wide tray size		88	.182		39	9.55		48.55	57
3740	24" wide tray size		80	.200		43	10.50		53.50	63
3760	30" wide tray size		70	.229		54.50	12		66.50	78
3770	36" wide tray size		64	.250		65.50	13.10		78.60	91.50
3800	24" radius, 6" wide tray size		136	.118		34	6.15		40.15	46.50
3810	9" wide tray size		116	.138		43	7.25		50.25	58.50
3820	12" wide tray size		96	.167		45	8.75		53.75	62.50
3830	18" wide tray size		80	.200		54.50	10.50		65	75.50
3840	24" wide tray size		72	.222		67.50	11.65		79.15	91.50
3860	30" wide tray size		64	.250		77.50	13.10		90.60	105
3870	36" wide tray size		56	.286		89.50	14.95		104.45	121
3900	36" radius, 6" wide tray size		120	.133		58.50	7		65.50	75
3910	9" wide tray size		104	.154		64	8.05		72.05	82
3920	12" wide tray size		84	.190		67.50	10		77.50	89
3930	18" wide tray size		76	.211		81	11.05		92.05	106
3940	24" wide tray size		64	.250		98	13.10		111.10	128
3960	30" wide tray size		56	.286		112	14.95		126.95	146
3970	36" wide tray size		50	.320		128	16.75		144.75	165
4000	Elbow vertical 90°, 12" radius, 6" wide tray size		150	.107		34	5.60		39.60	45.50
4010	9" wide tray size		128	.125		34	6.55		40.55	47.50
4020	12" wide tray size		108	.148		37	7.75		44.75	52.50
4030	18" wide tray size		88	.182		43	9.55		52.55	62

26 05 36.40 Cable Tray, Covers and Dividers		Crew	Daily Output	Labor-Hours	Unit	Material	2013 Bare Costs Labor	Equipment	Total	Total Incl O&P
4040	24" wide tray size	2 Elec	70	.229	Ea.	45	12		57	67.50
4060	30" wide tray size		64	.250		45.50	13.10		58.60	70
4070	36" wide tray size		54	.296		55	15.55		70.55	83.50
4100	24" radius, 6" wide tray size		136	.118		39	6.15		45.15	51.50
4110	9" wide tray size		116	.138		42.50	7.25		49.75	57.50
4120	12" wide tray size		96	.167		45.50	8.75		54.25	63.50
4130	18" wide tray size		80	.200		55	10.50		65.50	76
4140	24" wide tray size		64	.250		60.50	13.10		73.60	86
4160	30" wide tray size		56	.286		77.50	14.95		92.45	108
4170	36" wide tray size		48	.333		84.50	17.45		101.95	119
4200	36" radius, 6" wide tray size		120	.133		45	7		52	60
4210	9" wide tray size		104	.154		55	8.05		63.05	72.50
4220	12" wide tray size		84	.190		64	10		74	85
4230	18" wide tray size		76	.211		77.50	11.05		88.55	102
4240	24" wide tray size		56	.286		91.50	14.95		106.45	123
4260	30" wide tray size		50	.320		115	16.75		131.75	152
4270	36" wide tray size		44	.364		124	19.05		143.05	166
4300	Tee horizontal, 12" radius, 6" wide tray size		108	.148		55	7.75		62.75	72
4310	9" wide tray size		88	.182		58.50	9.55		68.05	79
4320	12" wide tray size		80	.200		65.50	10.50		76	87.50
4330	18" wide tray size		68	.235		77.50	12.35		89.85	104
4340	24" wide tray size		56	.286		98	14.95		112.95	131
4360	30" wide tray size		44	.364		115	19.05		134.05	156
4370	36" wide tray size		36	.444		139	23.50		162.50	188
4400	24" radius, 6" wide tray size		96	.167		89.50	8.75		98.25	112
4410	9" wide tray size		80	.200		100	10.50		110.50	126
4420	12" wide tray size		72	.222		112	11.65		123.65	140
4430	18" wide tray size		60	.267		131	13.95		144.95	165
4440	24" wide tray size		48	.333		209	17.45		226.45	256
4460	30" wide tray size		40	.400		230	21		251	285
4470	36" wide tray size		32	.500		257	26		283	320
4500	36" radius, 6" wide tray size		88	.182		164	9.55		173.55	194
4510	9" wide tray size		72	.222		167	11.65		178.65	201
4520	12" wide tray size		64	.250		184	13.10		197.10	222
4530	18" wide tray size		56	.286		209	14.95		223.95	253
4540	24" wide tray size		44	.364		265	19.05		284.05	320
4560	30" wide tray size		36	.444		298	23.50		321.50	365
4570	36" wide tray size		28	.571		335	30		365	415
4600	Cross horizontal, 12" radius, 6" wide tray size		80	.200		84.50	10.50		95	109
4610	9" wide tray size		72	.222		89.50	11.65		101.15	116
4620	12" wide tray size		64	.250		99.50	13.10		112.60	129
4630	18" wide tray size		56	.286		121	14.95		135.95	156
4640	24" wide tray size		48	.333		139	17.45		156.45	179
4660	30" wide tray size		40	.400		166	21		187	214
4670	36" wide tray size		32	.500		191	26		217	249
4700	24" radius, 6" wide tray size		72	.222		166	11.65		177.65	199
4710	9" wide tray size		64	.250		177	13.10		190.10	215
4720	12" wide tray size		56	.286		191	14.95		205.95	233
4730	18" wide tray size		48	.333		222	17.45		239.45	270
4740	24" wide tray size		40	.400		265	21		286	325
4760	30" wide tray size		32	.500		310	26		336	380
4770	36" wide tray size		24	.667		335	35		370	425
4800	36" radius, 6" wide tray size		64	.250		265	13.10		278.10	310

26 05 36.40 Cable Tray, Covers and Dividers		Crew	Daily Output	Labor-Hours	Unit	2013 Bare Costs Material	Labor	Equipment	Total	Total Incl O&P
4810	9″ wide tray size	2 Elec	56	.286	Ea.	277	14.95		291.95	330
4820	12″ wide tray size		50	.320		298	16.75		314.75	355
4830	18″ wide tray size		44	.364		330	19.05		349.05	390
4840	24″ wide tray size		36	.444		405	23.50		428.50	480
4860	30″ wide tray size		28	.571		465	30		495	555
4870	36″ wide tray size		22	.727		510	38		548	615
4900	Reducer, 9″ to 6″ wide tray size		128	.125		42.50	6.55		49.05	56.50
4910	12″ to 6″ wide tray size		108	.148		44	7.75		51.75	60
4920	12″ to 9″ wide tray size		108	.148		44	7.75		51.75	60
4930	18″ to 12″ wide tray size		88	.182		48	9.55		57.55	67.50
4950	18″ to 6″ wide tray size		88	.182		48	9.55		57.55	67.50
4960	24″ to 18″ wide tray size		80	.200		62	10.50		72.50	83.50
4970	24″ to 12″ wide tray size		80	.200		53.50	10.50		64	74
4990	30″ to 24″ wide tray size		70	.229		64.50	12		76.50	89
5000	30″ to 18″ wide tray size		70	.229		64.50	12		76.50	89
5010	30″ to 12″ wide tray size		70	.229		64.50	12		76.50	89
5040	36″ to 30″ wide tray size		64	.250		71.50	13.10		84.60	98
5050	36″ to 24″ wide tray size		64	.250		71.50	13.10		84.60	98
5060	36″ to 18″ wide tray size		64	.250		71.50	13.10		84.60	98
5070	36″ to 12″ wide tray size		64	.250		71.50	13.10		84.60	98
5710	Tray cover hold down clamp	1 Elec	60	.133		11.45	7		18.45	23
8000	Divider strip, straight, galvanized, 3″ deep		200	.040	L.F.	6.10	2.10		8.20	9.85
8020	4″ deep		180	.044		7.45	2.33		9.78	11.70
8040	6″ deep		160	.050		9.75	2.62		12.37	14.65
8060	Aluminum, straight, 3″ deep		210	.038		6.10	2		8.10	9.70
8080	4″ deep		190	.042		7.55	2.21		9.76	11.60
8100	6″ deep		170	.047		9.65	2.47		12.12	14.30
8110	Divider strip vertical fitting 3″ deep									
8120	12″ radius, galvanized, 30°	1 Elec	28	.286	Ea.	29.50	14.95		44.45	55
8140	45°		27	.296		37	15.55		52.55	63.50
8160	60°		26	.308		39	16.10		55.10	67
8180	90°		25	.320		49	16.75		65.75	79
8200	Aluminum, 30°		29	.276		19.70	14.45		34.15	43
8220	45°		28	.286		23.50	14.95		38.45	48
8240	60°		27	.296		27	15.55		42.55	53
8260	90°		26	.308		34.50	16.10		50.60	62
8280	24″ radius, galvanized, 30°		25	.320		44.50	16.75		61.25	74
8300	45°		24	.333		50	17.45		67.45	81
8320	60°		23	.348		63.50	18.25		81.75	97
8340	90°		22	.364		86	19.05		105.05	124
8360	Aluminum, 30°		26	.308		33	16.10		49.10	60
8380	45°		25	.320		38	16.75		54.75	66.50
8400	60°		24	.333		47.50	17.45		64.95	78
8420	90°		23	.348		64.50	18.25		82.75	98.50
8440	36″ radius, galvanized, 30°		22	.364		60	19.05		79.05	94.50
8460	45°		21	.381		70	19.95		89.95	107
8480	60°		20	.400		82	21		103	122
8500	90°		19	.421		114	22		136	158
8520	Aluminum, 30°		23	.348		50	18.25		68.25	82.50
8540	45°		22	.364		64.50	19.05		83.55	99.50
8560	60°		21	.381		83.50	19.95		103.45	122
8570	90°		20	.400		109	21		130	152
8590	Divider strip vertical fitting 4″ deep									

26 05 36.40 Cable Tray, Covers and Dividers		Crew	Daily Output	Labor-Hours	Unit	Material	Labor	Equipment	Total	Total Incl O&P
						2013 Bare Costs				
8600	12" radius, galvanized, 30°	1 Elec	27	.296	Ea.	37.50	15.55		53.05	64.50
8610	45°		26	.308		43.50	16.10		59.60	72
8620	60°		25	.320		49	16.75		65.75	79
8630	90°		24	.333		60	17.45		77.45	92
8640	Aluminum, 30°		28	.286		28	14.95		42.95	53
8650	45°		27	.296		32.50	15.55		48.05	58.50
8660	60°		26	.308		37	16.10		53.10	65
8670	90°		25	.320		44	16.75		60.75	73.50
8680	24" radius, galvanized, 30°		24	.333		60	17.45		77.45	92
8690	45°		23	.348		75.50	18.25		93.75	111
8700	60°		22	.364		85	19.05		104.05	122
8710	90°		21	.381		114	19.95		133.95	155
8720	Aluminum, 30°		25	.320		44	16.75		60.75	73.50
8730	45°		24	.333		53.50	17.45		70.95	84.50
8740	60°		23	.348		63	18.25		81.25	97
8750	90°		22	.364		86	19.05		105.05	124
8760	36" radius, galvanized, 30°		23	.348		70	18.25		88.25	105
8770	45°		22	.364		79.50	19.05		98.55	116
8780	60°		21	.381		100	19.95		119.95	140
8790	90°		20	.400		136	21		157	182
8800	Aluminum, 30°		24	.333		67.50	17.45		84.95	100
8810	45°		23	.348		86	18.25		104.25	123
8820	60°		22	.364		105	19.05		124.05	145
8830	90°	↓	21	.381	↓	133	19.95		152.95	176
8840	Divider strip vertical fitting 6" deep									
8850	12" radius, galvanized, 30°	1 Elec	24	.333	Ea.	41	17.45		58.45	71
8860	45°		23	.348		46	18.25		64.25	78
8870	60°		22	.364		52.50	19.05		71.55	86
8880	90°		21	.381		65.50	19.95		85.45	102
8890	Aluminum, 30°		25	.320		30.50	16.75		47.25	58.50
8900	45°		24	.333		36	17.45		53.45	65.50
8910	60°		23	.348		38	18.25		56.25	69
8920	90°		22	.364		45	19.05		64.05	78
8930	24" radius, galvanized, 30°		23	.348		60	18.25		78.25	93.50
8940	45°		22	.364		75.50	19.05		94.55	112
8950	60°		21	.381		86	19.95		105.95	125
8960	90°		20	.400		114	21		135	157
8970	Aluminum, 30°		24	.333		45	17.45		62.45	75.50
8980	45°		23	.348		61	18.25		79.25	95
8990	60°		22	.364		67.50	19.05		86.55	103
9000	90°		21	.381		91.50	19.95		111.45	130
9010	36" radius, galvanized, 30°		22	.364		70	19.05		89.05	106
9020	45°		21	.381		86	19.95		105.95	125
9030	60°		20	.400		114	21		135	157
9040	90°		19	.421		146	22		168	194
9050	Aluminum, 30°		23	.348		68	18.25		86.25	103
9060	45°		22	.364		91.50	19.05		110.55	129
9070	60°		21	.381		108	19.95		127.95	148
9080	90°		20	.400		129	21		150	174
9120	Divider strip, horizontal fitting, galvanized, 3" deep		33	.242		42	12.70		54.70	65
9130	4" deep		30	.267		47	13.95		60.95	72.50
9140	6" deep		27	.296		60	15.55		75.55	89
9150	Aluminum, 3" deep		35	.229		31.50	12		43.50	52.50

26 05 36 – Cable Trays for Electrical Systems

26 05 36.40 Cable Tray, Covers and Dividers		Crew	Daily Output	Labor-Hours	Unit	Material	Labor	Equipment	Total	Total Incl O&P
						2013 Bare Costs				
9160	4″ deep	1 Elec	32	.250	Ea.	34.50	13.10		47.60	57.50
9170	6″ deep		29	.276		45.50	14.45		59.95	72
9300	Divider strip protector		300	.027	L.F.	3.75	1.40		5.15	6.20
9310	Fastener, ladder tray				Ea.	.68			.68	.75
9320	Trough or solid bottom tray				"	.46			.46	.51
9900	Add to labor for higher elevated installation									
9910	15′ to 20′ high add						10%			
9920	20′ to 25′ high add						20%			
9930	25′ to 30′ high add						25%			
9940	30′ to 35′ high add						30%			
9960	Over 40′ high add						40%			

26 05 39 – Underfloor Raceways for Electrical Systems

26 05 39.30 Conduit In Concrete Slab

Line	Description	Crew	Daily Output	Labor-Hours	Unit	Material	Labor	Equipment	Total	Total Incl O&P
0010	**CONDUIT IN CONCRETE SLAB** Including terminations, R260539-30									
0020	fittings and supports									
3230	PVC, schedule 40, 1/2″ diameter	1 Elec	270	.030	L.F.	.71	1.55		2.26	3.10
3250	3/4″ diameter		230	.035		.85	1.82		2.67	3.66
3270	1″ diameter		200	.040		1.10	2.10		3.20	4.35
3300	1-1/4″ diameter		170	.047		1.57	2.47		4.04	5.40
3330	1-1/2″ diameter		140	.057		1.88	2.99		4.87	6.55
3350	2″ diameter		120	.067		2.40	3.49		5.89	7.90
3370	2-1/2″ diameter		90	.089		4.08	4.66		8.74	11.45
3400	3″ diameter	2 Elec	160	.100		5.10	5.25		10.35	13.45
3430	3-1/2″ diameter		120	.133		6.55	7		13.55	17.65
3440	4″ diameter		100	.160		7.20	8.40		15.60	20.50
3450	5″ diameter		80	.200		10.95	10.50		21.45	28
3460	6″ diameter		60	.267		15.45	13.95		29.40	38
3530	Sweeps, 1″ diameter, 30″ radius	1 Elec	32	.250	Ea.	32	13.10		45.10	55
3550	1-1/4″ diameter		24	.333		40.50	17.45		57.95	71
3570	1-1/2″ diameter		21	.381		42	19.95		61.95	76.50
3600	2″ diameter		18	.444		43	23.50		66.50	82.50
3630	2-1/2″ diameter		14	.571		66.50	30		96.50	118
3650	3″ diameter		10	.800		85	42		127	156
3670	3-1/2″ diameter		8	1		103	52.50		155.50	193
3700	4″ diameter		7	1.143		123	60		183	225
3710	5″ diameter		6	1.333		165	70		235	286
3730	Couplings, 1/2″ diameter					.29			.29	.32
3750	3/4″ diameter					.33			.33	.36
3770	1″ diameter					.50			.50	.55
3800	1-1/4″ diameter					.76			.76	.84
3830	1-1/2″ diameter					.94			.94	1.03
3850	2″ diameter					1.22			1.22	1.34
3870	2-1/2″ diameter					2.18			2.18	2.40
3900	3″ diameter					3.53			3.53	3.88
3930	3-1/2″ diameter					4.43			4.43	4.87
3950	4″ diameter					4.96			4.96	5.45
3960	5″ diameter					12.60			12.60	13.85
3970	6″ diameter					17.15			17.15	18.85
4030	End bells 1″ diameter, PVC	1 Elec	60	.133		5.40	7		12.40	16.40
4050	1-1/4″ diameter		53	.151		6.65	7.90		14.55	19.15
4100	1-1/2″ diameter		48	.167		6.65	8.75		15.40	20.50
4150	2″ diameter		34	.235		9.85	12.35		22.20	29.50

26 05 39 – Underfloor Raceways for Electrical Systems

26 05 39.30 Conduit In Concrete Slab		Crew	Daily Output	Labor-Hours	Unit	Material (2013 Bare Costs)	Labor (2013 Bare Costs)	Equipment (2013 Bare Costs)	Total (2013 Bare Costs)	Total Incl O&P
4170	2-1/2" diameter	1 Elec	27	.296	Ea.	10.95	15.55		26.50	35
4200	3" diameter		20	.400		11.55	21		32.55	44
4250	3-1/2" diameter		16	.500		13.70	26		39.70	54
4300	4" diameter		14	.571		14.95	30		44.95	61.50
4310	5" diameter		12	.667		15.75	35		50.75	70
4320	6" diameter		9	.889		18.70	46.50		65.20	90
4350	Rigid galvanized steel, 1/2" diameter		200	.040	L.F.	2.40	2.10		4.50	5.80
4400	3/4" diameter		170	.047		2.53	2.47		5	6.45
4450	1" diameter		130	.062		3.60	3.22		6.82	8.80
4500	1-1/4" diameter		110	.073		4.94	3.81		8.75	11.15
4600	1-1/2" diameter		100	.080		5.75	4.19		9.94	12.60
4800	2" diameter		90	.089		7	4.66		11.66	14.65

26 05 39.40 Conduit In Trench		Crew	Daily Output	Labor-Hours	Unit	Material	Labor	Equipment	Total	Total Incl O&P
0010	**CONDUIT IN TRENCH** Includes terminations and fittings R260539-40									
0020	Does not include excavation or backfill, see Section 31 23 16.00									
0200	Rigid galvanized steel, 2" diameter	1 Elec	150	.053	L.F.	6.60	2.79		9.39	11.50
0400	2-1/2" diameter	"	100	.080		13.15	4.19		17.34	20.50
0600	3" diameter	2 Elec	160	.100		16.40	5.25		21.65	26
0800	3-1/2" diameter		140	.114		21.50	6		27.50	32.50
1000	4" diameter		100	.160		24	8.40		32.40	39
1200	5" diameter		80	.200		49	10.50		59.50	69.50
1400	6" diameter		60	.267		73	13.95		86.95	101

26 05 43 – Underground Ducts and Raceways for Electrical Systems

26 05 43.10 Trench Duct		Crew	Daily Output	Labor-Hours	Unit	Material	Labor	Equipment	Total	Total Incl O&P
0010	**TRENCH DUCT** Steel with cover									
0020	Standard adjustable, depths to 4"									
0100	Straight, single compartment, 9" wide	2 Elec	40	.400	L.F.	114	21		135	158
0200	12" wide		32	.500		130	26		156	182
0400	18" wide		26	.615		173	32.50		205.50	239
0600	24" wide		22	.727		222	38		260	300
0700	27" wide		21	.762		238	40		278	320
0800	30" wide		20	.800		268	42		310	360
1000	36" wide		16	1		320	52.50		372.50	435
1020	Two compartment, 9" wide		38	.421		130	22		152	176
1030	12" wide		30	.533		148	28		176	205
1040	18" wide		24	.667		181	35		216	252
1050	24" wide		20	.800		237	42		279	325
1060	30" wide		18	.889		289	46.50		335.50	390
1070	36" wide		14	1.143		335	60		395	460
1090	Three compartment, 9" wide		36	.444		148	23.50		171.50	198
1100	12" wide		28	.571		163	30		193	224
1110	18" wide		22	.727		205	38		243	283
1120	24" wide		18	.889		258	46.50		304.50	355
1130	30" wide		16	1		310	52.50		362.50	420
1140	36" wide		12	1.333		365	70		435	505
1200	Horizontal elbow, 9" wide		5.40	2.963	Ea.	425	155		580	695
1400	12" wide		4.60	3.478		490	182		672	810
1600	18" wide		4	4		625	210		835	1,000
1800	24" wide		3.20	5		880	262		1,142	1,350
1900	27" wide		3	5.333		995	279		1,274	1,525
2000	30" wide		2.60	6.154		1,175	320		1,495	1,775
2200	36" wide		2.40	6.667		1,525	350		1,875	2,225

26 05 43.10 Trench Duct		Crew	Daily Output	Labor-Hours	Unit	Material	Labor	Equipment	Total	Total Incl O&P
						2013 Bare Costs				
2220	Two compartment, 9" wide	2 Elec	3.80	4.211	Ea.	685	221		906	1,075
2230	12" wide		3	5.333		755	279		1,034	1,250
2240	18" wide		2.40	6.667		910	350		1,260	1,525
2250	24" wide		2	8		1,150	420		1,570	1,900
2260	30" wide		1.80	8.889		1,550	465		2,015	2,400
2270	36" wide		1.60	10		1,875	525		2,400	2,825
2290	Three compartment, 9" wide		3.60	4.444		715	233		948	1,125
2300	12" wide		2.80	5.714		800	299		1,099	1,325
2310	18" wide		2.20	7.273		975	380		1,355	1,650
2320	24" wide		1.80	8.889		1,225	465		1,690	2,050
2330	30" wide		1.60	10		1,600	525		2,125	2,550
2350	36" wide		1.40	11.429		1,950	600		2,550	3,050
2400	Vertical elbow, 9" wide		5.40	2.963		148	155		303	395
2600	12" wide		4.60	3.478		160	182		342	450
2800	18" wide		4	4		184	210		394	520
3000	24" wide		3.20	5		229	262		491	640
3100	27" wide		3	5.333		238	279		517	680
3200	30" wide		2.60	6.154		252	320		572	760
3400	36" wide		2.40	6.667		278	350		628	830
3600	Cross, 9" wide		4	4		695	210		905	1,075
3800	12" wide		3.20	5		735	262		997	1,200
4000	18" wide		2.60	6.154		880	320		1,200	1,450
4200	24" wide		2.20	7.273		1,125	380		1,505	1,800
4300	27" wide		2.20	7.273		1,300	380		1,680	2,025
4400	30" wide		2	8		1,450	420		1,870	2,225
4600	36" wide		1.80	8.889		1,800	465		2,265	2,675
4620	Two compartment, 9" wide		3.80	4.211		725	221		946	1,125
4630	12" wide		3	5.333		760	279		1,039	1,250
4640	18" wide		2.40	6.667		920	350		1,270	1,525
4650	24" wide		2	8		1,175	420		1,595	1,900
4660	30" wide		1.80	8.889		1,550	465		2,015	2,400
4670	36" wide		1.60	10		1,875	525		2,400	2,825
4690	Three compartment, 9" wide		3.60	4.444		735	233		968	1,150
4700	12" wide		2.80	5.714		840	299		1,139	1,375
4710	18" wide		2.20	7.273		995	380		1,375	1,675
4720	24" wide		1.80	8.889		1,250	465		1,715	2,075
4730	30" wide		1.60	10		1,625	525		2,150	2,550
4740	36" wide		1.40	11.429		2,000	600		2,600	3,100
4800	End closure, 9" wide		14.40	1.111		43.50	58		101.50	135
5000	12" wide		12	1.333		49.50	70		119.50	160
5200	18" wide		10	1.600		76	84		160	209
5400	24" wide		8	2		100	105		205	267
5500	27" wide		7	2.286		116	120		236	305
5600	30" wide		6.60	2.424		126	127		253	330
5800	36" wide		5.80	2.759		149	145		294	380
6000	Tees, 9" wide		4	4		420	210		630	780
6200	12" wide		3.60	4.444		490	233		723	890
6400	18" wide		3.20	5		625	262		887	1,075
6600	24" wide		3	5.333		900	279		1,179	1,400
6700	27" wide		2.80	5.714		995	299		1,294	1,550
6800	30" wide		2.60	6.154		1,175	320		1,495	1,775
7000	36" wide		2	8		1,525	420		1,945	2,325
7020	Two compartment, 9" wide		3.80	4.211		490	221		711	870

26 05 Common Work Results for Electrical

26 05 43 – Underground Ducts and Raceways for Electrical Systems

26 05 43.10 Trench Duct		Crew	Daily Output	Labor-Hours	Unit	Material	Labor	Equipment	Total	Total Incl O&P
						2013 Bare Costs				
7030	12″ wide	2 Elec	3.40	4.706	Ea.	525	247		772	950
7040	18″ wide		3	5.333		705	279		984	1,200
7050	24″ wide		2.80	5.714		950	299		1,249	1,500
7060	30″ wide		2.40	6.667		1,300	350		1,650	1,950
7070	36″ wide		1.90	8.421		1,600	440		2,040	2,425
7090	Three compartment, 9″ wide		3.60	4.444		560	233		793	965
7100	12″ wide		3.20	5		585	262		847	1,025
7110	18″ wide		2.80	5.714		745	299		1,044	1,275
7120	24″ wide		2.60	6.154		1,025	320		1,345	1,600
7130	30″ wide		2.20	7.273		1,350	380		1,730	2,050
7140	36″ wide		1.80	8.889		1,700	465		2,165	2,575
7200	Riser, and cabinet connector, 9″ wide		5.40	2.963		184	155		339	435
7400	12″ wide		4.60	3.478		215	182		397	510
7600	18″ wide		4	4		264	210		474	605
7800	24″ wide		3.20	5		320	262		582	740
7900	27″ wide		3	5.333		330	279		609	780
8000	30″ wide		2.60	6.154		365	320		685	885
8200	36″ wide		2	8		425	420		845	1,100
8400	Insert assembly, cell to conduit adapter, 1-1/4″	1 Elec	16	.500		73	26		99	119
8500	Adjustable partition	″	320	.025	L.F.	26	1.31		27.31	30.50
8600	Depth of duct over 4″, per 1″, add					11.30			11.30	12.45
8700	Support post	1 Elec	240	.033		26	1.75		27.75	31
8800	Cover double tile trim, 2 sides					40			40	44
8900	4 sides					119			119	131
9160	Trench duct 3-1/2″ x 4-1/2″, add					10.75			10.75	11.80
9170	Trench duct 4″ x 5″, add					10.75			10.75	11.80
9200	For carpet trim, add					36.50			36.50	40
9210	For double carpet trim, add					111			111	122

26 05 43.20 Underfloor Duct

Line	Description	Crew	Daily Output	Labor-Hours	Unit	Material	Labor	Equipment	Total	Total Incl O&P
0010	**UNDERFLOOR DUCT** R260543-50									
0100	Duct, 1-3/8″ x 3-1/8″ blank, standard	2 Elec	160	.100	L.F.	13.95	5.25		19.20	23
0200	1-3/8″ x 7-1/4″ blank, super duct		120	.133		28	7		35	41
0400	7/8″ or 1-3/8″ insert type, 24″ O.C., 1-3/8″ x 3-1/8″, std.		140	.114		18.70	6		24.70	29.50
0600	1-3/8″ x 7-1/4″, super duct		100	.160		32.50	8.40		40.90	48.50
0800	Junction box, single duct, 1 level, 3-1/8″	1 Elec	4	2	Ea.	420	105		525	615
0820	3-1/8″ x 7-1/4″		4	2		490	105		595	690
0840	2 level, 3-1/8″ upper & lower		3.20	2.500		490	131		621	730
0860	3-1/8″ upper, 7-1/4″ lower		2.70	2.963		490	155		645	765
0880	Carpet pan for above		80	.100		345	5.25		350.25	390
0900	Terrazzo pan for above		67	.119		810	6.25		816.25	905
1000	Junction box, single duct, 1 level, 7-1/4″		2.70	2.963		490	155		645	765
1020	2 level, 7-1/4″ upper & lower		2.70	2.963		560	155		715	845
1040	2 duct, two 3-1/8″ upper & lower		3.20	2.500		740	131		871	1,000
1200	1 level, 2 duct, 3-1/8″		3.20	2.500		560	131		691	810
1220	Carpet pan for above boxes		80	.100		345	5.25		350.25	390
1240	Terrazzo pan for above boxes		67	.119		810	6.25		816.25	905
1260	Junction box, 1 level, two 3-1/8″ x one 3-1/8″ + one 7-1/4″		2.30	3.478		950	182		1,132	1,325
1280	2 level, two 3-1/8″ upper, one 3-1/8″ + one 7-1/4″ lower		2	4		1,050	210		1,260	1,475
1300	Carpet pan for above boxes		80	.100		345	5.25		350.25	390
1320	Terrazzo pan for above boxes		67	.119		810	6.25		816.25	905
1400	Junction box, 1 level, 2 duct, 7-1/4″		2.30	3.478		1,450	182		1,632	1,850
1420	Two 3-1/8″ + one 7-1/4″		2	4		1,450	210		1,660	1,900

26 05 43 – Underground Ducts and Raceways for Electrical Systems

26 05 43.20 Underfloor Duct		Crew	Daily Output	Labor-Hours	Unit	2013 Bare Costs Material	Labor	Equipment	Total	Total Incl O&P
1440	Carpet pan for above	1 Elec	80	.100	Ea.	345	5.25		350.25	390
1460	Terrazzo pan for above		67	.119		810	6.25		816.25	905
1580	Junction box, 1 level, one 3-1/8" + one 7-1/4" x same		2.30	3.478		950	182		1,132	1,325
1600	Triple duct, 3-1/8"		2.30	3.478		950	182		1,132	1,325
1700	Junction box, 1 level, one 3-1/8" + two 7-1/4"		2	4		1,600	210		1,810	2,100
1720	Carpet pan for above		80	.100		345	5.25		350.25	390
1740	Terrazzo pan for above		67	.119		810	6.25		816.25	905
1800	Insert to conduit adapter, 3/4" & 1"		32	.250		34.50	13.10		47.60	57
2000	Support, single cell		27	.296		51.50	15.55		67.05	80
2200	Super duct		16	.500		51.50	26		77.50	96
2400	Double cell		16	.500		51.50	26		77.50	96
2600	Triple cell		11	.727		51.50	38		89.50	114
2800	Vertical elbow, standard duct		10	.800		92.50	42		134.50	165
3000	Super duct		8	1		92.50	52.50		145	181
3200	Cabinet connector, standard duct		32	.250		69.50	13.10		82.60	96
3400	Super duct		27	.296		69.50	15.55		85.05	99.50
3600	Conduit adapter, 1" to 1-1/4"		32	.250		69.50	13.10		82.60	96
3800	2" to 1-1/4"		27	.296		83.50	15.55		99.05	115
4000	Outlet, low tension (tele, computer, etc.)		8	1		97.50	52.50		150	186
4200	High tension, receptacle (120 volt)		8	1		97.50	52.50		150	186
4300	End closure, standard duct		160	.050		3.90	2.62		6.52	8.20
4310	Super duct		160	.050		7.40	2.62		10.02	12
4350	Elbow, horiz., standard duct		26	.308		233	16.10		249.10	280
4360	Super duct		26	.308		233	16.10		249.10	280
4380	Elbow, offset, standard duct		26	.308		92.50	16.10		108.60	126
4390	Super duct		26	.308		92.50	16.10		108.60	126
4400	Marker screw assembly for inserts		50	.160		16.45	8.40		24.85	30.50
4410	Y take off, standard duct		26	.308		139	16.10		155.10	177
4420	Super duct		26	.308		139	16.10		155.10	177
4430	Box opening plug, standard duct		160	.050		13.05	2.62		15.67	18.25
4440	Super duct		160	.050		13.05	2.62		15.67	18.25
4450	Sleeve coupling, standard duct		160	.050		35	2.62		37.62	42.50
4460	Super duct		160	.050		43	2.62		45.62	51
4470	Conduit adapter, standard duct, 3/4"		32	.250		69.50	13.10		82.60	96
4480	1" or 1-1/4"		32	.250		69.50	13.10		82.60	96
4500	1-1/2"		32	.250		69.50	13.10		82.60	96

26 05 80 – Wiring Connections

26 05 80.10 Motor Connections		Crew	Daily Output	Labor-Hours	Unit	Material	Labor	Equipment	Total	Total Incl O&P
0010	**MOTOR CONNECTIONS**									
0020	Flexible conduit and fittings, 115 volt, 1 phase, up to 1 HP motor	1 Elec	8	1	Ea.	8.45	52.50		60.95	88
0050	2 HP motor R260580-75		6.50	1.231		15.95	64.50		80.45	114
0100	3 HP motor		5.50	1.455		14.20	76		90.20	130
0110	230 volt, 3 phase, 3 HP motor		6.78	1.180		10	62		72	104
0112	5 HP motor		5.47	1.463		8.60	76.50		85.10	124
0114	7-1/2 HP motor		4.61	1.735		12.15	91		103.15	149
0120	10 HP motor		4.20	1.905		27.50	100		127.50	179
0150	15 HP motor		3.30	2.424		27.50	127		154.50	220
0200	25 HP motor		2.70	2.963		38.50	155		193.50	275
0400	50 HP motor		2.20	3.636		77.50	191		268.50	370
0600	100 HP motor		1.50	5.333		182	279		461	620
1500	460 volt, 5 HP motor, 3 phase		8	1		9	52.50		61.50	88.50
1520	10 HP motor		8	1		9	52.50		61.50	88.50

26 05 Common Work Results for Electrical

26 05 80 – Wiring Connections

26 05 80.10 Motor Connections		Crew	Daily Output	Labor-Hours	Unit	Material	2013 Bare Costs Labor	Equipment	Total	Total Incl O&P
1530	25 HP motor	1 Elec	6	1.333	Ea.	17	70		87	124
1540	30 HP motor		6	1.333		17	70		87	124
1550	40 HP motor		5	1.600		26	84		110	154
1560	50 HP motor		5	1.600		30	84		114	158
1570	60 HP motor		3.80	2.105		33.50	110		143.50	202
1580	75 HP motor		3.50	2.286		45	120		165	228
1590	100 HP motor		2.50	3.200		67	168		235	325
1600	125 HP motor		2	4		88	210		298	410
1610	150 HP motor		1.80	4.444		92.50	233		325.50	450
1620	200 HP motor		1.50	5.333		149	279		428	585
2005	460 Volt, 5 HP motor, 3 Phase, w/sealtite		8	1		14.50	52.50		67	94.50
2010	10 HP motor		8	1		14.50	52.50		67	94.50
2015	25 HP motor		6	1.333		27	70		97	135
2020	30 HP motor		6	1.333		27	70		97	135
2025	40 HP motor		5	1.600		51	84		135	181
2030	50 HP motor		5	1.600		51.50	84		135.50	182
2035	60 HP motor		3.80	2.105		79.50	110		189.50	253
2040	75 HP motor		3.50	2.286		83	120		203	271
2045	100 HP motor		2.50	3.200		112	168		280	375
2055	150 HP motor		1.80	4.444		118	233		351	480
2060	200 HP motor		1.50	5.333		830	279		1,109	1,325

26 05 90 – Residential Applications

26 05 90.10 Residential Wiring		Crew	Daily Output	Labor-Hours	Unit	Material	Labor	Equipment	Total	Total Incl O&P
0010	**RESIDENTIAL WIRING**									
0020	20' avg. runs and #14/2 wiring incl. unless otherwise noted									
1000	Service & panel, includes 24' SE-AL cable, service eye, meter,									
1010	Socket, panel board, main bkr., ground rod, 15 or 20 amp									
1020	1-pole circuit breakers, and misc. hardware									
1100	100 amp, with 10 branch breakers	1 Elec	1.19	6.723	Ea.	555	350		905	1,150
1110	With PVC conduit and wire		.92	8.696		625	455		1,080	1,375
1120	With RGS conduit and wire		.73	10.959		780	575		1,355	1,725
1150	150 amp, with 14 branch breakers		1.03	7.767		860	405		1,265	1,550
1170	With PVC conduit and wire		.82	9.756		995	510		1,505	1,875
1180	With RGS conduit and wire		.67	11.940		1,300	625		1,925	2,350
1200	200 amp, with 18 branch breakers	2 Elec	1.80	8.889		1,125	465		1,590	1,925
1220	With PVC conduit and wire		1.46	10.959		1,250	575		1,825	2,225
1230	With RGS conduit and wire		1.24	12.903		1,675	675		2,350	2,825
1800	Lightning surge suppressor for above services, add	1 Elec	32	.250		50	13.10		63.10	74.50
2000	Switch devices									
2100	Single pole, 15 amp, Ivory, with a 1-gang box, cover plate,									
2110	Type NM (Romex) cable	1 Elec	17.10	.468	Ea.	12.20	24.50		36.70	50
2120	Type MC (BX) cable		14.30	.559		26	29.50		55.50	73
2130	EMT & wire		5.71	1.401		33.50	73.50		107	147
2150	3-way, #14/3, type NM cable		14.55	.550		16.20	29		45.20	61
2170	Type MC cable		12.31	.650		35	34		69	89
2180	EMT & wire		5	1.600		37	84		121	166
2200	4-way, #14/3, type NM cable		14.55	.550		24.50	29		53.50	69.50
2220	Type MC cable		12.31	.650		43	34		77	98
2230	EMT & wire		5	1.600		45	84		129	175
2250	S.P., 20 amp, #12/2, type NM cable		13.33	.600		21	31.50		52.50	70
2270	Type MC cable		11.43	.700		32.50	36.50		69	90.50
2280	EMT & wire		4.85	1.649		43.50	86.50		130	177

26 05 90.10 Residential Wiring		Crew	Daily Output	Labor-Hours	Unit	Material	2013 Bare Costs Labor	Equipment	Total	Total Incl O&P
2290	S.P. rotary dimmer, 600W, no wiring	1 Elec	17	.471	Ea.	25	24.50		49.50	64
2300	S.P. rotary dimmer, 600W, type NM cable		14.55	.550		30	29		59	76
2320	Type MC cable		12.31	.650		43.50	34		77.50	99
2330	EMT & wire		5	1.600		52.50	84		136.50	183
2350	3-way rotary dimmer, type NM cable		13.33	.600		34.50	31.50		66	85
2370	Type MC cable		11.43	.700		48.50	36.50		85	109
2380	EMT & wire		4.85	1.649		57.50	86.50		144	192
2400	Interval timer wall switch, 20 amp, 1-30 min., #12/2									
2410	Type NM cable	1 Elec	14.55	.550	Ea.	54	29		83	103
2420	Type MC cable		12.31	.650		60.50	34		94.50	118
2430	EMT & wire		5	1.600		76.50	84		160.50	210
2500	Decorator style									
2510	S.P., 15 amp, type NM cable	1 Elec	17.10	.468	Ea.	16.30	24.50		40.80	54.50
2520	Type MC cable		14.30	.559		30	29.50		59.50	77
2530	EMT & wire		5.71	1.401		37.50	73.50		111	151
2550	3-way, #14/3, type NM cable		14.55	.550		20.50	29		49.50	65.50
2570	Type MC cable		12.31	.650		39	34		73	93.50
2580	EMT & wire		5	1.600		41	84		125	170
2600	4-way, #14/3, type NM cable		14.55	.550		28.50	29		57.50	74
2620	Type MC cable		12.31	.650		47	34		81	103
2630	EMT & wire		5	1.600		49	84		133	179
2650	S.P., 20 amp, #12/2, type NM cable		13.33	.600		25	31.50		56.50	74.50
2670	Type MC cable		11.43	.700		36.50	36.50		73	95
2680	EMT & wire		4.85	1.649		47.50	86.50		134	182
2700	S.P., slide dimmer, type NM cable		17.10	.468		32.50	24.50		57	72
2720	Type MC cable		14.30	.559		46.50	29.50		76	95
2730	EMT & wire		5.71	1.401		55	73.50		128.50	171
2750	S.P., touch dimmer, type NM cable		17.10	.468		29	24.50		53.50	68.50
2770	Type MC cable		14.30	.559		43	29.50		72.50	91.50
2780	EMT & wire		5.71	1.401		52	73.50		125.50	167
2800	3-way touch dimmer, type NM cable		13.33	.600		51	31.50		82.50	103
2820	Type MC cable		11.43	.700		65	36.50		101.50	127
2830	EMT & wire		4.85	1.649		74	86.50		160.50	210
3000	Combination devices									
3100	S.P. switch/15 amp recpt., Ivory, 1-gang box, plate									
3110	Type NM cable	1 Elec	11.43	.700	Ea.	24	36.50		60.50	81.50
3120	Type MC cable		10	.800		38	42		80	105
3130	EMT & wire		4.40	1.818		46.50	95.50		142	195
3150	S.P. switch/pilot light, type NM cable		11.43	.700		24	36.50		60.50	81.50
3170	Type MC cable		10	.800		38	42		80	105
3180	EMT & wire		4.43	1.806		47	94.50		141.50	194
3190	2-S.P. switches, 2-#14/2, no wiring		14	.571		7.60	30		37.60	53.50
3200	2-S.P. switches, 2-#14/2, type NM cables		10	.800		27.50	42		69.50	92.50
3220	Type MC cable		8.89	.900		48.50	47		95.50	124
3230	EMT & wire		4.10	1.951		50	102		152	208
3250	3-way switch/15 amp recpt., #14/3, type NM cable		10	.800		31.50	42		73.50	97
3270	Type MC cable		8.89	.900		50	47		97	126
3280	EMT & wire		4.10	1.951		52	102		154	210
3300	2-3 way switches, 2-#14/3, type NM cables		8.89	.900		41	47		88	116
3320	Type MC cable		8	1		71.50	52.50		124	157
3330	EMT & wire		4	2		59.50	105		164.50	223
3350	S.P. switch/20 amp recpt., #12/2, type NM cable		10	.800		34	42		76	100
3370	Type MC cable		8.89	.900		40	47		87	115

26 05 90 – Residential Applications

26 05 90.10 Residential Wiring		Crew	Daily Output	Labor-Hours	Unit	Material	2013 Bare Costs Labor	Equipment	Total	Total Incl O&P
3380	EMT & wire	1 Elec	4.10	1.951	Ea.	56.50	102		158.50	216
3400	Decorator style									
3410	S.P. switch/15 amp recpt., type NM cable	1 Elec	11.43	.700	Ea.	28	36.50		64.50	86
3420	Type MC cable		10	.800		42	42		84	109
3430	EMT & wire		4.40	1.818		51	95.50		146.50	199
3450	S.P. switch/pilot light, type NM cable		11.43	.700		28	36.50		64.50	86
3470	Type MC cable		10	.800		42	42		84	109
3480	EMT & wire		4.40	1.818		51	95.50		146.50	199
3500	2-S.P. switches, 2-#14/2, type NM cables		10	.800		31.50	42		73.50	97
3520	Type MC cable		8.89	.900		52.50	47		99.50	129
3530	EMT & wire		4.10	1.951		54	102		156	213
3550	3-way/15 amp recpt., #14/3, type NM cable		10	.800		35.50	42		77.50	102
3570	Type MC cable		8.89	.900		54	47		101	130
3580	EMT & wire		4.10	1.951		56	102		158	215
3650	2-3 way switches, 2-#14/3, type NM cables		8.89	.900		45	47		92	120
3670	Type MC cable		8	1		75.50	52.50		128	162
3680	EMT & wire		4	2		63.50	105		168.50	227
3700	S.P. switch/20 amp recpt., #12/2, type NM cable		10	.800		38	42		80	105
3720	Type MC cable		8.89	.900		44.50	47		91.50	119
3730	EMT & wire	↓	4.10	1.951	↓	61	102		163	220
4000	Receptacle devices									
4010	Duplex outlet, 15 amp recpt., Ivory, 1-gang box, plate									
4015	Type NM cable	1 Elec	14.55	.550	Ea.	10.95	29		39.95	55
4020	Type MC cable		12.31	.650		25	34		59	78.50
4030	EMT & wire		5.33	1.501		32	78.50		110.50	153
4050	With #12/2, type NM cable		12.31	.650		13.55	34		47.55	66
4070	Type MC cable		10.67	.750		25	39.50		64.50	86.50
4080	EMT & wire		4.71	1.699		36	89		125	173
4100	20 amp recpt., #12/2, type NM cable		12.31	.650		22	34		56	75
4120	Type MC cable		10.67	.750		33.50	39.50		73	95.50
4130	EMT & wire	↓	4.71	1.699	↓	44.50	89		133.50	182
4140	For GFI see Section 26 05 90.10 line 4300 below									
4150	Decorator style, 15 amp recpt., type NM cable	1 Elec	14.55	.550	Ea.	15	29		44	59.50
4170	Type MC cable		12.31	.650		29	34		63	83
4180	EMT & wire		5.33	1.501		36	78.50		114.50	158
4200	With #12/2, type NM cable		12.31	.650		17.60	34		51.60	70.50
4220	Type MC cable		10.67	.750		29	39.50		68.50	91
4230	EMT & wire		4.71	1.699		40	89		129	177
4250	20 amp recpt. #12/2, type NM cable		12.31	.650		26	34		60	79.50
4270	Type MC cable		10.67	.750		37.50	39.50		77	100
4280	EMT & wire		4.71	1.699		48.50	89		137.50	187
4300	GFI, 15 amp recpt., type NM cable		12.31	.650		47	34		81	103
4320	Type MC cable		10.67	.750		61	39.50		100.50	126
4330	EMT & wire		4.71	1.699		68	89		157	208
4350	GFI with #12/2, type NM cable		10.67	.750		49.50	39.50		89	114
4370	Type MC cable		9.20	.870		61	45.50		106.50	135
4380	EMT & wire		4.21	1.900		72	99.50		171.50	229
4400	20 amp recpt., #12/2 type NM cable		10.67	.750		51.50	39.50		91	116
4420	Type MC cable		9.20	.870		63	45.50		108.50	138
4430	EMT & wire		4.21	1.900		74.50	99.50		174	231
4500	Weather-proof cover for above receptacles, add	↓	32	.250	↓	4.79	13.10		17.89	25
4550	Air conditioner outlet, 20 amp-240 volt recpt.									
4560	30′ of #12/2, 2 pole circuit breaker									

26 05 90.10 Residential Wiring		Crew	Daily Output	Labor-Hours	Unit	2013 Bare Costs Material	Labor	Equipment	Total	Total Incl O&P
4570	Type NM cable	1 Elec	10	.800	Ea.	62	42		104	131
4580	Type MC cable		9	.889		75.50	46.50		122	153
4590	EMT & wire		4	2		84.50	105		189.50	250
4600	Decorator style, type NM cable		10	.800		66.50	42		108.50	136
4620	Type MC cable		9	.889		80.50	46.50		127	158
4630	EMT & wire		4	2		89	105		194	255
4650	Dryer outlet, 30 amp-240 volt recpt., 20′ of #10/3									
4660	2 pole circuit breaker									
4670	Type NM cable	1 Elec	6.41	1.248	Ea.	65	65.50		130.50	170
4680	Type MC cable		5.71	1.401		73	73.50		146.50	190
4690	EMT & wire		3.48	2.299		79.50	120		199.50	267
4700	Range outlet, 50 amp-240 volt recpt., 30′ of #8/3									
4710	Type NM cable	1 Elec	4.21	1.900	Ea.	93	99.50		192.50	251
4720	Type MC cable		4	2		147	105		252	320
4730	EMT & wire		2.96	2.703		110	142		252	335
4750	Central vacuum outlet, Type NM cable		6.40	1.250		61.50	65.50		127	166
4770	Type MC cable		5.71	1.401		82	73.50		155.50	200
4780	EMT & wire		3.48	2.299		87.50	120		207.50	277
4800	30 amp-110 volt locking recpt., #10/2 circ. bkr.									
4810	Type NM cable	1 Elec	6.20	1.290	Ea.	71.50	67.50		139	180
4820	Type MC cable		5.40	1.481		94	77.50		171.50	219
4830	EMT & wire		3.20	2.500		98	131		229	305
4900	Low voltage outlets									
4910	Telephone recpt., 20′ of 4/C phone wire	1 Elec	26	.308	Ea.	11.15	16.10		27.25	36.50
4920	TV recpt., 20′ of RG59U coax wire, F type connector	"	16	.500	"	19.75	26		45.75	60.50
4950	Door bell chime, transformer, 2 buttons, 60′ of bellwire									
4970	Economy model	1 Elec	11.50	.696	Ea.	64.50	36.50		101	126
4980	Custom model		11.50	.696		118	36.50		154.50	185
4990	Luxury model, 3 buttons		9.50	.842		310	44		354	405
6000	Lighting outlets									
6050	Wire only (for fixture), type NM cable	1 Elec	32	.250	Ea.	7.10	13.10		20.20	27.50
6070	Type MC cable		24	.333		15.10	17.45		32.55	42.50
6080	EMT & wire		10	.800		21	42		63	85.50
6100	Box (4″), and wire (for fixture), type NM cable		25	.320		17.60	16.75		34.35	44.50
6120	Type MC cable		20	.400		25.50	21		46.50	59.50
6130	EMT & wire		11	.727		31.50	38		69.50	91.50
6200	Fixtures (use with lines 6050 or 6100 above)									
6210	Canopy style, economy grade	1 Elec	40	.200	Ea.	37	10.50		47.50	56.50
6220	Custom grade		40	.200		57	10.50		67.50	78
6250	Dining room chandelier, economy grade		19	.421		92.50	22		114.50	135
6260	Custom grade		19	.421		272	22		294	330
6270	Luxury grade		15	.533		610	28		638	715
6310	Kitchen fixture (fluorescent), economy grade		30	.267		68.50	13.95		82.45	96.50
6320	Custom grade		25	.320		188	16.75		204.75	232
6350	Outdoor, wall mounted, economy grade		30	.267		35	13.95		48.95	59.50
6360	Custom grade		30	.267		120	13.95		133.95	153
6370	Luxury grade		25	.320		272	16.75		288.75	325
6410	Outdoor PAR floodlights, 1 lamp, 150 watt		20	.400		37.50	21		58.50	73
6420	2 lamp, 150 watt each		20	.400		62.50	21		83.50	100
6425	Motion sensing, 2 lamp, 150 watt each		20	.400		75	21		96	114
6430	For infrared security sensor, add		32	.250		126	13.10		139.10	158
6450	Outdoor, quartz-halogen, 300 watt flood		20	.400		45.50	21		66.50	81.50
6600	Recessed downlight, round, pre-wired, 50 or 75 watt trim		30	.267		46.50	13.95		60.45	72

26 05 Common Work Results for Electrical

26 05 90 – Residential Applications

26 05 90.10 Residential Wiring		Crew	Daily Output	Labor-Hours	Unit	2013 Bare Costs Material	Labor	Equipment	Total	Total Incl O&P
6610	With shower light trim	1 Elec	30	.267	Ea.	57.50	13.95		71.45	84
6620	With wall washer trim		28	.286		69.50	14.95		84.45	98.50
6630	With eye-ball trim		28	.286		69.50	14.95		84.45	98.50
6640	For direct contact with insulation, add					2.47			2.47	2.72
6700	Porcelain lamp holder	1 Elec	40	.200		4.12	10.50		14.62	20
6710	With pull switch		40	.200		5.65	10.50		16.15	22
6750	Fluorescent strip, 1-20 watt tube, wrap around diffuser, 24"		24	.333		58	17.45		75.45	90
6760	1-34 watt tube, 48"		24	.333		74.50	17.45		91.95	108
6770	2-34 watt tubes, 48"		20	.400		88.50	21		109.50	129
6780	With residential ballast		20	.400		99.50	21		120.50	141
6800	Bathroom heat lamp, 1-250 watt		28	.286		51	14.95		65.95	78.50
6810	2-250 watt lamps		28	.286		81.50	14.95		96.45	112
6820	For timer switch, see Section 26 05 90.10 line 2400									
6900	Outdoor post lamp, incl. post, fixture, 35' of #14/2									
6910	Type NMC cable	1 Elec	3.50	2.286	Ea.	203	120		323	400
6920	Photo-eye, add		27	.296		37.50	15.55		53.05	64.50
6950	Clock dial time switch, 24 hr., w/enclosure, type NM cable		11.43	.700		70.50	36.50		107	133
6970	Type MC cable		11	.727		84.50	38		122.50	150
6980	EMT & wire		4.85	1.649		91.50	86.50		178	230
7000	Alarm systems									
7050	Smoke detectors, box, #14/3, type NM cable	1 Elec	14.55	.550	Ea.	35.50	29		64.50	82
7070	Type MC cable		12.31	.650		49	34		83	105
7080	EMT & wire		5	1.600		51	84		135	181
7090	For relay output to security system, add					14			14	15.40
8000	Residential equipment									
8050	Disposal hook-up, incl. switch, outlet box, 3' of flex									
8060	20 amp-1 pole circ. bkr., and 25' of #12/2									
8070	Type NM cable	1 Elec	10	.800	Ea.	32.50	42		74.50	98.50
8080	Type MC cable		8	1		45	52.50		97.50	128
8090	EMT & wire		5	1.600		57.50	84		141.50	189
8100	Trash compactor or dishwasher hook-up, incl. outlet box,									
8110	3' of flex, 15 amp-1 pole circ. bkr., and 25' of #14/2									
8120	Type NM cable	1 Elec	10	.800	Ea.	24	42		66	89
8130	Type MC cable		8	1		40	52.50		92.50	123
8140	EMT & wire		5	1.600		49.50	84		133.50	179
8150	Hot water sink dispensor hook-up, use line 8100									
8200	Vent/exhaust fan hook-up, type NM cable	1 Elec	32	.250	Ea.	7.10	13.10		20.20	27.50
8220	Type MC cable		24	.333		15.10	17.45		32.55	42.50
8230	EMT & wire		10	.800		21	42		63	85.50
8250	Bathroom vent fan, 50 CFM (use with above hook-up)									
8260	Economy model	1 Elec	15	.533	Ea.	27	28		55	71.50
8270	Low noise model		15	.533		36.50	28		64.50	82
8280	Custom model		12	.667		126	35		161	191
8300	Bathroom or kitchen vent fan, 110 CFM									
8310	Economy model	1 Elec	15	.533	Ea.	74.50	28		102.50	124
8320	Low noise model	"	15	.533	"	92.50	28		120.50	144
8350	Paddle fan, variable speed (w/o lights)									
8360	Economy model (AC motor)	1 Elec	10	.800	Ea.	124	42		166	199
8362	With light kit		10	.800		164	42		206	244
8370	Custom model (AC motor)		10	.800		195	42		237	278
8372	With light kit		10	.800		236	42		278	325
8380	Luxury model (DC motor)		8	1		385	52.50		437.50	505
8382	With light kit		8	1		425	52.50		477.50	550

26 05 Common Work Results for Electrical

26 05 90 – Residential Applications

	26 05 90.10 Residential Wiring	Crew	Daily Output	Labor-Hours	Unit	2013 Bare Costs Material	Labor	Equipment	Total	Total Incl O&P
8390	Remote speed switch for above, add	1 Elec	12	.667	Ea.	32	35		67	88
8500	Whole house exhaust fan, ceiling mount, 36", variable speed									
8510	Remote switch, incl. shutters, 20 amp-1 pole circ. bkr.									
8520	30' of #12/2, type NM cable	1 Elec	4	2	Ea.	1,225	105		1,330	1,500
8530	Type MC cable		3.50	2.286		1,225	120		1,345	1,525
8540	EMT & wire	↓	3	2.667	↓	1,250	140		1,390	1,575
8600	Whirlpool tub hook-up, incl. timer switch, outlet box									
8610	3' of flex, 20 amp-1 pole GFI circ. bkr.									
8620	30' of #12/2, type NM cable	1 Elec	5	1.600	Ea.	124	84		208	262
8630	Type MC cable		4.20	1.905		132	100		232	294
8640	EMT & wire	↓	3.40	2.353	↓	143	123		266	340
8650	Hot water heater hook-up, incl. 1-2 pole circ. bkr., box;									
8660	3' of flex, 20' of #10/2, type NM cable	1 Elec	5	1.600	Ea.	34	84		118	163
8670	Type MC cable		4.20	1.905		51	100		151	205
8680	EMT & wire	↓	3.40	2.353	↓	52	123		175	241
9000	Heating/air conditioning									
9050	Furnace/boiler hook-up, incl. firestat, local on-off switch									
9060	Emergency switch, and 40' of type NM cable	1 Elec	4	2	Ea.	56	105		161	219
9070	Type MC cable		3.50	2.286		77	120		197	264
9080	EMT & wire	↓	1.50	5.333	↓	90	279		369	520
9100	Air conditioner hook-up, incl. local 60 amp disc. switch									
9110	3' sealtite, 40 amp, 2 pole circuit breaker									
9130	40' of #8/2, type NM cable	1 Elec	3.50	2.286	Ea.	202	120		322	400
9140	Type MC cable		3	2.667		283	140		423	520
9150	EMT & wire	↓	1.30	6.154	↓	242	320		562	745
9200	Heat pump hook-up, 1-40 & 1-100 amp 2 pole circ. bkr.									
9210	Local disconnect switch, 3' sealtite									
9220	40' of #8/2 & 30' of #3/2									
9230	Type NM cable	1 Elec	1.30	6.154	Ea.	540	320		860	1,075
9240	Type MC cable		1.08	7.407		660	390		1,050	1,300
9250	EMT & wire	↓	.94	8.511	↓	620	445		1,065	1,350
9500	Thermostat hook-up, using low voltage wire									
9520	Heating only, 25' of #18-3	1 Elec	24	.333	Ea.	10.20	17.45		27.65	37.50
9530	Heating/cooling, 25' of #18-4	"	20	.400	"	12.35	21		33.35	45

26 09 Instrumentation and Control for Electrical Systems

26 09 13 – Electrical Power Monitoring

	26 09 13.10 Switchboard Instruments	Crew	Daily Output	Labor-Hours	Unit	Material	Labor	Equipment	Total	Total Incl O&P
0010	**SWITCHBOARD INSTRUMENTS** 3 phase, 4 wire R260913-80									
0100	AC indicating, ammeter & switch	1 Elec	8	1	Ea.	2,150	52.50		2,202.50	2,425
0200	Voltmeter & switch		8	1		2,150	52.50		2,202.50	2,425
0300	Wattmeter		8	1		4,250	52.50		4,302.50	4,750
0400	AC recording, ammeter		4	2		7,550	105		7,655	8,475
0500	Voltmeter		4	2		7,550	105		7,655	8,475
0600	Ground fault protection, zero sequence		2.70	2.963		6,675	155		6,830	7,575
0700	Ground return path		2.70	2.963		6,675	155		6,830	7,575
0800	3 current transformers, 5 to 800 amp		2	4		3,100	210		3,310	3,750
0900	1000 to 1500 amp		1.30	6.154		4,475	320		4,795	5,400
1200	2000 to 4000 amp		1	8		5,275	420		5,695	6,425
1300	Fused potential transformer, maximum 600 volt	↓	8	1	↓	1,175	52.50		1,227.50	1,350

26 09 Instrumentation and Control for Electrical Systems

26 09 13 – Electrical Power Monitoring

26 09 13.20 Voltage Monitor Systems

	26 09 13.20 Voltage Monitor Systems	Crew	Daily Output	Labor-Hours	Unit	2013 Bare Costs Material	Labor	Equipment	Total	Total Incl O&P
0010	**VOLTAGE MONITOR SYSTEMS** (test equipment)									
0100	AC voltage monitor system, 120/240 V, one-channel				Ea.	2,675			2,675	2,925
0110	Modem adapter					335			335	365
0120	Add-on detector only					1,400			1,400	1,550
0150	AC voltage remote monitor sys., 3 channel, 120, 230, or 480 V					4,850			4,850	5,325
0160	With internal modem					5,125			5,125	5,625
0170	Combination temperature and humidity probe					750			750	825
0180	Add-on detector only					3,525			3,525	3,875
0190	With internal modem				↓	3,825			3,825	4,200

26 09 13.30 Smart Metering

	26 09 13.30 Smart Metering		Crew	Daily Output	Labor-Hours	Unit	Material	Labor	Equipment	Total	Total Incl O&P
0010	**SMART METERING**, In panel										
0100	Single phase, 120/208 volt, 100 amp	G	1 Elec	8.78	.911	Ea.	410	47.50		457.50	520
0120	200 amp	G		8.78	.911		410	47.50		457.50	520
0200	277 volt, 100 amp	G		8.78	.911		425	47.50		472.50	540
0220	200 amp	G		8.78	.911		425	47.50		472.50	540
1100	Three phase, 120/208 volt, 100 amp	G		4.69	1.706		600	89.50		689.50	790
1120	200 amp	G		4.69	1.706		600	89.50		689.50	790
1130	400 amp	G		4.69	1.706		600	89.50		689.50	790
1140	800 amp	G		4.69	1.706		710	89.50		799.50	915
1150	1600 amp	G		4.69	1.706		710	89.50		799.50	915
1200	277/480 volt, 100 amp	G		4.69	1.706		665	89.50		754.50	865
1220	200 amp	G		4.69	1.706		665	89.50		754.50	865
1230	400 amp	G		4.69	1.706		665	89.50		754.50	865
1240	800 amp	G		4.69	1.706		665	89.50		754.50	865
1250	1600 amp	G		4.69	1.706		775	89.50		864.50	985
2000	Data recorder, 8 meters	G		10.97	.729		1,350	38		1,388	1,550
2100	16 meters	G	↓	8.53	.938		1,925	49		1,974	2,175
3000	Software package, per meter, basic	G					232			232	255
3100	premium	G				↓	600			600	660

26 09 23 – Lighting Control Devices

26 09 23.10 Energy Saving Lighting Devices

	26 09 23.10 Energy Saving Lighting Devices		Crew	Daily Output	Labor-Hours	Unit	Material	Labor	Equipment	Total	Total Incl O&P
0010	**ENERGY SAVING LIGHTING DEVICES**										
0100	Occupancy sensors, passive infrared ceiling mounted	G	1 Elec	7	1.143	Ea.	112	60		172	214
0110	Ultrasonic ceiling mounted	G		7	1.143		124	60		184	226
0120	Dual technology ceiling mounted	G		6.50	1.231		179	64.50		243.50	294
0150	Automatic wall switches	G		24	.333		66.50	17.45		83.95	99
0160	Daylighting sensor, manual control, ceiling mounted	G		7	1.143		112	60		172	214
0170	Remote and dimming control with remote controller	G		6.50	1.231		204	64.50		268.50	320
0200	Remote power pack	G		10	.800		37	42		79	104
0250	Photoelectric control, S.P.S.T. 120 V	G		8	1		16.90	52.50		69.40	97
0300	S.P.S.T. 208 V/277 V	G		8	1		24	52.50		76.50	105
0350	D.P.S.T. 120 V	G		6	1.333		171	70		241	293
0400	D.P.S.T. 208 V/277 V	G		6	1.333		176	70		246	299
0450	S.P.D.T. 208 V/277 V	G	↓	6	1.333	↓	210	70		280	335

26 12 Medium-Voltage Transformers

26 12 19 – Pad-Mounted, Liquid-Filled, Medium-Voltage Transformers

26 12 19.10 Transformer, Oil-Filled

	26 12 19.10 Transformer, Oil-Filled	Crew	Daily Output	Labor-Hours	Unit	2013 Bare Costs Material	Labor	Equipment	Total	Total Incl O&P
0010	**TRANSFORMER, OIL-FILLED** primary delta or Y, R262213-60									
0050	Pad mounted 5 kV or 15 kV, with taps, 277/480 V secondary, 3 phase									
0100	150 kVA	R-3	.65	30.769	Ea.	9,875	1,600	211	11,686	13,500
0110	225 kVA		.55	36.364		11,100	1,875	249	13,224	15,300
0200	300 kVA		.45	44.444		12,300	2,300	305	14,905	17,300
0300	500 kVA		.40	50		19,700	2,600	345	22,645	26,000
0400	750 kVA		.38	52.632		22,100	2,725	360	25,185	28,800
0500	1000 kVA		.26	76.923		26,200	4,000	530	30,730	35,500
0600	1500 kVA		.23	86.957		31,200	4,500	595	36,295	41,700
0700	2000 kVA		.20	100		39,400	5,200	685	45,285	52,000
0710	2500 kVA		.19	105		47,500	5,450	720	53,670	61,500
0720	3000 kVA		.17	117		57,500	6,100	805	64,405	73,000
0800	3750 kVA		.16	125		74,000	6,475	860	81,335	91,500
1990	Pole mounted distribution type, single phase									
2000	13.8 kV primary, 120/240 V secondary, 10 kVA	R-15	7.45	6.443	Ea.	930	330	44.50	1,304.50	1,575
2010	50 kVA		3.70	12.973		1,825	665	90	2,580	3,125
2020	100 kVA		2.75	17.455		3,225	895	121	4,241	5,025
2030	167 kVA		2.15	22.326		4,750	1,150	154	6,054	7,125
2900	2400 V primary, 120/240 V secondary, 10 kVA		7.45	6.443		945	330	44.50	1,319.50	1,600
2910	15 kVA		6.70	7.164		1,050	370	49.50	1,469.50	1,750
2920	25 kVA		6	8		1,275	410	55.50	1,740.50	2,075
2930	37.5 kVA		4.30	11.163		1,550	575	77.50	2,202.50	2,675
2940	50 kVA		3.70	12.973		1,875	665	90	2,630	3,150
2950	75 kVA		3	16		2,675	820	111	3,606	4,300
2960	100 kVA		2.75	17.455		3,250	895	121	4,266	5,050

26 12 19.20 Transformer, Liquid-Filled

	26 12 19.20 Transformer, Liquid-Filled	Crew	Daily Output	Labor-Hours	Unit	Material	Labor	Equipment	Total	Total Incl O&P
0010	**TRANSFORMER, LIQUID-FILLED** Pad mounted									
0020	5 kV or 15 kV primary, 277/480 volt secondary, 3 phase									
0050	225 kVA	R-3	.55	36.364	Ea.	13,100	1,875	249	15,224	17,500
0100	300 kVA		.45	44.444		15,600	2,300	305	18,205	20,900
0200	500 kVA		.40	50		19,700	2,600	345	22,645	25,900
0250	750 kVA		.38	52.632		25,400	2,725	360	28,485	32,500
0300	1000 kVA		.26	76.923		29,500	4,000	530	34,030	39,100
0350	1500 kVA		.23	86.957		34,400	4,500	595	39,495	45,300
0400	2000 kVA		.20	100		42,600	5,200	685	48,485	55,500
0450	2500 kVA		.19	105		51,000	5,450	720	57,170	65,000

26 13 Medium-Voltage Switchgear

26 13 16 – Medium-Voltage Fusible Interrupter Switchgear

26 13 16.10 Switchgear

	26 13 16.10 Switchgear	Crew	Daily Output	Labor-Hours	Unit	Material	Labor	Equipment	Total	Total Incl O&P
0010	**SWITCHGEAR**, Incorporate switch with cable connections, transformer,									
0100	& Low Voltage section									
0200	Load interrupter switch, 600 amp, 2 position									
0300	NEMA 1, 4.8 kV, 300 kVA & below w/CLF fuses	R-3	.40	50	Ea.	20,800	2,600	345	23,745	27,200
0400	400 kVA & above w/CLF fuses		.38	52.632		23,200	2,725	360	26,285	30,000
0500	Non fusible		.41	48.780		16,900	2,525	335	19,760	22,800
0600	13.8 kV, 300 kVA & below w/CLF fuses		.38	52.632		26,100	2,725	360	29,185	33,200
0700	400 kVA & above w/CLF fuses		.36	55.556		26,100	2,875	380	29,355	33,400
0800	Non fusible		.40	50		19,800	2,600	345	22,745	26,100
0900	Cable lugs for 2 feeders 4.8 kV or 13.8 kV	1 Elec	8	1		635	52.50		687.50	775
1000	Pothead, one 3 conductor or three 1 conductor		4	2		3,050	105		3,155	3,500

26 13 Medium-Voltage Switchgear

26 13 16 – Medium-Voltage Fusible Interrupter Switchgear

26 13 16.10 Switchgear		Crew	Daily Output	Labor-Hours	Unit	2013 Bare Costs Material	Labor	Equipment	Total	Total Incl O&P
1100	Two 3 conductor or six 1 conductor	1 Elec	2	4	Ea.	6,000	210		6,210	6,925
1200	Key interlocks		8	1		705	52.50		757.50	855
1300	Lightning arresters, Distribution class (no charge)									
1400	Intermediate class or line type 4.8 kV	1 Elec	2.70	2.963	Ea.	3,425	155		3,580	3,975
1500	13.8 kV		2	4		4,525	210		4,735	5,300
1600	Station class, 4.8 kV		2.70	2.963		5,850	155		6,005	6,675
1700	13.8 kV		2	4		10,100	210		10,310	11,400
1800	Transformers, 4800 volts to 480/277 volts, 75 kVA	R-3	.68	29.412		17,800	1,525	202	19,527	22,100
1900	112.5 kVA		.65	30.769		21,700	1,600	211	23,511	26,500
2000	150 kVA		.57	35.088		24,700	1,825	241	26,766	30,200
2100	225 kVA		.48	41.667		28,400	2,150	286	30,836	34,900
2200	300 kVA		.41	48.780		31,800	2,525	335	34,660	39,100
2300	500 kVA		.36	55.556		41,800	2,875	380	45,055	50,500
2400	750 kVA		.29	68.966		47,400	3,575	475	51,450	58,000
2500	13,800 volts to 480/277 volts, 75 kVA		.61	32.787		25,100	1,700	225	27,025	30,400
2600	112.5 kVA		.55	36.364		33,300	1,875	249	35,424	39,700
2700	150 kVA		.49	40.816		33,600	2,125	280	36,005	40,500
2800	225 kVA		.41	48.780		38,800	2,525	335	41,660	46,900
2900	300 kVA		.37	54.054		39,600	2,800	370	42,770	48,100
3000	500 kVA		.31	64.516		43,800	3,350	445	47,595	53,500
3100	750 kVA		.26	76.923		48,200	4,000	530	52,730	59,500
3200	Forced air cooling & temperature alarm	1 Elec	1	8		3,875	420		4,295	4,875
3300	Low voltage components									
3400	Maximum panel height 49-1/2", single or twin row									
3500	Breaker heights, type FA or FH, 6"									
3600	type KA or KH, 8"									
3700	type LA, 11"									
3800	type MA, 14"									
3900	Breakers, 2 pole, 15 to 60 amp, type FA	1 Elec	5.60	1.429	Ea.	350	75		425	495
4000	70 to 100 amp, type FA		4.20	1.905		445	100		545	640
4100	15 to 60 amp, type FH		5.60	1.429		580	75		655	750
4200	70 to 100 amp, type FH		4.20	1.905		675	100		775	895
4300	125 to 225 amp, type KA		3.40	2.353		1,025	123		1,148	1,325
4400	125 to 225 amp, type KH		3.40	2.353		2,425	123		2,548	2,825
4500	125 to 400 amp, type LA		2.50	3.200		2,675	168		2,843	3,200
4600	125 to 600 amp, type MA		1.80	4.444		2,975	233		3,208	3,625
4700	700 & 800 amp, type MA		1.50	5.333		3,850	279		4,129	4,675
4800	3 pole, 15 to 60 amp, type FA		5.30	1.509		455	79		534	620
4900	70 to 100 amp, type FA		4	2		560	105		665	770
5000	15 to 60 amp, type FH		5.30	1.509		680	79		759	865
5100	70 to 100 amp, type FH		4	2		770	105		875	1,000
5200	125 to 225 amp, type KA		3.20	2.500		1,300	131		1,431	1,625
5300	125 to 225 amp, type KH		3.20	2.500		2,925	131		3,056	3,425
5400	125 to 400 amp, type LA		2.30	3.478		3,325	182		3,507	3,925
5500	125 to 600 amp, type MA		1.60	5		4,375	262		4,637	5,200
5600	700 & 800 amp, type MA		1.30	6.154		4,975	320		5,295	5,925

26 22 13 – Low-Voltage Distribution Transformers

26 22 13.10 Transformer, Dry-Type		Crew	Daily Output	Labor-Hours	Unit	Material	Labor	Equipment	Total	Total Incl O&P
						2013 Bare Costs				
0010	**TRANSFORMER, DRY-TYPE** R262213-10									
0050	Single phase, 240/480 volt primary, 120/240 volt secondary									
0100	1 kVA R262213-60	1 Elec	2	4	Ea.	282	210		492	625
0300	2 kVA		1.60	5		425	262		687	855
0500	3 kVA		1.40	5.714		500	299		799	1,000
0700	5 kVA		1.20	6.667		690	350		1,040	1,275
0900	7.5 kVA	2 Elec	2.20	7.273		955	380		1,335	1,625
1100	10 kVA		1.60	10		1,275	525		1,800	2,175
1300	15 kVA R262213-65		1.20	13.333		1,450	700		2,150	2,650
1500	25 kVA		1	16		2,000	840		2,840	3,450
1700	37.5 kVA		.80	20		2,575	1,050		3,625	4,425
1900	50 kVA		.70	22.857		3,075	1,200		4,275	5,175
2100	75 kVA		.65	24.615		4,050	1,300		5,350	6,375
2110	100 kVA	R-3	.90	22.222		5,275	1,150	152	6,577	7,700
2120	167 kVA	″	.80	25		8,775	1,300	172	10,247	11,800
2190	480 V primary 120/240 V secondary, nonvent., 15 kVA	2 Elec	1.20	13.333		1,475	700		2,175	2,675
2200	25 kVA		.90	17.778		2,175	930		3,105	3,800
2210	37 kVA		.75	21.333		2,575	1,125		3,700	4,525
2220	50 kVA		.65	24.615		3,075	1,300		4,375	5,300
2230	75 kVA		.60	26.667		4,050	1,400		5,450	6,550
2240	100 kVA		.50	32		5,275	1,675		6,950	8,300
2250	Low operating temperature(80°C), 25 kVA		1	16		3,950	840		4,790	5,600
2260	37 kVA		.80	20		4,250	1,050		5,300	6,250
2270	50 kVA		.70	22.857		5,550	1,200		6,750	7,900
2280	75 kVA		.65	24.615		8,850	1,300		10,150	11,700
2290	100 kVA		.55	29.091		9,200	1,525		10,725	12,400
2300	3 phase, 480 volt primary 120/208 volt secondary									
2310	Ventilated, 3 kVA	1 Elec	1	8	Ea.	925	420		1,345	1,650
2700	6 kVA		.80	10		1,000	525		1,525	1,875
2900	9 kVA		.70	11.429		1,075	600		1,675	2,075
3100	15 kVA	2 Elec	1.10	14.545		1,150	760		1,910	2,400
3300	30 kVA		.90	17.778		1,200	930		2,130	2,725
3500	45 kVA		.80	20		1,350	1,050		2,400	3,075
3700	75 kVA		.70	22.857		2,050	1,200		3,250	4,050
3900	112.5 kVA	R-3	.90	22.222		2,725	1,150	152	4,027	4,875
4100	150 kVA		.85	23.529		3,550	1,225	161	4,936	5,900
4300	225 kVA		.65	30.769		4,800	1,600	211	6,611	7,900
4500	300 kVA		.55	36.364		6,075	1,875	249	8,199	9,775
4700	500 kVA		.45	44.444		10,000	2,300	305	12,605	14,800
4800	750 kVA		.35	57.143		17,600	2,975	390	20,965	24,300
4820	1000 kVA		.32	62.500		20,200	3,250	430	23,880	27,500
4850	K-4 rated, 15 kVA	2 Elec	1.10	14.545		3,125	760		3,885	4,600
4855	30 kVA		.90	17.778		5,000	930		5,930	6,875
4860	45 kVA		.80	20		6,000	1,050		7,050	8,175
4865	75 kVA		.70	22.857		8,225	1,200		9,425	10,900
4870	112.5 kVA	R-3	.90	22.222		11,000	1,150	152	12,302	14,000
4875	150 kVA		.85	23.529		14,300	1,225	161	15,686	17,700
4880	225 kVA		.65	30.769		20,600	1,600	211	22,411	25,200
4885	300 kVA		.55	36.364		28,700	1,875	249	30,824	34,700
4890	500 kVA		.45	44.444		39,400	2,300	305	42,005	47,100
4900	K-13 rated, 15 kVA	2 Elec	1.10	14.545		3,450	760		4,210	4,925
4905	30 kVA		.90	17.778		5,175	930		6,105	7,075
4910	45 kVA		.80	20		6,225	1,050		7,275	8,400

26 22 Low-Voltage Transformers

26 22 13 – Low-Voltage Distribution Transformers

26 22 13.10 Transformer, Dry-Type		Crew	Daily Output	Labor-Hours	Unit	Material	2013 Bare Costs Labor	Equipment	Total	Total Incl O&P
4915	75 kVA	2 Elec	.70	22.857	Ea.	10,300	1,200		11,500	13,100
4920	112.5 kVA	R-3	.90	22.222		13,700	1,150	152	15,002	17,000
4925	150 kVA		.85	23.529		17,900	1,225	161	19,286	21,700
4930	225 kVA		.65	30.769		24,400	1,600	211	26,211	29,400
4935	300 kVA		.55	36.364		29,500	1,875	249	31,624	35,600
4940	500 kVA		.45	44.444		46,900	2,300	305	49,505	55,500
5020	480 volt primary 120/208 volt secondary									
5030	Nonventilated, 15 kVA	2 Elec	1.10	14.545	Ea.	3,575	760		4,335	5,075
5040	30 kVA		.80	20		4,000	1,050		5,050	5,975
5050	45 kVA		.70	22.857		5,450	1,200		6,650	7,800
5060	75 kVA		.65	24.615		7,950	1,300		9,250	10,700
5070	112.5 kVA	R-3	.85	23.529		10,800	1,225	161	12,186	13,900
5081	150 kVA		.85	23.529		13,100	1,225	161	14,486	16,400
5090	225 kVA		.60	33.333		16,000	1,725	229	17,954	20,500
5100	300 kVA		.50	40		19,100	2,075	274	21,449	24,400
5200	Low operating temperature (80°C), 30 kVA	2 Elec	.90	17.778		3,275	930		4,205	5,000
5210	45 kVA		.80	20		4,900	1,050		5,950	6,975
5220	75 kVA		.70	22.857		6,525	1,200		7,725	9,000
5230	112.5 kVA	R-3	.90	22.222		11,600	1,150	152	12,902	14,600
5240	150 kVA		.85	23.529		14,600	1,225	161	15,986	18,100
5250	225 kVA		.65	30.769		20,200	1,600	211	22,011	24,800
5260	300 kVA		.55	36.364		24,200	1,875	249	26,324	29,700
5270	500 kVA		.45	44.444		36,400	2,300	305	39,005	43,800
5380	3 phase, 5 kV primary 277/480 volt secondary									
5400	High voltage, 112.5 kVA	R-3	.85	23.529	Ea.	15,900	1,225	161	17,286	19,500
5410	150 kVA		.65	30.769		16,700	1,600	211	18,511	21,000
5420	225 kVA		.55	36.364		19,700	1,875	249	21,824	24,800
5430	300 kVA		.45	44.444		25,200	2,300	305	27,805	31,500
5440	500 kVA		.35	57.143		32,500	2,975	390	35,865	40,600
5450	750 kVA		.32	62.500		50,000	3,250	430	53,680	60,500
5460	1000 kVA		.30	66.667		58,500	3,450	455	62,405	70,000
5470	1500 kVA		.27	74.074		68,500	3,850	510	72,860	81,500
5480	2000 kVA		.25	80		80,000	4,150	550	84,700	95,000
5490	2500 kVA		.20	100		90,000	5,200	685	95,885	107,500
5500	3000 kVA		.18	111		118,500	5,775	760	125,035	139,500
5590	15 kV primary 277/480 volt secondary									
5600	High voltage, 112.5 kVA	R-3	.85	23.529	Ea.	23,400	1,225	161	24,786	27,800
5610	150 kVA		.65	30.769		27,300	1,600	211	29,111	32,700
5620	225 kVA		.55	36.364		30,200	1,875	249	32,324	36,300
5630	300 kVA		.45	44.444		35,600	2,300	305	38,205	42,900
5640	500 kVA		.35	57.143		45,000	2,975	390	48,365	54,500
5650	750 kVA		.32	62.500		58,000	3,250	430	61,680	69,500
5660	1000 kVA		.30	66.667		66,000	3,450	455	69,905	78,500
5670	1500 kVA		.27	74.074		76,500	3,850	510	80,860	90,500
5680	2000 kVA		.25	80		85,000	4,150	550	89,700	100,500
5690	2500 kVA		.20	100		98,000	5,200	685	103,885	116,500
5700	3000 kVA		.18	111		117,000	5,775	760	123,535	138,000
6000	2400 volt primary, 480 volt secondary, 300 kVA		.45	44.444		25,200	2,300	305	27,805	31,500
6010	500 kVA		.35	57.143		32,500	2,975	390	35,865	40,600
6020	750 kVA		.32	62.500		46,300	3,250	430	49,980	56,500

26 22 13.20 Isolating Panels

0010	**ISOLATING PANELS** used with isolating transformers									

26 22 Low-Voltage Transformers

26 22 13 – Low-Voltage Distribution Transformers

26 22 13.20 Isolating Panels		Crew	Daily Output	Labor-Hours	Unit	Material	2013 Bare Costs Labor	Equipment	Total	Total Incl O&P
0020	For hospital applications									
0100	Critical care area, 8 circuit, 3 kVA	1 Elec	.58	13.793	Ea.	7,950	725		8,675	9,800
0200	5 kVA		.54	14.815		8,175	775		8,950	10,100
0400	7.5 kVA		.52	15.385		8,325	805		9,130	10,400
0600	10 kVA		.44	18.182		8,600	955		9,555	10,900
0800	Operating room power & lighting, 8 circuit, 3 kVA		.58	13.793		6,650	725		7,375	8,375
1000	5 kVA		.54	14.815		7,000	775		7,775	8,850
1200	7.5 kVA		.52	15.385		7,650	805		8,455	9,625
1400	10 kVA		.44	18.182		8,150	955		9,105	10,400
1600	X-ray systems, 15 kVA, 90 amp		.44	18.182		17,000	955		17,955	20,100
1800	25 kVA, 125 amp		.36	22.222		17,400	1,175		18,575	20,900

26 22 13.30 Isolating Transformer

		Crew	Daily Output	Labor-Hours	Unit	Material	Labor	Equipment	Total	Total Incl O&P
0010	**ISOLATING TRANSFORMER**									
0100	Single phase, 120/240 volt primary, 120/240 volt secondary									
0200	0.50 kVA R262213-60	1 Elec	4	2	Ea.	365	105		470	560
0400	1 kVA		2	4		460	210		670	820
0600	2 kVA		1.60	5		685	262		947	1,150
0800	3 kVA		1.40	5.714		785	299		1,084	1,325
1000	5 kVA		1.20	6.667		1,025	350		1,375	1,675
1200	7.5 kVA		1.10	7.273		1,300	380		1,680	2,025
1400	10 kVA		.80	10		1,675	525		2,200	2,625
1600	15 kVA		.60	13.333		2,150	700		2,850	3,400
1800	25 kVA		.50	16		3,075	840		3,915	4,625
1810	37.5 kVA	2 Elec	.80	20		5,075	1,050		6,125	7,175
1820	75 kVA	"	.65	24.615		7,650	1,300		8,950	10,300
1830	3 phase, 120/240 V primary, 120/240 V secondary, 112.5 kVA	R-3	.90	22.222		9,650	1,150	152	10,952	12,500
1840	150 kVA		.85	23.529		12,300	1,225	161	13,686	15,500
1850	225 kVA		.65	30.769		17,100	1,600	211	18,911	21,400
1860	300 kVA		.55	36.364		22,700	1,875	249	24,824	28,100
1870	500 kVA		.45	44.444		37,800	2,300	305	40,405	45,400
1880	750 kVA		.35	57.143		38,200	2,975	390	41,565	46,900

26 22 13.90 Transformer Handling

		Crew	Daily Output	Labor-Hours	Unit	Material	Labor	Equipment	Total	Total Incl O&P
0010	**TRANSFORMER HANDLING** Add to normal labor cost in restricted areas									
5000	Transformers									
5150	15 kVA, approximately 200 pounds	2 Elec	2.70	5.926	Ea.		310		310	465
5160	25 kVA, approximately 300 pounds		2.50	6.400			335		335	500
5170	37.5 kVA, approximately 400 pounds		2.30	6.957			365		365	545
5180	50 kVA, approximately 500 pounds		2	8			420		420	625
5190	75 kVA, approximately 600 pounds		1.80	8.889			465		465	695
5200	100 kVA, approximately 700 pounds		1.60	10			525		525	785
5210	112.5 kVA, approximately 800 pounds	3 Elec	2.20	10.909			570		570	855
5220	125 kVA, approximately 900 pounds		2	12			630		630	940
5230	150 kVA, approximately 1000 pounds		1.80	13.333			700		700	1,050
5240	167 kVA, approximately 1200 pounds		1.60	15			785		785	1,175
5250	200 kVA, approximately 1400 pounds		1.40	17.143			900		900	1,350
5260	225 kVA, approximately 1600 pounds		1.30	18.462			965		965	1,450
5270	250 kVA, approximately 1800 pounds		1.10	21.818			1,150		1,150	1,700
5280	300 kVA, approximately 2000 pounds		1	24			1,250		1,250	1,875
5290	500 kVA, approximately 3000 pounds		.75	32			1,675		1,675	2,500
5300	600 kVA, approximately 3500 pounds		.67	35.821			1,875		1,875	2,800
5310	750 kVA, approximately 4000 pounds		.60	40			2,100		2,100	3,125
5320	1000 kVA, approximately 5000 pounds		.50	48			2,525		2,525	3,775

26 22 Low-Voltage Transformers

26 22 16 – Low-Voltage Buck-Boost Transformers

26 22 16.10 Buck-Boost Transformer		Crew	Daily Output	Labor-Hours	Unit	2013 Bare Costs Material	Labor	Equipment	Total	Total Incl O&P
0010	**BUCK-BOOST TRANSFORMER** R262213-60									
0100	Single phase, 120/240 V primary, 12/24 V secondary									
0200	0.10 kVA	1 Elec	8	1	Ea.	92.50	52.50		145	181
0400	0.25 kVA		5.70	1.404		136	73.50		209.50	260
0600	0.50 kVA		4	2		186	105		291	360
0800	0.75 kVA		3.10	2.581		239	135		374	465
1000	1.0 kVA		2	4		297	210		507	640
1200	1.5 kVA		1.80	4.444		365	233		598	750
1400	2.0 kVA		1.60	5		440	262		702	875
1600	3.0 kVA		1.40	5.714		595	299		894	1,100
1800	5.0 kVA		1.20	6.667		780	350		1,130	1,375
2000	3 phase, 240 V primary, 208/120 V secondary, 15 kVA	2 Elec	2.40	6.667		2,050	350		2,400	2,775
2200	30 kVA		1.60	10		2,275	525		2,800	3,275
2400	45 kVA		1.40	11.429		2,725	600		3,325	3,900
2600	75 kVA		1.20	13.333		3,325	700		4,025	4,700
2800	112.5 kVA	R-3	1.40	14.286		4,125	740	98	4,963	5,725
3000	150 kVA		1.10	18.182		5,475	945	125	6,545	7,575
3200	225 kVA		1	20		7,150	1,050	137	8,337	9,550
3400	300 kVA		.90	22.222		9,700	1,150	152	11,002	12,600

26 24 Switchboards and Panelboards

26 24 13 – Switchboards

26 24 13.10 Incoming Switchboards

		Crew	Daily Output	Labor-Hours	Unit	Material	Labor	Equipment	Total	Total Incl O&P
0010	**INCOMING SWITCHBOARDS** main service section R262419-84									
0100	Aluminum bus bars, not including CT's or PT's									
0200	No main disconnect, includes CT compartment									
0300	120/208 volt, 4 wire, 600 amp	2 Elec	1	16	Ea.	3,825	840		4,665	5,450
0400	800 amp		.88	18.182		3,825	955		4,780	5,625
0500	1000 amp		.80	20		4,600	1,050		5,650	6,625
0600	1200 amp		.72	22.222		4,600	1,175		5,775	6,800
0700	1600 amp		.66	24.242		4,600	1,275		5,875	6,950
0800	2000 amp		.62	25.806		4,950	1,350		6,300	7,450
1000	3000 amp		.56	28.571		6,525	1,500		8,025	9,425
1200	277/480 volt, 4 wire, 600 amp		1	16		3,825	840		4,665	5,450
1300	800 amp		.88	18.182		3,825	955		4,780	5,625
1400	1000 amp		.80	20		4,850	1,050		5,900	6,900
1500	1200 amp		.72	22.222		4,850	1,175		6,025	7,075
1600	1600 amp		.66	24.242		4,850	1,275		6,125	7,225
1700	2000 amp		.62	25.806		4,950	1,350		6,300	7,450
1800	3000 amp		.56	28.571		6,525	1,500		8,025	9,425
1900	4000 amp		.52	30.769		8,075	1,600		9,675	11,300
2000	Fused switch & CT compartment									
2100	120/208 volt, 4 wire, 400 amp	2 Elec	1.12	14.286	Ea.	2,550	750		3,300	3,925
2200	600 amp		.94	17.021		3,025	890		3,915	4,650
2300	800 amp		.84	19.048		10,300	1,000		11,300	12,900
2400	1200 amp		.68	23.529		13,400	1,225		14,625	16,700
2500	277/480 volt, 4 wire, 400 amp		1.14	14.035		2,675	735		3,410	4,050
2600	600 amp		.94	17.021		3,075	890		3,965	4,700
2700	800 amp		.84	19.048		10,300	1,000		11,300	12,900
2800	1200 amp		.68	23.529		13,400	1,225		14,625	16,700

26 24 13.10 Incoming Switchboards

26 24 13.10 Incoming Switchboards		Crew	Daily Output	Labor-Hours	Unit	Material	2013 Bare Costs Labor	Equipment	Total	Total Incl O&P
2900	Pressure switch & CT compartment									
3000	120/208 volt, 4 wire, 800 amp	2 Elec	.80	20	Ea.	9,275	1,050		10,325	11,800
3100	1200 amp		.66	24.242		18,000	1,275		19,275	21,700
3200	1600 amp		.62	25.806		19,100	1,350		20,450	23,000
3300	2000 amp		.56	28.571		20,300	1,500		21,800	24,600
3310	2500 amp		.50	32		25,000	1,675		26,675	30,000
3320	3000 amp		.44	36.364		33,700	1,900		35,600	39,900
3330	4000 amp		.40	40		43,200	2,100		45,300	50,500
3340	120/208 volt, 4 wire, 800 amp, with ground fault		.80	20		15,300	1,050		16,350	18,500
3350	1200 amp, with ground fault		.66	24.242		19,800	1,275		21,075	23,700
3360	1600 amp, with ground fault		.62	25.806		21,500	1,350		22,850	25,700
3370	2000 amp, with ground fault		.56	28.571		23,200	1,500		24,700	27,800
3400	277/480 volt, 4 wire, 800 amp, with ground fault		.80	20		15,300	1,050		16,350	18,500
3600	1200 amp, with ground fault		.66	24.242		19,800	1,275		21,075	23,700
4000	1600 amp, with ground fault		.62	25.806		21,500	1,350		22,850	25,700
4200	2000 amp, with ground fault	↓	.56	28.571	↓	23,200	1,500		24,700	27,800
4400	Circuit breaker, molded case & CT compartment									
4600	3 pole, 4 wire, 600 amp	2 Elec	.94	17.021	Ea.	8,000	890		8,890	10,100
4800	800 amp		.84	19.048		9,550	1,000		10,550	12,000
5000	1200 amp	↓	.68	23.529	↓	13,000	1,225		14,225	16,200
5100	Copper bus bars, not incl. CT's or PT's, add, minimum					15%				

26 24 13.20 In Plant Distribution Switchboards

26 24 13.20 In Plant Distribution Switchboards		Crew	Daily Output	Labor-Hours	Unit	Material	Labor	Equipment	Total	Total Incl O&P
0010	**IN PLANT DISTRIBUTION SWITCHBOARDS**									
0100	Main lugs only, to 600 volt, 3 pole, 3 wire, 200 amp	2 Elec	1.20	13.333	Ea.	1,200	700		1,900	2,375
0110	400 amp		1.20	13.333		1,200	700		1,900	2,375
0120	600 amp		1.20	13.333		1,250	700		1,950	2,425
0130	800 amp		1.08	14.815		1,350	775		2,125	2,625
0140	1200 amp		.92	17.391		1,625	910		2,535	3,175
0150	1600 amp		.86	18.605		2,150	975		3,125	3,825
0160	2000 amp		.82	19.512		2,375	1,025		3,400	4,150
0250	To 480 volt, 3 pole, 4 wire, 200 amp		1.20	13.333		1,050	700		1,750	2,200
0260	400 amp		1.20	13.333		1,200	700		1,900	2,375
0270	600 amp		1.20	13.333		1,325	700		2,025	2,500
0280	800 amp		1.08	14.815		1,450	775		2,225	2,725
0290	1200 amp		.92	17.391		1,775	910		2,685	3,325
0300	1600 amp		.86	18.605		2,025	975		3,000	3,675
0310	2000 amp		.82	19.512		2,350	1,025		3,375	4,125
0400	Main circuit breaker, to 600 volt, 3 pole, 3 wire, 200 amp		1.20	13.333		2,925	700		3,625	4,275
0410	400 amp		1.14	14.035		2,925	735		3,660	4,325
0420	600 amp		1.10	14.545		3,700	760		4,460	5,225
0430	800 amp		1.04	15.385		6,200	805		7,005	8,000
0440	1200 amp		.88	18.182		8,100	955		9,055	10,300
0450	1600 amp		.84	19.048		13,000	1,000		14,000	15,800
0460	2000 amp		.80	20		13,900	1,050		14,950	16,900
0550	277/480 volt, 3 pole, 4 wire, 200 amp		1.20	13.333		3,075	700		3,775	4,425
0560	400 amp		1.14	14.035		3,075	735		3,810	4,475
0570	600 amp		1.10	14.545		3,850	760		4,610	5,400
0580	800 amp		1.04	15.385		6,450	805		7,255	8,275
0590	1200 amp		.88	18.182		8,325	955		9,280	10,600
0600	1600 amp		.84	19.048		12,900	1,000		13,900	15,700
0610	2000 amp		.80	20		13,800	1,050		14,850	16,700
0700	Main fusible switch w/fuse, 208/240 volt, 3 pole, 3 wire, 200 amp		1.20	13.333		3,225	700		3,925	4,600

26 24 13.20 In Plant Distribution Switchboards		Crew	Daily Output	Labor-Hours	Unit	Material	Labor (2013 Bare Costs)	Equipment	Total	Total Incl O&P
0710	400 amp	2 Elec	1.14	14.035	Ea.	3,225	735		3,960	4,650
0720	600 amp		1.10	14.545		3,925	760		4,685	5,450
0730	800 amp		1.04	15.385		8,025	805		8,830	10,000
0740	1200 amp		.88	18.182		9,325	955		10,280	11,700
0800	120/208, 120/240 volt, 3 pole, 4 wire, 200 amp		1.20	13.333		2,850	700		3,550	4,200
0810	400 amp		1.14	14.035		2,850	735		3,585	4,250
0820	600 amp		1.10	14.545		3,725	760		4,485	5,250
0830	800 amp		1.04	15.385		5,950	805		6,755	7,725
0840	1200 amp		.88	18.182		6,900	955		7,855	9,025
0900	480 or 600 volt, 3 pole, 3 wire, 200 amp		1.20	13.333		3,100	700		3,800	4,450
0910	400 amp		1.14	14.035		3,100	735		3,835	4,500
0920	600 amp		1.10	14.545		3,675	760		4,435	5,200
0930	800 amp		1.04	15.385		5,675	805		6,480	7,450
0940	1200 amp		.88	18.182		6,600	955		7,555	8,675
1000	277 or 480 volt, 3 pole, 4 wire, 200 amp		1.20	13.333		3,225	700		3,925	4,600
1010	400 amp		1.14	14.035		3,225	735		3,960	4,650
1020	600 amp		1.10	14.545		3,875	760		4,635	5,425
1030	800 amp		1.04	15.385		5,950	805		6,755	7,750
1040	1200 amp		.88	18.182		6,900	955		7,855	9,025
1120	1600 amp		.76	21.053		12,600	1,100		13,700	15,600
1130	2000 amp		.68	23.529		16,600	1,225		17,825	20,200
1150	Pressure switch, bolted, 3 pole, 208/240 volt, 3 wire, 800 amp		.96	16.667		9,350	875		10,225	11,600
1160	1200 amp		.80	20		11,900	1,050		12,950	14,700
1170	1600 amp		.76	21.053		13,600	1,100		14,700	16,700
1180	2000 amp		.68	23.529		15,500	1,225		16,725	19,000
1200	120/208 or 120/240 volt, 3 pole, 4 wire, 800 amp		.96	16.667		7,375	875		8,250	9,425
1210	1200 amp		.80	20		8,625	1,050		9,675	11,100
1220	1600 amp		.76	21.053		13,600	1,100		14,700	16,700
1230	2000 amp		.68	23.529		15,500	1,225		16,725	19,000
1300	480 or 600 volt, 3 wire, 800 amp		.96	16.667		9,350	875		10,225	11,600
1310	1200 amp		.80	20		13,100	1,050		14,150	16,000
1320	1600 amp		.76	21.053		14,700	1,100		15,800	17,900
1330	2000 amp		.68	23.529		16,600	1,225		17,825	20,200
1400	277-480 volt, 4 wire, 800 amp		.96	16.667		9,350	875		10,225	11,600
1410	1200 amp		.80	20		13,100	1,050		14,150	16,000
1420	1600 amp		.76	21.053		14,700	1,100		15,800	17,900
1430	2000 amp		.68	23.529		16,600	1,225		17,825	20,200
1500	Main ground fault protector, 1200-2000 amp		5.40	2.963		3,125	155		3,280	3,650
1600	Busway connection, 200 amp		5.40	2.963		470	155		625	745
1610	400 amp		4.60	3.478		470	182		652	790
1620	600 amp		4	4		470	210		680	830
1630	800 amp		3.20	5		470	262		732	905
1640	1200 amp		2.60	6.154		470	320		790	995
1650	1600 amp		2.40	6.667		970	350		1,320	1,600
1660	2000 amp		2	8		970	420		1,390	1,700
1700	Shunt trip for remote operation 200 amp		8	2		620	105		725	835
1710	400 amp		8	2		990	105		1,095	1,250
1720	600 amp		8	2		1,175	105		1,280	1,425
1730	800 amp		8	2		1,500	105		1,605	1,800
1740	1200-2000 amp		8	2		3,175	105		3,280	3,650
1800	Motor operated main breaker 200 amp		8	2		3,075	105		3,180	3,525
1810	400 amp		8	2		3,075	105		3,180	3,525
1820	600 amp		8	2		3,150	105		3,255	3,625

26 24 13.20 In Plant Distribution Switchboards

	26 24 13.20 In Plant Distribution Switchboards	Crew	Daily Output	Labor-Hours	Unit	2013 Bare Costs Material	Labor	Equipment	Total	Total Incl O&P
1830	800 amp	2 Elec	8	2	Ea.	3,150	105		3,255	3,625
1840	1200-2000 amp		8	2		3,225	105		3,330	3,700
1900	Current/potential transformer metering compartment 200-800 amp		5.40	2.963		2,425	155		2,580	2,875
1940	1200 amp		5.40	2.963		4,000	155		4,155	4,625
1950	1600-2000 amp		5.40	2.963		5,150	155		5,305	5,900
2000	With watt meter 200-800 amp		4	4		7,900	210		8,110	9,000
2040	1200 amp		4	4		9,750	210		9,960	11,000
2050	1600-2000 amp		4	4		9,750	210		9,960	11,000
2100	Split bus 60-200 amp	1 Elec	5.30	1.509		187	79		266	325
2130	400 amp	2 Elec	4.60	3.478		325	182		507	630
2140	600 amp		3.60	4.444		395	233		628	785
2150	800 amp		2.60	6.154		505	320		825	1,025
2170	1200 amp		2	8		575	420		995	1,250
2250	Contactor control 60 amp	1 Elec	2	4		1,225	210		1,435	1,675
2260	100 amp		1.50	5.333		1,375	279		1,654	1,950
2270	200 amp		1	8		2,075	420		2,495	2,900
2280	400 amp	2 Elec	1	16		6,325	840		7,165	8,200
2290	600 amp		.84	19.048		7,075	1,000		8,075	9,275
2300	800 amp		.72	22.222		8,375	1,175		9,550	11,000
2500	Modifier, two distribution sections, add		.80	20		3,225	1,050		4,275	5,125
2520	Three distribution sections, add		.40	40		7,200	2,100		9,300	11,000
2560	Auxiliary pull section, 20", add		2	8		1,050	420		1,470	1,775
2580	24", add		1.80	8.889		1,050	465		1,515	1,850
2600	30", add		1.60	10		1,050	525		1,575	1,925
2620	36", add		1.40	11.429		1,250	600		1,850	2,275
2640	Dog house, 12", add		2.40	6.667		216	350		566	765
2660	18", add		2	8		430	420		850	1,100
3000	Transition section between switchboard and transformer									
3050	or motor control center, 4 wire alum. bus, 600 amp	2 Elec	1.14	14.035	Ea.	2,125	735		2,860	3,450
3100	800 amp		1	16		2,375	840		3,215	3,875
3150	1000 amp		.88	18.182		2,675	955		3,630	4,375
3200	1200 amp		.80	20		2,950	1,050		4,000	4,825
3250	1600 amp		.72	22.222		3,475	1,175		4,650	5,575
3300	2000 amp		.66	24.242		4,000	1,275		5,275	6,300
3350	2500 amp		.62	25.806		4,675	1,350		6,025	7,150
3400	3000 amp		.56	28.571		5,375	1,500		6,875	8,150
4000	Weatherproof construction, per vertical section		1.76	9.091		2,450	475		2,925	3,425

26 24 13.30 Distribution Switchboards Section

	26 24 13.30 Distribution Switchboards Section	Crew	Daily Output	Labor-Hours	Unit	Material	Labor	Equipment	Total	Total Incl O&P
0010	**DISTRIBUTION SWITCHBOARDS SECTION** R262419-80									
0100	Aluminum bus bars, not including breakers									
0160	Subfeed lug-rated at 60 amp	2 Elec	1.30	12.308	Ea.	1,150	645		1,795	2,225
0170	100 amp		1.26	12.698		1,400	665		2,065	2,550
0180	200 amp		1.20	13.333		1,425	700		2,125	2,625
0190	400 amp		1.10	14.545		1,450	760		2,210	2,750
0200	120/208 or 277/480 volt, 4 wire, 600 amp		1	16		1,900	840		2,740	3,325
0300	800 amp		.88	18.182		2,475	955		3,430	4,150
0400	1000 amp		.80	20		3,050	1,050		4,100	4,925
0500	1200 amp		.72	22.222		3,625	1,175		4,800	5,750
0600	1600 amp		.66	24.242		4,175	1,275		5,450	6,475
0700	2000 amp		.62	25.806		4,875	1,350		6,225	7,400
0800	2500 amp		.60	26.667		5,500	1,400		6,900	8,150
0900	3000 amp		.56	28.571		6,675	1,500		8,175	9,600

26 24 13.30 Distribution Switchboards Section		Crew	Daily Output	Labor-Hours	Unit	Material	2013 Bare Costs Labor	Equipment	Total	Total Incl O&P
0950	4000 amp	2 Elec	.52	30.769	Ea.	9,750	1,600		11,350	13,100

26 24 13.40 Switchboards Feeder Section		Crew	Daily Output	Labor-Hours	Unit	Material	2013 Bare Costs Labor	Equipment	Total	Total Incl O&P
0010	**SWITCHBOARDS FEEDER SECTION** group mounted devices R262419-82									
0030	Circuit breakers									
0160	FA frame, 15 to 60 amp, 240 volt, 1 pole	1 Elec	8	1	Ea.	114	52.50		166.50	204
0170	2 pole		7	1.143		239	60		299	355
0180	3 pole		5.30	1.509		355	79		434	510
0210	480 volt, 1 pole		8	1		151	52.50		203.50	245
0220	2 pole		7	1.143		435	60		495	570
0230	3 pole		5.30	1.509		560	79		639	735
0260	600 volt, 2 pole		7	1.143		505	60		565	645
0270	3 pole		5.30	1.509		650	79		729	835
0280	FA frame, 70 to 100 amp, 240 volt, 1 pole		7	1.143		187	60		247	296
0310	2 pole		5	1.600		390	84		474	555
0320	3 pole		4	2		505	105		610	710
0330	480 volt, 1 pole		7	1.143		224	60		284	335
0360	2 pole		5	1.600		565	84		649	745
0370	3 pole		4	2		665	105		770	885
0380	600 volt, 2 pole		5	1.600		640	84		724	825
0410	3 pole		4	2		800	105		905	1,025
0420	KA frame, 70 to 225 amp		3.20	2.500		1,200	131		1,331	1,500
0430	LA frame, 125 to 400 amp		2.30	3.478		2,675	182		2,857	3,225
0460	MA frame, 450 to 600 amp		1.60	5		4,775	262		5,037	5,650
0470	700 to 800 amp		1.30	6.154		6,200	320		6,520	7,300
0480	MAL frame, 1000 amp		1	8		6,450	420		6,870	7,725
0490	PA frame, 1200 amp		.80	10		13,100	525		13,625	15,200
0500	Branch circuit, fusible switch, 600 volt, double 30/30 amp		4	2		845	105		950	1,075
0550	60/60 amp		3.20	2.500		865	131		996	1,150
0600	100/100 amp		2.70	2.963		1,100	155		1,255	1,425
0650	Single, 30 amp		5.30	1.509		700	79		779	890
0700	60 amp		4.70	1.702		700	89		789	905
0750	100 amp		4	2		925	105		1,030	1,175
0800	200 amp		2.70	2.963		1,350	155		1,505	1,700
0850	400 amp		2.30	3.478		2,475	182		2,657	3,000
0900	600 amp		1.80	4.444		3,025	233		3,258	3,675
0950	800 amp		1.30	6.154		5,075	320		5,395	6,050
1000	1200 amp		.80	10		5,800	525		6,325	7,150
1080	Branch circuit, circuit breakers, high interrupting capacity									
1100	60 amp, 240, 480 or 600 volt, 1 pole	1 Elec	8	1	Ea.	248	52.50		300.50	350
1120	2 pole		7	1.143		640	60		700	795
1140	3 pole		5.30	1.509		660	79		739	845
1150	100 amp, 240, 480 or 600 volt, 1 pole		7	1.143		280	60		340	395
1160	2 pole		5	1.600		745	84		829	940
1180	3 pole		4	2		690	105		795	915
1200	225 amp, 240, 480 or 600 volt, 2 pole		3.50	2.286		2,050	120		2,170	2,425
1220	3 pole		3.20	2.500		2,350	131		2,481	2,775
1240	400 amp, 240, 480 or 600 volt, 2 pole		2.50	3.200		3,250	168		3,418	3,825
1260	3 pole		2.30	3.478		2,900	182		3,082	3,450
1280	600 amp, 240, 480 or 600 volt, 2 pole		1.80	4.444		3,775	233		4,008	4,525
1300	3 pole		1.60	5		3,800	262		4,062	4,600
1320	800 amp, 240, 480 or 600 volt, 2 pole		1.50	5.333		4,850	279		5,129	5,750
1340	3 pole		1.30	6.154		5,075	320		5,395	6,075

26 24 Switchboards and Panelboards

26 24 13 – Switchboards

26 24 13.40 Switchboards Feeder Section		Crew	Daily Output	Labor-Hours	Unit	Material	Labor	Equipment	Total	Total Incl O&P
						2013 Bare Costs				
1360	1000 amp, 240, 480 or 600 volt, 2 pole	1 Elec	1.10	7.273	Ea.	6,100	380		6,480	7,300
1380	3 pole		1	8		8,500	420		8,920	9,975
1400	1200 amp, 240, 480 or 600 volt, 2 pole		.90	8.889		6,600	465		7,065	7,950
1420	3 pole		.80	10		7,475	525		8,000	9,000
1700	Fusible switch, 240 V, 60 amp, 2 pole		3.20	2.500		405	131		536	640
1720	3 pole		3	2.667		560	140		700	825
1740	100 amp, 2 pole		2.70	2.963		415	155		570	685
1760	3 pole		2.50	3.200		625	168		793	940
1780	200 amp, 2 pole		2	4		865	210		1,075	1,275
1800	3 pole		1.90	4.211		1,000	221		1,221	1,425
1820	400 amp, 2 pole		1.50	5.333		1,625	279		1,904	2,200
1840	3 pole		1.30	6.154		2,025	320		2,345	2,700
1860	600 amp, 2 pole		1	8		2,300	420		2,720	3,150
1880	3 pole		.90	8.889		2,825	465		3,290	3,800
1900	240-600 V, 800 amp, 2 pole		.70	11.429		5,125	600		5,725	6,525
1920	3 pole		.60	13.333		6,275	700		6,975	7,950
2000	600 V, 60 amp, 2 pole		3.20	2.500		615	131		746	870
2040	100 amp, 2 pole		2.70	2.963		635	155		790	925
2080	200 amp, 2 pole		2	4		1,100	210		1,310	1,525
2120	400 amp, 2 pole		1.50	5.333		2,175	279		2,454	2,825
2160	600 amp, 2 pole		1	8		2,650	420		3,070	3,550
2500	Branch circuit, circuit breakers, 60 amp, 600 volt, 3 pole		5.30	1.509		590	79		669	765
2520	240, 480 or 600 volt, 1 pole		8	1		95	52.50		147.50	184
2540	240 volt, 2 pole		7	1.143		163	60		223	270
2560	480 or 600 volt, 2 pole		7	1.143		345	60		405	470
2580	240 volt, 3 pole		5.30	1.509		265	79		344	410
2600	480 volt, 3 pole		5.30	1.509		460	79		539	625
2620	100 amp, 600 volt, 2 pole		5	1.600		440	84		524	610
2640	3 pole		4	2		560	105		665	770
2660	480 volt, 2 pole		5	1.600		380	84		464	545
2680	240 volt, 2 pole		5	1.600		218	84		302	365
2700	3 pole		4	2		350	105		455	540
2720	480 volt, 3 pole		4	2		510	105		615	720
2740	225 amp, 240, 480 or 600 volt, 2 pole		3.50	2.286		580	120		700	820
2760	3 pole		3.20	2.500		665	131		796	930
2780	400 amp, 240, 480 or 600 volt, 2 pole		2.50	3.200		1,325	168		1,493	1,725
2800	3 pole		2.30	3.478		1,550	182		1,732	1,975
2820	600 amp, 240 or 480 volt, 2 pole		1.80	4.444		2,175	233		2,408	2,750
2840	3 pole		1.60	5		2,675	262		2,937	3,350
2860	800 amp, 240, 480 volt or 600 volt, 2 pole		1.50	5.333		3,275	279		3,554	4,025
2880	3 pole		1.30	6.154		3,825	320		4,145	4,675
2900	1000 amp, 240, 480 or 600 volt, 2 pole		1.10	7.273		4,025	380		4,405	5,000
2920	480 volt, 600 volt, 3 pole		1	8		4,600	420		5,020	5,675
2940	1200 amp, 240, 480 or 600 volt, 2 pole		.90	8.889		5,650	465		6,115	6,925
2960	3 pole		.80	10		6,175	525		6,700	7,575
2980	600 volt, 3 pole		.80	10		6,175	525		6,700	7,575

26 24 16 – Panelboards

26 24 16.10 Load Centers

		Crew	Daily Output	Labor-Hours	Unit	Material	Labor	Equipment	Total	Total Incl O&P
0010	**LOAD CENTERS** (residential type) R262416-50									
0100	3 wire, 120/240 V, 1 phase, including 1 pole plug-in breakers									
0200	100 amp main lugs, indoor, 8 circuits	1 Elec	1.40	5.714	Ea.	164	299		463	630
0300	12 circuits		1.20	6.667		229	350		579	775

26 24 16.10 Load Centers		Crew	Daily Output	Labor-Hours	Unit	2013 Bare Costs Material	Labor	Equipment	Total	Total Incl O&P
0400	Rainproof, 8 circuits	1 Elec	1.40	5.714	Ea.	197	299		496	665
0500	12 circuits		1.20	6.667		277	350		627	830
0600	200 amp main lugs, indoor, 16 circuits	R-1A	1.80	8.889		315	385		700	925
0700	20 circuits		1.50	10.667		395	460		855	1,125
0800	24 circuits		1.30	12.308		515	530		1,045	1,375
0900	30 circuits		1.20	13.333		595	575		1,170	1,525
1000	40 circuits		.80	20		825	860		1,685	2,200
1200	Rainproof, 16 circuits		1.80	8.889		375	385		760	990
1300	20 circuits		1.50	10.667		450	460		910	1,200
1400	24 circuits		1.30	12.308		670	530		1,200	1,550
1500	30 circuits		1.20	13.333		750	575		1,325	1,700
1600	40 circuits		.80	20		980	860		1,840	2,375
1800	400 amp main lugs, indoor, 42 circuits		.72	22.222		1,150	955		2,105	2,725
1900	Rainproof, 42 circuit		.72	22.222		1,375	955		2,330	2,950
2200	Plug in breakers, 20 amp, 1 pole, 4 wire, 120/208 volts									
2210	125 amp main lugs, indoor, 12 circuits	1 Elec	1.20	6.667	Ea.	296	350		646	850
2300	18 circuits		.80	10		420	525		945	1,250
2400	Rainproof, 12 circuits		1.20	6.667		340	350		690	900
2500	18 circuits		.80	10		455	525		980	1,275
2600	200 amp main lugs, indoor, 24 circuits	R-1A	1.30	12.308		565	530		1,095	1,425
2700	30 circuits		1.20	13.333		645	575		1,220	1,575
2800	36 circuits		1	16		775	690		1,465	1,900
2900	42 circuits		.80	20		850	860		1,710	2,225
3000	Rainproof, 24 circuits		1.30	12.308		625	530		1,155	1,500
3100	30 circuits		1.20	13.333		705	575		1,280	1,650
3200	36 circuits		1	16		975	690		1,665	2,125
3300	42 circuits		.80	20		1,050	860		1,910	2,450
3500	400 amp main lugs, indoor, 42 circuits		.72	22.222		1,275	955		2,230	2,875
3600	Rainproof, 42 circuits		.72	22.222		1,525	955		2,480	3,125
3700	Plug-in breakers, 20 amp, 1 pole, 3 wire, 120/240 volts									
3800	100 amp main breaker, indoor, 12 circuits	1 Elec	1.20	6.667	Ea.	315	350		665	870
3900	18 circuits	″	.80	10		425	525		950	1,250
4000	200 amp main breaker, indoor, 20 circuits	R-1A	1.50	10.667		575	460		1,035	1,325
4200	24 circuits		1.30	12.308		695	530		1,225	1,550
4300	30 circuits		1.20	13.333		770	575		1,345	1,725
4400	40 circuits		.90	17.778		960	765		1,725	2,200
4500	Rainproof, 20 circuits		1.50	10.667		625	460		1,085	1,375
4600	24 circuits		1.30	12.308		775	530		1,305	1,650
4700	30 circuits		1.20	13.333		850	575		1,425	1,800
4800	40 circuits		.90	17.778		1,050	765		1,815	2,300
5000	400 amp main breaker, indoor, 42 circuits		.72	22.222		2,750	955		3,705	4,475
5100	Rainproof, 42 circuits		.72	22.222		3,075	955		4,030	4,850
5300	Plug in breakers, 20 amp, 1 pole, 4 wire, 120/208 volts									
5400	200 amp main breaker, indoor, 30 circuits	R-1A	1.20	13.333	Ea.	1,175	575		1,750	2,150
5500	42 circuits		.80	20		1,425	860		2,285	2,875
5600	Rainproof, 30 circuits		1.20	13.333		1,250	575		1,825	2,250
5700	42 circuits		.80	20		1,525	860		2,385	2,975

26 24 16.20 Panelboard and Load Center Circuit Breakers

		Crew	Daily Output	Labor-Hours	Unit	Material	Labor	Equipment	Total	Total Incl O&P
0010	**PANELBOARD AND LOAD CENTER CIRCUIT BREAKERS** R262419-82									
0050	Bolt-on, 10,000 amp I.C., 120 volt, 1 pole									
0100	15 to 50 amp	1 Elec	10	.800	Ea.	14	42		56	78
0200	60 amp		8	1		18.20	52.50		70.70	98.50

26 24 16.20 Panelboard and Load Center Circuit Breakers		Crew	Daily Output	Labor-Hours	Unit	Material	2013 Bare Costs Labor	Equipment	Total	Total Incl O&P
0300	70 amp	1 Elec	8	1	Ea.	26.50	52.50		79	108
0350	240 volt, 2 pole									
0400	15 to 50 amp	1 Elec	8	1	Ea.	31	52.50		83.50	113
0500	60 amp		7.50	1.067		49.50	56		105.50	138
0600	80 to 100 amp		5	1.600		79	84		163	212
0700	3 pole, 15 to 60 amp		6.20	1.290		97.50	67.50		165	208
0800	70 amp		5	1.600		123	84		207	260
0900	80 to 100 amp		3.60	2.222		139	116		255	325
1000	22,000 amp I.C., 240 volt, 2 pole, 70 – 225 amp		2.70	2.963		600	155		755	890
1100	3 pole, 70 – 225 amp		2.30	3.478		665	182		847	1,000
1200	14,000 amp I.C., 277 volts, 1 pole, 15 – 30 amp		8	1		37	52.50		89.50	120
1300	22,000 amp I.C., 480 volts, 2 pole, 70 – 225 amp		2.70	2.963		600	155		755	890
1400	3 pole, 70 – 225 amp		2.30	3.478		740	182		922	1,100
2000	Plug-in panel or load center, 120/240 volt, to 60 amp, 1 pole		12	.667		10.85	35		45.85	64.50
2010	2 pole		9	.889		24.50	46.50		71	96.50
2020	3 pole		7.50	1.067		84.50	56		140.50	177
2030	100 amp, 2 pole		6	1.333		113	70		183	230
2040	3 pole		4.50	1.778		126	93		219	278
2050	150 to 200 amp, 2 pole		3	2.667		207	140		347	435
2060	Plug-in tandem, 120/240 V, 2-15 A, 1 pole		11	.727		28.50	38		66.50	88.50
2070	1-15 A & 1-20 A		11	.727		28.50	38		66.50	88.50
2080	2-20 A		11	.727		28.50	38		66.50	88.50
2082	Arc fault circuit interrupter, 120/240 V, 1-15 A & 1-20 A, 1 pole		11	.727		90	38		128	156
2100	High interrupting capacity, 120/240 volt, plug-in, 30 amp, 1 pole		12	.667		21.50	35		56.50	76
2110	60 amp, 2 pole		9	.889		49.50	46.50		96	124
2120	3 pole		7.50	1.067		129	56		185	226
2130	100 amp, 2 pole		6	1.333		143	70		213	262
2140	3 pole		4.50	1.778		185	93		278	340
2150	125 amp, 2 pole		3	2.667		380	140		520	630
2200	Bolt-on, 30 amp, 1 pole		10	.800		27.50	42		69.50	92.50
2210	60 amp, 2 pole		7.50	1.067		58	56		114	147
2220	3 pole		6.20	1.290		148	67.50		215.50	264
2230	100 amp, 2 pole		5	1.600		165	84		249	305
2240	3 pole		3.60	2.222		215	116		331	410
2300	Ground fault, 240 volt, 30 amp, 1 pole		7	1.143		84	60		144	182
2310	2 pole		6	1.333		151	70		221	271
2350	Key operated, 240 volt, 1 pole, 30 amp		7	1.143		76.50	60		136.50	174
2360	Switched neutral, 240 volt, 30 amp, 2 pole		6	1.333		35	70		105	144
2370	3 pole		5.50	1.455		52.50	76		128.50	172
2400	Shunt trip, for 240 volt breaker, 60 amp, 1 pole		4	2		68.50	105		173.50	233
2410	2 pole		3.50	2.286		68.50	120		188.50	255
2420	3 pole		3	2.667		68.50	140		208.50	285
2430	100 amp, 2 pole		3	2.667		68.50	140		208.50	285
2440	3 pole		2.50	3.200		68.50	168		236.50	325
2450	150 amp, 2 pole		2	4		163	210		373	495
2500	Auxiliary switch, for 240 volt breaker, 60 amp, 1 pole		4	2		62.50	105		167.50	226
2510	2 pole		3.50	2.286		62.50	120		182.50	248
2520	3 pole		3	2.667		62.50	140		202.50	278
2530	100 amp, 2 pole		3	2.667		62.50	140		202.50	278
2540	3 pole		2.50	3.200		62.50	168		230.50	320
2550	150 amp, 2 pole		2	4		104	210		314	430
2600	Panel or load center, 277/480 volt, plug-in, 30 amp, 1 pole		12	.667		50.50	35		85.50	108
2610	60 amp, 2 pole		9	.889		151	46.50		197.50	236

26 24 16.20 Panelboard and Load Center Circuit Breakers		Crew	Daily Output	Labor-Hours	Unit	2013 Bare Costs Material	Labor	Equipment	Total	Total Incl O&P
2620	3 pole	1 Elec	7.50	1.067	Ea.	221	56		277	325
2650	Bolt-on, 60 amp, 2 pole		7.50	1.067		151	56		207	250
2660	3 pole		6.20	1.290		221	67.50		288.50	345
2700	I-line, 277/480 volt, 30 amp, 1 pole		8	1		58	52.50		110.50	142
2710	60 amp, 2 pole		7.50	1.067		203	56		259	305
2720	3 pole		6.20	1.290		264	67.50		331.50	390
2730	100 amp, 1 pole		7.50	1.067		111	56		167	206
2740	2 pole		5	1.600		264	84		348	415
2750	3 pole		3.50	2.286		310	120		430	525
2800	High interrupting capacity, 277/480 volt, plug-in, 30 amp, 1 pole		12	.667		291	35		326	375
2810	60 amp, 2 pole		9	.889		450	46.50		496.50	565
2820	3 pole		7	1.143		525	60		585	665
2830	Bolt-on, 30 amp, 1 pole		8	1		291	52.50		343.50	400
2840	60 amp, 2 pole		7.50	1.067		450	56		506	580
2850	3 pole		6.20	1.290		525	67.50		592.50	675
2900	I-line, 30 amp, 1 pole		8	1		291	52.50		343.50	400
2910	60 amp, 2 pole		7.50	1.067		415	56		471	540
2920	3 pole		6.20	1.290		460	67.50		527.50	605
2930	100 amp, 1 pole		7.50	1.067		470	56		526	600
2940	2 pole		5	1.600		470	84		554	640
2950	3 pole		3.60	2.222		520	116		636	745
2960	Shunt trip, 277/480 volt breaker, remote oper., 30 amp, 1 pole		4	2		310	105		415	500
2970	60 amp, 2 pole		3.50	2.286		455	120		575	685
2980	3 pole		3	2.667		520	140		660	780
2990	100 amp, 1 pole		3.50	2.286		365	120		485	580
3000	2 pole		3	2.667		520	140		660	780
3010	3 pole		2.50	3.200		565	168		733	870
3050	Under voltage trip, 277/480 volt breaker, 30 amp, 1 pole		4	2		310	105		415	500
3060	60 amp, 2 pole		3.50	2.286		455	120		575	685
3070	3 pole		3	2.667		520	140		660	780
3080	100 amp, 1 pole		3.50	2.286		365	120		485	580
3090	2 pole		3	2.667		520	140		660	780
3100	3 pole		2.50	3.200		565	168		733	870
3150	Motor operated, 277/480 volt breaker, 30 amp, 1 pole		4	2		555	105		660	765
3160	60 amp, 2 pole		3.50	2.286		700	120		820	950
3170	3 pole		3	2.667		760	140		900	1,050
3180	100 amp, 1 pole		3.50	2.286		610	120		730	850
3190	2 pole		3	2.667		760	140		900	1,050
3200	3 pole		2.50	3.200		810	168		978	1,150
3250	Panelboard spacers, per pole		40	.200		4.20	10.50		14.70	20.50

26 24 16.30 Panelboards Commercial Applications		Crew	Daily Output	Labor-Hours	Unit	Material	Labor	Equipment	Total	Total Incl O&P
0010	**PANELBOARDS COMMERCIAL APPLICATIONS** R262416-50									
0050	NQOD, w/20 amp 1 pole bolt-on circuit breakers									
0100	3 wire, 120/240 volts, 100 amp main lugs									
0150	10 circuits	1 Elec	1	8	Ea.	465	420		885	1,150
0200	14 circuits		.88	9.091		570	475		1,045	1,350
0250	18 circuits		.75	10.667		620	560		1,180	1,525
0300	20 circuits		.65	12.308		695	645		1,340	1,725
0350	225 amp main lugs, 24 circuits	2 Elec	1.20	13.333		780	700		1,480	1,900
0400	30 circuits		.90	17.778		905	930		1,835	2,400
0450	36 circuits		.80	20		1,025	1,050		2,075	2,700
0500	38 circuits		.72	22.222		1,100	1,175		2,275	2,975

26 24 16.30 Panelboards Commercial Applications		Crew	Daily Output	Labor-Hours	Unit	Material	Labor	Equipment	Total	Total Incl O&P
						2013 Bare Costs				
0550	42 circuits	2 Elec	.66	24.242	Ea.	1,150	1,275		2,425	3,175
0600	4 wire, 120/208 volts, 100 amp main lugs, 12 circuits	1 Elec	1	8		550	420		970	1,225
0650	16 circuits		.75	10.667		630	560		1,190	1,525
0700	20 circuits		.65	12.308		730	645		1,375	1,775
0750	24 circuits		.60	13.333		785	700		1,485	1,925
0800	30 circuits		.53	15.094		900	790		1,690	2,175
0850	225 amp main lugs, 32 circuits	2 Elec	.90	17.778		1,000	930		1,930	2,500
0900	34 circuits		.84	19.048		1,025	1,000		2,025	2,650
0950	36 circuits		.80	20		1,050	1,050		2,100	2,750
1000	42 circuits		.68	23.529		1,175	1,225		2,400	3,150
1040	225 amp main lugs, NEMA 7, 12 circuits		1	16		3,200	840		4,040	4,750
1100	24 circuits		.40	40		3,825	2,100		5,925	7,325
1200	NEHB,w/20 amp, 1 pole bolt-on circuit breakers									
1250	4 wire, 277/480 volts, 100 amp main lugs, 12 circuits	1 Elec	.88	9.091	Ea.	1,200	475		1,675	2,050
1300	20 circuits	″	.60	13.333		1,775	700		2,475	3,025
1350	225 amp main lugs, 24 circuits	2 Elec	.90	17.778		2,025	930		2,955	3,625
1400	30 circuits		.80	20		2,425	1,050		3,475	4,250
1450	36 circuits		.72	22.222		2,825	1,175		4,000	4,850
1500	42 circuits		.60	26.667		3,225	1,400		4,625	5,650
1510	225 amp main lugs, NEMA 7, 12 circuits		.90	17.778		5,400	930		6,330	7,350
1590	24 circuits		.30	53.333		7,275	2,800		10,075	12,200
1600	NQOD panel, w/20 amp, 1 pole, circuit breakers									
1650	3 wire, 120/240 volt with main circuit breaker									
1700	100 amp main, 12 circuits	1 Elec	.80	10	Ea.	785	525		1,310	1,650
1750	20 circuits	″	.60	13.333		995	700		1,695	2,150
1800	225 amp main, 30 circuits	2 Elec	.68	23.529		1,875	1,225		3,100	3,900
1850	42 circuits		.52	30.769		2,150	1,600		3,750	4,775
1900	400 amp main, 30 circuits		.54	29.630		2,575	1,550		4,125	5,150
1950	42 circuits		.50	32		2,875	1,675		4,550	5,650
2000	4 wire, 120/208 volts with main circuit breaker									
2050	100 amp main, 24 circuits	1 Elec	.47	17.021	Ea.	1,150	890		2,040	2,600
2100	30 circuits	″	.40	20		1,300	1,050		2,350	3,000
2200	225 amp main, 32 circuits	2 Elec	.72	22.222		2,175	1,175		3,350	4,125
2250	42 circuits		.56	28.571		2,375	1,500		3,875	4,850
2300	400 amp main, 42 circuits		.48	33.333		3,200	1,750		4,950	6,125
2350	600 amp main, 42 circuits		.40	40		4,725	2,100		6,825	8,325
2400	NEHB, with 20 amp, 1 pole circuit breaker									
2450	4 wire, 277/480 volts with main circuit breaker									
2500	100 amp main, 24 circuits	1 Elec	.42	19.048	Ea.	2,325	1,000		3,325	4,075
2550	30 circuits	″	.38	21.053		2,725	1,100		3,825	4,650
2600	225 amp main, 30 circuits	2 Elec	.72	22.222		3,425	1,175		4,600	5,525
2650	42 circuits		.56	28.571		4,225	1,500		5,725	6,900
2700	400 amp main, 42 circuits		.46	34.783		5,075	1,825		6,900	8,300
2750	600 amp main, 42 circuits		.38	42.105		6,950	2,200		9,150	11,000
2900	Note: the following line items don't include branch circuit breakers									
2910	For branch circuit breakers information, see Section 26 24 16.20									
3010	Main lug, no main breaker, 240 volt, 1 pole, 3 wire, 100 amp	1 Elec	2.30	3.478	Ea.	525	182		707	850
3020	225 amp	2 Elec	2.40	6.667		630	350		980	1,225
3030	400 amp	″	1.80	8.889		920	465		1,385	1,725
3060	3 pole, 3 wire, 100 amp	1 Elec	2.30	3.478		565	182		747	895
3070	225 amp	2 Elec	2.40	6.667		665	350		1,015	1,250
3080	400 amp		1.80	8.889		1,000	465		1,465	1,800
3090	600 amp		1.60	10		1,150	525		1,675	2,050

26 24 16 – Panelboards

26 24 16.30 Panelboards Commercial Applications		Crew	Daily Output	Labor-Hours	Unit	Material	Labor (2013 Bare Costs)	Equipment	Total	Total Incl O&P
3110	3 pole, 4 wire, 100 amp	1 Elec	2.30	3.478	Ea.	625	182		807	960
3120	225 amp	2 Elec	2.40	6.667		780	350		1,130	1,375
3130	400 amp		1.80	8.889		1,000	465		1,465	1,800
3140	600 amp		1.60	10		1,150	525		1,675	2,050
3160	480 volt, 3 pole, 3 wire, 100 amp	1 Elec	2.30	3.478		745	182		927	1,100
3170	225 amp	2 Elec	2.40	6.667		900	350		1,250	1,525
3180	400 amp		1.80	8.889		1,250	465		1,715	2,075
3190	600 amp		1.60	10		1,400	525		1,925	2,300
3210	277/480 volt, 3 pole, 4 wire, 100 amp	1 Elec	2.30	3.478		725	182		907	1,075
3220	225 amp	2 Elec	2.40	6.667		880	350		1,230	1,500
3230	400 amp		1.80	8.889		1,225	465		1,690	2,050
3240	600 amp		1.60	10		1,375	525		1,900	2,275
3260	Main circuit breaker, 240 volt, 1 pole, 3 wire, 100 amp	1 Elec	2	4		705	210		915	1,100
3270	225 amp	2 Elec	2	8		1,425	420		1,845	2,200
3280	400 amp	"	1.60	10		2,250	525		2,775	3,250
3310	3 pole, 3 wire, 100 amp	1 Elec	2	4		805	210		1,015	1,200
3320	225 amp	2 Elec	2	8		1,650	420		2,070	2,425
3330	400 amp		1.60	10		2,575	525		3,100	3,600
3360	120/208 volt, 3 pole, 4 wire, 100 amp		4	4		805	210		1,015	1,200
3370	225 amp		2	8		1,650	420		2,070	2,425
3380	400 amp		1.60	10		2,575	525		3,100	3,600
3410	480 volt, 3 pole, 3 wire, 100 amp	1 Elec	2	4		1,125	210		1,335	1,575
3420	225 amp	2 Elec	2	8		1,925	420		2,345	2,750
3430	400 amp		1.60	10		2,900	525		3,425	3,950
3460	277/480 volt, 3 pole, 4 wire, 100 amp		4	4		1,100	210		1,310	1,525
3470	225 amp		2	8		1,875	420		2,295	2,700
3480	400 amp		1.60	10		2,925	525		3,450	3,975
3510	Main circuit breaker, HIC, 240 volt, 1 pole, 3 wire, 100 amp	1 Elec	2	4		1,075	210		1,285	1,500
3520	225 amp	2 Elec	2	8		2,800	420		3,220	3,725
3530	400 amp	"	1.60	10		3,775	525		4,300	4,925
3560	3 pole, 3 wire, 100 amp	1 Elec	2	4		1,200	210		1,410	1,650
3570	225 amp	2 Elec	2	8		3,175	420		3,595	4,125
3580	400 amp	"	1.60	10		4,200	525		4,725	5,400
3610	120/208 volt, 3 pole, 4 wire, 100 amp	1 Elec	2	4		1,200	210		1,410	1,650
3620	225 amp	2 Elec	2	8		3,175	420		3,595	4,125
3630	400 amp	"	1.60	10		4,200	525		4,725	5,400
3660	480 volt, 3 pole, 3 wire, 100 amp	1 Elec	2	4		1,775	210		1,985	2,275
3670	225 amp	2 Elec	2	8		3,525	420		3,945	4,500
3680	400 amp	"	1.60	10		4,475	525		5,000	5,700
3710	277/480 volt, 3 pole, 4 wire, 100 amp	1 Elec	2	4		1,700	210		1,910	2,200
3720	225 amp	2 Elec	2	8		3,350	420		3,770	4,325
3730	400 amp	"	1.60	10		4,450	525		4,975	5,675
3760	Main circuit breaker, shunt trip, 100 amp	1 Elec	1.20	6.667		900	350		1,250	1,525
3770	225 amp	2 Elec	1.60	10		2,075	525		2,600	3,050
3780	400 amp	"	1.40	11.429		3,000	600		3,600	4,225

26 24 19 – Motor-Control Centers

26 24 19.20 Motor Control Center Components		Crew	Daily Output	Labor-Hours	Unit	Material	Labor	Equipment	Total	Total Incl O&P
0010	**MOTOR CONTROL CENTER COMPONENTS** R262419-60									
0100	Starter, size 1, FVNR, NEMA 1, type A, fusible	1 Elec	2.70	2.963	Ea.	1,525	155		1,680	1,900
0120	Circuit breaker		2.70	2.963		1,650	155		1,805	2,050
0140	Type B, fusible		2.70	2.963		1,675	155		1,830	2,050
0160	Circuit breaker		2.70	2.963		1,825	155		1,980	2,225

26 24 19.20 Motor Control Center Components		Crew	Daily Output	Labor-Hours	Unit	2013 Bare Costs Material	Labor	Equipment	Total	Total Incl O&P
0180	NEMA 12, type A, fusible	1 Elec	2.60	3.077	Ea.	1,550	161		1,711	1,950
0200	Circuit breaker		2.60	3.077		1,675	161		1,836	2,100
0220	Type B, fusible		2.60	3.077		1,700	161		1,861	2,125
0240	Circuit breaker		2.60	3.077		1,850	161		2,011	2,275
0300	Starter, size 1, FVR, NEMA 1, type A, fusible		2	4		2,175	210		2,385	2,725
0320	Circuit breaker		2	4		2,175	210		2,385	2,725
0340	Type B, fusible		2	4		2,400	210		2,610	2,950
0360	Circuit breaker		2	4		2,400	210		2,610	2,950
0380	NEMA 12, type A, fusible		1.90	4.211		2,200	221		2,421	2,750
0400	Circuit breaker		1.90	4.211		2,200	221		2,421	2,750
0420	Type B, fusible		1.90	4.211		2,425	221		2,646	3,000
0440	Circuit breaker		1.90	4.211		2,425	221		2,646	3,000
0490	Starter size 1, 2 speed, separate winding									
0500	NEMA 1, type A, fusible	1 Elec	2.60	3.077	Ea.	2,875	161		3,036	3,400
0520	Circuit breaker		2.60	3.077		2,875	161		3,036	3,400
0540	Type B, fusible		2.60	3.077		3,150	161		3,311	3,725
0560	Circuit breaker		2.60	3.077		3,150	161		3,311	3,725
0580	NEMA 12, type A, fusible		2.50	3.200		2,925	168		3,093	3,475
0600	Circuit breaker		2.50	3.200		2,925	168		3,093	3,475
0620	Type B, fusible		2.50	3.200		3,200	168		3,368	3,775
0640	Circuit breaker		2.50	3.200		3,200	168		3,368	3,775
0650	Starter size 1, 2 speed, consequent pole									
0660	NEMA 1, type A, fusible	1 Elec	2.60	3.077	Ea.	2,875	161		3,036	3,400
0680	Circuit breaker		2.60	3.077		2,875	161		3,036	3,400
0700	Type B, fusible		2.60	3.077		3,150	161		3,311	3,725
0720	Circuit breaker		2.60	3.077		3,150	161		3,311	3,725
0740	NEMA 12, type A, fusible		2.50	3.200		2,925	168		3,093	3,475
0760	Circuit breaker		2.50	3.200		2,925	168		3,093	3,475
0780	Type B, fusible		2.50	3.200		3,200	168		3,368	3,750
0800	Circuit breaker		2.50	3.200		3,200	168		3,368	3,775
0810	Starter size 1, 2 speed, space only									
0820	NEMA 1, type A, fusible	1 Elec	16	.500	Ea.	645	26		671	750
0840	Circuit breaker		16	.500		645	26		671	750
0860	Type B, fusible		16	.500		645	26		671	750
0880	Circuit breaker		16	.500		645	26		671	750
0900	NEMA 12, type A, fusible		15	.533		670	28		698	775
0920	Circuit breaker		15	.533		670	28		698	775
0940	Type B, fusible		15	.533		670	28		698	775
0960	Circuit breaker		15	.533		670	28		698	775
1100	Starter size 2, FVNR, NEMA 1, type A, fusible	2 Elec	4	4		1,725	210		1,935	2,225
1120	Circuit breaker		4	4		1,875	210		2,085	2,400
1140	Type B, fusible		4	4		1,900	210		2,110	2,400
1160	Circuit breaker		4	4		2,050	210		2,260	2,575
1180	NEMA 12, type A, fusible		3.80	4.211		1,750	221		1,971	2,250
1200	Circuit breaker		3.80	4.211		1,900	221		2,121	2,425
1220	Type B, fusible		3.80	4.211		1,900	221		2,121	2,425
1240	Circuit breaker		3.80	4.211		2,075	221		2,296	2,625
1300	FVR, NEMA 1, type A, fusible		3.20	5		2,900	262		3,162	3,600
1320	Circuit breaker		3.20	5		2,900	262		3,162	3,600
1340	Type B, fusible		3.20	5		3,200	262		3,462	3,925
1360	Circuit breaker		3.20	5		3,200	262		3,462	3,925
1380	NEMA type 12, type A, fusible		3	5.333		2,950	279		3,229	3,675
1400	Circuit breaker		3	5.333		2,950	279		3,229	3,675

26 24 Switchboards and Panelboards

26 24 19 – Motor-Control Centers

26 24 19.20 Motor Control Center Components		Crew	Daily Output	Labor-Hours	Unit	Material	2013 Bare Costs Labor	Equipment	Total	Total Incl O&P
1420	Type B, fusible	2 Elec	3	5.333	Ea.	3,250	279		3,529	4,000
1440	Circuit breaker		3	5.333		3,250	279		3,529	4,000
1490	Starter size 2, 2 speed, separate winding									
1500	NEMA 1, type A, fusible	2 Elec	3.80	4.211	Ea.	3,250	221		3,471	3,900
1520	Circuit breaker		3.80	4.211		3,250	221		3,471	3,900
1540	Type B, fusible		3.80	4.211		3,575	221		3,796	4,250
1560	Circuit breaker		3.80	4.211		3,575	221		3,796	4,250
1570	NEMA 12, type A, fusible		3.60	4.444		3,300	233		3,533	3,975
1580	Circuit breaker		3.60	4.444		3,300	233		3,533	3,975
1600	Type B, fusible		3.60	4.444		3,625	233		3,858	4,325
1620	Circuit breaker		3.60	4.444		3,625	233		3,858	4,325
1630	Starter size 2, 2 speed, consequent pole									
1640	NEMA 1, type A, fusible	2 Elec	3.80	4.211	Ea.	3,675	221		3,896	4,375
1660	Circuit breaker		3.80	4.211		3,675	221		3,896	4,375
1680	Type B, fusible		3.80	4.211		3,900	221		4,121	4,600
1700	Circuit breaker		3.80	4.211		3,900	221		4,121	4,600
1720	NEMA 12, type A, fusible		3.80	4.211		3,700	221		3,921	4,375
1740	Circuit breaker		3.60	4.444		3,725	233		3,958	4,450
1760	Type B, fusible		3.60	4.444		3,950	233		4,183	4,700
1780	Circuit breaker		3.60	4.444		3,950	233		4,183	4,700
1830	Starter size 2, autotransformer									
1840	NEMA 1, type A, fusible	2 Elec	3.40	4.706	Ea.	5,750	247		5,997	6,700
1860	Circuit breaker		3.40	4.706		5,875	247		6,122	6,825
1880	Type B, fusible		3.40	4.706		6,275	247		6,522	7,275
1900	Circuit breaker		3.40	4.706		6,275	247		6,522	7,275
1920	NEMA 12, type A, fusible		3.20	5		5,850	262		6,112	6,825
1940	Circuit breaker		3.20	5		5,850	262		6,112	6,825
1960	Type B, fusible		3.20	5		6,375	262		6,637	7,425
1980	Circuit breaker		3.20	5		6,375	262		6,637	7,425
2030	Starter size 2, space only									
2040	NEMA 1, type A, fusible	1 Elec	16	.500	Ea.	645	26		671	750
2060	Circuit breaker		16	.500		645	26		671	750
2080	Type B, fusible		16	.500		645	26		671	750
2100	Circuit breaker		16	.500		645	26		671	750
2120	NEMA 12, type A, fusible		15	.533		670	28		698	775
2140	Circuit breaker		15	.533		670	28		698	775
2160	Type B, fusible		15	.533		670	28		698	775
2180	Circuit breaker		15	.533		670	28		698	775
2300	Starter size 3, FVNR, NEMA 1, type A, fusible	2 Elec	2	8		3,275	420		3,695	4,225
2320	Circuit breaker		2	8		2,900	420		3,320	3,825
2340	Type B, fusible		2	8		3,600	420		4,020	4,575
2360	Circuit breaker		2	8		3,200	420		3,620	4,150
2380	NEMA 12, type A, fusible		1.90	8.421		3,325	440		3,765	4,325
2400	Circuit breaker		1.90	8.421		2,950	440		3,390	3,900
2420	Type B, fusible		1.90	8.421		3,725	440		4,165	4,725
2440	Circuit breaker		1.90	8.421		3,250	440		3,690	4,225
2500	Starter size 3, FVR, NEMA 1, type A, fusible		1.60	10		4,450	525		4,975	5,675
2520	Circuit breaker		1.60	10		4,250	525		4,775	5,450
2540	Type B, fusible		1.60	10		4,850	525		5,375	6,100
2560	Circuit breaker		1.60	10		4,650	525		5,175	5,900
2580	NEMA 12, type A, fusible		1.50	10.667		4,550	560		5,110	5,825
2600	Circuit breaker		1.50	10.667		4,325	560		4,885	5,600
2620	Type B, fusible		1.50	10.667		4,950	560		5,510	6,275

26 24 19.20 Motor Control Center Components		Crew	Daily Output	Labor-Hours	Unit	Material	2013 Bare Costs Labor	Equipment	Total	Total Incl O&P
2640	Circuit breaker	2 Elec	1.50	10.667	Ea.	5,125	560		5,685	6,475
2690	Starter size 3, 2 speed, separate winding									
2700	NEMA 1, type A, fusible	2 Elec	2	8	Ea.	5,125	420		5,545	6,250
2720	Circuit breaker		2	8		4,525	420		4,945	5,600
2740	Type B, fusible		2	8		5,600	420		6,020	6,800
2760	Circuit breaker		2	8		4,950	420		5,370	6,075
2780	NEMA 12, type A, fusible		1.90	8.421		5,225	440		5,665	6,400
2800	Circuit breaker		1.90	8.421		4,625	440		5,065	5,725
2820	Type B, fusible		1.90	8.421		5,725	440		6,165	6,925
2840	Circuit breaker		1.90	8.421		5,050	440		5,490	6,200
2850	Starter size 3, 2 speed, consequent pole									
2860	NEMA 1, type A, fusible	2 Elec	2	8	Ea.	5,725	420		6,145	6,925
2880	Circuit breaker		2	8		5,125	420		5,545	6,250
2900	Type B, fusible		2	8		6,200	420		6,620	7,450
2920	Circuit breaker		2	8		5,550	420		5,970	6,725
2940	NEMA 12, type A, fusible		1.90	8.421		5,825	440		6,265	7,050
2960	Circuit breaker		1.90	8.421		5,200	440		5,640	6,375
2980	Type B, fusible		1.90	8.421		6,300	440		6,740	7,600
3000	Circuit breaker		1.90	8.421		5,625	440		6,065	6,850
3100	Starter size 3, autotransformer, NEMA 1, type A, fusible		1.60	10		7,300	525		7,825	8,800
3120	Circuit breaker		1.60	10		7,125	525		7,650	8,600
3140	Type B, fusible		1.60	10		7,775	525		8,300	9,325
3160	Circuit breaker		1.60	10		7,775	525		8,300	9,325
3180	NEMA 12, type A, fusible		1.50	10.667		7,250	560		7,810	8,800
3200	Circuit breaker		1.50	10.667		7,250	560		7,810	8,800
3220	Type B, fusible		1.50	10.667		7,900	560		8,460	9,525
3240	Circuit breaker		1.50	10.667		7,900	560		8,460	9,525
3260	Starter size 3, space only, NEMA 1, type A, fusible	1 Elec	15	.533		1,100	28		1,128	1,275
3280	Circuit breaker		15	.533		860	28		888	985
3300	Type B, fusible		15	.533		1,100	28		1,128	1,275
3320	Circuit breaker		15	.533		860	28		888	985
3340	NEMA 12, type A, fusible		14	.571		1,175	30		1,205	1,350
3360	Circuit breaker		14	.571		910	30		940	1,050
3380	Type B, fusible		14	.571		1,175	30		1,205	1,350
3400	Circuit breaker		14	.571		910	30		940	1,050
3500	Starter size 4, FVNR, NEMA 1, type A, fusible	2 Elec	1.60	10		4,250	525		4,775	5,450
3520	Circuit breaker		1.60	10		3,850	525		4,375	5,025
3540	Type B, fusible		1.60	10		4,650	525		5,175	5,900
3560	Circuit breaker		1.60	10		4,250	525		4,775	5,450
3580	NEMA 12, type A, fusible		1.50	10.667		4,375	560		4,935	5,625
3600	Circuit breaker		1.50	10.667		3,925	560		4,485	5,150
3620	Type B, fusible		1.50	10.667		4,750	560		5,310	6,050
3640	Circuit breaker		1.50	10.667		4,325	560		4,885	5,575
3700	Starter size 4, FVR, NEMA 1, type A, fusible		1.20	13.333		5,825	700		6,525	7,450
3720	Circuit breaker		1.20	13.333		5,250	700		5,950	6,825
3740	Type B, fusible		1.20	13.333		6,350	700		7,050	8,025
3760	Circuit breaker		1.20	13.333		5,775	700		6,475	7,400
3780	NEMA 12, type A, fusible		1.16	13.793		5,950	725		6,675	7,625
3800	Circuit breaker		1.16	13.793		5,325	725		6,050	6,925
3820	Type B, fusible		1.16	13.793		5,375	725		6,100	7,000
3840	Circuit breaker		1.16	13.793		5,850	725		6,575	7,525
3890	Starter size 4, 2 speed, separate windings									
3900	NEMA 1, type A, fusible	2 Elec	1.60	10	Ea.	7,350	525		7,875	8,850

26 24 19 – Motor-Control Centers

26 24 19.20 Motor Control Center Components		Crew	Daily Output	Labor-Hours	Unit	2013 Bare Costs Material	Labor	Equipment	Total	Total Incl O&P
3920	Circuit breaker	2 Elec	1.60	10	Ea.	5,525	525		6,050	6,850
3940	Type B, fusible		1.60	10		8,075	525		8,600	9,650
3960	Circuit breaker		1.60	10		6,050	525		6,575	7,425
3980	NEMA 12, type A, fusible		1.50	10.667		7,475	560		8,035	9,050
4000	Circuit breaker		1.50	10.667		5,600	560		6,160	7,000
4020	Type B, fusible		1.50	10.667		8,200	560		8,760	9,850
4040	Circuit breaker		1.50	10.667		6,150	560		6,710	7,575
4050	Starter size 4, 2 speed, consequent pole									
4060	NEMA 1, type A, fusible	2 Elec	1.60	10	Ea.	8,575	525		9,100	10,200
4080	Circuit breaker		1.60	10		6,375	525		6,900	7,800
4100	Type B, fusible		1.60	10		9,450	525		9,975	11,200
4120	Circuit breaker		1.60	10		7,000	525		7,525	8,475
4140	NEMA 12, type A, fusible		1.50	10.667		8,700	560		9,260	10,400
4160	Circuit breaker		1.50	10.667		6,475	560		7,035	7,950
4180	Type B, fusible		1.50	10.667		9,575	560		10,135	11,300
4200	Circuit breaker		1.50	10.667		7,100	560		7,660	8,625
4300	Starter size 4, autotransformer, NEMA 1, type A, fusible		1.30	12.308		8,450	645		9,095	10,300
4320	Circuit breaker		1.30	12.308		8,525	645		9,170	10,300
4340	Type B, fusible		1.30	12.308		9,300	645		9,945	11,200
4360	Circuit breaker		1.30	12.308		9,300	645		9,945	11,200
4380	NEMA 12, type A, fusible		1.24	12.903		8,600	675		9,275	10,500
4400	Circuit breaker		1.24	12.903		8,625	675		9,300	10,500
4420	Type B, fusible		1.24	12.903		9,450	675		10,125	11,400
4440	Circuit breaker		1.24	12.903		9,425	675		10,100	11,400
4500	Starter size 4, space only, NEMA 1, type A, fusible	1 Elec	14	.571		1,525	30		1,555	1,750
4520	Circuit breaker		14	.571		1,100	30		1,130	1,275
4540	Type B, fusible		14	.571		1,525	30		1,555	1,750
4560	Circuit breaker		14	.571		1,100	30		1,130	1,275
4580	NEMA 12, type A, fusible		13	.615		1,625	32.50		1,657.50	1,850
4600	Circuit breaker		13	.615		1,175	32.50		1,207.50	1,350
4620	Type B, fusible		13	.615		1,625	32.50		1,657.50	1,850
4640	Circuit breaker		13	.615		1,175	32.50		1,207.50	1,350
4800	Starter size 5, FVNR, NEMA 1, type A, fusible	2 Elec	1	16		8,800	840		9,640	10,900
4820	Circuit breaker		1	16		6,225	840		7,065	8,100
4840	Type B, fusible		1	16		9,650	840		10,490	11,900
4860	Circuit breaker		1	16		6,875	840		7,715	8,800
4880	NEMA 12, type A, fusible		.96	16.667		8,900	875		9,775	11,100
4900	Circuit breaker		.96	16.667		6,350	875		7,225	8,300
4920	Type B, fusible		.96	16.667		9,775	875		10,650	12,000
4940	Circuit breaker		.96	16.667		6,975	875		7,850	8,975
5000	Starter size 5, FVR, NEMA 1, type A, fusible		.80	20		13,000	1,050		14,050	15,800
5020	Circuit breaker		.80	20		10,200	1,050		11,250	12,800
5040	Type B, fusible		.80	20		14,200	1,050		15,250	17,200
5060	Circuit breaker		.80	20		11,200	1,050		12,250	13,900
5080	NEMA 12, type A, fusible		.76	21.053		13,200	1,100		14,300	16,200
5100	Circuit breaker		.76	21.053		10,300	1,100		11,400	13,100
5120	Type B, fusible		.76	21.053		14,400	1,100		15,500	17,600
5140	Circuit breaker		.76	21.053		11,300	1,100		12,400	14,100
5190	Starter size 5, 2 speed, separate windings									
5200	NEMA 1, type A, fusible	2 Elec	1	16	Ea.	17,800	840		18,640	20,900
5220	Circuit breaker		1	16		13,300	840		14,140	15,900
5240	Type B, fusible		1	16		19,600	840		20,440	22,900
5260	Circuit breaker		1	16		14,600	840		15,440	17,400

26 24 19.20 Motor Control Center Components		Crew	Daily Output	Labor-Hours	Unit	2013 Bare Costs Material	Labor	Equipment	Total	Total Incl O&P
5280	NEMA 12, type A, fusible	2 Elec	.96	16.667	Ea.	18,100	875		18,975	21,200
5300	Circuit breaker		.96	16.667		13,500	875		14,375	16,100
5320	Type B, fusible		.96	16.667		19,900	875		20,775	23,100
5340	Circuit breaker		.96	16.667		14,800	875		15,675	17,600
5400	Starter size 5, autotransformer, NEMA 1, type A, fusible		.70	22.857		14,600	1,200		15,800	17,900
5420	Circuit breaker		.70	22.857		12,100	1,200		13,300	15,200
5440	Type B, fusible		.70	22.857		16,000	1,200		17,200	19,400
5460	Circuit breaker		.70	22.857		13,300	1,200		14,500	16,500
5480	NEMA 12, type A, fusible		.68	23.529		14,800	1,225		16,025	18,200
5500	Circuit breaker		.68	23.529		12,300	1,225		13,525	15,400
5520	Type B, fusible		.68	23.529		16,200	1,225		17,425	19,700
5540	Circuit breakers		.68	23.529		13,500	1,225		14,725	16,700
5600	Starter size 5, space only, NEMA 1, type A, fusible	1 Elec	12	.667		1,975	35		2,010	2,200
5620	Circuit breaker		12	.667		1,975	35		2,010	2,200
5640	Type B, fusible		12	.667		1,975	35		2,010	2,200
5660	Circuit breaker		12	.667		1,275	35		1,310	1,475
5680	NEMA 12, type A, fusible		11	.727		2,075	38		2,113	2,350
5700	Circuit breaker		11	.727		1,375	38		1,413	1,550
5720	Type B, fusible		11	.727		2,075	38		2,113	2,350
5740	Circuit breaker		11	.727		1,375	38		1,413	1,550
5800	Fuse, light contactor NEMA 1, type A, 30 amp		2.70	2.963		1,525	155		1,680	1,900
5820	60 amp		2	4		1,725	210		1,935	2,225
5840	100 amp		1	8		3,275	420		3,695	4,225
5860	200 amp		.80	10		7,600	525		8,125	9,125
5880	Type B, 30 amp		2.70	2.963		1,650	155		1,805	2,050
5900	60 amp		2	4		1,850	210		2,060	2,350
5920	100 amp		1	8		3,600	420		4,020	4,575
5940	200 amp	2 Elec	1.60	10		8,325	525		8,850	9,925
5960	NEMA 12, type A, 30 amp	1 Elec	2.60	3.077		1,550	161		1,711	1,950
5980	60 amp		1.90	4.211		1,750	221		1,971	2,250
6000	100 amp		.95	8.421		3,325	440		3,765	4,325
6020	200 amp	2 Elec	1.50	10.667		7,700	560		8,260	9,300
6040	Type B, 30 amp	1 Elec	2.60	3.077		1,675	161		1,836	2,100
6060	60 amp		1.90	4.211		1,875	221		2,096	2,375
6080	100 amp		.95	8.421		3,675	440		4,115	4,675
6100	200 amp	2 Elec	1.50	10.667		8,450	560		9,010	10,100
6200	Circuit breaker, light contactor NEMA 1, type A, 30 amp	1 Elec	2.70	2.963		1,650	155		1,805	2,050
6220	60 amp		2	4		1,875	210		2,085	2,400
6240	100 amp		1	8		2,900	420		3,320	3,825
6260	200 amp	2 Elec	1.60	10		6,300	525		6,825	7,700
6280	Type B, 30 amp	1 Elec	2.70	2.963		1,825	155		1,980	2,225
6300	60 amp		2	4		2,075	210		2,285	2,600
6320	100 amp		1	8		3,175	420		3,595	4,100
6340	200 amp	2 Elec	1.60	10		6,875	525		7,400	8,325
6360	NEMA 12, type A, 30 amp	1 Elec	2.60	3.077		1,675	161		1,836	2,100
6380	60 amp		1.90	4.211		1,900	221		2,121	2,425
6400	100 amp		.95	8.421		2,950	440		3,390	3,900
6420	200 amp	2 Elec	1.50	10.667		6,325	560		6,885	7,775
6440	Type B, 30 amp	1 Elec	2.60	3.077		1,975	161		2,136	2,425
6460	60 amp		1.90	4.211		2,100	221		2,321	2,650
6480	100 amp		.95	8.421		3,225	440		3,665	4,200
6500	200 amp	2 Elec	1.50	10.667		6,950	560		7,510	8,475
6600	Fusible switch, NEMA 1, type A, 30 amp	1 Elec	5.30	1.509		1,000	79		1,079	1,225

26 24 19.20 Motor Control Center Components		Crew	Daily Output	Labor-Hours	Unit	2013 Bare Costs Material	Labor	Equipment	Total	Total Incl O&P
6620	60 amp	1 Elec	5	1.600	Ea.	1,075	84		1,159	1,300
6640	100 amp		4	2		1,200	105		1,305	1,450
6660	200 amp		3.20	2.500		2,050	131		2,181	2,450
6680	400 amp	2 Elec	4.60	3.478		4,825	182		5,007	5,575
6700	600 amp		3.20	5		5,225	262		5,487	6,125
6720	800 amp		2.60	6.154		14,300	320		14,620	16,200
6740	NEMA 12, type A, 30 amp	1 Elec	5.20	1.538		1,025	80.50		1,105.50	1,250
6760	60 amp		4.90	1.633		1,100	85.50		1,185.50	1,325
6780	100 amp		3.90	2.051		1,225	107		1,332	1,475
6800	200 amp		3.10	2.581		2,100	135		2,235	2,500
6820	400 amp	2 Elec	4.40	3.636		4,925	191		5,116	5,700
6840	600 amp		3	5.333		5,375	279		5,654	6,325
6860	800 amp		2.40	6.667		14,500	350		14,850	16,400
6900	Circuit breaker, NEMA 1, type A, 30 amp	1 Elec	5.30	1.509		925	79		1,004	1,150
6920	60 amp		5	1.600		925	84		1,009	1,150
6940	100 amp		4	2		925	105		1,030	1,175
6960	225 amp		3.20	2.500		1,650	131		1,781	2,025
6980	400 amp	2 Elec	4.60	3.478		3,125	182		3,307	3,725
7000	600 amp		3.20	5		3,675	262		3,937	4,425
7020	800 amp		2.60	6.154		7,650	320		7,970	8,900
7040	NEMA 12, type A, 30 amp	1 Elec	5.20	1.538		950	80.50		1,030.50	1,175
7060	60 amp		4.90	1.633		950	85.50		1,035.50	1,175
7080	100 amp		3.90	2.051		950	107		1,057	1,200
7100	225 amp		3.10	2.581		1,675	135		1,810	2,050
7120	400 amp	2 Elec	4.40	3.636		3,200	191		3,391	3,775
7140	600 amp		3	5.333		3,725	279		4,004	4,500
7160	800 amp		2.40	6.667		7,800	350		8,150	9,100
7300	Incoming line, main lug only, 600 amp, alum, NEMA 1		1.60	10		1,175	525		1,700	2,050
7320	NEMA 12		1.50	10.667		1,200	560		1,760	2,150
7340	Copper, NEMA 1		1.60	10		1,225	525		1,750	2,125
7360	800 amp, alum., NEMA 1		1.50	10.667		2,925	560		3,485	4,050
7380	NEMA 12		1.40	11.429		2,975	600		3,575	4,175
7400	Copper, NEMA 1		1.50	10.667		3,050	560		3,610	4,175
7420	1200 amp, copper, NEMA 1		1.40	11.429		3,150	600		3,750	4,375
7440	Incoming line, fusible switch, 400 amp, alum., NEMA 1		1.20	13.333		4,025	700		4,725	5,475
7460	NEMA 12		1.10	14.545		4,100	760		4,860	5,675
7480	Copper, NEMA 1		1.20	13.333		4,100	700		4,800	5,550
7500	600 amp, alum., NEMA 1		1.10	14.545		4,925	760		5,685	6,575
7520	NEMA 12		1	16		5,000	840		5,840	6,750
7540	Copper, NEMA 1		1.10	14.545		4,975	760		5,735	6,625
7560	Incoming line, circuit breaker, 225 amp, alum., NEMA 1		1.20	13.333		2,100	700		2,800	3,350
7580	NEMA 12		1.10	14.545		2,150	760		2,910	3,500
7600	Copper, NEMA 1		1.20	13.333		2,150	700		2,850	3,425
7620	400 amp, alum., NEMA 1		1.20	13.333		3,125	700		3,825	4,500
7640	NEMA 12		1.10	14.545		3,200	760		3,960	4,650
7660	Copper, NEMA 1		1.20	13.333		3,200	700		3,900	4,575
7680	600 amp, alum., NEMA 1		1.10	14.545		3,675	760		4,435	5,175
7700	NEMA 12		1	16		3,725	840		4,565	5,325
7720	Copper, NEMA 1		1.10	14.545		3,725	760		4,485	5,250
7740	800 amp, copper, NEMA 1		.90	17.778		7,650	930		8,580	9,825
7760	Incoming line, for copper bus, add					116			116	127
7780	For 65000 amp bus bracing, add					172			172	189
7800	For NEMA 3R enclosure, add					5,400			5,400	5,925

26 24 19.20 Motor Control Center Components		Crew	Daily Output	Labor-Hours	Unit	Material	Labor	Equipment	Total	Total Incl O&P
						2013 Bare Costs				
7820	For NEMA 12 enclosure, add				Ea.	149			149	163
7840	For 1/4" x 1" ground bus, add	1 Elec	16	.500		95.50	26		121.50	144
7860	For 1/4" x 2" ground bus, add	"	12	.667		95.50	35		130.50	158
7900	Main rating basic section, alum., NEMA 1, 800 amp	2 Elec	1.40	11.429		315	600		915	1,250
7920	1200 amp	"	1.20	13.333		615	700		1,315	1,725
7940	For copper bus, add					460			460	510
7960	For 65000 amp bus bracing, add					310			310	340
7980	For NEMA 3R enclosure, add					5,400			5,400	5,925
8000	For NEMA 12, enclosure, add					149			149	163
8020	For 1/4" x 1" ground bus, add	1 Elec	16	.500		95.50	26		121.50	144
8040	For 1/4" x 2" ground bus, add		12	.667		95.50	35		130.50	158
8060	Unit devices, pilot light, standard		16	.500		82.50	26		108.50	130
8080	Pilot light, push to test		16	.500		116	26		142	166
8100	Pilot light, standard, and push button		12	.667		198	35		233	271
8120	Pilot light, push to test, and push button		12	.667		231	35		266	305
8140	Pilot light, standard, and select switch		12	.667		198	35		233	271
8160	Pilot light, push to test, and select switch		12	.667		231	35		266	305

26 24 19.30 Motor Control Center

		Crew	Daily Output	Labor-Hours	Unit	Material	Labor	Equipment	Total	Total Incl O&P
0010	**MOTOR CONTROL CENTER** Consists of starters & structures R262419-60									
0050	Starters, class 1, type B, comb. MCP, FVNR, with									
0100	control transformer, 10 HP, size 1, 12" high	1 Elec	2.70	2.963	Ea.	1,650	155		1,805	2,050
0200	25 HP, size 2, 18" high	2 Elec	4	4		1,875	210		2,085	2,400
0300	50 HP, size 3, 24" high		2	8		2,900	420		3,320	3,825
0350	75 HP, size 4, 24" high		1.60	10		3,850	525		4,375	5,025
0400	100 HP, size 4, 30" high		1.40	11.429		5,100	600		5,700	6,500
0500	200 HP, size 5, 48" high		1	16		7,600	840		8,440	9,625
0600	400 HP, size 6, 72" high		.80	20		15,600	1,050		16,650	18,800
0800	Structures, 600 amp, 22,000 rms, takes any									
0900	combination of starters up to 72" high	2 Elec	1.60	10	Ea.	2,250	525		2,775	3,250
1000	Back to back, 72" front & 66" back	"	1.20	13.333		3,025	700		3,725	4,375
1100	For copper bus add per structure					257			257	283
1200	For NEMA 12, add per structure					149			149	163
1300	For 42,000 rms, add per structure					194			194	213
1400	For 100,000 rms, size 1 & 2, add					660			660	725
1500	Size 3, add					1,075			1,075	1,175
1600	Size 4, add					860			860	945
1700	For pilot lights, add per starter	1 Elec	16	.500		116	26		142	166
1800	For push button, add per starter		16	.500		116	26		142	166
1900	For auxiliary contacts, add per starter		16	.500		194	26		220	252

26 24 19.40 Motor Starters and Controls

		Crew	Daily Output	Labor-Hours	Unit	Material	Labor	Equipment	Total	Total Incl O&P
0010	**MOTOR STARTERS AND CONTROLS** R262419-65									
0050	Magnetic, FVNR, with enclosure and heaters, 480 volt									
0080	2 HP, size 00	1 Elec	3.50	2.286	Ea.	193	120		313	390
0100	5 HP, size 0		2.30	3.478		243	182		425	540
0200	10 HP, size 1		1.60	5		273	262		535	690
0300	25 HP, size 2 R263413-33	2 Elec	2.20	7.273		515	380		895	1,125
0400	50 HP, size 3		1.80	8.889		835	465		1,300	1,625
0500	100 HP, size 4		1.20	13.333		1,850	700		2,550	3,100
0600	200 HP, size 5		.90	17.778		4,350	930		5,280	6,175
0610	400 HP, size 6		.80	20		19,500	1,050		20,550	23,000
0620	NEMA 7, 5 HP, size 0	1 Elec	1.60	5		1,450	262		1,712	2,000
0630	10 HP, size 1	"	1.10	7.273		1,500	380		1,880	2,225

26 24 19.40 Motor Starters and Controls		Crew	Daily Output	Labor-Hours	Unit	Material	Labor	Equipment	Total	Total Incl O&P
						2013 Bare Costs				
0640	25 HP, size 2	2 Elec	1.80	8.889	Ea.	2,450	465		2,915	3,375
0650	50 HP, size 3		1.20	13.333		3,675	700		4,375	5,075
0660	100 HP, size 4		.90	17.778		5,925	930		6,855	7,900
0670	200 HP, size 5		.50	32		14,100	1,675		15,775	18,000
0700	Combination, with motor circuit protectors, 5 HP, size 0	1 Elec	1.80	4.444		755	233		988	1,175
0800	10 HP, size 1	"	1.30	6.154		785	320		1,105	1,350
0900	25 HP, size 2	2 Elec	2	8		1,100	420		1,520	1,850
1000	50 HP, size 3		1.32	12.121		1,600	635		2,235	2,700
1200	100 HP, size 4		.80	20		3,475	1,050		4,525	5,375
1220	NEMA 7, 5 HP, size 0	1 Elec	1.30	6.154		2,625	320		2,945	3,350
1230	10 HP, size 1	"	1	8		2,675	420		3,095	3,575
1240	25 HP, size 2	2 Elec	1.32	12.121		3,575	635		4,210	4,875
1250	50 HP, size 3		.80	20		5,900	1,050		6,950	8,075
1260	100 HP, size 4		.60	26.667		9,200	1,400		10,600	12,200
1270	200 HP, size 5		.40	40		20,000	2,100		22,100	25,100
1400	Combination, with fused switch, 5 HP, size 0	1 Elec	1.80	4.444		580	233		813	985
1600	10 HP, size 1	"	1.30	6.154		620	320		940	1,150
1800	25 HP, size 2	2 Elec	2	8		1,000	420		1,420	1,725
2000	50 HP, size 3		1.32	12.121		1,700	635		2,335	2,825
2200	100 HP, size 4		.80	20		2,975	1,050		4,025	4,850
2610	NEMA 4, with start-stop pushbutton size 1	1 Elec	1.30	6.154		1,550	320		1,870	2,200
2620	Size 2	2 Elec	2	8		2,075	420		2,495	2,925
2630	Size 3		1.32	12.121		3,300	635		3,935	4,575
2640	Size 4		.80	20		5,075	1,050		6,125	7,150
2650	NEMA 4, FVNR, including control transformer									
2660	Size 1	2 Elec	2.60	6.154	Ea.	1,425	320		1,745	2,050
2670	Size 2		2	8		2,025	420		2,445	2,875
2680	Size 3		1.32	12.121		3,250	635		3,885	4,525
2690	Size 4		.80	20		5,025	1,050		6,075	7,100
2710	Magnetic, FVR, control circuit transformer, NEMA 1, size 1	1 Elec	1.30	6.154		845	320		1,165	1,400
2720	Size 2	2 Elec	2	8		1,350	420		1,770	2,100
2730	Size 3		1.32	12.121		2,050	635		2,685	3,200
2740	Size 4		.80	20		4,450	1,050		5,500	6,475
2760	NEMA 4, size 1	1 Elec	1.10	7.273		1,200	380		1,580	1,900
2770	Size 2	2 Elec	1.60	10		1,950	525		2,475	2,925
2780	Size 3		1.20	13.333		2,925	700		3,625	4,275
2790	Size 4		.70	22.857		6,025	1,200		7,225	8,425
2820	NEMA 12, size 1	1 Elec	1.10	7.273		995	380		1,375	1,675
2830	Size 2	2 Elec	1.60	10		1,575	525		2,100	2,500
2840	Size 3		1.20	13.333		2,475	700		3,175	3,775
2850	Size 4		.70	22.857		5,100	1,200		6,300	7,400
2870	Combination FVR, fused, w/control XFMR & PB, NEMA 1, size 1	1 Elec	1	8		1,450	420		1,870	2,225
2880	Size 2	2 Elec	1.50	10.667		2,075	560		2,635	3,100
2890	Size 3		1.10	14.545		3,075	760		3,835	4,525
2900	Size 4		.70	22.857		6,250	1,200		7,450	8,675
2910	NEMA 4, size 1	1 Elec	.90	8.889		2,150	465		2,615	3,075
2920	Size 2	2 Elec	1.40	11.429		3,150	600		3,750	4,350
2930	Size 3		1	16		4,950	840		5,790	6,700
2940	Size 4		.60	26.667		8,225	1,400		9,625	11,200
2950	NEMA 12, size 1	1 Elec	1	8		1,650	420		2,070	2,450
2960	Size 2	2 Elec	1.40	11.429		2,325	600		2,925	3,450
2970	Size 3		1	16		3,425	840		4,265	5,000
2980	Size 4		.60	26.667		6,775	1,400		8,175	9,550

26 24 19.40	Motor Starters and Controls	Crew	Daily Output	Labor-Hours	Unit	2013 Bare Costs Material	Labor	Equipment	Total	Total Incl O&P
3010	Manual, single phase, w/pilot, 1 pole 120 V NEMA 1	1 Elec	6.40	1.250	Ea.	60.50	65.50		126	165
3020	NEMA 4		4	2		340	105		445	530
3030	2 pole, 120/240 V, NEMA 1		6.40	1.250		71	65.50		136.50	176
3040	NEMA 4		4	2		269	105		374	450
3041	3 phase, 3 pole 600 V, NEMA 1		5.50	1.455		236	76		312	375
3042	NEMA 4		3.50	2.286		460	120		580	685
3043	NEMA 12		3.50	2.286		276	120		396	485
3070	Auxiliary contact, normally open					79			79	87
3500	Magnetic FVNR with NEMA 12, enclosure & heaters, 480 volt									
3600	5 HP, size 0	1 Elec	2.20	3.636	Ea.	220	191		411	525
3700	10 HP, size 1	"	1.50	5.333		330	279		609	785
3800	25 HP, size 2	2 Elec	2	8		620	420		1,040	1,300
3900	50 HP, size 3		1.60	10		955	525		1,480	1,825
4000	100 HP, size 4		1	16		2,275	840		3,115	3,750
4100	200 HP, size 5		.80	20		5,475	1,050		6,525	7,600
4200	Combination, with motor circuit protectors, 5 HP, size 0	1 Elec	1.70	4.706		730	247		977	1,175
4300	10 HP, size 1	"	1.20	6.667		755	350		1,105	1,350
4400	25 HP, size 2	2 Elec	1.80	8.889		1,125	465		1,590	1,950
4500	50 HP, size 3		1.20	13.333		1,850	700		2,550	3,075
4600	100 HP, size 4		.74	21.622		4,175	1,125		5,300	6,275
4700	Combination, with fused switch, 5 HP, size 0	1 Elec	1.70	4.706		705	247		952	1,150
4800	10 HP, size 1	"	1.20	6.667		735	350		1,085	1,325
4900	25 HP, size 2	2 Elec	1.80	8.889		1,125	465		1,590	1,925
5000	50 HP, size 3		1.20	13.333		1,800	700		2,500	3,025
5100	100 HP, size 4		.74	21.622		3,625	1,125		4,750	5,700
5200	Factory installed controls, adders to size 0 thru 5									
5300	Start-stop push button	1 Elec	32	.250	Ea.	46	13.10		59.10	70
5400	Hand-off-auto-selector switch		32	.250		46	13.10		59.10	70
5500	Pilot light		32	.250		86.50	13.10		99.60	115
5600	Start-stop-pilot		32	.250		132	13.10		145.10	166
5700	Auxiliary contact, NO or NC		32	.250		63.50	13.10		76.60	89
5800	NO-NC		32	.250		127	13.10		140.10	159
5810	Magnetic FVR, NEMA 7, w/heaters, size 1		.66	12.121		2,500	635		3,135	3,700
5830	Size 2	2 Elec	1.10	14.545		4,150	760		4,910	5,725
5840	Size 3		.70	22.857		6,675	1,200		7,875	9,150
5850	Size 4		.60	26.667		7,600	1,400		9,000	10,500
5860	Combination w/circuit breakers, heaters, control XFMR PB, size 1	1 Elec	.60	13.333		1,625	700		2,325	2,825
5870	Size 2	2 Elec	.80	20		2,075	1,050		3,125	3,850
5880	Size 3		.50	32		2,775	1,675		4,450	5,550
5890	Size 4		.40	40		5,475	2,100		7,575	9,150
5900	Manual, 240 volt, .75 HP motor	1 Elec	4	2		49	105		154	211
5910	2 HP motor		4	2		132	105		237	305
6000	Magnetic, 240 volt, 1 or 2 pole, .75 HP motor		4	2		193	105		298	370
6020	2 HP motor		4	2		213	105		318	390
6040	5 HP motor		3	2.667		305	140		445	545
6060	10 HP motor		2.30	3.478		755	182		937	1,100
6100	3 pole, .75 HP motor		3	2.667		193	140		333	420
6120	5 HP motor		2.30	3.478		262	182		444	560
6140	10 HP motor		1.60	5		490	262		752	930
6160	15 HP motor		1.60	5		490	262		752	930
6180	20 HP motor		1.10	7.273		805	380		1,185	1,450
6200	25 HP motor	2 Elec	2.20	7.273		805	380		1,185	1,450
6210	30 HP motor		1.80	8.889		805	465		1,270	1,575

26 24 19.40	Motor Starters and Controls	Crew	Daily Output	Labor-Hours	Unit	2013 Bare Costs Material	Labor	Equipment	Total	Total Incl O&P
6220	40 HP motor	2 Elec	1.80	8.889	Ea.	1,775	465		2,240	2,650
6230	50 HP motor		1.80	8.889		1,775	465		2,240	2,650
6240	60 HP motor		1.20	13.333		4,175	700		4,875	5,625
6250	75 HP motor		1.20	13.333		4,175	700		4,875	5,625
6260	100 HP motor		1.20	13.333		4,175	700		4,875	5,625
6270	125 HP motor		.90	17.778		11,700	930		12,630	14,300
6280	150 HP motor		.90	17.778		11,700	930		12,630	14,300
6290	200 HP motor		.90	17.778		11,700	930		12,630	14,300
6400	Starter & nonfused disconnect, 240 volt, 1-2 pole, .75 HP motor	1 Elec	2	4		275	210		485	615
6410	2 HP motor		2	4		295	210		505	640
6420	5 HP motor		1.80	4.444		385	233		618	775
6430	10 HP motor		1.40	5.714		865	299		1,164	1,400
6440	3 pole, .75 HP motor		1.60	5		275	262		537	690
6450	5 HP motor		1.40	5.714		345	299		644	830
6460	10 HP motor		1.10	7.273		600	380		980	1,225
6470	15 HP motor		1	8		600	420		1,020	1,275
6480	20 HP motor	2 Elec	1.50	10.667		1,050	560		1,610	1,975
6490	25 HP motor		1.50	10.667		1,050	560		1,610	1,975
6500	30 HP motor		1.30	12.308		1,050	645		1,695	2,125
6510	40 HP motor		1.24	12.903		2,250	675		2,925	3,475
6520	50 HP motor		1.12	14.286		2,250	750		3,000	3,600
6530	60 HP motor		.90	17.778		4,625	930		5,555	6,500
6540	75 HP motor		.76	21.053		4,625	1,100		5,725	6,750
6550	100 HP motor		.70	22.857		4,625	1,200		5,825	6,900
6560	125 HP motor		.60	26.667		12,800	1,400		14,200	16,200
6570	150 HP motor		.52	30.769		12,800	1,600		14,400	16,500
6580	200 HP motor		.50	32		12,800	1,675		14,475	16,600
6600	Starter & fused disconnect, 240 volt, 1-2 pole, .75 HP motor	1 Elec	2	4		295	210		505	640
6610	2 HP motor		2	4		315	210		525	660
6620	5 HP motor		1.80	4.444		405	233		638	800
6630	10 HP motor		1.40	5.714		930	299		1,229	1,475
6640	3 pole, .75 HP motor		1.60	5		295	262		557	715
6650	5 HP motor		1.40	5.714		365	299		664	850
6660	10 HP motor		1.10	7.273		665	380		1,045	1,300
6690	15 HP motor		1	8		665	420		1,085	1,350
6700	20 HP motor	2 Elec	1.60	10		1,100	525		1,625	1,975
6710	25 HP motor		1.60	10		1,100	525		1,625	1,975
6720	30 HP motor		1.40	11.429		1,100	600		1,700	2,100
6730	40 HP motor		1.20	13.333		2,425	700		3,125	3,700
6740	50 HP motor		1.20	13.333		2,425	700		3,125	3,700
6750	60 HP motor		.90	17.778		4,800	930		5,730	6,675
6760	75 HP motor		.90	17.778		4,800	930		5,730	6,675
6770	100 HP motor		.70	22.857		4,800	1,200		6,000	7,075
6780	125 HP motor		.54	29.630		13,300	1,550		14,850	16,900
6790	Combination starter & nonfusible disconnect									
6800	240 volt, 1-2 pole, .75 HP motor	1 Elec	2	4	Ea.	550	210		760	920
6810	2 HP motor		2	4		550	210		760	920
6820	5 HP motor		1.50	5.333		580	279		859	1,050
6830	10 HP motor		1.20	6.667		900	350		1,250	1,525
6840	3 pole, .75 HP motor		1.80	4.444		550	233		783	955
6850	5 HP motor		1.30	6.154		580	320		900	1,125
6860	10 HP motor		1	8		900	420		1,320	1,625
6870	15 HP motor		1	8		900	420		1,320	1,625

26 24 19.40 Motor Starters and Controls		Crew	Daily Output	Labor-Hours	Unit	Material	Labor	Equipment	Total	Total Incl O&P
						2013 Bare Costs				
6880	20 HP motor	2 Elec	1.32	12.121	Ea.	1,475	635		2,110	2,575
6890	25 HP motor		1.32	12.121		1,475	635		2,110	2,575
6900	30 HP motor		1.32	12.121		1,475	635		2,110	2,575
6910	40 HP motor		.80	20		2,825	1,050		3,875	4,700
6920	50 HP motor		.80	20		2,825	1,050		3,875	4,700
6930	60 HP motor		.70	22.857		6,325	1,200		7,525	8,775
6940	75 HP motor		.70	22.857		6,325	1,200		7,525	8,775
6950	100 HP motor		.70	22.857		6,325	1,200		7,525	8,775
6960	125 HP motor		.60	26.667		16,700	1,400		18,100	20,400
6970	150 HP motor		.60	26.667		16,700	1,400		18,100	20,400
6980	200 HP motor		.60	26.667		16,700	1,400		18,100	20,400
6990	Combination starter and fused disconnect									
7000	240 volt, 1-2 pole, .75 HP motor	1 Elec	2	4	Ea.	565	210		775	940
7010	2 HP motor		2	4		565	210		775	940
7020	5 HP motor		1.50	5.333		595	279		874	1,075
7030	10 HP motor		1.20	6.667		925	350		1,275	1,550
7040	3 pole, .75 HP motor		1.80	4.444		565	233		798	975
7050	5 HP motor		1.30	6.154		595	320		915	1,125
7060	10 HP motor		1	8		925	420		1,345	1,650
7070	15 HP motor		1	8		925	420		1,345	1,650
7080	20 HP motor	2 Elec	1.32	12.121		1,550	635		2,185	2,650
7090	25 HP motor		1.32	12.121		1,550	635		2,185	2,650
7100	30 HP motor		1.32	12.121		1,550	635		2,185	2,650
7110	40 HP motor		.80	20		2,925	1,050		3,975	4,800
7120	50 HP motor		.80	20		2,925	1,050		3,975	4,800
7130	60 HP motor		.80	20		6,550	1,050		7,600	8,775
7140	75 HP motor		.70	22.857		6,550	1,200		7,750	9,000
7150	100 HP motor		.70	22.857		6,550	1,200		7,750	9,000
7160	125 HP motor		.70	22.857		17,300	1,200		18,500	20,800
7170	150 HP motor		.60	26.667		17,300	1,400		18,700	21,100
7180	200 HP motor		.60	26.667		17,300	1,400		18,700	21,100
7190	Combination starter & circuit breaker disconnect									
7200	240 volt, 1-2 pole, .75 HP motor	1 Elec	2	4	Ea.	575	210		785	945
7210	2 HP motor		2	4		575	210		785	945
7220	5 HP motor		1.50	5.333		600	279		879	1,075
7230	10 HP motor		1.20	6.667		920	350		1,270	1,525
7240	3 pole, .75 HP motor		1.80	4.444		590	233		823	1,000
7250	5 HP motor		1.30	6.154		620	320		940	1,150
7260	10 HP motor		1	8		935	420		1,355	1,650
7270	15 HP motor		1	8		935	420		1,355	1,650
7280	20 HP motor	2 Elec	1.32	12.121		1,600	635		2,235	2,700
7290	25 HP motor		1.32	12.121		1,600	635		2,235	2,700
7300	30 HP motor		1.32	12.121		1,600	635		2,235	2,700
7310	40 HP motor		.80	20		3,475	1,050		4,525	5,375
7320	50 HP motor		.80	20		3,475	1,050		4,525	5,375
7330	60 HP motor		.80	20		8,000	1,050		9,050	10,400
7340	75 HP motor		.70	22.857		8,000	1,200		9,200	10,600
7350	100 HP motor		.70	22.857		8,000	1,200		9,200	10,600
7360	125 HP motor		.70	22.857		17,300	1,200		18,500	20,900
7370	150 HP motor		.60	26.667		17,300	1,400		18,700	21,200
7380	200 HP motor		.60	26.667		17,300	1,400		18,700	21,200
7400	Magnetic FVNR with enclosure & heaters, 2 pole,									
7410	230 volt, 1 HP size 00	1 Elec	4	2	Ea.	176	105		281	350

26 24 Switchboards and Panelboards

26 24 19 – Motor-Control Centers

26 24 19.40 Motor Starters and Controls		Crew	Daily Output	Labor-Hours	Unit	Material	Labor	Equipment	Total	Total Incl O&P
						2013 Bare Costs				
7420	2 HP, size 0	1 Elec	4	2	Ea.	196	105		301	370
7430	3 HP, size 1		3	2.667		225	140		365	455
7440	5 HP, size 1p		3	2.667		288	140		428	525
7450	115 volt, 1/3 HP, size 00		4	2		176	105		281	350
7460	1 HP, size 0		4	2		196	105		301	370
7470	2 HP, size 1		3	2.667		225	140		365	455
7480	3 HP, size 1P		3	2.667		279	140		419	515
7500	3 pole, 480 volt, 600 HP, size 7	2 Elec	.70	22.857		15,500	1,200		16,700	18,900
7590	Magnetic FVNR with heater, NEMA 1									
7600	600 volt, 3 pole, 5 HP motor	1 Elec	2.30	3.478	Ea.	233	182		415	530
7610	10 HP motor	"	1.60	5		262	262		524	680
7620	25 HP motor	2 Elec	2.20	7.273		490	380		870	1,100
7630	30 HP motor		1.80	8.889		805	465		1,270	1,575
7640	40 HP motor		1.80	8.889		805	465		1,270	1,575
7650	50 HP motor		1.80	8.889		805	465		1,270	1,575
7660	60 HP motor		1.20	13.333		1,775	700		2,475	3,000
7670	75 HP motor		1.20	13.333		1,775	700		2,475	3,000
7680	100 HP motor		1.20	13.333		2,725	700		3,425	4,025
7690	125 HP motor		.90	17.778		4,175	930		5,105	5,975
7700	150 HP motor		.90	17.778		4,175	930		5,105	5,975
7710	200 HP motor		.90	17.778		4,175	930		5,105	5,975
7750	Starter & nonfused disconnect, 600 volt, 3 pole, 5 HP motor	1 Elec	1.40	5.714		375	299		674	865
7760	10 HP motor	"	1.10	7.273		405	380		785	1,025
7770	25 HP motor	2 Elec	1.50	10.667		635	560		1,195	1,525
7780	30 HP motor		1.30	12.308		1,075	645		1,720	2,150
7790	40 HP motor		1.30	12.308		1,075	645		1,720	2,150
7800	50 HP motor		1.30	12.308		1,075	645		1,720	2,150
7810	60 HP motor		.92	17.391		2,200	910		3,110	3,800
7820	75 HP motor		.92	17.391		2,200	910		3,110	3,800
7830	100 HP motor		.84	19.048		3,125	1,000		4,125	4,925
7840	125 HP motor		.70	22.857		4,800	1,200		6,000	7,075
7850	150 HP motor		.70	22.857		4,800	1,200		6,000	7,075
7860	200 HP motor		.60	26.667		4,800	1,400		6,200	7,375
7870	Starter & fused disconnect, 600 volt, 3 pole, 5 HP motor	1 Elec	1.40	5.714		510	299		809	1,000
7880	10 HP motor	"	1.10	7.273		540	380		920	1,175
7890	25 HP motor	2 Elec	1.50	10.667		770	560		1,330	1,675
7900	30 HP motor		1.30	12.308		1,150	645		1,795	2,225
7910	40 HP motor		1.30	12.308		1,150	645		1,795	2,225
7920	50 HP motor		1.30	12.308		1,150	645		1,795	2,225
7930	60 HP motor		.92	17.391		2,400	910		3,310	4,000
7940	75 HP motor		.92	17.391		2,400	910		3,310	4,000
7950	100 HP motor		.84	19.048		3,325	1,000		4,325	5,150
7960	125 HP motor		.70	22.857		5,050	1,200		6,250	7,350
7970	150 HP motor		.70	22.857		5,050	1,200		6,250	7,350
7980	200 HP motor		.60	26.667		5,050	1,400		6,450	7,650
7990	Combination starter and nonfusible disconnect									
8000	600 volt, 3 pole, 5 HP motor	1 Elec	1.80	4.444	Ea.	575	233		808	980
8010	10 HP motor	"	1.30	6.154		605	320		925	1,150
8020	25 HP motor	2 Elec	2	8		940	420		1,360	1,650
8030	30 HP motor		1.32	12.121		1,550	635		2,185	2,650
8040	40 HP motor		1.32	12.121		1,550	635		2,185	2,650
8050	50 HP motor		1.32	12.121		2,050	635		2,685	3,200
8060	60 HP motor		.80	20		2,925	1,050		3,975	4,800

26 24 19 – Motor-Control Centers

26 24 19.40	Motor Starters and Controls	Crew	Daily Output	Labor-Hours	Unit	2013 Bare Costs Material	Labor	Equipment	Total	Total Incl O&P
8070	75 HP motor	2 Elec	.80	20	Ea.	2,925	1,050		3,975	4,800
8080	100 HP motor		.80	20		4,325	1,050		5,375	6,325
8090	125 HP motor		.70	22.857		6,550	1,200		7,750	9,000
8100	150 HP motor		.70	22.857		6,550	1,200		7,750	9,000
8110	200 HP motor		.70	22.857		7,600	1,200		8,800	10,200
8140	Combination starter and fused disconnect									
8150	600 volt, 3 pole, 5 HP motor	1 Elec	1.80	4.444	Ea.	565	233		798	975
8160	10 HP motor	"	1.30	6.154		620	320		940	1,150
8170	25 HP motor	2 Elec	2	8		1,000	420		1,420	1,725
8180	30 HP motor		1.32	12.121		1,575	635		2,210	2,675
8190	40 HP motor		1.32	12.121		1,700	635		2,335	2,800
8200	50 HP motor		1.32	12.121		1,700	635		2,335	2,800
8210	60 HP motor		.80	20		2,950	1,050		4,000	4,825
8220	75 HP motor		.80	20		2,950	1,050		4,000	4,825
8230	100 HP motor		.80	20		2,950	1,050		4,000	4,825
8240	125 HP motor		.70	22.857		6,550	1,200		7,750	9,000
8250	150 HP motor		.70	22.857		6,550	1,200		7,750	9,000
8260	200 HP motor		.70	22.857		6,550	1,200		7,750	9,000
8290	Combination starter & circuit breaker disconnect									
8300	600 volt, 3 pole, 5 HP motor	1 Elec	1.80	4.444	Ea.	755	233		988	1,175
8310	10 HP motor	"	1.30	6.154		785	320		1,105	1,350
8320	25 HP motor	2 Elec	2	8		1,100	420		1,520	1,850
8330	30 HP motor		1.32	12.121		1,600	635		2,235	2,700
8340	40 HP motor		1.32	12.121		1,600	635		2,235	2,700
8350	50 HP motor		1.32	12.121		1,600	635		2,235	2,700
8360	60 HP motor		.80	20		3,475	1,050		4,525	5,375
8370	75 HP motor		.80	20		3,475	1,050		4,525	5,375
8380	100 HP motor		.80	20		3,475	1,050		4,525	5,375
8390	125 HP motor		.70	22.857		8,000	1,200		9,200	10,600
8400	150 HP motor		.70	22.857		8,000	1,200		9,200	10,600
8410	200 HP motor		.70	22.857		8,000	1,200		9,200	10,600
8430	Starter & circuit breaker disconnect									
8440	600 volt, 3 pole, 5 HP motor	1 Elec	1.40	5.714	Ea.	710	299		1,009	1,225
8450	10 HP motor	"	1.10	7.273		740	380		1,120	1,375
8460	25 HP motor	2 Elec	1.50	10.667		970	560		1,530	1,900
8470	30 HP motor		1.30	12.308		1,400	645		2,045	2,500
8480	40 HP motor		1.30	12.308		1,400	645		2,045	2,500
8490	50 HP motor		1.30	12.308		1,400	645		2,045	2,500
8500	60 HP motor		.92	17.391		3,325	910		4,235	5,025
8510	75 HP motor		.92	17.391		3,325	910		4,235	5,025
8520	100 HP motor		.84	19.048		4,250	1,000		5,250	6,175
8530	125 HP motor		.70	22.857		5,725	1,200		6,925	8,100
8540	150 HP motor		.70	22.857		5,725	1,200		6,925	8,100
8550	200 HP motor		.60	26.667		6,800	1,400		8,200	9,600
8900	240 volt, 1-2 pole, .75 HP motor	1 Elec	2	4		670	210		880	1,050
8910	2 HP motor		2	4		690	210		900	1,075
8920	5 HP motor		1.80	4.444		780	233		1,013	1,200
8930	10 HP motor		1.40	5.714		1,350	299		1,649	1,925
8950	3 pole, .75 HP motor		1.60	5		670	262		932	1,125
8970	5 HP motor		1.40	5.714		740	299		1,039	1,250
8980	10 HP motor		1.10	7.273		1,075	380		1,455	1,750
8990	15 HP motor		1	8		1,075	420		1,495	1,800
9100	20 HP motor	2 Elec	1.50	10.667		1,475	560		2,035	2,450

26 24 Switchboards and Panelboards

26 24 19 – Motor-Control Centers

26 24 19.40 Motor Starters and Controls		Crew	Daily Output	Labor-Hours	Unit	Material	Labor	Equipment	Total	Total Incl O&P
						2013 Bare Costs				
9110	25 HP motor	2 Elec	1.50	10.667	Ea.	1,475	560		2,035	2,450
9120	30 HP motor		1.30	12.308		1,475	645		2,120	2,600
9130	40 HP motor		1.24	12.903		3,325	675		4,000	4,650
9140	50 HP motor		1.12	14.286		3,325	750		4,075	4,775
9150	60 HP motor		.90	17.778		6,800	930		7,730	8,900
9160	75 HP motor		.76	21.053		6,800	1,100		7,900	9,150
9170	100 HP motor		.70	22.857		6,800	1,200		8,000	9,300
9180	125 HP motor		.60	26.667		15,500	1,400		16,900	19,200
9190	150 HP motor		.52	30.769		15,500	1,600		17,100	19,500
9200	200 HP motor		.50	32		15,500	1,675		17,175	19,600

26 25 Enclosed Bus Assemblies

26 25 13 – Bus Duct/Busway and Fittings

26 25 13.10 Aluminum Bus Duct

		Crew	Daily Output	Labor-Hours	Unit	Material	Labor	Equipment	Total	Total Incl O&P
0010	**ALUMINUM BUS DUCT** 10 ft. long R262513-10									
0050	Indoor 3 pole 4 wire, plug-in, straight section, 225 amp	2 Elec	44	.364	L.F.	109	19.05		128.05	148
0100	400 amp		36	.444		128	23.50		151.50	176
0150	600 amp		32	.500		144	26		170	197
0200	800 amp R262513-15		26	.615		162	32.50		194.50	228
0250	1000 amp		24	.667		186	35		221	258
0300	1350 amp		22	.727		239	38		277	320
0310	1600 amp		18	.889		254	46.50		300.50	350
0320	2000 amp		16	1		335	52.50		387.50	445
0330	2500 amp		14	1.143		380	60		440	510
0340	3000 amp		12	1.333		460	70		530	615
0350	Feeder, 600 amp		34	.471		104	24.50		128.50	152
0400	800 amp		28	.571		121	30		151	179
0450	1000 amp		26	.615		139	32.50		171.50	202
0455	1200 amp		25	.640		139	33.50		172.50	203
0500	1350 amp		24	.667		217	35		252	292
0550	1600 amp		20	.800		252	42		294	340
0600	2000 amp		18	.889		295	46.50		341.50	395
0620	2500 amp		14	1.143		380	60		440	510
0630	3000 amp		12	1.333		435	70		505	580
0640	4000 amp		10	1.600		625	84		709	810
0650	Elbow, 225 amp		4.40	3.636	Ea.	615	191		806	960
0700	400 amp		3.80	4.211		615	221		836	1,000
0750	600 amp		3.40	4.706		615	247		862	1,050
0800	800 amp		3	5.333		640	279		919	1,125
0850	1000 amp		2.80	5.714		830	299		1,129	1,375
0870	1200 amp		2.70	5.926		1,025	310		1,335	1,600
0900	1350 amp		2.60	6.154		1,100	320		1,420	1,700
0950	1600 amp		2.40	6.667		1,300	350		1,650	1,975
1000	2000 amp		2	8		1,450	420		1,870	2,200
1020	2500 amp		1.80	8.889		1,700	465		2,165	2,575
1030	3000 amp		1.60	10		1,950	525		2,475	2,925
1040	4000 amp		1.40	11.429		3,175	600		3,775	4,375
1100	Cable tap box end, 225 amp		3.60	4.444		860	233		1,093	1,300
1150	400 amp		3.20	5		860	262		1,122	1,325
1200	600 amp		2.60	6.154		860	320		1,180	1,425
1250	800 amp		2.20	7.273		935	380		1,315	1,600

26 25 Enclosed Bus Assemblies

26 25 13 – Bus Duct/Busway and Fittings

26 25 13.10 Aluminum Bus Duct		Crew	Daily Output	Labor-Hours	Unit	Material	Labor	Equipment	Total	Total Incl O&P
						2013 Bare Costs				
1300	1000 amp	2 Elec	2	8	Ea.	1,025	420		1,445	1,750
1320	1200 amp		2	8		1,225	420		1,645	1,975
1350	1350 amp		1.60	10		1,325	525		1,850	2,250
1400	1600 amp		1.40	11.429		1,500	600		2,100	2,550
1450	2000 amp		1.20	13.333		1,725	700		2,425	2,950
1460	2500 amp		1	16		2,075	840		2,915	3,550
1470	3000 amp		.80	20		2,300	1,050		3,350	4,125
1480	4000 amp		.60	26.667		3,075	1,400		4,475	5,475
1500	Switchboard stub, 225 amp		5.80	2.759		755	145		900	1,050
1550	400 amp		5.40	2.963		755	155		910	1,075
1600	600 amp		4.60	3.478		755	182		937	1,100
1650	800 amp		4	4		860	210		1,070	1,250
1700	1000 amp		3.20	5		960	262		1,222	1,450
1720	1200 amp		3.10	5.161		1,050	270		1,320	1,575
1750	1350 amp		3	5.333		1,175	279		1,454	1,725
1800	1600 amp		2.60	6.154		1,325	320		1,645	1,925
1850	2000 amp		2.40	6.667		1,500	350		1,850	2,175
1860	2500 amp		2.20	7.273		1,800	380		2,180	2,550
1870	3000 amp		2	8		2,100	420		2,520	2,925
1880	4000 amp		1.80	8.889		2,625	465		3,090	3,575
1890	Tee fittings, 225 amp		3.20	5		820	262		1,082	1,300
1900	400 amp		2.80	5.714		820	299		1,119	1,350
1950	600 amp		2.60	6.154		820	320		1,140	1,375
2000	800 amp		2.40	6.667		875	350		1,225	1,475
2050	1000 amp		2.20	7.273		925	380		1,305	1,600
2070	1200 amp		2.10	7.619		1,075	400		1,475	1,800
2100	1350 amp		2	8		1,500	420		1,920	2,275
2150	1600 amp		1.60	10		1,800	525		2,325	2,750
2200	2000 amp		1.20	13.333		2,000	700		2,700	3,250
2220	2500 amp		1	16		2,400	840		3,240	3,875
2230	3000 amp		.80	20		2,725	1,050		3,775	4,575
2240	4000 amp		.60	26.667		4,525	1,400		5,925	7,100
2300	Wall flange, 600 amp		20	.800		172	42		214	252
2310	800 amp		16	1		172	52.50		224.50	268
2320	1000 amp		13	1.231		172	64.50		236.50	286
2325	1200 amp		12	1.333		172	70		242	294
2330	1350 amp		10.80	1.481		172	77.50		249.50	305
2340	1600 amp		9	1.778		172	93		265	330
2350	2000 amp		8	2		172	105		277	345
2360	2500 amp		6.60	2.424		172	127		299	380
2370	3000 amp		5.40	2.963		172	155		327	420
2380	4000 amp		4	4		249	210		459	590
2390	5000 amp		3	5.333		249	279		528	695
2400	Vapor barrier		8	2		315	105		420	500
2420	Roof flange kit		4	4		560	210		770	930
2600	Expansion fitting, 225 amp		10	1.600		1,000	84		1,084	1,225
2610	400 amp		8	2		1,000	105		1,105	1,250
2620	600 amp		6	2.667		1,000	140		1,140	1,300
2630	800 amp		4.60	3.478		1,175	182		1,357	1,575
2640	1000 amp		4	4		1,300	210		1,510	1,775
2650	1350 amp		3.60	4.444		1,925	233		2,158	2,450
2660	1600 amp		3.20	5		2,300	262		2,562	2,925
2670	2000 amp		2.80	5.714		2,550	299		2,849	3,250

26 25 13 – Bus Duct/Busway and Fittings

26 25 13.10 Aluminum Bus Duct		Crew	Daily Output	Labor-Hours	Unit	Material	Labor	Equipment	Total	Total Incl O&P
						2013 Bare Costs				
2680	2500 amp	2 Elec	2.40	6.667	Ea.	3,075	350		3,425	3,925
2690	3000 amp		2	8		3,550	420		3,970	4,525
2700	4000 amp		1.60	10		4,675	525		5,200	5,925
2800	Reducer nonfused, 400 amp		8	2		825	105		930	1,075
2810	600 amp		6	2.667		825	140		965	1,125
2820	800 amp		4.60	3.478		995	182		1,177	1,375
2830	1000 amp		4	4		1,150	210		1,360	1,600
2840	1350 amp		3.60	4.444		1,525	233		1,758	2,025
2850	1600 amp		3.20	5		2,050	262		2,312	2,650
2860	2000 amp		2.80	5.714		2,350	299		2,649	3,050
2870	2500 amp		2.40	6.667		2,975	350		3,325	3,775
2880	3000 amp		2	8		3,425	420		3,845	4,400
2890	4000 amp		1.60	10		4,575	525		5,100	5,825
2950	Reducer fuse included, 225 amp		4.40	3.636		2,425	191		2,616	2,950
2960	400 amp		4.20	3.810		2,475	200		2,675	3,000
2970	600 amp		3.60	4.444		2,900	233		3,133	3,550
2980	800 amp		3.20	5		4,625	262		4,887	5,500
2990	1000 amp		3	5.333		5,300	279		5,579	6,250
3000	1200 amp		2.80	5.714		5,300	299		5,599	6,275
3010	1600 amp		2.20	7.273		12,100	380		12,480	13,900
3020	2000 amp		1.80	8.889		13,400	465		13,865	15,400
3100	Reducer circuit breaker, 225 amp		4.40	3.636		2,375	191		2,566	2,900
3110	400 amp		4.20	3.810		2,900	200		3,100	3,500
3120	600 amp		3.60	4.444		4,125	233		4,358	4,900
3130	800 amp		3.20	5		4,850	262		5,112	5,725
3140	1000 amp		3	5.333		5,500	279		5,779	6,475
3150	1200 amp		2.80	5.714		6,600	299		6,899	7,700
3160	1600 amp		2.20	7.273		9,750	380		10,130	11,300
3170	2000 amp		1.80	8.889		10,700	465		11,165	12,500
3250	Reducer circuit breaker, 75,000 AIC, 225 amp		4.40	3.636		3,725	191		3,916	4,375
3260	400 amp		4.20	3.810		3,725	200		3,925	4,400
3270	600 amp		3.60	4.444		5,000	233		5,233	5,850
3280	800 amp		3.20	5		5,475	262		5,737	6,425
3290	1000 amp		3	5.333		8,825	279		9,104	10,100
3300	1200 amp		2.80	5.714		8,825	299		9,124	10,200
3310	1600 amp		2.20	7.273		9,750	380		10,130	11,300
3320	2000 amp		1.80	8.889		10,700	465		11,165	12,500
3400	Reducer circuit breaker CLF 225 amp		4.40	3.636		3,850	191		4,041	4,500
3410	400 amp		4.20	3.810		4,550	200		4,750	5,325
3420	600 amp		3.60	4.444		6,825	233		7,058	7,850
3430	800 amp		3.20	5		7,125	262		7,387	8,225
3440	1000 amp		3	5.333		7,425	279		7,704	8,600
3450	1200 amp		2.80	5.714		9,700	299		9,999	11,200
3460	1600 amp		2.20	7.273		9,750	380		10,130	11,300
3470	2000 amp		1.80	8.889		10,700	465		11,165	12,500
3550	Ground bus added to bus duct, 225 amp		320	.050	L.F.	31	2.62		33.62	38
3560	400 amp		320	.050		31	2.62		33.62	38
3570	600 amp		280	.057		31	2.99		33.99	38.50
3580	800 amp		240	.067		31	3.49		34.49	39.50
3590	1000 amp		200	.080		31	4.19		35.19	40.50
3600	1350 amp		180	.089		31	4.66		35.66	41
3610	1600 amp		160	.100		31	5.25		36.25	42
3620	2000 amp		160	.100		31	5.25		36.25	42

26 25 13.10 Aluminum Bus Duct		Crew	Daily Output	Labor-Hours	Unit	Material	2013 Bare Costs Labor	Equipment	Total	Total Incl O&P
3630	2500 amp	2 Elec	140	.114	L.F.	31	6		37	43
3640	3000 amp		120	.133		31	7		38	44.50
3650	4000 amp		100	.160		31	8.40		39.40	46.50
3810	High short circuit, 400 amp		36	.444		104	23.50		127.50	150
3820	600 amp		32	.500		104	26		130	154
3830	800 amp		26	.615		121	32.50		153.50	183
3840	1000 amp		24	.667		130	35		165	196
3850	1350 amp		22	.727		173	38		211	248
3860	1600 amp		18	.889		200	46.50		246.50	289
3870	2000 amp		16	1		234	52.50		286.50	335
3880	2500 amp		14	1.143		390	60		450	520
3890	3000 amp		12	1.333	↓	440	70		510	590
3920	Cross, 225 amp		5.60	2.857	Ea.	1,225	150		1,375	1,575
3930	400 amp		4.60	3.478		1,225	182		1,407	1,625
3940	600 amp		4	4		1,225	210		1,435	1,675
3950	800 amp		3.40	4.706		1,300	247		1,547	1,800
3960	1000 amp		3	5.333		1,375	279		1,654	1,925
3970	1350 amp		2.80	5.714		2,200	299		2,499	2,875
3980	1600 amp		2.20	7.273		2,625	380		3,005	3,450
3990	2000 amp		1.80	8.889		2,875	465		3,340	3,875
4000	2500 amp		1.60	10		3,400	525		3,925	4,525
4010	3000 amp		1.20	13.333		3,925	700		4,625	5,375
4020	4000 amp		1	16		6,000	840		6,840	7,850
4040	Cable tap box center, 225 amp		3.60	4.444		860	233		1,093	1,300
4050	400 amp		3.20	5		860	262		1,122	1,325
4060	600 amp		2.60	6.154		860	320		1,180	1,425
4070	800 amp		2.20	7.273		935	380		1,315	1,600
4080	1000 amp		2	8		1,025	420		1,445	1,750
4090	1350 amp		1.60	10		1,325	525		1,850	2,250
4100	1600 amp		1.40	11.429		1,500	600		2,100	2,550
4110	2000 amp		1.20	13.333		1,725	700		2,425	2,950
4120	2500 amp		1	16		2,075	840		2,915	3,550
4130	3000 amp		.80	20		2,300	1,050		3,350	4,125
4140	4000 amp		.60	26.667	↓	3,075	1,400		4,475	5,475
4500	Weatherproof 3 pole 4 wire, feeder, 600 amp		30	.533	L.F.	125	28		153	179
4520	800 amp		24	.667		146	35		181	213
4540	1000 amp		22	.727		166	38		204	240
4550	1200 amp		21	.762		230	40		270	315
4560	1350 amp		20	.800		260	42		302	350
4580	1600 amp		17	.941		300	49.50		349.50	405
4600	2000 amp		16	1		355	52.50		407.50	470
4620	2500 amp		12	1.333		460	70		530	610
4640	3000 amp		10	1.600		520	84		604	700
4660	4000 amp		8	2		750	105		855	980
5000	Indoor 3 pole, 3 wire, feeder, 600 amp		40	.400		95.50	21		116.50	137
5010	800 amp		32	.500		113	26		139	163
5020	1000 amp		30	.533		121	28		149	176
5025	1200 amp		29	.552		121	29		150	178
5030	1350 amp		28	.571		165	30		195	226
5040	1600 amp		24	.667		191	35		226	263
5050	2000 amp		20	.800		226	42		268	310
5060	2500 amp		16	1		310	52.50		362.50	425
5070	3000 amp		14	1.143		375	60		435	500

26 25 13 – Bus Duct/Busway and Fittings

26 25 13.10 Aluminum Bus Duct		Crew	Daily Output	Labor-Hours	Unit	Material	Labor	Equipment	Total	Total Incl O&P
						2013 Bare Costs				
5080	4000 amp	2 Elec	12	1.333	L.F.	450	70		520	600
5200	Plug-in type, 225 amp		50	.320		104	16.75		120.75	140
5210	400 amp		42	.381		104	19.95		123.95	145
5220	600 amp		36	.444		104	23.50		127.50	150
5230	800 amp		30	.533		121	28		149	176
5240	1000 amp		28	.571		130	30		160	188
5245	1200 amp		27	.593		147	31		178	209
5250	1350 amp		26	.615		173	32.50		205.50	240
5260	1600 amp		20	.800		200	42		242	282
5270	2000 amp		18	.889		234	46.50		280.50	330
5280	2500 amp		16	1		320	52.50		372.50	435
5290	3000 amp		14	1.143		380	60		440	510
5300	4000 amp		12	1.333		460	70		530	610
5330	High short circuit, 400 amp		42	.381		104	19.95		123.95	145
5340	600 amp		36	.444		104	23.50		127.50	150
5350	800 amp		30	.533		121	28		149	176
5360	1000 amp		28	.571		130	30		160	188
5370	1350 amp		26	.615		173	32.50		205.50	240
5380	1600 amp		20	.800		200	42		242	282
5390	2000 amp		18	.889		234	46.50		280.50	330
5400	2500 amp		16	1		370	52.50		422.50	485
5410	3000 amp		14	1.143		380	60		440	510
5440	Elbow, 225 amp		5	3.200	Ea.	530	168		698	835
5450	400 amp		4.40	3.636		530	191		721	870
5460	600 amp		4	4		530	210		740	900
5470	800 amp		3.40	4.706		565	247		812	990
5480	1000 amp		3.20	5		580	262		842	1,025
5485	1200 amp		3.10	5.161		615	270		885	1,075
5490	1350 amp		3	5.333		670	279		949	1,150
5500	1600 amp		2.80	5.714		1,025	299		1,324	1,575
5510	2000 amp		2.40	6.667		1,125	350		1,475	1,775
5520	2500 amp		2	8		1,400	420		1,820	2,150
5530	3000 amp		1.80	8.889		1,650	465		2,115	2,525
5540	4000 amp		1.60	10		2,200	525		2,725	3,200
5560	Tee fittings, 225 amp		3.60	4.444		735	233		968	1,150
5570	400 amp		3.20	5		735	262		997	1,200
5580	600 amp		3	5.333		735	279		1,014	1,225
5590	800 amp		2.80	5.714		790	299		1,089	1,325
5600	1000 amp		2.60	6.154		815	320		1,135	1,375
5605	1200 amp		2.50	6.400		870	335		1,205	1,450
5610	1350 amp		2.40	6.667		1,200	350		1,550	1,850
5620	1600 amp		1.80	8.889		1,425	465		1,890	2,250
5630	2000 amp		1.40	11.429		1,575	600		2,175	2,650
5640	2500 amp		1.20	13.333		1,975	700		2,675	3,225
5650	3000 amp		1	16		2,325	840		3,165	3,800
5660	4000 amp		.70	22.857		3,400	1,200		4,600	5,525
5680	Cross, 225 amp		6.40	2.500		1,200	131		1,331	1,500
5690	400 amp		5.40	2.963		1,200	155		1,355	1,525
5700	600 amp		4.60	3.478		1,200	182		1,382	1,575
5710	800 amp		4	4		1,250	210		1,460	1,725
5720	1000 amp		3.60	4.444		1,300	233		1,533	1,775
5730	1350 amp		3.20	5		1,925	262		2,187	2,525
5740	1600 amp		2.60	6.154		2,250	320		2,570	2,950

26 25 13.10 Aluminum Bus Duct		Crew	Daily Output	Labor-Hours	Unit	2013 Bare Costs Material	Labor	Equipment	Total	Total Incl O&P
5750	2000 amp	2 Elec	2.20	7.273	Ea.	2,475	380		2,855	3,275
5760	2500 amp		1.80	8.889		2,975	465		3,440	3,975
5770	3000 amp		1.40	11.429		3,550	600		4,150	4,800
5780	4000 amp		1.20	13.333		4,950	700		5,650	6,500
5800	Expansion fitting, 225 amp		11.60	1.379		865	72.50		937.50	1,050
5810	400 amp		9.20	1.739		865	91		956	1,075
5820	600 amp		7	2.286		865	120		985	1,125
5830	800 amp		5.20	3.077		1,025	161		1,186	1,375
5840	1000 amp		4.60	3.478		1,125	182		1,307	1,500
5850	1350 amp		4.20	3.810		1,350	200		1,550	1,800
5860	1600 amp		3.60	4.444		1,675	233		1,908	2,200
5870	2000 amp		3.20	5		2,000	262		2,262	2,600
5880	2500 amp		2.80	5.714		2,250	299		2,549	2,925
5890	3000 amp		2.40	6.667		2,650	350		3,000	3,425
5900	4000 amp		1.80	8.889		3,625	465		4,090	4,675
5940	Reducer, nonfused, 400 amp		9.20	1.739		710	91		801	915
5950	600 amp		7	2.286		710	120		830	960
5960	800 amp		5.20	3.077		745	161		906	1,050
5970	1000 amp		4.60	3.478		915	182		1,097	1,275
5980	1350 amp		4.20	3.810		1,350	200		1,550	1,800
5990	1600 amp		3.60	4.444		1,550	233		1,783	2,050
6000	2000 amp		3.20	5		1,800	262		2,062	2,400
6010	2500 amp		2.80	5.714		2,300	299		2,599	3,000
6020	3000 amp		2.20	7.273		2,700	380		3,080	3,550
6030	4000 amp		1.80	8.889		3,700	465		4,165	4,775
6050	Reducer, fuse included, 225 amp		5	3.200		1,850	168		2,018	2,300
6060	400 amp		4.80	3.333		2,475	175		2,650	2,950
6070	600 amp		4.20	3.810		3,150	200		3,350	3,775
6080	800 amp		3.60	4.444		4,900	233		5,133	5,725
6090	1000 amp		3.40	4.706		5,325	247		5,572	6,250
6100	1350 amp		3.20	5		9,750	262		10,012	11,100
6110	1600 amp		2.60	6.154		11,600	320		11,920	13,200
6120	2000 amp		2	8		13,400	420		13,820	15,300
6160	Reducer, circuit breaker, 225 amp		5	3.200		2,275	168		2,443	2,775
6170	400 amp		4.80	3.333		2,800	175		2,975	3,325
6180	600 amp		4.20	3.810		4,025	200		4,225	4,725
6190	800 amp		3.60	4.444		4,700	233		4,933	5,525
6200	1000 amp		3.40	4.706		5,325	247		5,572	6,250
6210	1350 amp		3.20	5		6,450	262		6,712	7,500
6220	1600 amp		2.60	6.154		9,625	320		9,945	11,100
6230	2000 amp		2	8		10,500	420		10,920	12,200
6270	Cable tap box center, 225 amp		4.20	3.810		830	200		1,030	1,225
6280	400 amp		3.60	4.444		830	233		1,063	1,275
6290	600 amp		3	5.333		830	279		1,109	1,325
6300	800 amp		2.60	6.154		905	320		1,225	1,475
6310	1000 amp		2.40	6.667		960	350		1,310	1,575
6320	1350 amp		1.80	8.889		1,150	465		1,615	1,975
6330	1600 amp		1.60	10		1,300	525		1,825	2,200
6340	2000 amp		1.40	11.429		1,500	600		2,100	2,550
6350	2500 amp		1.20	13.333		1,850	700		2,550	3,100
6360	3000 amp		1	16		2,100	840		2,940	3,575
6370	4000 amp		.70	22.857		2,500	1,200		3,700	4,550
6390	Cable tap box end, 225 amp		4.20	3.810		505	200		705	860

26 25 Enclosed Bus Assemblies

26 25 13 – Bus Duct/Busway and Fittings

26 25 13.10 Aluminum Bus Duct		Crew	Daily Output	Labor-Hours	Unit	2013 Bare Costs Material	Labor	Equipment	Total	Total Incl O&P
6400	400 amp	2 Elec	3.60	4.444	Ea.	505	233		738	910
6410	600 amp		3	5.333		505	279		784	980
6420	800 amp		2.60	6.154		555	320		875	1,100
6430	1000 amp		2.40	6.667		600	350		950	1,175
6435	1200 amp		2.10	7.619		650	400		1,050	1,300
6440	1350 amp		1.80	8.889		720	465		1,185	1,500
6450	1600 amp		1.60	10		820	525		1,345	1,675
6460	2000 amp		1.40	11.429		925	600		1,525	1,925
6470	2500 amp		1.20	13.333		1,075	700		1,775	2,225
6480	3000 amp		1	16		1,275	840		2,115	2,650
6490	4000 amp		.70	22.857		1,525	1,200		2,725	3,475
7000	Weatherproof 3 pole 3 wire, feeder, 600 amp		34	.471	L.F.	123	24.50		147.50	172
7020	800 amp		28	.571		136	30		166	194
7040	1000 amp		26	.615		146	32.50		178.50	209
7050	1200 amp		25	.640		170	33.50		203.50	237
7060	1350 amp		24	.667		198	35		233	271
7080	1600 amp		20	.800		229	42		271	315
7100	2000 amp		18	.889		270	46.50		316.50	370
7120	2500 amp		14	1.143		375	60		435	500
7140	3000 amp		12	1.333		450	70		520	600
7160	4000 amp		10	1.600		540	84		624	720

26 25 13.20 Bus Duct		Crew	Daily Output	Labor-Hours	Unit	Material	Labor	Equipment	Total	Total Incl O&P
0010	**BUS DUCT** 100 amp and less, aluminum or copper, plug-in									
0080	Bus duct, 3 pole 3 wire, 100 amp	1 Elec	42	.190	L.F.	19	10		29	36
0110	Elbow		4	2	Ea.	49	105		154	211
0120	Tee		2	4		71	210		281	395
0130	Wall flange		8	1		6.35	52.50		58.85	85.50
0140	Ground kit		16	.500		13.45	26		39.45	54
0180	3 pole 4 wire, 100 amp		40	.200	L.F.	22	10.50		32.50	39.50
0200	Cable tap box		3.10	2.581	Ea.	59.50	135		194.50	268
0300	End closure		16	.500		7.90	26		33.90	47.50
0400	Elbow		4	2		49	105		154	211
0500	Tee		2	4		71	210		281	395
0600	Hangers		10	.800		4.75	42		46.75	68
0700	Circuit breakers, 15 to 50 amp, 1 pole		8	1		130	52.50		182.50	222
0800	15 to 60 amp, 2 pole		6.70	1.194		505	62.50		567.50	650
0900	3 pole		5.30	1.509		505	79		584	675
1000	60 to 100 amp, 1 pole		6.70	1.194		505	62.50		567.50	650
1100	70 to 100 amp, 2 pole		5.30	1.509		505	79		584	675
1200	3 pole		4.50	1.778		505	93		598	695
1220	Switch, nonfused, 3 pole, 4 wire		8	1		69	52.50		121.50	155
1240	Fused, 3 fuses, 4 wire, 30 amp		8	1		250	52.50		302.50	355
1260	60 amp		5.30	1.509		262	79		341	405
1280	100 amp		4.50	1.778		435	93		528	620
1300	Plug, fusible, 3 pole 250 volt, 30 amp		5.30	1.509		250	79		329	395
1310	60 amp		5.30	1.509		296	79		375	445
1320	100 amp		4.50	1.778		435	93		528	620
1330	3 pole 480 volt, 30 amp		5.30	1.509		285	79		364	435
1340	60 amp		5.30	1.509		310	79		389	460
1350	100 amp		4.50	1.778		445	93		538	630
1360	Circuit breaker, 3 pole 250 volt, 60 amp		5.30	1.509		505	79		584	675
1370	3 pole 480 volt, 100 amp		4.50	1.778		505	93		598	695

26 25 13 – Bus Duct/Busway and Fittings

26 25 13.20 Bus Duct		Crew	Daily Output	Labor-Hours	Unit	Material	2013 Bare Costs Labor	Equipment	Total	Total Incl O&P
2000	Bus duct, 2 wire, 250 volt, 30 amp	1 Elec	60	.133	L.F.	2.75	7		9.75	13.50
2100	60 amp		50	.160		2.75	8.40		11.15	15.60
2200	300 volt, 30 amp		60	.133		2.75	7		9.75	13.50
2300	60 amp		50	.160		2.75	8.40		11.15	15.60
2400	3 wire, 250 volt, 30 amp		60	.133		2.75	7		9.75	13.50
2500	60 amp		50	.160		2.75	8.40		11.15	15.60
2600	480/277 volt, 30 amp		60	.133		2.75	7		9.75	13.50
2700	60 amp		50	.160		2.75	8.40		11.15	15.60
2750	End feed, 300 volt 2 wire max. 30 amp		6	1.333	Ea.	41	70		111	150
2800	60 amp		5.50	1.455		41	76		117	159
2850	30 amp miniature		6	1.333		41	70		111	150
2900	3 wire, 30 amp		6	1.333		52	70		122	162
2950	60 amp		5.50	1.455		52	76		128	171
3000	30 amp miniature		6	1.333		52	70		122	162
3050	Center feed, 300 volt 2 wire, 30 amp		6	1.333		57	70		127	168
3100	60 amp		5.50	1.455		57	76		133	177
3150	3 wire, 30 amp		6	1.333		65	70		135	177
3200	60 amp		5.50	1.455		65	76		141	186
3220	Elbow, 30 amp		6	1.333		39.50	70		109.50	149
3240	60 amp		5.50	1.455		39.50	76		115.50	158
3260	End cap		40	.200		7.90	10.50		18.40	24.50
3280	Strength beam, 10 ft.		15	.533		22	28		50	66.50
3300	Hanger		24	.333		4.75	17.45		22.20	31.50
3320	Tap box, nonfusible		6.30	1.270		67	66.50		133.50	174
3340	Fusible switch 30 amp, 1 fuse		6	1.333		293	70		363	425
3360	2 fuse		6	1.333		297	70		367	430
3380	3 fuse		6	1.333		300	70		370	435
3400	Circuit breaker handle on cover, 1 pole		6	1.333		46.50	70		116.50	156
3420	2 pole		6	1.333		57	70		127	168
3440	3 pole		6	1.333		73	70		143	185
3460	Circuit breaker external operhandle, 1 pole		6	1.333		48	70		118	158
3480	2 pole		6	1.333		57.50	70		127.50	169
3500	3 pole		6	1.333		84	70		154	198
3520	Terminal plug only		16	.500		8.70	26		34.70	48.50
3540	Terminal with receptacle		16	.500		11.25	26		37.25	51.50
3560	Fixture plug		16	.500		7.75	26		33.75	47.50
4000	Copper bus duct, lighting, 2 wire 300 volt, 20 amp		70	.114	L.F.	5.10	6		11.10	14.55
4020	35 amp		60	.133		5.10	7		12.10	16.05
4040	50 amp		55	.145		5.10	7.60		12.70	17
4060	60 amp		50	.160		5.10	8.40		13.50	18.15
4080	3 wire 300 volt, 20 amp		70	.114		5.55	6		11.55	15.05
4100	35 amp		60	.133		5.55	7		12.55	16.55
4120	50 amp		55	.145		5.55	7.60		13.15	17.50
4140	60 amp		50	.160		5.55	8.40		13.95	18.65
4160	Feeder in box, end, 1 circuit		6	1.333	Ea.	76	70		146	189
4180	2 circuit		5.50	1.455		79	76		155	201
4200	Center, 1 circuit		6	1.333		104	70		174	220
4220	2 circuit		5.50	1.455		108	76		184	233
4240	End cap		40	.200		12.70	10.50		23.20	29.50
4260	Hanger, surface mount		24	.333		7.60	17.45		25.05	34.50
4280	Coupling		40	.200		9.65	10.50		20.15	26.50

26 25 13 – Bus Duct/Busway and Fittings

26 25 13.30 Copper Bus Duct		Crew	Daily Output	Labor-Hours	Unit	Material	Labor	Equipment	Total	Total Incl O&P
						2013 Bare Costs				
0010	**COPPER BUS DUCT** R262513-15									
0100	Weatherproof 3 pole 4 wire, feeder duct, 600 amp	2 Elec	24	.667	L.F.	239	35		274	315
0110	800 amp		18	.889		325	46.50		371.50	430
0120	1000 amp		17	.941		365	49.50		414.50	475
0125	1200 amp		16.50	.970		425	51		476	545
0130	1350 amp		16	1		445	52.50		497.50	570
0140	1600 amp		12	1.333		480	70		550	630
0150	2000 amp		10	1.600		535	84		619	710
0160	2500 amp		7	2.286		580	120		700	820
0170	3000 amp		5	3.200		805	168		973	1,125
0180	4000 amp		3.60	4.444		1,250	233		1,483	1,750
0200	Indoor 3 pole 4 wire, plug-in, bus duct high short circuit, 400 amp		32	.500		227	26		253	289
0210	600 amp		26	.615		227	32.50		259.50	299
0220	800 amp		20	.800		270	42		312	360
0230	1000 amp		18	.889		298	46.50		344.50	400
0240	1350 amp		16	1		405	52.50		457.50	525
0250	1600 amp		12	1.333		460	70		530	610
0260	2000 amp		10	1.600		580	84		664	765
0270	2500 amp		8	2		715	105		820	940
0280	3000 amp		6	2.667	▼	820	140		960	1,100
0310	Cross, 225 amp		3	5.333	Ea.	2,700	279		2,979	3,400
0320	400 amp		2.80	5.714		2,700	299		2,999	3,425
0330	600 amp		2.60	6.154		2,700	320		3,020	3,450
0340	800 amp		2.20	7.273		2,925	380		3,305	3,800
0350	1000 amp		2	8		3,250	420		3,670	4,225
0360	1350 amp		1.80	8.889		3,500	465		3,965	4,550
0370	1600 amp		1.70	9.412		3,900	495		4,395	5,025
0380	2000 amp		1.60	10		6,400	525		6,925	7,825
0390	2500 amp		1.40	11.429		7,750	600		8,350	9,425
0400	3000 amp		1.20	13.333		8,475	700		9,175	10,400
0410	4000 amp		1	16		11,000	840		11,840	13,400
0430	Expansion fitting, 225 amp		5.40	2.963		1,425	155		1,580	1,775
0440	400 amp		4.60	3.478		1,600	182		1,782	2,025
0450	600 amp		4	4		1,975	210		2,185	2,500
0460	800 amp		3.40	4.706		2,325	247		2,572	2,925
0470	1000 amp		3	5.333		2,675	279		2,954	3,375
0480	1350 amp		2.80	5.714		3,200	299		3,499	3,950
0490	1600 amp		2.60	6.154		4,450	320		4,770	5,375
0500	2000 amp		2.20	7.273		5,125	380		5,505	6,200
0510	2500 amp		1.80	8.889		6,225	465		6,690	7,550
0520	3000 amp		1.60	10		6,975	525		7,500	8,450
0530	4000 amp		1.20	13.333		9,000	700		9,700	11,000
0550	Reducer nonfused, 225 amp		5.40	2.963		1,650	155		1,805	2,050
0560	400 amp		4.60	3.478		1,650	182		1,832	2,100
0570	600 amp		4	4		1,650	210		1,860	2,150
0580	800 amp		3.40	4.706		2,000	247		2,247	2,575
0590	1000 amp		3	5.333		2,400	279		2,679	3,075
0600	1350 amp		2.80	5.714		3,475	299		3,774	4,275
0610	1600 amp		2.60	6.154		4,000	320		4,320	4,875
0620	2000 amp		2.20	7.273		4,800	380		5,180	5,850
0630	2500 amp		1.80	8.889		6,050	465		6,515	7,350
0640	3000 amp		1.60	10		6,825	525		7,350	8,275

26 25 Enclosed Bus Assemblies

26 25 13 – Bus Duct/Busway and Fittings

26 25 13.30 Copper Bus Duct		Crew	Daily Output	Labor-Hours	Unit	2013 Bare Costs Material	Labor	Equipment	Total	Total Incl O&P
0650	4000 amp	2 Elec	1.20	13.333	Ea.	8,875	700		9,575	10,800
0670	Reducer fuse included, 225 amp		4.40	3.636		3,550	191		3,741	4,175
0680	400 amp		4.20	3.810		4,450	200		4,650	5,175
0690	600 amp		3.60	4.444		5,475	233		5,708	6,375
0700	800 amp		3.20	5		7,725	262		7,987	8,900
0710	1000 amp		3	5.333		9,675	279		9,954	11,000
0720	1350 amp		2.80	5.714		15,000	299		15,299	17,000
0730	1600 amp		2.20	7.273		20,400	380		20,780	23,000
0740	2000 amp		1.80	8.889		23,000	465		23,465	25,900
0790	Reducer, circuit breaker, 225 amp		4.40	3.636		4,500	191		4,691	5,225
0800	400 amp		4.20	3.810		5,250	200		5,450	6,075
0810	600 amp		3.60	4.444		7,425	233		7,658	8,525
0820	800 amp		3.20	5		8,725	262		8,987	10,000
0830	1000 amp		3	5.333		9,900	279		10,179	11,300
0840	1350 amp		2.80	5.714		11,300	299		11,599	12,900
0850	1600 amp		2.20	7.273		16,700	380		17,080	19,000
0860	2000 amp		1.80	8.889		18,300	465		18,765	20,800
0910	Cable tap box, center, 225 amp		3.20	5		1,625	262		1,887	2,200
0920	400 amp		2.60	6.154		1,625	320		1,945	2,275
0930	600 amp		2.20	7.273		1,625	380		2,005	2,375
0940	800 amp		2	8		1,800	420		2,220	2,600
0950	1000 amp		1.60	10		1,950	525		2,475	2,925
0960	1350 amp		1.40	11.429		2,350	600		2,950	3,475
0970	1600 amp		1.20	13.333		2,625	700		3,325	3,925
0980	2000 amp		1	16		3,150	840		3,990	4,725
1040	2500 amp		.80	20		3,725	1,050		4,775	5,675
1060	3000 amp		.60	26.667		4,075	1,400		5,475	6,575
1080	4000 amp		.40	40		5,150	2,100		7,250	8,775
1800	Weatherproof 3 pole 3 wire, feeder duct, 600 amp		28	.571	L.F.	239	30		269	310
1820	800 amp		22	.727		290	38		328	375
1840	1000 amp		20	.800		325	42		367	420
1850	1200 amp		19	.842		335	44		379	435
1860	1350 amp		18	.889		455	46.50		501.50	570
1880	1600 amp		14	1.143		520	60		580	660
1900	2000 amp		12	1.333		665	70		735	835
1920	2500 amp		8	2		825	105		930	1,075
1940	3000 amp		6	2.667		950	140		1,090	1,250
1960	4000 amp		4	4		1,250	210		1,460	1,700
2000	Indoor 3 pole 3 wire, feeder duct, 600 amp		32	.500		199	26		225	258
2010	800 amp		26	.615		242	32.50		274.50	315
2020	1000 amp		24	.667		270	35		305	350
2025	1200 amp		22	.727		355	38		393	445
2030	1350 amp		20	.800		380	42		422	480
2040	1600 amp		16	1		430	52.50		482.50	555
2050	2000 amp		14	1.143		555	60		615	700
2060	2500 amp		10	1.600		690	84		774	880
2070	3000 amp		8	2		795	105		900	1,025
2080	4000 amp		6	2.667		1,050	140		1,190	1,350
2090	5000 amp		5	3.200		1,275	168		1,443	1,650
2200	Indoor 3 pole 3 wire, bus duct plug-in, 225 amp		46	.348		227	18.25		245.25	278
2210	400 amp		36	.444		227	23.50		250.50	285
2220	600 amp		30	.533		227	28		255	292
2230	800 amp		24	.667		270	35		305	350

26 25 Enclosed Bus Assemblies

26 25 13 – Bus Duct/Busway and Fittings

26 25 13.30 Copper Bus Duct		Crew	Daily Output	Labor-Hours	Unit	2013 Bare Costs Material	Labor	Equipment	Total	Total Incl O&P
2240	1000 amp	2 Elec	20	.800	L.F.	298	42		340	395
2250	1350 amp		18	.889		405	46.50		451.50	515
2260	1600 amp		14	1.143		460	60		520	595
2270	2000 amp		12	1.333		580	70		650	745
2280	2500 amp		10	1.600		715	84		799	910
2290	3000 amp		8	2		820	105		925	1,050
2330	High short circuit, 400 amp		36	.444		227	23.50		250.50	285
2340	600 amp		30	.533		227	28		255	292
2350	800 amp		24	.667		270	35		305	350
2360	1000 amp		20	.800		298	42		340	395
2370	1350 amp		18	.889		405	46.50		451.50	515
2380	1600 amp		14	1.143		460	60		520	595
2390	2000 amp		12	1.333		580	70		650	745
2400	2500 amp		10	1.600		715	84		799	910
2410	3000 amp		8	2		820	105		925	1,050
2440	Elbows, 225 amp		4.60	3.478	Ea.	1,125	182		1,307	1,525
2450	400 amp		4.20	3.810		1,125	200		1,325	1,550
2460	600 amp		3.60	4.444		1,125	233		1,358	1,600
2470	800 amp		3.20	5		1,225	262		1,487	1,750
2480	1000 amp		3	5.333		1,275	279		1,554	1,825
2485	1200 amp		2.90	5.517		1,375	289		1,664	1,950
2490	1350 amp		2.80	5.714		1,450	299		1,749	2,050
2500	1600 amp		2.60	6.154		1,575	320		1,895	2,200
2510	2000 amp		2	8		1,900	420		2,320	2,725
2520	2500 amp		1.80	8.889		2,875	465		3,340	3,875
2530	3000 amp		1.60	10		3,150	525		3,675	4,250
2540	4000 amp		1.40	11.429		4,050	600		4,650	5,350
2560	Tee fittings, 225 amp		2.80	5.714		1,350	299		1,649	1,925
2570	400 amp		2.40	6.667		1,350	350		1,700	2,000
2580	600 amp		2	8		1,350	420		1,770	2,100
2590	800 amp		1.80	8.889		1,450	465		1,915	2,300
2600	1000 amp		1.60	10		1,625	525		2,150	2,575
2605	1200 amp		1.50	10.667		1,750	560		2,310	2,750
2610	1350 amp		1.40	11.429		1,825	600		2,425	2,900
2620	1600 amp		1.20	13.333		2,125	700		2,825	3,375
2630	2000 amp		1	16		3,525	840		4,365	5,125
2640	2500 amp		.70	22.857		4,125	1,200		5,325	6,325
2650	3000 amp		.60	26.667		4,550	1,400		5,950	7,100
2660	4000 amp		.50	32		5,875	1,675		7,550	8,950
2680	Cross, 225 amp		3.60	4.444		2,050	233		2,283	2,600
2690	400 amp		3.20	5		2,050	262		2,312	2,650
2700	600 amp		3	5.333		2,050	279		2,329	2,675
2710	800 amp		2.60	6.154		2,450	320		2,770	3,175
2720	1000 amp		2.40	6.667		2,550	350		2,900	3,350
2730	1350 amp		2.20	7.273		2,925	380		3,305	3,775
2740	1600 amp		2	8		3,125	420		3,545	4,075
2750	2000 amp		1.80	8.889		4,950	465		5,415	6,150
2760	2500 amp		1.60	10		5,775	525		6,300	7,125
2770	3000 amp		1.40	11.429		6,325	600		6,925	7,850
2780	4000 amp		1	16		8,075	840		8,915	10,100
2800	Expansion fitting, 225 amp		6.40	2.500		1,625	131		1,756	2,000
2810	400 amp		5.40	2.963		1,625	155		1,780	2,025
2820	600 amp		4.60	3.478		1,625	182		1,807	2,075

26 25 13 – Bus Duct/Busway and Fittings

26 25 13.30 Copper Bus Duct		Crew	Daily Output	Labor-Hours	Unit	Material	Labor	Equipment	Total	Total Incl O&P
						2013 Bare Costs				
2830	800 amp	2 Elec	4	4	Ea.	1,950	210		2,160	2,475
2840	1000 amp		3.60	4.444		2,150	233		2,383	2,700
2850	1350 amp		3.20	5		2,550	262		2,812	3,200
2860	1600 amp		3	5.333		2,800	279		3,079	3,500
2870	2000 amp		2.60	6.154		3,375	320		3,695	4,175
2880	2500 amp		2.20	7.273		4,700	380		5,080	5,750
2890	3000 amp		1.80	8.889		5,250	465		5,715	6,475
2900	4000 amp		1.40	11.429		6,700	600		7,300	8,275
2920	Reducer nonfused, 225 amp		6.40	2.500		1,350	131		1,481	1,700
2930	400 amp		5.40	2.963		1,350	155		1,505	1,725
2940	600 amp		4.60	3.478		1,350	182		1,532	1,775
2950	800 amp		4	4		1,575	210		1,785	2,075
2960	1000 amp		3.60	4.444		1,775	233		2,008	2,300
2970	1350 amp		3.20	5		2,250	262		2,512	2,875
2980	1600 amp		3	5.333		2,550	279		2,829	3,225
2990	2000 amp		2.60	6.154		3,050	320		3,370	3,825
3000	2500 amp		2.20	7.273		4,475	380		4,855	5,500
3010	3000 amp		1.80	8.889		5,000	465		5,465	6,200
3020	4000 amp		1.40	11.429		6,500	600		7,100	8,050
3040	Reducer fuse included, 225 amp		5	3.200		3,225	168		3,393	3,800
3050	400 amp		4.80	3.333		4,275	175		4,450	4,975
3060	600 amp		4.20	3.810		5,100	200		5,300	5,925
3070	800 amp		3.60	4.444		7,275	233		7,508	8,350
3080	1000 amp		3.40	4.706		8,625	247		8,872	9,850
3090	1350 amp		3.20	5		8,350	262		8,612	9,575
3100	1600 amp		2.60	6.154		19,400	320		19,720	21,900
3110	2000 amp		2	8		22,000	420		22,420	24,800
3160	Reducer circuit breaker, 225 amp		5	3.200		4,125	168		4,293	4,775
3170	400 amp		4.80	3.333		5,050	175		5,225	5,800
3180	600 amp		4.20	3.810		7,250	200		7,450	8,275
3190	800 amp		3.60	4.444		8,450	233		8,683	9,650
3200	1000 amp		3.40	4.706		9,625	247		9,872	11,000
3210	1350 amp		3.20	5		11,000	262		11,262	12,500
3220	1600 amp		2.60	6.154		16,500	320		16,820	18,600
3230	2000 amp		2	8		18,000	420		18,420	20,400
3280	3 pole, 3 wire, cable tap box center, 225 amp		3.60	4.444		1,850	233		2,083	2,400
3290	400 amp		3	5.333		1,850	279		2,129	2,475
3300	600 amp		2.60	6.154		1,850	320		2,170	2,525
3310	800 amp		2.40	6.667		2,075	350		2,425	2,800
3320	1000 amp		1.80	8.889		2,250	465		2,715	3,175
3330	1350 amp		1.60	10		2,750	525		3,275	3,800
3340	1600 amp		1.40	11.429		3,075	600		3,675	4,275
3350	2000 amp		1.20	13.333		3,725	700		4,425	5,150
3360	2500 amp		1	16		4,450	840		5,290	6,125
3370	3000 amp		.70	22.857		4,900	1,200		6,100	7,175
3380	4000 amp		.50	32		6,225	1,675		7,900	9,325
3400	Cable tap box end, 225 amp		3.60	4.444		1,000	233		1,233	1,450
3410	400 amp		3	5.333		1,000	279		1,279	1,525
3420	600 amp		2.60	6.154		1,125	320		1,445	1,725
3430	800 amp		2.40	6.667		1,125	350		1,475	1,775
3440	1000 amp		1.80	8.889		1,250	465		1,715	2,075
3445	1200 amp		1.70	9.412		1,400	495		1,895	2,275
3450	1350 amp		1.60	10		1,475	525		2,000	2,400

26 25 Enclosed Bus Assemblies

26 25 13 – Bus Duct/Busway and Fittings

26 25 13.30 Copper Bus Duct		Crew	Daily Output	Labor-Hours	Unit	Material	2013 Bare Costs Labor	Equipment	Total	Total Incl O&P
3460	1600 amp	2 Elec	1.40	11.429	Ea.	1,700	600		2,300	2,750
3470	2000 amp		1.20	13.333		1,925	700		2,625	3,175
3480	2500 amp		1	16		2,300	840		3,140	3,800
3490	3000 amp		.70	22.857		2,525	1,200		3,725	4,575
3500	4000 amp		.50	32		3,200	1,675		4,875	6,025
4600	Plug-in, fusible switch w/3 fuses, 3 pole, 250 volt, 30 amp	1 Elec	4	2		405	105		510	600
4610	60 amp		3.60	2.222		525	116		641	755
4620	100 amp		2.70	2.963		755	155		910	1,050
4630	200 amp	2 Elec	3.20	5		1,275	262		1,537	1,800
4640	400 amp		1.40	11.429		3,300	600		3,900	4,525
4650	600 amp		.90	17.778		4,575	930		5,505	6,425
4700	4 pole, 120/208 volt, 30 amp	1 Elec	3.90	2.051		550	107		657	765
4710	60 amp		3.50	2.286		595	120		715	835
4720	100 amp		2.60	3.077		825	161		986	1,150
4730	200 amp	2 Elec	3	5.333		1,400	279		1,679	1,950
4740	400 amp		1.30	12.308		3,250	645		3,895	4,550
4750	600 amp		.80	20		4,575	1,050		5,625	6,600
4800	3 pole, 480 volt, 30 amp	1 Elec	4	2		415	105		520	610
4810	60 amp		3.60	2.222		435	116		551	655
4820	100 amp		2.70	2.963		735	155		890	1,050
4830	200 amp	2 Elec	3.20	5		1,275	262		1,537	1,800
4840	400 amp		1.40	11.429		2,950	600		3,550	4,150
4850	600 amp		.90	17.778		4,200	930		5,130	6,025
4860	800 amp		.66	24.242		13,800	1,275		15,075	17,100
4870	1000 amp		.60	26.667		14,600	1,400		16,000	18,200
4880	1200 amp		.50	32		15,400	1,675		17,075	19,400
4890	1600 amp		.44	36.364		16,900	1,900		18,800	21,500
4900	4 pole, 277/480 volt, 30 amp	1 Elec	3.90	2.051		595	107		702	815
4910	60 amp		3.50	2.286		635	120		755	880
4920	100 amp		2.60	3.077		930	161		1,091	1,275
4930	200 amp	2 Elec	3	5.333		1,850	279		2,129	2,450
4940	400 amp		1.30	12.308		3,475	645		4,120	4,800
4950	600 amp		.80	20		4,775	1,050		5,825	6,825
5050	800 amp		.60	26.667		15,700	1,400		17,100	19,400
5060	1000 amp		.56	28.571		18,100	1,500		19,600	22,200
5070	1200 amp		.48	33.333		18,200	1,750		19,950	22,600
5080	1600 amp		.42	38.095		20,100	2,000		22,100	25,100
5150	Fusible with starter, 3 pole 250 volt, 30 amp	1 Elec	3.50	2.286		2,125	120		2,245	2,525
5160	60 amp		3.20	2.500		2,250	131		2,381	2,675
5170	100 amp		2.50	3.200		2,550	168		2,718	3,050
5180	200 amp	2 Elec	2.80	5.714		4,200	299		4,499	5,075
5200	3 pole 480 volt, 30 amp	1 Elec	3.50	2.286		2,125	120		2,245	2,525
5210	60 amp		3.20	2.500		2,250	131		2,381	2,675
5220	100 amp		2.50	3.200		2,550	168		2,718	3,050
5230	200 amp	2 Elec	2.80	5.714		4,200	299		4,499	5,075
5300	Fusible with contactor, 3 pole 250 volt, 30 amp	1 Elec	3.50	2.286		2,075	120		2,195	2,450
5310	60 amp		3.20	2.500		2,650	131		2,781	3,100
5320	100 amp		2.50	3.200		3,700	168		3,868	4,325
5330	200 amp	2 Elec	2.80	5.714		4,225	299		4,524	5,100
5400	3 pole 480 volt, 30 amp	1 Elec	3.50	2.286		2,225	120		2,345	2,625
5410	60 amp		3.20	2.500		3,150	131		3,281	3,650
5420	100 amp		2.50	3.200		4,325	168		4,493	5,000
5430	200 amp	2 Elec	2.80	5.714		4,425	299		4,724	5,325

26 25 Enclosed Bus Assemblies

26 25 13 – Bus Duct/Busway and Fittings

26 25 13.30 Copper Bus Duct		Crew	Daily Output	Labor-Hours	Unit	2013 Bare Costs Material	Labor	Equipment	Total	Total Incl O&P
5450	Fusible with capacitor, 3 pole 250 volt, 30 amp	1 Elec	3	2.667	Ea.	5,375	140		5,515	6,125
5460	60 amp		2	4		6,275	210		6,485	7,225
5500	3 pole 480 volt, 30 amp		3	2.667		4,525	140		4,665	5,175
5510	60 amp		2	4		5,650	210		5,860	6,525
5600	Circuit breaker, 3 pole, 250 volt, 60 amp		4.50	1.778		565	93		658	765
5610	100 amp		3.20	2.500		695	131		826	960
5650	4 pole, 120/208 volt, 60 amp		4.40	1.818		640	95.50		735.50	850
5660	100 amp		3.10	2.581		760	135		895	1,025
5700	3 pole, 4 wire 277/480 volt, 60 amp		4.30	1.860		860	97.50		957.50	1,100
5710	100 amp		3	2.667		960	140		1,100	1,250
5720	225 amp	2 Elec	3.20	5		2,125	262		2,387	2,750
5730	400 amp		1.20	13.333		4,400	700		5,100	5,900
5740	600 amp		.96	16.667		5,925	875		6,800	7,825
5750	700 amp		.60	26.667		7,500	1,400		8,900	10,400
5760	800 amp		.60	26.667		7,500	1,400		8,900	10,400
5770	900 amp		.54	29.630		9,900	1,550		11,450	13,200
5780	1000 amp		.54	29.630		9,900	1,550		11,450	13,200
5790	1200 amp		.42	38.095		11,900	2,000		13,900	16,100
5810	Circuit breaker w/HIC fuses, 3 pole 480 volt, 60 amp	1 Elec	4.40	1.818		1,075	95.50		1,170.50	1,350
5820	100 amp	"	3.10	2.581		1,200	135		1,335	1,500
5830	225 amp	2 Elec	3.40	4.706		3,800	247		4,047	4,550
5840	400 amp		1.40	11.429		6,025	600		6,625	7,525
5850	600 amp		1	16		6,100	840		6,940	7,950
5860	700 amp		.64	25		8,100	1,300		9,400	10,900
5870	800 amp		.64	25		8,100	1,300		9,400	10,900
5880	900 amp		.56	28.571		17,500	1,500		19,000	21,500
5890	1000 amp		.56	28.571		17,500	1,500		19,000	21,500
5950	3 pole 4 wire 277/480 volt, 60 amp	1 Elec	4.30	1.860		1,075	97.50		1,172.50	1,350
5960	100 amp	"	3	2.667		1,200	140		1,340	1,500
5970	225 amp	2 Elec	3	5.333		3,800	279		4,079	4,600
5980	400 amp		1.10	14.545		6,025	760		6,785	7,775
5990	600 amp		.94	17.021		6,100	890		6,990	8,025
6000	700 amp		.58	27.586		8,100	1,450		9,550	11,100
6010	800 amp		.58	27.586		8,100	1,450		9,550	11,100
6020	900 amp		.52	30.769		17,500	1,600		19,100	21,600
6030	1000 amp		.52	30.769		17,500	1,600		19,100	21,600
6040	1200 amp		.40	40		17,500	2,100		19,600	22,300
6100	Circuit breaker with starter, 3 pole 250 volt, 60 amp	1 Elec	3.20	2.500		1,575	131		1,706	1,925
6110	100 amp	"	2.50	3.200		2,175	168		2,343	2,650
6120	225 amp	2 Elec	3	5.333		2,550	279		2,829	3,225
6130	3 pole 480 volt, 60 amp	1 Elec	3.20	2.500		1,575	131		1,706	1,925
6140	100 amp	"	2.50	3.200		2,175	168		2,343	2,650
6150	225 amp	2 Elec	3	5.333		2,550	279		2,829	3,225
6200	Circuit breaker with contactor, 3 pole 250 volt, 60 amp	1 Elec	3.20	2.500		1,475	131		1,606	1,825
6210	100 amp	"	2.50	3.200		2,025	168		2,193	2,475
6220	225 amp	2 Elec	3	5.333		2,375	279		2,654	3,025
6250	3 pole 480 volt, 60 amp	1 Elec	3.20	2.500		1,475	131		1,606	1,825
6260	100 amp	"	2.50	3.200		2,025	168		2,193	2,475
6270	225 amp	2 Elec	3	5.333		2,375	279		2,654	3,025
6300	Circuit breaker with capacitor, 3 pole 250 volt, 60 amp	1 Elec	2	4		6,850	210		7,060	7,850
6310	3 pole 480 volt, 60 amp		2	4		6,975	210		7,185	8,000
6400	Add control transformer with pilot light to above starter		16	.500		435	26		461	515
6410	Switch, fusible, mechanically held contactor optional		16	.500		1,150	26		1,176	1,300

26 25 13 – Bus Duct/Busway and Fittings

26 25 13.30 Copper Bus Duct		Crew	Daily Output	Labor-Hours	Unit	Material	2013 Bare Costs Labor	Equipment	Total	Total Incl O&P
6430	Circuit breaker, mechanically held contactor optional	1 Elec	16	.500	Ea.	1,150	26		1,176	1,300
6450	Ground neutralizer, 3 pole	↓	16	.500	↓	53	26		79	97
26 25 13.40 Copper Bus Duct										
0010	**COPPER BUS DUCT** 10 ft. long									
0050	Indoor 3 pole 4 wire, plug-in, straight section, 225 amp	2 Elec	40	.400	L.F.	228	21		249	283
1000	400 amp		32	.500		228	26		254	290
1500	600 amp		26	.615		228	32.50		260.50	300
2400	800 amp		20	.800		271	42		313	360
2450	1000 amp		18	.889		299	46.50		345.50	400
2470	1200 amp		17	.941		385	49.50		434.50	500
2500	1350 amp		16	1		405	52.50		457.50	525
2510	1600 amp		12	1.333		460	70		530	610
2520	2000 amp		10	1.600		580	84		664	765
2530	2500 amp		8	2		720	105		825	945
2540	3000 amp		6	2.667		820	140		960	1,125
2550	Feeder, 600 amp		28	.571		200	30		230	264
2600	800 amp		22	.727		242	38		280	325
2700	1000 amp		20	.800		271	42		313	360
2750	1200 amp		19	.842		355	44		399	455
2800	1350 amp		18	.889		380	46.50		426.50	485
2900	1600 amp		14	1.143		435	60		495	565
3000	2000 amp		12	1.333		555	70		625	715
3010	2500 amp		8	2		690	105		795	915
3020	3000 amp		6	2.667		795	140		935	1,075
3030	4000 amp		4	4		1,050	210		1,260	1,475
3040	5000 amp		2	8	↓	1,275	420		1,695	2,025
3100	Elbows, 225 amp		4	4	Ea.	1,350	210		1,560	1,800
3200	400 amp		3.60	4.444		1,350	233		1,583	1,825
3300	600 amp		3.20	5		1,350	262		1,612	1,875
3400	800 amp		2.80	5.714		1,475	299		1,774	2,050
3500	1000 amp		2.60	6.154		1,625	320		1,945	2,275
3550	1200 amp		2.50	6.400		1,800	335		2,135	2,500
3600	1350 amp		2.40	6.667		1,825	350		2,175	2,525
3700	1600 amp		2.20	7.273		2,000	380		2,380	2,750
3800	2000 amp		1.80	8.889		2,450	465		2,915	3,400
3810	2500 amp		1.60	10		3,875	525		4,400	5,025
3820	3000 amp		1.40	11.429		4,250	600		4,850	5,575
3830	4000 amp		1.20	13.333		5,500	700		6,200	7,100
3840	5000 amp		1	16		8,875	840		9,715	11,000
4000	End box, 225 amp		34	.471		182	24.50		206.50	238
4100	400 amp		32	.500		182	26		208	240
4200	600 amp		28	.571		182	30		212	246
4300	800 amp		26	.615		182	32.50		214.50	250
4400	1000 amp		24	.667		182	35		217	254
4410	1200 amp		23	.696		182	36.50		218.50	256
4500	1350 amp		22	.727		173	38		211	248
4600	1600 amp		20	.800		173	42		215	254
4700	2000 amp		18	.889		213	46.50		259.50	305
4710	2500 amp		16	1		213	52.50		265.50	315
4720	3000 amp		14	1.143		201	60		261	310
4730	4000 amp		12	1.333		244	70		314	375
4740	5000 amp	↓	10	1.600	↓	244	84		328	395

26 25 13 – Bus Duct/Busway and Fittings

26 25 13.40 Copper Bus Duct		Crew	Daily Output	Labor-Hours	Unit	Material	Labor	Equipment	Total	Total Incl O&P
						2013 Bare Costs				
4800	Cable tap box end, 225 amp	2 Elec	3.20	5	Ea.	1,475	262		1,737	2,025
5000	400 amp		2.60	6.154		1,475	320		1,795	2,100
5100	600 amp		2.20	7.273		1,475	380		1,855	2,200
5200	800 amp		2	8		1,700	420		2,120	2,500
5300	1000 amp		1.60	10		1,825	525		2,350	2,800
5350	1200 amp		1.50	10.667		1,925	560		2,485	2,950
5400	1350 amp		1.40	11.429		2,175	600		2,775	3,275
5500	1600 amp		1.20	13.333		2,425	700		3,125	3,700
5600	2000 amp		1	16		2,700	840		3,540	4,225
5610	2500 amp		.80	20		3,000	1,050		4,050	4,875
5620	3000 amp		.60	26.667		3,500	1,400		4,900	5,950
5630	4000 amp		.40	40		4,025	2,100		6,125	7,550
5640	5000 amp		.20	80		4,750	4,200		8,950	11,500
5700	Switchboard stub, 225 amp		5.40	2.963		1,375	155		1,530	1,725
5800	400 amp		4.60	3.478		1,375	182		1,557	1,775
5900	600 amp		4	4		1,375	210		1,585	1,825
6000	800 amp		3.20	5		1,650	262		1,912	2,200
6100	1000 amp		3	5.333		1,925	279		2,204	2,525
6150	1200 amp		2.80	5.714		2,225	299		2,524	2,900
6200	1350 amp		2.60	6.154		2,350	320		2,670	3,050
6300	1600 amp		2.40	6.667		2,650	350		3,000	3,425
6400	2000 amp		2	8		3,200	420		3,620	4,150
6410	2500 amp		1.80	8.889		3,900	465		4,365	4,975
6420	3000 amp		1.60	10		4,325	525		4,850	5,525
6430	4000 amp		1.40	11.429		5,625	600		6,225	7,100
6440	5000 amp		1.20	13.333		6,900	700		7,600	8,650
6490	Tee fittings, 225 amp		2.40	6.667		1,850	350		2,200	2,575
6500	400 amp		2	8		1,850	420		2,270	2,675
6600	600 amp		1.80	8.889		1,850	465		2,315	2,750
6700	800 amp		1.60	10		2,150	525		2,675	3,125
6750	1000 amp		1.40	11.429		2,475	600		3,075	3,625
6770	1200 amp		1.30	12.308		2,775	645		3,420	4,025
6800	1350 amp		1.20	13.333		2,925	700		3,625	4,275
7000	1600 amp		1	16		3,350	840		4,190	4,925
7100	2000 amp		.80	20		3,950	1,050		5,000	5,925
7110	2500 amp		.60	26.667		4,850	1,400		6,250	7,450
7120	3000 amp		.50	32		5,350	1,675		7,025	8,400
7130	4000 amp		.40	40		6,925	2,100		9,025	10,800
7140	5000 amp		.20	80		8,200	4,200		12,400	15,300
7200	Plug-in fusible switches w/3 fuses, 600 volt, 3 pole, 30 amp	1 Elec	4	2		835	105		940	1,075
7300	60 amp		3.60	2.222		935	116		1,051	1,200
7400	100 amp		2.70	2.963		1,425	155		1,580	1,800
7500	200 amp	2 Elec	3.20	5		2,550	262		2,812	3,200
7600	400 amp		1.40	11.429		7,475	600		8,075	9,125
7700	600 amp		.90	17.778		8,475	930		9,405	10,700
7800	800 amp		.66	24.242		11,900	1,275		13,175	15,000
7900	1200 amp		.50	32		22,300	1,675		23,975	27,000
7910	1600 amp		.44	36.364		21,200	1,900		23,100	26,200
8000	Plug-in circuit breakers, molded case, 15 to 50 amp	1 Elec	4.40	1.818		790	95.50		885.50	1,025
8100	70 to 100 amp	"	3.10	2.581		875	135		1,010	1,175
8200	150 to 225 amp	2 Elec	3.40	4.706		2,375	247		2,622	2,975
8300	250 to 400 amp		1.40	11.429		4,175	600		4,775	5,475
8400	500 to 600 amp		1	16		5,625	840		6,465	7,425

26 25 13.40 Copper Bus Duct		Crew	Daily Output	Labor-Hours	Unit	Material	2013 Bare Costs Labor	Equipment	Total	Total Incl O&P
8500	700 to 800 amp	2 Elec	.64	25	Ea.	6,950	1,300		8,250	9,575
8600	900 to 1000 amp		.56	28.571		9,925	1,500		11,425	13,200
8700	1200 amp		.44	36.364		11,900	1,900		13,800	16,000
8720	1400 amp		.40	40		16,700	2,100		18,800	21,500
8730	1600 amp		.40	40		18,400	2,100		20,500	23,300
8750	Circuit breakers, with current limiting fuse, 15 to 50 amp	1 Elec	4.40	1.818		1,575	95.50		1,670.50	1,875
8760	70 to 100 amp	"	3.10	2.581		1,875	135		2,010	2,250
8770	150 to 225 amp	2 Elec	3.40	4.706		4,025	247		4,272	4,800
8780	250 to 400 amp		1.40	11.429		6,200	600		6,800	7,725
8790	500 to 600 amp		1	16		7,150	840		7,990	9,125
8800	700 to 800 amp		.64	25		11,800	1,300		13,100	14,900
8810	900 to 1000 amp		.56	28.571		13,400	1,500		14,900	17,100
8850	Combination starter FVNR, fusible switch, NEMA size 0, 30 amp	1 Elec	2	4		2,125	210		2,335	2,675
8860	NEMA size 1, 60 amp		1.80	4.444		2,250	233		2,483	2,825
8870	NEMA size 2, 100 amp		1.30	6.154		2,850	320		3,170	3,600
8880	NEMA size 3, 200 amp	2 Elec	2	8		4,525	420		4,945	5,600
8900	Circuit breaker, NEMA size 0, 30 amp	1 Elec	2	4		2,200	210		2,410	2,725
8910	NEMA size 1, 60 amp		1.80	4.444		2,275	233		2,508	2,850
8920	NEMA size 2, 100 amp		1.30	6.154		3,275	320		3,595	4,075
8930	NEMA size 3, 200 amp	2 Elec	2	8		4,150	420		4,570	5,175
8950	Combination contactor, fusible switch, NEMA size 0, 30 amp	1 Elec	2	4		1,250	210		1,460	1,700
8960	NEMA size 1, 60 amp		1.80	4.444		1,275	233		1,508	1,750
8970	NEMA size 2, 100 amp		1.30	6.154		1,900	320		2,220	2,575
8980	NEMA size 3, 200 amp	2 Elec	2	8		2,200	420		2,620	3,025
9000	Circuit breaker, NEMA size 0, 30 amp	1 Elec	2	4		1,425	210		1,635	1,900
9010	NEMA size 1, 60 amp		1.80	4.444		1,450	233		1,683	1,950
9020	NEMA size 2, 100 amp		1.30	6.154		2,250	320		2,570	2,950
9030	NEMA size 3, 200 amp	2 Elec	2	8		2,750	420		3,170	3,650
9050	Control transformer for above, NEMA size 0, 30 amp	1 Elec	8	1		241	52.50		293.50	345
9060	NEMA size 1, 60 amp		8	1		241	52.50		293.50	345
9070	NEMA size 2, 100 amp		7	1.143		335	60		395	460
9080	NEMA size 3, 200 amp	2 Elec	14	1.143		470	60		530	605
9100	Comb. fusible switch & lighting control, electrically held, 30 amp	1 Elec	2	4		985	210		1,195	1,400
9110	60 amp		1.80	4.444		1,425	233		1,658	1,900
9120	100 amp		1.30	6.154		1,825	320		2,145	2,475
9130	200 amp	2 Elec	2	8		4,500	420		4,920	5,575
9150	Mechanically held, 30 amp	1 Elec	2	4		1,225	210		1,435	1,675
9160	60 amp		1.80	4.444		1,850	233		2,083	2,375
9170	100 amp		1.30	6.154		2,375	320		2,695	3,075
9180	200 amp	2 Elec	2	8		4,825	420		5,245	5,925
9200	Ground bus added to bus duct, 225 amp		320	.050	L.F.	43	2.62		45.62	51
9210	400 amp		240	.067		43	3.49		46.49	52.50
9220	600 amp		240	.067		43	3.49		46.49	52.50
9230	800 amp		160	.100		50	5.25		55.25	63
9240	1000 amp		160	.100		57	5.25		62.25	70.50
9250	1350 amp		140	.114		81	6		87	98.50
9260	1600 amp		120	.133		88	7		95	107
9270	2000 amp		110	.145		115	7.60		122.60	138
9280	2500 amp		100	.160		142	8.40		150.40	169
9290	3000 amp		90	.178		160	9.30		169.30	190
9300	4000 amp		80	.200		212	10.50		222.50	249
9310	5000 amp		70	.229		257	12		269	300
9320	High short circuit bracing, add					17.90			17.90	19.70

26 25 13.60 Copper or Aluminum Bus Duct Fittings		Crew	Daily Output	Labor-Hours	Unit	Material	2013 Bare Costs Labor	Equipment	Total	Total Incl O&P
0010	**COPPER OR ALUMINUM BUS DUCT FITTINGS**									
0100	Flange, wall, with vapor barrier, 225 amp	2 Elec	6.20	2.581	Ea.	780	135		915	1,050
0110	400 amp		6	2.667		780	140		920	1,075
0120	600 amp		5.80	2.759		780	145		925	1,075
0130	800 amp		5.40	2.963		780	155		935	1,100
0140	1000 amp		5	3.200		780	168		948	1,100
0145	1200 amp		4.80	3.333		780	175		955	1,125
0150	1350 amp		4.60	3.478		780	182		962	1,125
0160	1600 amp		4.20	3.810		780	200		980	1,150
0170	2000 amp		4	4		780	210		990	1,175
0180	2500 amp		3.60	4.444		780	233		1,013	1,200
0190	3000 amp		3.20	5		780	262		1,042	1,250
0200	4000 amp		2.60	6.154		780	320		1,100	1,350
0300	Roof, 225 amp		6.20	2.581		900	135		1,035	1,200
0310	400 amp		6	2.667		900	140		1,040	1,200
0320	600 amp		5.80	2.759		900	145		1,045	1,200
0330	800 amp		5.40	2.963		900	155		1,055	1,225
0340	1000 amp		5	3.200		900	168		1,068	1,250
0345	1200 amp		4.80	3.333		900	175		1,075	1,250
0350	1350 amp		4.60	3.478		900	182		1,082	1,275
0360	1600 amp		4.20	3.810		900	200		1,100	1,300
0370	2000 amp		4	4		900	210		1,110	1,300
0380	2500 amp		3.60	4.444		900	233		1,133	1,350
0390	3000 amp		3.20	5		900	262		1,162	1,375
0400	4000 amp		2.60	6.154		900	320		1,220	1,475
0420	Support, floor mounted, 225 amp		20	.800		163	42		205	243
0430	400 amp		20	.800		163	42		205	243
0440	600 amp		18	.889		163	46.50		209.50	250
0450	800 amp		16	1		163	52.50		215.50	259
0460	1000 amp		13	1.231		163	64.50		227.50	277
0465	1200 amp		11.80	1.356		163	71		234	286
0470	1350 amp		10.60	1.509		163	79		242	298
0480	1600 amp		9.20	1.739		163	91		254	315
0490	2000 amp		8	2		163	105		268	335
0500	2500 amp		6.40	2.500		163	131		294	375
0510	3000 amp		5.40	2.963		163	155		318	410
0520	4000 amp		4	4		163	210		373	495
0540	Weather stop, 225 amp		12	1.333		505	70		575	660
0550	400 amp		10	1.600		505	84		589	680
0560	600 amp		9	1.778		505	93		598	695
0570	800 amp		8	2		505	105		610	710
0580	1000 amp		6.40	2.500		505	131		636	750
0585	1200 amp		5.90	2.712		505	142		647	770
0590	1350 amp		5.40	2.963		505	155		660	785
0600	1600 amp		4.60	3.478		505	182		687	830
0610	2000 amp		4	4		505	210		715	870
0620	2500 amp		3.20	5		505	262		767	945
0630	3000 amp		2.60	6.154		505	320		825	1,025
0640	4000 amp		2	8		505	420		925	1,175
0660	End closure, 225 amp		34	.471		162	24.50		186.50	215
0670	400 amp		32	.500		162	26		188	217
0680	600 amp		28	.571		162	30		192	223

26 25 13 – Bus Duct/Busway and Fittings

26 25 13.60 Copper or Aluminum Bus Duct Fittings		Crew	Daily Output	Labor-Hours	Unit	Material	Labor	Equipment	Total	Total Incl O&P
						2013 Bare Costs				
0690	800 amp	2 Elec	26	.615	Ea.	162	32.50		194.50	227
0700	1000 amp		24	.667		162	35		197	231
0705	1200 amp		23	.696		162	36.50		198.50	233
0710	1350 amp		22	.727		162	38		200	235
0720	1600 amp		20	.800		162	42		204	241
0730	2000 amp		18	.889		199	46.50		245.50	289
0740	2500 amp		16	1		199	52.50		251.50	298
0750	3000 amp		14	1.143		199	60		259	310
0760	4000 amp		12	1.333		241	70		311	370
0780	Switchboard stub, 3 pole 3 wire, 225 amp		6	2.667		975	140		1,115	1,275
0790	400 amp		5.20	3.077		975	161		1,136	1,325
0800	600 amp		4.60	3.478		975	182		1,157	1,350
0810	800 amp		3.60	4.444		1,175	233		1,408	1,650
0820	1000 amp		3.40	4.706		1,375	247		1,622	1,900
0825	1200 amp		3.20	5		1,375	262		1,637	1,925
0830	1350 amp		3	5.333		1,650	279		1,929	2,250
0840	1600 amp		2.80	5.714		1,950	299		2,249	2,600
0850	2000 amp		2.40	6.667		2,275	350		2,625	3,025
0860	2500 amp		2	8		2,775	420		3,195	3,675
0870	3000 amp		1.80	8.889		3,150	465		3,615	4,175
0880	4000 amp		1.60	10		4,075	525		4,600	5,275
0890	5000 amp		1.40	11.429		5,025	600		5,625	6,425
0900	3 pole 4 wire, 225 amp		5.40	2.963		1,225	155		1,380	1,550
0910	400 amp		4.60	3.478		1,225	182		1,407	1,600
0920	600 amp		4	4		1,225	210		1,435	1,650
0930	800 amp		3.20	5		1,475	262		1,737	2,000
0940	1000 amp		3	5.333		1,700	279		1,979	2,300
0950	1350 amp		2.60	6.154		2,200	320		2,520	2,875
0960	1600 amp		2.40	6.667		2,475	350		2,825	3,250
0970	2000 amp		2	8		3,000	420		3,420	3,925
0980	2500 amp		1.80	8.889		3,650	465		4,115	4,725
0990	3000 amp		1.60	10		4,275	525		4,800	5,475
1000	4000 amp		1.40	11.429		5,550	600		6,150	7,000
1050	Service head, weatherproof, 3 pole 3 wire, 225 amp		3	5.333		1,650	279		1,929	2,250
1060	400 amp		2.80	5.714		1,650	299		1,949	2,275
1070	600 amp		2.60	6.154		1,650	320		1,970	2,300
1080	800 amp		2.40	6.667		1,850	350		2,200	2,575
1090	1000 amp		2	8		2,000	420		2,420	2,850
1100	1350 amp		1.80	8.889		2,600	465		3,065	3,575
1110	1600 amp		1.60	10		2,925	525		3,450	3,975
1120	2000 amp		1.40	11.429		3,550	600		4,150	4,800
1130	2500 amp		1.20	13.333		4,250	700		4,950	5,700
1140	3000 amp		.90	17.778		4,950	930		5,880	6,825
1150	4000 amp		.70	22.857		6,275	1,200		7,475	8,700
1200	3 pole 4 wire, 225 amp		2.60	6.154		1,850	320		2,170	2,525
1210	400 amp		2.40	6.667		1,850	350		2,200	2,575
1220	600 amp		2.20	7.273		1,850	380		2,230	2,625
1230	800 amp		2	8		2,150	420		2,570	3,000
1240	1000 amp		1.70	9.412		2,475	495		2,970	3,475
1250	1350 amp		1.50	10.667		2,950	560		3,510	4,075
1260	1600 amp		1.40	11.429		3,200	600		3,800	4,425
1270	2000 amp		1.20	13.333		4,300	700		5,000	5,775
1280	2500 amp		1	16		5,275	840		6,115	7,075

26 25 13 – Bus Duct/Busway and Fittings

26 25 13.60 Copper or Aluminum Bus Duct Fittings		Crew	Daily Output	Labor-Hours	Unit	Material	Labor	Equipment	Total	Total Incl O&P
						2013 Bare Costs				
1290	3000 amp	2 Elec	.80	20	Ea.	6,200	1,050		7,250	8,375
1300	4000 amp		.60	26.667		8,025	1,400		9,425	10,900
1350	Flanged end, 3 pole 3 wire, 225 amp		6	2.667		905	140		1,045	1,200
1360	400 amp		5.20	3.077		905	161		1,066	1,225
1370	600 amp		4.60	3.478		905	182		1,087	1,275
1380	800 amp		3.60	4.444		1,000	233		1,233	1,450
1390	1000 amp		3.40	4.706		1,125	247		1,372	1,600
1395	1200 amp		3.20	5		1,250	262		1,512	1,775
1400	1350 amp		3	5.333		1,350	279		1,629	1,900
1410	1600 amp		2.80	5.714		1,525	299		1,824	2,125
1420	2000 amp		2.40	6.667		1,800	350		2,150	2,500
1430	2500 amp		2	8		2,100	420		2,520	2,950
1440	3000 amp		1.80	8.889		2,400	465		2,865	3,325
1450	4000 amp		1.60	10		2,975	525		3,500	4,050
1500	3 pole 4 wire, 225 amp		5.40	2.963		1,025	155		1,180	1,375
1510	400 amp		4.60	3.478		1,025	182		1,207	1,425
1520	600 amp		4	4		1,025	210		1,235	1,475
1530	800 amp		3.20	5		1,225	262		1,487	1,750
1540	1000 amp		3	5.333		1,375	279		1,654	1,925
1545	1200 amp		2.80	5.714		1,550	299		1,849	2,150
1550	1350 amp		2.60	6.154		1,675	320		1,995	2,300
1560	1600 amp		2.40	6.667		1,900	350		2,250	2,625
1570	2000 amp		2	8		2,250	420		2,670	3,100
1580	2500 amp		1.80	8.889		2,675	465		3,140	3,650
1590	3000 amp		1.60	10		3,050	525		3,575	4,125
1600	4000 amp		1.40	11.429		3,950	600		4,550	5,250
1650	Hanger, standard, 225 amp		64	.250		19	13.10		32.10	40.50
1660	400 amp		48	.333		19	17.45		36.45	47
1670	600 amp		40	.400		19	21		40	52.50
1680	800 amp		32	.500		19	26		45	60
1690	1000 amp		24	.667		19	35		54	73.50
1695	1200 amp		22	.727		19	38		57	78
1700	1350 amp		20	.800		19	42		61	83.50
1710	1600 amp		20	.800		19	42		61	83.50
1720	2000 amp		18	.889		19	46.50		65.50	90.50
1730	2500 amp		16	1		19	52.50		71.50	99.50
1740	3000 amp		16	1		19	52.50		71.50	99.50
1750	4000 amp		16	1		19	52.50		71.50	99.50
1800	Spring type, 225 amp		16	1		85	52.50		137.50	172
1810	400 amp		14	1.143		85	60		145	183
1820	600 amp		14	1.143		85	60		145	183
1830	800 amp		14	1.143		85	60		145	183
1840	1000 amp		14	1.143		85	60		145	183
1845	1200 amp		14	1.143		85	60		145	183
1850	1350 amp		14	1.143		85	60		145	183
1860	1600 amp		12	1.333		85	70		155	199
1870	2000 amp		12	1.333		85	70		155	199
1880	2500 amp		12	1.333		85	70		155	199
1890	3000 amp		10	1.600		85	84		169	219
1900	4000 amp	▼	10	1.600	▼	85	84		169	219

26 25 Enclosed Bus Assemblies

26 25 13 – Bus Duct/Busway and Fittings

26 25 13.70 Feedrail		Crew	Daily Output	Labor-Hours	Unit	Material	Labor	Equipment	Total	Total Incl O&P
						2013 Bare Costs				
0010	**FEEDRAIL**, 12 foot mounting									
0050	Trolley busway, 3 pole									
0100	300 volt 60 amp, plain, 10 ft. lengths	1 Elec	50	.160	L.F.	26.50	8.40		34.90	41.50
0300	Door track		50	.160		33.50	8.40		41.90	49
0500	Curved track		30	.267		385	13.95		398.95	445
0700	Coupling				Ea.	17.40			17.40	19.15
0900	Center feed	1 Elec	5.30	1.509		65.50	79		144.50	190
1100	End feed		5.30	1.509		57.50	79		136.50	182
1300	Hanger set		24	.333		5.20	17.45		22.65	31.50
3000	600 volt 100 amp, plain, 10 ft. lengths		35	.229	L.F.	60.50	12		72.50	84.50
3300	Door track		35	.229	"	74	12		86	99.50
3700	Coupling				Ea.	52.50			52.50	58
4000	End cap	1 Elec	40	.200		42	10.50		52.50	62
4200	End feed		4	2		253	105		358	435
4500	Trolley, 600 volt, 20 amp		5.30	1.509		345	79		424	500
4700	30 amp		5.30	1.509		345	79		424	500
4900	Duplex, 40 amp		4	2		835	105		940	1,075
5000	60 amp		4	2		835	105		940	1,075
5300	Fusible, 20 amp		4	2		945	105		1,050	1,200
5500	30 amp		4	2		945	105		1,050	1,200
5900	300 volt, 20 amp		5.30	1.509		293	79		372	440
6000	30 amp		5.30	1.509		325	79		404	480
6300	Fusible, 20 amp		4.70	1.702		390	89		479	560
6500	30 amp		4.70	1.702		555	89		644	745
7300	Busway, 250 volt 50 amp, 2 wire		70	.114	L.F.	19.35	6		25.35	30.50
7330	Coupling				Ea.	35.50			35.50	39
7340	Center feed	1 Elec	6	1.333		415	70		485	560
7350	End feed		6	1.333		88	70		158	202
7360	End cap		40	.200		26.50	10.50		37	45
7370	Hanger set		24	.333		3.04	17.45		20.49	29.50
7400	125/250 volt 50 amp, 3 wire		60	.133	L.F.	19.60	7		26.60	32
7430	Coupling		6	1.333	Ea.	49.50	70		119.50	160
7440	Center feed		6	1.333		440	70		510	590
7450	End feed		6	1.333		88	70		158	202
7460	End cap		40	.200		33.50	10.50		44	52.50
7470	Hanger set		24	.333		3.04	17.45		20.49	29.50
7480	Trolley, 250 volt, 2 pole, 20 amp		6	1.333		45.50	70		115.50	155
7490	30 amp		6	1.333		45.50	70		115.50	155
7500	125/250 volt, 3 pole, 20 amp		6	1.333		45.50	70		115.50	155
7510	30 amp		6	1.333		45.50	70		115.50	155
8000	Cleaning tools, 300 volt, dust remover					92			92	101
8100	Bus bar cleaner					164			164	180
8300	600 volt, dust remover, 60 amp					345			345	380
8400	100 amp					475			475	520
8600	Bus bar cleaner, 60 amp					495			495	545
8700	100 amp					585			585	640

26 27 13.10 Meter Centers and Sockets		Crew	Daily Output	Labor-Hours	Unit	Material	2013 Bare Costs Labor	Equipment	Total	Total Incl O&P
0010	**METER CENTERS AND SOCKETS**									
0100	Sockets, single position, 4 terminal, 100 amp	1 Elec	3.20	2.500	Ea.	46.50	131		177.50	247
0200	150 amp		2.30	3.478		64	182		246	345
0300	200 amp		1.90	4.211		84	221		305	420
0400	Transformer rated, 20 amp		3.20	2.500		139	131		270	350
0500	Double position, 4 terminal, 100 amp		2.80	2.857		192	150		342	435
0600	150 amp		2.10	3.810		230	200		430	550
0700	200 amp		1.70	4.706		465	247		712	880
0800	Trans-socket, 13 terminal, 3 CT mounts, 400 amp		1	8		1,325	420		1,745	2,075
0900	800 amp	2 Elec	1.20	13.333		1,500	700		2,200	2,700
1100	Meter centers and sockets, three phase, single pos, 7 terminal, 100 amp	1 Elec	2.80	2.857		125	150		275	360
1200	200 amp		2.10	3.810		330	200		530	660
1400	400 amp		1.70	4.706		840	247		1,087	1,300
2000	Meter center, main fusible switch, 1P 3W 120/240 volt									
2030	400 amp	2 Elec	1.60	10	Ea.	1,525	525		2,050	2,450
2040	600 amp		1.10	14.545		2,650	760		3,410	4,075
2050	800 amp		.90	17.778		4,150	930		5,080	5,975
2060	Rainproof 1P 3W 120/240 volt, 400 amp		1.60	10		1,525	525		2,050	2,450
2070	600 amp		1.10	14.545		2,650	760		3,410	4,075
2080	800 amp		.90	17.778		4,150	930		5,080	5,975
2100	3P 4W 120/208 V, 400 amp		1.60	10		1,750	525		2,275	2,700
2110	600 amp		1.10	14.545		3,250	760		4,010	4,725
2120	800 amp		.90	17.778		6,050	930		6,980	8,050
2130	Rainproof 3P 4W 120/208 V, 400 amp		1.60	10		1,750	525		2,275	2,700
2140	600 amp		1.10	14.545		3,250	760		4,010	4,725
2150	800 amp		.90	17.778		6,050	930		6,980	8,050
2170	Main circuit breaker, 1P 3W 120/240 V									
2180	400 amp	2 Elec	1.60	10	Ea.	2,725	525		3,250	3,775
2190	600 amp		1.10	14.545		3,675	760		4,435	5,175
2200	800 amp		.90	17.778		4,275	930		5,205	6,100
2210	1000 amp		.80	20		5,900	1,050		6,950	8,075
2220	1200 amp		.76	21.053		7,975	1,100		9,075	10,400
2230	1600 amp		.68	23.529		17,800	1,225		19,025	21,500
2240	Rainproof 1P 3W 120/240 V, 400 amp		1.60	10		2,725	525		3,250	3,775
2250	600 amp		1.10	14.545		3,675	760		4,435	5,175
2260	800 amp		.90	17.778		4,275	930		5,205	6,100
2270	1000 amp		.80	20		5,900	1,050		6,950	8,075
2280	1200 amp		.76	21.053		7,975	1,100		9,075	10,400
2300	3P 4W 120/208 V, 400 amp		1.60	10		3,100	525		3,625	4,200
2310	600 amp		1.10	14.545		4,375	760		5,135	5,950
2320	800 amp		.90	17.778		5,200	930		6,130	7,125
2330	1000 amp		.80	20		6,825	1,050		7,875	9,075
2340	1200 amp		.76	21.053		8,725	1,100		9,825	11,300
2350	1600 amp		.68	23.529		17,800	1,225		19,025	21,500
2360	Rainproof 3P 4W 120/208 V, 400 amp		1.60	10		3,100	525		3,625	4,200
2370	600 amp		1.10	14.545		4,375	760		5,135	5,950
2380	800 amp		.90	17.778		5,200	930		6,130	7,125
2390	1000 amp		.76	21.053		6,825	1,100		7,925	9,150
2400	1200 amp		.68	23.529		8,725	1,225		9,950	11,500
2420	Main lugs terminal box, 1P 3W 120/240 V									
2430	800 amp	2 Elec	.94	17.021	Ea.	510	890		1,400	1,900
2440	1200 amp		.72	22.222		1,025	1,175		2,200	2,875
2450	Rainproof 1P 3W 120/240 V, 225 amp		2.40	6.667		350	350		700	910

26 27 13.10 Meter Centers and Sockets		Crew	Daily Output	Labor-Hours	Unit	2013 Bare Costs Material	Labor	Equipment	Total	Total Incl O&P
2460	800 amp	2 Elec	.94	17.021	Ea.	510	890		1,400	1,900
2470	1200 amp		.72	22.222		1,025	1,175		2,200	2,875
2500	3P 4W 120/208 V, 800 amp		.94	17.021		570	890		1,460	1,950
2510	1200 amp		.72	22.222		1,150	1,175		2,325	3,000
2520	Rainproof 3P 4W 120/208 V, 225 amp		2.40	6.667		350	350		700	910
2530	800 amp		.94	17.021		570	890		1,460	1,950
2540	1200 amp		.72	22.222		1,150	1,175		2,325	3,000
2590	Basic meter device									
2600	1P 3W 120/240 V 4 jaw 125A sockets, 3 meter	2 Elec	1	16	Ea.	585	840		1,425	1,900
2610	4 meter		.90	17.778		700	930		1,630	2,175
2620	5 meter		.80	20		880	1,050		1,930	2,550
2630	6 meter		.60	26.667		1,000	1,400		2,400	3,200
2640	7 meter		.56	28.571		1,300	1,500		2,800	3,675
2650	8 meter		.52	30.769		1,400	1,600		3,000	3,950
2660	10 meter		.48	33.333		1,750	1,750		3,500	4,550
2680	Rainproof 1P 3W 120/240 V 4 jaw 125A sockets									
2690	3 meter	2 Elec	1	16	Ea.	585	840		1,425	1,900
2700	4 meter		.90	17.778		700	930		1,630	2,175
2710	6 meter		.60	26.667		1,000	1,400		2,400	3,200
2720	7 meter		.56	28.571		1,300	1,500		2,800	3,675
2730	8 meter		.52	30.769		1,400	1,600		3,000	3,950
2750	1P 3W 120/240 V 4 jaw sockets									
2760	with 125A circuit breaker, 3 meter	2 Elec	1	16	Ea.	1,100	840		1,940	2,450
2770	4 meter		.90	17.778		1,375	930		2,305	2,925
2780	5 meter		.80	20		1,725	1,050		2,775	3,475
2790	6 meter		.60	26.667		2,025	1,400		3,425	4,325
2800	7 meter		.56	28.571		2,475	1,500		3,975	4,975
2810	8 meter		.52	30.769		2,775	1,600		4,375	5,450
2820	10 meter		.48	33.333		3,450	1,750		5,200	6,425
2830	Rainproof 1P 3W 120/240 V 4 jaw sockets									
2840	with 125A circuit breaker, 3 meter	2 Elec	1	16	Ea.	1,100	840		1,940	2,450
2850	4 meter		.90	17.778		1,375	930		2,305	2,925
2870	6 meter		.60	26.667		2,025	1,400		3,425	4,325
2880	7 meter		.56	28.571		2,475	1,500		3,975	4,975
2890	8 meter		.52	30.769		2,775	1,600		4,375	5,450
2920	1P 3W on 3P 4W 120/208 V system 5 jaw									
2930	125A sockets, 3 meter	2 Elec	1	16	Ea.	585	840		1,425	1,900
2940	4 meter		.90	17.778		700	930		1,630	2,175
2950	5 meter		.80	20		880	1,050		1,930	2,550
2960	6 meter		.60	26.667		1,000	1,400		2,400	3,200
2970	7 meter		.56	28.571		1,300	1,500		2,800	3,675
2980	8 meter		.52	30.769		1,400	1,600		3,000	3,950
2990	10 meter		.48	33.333		1,750	1,750		3,500	4,550
3000	Rainproof 1P 3W on 3P 4W 120/208 V system									
3020	5 jaw 125A sockets, 3 meter	2 Elec	1	16	Ea.	585	840		1,425	1,900
3030	4 meter		.90	17.778		700	930		1,630	2,175
3050	6 meter		.60	26.667		1,000	1,400		2,400	3,200
3060	7 meter		.56	28.571		1,300	1,500		2,800	3,675
3070	8 meter		.52	30.769		1,400	1,600		3,000	3,950
3090	1P 3W on 3P 4W 120/208 V system 5 jaw sockets									
3100	With 125A circuit breaker, 3 meter	2 Elec	1	16	Ea.	1,100	840		1,940	2,450
3110	4 meter		.90	17.778		1,375	930		2,305	2,925
3120	5 meter		.80	20		1,725	1,050		2,775	3,475

26 27 13.10 Meter Centers and Sockets		Crew	Daily Output	Labor-Hours	Unit	2013 Bare Costs Material	Labor	Equipment	Total	Total Incl O&P
3130	6 meter	2 Elec	.60	26.667	Ea.	2,025	1,400		3,425	4,325
3140	7 meter		.56	28.571		2,475	1,500		3,975	4,975
3150	8 meter		.52	30.769		2,775	1,600		4,375	5,450
3160	10 meter		.48	33.333		3,450	1,750		5,200	6,425
3170	Rainproof 1P 3W on 3P 4W 120/208 V system									
3180	5 jaw sockets w/125A circuit breaker, 3 meter	2 Elec	1	16	Ea.	1,100	840		1,940	2,450
3190	4 meter		.90	17.778		1,375	930		2,305	2,925
3210	6 meter		.60	26.667		2,025	1,400		3,425	4,325
3220	7 meter		.56	28.571		2,475	1,500		3,975	4,975
3230	8 meter		.52	30.769		2,775	1,600		4,375	5,450
3250	1P 3W 120/240 V 4 jaw sockets									
3260	with 200A circuit breaker, 3 meter	2 Elec	1	16	Ea.	1,650	840		2,490	3,050
3270	4 meter		.90	17.778		2,225	930		3,155	3,850
3290	6 meter		.60	26.667		3,300	1,400		4,700	5,725
3300	7 meter		.56	28.571		3,875	1,500		5,375	6,500
3310	8 meter		.56	28.571		4,450	1,500		5,950	7,150
3330	Rainproof 1P 3W 120/240 V 4 jaw sockets									
3350	with 200A circuit breaker, 3 meter	2 Elec	1	16	Ea.	1,650	840		2,490	3,050
3360	4 meter		.90	17.778		2,225	930		3,155	3,850
3380	6 meter		.60	26.667		3,300	1,400		4,700	5,725
3390	7 meter		.56	28.571		3,875	1,500		5,375	6,500
3400	8 meter		.52	30.769		4,450	1,600		6,050	7,300
3420	1P 3W on 3P 4W 120/208 V 5 jaw sockets									
3430	with 200A circuit breaker, 3 meter	2 Elec	1	16	Ea.	1,650	840		2,490	3,050
3440	4 meter		.90	17.778		2,225	930		3,155	3,850
3460	6 meter		.60	26.667		3,300	1,400		4,700	5,725
3470	7 meter		.56	28.571		3,875	1,500		5,375	6,500
3480	8 meter		.52	30.769		4,450	1,600		6,050	7,300
3500	Rainproof 1P 3W on 3P 4W 120/208 V 5 jaw socket									
3510	with 200A circuit breaker, 3 meter	2 Elec	1	16	Ea.	1,650	840		2,490	3,050
3520	4 meter		.90	17.778		2,225	930		3,155	3,850
3540	6 meter		.60	26.667		3,300	1,400		4,700	5,725
3550	7 meter		.56	28.571		3,875	1,500		5,375	6,500
3560	8 meter		.52	30.769		4,450	1,600		6,050	7,300
3600	Automatic circuit closing, add					63.50			63.50	70
3610	Manual circuit closing, add					72.50			72.50	80
3650	Branch meter device									
3660	3P 4W 208/120 or 240/120 V 7 jaw sockets									
3670	with 200A circuit breaker, 2 meter	2 Elec	.90	17.778	Ea.	2,750	930		3,680	4,425
3680	3 meter		.80	20		4,125	1,050		5,175	6,100
3690	4 meter		.70	22.857		5,500	1,200		6,700	7,850
3700	Main circuit breaker 42,000 rms, 400 amp		1.60	10		2,400	525		2,925	3,400
3710	600 amp		1.10	14.545		3,875	760		4,635	5,400
3720	800 amp		.90	17.778		5,200	930		6,130	7,125
3730	Rainproof main circ. breaker 42,000 rms, 400 amp		1.60	10		2,400	525		2,925	3,400
3740	600 amp		1.10	14.545		3,875	760		4,635	5,400
3750	800 amp		.90	17.778		5,200	930		6,130	7,125
3760	Main circuit breaker 65,000 rms, 400 amp		1.60	10		3,250	525		3,775	4,350
3770	600 amp		1.10	14.545		4,600	760		5,360	6,200
3780	800 amp		.90	17.778		5,200	930		6,130	7,125
3790	1000 amp		.80	20		6,825	1,050		7,875	9,075
3800	1200 amp		.76	21.053		8,725	1,100		9,825	11,300
3810	1600 amp		.68	23.529		17,800	1,225		19,025	21,500

26 27 Low-Voltage Distribution Equipment

26 27 13 – Electricity Metering

26 27 13.10 Meter Centers and Sockets		Crew	Daily Output	Labor-Hours	Unit	Material	2013 Bare Costs Labor	Equipment	Total	Total Incl O&P
3820	Rainproof main circ. breaker 65,000 rms, 400 amp	2 Elec	1.60	10	Ea.	3,250	525		3,775	4,350
3830	600 amp		1.10	14.545		4,600	760		5,360	6,200
3840	800 amp		.90	17.778		5,200	930		6,130	7,100
3850	1000 amp		.80	20		6,825	1,050		7,875	9,075
3860	1200 amp		.76	21.053		8,725	1,100		9,825	11,300
3880	Main circuit breaker 100,000 rms, 400 amp		1.60	10		3,250	525		3,775	4,350
3890	600 amp		1.10	14.545		4,600	760		5,360	6,200
3900	800 amp		.90	17.778		5,375	930		6,305	7,325
3910	Rainproof main circ. breaker 100,000 rms, 400 amp		1.60	10		3,250	525		3,775	4,350
3920	600 amp		1.10	14.545		4,600	760		5,360	6,200
3930	800 amp		.90	17.778		5,375	930		6,305	7,325
3940	Main lugs terminal box, 800 amp		.94	17.021		570	890		1,460	1,950
3950	1600 amp		.72	22.222		1,925	1,175		3,100	3,850
3960	Rainproof, 800 amp		.94	17.021		570	890		1,460	1,950
3970	1600 amp		.72	22.222		1,925	1,175		3,100	3,850

26 27 16 – Electrical Cabinets and Enclosures

26 27 16.10 Cabinets		Crew	Daily Output	Labor-Hours	Unit	Material	Labor	Equipment	Total	Total Incl O&P
0010	CABINETS R260533-70									
7000	Cabinets, current transformer									
7050	Single door, 24" H x 24" W x 10" D R262716-40	1 Elec	1.60	5	Ea.	154	262		416	560
7100	30" H x 24" W x 10" D		1.30	6.154		156	320		476	650
7150	36" H x 24" W x 10" D		1.10	7.273		161	380		541	745
7200	30" H x 30" W x 10" D		1	8		168	420		588	810
7250	36" H x 30" W x 10" D		.90	8.889		198	465		663	915
7300	36" H x 36" W x 10" D		.80	10		204	525		729	1,000
7500	Double door, 48" H x 36" W x 10" D		.60	13.333		505	700		1,205	1,600
7550	24" H x 24" W x 12" D		1	8		157	420		577	795
7600	Telephone with wood backboard									
7620	Single door, 12" H x 12" W x 4" D	1 Elec	5.30	1.509	Ea.	81.50	79		160.50	208
7650	18" H x 12" W x 4" D		4.70	1.702		95.50	89		184.50	238
7700	24" H x 12" W x 4" D		4.20	1.905		105	100		205	265
7720	18" H x 18" W x 4" D		4.20	1.905		114	100		214	274
7750	24" H x 18" W x 4" D		4	2		162	105		267	335
7780	36" H x 36" W x 4" D		3.60	2.222		206	116		322	400
7800	24" H x 24" W x 6" D		3.60	2.222		201	116		317	395
7820	30" H x 24" W x 6" D		3.20	2.500		253	131		384	475
7850	30" H x 30" W x 6" D		2.70	2.963		293	155		448	555
7880	36" H x 30" W x 6" D		2.50	3.200		340	168		508	625
7900	48" H x 36" W x 6" D		2.20	3.636		560	191		751	900
7920	48" H x 36" W x 6" D		2	4		775	210		985	1,175
8000	NEMA 12, double door, floor mounted									
8020	54" H x 42" W x 8" D	2 Elec	6	2.667	Ea.	1,225	140		1,365	1,550
8040	60" H x 48" W x 8" D		5.40	2.963		1,650	155		1,805	2,025
8060	60" H x 48" W x 10" D		5.40	2.963		1,700	155		1,855	2,075
8080	60" H x 60" W x 10" D		5	3.200		1,900	168		2,068	2,325
8100	72" H x 60" W x 10" D		4	4		2,225	210		2,435	2,775
8120	72" H x 72" W x 10" D		3.40	4.706		2,800	247		3,047	3,450
8140	60" H x 48" W x 12" D		3.40	4.706		1,725	247		1,972	2,275
8160	60" H x 60" W x 12" D		3.20	5		1,950	262		2,212	2,550
8180	72" H x 60" W x 12" D		3	5.333		2,200	279		2,479	2,850
8200	72" H x 72" W x 12" D		3	5.333		2,475	279		2,754	3,125
8220	60" H x 48" W x 16" D		3.20	5		1,800	262		2,062	2,375

26 27 16.10 Cabinets		Crew	Daily Output	Labor-Hours	Unit	Material	Labor	Equipment	Total	Total Incl O&P
						2013 Bare Costs				
8240	72" H x 72" W x 16" D	2 Elec	2.60	6.154	Ea.	2,600	320		2,920	3,325
8260	60" H x 48" W x 20" D		3	5.333		1,975	279		2,254	2,575
8280	72" H x 72" W x 20" D		2.20	7.273		2,800	380		3,180	3,650
8300	60" H x 48" W x 24" D		2.60	6.154		2,075	320		2,395	2,775
8320	72" H x 72" W x 24" D		2	8		2,950	420		3,370	3,875
8340	Pushbutton enclosure, oiltight									
8360	3-1/2" H x 3-1/4" W x 2-3/4" D, for 1 P.B.	1 Elec	12	.667	Ea.	50	35		85	108
8380	5-3/4" H x 3-1/4" W x 2-3/4" D, for 2 P.B.		11	.727		71	38		109	135
8400	8" H x 3-1/4" W x 2-3/4" D, for 3 P.B.		10.50	.762		79	40		119	147
8420	10-1/4" H x 3-1/4" W x 2-3/4" D, for 4 P.B.		10.50	.762		77.50	40		117.50	145
8460	12-1/2" H x 3-1/4" W x 3" D, for 5 P.B.		9	.889		86.50	46.50		133	165
8480	9-1/2" H x 6-1/4" W x 3" D, for 6 P.B.		8.50	.941		107	49.50		156.50	191
8500	9-1/2" H x 8-1/2" W x 3" D, for 9 P.B.		8	1		106	52.50		158.50	196
8510	11-3/4" H x 8-1/2" W x 3" D, for 12 P.B.		7	1.143		118	60		178	220
8520	11-3/4" H x 10-3/4" W x 3" D, for 16 P.B.		6.50	1.231		152	64.50		216.50	264
8540	14" H x 10-3/4" W x 3" D, for 20 P.B.		5	1.600		172	84		256	315
8560	14" H x 13" W x 3" D, for 25 P.B.		4.50	1.778		197	93		290	355
8580	Sloping front pushbutton enclosures									
8600	3-1/2" H x 7-3/4" W x 4-7/8" D, for 3 P.B.	1 Elec	10	.800	Ea.	90.50	42		132.50	163
8620	7-1/4" H x 8-1/2" W x 6-3/4" D, for 6 P.B.		8	1		132	52.50		184.50	225
8640	9-1/2" H x 8-1/2" W x 7-7/8" D, for 9 P.B.		7	1.143		157	60		217	263
8660	11-1/4" H x 8-1/2" W x 9" D, for 12 P.B.		5	1.600		182	84		266	325
8680	11-3/4" H x 10" W x 9" D, for 16 P.B.		5	1.600		195	84		279	340
8700	11-3/4" H x 13" W x 9" D, for 20 P.B.		5	1.600		226	84		310	375
8720	14" H x 13" W x 10-1/8" D, for 25 P.B.		4.50	1.778		256	93		349	420
8740	Pedestals, not including P.B. enclosure or base									
8760	Straight column 4" x 4"	1 Elec	4.50	1.778	Ea.	254	93		347	420
8780	6" x 6"		4	2		395	105		500	590
8800	Angled column 4" x 4"		4.50	1.778		293	93		386	460
8820	6" x 6"		4	2		440	105		545	640
8840	Pedestal, base 18" x 18"		10	.800		150	42		192	228
8860	24" x 24"		9	.889		345	46.50		391.50	450
8900	Electronic rack enclosures									
8920	72" H x 25" W x 24" D	1 Elec	1.50	5.333	Ea.	2,200	279		2,479	2,850
8940	72" H x 30" W x 24" D		1.50	5.333		2,325	279		2,604	3,000
8960	72" H x 25" W x 30" D		1.30	6.154		2,375	320		2,695	3,075
8980	72" H x 25" W x 36" D		1.20	6.667		2,550	350		2,900	3,350
9000	72" H x 30" W x 36" D		1.20	6.667		2,675	350		3,025	3,475
9020	NEMA 12 & 4 enclosure panels									
9040	12" x 24"	1 Elec	20	.400	Ea.	33	21		54	67.50
9060	16" x 12"		20	.400		24	21		45	58
9080	20" x 16"		20	.400		34.50	21		55.50	69.50
9100	20" x 20"		19	.421		41.50	22		63.50	79
9120	24" x 20"		18	.444		54	23.50		77.50	94.50
9140	24" x 24"		17	.471		62	24.50		86.50	105
9160	30" x 20"		16	.500		76.50	26		102.50	124
9180	30" x 24"		16	.500		75.50	26		101.50	122
9200	36" x 24"		15	.533		89	28		117	140
9220	36" x 30"		15	.533		120	28		148	173
9240	42" x 24"		15	.533		102	28		130	154
9260	42" x 30"		14	.571		135	30		165	193
9280	42" x 36"		14	.571		158	30		188	219
9300	48" x 24"		14	.571		121	30		151	179

26 27 Low-Voltage Distribution Equipment

26 27 16 – Electrical Cabinets and Enclosures

26 27 16.10 Cabinets		Crew	Daily Output	Labor-Hours	Unit	Material	2013 Bare Costs Labor	Equipment	Total	Total Incl O&P
9320	48″ x 30″	1 Elec	14	.571	Ea.	157	30		187	218
9340	48″ x 36″		13	.615		182	32.50		214.50	250
9360	60″ x 36″		12	.667		222	35		257	297
9400	Wiring trough steel JIC, clamp cover									
9490	4″ x 4″, 12″ long	1 Elec	12	.667	Ea.	85.50	35		120.50	147
9510	24″ long		10	.800		108	42		150	181
9530	36″ long		8	1		135	52.50		187.50	227
9540	48″ long		7	1.143		156	60		216	262
9550	60″ long		6	1.333		183	70		253	305
9560	6″ x 6″, 12″ long		11	.727		113	38		151	181
9580	24″ long		9	.889		150	46.50		196.50	235
9600	36″ long		7	1.143		184	60		244	292
9610	48″ long		6	1.333		223	70		293	350
9620	60″ long		5	1.600		263	84		347	415

26 27 16.20 Cabinets and Enclosures		Crew	Daily Output	Labor-Hours	Unit	Material	Labor	Equipment	Total	Total Incl O&P
0010	**CABINETS AND ENCLOSURES** Nonmetallic									
0080	Enclosures fiberglass NEMA 4X									
0100	Wall mount, quick release latch door, 20″H x 16″W x 6″D	1 Elec	4.80	1.667	Ea.	565	87.50		652.50	755
0110	20″H x 20″W x 6″D		4.50	1.778		680	93		773	890
0120	24″H x 20″W x 6″D		4.20	1.905		725	100		825	950
0130	20″H x 16″W x 8″D		4.50	1.778		645	93		738	850
0140	20″H x 20″W x 8″D		4.20	1.905		725	100		825	945
0150	24″H x 24″W x 8″D		3.80	2.105		820	110		930	1,075
0160	30″H x 24″W x 8″D		3.20	2.500		885	131		1,016	1,175
0170	36″H x 30″W x 8″D		3	2.667		1,250	140		1,390	1,575
0180	20″H x 16″W x 10″D		3.50	2.286		755	120		875	1,000
0190	20″H x 20″W x 10″D		3.20	2.500		805	131		936	1,075
0200	24″H x 20″W x 10″D		3	2.667		855	140		995	1,150
0210	30″H x 24″W x 10″D		2.80	2.857		970	150		1,120	1,300
0220	20″H x 16″W x 12″D		3	2.667		800	140		940	1,100
0230	20″H x 20″W x 12″D		2.80	2.857		860	150		1,010	1,175
0240	24″H x 24″W x 12″D		2.60	3.077		950	161		1,111	1,300
0250	30″H x 24″W x 12″D		2.40	3.333		1,050	175		1,225	1,425
0260	36″H x 30″W x 12″D		2.20	3.636		1,475	191		1,666	1,875
0270	36″H x 36″W x 12″D		2.10	3.810		1,625	200		1,825	2,100
0280	48″H x 36″W x 12″D		2	4		1,850	210		2,060	2,375
0290	60″H x 36″W x 12″D		1.80	4.444		2,100	233		2,333	2,675
0300	30″H x 24″W x 16″D		1.40	5.714		1,225	299		1,524	1,800
0310	48″H x 36″W x 16″D		1.20	6.667		2,050	350		2,400	2,775
0320	60″H x 36″W x 16″D		1	8		2,325	420		2,745	3,175
0480	Freestanding, one door, 72″H x 25″W x 25″D		.80	10		3,850	525		4,375	5,000
0490	Two doors with two panels, 72″H x 49″W x 24″D		.50	16		9,575	840		10,415	11,800
0500	Floor stand kits, for NEMA 4 & 12, 20″W or more, 6″H x 10″D		24	.333		140	17.45		157.45	180
0510	6″H x 10″D		24	.333		153	17.45		170.45	195
0520	6″H x 12″D		24	.333		171	17.45		188.45	215
0530	6″H x 18″D		24	.333		199	17.45		216.45	245
0540	12″H x 8″D		22	.364		177	19.05		196.05	224
0550	12″H x 10″D		22	.364		188	19.05		207.05	236
0560	12″H x 12″D		22	.364		200	19.05		219.05	249
0570	12″H x 16″D		22	.364		233	19.05		252.05	285
0580	12″H x 18″D		22	.364		239	19.05		258.05	292
0590	12″H x 20″D		22	.364		254	19.05		273.05	310

26 27 16 – Electrical Cabinets and Enclosures

26 27 16.20 Cabinets and Enclosures		Crew	Daily Output	Labor-Hours	Unit	Material	Labor (2013 Bare Costs)	Equipment	Total	Total Incl O&P
0600	18"H x 8"D	1 Elec	20	.400	Ea.	211	21		232	264
0610	18"H x 10"D		20	.400		227	21		248	281
0620	18"H x 12"D		20	.400		243	21		264	299
0630	18"H x 16"D		20	.400		270	21		291	330
0640	24"H x 8"D		16	.500		261	26		287	325
0650	24"H x 10"D		16	.500		270	26		296	335
0660	24"H x 12"D		16	.500		291	26		317	360
0670	24"H x 16"D		16	.500		305	26		331	375
0680	Small, screw cover, 5-1/2"H x 4"W x 4-15/16"D		12	.667		69.50	35		104.50	129
0690	7-1/2"H x 4"W x 4-15/16"D		12	.667		74	35		109	134
0700	7-1/2"H x 6"W x 5-3/16"D		10	.800		86.50	42		128.50	158
0710	9-1/2"H x 6"W x 5-11/16"D		10	.800		88.50	42		130.50	160
0720	11-1/2"H x 8"W x 6-11/16"D		8	1		131	52.50		183.50	224
0730	13-1/2"H x 10"W x 7-3/16"D		7	1.143		158	60		218	264
0740	15-1/2"H x 12"W x 8-3/16"D		6	1.333		207	70		277	335
0750	17-1/2"H x 14"W x 8-11/16"D		5	1.600		246	84		330	395
0760	Screw cover with window, 6"H x 4"W x 5"D		12	.667		149	35		184	217
0770	8"H x 4"W x 5"D		11	.727		157	38		195	229
0780	8"H x 6"W x 5"D		11	.727		178	38		216	253
0790	10"H x 6"W x 6"D		10	.800		192	42		234	274
0800	12"H x 8"W x 7"D		8	1		258	52.50		310.50	365
0810	14"H x 10"W x 7"D		7	1.143		281	60		341	400
0820	16"H x 12"W x 8"D		6	1.333		355	70		425	495
0830	18"H x 14"W x 9"D		5	1.600		425	84		509	590
0840	Quick-release latch cover, 5-1/2"H x 4"W x 5"D		12	.667		95	35		130	157
0850	7-1/2"H x 4"W x 5"D		12	.667		102	35		137	165
0860	7-1/2"H x 6"W x 5-1/4"D		10	.800		110	42		152	184
0870	9-1/2"H x 6"W x 5-3/4"D		10	.800		121	42		163	196
0880	11-1/2"H x 8"W x 6-3/4"D		8	1		172	52.50		224.50	268
0890	13-1/2"H x 10"W x 7-1/4"D		7	1.143		215	60		275	325
0900	15-1/2"H x 12"W x 8-1/4"D		6	1.333		265	70		335	395
0910	17-1/2"H x 14"W x 8-3/4"D		5	1.600		320	84		404	475
0920	Pushbutton, 1 hole 5-1/2"H x 4"W x 4-15/16"D		12	.667		76	35		111	136
0930	2 hole 7-1/2"H x 4"W x 4-15/16"D		11	.727		86.50	38		124.50	152
0940	4 hole 7-1/2"H x 6"W x 5-3/16"D		10.50	.762		102	40		142	172
0950	6 hole 9-1/2"H x 6"W x 5-11/16"D		9	.889		125	46.50		171.50	207
0960	8 hole 11-1/2"H x 8"W x 6-11/16"D		8.50	.941		159	49.50		208.50	249
0970	12 hole 13-1/2"H x 10"W x 7-3/16"D		8	1		204	52.50		256.50	305
0980	20 hole 15-1/2"H x 12"W x 8-3/16"D		5	1.600		279	84		363	430
0990	30 hole 17-1/2"H x 14"W x 8-11/16"D	↓	4.50	1.778	↓	310	93		403	480
1450	Enclosures polyester NEMA 4X									
1460	Small, screw cover,									
1500	3-15/16"H x 3-15/16"W x 3-1/16"D	1 Elec	12	.667	Ea.	64	35		99	123
1510	5-3/16"H x 3-5/16"W x 3-1/16"D		12	.667		63	35		98	122
1520	5-7/8"H x 3-7/8"W x 4-3/16"D		12	.667		66.50	35		101.50	126
1530	5-7/8"H x 5-7/8"W x 4-3/16"D		12	.667		74.50	35		109.50	135
1540	7-5/8"H x 3-5/16"W x 3-1/16"D		12	.667		69	35		104	128
1550	10-3/16"H x 3-5/16"W x 3-1/16"D		10	.800		81.50	42		123.50	153
1560	Clear cover, 3-15/16"H x 3-15/16"W x 2-7/8"D		12	.667		72.50	35		107.50	132
1570	5-3/16"H x 3-5/16"W x 2-7/8"D		12	.667		77	35		112	137
1580	5-7/8"H x 3-7/8"W x 4"D		12	.667		95.50	35		130.50	158
1590	5-7/8"H x 5-7/8"W x 4"D		12	.667		118	35		153	183
1600	7-5/8"H x 3-5/16"W x 2-7/8"D		12	.667		86.50	35		121.50	148

26 27 16 – Electrical Cabinets and Enclosures

26 27 16.20 Cabinets and Enclosures		Crew	Daily Output	Labor-Hours	Unit	Material	Labor	Equipment	Total	Total Incl O&P
						2013 Bare Costs				
1610	10-3/16"H x 3-5/16"W x 2-7/8"D	1 Elec	10	.800	Ea.	109	42		151	182
1620	Pushbutton, 1 hole, 5-5/16"H x 3-5/16"W x 3-1/16"D		12	.667		56	35		91	114
1630	2 hole, 7-5/8"H x 3-5/16"W x 3-1/8"D		11	.727		63	38		101	127
1640	3 hole, 10-3/16"H x 3-5/16"W x 3-1/16"D		10.50	.762		74.50	40		114.50	142
8000	Wireway fiberglass, straight sect. screwcover, 12" L, 4" W x 4" D		40	.200		176	10.50		186.50	210
8010	6" W x 6" D		30	.267		234	13.95		247.95	279
8020	24" L, 4" W x 4" D		20	.400		214	21		235	267
8030	6" W x 6" D		15	.533		355	28		383	430
8040	36" L, 4" W x 4" D		13.30	.602		268	31.50		299.50	340
8050	6" W x 6" D		10	.800		420	42		462	525
8060	48" L, 4" W x 4" D		10	.800		320	42		362	415
8070	6" W x 6" D		7.50	1.067		525	56		581	665
8080	60" L, 4" W x 4" D		8	1		465	52.50		517.50	590
8090	6" W x 6" D		6	1.333		610	70		680	780
8100	Elbow, 90°, 4" W x 4" D		20	.400		181	21		202	231
8110	6" W x 6" D		18	.444		360	23.50		383.50	435
8120	Elbow, 45°, 4" W x 4" D		20	.400		172	21		193	221
8130	6" W x 6" D		18	.444		340	23.50		363.50	410
8140	Tee, 4" W x 4" D		16	.500		229	26		255	291
8150	6" W x 6" D		14	.571		415	30		445	500
8160	Cross, 4" W x 4" D		14	.571		325	30		355	405
8170	6" W x 6" D		12	.667		705	35		740	830
8180	Cut-off fitting, w/flange & adhesive, 4" W x 4" D		18	.444		128	23.50		151.50	176
8190	6" W x 6" D		16	.500		278	26		304	345
8200	Flexible ftng., hvy. neoprene coated nylon, 4" W x 4" D		20	.400		248	21		269	305
8210	6" W x 6" D		18	.444		365	23.50		388.50	440
8220	Closure plate, fiberglass, 4" W x 4" D		20	.400		49.50	21		70.50	86
8230	6" W x 6" D		18	.444		54	23.50		77.50	94.50
8240	Box connector, stainless steel type 304, 4" W x 4" D		20	.400		80.50	21		101.50	120
8250	6" W x 6" D		18	.444		94.50	23.50		118	139
8260	Hanger, 4" W x 4" D		100	.080		21	4.19		25.19	29.50
8270	6" W x 6" D		80	.100		26.50	5.25		31.75	37
8280	Straight tube section fiberglass, 4"W x 4"D, 12" long		40	.200		152	10.50		162.50	183
8290	24" long		20	.400		183	21		204	234
8300	36" long		13.30	.602		229	31.50		260.50	299
8310	48" long		10	.800		222	42		264	305
8320	60" long		8	1		274	52.50		326.50	380
8330	120" long		4	2		420	105		525	615

26 27 19 – Multi-Outlet Assemblies

26 27 19.10 Wiring Duct		Crew	Daily Output	Labor-Hours	Unit	Material	Labor	Equipment	Total	Total Incl O&P
0010	**WIRING DUCT** Plastic									
1250	PVC, snap-in slots, adhesive backed									
1270	1-1/2"W x 2"H	2 Elec	120	.133	L.F.	4.86	7		11.86	15.80
1280	1-1/2"W x 3"H		120	.133		5.85	7		12.85	16.85
1290	1-1/2"W x 4"H		120	.133		6.90	7		13.90	18.05
1300	2"W x 1"H		120	.133		4.36	7		11.36	15.25
1310	2"W x 1-1/2"H		120	.133		4.73	7		11.73	15.65
1320	2"W x 2"H		120	.133		4.75	7		11.75	15.70
1340	2"W x 3"H		120	.133		6.20	7		13.20	17.25
1350	2"W x 4"H		120	.133		7.50	7		14.50	18.70
1360	2-1/2"W x 3"H		120	.133		6.65	7		13.65	17.80
1370	3"W x 1"H		110	.145		4.95	7.60		12.55	16.85

26 27 19 – Multi-Outlet Assemblies

26 27 19.10 Wiring Duct		Crew	Daily Output	Labor-Hours	Unit	Material	2013 Bare Costs Labor	Equipment	Total	Total Incl O&P
1390	3"W x 2"H	2 Elec	110	.145	L.F.	5.90	7.60		13.50	17.90
1400	3"W x 3"H		110	.145		7.15	7.60		14.75	19.25
1410	3"W x 4"H		110	.145		8.95	7.60		16.55	21
1420	3"W x 5"H		110	.145		11.75	7.60		19.35	24.50
1430	4"W x 1-1/2"H		100	.160		5.90	8.40		14.30	19.05
1440	4"W x 2"H		100	.160		6.90	8.40		15.30	20
1450	4"W x 3"H		100	.160		7.80	8.40		16.20	21
1460	4"W x 4"H		100	.160		9.45	8.40		17.85	23
1470	4"W x 5"H		100	.160		13.30	8.40		21.70	27
1550	Cover, 1-1/2"W		200	.080		.99	4.19		5.18	7.35
1560	2"W		200	.080		1.22	4.19		5.41	7.60
1570	2-1/2"W		200	.080		1.53	4.19		5.72	7.95
1580	3"W		200	.080		1.86	4.19		6.05	8.30
1590	4"W		200	.080		2.24	4.19		6.43	8.70

26 27 23 – Indoor Service Poles

26 27 23.40 Surface Raceway		Crew	Daily Output	Labor-Hours	Unit	Material	Labor	Equipment	Total	Total Incl O&P
0010	**SURFACE RACEWAY**									
0090	Metal, straight section									
0100	No. 500	1 Elec	100	.080	L.F.	1.04	4.19		5.23	7.40
0110	No. 700		100	.080		1.17	4.19		5.36	7.55
0400	No. 1500, small pancake		90	.089		2.15	4.66		6.81	9.30
0600	No. 2000, base & cover, blank		90	.089		2.19	4.66		6.85	9.35
0610	Receptacle, 6" O.C.		40	.200		13.55	10.50		24.05	30.50
0620	12" O.C.		44	.182		8.65	9.55		18.20	24
0630	18" O.C.		46	.174		7.45	9.10		16.55	22
0650	30" O.C.		50	.160		4.84	8.40		13.24	17.85
0670	No. 2400, base & cover, blank		80	.100		2.49	5.25		7.74	10.60
0680	Receptacle, 6" O.C.		42	.190		32.50	10		42.50	50.50
0690	12" O.C.		53	.151		17.85	7.90		25.75	31.50
0700	18" O.C.		55	.145		13.15	7.60		20.75	26
0710	24" O.C.		57	.140		9.30	7.35		16.65	21.50
0720	30" O.C.		59	.136		6.80	7.10		13.90	18.15
0730	60" O.C.		61	.131		5.15	6.85		12	15.95
0800	No. 3000, base & cover, blank		75	.107		4.92	5.60		10.52	13.75
0810	Receptacle, 6" O.C.		45	.178		37	9.30		46.30	54.50
0820	12" O.C.		62	.129		21	6.75		27.75	33
0830	18" O.C.		64	.125		16.80	6.55		23.35	28.50
0840	24" O.C.		66	.121		12.60	6.35		18.95	23.50
0850	30" O.C.		68	.118		11.35	6.15		17.50	21.50
0860	60" O.C.		70	.114		8.30	6		14.30	18.10
1000	No. 4000, base & cover, blank		65	.123		8.20	6.45		14.65	18.65
1010	Receptacle, 6" O.C.		41	.195		49.50	10.20		59.70	70
1020	12" O.C.		52	.154		30.50	8.05		38.55	46
1030	18" O.C.		54	.148		25	7.75		32.75	39
1040	24" O.C.		56	.143		20.50	7.50		28	33.50
1050	30" O.C.		58	.138		18.60	7.25		25.85	31.50
1060	60" O.C.		60	.133		14.30	7		21.30	26
1200	No. 6000, base & cover, blank		50	.160		12.30	8.40		20.70	26
1210	Receptacle, 6" O.C.		30	.267		60.50	13.95		74.45	87.50
1220	12" O.C.		37	.216		40	11.35		51.35	61
1230	18" O.C.		39	.205		34	10.75		44.75	53
1240	24" O.C.		41	.195		27.50	10.20		37.70	46

26 27 Low-Voltage Distribution Equipment

26 27 23 – Indoor Service Poles

26 27 23.40 Surface Raceway		Crew	Daily Output	Labor-Hours	Unit	2013 Bare Costs Material	Labor	Equipment	Total	Total Incl O&P
1250	30″ O.C.	1 Elec	43	.186	L.F.	26.50	9.75		36.25	43.50
1260	60″ O.C.		45	.178		20.50	9.30		29.80	36.50
2400	Fittings, elbows, No. 500		40	.200	Ea.	1.90	10.50		12.40	17.80
2800	Elbow cover, No. 2000		40	.200		3.57	10.50		14.07	19.65
2880	Tee, No. 500		42	.190		3.66	10		13.66	19
2900	No. 2000		27	.296		11.85	15.55		27.40	36
3000	Switch box, No. 500		16	.500		11	26		37	51
3400	Telephone outlet, No. 1500		16	.500		14.10	26		40.10	54.50
3600	Junction box, No. 1500		16	.500		9.65	26		35.65	49.50
3800	Plugmold wired sections, No. 2000									
4000	1 circuit, 6 outlets, 3 ft. long	1 Elec	8	1	Ea.	36	52.50		88.50	118
4100	2 circuits, 8 outlets, 6 ft. long		5.30	1.509		53	79		132	177
4110	Tele-power pole, alum, w/2 recept, 10′		4	2		205	105		310	385
4120	12′		3.85	2.078		215	109		324	400
4130	15′		3.70	2.162		310	113		423	510
4140	Steel, w/2 recept, 10′		4	2		130	105		235	299
4150	One phone fitting, 10′		4	2		173	105		278	345
4160	Alum, 4 outlets, 10′		3.70	2.162		262	113		375	460
4300	Overhead distribution systems, 125 volt									
4800	No. 2000, entrance end fitting	1 Elec	20	.400	Ea.	5.50	21		26.50	37.50
5000	Blank end fitting		40	.200		2.21	10.50		12.71	18.15
5200	Supporting clip		40	.200		1.31	10.50		11.81	17.15
5800	No. 3000, entrance end fitting		20	.400		9.05	21		30.05	41.50
6000	Blank end fitting		40	.200		2.56	10.50		13.06	18.50
6020	Internal elbow		20	.400		11.85	21		32.85	44.50
6030	External elbow		20	.400		16.75	21		37.75	50
6040	Device bracket		53	.151		4.02	7.90		11.92	16.25
6400	Hanger clamp		32	.250		6.10	13.10		19.20	26.50
7000	No. 4000 Base		90	.089	L.F.	4.32	4.66		8.98	11.70
7200	Divider		100	.080	″	.83	4.19		5.02	7.15
7400	Entrance end fitting		16	.500	Ea.	22	26		48	63
7600	Blank end fitting		40	.200		6.05	10.50		16.55	22.50
7610	Recpt. & tele. cover		53	.151		10.35	7.90		18.25	23
7620	External elbow		16	.500		34	26		60	76
7630	Coupling		53	.151		5.15	7.90		13.05	17.55
7640	Divider clip & coupling		80	.100		1.02	5.25		6.27	8.95
7650	Panel connector		16	.500		21.50	26		47.50	62.50
7800	Take off connector		16	.500		71.50	26		97.50	118
8000	No. 6000, take off connector		16	.500		84.50	26		110.50	132
8100	Take off fitting		16	.500		63	26		89	108
8200	Hanger clamp		32	.250		15.25	13.10		28.35	36.50
8230	Coupling					8.35			8.35	9.15
8240	One gang device plate	1 Elec	53	.151		8.15	7.90		16.05	21
8250	Two gang device plate		40	.200		9.95	10.50		20.45	26.50
8260	Blank end fitting		40	.200		8.65	10.50		19.15	25
8270	Combination elbow		14	.571		35	30		65	83.50
8300	Panel connector		16	.500		15.45	26		41.45	56
8500	Chan-L-Wire system installed in 1-5/8″ x 1-5/8″ strut. Strut									
8600	not incl., 30 amp, 4 wire, 3 phase	1 Elec	200	.040	L.F.	4.85	2.10		6.95	8.50
8700	Junction box		8	1	Ea.	32.50	52.50		85	114
8800	Insulating end cap		40	.200		8.60	10.50		19.10	25
8900	Strut splice plate		40	.200		11.40	10.50		21.90	28.50
9000	Tap		40	.200		22.50	10.50		33	40.50

26 27 23 – Indoor Service Poles

26 27 23.40 Surface Raceway		Crew	Daily Output	Labor-Hours	Unit	Material	Labor	Equipment	Total	Total Incl O&P
						2013 Bare Costs				
9100	Fixture hanger	1 Elec	60	.133	Ea.	9.80	7		16.80	21.50
9200	Pulling tool					87			87	95.50
9300	Non-metallic, straight section									
9310	7/16" x 7/8", base & cover, blank	1 Elec	160	.050	L.F.	1.68	2.62		4.30	5.75
9320	Base & cover w/adhesive		160	.050		1.56	2.62		4.18	5.65
9340	7/16" x 1-5/16", base & cover, blank		145	.055		1.84	2.89		4.73	6.35
9350	Base & cover w/adhesive		145	.055		2.17	2.89		5.06	6.70
9370	11/16" x 2-1/4", base & cover, blank		130	.062		2.60	3.22		5.82	7.70
9380	Base & cover w/adhesive		130	.062		3.01	3.22		6.23	8.15
9385	1-11/16" x 5-1/4", two compartment base & cover w/screws		80	.100		7.40	5.25		12.65	15.95
9400	Fittings, elbows, 7/16" x 7/8"		50	.160	Ea.	1.88	8.40		10.28	14.60
9410	7/16" x 1-5/16"		45	.178		1.95	9.30		11.25	16.10
9420	11/16" x 2-1/4"		40	.200		2.11	10.50		12.61	18
9425	1-11/16" x 5-1/4"		28	.286		11.05	14.95		26	34.50
9430	Tees, 7/16" x 7/8"		35	.229		2.42	12		14.42	20.50
9440	7/16" x 1-5/16"		32	.250		2.49	13.10		15.59	22.50
9450	11/16" x 2-1/4"		30	.267		2.56	13.95		16.51	24
9455	1-11/16" x 5-1/4"		24	.333		17.75	17.45		35.20	45.50
9460	Cover clip, 7/16" x 7/8"		80	.100		.49	5.25		5.74	8.40
9470	7/16" x 1-5/16"		72	.111		.44	5.80		6.24	9.20
9480	11/16" x 2-1/4"		64	.125		.75	6.55		7.30	10.65
9484	1-11/16" x 5-1/4"		42	.190		2.61	10		12.61	17.80
9486	Wire clip, 1-11/16" x 5-1/4"		68	.118		.45	6.15		6.60	9.70
9490	Blank end, 7/16" x 7/8"		50	.160		.70	8.40		9.10	13.30
9500	7/16" x 1-5/16"		45	.178		.78	9.30		10.08	14.80
9510	11/16" x 2-1/4"		40	.200		1.18	10.50		11.68	17
9515	1-11/16" x 5-1/4"		38	.211		5.45	11.05		16.50	22.50
9520	Round fixture box, 5.5" dia x 1"		25	.320		11.10	16.75		27.85	37
9530	Device box, 1 gang		30	.267		5	13.95		18.95	26.50
9540	2 gang		25	.320		7.30	16.75		24.05	33

26 27 26 – Wiring Devices

26 27 26.10 Low Voltage Switching		Crew	Daily Output	Labor-Hours	Unit	Material	Labor	Equipment	Total	Total Incl O&P
0010	**LOW VOLTAGE SWITCHING**									
3600	Relays, 120 V or 277 V standard	1 Elec	12	.667	Ea.	41	35		76	97.50
3800	Flush switch, standard		40	.200		11.30	10.50		21.80	28
4000	Interchangeable		40	.200		14.75	10.50		25.25	32
4100	Surface switch, standard		40	.200		7.95	10.50		18.45	24.50
4200	Transformer 115 V to 25 V		12	.667		128	35		163	193
4400	Master control, 12 circuit, manual		4	2		124	105		229	293
4500	25 circuit, motorized		4	2		137	105		242	310
4600	Rectifier, silicon		12	.667		44.50	35		79.50	102
4800	Switchplates, 1 gang, 1, 2 or 3 switch, plastic		80	.100		4.90	5.25		10.15	13.25
5000	Stainless steel		80	.100		11.15	5.25		16.40	20
5400	2 gang, 3 switch, stainless steel		53	.151		22.50	7.90		30.40	36.50
5500	4 switch, plastic		53	.151		10.10	7.90		18	23
5600	2 gang, 4 switch, stainless steel		53	.151		21	7.90		28.90	35.50
5700	6 switch, stainless steel		53	.151		42.50	7.90		50.40	58.50
5800	3 gang, 9 switch, stainless steel		32	.250		63.50	13.10		76.60	89.50
5900	Receptacle, triple, 1 return, 1 feed		26	.308		42	16.10		58.10	70
6000	2 feed		20	.400		42	21		63	77.50
6100	Relay gang boxes, flush or surface, 6 gang		5.30	1.509		88	79		167	215
6200	12 gang		4.70	1.702		99	89		188	242

26 27 26 – Wiring Devices

26 27 26.10 Low Voltage Switching		Crew	Daily Output	Labor-Hours	Unit	Material	Labor	Equipment	Total	Total Incl O&P
6400	18 gang	1 Elec	4	2	Ea.	110	105		215	278
6500	Frame, to hold up to 6 relays		12	.667		81.50	35		116.50	142
7200	Control wire, 2 conductor		6.30	1.270	C.L.F.	23	66.50		89.50	125
7400	3 conductor		5	1.600		32	84		116	160
7600	19 conductor		2.50	3.200		275	168		443	555
7800	26 conductor		2	4		375	210		585	725
8000	Weatherproof, 3 conductor		5	1.600		67	84		151	199

26 27 26.20 Wiring Devices Elements		Crew	Daily Output	Labor-Hours	Unit	Material	Labor	Equipment	Total	Total Incl O&P
0010	**WIRING DEVICES ELEMENTS** R262726-90									
0200	Toggle switch, quiet type, single pole, 15 amp	1 Elec	40	.200	Ea.	5.45	10.50		15.95	21.50
0500	20 amp		27	.296		8.70	15.55		24.25	32.50
0510	30 amp		23	.348		23	18.25		41.25	53
0530	Lock handle, 20 amp		27	.296		23	15.55		38.55	48.50
0540	Security key, 20 amp		26	.308		89	16.10		105.10	122
0550	Rocker, 15 amp		40	.200		5.70	10.50		16.20	22
0560	20 amp		27	.296		13.20	15.55		28.75	37.50
0600	3 way, 15 amp		23	.348		7	18.25		25.25	35
0800	20 amp		18	.444		8.45	23.50		31.95	44.50
0810	30 amp		9	.889		28.50	46.50		75	101
0830	Lock handle, 20 amp		18	.444		25.50	23.50		49	63
0840	Security key, 20 amp		17	.471		91.50	24.50		116	138
0850	Rocker, 15 amp		23	.348		8	18.25		26.25	36.50
0860	20 amp		18	.444		18.95	23.50		42.45	56
0900	4 way, 15 amp		15	.533		12.60	28		40.60	56
1000	20 amp		11	.727		41.50	38		79.50	103
1020	Lock handle, 20 amp		11	.727		55	38		93	118
1030	Rocker, 15 amp		15	.533		37	28		65	82.50
1040	20 amp		11	.727		46.50	38		84.50	109
1100	Toggle switch, quiet type, double pole, 15 amp		15	.533		14.50	28		42.50	58
1200	20 amp		11	.727		16.75	38		54.75	75.50
1210	30 amp		9	.889		29	46.50		75.50	102
1230	Lock handle, 20 amp		11	.727		25.50	38		63.50	85
1250	Security key, 20 amp		10	.800		94.50	42		136.50	167
1420	Toggle switch quiet type, 1 pole, 2 throw center off, 15 amp		23	.348		56.50	18.25		74.75	90
1440	20 amp		18	.444		54.50	23.50		78	95
1460	2 pole, 2 throw center off, lock handle, 20 amp		11	.727		71.50	38		109.50	136
1480	1 pole, momentary contact, 15 amp		23	.348		21	18.25		39.25	50.50
1500	20 amp		18	.444		34	23.50		57.50	72.50
1520	Momentary contact, lock handle, 20 amp		18	.444		35.50	23.50		59	74.50
1650	Dimmer switch, 120 volt, incandescent, 600 watt, 1 pole G		16	.500		16	26		42	56.50
1700	600 watt, 3 way G		12	.667		21	35		56	75.50
1750	1000 watt, 1 pole G		16	.500		43.50	26		69.50	87
1800	1000 watt, 3 way G		12	.667		71.50	35		106.50	132
2000	1500 watt, 1 pole G		11	.727		82.50	38		120.50	148
2100	2000 watt, 1 pole G		8	1		126	52.50		178.50	218
2110	Fluorescent, 600 watt G		15	.533		107	28		135	160
2120	1000 watt G		15	.533		130	28		158	185
2130	1500 watt G		10	.800		241	42		283	330
2160	Explosionproof, toggle switch, wall, single pole 20 amp		5.30	1.509		182	79		261	320
2180	Receptacle, single outlet, 20 amp		5.30	1.509		315	79		394	465
2190	30 amp		4	2		680	105		785	905
2290	60 amp		2.50	3.200		760	168		928	1,075

26 27 26.20 Wiring Devices Elements		Crew	Daily Output	Labor-Hours	Unit	Material	Labor	Equipment	Total	Total Incl O&P
						2013 Bare Costs				
2360	Plug, 20 amp	1 Elec	16	.500	Ea.	167	26		193	222
2370	30 amp		12	.667		283	35		318	365
2380	60 amp		8	1		375	52.50		427.50	490
2410	Furnace, thermal cutoff switch with plate		26	.308		15.45	16.10		31.55	41
2460	Receptacle, duplex, 120 volt, grounded, 15 amp		40	.200		1.22	10.50		11.72	17.05
2470	20 amp		27	.296		9.60	15.55		25.15	33.50
2480	Ground fault interrupting, 15 amp		27	.296		37	15.55		52.55	64
2482	20 amp		27	.296		39.50	15.55		55.05	66.50
2486	Clock receptacle, 15 amp		40	.200		27	10.50		37.50	45
2490	Dryer, 30 amp		15	.533		5.75	28		33.75	48.50
2500	Range, 50 amp		11	.727		12.80	38		50.80	71
2530	Surge suppresser receptacle, duplex, 20 amp		27	.296		44	15.55		59.55	71.50
2532	Quad, 20 amp		20	.400		79	21		100	119
2540	Isolated ground receptacle, duplex, 20 amp		27	.296		21.50	15.55		37.05	47
2542	Quad, 20 amp		20	.400		43.50	21		64.50	79
2550	Simplex, 20 amp		27	.296		24.50	15.55		40.05	50
2560	Simplex, 30 amp		15	.533		28.50	28		56.50	73
2570	Cable reel w/receptacle 50′ w/3#12, 120 V, 20 A	2 Elec	2.67	5.999		870	315		1,185	1,425
2600	Wall plates, stainless steel, 1 gang	1 Elec	80	.100		2.50	5.25		7.75	10.60
2800	2 gang		53	.151		4.25	7.90		12.15	16.55
3000	3 gang		32	.250		8.30	13.10		21.40	29
3100	4 gang		27	.296		10.80	15.55		26.35	35
3110	Brown plastic, 1 gang		80	.100		.38	5.25		5.63	8.25
3120	2 gang		53	.151		.75	7.90		8.65	12.70
3130	3 gang		32	.250		1.11	13.10		14.21	21
3140	4 gang		27	.296		2.86	15.55		18.41	26
3150	Brushed brass, 1 gang		80	.100		4.44	5.25		9.69	12.75
3160	Anodized aluminum, 1 gang		80	.100		2.95	5.25		8.20	11.10
3170	Switch cover, weatherproof, 1 gang		60	.133		9.70	7		16.70	21
3180	Vandal proof lock, 1 gang		60	.133		13.75	7		20.75	25.50
3200	Lampholder, keyless		26	.308		11.55	16.10		27.65	36.50
3400	Pullchain with receptacle		22	.364		20	19.05		39.05	50.50
3500	Pilot light, neon with jewel		27	.296		13.25	15.55		28.80	37.50
3600	Receptacle, 20 amp, 250 volt, NEMA 6		27	.296		21	15.55		36.55	46
3620	277 volt NEMA 7		27	.296		18.40	15.55		33.95	43
3640	125/250 volt NEMA 10		27	.296		19.25	15.55		34.80	44
3680	125/250 volt NEMA 14		25	.320		24.50	16.75		41.25	52
3700	3 pole, 250 volt NEMA 15		25	.320		25.50	16.75		42.25	53
3720	120/208 volt NEMA 18		25	.320		26.50	16.75		43.25	54.50
3740	30 amp, 125 volt NEMA 5		15	.533		17.35	28		45.35	61
3760	250 volt NEMA 6		15	.533		19.80	28		47.80	64
3780	277 volt NEMA 7		15	.533		26	28		54	70.50
3820	125/250 volt NEMA 14		14	.571		53.50	30		83.50	104
3840	3 pole, 250 volt NEMA 15		14	.571		50	30		80	100
3880	50 amp, 125 volt NEMA 5		11	.727		22.50	38		60.50	82
3900	250 volt NEMA 6		11	.727		22.50	38		60.50	82
3920	277 volt NEMA 7		11	.727		29	38		67	89
3960	125/250 volt NEMA 14		10	.800		69.50	42		111.50	139
3980	3 pole 250 volt NEMA 15		10	.800		68.50	42		110.50	138
4020	60 amp, 125/250 volt, NEMA 14		8	1		76.50	52.50		129	163
4040	3 pole, 250 volt NEMA 15		8	1		80.50	52.50		133	167
4060	120/208 volt NEMA 18		8	1		102	52.50		154.50	191
4100	Receptacle locking, 20 amp, 125 volt NEMA L5		27	.296		17.30	15.55		32.85	42

26 27 26.20 Wiring Devices Elements		Crew	Daily Output	Labor-Hours	Unit	Material	Labor	Equipment	Total	Total Incl O&P
						2013 Bare Costs				
4120	250 volt NEMA L6	1 Elec	27	.296	Ea.	17.40	15.55		32.95	42
4140	277 volt NEMA L7		27	.296		17.35	15.55		32.90	42
4150	3 pole, 250 volt, NEMA L11		27	.296		27	15.55		42.55	52.50
4160	20 amp, 480 volt NEMA L8		27	.296		22	15.55		37.55	47
4180	600 volt NEMA L9		27	.296		33	15.55		48.55	59
4200	125/250 volt NEMA L10		27	.296		27	15.55		42.55	52.50
4230	125/250 volt NEMA L14		25	.320		26.50	16.75		43.25	54
4280	250 volt NEMA L15		25	.320		28.50	16.75		45.25	56
4300	480 volt NEMA L16		25	.320		28.50	16.75		45.25	56.50
4320	3 phase, 120/208 volt NEMA L18		25	.320		37	16.75		53.75	65.50
4340	277/480 volt NEMA L19		25	.320		37.50	16.75		54.25	66
4360	347/600 volt NEMA L20		25	.320		37	16.75		53.75	66
4380	120/208 volt NEMA L21		23	.348		31.50	18.25		49.75	62.50
4400	277/480 volt NEMA L22		23	.348		35.50	18.25		53.75	66.50
4420	347/600 volt NEMA L23		23	.348		35	18.25		53.25	66
4440	30 amp, 125 volt NEMA L5		15	.533		24.50	28		52.50	69
4460	250 volt NEMA L6		15	.533		26.50	28		54.50	71
4480	277 volt NEMA L7		15	.533		26.50	28		54.50	71
4500	480 volt NEMA L8		15	.533		31.50	28		59.50	76.50
4520	600 volt NEMA L9		15	.533		31.50	28		59.50	76.50
4540	125/250 volt NEMA L10		15	.533		29.50	28		57.50	74.50
4560	3 phase, 250 volt NEMA L11		15	.533		29	28		57	74
4620	125/250 volt NEMA L14		14	.571		39	30		69	88
4640	250 volt NEMA L15		14	.571		38.50	30		68.50	87
4660	480 volt NEMA L16		14	.571		45	30		75	94.50
4680	600 volt NEMA L17		14	.571		43.50	30		73.50	93
4700	120/208 volt NEMA L18		14	.571		54.50	30		84.50	105
4720	277/480 volt NEMA L19		14	.571		54.50	30		84.50	105
4740	347/600 volt NEMA L20		14	.571		54.50	30		84.50	105
4760	120/208 volt NEMA L21		13	.615		40.50	32.50		73	93.50
4780	277/480 volt NEMA L22		13	.615		48.50	32.50		81	102
4800	347/600 volt NEMA L23		13	.615		60.50	32.50		93	115
4840	Receptacle, corrosion resistant, 15 or 20 amp, 125 volt NEMA L5		27	.296		26	15.55		41.55	51.50
4860	250 volt NEMA L6		27	.296		25	15.55		40.55	50
4900	Receptacle, cover plate, phenolic plastic, NEMA 5 & 6		80	.100		.61	5.25		5.86	8.50
4910	NEMA 7-23		80	.100		.73	5.25		5.98	8.65
4920	Stainless steel, NEMA 5 & 6		80	.100		2.65	5.25		7.90	10.75
4930	NEMA 7-23		80	.100		2.81	5.25		8.06	10.95
4940	Brushed brass NEMA 5 & 6		80	.100		5.30	5.25		10.55	13.70
4950	NEMA 7-23		80	.100		5.75	5.25		11	14.20
4960	Anodized aluminum, NEMA 5 & 6		80	.100		3.45	5.25		8.70	11.65
4970	NEMA 7-23		80	.100		5.70	5.25		10.95	14.10
4980	Weatherproof NEMA 7-23		60	.133		32	7		39	45.50
5100	Plug, 20 amp, 250 volt, NEMA 6		30	.267		15.25	13.95		29.20	38
5110	277 volt NEMA 7		30	.267		18.80	13.95		32.75	41.50
5120	3 pole, 120/250 volt, NEMA 10		26	.308		20.50	16.10		36.60	47
5130	125/250 volt NEMA 14		26	.308		41	16.10		57.10	69
5140	250 volt NEMA 15		26	.308		43	16.10		59.10	71
5150	120/208 volt NEMA 8		26	.308		44	16.10		60.10	72.50
5160	30 amp, 125 volt NEMA 5		13	.615		55	32.50		87.50	109
5170	250 volt NEMA 6		13	.615		52	32.50		84.50	106
5180	277 volt NEMA 7		13	.615		60.50	32.50		93	115
5190	125/250 volt NEMA 14		13	.615		47.50	32.50		80	101

26 27 26.20 Wiring Devices Elements		Crew	Daily Output	Labor-Hours	Unit	Material	Labor	Equipment	Total	Total Incl O&P
						2013 Bare Costs				
5200	3 pole, 250 volt NEMA 15	1 Elec	12	.667	Ea.	49.50	35		84.50	107
5210	50 amp, 125 volt NEMA 5		9	.889		60.50	46.50		107	136
5220	250 volt NEMA 6		9	.889		63.50	46.50		110	139
5230	277 volt NEMA 7		9	.889		67.50	46.50		114	144
5240	125/250 volt NEMA 14		9	.889		58	46.50		104.50	133
5250	3 pole, 250 volt NEMA 15		8	1		60.50	52.50		113	145
5260	60 amp, 125/250 volt NEMA 14		7	1.143		66.50	60		126.50	163
5270	3 pole, 250 volt NEMA 15		7	1.143		71	60		131	168
5280	120/208 volt NEMA 18		7	1.143		81	60		141	179
5300	Plug angle, 20 amp, 250 volt NEMA 6		30	.267		23.50	13.95		37.45	47
5310	30 amp, 125 volt NEMA 5		13	.615		42	32.50		74.50	94.50
5320	250 volt NEMA 6		13	.615		43.50	32.50		76	96
5330	277 volt NEMA 7		13	.615		53	32.50		85.50	107
5340	125/250 volt NEMA 14		13	.615		52.50	32.50		85	107
5350	3 pole, 250 volt NEMA 15		12	.667		55.50	35		90.50	114
5360	50 amp, 125 volt NEMA 5		9	.889		45.50	46.50		92	120
5370	250 volt NEMA 6		9	.889		45	46.50		91.50	119
5380	277 volt NEMA 7		9	.889		58	46.50		104.50	133
5390	125/250 volt NEMA 14		9	.889		63.50	46.50		110	140
5400	3 pole, 250 volt NEMA 15		8	1		66.50	52.50		119	152
5410	60 amp, 125/250 volt NEMA 14		7	1.143		74.50	60		134.50	172
5420	3 pole, 250 volt NEMA 15		7	1.143		77.50	60		137.50	175
5430	120/208 volt NEMA 18		7	1.143		77.50	60		137.50	175
5500	Plug, locking, 20 amp, 125 volt NEMA L5		30	.267		13.60	13.95		27.55	36
5510	250 volt NEMA L6		30	.267		13.55	13.95		27.50	36
5520	277 volt NEMA L7		30	.267		13.50	13.95		27.45	36
5530	480 volt NEMA L8		30	.267		16.95	13.95		30.90	39.50
5540	600 volt NEMA L9		30	.267		20.50	13.95		34.45	43.50
5550	3 pole, 125/250 volt NEMA L10		26	.308		21.50	16.10		37.60	47.50
5560	250 volt NEMA L11		26	.308		21.50	16.10		37.60	47.50
5570	480 volt NEMA L12		26	.308		25	16.10		41.10	51.50
5580	125/250 volt NEMA L14		26	.308		20.50	16.10		36.60	46.50
5590	250 volt NEMA L15		26	.308		22	16.10		38.10	48
5600	480 volt NEMA L16		26	.308		25.50	16.10		41.60	52.50
5610	4 pole, 120/208 volt NEMA L18		24	.333		33	17.45		50.45	62.50
5620	277/480 volt NEMA L19		24	.333		33	17.45		50.45	62
5630	347/600 volt NEMA L20		24	.333		33	17.45		50.45	62
5640	120/208 volt NEMA L21		24	.333		26.50	17.45		43.95	55
5650	277/480 volt NEMA L22		24	.333		31.50	17.45		48.95	60.50
5660	347/600 volt NEMA L23		24	.333		31.50	17.45		48.95	60.50
5670	30 amp, 125 volt NEMA L5		13	.615		21	32.50		53.50	71.50
5680	250 volt NEMA L6		13	.615		21.50	32.50		54	72
5690	277 volt NEMA L7		13	.615		21	32.50		53.50	71.50
5700	480 volt NEMA L8		13	.615		25	32.50		57.50	76
5710	600 volt NEMA L9		13	.615		24.50	32.50		57	75.50
5720	3 pole, 125/250 volt NEMA L10		11	.727		23.50	38		61.50	82.50
5730	250 volt NEMA L11		11	.727		23.50	38		61.50	82.50
5760	125/250 volt NEMA L14		11	.727		28	38		66	88
5770	250 volt NEMA L15		11	.727		28.50	38		66.50	88
5780	480 volt NEMA L16		11	.727		33	38		71	93.50
5790	600 volt NEMA L17		11	.727		33.50	38		71.50	94
5800	4 pole, 120/208 volt NEMA L18		10	.800		41	42		83	108
5810	120/208 volt NEMA L19		10	.800		41.50	42		83.50	108

26 27 26.20 Wiring Devices Elements		Crew	Daily Output	Labor-Hours	Unit	Material	Labor	Equipment	Total	Total Incl O&P
5820	347/600 volt NEMA L20	1 Elec	10	.800	Ea.	41.50	42		83.50	108
5830	120/208 volt NEMA L21		10	.800		31	42		73	96.50
5840	277/480 volt NEMA L22		10	.800		37.50	42		79.50	104
5850	347/600 volt NEMA L23		10	.800		46.50	42		88.50	114
6000	Connector, 20 amp, 250 volt NEMA 6		30	.267		23.50	13.95		37.45	47
6010	277 volt NEMA 7		30	.267		30	13.95		43.95	54
6020	3 pole, 120/250 volt NEMA 10		26	.308		34	16.10		50.10	61.50
6030	125/250 volt NEMA 14		26	.308		34	16.10		50.10	61.50
6040	250 volt NEMA 15		26	.308		34.50	16.10		50.60	61.50
6050	120/208 volt NEMA 18		26	.308		37.50	16.10		53.60	65
6060	30 amp, 125 volt NEMA 5		13	.615		59	32.50		91.50	114
6070	250 volt NEMA 6		13	.615		59	32.50		91.50	114
6080	277 volt NEMA 7		13	.615		59	32.50		91.50	114
6110	50 amp, 125 volt NEMA 5		9	.889		81	46.50		127.50	159
6120	250 volt NEMA 6		9	.889		81	46.50		127.50	159
6130	277 volt NEMA 7		9	.889		81	46.50		127.50	159
6200	Connector, locking, 20 amp, 125 volt NEMA L5		30	.267		20.50	13.95		34.45	43.50
6210	250 volt NEMA L6		30	.267		20.50	13.95		34.45	43.50
6220	277 volt NEMA L7		30	.267		20.50	13.95		34.45	43.50
6230	480 volt NEMA L8		30	.267		30.50	13.95		44.45	54.50
6240	600 volt NEMA L9		30	.267		30.50	13.95		44.45	54.50
6250	3 pole, 125/250 volt NEMA L10		26	.308		33	16.10		49.10	60
6260	250 volt NEMA L11		26	.308		33	16.10		49.10	60
6280	125/250 volt NEMA L14		26	.308		28	16.10		44.10	54.50
6290	250 volt NEMA L15		26	.308		28	16.10		44.10	55
6300	480 volt NEMA L16		26	.308		33.50	16.10		49.60	61
6310	4 pole, 120/208 volt NEMA L18		24	.333		44.50	17.45		61.95	74.50
6320	277/480 volt NEMA L19		24	.333		44	17.45		61.45	74.50
6330	347/600 volt NEMA L20		24	.333		44.50	17.45		61.95	74.50
6340	120/208 volt NEMA L21		24	.333		42.50	17.45		59.95	73
6350	277/480 volt NEMA L22		24	.333		52.50	17.45		69.95	84
6360	347/600 volt NEMA L23		24	.333		53	17.45		70.45	84
6370	30 amp, 125 volt NEMA L5		13	.615		40.50	32.50		73	93
6380	250 volt NEMA L6		13	.615		40.50	32.50		73	93.50
6390	277 volt NEMA L7		13	.615		41	32.50		73.50	93.50
6400	480 volt NEMA L8		13	.615		48.50	32.50		81	102
6410	600 volt NEMA L9		13	.615		48.50	32.50		81	102
6420	3 pole, 125/250 volt NEMA L10		11	.727		63.50	38		101.50	127
6430	250 volt NEMA L11		11	.727		63.50	38		101.50	127
6460	125/250 volt NEMA L14		11	.727		56	38		94	119
6470	250 volt NEMA L15		11	.727		56	38		94	119
6480	480 volt NEMA L16		11	.727		67.50	38		105.50	131
6490	600 volt NEMA L17		11	.727		68	38		106	132
6500	4 pole, 120/208 volt NEMA L18		10	.800		84	42		126	155
6510	120/208 volt NEMA L19		10	.800		84.50	42		126.50	156
6520	347/600 volt NEMA L20		10	.800		84.50	42		126.50	156
6530	120/208 volt NEMA L21		10	.800		56	42		98	124
6540	277/480 volt NEMA L22		10	.800		67	42		109	136
6550	347/600 volt NEMA L23		10	.800		83	42		125	154
7000	Receptacle computer, 250 volt, 15 amp, 3 pole 4 wire		8	1		81	52.50		133.50	168
7010	20 amp, 2 pole 3 wire		8	1		82	52.50		134.50	169
7020	30 amp, 2 pole 3 wire		6.50	1.231		133	64.50		197.50	243
7030	30 amp, 3 pole 4 wire		6.50	1.231		144	64.50		208.50	255

Table column group "2013 Bare Costs" covers Material, Labor, Equipment, Total.

26 27 Low-Voltage Distribution Equipment

26 27 26 – Wiring Devices

26 27 26.20 Wiring Devices Elements		Crew	Daily Output	Labor-Hours	Unit	Material	Labor	Equipment	Total	Total Incl O&P
						2013 Bare Costs				
7040	60 amp, 3 pole 4 wire	1 Elec	4.50	1.778	Ea.	243	93		336	405
7050	100 amp, 3 pole 4 wire		3	2.667		300	140		440	540
7100	Connector computer, 250 volt, 15 amp, 3 pole 4 wire		27	.296		127	15.55		142.55	163
7110	20 amp, 2 pole 3 wire		27	.296		114	15.55		129.55	149
7120	30 amp, 2 pole 3 wire		15	.533		175	28		203	235
7130	30 amp, 3 pole 4 wire		15	.533		181	28		209	241
7140	60 amp, 3 pole 4 wire		8	1		300	52.50		352.50	410
7150	100 amp, 3 pole 4 wire		4	2		415	105		520	610
7200	Plug, computer, 250 volt, 15 amp, 3 pole 4 wire		27	.296		114	15.55		129.55	148
7210	20 amp, 2 pole, 3 wire		27	.296		106	15.55		121.55	140
7220	30 amp, 2 pole, 3 wire		15	.533		186	28		214	247
7230	30 amp, 3 pole, 4 wire		15	.533		180	28		208	240
7240	60 amp, 3 pole, 4 wire		8	1		271	52.50		323.50	375
7250	100 amp, 3 pole, 4 wire		4	2		340	105		445	525
7300	Connector adapter to flexible conduit, 1/2"		60	.133		3.82	7		10.82	14.65
7310	3/4"		50	.160		5.45	8.40		13.85	18.55
7320	1-1/4"		30	.267		17.50	13.95		31.45	40.50
7330	1-1/2"	↓	23	.348	↓	25	18.25		43.25	55

26 27 73 – Door Chimes

26 27 73.10 Doorbell System		Crew	Daily Output	Labor-Hours	Unit	Material	Labor	Equipment	Total	Total Incl O&P
0010	**DOORBELL SYSTEM**, incl. transformer, button & signal									
0100	6" bell	1 Elec	4	2	Ea.	110	105		215	278
0200	Buzzer		4	2		100	105		205	267
1000	Door chimes, 2 notes, minimum		16	.500		25.50	26		51.50	67
1020	Maximum		12	.667		134	35		169	201
1100	Tube type, 3 tube system		12	.667		198	35		233	271
1180	4 tube system		10	.800		385	42		427	485
1900	For transformer & button, minimum add		5	1.600		17.65	84		101.65	144
1960	Maximum, add		4.50	1.778		79	93		172	226
3000	For push button only, minimum		24	.333		3.18	17.45		20.63	29.50
3100	Maximum		20	.400		23	21		44	56.50
3200	Bell transformer	↓	16	.500	↓	22	26		48	63.50

26 28 Low-Voltage Circuit Protective Devices

26 28 13 – Fuses

26 28 13.10 Fuse Elements		Crew	Daily Output	Labor-Hours	Unit	Material	Labor	Equipment	Total	Total Incl O&P
0010	**FUSE ELEMENTS**									
0020	Cartridge, nonrenewable									
0050	250 volt, 30 amp	1 Elec	50	.160	Ea.	2.01	8.40		10.41	14.75
0100	60 amp		50	.160		3.26	8.40		11.66	16.15
0150	100 amp		40	.200		13.65	10.50		24.15	31
0200	200 amp		36	.222		33	11.65		44.65	54
0250	400 amp		30	.267		74	13.95		87.95	103
0300	600 amp		24	.333		120	17.45		137.45	158
0400	600 volt, 30 amp		40	.200		9.80	10.50		20.30	26.50
0450	60 amp		40	.200		15.60	10.50		26.10	33
0500	100 amp		36	.222		30	11.65		41.65	50.50
0550	200 amp		30	.267		80	13.95		93.95	109
0600	400 amp		24	.333		165	17.45		182.45	208
0650	600 amp		20	.400		247	21		268	305

26 28 Low-Voltage Circuit Protective Devices

26 28 13 – Fuses

26 28 13.10 Fuse Elements		Crew	Daily Output	Labor-Hours	Unit	2013 Bare Costs Material	Labor	Equipment	Total	Total Incl O&P
0800	Dual element, time delay, 250 volt, 30 amp	1 Elec	50	.160	Ea.	7.15	8.40		15.55	20.50
0850	60 amp		50	.160		11.60	8.40		20	25.50
0900	100 amp		40	.200		26	10.50		36.50	44.50
0950	200 amp		36	.222		57.50	11.65		69.15	81
1000	400 amp		30	.267		105	13.95		118.95	137
1050	600 amp		24	.333		172	17.45		189.45	215
1300	600 volt, 15 to 30 amp		40	.200		14.25	10.50		24.75	31.50
1350	35 to 60 amp		40	.200		24.50	10.50		35	42
1400	70 to 100 amp		36	.222		50.50	11.65		62.15	73
1450	110 to 200 amp		30	.267		101	13.95		114.95	132
1500	225 to 400 amp		24	.333		202	17.45		219.45	248
1550	600 amp		20	.400		293	21		314	350
1800	Class RK1, high capacity, 250 volt, 30 amp		50	.160		6.70	8.40		15.10	19.90
1850	60 amp		50	.160		12.45	8.40		20.85	26.50
1900	100 amp		40	.200		27.50	10.50		38	46
1950	200 amp		36	.222		61	11.65		72.65	84.50
2000	400 amp		30	.267		110	13.95		123.95	142
2050	600 amp		24	.333		181	17.45		198.45	225
2200	600 volt, 30 amp		40	.200		15.45	10.50		25.95	32.50
2250	60 amp		40	.200		26.50	10.50		37	44.50
2300	100 amp		36	.222		56.50	11.65		68.15	79.50
2350	200 amp		30	.267		110	13.95		123.95	142
2400	400 amp		24	.333		218	17.45		235.45	266
2450	600 amp		20	.400		320	21		341	380
2700	Class J, current limiting, 250 or 600 volt, 30 amp		40	.200		18.75	10.50		29.25	36
2750	60 amp		40	.200		30.50	10.50		41	49
2800	100 amp		36	.222		46.50	11.65		58.15	69
2850	200 amp		30	.267		87.50	13.95		101.45	117
2900	400 amp		24	.333		206	17.45		223.45	252
2950	600 amp		20	.400		310	21		331	370
3100	Class L, current limiting, 250 or 600 volt, 601 to 1200 amp		16	.500		525	26		551	620
3150	1500-1600 amp		13	.615		685	32.50		717.50	805
3200	1800-2000 amp		10	.800		900	42		942	1,050
3250	2500 amp		10	.800		1,200	42		1,242	1,400
3300	3000 amp		8	1		1,375	52.50		1,427.50	1,600
3350	3500-4000 amp		8	1		1,900	52.50		1,952.50	2,175
3400	4500-5000 amp		6.70	1.194		3,675	62.50		3,737.50	4,125
3450	6000 amp		5.70	1.404		4,925	73.50		4,998.50	5,525
3600	Plug, 120 volt, 1 to 10 amp		50	.160		2.37	8.40		10.77	15.15
3650	15 to 30 amp		50	.160		6.70	8.40		15.10	19.95
3700	Dual element 0.3 to 14 amp		50	.160		3.50	8.40		11.90	16.40
3750	15 to 30 amp		50	.160		9.15	8.40		17.55	22.50
3800	Fustat, 120 volt, 15 to 30 amp		50	.160		4.66	8.40		13.06	17.70
3850	0.3 to 14 amp		50	.160		9.80	8.40		18.20	23.50
3900	Adapters 0.3 to 10 amp		50	.160		5.05	8.40		13.45	18.10
3950	15 to 30 amp		50	.160		10.95	8.40		19.35	24.50

26 28 16 – Enclosed Switches and Circuit Breakers

26 28 16.10 Circuit Breakers

		Crew	Daily Output	Labor-Hours	Unit	Material	Labor	Equipment	Total	Total Incl O&P
0010	**CIRCUIT BREAKERS** (in enclosure)									
0100	Enclosed (NEMA 1), 600 volt, 3 pole, 30 amp	1 Elec	3.20	2.500	Ea.	475	131		606	720
0200	60 amp		2.80	2.857		585	150		735	870
0400	100 amp		2.30	3.478		670	182		852	1,000

26 28 16 – Enclosed Switches and Circuit Breakers

26 28 16.10 Circuit Breakers		Crew	Daily Output	Labor-Hours	Unit	Material	2013 Bare Costs Labor	Equipment	Total	Total Incl O&P
0500	200 amp	1 Elec	1.50	5.333	Ea.	1,400	279		1,679	1,975
0600	225 amp		1.50	5.333		1,550	279		1,829	2,125
0700	400 amp	2 Elec	1.60	10		2,650	525		3,175	3,675
0800	600 amp		1.20	13.333		3,825	700		4,525	5,250
1000	800 amp		.94	17.021		4,975	890		5,865	6,800
1200	1000 amp		.84	19.048		6,275	1,000		7,275	8,425
1220	1200 amp		.80	20		8,050	1,050		9,100	10,400
1240	1600 amp		.72	22.222		15,500	1,175		16,675	18,800
1260	2000 amp		.64	25		16,800	1,300		18,100	20,400
1400	1200 amp with ground fault		.80	20		13,400	1,050		14,450	16,300
1600	1600 amp with ground fault		.72	22.222		17,500	1,175		18,675	21,100
1800	2000 amp with ground fault		.64	25		18,800	1,300		20,100	22,700
2000	Disconnect, 240 volt 3 pole, 5 HP motor	1 Elec	3.20	2.500		410	131		541	645
2020	10 HP motor		3.20	2.500		410	131		541	645
2040	15 HP motor		2.80	2.857		410	150		560	675
2060	20 HP motor		2.30	3.478		495	182		677	820
2080	25 HP motor		2.30	3.478		495	182		677	820
2100	30 HP motor		2.30	3.478		495	182		677	820
2120	40 HP motor		2	4		850	210		1,060	1,250
2140	50 HP motor		1.50	5.333		850	279		1,129	1,350
2160	60 HP motor		1.50	5.333		1,950	279		2,229	2,575
2180	75 HP motor	2 Elec	2	8		1,950	420		2,370	2,775
2200	100 HP motor		1.60	10		1,950	525		2,475	2,925
2220	125 HP motor		1.60	10		1,950	525		2,475	2,925
2240	150 HP motor		1.20	13.333		3,825	700		4,525	5,250
2260	200 HP motor		1.20	13.333		4,975	700		5,675	6,525
2300	Enclosed (NEMA 7), explosion proof, 600 volt 3 pole, 50 amp	1 Elec	2.30	3.478		1,400	182		1,582	1,825
2350	100 amp		1.50	5.333		1,475	279		1,754	2,025
2400	150 amp		1	8		3,525	420		3,945	4,500
2450	250 amp	2 Elec	1.60	10		4,400	525		4,925	5,625
2500	400 amp	"	1.20	13.333		4,875	700		5,575	6,425

26 28 16.20 Safety Switches		Crew	Daily Output	Labor-Hours	Unit	Material	Labor	Equipment	Total	Total Incl O&P
0010	**SAFETY SWITCHES** R262816-80									
0100	General duty 240 volt, 3 pole NEMA 1, fusible, 30 amp	1 Elec	3.20	2.500	Ea.	102	131		233	310
0200	60 amp		2.30	3.478		172	182		354	460
0300	100 amp		1.90	4.211		294	221		515	655
0400	200 amp		1.30	6.154		630	320		950	1,175
0500	400 amp	2 Elec	1.80	8.889		1,600	465		2,065	2,450
0600	600 amp	"	1.20	13.333		3,000	700		3,700	4,325
0610	Nonfusible, 30 amp	1 Elec	3.20	2.500		82	131		213	286
0650	60 amp		2.30	3.478		109	182		291	390
0700	100 amp		1.90	4.211		252	221		473	610
0750	200 amp		1.30	6.154		465	320		785	995
0800	400 amp	2 Elec	1.80	8.889		1,125	465		1,590	1,925
0850	600 amp	"	1.20	13.333		2,150	700		2,850	3,400
1100	Heavy duty, 600 volt, 3 pole NEMA 1 nonfused									
1110	30 amp	1 Elec	3.20	2.500	Ea.	144	131		275	355
1500	60 amp		2.30	3.478		260	182		442	560
1700	100 amp		1.90	4.211		410	221		631	780
1900	200 amp		1.30	6.154		625	320		945	1,175
2100	400 amp	2 Elec	1.80	8.889		1,400	465		1,865	2,225
2300	600 amp		1.20	13.333		2,500	700		3,200	3,800

26 28 Low-Voltage Circuit Protective Devices

26 28 16 – Enclosed Switches and Circuit Breakers

26 28 16.20 Safety Switches		Crew	Daily Output	Labor-Hours	Unit	2013 Bare Costs Material	Labor	Equipment	Total	Total Incl O&P
2500	800 amp	2 Elec	.94	17.021	Ea.	5,075	890		5,965	6,925
2700	1200 amp		.80	20		6,800	1,050		7,850	9,050
2900	Heavy duty, 240 volt, 3 pole NEMA 1 fusible									
2910	30 amp	1 Elec	3.20	2.500	Ea.	163	131		294	375
3000	60 amp		2.30	3.478		276	182		458	580
3300	100 amp		1.90	4.211		435	221		656	810
3500	200 amp		1.30	6.154		750	320		1,070	1,300
3700	400 amp	2 Elec	1.80	8.889		1,925	465		2,390	2,800
3900	600 amp		1.20	13.333		3,325	700		4,025	4,700
4100	800 amp		.94	17.021		7,750	890		8,640	9,850
4300	1200 amp		.80	20		10,200	1,050		11,250	12,800
4340	2 pole fusible, 30 amp	1 Elec	3.50	2.286		123	120		243	315
4350	600 volt, 3 pole, fusible, 30 amp		3.20	2.500		277	131		408	500
4380	60 amp		2.30	3.478		335	182		517	645
4400	100 amp		1.90	4.211		610	221		831	1,000
4420	200 amp		1.30	6.154		885	320		1,205	1,450
4440	400 amp	2 Elec	1.80	8.889		2,325	465		2,790	3,250
4450	600 amp		1.20	13.333		3,875	700		4,575	5,325
4460	800 amp		.94	17.021		7,750	890		8,640	9,850
4480	1200 amp		.80	20		10,200	1,050		11,250	12,800
4500	240 volt 3 pole NEMA 3R (no hubs), fusible									
4510	30 amp	1 Elec	3.10	2.581	Ea.	292	135		427	520
4700	60 amp		2.20	3.636		460	191		651	790
4900	100 amp		1.80	4.444		660	233		893	1,075
5100	200 amp		1.20	6.667		910	350		1,260	1,525
5300	400 amp	2 Elec	1.60	10		2,000	525		2,525	2,975
5500	600 amp	″	1	16		4,200	840		5,040	5,875
5510	Heavy duty, 600 volt, 3 pole 3ph. NEMA 3R fusible, 30 amp	1 Elec	3.10	2.581		465	135		600	710
5520	60 amp		2.20	3.636		545	191		736	885
5530	100 amp		1.80	4.444		850	233		1,083	1,300
5540	200 amp		1.20	6.667		1,175	350		1,525	1,825
5550	400 amp	2 Elec	1.60	10		2,725	525		3,250	3,775
5700	600 volt, 3 pole NEMA 3R nonfused									
5710	30 amp	1 Elec	3.10	2.581	Ea.	259	135		394	485
5900	60 amp		2.20	3.636		455	191		646	785
6100	100 amp		1.80	4.444		630	233		863	1,050
6300	200 amp		1.20	6.667		765	350		1,115	1,375
6500	400 amp	2 Elec	1.60	10		1,925	525		2,450	2,875
6700	600 amp	″	1	16		3,825	840		4,665	5,475
6900	600 volt, 6 pole NEMA 3R nonfused, 30 amp	1 Elec	2.70	2.963		1,650	155		1,805	2,050
7100	60 amp		2	4		1,900	210		2,110	2,425
7300	100 amp		1.50	5.333		2,350	279		2,629	3,000
7500	200 amp		1.20	6.667		6,750	350		7,100	7,950
7600	600 volt, 3 pole NEMA 7 explosion proof nonfused									
7610	30 amp	1 Elec	2.20	3.636	Ea.	1,250	191		1,441	1,675
7620	60 amp		1.80	4.444		1,525	233		1,758	2,025
7630	100 amp		1.20	6.667		1,850	350		2,200	2,550
7640	200 amp		.80	10		3,825	525		4,350	4,975
7710	600 volt 6 pole, NEMA 3R fusible, 30 amp		2.70	2.963		1,850	155		2,005	2,250
7900	60 amp		2	4		2,225	210		2,435	2,775
8100	100 amp		1.50	5.333		2,700	279		2,979	3,400
8110	240 volt 3 pole, NEMA 12 fusible, 30 amp		3.10	2.581		360	135		495	595
8120	60 amp		2.20	3.636		480	191		671	815

26 28 16 – Enclosed Switches and Circuit Breakers

26 28 16.20 Safety Switches		Crew	Daily Output	Labor-Hours	Unit	Material	2013 Bare Costs Labor	Equipment	Total	Total Incl O&P
8130	100 amp	1 Elec	1.80	4.444	Ea.	710	233		943	1,125
8140	200 amp		1.20	6.667		1,025	350		1,375	1,650
8150	400 amp	2 Elec	1.60	10		2,325	525		2,850	3,325
8160	600 amp	"	1	16		3,725	840		4,565	5,350
8180	600 volt 3 pole, NEMA 12 fusible, 30 amp	1 Elec	3.10	2.581		470	135		605	720
8190	60 amp		2.20	3.636		485	191		676	820
8200	100 amp		1.80	4.444		780	233		1,013	1,200
8210	200 amp		1.20	6.667		1,200	350		1,550	1,850
8220	400 amp	2 Elec	1.60	10		2,675	525		3,200	3,700
8230	600 amp	"	1	16		4,475	840		5,315	6,175
8240	600 volt 3 pole, NEMA 12 nonfused, 30 amp	1 Elec	3.10	2.581		330	135		465	565
8250	60 amp		2.20	3.636		430	191		621	755
8260	100 amp		1.80	4.444		610	233		843	1,025
8270	200 amp		1.20	6.667		810	350		1,160	1,425
8280	400 amp	2 Elec	1.60	10		2,000	525		2,525	2,975
8290	600 amp	"	1	16		3,275	840		4,115	4,850
8310	600 volt, 3 pole NEMA 4 fusible, 30 amp	1 Elec	3	2.667		1,300	140		1,440	1,625
8320	60 amp		2.20	3.636		1,425	191		1,616	1,850
8330	100 amp		1.80	4.444		3,175	233		3,408	3,850
8340	200 amp		1.20	6.667		4,025	350		4,375	4,950
8350	400 amp	2 Elec	1.60	10		7,925	525		8,450	9,500
8360	600 volt 3 pole NEMA 4 nonfused, 30 amp	1 Elec	3	2.667		1,075	140		1,215	1,375
8370	60 amp		2.20	3.636		1,275	191		1,466	1,675
8380	100 amp		1.80	4.444		2,600	233		2,833	3,200
8390	200 amp		1.20	6.667		3,525	350		3,875	4,400
8400	400 amp	2 Elec	1.60	10		7,125	525		7,650	8,625
8490	Motor starters, manual, single phase, NEMA 1	1 Elec	6.40	1.250		65	65.50		130.50	170
8500	NEMA 4		4	2		182	105		287	360
8700	NEMA 7		4	2		200	105		305	375
8900	NEMA 1 with pilot		6.40	1.250		90	65.50		155.50	197
8920	3 pole, NEMA 1, 230/460 volt, 5 HP, size 0		3.50	2.286		203	120		323	400
8940	10 HP, size 1		2	4		241	210		451	580
9010	Disc. switch, 600 V 3 pole fusible, 30 amp, to 10 HP motor		3.20	2.500		345	131		476	575
9050	60 amp, to 30 HP motor		2.30	3.478		800	182		982	1,150
9070	100 amp, to 60 HP motor		1.90	4.211		800	221		1,021	1,200
9100	200 amp, to 125 HP motor		1.30	6.154		1,200	320		1,520	1,800
9110	400 amp, to 200 HP motor	2 Elec	1.80	8.889		3,025	465		3,490	4,025

26 28 16.40 Time Switches		Crew	Daily Output	Labor-Hours	Unit	Material	Labor	Equipment	Total	Total Incl O&P
0010	**TIME SWITCHES**									
0100	Single pole, single throw, 24 hour dial	1 Elec	4	2	Ea.	137	105		242	310
0200	24 hour dial with reserve power		3.60	2.222		580	116		696	815
0300	Astronomic dial		3.60	2.222		212	116		328	405
0400	Astronomic dial with reserve power		3.30	2.424		665	127		792	920
0500	7 day calendar dial		3.30	2.424		158	127		285	365
0600	7 day calendar dial with reserve power		3.20	2.500		465	131		596	705
0700	Photo cell 2000 watt		8	1		24	52.50		76.50	105
1080	Load management device, 4 loads		2	4		760	210		970	1,150
1100	8 loads		1	8		1,925	420		2,345	2,750

26 29 Low-Voltage Controllers

26 29 13 – Enclosed Controllers

26 29 13.10 Contactors, AC		Crew	Daily Output	Labor-Hours	Unit	Material	2013 Bare Costs Labor	Equipment	Total	Total Incl O&P
0010	**CONTACTORS, AC** Enclosed (NEMA 1)									
0050	Lighting, 600 volt 3 pole, electrically held									
0100	20 amp	1 Elec	4	2	Ea.	310	105		415	495
0200	30 amp		3.60	2.222		345	116		461	555
0300	60 amp		3	2.667		650	140		790	925
0400	100 amp		2.50	3.200		975	168		1,143	1,325
0500	200 amp		1.40	5.714		2,425	299		2,724	3,125
0600	300 amp	2 Elec	1.60	10		6,325	525		6,850	7,725
0800	600 volt 3 pole, mechanically held, 30 amp	1 Elec	3.60	2.222		445	116		561	665
0900	60 amp		3	2.667		880	140		1,020	1,175
1000	75 amp		2.80	2.857		1,200	150		1,350	1,525
1100	100 amp		2.50	3.200		1,250	168		1,418	1,625
1200	150 amp		2	4		3,275	210		3,485	3,925
1300	200 amp		1.40	5.714		3,425	299		3,724	4,225
1500	Magnetic with auxiliary contact, size 00, 9 amp		4	2		179	105		284	355
1600	Size 0, 18 amp		4	2		213	105		318	390
1700	Size 1, 27 amp		3.60	2.222		242	116		358	440
1800	Size 2, 45 amp		3	2.667		450	140		590	705
1900	Size 3, 90 amp		2.50	3.200		725	168		893	1,050
2000	Size 4, 135 amp		2.30	3.478		1,650	182		1,832	2,100
2100	Size 5, 270 amp	2 Elec	1.80	8.889		3,500	465		3,965	4,550
2200	Size 6, 540 amp		1.20	13.333		10,200	700		10,900	12,300
2300	Size 7, 810 amp		1	16		13,700	840		14,540	16,400
2310	Size 8, 1215 amp		.80	20		21,400	1,050		22,450	25,100
2500	Magnetic, 240 volt, 1-2 pole, .75 HP motor	1 Elec	4	2		144	105		249	315
2520	2 HP motor		3.60	2.222		161	116		277	350
2540	5 HP motor		2.50	3.200		390	168		558	680
2560	10 HP motor		1.40	5.714		645	299		944	1,150
2600	240 volt or less, 3 pole, .75 HP motor		4	2		144	105		249	315
2620	5 HP motor		3.60	2.222		179	116		295	370
2640	10 HP motor		3.60	2.222		207	116		323	400
2660	15 HP motor		2.50	3.200		415	168		583	705
2700	25 HP motor		2.50	3.200		415	168		583	705
2720	30 HP motor	2 Elec	2.80	5.714		690	299		989	1,200
2740	40 HP motor		2.80	5.714		690	299		989	1,200
2760	50 HP motor		1.60	10		690	525		1,215	1,550
2800	75 HP motor		1.60	10		1,625	525		2,150	2,550
2820	100 HP motor		1	16		1,625	840		2,465	3,025
2860	150 HP motor		1	16		3,450	840		4,290	5,050
2880	200 HP motor		1	16		3,450	840		4,290	5,050
3000	600 volt, 3 pole, 5 HP motor	1 Elec	4	2		179	105		284	355
3020	10 HP motor		3.60	2.222		207	116		323	400
3040	25 HP motor		3	2.667		415	140		555	665
3100	50 HP motor		2.50	3.200		690	168		858	1,000
3160	100 HP motor	2 Elec	2.80	5.714		1,625	299		1,924	2,225
3220	200 HP motor	"	1.60	10		3,450	525		3,975	4,575

26 29 13.20 Control Stations		Crew	Daily Output	Labor-Hours	Unit	Material	Labor	Equipment	Total	Total Incl O&P
0010	**CONTROL STATIONS**									
0050	NEMA 1, heavy duty, stop/start	1 Elec	8	1	Ea.	132	52.50		184.50	225
0100	Stop/start, pilot light		6.20	1.290		180	67.50		247.50	300
0200	Hand/off/automatic		6.20	1.290		98	67.50		165.50	209
0400	Stop/start/reverse		5.30	1.509		179	79		258	315

26 29 Low-Voltage Controllers

26 29 13 – Enclosed Controllers

26 29 13.20 Control Stations		Crew	Daily Output	Labor-Hours	Unit	Material	2013 Bare Costs Labor	Equipment	Total	Total Incl O&P
0500	NEMA 7, heavy duty, stop/start	1 Elec	6	1.333	Ea.	390	70		460	535
0600	Stop/start, pilot light		4	2		480	105		585	680
0700	NEMA 7 or 9, 1 element		6	1.333		320	70		390	455
0800	2 element		6	1.333		410	70		480	560
0900	3 element		4	2		820	105		925	1,050
0910	Selector switch, 2 position		6	1.333		320	70		390	455
0920	3 position		4	2		320	105		425	505
0930	Oiltight, 1 element		8	1		92	52.50		144.50	180
0940	2 element		6.20	1.290		132	67.50		199.50	247
0950	3 element		5.30	1.509		179	79		258	315
0960	Selector switch, 2 position		6.20	1.290		98	67.50		165.50	209
0970	3 position	↓	5.30	1.509	↓	98	79		177	226

26 29 13.30 Control Switches		Crew	Daily Output	Labor-Hours	Unit	Material	Labor	Equipment	Total	Total Incl O&P
0010	**CONTROL SWITCHES** Field installed									
6000	Push button 600 V 10A, momentary contact									
6150	Standard operator with colored button	1 Elec	34	.235	Ea.	15.55	12.35		27.90	35.50
6160	With single block 1NO 1NC		18	.444		33	23.50		56.50	71
6170	With double block 2NO 2NC		15	.533		50	28		78	97
6180	Std operator w/mushroom button 1-9/16" diam.	↓	34	.235	↓	33	12.35		45.35	54.50
6190	Std operator w/mushroom button 2-1/4" diam.									
6200	With single block 1NO 1NC	1 Elec	18	.444	Ea.	50	23.50		73.50	90
6210	With double block 2NO 2NC		15	.533		67.50	28		95.50	116
6500	Maintained contact, selector operator		34	.235		50	12.35		62.35	73.50
6510	With single block 1NO 1NC		18	.444		67.50	23.50		91	109
6520	With double block 2NO 2NC		15	.533		84.50	28		112.50	135
6560	Spring-return selector operator		34	.235		50	12.35		62.35	73.50
6570	With single block 1NO 1NC		18	.444		67.50	23.50		91	109
6580	With double block 2NO 2NC	↓	15	.533	↓	84.50	28		112.50	135
6620	Transformer operator w/illuminated									
6630	button 6 V #12 lamp	1 Elec	32	.250	Ea.	84.50	13.10		97.60	113
6640	With single block 1NO 1NC w/guard		16	.500		102	26		128	151
6650	With double block 2NO 2NC w/guard		13	.615		119	32.50		151.50	180
6690	Combination operator		34	.235		50	12.35		62.35	73.50
6700	With single block 1NO 1NC		18	.444		67.50	23.50		91	109
6710	With double block 2NO 2NC	↓	15	.533	↓	84.50	28		112.50	135
9000	Indicating light unit, full voltage									
9010	110-125 V front mount	1 Elec	32	.250	Ea.	67.50	13.10		80.60	93.50
9020	130 V resistor type		32	.250		50	13.10		63.10	74.50
9030	6 V transformer type	↓	32	.250	↓	61.50	13.10		74.60	87.50

26 29 13.40 Relays		Crew	Daily Output	Labor-Hours	Unit	Material	Labor	Equipment	Total	Total Incl O&P
0010	**RELAYS** Enclosed (NEMA 1)									
0050	600 volt AC, 1 pole, 12 amp	1 Elec	5.30	1.509	Ea.	86.50	79		165.50	213
0100	2 pole, 12 amp		5	1.600		86.50	84		170.50	220
0200	4 pole, 10 amp		4.50	1.778		115	93		208	266
0500	250 volt DC, 1 pole, 15 amp		5.30	1.509		113	79		192	243
0600	2 pole, 10 amp		5	1.600		108	84		192	244
0700	4 pole, 4 amp	↓	4.50	1.778	↓	143	93		236	296

26 29 23 – Variable-Frequency Motor Controllers

26 29 23.10 Variable Frequency Drives/Adj. Frequency Drives		Crew	Daily Output	Labor-Hours	Unit	Material	Labor	Equipment	Total	Total Incl O&P
0010	**VARIABLE FREQUENCY DRIVES/ADJ. FREQUENCY DRIVES**									
0100	Enclosed (NEMA 1), 460 volt, for 3 HP motor size G	1 Elec	.80	10	Ea.	1,275	525		1,800	2,200
0110	5 HP motor size G		.80	10		1,450	525		1,975	2,375

26 29 Low-Voltage Controllers

26 29 23 – Variable-Frequency Motor Controllers

26 29 23.10 Variable Frequency Drives/Adj. Frequency Drives		Crew	Daily Output	Labor-Hours	Unit	2013 Bare Costs Material	Labor	Equipment	Total	Total Incl O&P
0120	7.5 HP motor size G	1 Elec	.67	11.940	Ea.	1,725	625		2,350	2,825
0130	10 HP motor size G		.67	11.940		2,025	625		2,650	3,175
0140	15 HP motor size G	2 Elec	.89	17.978		2,450	940		3,390	4,075
0150	20 HP motor size G		.89	17.978		2,900	940		3,840	4,575
0160	25 HP motor size G		.67	23.881		3,600	1,250		4,850	5,825
0170	30 HP motor size G		.67	23.881		4,425	1,250		5,675	6,750
0180	40 HP motor size G		.67	23.881		5,350	1,250		6,600	7,750
0190	50 HP motor size G		.53	30.189		6,650	1,575		8,225	9,675
0200	60 HP motor size G	R-3	.56	35.714		7,950	1,850	245	10,045	11,800
0210	75 HP motor size G		.56	35.714		9,200	1,850	245	11,295	13,100
0220	100 HP motor size G		.50	40		10,600	2,075	274	12,949	15,100
0230	125 HP motor size G		.50	40		11,900	2,075	274	14,249	16,500
0240	150 HP motor size G		.50	40		13,600	2,075	274	15,949	18,300
0250	200 HP motor size G		.42	47.619		17,800	2,475	325	20,600	23,700
1100	Custom-engineered, 460 volt, for 3 HP motor size G	1 Elec	.56	14.286		2,575	750		3,325	3,950
1110	5 HP motor size G		.56	14.286		2,575	750		3,325	3,950
1120	7.5 HP motor size G		.47	17.021		2,700	890		3,590	4,300
1130	10 HP motor size G		.47	17.021		2,825	890		3,715	4,425
1140	15 HP motor size G	2 Elec	.62	25.806		3,525	1,350		4,875	5,925
1150	20 HP motor size G		.62	25.806		3,975	1,350		5,325	6,400
1160	25 HP motor size G		.47	34.043		4,650	1,775		6,425	7,800
1170	30 HP motor size G		.47	34.043		5,825	1,775		7,600	9,075
1180	40 HP motor size G		.47	34.043		6,875	1,775		8,650	10,300
1190	50 HP motor size G		.37	43.243		7,900	2,275		10,175	12,100
1200	60 HP motor size G	R-3	.39	51.282		11,900	2,650	350	14,900	17,500
1210	75 HP motor size G		.39	51.282		12,700	2,650	350	15,700	18,300
1220	100 HP motor size G		.35	57.143		13,900	2,975	390	17,265	20,200
1230	125 HP motor size G		.35	57.143		14,800	2,975	390	18,165	21,200
1240	150 HP motor size G		.35	57.143		17,000	2,975	390	20,365	23,600
1250	200 HP motor size G		.29	68.966		22,200	3,575	475	26,250	30,300
2000	For complex & special design systems to meet specific									
2010	requirements, obtain quote from vendor.									

26 31 Photovoltaic Collectors

26 31 13 – Photovoltaics

26 31 13.50 Solar Energy - Photovoltaics		Crew	Daily Output	Labor-Hours	Unit	Material	Labor	Equipment	Total	Total Incl O&P
0010	**SOLAR ENERGY - PHOTOVOLTAICS**									
0220	Alt energy source, photovoltaic module, 6 watt, 15 volts G	1 Elec	8	1	Ea.	66.50	52.50		119	152
0230	10 watt, 16.3 volts G		8	1		106	52.50		158.50	196
0240	20 watt, 14.5 volts G		8	1		160	52.50		212.50	254
0250	36 watt, 17 volts G		8	1		215	52.50		267.50	315
0260	55 watt, 17 volts G		8	1		298	52.50		350.50	405
0270	75 watt, 17 volts G		8	1		385	52.50		437.50	505
0280	130 watt, 33 volts G		8	1		555	52.50		607.50	690
0290	140 watt, 33 volts G		8	1		580	52.50		632.50	720
0300	150 watt, 33 volts G		8	1		640	52.50		692.50	785
0310	DC to AC inverter for, 12 V 2,000 watt G		4	2		1,150	105		1,255	1,425
0320	12 V, 2,500 watt G		4	2		1,400	105		1,505	1,675
0330	24 V, 2,500 watt G		4	2		1,975	105		2,080	2,325
0340	12 V, 3,000 watt G		3	2.667		1,625	140		1,765	2,000
0350	24 V, 3,000 watt G		3	2.667		2,375	140		2,515	2,825

26 31 Photovoltaic Collectors

26 31 13 – Photovoltaics

26 31 13.50 Solar Energy - Photovoltaics			Crew	Daily Output	Labor-Hours	Unit	2013 Bare Costs Material	Labor	Equipment	Total	Total Incl O&P
0360	24 V, 4,000 watt	G	1 Elec	2	4	Ea.	3,250	210		3,460	3,900
0370	48 V, 4,000 watt	G		2	4		3,250	210		3,460	3,900
0380	48 V, 5,500 watt	G		2	4		3,700	210		3,910	4,400
0390	PV components, combiner box, 10 lug, NEMA 3R enclosure	G		4	2		187	105		292	365
0400	Fuse, 15 A for combiner box	G		40	.200		14.25	10.50		24.75	31.50
0410	Battery charger controller w/temperature sensor	G		4	2		480	105		585	685
0420	Digital readout panel, displays hours, volts, amps, etc.	G		4	2		127	105		232	297
0430	Deep cycle solar battery, 6 V, 180 Ah (C/20)	G		8	1		230	52.50		282.50	330
0440	Battery interconn, 15" AWG #2/0, sealed w/copper ring lugs	G		16	.500		12.55	26		38.55	53
0442	Battery interconn, 24" AWG #2/0, sealed w/copper ring lugs	G		16	.500		18	26		44	59
0444	Battery interconn, 60" AWG #2/0, sealed w/copper ring lugs	G		16	.500		43	26		69	86
0446	Batt temp computer probe, RJ11 jack, 15' cord	G		16	.500		19.30	26		45.30	60
0450	System disconnect, DC 175 amp circuit breaker	G		8	1		137	52.50		189.50	230
0460	Conduit box for inverter	G		8	1		43.50	52.50		96	126
0470	Low voltage disconnect	G		8	1		39	52.50		91.50	122
0480	Vented battery enclosure, wood	G	1 Carp	2	4		171	180		351	465
0490	Roof mounting frame, metal, for 6 modules	G	1 Rofc	4	2		875	76.50		951.50	1,100

26 32 Packaged Generator Assemblies

26 32 13 – Engine Generators

26 32 13.13 Diesel-Engine-Driven Generator Sets

Line	Description	Crew	Daily Output	Labor-Hours	Unit	Material	Labor	Equipment	Total	Total Incl O&P
0010	**DIESEL-ENGINE-DRIVEN GENERATOR SETS**									
2000	Diesel engine, including battery, charger,									
2010	muffler, automatic transfer switch & day tank, 30 kW	R-3	.55	36.364	Ea.	19,300	1,875	249	21,424	24,300
2100	50 kW		.42	47.619		22,900	2,475	325	25,700	29,300
2110	60 kW		.39	51.282		25,200	2,650	350	28,200	32,100
2200	75 kW		.35	57.143		29,700	2,975	390	33,065	37,600
2300	100 kW		.31	64.516		33,000	3,350	445	36,795	41,800
2400	125 kW		.29	68.966		34,800	3,575	475	38,850	44,200
2500	150 kW		.26	76.923		39,900	4,000	530	44,430	50,500
2600	175 kW		.25	80		43,400	4,150	550	48,100	54,500
2700	200 kW		.24	83.333		44,800	4,325	570	49,695	56,500
2800	250 kW		.23	86.957		53,000	4,500	595	58,095	65,500
2850	275 kW		.22	90.909		55,000	4,725	625	60,350	68,500
2900	300 kW		.22	90.909		57,000	4,725	625	62,350	70,500
3000	350 kW		.20	100		64,500	5,200	685	70,385	79,500
3100	400 kW		.19	105		79,500	5,450	720	85,670	96,500
3200	500 kW		.18	111		100,000	5,775	760	106,535	120,000
3220	600 kW		.17	117		131,500	6,100	805	138,405	154,500
3230	650 kW	R-13	.38	110		160,500	5,575	490	166,565	186,000
3240	750 kW		.38	110		165,000	5,575	490	171,065	190,500
3250	800 kW		.36	116		171,000	5,875	515	177,390	198,000
3260	900 kW		.31	135		198,000	6,825	600	205,425	228,500
3270	1000 kW		.27	155		204,000	7,850	690	212,540	236,500

26 32 13.16 Gas-Engine-Driven Generator Sets

Line	Description	Crew	Daily Output	Labor-Hours	Unit	Material	Labor	Equipment	Total	Total Incl O&P
0010	**GAS-ENGINE-DRIVEN GENERATOR SETS** R263213-45									
0020	Gas or gasoline operated, includes battery,									
0050	charger, muffler & transfer switch									
0200	3 phase 4 wire, 277/480 volt, 7.5 kW	R-3	.83	24.096	Ea.	8,550	1,250	165	9,965	11,500
0300	11.5 kW		.71	28.169		12,100	1,450	193	13,743	15,700
0400	20 kW		.63	31.746		14,300	1,650	218	16,168	18,400

26 32 Packaged Generator Assemblies

26 32 13 – Engine Generators

26 32 13.16 Gas-Engine-Driven Generator Sets		Crew	Daily Output	Labor-Hours	Unit	Material	Labor	Equipment	Total	Total Incl O&P
0500	35 kW	R-3	.55	36.364	Ea.	17,000	1,875	249	19,124	21,800
0520	60 kW		.50	40		22,400	2,075	274	24,749	28,100
0600	80 kW		.40	50		27,900	2,600	345	30,845	35,000
0700	100 kW		.33	60.606		30,600	3,150	415	34,165	38,900
0800	125 kW		.28	71.429		62,500	3,700	490	66,690	75,000
0900	185 kW	↓	.25	80	↓	83,000	4,150	550	87,700	98,000

26 33 Battery Equipment

26 33 43 – Battery Chargers

26 33 43.55 Electric Vehicle Charging

			Crew	Daily Output	Labor-Hours	Unit	Material	Labor	Equipment	Total	Total Incl O&P
0010	**ELECTRIC VEHICLE CHARGING**										
0020	Level 2, wall mounted										
2200	Heavy duty	G	R-1A	15.36	1.042	Ea.	3,500	45		3,545	3,925
2210	with RFID	G		12.29	1.302		3,775	56		3,831	4,225
2300	Free standing, single connector	G		10.24	1.563		4,000	67.50		4,067.50	4,500
2310	with RFID	G		8.78	1.822		4,600	78.50		4,678.50	5,175
2320	Double connector	G		7.68	2.083		6,000	90		6,090	6,725
2330	with RFID	G	↓	6.83	2.343	↓	6,600	101		6,701	7,400

26 33 53 – Static Uninterruptible Power Supply

26 33 53.10 Uninterruptible Power Supply/Conditioner Transformers

			Crew	Daily Output	Labor-Hours	Unit	Material	Labor	Equipment	Total	Total Incl O&P
0010	**UNINTERRUPTIBLE POWER SUPPLY/CONDITIONER TRANSFORMERS**										
0100	Volt. regulating, isolating transf., w/invert. & 10 min. battery pack										
0110	Single-phase, 120 V, 0.35 kVA	R263353-80	1 Elec	2.29	3.493	Ea.	1,000	183		1,183	1,375
0120	0.5 kVA			2	4		1,050	210		1,260	1,475
0130	For additional 55 min. battery, add to 0.35 kVA			2.29	3.493		590	183		773	925
0140	Add to 0.5 kVA			1.14	7.018		620	370		990	1,225
0150	Single-phase, 120 V, 0.75 kVA			.80	10		1,325	525		1,850	2,225
0160	1.0 kVA		↓	.80	10		1,900	525		2,425	2,875
0170	1.5 kVA		2 Elec	1.14	14.035		3,275	735		4,010	4,700
0180	2 kVA		"	.89	17.978		3,550	940		4,490	5,300
0190	3 kVA		R-3	.63	31.746		4,300	1,650	218	6,168	7,450
0200	5 kVA			.42	47.619		6,225	2,475	325	9,025	10,900
0210	7.5 kVA			.33	60.606		7,950	3,150	415	11,515	13,900
0220	10 kVA			.28	71.429		10,300	3,700	490	14,490	17,400
0230	15 kVA			.22	90.909		13,100	4,725	625	18,450	22,200
0240	3 phase, 120/208 V input 120/208 V output, 20 kVA, incl 17 min. battery			.21	95.238		24,100	4,950	655	29,705	34,600
0242	30 kVA, incl 11 min. battery			.20	100		27,000	5,200	685	32,885	38,200
0250	40 kVA, incl 15 min. battery			.20	100		36,000	5,200	685	41,885	48,100
0260	480 V input 277/480 V output, 60 kVA, incl 6 min. battery			.19	105		40,500	5,450	720	46,670	53,500
0262	80 kVA, incl 4 min. battery			.19	105		48,300	5,450	720	54,470	62,000
0400	For additional 34 min./15 min. battery, add to 40 kVA		↓	.71	28.011	↓	12,600	1,450	192	14,242	16,200
0500	For options & accessories add to above, minimum										10%
0520	Maximum										35%
0600	For complex & special design systems to meet specific										
0610	requirements, obtain quote from vendor										

26 35 Power Filters and Conditioners

26 35 13 – Capacitors

26 35 13.10 Capacitors Indoor		Crew	Daily Output	Labor-Hours	Unit	Material	Labor	Equipment	Total	Total Incl O&P
						2013 Bare Costs				
0010	**CAPACITORS INDOOR**									
0020	240 volts, single & 3 phase, 0.5 kVAR	1 Elec	2.70	2.963	Ea.	400	155		555	670
0100	1.0 kVAR		2.70	2.963		480	155		635	760
0150	2.5 kVAR		2	4		540	210		750	910
0200	5.0 kVAR		1.80	4.444		650	233		883	1,075
0250	7.5 kVAR		1.60	5		755	262		1,017	1,225
0300	10 kVAR		1.50	5.333		910	279		1,189	1,425
0350	15 kVAR		1.30	6.154		1,250	320		1,570	1,850
0400	20 kVAR		1.10	7.273		1,550	380		1,930	2,275
0450	25 kVAR		1	8		1,825	420		2,245	2,650
1000	480 volts, single & 3 phase, 1 kVAR		2.70	2.963		365	155		520	630
1050	2 kVAR		2.70	2.963		420	155		575	690
1100	5 kVAR		2	4		525	210		735	895
1150	7.5 kVAR		2	4		570	210		780	940
1200	10 kVAR		2	4		635	210		845	1,025
1250	15 kVAR		2	4		785	210		995	1,175
1300	20 kVAR		1.60	5		875	262		1,137	1,350
1350	30 kVAR		1.50	5.333		1,100	279		1,379	1,625
1400	40 kVAR		1.20	6.667		1,400	350		1,750	2,050
1450	50 kVAR		1.10	7.273		1,575	380		1,955	2,325
2000	600 volts, single & 3 phase, 1 kVAR		2.70	2.963		375	155		530	645
2050	2 kVAR		2.70	2.963		430	155		585	705
2100	5 kVAR		2	4		525	210		735	895
2150	7.5 kVAR		2	4		570	210		780	940
2200	10 kVAR		2	4		635	210		845	1,025
2250	15 kVAR		1.60	5		785	262		1,047	1,250
2300	20 kVAR		1.60	5		875	262		1,137	1,350
2350	25 kVAR		1.50	5.333		980	279		1,259	1,500
2400	35 kVAR		1.40	5.714		1,250	299		1,549	1,825
2450	50 kVAR		1.30	6.154		1,775	320		2,095	2,425

26 35 26 – Harmonic Filters

26 35 26.10 Computer Isolation Transformer		Crew	Daily Output	Labor-Hours	Unit	Material	Labor	Equipment	Total	Total Incl O&P
0010	**COMPUTER ISOLATION TRANSFORMER**									
0100	Computer grade									
0110	Single-phase, 120/240 V, 0.5 kVA	1 Elec	4	2	Ea.	365	105		470	555
0120	1.0 kVA		2.67	2.996		515	157		672	800
0130	2.5 kVA		2	4		785	210		995	1,175
0140	5 kVA		1.14	7.018		895	370		1,265	1,525

26 35 26.20 Computer Regulator Transformer		Crew	Daily Output	Labor-Hours	Unit	Material	Labor	Equipment	Total	Total Incl O&P
0010	**COMPUTER REGULATOR TRANSFORMER**									
0100	Ferro-resonant, constant voltage, variable transformer									
0110	Single-phase, 240 V, 0.5 kVA	1 Elec	2.67	2.996	Ea.	375	157		532	645
0120	1.0 kVA		2	4		515	210		725	880
0130	2.0 kVA		1	8		875	420		1,295	1,600
0210	Plug-in unit 120 V, 0.14 kVA		8	1		216	52.50		268.50	315
0220	0.25 kVA		8	1		252	52.50		304.50	355
0230	0.5 kVA		8	1		375	52.50		427.50	490
0240	1.0 kVA		5.33	1.501		515	78.50		593.50	685
0250	2.0 kVA		4	2		875	105		980	1,125

26 35 Power Filters and Conditioners

26 35 26 – Harmonic Filters

26 35 26.30 Power Conditioner Transformer

	26 35 26.30 Power Conditioner Transformer	Crew	Daily Output	Labor-Hours	Unit	Material	Labor	Equipment	Total	Total Incl O&P
						2013 Bare Costs				
0010	**POWER CONDITIONER TRANSFORMER**									
0100	Electronic solid state, buck-boost, transformer, w/tap switch									
0110	Single-phase, 115 V, 3.0 kVA, + or - 3% accuracy	2 Elec	1.60	10	Ea.	2,025	525		2,550	3,000
0120	208, 220, 230, or 240 V, 5.0 kVA, + or - 1.5% accuracy	3 Elec	1.60	15		2,625	785		3,410	4,050
0130	5.0 kVA, + or - 6% accuracy	2 Elec	1.14	14.035		2,350	735		3,085	3,675
0140	7.5 kVA, + or - 1.5% accuracy	3 Elec	1.50	16		3,325	840		4,165	4,900
0150	7.5 kVA, + or - 6% accuracy		1.60	15		2,775	785		3,560	4,225
0160	10.0 kVA, + or - 1.5% accuracy		1.33	18.045		4,425	945		5,370	6,275
0170	10.0 kVA, + or - 6% accuracy		1.41	17.021		3,750	890		4,640	5,450

26 35 26.40 Transient Voltage Suppressor Transformer

	26 35 26.40 Transient Voltage Suppressor Transformer	Crew	Daily Output	Labor-Hours	Unit	Material	Labor	Equipment	Total	Total Incl O&P
0010	**TRANSIENT VOLTAGE SUPPRESSOR TRANSFORMER**									
0110	Single-phase, 120 V, 1.8 kVA	1 Elec	4	2	Ea.	750	105		855	980
0120	3.6 kVA		4	2		1,525	105		1,630	1,825
0130	7.2 kVA		3.20	2.500		2,650	131		2,781	3,125
0150	240 V, 3.6 kVA		4	2		1,525	105		1,630	1,825
0160	7.2 kVA		4	2		1,950	105		2,055	2,300
0170	14.4 kVA		3.20	2.500		3,400	131		3,531	3,950
0210	Plug-in unit, 120 V, 1.8 kVA		8	1		565	52.50		617.50	700

26 35 53 – Voltage Regulators

26 35 53.10 Automatic Voltage Regulators

	26 35 53.10 Automatic Voltage Regulators	Crew	Daily Output	Labor-Hours	Unit	Material	Labor	Equipment	Total	Total Incl O&P
0010	**AUTOMATIC VOLTAGE REGULATORS**									
0100	Computer grade, solid state, variable transf. volt. regulator									
0110	Single-phase, 120 V, 8.6 kVA	2 Elec	1.33	12.030	Ea.	4,675	630		5,305	6,100
0120	17.3 kVA		1.14	14.035		5,525	735		6,260	7,175
0130	208/240 V, 7.5/8.6 kVA		1.33	12.030		4,675	630		5,305	6,100
0140	13.5/15.6 kVA		1.33	12.030		5,525	630		6,155	7,025
0150	27.0/31.2 kVA		1.14	14.035		6,975	735		7,710	8,775
0210	Two-phase, single control, 208/240 V, 15.0/17.3 kVA		1.14	14.035		5,525	735		6,260	7,175
0220	Individual phase control, 15.0/17.3 kVA		1.14	14.035		5,525	735		6,260	7,175
0230	30.0/34.6 kVA	3 Elec	1.33	18.045		6,975	945		7,920	9,100
0310	Three-phase single control, 208/240 V, 26/30 kVA	2 Elec	1	16		5,525	840		6,365	7,325
0320	380/480 V, 24/30 kVA	″	1	16		5,525	840		6,365	7,325
0330	43/54 kVA	3 Elec	1.33	18.045		10,000	945		10,945	12,400
0340	Individual phase control, 208 V, 26 kVA	″	1.33	18.045		5,525	945		6,470	7,500
0350	52 kVA	R-3	.91	21.978		6,975	1,150	151	8,276	9,550
0360	340/480 V, 24/30 kVA	2 Elec	1	16		5,525	840		6,365	7,325
0370	43/54 kVA	″	1	16		6,975	840		7,815	8,925
0380	48/60 kVA	3 Elec	1.33	18.045		10,100	945		11,045	12,500
0390	86/108 kVA	R-3	.91	21.978		11,200	1,150	151	12,501	14,200
0500	Standard grade, solid state, variable transformer volt. regulator									
0510	Single-phase, 115 V, 2.3 kVA	1 Elec	2.29	3.493	Ea.	1,825	183		2,008	2,275
0520	4.2 kVA		2	4		2,925	210		3,135	3,550
0530	6.6 kVA		1.14	7.018		3,600	370		3,970	4,500
0540	13.0 kVA		1.14	7.018		6,200	370		6,570	7,350
0550	16.6 kVA	2 Elec	1.23	13.008		7,325	680		8,005	9,075
0610	230 V, 8.3 kVA		1.33	12.030		6,200	630		6,830	7,750
0620	21.4 kVA		1.23	13.008		7,325	680		8,005	9,075
0630	29.9 kVA		1.23	13.008		7,325	680		8,005	9,075
0710	460 V, 9.2 kVA		1.33	12.030		6,200	630		6,830	7,750
0720	20.7 kVA		1.23	13.008		7,325	680		8,005	9,075
0810	Three-phase, 230 V, 13.1 kVA	3 Elec	1.41	17.021		6,200	890		7,090	8,125

26 35 Power Filters and Conditioners

26 35 53 – Voltage Regulators

26 35 53.10 Automatic Voltage Regulators		Crew	Daily Output	Labor-Hours	Unit	2013 Bare Costs Material	Labor	Equipment	Total	Total Incl O&P
0820	19.1 kVA	3 Elec	1.41	17.021	Ea.	7,325	890		8,215	9,375
0830	25.1 kVA	↓	1.23	19.512		7,325	1,025		8,350	9,575
0840	57.8 kVA	R-3	.95	21.053		13,300	1,100	144	14,544	16,400
0850	74.9 kVA	″	.91	21.978		13,300	1,150	151	14,601	16,500
0910	460 V, 14.3 kVA	3 Elec	1.41	17.021		6,200	890		7,090	8,125
0920	19.1 kVA		1.41	17.021		7,325	890		8,215	9,375
0930	27.9 kVA	↓	1.23	19.512		7,325	1,025		8,350	9,575
0940	59.8 kVA	R-3	1	20		13,300	1,050	137	14,487	16,300
0950	79.7 kVA		.95	21.053		14,900	1,100	144	16,144	18,100
0960	118 kVA	↓	.95	21.053	↓	15,600	1,100	144	16,844	19,000
1000	Laboratory grade, precision, electronic voltage regulator									
1110	Single-phase, 115 V, 0.5 kVA	1 Elec	2.29	3.493	Ea.	1,150	183		1,333	1,525
1120	1.0 kVA		2	4		1,225	210		1,435	1,675
1130	3.0 kVA	↓	.80	10		1,700	525		2,225	2,650
1140	6.0 kVA	2 Elec	1.46	10.959		3,150	575		3,725	4,325
1150	10.0 kVA	3 Elec	1	24		4,100	1,250		5,350	6,400
1160	15.0 kVA	″	1.50	16		4,700	840		5,540	6,400
1210	230 V, 3.0 kVA	1 Elec	.80	10		1,925	525		2,450	2,900
1220	6.0 kVA	2 Elec	1.46	10.959		3,225	575		3,800	4,400
1230	10.0 kVA	3 Elec	1.71	14.035		4,300	735		5,035	5,825
1240	15.0 kVA	″	1.60	15	↓	4,850	785		5,635	6,525

26 35 53.30 Transient Suppressor/Voltage Regulator		Crew	Daily Output	Labor-Hours	Unit	Material	Labor	Equipment	Total	Total Incl O&P
0010	**TRANSIENT SUPPRESSOR/VOLTAGE REGULATOR** (without isolation)									
0110	Single-phase, 115 V, 1.0 kVA	1 Elec	2.67	2.996	Ea.	980	157		1,137	1,300
0120	2.0 kVA		2.29	3.493		1,325	183		1,508	1,750
0130	4.0 kVA		2.13	3.756		1,650	197		1,847	2,125
0140	220 V, 1.0 kVA		2.67	2.996		980	157		1,137	1,300
0150	2.0 kVA		2.29	3.493		1,325	183		1,508	1,750
0160	4.0 kVA		2.13	3.756		1,750	197		1,947	2,200
0210	Plug-in unit, 120 V, 1.0 kVA		8	1		935	52.50		987.50	1,100
0220	2.0 kVA	↓	8	1	↓	1,325	52.50		1,377.50	1,525

26 36 Transfer Switches

26 36 13 – Manual Transfer Switches

26 36 13.10 Non-Automatic Transfer Switches		Crew	Daily Output	Labor-Hours	Unit	Material	Labor	Equipment	Total	Total Incl O&P
0010	**NON-AUTOMATIC TRANSFER SWITCHES** enclosed									
0100	Manual operated, 480 volt 3 pole, 30 amp	1 Elec	2.30	3.478	Ea.	1,400	182		1,582	1,825
0150	60 amp		1.90	4.211		1,400	221		1,621	1,850
0200	100 amp	↓	1.30	6.154		1,425	320		1,745	2,025
0250	200 amp	2 Elec	2	8		2,775	420		3,195	3,675
0300	400 amp		1.60	10		4,025	525		4,550	5,200
0350	600 amp	↓	1	16		6,150	840		6,990	8,000
1000	250 volt 3 pole, 30 amp	1 Elec	2.30	3.478		1,150	182		1,332	1,550
1100	60 amp		1.90	4.211		1,400	221		1,621	1,850
1150	100 amp	↓	1.30	6.154		1,400	320		1,720	2,000
1200	200 amp	2 Elec	2	8		2,800	420		3,220	3,700
1300	600 amp	″	1	16		6,175	840		7,015	8,025
1500	Electrically operated, 480 volt 3 pole, 60 amp	1 Elec	1.90	4.211		2,325	221		2,546	2,875
1600	100 amp	″	1.30	6.154		2,325	320		2,645	3,025
1650	200 amp	2 Elec	2	8		3,825	420		4,245	4,850
1700	400 amp		1.60	10		5,350	525		5,875	6,675

26 36 Transfer Switches

26 36 13 – Manual Transfer Switches

26 36 13.10 Non-Automatic Transfer Switches		Crew	Daily Output	Labor-Hours	Unit	2013 Bare Costs Material	Labor	Equipment	Total	Total Incl O&P
1750	600 amp	2 Elec	1	16	Ea.	7,725	840		8,565	9,750
2000	250 volt 3 pole, 30 amp	1 Elec	2.30	3.478		2,325	182		2,507	2,825
2050	60 amp	″	1.90	4.211		2,325	221		2,546	2,875
2150	200 amp	2 Elec	2	8		3,825	420		4,245	4,850
2200	400 amp		1.60	10		5,350	525		5,875	6,675
2250	600 amp		1	16		7,725	840		8,565	9,750
2500	NEMA 3R, 480 volt 3 pole, 60 amp	1 Elec	1.80	4.444		2,725	233		2,958	3,350
2550	100 amp	″	1.20	6.667		2,725	350		3,075	3,525
2600	200 amp	2 Elec	1.80	8.889		4,350	465		4,815	5,475
2650	400 amp	″	1.40	11.429		5,875	600		6,475	7,350
2800	NEMA 3R, 250 volt 3 pole solid state, 100 amp	1 Elec	1.20	6.667		2,725	350		3,075	3,525
2850	150 amp	2 Elec	1.80	8.889		3,650	465		4,115	4,700
2900	250 volt 2 pole solid state, 100 amp	1 Elec	1.30	6.154		2,675	320		2,995	3,400
2950	150 amp	2 Elec	2	8		3,550	420		3,970	4,525

26 36 23 – Automatic Transfer Switches

26 36 23.10 Automatic Transfer Switch Devices

Line	Description	Crew	Daily Output	Labor-Hours	Unit	Material	Labor	Equipment	Total	Total Incl O&P
0010	**AUTOMATIC TRANSFER SWITCH DEVICES** R263623-60									
0015	Switches, enclosed 120/240 volt, 2 pole, 30 amp	1 Elec	2.40	3.333	Ea.	2,325	175		2,500	2,800
0020	70 amp		2	4		2,325	210		2,535	2,875
0030	100 amp		1.35	5.926		2,325	310		2,635	3,025
0040	225 amp	2 Elec	2.10	7.619		3,425	400		3,825	4,350
0050	400 amp		1.70	9.412		5,200	495		5,695	6,475
0060	600 amp		1.06	15.094		7,625	790		8,415	9,575
0070	800 amp		.84	19.048		8,950	1,000		9,950	11,400
0100	Switches, enclosed 480 volt, 3 pole, 30 amp	1 Elec	2.30	3.478		2,575	182		2,757	3,100
0200	60 amp		1.90	4.211		2,575	221		2,796	3,150
0300	100 amp		1.30	6.154		2,575	320		2,895	3,300
0400	150 amp	2 Elec	2.40	6.667		3,175	350		3,525	4,025
0500	225 amp		2	8		4,050	420		4,470	5,100
0600	260 amp		2	8		4,700	420		5,120	5,775
0700	400 amp		1.60	10		6,125	525		6,650	7,500
0800	600 amp		1	16		8,475	840		9,315	10,600
0900	800 amp		.80	20		9,950	1,050		11,000	12,500
1000	1000 amp		.76	21.053		13,100	1,100		14,200	16,100
1100	1200 amp		.70	22.857		18,000	1,200		19,200	21,600
1200	1600 amp		.60	26.667		20,500	1,400		21,900	24,600
1300	2000 amp		.50	32		22,900	1,675		24,575	27,700
1600	Accessories, time delay on engine starting					201			201	221
1700	Adjustable time delay on retransfer					201			201	221
1800	Shunt trips for customer connections					360			360	395
1900	Maintenance select switch					81.50			81.50	90
2000	Auxiliary contact when normal fails					94.50			94.50	104
2100	Pilot light-emergency					81.50			81.50	90
2200	Pilot light-normal					81.50			81.50	90
2300	Auxiliary contact-closed on normal					94.50			94.50	104
2400	Auxiliary contact-closed on emergency					94.50			94.50	104
2500	Emergency source sensing, frequency relay					415			415	455

26 41 Facility Lightning Protection

26 41 13 – Lightning Protection for Structures

26 41 13.13 Lightning Protection for Buildings		Crew	Daily Output	Labor-Hours	Unit	2013 Bare Costs Material	Labor	Equipment	Total	Total Incl O&P
0010	**LIGHTNING PROTECTION FOR BUILDINGS**									
0200	Air terminals & base, copper									
0400	3/8″ diameter x 10″ (to 75′ high)	1 Elec	8	1	Ea.	29	52.50		81.50	110
0500	1/2″ diameter x 12″ (over 75′ high)		8	1		33.50	52.50		86	115
0520	1/2″ diameter x 24″		7.30	1.096		44.50	57.50		102	135
0540	1/2″ diameter x 60″		6.70	1.194		76.50	62.50		139	178
1000	Aluminum, 1/2″ diameter x 12″ (to 75′ high)		8	1		15	52.50		67.50	95
1020	1/2″ diameter x 24″		7.30	1.096		17.95	57.50		75.45	106
1040	1/2″ diameter x 60″		6.70	1.194		26.50	62.50		89	123
1100	5/8″ diameter x 12″ (over 75′ high)		8	1		16.75	52.50		69.25	97
2000	Cable, copper, 220 lb. per thousand ft. (to 75′ high)		320	.025	L.F.	3.54	1.31		4.85	5.85
2100	375 lb. per thousand ft. (over 75′ high)		230	.035		6.40	1.82		8.22	9.80
2500	Aluminum, 101 lb. per thousand ft. (to 75′ high)		280	.029		1.10	1.50		2.60	3.45
2600	199 lb. per thousand ft. (over 75′ high)		240	.033		1.62	1.75		3.37	4.39
3000	Arrester, 175 volt AC to ground		8	1	Ea.	78.50	52.50		131	165
3100	650 volt AC to ground		6.70	1.194	″	68	62.50		130.50	169
4400	Connector cable, copper									
4410	Through wall connector	1 Elec	7	1.143	Ea.	57.50	60		117.50	153
4420	Through roof connector		5	1.600		67	84		151	199
4430	Beam connector		7	1.143		33	60		93	126
4440	Double bolt connector		38	.211		12.10	11.05		23.15	30
4450	Bar, copper		40	.200		11.90	10.50		22.40	29
4500	Connector cable, aluminum									
4510	Through wall connector	1 Elec	7	1.143	Ea.	27.50	60		87.50	120
4520	Through roof connector		5	1.600		33	84		117	162
4530	Beam connector		7	1.143		19.20	60		79.20	111
4540	Double bolt connector		38	.211		8	11.05		19.05	25.50
4600	Bonding plates, copper									
4610	I Beam, 8″ square	1 Elec	7	1.143	Ea.	25.50	60		85.50	118
4620	Purlin, 8″ square		7	1.143		27	60		87	119
4630	Large heavy duty, 16″ square		5	1.600		27	84		111	155
4640	Aluminum, I Beam, 8″ square		7	1.143		14.90	60		74.90	106
4650	Purlin, 8″ square		7	1.143		13.70	60		73.70	105
4660	Large heavy duty, 16″ square		5	1.600		17.80	84		101.80	145
4700	Cable support, copper, loop fastener		38	.211		.72	11.05		11.77	17.30
4710	Adhesive cable holder		38	.211		1.43	11.05		12.48	18.05
4720	Aluminum, loop fastener		38	.211		.28	11.05		11.33	16.80
4730	Adhesive cable holder		38	.211		.79	11.05		11.84	17.35
4740	Bonding strap, copper, 3/4″ x 9-1/2″		15	.533		8.85	28		36.85	52
4750	Aluminum, 3/4″ x 9-1/2″		15	.533		5	28		33	47.50
4760	Swivel adapter, copper, 1/2″		73	.110		16.65	5.75		22.40	27
4770	Aluminum, 1/2″		73	.110		10.35	5.75		16.10	20

26 42 Cathodic Protection

26 42 16 – Passive Cathodic Protection for Underground Storage Tank

26 42 16.50 Cathodic Protection Wiring Methods		Crew	Daily Output	Labor-Hours	Unit	Material	Labor	Equipment	Total	Total Incl O&P
						2013 Bare Costs				
0010	**CATHODIC PROTECTION WIRING METHODS**									
1000	Anodes, magnesium type, 9 #	R-15	18.50	2.595	Ea.	38.50	133	17.95	189.45	262
1010	17 #		13	3.692		70	190	25.50	285.50	390
1020	32 #		10	4.800		122	247	33	402	540
1030	48 #		7.20	6.667		160	345	46	551	740
1100	Graphite type w/epoxy cap, 3" x 60" (32 #)	R-22	8.40	4.438		125	197		322	435
1110	4" x 80" (68 #)		6	6.213		237	276		513	675
1120	6" x 72" (80 #)		5.20	7.169		1,475	320		1,795	2,100
1130	6" x 36" (45 #)		9.60	3.883		745	173		918	1,075
2000	Rectifiers, silicon type, air cooled, 28 V/10 A	R-19	3.50	5.714		2,275	300		2,575	2,950
2010	20 V/20 A		3.50	5.714		2,325	300		2,625	3,000
2100	Oil immersed, 28 V/10 A		3	6.667		3,100	350		3,450	3,925
2110	20 V/20 A		3	6.667		3,125	350		3,475	3,950
3000	Anode backfill, coke breeze	R-22	3850	.010	Lb.	.22	.43		.65	.89
4000	Cable, HMWPE, No. 8		2.40	15.533	M.L.F.	490	690		1,180	1,600
4010	No. 6		2.40	15.533		730	690		1,420	1,850
4020	No. 4		2.40	15.533		1,100	690		1,790	2,275
4030	No. 2		2.40	15.533		1,725	690		2,415	2,950
4040	No. 1		2.20	16.945		2,350	755		3,105	3,750
4050	No. 1/0		2.20	16.945		2,925	755		3,680	4,350
4060	No. 2/0		2.20	16.945		4,525	755		5,280	6,150
4070	No. 4/0		2	18.640		5,875	830		6,705	7,725
5000	Test station, 7 terminal box, flush curb type w/lockable cover	R-19	12	1.667	Ea.	70.50	87.50		158	209
5010	Reference cell, 2" dia PVC conduit, cplg., plug, set flush	"	4.80	4.167	"	148	219		367	490

26 51 Interior Lighting

26 51 13 – Interior Lighting Fixtures, Lamps, and Ballasts

26 51 13.10 Fixture Hangers		Crew	Daily Output	Labor-Hours	Unit	Material	Labor	Equipment	Total	Total Incl O&P
0010	**FIXTURE HANGERS**									
0220	Box hub cover	1 Elec	32	.250	Ea.	3.81	13.10		16.91	24
0240	Canopy		12	.667		9.10	35		44.10	62.50
0260	Connecting block		40	.200		3.49	10.50		13.99	19.55
0280	Cushion hanger		16	.500		22.50	26		48.50	64
0300	Box hanger, with mounting strap		8	1		9	52.50		61.50	88.50
0320	Connecting block		40	.200		2.13	10.50		12.63	18.05
0340	Flexible, 1/2" diameter, 4" long		12	.667		11.70	35		46.70	65.50
0360	6" long		12	.667		12.55	35		47.55	66.50
0380	8" long		12	.667		13.85	35		48.85	68
0400	10" long		12	.667		14.60	35		49.60	68.50
0420	12" long		12	.667		15.75	35		50.75	70
0440	15" long		12	.667		16.55	35		51.55	70.50
0460	18" long		12	.667		18.75	35		53.75	73
0480	3/4" diameter, 4" long		10	.800		13.95	42		55.95	78
0500	6" long		10	.800		15.55	42		57.55	79.50
0520	8" long		10	.800		16.85	42		58.85	81
0540	10" long		10	.800		17.80	42		59.80	82
0560	12" long		10	.800		19.30	42		61.30	83.50
0580	15" long		10	.800		21	42		63	86
0600	18" long		10	.800		23.50	42		65.50	88.50

26 51 Interior Lighting

26 51 13 – Interior Lighting Fixtures, Lamps, and Ballasts

	26 51 13.40 Interior HID Fixtures	Crew	Daily Output	Labor-Hours	Unit	Material	Labor	Equipment	Total	Total Incl O&P
						2013 Bare Costs				
0010	**INTERIOR HID FIXTURES** Incl. lamps, and mounting hardware									
0700	High pressure sodium, recessed, round, 70 watt	1 Elec	3.50	2.286	Ea.	425	120		545	645
0720	100 watt		3.50	2.286		445	120		565	670
0740	150 watt		3.20	2.500		500	131		631	745
0760	Square, 70 watt		3.60	2.222		420	116		536	640
0780	100 watt		3.60	2.222		445	116		561	665
0820	250 watt		3	2.667		640	140		780	915
0840	1000 watt	2 Elec	4.80	3.333		1,200	175		1,375	1,575
0860	Surface, round, 70 watt	1 Elec	3	2.667		600	140		740	870
0880	100 watt		3	2.667		615	140		755	885
0900	150 watt		2.70	2.963		640	155		795	935
0920	Square, 70 watt		3	2.667		580	140		720	850
0940	100 watt		3	2.667		615	140		755	885
0980	250 watt		2.50	3.200		650	168		818	965
1040	Pendent, round, 70 watt		3	2.667		635	140		775	910
1060	100 watt		3	2.667		610	140		750	880
1080	150 watt		2.70	2.963		625	155		780	915
1100	Square, 70 watt		3	2.667		650	140		790	925
1120	100 watt		3	2.667		660	140		800	935
1140	150 watt		2.70	2.963		680	155		835	975
1160	250 watt		2.50	3.200		915	168		1,083	1,250
1180	400 watt		2.40	3.333		965	175		1,140	1,300
1220	Wall, round, 70 watt		3	2.667		545	140		685	810
1240	100 watt		3	2.667		560	140		700	825
1260	150 watt		2.70	2.963		575	155		730	860
1300	Square, 70 watt		3	2.667		580	140		720	850
1320	100 watt		3	2.667		610	140		750	880
1340	150 watt		2.70	2.963		645	155		800	940
1360	250 watt		2.50	3.200		690	168		858	1,000
1380	400 watt	2 Elec	4.80	3.333		850	175		1,025	1,200
1400	1000 watt	"	3.60	4.444		1,200	233		1,433	1,675
1500	Metal halide, recessed, round, 175 watt	1 Elec	3.40	2.353		420	123		543	645
1520	250 watt	"	3.20	2.500		465	131		596	710
1540	400 watt	2 Elec	5.80	2.759		660	145		805	945
1580	Square, 175 watt	1 Elec	3.40	2.353		370	123		493	595
1640	Surface, round, 175 watt		2.90	2.759		515	145		660	785
1660	250 watt		2.70	2.963		815	155		970	1,125
1680	400 watt	2 Elec	4.80	3.333		915	175		1,090	1,250
1720	Square, 175 watt	1 Elec	2.90	2.759		560	145		705	830
1800	Pendent, round, 175 watt		2.90	2.759		610	145		755	885
1820	250 watt		2.70	2.963		880	155		1,035	1,200
1840	400 watt	2 Elec	4.80	3.333		980	175		1,155	1,325
1880	Square, 175 watt	1 Elec	2.90	2.759		590	145		735	865
1900	250 watt	"	2.70	2.963		600	155		755	890
1920	400 watt	2 Elec	4.80	3.333		965	175		1,140	1,325
1980	Wall, round, 175 watt	1 Elec	2.90	2.759		515	145		660	785
2000	250 watt	"	2.70	2.963		810	155		965	1,125
2020	400 watt	2 Elec	4.80	3.333		835	175		1,010	1,175
2060	Square, 175 watt	1 Elec	2.90	2.759		545	145		690	810
2080	250 watt	"	2.70	2.963		590	155		745	880
2100	400 watt	2 Elec	4.80	3.333		830	175		1,005	1,175
2800	High pressure sodium, vaporproof, recessed, 70 watt	1 Elec	3.50	2.286		565	120		685	805

26 51 13 – Interior Lighting Fixtures, Lamps, and Ballasts

26 51 13.40 Interior HID Fixtures		Crew	Daily Output	Labor-Hours	Unit	2013 Bare Costs Material	Labor	Equipment	Total	Total Incl O&P
2820	100 watt	1 Elec	3.50	2.286	Ea.	580	120		700	815
2840	150 watt		3.20	2.500		595	131		726	850
2900	Surface, 70 watt		3	2.667		640	140		780	915
2920	100 watt		3	2.667		670	140		810	945
2940	150 watt		2.70	2.963		695	155		850	995
3000	Pendent, 70 watt		3	2.667		630	140		770	905
3020	100 watt		3	2.667		650	140		790	925
3040	150 watt		2.70	2.963		690	155		845	990
3100	Wall, 70 watt		3	2.667		680	140		820	960
3120	100 watt		3	2.667		710	140		850	995
3140	150 watt		2.70	2.963		740	155		895	1,050
3200	Metal halide, vaporproof, recessed, 175 watt		3.40	2.353		545	123		668	785
3220	250 watt		3.20	2.500		585	131		716	835
3240	400 watt	2 Elec	5.80	2.759		730	145		875	1,025
3260	1000 watt	"	4.80	3.333		1,325	175		1,500	1,725
3280	Surface, 175 watt	1 Elec	2.90	2.759		530	145		675	800
3300	250 watt	"	2.70	2.963		830	155		985	1,150
3320	400 watt	2 Elec	4.80	3.333		1,025	175		1,200	1,375
3340	1000 watt	"	3.60	4.444		1,500	233		1,733	2,000
3360	Pendent, 175 watt	1 Elec	2.90	2.759		555	145		700	825
3380	250 watt	"	2.70	2.963		855	155		1,010	1,175
3400	400 watt	2 Elec	4.80	3.333		1,025	175		1,200	1,375
3420	1000 watt	"	3.60	4.444		1,650	233		1,883	2,175
3440	Wall, 175 watt	1 Elec	2.90	2.759		605	145		750	880
3460	250 watt	"	2.70	2.963		905	155		1,060	1,225
3480	400 watt	2 Elec	4.80	3.333		1,075	175		1,250	1,425
3500	1000 watt	"	3.60	4.444		1,725	233		1,958	2,225

26 51 13.50 Interior Lighting Fixtures		Crew	Daily Output	Labor-Hours	Unit	Material	Labor	Equipment	Total	Total Incl O&P
0010	**INTERIOR LIGHTING FIXTURES** Including lamps, mounting R265113-40									
0030	hardware and connections									
0100	Fluorescent, C.W. lamps, troffer, recess mounted in grid, RS									
0130	Grid ceiling mount									
0200	Acrylic lens, 1'W x 4'L, two 40 watt	1 Elec	5.70	1.404	Ea.	46	73.50		119.50	161
0210	1'W x 4'L, three 40 watt		5.40	1.481		53.50	77.50		131	175
0300	2'W x 2'L, two U40 watt		5.70	1.404		50.50	73.50		124	166
0400	2'W x 4'L, two 40 watt		5.30	1.509		49.50	79		128.50	173
0500	2'W x 4'L, three 40 watt		5	1.600		54.50	84		138.50	185
0600	2'W x 4'L, four 40 watt		4.70	1.702		57	89		146	196
0700	4'W x 4'L, four 40 watt	2 Elec	6.40	2.500		290	131		421	515
0800	4'W x 4'L, six 40 watt		6.20	2.581		300	135		435	530
0900	4'W x 4'L, eight 40 watt		5.80	2.759		310	145		455	555
0910	Acrylic lens, 1'W x 4'L, two 32 watt T8 G	1 Elec	5.70	1.404		57	73.50		130.50	173
0930	2'W x 2'L, two U32 watt T8 G		5.70	1.404		68	73.50		141.50	185
0940	2'W x 4'L, two 32 watt T8 G		5.30	1.509		62	79		141	186
0950	2'W x 4'L, three 32 watt T8 G		5	1.600		65	84		149	197
0960	2'W x 4'L, four 32 watt T8 G		4.70	1.702		66.50	89		155.50	207
1000	Surface mounted, RS									
1030	Acrylic lens with hinged & latched door frame									
1100	1'W x 4'L, two 40 watt	1 Elec	7	1.143	Ea.	63	60		123	159
1110	1'W x 4'L, three 40 watt		6.70	1.194		65	62.50		127.50	165
1200	2'W x 2'L, two U40 watt		7	1.143		67.50	60		127.50	164
1300	2'W x 4'L, two 40 watt		6.20	1.290		77	67.50		144.50	186

26 51 13 – Interior Lighting Fixtures, Lamps, and Ballasts

26 51 13.50 Interior Lighting Fixtures		Crew	Daily Output	Labor-Hours	Unit	2013 Bare Costs Material	Labor	Equipment	Total	Total Incl O&P
1400	2′W x 4′L, three 40 watt	1 Elec	5.70	1.404	Ea.	78	73.50		151.50	196
1500	2′W x 4′L, four 40 watt		5.30	1.509		80.50	79		159.50	207
1600	4′W x 4′L, four 40 watt	2 Elec	7.20	2.222		395	116		511	610
1700	4′W x 4′L, six 40 watt		6.60	2.424		430	127		557	660
1800	4′W x 4′L, eight 40 watt		6.20	2.581		445	135		580	690
1900	2′W x 8′L, four 40 watt		6.40	2.500		157	131		288	370
2000	2′W x 8′L, eight 40 watt		6.20	2.581		169	135		304	390
2010	Acrylic wrap around lens									
2020	6″W x 4′L, one 40 watt	1 Elec	8	1	Ea.	64	52.50		116.50	149
2030	6″W x 8′L, two 40 watt	2 Elec	8	2		70.50	105		175.50	235
2040	11″W x 4′L, two 40 watt	1 Elec	7	1.143		43	60		103	137
2050	11″W x 8′L, four 40 watt	2 Elec	6.60	2.424		71	127		198	268
2060	16″W x 4′L, four 40 watt	1 Elec	5.30	1.509		71	79		150	196
2070	16″W x 8′L, eight 40 watt	2 Elec	6.40	2.500		155	131		286	365
2080	2′W x 2′L, two U40 watt	1 Elec	7	1.143		86.50	60		146.50	185
2100	Strip fixture									
2130	Surface mounted									
2200	4′ long, one 40 watt, RS	1 Elec	8.50	.941	Ea.	27	49.50		76.50	104
2300	4′ long, two 40 watt, RS		8	1		38	52.50		90.50	120
2310	4′ long, two 32 watt T8, RS G		8	1		68.50	52.50		121	154
2400	4′ long, one 40 watt, SL		8	1		46	52.50		98.50	129
2500	4′ long, two 40 watt, SL		7	1.143		62.50	60		122.50	159
2580	8′ long, one 60 watt T8, SL G	2 Elec	13.40	1.194		82	62.50		144.50	184
2590	8′ long, two 60 watt T8,.SL G		12.40	1.290		85	67.50		152.50	195
2600	8′ long, one 75 watt, SL		13.40	1.194		47	62.50		109.50	146
2700	8′ long, two 75 watt, SL		12.40	1.290		57	67.50		124.50	164
2800	4′ long, two 60 watt, HO	1 Elec	6.70	1.194		91.50	62.50		154	195
2810	4′ long, two 54 watt, T5HO G	"	6.70	1.194		158	62.50		220.50	268
2900	8′ long, two 110 watt, HO	2 Elec	10.60	1.509		96.50	79		175.50	224
2910	4′ long, two 115 watt, VHO	1 Elec	6.50	1.231		127	64.50		191.50	237
2920	8′ long, two 215 watt, VHO	2 Elec	10.40	1.538		137	80.50		217.50	272
2950	High bay pendent mounted, 16″ W x 4′ L, four 54 watt, T5HO G		8.90	1.798		224	94		318	385
2952	2′ W x 4′ L, six 54 watt, T5HO G		8.50	1.882		300	98.50		398.50	480
2954	2′ W x 4′ L, six 32 watt, T8 G		8.50	1.882		175	98.50		273.50	340
3000	Strip, pendent mounted, industrial, white porcelain enamel									
3100	4′ long, two 40 watt, RS	1 Elec	5.70	1.404	Ea.	52	73.50		125.50	167
3110	4′ long, two 32 watt T8, RS G		5.70	1.404		72.50	73.50		146	190
3200	4′ long, two 60 watt, HO		5	1.600		81	84		165	214
3290	8′ long, two 60 watt T8, SL G	2 Elec	8.80	1.818		113	95.50		208.50	267
3300	8′ long, two 75 watt, SL		8.80	1.818		96	95.50		191.50	249
3400	8′ long, two 110 watt, HO		8	2		123	105		228	292
3410	Acrylic finish, 4′ long, two 40 watt, RS	1 Elec	5.70	1.404		85	73.50		158.50	204
3420	4′ long, two 60 watt, HO		5	1.600		157	84		241	298
3430	4′ long, two 115 watt, VHO		4.80	1.667		203	87.50		290.50	355
3440	8′ long, two 75 watt, SL	2 Elec	8.80	1.818		166	95.50		261.50	325
3450	8′ long, two 110 watt, HO		8	2		188	105		293	365
3460	8′ long, two 215 watt, VHO		7.60	2.105		266	110		376	460
3470	Troffer, air handling, 2′W x 4′L with four 32 watt T8 G	1 Elec	4	2		110	105		215	278
3480	2′W x 2′L with two U32 watt T8 G		5.50	1.455		106	76		182	231
3490	Air connector insulated, 5″ diameter		20	.400		66	21		87	105
3500	6″ diameter		20	.400		67.50	21		88.50	106
3502	Troffer, direct/indirect, 2′W x 4′L with two 32 W T8 G		5.30	1.509		268	79		347	415
3510	Troffer parabolic lay-in, 1′W x 4′L with one 32 W T8 G		5.70	1.404		112	73.50		185.50	233

26 51 13.50 Interior Lighting Fixtures		Crew	Daily Output	Labor-Hours	Unit	2013 Bare Costs Material	Labor	Equipment	Total	Total Incl O&P
3520	1'W x 4'L with two 32 W T8 G	1 Elec	5.30	1.509	Ea.	134	79		213	265
3525	2'W x 2'L with two U32 W T8 G		5.70	1.404		113	73.50		186.50	234
3530	2'W x 4'L with three 32 W T8 G		5	1.600		125	84		209	263
3535	Downlight, recess mounted G		8	1		145	52.50		197.50	238
3540	Wall wash reflector, recess mounted G		8	1		103	52.50		155.50	192
3550	Direct/indirect, 4' long, stl., pendent mtd. G		5	1.600		159	84		243	300
3560	4' long, alum., pendent mtd. G		5	1.600		330	84		414	485
3565	Prefabricated cove, 4' long, stl. continuous row G		5	1.600		199	84		283	345
3570	4' long, alum. continuous row G		5	1.600		340	84		424	500
3580	Wet location, recess mounted, 2'W x 4'L with two 32 watt T8 G		5.30	1.509		232	79		311	375
3590	Pendent mounted, 2'W x 4'L with two 32 watt T8 G		5.70	1.404		360	73.50		433.50	505
4000	Induction lamp, integral ballast, ceiling mounted									
4110	High bay, aluminum reflector, 160 watt	1 Elec	3.20	2.500	Ea.	935	131		1,066	1,225
4120	320 watt	″	3	2.667		1,675	140		1,815	2,050
4130	480 watt	2 Elec	5.80	2.759		2,550	145		2,695	3,025
4150	Low bay, aluminum reflector, 250 watt	1 Elec	3.20	2.500		740	131		871	1,000
4170	Garage, aluminum reflector, 80 watt		3.60	2.222		740	116		856	990
4180	Vandalproof, aluminum reflector, 100 watt		3.20	2.500		525	131		656	770
4220	Metal halide, integral ballast, ceiling, recess mounted									
4230	prismatic glass lens, floating door									
4240	2'W x 2'L, 250 watt	1 Elec	3.20	2.500	Ea.	310	131		441	535
4250	2'W x 2'L, 400 watt	2 Elec	5.80	2.759		360	145		505	610
4260	Surface mounted, 2'W x 2'L, 250 watt	1 Elec	2.70	2.963		335	155		490	600
4270	400 watt	2 Elec	4.80	3.333		395	175		570	695
4280	High bay, aluminum reflector,									
4290	Single unit, 400 watt	2 Elec	4.60	3.478	Ea.	405	182		587	720
4300	Single unit, 1000 watt		4	4		580	210		790	955
4310	Twin unit, 400 watt		3.20	5		805	262		1,067	1,275
4320	Low bay, aluminum reflector, 250W DX lamp	1 Elec	3.20	2.500		355	131		486	585
4330	400 watt lamp	2 Elec	5	3.200		515	168		683	815
4340	High pressure sodium integral ballast ceiling, recess mounted									
4350	prismatic glass lens, floating door									
4360	2'W x 2'L, 150 watt lamp	1 Elec	3.20	2.500	Ea.	380	131		511	610
4370	2'W x 2'L, 400 watt lamp	2 Elec	5.80	2.759		450	145		595	710
4380	Surface mounted, 2'W x 2'L, 150 watt lamp	1 Elec	2.70	2.963		460	155		615	740
4390	400 watt lamp	2 Elec	4.80	3.333		515	175		690	830
4400	High bay, aluminum reflector,									
4410	Single unit, 400 watt lamp	2 Elec	4.60	3.478	Ea.	370	182		552	685
4430	Single unit, 1000 watt lamp	″	4	4		535	210		745	905
4440	Low bay, aluminum reflector, 150 watt lamp	1 Elec	3.20	2.500		320	131		451	550
4445	High bay H.I.D. quartz restrike	″	16	.500		160	26		186	215
4450	Incandescent, high hat can, round alzak reflector, prewired									
4470	100 watt	1 Elec	8	1	Ea.	70	52.50		122.50	156
4480	150 watt		8	1		102	52.50		154.50	191
4500	300 watt		6.70	1.194		236	62.50		298.50	355
4520	Round with reflector and baffles, 150 watt		8	1		50	52.50		102.50	134
4540	Round with concentric louver, 150 watt PAR		8	1		77	52.50		129.50	163
4600	Square glass lens with metal trim, prewired									
4630	100 watt	1 Elec	6.70	1.194	Ea.	54	62.50		116.50	153
4700	200 watt		6.70	1.194		95	62.50		157.50	199
4800	300 watt		5.70	1.404		142	73.50		215.50	266
4810	500 watt		5	1.600		279	84		363	430
4900	Ceiling/wall, surface mounted, metal cylinder, 75 watt		10	.800		55	42		97	123

26 51 13 – Interior Lighting Fixtures, Lamps, and Ballasts

26 51 13.50 Interior Lighting Fixtures		Crew	Daily Output	Labor-Hours	Unit	2013 Bare Costs Material	Labor	Equipment	Total	Total Incl O&P
4920	150 watt	1 Elec	10	.800	Ea.	79	42		121	150
4930	300 watt		8	1		164	52.50		216.50	259
5000	500 watt		6.70	1.194		355	62.50		417.50	485
5010	Square, 100 watt		8	1		110	52.50		162.50	200
5020	150 watt		8	1		121	52.50		173.50	212
5030	300 watt		7	1.143		330	60		390	455
5040	500 watt		6	1.333		335	70		405	475
5200	Ceiling, surface mounted, opal glass drum									
5300	8", one 60 watt lamp	1 Elec	10	.800	Ea.	43.50	42		85.50	111
5400	10", two 60 watt lamps		8	1		49	52.50		101.50	133
5500	12", four 60 watt lamps		6.70	1.194		69	62.50		131.50	170
5510	Pendent, round, 100 watt		8	1		110	52.50		162.50	200
5520	150 watt		8	1		120	52.50		172.50	211
5530	300 watt		6.70	1.194		167	62.50		229.50	278
5540	500 watt		5.50	1.455		315	76		391	460
5550	Square, 100 watt		6.70	1.194		148	62.50		210.50	257
5560	150 watt		6.70	1.194		154	62.50		216.50	263
5570	300 watt		5.70	1.404		225	73.50		298.50	360
5580	500 watt		5	1.600		305	84		389	460
5600	Wall, round, 100 watt		8	1		64	52.50		116.50	149
5620	300 watt		8	1		112	52.50		164.50	202
5630	500 watt		6.70	1.194		370	62.50		432.50	500
5640	Square, 100 watt		8	1		101	52.50		153.50	190
5650	150 watt		8	1		103	52.50		155.50	192
5660	300 watt		7	1.143		162	60		222	268
5670	500 watt		6	1.333		285	70		355	420
6010	Vapor tight, incandescent, ceiling mounted, 200 watt		6.20	1.290		81	67.50		148.50	190
6020	Recessed, 200 watt		6.70	1.194		121	62.50		183.50	227
6030	Pendent, 200 watt		6.70	1.194		81	62.50		143.50	183
6040	Wall, 200 watt		8	1		78	52.50		130.50	165
6100	Fluorescent, surface mounted, 2 lamps, 4'L, RS, 40 watt		3.20	2.500		113	131		244	320
6110	Industrial, 2 lamps 4' long in tandem, 430 MA		2.20	3.636		208	191		399	515
6130	2 lamps 4' long, 800 MA		1.90	4.211		178	221		399	525
6160	Pendent, indust, 2 lamps 4'L in tandem, 430 MA		1.90	4.211		243	221		464	595
6170	2 lamps 4' long, 430 MA		2.30	3.478		163	182		345	450
6180	2 lamps 4' long, 800 MA		1.70	4.706		205	247		452	595
6850	Vandalproof, surface mounted, fluorescent, two 32 watt T8 G		3.20	2.500		248	131		379	470
6860	Incandescent, one 150 watt		8	1		84	52.50		136.50	171
6900	Mirror light, fluorescent, RS, acrylic enclosure, two 40 watt		8	1		125	52.50		177.50	217
6910	One 40 watt		8	1		104	52.50		156.50	193
6920	One 20 watt		12	.667		80	35		115	141
7000	Low bay, aluminum reflector. 70 watt, high pressure sodium		4	2		270	105		375	455
7010	250 watt		3.20	2.500		360	131		491	590
7020	400 watt	2 Elec	5	3.200		375	168		543	665
7500	Ballast replacement, by weight of ballast, to 15' high									
7520	Indoor fluorescent, less than 2 lb.	1 Elec	10	.800	Ea.	25.50	42		67.50	90.50
7540	Two 40W, watt reducer, 2 to 5 lb.		9.40	.851		40.50	44.50		85	111
7560	Two F96 slimline, over 5 lb.		8	1		75.50	52.50		128	162
7580	Vaportite ballast, less than 2 lb.		9.40	.851		25.50	44.50		70	94.50
7600	2 lb. to 5 lb.		8.90	.899		40.50	47		87.50	115
7620	Over 5 lb.		7.60	1.053		75.50	55		130.50	166
7630	Electronic ballast for two tubes		8	1		36.50	52.50		89	119
7640	Dimmable ballast one lamp G		8	1		104	52.50		156.50	193

26 51 13.50 Interior Lighting Fixtures		Crew	Daily Output	Labor-Hours	Unit	2013 Bare Costs Material	Labor	Equipment	Total	Total Incl O&P
7650	Dimmable ballast two-lamp G	1 Elec	7.60	1.053	Ea.	102	55		157	195
7690	Emergency ballast (factory installed in fixture)					154			154	169
7990	Decorator									
8000	Pendent RLM in colors, shallow dome, 12" diam. 100 W	1 Elec	8	1	Ea.	79	52.50		131.50	166
8010	Regular dome, 12" diam., 100 watt		8	1		81.50	52.50		134	168
8020	16" diam., 200 watt		7	1.143		83	60		143	181
8030	18" diam., 300 watt		6	1.333		91	70		161	205
8100	Picture framing light, minimum		16	.500		92	26		118	140
8110	Maximum		16	.500		102	26		128	151
8150	Miniature low voltage, recessed, pinhole		8	1		159	52.50		211.50	254
8160	Star		8	1		131	52.50		183.50	223
8170	Adjustable cone		8	1		178	52.50		230.50	274
8180	Eyeball		8	1		145	52.50		197.50	238
8190	Cone		8	1		127	52.50		179.50	219
8200	Coilex baffle		8	1		125	52.50		177.50	216
8210	Surface mounted, adjustable cylinder		8	1		129	52.50		181.50	220
8250	Chandeliers, incandescent									
8260	24" diam. x 42" high, 6 light candle	1 Elec	6	1.333	Ea.	435	70		505	585
8270	24" diam. x 42" high, 6 light candle w/glass shade		6	1.333		445	70		515	595
8280	17" diam. x 12" high, 8 light w/glass panels		8	1		276	52.50		328.50	385
8300	27" diam. x 29"H, 10 light bohemian lead crystal		4	2		580	105		685	790
8310	21" diam. x 9" high 6 light sculptured ice crystal		8	1		415	52.50		467.50	535
8500	Accent lights, on floor or edge, 0.5 W low volt incandescent									
8520	incl. transformer & fastenings, based on 100' lengths									
8550	Lights in clear tubing, 12" on center	1 Elec	230	.035	L.F.	8.65	1.82		10.47	12.25
8560	6" on center		160	.050		11.25	2.62		13.87	16.30
8570	4" on center		130	.062		17.20	3.22		20.42	24
8580	3" on center		125	.064		19.10	3.35		22.45	26
8590	2" on center		100	.080		28	4.19		32.19	37
8600	Carpet, lights both sides 6" OC, in alum. extrusion		270	.030		26	1.55		27.55	31
8610	In bronze extrusion		270	.030		29.50	1.55		31.05	35
8620	Carpet-bare floor, lights 18" OC, in alum. extrusion		270	.030		21	1.55		22.55	25.50
8630	In bronze extrusion		270	.030		24.50	1.55		26.05	29
8640	Carpet edge-wall, lights 6" OC in alum. extrusion		270	.030		26	1.55		27.55	31
8650	In bronze extrusion		270	.030		29.50	1.55		31.05	35
8660	Bare floor, lights 18" OC, in aluminum extrusion		300	.027		21	1.40		22.40	25
8670	In bronze extrusion		300	.027		24.50	1.40		25.90	28.50
8680	Bare floor conduit, aluminum extrusion		300	.027		6.90	1.40		8.30	9.70
8690	In bronze extrusion		300	.027		13.75	1.40		15.15	17.25
8700	Step edge to 36", lights 6" OC, in alum. extrusion		100	.080	Ea.	69.50	4.19		73.69	83
8710	In bronze extrusion		100	.080		72	4.19		76.19	86
8720	Step edge to 54", lights 6" OC, in alum. extrusion		100	.080		104	4.19		108.19	120
8730	In bronze extrusion		100	.080		110	4.19		114.19	127
8740	Step edge to 72", lights 6" OC, in alum. extrusion		100	.080		139	4.19		143.19	159
8750	In bronze extrusion		100	.080		152	4.19		156.19	173
8760	Connector, male		32	.250		2.57	13.10		15.67	22.50
8770	Female with pigtail		32	.250		5.40	13.10		18.50	25.50
8780	Clamps		400	.020		.51	1.05		1.56	2.13
8790	Transformers, 50 watt		8	1		77.50	52.50		130	164
8800	250 watt		4	2		257	105		362	440
8810	1000 watt		2.70	2.963		465	155		620	740

26 51 Interior Lighting

26 51 13 – Interior Lighting Fixtures, Lamps, and Ballasts

26 51 13.55 Interior LED Fixtures			Crew	Daily Output	Labor-Hours	Unit	Material	2013 Bare Costs Labor	Equipment	Total	Total Incl O&P
0010	**INTERIOR LED FIXTURES** Incl. lamps, and mounting hardware										
0100	Downlight, recess mounted, 7.5" diameter, 25 watt	G	1 Elec	8	1	Ea.	400	52.50		452.50	520
0120	10" diameter, 36 watt	G		8	1		580	52.50		632.50	720
0140	12" x 12", 40 watt	G		6.70	1.194		675	62.50		737.50	840
0160	cylinder, 7.8 watts			8	1		395	52.50		447.50	515
0180	13.3 watts			8	1		520	52.50		572.50	650
1000	Troffer, recess mounted, 2′ x 4′, 37 watt	G		5.30	1.509		590	79		669	770
1010	60 watt	G		5	1.600		755	84		839	955
1020	74 watt	G		4.70	1.702		755	89		844	965
1100	Troffer retrofit kit, 46", 37 watt	G		21	.381		475	19.95		494.95	555
1110	55 watt	G		20	.400		655	21		676	755
1120	74 watt	G		18	.444		655	23.50		678.50	760
1200	Troffer, volumetric recess mounted, 2′ x 2′	G		5.70	1.404		655	73.50		728.50	830
2000	Strip, surface mounted, one light bar 4′ long, 40 watt	G		8.50	.941		845	49.50		894.50	1,000
2010	60 watt	G		8	1		880	52.50		932.50	1,050
2020	Two light bar 4′ long, 80 watt	G		7	1.143		1,150	60		1,210	1,375
3000	Linear, suspended mounted, one light bar 4′ long, 37 watt	G		6.70	1.194		390	62.50		452.50	525
3010	One light bar 8′ long, 74 watt	G	2 Elec	12.20	1.311		725	68.50		793.50	900
3020	Two light bar 4′ long, 74 watt	G	1 Elec	5.70	1.404		775	73.50		848.50	965
3030	Two light bar 8′ long, 148 watt	G	2 Elec	8.80	1.818		1,450	95.50		1,545.50	1,725
4000	High bay, surface mounted, round, 150 watts	G		5.41	2.959		1,625	155		1,780	2,000
4010	2 bars,164 watts			5.41	2.959		1,150	155		1,305	1,475
4020	3 bars, 246 watts			5.01	3.197		1,475	168		1,643	1,850
4030	4 bars, 328 watts			4.60	3.478		1,800	182		1,982	2,250
4040	5 bars, 410 watts		3 Elec	4.20	5.716		2,100	300		2,400	2,750
4050	6 bars, 492 watts			3.80	6.324		2,375	330		2,705	3,125
4060	7 bars, 574 watts			3.39	7.075		2,700	370		3,070	3,525
4070	8 bars, 656 watts			2.99	8.029		3,000	420		3,420	3,900
5000	track, lighthead, 6 watt		1 Elec	32	.250		77	13.10		90.10	105
5010	9 watt		"	32	.250		100	13.10		113.10	130
6000	Garage, surface mount, 103 watts		2 Elec	6.50	2.462		1,375	129		1,504	1,700
6100	pendent mount, 80 watts			6.50	2.462		975	129		1,104	1,275
6200	95 watts			6.50	2.462		1,100	129		1,229	1,425
6300	125 watts			6.50	2.462		1,225	129		1,354	1,550

26 51 13.70 Residential Fixtures		Crew	Daily Output	Labor-Hours	Unit	Material	Labor	Equipment	Total	Total Incl O&P
0010	**RESIDENTIAL FIXTURES**									
0400	Fluorescent, interior, surface, circline, 32 watt & 40 watt	1 Elec	20	.400	Ea.	103	21		124	145
0700	Shallow under cabinet, two 20 watt		16	.500		47.50	26		73.50	91.50
0900	Wall mounted, 4′L, two 32 watt T8, with baffle		10	.800		161	42		203	240
2000	Incandescent, exterior lantern, wall mounted, 60 watt		16	.500		42	26		68	85
2100	Post light, 150W, with 7′ post		4	2		117	105		222	286
2500	Lamp holder, weatherproof with 150W PAR		16	.500		32	26		58	74
2550	With reflector and guard		12	.667		64	35		99	123
2600	Interior pendent, globe with shade, 150 watt		20	.400		209	21		230	262

26 51 13.90 Ballast, Replacement HID		Crew	Daily Output	Labor-Hours	Unit	Material	Labor	Equipment	Total	Total Incl O&P
0010	**BALLAST, REPLACEMENT HID**									
7510	Multi-tap 120/208/240/277 volt									
7550	High pressure sodium, 70 watt	1 Elec	10	.800	Ea.	188	42		230	270
7560	100 watt		9.40	.851		197	44.50		241.50	283
7570	150 watt		9	.889		212	46.50		258.50	305
7580	250 watt		8.50	.941		315	49.50		364.50	420
7590	400 watt		7	1.143		355	60		415	480

26 51 Interior Lighting

26 51 13 – Interior Lighting Fixtures, Lamps, and Ballasts

26 51 13.90 Ballast, Replacement HID		Crew	Daily Output	Labor-Hours	Unit	2013 Bare Costs Material	Labor	Equipment	Total	Total Incl O&P
7600	1000 watt	1 Elec	6	1.333	Ea.	490	70		560	645
7610	Metal halide, 175 watt		8	1		115	52.50		167.50	206
7620	250 watt		8	1		149	52.50		201.50	243
7630	400 watt		7	1.143		186	60		246	295
7640	1000 watt		6	1.333		320	70		390	455
7650	1500 watt		5	1.600		395	84		479	560

26 52 Emergency Lighting

26 52 13 – Emergency Lighting Equipments

26 52 13.10 Emergency Lighting and Battery Units

		Crew	Daily Output	Labor-Hours	Unit	Material	Labor	Equipment	Total	Total Incl O&P
0010	**EMERGENCY LIGHTING AND BATTERY UNITS**									
0300	Emergency light units, battery operated									
0350	Twin sealed beam light, 25 watt, 6 volt each									
0500	Lead battery operated	1 Elec	4	2	Ea.	136	105		241	305
0700	Nickel cadmium battery operated		4	2		630	105		735	850
0780	Additional remote mount, sealed beam, 25 W 6 V		26.70	.300		28.50	15.70		44.20	55
0790	Twin sealed beam light, 25 W 6 V each		26.70	.300		54.50	15.70		70.20	83.50
0900	Self-contained fluorescent lamp pack		10	.800		157	42		199	236

26 53 Exit Signs

26 53 13 – Exit Lighting

26 53 13.10 Exit Lighting Fixtures

		Crew	Daily Output	Labor-Hours	Unit	Material	Labor	Equipment	Total	Total Incl O&P
0010	**EXIT LIGHTING FIXTURES**									
0080	Exit light ceiling or wall mount, incandescent, single face	1 Elec	8	1	Ea.	38.50	52.50		91	121
0100	Double face		6.70	1.194		38.50	62.50		101	136
0120	Explosion proof		3.80	2.105		585	110		695	805
0150	Fluorescent, single face		8	1		65.50	52.50		118	151
0160	Double face		6.70	1.194		73.50	62.50		136	175
0200	LED standard, single face G		8	1		75.50	52.50		128	162
0220	Double face G		6.70	1.194		75.50	62.50		138	177
0230	LED vandal-resistant, single face G		7.27	1.100		225	57.50		282.50	335
0240	LED w/battery unit, single face G		4.40	1.818		137	95.50		232.50	294
0260	Double face G		4	2		139	105		244	310
0262	LED w/battery unit, vandal-resistant, single face G		4.40	1.818		375	95.50		470.50	555
0270	Combination emergency light units and exit sign		4	2		188	105		293	365
0290	LED retrofit kits G		60	.133		50	7		57	65.50
1780	With emergency battery, explosion proof	R-19	7.70	2.597		3,100	136		3,236	3,625

26 54 Classified Location Lighting

26 54 13 – Classified Lighting

26 54 13.20 Explosionproof		Crew	Daily Output	Labor-Hours	Unit	Material	Labor	Equipment	Total	Total Incl O&P
0010	**EXPLOSIONPROOF**, incl lamps, mounting hardware and connections									
6310	Metal halide with ballast, ceiling, surface mounted, 175 watt	1 Elec	2.90	2.759	Ea.	895	145		1,040	1,200
6320	250 watt	"	2.70	2.963		1,075	155		1,230	1,400
6330	400 watt	2 Elec	4.80	3.333		1,150	175		1,325	1,525
6340	Ceiling, pendent mounted, 175 watt	1 Elec	2.60	3.077		850	161		1,011	1,175
6350	250 watt	"	2.40	3.333		1,025	175		1,200	1,375
6360	400 watt	2 Elec	4.20	3.810		1,100	200		1,300	1,525
6370	Wall, surface mounted, 175 watt	1 Elec	2.90	2.759		960	145		1,105	1,275
6380	250 watt	"	2.70	2.963		1,150	155		1,305	1,475
6390	400 watt	2 Elec	4.80	3.333		1,225	175		1,400	1,600
6400	High pressure sodium, ceiling surface mounted, 70 watt	1 Elec	3	2.667		1,250	140		1,390	1,575
6410	100 watt		3	2.667		1,300	140		1,440	1,625
6420	150 watt		2.70	2.963		1,350	155		1,505	1,700
6430	Pendent mounted, 70 watt		2.70	2.963		1,200	155		1,355	1,550
6440	100 watt		2.70	2.963		1,250	155		1,405	1,600
6450	150 watt		2.40	3.333		1,275	175		1,450	1,650
6460	Wall mounted, 70 watt		3	2.667		1,350	140		1,490	1,675
6470	100 watt		3	2.667		1,375	140		1,515	1,725
6480	150 watt		2.70	2.963		1,425	155		1,580	1,800
6510	Incandescent, ceiling mounted, 200 watt		4	2		1,050	105		1,155	1,300
6520	Pendent mounted, 200 watt		3.50	2.286		905	120		1,025	1,175
6530	Wall mounted, 200 watt		4	2		1,050	105		1,155	1,300
6600	Fluorescent, RS, 4′ long, ceiling mounted, two 40 watt		2.70	2.963		2,800	155		2,955	3,300
6610	Three 40 watt		2.20	3.636		4,050	191		4,241	4,725
6620	Four 40 watt		1.90	4.211		5,200	221		5,421	6,050
6630	Pendent mounted, two 40 watt		2.30	3.478		3,250	182		3,432	3,850
6640	Three 40 watt		1.90	4.211		4,625	221		4,846	5,400
6650	Four 40 watt	↓	1.70	4.706	↓	6,075	247		6,322	7,050

26 55 Special Purpose Lighting

26 55 59 – Display Lighting

26 55 59.10 Track Lighting		Crew	Daily Output	Labor-Hours	Unit	Material	Labor	Equipment	Total	Total Incl O&P
0010	**TRACK LIGHTING**									
0080	Track, 1 circuit, 4′ section	1 Elec	6.70	1.194	Ea.	48	62.50		110.50	147
0100	8′ section	2 Elec	10.60	1.509		74.50	79		153.50	200
0200	12′ section	"	8.80	1.818		111	95.50		206.50	265
0300	3 circuits, 4′ section	1 Elec	6.70	1.194		78	62.50		140.50	180
0400	8′ section	2 Elec	10.60	1.509		109	79		188	238
0500	12′ section	"	8.80	1.818		150	95.50		245.50	310
1000	Feed kit, surface mounting	1 Elec	16	.500		13	26		39	53.50
1100	End cover		24	.333		5.55	17.45		23	32
1200	Feed kit, stem mounting, 1 circuit		16	.500		42	26		68	85
1300	3 circuit		16	.500		42	26		68	85
2000	Electrical joiner, for continuous runs, 1 circuit		32	.250		18.05	13.10		31.15	39.50
2100	3 circuit		32	.250		48.50	13.10		61.60	73
2200	Fixtures, spotlight, 75W PAR halogen		16	.500		91	26		117	139
2210	50W MR16 halogen		16	.500		151	26		177	205
3000	Wall washer, 250 watt tungsten halogen		16	.500		124	26		150	176
3100	Low voltage, 25/50 watt, 1 circuit		16	.500		125	26		151	177
3120	3 circuit	↓	16	.500	↓	149	26		175	203

26 55 Special Purpose Lighting

26 55 61 – Theatrical Lighting

26 55 61.10 Lights		Crew	Daily Output	Labor-Hours	Unit	2013 Bare Costs Material	Labor	Equipment	Total	Total Incl O&P
0010	**LIGHTS**									
2000	Lights, border, quartz, reflector, vented,									
2100	colored or white	1 Elec	20	.400	L.F.	173	21		194	222
2500	Spotlight, follow spot, with transformer, 2,100 watt	″	4	2	Ea.	3,075	105		3,180	3,525
2600	For no transformer, deduct					910			910	1,000
3000	Stationary spot, fresnel quartz, 6″ lens	1 Elec	4	2		131	105		236	300
3100	8″ lens		4	2		231	105		336	410
3500	Ellipsoidal quartz, 1,000W, 6″ lens		4	2		335	105		440	525
3600	12″ lens		4	2		600	105		705	815
4000	Strobe light, 1 to 15 flashes per second, quartz		3	2.667		750	140		890	1,050
4500	Color wheel, portable, five hole, motorized		4	2		194	105		299	370

26 56 Exterior Lighting

26 56 13 – Lighting Poles and Standards

26 56 13.10 Lighting Poles		Crew	Daily Output	Labor-Hours	Unit	Material	Labor	Equipment	Total	Total Incl O&P
0010	**LIGHTING POLES**									
2800	Light poles, anchor base									
2820	not including concrete bases									
2840	Aluminum pole, 8′ high	1 Elec	4	2	Ea.	685	105		790	910
2850	10′ high		4	2		720	105		825	950
2860	12′ high		3.80	2.105		755	110		865	995
2870	14′ high		3.40	2.353		780	123		903	1,050
2880	16′ high		3	2.667		865	140		1,005	1,150
3000	20′ high	R-3	2.90	6.897		910	360	47.50	1,317.50	1,575
3200	30′ high		2.60	7.692		1,750	400	53	2,203	2,550
3400	35′ high		2.30	8.696		1,900	450	59.50	2,409.50	2,825
3600	40′ high		2	10		2,150	520	68.50	2,738.50	3,225
3800	Bracket arms, 1 arm	1 Elec	8	1		119	52.50		171.50	209
4000	2 arms		8	1		238	52.50		290.50	340
4200	3 arms		5.30	1.509		360	79		439	515
4400	4 arms		5.30	1.509		475	79		554	645
4500	Steel pole, galvanized, 8′ high		3.80	2.105		610	110		720	835
4510	10′ high		3.70	2.162		635	113		748	870
4520	12′ high		3.40	2.353		690	123		813	940
4530	14′ high		3.10	2.581		730	135		865	1,000
4540	16′ high		2.90	2.759		775	145		920	1,075
4550	18′ high		2.70	2.963		820	155		975	1,125
4600	20′ high	R-3	2.60	7.692		1,075	400	53	1,528	1,825
4800	30′ high		2.30	8.696		1,275	450	59.50	1,784.50	2,150
5000	35′ high		2.20	9.091		1,400	470	62.50	1,932.50	2,300
5200	40′ high		1.70	11.765		1,725	610	80.50	2,415.50	2,900
5400	Bracket arms, 1 arm	1 Elec	8	1		177	52.50		229.50	273
5600	2 arms		8	1		274	52.50		326.50	380
5800	3 arms		5.30	1.509		297	79		376	445
6000	4 arms		5.30	1.509		415	79		494	575
6100	Fiberglass pole, 1 or 2 fixtures, 20′ high	R-3	4	5		660	259	34.50	953.50	1,150
6200	30′ high		3.60	5.556		820	288	38	1,146	1,375
6300	35′ high		3.20	6.250		1,275	325	43	1,643	1,925
6400	40′ high		2.80	7.143		1,500	370	49	1,919	2,250
6420	Wood pole, 4-1/2″ x 5-1/8″, 8′ high	1 Elec	6	1.333		325	70		395	460

26 56 Exterior Lighting

26 56 13 – Lighting Poles and Standards

26 56 13.10 Lighting Poles			Crew	Daily Output	Labor-Hours	Unit	2013 Bare Costs Material	Labor	Equipment	Total	Total Incl O&P
6430	10′ high		1 Elec	6	1.333	Ea.	365	70		435	505
6440	12′ high			5.70	1.404		465	73.50		538.50	620
6450	15′ high			5	1.600		540	84		624	720
6460	20′ high			4	2		655	105		760	875
7300	Transformer bases, not including concrete bases										
7320	Maximum pole size, steel, 40′ high		1 Elec	2	4	Ea.	1,375	210		1,585	1,825
7340	Cast aluminum, 30′ high			3	2.667		735	140		875	1,025
7350	40′ high			2.50	3.200		1,100	168		1,268	1,475

26 56 16 – Parking Lighting

26 56 16.55 Parking Led Lighting			Crew	Daily Output	Labor-Hours	Unit	Material	Labor	Equipment	Total	Total Incl O&P
0010	**PARKING LED LIGHTING**										
0100	Round pole mounting, 88 lamp watts	G	1 Elec	2	4	Ea.	1,950	210		2,160	2,475
0110	Square pole mounting, 223 lamp watts	G	″	2	4	″	3,425	210		3,635	4,100

26 56 19 – Roadway Lighting

26 56 19.20 Roadway Luminaire			Crew	Daily Output	Labor-Hours	Unit	Material	Labor	Equipment	Total	Total Incl O&P
0010	**ROADWAY LUMINAIRE**										
2650	Roadway area luminaire, low pressure sodium, 135 watt		1 Elec	2	4	Ea.	640	210		850	1,025
2700	180 watt		″	2	4		685	210		895	1,075
2750	Metal halide, 400 watt		2 Elec	4.40	3.636		545	191		736	885
2760	1000 watt			4	4		615	210		825	990
2780	High pressure sodium, 400 watt			4.40	3.636		570	191		761	910
2790	1000 watt			4	4		645	210		855	1,025

26 56 19.55 Roadway Led Luminaire			Crew	Daily Output	Labor-Hours	Unit	Material	Labor	Equipment	Total	Total Incl O&P
0010	**ROADWAY LED LUMINAIRE**										
0100	LED fixture, 72 LEDs, 120 V AC or 12 V DC, equal to 60 watt	G	1 Elec	2.70	2.963	Ea.	580	155		735	870
0110	108 LEDs, 120 V AC or 12 V DC, equal to 90 watt	G		2.70	2.963		680	155		835	980
0120	144 LEDs, 120 V AC or 12 V DC, equal to 120 watt	G		2.70	2.963		840	155		995	1,150
0130	252 LEDs, 120 V AC or 12 V DC, equal to 210 watt	G	2 Elec	4.40	3.636		1,150	191		1,341	1,550
0140	Replaces high pressure sodium fixture, 75 watt	G	1 Elec	2.70	2.963		705	155		860	1,000
0150	125 watt	G		2.70	2.963		810	155		965	1,125
0160	150 watt	G		2.70	2.963		1,000	155		1,155	1,325
0170	175 watt	G		2.70	2.963		1,200	155		1,355	1,550
0180	200 watt	G		2.70	2.963		1,425	155		1,580	1,775
0190	250 watt	G	2 Elec	4.40	3.636		1,625	191		1,816	2,050
0200	320 watt	G	″	4.40	3.636		1,775	191		1,966	2,225

26 56 23 – Area Lighting

26 56 23.10 Exterior Fixtures			Crew	Daily Output	Labor-Hours	Unit	Material	Labor	Equipment	Total	Total Incl O&P
0010	**EXTERIOR FIXTURES** With lamps										
0200	Wall mounted, incandescent, 100 watt		1 Elec	8	1	Ea.	34	52.50		86.50	116
0400	Quartz, 500 watt			5.30	1.509		58	79		137	182
0420	1500 watt			4.20	1.905		120	100		220	281
1100	Wall pack, low pressure sodium, 35 watt			4	2		223	105		328	400
1150	55 watt			4	2		265	105		370	450
1160	High pressure sodium, 70 watt			4	2		219	105		324	400
1170	150 watt			4	2		239	105		344	420
1180	Metal Halide, 175 watt			4	2		243	105		348	425
1190	250 watt			4	2		276	105		381	460
1195	400 watt			4	2		325	105		430	515
1250	Induction lamp, 40 watt			4	2		335	105		440	525
1260	80 watt			4	2		570	105		675	780
1278	LED, poly lens, 26 watt			4	2		320	105		425	510

26 56 Exterior Lighting

26 56 23 – Area Lighting

26 56 23.10 Exterior Fixtures		Crew	Daily Output	Labor-Hours	Unit	2013 Bare Costs Material	Labor	Equipment	Total	Total Incl O&P
1280	110 watt	1 Elec	4	2	Ea.	1,275	105		1,380	1,575
1500	LED, glass lens, 13 watt		4	2		320	105		425	510

26 56 23.55 Exterior Led Fixtures

			Crew	Daily Output	Labor-Hours	Unit	Material	Labor	Equipment	Total	Total Incl O&P
0010	**EXTERIOR LED FIXTURES**										
0100	Wall mounted, indoor/outdoor, 12 watt	G	1 Elec	10	.800	Ea.	365	42		407	465
0110	32 watt	G		10	.800		490	42		532	605
0120	66 watt	G		10	.800		710	42		752	845
0200	outdoor, 110 watt	G		10	.800		1,075	42		1,117	1,275
0210	220 watt	G		10	.800		1,900	42		1,942	2,175
0300	modular, type IV, 120 V, 50 lamp watts	G		9	.889		1,325	46.50		1,371.50	1,525
0310	101 lamp watts	G		9	.889		1,500	46.50		1,546.50	1,700
0320	126 lamp watts	G		9	.889		1,875	46.50		1,921.50	2,125
0330	202 lamp watts	G		9	.889		2,125	46.50		2,171.50	2,400
0340	240 V, 50 lamp watts	G		8	1		1,375	52.50		1,427.50	1,575
0350	101 lamp watts	G		8	1		1,550	52.50		1,602.50	1,775
0360	126 lamp watts	G		8	1		1,925	52.50		1,977.50	2,200
0370	202 lamp watts	G		8	1		2,175	52.50		2,227.50	2,475
0400	wall pack, glass, 13 lamp watts	G		4	2		510	105		615	715
0410	poly w/photocell, 26 lamp watts	G		4	2		320	105		425	505
0420	50 lamp watts	G		4	2		750	105		855	980
0430	replacement, 40 watts	G		4	2		655	105		760	875
0440	60 watts	G		4	2		820	105		925	1,050

26 56 26 – Landscape Lighting

26 56 26.20 Landscape Fixtures

		Crew	Daily Output	Labor-Hours	Unit	Material	Labor	Equipment	Total	Total Incl O&P
0010	**LANDSCAPE FIXTURES**									
7380	Landscape recessed uplight, incl. housing, ballast, transformer									
7390	& reflector, not incl. conduit, wire, trench									
7420	Incandescent, 250 watt	1 Elec	5	1.600	Ea.	575	84		659	760
7440	Quartz, 250 watt		5	1.600		545	84		629	725
7460	500 watt		4	2		560	105		665	775

26 56 33 – Walkway Lighting

26 56 33.10 Walkway Luminaire

		Crew	Daily Output	Labor-Hours	Unit	Material	Labor	Equipment	Total	Total Incl O&P
0010	**WALKWAY LUMINAIRE**									
6500	Bollard light, lamp & ballast, 42" high with polycarbonate lens									
6800	Metal halide, 175 watt	1 Elec	3	2.667	Ea.	785	140		925	1,075
6900	High pressure sodium, 70 watt		3	2.667		805	140		945	1,100
7000	100 watt		3	2.667		805	140		945	1,100
7100	150 watt		3	2.667		785	140		925	1,075
7200	Incandescent, 150 watt		3	2.667		575	140		715	840
7810	Walkway luminaire, square 16", metal halide 250 watt		2.70	2.963		610	155		765	900
7820	High pressure sodium, 70 watt		3	2.667		700	140		840	980
7830	100 watt		3	2.667		715	140		855	995
7840	150 watt		3	2.667		715	140		855	995
7850	200 watt		3	2.667		720	140		860	1,000
7910	Round 19", metal halide, 250 watt		2.70	2.963		890	155		1,045	1,200
7920	High pressure sodium, 70 watt		3	2.667		980	140		1,120	1,275
7930	100 watt		3	2.667		980	140		1,120	1,275
7940	150 watt		3	2.667		985	140		1,125	1,275
7950	250 watt		2.70	2.963		1,025	155		1,180	1,350
8000	Sphere 14" opal, incandescent, 200 watt		4	2		288	105		393	470
8020	Sphere 18" opal, incandescent, 300 watt		3.50	2.286		345	120		465	560

26 56 Exterior Lighting

26 56 33 – Walkway Lighting

26 56 33.10 Walkway Luminaire		Crew	Daily Output	Labor-Hours	Unit	2013 Bare Costs Material	Labor	Equipment	Total	Total Incl O&P
8040	Sphere 16″ clear, high pressure sodium, 70 watt	1 Elec	3	2.667	Ea.	600	140		740	870
8050	100 watt		3	2.667		640	140		780	915
8100	Cube 16″ opal, incandescent, 300 watt		3.50	2.286		385	120		505	605
8120	High pressure sodium, 70 watt		3	2.667		560	140		700	825
8130	100 watt		3	2.667		575	140		715	845
8230	Lantern, high pressure sodium, 70 watt		3	2.667		495	140		635	755
8240	100 watt		3	2.667		535	140		675	800
8250	150 watt		3	2.667		505	140		645	765
8260	250 watt		2.70	2.963		700	155		855	1,000
8270	Incandescent, 300 watt		3.50	2.286		370	120		490	585
8330	Reflector 22″ w/globe, high pressure sodium, 70 watt		3	2.667		480	140		620	740
8340	100 watt		3	2.667		490	140		630	750
8350	150 watt		3	2.667		495	140		635	755
8360	250 watt		2.70	2.963		635	155		790	925

26 56 33.55 Walkway Led Luminaire		Crew	Daily Output	Labor-Hours	Unit	Material	Labor	Equipment	Total	Total Incl O&P
0010	**WALKWAY LED LUMINAIRE** G									
0100	Pole mounted, 86 watts, 4350 lumens	1 Elec	3	2.667	Ea.	1,350	140		1,490	1,675
0110	4630 lumens G		3	2.667		2,250	140		2,390	2,675
0120	80 watts, 4000 lumens G		3	2.667		2,125	140		2,265	2,525

26 56 36 – Flood Lighting

26 56 36.20 Floodlights		Crew	Daily Output	Labor-Hours	Unit	Material	Labor	Equipment	Total	Total Incl O&P
0010	**FLOODLIGHTS** with ballast and lamp,									
1290	floor mtd, mount with swivel bracket									
1300	Induction lamp, 40 watt	1 Elec	3	2.667	Ea.	435	140		575	690
1310	80 watt		3	2.667		665	140		805	940
1320	150 watt		3	2.667		1,200	140		1,340	1,525
1400	Pole mounted, pole not included									
1950	Metal halide, 175 watt	1 Elec	2.70	2.963	Ea.	330	155		485	590
2000	400 watt	2 Elec	4.40	3.636		410	191		601	735
2200	1000 watt		4	4		560	210		770	935
2210	1500 watt		3.70	4.324		585	227		812	985
2250	Low pressure sodium, 55 watt	1 Elec	2.70	2.963		570	155		725	855
2270	90 watt		2	4		625	210		835	1,000
2290	180 watt		2	4		795	210		1,005	1,200
2340	High pressure sodium, 70 watt		2.70	2.963		238	155		393	495
2360	100 watt		2.70	2.963		244	155		399	500
2380	150 watt		2.70	2.963		285	155		440	545
2400	400 watt	2 Elec	4.40	3.636		370	191		561	690
2600	1000 watt	″	4	4		630	210		840	1,000

26 56 36.55 Led Floodlights		Crew	Daily Output	Labor-Hours	Unit	Material	Labor	Equipment	Total	Total Incl O&P
0010	**LED FLOODLIGHTS** with ballast and lamp,									
0020	Pole mounted, pole not included									
0100	11 watt G	1 Elec	4	2	Ea.	340	105		445	530
0110	46 watt G		4	2		1,025	105		1,130	1,275
0120	90 watt G		4	2		1,500	105		1,605	1,775
0130	288 watt G		4	2		2,300	105		2,405	2,675

26 61 13 – Lighting Accessories

26 61 13.30 Fixture Whips		Crew	Daily Output	Labor-Hours	Unit	2013 Bare Costs Material	Labor	Equipment	Total	Total Incl O&P
0010	**FIXTURE WHIPS**									
0080	3/8" Greenfield, 2 connectors, 6' long									
0100	TFFN wire, three #18	1 Elec	32	.250	Ea.	10.10	13.10		23.20	31
0150	Four #18		28	.286		10.70	14.95		25.65	34.50
0200	Three #16		32	.250		10.20	13.10		23.30	31
0250	Four #16		28	.286		10.90	14.95		25.85	34.50
0300	THHN wire, three #14		32	.250		11.85	13.10		24.95	32.50
0350	Four #14		28	.286		13.55	14.95		28.50	37.50
0360	Three #12	↓	32	.250	↓	15.35	13.10		28.45	36.50

26 61 23 – Lamps Applications

26 61 23.10 Lamps

26 61 23.10 Lamps		Crew	Daily Output	Labor-Hours	Unit	Material	Labor	Equipment	Total	Total Incl O&P
0010	**LAMPS** R265723-05									
0080	Fluorescent, rapid start, cool white, 2' long, 20 watt	1 Elec	1	8	C	300	420		720	955
0100	4' long, 40 watt		.90	8.889		267	465		732	990
0120	3' long, 30 watt		.90	8.889		410	465		875	1,150
0125	3' long, 25 watt energy saver R265723-10 (G)		.90	8.889		1,050	465		1,515	1,850
0150	U-40 watt		.80	10		1,175	525		1,700	2,075
0155	U-34 watt energy saver (G)		.80	10		1,175	525		1,700	2,075
0170	4' long, 34 watt energy saver (G)		.90	8.889		425	465		890	1,150
0176	2' long, T8, 17 W energy saver R265723-20 (G)		1	8		365	420		785	1,025
0178	3' long, T8, 25 W energy saver (G)		.90	8.889		415	465		880	1,150
0180	4' long, T8, 32 watt energy saver (G)		.90	8.889		425	465		890	1,175
0200	Slimline, 4' long, 40 watt		.90	8.889		1,025	465		1,490	1,825
0210	4' long, 30 watt energy saver R265723-25 (G)		.90	8.889		1,025	465		1,490	1,825
0300	8' long, 75 watt		.80	10		1,050	525		1,575	1,925
0350	8' long, 60 watt energy saver (G)		.80	10		505	525		1,030	1,350
0400	High output, 4' long, 60 watt		.90	8.889		870	465		1,335	1,650
0410	8' long, 95 watt energy saver (G)		.80	10		885	525		1,410	1,750
0500	8' long, 110 watt		.80	10		885	525		1,410	1,750
0512	2' long, T5, 14 watt energy saver (G)		1	8		1,200	420		1,620	1,950
0514	3' long, T5, 21 watt energy saver (G)		.90	8.889		1,200	465		1,665	2,025
0516	4' long, T5, 28 watt energy saver (G)		.90	8.889		1,025	465		1,490	1,825
0517	4' long, T5, 54 watt energy saver (G)		.90	8.889		1,575	465		2,040	2,425
0520	Very high output, 4' long, 110 watt		.90	8.889		2,175	465		2,640	3,100
0525	8' long, 195 watt energy saver (G)		.70	11.429		2,175	600		2,775	3,300
0550	8' long, 215 watt		.70	11.429		2,175	600		2,775	3,300
0554	Full spectrum, 4' long, 60 watt		.90	8.889		1,650	465		2,115	2,500
0556	6' long, 85 watt		.90	8.889		1,775	465		2,240	2,650
0558	8' long, 110 watt		.80	10		1,800	525		2,325	2,750
0560	Twin tube compact lamp (G)		.90	8.889		500	465		965	1,250
0570	Double twin tube compact lamp (G)		.80	10		1,125	525		1,650	2,000
0600	Mercury vapor, mogul base, deluxe white, 100 watt		.30	26.667		3,625	1,400		5,025	6,100
0650	175 watt		.30	26.667		1,900	1,400		3,300	4,200
0700	250 watt		.30	26.667		4,175	1,400		5,575	6,675
0800	400 watt		.30	26.667		6,000	1,400		7,400	8,675
0900	1000 watt		.20	40		7,800	2,100		9,900	11,700
1000	Metal halide, mogul base, 175 watt		.30	26.667		3,550	1,400		4,950	6,000
1100	250 watt		.30	26.667		4,025	1,400		5,425	6,525
1200	400 watt		.30	26.667		6,600	1,400		8,000	9,350
1300	1000 watt		.20	40		9,175	2,100		11,275	13,200
1320	1000 watt, 125,000 initial lumens		.20	40		22,100	2,100		24,200	27,400
1330	1500 watt		.20	40		20,200	2,100		22,300	25,400

26 61 Lighting Systems and Accessories

26 61 23 – Lamps Applications

26 61 23.10 Lamps		Crew	Daily Output	Labor-Hours	Unit	2013 Bare Costs Material	Labor	Equipment	Total	Total Incl O&P
1350	High pressure sodium, 70 watt	1 Elec	.30	26.667	C	4,300	1,400		5,700	6,825
1360	100 watt		.30	26.667		4,475	1,400		5,875	7,025
1370	150 watt		.30	26.667		4,600	1,400		6,000	7,175
1380	250 watt		.30	26.667		5,750	1,400		7,150	8,425
1400	400 watt		.30	26.667		5,025	1,400		6,425	7,625
1450	1000 watt		.20	40		13,900	2,100		16,000	18,400
1500	Low pressure sodium, 35 watt		.30	26.667		12,000	1,400		13,400	15,300
1550	55 watt		.30	26.667		11,900	1,400		13,300	15,200
1600	90 watt		.30	26.667		13,200	1,400		14,600	16,600
1650	135 watt		.20	40		20,000	2,100		22,100	25,100
1700	180 watt		.20	40		22,800	2,100		24,900	28,200
1750	Quartz line, clear, 500 watt		1.10	7.273		1,375	380		1,755	2,075
1760	1500 watt		.20	40		4,225	2,100		6,325	7,775
1762	Spot, MR 16, 50 watt		1.30	6.154		1,050	320		1,370	1,625
1770	Tungsten halogen, T4, 400 watt		1.10	7.273		3,900	380		4,280	4,875
1775	T3, 1200 watt		.30	26.667		4,825	1,400		6,225	7,400
1778	PAR 30, 50 watt		1.30	6.154		1,125	320		1,445	1,700
1780	PAR 38, 90 watt		1.30	6.154		940	320		1,260	1,500
1800	Incandescent, interior, A21, 100 watt		1.60	5		231	262		493	645
1900	A21, 150 watt		1.60	5		166	262		428	575
2000	A23, 200 watt		1.60	5		310	262		572	730
2200	PS 35, 300 watt		1.60	5		860	262		1,122	1,325
2210	PS 35, 500 watt		1.60	5		1,425	262		1,687	1,950
2230	PS 52, 1000 watt		1.30	6.154		2,500	320		2,820	3,225
2240	PS 52, 1500 watt		1.30	6.154		6,700	320		7,020	7,850
2300	R30, 75 watt		1.30	6.154		605	320		925	1,150
2400	R40, 100 watt		1.30	6.154		605	320		925	1,150
2500	Exterior, PAR 38, 75 watt		1.30	6.154		1,675	320		1,995	2,325
2600	PAR 38, 150 watt		1.30	6.154		2,000	320		2,320	2,675
2700	PAR 46, 200 watt		1.10	7.273		3,700	380		4,080	4,650
2800	PAR 56, 300 watt		1.10	7.273		4,425	380		4,805	5,450
3000	Guards, fluorescent lamp, 4' long		1	8		1,325	420		1,745	2,100
3200	8' long		.90	8.889		2,675	465		3,140	3,625

26 61 23.55 Led Lamps		Crew	Daily Output	Labor-Hours	Unit	Material	Labor	Equipment	Total	Total Incl O&P
0010	**LED LAMPS**									
0100	LED lamp, interior, shape A60, equal to 60 watt G	1 Elec	160	.050	Ea.	22	2.62		24.62	28
0200	Globe frosted A60, equal to 60 watt G		160	.050		22	2.62		24.62	28
0300	Globe earth, equal to 100 watt G		160	.050		72.50	2.62		75.12	84
1100	MR16, 3 watt, replacement of halogen lamp 25 watt G		130	.062		21	3.22		24.22	28
1200	6 watt replacement of halogen lamp 45 watt G		130	.062		46	3.22		49.22	55.50
2100	10 watt, PAR20, equal to 60 watt G		130	.062		47	3.22		50.22	56.50
2200	15 watt, PAR30, equal to 100 watt G		130	.062		90.50	3.22		93.72	105

26 71 13.10 Handling

26 71 13.10 Handling		Crew	Daily Output	Labor-Hours	Unit	Material (2013 Bare Costs)	Labor (2013 Bare Costs)	Equipment (2013 Bare Costs)	Total (2013 Bare Costs)	Total Incl O&P
0010	**HANDLING** Add to normal labor cost for restricted areas									
5000	Motors									
5100	1/2 HP, 23 pounds	1 Elec	4	2	Ea.		105		105	157
5110	3/4 HP, 28 pounds		4	2			105		105	157
5120	1 HP, 33 pounds		4	2			105		105	157
5130	1-1/2 HP, 44 pounds		3.20	2.500			131		131	196
5140	2 HP, 56 pounds		3	2.667			140		140	209
5150	3 HP, 71 pounds		2.30	3.478			182		182	273
5160	5 HP, 82 pounds		1.90	4.211			221		221	330
5170	7-1/2 HP, 124 pounds		1.50	5.333			279		279	420
5180	10 HP, 144 pounds		1.20	6.667			350		350	525
5190	15 HP, 185 pounds		1	8			420		420	625
5200	20 HP, 214 pounds	2 Elec	1.50	10.667			560		560	835
5210	25 HP, 266 pounds		1.40	11.429			600		600	895
5220	30 HP, 310 pounds		1.20	13.333			700		700	1,050
5230	40 HP, 400 pounds		1	16			840		840	1,250
5240	50 HP, 450 pounds		.90	17.778			930		930	1,400
5250	75 HP, 680 pounds		.80	20			1,050		1,050	1,575
5260	100 HP, 870 pounds	3 Elec	1	24			1,250		1,250	1,875
5270	125 HP, 940 pounds		.80	30			1,575		1,575	2,350
5280	150 HP, 1200 pounds		.70	34.286			1,800		1,800	2,700
5290	175 HP, 1300 pounds		.60	40			2,100		2,100	3,125
5300	200 HP, 1400 pounds		.50	48			2,525		2,525	3,775

26 71 13.20 Motors

26 71 13.20 Motors		Crew	Daily Output	Labor-Hours	Unit	Material	Labor	Equipment	Total	Total Incl O&P
0010	**MOTORS** 230/460 volts, 60 HZ R263413-30									
0050	Dripproof, premium efficiency, 1.15 service factor									
0060	1800 RPM, 1/4 HP	1 Elec	5.33	1.501	Ea.	167	78.50		245.50	300
0070	1/3 HP		5.33	1.501		172	78.50		250.50	305
0080	1/2 HP		5.33	1.501		245	78.50		323.50	385
0090	3/4 HP		5.33	1.501		271	78.50		349.50	415
0100	1 HP		4.50	1.778		283	93		376	450
0150	2 HP		4.50	1.778		325	93		418	495
0200	3 HP		4.50	1.778		440	93		533	620
0250	5 HP		4.50	1.778		525	93		618	715
0300	7.5 HP		4.20	1.905		610	100		710	820
0350	10 HP		4	2		745	105		850	975
0400	15 HP		3.20	2.500		1,050	131		1,181	1,350
0450	20 HP	2 Elec	5.20	3.077		1,325	161		1,486	1,725
0500	25 HP		5	3.200		1,675	168		1,843	2,075
0550	30 HP		4.80	3.333		1,925	175		2,100	2,375
0600	40 HP		4	4		2,550	210		2,760	3,150
0650	50 HP		3.20	5		2,425	262		2,687	3,075
0700	60 HP		2.80	5.714		3,600	299		3,899	4,400
0750	75 HP		2.40	6.667		3,150	350		3,500	3,975
0800	100 HP	3 Elec	2.70	8.889		4,050	465		4,515	5,150
0850	125 HP		2.10	11.429		4,650	600		5,250	6,025
0900	150 HP		1.80	13.333		6,700	700		7,400	8,425
0950	200 HP		1.50	16		8,175	840		9,015	10,300
1000	1200 RPM, 1 HP	1 Elec	4.50	1.778		365	93		458	545
1050	2 HP		4.50	1.778		425	93		518	605
1100	3 HP		4.50	1.778		590	93		683	790
1150	5 HP		4.50	1.778		800	93		893	1,025

26 71 13.20 Motors		Crew	Daily Output	Labor-Hours	Unit	Material	Labor	Equipment	Total	Total Incl O&P
						2013 Bare Costs				
1200	3600 RPM, 2 HP	1 Elec	4.50	1.778	Ea.	410	93		503	595
1250	3 HP		4.50	1.778		415	93		508	600
1300	5 HP		4.50	1.778		455	93		548	640
1350	Totally enclosed, premium efficiency 1.15 service factor									
1360	1800 RPM, 1/4 HP	1 Elec	5.33	1.501	Ea.	226	78.50		304.50	365
1370	1/3 HP		5.33	1.501		252	78.50		330.50	395
1380	1/2 HP		5.33	1.501		295	78.50		373.50	445
1390	3/4 HP		5.33	1.501		325	78.50		403.50	475
1400	1 HP		4.50	1.778		365	93		458	540
1450	2 HP		4.50	1.778		445	93		538	630
1500	3 HP		4.50	1.778		525	93		618	720
1550	5 HP		4.50	1.778		570	93		663	765
1600	7.5 HP		4.20	1.905		820	100		920	1,050
1650	10 HP		4	2		865	105		970	1,100
1700	15 HP		3.20	2.500		1,225	131		1,356	1,550
1750	20 HP	2 Elec	5.20	3.077		1,700	161		1,861	2,125
1800	25 HP		5	3.200		1,950	168		2,118	2,400
1850	30 HP		4.80	3.333		2,050	175		2,225	2,500
1900	40 HP		4	4		2,500	210		2,710	3,075
1950	50 HP		3.20	5		2,700	262		2,962	3,375
2000	60 HP		2.80	5.714		4,525	299		4,824	5,425
2050	75 HP		2.40	6.667		5,550	350		5,900	6,625
2100	100 HP	3 Elec	2.70	8.889		7,025	465		7,490	8,450
2150	125 HP		2.10	11.429		9,900	600		10,500	11,800
2200	150 HP		1.80	13.333		10,800	700		11,500	13,000
2250	200 HP		1.50	16		13,800	840		14,640	16,400
2300	1200 RPM, 1 HP	1 Elec	4.50	1.778		350	93		443	525
2350	2 HP		4.50	1.778		405	93		498	585
2400	3 HP		4.50	1.778		550	93		643	745
2450	5 HP		4.50	1.778		750	93		843	965
2500	3600 RPM, 2 HP		4.50	1.778		315	93		408	490
2550	3 HP		4.50	1.778		395	93		488	570
2600	5 HP		4.50	1.778		490	93		583	680

Division Notes

		CREW	DAILY OUTPUT	LABOR-HOURS	UNIT	2012 BARE COSTS MAT.	LABOR	EQUIP.	TOTAL	TOTAL INCL O&P

Estimating Tips

27 20 00 Data Communications
27 30 00 Voice Communications
27 40 00 Audio-Video Communications

When estimating material costs for special systems, it is always prudent to obtain manufacturers' quotations for equipment prices and special installation requirements which will affect the total costs.

Reference Numbers

Reference numbers are shown in shaded boxes at the beginning of some major classifications. These numbers refer to related items in the Reference Section. The reference information may be an estimating procedure, an alternate pricing method, or technical information.

Note: Not all subdivisions listed here necessarily appear in this publication.

Note: **Trade Service,** *in part, has been used as a reference source for some of the material prices used in Division 27.*

27 01 Operation and Maintenance of Communications Systems

27 01 30 – Operation and Maintenance of Voice Communications

27 01 30.51 Operation and Maintenance of Voice Equipment		Crew	Daily Output	Labor-Hours	Unit	2013 Bare Costs Material	Labor	Equipment	Total	Total Incl O&P
0010	**OPERATION AND MAINTENANCE OF VOICE EQUIPMENT**									
3400	Remove and replace (reinstall), speaker	1 Elec	6	1.333	Ea.		70		70	105

27 05 Common Work Results for Communications

27 05 05 – Selective Communications Demolition

27 05 05.20 Electrical Demolition, Communications

Line	Description	Crew	Daily Output	Labor-Hours	Unit	Material	Labor	Equipment	Total	Total Incl O&P
0010	**ELECTRICAL DEMOLITION, COMMUNICATIONS** R024119-10									
0100	Fiber optics									
0120	Cable R260105-30	1 Elec	2400	.003	L.F.		.17		.17	.26
0160	Multi-channel rack enclosure		6	1.333	Ea.		70		70	105
0180	Patch panel		18	.444	″		23.50		23.50	35
0200	Communication cables & fittings									
0220	Voice/data outlet	1 Elec	140	.057	Ea.		2.99		2.99	4.48
0240	Telephone cable		2800	.003	L.F.		.15		.15	.22
0260	Phone jack		135	.059	Ea.		3.11		3.11	4.65
0300	High performance cable, 2 pair		3000	.003	L.F.		.14		.14	.21
0320	4 pair		2100	.004			.20		.20	.30
0340	25 pair		900	.009			.47		.47	.70
0400	Terminal cabinet		5	1.600	Ea.		84		84	125
1000	Nurse call system									
1020	Station	1 Elec	24	.333	Ea.		17.45		17.45	26
1040	Standard call button		24	.333			17.45		17.45	26
1060	Corridor dome light or zone indictor		24	.333			17.45		17.45	26
1080	Master control station	2 Elec	2	8			420		420	625

27 05 05.30 Electrical Demolition, Sound and Video

Line	Description	Crew	Daily Output	Labor-Hours	Unit	Material	Labor	Equipment	Total	Total Incl O&P
0010	**ELECTRICAL DEMOLITION, SOUND & VIDEO** R024119-10									
0100	Cables									
0120	TV antenna lead-in cable R260105-30	1 Elec	2100	.004	L.F.		.20		.20	.30
0140	Sound cable		2400	.003			.17		.17	.26
0160	Microphone cable		2400	.003			.17		.17	.26
0180	Coaxial cable		2400	.003			.17		.17	.26
0200	Doorbell system, not including wires, cables, and conduit		16	.500	Ea.		26		26	39
0220	Door chime or devices		36	.222	″		11.65		11.65	17.40
0300	Public address system, not including wires, cables, and conduit									
0320	Conventional office	1 Elec	16	.500	Speaker		26		26	39
0340	Conventional industrial	″	8	1	″		52.50		52.50	78.50
0352	PA cabinet/panel	2 Elec	5	3.200	Ea.		168		168	251
0400	Sound system, not including wires, cables, and conduit									
0410	Components	1 Elec	24	.333	Ea.		17.45		17.45	26
0412	Speaker		24	.333			17.45		17.45	26
0416	Volume control		24	.333			17.45		17.45	26
0420	Intercom, master station		6	1.333			70		70	105
0440	Remote station		24	.333			17.45		17.45	26
0460	Intercom outlets		24	.333			17.45		17.45	26
0480	Handset		12	.667			35		35	52.50
0500	Emergency call system, not including wires, cables, and conduit									
0520	Annunciator	1 Elec	4	2	Ea.		105		105	157
0540	Devices	″	16	.500			26		26	39
0600	Master door, buzzer type unit	2 Elec	1.60	10			525		525	785
0800	TV System, not including wires, cables, and conduit									
0820	Master TV antenna system, per outlet	1 Elec	39	.205	Outlet		10.75		10.75	16.10

27 05 Common Work Results for Communications

27 05 05 – Selective Communications Demolition

27 05 05.30 Electrical Demolition, Sound and Video		Crew	Daily Output	Labor-Hours	Unit	2013 Bare Costs Material	Labor	Equipment	Total	Total Incl O&P
0840	School application, per outlet	1 Elec	16	.500	Outlet		26		26	39
0860	Amplifier		12	.667	Ea.		35		35	52.50
0880	Antenna		6	1.333	″		70		70	105
0900	One camera & one monitor	2 Elec	7.80	2.051	Total		107		107	161
0920	One camera	1 Elec	8	1	Ea.		52.50		52.50	78.50

27 11 Communications Equipment Room Fittings

27 11 19 – Communications Termination Blocks and Patch Panels

27 11 19.10 Termination Blocks and Patch Panels

		Crew	Daily Output	Labor-Hours	Unit	Material	Labor	Equipment	Total	Total Incl O&P
0010	TERMINATION BLOCKS AND PATCH PANELS									
2960	Patch panel, RJ-45/110 type, 24 ports	2 Elec	6	2.667	Ea.	188	140		328	415
3000	48 ports	3 Elec	6	4		375	210		585	725
3040	96 ports	″	4	6		620	315		935	1,150
3100	Punch down termination per port	1 Elec	107	.075			3.92		3.92	5.85

27 13 Communications Backbone Cabling

27 13 23 – Communications Optical Fiber Backbone Cabling

27 13 23.13 Communications Optical Fiber

		Crew	Daily Output	Labor-Hours	Unit	Material	Labor	Equipment	Total	Total Incl O&P
0010	COMMUNICATIONS OPTICAL FIBER									
0040	Specialized tools & techniques cause installation costs to vary.									
0070	Fiber optic, cable, bulk simplex, single mode R271323-40	1 Elec	8	1	C.L.F.	22.50	52.50		75	104
0080	Multi mode		2.29	3.493		57.50	183		240.50	340
0090	4 strand, single mode		8	1		51	52.50		103.50	135
0095	Multi mode		8	1		51	52.50		103.50	135
0100	12 strand, single mode		6.67	1.199		123	63		186	229
0105	Multi mode		6.67	1.199		123	63		186	230
0150	Jumper				Ea.	33			33	36.50
0200	Pigtail					31			31	34
0300	Connector	1 Elec	24	.333		20	17.45		37.45	48
0350	Finger splice		32	.250		32	13.10		45.10	54.50
0400	Transceiver (low cost bi-directional)		8	1		410	52.50		462.50	535
0450	Rack housing, 4 rack spaces, 12 panels (144 fibers)		2	4		490	210		700	855
0500	Patch panel, 12 ports		6	1.333		242	70		312	370

27 15 Communications Horizontal Cabling

27 15 10 – Special Communications Cabling

27 15 10.23 Sound and Video Cables and Fittings

		Crew	Daily Output	Labor-Hours	Unit	Material	Labor	Equipment	Total	Total Incl O&P
0010	SOUND AND VIDEO CABLES & FITTINGS									
0900	TV antenna lead-in, 300 ohm, #20-2 conductor	1 Elec	7	1.143	C.L.F.	23	60		83	115
0950	Coaxial, feeder outlet		7	1.143		20.50	60		80.50	112
1000	Coaxial, main riser		6	1.333		31	70		101	139
1100	Sound, shielded with drain, #22-2 conductor		8	1		18	52.50		70.50	98.50
1150	#22-3 conductor		7.50	1.067		23.50	56		79.50	109
1200	#22-4 conductor		6.50	1.231		29	64.50		93.50	129
1250	Nonshielded, #22-2 conductor		10	.800		14.25	42		56.25	78
1300	#22-3 conductor		9	.889		21.50	46.50		68	93
1350	#22-4 conductor		8	1		28.50	52.50		81	110

27 15 Communications Horizontal Cabling

27 15 10 – Special Communications Cabling

27 15 10.23 Sound and Video Cables and Fittings		Crew	Daily Output	Labor-Hours	Unit	Material	Labor	Equipment	Total	Total Incl O&P
						2013 Bare Costs				
1400	Microphone cable	1 Elec	8	1	C.L.F.	62.50	52.50		115	148

27 15 13 – Communications Copper Horizontal Cabling

27 15 13.13 Communication Cables		Crew	Daily Output	Labor-Hours	Unit	Material	Labor	Equipment	Total	Total Incl O&P
						2013 Bare Costs				
0010	**COMMUNICATION CABLES**									
2200	Telephone twisted, PVC insulation, #22-2 conductor	1 Elec	10	.800	C.L.F.	9.30	42		51.30	73
2250	#22-3 conductor		9	.889		12.20	46.50		58.70	83
2300	#22-4 conductor		8	1		16.25	52.50		68.75	96.50
2350	#18-2 conductor		9	.889		12.55	46.50		59.05	83.50
2370	Telephone jack, eight pins		32	.250	Ea.	4.79	13.10		17.89	25
5000	High performance unshielded twisted pair (UTP)									
5100	Cable, category 3, #24, 2 pair solid, PVC jacket R271513-75	1 Elec	10	.800	C.L.F.	8	42		50	71.50
5200	4 pair solid, PVC jacket		7	1.143		13	60		73	104
5300	25 pair solid, PVC jacket		3	2.667		68	140		208	284
5400	2 pair solid, plenum		10	.800		11.25	42		53.25	75
5500	4 pair solid, plenum		7	1.143		15.95	60		75.95	107
5600	25 pair solid, plenum		3	2.667		129	140		269	350
5700	4 pair stranded, PVC jacket		7	1.143		30	60		90	123
7000	Category 5, #24, 4 pair solid, PVC jacket		7	1.143		14.65	60		74.65	106
7100	4 pair solid, plenum		7	1.143		45	60		105	139
7200	4 pair stranded, PVC jacket		7	1.143		22	60		82	114
7210	Category 5e, #24, 4 pair solid, PVC jacket		7	1.143		14.05	60		74.05	105
7212	4 pair solid, plenum		7	1.143		23	60		83	115
7214	4 pair stranded, PVC jacket		7	1.143		25	60		85	118
7240	Category 6, #24, 4 pair solid, PVC jacket		7	1.143		20	60		80	112
7242	4 pair solid, plenum		7	1.143		24	60		84	116
7244	4 pair stranded, PVC jacket		7	1.143		25	60		85	118
7300	Connector, RJ-45, category 5		80	.100	Ea.	1.31	5.25		6.56	9.30
7302	Shielded RJ-45, category 5		72	.111		3.33	5.80		9.13	12.35
7310	Jack, UTP RJ-45, category 3		72	.111		2.31	5.80		8.11	11.25
7312	Category 5		65	.123		4.95	6.45		11.40	15.10
7314	Category 5e		65	.123		3.23	6.45		9.68	13.20
7316	Category 6		65	.123		3.28	6.45		9.73	13.25
7322	Jack, shielded RJ-45, category 5		60	.133		5.95	7		12.95	17
7324	Category 5e		60	.133		5.95	7		12.95	17
7326	Category 6		60	.133		5.95	7		12.95	17
7400	Voice/data expansion module, category 5e		8	1		45.50	52.50		98	129
8000	Multipair unshielded non-plenum cable, 150 V PVC jacket									
8002	#22, 2 pair	1 Elec	8.40	.952	C.L.F.	16.45	50		66.45	92.50
8003	3 pair		8	1		21.50	52.50		74	103
8004	4 pair		7.30	1.096		31.50	57.50		89	121
8006	6 pair		6.20	1.290		41.50	67.50		109	147
8008	8 pair		5.70	1.404		54.50	73.50		128	170
8010	10 pair		5.30	1.509		63	79		142	188
8012	12 pair		4.20	1.905		106	100		206	265
8015	15 pair		3.80	2.105		144	110		254	325
8020	20 pair	2 Elec	6.60	2.424		191	127		318	400
8025	25 pair		6	2.667		222	140		362	455
8030	30 pair		5.60	2.857		370	150		520	630
8040	40 pair		5.20	3.077		485	161		646	770
8050	50 pair		4.90	3.265		605	171		776	920
8100	Multipair unshielded non-plenum cable, 300 V PVC jacket									
8102	#20, 2 pair	1 Elec	7.30	1.096	C.L.F.	25.50	57.50		83	114

27 15 Communications Horizontal Cabling

27 15 13 – Communications Copper Horizontal Cabling

27 15 13.13 Communication Cables		Crew	Daily Output	Labor-Hours	Unit	Material	2013 Bare Costs Labor	Equipment	Total	Total Incl O&P
8103	3 pair	1 Elec	6.70	1.194	C.L.F.	33	62.50		95.50	130
8104	4 pair		6	1.333		37	70		107	146
8106	6 pair		5.30	1.509		58	79		137	182
8108	8 pair	↓	4.40	1.818		95	95.50		190.50	247
8110	10 pair	2 Elec	7.30	2.192		105	115		220	287
8112	12 pair		6.20	2.581		132	135		267	345
8115	15 pair	↓	5.90	2.712		166	142		308	395
8201	#18, 1 pair	1 Elec	8	1		62	52.50		114.50	147
8202	2 pair		6.50	1.231		78.50	64.50		143	183
8203	3 pair		5.60	1.429		119	75		194	243
8204	4 pair		5	1.600		155	84		239	296
8206	6 pair		4.40	1.818		261	95.50		356.50	430
8208	8 pair	↓	4	2		310	105		415	495
8215	15 pair	2 Elec	5	3.200	↓	640	168		808	955
8300	Multipair shielded non-plenum cable, 300 V PVC jacket									
8303	#22, 3 pair	1 Elec	6.70	1.194	C.L.F.	61	62.50		123.50	161
8306	6 pair		5.70	1.404		122	73.50		195.50	244
8309	9 pair	↓	5	1.600		270	84		354	420
8312	12 pair	2 Elec	8	2		480	105		585	680
8315	15 pair		7.30	2.192		495	115		610	715
8317	17 pair		6.80	2.353		675	123		798	925
8319	19 pair		6.40	2.500		675	131		806	940
8327	27 pair	↓	5.90	2.712		1,225	142		1,367	1,575
8402	#20, 2 pair	1 Elec	6.70	1.194		76	62.50		138.50	177
8403	3 pair		6.20	1.290		95	67.50		162.50	205
8406	6 pair		4.40	1.818		239	95.50		334.50	405
8409	9 pair	↓	3.60	2.222		355	116		471	565
8412	12 pair	2 Elec	5.90	2.712		555	142		697	825
8415	15 pair	"	5.70	2.807		650	147		797	935
8502	#18, 2 pair	1 Elec	6.20	1.290		130	67.50		197.50	244
8503	3 pair		5.30	1.509		203	79		282	340
8504	4 pair		4.70	1.702		242	89		331	400
8506	6 pair		4	2		380	105		485	570
8509	9 pair	↓	3.40	2.353		575	123		698	820
8515	15 pair	2 Elec	4.80	3.333	↓	970	175		1,145	1,325

27 15 33 – Communications Coaxial Horizontal Cabling

27 15 33.10 Coaxial Cable and Fittings		Crew	Daily Output	Labor-Hours	Unit	Material	Labor	Equipment	Total	Total Incl O&P
0010	**COAXIAL CABLE & FITTINGS**									
3500	Coaxial connectors, 50 ohm impedance quick disconnect									
3540	BNC plug, for RG A/U #58 cable	1 Elec	42	.190	Ea.	4.55	10		14.55	19.95
3550	RG A/U #59 cable		42	.190		4.55	10		14.55	19.95
3560	RG A/U #62 cable		42	.190		4.55	10		14.55	19.95
3600	BNC jack, for RG A/U #58 cable		42	.190		4.80	10		14.80	20.50
3610	RG A/U #59 cable		42	.190		4.80	10		14.80	20.50
3620	RG A/U #62 cable		42	.190		4.80	10		14.80	20.50
3660	BNC panel jack, for RG A/U #58 cable		40	.200		7.45	10.50		17.95	24
3670	RG A/U #59 cable		40	.200		7.45	10.50		17.95	24
3680	RG A/U #62 cable		40	.200		7.45	10.50		17.95	24
3720	BNC bulkhead jack, for RG A/U #58 cable		40	.200		7.70	10.50		18.20	24
3730	RG A/U #59 cable		40	.200		7.70	10.50		18.20	24
3740	RG A/U #62 cable		40	.200	↓	7.70	10.50		18.20	24
3850	Coaxial cable, RG A/U 58, 50 ohm		8	1	C.L.F.	48.50	52.50		101	132

27 15 Communications Horizontal Cabling

27 15 33 – Communications Coaxial Horizontal Cabling

27 15 33.10 Coaxial Cable and Fittings		Crew	Daily Output	Labor-Hours	Unit	2013 Bare Costs Material	Labor	Equipment	Total	Total Incl O&P
3860	RG A/U 59, 75 ohm	1 Elec	8	1	C.L.F.	38.50	52.50		91	121
3870	RG A/U 62, 93 ohm		8	1		45.50	52.50		98	129
3875	RG 6/U, 75 ohm		8	1		28	52.50		80.50	110
3950	Fire rated, RG A/U 58, 50 ohm		8	1		92.50	52.50		145	181
3960	RG A/U 59, 75 ohm		8	1		122	52.50		174.50	214
3970	RG A/U 62, 93 ohm		8	1		108	52.50		160.50	198

27 15 43 – Communications Faceplates and Connectors

27 15 43.13 Communication Outlets

Line	Description	Crew	Daily Output	Labor-Hours	Unit	Material	Labor	Equipment	Total	Total Incl O&P
0010	**COMMUNICATION OUTLETS**									
0100	Voice/data devices not included									
0120	Voice/Data outlets, single opening	1 Elec	48	.167	Ea.	7.85	8.75		16.60	21.50
0140	Two jack openings		48	.167		2.56	8.75		11.31	15.85
0160	One jack & one 3/4" round opening		48	.167		6.95	8.75		15.70	20.50
0180	One jack & one twinaxial opening		48	.167		7.50	8.75		16.25	21.50
0200	One jack & one connector cabling opening		48	.167		6.95	8.75		15.70	20.50
0220	Two 3/8" coaxial openings		48	.167		6.95	8.75		15.70	20.50
0300	Data outlets, single opening		48	.167		6.95	8.75		15.70	20.50
0320	One 25-pin subminiature opening		48	.167		6.95	8.75		15.70	20.50
1000	Voice/Data wall plate plastic, 1 gang, 1-port		72	.111		2.11	5.80		7.91	11
1020	2-port		72	.111		2.11	5.80		7.91	11
1040	3-port		72	.111		2.11	5.80		7.91	11
1060	4-port		72	.111		2.11	5.80		7.91	11
1080	6-port		72	.111		2.11	5.80		7.91	11
1100	2 gang, 6-port		48	.167		6	8.75		14.75	19.65
1120	Voice/Data wall plate stainless steel, 1 gang, 1-port		72	.111		6.40	5.80		12.20	15.75
1140	2-port		72	.111		6.40	5.80		12.20	15.75
1160	3-port		72	.111		6.40	5.80		12.20	15.75
1180	4-port		72	.111		6.85	5.80		12.65	16.20
1200	2 gang, 6-port		48	.167		11.95	8.75		20.70	26

27 21 Data Communications Network Equipment

27 21 23 – Data Communications Switches and Hubs

27 21 23.10 Switching and Routing Equipment

Line	Description	Crew	Daily Output	Labor-Hours	Unit	Material	Labor	Equipment	Total	Total Incl O&P
0010	**SWITCHING AND ROUTING EQUIPMENT**									
1100	Network hub, dual speed, 24 ports, includes cabinet	3 Elec	.66	36.364	Ea.	1,675	1,900		3,575	4,700
2000	Network switch, 10/100/1000 Mbps, 24 ports		.75	32		1,825	1,675		3,500	4,525
2040	48 ports		.66	36.364		2,300	1,900		4,200	5,375

27 32 Voice Communications Terminal Equipment

27 32 36 – TTY Equipment

27 32 36.10 TTY Telephone Equipment

Line	Description	Crew	Daily Output	Labor-Hours	Unit	Material	Labor	Equipment	Total	Total Incl O&P
0010	**TTY TELEPHONE EQUIPMENT**									
1620	Telephone, TTY, compact, pocket type				Ea.	248			248	273
1630	Advanced, desk type	2 Elec	20	.800		500	42		542	615
1640	Full-featured public, wall type	"	4	4		715	210		925	1,100

27 41 Audio-Video Systems

27 41 19 – Portable Audio-Video Equipment

27 41 19.10 T.V. Systems		Crew	Daily Output	Labor-Hours	Unit	Material	2013 Bare Costs Labor	Equipment	Total	Total Incl O&P
0010	**T.V. SYSTEMS**, not including rough-in wires, cables & conduits									
0100	Master TV antenna system									
0200	VHF reception & distribution, 12 outlets	1 Elec	6	1.333	Outlet	156	70		226	277
0400	30 outlets		10	.800		139	42		181	216
0600	100 outlets		13	.615		141	32.50		173.50	205
0800	VHF & UHF reception & distribution, 12 outlets		6	1.333		209	70		279	335
1000	30 outlets		10	.800		139	42		181	216
1200	100 outlets		13	.615		141	32.50		173.50	205
1400	School and deluxe systems, 12 outlets		2.40	3.333		276	175		451	565
1600	30 outlets		4	2		242	105		347	425
1800	80 outlets		5.30	1.509		233	79		312	375
1900	Amplifier		4	2	Ea.	695	105		800	920
1910	Antenna		2	4	"	335	210		545	680

27 51 Distributed Audio-Video Communications Systems

27 51 16 – Public Address and Mass Notification Systems

27 51 16.10 Public Address System

		Crew	Daily Output	Labor-Hours	Unit	Material	Labor	Equipment	Total	Total Incl O&P
0010	**PUBLIC ADDRESS SYSTEM**									
0100	Conventional, office	1 Elec	5.33	1.501	Speaker	132	78.50		210.50	264
0200	Industrial	"	2.70	2.963	"	256	155		411	515

27 51 19 – Sound Masking Systems

27 51 19.10 Sound System

		Crew	Daily Output	Labor-Hours	Unit	Material	Labor	Equipment	Total	Total Incl O&P
0010	**SOUND SYSTEM**, not including rough-in wires, cables & conduits									
0100	Components, projector outlet	1 Elec	8	1	Ea.	49	52.50		101.50	133
0200	Microphone		4	2		79	105		184	244
0400	Speakers, ceiling or wall		8	1		115	52.50		167.50	206
0600	Trumpets		4	2		215	105		320	395
0800	Privacy switch		8	1		85.50	52.50		138	173
1000	Monitor panel		4	2		380	105		485	575
1200	Antenna, AM/FM		4	2		133	105		238	305
1400	Volume control		8	1		87	52.50		139.50	174
1600	Amplifier, 250 watts		1	8		1,225	420		1,645	1,975
1800	Cabinets		1	8		830	420		1,250	1,550
2000	Intercom, 30 station capacity, master station	2 Elec	2	8		2,275	420		2,695	3,125
2020	10 station capacity	"	4	4		1,275	210		1,485	1,725
2200	Remote station	1 Elec	8	1		160	52.50		212.50	256
2400	Intercom outlets		8	1		94.50	52.50		147	183
2600	Handset		4	2		310	105		415	500
2800	Emergency call system, 12 zones, annunciator		1.30	6.154		940	320		1,260	1,500
3000	Bell		5.30	1.509		97	79		176	225
3200	Light or relay		8	1		48.50	52.50		101	132
3400	Transformer		4	2		214	105		319	390
3600	House telephone, talking station		1.60	5		460	262		722	895
3800	Press to talk, release to listen		5.30	1.509		107	79		186	235
4000	System-on button					64			64	70.50
4200	Door release	1 Elec	4	2		114	105		219	283
4400	Combination speaker and microphone		8	1		194	52.50		246.50	293
4600	Termination box		3.20	2.500		61	131		192	264
4800	Amplifier or power supply		5.30	1.509		705	79		784	895
5000	Vestibule door unit		16	.500	Name	129	26		155	181

27 51 Distributed Audio-Video Communications Systems

27 51 19 – Sound Masking Systems

27 51 19.10 Sound System		Crew	Daily Output	Labor-Hours	Unit	Material	Labor	Equipment	Total	Total Incl O&P
						2013 Bare Costs				
5200	Strip cabinet	1 Elec	27	.296	Ea.	244	15.55		259.55	292
5400	Directory		16	.500		115	26		141	165
6000	Master door, button buzzer type, 100 unit	2 Elec	.54	29.630		1,175	1,550		2,725	3,625
6020	200 unit		.30	53.333		2,200	2,800		5,000	6,600
6040	300 unit		.20	80		3,350	4,200		7,550	9,975
6060	Transformer	1 Elec	8	1		30	52.50		82.50	112
6080	Door opener		5.30	1.509		43	79		122	165
6100	Buzzer with door release and plate		4	2		43	105		148	204
6200	Intercom type, 100 unit	2 Elec	.54	29.630		1,450	1,550		3,000	3,925
6220	200 unit		.30	53.333		2,850	2,800		5,650	7,300
6240	300 unit		.20	80		4,275	4,200		8,475	11,000
6260	Amplifier	1 Elec	2	4		214	210		424	550
6280	Speaker with door release	″	4	2		64	105		169	228

27 52 Healthcare Communications and Monitoring Systems

27 52 23 – Nurse Call/Code Blue Systems

27 52 23.10 Nurse Call Systems

		Crew	Daily Output	Labor-Hours	Unit	Material	Labor	Equipment	Total	Total Incl O&P
0010	**NURSE CALL SYSTEMS**									
0100	Single bedside call station	1 Elec	8	1	Ea.	228	52.50		280.50	330
0200	Ceiling speaker station		8	1		66.50	52.50		119	152
0400	Emergency call station		8	1		71.50	52.50		124	157
0600	Pillow speaker		8	1		176	52.50		228.50	272
0800	Double bedside call station		4	2		141	105		246	310
1000	Duty station		4	2		146	105		251	320
1200	Standard call button		8	1		86	52.50		138.50	173
1400	Lights, corridor, dome or zone indicator		8	1		48.50	52.50		101	132
1600	Master control station for 20 stations	2 Elec	.65	24.615	Total	3,950	1,300		5,250	6,250

27 53 Distributed Systems

27 53 13 – Clock Systems

27 53 13.50 Clock Equipments

		Crew	Daily Output	Labor-Hours	Unit	Material	Labor	Equipment	Total	Total Incl O&P
0010	**CLOCK EQUIPMENTS**, not including wires & conduits									
0100	Time system components, master controller	1 Elec	.33	24.242	Ea.	1,825	1,275		3,100	3,900
0200	Program bell		8	1		85	52.50		137.50	172
0400	Combination clock & speaker		3.20	2.500		210	131		341	425
0600	Frequency generator		2	4		2,350	210		2,560	2,925
0800	Job time automatic stamp recorder		4	2		530	105		635	740
1600	Master time clock system, clocks & bells, 20 room	4 Elec	.20	160		6,050	8,375		14,425	19,200
1800	50 room	″	.08	400		12,400	21,000		33,400	45,000
1900	Time clock	1 Elec	3.20	2.500		435	131		566	675
2000	100 cards in & out, 1 color					9			9	9.90
2200	2 colors					9			9	9.90
2800	Metal rack for 25 cards	1 Elec	7	1.143		43	60		103	137
4000	Wireless time systems component, master controller	″	2	4		595	210		805	970
4010	For transceiver and antenna, see Section 28 39 10.10									
4100	Wireless analog clock w/battery operated	1 Elec	8	1	Ea.	111	52.50		163.50	201
4200	Wireless digital clock w/battery operated	″	8	1	″	360	52.50		412.50	480

Division 28 - Electronic Safety and Security

Estimating Tips

- When estimating material costs for electronic safety and security systems, it is always prudent to obtain manufacturers' quotations for equipment prices and special installation requirements that affect the total cost.
- Fire alarm systems consist of control panels, annunciator panels, battery with rack, charger, and fire alarm actuating and indicating devices. Some fire alarm systems include speakers, telephone lines, door closer controls, and other components. Be careful not to overlook the costs related to installation for these items.

  Also be aware of costs for integrated automation instrumentation and terminal devices, control equipment, control wiring, and programming.
- Security equipment includes items such as CCTV, access control, and other detection and identification systems to perform alert and alarm functions. Be sure to consider the costs related to installation for this security equipment, such as for integrated automation instrumentation and terminal devices, control equipment, control wiring, and programming.

Reference Numbers

Reference numbers are shown in shaded boxes at the beginning of some major classifications. These numbers refer to related items in the Reference Section. The reference information may be an estimating procedure, an alternate pricing method, or technical information.

Note: Not all subdivisions listed here necessarily appear in this publication.

28 01 Operation and Maint. of Electronic Safety and Security

28 01 30 – Operation and Maint. of Electronic Detection and Alarm

28 01 30.51 Maint. and Admin. of Elec. Detection and Alarm		Crew	Daily Output	Labor-Hours	Unit	2013 Bare Costs Material	Labor	Equipment	Total	Total Incl O&P
0010	MAINT. AND ADMIN. OF ELEC. DETECTION AND ALARM									
3300	Remove and replace (reinstall), fire alarm device	1 Elec	5.33	1.501	Ea.		78.50		78.50	118

28 05 Common Work Results for Electronic Safety and Security

28 05 05 – Selective Safety and Security Demolition

28 05 05.10 Safety and Security Demolition

		Crew	Daily Output	Labor-Hours	Unit	Material	Labor	Equipment	Total	Total Incl O&P
0010	SAFETY AND SECURITY DEMOLITION									
1050	Finger print/card reader	1 Elec	8	1	Ea.		52.50		52.50	78.50
1060	Video camera		8	1			52.50		52.50	78.50
1070	Motion detector, multi-channel		7	1.143			60		60	89.50
1090	Infrared detector		7	1.143			60		60	89.50
1130	Video monitor, 19"		8	1			52.50		52.50	78.50
1210	Fire alarm horn and strobe light		16	.500			26		26	39
1220	Flame detector		23	.348			18.25		18.25	27.50
1230	Duct detector		10	.800			42		42	62.50
1240	Smoke detector		23	.348			18.25		18.25	27.50
1250	Fire alarm pull station		23	.348			18.25		18.25	27.50
1260	Fire alarm control panel, 4 to 8 zone	2 Elec	4	4			210		210	315
1270	12 to 16 zone	"	2.67	5.993			315		315	470
1280	Fire alarm annunciation panel, 4 to 8 zone	1 Elec	6	1.333			70		70	105
1290	12 to 16 zone	2 Elec	6	2.667			140		140	209

28 13 Access Control

28 13 53 – Security Access Detection

28 13 53.13 Security Access Metal Detectors

		Crew	Daily Output	Labor-Hours	Unit	Material	Labor	Equipment	Total	Total Incl O&P
0010	SECURITY ACCESS METAL DETECTORS									
0240	Metal detector, hand-held, wand type, unit only				Ea.	89			89	98
0250	Metal detector, walk through portal type, single zone	1 Elec	2	4		3,625	210		3,835	4,325
0260	Multi-zone	"	2	4		4,625	210		4,835	5,400

28 13 53.16 Security Access X-Ray Equipment

		Crew	Daily Output	Labor-Hours	Unit	Material	Labor	Equipment	Total	Total Incl O&P
0010	SECURITY ACCESS X-RAY EQUIPMENT									
0290	X-ray machine, desk top, for mail/small packages/letters	1 Elec	4	2	Ea.	3,350	105		3,455	3,850
0300	Conveyor type, incl monitor, minimum		2	4		15,700	210		15,910	17,500
0310	Maximum		2	4		28,000	210		28,210	31,100
0320	X-ray machine, large unit, for airports, incl monitor, min.	2 Elec	1	16		39,200	840		40,040	44,400
0330	Maximum	"	.50	32		67,000	1,675		68,675	76,500

28 13 53.23 Security Access Explosive Detection Equipment

		Crew	Daily Output	Labor-Hours	Unit	Material	Labor	Equipment	Total	Total Incl O&P
0010	SECURITY ACCESS EXPLOSIVE DETECTION EQUIPMENT									
0270	Explosives detector, walk through portal type	1 Elec	2	4	Ea.	43,300	210		43,510	47,900
0280	Hand-held, battery operated				"				25,500	28,100

28 16 Intrusion Detection

28 16 16 – Intrusion Detection Systems Infrastructure

28 16 16.50 Intrusion Detection		Crew	Daily Output	Labor-Hours	Unit	2013 Bare Costs Material	Labor	Equipment	Total	Total Incl O&P
0010	**INTRUSION DETECTION**, not including wires & conduits									
0100	Burglar alarm, battery operated, mechanical trigger	1 Elec	4	2	Ea.	272	105		377	455
0200	Electrical trigger		4	2		325	105		430	510
0400	For outside key control, add		8	1		82	52.50		134.50	169
0600	For remote signaling circuitry, add		8	1		122	52.50		174.50	214
0800	Card reader, flush type, standard		2.70	2.963		910	155		1,065	1,225
1000	Multi-code		2.70	2.963		1,175	155		1,330	1,525
1010	Card reader, proximity type		2.70	2.963		310	155		465	570
1200	Door switches, hinge switch		5.30	1.509		57.50	79		136.50	181
1400	Magnetic switch		5.30	1.509		67.50	79		146.50	193
1600	Exit control locks, horn alarm		4	2		272	105		377	455
1800	Flashing light alarm		4	2		299	105		404	485
2000	Indicating panels, 1 channel		2.70	2.963		360	155		515	625
2200	10 channel	2 Elec	3.20	5		1,225	262		1,487	1,750
2400	20 channel		2	8		2,400	420		2,820	3,275
2600	40 channel		1.14	14.035		4,375	735		5,110	5,900
2800	Ultrasonic motion detector, 12 volt	1 Elec	2.30	3.478		225	182		407	520
3000	Infrared photoelectric detector		4	2		186	105		291	360
3200	Passive infrared detector		4	2		278	105		383	460
3400	Glass break alarm switch		8	1		91.50	52.50		144	179
3420	Switchmats, 30″ x 5′		5.30	1.509		83	79		162	210
3440	30″ x 25′		4	2		199	105		304	375
3460	Police connect panel		4	2		239	105		344	420
3480	Telephone dialer		5.30	1.509		375	79		454	535
3500	Alarm bell		4	2		102	105		207	270
3520	Siren		4	2		143	105		248	315
3540	Microwave detector, 10′ to 200′		2	4		655	210		865	1,025
3560	10′ to 350′		2	4		1,900	210		2,110	2,425

28 23 Video Surveillance

28 23 13 – Video Surveillance Control and Management Systems

28 23 13.10 Closed Circuit Television System		Crew	Daily Output	Labor-Hours	Unit	Material	Labor	Equipment	Total	Total Incl O&P
0010	**CLOSED CIRCUIT TELEVISION SYSTEM**									
2000	Surveillance, one station (camera & monitor)	2 Elec	2.60	6.154	Total	1,350	320		1,670	1,950
2200	For additional camera stations, add	1 Elec	2.70	2.963	Ea.	750	155		905	1,050
2400	Industrial quality, one station (camera & monitor)	2 Elec	2.60	6.154	Total	2,775	320		3,095	3,525
2600	For additional camera stations, add	1 Elec	2.70	2.963	Ea.	1,700	155		1,855	2,100
2610	For low light, add		2.70	2.963		1,375	155		1,530	1,725
2620	For very low light, add		2.70	2.963		10,100	155		10,255	11,300
2800	For weatherproof camera station, add		1.30	6.154		1,050	320		1,370	1,625
3000	For pan and tilt, add		1.30	6.154		2,700	320		3,020	3,450
3200	For zoom lens - remote control, add, minimum		2	4		2,500	210		2,710	3,075
3400	Maximum		2	4		9,125	210		9,335	10,300
3410	For automatic iris for low light, add		2	4		2,175	210		2,385	2,725
3600	Educational T.V. studio, basic 3 camera system, black & white,									
3800	electrical & electronic equip. only, minimum	4 Elec	.80	40	Total	13,000	2,100		15,100	17,400
4000	Maximum (full console)		.28	114		55,000	6,000		61,000	69,500
4100	As above, but color system, minimum		.28	114		73,000	6,000		79,000	89,500
4120	Maximum		.12	266		317,500	14,000		331,500	370,500
4200	For film chain, black & white, add	1 Elec	1	8	Ea.	14,800	420		15,220	16,900
4250	Color, add		.25	32		18,000	1,675		19,675	22,300

28 23 Video Surveillance

28 23 13 – Video Surveillance Control and Management Systems

28 23 13.10 Closed Circuit Television System		Crew	Daily Output	Labor-Hours	Unit	2013 Bare Costs Material	Labor	Equipment	Total	Total Incl O&P
4400	For video recorders, add, minimum	1 Elec	1	8	Ea.	3,125	420		3,545	4,050
4600	Maximum	4 Elec	.40	80		25,900	4,200		30,100	34,800

28 23 19 – Digital Video Recorders and Analog Recording Devices

28 23 19.10 Digital Video Recorder (DVR)		Crew	Daily Output	Labor-Hours	Unit	Material	Labor	Equipment	Total	Total Incl O&P
0010	**DIGITAL VIDEO RECORDER (DVR)**									
0100	Pentaplex hybrid, internet protocol, and hard drive									
0200	4 channel	1 Elec	1.33	6.015	Ea.	1,425	315		1,740	2,050
0300	8 channel		1	8		2,225	420		2,645	3,075
0400	16 channel		1	8		2,150	420		2,570	3,000

28 23 23 – Video Surveillance Systems Infrastructure

28 23 23.50 Video Surveillance Equipments		Crew	Daily Output	Labor-Hours	Unit	Material	Labor	Equipment	Total	Total Incl O&P
0010	**VIDEO SURVEILLANCE EQUIPMENTS**									
0200	Video cameras, wireless, hidden in exit signs, clocks, etc., incl. receiver	1 Elec	3	2.667	Ea.	106	140		246	325
0210	Accessories for video recorder, single camera		3	2.667		179	140		319	405
0220	For multiple cameras		3	2.667		1,725	140		1,865	2,100
0230	Video cameras, wireless, for under vehicle searching, complete		2	4		9,975	210		10,185	11,300
0400	Internet protocol network camera, day/night, color & power supply		2.60	3.077		1,000	161		1,161	1,350
0500	Monitor, color flat screen, liquid crystal display (LCD), 15″		2.70	2.963		685	155		840	980
0520	17″		2.70	2.963		805	155		960	1,125
0540	19″		2.70	2.963		905	155		1,060	1,225

28 31 Fire Detection and Alarm

28 31 23 – Fire Detection and Alarm Annunciation Panels and Fire Stations

28 31 23.50 Alarm Panels and Devices		Crew	Daily Output	Labor-Hours	Unit	Material	Labor	Equipment	Total	Total Incl O&P
0010	**ALARM PANELS AND DEVICES**, not including wires & conduits									
3594	Fire, alarm control panel									
3600	4 zone	2 Elec	2	8	Ea.	390	420		810	1,050
3800	8 zone		1	16		765	840		1,605	2,100
4000	12 zone		.67	23.988		2,350	1,250		3,600	4,450
4020	Alarm device	1 Elec	8	1		233	52.50		285.50	335
4050	Actuating device	″	8	1		330	52.50		382.50	440
4160	Alarm control panel, addressable w/o voice, up to 200 points	2 Elec	1.14	13.998		4,400	735		5,135	5,925
4170	addressable w/voice, up to 400 points	″	.73	22.008		9,225	1,150		10,375	11,900
4175	Addressable interface device	1 Elec	7.25	1.103		132	58		190	232
4200	Battery and rack		4	2		405	105		510	600
4400	Automatic charger		8	1		570	52.50		622.50	710
4600	Signal bell		8	1		77	52.50		129.50	164
4800	Trouble buzzer or manual station		8	1		81.50	52.50		134	169
5600	Strobe and horn		5.30	1.509		149	79		228	282
5610	Strobe and horn (ADA type)		5.30	1.509		149	79		228	282
5620	Visual alarm (ADA type)		6.70	1.194		101	62.50		163.50	205
5800	Fire alarm horn		6.70	1.194		59.50	62.50		122	159
6000	Door holder, electro-magnetic		4	2		101	105		206	268
6200	Combination holder and closer		3.20	2.500		121	131		252	330
6600	Drill switch		8	1		365	52.50		417.50	480
6800	Master box		2.70	2.963		6,275	155		6,430	7,125
7000	Break glass station		8	1		54.50	52.50		107	139
7010	Break glass station, addressable		7.25	1.103		145	58		203	246
7800	Remote annunciator, 8 zone lamp		1.80	4.444		205	233		438	575
8000	12 zone lamp	2 Elec	2.60	6.154		330	320		650	840

28 31 Fire Detection and Alarm

28 31 23 – Fire Detection and Alarm Annunciation Panels and Fire Stations

28 31 23.50 Alarm Panels and Devices		Crew	Daily Output	Labor-Hours	Unit	2013 Bare Costs Material	Labor	Equipment	Total	Total Incl O&P
8200	16 zone lamp	2 Elec	2.20	7.273	Ea.	410	380		790	1,025

28 31 43 – Fire Detection Sensors

28 31 43.50 Fire and Heat Detectors		Crew	Daily Output	Labor-Hours	Unit	Material	Labor	Equipment	Total	Total Incl O&P
0010	**FIRE & HEAT DETECTORS**									
5000	Detector, rate of rise	1 Elec	8	1	Ea.	50	52.50		102.50	134
5010	Detector, heat addressable type		7.25	1.103		252	58		310	365
5100	Fixed temperature		8	1		50	52.50		102.50	134

28 31 46 – Smoke Detection Sensors

28 31 46.50 Smoke Detectors		Crew	Daily Output	Labor-Hours	Unit	Material	Labor	Equipment	Total	Total Incl O&P
0010	**SMOKE DETECTORS**									
5200	Smoke detector, ceiling type	1 Elec	6.20	1.290	Ea.	108	67.50		175.50	220
5240	Smoke detector addressable type		6	1.333		217	70		287	345
5400	Duct type		3.20	2.500		320	131		451	550
5420	Duct addressable type		3.20	2.500		500	131		631	745

28 39 Mass Notification Systems

28 39 10 – Notification Systems

28 39 10.10 Mass Notification System		Crew	Daily Output	Labor-Hours	Unit	Material	Labor	Equipment	Total	Total Incl O&P
0010	**MASS NOTIFICATION SYSTEM**									
0100	Wireless command center, 10,000 devices	2 Elec	1.33	12.030	Ea.	3,525	630		4,155	4,825
0200	Option, email notification					1,725			1,725	1,900
0210	Remote device supervision & monitor					2,400			2,400	2,625
0300	Antenna VHF or UHF, for medium range	1 Elec	4	2		127	105		232	296
0310	For high-power transmitter		2	4		660	210		870	1,050
0400	Transmitter, 25 watt		4	2		2,000	105		2,105	2,350
0410	40 watt		2.66	3.008		2,650	158		2,808	3,150
0420	100 watt		1.33	6.015		6,650	315		6,965	7,775
0500	Wireless receiver/control module for speaker		8	1		260	52.50		312.50	365
0600	Desktop paging controller, stand alone		4	2		365	105		470	555

Division Notes

		CREW	DAILY OUTPUT	LABOR-HOURS	UNIT	2012 BARE COSTS MAT.	LABOR	EQUIP.	TOTAL	TOTAL INCL O&P

Estimating Tips

31 05 00 Common Work Results for Earthwork

- Estimating the actual cost of performing earthwork requires careful consideration of the variables involved. This includes items such as type of soil, whether water will be encountered, dewatering, whether banks need bracing, disposal of excavated earth, and length of haul to fill or spoil sites, etc. If the project has large quantities of cut or fill, consider raising or lowering the site to reduce costs, while paying close attention to the effect on site drainage and utilities.
- If the project has large quantities of fill, creating a borrow pit on the site can significantly lower the costs.
- It is very important to consider what time of year the project is scheduled for completion. Bad weather can create large cost overruns from dewatering, site repair, and lost productivity from cold weather.
- New lines have been added for marine piling and rough grading.

Reference Numbers

Reference numbers are shown in shaded boxes at the beginning of some major classifications. These numbers refer to related items in the Reference Section. The reference information may be an estimating procedure, an alternate pricing method, or technical information.

Note: Not all subdivisions listed here necessarily appear in this publication.

31 23 16.13 Excavating, Trench

	31 23 16.13 Excavating, Trench	Crew	Daily Output	Labor-Hours	Unit	Material	2013 Bare Costs Labor	Equipment	Total	Total Incl O&P
0010	**EXCAVATING, TRENCH**									
0011	Or continuous footing									
0020	Common earth with no sheeting or dewatering included									
0050	1′ to 4′ deep, 3/8 C.Y. excavator	B-11C	150	.107	B.C.Y.		4.42	2.45	6.87	9.45
0060	1/2 C.Y. excavator	B-11M	200	.080			3.32	2	5.32	7.25
0062	3/4 C.Y. excavator	B-12F	270	.059			2.50	2.57	5.07	6.65
1400	By hand with pick and shovel 2′ to 6′ deep, light soil	1 Clab	8	1			35.50		35.50	54.50
1500	Heavy soil	"	4	2			71		71	109
3000	Backfill trench, F.E. loader, wheel mtd., 1 C.Y. bucket									
3020	Minimal haul	B-10R	400	.030	L.C.Y.		1.30	.74	2.04	2.79
5020	Loam & Sandy clay with no sheeting or dewatering included									
5050	1′ to 4′ deep, 3/8 C.Y. tractor loader/backhoe	B-11C	162	.099	B.C.Y.		4.10	2.27	6.37	8.75
5060	1/2 C.Y. excavator	B-11M	216	.074			3.07	1.85	4.92	6.70
5070	3/4 C.Y. excavator	B-12F	292	.055			2.31	2.38	4.69	6.15
6020	Sand & gravel with no sheeting or dewatering included									
6050	1′ to 4′ deep, 3/8 C.Y. excavator	B-11C	165	.097	B.C.Y.		4.02	2.23	6.25	8.60
6060	1/2 C.Y. excavator	B-11M	220	.073			3.02	1.82	4.84	6.60
6070	3/4 C.Y. excavator	B-12F	297	.054			2.27	2.34	4.61	6.05
7020	Dense hard clay with no sheeting or dewatering included									
7050	1′ to 4′ deep, 3/8 C.Y. excavator	B-11C	132	.121	B.C.Y.		5.05	2.78	7.83	10.70
7060	1/2 C.Y. excavator	B-11M	176	.091			3.77	2.28	6.05	8.25
7070	3/4 C.Y. excavator	B-12F	238	.067			2.83	2.92	5.75	7.50

31 23 16.14 Excavating, Utility Trench

	31 23 16.14 Excavating, Utility Trench	Crew	Daily Output	Labor-Hours	Unit	Material	Labor	Equipment	Total	Total Incl O&P
0010	**EXCAVATING, UTILITY TRENCH**									
0011	Common earth									
0050	Trenching with chain trencher, 12 H.P., operator walking									
0100	4″ wide trench, 12″ deep	B-53	800	.010	L.F.		.46	.09	.55	.79
0150	18″ deep		750	.011			.49	.09	.58	.84
0200	24″ deep		700	.011			.52	.10	.62	.90
0300	6″ wide trench, 12″ deep		650	.012			.56	.11	.67	.97
0350	18″ deep		600	.013			.61	.12	.73	1.05
0400	24″ deep		550	.015			.67	.13	.80	1.15
0450	36″ deep		450	.018			.81	.16	.97	1.40
0600	8″ wide trench, 12″ deep		475	.017			.77	.15	.92	1.33
0650	18″ deep		400	.020			.92	.18	1.10	1.57
0700	24″ deep		350	.023			1.05	.20	1.25	1.80
0750	36″ deep		300	.027			1.22	.23	1.45	2.11
1000	Backfill by hand including compaction, add									
1050	4″ wide trench, 12″ deep	A-1G	800	.010	L.F.		.35	.07	.42	.62
1100	18″ deep		530	.015			.53	.10	.63	.93
1150	24″ deep		400	.020			.71	.13	.84	1.24
1300	6″ wide trench, 12″ deep		540	.015			.53	.10	.63	.92
1350	18″ deep		405	.020			.70	.13	.83	1.23
1400	24″ deep		270	.030			1.05	.20	1.25	1.84
1450	36″ deep		180	.044			1.58	.30	1.88	2.76
1600	8″ wide trench, 12″ deep		400	.020			.71	.13	.84	1.24
1650	18″ deep		265	.030			1.07	.20	1.27	1.87
1700	24″ deep		200	.040			1.42	.27	1.69	2.47
1750	36″ deep		135	.059			2.10	.40	2.50	3.68
2000	Chain trencher, 40 H.P. operator riding									
2050	6″ wide trench and backfill, 12″ deep	B-54	1200	.007	L.F.		.31	.28	.59	.77
2100	18″ deep		1000	.008			.37	.34	.71	.92

31 23 16 – Excavation

31 23 16.14 Excavating, Utility Trench

	31 23 16.14 Excavating, Utility Trench	Crew	Daily Output	Labor-Hours	Unit	Material	2013 Bare Costs Labor	Equipment	Total	Total Incl O&P
2150	24″ deep	B-54	975	.008	L.F.		.38	.35	.73	.95
2200	36″ deep		900	.009			.41	.38	.79	1.03
2250	48″ deep		750	.011			.49	.45	.94	1.24
2300	60″ deep		650	.012			.56	.52	1.08	1.42
2400	8″ wide trench and backfill, 12″ deep		1000	.008			.37	.34	.71	.92
2450	18″ deep		950	.008			.39	.36	.75	.97
2500	24″ deep		900	.009			.41	.38	.79	1.03
2550	36″ deep		800	.010			.46	.42	.88	1.16
2600	48″ deep		650	.012			.56	.52	1.08	1.42
2700	12″ wide trench and backfill, 12″ deep		975	.008			.38	.35	.73	.95
2750	18″ deep		860	.009			.43	.39	.82	1.07
2800	24″ deep		800	.010			.46	.42	.88	1.16
2850	36″ deep		725	.011			.51	.47	.98	1.27
3000	16″ wide trench and backfill, 12″ deep		835	.010			.44	.41	.85	1.11
3050	18″ deep		750	.011			.49	.45	.94	1.24
3100	24″ deep		700	.011			.52	.48	1	1.32
3200	Compaction with vibratory plate, add								35%	35%
5100	Hand excavate and trim for pipe bells after trench excavation									
5200	8″ pipe	1 Clab	155	.052	L.F.		1.83		1.83	2.82
5300	18″ pipe	″	130	.062	″		2.18		2.18	3.36

31 23 16.15 Excavating, Utility Trench, Plow

	31 23 16.15 Excavating, Utility Trench, Plow	Crew	Daily Output	Labor-Hours	Unit	Material	Labor	Equipment	Total	Total Incl O&P
0010	**EXCAVATING, UTILITY TRENCH, PLOW**									
0100	Single cable, plowed into fine material	B-54C	3800	.004	L.F.		.17	.32	.49	.62
0200	Two cable		3200	.005			.21	.37	.58	.73
0300	Single cable, plowed into coarse material		2000	.008			.33	.60	.93	1.17

31 23 16.16 Structural Excavation for Minor Structures

	31 23 16.16 Structural Excavation for Minor Structures	Crew	Daily Output	Labor-Hours	Unit	Material	Labor	Equipment	Total	Total Incl O&P
0010	**STRUCTURAL EXCAVATION FOR MINOR STRUCTURES**									
0015	Hand, pits to 6′ deep, sandy soil	1 Clab	8	1	B.C.Y.		35.50		35.50	54.50
0100	Heavy soil or clay		4	2			71		71	109
0300	Pits 6′ to 12′ deep, sandy soil		5	1.600			56.50		56.50	87.50
0500	Heavy soil or clay		3	2.667			94.50		94.50	146
0700	Pits 12′ to 18′ deep, sandy soil		4	2			71		71	109
0900	Heavy soil or clay		2	4			142		142	218
6030	Common earth, hydraulic backhoe, 1/2 C.Y. bucket	B-12E	55	.291			12.25	7.30	19.55	26.50
6035	3/4 C.Y. bucket	B-12F	90	.178			7.50	7.70	15.20	19.90
6040	1 C.Y. bucket	B-12A	108	.148			6.25	7.55	13.80	17.80
6050	1-1/2 C.Y. bucket	B-12B	144	.111			4.68	7.20	11.88	15.05
6060	2 C.Y. bucket	B-12C	200	.080			3.37	6.80	10.17	12.65
6070	Sand and gravel, 3/4 C.Y. bucket	B-12F	100	.160			6.75	6.95	13.70	17.90
6080	1 C.Y. bucket	B-12A	120	.133			5.60	6.80	12.40	16.05
6090	1-1/2 C.Y. bucket	B-12B	160	.100			4.21	6.50	10.71	13.50
6100	2 C.Y. bucket	B-12C	220	.073			3.06	6.20	9.26	11.45
6110	Clay, till, or blasted rock, 3/4 C.Y. bucket	B-12F	80	.200			8.45	8.70	17.15	22.50
6120	1 C.Y. bucket	B-12A	95	.168			7.10	8.60	15.70	20.50
6130	1-1/2 C.Y. bucket	B-12B	130	.123			5.20	7.95	13.15	16.65
6140	2 C.Y. bucket	B-12C	175	.091			3.85	7.80	11.65	14.40
6230	Sandy clay & loam, hydraulic backhoe, 1/2 C.Y. bucket	B-12E	60	.267			11.25	6.65	17.90	24.50
6235	3/4 C.Y. bucket	B-12F	98	.163			6.90	7.10	14	18.30
6240	1 C.Y. bucket	B-12A	116	.138			5.80	7.05	12.85	16.60
6250	1-1/2 C.Y. bucket	B-12B	156	.103			4.32	6.65	10.97	13.90

31 23 Excavation and Fill

31 23 19 – Dewatering

31 23 19.20 Dewatering Systems

	31 23 19.20 Dewatering Systems	Crew	Daily Output	Labor-Hours	Unit	Material	2013 Bare Costs Labor	Equipment	Total	Total Incl O&P
0010	**DEWATERING SYSTEMS**									
0020	Excavate drainage trench, 2′ wide, 2′ deep	B-11C	90	.178	C.Y.		7.35	4.08	11.43	15.75
0100	2′ wide, 3′ deep, with backhoe loader	"	135	.119			4.92	2.72	7.64	10.50
0200	Excavate sump pits by hand, light soil	1 Clab	7.10	1.127			40		40	61.50
0300	Heavy soil	"	3.50	2.286	↓		81		81	125
0500	Pumping 8 hr., attended 2 hrs. per day, including 20 L.F.									
0550	of suction hose & 100 L.F. discharge hose									
0600	2″ diaphragm pump used for 8 hours	B-10H	4	3	Day		130	19.25	149.25	219
0650	4″ diaphragm pump used for 8 hours	B-10I	4	3			130	32	162	233
0800	8 hrs. attended, 2″ diaphragm pump	B-10H	1	12			520	77	597	875
0900	3″ centrifugal pump	B-10J	1	12			520	87.50	607.50	885
1000	4″ diaphragm pump	B-10I	1	12			520	127	647	930
1100	6″ centrifugal pump	B-10K	1	12	↓		520	375	895	1,200

31 23 23 – Fill

31 23 23.13 Backfill

	31 23 23.13 Backfill	Crew	Daily Output	Labor-Hours	Unit	Material	Labor	Equipment	Total	Total Incl O&P
0010	**BACKFILL**									
0015	By hand, no compaction, light soil	1 Clab	14	.571	L.C.Y.		20.50		20.50	31
0100	Heavy soil		11	.727	"		26		26	39.50
0300	Compaction in 6″ layers, hand tamp, add to above	↓	20.60	.388	E.C.Y.		13.75		13.75	21
0400	Roller compaction operator walking, add	B-10A	100	.120			5.20	1.74	6.94	9.80
0500	Air tamp, add	B-9D	190	.211			7.55	1.37	8.92	13.10
0600	Vibrating plate, add	A-1D	60	.133			4.73	.58	5.31	7.95
0800	Compaction in 12″ layers, hand tamp, add to above	1 Clab	34	.235			8.35		8.35	12.85
0900	Roller compaction operator walking, add	B-10A	150	.080			3.48	1.16	4.64	6.60
1000	Air tamp, add	B-9	285	.140			5.05	.80	5.85	8.65
1100	Vibrating plate, add	A-1E	90	.089	↓		3.15	.49	3.64	5.40
1300	Dozer backfilling, bulk, up to 300′ haul, no compaction	B-10B	1200	.010	L.C.Y.		.43	1.11	1.54	1.88
1400	Air tamped, add	B-11B	80	.200	E.C.Y.		8.15	3.67	11.82	16.45
1600	Compacting backfill, 6″ to 12″ lifts, vibrating roller	B-10C	800	.015			.65	2.18	2.83	3.38
1700	Sheepsfoot roller	B-10D	750	.016	↓		.70	2.36	3.06	3.66
1900	Dozer backfilling, trench, up to 300′ haul, no compaction	B-10B	900	.013	L.C.Y.		.58	1.48	2.06	2.51
2000	Air tamped, add	B-11B	80	.200	E.C.Y.		8.15	3.67	11.82	16.45
2200	Compacting backfill, 6″ to 12″ lifts, vibrating roller	B-10C	700	.017			.75	2.49	3.24	3.86
2300	Sheepsfoot roller	B-10D	650	.018	↓		.80	2.73	3.53	4.22

31 23 23.16 Fill By Borrow and Utility Bedding

	31 23 23.16 Fill By Borrow and Utility Bedding	Crew	Daily Output	Labor-Hours	Unit	Material	Labor	Equipment	Total	Total Incl O&P
0010	**FILL BY BORROW AND UTILITY BEDDING**									
0049	Utility bedding, for pipe & conduit, not incl. compaction									
0050	Crushed or screened bank run gravel	B-6	150	.160	L.C.Y.	27	6.20	2.45	35.65	42
0100	Crushed stone 3/4″ to 1/2″		150	.160		30.50	6.20	2.45	39.15	46
0200	Sand, dead or bank	↓	150	.160	↓	18	6.20	2.45	26.65	32
0500	Compacting bedding in trench	A-1D	90	.089	E.C.Y.		3.15	.38	3.53	5.25
0610	See Section 31 23 23.20 for hauling mileage add.									

31 23 23.20 Hauling

	31 23 23.20 Hauling	Crew	Daily Output	Labor-Hours	Unit	Material	Labor	Equipment	Total	Total Incl O&P
0010	**HAULING**									
0011	Excavated or borrow, loose cubic yards									
0012	no loading equipment, including hauling,waiting, loading/dumping									
0013	time per cycle (wait, load, travel, unload or dump & return)									
0014	8 C.Y. truck, 15 MPH ave, cycle 0.5 miles, 10 min. wait/Ld./Uld.	B-34A	320	.025	L.C.Y.		.92	1.30	2.22	2.83
0016	cycle 1 mile		272	.029			1.08	1.53	2.61	3.33
0018	cycle 2 miles		208	.038			1.41	2	3.41	4.36
0020	cycle 4 miles		144	.056			2.03	2.89	4.92	6.30

31 23 23 – Fill

31 23 23.20 Hauling		Crew	Daily Output	Labor-Hours	Unit	Material	2013 Bare Costs Labor	Equipment	Total	Total Incl O&P
0022	cycle 6 miles	B-34A	112	.071	L.C.Y.		2.61	3.72	6.33	8.10
0024	cycle 8 miles		88	.091			3.33	4.73	8.06	10.30
0026	20 MPH ave,cycle 0.5 mile		336	.024			.87	1.24	2.11	2.69
0028	cycle 1 mile		296	.027			.99	1.41	2.40	3.07
0030	cycle 2 miles		240	.033			1.22	1.74	2.96	3.78
0032	cycle 4 miles		176	.045			1.66	2.37	4.03	5.15
0034	cycle 6 miles		136	.059			2.15	3.06	5.21	6.65
0036	cycle 8 miles		112	.071			2.61	3.72	6.33	8.10
0044	25 MPH ave, cycle 4 miles		192	.042			1.53	2.17	3.70	4.73
0046	cycle 6 miles		160	.050			1.83	2.60	4.43	5.65
0048	cycle 8 miles		128	.063			2.29	3.26	5.55	7.10
0050	30 MPH ave, cycle 4 miles		216	.037			1.36	1.93	3.29	4.20
0052	cycle 6 miles		176	.045			1.66	2.37	4.03	5.15
0054	cycle 8 miles		144	.056			2.03	2.89	4.92	6.30
0114	15 MPH ave, cycle 0.5 mile, 15 min. wait/Ld./Uld.		224	.036			1.31	1.86	3.17	4.05
0116	cycle 1 mile		200	.040			1.46	2.08	3.54	4.53
0118	cycle 2 miles		168	.048			1.74	2.48	4.22	5.40
0120	cycle 4 miles		120	.067			2.44	3.47	5.91	7.55
0122	cycle 6 miles		96	.083			3.05	4.34	7.39	9.45
0124	cycle 8 miles		80	.100			3.66	5.20	8.86	11.35
0126	20 MPH ave, cycle 0.5 mile		232	.034			1.26	1.80	3.06	3.91
0128	cycle 1 mile		208	.038			1.41	2	3.41	4.36
0130	cycle 2 miles		184	.043			1.59	2.26	3.85	4.93
0132	cycle 4 miles		144	.056			2.03	2.89	4.92	6.30
0134	cycle 6 miles		112	.071			2.61	3.72	6.33	8.10
0136	cycle 8 miles		96	.083			3.05	4.34	7.39	9.45
0144	25 MPH ave, cycle 4 miles		152	.053			1.93	2.74	4.67	5.95
0146	cycle 6 miles		128	.063			2.29	3.26	5.55	7.10
0148	cycle 8 miles		112	.071			2.61	3.72	6.33	8.10
0150	30 MPH ave, cycle 4 miles		168	.048			1.74	2.48	4.22	5.40
0152	cycle 6 miles		144	.056			2.03	2.89	4.92	6.30
0154	cycle 8 miles		120	.067			2.44	3.47	5.91	7.55
0214	15 MPH ave, cycle 0.5 mile, 20 min wait/Ld./Uld.		176	.045			1.66	2.37	4.03	5.15
0216	cycle 1 mile		160	.050			1.83	2.60	4.43	5.65
0218	cycle 2 miles		136	.059			2.15	3.06	5.21	6.65
0220	cycle 4 miles		104	.077			2.82	4.01	6.83	8.70
0222	cycle 6 miles		88	.091			3.33	4.73	8.06	10.30
0224	cycle 8 miles		72	.111			4.07	5.80	9.87	12.60
0226	20 MPH ave, cycle 0.5 mile		176	.045			1.66	2.37	4.03	5.15
0228	cycle 1 mile		168	.048			1.74	2.48	4.22	5.40
0230	cycle 2 miles		144	.056			2.03	2.89	4.92	6.30
0232	cycle 4 miles		120	.067			2.44	3.47	5.91	7.55
0234	cycle 6 miles		96	.083			3.05	4.34	7.39	9.45
0236	cycle 8 miles		88	.091			3.33	4.73	8.06	10.30
0244	25 MPH ave, cycle 4 miles		128	.063			2.29	3.26	5.55	7.10
0246	cycle 6 miles		112	.071			2.61	3.72	6.33	8.10
0248	cycle 8 miles		96	.083			3.05	4.34	7.39	9.45
0250	30 MPH ave, cycle 4 miles		136	.059			2.15	3.06	5.21	6.65
0252	cycle 6 miles		120	.067			2.44	3.47	5.91	7.55
0254	cycle 8 miles		104	.077			2.82	4.01	6.83	8.70
0314	15 MPH ave, cycle 0.5 mile, 25 min wait/Ld./Uld.		144	.056			2.03	2.89	4.92	6.30
0316	cycle 1 mile		128	.063			2.29	3.26	5.55	7.10
0318	cycle 2 miles		112	.071			2.61	3.72	6.33	8.10

31 23 23.20 Hauling		Crew	Daily Output	Labor-Hours	Unit	Material	2013 Bare Costs Labor	Equipment	Total	Total Incl O&P
0320	cycle 4 miles	B-34A	96	.083	L.C.Y.		3.05	4.34	7.39	9.45
0322	cycle 6 miles		80	.100			3.66	5.20	8.86	11.35
0324	cycle 8 miles		64	.125			4.58	6.50	11.08	14.15
0326	20 MPH ave, cycle 0.5 mile		144	.056			2.03	2.89	4.92	6.30
0328	cycle 1 mile		136	.059			2.15	3.06	5.21	6.65
0330	cycle 2 miles		120	.067			2.44	3.47	5.91	7.55
0332	cycle 4 miles		104	.077			2.82	4.01	6.83	8.70
0334	cycle 6 miles		88	.091			3.33	4.73	8.06	10.30
0336	cycle 8 miles		80	.100			3.66	5.20	8.86	11.35
0344	25 MPH ave, cycle 4 miles		112	.071			2.61	3.72	6.33	8.10
0346	cycle 6 miles		96	.083			3.05	4.34	7.39	9.45
0348	cycle 8 miles		88	.091			3.33	4.73	8.06	10.30
0350	30 MPH ave, cycle 4 miles		112	.071			2.61	3.72	6.33	8.10
0352	cycle 6 miles		104	.077			2.82	4.01	6.83	8.70
0354	cycle 8 miles		96	.083			3.05	4.34	7.39	9.45
0414	15 MPH ave, cycle 0.5 mile, 30 min wait/Ld./Uld.		120	.067			2.44	3.47	5.91	7.55
0416	cycle 1 mile		112	.071			2.61	3.72	6.33	8.10
0418	cycle 2 miles		96	.083			3.05	4.34	7.39	9.45
0420	cycle 4 miles		80	.100			3.66	5.20	8.86	11.35
0422	cycle 6 miles		72	.111			4.07	5.80	9.87	12.60
0424	cycle 8 miles		64	.125			4.58	6.50	11.08	14.15
0426	20 MPH ave, cycle 0.5 mile		120	.067			2.44	3.47	5.91	7.55
0428	cycle 1 mile		112	.071			2.61	3.72	6.33	8.10
0430	cycle 2 miles		104	.077			2.82	4.01	6.83	8.70
0432	cycle 4 miles		88	.091			3.33	4.73	8.06	10.30
0434	cycle 6 miles		80	.100			3.66	5.20	8.86	11.35
0436	cycle 8 miles		72	.111			4.07	5.80	9.87	12.60
0444	25 MPH ave, cycle 4 miles		96	.083			3.05	4.34	7.39	9.45
0446	cycle 6 miles		88	.091			3.33	4.73	8.06	10.30
0448	cycle 8 miles		80	.100			3.66	5.20	8.86	11.35
0450	30 MPH ave, cycle 4 miles		96	.083			3.05	4.34	7.39	9.45
0452	cycle 6 miles		88	.091			3.33	4.73	8.06	10.30
0454	cycle 8 miles		80	.100			3.66	5.20	8.86	11.35
0514	15 MPH ave, cycle 0.5 mile, 35 min wait/Ld./Uld.		104	.077			2.82	4.01	6.83	8.70
0516	cycle 1 mile		96	.083			3.05	4.34	7.39	9.45
0518	cycle 2 miles		88	.091			3.33	4.73	8.06	10.30
0520	cycle 4 miles		72	.111			4.07	5.80	9.87	12.60
0522	cycle 6 miles		64	.125			4.58	6.50	11.08	14.15
0524	cycle 8 miles		56	.143			5.25	7.45	12.70	16.20
0526	20 MPH ave, cycle 0.5 mile		104	.077			2.82	4.01	6.83	8.70
0528	cycle 1 mile		96	.083			3.05	4.34	7.39	9.45
0530	cycle 2 miles		96	.083			3.05	4.34	7.39	9.45
0532	cycle 4 miles		80	.100			3.66	5.20	8.86	11.35
0534	cycle 6 miles		72	.111			4.07	5.80	9.87	12.60
0536	cycle 8 miles		64	.125			4.58	6.50	11.08	14.15
0544	25 MPH ave, cycle 4 miles		88	.091			3.33	4.73	8.06	10.30
0546	cycle 6 miles		80	.100			3.66	5.20	8.86	11.35
0548	cycle 8 miles		72	.111			4.07	5.80	9.87	12.60
0550	30 MPH ave, cycle 4 miles		88	.091			3.33	4.73	8.06	10.30
0552	cycle 6 miles		80	.100			3.66	5.20	8.86	11.35
0554	cycle 8 miles		72	.111			4.07	5.80	9.87	12.60
1014	12 C.Y. truck, cycle 0.5 mile, 15 MPH ave, 15 min. wait/Ld./Uld.	B-34B	336	.024			.87	2.06	2.93	3.60
1016	cycle 1 mile		300	.027			.98	2.31	3.29	4.03

31 23 23.20 Hauling		Crew	Daily Output	Labor-Hours	Unit	Material	2013 Bare Costs Labor	Equipment	Total	Total Incl O&P
1018	cycle 2 miles	B-34B	252	.032	L.C.Y.		1.16	2.75	3.91	4.80
1020	cycle 4 miles		180	.044			1.63	3.85	5.48	6.70
1022	cycle 6 miles		144	.056			2.03	4.81	6.84	8.40
1024	cycle 8 miles		120	.067			2.44	5.75	8.19	10.10
1025	cycle 10 miles		96	.083			3.05	7.20	10.25	12.60
1026	20 MPH ave, cycle 0.5 mile		348	.023			.84	1.99	2.83	3.48
1028	cycle 1 mile		312	.026			.94	2.22	3.16	3.88
1030	cycle 2 miles		276	.029			1.06	2.51	3.57	4.38
1032	cycle 4 miles		216	.037			1.36	3.21	4.57	5.60
1034	cycle 6 miles		168	.048			1.74	4.12	5.86	7.20
1036	cycle 8 miles		144	.056			2.03	4.81	6.84	8.40
1038	cycle 10 miles		120	.067			2.44	5.75	8.19	10.10
1040	25 MPH ave, cycle 4 miles		228	.035			1.28	3.04	4.32	5.30
1042	cycle 6 miles		192	.042			1.53	3.61	5.14	6.30
1044	cycle 8 miles		168	.048			1.74	4.12	5.86	7.20
1046	cycle 10 miles		144	.056			2.03	4.81	6.84	8.40
1050	30 MPH ave, cycle 4 miles		252	.032			1.16	2.75	3.91	4.80
1052	cycle 6 miles		216	.037			1.36	3.21	4.57	5.60
1054	cycle 8 miles		180	.044			1.63	3.85	5.48	6.70
1056	cycle 10 miles		156	.051			1.88	4.44	6.32	7.75
1060	35 MPH ave, cycle 4 miles		264	.030			1.11	2.62	3.73	4.58
1062	cycle 6 miles		228	.035			1.28	3.04	4.32	5.30
1064	cycle 8 miles		204	.039			1.44	3.39	4.83	5.95
1066	cycle 10 miles		180	.044			1.63	3.85	5.48	6.70
1068	cycle 20 miles		120	.067			2.44	5.75	8.19	10.10
1069	cycle 30 miles		84	.095			3.49	8.25	11.74	14.40
1070	cycle 40 miles		72	.111			4.07	9.60	13.67	16.85
1072	40 MPH ave, cycle 6 miles		240	.033			1.22	2.88	4.10	5.05
1074	cycle 8 miles		216	.037			1.36	3.21	4.57	5.60
1076	cycle 10 miles		192	.042			1.53	3.61	5.14	6.30
1078	cycle 20 miles		120	.067			2.44	5.75	8.19	10.10
1080	cycle 30 miles		96	.083			3.05	7.20	10.25	12.60
1082	cycle 40 miles		72	.111			4.07	9.60	13.67	16.85
1084	cycle 50 miles		60	.133			4.88	11.55	16.43	20
1094	45 MPH ave, cycle 8 miles		216	.037			1.36	3.21	4.57	5.60
1096	cycle 10 miles		204	.039			1.44	3.39	4.83	5.95
1098	cycle 20 miles		132	.061			2.22	5.25	7.47	9.15
1100	cycle 30 miles		108	.074			2.71	6.40	9.11	11.20
1102	cycle 40 miles		84	.095			3.49	8.25	11.74	14.40
1104	cycle 50 miles		72	.111			4.07	9.60	13.67	16.85
1106	50 MPH ave, cycle 10 miles		216	.037			1.36	3.21	4.57	5.60
1108	cycle 20 miles		144	.056			2.03	4.81	6.84	8.40
1110	cycle 30 miles		108	.074			2.71	6.40	9.11	11.20
1112	cycle 40 miles		84	.095			3.49	8.25	11.74	14.40
1114	cycle 50 miles		72	.111			4.07	9.60	13.67	16.85
1214	15 MPH ave, cycle 0.5 mile, 20 min. wait/Ld./Uld.		264	.030			1.11	2.62	3.73	4.58
1216	cycle 1 mile		240	.033			1.22	2.88	4.10	5.05
1218	cycle 2 miles		204	.039			1.44	3.39	4.83	5.95
1220	cycle 4 miles		156	.051			1.88	4.44	6.32	7.75
1222	cycle 6 miles		132	.061			2.22	5.25	7.47	9.15
1224	cycle 8 miles		108	.074			2.71	6.40	9.11	11.20
1225	cycle 10 miles		96	.083			3.05	7.20	10.25	12.60
1226	20 MPH ave, cycle 0.5 mile		264	.030			1.11	2.62	3.73	4.58

31 23 23.20 Hauling		Crew	Daily Output	Labor-Hours	Unit	Material	2013 Bare Costs Labor	Equipment	Total	Total Incl O&P
1228	cycle 1 mile	B-34B	252	.032	L.C.Y.		1.16	2.75	3.91	4.80
1230	cycle 2 miles		216	.037			1.36	3.21	4.57	5.60
1232	cycle 4 miles		180	.044			1.63	3.85	5.48	6.70
1234	cycle 6 miles		144	.056			2.03	4.81	6.84	8.40
1236	cycle 8 miles		132	.061			2.22	5.25	7.47	9.15
1238	cycle 10 miles		108	.074			2.71	6.40	9.11	11.20
1240	25 MPH ave, cycle 4 miles		192	.042			1.53	3.61	5.14	6.30
1242	cycle 6 miles		168	.048			1.74	4.12	5.86	7.20
1244	cycle 8 miles		144	.056			2.03	4.81	6.84	8.40
1246	cycle 10 miles		132	.061			2.22	5.25	7.47	9.15
1250	30 MPH ave, cycle 4 miles		204	.039			1.44	3.39	4.83	5.95
1252	cycle 6 miles		180	.044			1.63	3.85	5.48	6.70
1254	cycle 8 miles		156	.051			1.88	4.44	6.32	7.75
1256	cycle 10 miles		144	.056			2.03	4.81	6.84	8.40
1260	35 MPH ave, cycle 4 miles		216	.037			1.36	3.21	4.57	5.60
1262	cycle 6 miles		192	.042			1.53	3.61	5.14	6.30
1264	cycle 8 miles		168	.048			1.74	4.12	5.86	7.20
1266	cycle 10 miles		156	.051			1.88	4.44	6.32	7.75
1268	cycle 20 miles		108	.074			2.71	6.40	9.11	11.20
1269	cycle 30 miles		72	.111			4.07	9.60	13.67	16.85
1270	cycle 40 miles		60	.133			4.88	11.55	16.43	20
1272	40 MPH ave, cycle 6 miles		192	.042			1.53	3.61	5.14	6.30
1274	cycle 8 miles		180	.044			1.63	3.85	5.48	6.70
1276	cycle 10 miles		156	.051			1.88	4.44	6.32	7.75
1278	cycle 20 miles		108	.074			2.71	6.40	9.11	11.20
1280	cycle 30 miles		84	.095			3.49	8.25	11.74	14.40
1282	cycle 40 miles		72	.111			4.07	9.60	13.67	16.85
1284	cycle 50 miles		60	.133			4.88	11.55	16.43	20
1294	45 MPH ave, cycle 8 miles		180	.044			1.63	3.85	5.48	6.70
1296	cycle 10 miles		168	.048			1.74	4.12	5.86	7.20
1298	cycle 20 miles		120	.067			2.44	5.75	8.19	10.10
1300	cycle 30 miles		96	.083			3.05	7.20	10.25	12.60
1302	cycle 40 miles		72	.111			4.07	9.60	13.67	16.85
1304	cycle 50 miles		60	.133			4.88	11.55	16.43	20
1306	50 MPH ave, cycle 10 miles		180	.044			1.63	3.85	5.48	6.70
1308	cycle 20 miles		132	.061			2.22	5.25	7.47	9.15
1310	cycle 30 miles		96	.083			3.05	7.20	10.25	12.60
1312	cycle 40 miles		84	.095			3.49	8.25	11.74	14.40
1314	cycle 50 miles		72	.111			4.07	9.60	13.67	16.85
1414	15 MPH ave, cycle 0.5 mile, 25 min. wait/Ld./Uld.		204	.039			1.44	3.39	4.83	5.95
1416	cycle 1 mile		192	.042			1.53	3.61	5.14	6.30
1418	cycle 2 miles		168	.048			1.74	4.12	5.86	7.20
1420	cycle 4 miles		132	.061			2.22	5.25	7.47	9.15
1422	cycle 6 miles		120	.067			2.44	5.75	8.19	10.10
1424	cycle 8 miles		96	.083			3.05	7.20	10.25	12.60
1425	cycle 10 miles		84	.095			3.49	8.25	11.74	14.40
1426	20 MPH ave, cycle 0.5 mile		216	.037			1.36	3.21	4.57	5.60
1428	cycle 1 mile		204	.039			1.44	3.39	4.83	5.95
1430	cycle 2 miles		180	.044			1.63	3.85	5.48	6.70
1432	cycle 4 miles		156	.051			1.88	4.44	6.32	7.75
1434	cycle 6 miles		132	.061			2.22	5.25	7.47	9.15
1436	cycle 8 miles		120	.067			2.44	5.75	8.19	10.10
1438	cycle 10 miles		96	.083			3.05	7.20	10.25	12.60

31 23 23.20 Hauling		Crew	Daily Output	Labor-Hours	Unit	Material	2013 Bare Costs Labor	Equipment	Total	Total Incl O&P
1440	25 MPH ave, cycle 4 miles	B-34B	168	.048	L.C.Y.		1.74	4.12	5.86	7.20
1442	cycle 6 miles		144	.056			2.03	4.81	6.84	8.40
1444	cycle 8 miles		132	.061			2.22	5.25	7.47	9.15
1446	cycle 10 miles		108	.074			2.71	6.40	9.11	11.20
1450	30 MPH ave, cycle 4 miles		168	.048			1.74	4.12	5.86	7.20
1452	cycle 6 miles		156	.051			1.88	4.44	6.32	7.75
1454	cycle 8 miles		132	.061			2.22	5.25	7.47	9.15
1456	cycle 10 miles		120	.067			2.44	5.75	8.19	10.10
1460	35 MPH ave, cycle 4 miles		180	.044			1.63	3.85	5.48	6.70
1462	cycle 6 miles		156	.051			1.88	4.44	6.32	7.75
1464	cycle 8 miles		144	.056			2.03	4.81	6.84	8.40
1466	cycle 10 miles		132	.061			2.22	5.25	7.47	9.15
1468	cycle 20 miles		96	.083			3.05	7.20	10.25	12.60
1469	cycle 30 miles		72	.111			4.07	9.60	13.67	16.85
1470	cycle 40 miles		60	.133			4.88	11.55	16.43	20
1472	40 MPH ave, cycle 6 miles		168	.048			1.74	4.12	5.86	7.20
1474	cycle 8 miles		156	.051			1.88	4.44	6.32	7.75
1476	cycle 10 miles		144	.056			2.03	4.81	6.84	8.40
1478	cycle 20 miles		96	.083			3.05	7.20	10.25	12.60
1480	cycle 30 miles		84	.095			3.49	8.25	11.74	14.40
1482	cycle 40 miles		60	.133			4.88	11.55	16.43	20
1484	cycle 50 miles		60	.133			4.88	11.55	16.43	20
1494	45 MPH ave, cycle 8 miles		156	.051			1.88	4.44	6.32	7.75
1496	cycle 10 miles		144	.056			2.03	4.81	6.84	8.40
1498	cycle 20 miles		108	.074			2.71	6.40	9.11	11.20
1500	cycle 30 miles		84	.095			3.49	8.25	11.74	14.40
1502	cycle 40 miles		72	.111			4.07	9.60	13.67	16.85
1504	cycle 50 miles		60	.133			4.88	11.55	16.43	20
1506	50 MPH ave, cycle 10 miles		156	.051			1.88	4.44	6.32	7.75
1508	cycle 20 miles		120	.067			2.44	5.75	8.19	10.10
1510	cycle 30 miles		96	.083			3.05	7.20	10.25	12.60
1512	cycle 40 miles		72	.111			4.07	9.60	13.67	16.85
1514	cycle 50 miles		60	.133			4.88	11.55	16.43	20
1614	15 MPH, cycle 0.5 mile, 30 min. wait/Ld./Uld.		180	.044			1.63	3.85	5.48	6.70
1616	cycle 1 mile		168	.048			1.74	4.12	5.86	7.20
1618	cycle 2 miles		144	.056			2.03	4.81	6.84	8.40
1620	cycle 4 miles		120	.067			2.44	5.75	8.19	10.10
1622	cycle 6 miles		108	.074			2.71	6.40	9.11	11.20
1624	cycle 8 miles		84	.095			3.49	8.25	11.74	14.40
1625	cycle 10 miles		84	.095			3.49	8.25	11.74	14.40
1626	20 MPH ave, cycle 0.5 mile		180	.044			1.63	3.85	5.48	6.70
1628	cycle 1 mile		168	.048			1.74	4.12	5.86	7.20
1630	cycle 2 miles		156	.051			1.88	4.44	6.32	7.75
1632	cycle 4 miles		132	.061			2.22	5.25	7.47	9.15
1634	cycle 6 miles		120	.067			2.44	5.75	8.19	10.10
1636	cycle 8 miles		108	.074			2.71	6.40	9.11	11.20
1638	cycle 10 miles		96	.083			3.05	7.20	10.25	12.60
1640	25 MPH ave, cycle 4 miles		144	.056			2.03	4.81	6.84	8.40
1642	cycle 6 miles		132	.061			2.22	5.25	7.47	9.15
1644	cycle 8 miles		108	.074			2.71	6.40	9.11	11.20
1646	cycle 10 miles		108	.074			2.71	6.40	9.11	11.20
1650	30 MPH ave, cycle 4 miles		144	.056			2.03	4.81	6.84	8.40
1652	cycle 6 miles		132	.061			2.22	5.25	7.47	9.15

31 23 23.20 Hauling		Crew	Daily Output	Labor-Hours	Unit	2013 Bare Costs Material	Labor	Equipment	Total	Total Incl O&P
1654	cycle 8 miles	B-34B	120	.067	L.C.Y.		2.44	5.75	8.19	10.10
1656	cycle 10 miles		108	.074			2.71	6.40	9.11	11.20
1660	35 MPH ave, cycle 4 miles		156	.051			1.88	4.44	6.32	7.75
1662	cycle 6 miles		144	.056			2.03	4.81	6.84	8.40
1664	cycle 8 miles		132	.061			2.22	5.25	7.47	9.15
1666	cycle 10 miles		120	.067			2.44	5.75	8.19	10.10
1668	cycle 20 miles		84	.095			3.49	8.25	11.74	14.40
1669	cycle 30 miles		72	.111			4.07	9.60	13.67	16.85
1670	cycle 40 miles		60	.133			4.88	11.55	16.43	20
1672	40 MPH, cycle 6 miles		144	.056			2.03	4.81	6.84	8.40
1674	cycle 8 miles		132	.061			2.22	5.25	7.47	9.15
1676	cycle 10 miles		120	.067			2.44	5.75	8.19	10.10
1678	cycle 20 miles		96	.083			3.05	7.20	10.25	12.60
1680	cycle 30 miles		72	.111			4.07	9.60	13.67	16.85
1682	cycle 40 miles		60	.133			4.88	11.55	16.43	20
1684	cycle 50 miles		48	.167			6.10	14.45	20.55	25
1694	45 MPH ave, cycle 8 miles		144	.056			2.03	4.81	6.84	8.40
1696	cycle 10 miles		132	.061			2.22	5.25	7.47	9.15
1698	cycle 20 miles		96	.083			3.05	7.20	10.25	12.60
1700	cycle 30 miles		84	.095			3.49	8.25	11.74	14.40
1702	cycle 40 miles		60	.133			4.88	11.55	16.43	20
1704	cycle 50 miles		60	.133			4.88	11.55	16.43	20
1706	50 MPH ave, cycle 10 miles		132	.061			2.22	5.25	7.47	9.15
1708	cycle 20 miles		108	.074			2.71	6.40	9.11	11.20
1710	cycle 30 miles		84	.095			3.49	8.25	11.74	14.40
1712	cycle 40 miles		72	.111			4.07	9.60	13.67	16.85
1714	cycle 50 miles		60	.133			4.88	11.55	16.43	20
2000	Hauling, 8 C.Y. truck, small project cost per hour	B-34A	8	1	Hr.		36.50	52	88.50	114
2100	12 C.Y. Truck	B-34B	8	1			36.50	86.50	123	151
2150	16.5 C.Y. Truck	B-34C	8	1			36.50	92.50	129	158
2175	18 C.Y. 8 wheel Truck	B-34I	8	1			36.50	108	144.50	175
2200	20 C.Y. Truck	B-34D	8	1			36.50	94	130.50	160
2300	Grading at dump, or embankment if required, by dozer	B-10B	1000	.012	L.C.Y.		.52	1.33	1.85	2.26
2310	Spotter at fill or cut, if required	1 Clab	8	1	Hr.		35.50		35.50	54.50
9014	18 C.Y. truck, 8 wheels,15 min. wait/Ld./Uld.,15 MPH, cycle 0.5 mi.	B-34I	504	.016	L.C.Y.		.58	1.72	2.30	2.78
9016	cycle 1 mile		450	.018			.65	1.93	2.58	3.12
9018	cycle 2 miles		378	.021			.77	2.29	3.06	3.71
9020	cycle 4 miles		270	.030			1.08	3.21	4.29	5.20
9022	cycle 6 miles		216	.037			1.36	4.01	5.37	6.50
9024	cycle 8 miles		180	.044			1.63	4.82	6.45	7.80
9025	cycle 10 miles		144	.056			2.03	6	8.03	9.70
9026	20 MPH ave, cycle 0.5 mile		522	.015			.56	1.66	2.22	2.69
9028	cycle 1 mile		468	.017			.63	1.85	2.48	3
9030	cycle 2 miles		414	.019			.71	2.09	2.80	3.38
9032	cycle 4 miles		324	.025			.90	2.68	3.58	4.32
9034	cycle 6 miles		252	.032			1.16	3.44	4.60	5.55
9036	cycle 8 miles		216	.037			1.36	4.01	5.37	6.50
9038	cycle 10 miles		180	.044			1.63	4.82	6.45	7.80
9040	25 MPH ave, cycle 4 miles		342	.023			.86	2.54	3.40	4.10
9042	cycle 6 miles		288	.028			1.02	3.01	4.03	4.87
9044	cycle 8 miles		252	.032			1.16	3.44	4.60	5.55
9046	cycle 10 miles		216	.037			1.36	4.01	5.37	6.50
9050	30 MPH ave, cycle 4 miles		378	.021			.77	2.29	3.06	3.71

31 23 23.20 Hauling		Crew	Daily Output	Labor-Hours	Unit	Material	2013 Bare Costs Labor	Equipment	Total	Total Incl O&P
9052	cycle 6 miles	B-34I	324	.025	L.C.Y.		.90	2.68	3.58	4.32
9054	cycle 8 miles		270	.030			1.08	3.21	4.29	5.20
9056	cycle 10 miles		234	.034			1.25	3.71	4.96	6
9060	35 MPH ave, cycle 4 miles		396	.020			.74	2.19	2.93	3.54
9062	cycle 6 miles		342	.023			.86	2.54	3.40	4.10
9064	cycle 8 miles		288	.028			1.02	3.01	4.03	4.87
9066	cycle 10 miles		270	.030			1.08	3.21	4.29	5.20
9068	cycle 20 miles		162	.049			1.81	5.35	7.16	8.65
9070	cycle 30 miles		126	.063			2.32	6.90	9.22	11.10
9072	cycle 40 miles		90	.089			3.25	9.65	12.90	15.60
9074	40 MPH ave, cycle 6 miles		360	.022			.81	2.41	3.22	3.90
9076	cycle 8 miles		324	.025			.90	2.68	3.58	4.32
9078	cycle 10 miles		288	.028			1.02	3.01	4.03	4.87
9080	cycle 20 miles		180	.044			1.63	4.82	6.45	7.80
9082	cycle 30 miles		144	.056			2.03	6	8.03	9.70
9084	cycle 40 miles		108	.074			2.71	8.05	10.76	13
9086	cycle 50 miles		90	.089			3.25	9.65	12.90	15.60
9094	45 MPH ave, cycle 8 miles		324	.025			.90	2.68	3.58	4.32
9096	cycle 10 miles		306	.026			.96	2.83	3.79	4.59
9098	cycle 20 miles		198	.040			1.48	4.38	5.86	7.10
9100	cycle 30 miles		144	.056			2.03	6	8.03	9.70
9102	cycle 40 miles		126	.063			2.32	6.90	9.22	11.10
9104	cycle 50 miles		108	.074			2.71	8.05	10.76	13
9106	50 MPH ave, cycle 10 miles		324	.025			.90	2.68	3.58	4.32
9108	cycle 20 miles		216	.037			1.36	4.01	5.37	6.50
9110	cycle 30 miles		162	.049			1.81	5.35	7.16	8.65
9112	cycle 40 miles		126	.063			2.32	6.90	9.22	11.10
9114	cycle 50 miles		108	.074			2.71	8.05	10.76	13
9214	20 min. wait/Ld./Uld.,15 MPH, cycle 0.5 mi.		396	.020			.74	2.19	2.93	3.54
9216	cycle 1 mile		360	.022			.81	2.41	3.22	3.90
9218	cycle 2 miles		306	.026			.96	2.83	3.79	4.59
9220	cycle 4 miles		234	.034			1.25	3.71	4.96	6
9222	cycle 6 miles		198	.040			1.48	4.38	5.86	7.10
9224	cycle 8 miles		162	.049			1.81	5.35	7.16	8.65
9225	cycle 10 miles		144	.056			2.03	6	8.03	9.70
9226	20 MPH ave, cycle 0.5 mile		396	.020			.74	2.19	2.93	3.54
9228	cycle 1 mile		378	.021			.77	2.29	3.06	3.71
9230	cycle 2 miles		324	.025			.90	2.68	3.58	4.32
9232	cycle 4 miles		270	.030			1.08	3.21	4.29	5.20
9234	cycle 6 miles		216	.037			1.36	4.01	5.37	6.50
9236	cycle 8 miles		198	.040			1.48	4.38	5.86	7.10
9238	cycle 10 miles		162	.049			1.81	5.35	7.16	8.65
9240	25 MPH ave, cycle 4 miles		288	.028			1.02	3.01	4.03	4.87
9242	cycle 6 miles		252	.032			1.16	3.44	4.60	5.55
9244	cycle 8 miles		216	.037			1.36	4.01	5.37	6.50
9246	cycle 10 miles		198	.040			1.48	4.38	5.86	7.10
9250	30 MPH ave, cycle 4 miles		306	.026			.96	2.83	3.79	4.59
9252	cycle 6 miles		270	.030			1.08	3.21	4.29	5.20
9254	cycle 8 miles		234	.034			1.25	3.71	4.96	6
9256	cycle 10 miles		216	.037			1.36	4.01	5.37	6.50
9260	35 MPH ave, cycle 4 miles		324	.025			.90	2.68	3.58	4.32
9262	cycle 6 miles		288	.028			1.02	3.01	4.03	4.87
9264	cycle 8 miles		252	.032			1.16	3.44	4.60	5.55

31 23 23 – Fill

31 23 23.20 Hauling		Crew	Daily Output	Labor-Hours	Unit	Material	2013 Bare Costs Labor	Equipment	Total	Total Incl O&P
9266	cycle 10 miles	B-34I	234	.034	L.C.Y.		1.25	3.71	4.96	6
9268	cycle 20 miles		162	.049			1.81	5.35	7.16	8.65
9270	cycle 30 miles		108	.074			2.71	8.05	10.76	13
9272	cycle 40 miles		90	.089			3.25	9.65	12.90	15.60
9274	40 MPH ave, cycle 6 miles		288	.028			1.02	3.01	4.03	4.87
9276	cycle 8 miles		270	.030			1.08	3.21	4.29	5.20
9278	cycle 10 miles		234	.034			1.25	3.71	4.96	6
9280	cycle 20 miles		162	.049			1.81	5.35	7.16	8.65
9282	cycle 30 miles		126	.063			2.32	6.90	9.22	11.10
9284	cycle 40 miles		108	.074			2.71	8.05	10.76	13
9286	cycle 50 miles		90	.089			3.25	9.65	12.90	15.60
9294	45 MPH ave, cycle 8 miles		270	.030			1.08	3.21	4.29	5.20
9296	cycle 10 miles		252	.032			1.16	3.44	4.60	5.55
9298	cycle 20 miles		180	.044			1.63	4.82	6.45	7.80
9300	cycle 30 miles		144	.056			2.03	6	8.03	9.70
9302	cycle 40 miles		108	.074			2.71	8.05	10.76	13
9304	cycle 50 miles		90	.089			3.25	9.65	12.90	15.60
9306	50 MPH ave, cycle 10 miles		270	.030			1.08	3.21	4.29	5.20
9308	cycle 20 miles		198	.040			1.48	4.38	5.86	7.10
9310	cycle 30 miles		144	.056			2.03	6	8.03	9.70
9312	cycle 40 miles		126	.063			2.32	6.90	9.22	11.10
9314	cycle 50 miles		108	.074			2.71	8.05	10.76	13
9414	25 min. wait/Ld./Uld.,15 MPH, cycle 0.5 mi.		306	.026			.96	2.83	3.79	4.59
9416	cycle 1 mile		288	.028			1.02	3.01	4.03	4.87
9418	cycle 2 miles		252	.032			1.16	3.44	4.60	5.55
9420	cycle 4 miles		198	.040			1.48	4.38	5.86	7.10
9422	cycle 6 miles		180	.044			1.63	4.82	6.45	7.80
9424	cycle 8 miles		144	.056			2.03	6	8.03	9.70
9425	cycle 10 miles		126	.063			2.32	6.90	9.22	11.10
9426	20 MPH ave, cycle 0.5 mile		324	.025			.90	2.68	3.58	4.32
9428	cycle 1 mile		306	.026			.96	2.83	3.79	4.59
9430	cycle 2 miles		270	.030			1.08	3.21	4.29	5.20
9432	cycle 4 miles		234	.034			1.25	3.71	4.96	6
9434	cycle 6 miles		198	.040			1.48	4.38	5.86	7.10
9436	cycle 8 miles		180	.044			1.63	4.82	6.45	7.80
9438	cycle 10 miles		144	.056			2.03	6	8.03	9.70
9440	25 MPH ave, cycle 4 miles		252	.032			1.16	3.44	4.60	5.55
9442	cycle 6 miles		216	.037			1.36	4.01	5.37	6.50
9444	cycle 8 miles		198	.040			1.48	4.38	5.86	7.10
9446	cycle 10 miles		180	.044			1.63	4.82	6.45	7.80
9450	30 MPH ave, cycle 4 miles		252	.032			1.16	3.44	4.60	5.55
9452	cycle 6 miles		234	.034			1.25	3.71	4.96	6
9454	cycle 8 miles		198	.040			1.48	4.38	5.86	7.10
9456	cycle 10 miles		180	.044			1.63	4.82	6.45	7.80
9460	35 MPH ave, cycle 4 miles		270	.030			1.08	3.21	4.29	5.20
9462	cycle 6 miles		234	.034			1.25	3.71	4.96	6
9464	cycle 8 miles		216	.037			1.36	4.01	5.37	6.50
9466	cycle 10 miles		198	.040			1.48	4.38	5.86	7.10
9468	cycle 20 miles		144	.056			2.03	6	8.03	9.70
9470	cycle 30 miles		108	.074			2.71	8.05	10.76	13
9472	cycle 40 miles		90	.089			3.25	9.65	12.90	15.60
9474	40 MPH ave, cycle 6 miles		252	.032			1.16	3.44	4.60	5.55
9476	cycle 8 miles		234	.034			1.25	3.71	4.96	6

31 23 23.20 Hauling		Crew	Daily Output	Labor-Hours	Unit	Material	2013 Bare Costs Labor	Equipment	Total	Total Incl O&P
9478	cycle 10 miles	B-34I	216	.037	L.C.Y.		1.36	4.01	5.37	6.50
9480	cycle 20 miles		144	.056			2.03	6	8.03	9.70
9482	cycle 30 miles		126	.063			2.32	6.90	9.22	11.10
9484	cycle 40 miles		90	.089			3.25	9.65	12.90	15.60
9486	cycle 50 miles		90	.089			3.25	9.65	12.90	15.60
9494	45 MPH ave, cycle 8 miles		234	.034			1.25	3.71	4.96	6
9496	cycle 10 miles		216	.037			1.36	4.01	5.37	6.50
9498	cycle 20 miles		162	.049			1.81	5.35	7.16	8.65
9500	cycle 30 miles		126	.063			2.32	6.90	9.22	11.10
9502	cycle 40 miles		108	.074			2.71	8.05	10.76	13
9504	cycle 50 miles		90	.089			3.25	9.65	12.90	15.60
9506	50 MPH ave, cycle 10 miles		234	.034			1.25	3.71	4.96	6
9508	cycle 20 miles		180	.044			1.63	4.82	6.45	7.80
9510	cycle 30 miles		144	.056			2.03	6	8.03	9.70
9512	cycle 40 miles		108	.074			2.71	8.05	10.76	13
9514	cycle 50 miles		90	.089			3.25	9.65	12.90	15.60
9614	30 min. wait/Ld./Uld.,15 MPH, cycle 0.5 mi.		270	.030			1.08	3.21	4.29	5.20
9616	cycle 1 mile		252	.032			1.16	3.44	4.60	5.55
9618	cycle 2 miles		216	.037			1.36	4.01	5.37	6.50
9620	cycle 4 miles		180	.044			1.63	4.82	6.45	7.80
9622	cycle 6 miles		162	.049			1.81	5.35	7.16	8.65
9624	cycle 8 miles		126	.063			2.32	6.90	9.22	11.10
9625	cycle 10 miles		126	.063			2.32	6.90	9.22	11.10
9626	20 MPH ave, cycle 0.5 mile		270	.030			1.08	3.21	4.29	5.20
9628	cycle 1 mile		252	.032			1.16	3.44	4.60	5.55
9630	cycle 2 miles		234	.034			1.25	3.71	4.96	6
9632	cycle 4 miles		198	.040			1.48	4.38	5.86	7.10
9634	cycle 6 miles		180	.044			1.63	4.82	6.45	7.80
9636	cycle 8 miles		162	.049			1.81	5.35	7.16	8.65
9638	cycle 10 miles		144	.056			2.03	6	8.03	9.70
9640	25 MPH ave, cycle 4 miles		216	.037			1.36	4.01	5.37	6.50
9642	cycle 6 miles		198	.040			1.48	4.38	5.86	7.10
9644	cycle 8 miles		180	.044			1.63	4.82	6.45	7.80
9646	cycle 10 miles		162	.049			1.81	5.35	7.16	8.65
9650	30 MPH ave, cycle 4 miles		216	.037			1.36	4.01	5.37	6.50
9652	cycle 6 miles		198	.040			1.48	4.38	5.86	7.10
9654	cycle 8 miles		180	.044			1.63	4.82	6.45	7.80
9656	cycle 10 miles		162	.049			1.81	5.35	7.16	8.65
9660	35 MPH ave, cycle 4 miles		234	.034			1.25	3.71	4.96	6
9662	cycle 6 miles		216	.037			1.36	4.01	5.37	6.50
9664	cycle 8 miles		198	.040			1.48	4.38	5.86	7.10
9666	cycle 10 miles		180	.044			1.63	4.82	6.45	7.80
9668	cycle 20 miles		126	.063			2.32	6.90	9.22	11.10
9670	cycle 30 miles		108	.074			2.71	8.05	10.76	13
9672	cycle 40 miles		90	.089			3.25	9.65	12.90	15.60
9674	40 MPH ave, cycle 6 miles		216	.037			1.36	4.01	5.37	6.50
9676	cycle 8 miles		198	.040			1.48	4.38	5.86	7.10
9678	cycle 10 miles		180	.044			1.63	4.82	6.45	7.80
9680	cycle 20 miles		144	.056			2.03	6	8.03	9.70
9682	cycle 30 miles		108	.074			2.71	8.05	10.76	13
9684	cycle 40 miles		90	.089			3.25	9.65	12.90	15.60
9686	cycle 50 miles		72	.111			4.07	12.05	16.12	19.50
9694	45 MPH ave, cycle 8 miles		216	.037			1.36	4.01	5.37	6.50

31 23 23 – Fill

31 23 23.20 Hauling		Crew	Daily Output	Labor-Hours	Unit	Material	2013 Bare Costs Labor	Equipment	Total	Total Incl O&P
9696	cycle 10 miles	B-34I	198	.040	L.C.Y.		1.48	4.38	5.86	7.10
9698	cycle 20 miles		144	.056			2.03	6	8.03	9.70
9700	cycle 30 miles		126	.063			2.32	6.90	9.22	11.10
9702	cycle 40 miles		108	.074			2.71	8.05	10.76	13
9704	cycle 50 miles		90	.089			3.25	9.65	12.90	15.60
9706	50 MPH ave, cycle 10 miles		198	.040			1.48	4.38	5.86	7.10
9708	cycle 20 miles		162	.049			1.81	5.35	7.16	8.65
9710	cycle 30 miles		126	.063			2.32	6.90	9.22	11.10
9712	cycle 40 miles		108	.074			2.71	8.05	10.76	13
9714	cycle 50 miles	↓	90	.089	↓		3.25	9.65	12.90	15.60

31 23 23.23 Compaction		Crew	Daily Output	Labor-Hours	Unit	Material	Labor	Equipment	Total	Total Incl O&P
0010	**COMPACTION**									
7000	Walk behind, vibrating plate 18" wide, 6" lifts, 2 passes	A-1D	200	.040	E.C.Y.		1.42	.17	1.59	2.37
7020	3 passes		185	.043			1.53	.19	1.72	2.57
7040	4 passes	↓	140	.057	↓		2.03	.25	2.28	3.39

Estimating Tips

33 10 00 Water Utilities
33 30 00 Sanitary Sewerage Utilities
33 40 00 Storm Drainage Utilities

- Never assume that the water, sewer, and drainage lines will go in at the early stages of the project. Consider the site access needs before dividing the site in half with open trenches, loose pipe, and machinery obstructions. Always inspect the site to establish that the site drawings are complete. Check off all existing utilities on your drawings as you locate them. Be especially careful with underground utilities because appurtenances are sometimes buried during regrading or repaving operations. If you find any discrepancies, mark up the site plan for further research. Differing site conditions can be very costly if discovered later in the project.
- See also Section 33 01 00 for restoration of pipe where removal/replacement may be undesirable. Use of new types of piping materials can reduce the overall project cost. Owners/design engineers should consider the installing contractor as a valuable source of current information on utility products and local conditions that could lead to significant cost savings.

Reference Numbers

Reference numbers are shown in shaded boxes at the beginning of some major classifications. These numbers refer to related items in the Reference Section. The reference information may be an estimating procedure, an alternate pricing method, or technical information.

Note: Not all subdivisions listed here necessarily appear in this publication.

Note: **Trade Service,** *in part, has been used as a reference source for some of the material prices used in Division 33.*

33 05 Common Work Results for Utilities

33 05 16 – Utility Structures

	33 05 16.13 Precast Concrete Utility Boxes	Crew	Daily Output	Labor-Hours	Unit	Material	Labor	Equipment	Total	Total Incl O&P
						2013 Bare Costs				
0010	**PRECAST CONCRETE UTILITY BOXES**, 6" thick									
0050	5' x 10' x 6' high, I.D.	B-13	2	28	Ea.	3,600	1,075	365	5,040	6,000
0100	6' x 10' x 6' high, I.D.		2	28		3,725	1,075	365	5,165	6,150
0150	5' x 12' x 6' high, I.D.		2	28		3,950	1,075	365	5,390	6,400
0200	6' x 12' x 6' high, I.D.		1.80	31.111		4,425	1,200	405	6,030	7,175
0250	6' x 13' x 6' high, I.D.		1.50	37.333		5,800	1,450	485	7,735	9,100
0300	8' x 14' x 7' high, I.D.		1	56		6,275	2,150	730	9,155	11,000

33 05 23 – Trenchless Utility Installation

33 05 23.19 Microtunneling

	33 05 23.19 Microtunneling	Crew	Daily Output	Labor-Hours	Unit	Material	Labor	Equipment	Total	Total Incl O&P
0010	**MICROTUNNELING**									
0011	Not including excavation, backfill, shoring,									
0020	or dewatering, average 50'/day, slurry method									
0100	24" to 48" outside diameter, minimum				L.F.				850	935
0110	Adverse conditions, add				%				50%	50%
1000	Rent microtunneling machine, average monthly lease				Month				94,500	104,000
1010	Operating technician				Day				620	690
1100	Mobilization and demobilization, minimum				Job				40,000	45,000
1110	Maximum				"				432,500	476,000

33 05 23.20 Horizontal Boring

	33 05 23.20 Horizontal Boring	Crew	Daily Output	Labor-Hours	Unit	Material	Labor	Equipment	Total	Total Incl O&P
0010	**HORIZONTAL BORING**									
0011	Casing only, 100' minimum,									
0020	not incl. jacking pits or dewatering									
0100	Roadwork, 1/2" thick wall, 24" diameter casing	B-42	20	3.200	L.F.	132	128	67	327	420
0200	36" diameter		16	4		242	160	83.50	485.50	610
0300	48" diameter		15	4.267		340	171	89	600	735
0500	Railroad work, 24" diameter		15	4.267		132	171	89	392	510
0600	36" diameter		14	4.571		242	183	95.50	520.50	660
0700	48" diameter		12	5.333		340	214	111	665	830
0900	For ledge, add								20%	20%
1000	Small diameter boring, 3", sandy soil	B-82	900	.018		24	.72	.06	24.78	27.50
1040	Rocky soil	"	500	.032		24	1.30	.12	25.42	28.50
1100	Prepare jacking pits, incl. mobilization & demobilization, minimum				Ea.				3,150	3,625
1101	Maximum				"				21,600	25,000

33 05 23.22 Directional Drilling

	33 05 23.22 Directional Drilling	Crew	Daily Output	Labor-Hours	Unit	Material	Labor	Equipment	Total	Total Incl O&P
0010	**DIRECTIONAL DRILLING**									
0011	Excluding cost of conduit, 100' minimum									
0100	Small equipment to 300', not to exceed 12" diam.									
0102	small unit mobilization to site	B-82A	2	8	Ea.		325	540	865	1,075
0105	small unit setup per drill		4	4			163	269	432	545
0109	minimum charge gravel, sand & silt, up to 12" diam.		3.20	5		9.85	203	335	547.85	690
0110	gravel, sand & silt, up to 12" diam.		300	.053	L.F.	.07	2.17	3.59	5.83	7.30
0118	min charge , clay & soft sandstone, up to 10" diam.		3.20	5	Ea.	19.70	203	335	557.70	700
0120	Clay & soft sandstone, up to 10" diam.		300	.053	L.F.	.13	2.17	3.59	5.89	7.40
0128	min charge , hard clay & cobble, up to 8" diam.		1	16	Ea.	52.50	650	1,075	1,777.50	2,225
0130	Hard clay & cobble, up to 8" diam.		100	.160	L.F.	.53	6.50	10.75	17.78	22.50
0200	Medium equipment to 600', not to exceed 12" diam.									
0202	medium unit mobilization to site	B-82B	2	12	Ea.		465	675	1,140	1,450
0205	medium unit setup per drill		4	6			233	335	568	725
0209	minimum charge gravel, sand & silt, up to 12" diam.		3.20	7.500		9.85	292	420	721.85	915
0210	gravel, sand & silt, up to 12" diam.		350	.069	L.F.	.07	2.67	3.84	6.58	8.40
0218	min charge , clay & soft sandstone, up to 10" diam.		3.20	7.500	Ea.	19.70	292	420	731.70	925
0220	Clay & soft sandstone, up to 10" diam.		300	.080	L.F.	.13	3.11	4.48	7.72	9.85

33 05 Common Work Results for Utilities

33 05 23 – Trenchless Utility Installation

33 05 23.22 Directional Drilling		Crew	Daily Output	Labor-Hours	Unit	2013 Bare Costs Material	Labor	Equipment	Total	Total Incl O&P
0228	min charge , hard clay & cobble, up to 8" diam.	B-82B	1.78	13.483	Ea.	52.50	525	755	1,332.50	1,700
0230	Hard clay & cobble, up to 8" diam.		150	.160	L.F.	.53	6.20	8.95	15.68	19.95
0300	Large equipment to 1000', not to exceed 12" diam.									
0302	large unit mobilization to site	B-82C	2	12	Ea.		465	820	1,285	1,625
0305	large unit setup per drill		4	6			233	410	643	805
0309	minimum charge gravel, sand & silt, up to 12" diam.		4	6		9.85	233	410	652.85	815
0310	gravel, sand & silt, up to 12" diam.		400	.060	L.F.	.07	2.33	4.11	6.51	8.15
0318	min charge , clay & soft sandstone, up to 12" diam.		3.20	7.500	Ea.	19.70	292	515	826.70	1,025
0320	Clay & soft sandstone, up to 12" diam.		350	.069	L.F.	.13	2.67	4.70	7.50	9.35
0328	min charge , hard clay & cobble, up to 10" diam.		1.78	13.483	Ea.	52.50	525	925	1,502.50	1,875
0330	Rock & cobble, up to 10" diam.		150	.160	L.F.	.53	6.20	10.95	17.68	22
0400	Directional drilling, mud trailer per day	B-82D	1	8	Day		365	370	735	960
1000	Additional charges for mobilization may apply at some locations									

33 05 26 – Utility Identification

33 05 26.10 Utility Accessories		Crew	Daily Output	Labor-Hours	Unit	Material	Labor	Equipment	Total	Total Incl O&P
0010	**UTILITY ACCESSORIES**									
0400	Underground tape, detectable, reinforced, alum. foil core, 2"	1 Clab	150	.053	C.L.F.	2.37	1.89		4.26	5.50
0500	6"	"	140	.057	"	12	2.03		14.03	16.30

33 44 Storm Utility Water Drains

33 44 13 – Utility Area Drains

33 44 13.13 Catchbasins		Crew	Daily Output	Labor-Hours	Unit	Material	Labor	Equipment	Total	Total Incl O&P
0010	**CATCHBASINS**									
0011	Not including footing & excavation									
1600	Frames & covers, C.I., 24" square, 500 lb.	B-6	7.80	3.077	Ea.	320	120	47	487	590
1700	26" D shape, 600 lb.		7	3.429		475	133	52.50	660.50	785
1800	Light traffic, 18" diameter, 100 lb.		10	2.400		115	93.50	36.50	245	310
1900	24" diameter, 300 lb.		8.70	2.759		185	107	42	334	415
2000	36" diameter, 900 lb.		5.80	4.138		535	161	63.50	759.50	905
2100	Heavy traffic, 24" diameter, 400 lb.		7.80	3.077		230	120	47	397	490
2200	36" diameter, 1150 lb.		3	8		750	310	122	1,182	1,425
2300	Mass. State standard, 26" diameter, 475 lb.		7	3.429		250	133	52.50	435.50	535
2400	30" diameter, 620 lb.		7	3.429		325	133	52.50	510.50	620
2500	Watertight, 24" diameter, 350 lb.		7.80	3.077		300	120	47	467	565
2600	26" diameter, 500 lb.		7	3.429		400	133	52.50	585.50	700
2700	32" diameter, 575 lb.		6	4		800	156	61	1,017	1,175
2800	3 piece cover & frame, 10" deep,									
2900	1200 lb., for heavy equipment	B-6	3	8	Ea.	995	310	122	1,427	1,700
3000	Raised for paving 1-1/4" to 2" high									
3100	4 piece expansion ring									
3200	20" to 26" diameter	1 Clab	3	2.667	Ea.	148	94.50		242.50	310
3300	30" to 36" diameter	"	3	2.667	"	204	94.50		298.50	370
3320	Frames and covers, existing, raised for paving, 2", including									
3340	row of brick, concrete collar, up to 12" wide frame	B-6	18	1.333	Ea.	45	52	20.50	117.50	152
3360	20" to 26" wide frame		11	2.182		67	85	33.50	185.50	241
3380	30" to 36" wide frame		9	2.667		83	104	41	228	296
3400	Inverts, single channel brick	D-1	3	5.333		97	215		312	430
3500	Concrete		5	3.200		102	129		231	310
3600	Triple channel, brick		2	8		148	325		473	655
3700	Concrete		3	5.333		137	215		352	475

33 49 Storm Drainage Structures

33 49 13 – Storm Drainage Manholes, Frames, and Covers

33 49 13.10 Storm Drainage Manholes, Frames and Covers		Crew	Daily Output	Labor-Hours	Unit	2013 Bare Costs Material	Labor	Equipment	Total	Total Incl O&P
0010	**STORM DRAINAGE MANHOLES, FRAMES & COVERS**									
0020	Excludes footing, excavation, backfill (See line items for frame & cover)									
0050	Brick, 4′ inside diameter, 4′ deep	D-1	1	16	Ea.	395	645		1,040	1,400
0100	6′ deep		.70	22.857		550	925		1,475	2,000
0150	8′ deep		.50	32		705	1,300		2,005	2,725
0200	For depths over 8′, add		4	4	V.L.F.	84	162		246	340
0400	Concrete blocks (radial), 4′ I.D., 4′ deep		1.50	10.667	Ea.	360	430		790	1,050
0500	6′ deep		1	16		485	645		1,130	1,500
0600	8′ deep		.70	22.857		605	925		1,530	2,075
0700	For depths over 8′, add		5.50	2.909	V.L.F.	63.50	117		180.50	248
0800	Concrete, cast in place, 4′ x 4′, 8″ thick, 4′ deep	C-14H	2	24	Ea.	500	1,050	16.55	1,566.55	2,200
0900	6′ deep		1.50	32		720	1,425	22	2,167	3,000
1000	8′ deep		1	48		1,025	2,125	33	3,183	4,400
1100	For depths over 8′, add		8	6	V.L.F.	117	265	4.14	386.14	540
1110	Precast, 4′ I.D., 4′ deep	B-22	4.10	7.317	Ea.	730	305	50	1,085	1,325
1120	6′ deep		3	10		930	415	68.50	1,413.50	1,725
1130	8′ deep		2	15		1,075	625	103	1,803	2,250
1140	For depths over 8′, add		16	1.875	V.L.F.	123	78	12.85	213.85	269
1150	5′ I.D., 4′ deep	B-6	3	8	Ea.	1,525	310	122	1,957	2,275
1160	6′ deep		2	12		1,850	465	184	2,499	2,950
1170	8′ deep		1.50	16		2,275	620	245	3,140	3,725
1180	For depths over 8′, add		12	2	V.L.F.	272	78	30.50	380.50	450
1190	6′ I.D., 4′ deep		2	12	Ea.	2,150	465	184	2,799	3,300
1200	6′ deep		1.50	16		2,575	620	245	3,440	4,075
1210	8′ deep		1	24		3,175	935	365	4,475	5,300
1220	For depths over 8′, add		8	3	V.L.F.	365	117	46	528	635
1250	Slab tops, precast, 8″ thick									
1300	4′ diameter manhole	B-6	8	3	Ea.	246	117	46	409	500
1400	5′ diameter manhole		7.50	3.200		405	124	49	578	690
1500	6′ diameter manhole		7	3.429		610	133	52.50	795.50	930
3800	Steps, heavyweight cast iron, 7″ x 9″	1 Bric	40	.200		18	8.95		26.95	33.50
3900	8″ x 9″		40	.200		21.50	8.95		30.45	37.50
3928	12″ x 10-1/2″		40	.200		25	8.95		33.95	41
4000	Standard sizes, galvanized steel		40	.200		24	8.95		32.95	39.50
4100	Aluminum		40	.200		24	8.95		32.95	39.50
4150	Polyethylene		40	.200		22	8.95		30.95	37.50

33 71 Electrical Utility Transmission and Distribution

33 71 13 – Electrical Utility Towers

33 71 13.23 Steel Electrical Utility Towers		Crew	Daily Output	Labor-Hours	Unit	Material	Labor	Equipment	Total	Total Incl O&P
0010	**STEEL ELECTRICAL UTILITY TOWERS**									
0100	Excavation and backfill, earth	R-5	135.38	.650	C.Y.		29.50	11.10	40.60	57
0105	Rock		21.46	4.101	″		187	70	257	360
0200	Steel footings (grillage) in earth		3.91	22.506	Ton	2,125	1,025	385	3,535	4,325
0205	In rock		3.20	27.500	″	2,125	1,250	470	3,845	4,775
0290	See also Section 33 05 23.19									
0300	Rock anchors	R-5	5.87	14.991	Ea.	585	685	256	1,526	1,950
0400	Concrete foundations	″	12.85	6.848	C.Y.	129	315	117	561	740
0490	See also Section 03 30 53.40									
0500	Towers-material handling and spotting	R-7	22.56	2.128	Ton		78.50	9.80	88.30	131
0540	Steel tower erection	R-5	7.65	11.503		2,075	525	196	2,796	3,300

33 71 Electrical Utility Transmission and Distribution

33 71 13 – Electrical Utility Towers

33 71 13.23 Steel Electrical Utility Towers		Crew	Daily Output	Labor-Hours	Unit	2013 Bare Costs Material	Labor	Equipment	Total	Total Incl O&P
0550	Lace and box	R-5	7.10	12.394	Ton	2,075	565	212	2,852	3,400
0560	Painting total structure	↓	1.47	59.864	Ea.	450	2,725	1,025	4,200	5,750
0570	Disposal of surplus material	R-7	20.87	2.300	Mile		85	10.60	95.60	142
0600	Special towers-material handling and spotting	"	12.31	3.899	Ton		144	18	162	240
0640	Special steel structure erection	R-6	6.52	13.497		2,600	615	520	3,735	4,350
0650	Special steel lace and box	"	6.29	13.990	↓	2,600	640	540	3,780	4,400
0670	Disposal of surplus material	R-7	7.87	6.099	Mile		225	28	253	375

33 71 13.80 Transmission Line Right of Way		Crew	Daily Output	Labor-Hours	Unit	Material	Labor	Equipment	Total	Total Incl O&P
0010	**TRANSMISSION LINE RIGHT OF WAY**									
0100	Clearing right of way	B-87	6.67	5.997	Acre		270	415	685	865
0200	Restoration & seeding	B-10D	4	3	"	765	130	445	1,340	1,525

33 71 16 – Electrical Utility Poles

33 71 16.23 Steel Electrical Utility Poles		Crew	Daily Output	Labor-Hours	Unit	Material	Labor	Equipment	Total	Total Incl O&P
0010	**STEEL ELECTRICAL UTILITY POLES**									
6000	Digging holes in earth, average	R-5	25.14	3.500	Ea.		160	60	220	305
6010	In rock, average	"	4.51	19.512	"		890	335	1,225	1,725
6020	Formed plate pole structure									
6030	Material handling and spotting	R-7	2.40	20	Ea.		740	92	832	1,225
6040	Erect steel plate pole	R-5	1.95	45.128		9,025	2,050	770	11,845	13,900
6050	Guys, anchors and hardware for pole, in earth		7.04	12.500		545	570	213	1,328	1,700
6060	In rock	↓	17.96	4.900	↓	655	224	83.50	962.50	1,150
6070	Foundations for line poles									
6080	Excavation, in earth	R-5	135.38	.650	C.Y.		29.50	11.10	40.60	57
6090	In rock		20	4.400			201	75	276	390
6110	Concrete foundations	↓	11	8	↓	129	365	137	631	840

33 71 16.33 Wood Electrical Utility Poles		Crew	Daily Output	Labor-Hours	Unit	Material	Labor	Equipment	Total	Total Incl O&P
0010	**WOOD ELECTRICAL UTILITY POLES**									
0011	Excludes excavation, backfill and cast-in-place concrete									
6200	Electric & tel sitework, 20′ high, treated wd., see Section 26 56 13.10	R-3	3.10	6.452	Ea.	273	335	44.50	652.50	850
6400	25′ high R337119-30		2.90	6.897		330	360	47.50	737.50	945
6600	30′ high		2.60	7.692		430	400	53	883	1,125
6800	35′ high		2.40	8.333		535	430	57	1,022	1,300
7000	40′ high		2.30	8.696		755	450	59.50	1,264.50	1,575
7200	45′ high	↓	1.70	11.765	↓	910	610	80.50	1,600.50	2,000
7400	Cross arms with hardware & insulators									
7600	4′ long	1 Elec	2.50	3.200	Ea.	147	168		315	415
7800	5′ long		2.40	3.333		169	175		344	445
8000	6′ long	↓	2.20	3.636	↓	194	191		385	500
9000	Disposal of pole & hardware surplus material	R-7	20.87	2.300	Mile		85	10.60	95.60	142
9100	Disposal of crossarms & hardware surplus material	"	40	1.200	"		44.50	5.55	50.05	74

33 71 19 – Electrical Underground Ducts and Manholes

33 71 19.15 Underground Ducts and Manholes		Crew	Daily Output	Labor-Hours	Unit	Material	Labor	Equipment	Total	Total Incl O&P
0010	**UNDERGROUND DUCTS AND MANHOLES**									
0011	Not incl. excavation, backfill and concrete, in slab or duct bank									
1000	Direct burial									
1010	PVC, schedule 40, w/coupling, 1/2″ diameter	1 Elec	340	.024	L.F.	.48	1.23		1.71	2.37
1020	3/4″ diameter		290	.028		.65	1.45		2.10	2.88
1030	1″ diameter		260	.031		.95	1.61		2.56	3.46
1040	1-1/2″ diameter		210	.038		1.56	2		3.56	4.71
1050	2″ diameter	↓	180	.044		2.05	2.33		4.38	5.75
1060	3″ diameter	2 Elec	240	.067		3.91	3.49		7.40	9.55

33 71 19 – Electrical Underground Ducts and Manholes

33 71 19.15 Underground Ducts and Manholes		Crew	Daily Output	Labor-Hours	Unit	Material	2013 Bare Costs Labor	Equipment	Total	Total Incl O&P
1070	4" diameter	2 Elec	160	.100	L.F.	5.50	5.25		10.75	13.90
1080	5" diameter		120	.133		7.95	7		14.95	19.20
1090	6" diameter		90	.178		10.55	9.30		19.85	25.50
1110	Elbows, 1/2" diameter	1 Elec	48	.167	Ea.	1.24	8.75		9.99	14.40
1120	3/4" diameter		38	.211		1.22	11.05		12.27	17.85
1130	1" diameter		32	.250		1.85	13.10		14.95	21.50
1140	1-1/2" diameter		21	.381		3.67	19.95		23.62	34
1150	2" diameter		16	.500		4.97	26		30.97	44.50
1160	3" diameter		12	.667		14.95	35		49.95	69
1170	4" diameter		9	.889		24	46.50		70.50	95.50
1180	5" diameter		8	1		39	52.50		91.50	122
1190	6" diameter		5	1.600		89.50	84		173.50	224
1210	Adapters, 1/2" diameter		52	.154		.35	8.05		8.40	12.45
1220	3/4" diameter		43	.186		.58	9.75		10.33	15.25
1230	1" diameter		39	.205		.77	10.75		11.52	16.95
1240	1-1/2" diameter		35	.229		1.34	12		13.34	19.35
1250	2" diameter		26	.308		1.69	16.10		17.79	26
1260	3" diameter		20	.400		4.11	21		25.11	36
1270	4" diameter		14	.571		7.25	30		37.25	53
1280	5" diameter		12	.667		15.55	35		50.55	69.50
1290	6" diameter		9	.889		26	46.50		72.50	98
1340	Bell end & cap, 1-1/2" diameter		35	.229		9.65	12		21.65	28.50
1350	Bell end & plug, 2" diameter		26	.308		12.80	16.10		28.90	38
1360	3" diameter		20	.400		15.55	21		36.55	48.50
1370	4" diameter		14	.571		19.65	30		49.65	66.50
1380	5" diameter		12	.667		22	35		57	77
1390	6" diameter		9	.889		26.50	46.50		73	98.50
1450	Base spacer, 2" diameter		56	.143		1.53	7.50		9.03	12.90
1460	3" diameter		46	.174		1.60	9.10		10.70	15.40
1470	4" diameter		41	.195		1.62	10.20		11.82	17.10
1480	5" diameter		37	.216		2.03	11.35		13.38	19.20
1490	6" diameter		34	.235		2.78	12.35		15.13	21.50
1550	Intermediate spacer, 2" diameter		60	.133		1.20	7		8.20	11.75
1560	3" diameter		46	.174		1.69	9.10		10.79	15.50
1570	4" diameter		41	.195		1.55	10.20		11.75	17
1580	5" diameter		37	.216		1.83	11.35		13.18	18.95
1590	6" diameter		34	.235		2.68	12.35		15.03	21.50
4010	PVC, schedule 80, w/coupling, 1/2" diameter		215	.037	L.F.	1.15	1.95		3.10	4.19
4020	3/4" diameter		180	.044		1.54	2.33		3.87	5.15
4030	1" diameter		145	.055		2.21	2.89		5.10	6.75
4040	1-1/2" diameter		120	.067		3.67	3.49		7.16	9.30
4050	2" diameter		100	.080		5.05	4.19		9.24	11.80
4060	3" diameter	2 Elec	130	.123		10.35	6.45		16.80	21
4070	4" diameter		90	.178		15	9.30		24.30	30.50
4080	5" diameter		70	.229		21.50	12		33.50	41.50
4090	6" diameter		50	.320		27	16.75		43.75	54.50
4110	Elbows, 1/2" diameter	1 Elec	29	.276	Ea.	1.65	14.45		16.10	23.50
4120	3/4" diameter		23	.348		3.20	18.25		21.45	31
4130	1" diameter		20	.400		3.98	21		24.98	36
4140	1-1/2" diameter		16	.500		8.65	26		34.65	48.50
4150	2" diameter		12	.667		10.35	35		45.35	64
4160	3" diameter		9	.889		29	46.50		75.50	102
4170	4" diameter		7	1.143		58.50	60		118.50	154

33 71 19.15 Underground Ducts and Manholes		Crew	Daily Output	Labor-Hours	Unit	Material	2013 Bare Costs Labor	Equipment	Total	Total Incl O&P
4180	5″ diameter	1 Elec	6	1.333	Ea.	244	70		314	375
4190	6″ diameter		4	2		289	105		394	475
4210	Adapter, 1/2″ diameter		39	.205		.35	10.75		11.10	16.50
4220	3/4″ diameter		33	.242		.58	12.70		13.28	19.65
4230	1″ diameter		29	.276		.77	14.45		15.22	22.50
4240	1-1/2″ diameter		26	.308		1.34	16.10		17.44	25.50
4250	2″ diameter		23	.348		1.69	18.25		19.94	29.50
4260	3″ diameter		18	.444		4.11	23.50		27.61	39.50
4270	4″ diameter		13	.615		7.25	32.50		39.75	56.50
4280	5″ diameter		11	.727		15.55	38		53.55	74
4290	6″ diameter		8	1		26	52.50		78.50	107
4310	Bell end & cap, 1-1/2″ diameter		26	.308		9.65	16.10		25.75	34.50
4320	Bell end & plug, 2″ diameter		23	.348		12.80	18.25		31.05	41.50
4330	3″ diameter		18	.444		15.55	23.50		39.05	52
4340	4″ diameter		13	.615		19.65	32.50		52.15	70
4350	5″ diameter		11	.727		22	38		60	81.50
4360	6″ diameter		8	1		26.50	52.50		79	108
4370	Base spacer, 2″ diameter		42	.190		1.53	10		11.53	16.65
4380	3″ diameter		33	.242		1.60	12.70		14.30	21
4390	4″ diameter		29	.276		1.62	14.45		16.07	23.50
4400	5″ diameter		26	.308		2.03	16.10		18.13	26
4410	6″ diameter		25	.320		2.78	16.75		19.53	28
4420	Intermediate spacer, 2″ diameter		45	.178		1.20	9.30		10.50	15.25
4430	3″ diameter		34	.235		1.69	12.35		14.04	20.50
4440	4″ diameter		31	.258		1.55	13.50		15.05	21.50
4450	5″ diameter		28	.286		1.83	14.95		16.78	24.50
4460	6″ diameter		25	.320		2.68	16.75		19.43	28

33 71 19.17 Electric and Telephone Underground		Crew	Daily Output	Labor-Hours	Unit	Material	Labor	Equipment	Total	Total Incl O&P
0010	**ELECTRIC AND TELEPHONE UNDERGROUND**									
0011	Not including excavation									
0200	backfill and cast in place concrete									
0250	For bedding, see Section 31 23 23.16 R337119-30									
0400	Hand holes, precast concrete, with concrete cover									
0600	2′ x 2′ x 3′ deep	R-3	2.40	8.333	Ea.	410	430	57	897	1,175
0800	3′ x 3′ x 3′ deep		1.90	10.526		530	545	72	1,147	1,475
1000	4′ x 4′ x 4′ deep		1.40	14.286		1,200	740	98	2,038	2,500
1200	Manholes, precast with iron racks & pulling irons, C.I. frame									
1400	and cover, 4′ x 6′ x 7′ deep	B-13	2	28	Ea.	5,800	1,075	365	7,240	8,450
1600	6′ x 8′ x 7′ deep		1.90	29.474		6,525	1,150	385	8,060	9,350
1800	6′ x 10′ x 7′ deep		1.80	31.111		7,325	1,200	405	8,930	10,300
4200	Underground duct, banks ready for concrete fill, min. of 7.5″									
4400	between conduits, center to center									
4580	PVC, type EB, 1 @ 2″ diameter	2 Elec	480	.033	L.F.	.46	1.75		2.21	3.12
4600	2 @ 2″ diameter		240	.067		.93	3.49		4.42	6.25
4800	4 @ 2″ diameter		120	.133		1.85	7		8.85	12.50
4900	1 @ 3″ diameter		400	.040		.60	2.10		2.70	3.80
5000	2 @ 3″ diameter		200	.080		1.20	4.19		5.39	7.55
5200	4 @ 3″ diameter		100	.160		2.40	8.40		10.80	15.20
5300	1 @ 4″ diameter		320	.050		.88	2.62		3.50	4.89
5400	2 @ 4″ diameter		160	.100		1.76	5.25		7.01	9.80
5600	4 @ 4″ diameter		80	.200		3.53	10.50		14.03	19.60
5800	6 @ 4″ diameter		54	.296		5.30	15.55		20.85	29

33 71 Electrical Utility Transmission and Distribution

33 71 19 – Electrical Underground Ducts and Manholes

33 71 19.17 Electric and Telephone Underground		Crew	Daily Output	Labor-Hours	Unit	Material	Labor	Equipment	Total	Total Incl O&P
						2013 Bare Costs				
5810	1 @ 5″ diameter	2 Elec	260	.062	L.F.	1.29	3.22		4.51	6.25
5820	2 @ 5″ diameter		130	.123		2.59	6.45		9.04	12.50
5840	4 @ 5″ diameter		70	.229		5.15	12		17.15	23.50
5860	6 @ 5″ diameter		50	.320		7.75	16.75		24.50	33.50
5870	1 @ 6″ diameter		200	.080		1.84	4.19		6.03	8.25
5880	2 @ 6″ diameter		100	.160		3.68	8.40		12.08	16.60
5900	4 @ 6″ diameter		50	.320		7.35	16.75		24.10	33
5920	6 @ 6″ diameter		30	.533		11.05	28		39.05	54
6200	Rigid galvanized steel, 2 @ 2″ diameter		180	.089		13.30	4.66		17.96	21.50
6400	4 @ 2″ diameter		90	.178		26.50	9.30		35.80	43.50
6800	2 @ 3″ diameter		100	.160		32.50	8.40		40.90	48
7000	4 @ 3″ diameter		50	.320		64.50	16.75		81.25	96
7200	2 @ 4″ diameter		70	.229		47	12		59	69.50
7400	4 @ 4″ diameter		34	.471		93.50	24.50		118	140
7600	6 @ 4″ diameter		22	.727		140	38		178	211
7620	2 @ 5″ diameter		60	.267		95.50	13.95		109.45	126
7640	4 @ 5″ diameter		30	.533		191	28		219	252
7660	6 @ 5″ diameter		18	.889		286	46.50		332.50	385
7680	2 @ 6″ diameter		40	.400		139	21		160	185
7700	4 @ 6″ diameter		20	.800		279	42		321	370
7720	6 @ 6″ diameter		14	1.143		420	60		480	550
7800	For Cast-in-place Concrete - Add									
7810	Under 1 C.Y.	C-6	16	3	C.Y.	155	111	4.14	270.14	345
7820	1 C.Y. - 5 C.Y.		19.20	2.500		141	92.50	3.45	236.95	300
7830	Over 5 C.Y.		24	2		116	74	2.76	192.76	244
7850	For Reinforcing Rods - Add									
7860	#4 to #7	2 Rodm	1.10	14.545	Ton	1,000	725		1,725	2,250
7870	#8 to #14	″	1.50	10.667	″	1,000	530		1,530	1,950
8000	Fittings, PVC type EB, elbow, 2″ diameter	1 Elec	16	.500	Ea.	13.60	26		39.60	54
8200	3″ diameter		14	.571		17.60	30		47.60	64.50
8400	4″ diameter		12	.667		24.50	35		59.50	79
8420	5″ diameter		10	.800		130	42		172	206
8440	6″ diameter		9	.889		148	46.50		194.50	233
8500	Coupling, 2″ diameter					.60			.60	.66
8600	3″ diameter					2.75			2.75	3.03
8700	4″ diameter					3.01			3.01	3.31
8720	5″ diameter					5.70			5.70	6.25
8740	6″ diameter					19.20			19.20	21
8800	Adapter, 2″ diameter	1 Elec	26	.308		.88	16.10		16.98	25
9000	3″ diameter		20	.400		2.46	21		23.46	34
9200	4″ diameter		16	.500		3.63	26		29.63	43
9220	5″ diameter		13	.615		8.75	32.50		41.25	58
9240	6″ diameter		10	.800		11.35	42		53.35	75
9400	End bell, 2″ diameter		16	.500		1.12	26		27.12	40
9600	3″ diameter		14	.571		5.70	30		35.70	51.50
9800	4″ diameter		12	.667		3.65	35		38.65	56.50
9810	5″ diameter		10	.800		11.15	42		53.15	75
9820	6″ diameter		8	1		56.50	52.50		109	141
9830	5° angle coupling, 2″ diameter		26	.308		16.75	16.10		32.85	42.50
9840	3″ diameter		20	.400		25	21		46	59
9850	4″ diameter		16	.500		13	26		39	53.50
9860	5″ diameter		13	.615		12.60	32.50		45.10	62.50
9870	6″ diameter		10	.800		19.50	42		61.50	84

33 71 Electrical Utility Transmission and Distribution

33 71 19 – Electrical Underground Ducts and Manholes

33 71 19.17	Electric and Telephone Underground	Crew	Daily Output	Labor-Hours	Unit	Material	2013 Bare Costs Labor	Equipment	Total	Total Incl O&P
9880	Expansion joint, 2" diameter	1 Elec	16	.500	Ea.	29	26		55	70.50
9890	3" diameter		18	.444		54.50	23.50		78	95
9900	4" diameter		12	.667		73	35		108	133
9910	5" diameter		10	.800		97.50	42		139.50	170
9920	6" diameter		8	1		122	52.50		174.50	213
9930	Heat bender, 2" diameter					465			465	510
9940	6" diameter					1,575			1,575	1,725
9950	Cement, quart					23			23	25.50
9960	Nylon polyethylene pull rope, 1/4"	2 Elec	2000	.008	L.F.	.21	.42		.63	.86

33 71 23 – Insulators and Fittings

33 71 23.16 Post Insulators

		Crew	Daily Output	Labor-Hours	Unit	Material	Labor	Equipment	Total	Total Incl O&P
0010	**POST INSULATORS**									
7400	Insulators, pedestal type	R-11	112	.500	Ea.		25	7.75	32.75	45.50
7490	See also line 33 71 39.13 1000									

33 71 26 – Transmission and Distribution Equipment

33 71 26.13 Capacitor Banks

		Crew	Daily Output	Labor-Hours	Unit	Material	Labor	Equipment	Total	Total Incl O&P
0010	**CAPACITOR BANKS**									
1300	Station capacitors									
1350	Synchronous, 13 to 26 kV	R-11	3.11	18.006	MVAR	6,650	890	279	7,819	8,975
1360	46 kV		3.33	16.817		8,500	835	260	9,595	10,900
1370	69 kV		3.81	14.698		8,350	730	228	9,308	10,600
1380	161 kV		6.51	8.602		7,800	425	133	8,358	9,375
1390	500 kV		10.37	5.400		6,800	268	83.50	7,151.50	7,975
1450	Static, 13 to 26 kV		3.11	18.006		5,650	890	279	6,819	7,850
1460	46 kV		3.01	18.605		7,125	920	288	8,333	9,550
1470	69 kV		3.81	14.698		6,925	730	228	7,883	8,975
1480	161 kV		6.51	8.602		6,425	425	133	6,983	7,850
1490	500 kV		10.37	5.400		5,875	268	83.50	6,226.50	6,950
1600	Voltage regulators, 13 to 26 kV		.75	74.667	Ea.	252,500	3,700	1,150	257,350	285,000

33 71 26.23 Current Transformers

		Crew	Daily Output	Labor-Hours	Unit	Material	Labor	Equipment	Total	Total Incl O&P
0010	**CURRENT TRANSFORMERS**									
4050	Current transformers, 13 to 26 kV	R-11	14	4	Ea.	3,100	198	62	3,360	3,775
4060	46 kV		9.33	6.002		9,075	297	93	9,465	10,500
4070	69 kV		7	8		9,400	395	124	9,919	11,000
4080	161 kV		1.87	29.947		30,500	1,475	465	32,440	36,200

33 71 26.26 Potential Transformers

		Crew	Daily Output	Labor-Hours	Unit	Material	Labor	Equipment	Total	Total Incl O&P
0010	**POTENTIAL TRANSFORMERS**									
4100	Potential transformers, 13 to 26 kV	R-11	11.20	5	Ea.	4,450	248	77.50	4,775.50	5,325
4110	46 kV		8	7		9,125	345	108	9,578	10,600
4120	69 kV		6.22	9.003		9,675	445	139	10,259	11,400
4130	161 kV		2.24	25		20,900	1,250	385	22,535	25,300
4140	500 kV		1.40	40		62,500	1,975	620	65,095	72,000

33 71 39 – High-Voltage Wiring

33 71 39.13 Overhead High-Voltage Wiring

		Crew	Daily Output	Labor-Hours	Unit	Material	Labor	Equipment	Total	Total Incl O&P
0010	**OVERHEAD HIGH-VOLTAGE WIRING** R337116-60									
0100	Conductors, primary circuits									
0110	Material handling and spotting	R-5	9.78	8.998	W.Mile		410	154	564	790
0120	For river crossing, add		11	8			365	137	502	700
0150	Conductors, per wire, 210 to 636 kcmil		1.96	44.898		10,300	2,050	765	13,115	15,200
0160	795 to 954 kcmil		1.87	47.059		20,500	2,150	805	23,455	26,600

33 71 39.13 Overhead High-Voltage Wiring		Crew	Daily Output	Labor-Hours	Unit	2013 Bare Costs Material	Labor	Equipment	Total	Total Incl O&P
0170	1000 to 1600 kcmil	R-5	1.47	59.864	W.Mile	35,100	2,725	1,025	38,850	43,900
0180	Over 1600 kcmil		1.35	65.185		49,100	2,975	1,125	53,200	59,500
0200	For river crossing, add, 210 to 636 kcmil		1.24	70.968			3,250	1,200	4,450	6,225
0220	795 to 954 kcmil		1.09	80.734			3,675	1,375	5,050	7,075
0230	1000 to 1600 kcmil		.97	90.722			4,150	1,550	5,700	7,950
0240	Over 1600 kcmil		.87	101			4,625	1,725	6,350	8,875
0300	Joints and dead ends	R-8	6	8	Ea.	1,400	370	62.50	1,832.50	2,175
0400	Sagging	R-5	7.33	12.001	W.Mile		550	205	755	1,050
0500	Clipping, per structure, 69 kV	R-10	9.60	5	Ea.		247	63.50	310.50	440
0510	161 kV		5.33	9.006			445	114	559	795
0520	345 to 500 kV		2.53	18.972			935	241	1,176	1,675
0600	Make and install jumpers, per structure, 69 kV	R-8	3.20	15		380	695	117	1,192	1,600
0620	161 kV		1.20	40		765	1,850	315	2,930	3,975
0640	345 to 500 kV		.32	150		1,275	6,950	1,175	9,400	13,200
0700	Spacers	R-10	68.57	.700		76	34.50	8.90	119.40	146
0720	For river crossings, add	"	60	.800			39.50	10.15	49.65	70.50
0800	Installing pulling line (500 kV only)	R-9	1.45	44.138	W.Mile	755	1,900	259	2,914	4,000
0810	Disposal of surplus material, high voltage conductors	R-7	6.96	6.897	Mile		255	32	287	425
0820	With trailer mounted reel stands	"	13.71	3.501	"		129	16.15	145.15	216
0900	Insulators and hardware, primary circuits									
0920	Material handling and spotting, 69 kV	R-7	480	.100	Ea.		3.69	.46	4.15	6.15
0930	161 kV		685.71	.070			2.59	.32	2.91	4.31
0950	345 to 500 kV		960	.050			1.85	.23	2.08	3.08
1000	Disk insulators, 69 kV	R-5	880	.100		77.50	4.57	1.71	83.78	94.50
1020	161 kV		977.78	.090		89	4.11	1.54	94.65	105
1040	345 to 500 kV		1100	.080		89	3.65	1.37	94.02	105
1060	See Section 33 71 23.16 for pin or pedestal insulator									
1100	Install disk insulator at river crossing, add									
1110	69 kV	R-5	586.67	.150	Ea.		6.85	2.56	9.41	13.15
1120	161 kV		880	.100			4.57	1.71	6.28	8.80
1140	345 to 500 kV		880	.100			4.57	1.71	6.28	8.80
1150	Disposal of surplus material, high voltage insulators	R-7	41.74	1.150	Mile		42.50	5.30	47.80	71
1300	Overhead ground wire installation									
1320	Material handling and spotting	R-7	5.65	8.496	W.Mile		315	39	354	525
1340	Overhead ground wire	R-5	1.76	50		2,975	2,275	855	6,105	7,675
1350	At river crossing, add		1.17	75.214			3,425	1,275	4,700	6,600
1360	Disposal of surplus material, grounding wire		41.74	2.108	Mile		96.50	36	132.50	185
1400	Installing conductors, underbuilt circuits									
1420	Material handling and spotting	R-7	5.65	8.496	W.Mile		315	39	354	525
1440	Conductors, per wire, 210 to 636 kcmil	R-5	1.96	44.898		10,300	2,050	765	13,115	15,200
1450	795 to 954 kcmil		1.87	47.059		20,500	2,150	805	23,455	26,600
1460	1000 to 1600 kcmil		1.47	59.864		35,100	2,725	1,025	38,850	43,900
1470	Over 1600 kcmil		1.35	65.185		49,100	2,975	1,125	53,200	59,500
1500	Joints and dead ends	R-8	6	8	Ea.	1,400	370	62.50	1,832.50	2,175
1550	Sagging	R-5	8.80	10	W.Mile		455	171	626	880
1600	Clipping, per structure, 69 kV	R-10	9.60	5	Ea.		247	63.50	310.50	440
1620	161 kV		5.33	9.006			445	114	559	795
1640	345 to 500 kV		2.53	18.972			935	241	1,176	1,675
1700	Making and installing jumpers, per structure, 69 kV	R-8	5.87	8.177		380	380	64	824	1,050
1720	161 kV		.96	50		765	2,325	390	3,480	4,750
1740	345 to 500 kV		.32	150		1,275	6,950	1,175	9,400	13,200
1800	Spacers	R-10	96	.500		76	24.50	6.35	106.85	128
1810	Disposal of surplus material, conductors & hardware	R-7	6.96	6.897	Mile		255	32	287	425

33 71 Electrical Utility Transmission and Distribution

33 71 39 – High-Voltage Wiring

33 71 39.13 Overhead High-Voltage Wiring		Crew	Daily Output	Labor-Hours	Unit	2013 Bare Costs Material	Labor	Equipment	Total	Total Incl O&P
2000	Insulators and hardware for underbuilt circuits									
2100	Material handling and spotting	R-7	1200	.040	Ea.		1.48	.18	1.66	2.46
2150	Disk insulators, 69 kV	R-8	600	.080		77.50	3.70	.63	81.83	92
2160	161 kV		686	.070		89	3.24	.55	92.79	103
2170	345 to 500 kV	↓	800	.060	↓	89	2.78	.47	92.25	102
2180	Disposal of surplus material, insulators & hardware	R-7	41.74	1.150	Mile		42.50	5.30	47.80	71
2300	Sectionalizing switches, 69 kV	R-5	1.26	69.841	Ea.	20,000	3,200	1,200	24,400	28,100
2310	161 kV		.80	110		22,600	5,025	1,875	29,500	34,600
2500	Protective devices	↓	5.50	16	↓	6,550	730	273	7,553	8,600
2600	Clearance poles, 8 poles per mile									
2650	In earth, 69 kV	R-5	1.16	75.862	Mile	5,400	3,475	1,300	10,175	12,600
2660	161 kV	"	.64	137		8,850	6,275	2,350	17,475	21,800
2670	345 to 500 kV	R-6	.48	183		10,600	8,375	7,050	26,025	32,100
2800	In rock, 69 kV	R-5	.69	127		5,400	5,825	2,175	13,400	17,100
2820	161 kV	"	.35	251		8,850	11,500	4,300	24,650	31,800
2840	345 to 500 kV	R-6	.24	366	↓	10,600	16,700	14,100	41,400	52,500

33 72 Utility Substations

33 72 26 – Substation Bus Assemblies

33 72 26.13 Aluminum Substation Bus Assemblies

		Crew	Daily Output	Labor-Hours	Unit	Material	Labor	Equipment	Total	Total Incl O&P
0010	**ALUMINUM SUBSTATION BUS ASSEMBLIES**									
7300	Bus	R-11	590	.095	Lb.	2.73	4.70	1.47	8.90	11.65

33 72 33 – Control House Equipment

33 72 33.33 Raceway/Boxes for Utility Substations

		Crew	Daily Output	Labor-Hours	Unit	Material	Labor	Equipment	Total	Total Incl O&P
0010	**RACEWAY/BOXES FOR UTILITY SUBSTATIONS**									
7000	Conduit, conductors, and insulators									
7100	Conduit, metallic	R-11	560	.100	Lb.	2.50	4.95	1.55	9	11.90
7110	Non-metallic	"	800	.070	"	1.75	3.47	1.08	6.30	8.30
7190	See Section 26 05 33									
7200	Wire and cable	R-11	700	.080	Lb.	8.20	3.96	1.24	13.40	16.30
7290	See Section 26 05 19									

33 72 33.36 Cable Trays for Utility Substations

		Crew	Daily Output	Labor-Hours	Unit	Material	Labor	Equipment	Total	Total Incl O&P
0010	**CABLE TRAYS FOR UTILITY SUBSTATIONS**									
7700	Cable tray	R-11	40	1.400	L.F.	23	69.50	21.50	114	154
7790	See Section 26 05 36									

33 72 33.43 Substation Backup Batteries

		Crew	Daily Output	Labor-Hours	Unit	Material	Labor	Equipment	Total	Total Incl O&P
0010	**SUBSTATION BACKUP BATTERIES**									
9120	Battery chargers	R-11	11.20	5	Ea.	3,775	248	77.50	4,100.50	4,625
9200	Control batteries	"	14	4	K.A.H.	87	198	62	347	460

33 72 33.46 Substation Converter Stations

		Crew	Daily Output	Labor-Hours	Unit	Material	Labor	Equipment	Total	Total Incl O&P
0010	**SUBSTATION CONVERTER STATIONS**									
9000	Station service equipment									
9100	Conversion equipment									
9110	Station service transformers	R-11	5.60	10	Ea.	91,500	495	155	92,150	101,500

33 73 Utility Transformers

33 73 23 – Dry-Type Utility Transformers

	33 73 23.20 Primary Transformers	Crew	Daily Output	Labor-Hours	Unit	2013 Bare Costs Material	Labor	Equipment	Total	Total Incl O&P
0010	**PRIMARY TRANSFORMERS**									
1000	Main conversion equipment									
1050	Power transformers, 13 to 26 kV	R-11	1.72	32.558	MVA	23,900	1,625	505	26,030	29,300
1060	46 kV		3.50	16		22,400	795	248	23,443	26,100
1070	69 kV		3.11	18.006		19,100	890	279	20,269	22,700
1080	110 kV		3.29	17.021		18,100	845	264	19,209	21,600
1090	161 kV		4.31	12.993		16,800	645	201	17,646	19,700
1100	500 kV		7	8		16,800	395	124	17,319	19,100
1200	Grounding transformers		3.11	18.006	Ea.	105,500	890	279	106,669	117,500

33 75 High-Voltage Switchgear and Protection Devices

33 75 13 – Air High-Voltage Circuit Breaker

33 75 13.13 High-Voltage Circuit Breaker, Air

		Crew	Daily Output	Labor-Hours	Unit	Material	Labor	Equipment	Total	Total Incl O&P
0010	**HIGH-VOLTAGE CIRCUIT BREAKER, AIR**									
2100	Air circuit breakers, 13 to 26 kV	R-11	.56	100	Ea.	59,000	4,950	1,550	65,500	74,000
2110	161 kV	"	.24	233	"	253,500	11,600	3,625	268,725	300,500

33 75 16 – Oil High-Voltage Circuit Breaker

33 75 16.13 Oil Circuit Breaker

		Crew	Daily Output	Labor-Hours	Unit	Material	Labor	Equipment	Total	Total Incl O&P
0010	**OIL CIRCUIT BREAKER**									
2000	Power circuit breakers									
2050	Oil circuit breakers, 13 to 26 kV	R-11	1.12	50	Ea.	56,500	2,475	775	59,750	66,500
2060	46 kV		.75	74.667		82,500	3,700	1,150	87,350	97,500
2070	69 kV		.45	124		191,500	6,175	1,925	199,600	222,000
2080	161 kV		.16	350		292,500	17,300	5,425	315,225	353,500
2090	500 kV		.06	933		1,094,500	46,200	14,500	1,155,200	1,289,500

33 75 19 – Gas High-Voltage Circuit Breaker

33 75 19.13 Gas Circuit Breaker

		Crew	Daily Output	Labor-Hours	Unit	Material	Labor	Equipment	Total	Total Incl O&P
0010	**GAS CIRCUIT BREAKER**									
2000	Power circuit breakers									
2150	Gas circuit breakers, 13 to 26 kV	R-11	.56	100	Ea.	221,500	4,950	1,550	228,000	252,500
2160	161 kV		.08	700		291,000	34,700	10,800	336,500	384,000
2170	500 kV		.04	1400		1,024,500	69,500	21,700	1,115,700	1,255,000

33 75 23 – Vacuum High-Voltage Circuit Breaker

33 75 23.13 Vacuum Circuit Breaker

		Crew	Daily Output	Labor-Hours	Unit	Material	Labor	Equipment	Total	Total Incl O&P
0010	**VACUUM CIRCUIT BREAKER**									
2000	Power circuit breakers									
2200	Vacuum circuit breakers, 13 to 26 kV	R-11	.56	100	Ea.	48,300	4,950	1,550	54,800	62,000

33 75 36 – High-Voltage Utility Fuses

33 75 36.13 Fuses

		Crew	Daily Output	Labor-Hours	Unit	Material	Labor	Equipment	Total	Total Incl O&P
0010	**FUSES**									
8000	Protective equipment									
8250	Fuses, 13 to 26 kV	R-11	18.67	2.999	Ea.	2,050	149	46.50	2,245.50	2,525
8260	46 kV		11.20	5		2,325	248	77.50	2,650.50	3,025
8270	69 kV		8	7		2,425	345	108	2,878	3,325
8280	161 kV		4.67	11.991		3,075	595	186	3,856	4,475

33 75 High-Voltage Switchgear and Protection Devices

33 75 39 – High-Voltage Surge Arresters

33 75 39.13 Surge Arresters		Crew	Daily Output	Labor-Hours	Unit	Material	Labor	Equipment	Total	Total Incl O&P
						2013 Bare Costs				
0010	**SURGE ARRESTERS**									
8000	Protective equipment									
8050	Lightning arresters, 13 to 26 kV	R-11	18.67	2.999	Ea.	1,525	149	46.50	1,720.50	1,950
8060	46 kV		14	4		4,125	198	62	4,385	4,925
8070	69 kV		11.20	5		5,300	248	77.50	5,625.50	6,275
8080	161 kV		5.60	10		7,300	495	155	7,950	8,975
8090	500 kV	↓	1.40	40	↓	24,800	1,975	620	27,395	30,900

33 75 43 – Shunt Reactors

33 75 43.13 Reactors		Crew	Daily Output	Labor-Hours	Unit	Material	Labor	Equipment	Total	Total Incl O&P
0010	**REACTORS**									
8000	Protective equipment									
8150	Reactors and resistors, 13 to 26 kV	R-11	28	2	Ea.	3,450	99	31	3,580	3,975
8160	46 kV		4.31	12.993		10,400	645	201	11,246	12,700
8170	69 kV		2.80	20		17,000	990	310	18,300	20,500
8180	161 kV		2.24	25		19,400	1,250	385	21,035	23,700
8190	500 kV	↓	.08	700	↓	78,000	34,700	10,800	123,500	149,500

33 75 53 – High-Voltage Switches

33 75 53.13 Switches		Crew	Daily Output	Labor-Hours	Unit	Material	Labor	Equipment	Total	Total Incl O&P
0010	**SWITCHES**									
3000	Disconnecting switches									
3050	Gang operated switches									
3060	Manual operation, 13 to 26 kV	R-11	1.65	33.939	Ea.	13,600	1,675	525	15,800	18,000
3070	46 kV		1.12	50		22,600	2,475	775	25,850	29,400
3080	69 kV		.80	70		25,300	3,475	1,075	29,850	34,300
3090	161 kV		.56	100		30,500	4,950	1,550	37,000	42,700
3100	500 kV		.14	400		83,000	19,800	6,200	109,000	128,000
3110	Motor operation, 161 kV		.51	109		45,900	5,450	1,700	53,050	60,500
3120	500 kV		.28	200		127,000	9,900	3,100	140,000	158,000
3250	Circuit switches, 161 kV	↓	.41	136	↓	91,500	6,775	2,125	100,400	113,000
3300	Single pole switches									
3350	Disconnecting switches, 13 to 26 kV	R-11	28	2	Ea.	11,900	99	31	12,030	13,300
3360	46 kV		8	7		21,000	345	108	21,453	23,700
3370	69 kV		5.60	10		23,300	495	155	23,950	26,600
3380	161 kV		2.80	20		89,000	990	310	90,300	99,500
3390	500 kV		.22	254		253,000	12,600	3,950	269,550	301,500
3450	Grounding switches, 46 kV		5.60	10		32,100	495	155	32,750	36,200
3460	69 kV		3.73	15.013		33,300	745	233	34,278	38,100
3470	161 kV		2.24	25		35,300	1,250	385	36,935	41,100
3480	500 kV	↓	.62	90.323	↓	43,800	4,475	1,400	49,675	56,500

33 79 Site Grounding

33 79 83 – Site Grounding Conductors

33 79 83.13 Grounding Wire, Bar, and Rod		Crew	Daily Output	Labor-Hours	Unit	2013 Bare Costs Material	Labor	Equipment	Total	Total Incl O&P
0010	**GROUNDING WIRE, BAR, AND ROD**									
7500	Grounding systems	R-11	280	.200	Lb.	11.25	9.90	3.10	24.25	30.50

33 81 Communications Structures

33 81 13 – Communications Transmission Towers

33 81 13.10 Radio Towers		Crew	Daily Output	Labor-Hours	Unit	2013 Bare Costs Material	Labor	Equipment	Total	Total Incl O&P
0010	**RADIO TOWERS**									
0020	Guyed, 50′ H, 40 lb. sect., 70 MPH basic wind spd.	2 Sswk	1	16	Ea.	2,925	800		3,725	4,650
0100	Wind load 90 MPH basic wind speed	"	1	16		3,475	800		4,275	5,250
0300	190′ high, 40 lb. section, wind load 70 MPH basic wind speed	K-2	.33	72.727		9,100	3,350	995	13,445	16,900
0400	200′ high, 70 lb. section, wind load 90 MPH basic wind speed		.33	72.727		15,200	3,350	995	19,545	23,600
0600	300′ high, 70 lb. section, wind load 70 MPH basic wind speed		.20	120		24,200	5,500	1,650	31,350	38,000
0700	270′ high, 90 lb. section, wind load 90 MPH basic wind speed		.20	120		26,300	5,500	1,650	33,450	40,200
0800	400′ high, 100 lb. section, wind load 70 MPH basic wind speed		.14	171		36,400	7,875	2,350	46,625	56,000
0900	Self-supporting, 60′ high, wind load 70 MPH basic wind speed		.80	30		4,550	1,375	410	6,335	7,825
0910	60′ high, wind load 90 MPH basic wind speed		.45	53.333		5,550	2,450	730	8,730	11,100
1000	120′ high, wind load 70 MPH basic wind speed		.40	60		10,600	2,750	820	14,170	17,300
1200	190′ high, wind load 90 MPH basic wind speed		.20	120		29,800	5,500	1,650	36,950	44,100
2000	For states west of Rocky Mountains, add for shipping					10%				

Estimating Tips

34 11 00 Rail Tracks

This subdivision includes items that may involve either repair of existing, or construction of new, railroad tracks. Additional preparation work, such as the roadbed earthwork, would be found in Division 31. Additional new construction siding and turnouts are found in Subdivision 34 72. Maintenance of railroads is found under 34 01 23 Operation and Maintenance of Railways.

34 40 00 Traffic Signals

This subdivision includes traffic signal systems. Other traffic control devices such as traffic signs are found in Subdivision 10 14 53 Traffic Signage.

34 70 00 Vehicle Barriers

This subdivision includes security vehicle barriers, guide and guard rails, crash barriers, and delineators. The actual maintenance and construction of concrete and asphalt pavement is found in Division 32.

Reference Numbers

Reference numbers are shown in shaded boxes at the beginning of some major classifications. These numbers refer to related items in the Reference Section. The reference information may be an estimating procedure, an alternate pricing method, or technical information.

Note: Not all subdivisions listed here necessarily appear in this publication.

34 43 Airfield Signaling and Control Equipment

34 43 13 – Airfield Signals

34 43 13.10 Airport Lighting		Crew	Daily Output	Labor-Hours	Unit	2013 Bare Costs Material	Labor	Equipment	Total	Total Incl O&P
0010	**AIRPORT LIGHTING**									
0100	Runway centerline, bidir., semi-flush, 200 W, w/shallow insert base	R-22	12.40	3.006	Ea.	1,525	134		1,659	1,875
0120	Flush, 200 W, w/shallow insert base		12.40	3.006		1,550	134		1,684	1,900
0130	for mounting in base housing		18.64	2		820	89		909	1,025
0150	Touchdown zone light, unidirectional, 200 W, w/shallow insert base		12.40	3.006		1,275	134		1,409	1,625
0160	115 W		12.40	3.006		1,275	134		1,409	1,600
0180	Unidirectional, 200 W, for mounting in base housing		18.60	2.004		575	89		664	765
0190	115 W		18.60	2.004		620	89		709	815
0210	Runway edge & threshold light, bidir., 200 W, for base housing		9.36	3.983		885	177		1,062	1,250
0240	Threshold & approach light, unidir., 200 W, for base housing		9.36	3.983		515	177		692	835
0260	Runway edge, bi-directional, 2-115 W, for base housing		12.40	3.006		1,350	134		1,484	1,700
0280	Runway threshold & end, bidir., 2-115 W, for base housing		12.40	3.006		1,500	134		1,634	1,850
0370	45 W, flush, for mounting in base housing		18.64	2		730	89		819	940
0380	115 W	▼	18.64	2	▼	750	89		839	960

34 43 23 – Weather Observation Equipment

34 43 23.16 Airfield Wind Cones

		Crew	Daily Output	Labor-Hours	Unit	Material	Labor	Equipment	Total	Total Incl O&P
0010	**AIRFIELD WIND CONES**									
1200	Wind cone, 12′ lighted assembly, rigid, w/obstruction light	R-21	1.36	24.118	Ea.	8,575	1,275	43.50	9,893.50	11,400
1210	Without obstruction light		1.52	21.579		7,200	1,125	39	8,364	9,650
1220	Unlighted assembly, w/obstruction light		1.68	19.524		7,850	1,025	35.50	8,910.50	10,200
1230	Without obstruction light		1.84	17.826		7,200	935	32.50	8,167.50	9,325
1240	Wind cone slip fitter, 2-1/2″ pipe		21.84	1.502		95	78.50	2.72	176.22	225
1250	Wind cone sock, 12′ x 3′, cotton		6.56	5		520	262	9.05	791.05	970
1260	Nylon	▼	6.56	5	▼	530	262	9.05	801.05	980

Estimating Tips

- When estimating costs for the installation of electrical power generation equipment, factors to review include access to the job site, access and setting at the installation site, required connections, uncrating pads, anchors, leveling, final assembly of the components, and temporary protection from physical damage, including from exposure to the environment.
- Be aware of the cost of equipment supports, concrete pads, and vibration isolators, and cross-reference to other trades' specifications. Also, review site and structural drawings for items that must be included in the estimates.
- It is important to include items that are not documented in the plans and specifications but must be priced. These items include, but are not limited to, testing, dust protection, roof penetration, core drilling concrete floors and walls, patching, cleanup, and final adjustments. Add a contingency or allowance for utility company fees for power hookups, if needed.
- The project size and scope of electrical power generation equipment will have a significant impact on cost. The intent of RSMeans cost data is to provide a benchmark cost so that owners, engineers, and electrical contractors will have a comfortable number with which to start a project. Additionally, there are many websites available to use for research and to obtain a vendor's quote to finalize costs.

Reference Numbers

Reference numbers are shown in shaded boxes at the beginning of some major classifications. These numbers refer to related items in the Reference Section. The reference information may be an estimating procedure, an alternate pricing method, or technical information.

Note: Not all subdivisions listed here necessarily appear in this publication.

48 15 Wind Energy Electrical Power Generation Equipment

48 15 13 – Wind Turbines

48 15 13.50 Wind Turbines and Components			Crew	Daily Output	Labor-Hours	Unit	Material	Labor	Equipment	Total	Total Incl O&P
							2013 Bare Costs				
0010	**WIND TURBINES & COMPONENTS**										
0500	Complete system, grid connected										
1000	20 kW, 31′ dia, incl labor & material, min	G				System					49,900
1010	Maximum	G									92,000
1500	10 kW, 23′ dia, incl labor & material	G									74,000
2000	2.4 kW, 12′ dia, incl labor & material	G				↓					18,000
2900	Component system										
3000	Turbine, 400 watt, 3′ dia	G	1 Elec	3.41	2.346	Ea.	750	123		873	1,000
3100	600 watt, 3′ dia	G		2.56	3.125		945	164		1,109	1,300
3200	1000 watt, 9′ dia	G	↓	2.05	3.902	↓	3,800	204		4,004	4,475
3400	Mounting hardware										
3500	30′ guyed tower kit	G	2 Clab	5.12	3.125	Ea.	450	111		561	665
3505	3′ galvanized helical earth screw	G	1 Clab	8	1		52.50	35.50		88	112
3510	Attic mount kit	G	1 Rofc	2.56	3.125		132	119		251	345
3520	Roof mount kit	G	1 Clab	3.41	2.346	↓	117	83		200	256
8900	Equipment										
9100	DC to AC inverter for, 48 V, 4,000 watt	G	1 Elec	2	4	Ea.	2,700	210		2,910	3,300

48 18 Fuel Cell Electrical Power Generation Equipment

48 18 13 – Electrical Power Generation Fuel Cells

48 18 13.10 Fuel Cells			Crew	Daily Output	Labor-Hours	Unit	Material	Labor	Equipment	Total	Total Incl O&P
0010	**FUEL CELLS**										
2001	Complete system, natural gas, installed, 200 kW	G				System					1,150,000
2002	400 kW	G									2,168,000
2003	600 kW	G									3,060,000
2004	800 kW	G									4,000,000
2005	1000 kW	G				↓					5,005,000
2010	Comm type, for battery charging, uses hydrogen, 100 watt, 12 V	G				Ea.	2,450			2,450	2,700
2020	500 watt, 12 V	G					5,050			5,050	5,550
2030	1000 watt, 48 V	G					8,225			8,225	9,050
2040	Small, for demonstration or battery charging units, 12 V	G					164			164	180
2050	Spare anode set for	G					50			50	55
2060	3 watts, uses hydrogen gas	G					510			510	560
2070	6 watts, uses hydrogen gas	G					920			920	1,000
2080	10 watts, uses hydrogen gas	G					1,325			1,325	1,450
2090	One month rental of 261 C.F. hydrogen gas cylinder for, min	G				Ea.	18.70			18.70	20.50
2100	Maximum	G				Ea.	690			690	760

Assemblies Section

Table of Contents

How RSMeans Assemblies Data Works

Assemblies estimating provides a fast and reasonably accurate way to develop construction costs. An assembly is the grouping of individual work items, with appropriate quantities, to provide a cost for a major construction component in a convenient unit of measure.

An assemblies estimate is often used during early stages of design development to compare the cost impact of various design alternatives on total building cost.

Assemblies estimates are also used as an efficient tool to verify construction estimates.

Assemblies estimates do not require a completed design or detailed drawings. Instead, they are based on the general size of the structure and other known parameters of the project. The degree of accuracy of an assemblies estimate is generally within +/- 15%.

D50 Electrical

D5020 Lighting and Branch Wiring

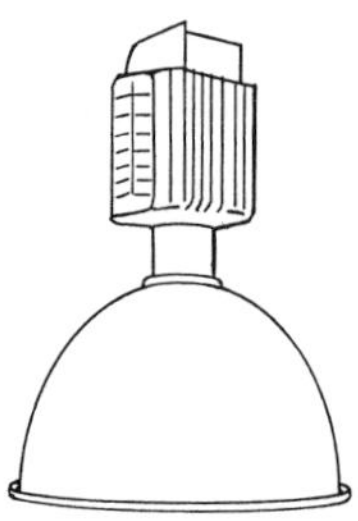

HIGH BAY FIXTURES
B. Metal halide 400 watt
C. High pressure sodium 400 watt
E. Metal halide 1000 watt
F. High pressure sodium 1000 watt
G. Metal halide 1000 watt
125,000 lumen lamp

System Components	QUANTITY	UNIT	COST PER S.F. MAT.	INST.	TOTAL
SYSTEM D5020 222 0240					
HIGH INTENSITY DISCHARGE FIXTURE, 8'-10' ABOVE WORK PLANE					
1 WATT/S.F., TYPE B, 29 FC, 2 FIXTURES/1000 S.F.					
Steel intermediate conduit, (IMC) 1/2" diam	.100	L.F.	.20	.63	.83
Wire, 600V, type THWN-THHN, copper, solid, #10	.002	C.L.F.	.04	.13	.17
Steel outlet box 4" concrete	.002	Ea.	.04	.06	.10
Steel outlet box plate with stud, 4" concrete	.002	Ea.	.02	.02	.04
Metal halide, hi bay, aluminum reflector, 400 W lamp	.002	Ea.	.89	.55	1.44
TOTAL			1.19	1.39	2.58

D5020 222	H.I.D. Fixture, High Bay, 8'-10' (by Wattage)	COST PER S.F. MAT.	INST.	TOTAL
0190	High intensity discharge fixture, 8'-10' above work plane			
0240	1 watt/S.F., type B, 29 FC, 2 fixtures/1000 S.F.	1.19	1.39	2.58
0280	Type C, 54 FC, 2 fixtures/1000 S.F.	[illegible]	1.05	2.06
0400	2 watt/S.F., type B, 59 FC, 4 fixtures/1000 S.F.	2.36	2.66	5.02
0440	Type C, 108 FC, 4 fixtures/1000 S.F. R265723-05	2.01	2.04	4.05
0560	3 watt/S.F., type B, 103 FC, 7 fixtures/1000 S.F.	4.01	4.25	8.26
0600	Type C, 189 FC, 6 fixtures/1000 S.F.	3.23	2.62	5.85
0720	4 watt/S.F., type B, 133 FC, 9 fixtures/1000 S.F.	5.20	5.60	10.80
0760	Type C, 243 FC, 9 fixtures/1000 S.F.	4.53	4.51	9.04
0880	5 watt/S.F., type B, 162 FC, 11 fixtures/1000 S.F.	6.35	6.90	13.25
0920	Type C, 297 FC, 11 fixtures/1000 S.F.	5.55	5.60	11.15

RSMeans assemblies are identified by a **unique 12-character identifier**. The assemblies are numbered using UNIFORMAT II, ASTM Standard E1557. The first 6 characters represent this system to Level 4. The last 6 characters represent further breakdown by RSMeans in order to arrange items in understandable groups of similar tasks. Line numbers are consistent across all RSMeans publications, so a line number in any RSMeans assemblies data set will always refer to the same work.

Information is available in the Reference Section to assist the estimator with estimating procedures, alternate pricing methods, and additional technical information.

The **Reference Box** indicates the exact location of this information in the Reference Section at the back of the book. The "R" stands for "reference," and the remaining characters are the RSMeans line numbers.

RSMeans assemblies **Narrative Descriptions** are shown in a hierarchical structure to make them readable. In order to read a complete description, read up through the indents to the top of the section. Include everything that is above and to the left that is not contradicted by information below.

Most assemblies consist of three major elements: a graphic, the system components, and the cost data itself. The **Graphic** is a visual representation showing the typical appearance of the assembly in question, frequently accompanied by additional explanatory technical information describing the class of items. The **System Components** is a listing of the individual tasks that make up the assembly, including the quantity and unit of measure for each item, along with the cost of material and installation. The **Assemblies Data** below lists prices for other similar systems with dimensional and/or size variations.

All RSMeans assemblies costs represent the cost for the installing contractor. An allowance for profit has been added to all material, labor, and equipment rental costs. A markup for labor burdens, including workers' compensation, fixed overhead, and business overhead, is included with installation costs.

The data published in RSMeans print books represents a "national average" cost. This data should be modified to the project location using the **City Cost Indexes** or **Location Factors** tables found in the Reference Section in the back of the book.

D50 Electrical

D5020 Lighting and Branch Wiring

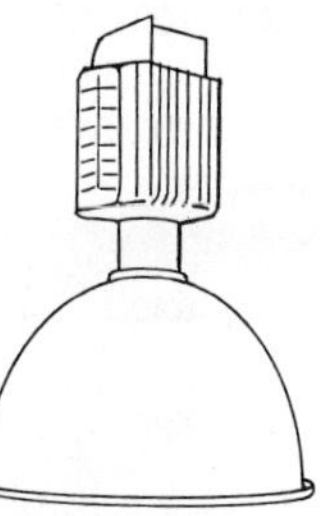

HIGH BAY FIXTURES
B. Metal halide 400 watt
C. High pressure sodium 400 watt
E. Metal halide 1000 watt
F. High pressure sodium 1000 watt
G. Metal halide 1000 watt
125,000 lumen lamp

All RSMeans assemblies data includes a typical **Unit of Measure** used for estimating that item. For instance,while the unit of measure for the graphic shown here is Each, some A/C systems are estimated by the square foot (S.F.). The estimator needs to take special care that the unit in the data matches the unit in the takeoff. Abbreviations can be found in the Reference Section.

System Components	QUANTITY	UNIT	COST PER S.F. MAT.	COST PER S.F. INST.	COST PER S.F. TOTAL
SYSTEM D5020 222 0240					
HIGH INTENSITY DISCHARGE FIXTURE, 8′-10′ ABOVE WORK PLANE					
1 WATT/S.F., TYPE B, 29 FC, 2 FIXTURES/1000 S.F.					
Steel intermediate conduit, (IMC) 1/2″ diam	.100	L.F.	.20	.63	.83
Wire, 600V, type THWN-THHN, copper, solid, #10	.002	C.L.F.	.04	.13	.17
Steel outlet box 4″ concrete	.002	Ea.	.04	.06	.10
Steel outlet box plate with stud, 4″ concrete	.002	Ea.	.02	.02	.04
Metal halide, hi bay, aluminum reflector, 400 W lamp	.002	Ea.	.89	.55	1.44
TOTAL			1.19	1.39	2.58

System components are listed separately to detail what is included in the development of the total system price.

D5020 222	H.I.D. Fixture, High Bay, 8′-10′ (by Wattage)	COST PER S.F. MAT.	COST PER S.F. INST.	COST PER S.F. TOTAL
0190	High intensity discharge fixture, 8′-10′ above work plane			
0240	1 watt/S.F., type B, 29 FC, 2 fixtures/1000 S.F.	1.19	1.39	2.58
0280	Type C, 54 FC, 2 fixtures/1000 S.F.	1.01	1.05	2.06
0400	2 watt/S.F., type B, 59 FC, 4 fixtures/1000 S.F.	2.36	2.66	5.02
0440	Type C, 108 FC, 4 fixtures/1000 S.F.	2.01	2.04	4.05
0560	3 watt/S.F., type B, 103 FC, 7 fixtures/1000 S.F. R265723-05	4.01	4.25	8.26
0600	Type C, 189 FC, 6 fixtures/1000 S.F.	3.23	2.62	5.85
0720	4 watt/S.F., type B, 133 FC, 9 fixtures/1000 S.F.	5.20	5.60	10.80
0760	Type C, 243 FC, 9 fixtures/1000 S.F.	4.53	4.51	9.04
0880	5 watt/S.F., type B, 162 FC, 11 fixtures/1000 S.F.	6.35	6.90	13.25
0920	Type C, 297 FC, 11 fixtures/1000 S.F.	5.55	5.60	11.15

How RSMeans Assemblies Data Works (continued)

Sample Estimate

This sample demonstrates the elements of an estimate, including a tally of the RSMeans data lines. Published assemblies costs include all markups for labor burden and profit for the installing contractor. This estimate adds a summary of the markups applied by a general contractor on the installing contractors' work. These figures represent the total cost to the owner. The RSMeans location factor is added at the bottom of the estimate to adjust the cost of the work to a specific location

Work performed: The body of the estimate shows the RSMeans data selected, including line numbers, a brief description of each item, its takeoff quantity and unit, and the total installed cost, including the installing contractor's overhead and profit.

Location Factor: RSMeans published data is based on national average costs. If necessary, adjust the total cost of the project using a location factor from the "Location Factor" table or the "City Cost Indexes" table, found in the Reference Section. Use location factors if the work is general, covering the work of multiple trades. If the work is by a single trade (e.g. masonry) use the more specific data found in the City Cost Indexes.

To adjust costs by location factors, multiply the base cost by the factor and divide by 100.

Contingency: A factor for contingency may be added to any estimate to represent the cost of unknowns that may occur between the time that the estimate is performed and the time the project is constructed. The amount of the allowance will depend on the stage of design at which the estimate is done, and the contractor's assessment of the risk invloved.

Project Name:	Interior Fit-out, ABC Office
Location:	Anywhere, USA
Assembly Number	**Description**
C1010 124 1200	Wood partition,2 x 4,@ 16" OC w/5/8" FR gypsum board
C1020 114 1800	Metal door & frame, flush-hollow core, 3'-0" x 7'-0"
C3010 230 0080	Painting, brushwork, primer & 2 coats
C3020 410 0140	Carpet, tufted, nylon, roll goods, 12' wide, 26 oz
C3030 210 6000	Acoustic ceilings, 24" x 48" tile, tee grid suspension
D5020 110 0360	Receptacles incl plate, box, conduit, wire, 5 per 1000 SF
D5020 210 0540	Fluorescent fixtures, recess mounted
	Assembly Subtotal
	Sales Tax @
	General Requirements @
	Subtotal A
	GC Overhead @
	Subtotal B
	GC Profit @
	Subtotal C
	Adjusted by Location Factor (Boston, MA)
	Architects Fee @
	Contingency @
	Project Total Cost

This estimate is based on an interactive spreadsheet.
A copy of this spreadsheet is located on the RSMeans website at **http://www.reedconstructiondata.com/rsmeans/extras/546011**.
You are free to download it and adjust it to your methodology.

Date:		
Qty.	**Unit**	**Subtotal**
600.000	S.F.	$2,682.00
2.000	Ea.	$2,550.00
600.000	S.F.	$696.00
240.000	S.F.	$1,708.80
200.000	S.F.	$870.00
6.000	S.F.	$14.76
200.000	S.F.	$1,580.00
		$10,101.56
5 %		**$ 252.54**
7 %		**$ 707.11**
		$11,061.21
5 %		**$ 553.06**
		$11,614.27
5 %		**$ 580.7**
		$12,194.98
118.1		**$ 14,402.27**
8 %		**$ 1,152.18**
15 %		**$ 2,160.34**
		$ 17,714.80

Sales Tax: If the work is subject to state or local sales taxes, the amount must be added to the estimate. In a conceptual estimate it can be assumed that one half of the total represents material costs. Therefore, apply the sales tax rate to 50% of the assembly subtotal.

General Requirements: This item covers project-wide needs provided by the General Contractor. These items vary by project, but may include temporary facilities and utilities, security, testing, project cleanup, etc. In assemblies estimates a percentage is used, typically between 5% and 15% of project cost.

General Contractor Overhead: This entry represents the General Contractor's markup on all work to cover project administration costs.

General Contractor Profit: This entry represents the GC's profit on all work performed. The value included here can vary widely by project, and is influenced by the GC's perception of the project's financial risk and market conditions.

Architects Fee: If appropriate, add the design cost to the project estimate. These fees vary based on project complexity and size. Typical design and engineering fees can be found in the reference section.

D Services

D2020 Domestic Water Distribution

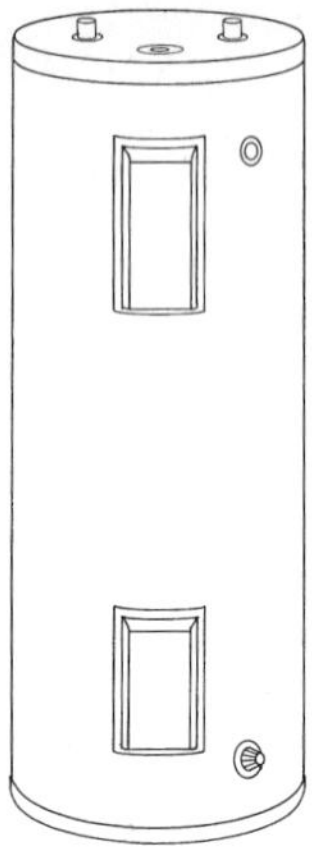

Installation includes piping and fittings within 10′ of heater. Electric water heaters do not require venting.

1 Kilowatt hour will raise:			
Gallons of Water	Degrees F	Gallons of Water	Degrees F
4.1	100°	6.8	60°
4.5	90°	8.2	50°
5.1	80°	10.0	40°
5.9	70°		

System Components	QUANTITY	UNIT	COST EACH MAT.	COST EACH INST.	COST EACH TOTAL
SYSTEM D2020 210 1780					
ELECTRIC WATER HEATER, RESIDENTIAL, 100°F RISE					
10 GALLON TANK, 7 GPH					
Water heater, residential electric, glass lined tank, 10 gal.	1.000	Ea.	335	292	627
Copper tubing, type L, solder joint, hanger 10′ OC 1/2″ diam.	30.000	L.F.	139.80	249	388.80
Wrought copper 90° elbow for solder joints 1/2″ diam.	4.000	Ea.	12.44	134	146.44
Wrought copper Tee for solder joints, 1/2″ diam.	2.000	Ea.	10.60	103	113.60
Union, wrought copper, 1/2″ diam.	2.000	Ea.	74	71	145
Valve, gate, bronze, 125 lb, NRS, soldered 1/2″ diam.	2.000	Ea.	97	56	153
Relief valve, bronze, press & temp, self-close, 3/4″ IPS	1.000	Ea.	171	24	195
Wrought copper adapter, CTS to MPT 3/4″ IPS	1.000	Ea.	10.95	39.50	50.45
Copper tubing, type L, solder joints, 3/4″ diam.	1.000	L.F.	7.20	8.85	16.05
Wrought copper 90° elbow for solder joints 3/4″ diam.	1.000	Ea.	7	35.50	42.50
TOTAL			864.99	1,012.85	1,877.84

D2020 210	Electric Water Heaters - Residential Systems	COST EACH MAT.	COST EACH INST.	COST EACH TOTAL
1760	Electric water heater, residential, 100°F rise			
1780	10 gallon tank, 7 GPH	865	1,025	1,890
1820	20 gallon tank, 7 GPH (R224000-20)	1,100	1,075	2,175
1860	30 gallon tank, 7 GPH	1,600	1,125	2,725
1900	40 gallon tank, 8 GPH	1,950	1,250	3,200
1940	52 gallon tank, 10 GPH	2,050	1,250	3,300
1980	66 gallon tank, 13 GPH	3,075	1,450	4,525
2020	80 gallon tank, 16 GPH	3,250	1,500	4,750
2060	120 gallon tank, 23 GPH	4,750	1,750	6,500

D2020 Domestic Water Distribution

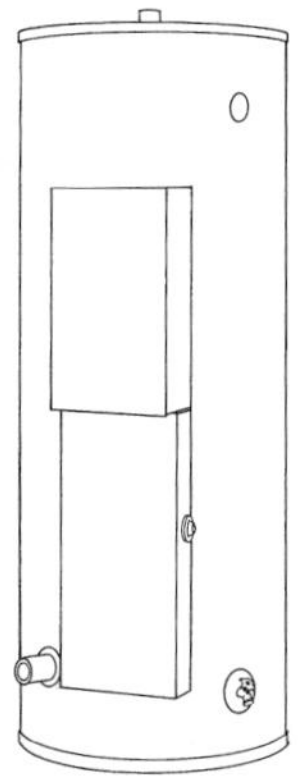

Systems below include piping and fittings within 10′ of heater. Electric water heaters do not require venting.

System Components	QUANTITY	UNIT	COST EACH MAT.	COST EACH INST.	COST EACH TOTAL
SYSTEM D2020 240 1820					
ELECTRIC WATER HEATER, COMMERCIAL, 100°F RISE					
50 GALLON TANK, 9 KW, 37 GPH					
Water heater, commercial, electric, 50 Gal, 9 KW, 37 GPH	1.000	Ea.	4,375	375	4,750
Copper tubing, type L, solder joint, hanger 10′ OC, 3/4″ diam	34.000	L.F.	244.80	300.90	545.70
Wrought copper 90° elbow for solder joints 3/4″ diam	5.000	Ea.	35	177.50	212.50
Wrought copper Tee for solder joints, 3/4″ diam	2.000	Ea.	25.70	112	137.70
Wrought copper union for soldered joints, 3/4″ diam.	2.000	Ea.	93	75	168
Valve, gate, bronze, 125 lb, NRS, soldered 3/4″ diam	2.000	Ea.	109	67	176
Relief valve, bronze, press & temp, self-close, 3/4″ IPS	1.000	Ea.	171	24	195
Wrought copper adapter, copper tubing to male, 3/4″ IPS	1.000	Ea.	10.95	39.50	50.45
TOTAL			5,064.45	1,170.90	6,235.35

D2020 240	Electric Water Heaters - Commercial Systems	COST EACH MAT.	COST EACH INST.	COST EACH TOTAL
1800	Electric water heater, commercial, 100°F rise			
1820	50 gallon tank, 9 KW 37 GPH	5,075	1,175	6,250
1860	80 gal, 12 KW 49 GPH (R224000-10)	6,800	1,450	8,250
1900	36 KW 147 GPH	9,225	1,550	10,775
1940	120 gal, 36 KW 147 GPH (R224000-20)	9,875	1,675	11,550
1980	150 gal, 120 KW 490 GPH	28,600	1,800	30,400
2020	200 gal, 120 KW 490 GPH	30,000	1,850	31,850
2060	250 gal, 150 KW 615 GPH	33,800	2,150	35,950
2100	300 gal, 180 KW 738 GPH	43,200	2,275	45,475
2140	350 gal, 30 KW 123 GPH	27,600	2,450	30,050
2180	180 KW 738 GPH	37,700	2,450	40,150
2220	500 gal, 30 KW 123 GPH	35,600	2,900	38,500
2260	240 KW 984 GPH	60,500	2,900	63,400
2300	700 gal, 30 KW 123 GPH	29,400	3,300	32,700
2340	300 KW 1230 GPH	42,600	3,300	45,900
2380	1000 gal, 60 KW 245 GPH	37,100	4,600	41,700
2420	480 KW 1970 GPH	57,500	4,625	62,125
2460	1500 gal, 60 KW 245 GPH	75,500	5,700	81,200
2500	480 KW 1970 GPH	102,500	5,700	108,200

D3020 Heat Generating Systems

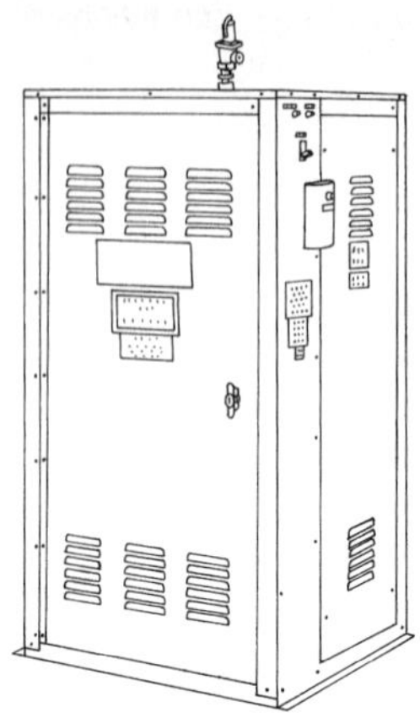

Boiler

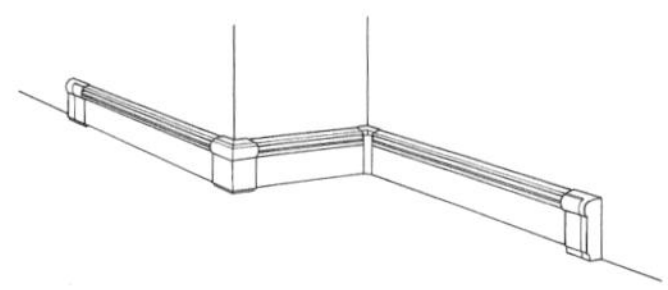

Baseboard Radiation

Small Electric Boiler
System Considerations:

1. Terminal units are fin tube baseboard radiation rated at 720 BTU/hr with 200° water temperature or 820 BTU/hr steam.
2. Primary use being for residential or smaller supplementary areas, the floor levels are based on 7-1/2′ ceiling heights.
3. All distribution piping is copper for boilers through 205 MBH. All piping for larger systems is steel pipe.

System Components	QUANTITY	UNIT	COST EACH MAT.	COST EACH INST.	COST EACH TOTAL
SYSTEM D3020 102 1120					
SMALL HEATING SYSTEM, HYDRONIC, ELECTRIC BOILER					
1,480 S.F., 61 MBH, STEAM, 1 FLOOR					
Boiler, electric steam, std cntrls, trim, ftngs and valves, 18 KW, 61.4 MBH	1.000	Ea.	4,867.50	1,705	6,572.50
Copper tubing type L, solder joint, hanger 10′OC, 1-1/4″ diam	160.000	L.F.	2,408	1,856	4,264
Radiation, 3/4″ copper tube w/alum fin baseboard pkg 7″ high	60.000	L.F.	555	1,260	1,815
Rough in baseboard panel or fin tube with valves & traps	10.000	Set	3,450	6,450	9,900
Pipe covering, calcium silicate w/cover, 1″ wall 1-1/4″ diam	160.000	L.F.	580.80	1,080	1,660.80
Low water cut-off, quick hookup, in gage glass tappings	1.000	Ea.	260	42.50	302.50
TOTAL			12,121.30	12,393.50	24,514.80
COST PER S.F.			8.19	8.37	16.56

D3020 102	Small Heating Systems, Hydronic, Electric Boilers	COST PER S.F. MAT.	COST PER S.F. INST.	COST PER S.F. TOTAL
1100	Small heating systems, hydronic, electric boilers			
1120	Steam, 1 floor, 1480 S.F., 61 M.B.H.	8.18	8.38	16.56
1160	3,000 S.F., 123 M.B.H.	6.45	7.35	13.80
1200	5,000 S.F., 205 M.B.H.	5.75	6.80	12.55
1240	2 floors, 12,400 S.F., 512 M.B.H.	4.58	6.75	11.33
1280	3 floors, 24,800 S.F., 1023 M.B.H.	5	6.65	11.65
1320	34,750 S.F., 1,433 M.B.H.	4.62	6.45	11.07
1360	Hot water, 1 floor, 1,000 S.F., 41 M.B.H.	12.90	4.64	17.54
1400	2,500 S.F., 103 M.B.H.	9.85	8.35	18.20
1440	2 floors, 4,850 S.F., 205 M.B.H.	9.90	10.05	19.95
1480	3 floors, 9,700 S.F., 410 M.B.H.	10.95	10.35	21.30

D3020 Heat Generating Systems

Boiler

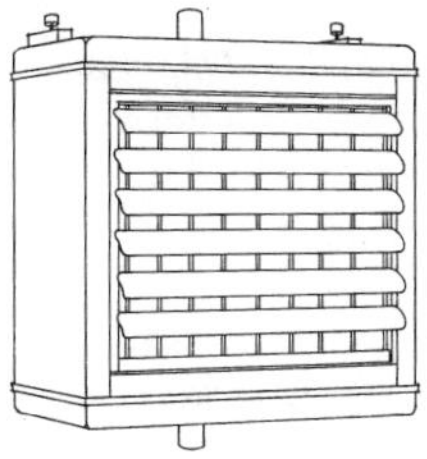

Unit Heater

Large Electric Boiler System Considerations:

1. Terminal units are all unit heaters of the same size. Quantities are varied to accommodate total requirements.
2. All air is circulated through the heaters a minimum of three times per hour.
3. As the capacities are adequate for commercial use, floor levels are based on 10′ ceiling heights.
4. All distribution piping is black steel pipe.

System Components	QUANTITY	UNIT	COST EACH MAT.	COST EACH INST.	COST EACH TOTAL
SYSTEM D3020 104 1240					
LARGE HEATING SYSTEM, HYDRONIC, ELECTRIC BOILER					
9,280 S.F., 135 KW, 461 MBH, 1 FLOOR					
Boiler, electric hot water, std ctrls, trim, ftngs, valves, 135 KW, 461 MBH	1.000	Ea.	11,850	3,712.50	15,562.50
Expansion tank, painted steel, 60 Gal capacity ASME	1.000	Ea.	4,025	205	4,230
Circulating pump, CI, close cpld, 50 GPM, 2 HP, 2″ pipe conn	1.000	Ea.	2,775	405	3,180
Unit heater, 1 speed propeller, horizontal, 200° EWT, 72.7 MBH	7.000	Ea.	6,125	1,561	7,686
Unit heater piping hookup with controls	7.000	Set	5,180	10,150	15,330
Pipe, steel, black, schedule 40, welded, 2-1/2″ diam	380.000	L.F.	5,947	10,351.20	16,298.20
Pipe covering, calcium silicate w/cover, 1″ wall, 2-1/2″ diam	380.000	L.F.	1,710	2,641	4,351
TOTAL			37,612	29,025.70	66,637.70
COST PER S.F.			4.05	3.13	7.18

D3020 104	Large Heating Systems, Hydronic, Electric Boilers	COST PER S.F. MAT.	COST PER S.F. INST.	COST PER S.F. TOTAL
1230	Large heating systems, hydronic, electric boilers			
1240	9,280 S.F., 135 K.W., 461 M.B.H., 1 floor	4.05	3.14	7.19
1280	14,900 S.F., 240 K.W., 820 M.B.H., 2 floors	5.40	5.05	10.45
1320	18,600 S.F., 296 K.W., 1,010 M.B.H., 3 floors	5.35	5.50	10.85
1360	26,100 S.F., 420 K.W., 1,432 M.B.H., 4 floors	5.25	5.40	10.65
1400	39,100 S.F., 666 K.W., 2,273 M.B.H., 4 floors	4.49	4.51	9
1440	57,700 S.F., 900 K.W., 3,071 M.B.H., 5 floors	4.34	4.42	8.76
1480	111,700 S.F., 1,800 K.W., 6,148 M.B.H., 6 floors	3.85	3.80	7.65
1520	149,000 S.F., 2,400 K.W., 8,191 M.B.H., 8 floors	3.80	3.80	7.60
1560	223,300 S.F., 3,600 K.W., 12,283 M.B.H., 14 floors	4.06	4.30	8.36

D4090 Other Fire Protection Systems

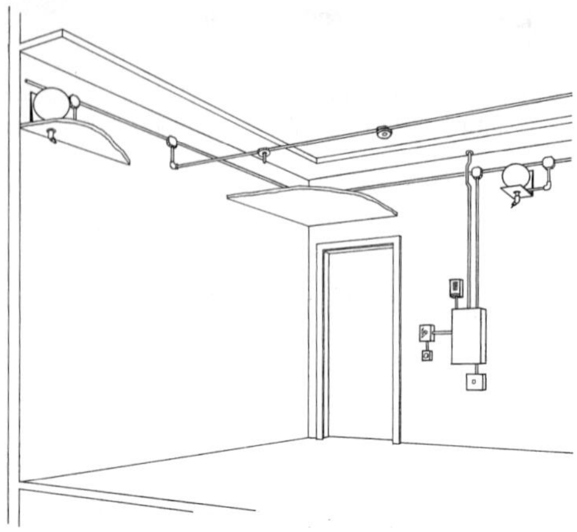

General: Automatic fire protection (suppression) systems other than water sprinklers may be desired for special environments, high risk areas, isolated locations or unusual hazards. Some typical applications would include:

Paint dip tanks
Securities vaults
Electronic data processing
Tape and data storage
Transformer rooms
Spray booths
Petroleum storage
High rack storage

Piping and wiring costs are dependent on the individual application and must be added to the component costs shown below.

All areas are assumed to be open.

D4090 910	Fire Suppression Unit Components	COST EACH		
		MAT.	INST.	TOTAL
0020	Detectors with brackets			
0040	Fixed temperature heat detector	55	78.50	133.50
0060	Rate of temperature rise detector	55	78.50	133.50
0080	Ion detector (smoke) detector	119	101	220
0200	Extinguisher agent			
0240	200 lb FM200, container	7,900	305	8,205
0280	75 lb carbon dioxide cylinder	1,300	202	1,502
0320	Dispersion nozzle			
0340	FM200 1-1/2" dispersion nozzle	68	48	116
0380	Carbon dioxide 3" x 5" dispersion nozzle	68	37.50	105.50
0420	Control station			
0440	Single zone control station with batteries	1,750	625	2,375
0470	Multizone (4) control station with batteries	3,350	1,250	4,600
0490				
0500	Electric mechanical release	170	325	495
0520				
0550	Manual pull station	61.50	112	173.50
0570				
0640	Battery standby power 10" x 10" x 17"	445	157	602
0700				
0740	Bell signalling device	85	78.50	163.50

D4090 920	FM200 Systems	COST PER C.F.		
		MAT.	INST.	TOTAL
0820	Average FM200 system, minimum			1.79
0840	Maximum			3.56

D5010 Electrical Service/Distribution

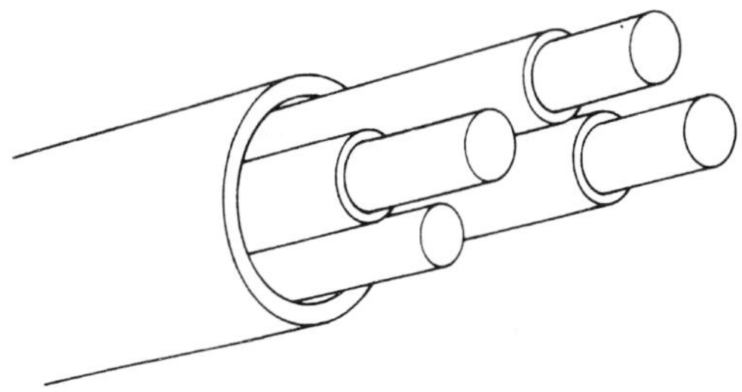

System Components	QUANTITY	UNIT	COST PER L.F. MAT.	INST.	TOTAL
SYSTEM D5010 110 0200					
HIGH VOLTAGE CABLE, NEUTRAL AND CONDUIT INCLUDED, COPPER #2, 5 kV					
Shielded cable, no splice/termn, copper, XLP shielding, 5 kV, #2	.030	C.L.F.	8.19	9.45	17.64
Wire, 600 volt, type THW, copper, stranded, #8	.010	C.L.F.	.41	.79	1.20
Rigid galv steel conduit to 15′ H, 2″ diam, w/term, ftng & support	1.000	L.F.	8.70	13.95	22.65
TOTAL			17.30	24.19	41.49

D5010 110	High Voltage Shielded Conductors	COST PER L.F. MAT.	INST.	TOTAL
0200	High voltage cable, neutral & conduit included, copper #2, 5 kV	17.30	24	41.30
0240	Copper #1, 5 kV	26.50	28.50	55
0280	15 kV	28.50	28.50	57
0320	Copper 1/0, 5 kV	31	36	67
0360	15 kV	34.50	36	70.50
0400	25 kV	39.50	36.50	76
0440	35 kV	40.50	37	77.50
0480	Copper 2/0, 5 kV	34	36.50	70.50
0520	15 kV	36.50	36.50	73
0560	25 kV	42	37.50	79.50
0600	35 kV	49.50	41.50	91
0640	Copper 4/0, 5 kV	43.50	41.50	85
0680	15 kV	47.50	41.50	89
0720	25 kV	53	42.50	95.50
0760	35 kV	56.50	43.50	100
0800	Copper 250 kcmil, 5 kV	47.50	42.50	90
0840	15 kV	49.50	42.50	92
0880	25 kV	59	43.50	102.50
0920	35 kV	65.50	47.50	113
0960	Copper 350 kcmil, 5 kV	58	47.50	105.50
1000	15 kV	60	47.50	107.50
1040	25 kV	69.50	48.50	118
1080	35 kV	72.50	50	122.50
1120	Copper 500 kcmil, 5 kV	93	59.50	152.50
1160	15 kV	95.50	59.50	155
1200	25 kV	104	61	165
1240	35 kV	109	62.50	171.50

D5010 Electrical Service/Distribution

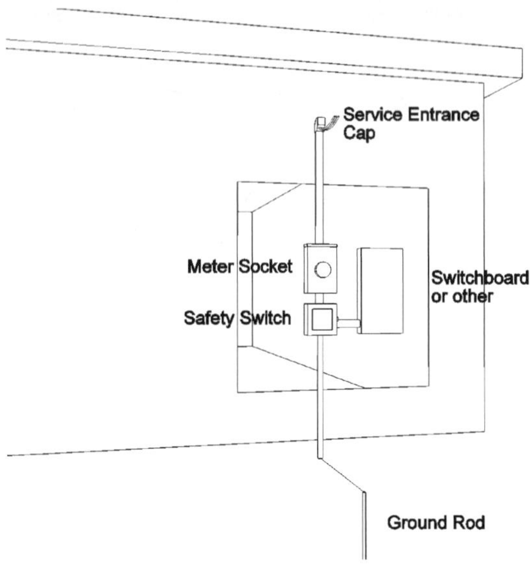

System Components	QUANTITY	UNIT	COST EACH MAT.	COST EACH INST.	COST EACH TOTAL
SYSTEM D5010 120 0220					
SERVICE INSTALLATION, INCLUDES BREAKERS, METERING, 20′ CONDUIT & WIRE					
3 PHASE, 4 WIRE, 60 A					
Wire, 600 volt, copper type XHHW, stranded #6	.600	C.L.F.	55.80	57.90	113.70
Rigid galvanized steel conduit, 3/4″ including fittings	20.000	L.F.	59.20	157	216.20
Service entrance cap, 3/4″ diameter	1.000	Ea.	9.90	48.50	58.40
Conduit LB fitting with cover, 3/4″ diameter	1.000	Ea.	12.20	48.50	60.70
Meter socket, three phase, 100 A	1.000	Ea.	138	224	362
Safety switches, heavy duty, 240 volt, 3 pole NEMA 1 fusible, 60 amp	1.000	Ea.	305	273	578
Grounding, wire ground bare armored, #8-1 conductor	.200	C.L.F.	23.20	63	86.20
Grounding, clamp, bronze, 3/4″ dia	1.000	Ea.	8.15	19.60	27.75
Grounding, rod, copper clad, 8′ long, 3/4″ diameter	1.000	Ea.	37	118	155
Wireway w/fittings, 2-1/2″ x 2-1/2″	1.000	L.F.	11.45	13.95	25.40
TOTAL			659.90	1,023.45	1,683.35

D5010 120	Overhead Electric Service, 3 Phase - 4 Wire	COST EACH MAT.	COST EACH INST.	COST EACH TOTAL
0200	Service installation, includes breakers, metering, 20′ conduit & wire			
0220	3 phase, 4 wire, 120/208 volts, 60 A	660	1,025	1,685
0240	3 phase, 4 wire, 120/208 volts, 100 A	920	1,150	2,070
0245	100 A w/circuit breaker	1,650	1,425	3,075
0280	200 A	1,850	1,600	3,450
0285	200 A w/circuit breaker	3,400	2,000	5,400
0320	400 A	3,725	3,175	6,900
0325	400 A w/circuit breaker	6,625	3,950	10,575
0360	600 A	6,200	4,700	10,900
0365	600 A, w/switchboard	10,400	5,950	16,350
0400	800 A	8,925	5,625	14,550
0405	800 A, w/switchboard	13,100	7,050	20,150
0440	1000 A	10,800	6,925	17,725
0445	1000 A, w/switchboard	15,900	8,500	24,400
0480	1200 A	14,300	7,900	22,200
0485	1200 A, w/groundfault switchboard	36,100	9,800	45,900
0520	1600 A	17,900	10,300	28,200
0525	1600 A, w/groundfault switchboard	41,600	12,400	54,000

D5010 Electrical Service/Distribution

D5010 120	Overhead Electric Service, 3 Phase - 4 Wire	COST EACH		
		MAT.	INST.	TOTAL
0560	2000 A	23,700	12,500	36,200
0565	2000 A, w/groundfault switchboard	51,500	15,500	67,000
0610	1 phase, 3 wire, 120/240 volts, 100 A (no safety switch)	204	565	769
0615	100 A w/load center	550	1,100	1,650
0620	200 A	465	775	1,240
0625	200 A w/load center	1,225	1,575	2,800

D5010 Underground Electric Service

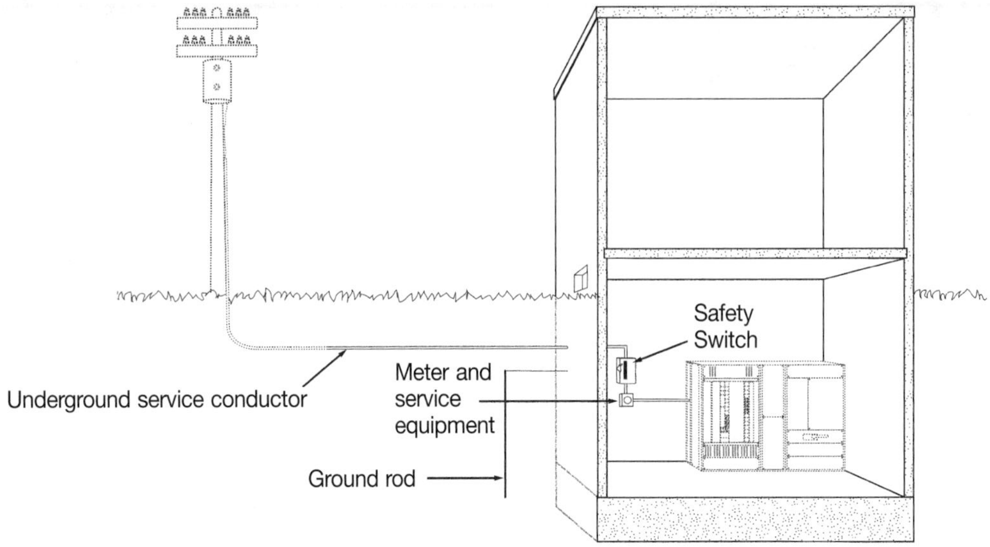

System Components	QUANTITY	UNIT	COST EACH MAT.	COST EACH INST.	COST EACH TOTAL
SYSTEM D5010 130 1000					
2000 AMP UNDERGROUND ELECTRIC SERVICE WITH GROUNDFAULT SWITCHBOARD					
INCLUDING EXCAVATION, BACKFILL, AND COMPACTION					
Excavate Trench	44.440	B.C.Y.		294.64	294.64
4 inch conduit bank	108.000	L.F.	626.40	2,484	3,110.40
4 inch fitting	8.000	Ea.	212	420	632
4 inch bells	4.000	Ea.	16.08	210	226.08
Concrete material	16.580	C.Y.	1,774.06		1,774.06
Concrete placement	16.580	C.Y.		383	383
Backfill trench	33.710	L.C.Y.		94.06	94.06
Compact fill material in trench	25.930	E.C.Y.		121.36	121.36
Dispose of excess fill material on-site	24.070	L.C.Y.		268.62	268.62
500 kcmil power cable	18.000	C.L.F.	22,950	7,020	29,970
Wire, 600 volt, type THW, copper, stranded, 1/0	6.000	C.L.F.	1,638	1,140	2,778
Saw cutting, concrete walls, plain, per inch of depth	64.000	L.F.	2.56	501.12	503.68
Meter centers and sockets, single pos, 4 terminal, 400 amp	5.000	Ea.	4,600	1,850	6,450
Safety switch, 400 amp	5.000	Ea.	10,500	3,475	13,975
Ground rod clamp	1.000	Ea.	8.15	19.60	27.75
Wireway	1.000	L.F.	11.45	13.95	25.40
600 volt stranded copper wire, 1/0	1.200	C.L.F.	327.60	228	555.60
Flexible metallic conduit	120.000	L.F.	121.20	470.40	591.60
Ground rod	6.000	Ea.	468	942	1,410
TOTAL			43,255.50	19,935.75	63,191.25

D5010 130	Underground Electric Service	COST EACH MAT.	COST EACH INST.	COST EACH TOTAL
0950	Underground electric service including excavation,backfill, and compaction			
1000	3 phase, 4 wire, 277/480 volts, 2000 A	43,300	19,900	63,200
1050	2000 A w/groundfault switch	69,000	22,300	91,300
1100	1600 A	34,400	16,100	50,500
1150	1600 A w/groundfault switchboard	58,000	18,200	76,200
1200	1200 A	26,200	12,800	39,000
1250	1200 A w/groundfault switchboard	48,000	14,700	62,700
1400	800 A	18,800	10,700	29,500
1450	800 A w/ switchboard	23,000	12,100	35,100
1500	600 A	11,900	9,800	21,700

D5010 Underground Electric Service

D5010 130	Underground Electric Service	COST EACH		
		MAT.	INST.	TOTAL
1550	600 A w/ switchboard	16,100	11,000	27,100
1600	1 phase, 3 wire, 120/240 volts, 200 A	3,375	3,400	6,775
1650	200 A w/load center	4,125	4,200	8,325
1700	100 A	2,250	2,775	5,025
1750	100 A w/load center	3,025	3,575	6,600

D5010 Electrical Service/Distribution

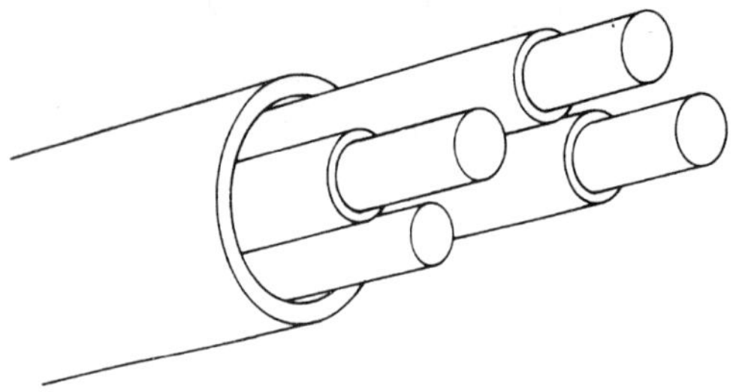

System Components	QUANTITY	UNIT	COST PER L.F. MAT.	COST PER L.F. INST.	COST PER L.F. TOTAL
SYSTEM D5010 230 0200					
FEEDERS, INCLUDING STEEL CONDUIT & WIRE, 60 A					
Rigid galvanized steel conduit, 3/4", including fittings	1.000	L.F.	2.96	7.85	10.81
Wire 600 volt, type XHHW copper stranded #6	.030	C.L.F.	2.79	2.90	5.69
Wire 600 volt, type XHHW copper stranded #8	.010	C.L.F.	.57	.79	1.36
TOTAL			6.32	11.54	17.86

D5010 230	Feeder Installation	COST PER L.F. MAT.	COST PER L.F. INST.	COST PER L.F. TOTAL
0200	Feeder installation 600 V, including RGS conduit and XHHW wire, 60 A	6.30	11.55	17.85
0240	100 A	11.75	15	26.75
0280	200 A	28	22.50	50.50
0320	400 A	56	45.50	101.50
0360	600 A	103	74	177
0400	800 A	135	87.50	222.50
0440	1000 A	168	115	283
0480	1200 A	207	148	355
0520	1600 A	271	175	446
0560	2000 A	335	230	565
1200	Branch installation 600 V, including EMT conduit and THW wire, 15 A	1.10	5.65	6.75
1240	20 A	1.32	5.95	7.27
1280	30 A	1.67	6.20	7.87
1320	50 A	3.50	8.60	12.10
1360	65 A	4.36	9.15	13.51
1400	85 A	6.80	10.60	17.40
1440	100 A	7.60	10.80	18.40
1480	130 A	11.30	12.65	23.95
1520	150 A	14.05	14.50	28.55
1560	200 A	19.15	16.55	35.70

D5010 Electrical Service/Distribution

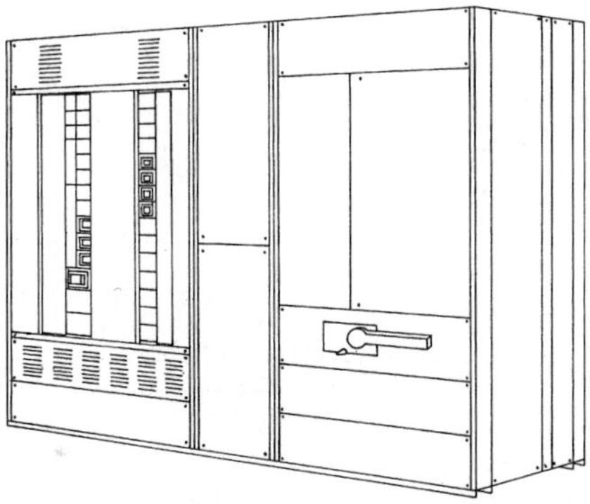

System Components	QUANTITY	UNIT	COST EACH MAT.	COST EACH INST.	COST EACH TOTAL
SYSTEM D5010 240 0240					
SWITCHGEAR INSTALLATION, INCL SWBD, PANELS & CIRC BREAKERS, 600 A					
Switchboards, 120/208 V, 600 amp	1.000	Ea.	4,200	1,250	5,450
Aluminum bus bars, 120/208 V, 600 amp	1.000	Ea.	2,075	1,250	3,325
Feeder section, circuit breakers, KA frame, 70 to 225 amp	1.000	Ea.	1,300	196	1,496
Feeder section, circuit breakers, LA frame, 125 to 400 amp	1.000	Ea.	2,950	273	3,223
TOTAL			10,525	2,969	13,494

D5010 240	Switchgear	COST EACH MAT.	COST EACH INST.	COST EACH TOTAL
0190	Switchgear installation, including switchboard, panels, & circuit breaker			
0200	120/208 V, 1 phase, 400 A	8,050	2,675	10,725
0240	600 A	10,500	2,975	13,475
0280	800 A	13,500	3,425	16,925
0300	1000 A	16,500	3,825	20,325
0320	1200 A	17,500	4,325	21,825
0360	1600 A	26,600	4,975	31,575
0400	2000 A	31,800	5,425	37,225
0500	277/480 V, 3 phase, 400 A	11,400	4,975	16,375
0520	600 A	16,000	5,675	21,675
0540	800 A	18,900	6,275	25,175
0560	1000 A	23,500	6,975	30,475
0580	1200 A	25,700	7,825	33,525
0600	1600 A	34,800	8,575	43,375
0620	2000 A	42,500	9,475	51,975

D5010 Electrical Service/Distribution

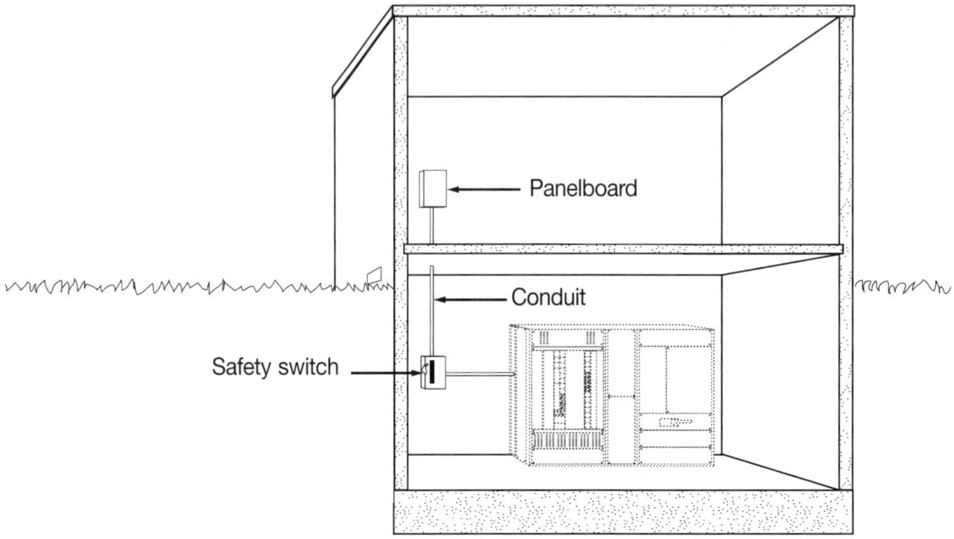

System Components	QUANTITY	UNIT	COST EACH MAT.	COST EACH INST.	COST EACH TOTAL
SYSTEM D5010 250 1020					
PANELBOARD INSTALLATION, INCLUDES PANELBOARD, CONDUCTOR, & CONDUIT					
Conduit, galvanized steel, 1-1/4"dia	37.000	L.F.	216.45	386.65	603.10
Panelboards, NQOD, 4 wire, 120/208 volts, 100 amp main, 24 circuits	1.000	Ea.	1,275	1,325	2,600
Wire, 600 volt, type THW, copper, stranded, #3	1.480	C.L.F.	202.76	185	387.76
TOTAL			1,694.21	1,896.65	3,590.86

D5010 250	Panelboard	COST EACH MAT.	COST EACH INST.	COST EACH TOTAL
0900	Panelboards, NQOD, 4 wire, 120/208 volts w/conductor & conduit			
1000	100 A, 0 stories, 0′ horizontal	1,275	1,325	2,600
1020	1 stories, 25′ horizontal	1,700	1,900	3,600
1040	5 stories, 50′ horizontal	2,525	3,025	5,550
1060	10 stories, 75′ horizontal	3,475	4,350	7,825
1080	225A, 0 stories, 0′ horizontal	2,375	1,750	4,125
2000	1 stories, 25′ horizontal	3,650	2,775	6,425
2020	5 stories, 50′ horizontal	6,175	4,800	10,975
2040	10 stories, 75′ horizontal	9,100	7,175	16,275
2060	400A, 0 stories, 0′ horizontal	3,500	2,625	6,125
2080	1 stories, 25′ horizontal	5,175	4,300	9,475
3000	5 stories, 50′ horizontal	8,500	7,600	16,100
3020	10 stories, 75′ horizontal	17,800	14,900	32,700
3040	600 A, 0 stories, 0′ horizontal	5,200	3,125	8,325
3060	1 stories, 25′ horizontal	8,875	5,850	14,725
3080	5 stories, 50′ horizontal	16,200	11,200	27,400
4000	10 stories, 75′ horizontal	36,200	22,100	58,300
4010	Panelboards, NEHB, 4 wire, 277/480 volts w/conductor, conduit, & safety switch			
4020	100 A, 0 stories, 0′ horizontal, includes safety switch	3,050	1,825	4,875
4040	1 stories, 25′ horizontal	3,475	2,400	5,875
4060	5 stories, 50′ horizontal	4,300	3,525	7,825
4080	10 stories, 75′ horizontal	5,275	4,850	10,125
5000	225 A, 0 stories, 0′ horizontal	4,600	2,225	6,825
5020	1 stories, 25′ horizontal	5,875	3,250	9,125
5040	5 stories, 50′ horizontal	8,400	5,300	13,700
5060	10 stories, 75′ horizontal	11,300	7,650	18,950

D5010 Electrical Service/Distribution

D5010 250	Panelboard	COST EACH		
		MAT.	INST.	TOTAL
5080	400 A, 0 stories, 0′ horizontal	7,675	3,425	11,100
6000	1 stories, 25′ horizontal	9,350	5,100	14,450
6020	5 stories, 50′ horizontal	12,700	8,400	21,100
6040	10 stories, 75′ horizontal	22,000	15,700	37,700
6060	600 A, 0 stories, 0′ horizontal	11,300	4,350	15,650
6080	1 stories, 25′ horizontal	15,000	7,075	22,075
7000	5 stories, 50′ horizontal	22,300	12,500	34,800
7020	10 stories, 75′ horizontal	42,300	23,400	65,700

D5020 Lighting and Branch Wiring

Duplex Receptacle

System Components	QUANTITY	UNIT	COST PER S.F. MAT.	COST PER S.F. INST.	COST PER S.F. TOTAL
SYSTEM D5020 110 0200					
RECEPTACLES INCL. PLATE, BOX, CONDUIT, WIRE & TRANS. WHEN REQUIRED					
2.5 PER 1000 S.F., .3 WATTS PER S.F.					
Steel intermediate conduit, (IMC) 1/2" diam	167.000	L.F.	.33	1.04	1.37
Wire 600V type THWN-THHN, copper solid #12	3.340	C.L.F.	.05	.19	.24
Wiring device, receptacle, duplex, 120V grounded, 15 amp	2.500	Ea.		.04	.04
Wall plate, 1 gang, brown plastic	2.500	Ea.		.02	.02
Steel outlet box 4" square	2.500	Ea.	.01	.08	.09
Steel outlet box 4" plaster rings	2.500	Ea.	.01	.02	.03
TOTAL			.40	1.39	1.79

D5020 110	Receptacle (by Wattage)	COST PER S.F. MAT.	COST PER S.F. INST.	COST PER S.F. TOTAL
0190	Receptacles include plate, box, conduit, wire & transformer when required			
0200	2.5 per 1000 S.F., .3 watts per S.F.	.40	1.39	1.79
0240	With transformer	.47	1.46	1.93
0280	4 per 1000 S.F., .5 watts per S.F.	.45	1.63	2.08
0320	With transformer	.56	1.74	2.30
0360	5 per 1000 S.F., .6 watts per S.F.	.53	1.93	2.46
0400	With transformer	.67	2.07	2.74
0440	8 per 1000 S.F., .9 watts per S.F.	.56	2.13	2.69
0480	With transformer	.78	2.35	3.13
0520	10 per 1000 S.F., 1.2 watts per S.F.	.59	2.33	2.92
0560	With transformer	.87	2.61	3.48
0600	16.5 per 1000 S.F., 2.0 watts per S.F.	.69	2.89	3.58
0640	With transformer	1.14	3.35	4.49
0680	20 per 1000 S.F., 2.4 watts per S.F.	.73	3.15	3.88
0720	With transformer	1.28	3.71	4.99

D5020 Lighting and Branch Wiring

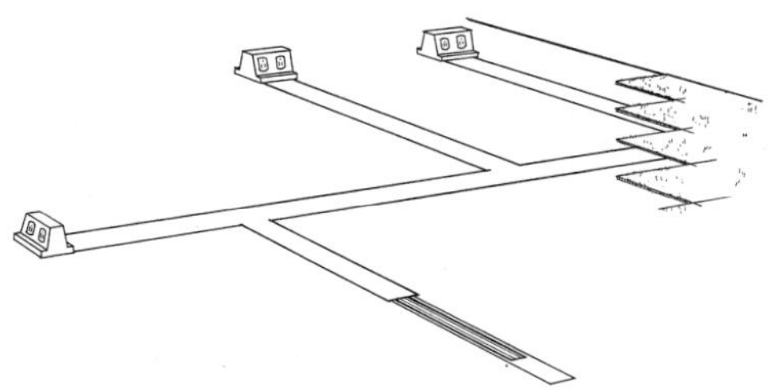

Underfloor Receptacle System

Description: Table D5020 115 includes installed costs of raceways and copper wire from panel to and including receptacle.

National Electrical Code prohibits use of undercarpet system in residential, school or hospital buildings. Can only be used with carpet squares.

Low density = (1) Outlet per 259 S.F. of floor area.

High density = (1) Outlet per 127 S.F. of floor area.

System Components	QUANTITY	UNIT	COST PER S.F. MAT.	INST.	TOTAL
SYSTEM D5020 115 0200					
RECEPTACLE SYSTEMS, UNDERFLOOR DUCT, 5′ ON CENTER, LOW DENSITY					
Underfloor duct 3-1/8″ x 7/8″ w/insert 24″ on center	.190	L.F.	3.90	1.70	5.60
Underfloor duct junction box, single duct, 3-1/8″	.003	Ea.	1.38	.47	1.85
Underfloor junction box carpet pan	.003	Ea.	1.14	.02	1.16
Underfloor duct outlet, high tension receptacle	.004	Ea.	.43	.31	.74
Wire 600V type THWN-THHN copper solid #12	.010	C.L.F.	.14	.57	.71
Vertical elbow for underfloor duct, 3-1/8″, included					
Underfloor duct conduit adapter, 2″ x 1-1/4″, included					
TOTAL			6.99	3.07	10.06

D5020 115	Receptacles, Floor	COST PER S.F. MAT.	INST.	TOTAL
0200	Receptacle systems, underfloor duct, 5′ on center, low density	7	3.07	10.07
0240	High density	7.55	3.96	11.51
0280	7′ on center, low density	5.55	2.64	8.19
0320	High density	6.10	3.53	9.63
0400	Poke thru fittings, low density	1.20	1.50	2.70
0440	High density	2.37	3	5.37
0520	Telepoles, using Romex, low density	1.19	.93	2.12
0560	High density	2.38	1.86	4.24
0600	Using EMT, low density	1.27	1.24	2.51
0640	High density	2.51	2.46	4.97
0720	Conduit system with floor boxes, low density	1.29	1.04	2.33
0760	High density	2.58	2.10	4.68
0840	Undercarpet power system, 3 conductor with 5 conductor feeder, low density	1.49	.38	1.87
0880	High density	2.95	.79	3.74

D5020 Lighting and Branch Wiring

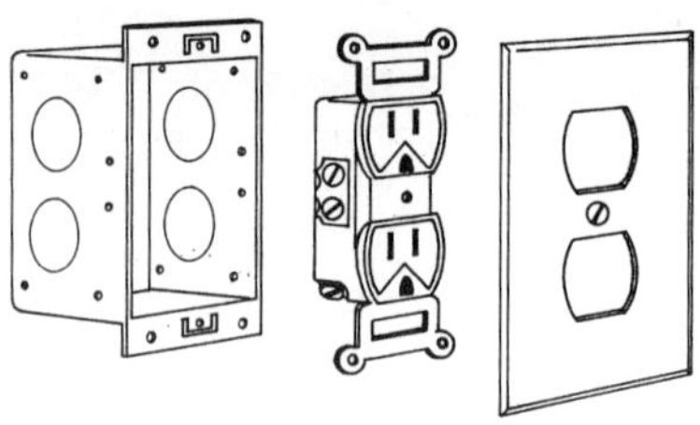

Duplex Receptacle

Wall Switch

System Components	QUANTITY	UNIT	COST PER S.F. MAT.	COST PER S.F. INST.	COST PER S.F. TOTAL
SYSTEM D5020 120 0520					
RECEPTACLES AND WALL SWITCHES					
4 RECEPTACLES PER 400 S.F.					
Steel intermediate conduit, (IMC), 1/2″ diam	.220	L.F.	.44	1.38	1.82
Wire, 600 volt, type THWN-THHN, copper, solid #12	.005	C.L.F.	.07	.29	.36
Steel outlet box 4″ square	.010	Ea.	.03	.32	.35
Steel outlet box, 4″ square, plaster rings	.010	Ea.	.04	.10	.14
Receptacle, duplex, 120 volt grounded, 15 amp	.010	Ea.	.01	.16	.17
Wall plate, 1 gang, brown plastic	.010	Ea.		.08	.08
TOTAL			.59	2.33	2.92

D5020 120	Receptacles & Wall Switches	COST PER S.F. MAT.	COST PER S.F. INST.	COST PER S.F. TOTAL
0520	Receptacles and wall switches, 400 S.F., 4 receptacles	.59	2.33	2.92
0560	6 receptacles	.65	2.71	3.36
0600	8 receptacles	.72	3.14	3.86
0640	1 switch	.14	.52	.66
0680	600 S.F., 6 receptacles	.59	2.33	2.92
0720	8 receptacles	.66	2.62	3.28
0760	10 receptacles	.70	2.95	3.65
0800	2 switches	.19	.68	.87
0840	1000 S.F., 10 receptacles	.58	2.31	2.89
0880	12 receptacles	.60	2.43	3.03
0920	14 receptacles	.64	2.63	3.27
0960	2 switches	.10	.38	.48
1000	1600 S.F., 12 receptacles	.52	2.05	2.57
1040	14 receptacles	.56	2.23	2.79
1080	16 receptacles	.58	2.31	2.89
1120	4 switches	.12	.44	.56
1160	2000 S.F., 14 receptacles	.56	2.10	2.66
1200	16 receptacles	.54	2.11	2.65
1240	18 receptacles	.56	2.23	2.79
1280	4 switches	.10	.38	.48
1320	3000 S.F., 12 receptacles	.44	1.61	2.05
1360	18 receptacles	.53	1.98	2.51
1400	24 receptacles	.54	2.11	2.65
1440	6 switches	.10	.38	.48
1480	3600 S.F., 20 receptacles	.51	1.92	2.43
1520	24 receptacles	.52	1.98	2.50
1560	28 receptacles	.54	2.04	2.58
1600	8 switches	.12	.43	.55
1640	4000 S.F., 16 receptacles	.44	1.61	2.05
1680	24 receptacles	.53	1.98	2.51
1720	30 receptacles	.52	2.05	2.57
1760	8 switches	.10	.38	.48

D5020 Lighting and Branch Wiring

D5020 120	Receptacles & Wall Switches	COST PER S.F.		
		MAT.	INST.	TOTAL
1800	5000 S.F., 20 receptacles	.44	1.61	2.05
1840	26 receptacles	.52	1.92	2.44
1880	30 receptacles	.53	1.98	2.51
1920	10 switches	.10	.38	.48

D5020 Lighting and Branch Wiring

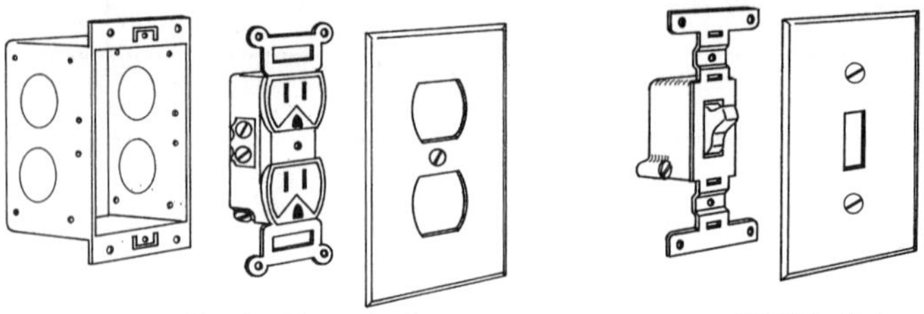

System Components	QUANTITY	UNIT	COST PER EACH MAT.	COST PER EACH INST.	COST PER EACH TOTAL
SYSTEM D5020 125 0520					
RECEPTACLES AND WALL SWITCHES, RECEPTICLE DUPLEX 120 V GROUNDED, 15 A					
Electric metallic tubing conduit, (EMT), 3/4" diam	22.000	L.F.	23.32	106.04	129.36
Wire, 600 volt, type THWN-THHN, copper, solid #12	.630	C.L.F.	8.63	35.91	44.54
Steel outlet box 4" square	1.000	Ea.	2.71	31.50	34.21
Steel outlet box, 4" square, plaster rings	1.000	Ea.	3.65	9.80	13.45
Receptacle, duplex, 120 volt grounded, 15 amp	1.000	Ea.	1.34	15.70	17.04
Wall plate, 1 gang, brown plastic	1.000	Ea.	.42	7.85	8.27
TOTAL			40.07	206.80	246.87

D5020 125	Receptacles & Switches by Each	COST PER EACH MAT.	COST PER EACH INST.	COST PER EACH TOTAL
0460	Receptacles & Switches, with box, plate, 3/4" EMT conduit & wire			
0520	Receptacle duplex 120 V grounded, 15 A	40	207	247
0560	20 A	49.50	214	263.50
0600	Receptacle duplex ground fault interrupting, 15 A	79.50	214	293.50
0640	20 A	82	214	296
0680	Toggle switch single, 15 A	44.50	207	251.50
0720	20 A	48.50	214	262.50
0760	3 way switch, 15 A	46.50	219	265.50
0800	20 A	48	226	274
0840	4 way switch, 15 A	52.50	233	285.50
0880	20 A	84.50	248	332.50

D5020 Lighting and Branch Wiring

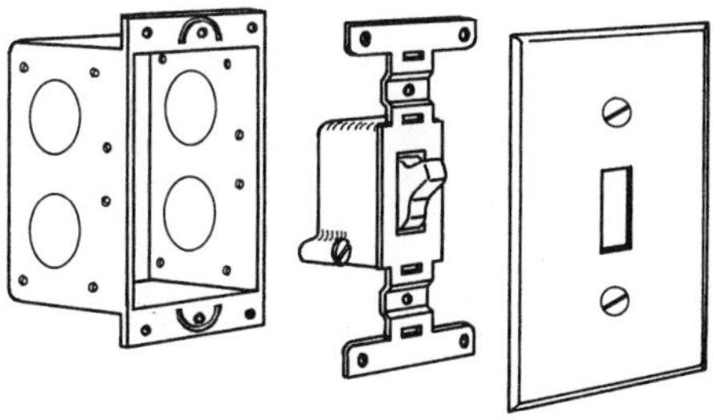

Description: Table D5020 130 includes the cost for switch, plate, box, conduit in slab or EMT exposed and copper wire. Add 20% for exposed conduit.

No power required for switches.

Federal energy guidelines recommend the maximum lighting area controlled per switch shall not exceed 1000 S.F. and that areas over 500 S.F. shall be so controlled that total illumination can be reduced by at least 50%.

System Components	QUANTITY	UNIT	COST PER S.F. MAT.	COST PER S.F. INST.	COST PER S.F. TOTAL
SYSTEM D5020 130 0360					
WALL SWITCHES, 5.0 PER 1000 S.F.					
Steel, intermediate conduit (IMC), 1/2" diameter	88.000	L.F.	.18	.55	.73
Wire, 600V type THWN-THHN, copper solid #12	1.710	C.L.F.	.02	.10	.12
Toggle switch, single pole, 15 amp	5.000	Ea.	.03	.08	.11
Wall plate, 1 gang, brown plastic	5.000	Ea.		.04	.04
Steel outlet box 4" plaster rings	5.000	Ea.	.01	.16	.17
Plaster rings	5.000	Ea.	.02	.05	.07
TOTAL			.26	.98	1.24

D5020 130	Wall Switch by Sq. Ft.	COST PER S.F. MAT.	COST PER S.F. INST.	COST PER S.F. TOTAL
0200	Wall switches, 1.0 per 1000 S.F.	.06	.23	.29
0240	1.2 per 1000 S.F.	.07	.25	.32
0280	2.0 per 1000 S.F.	.10	.36	.46
0320	2.5 per 1000 S.F.	.13	.45	.58
0360	5.0 per 1000 S.F.	.26	.98	1.24
0400	10.0 per 1000 S.F.	.54	1.98	2.52

D5020 Lighting and Branch Wiring

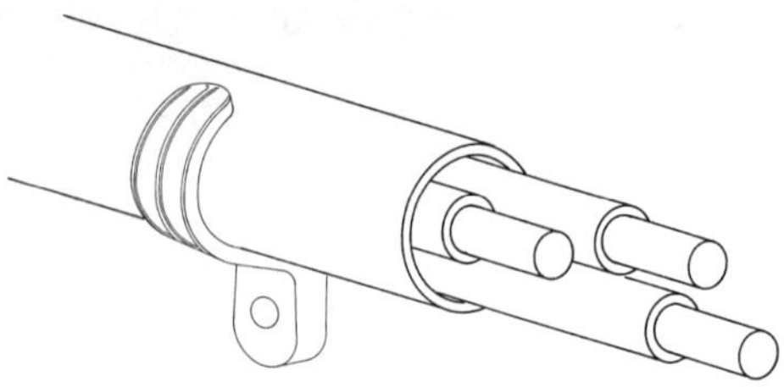

System D5020 135 includes all wiring and connections.

System Components	QUANTITY	UNIT	COST PER S.F. MAT.	INST.	TOTAL
SYSTEM D5020 135 0200					
MISCELLANEOUS POWER, TO .5 WATTS					
Steel intermediate conduit, (IMC) 1/2" diam	15.000	L.F.	.03	.09	.12
Wire 600V type THWN-THHN, copper solid #12	.325	C.L.F.		.02	.02
TOTAL			.03	.11	.14

D5020 135	Miscellaneous Power	COST PER S.F. MAT.	INST.	TOTAL
0200	Miscellaneous power, to .5 watts	.03	.11	.14
0240	.8 watts	.05	.16	.21
0280	1 watt	.07	.21	.28
0320	1.2 watts	.08	.25	.33
0360	1.5 watts	.09	.29	.38
0400	1.8 watts	.10	.34	.44
0440	2 watts	.12	.40	.52
0480	2.5 watts	.15	.50	.65
0520	3 watts	.18	.58	.76

D5020 Lighting and Branch Wiring

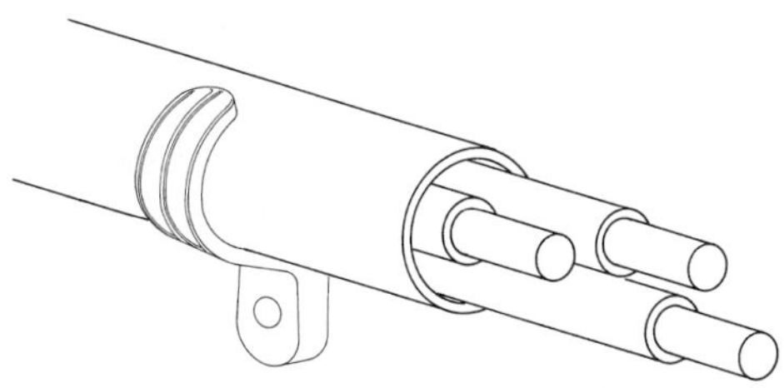

System D5020 140 includes all wiring and connections for central air conditioning units.

System Components	QUANTITY	UNIT	COST PER S.F. MAT.	INST.	TOTAL
SYSTEM D5020 140 0200					
CENTRAL AIR CONDITIONING POWER, 1 WATT					
Steel intermediate conduit, 1/2" diam.	.030	L.F.	.06	.19	.25
Wire 600V type THWN-THHN, copper solid #12	.001	C.L.F.	.01	.06	.07
TOTAL			.07	.25	.32

D5020 140	Central A. C. Power (by Wattage)	COST PER S.F. MAT.	INST.	TOTAL
0200	Central air conditioning power, 1 watt	.07	.25	.32
0220	2 watts	.08	.29	.37
0240	3 watts	.12	.32	.44
0280	4 watts	.18	.43	.61
0320	6 watts	.33	.56	.89
0360	8 watts	.43	.61	1.04
0400	10 watts	.61	.71	1.32

D5020 Lighting and Branch Wiring

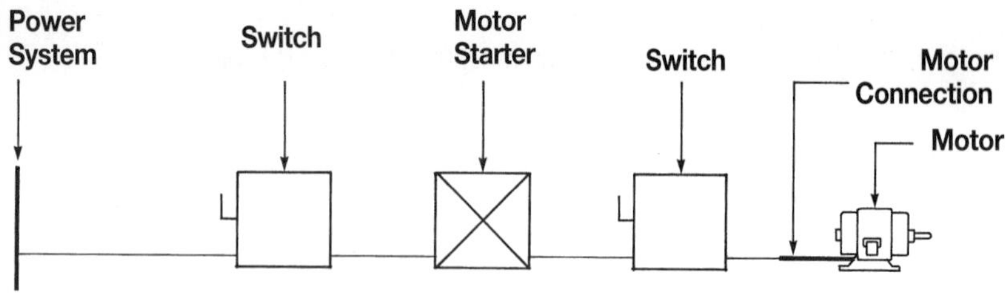

System D5020 145 installed cost of motor wiring using 50′ of rigid conduit and copper wire. **Cost and setting of motor not included.**

System Components	QUANTITY	UNIT	COST EACH MAT.	COST EACH INST.	COST EACH TOTAL
SYSTEM D5020 145 0200					
MOTOR INST., SINGLE PHASE, 115V, TO AND INCLUDING 1/3 HP MOTOR SIZE					
Wire 600V type THWN-THHN, copper solid #12	1.250	C.L.F.	17.13	71.25	88.38
Steel intermediate conduit, (IMC) 1/2″ diam.	50.000	L.F.	100	312.50	412.50
Magnetic FVNR, 115V, 1/3 HP, size 00 starter	1.000	Ea.	193	157	350
Safety switch, fused, heavy duty, 240V 2P 30 amp	1.000	Ea.	135	179	314
Safety switch, non fused, heavy duty, 600V, 3 phase, 30 A	1.000	Ea.	159	196	355
Flexible metallic conduit, Greenfield 1/2″ diam.	1.500	L.F.	1.11	4.71	5.82
Connectors for flexible metallic conduit Greenfield 1/2″ diam.	1.000	Ea.	3.45	7.85	11.30
Coupling for Greenfield to conduit 1/2″ diam. flexible metalic conduit	1.000	Ea.	2.26	12.55	14.81
Fuse cartridge nonrenewable, 250V 30 amp	1.000	Ea.	2.21	12.55	14.76
TOTAL			613.16	953.41	1,566.57

D5020 145	Motor Installation	COST EACH MAT.	COST EACH INST.	COST EACH TOTAL
0200	Motor installation, single phase, 115V, 1/3 HP motor size	615	955	1,570
0240	1 HP motor size	635	955	1,590
0280	2 HP motor size	685	1,000	1,685
0320	3 HP motor size	765	1,025	1,790
0360	230V, 1 HP motor size	615	965	1,580
0400	2 HP motor size	650	965	1,615
0440	3 HP motor size	730	1,050	1,780
0520	Three phase, 200V, 1-1/2 HP motor size	730	1,050	1,780
0560	3 HP motor size	785	1,150	1,935
0600	5 HP motor size	850	1,275	2,125
0640	7-1/2 HP motor size	885	1,300	2,185
0680	10 HP motor size	1,400	1,650	3,050
0720	15 HP motor size	1,800	1,825	3,625
0760	20 HP motor size	2,325	2,100	4,425
0800	25 HP motor size	2,375	2,125	4,500
0840	30 HP motor size	3,675	2,500	6,175
0880	40 HP motor size	4,700	2,950	7,650
0920	50 HP motor size	7,875	3,450	11,325
0960	60 HP motor size	8,250	3,625	11,875
1000	75 HP motor size	10,800	4,150	14,950
1040	100 HP motor size	28,800	4,900	33,700
1080	125 HP motor size	29,700	5,375	35,075
1120	150 HP motor size	34,600	6,325	40,925
1160	200 HP motor size	35,300	7,500	42,800
1240	230V, 1-1/2 HP motor size	690	1,050	1,740
1280	3 HP motor size	745	1,150	1,895
1320	5 HP motor size	810	1,275	2,085
1360	7-1/2 HP motor size	810	1,275	2,085
1400	10 HP motor size	1,250	1,550	2,800
1440	15 HP motor size	1,450	1,700	3,150
1480	20 HP motor size	2,175	2,050	4,225
1520	25 HP motor size	2,325	2,100	4,425

D5020 Lighting and Branch Wiring

D5020 145	Motor Installation	MAT.	INST.	TOTAL
		COST EACH		
1560	30 HP motor size	2,375	2,125	4,500
1600	40 HP motor size	4,400	2,875	7,275
1640	50 HP motor size	4,700	3,050	7,750
1680	60 HP motor size	7,875	3,450	11,325
1720	75 HP motor size	9,750	3,925	13,675
1760	100 HP motor size	11,200	4,375	15,575
1800	125 HP motor size	29,300	5,050	34,350
1840	150 HP motor size	31,400	5,750	37,150
1880	200 HP motor size	33,400	6,400	39,800
1960	460V, 2 HP motor size	855	1,050	1,905
2000	5 HP motor size	910	1,150	2,060
2040	10 HP motor size	940	1,275	2,215
2080	15 HP motor size	1,225	1,450	2,675
2120	20 HP motor size	1,300	1,575	2,875
2160	25 HP motor size	1,425	1,650	3,075
2200	30 HP motor size	1,775	1,775	3,550
2240	40 HP motor size	2,300	1,900	4,200
2280	50 HP motor size	2,600	2,100	4,700
2320	60 HP motor size	3,775	2,475	6,250
2360	75 HP motor size	4,450	2,750	7,200
2400	100 HP motor size	4,975	3,050	8,025
2440	125 HP motor size	8,175	3,475	11,650
2480	150 HP motor size	10,100	3,875	13,975
2520	200 HP motor size	12,000	4,375	16,375
2600	575V, 2 HP motor size	855	1,050	1,905
2640	5 HP motor size	910	1,150	2,060
2680	10 HP motor size	940	1,275	2,215
2720	20 HP motor size	1,225	1,450	2,675
2760	25 HP motor size	1,300	1,575	2,875
2800	30 HP motor size	1,775	1,775	3,550
2840	50 HP motor size	1,925	1,825	3,750
2880	60 HP motor size	3,725	2,475	6,200
2920	75 HP motor size	3,775	2,475	6,250
2960	100 HP motor size	4,450	2,750	7,200
3000	125 HP motor size	7,700	3,400	11,100
3040	150 HP motor size	8,175	3,475	11,650
3080	200 HP motor size	10,500	3,925	14,425

D5020 Lighting and Branch Wiring

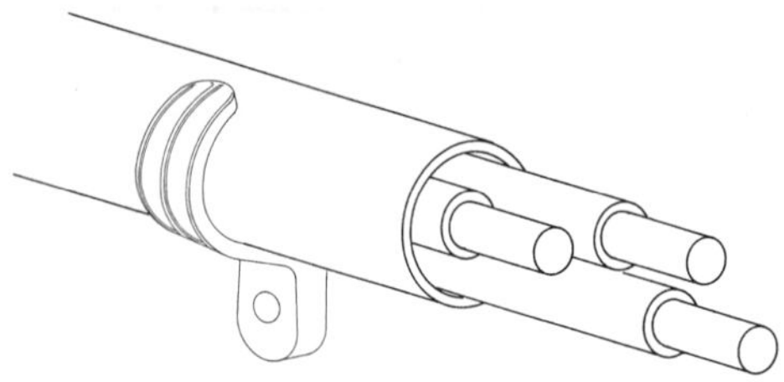

System Components	QUANTITY	UNIT	COST PER L.F. MAT.	INST.	TOTAL
SYSTEM D5020 155 0200					
MOTOR FEEDER SYSTEMS, SINGLE PHASE, UP TO 115V, 1HP OR 230V, 2HP					
Steel intermediate conduit, (IMC) 1/2" diam	1.000	L.F.	2	6.25	8.25
Wire 600V type THWN-THHN, copper solid #12	.020	C.L.F.	.27	1.14	1.41
TOTAL			2.27	7.39	9.66

D5020 155	Motor Feeder	COST PER L.F. MAT.	INST.	TOTAL
0200	Motor feeder systems, single phase, feed up to 115V 1HP or 230V 2 HP	2.27	7.40	9.67
0240	115V 2HP, 230V 3HP	2.43	7.50	9.93
0280	115V 3HP	2.81	7.80	10.61
0360	Three phase, feed to 200V 3HP, 230V 5HP, 460V 10HP, 575V 10HP	2.41	7.95	10.36
0440	200V 5HP, 230V 7.5HP, 460V 15HP, 575V 20HP	2.65	8.15	10.80
0520	200V 10HP, 230V 10HP, 460V 30HP, 575V 30HP	3.22	8.60	11.82
0600	200V 15HP, 230V 15HP, 460V 40HP, 575V 50HP	4.48	9.85	14.33
0680	200V 20HP, 230V 25HP, 460V 50HP, 575V 60HP	6.85	12.50	19.35
0760	200V 25HP, 230V 30HP, 460V 60HP, 575V 75HP	7.65	12.70	20.35
0840	200V 30HP	8.70	13.10	21.80
0920	230V 40HP, 460V 75HP, 575V 100HP	11.30	14.35	25.65
1000	200V 40HP	12.75	15.35	28.10
1080	230V 50HP, 460V 100HP, 575V 125HP	16	16.95	32.95
1160	200V 50HP, 230V 60HP, 460V 125HP, 575V 150HP	18.70	18	36.70
1240	200V 60HP, 460V 150HP	23	21	44
1320	230V 75HP, 575V 200HP	26.50	22	48.50
1400	200V 75HP	30	22.50	52.50
1480	230V 100HP, 460V 200HP	40.50	26	66.50
1560	200V 100HP	55	32.50	87.50
1640	230V 125HP	55	32.50	87.50
1720	200V 125HP, 230V 150HP	64	41	105
1800	200V 150HP	78	44	122
1880	200V 200HP	98	46.50	144.50
1960	230V 200HP	100	48.50	148.50

D5020 Lighting and Branch Wiring

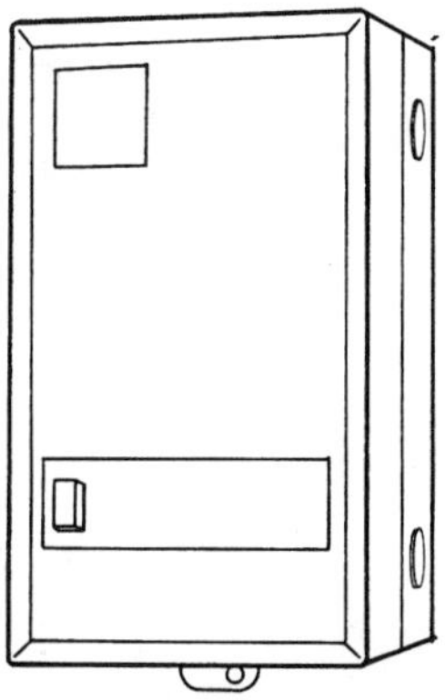

Starters are full voltage, type NEMA 1 for general purpose indoor application with motor overload protection and include mounting and wire connections.

System Components	QUANTITY	UNIT	COST EACH MAT.	COST EACH INST.	COST EACH TOTAL
SYSTEM D5020 160 0200					
MAGNETIC STARTER, SIZE 00 TO 1/3 HP, 1 PHASE 115V OR 1 HP 230V					
Magnetic starter, size 00, to 1/3 HP, 1 phase, 115V or 1 HP 230V	1.000	Ea.	193	157	350
TOTAL			193	157	350

D5020 160	Magnetic Starter	COST EACH MAT.	COST EACH INST.	COST EACH TOTAL
0200	Magnetic starter, size 00, to 1/3 HP, 1 phase, 115V or 1 HP 230V	193	157	350
0280	Size 00, to 1-1/2 HP, 3 phase, 200-230V or 2 HP 460-575V	212	179	391
0360	Size 0, to 1 HP, 1 phase, 115V or 2 HP 230V	215	157	372
0440	Size 0, to 3 HP, 3 phase, 200-230V or 5 HP 460-575V	267	273	540
0520	Size 1, to 2 HP, 1 phase, 115V or 3 HP 230V	247	209	456
0600	Size 1, to 7-1/2 HP, 3 phase, 200-230V or 10 HP 460-575V	300	390	690
0680	Size 2, to 10 HP, 3 phase, 200V, 15 HP-230V or 25 HP 460-575V	565	570	1,135
0760	Size 3, to 25 HP, 3 phase, 200V, 30 HP-230V or 50 HP 460-575V	920	695	1,615
0840	Size 4, to 40 HP, 3 phase, 200V, 50 HP-230V or 100 HP 460-575V	2,050	1,050	3,100
0920	Size 5, to 75 HP, 3 phase, 200V, 100 HP-230V or 200 HP 460-575V	4,775	1,400	6,175
1000	Size 6, to 150 HP, 3 phase, 200V, 200 HP-230V or 400 HP 460-575V	21,400	1,575	22,975

D5020 Lighting and Branch Wiring

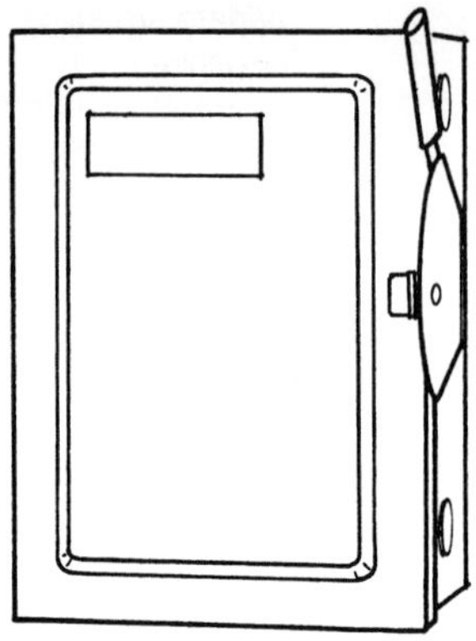

Safety switches are type NEMA 1 for general purpose indoor application, and include time delay fuses, insulation and wire terminations.

System Components	QUANTITY	UNIT	COST EACH MAT.	COST EACH INST.	COST EACH TOTAL
SYSTEM D5020 165 0200					
SAFETY SWITCH, 30A FUSED, 1 PHASE, 115V OR 230V					
Safety switch fused, hvy duty, 240V 2p 30 amp	1.000	Ea.	135	179	314
Fuse, dual element time delay 250V, 30 amp	2.000	Ea.	15.80	25.10	40.90
TOTAL			150.80	204.10	354.90

D5020 165	Safety Switches	COST EACH MAT.	COST EACH INST.	COST EACH TOTAL
0200	Safety switch, 30 A fused, 1 phase, 2 HP 115 V or 3 HP, 230 V	151	204	355
0280	3 phase, 5 HP, 200 V or 7 1/2 HP, 230 V	204	234	438
0360	15 HP, 460 V or 20 HP, 575 V	350	243	593
0440	60 A fused, 3 phase, 15 HP 200 V or 15 HP 230 V	345	310	655
0520	30 HP 460 V or 40 HP 575 V	450	320	770
0600	100 A fused, 3 phase, 20 HP 200 V or 25 HP 230 V	565	375	940
0680	50 HP 460 V or 60 HP 575 V	840	380	1,220
0760	200 A fused, 3 phase, 50 HP 200 V or 60 HP 230 V	1,025	530	1,555
0840	125 HP 460 V or 150 HP 575 V	1,300	545	1,845
0920	400 A fused, 3 phase, 100 HP 200 V or 125 HP 230 V	2,450	760	3,210
1000	250 HP 460 V or 350 HP 575 V	3,225	775	4,000
1020	600 A fused, 3 phase, 150 HP 200 V or 200 HP 230 V	4,225	1,125	5,350
1040	400 HP 460 V	5,225	1,150	6,375

D5020 Lighting and Branch Wiring

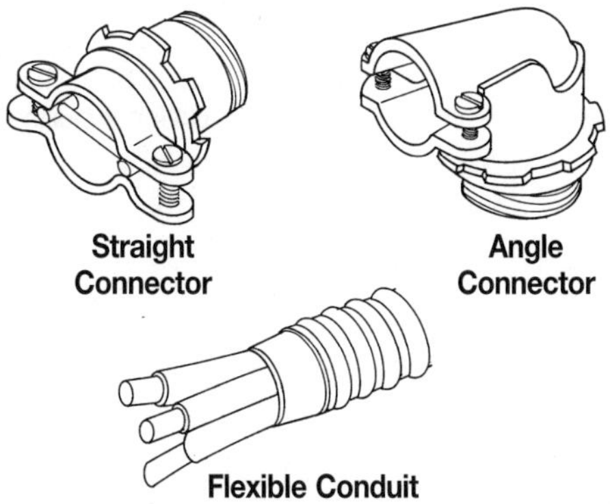

Table below includes costs for the flexible conduit. Not included are wire terminations and testing motor for correct rotation.

System Components	QUANTITY	UNIT	COST EACH MAT.	COST EACH INST.	COST EACH TOTAL
SYSTEM D5020 170 0200					
MOTOR CONNECTIONS, SINGLE PHASE, 115V/230V UP TO 1 HP					
Motor connection, flexible conduit & fittings, 1 HP motor 115V	1.000	Ea.	9.30	78.50	87.80
TOTAL			9.30	78.50	87.80

D5020 170	Motor Connections	COST EACH MAT.	COST EACH INST.	COST EACH TOTAL
0200	Motor connections, single phase, 115/230V, up to 1 HP	9.30	78.50	87.80
0240	Up to 3 HP	17.55	96.50	114.05
0280	Three phase, 200/230/460/575V, up to 3 HP	11	92.50	103.50
0320	Up to 5 HP	9.45	115	124.45
0360	Up to 7-1/2 HP	13.40	136	149.40
0400	Up to 10 HP	30	149	179
0440	Up to 15 HP	30	190	220
0480	Up to 25 HP	42.50	232	274.50
0520	Up to 50 HP	85	285	370
0560	Up to 100 HP	200	420	620

D5020 Lighting and Branch Wiring

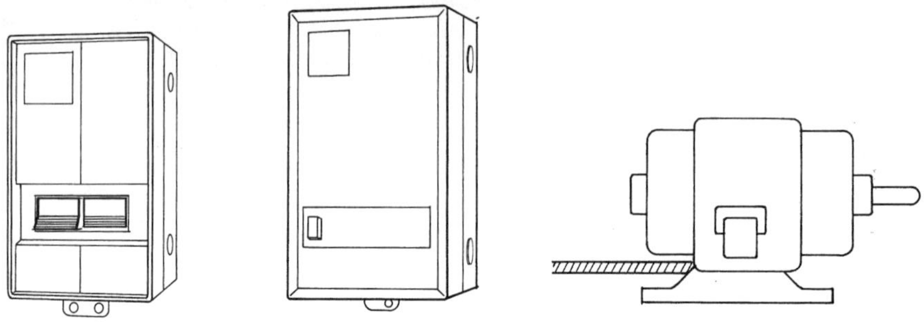

Manual Starter — Magnetic Starter — Induction Motor

For 230/460 Volt A.C., 3 phase, 60 cycle ball bearing squirrel cage induction motors, NEMA Class B standard line. Installation included.

No conduit, wire, or terminations included.

System Components	QUANTITY	UNIT	COST EACH MAT.	COST EACH INST.	COST EACH TOTAL
SYSTEM D5020 175 0220					
MOTOR, DRIPPROOF CLASS B INSULATION, 1.15 SERVICE FACTOR, WITH STARTER					
1 H.P., 1200 RPM WITH MANUAL STARTER					
Motor, dripproof, class B insul, 1.15 serv fact, 1200 RPM, 1 HP	1.000	Ea.	405	139	544
Motor starter, manual, 3 phase, 1 HP motor	1.000	Ea.	223	179	402
TOTAL			628	318	946

D5020 175	Motor & Starter	COST EACH MAT.	COST EACH INST.	COST EACH TOTAL
0190	Motor, dripproof, premium efficient, 1.15 service factor			
0200	1 HP, 1200 RPM, motor only	405	139	544
0220	With manual starter	630	320	950
0240	With magnetic starter	615	320	935
0260	1800 RPM, motor only	310	139	449
0280	With manual starter	535	320	855
0300	With magnetic starter	520	320	840
0320	2 HP, 1200 RPM, motor only	465	139	604
0340	With manual starter	690	320	1,010
0360	With magnetic starter	730	410	1,140
0380	1800 RPM, motor only	355	139	494
0400	With manual starter	580	320	900
0420	With magnetic starter	620	410	1,030
0440	3600 RPM, motor only	455	139	594
0460	With manual starter	680	320	1,000
0480	With magnetic starter	720	410	1,130
0500	3 HP, 1200 RPM, motor only	650	139	789
0520	With manual starter	875	320	1,195
0540	With magnetic starter	915	410	1,325
0560	1800 RPM, motor only	480	139	619
0580	With manual starter	705	320	1,025
0600	With magnetic starter	745	410	1,155
0620	3600 RPM, motor only	460	139	599
0640	With manual starter	685	320	1,005
0660	With magnetic starter	725	410	1,135
0680	5 HP, 1200 RPM, motor only	880	139	1,019
0700	With manual starter	1,150	455	1,605
0720	With magnetic starter	1,175	530	1,705
0740	1800 RPM, motor only	575	139	714
0760	With manual starter	840	455	1,295
0780	With magnetic starter	875	530	1,405
0800	3600 RPM, motor only	500	139	639

D5020 Lighting and Branch Wiring

D5020 175	Motor & Starter	COST EACH		
		MAT.	INST.	TOTAL
0820	With manual starter	765	455	1,220
0840	With magnetic starter	800	530	1,330
0860	7.5 HP, 1800 RPM, motor only	670	149	819
0880	With manual starter	935	465	1,400
0900	With magnetic starter	1,225	720	1,945
0920	10 HP, 1800 RPM, motor only	820	157	977
0940	With manual starter	1,075	470	1,545
0960	With magnetic starter	1,375	725	2,100
0980	15 HP, 1800 RPM, motor only	1,150	196	1,346
1000	With magnetic starter	1,725	765	2,490
1040	20 HP, 1800 RPM, motor only	1,475	241	1,716
1060	With magnetic starter	2,400	935	3,335
1100	25 HP, 1800 RPM, motor only	1,825	251	2,076
1120	With magnetic starter	2,750	945	3,695
1160	30 HP, 1800 RPM, motor only	2,125	261	2,386
1180	With magnetic starter	3,050	955	4,005
1220	40 HP, 1800 RPM, motor only	2,825	315	3,140
1240	With magnetic starter	4,875	1,375	6,250
1280	50 HP, 1800 RPM, motor only	2,675	390	3,065
1300	With magnetic starter	4,725	1,450	6,175
1340	60 HP, 1800 RPM, motor only	3,950	450	4,400
1360	With magnetic starter	8,725	1,850	10,575
1400	75 HP, 1800 RPM, motor only	3,450	525	3,975
1420	With magnetic starter	8,225	1,925	10,150
1460	100 HP, 1800 RPM, motor only	4,450	695	5,145
1480	With magnetic starter	9,225	2,100	11,325
1520	125 HP, 1800 RPM, motor only	5,125	895	6,020
1540	With magnetic starter	26,500	2,475	28,975
1580	150 HP, 1800 RPM, motor only	7,375	1,050	8,425
1600	With magnetic starter	28,800	2,625	31,425
1640	200 HP, 1800 RPM, motor only	9,000	1,250	10,250
1660	With magnetic starter	30,400	2,825	33,225
1680	Totally encl, premium efficient, 1.0 ser. fac., 1HP, 1200 RPM, motor only	385	139	524
1700	With manual starter	610	320	930
1720	With magnetic starter	595	320	915
1740	1800 RPM, motor only	400	139	539
1760	With manual starter	625	320	945
1780	With magnetic starter	610	320	930
1800	2 HP, 1200 RPM, motor only	445	139	584
1820	With manual starter	670	320	990
1840	With magnetic starter	710	410	1,120
1860	1800 RPM, motor only	490	139	629
1880	With manual starter	715	320	1,035
1900	With magnetic starter	755	410	1,165
1920	3600 RPM, motor only	350	139	489
1940	With manual starter	575	320	895
1960	With magnetic starter	615	410	1,025
1980	3 HP, 1200 RPM, motor only	605	139	744
2000	With manual starter	830	320	1,150
2020	With magnetic starter	870	410	1,280
2040	1800 RPM, motor only	580	139	719
2060	With manual starter	805	320	1,125
2080	With magnetic starter	845	410	1,255
2100	3600 RPM, motor only	430	139	569
2120	With manual starter	655	320	975
2140	With magnetic starter	695	410	1,105
2160	5 HP, 1200 RPM, motor only	825	139	964
2180	With manual starter	1,100	455	1,555

D5020 Lighting and Branch Wiring

D5020 175	Motor & Starter	COST EACH		
		MAT.	INST.	TOTAL
2200	With magnetic starter	1,125	530	1,655
2220	1800 RPM, motor only	625	139	764
2240	With manual starter	890	455	1,345
2260	With magnetic starter	925	530	1,455
2280	3600 RPM, motor only	540	139	679
2300	With manual starter	805	455	1,260
2320	With magnetic starter	840	530	1,370
2340	7.5 HP, 1800 RPM, motor only	900	149	1,049
2360	With manual starter	1,175	465	1,640
2380	With magnetic starter	1,475	720	2,195
2400	10 HP, 1800 RPM, motor only	955	157	1,112
2420	With manual starter	1,225	470	1,695
2440	With magnetic starter	1,525	725	2,250
2460	15 HP, 1800 RPM, motor only	1,350	196	1,546
2480	With magnetic starter	1,925	765	2,690
2500	20 HP, 1800 RPM, motor only	1,875	241	2,116
2520	With magnetic starter	2,800	935	3,735
2540	25 HP, 1800 RPM, motor only	2,150	251	2,401
2560	With magnetic starter	3,075	945	4,020
2580	30 HP, 1800 RPM, motor only	2,250	261	2,511
2600	With magnetic starter	3,175	955	4,130
2620	40 HP, 1800 RPM, motor only	2,750	315	3,065
2640	With magnetic starter	4,800	1,375	6,175
2660	50 HP, 1800 RPM, motor only	2,975	390	3,365
2680	With magnetic starter	5,025	1,450	6,475
2700	60 HP, 1800 RPM, motor only	4,975	450	5,425
2720	With magnetic starter	9,750	1,850	11,600
2740	75 HP, 1800 RPM, motor only	6,100	525	6,625
2760	With magnetic starter	10,900	1,925	12,825
2780	100 HP, 1800 RPM, motor only	7,750	695	8,445
2800	With magnetic starter	12,500	2,100	14,600
2820	125 HP, 1800 RPM, motor only	10,900	895	11,795
2840	With magnetic starter	32,300	2,475	34,775
2860	150 HP, 1800 RPM, motor only	11,900	1,050	12,950
2880	With magnetic starter	33,300	2,625	35,925
2900	200 HP, 1800 RPM, motor only	15,100	1,250	16,350
2920	With magnetic starter	36,500	2,825	39,325

D5020 Lighting and Branch Wiring

General: The cost of the lighting portion of the electrical costs is dependent upon:
1. The footcandle requirement of the proposed building.
2. The type of fixtures required.
3. The ceiling heights of the building.
4. Reflectance value of ceilings, walls and floors.
5. Fixture efficiencies and spacing vs. mounting height ratios.

Footcandle Requirements: See Table D5020-204 for Footcandle and Watts per S.F. determination.

Table D5020-201 IESNA* Recommended Illumination Levels in Footcandles

Commercial Buildings			Industrial Buildings		
Type	Description	Footcandles	Type	Description	Footcandles
Bank	Lobby	50	Assembly Areas	Rough bench & machine work	50
	Customer Areas	70		Medium bench & machine work	100
	Teller Stations	150		Fine bench & machine work	500
	Accounting Areas	150	Inspection Areas	Ordinary	50
Offices	Routine Work	100		Difficult	100
	Accounting	150		Highly Difficult	200
	Drafting	200	Material Handling	Loading	20
	Corridors, Halls, Washrooms	30		Stock Picking	30
Schools	Reading or Writing	70		Packing, Wrapping	50
	Drafting, Labs, Shops	100	Stairways	Service Areas	20
	Libraries	70	Washrooms	Service Areas	20
	Auditoriums, Assembly	15	Storage Areas	Inactive	5
	Auditoriums, Exhibition	30		Active, Rough, Bulky	10
Stores	Circulation Areas	30		Active, Medium	20
	Stock Rooms	30		Active, Fine	50
	Merchandise Areas, Service	100	Garages	Active Traffic Areas	20
	Self-Service Areas	200		Service & Repair	100

*IESNA - Illuminating Engineering Society of North America

D5020 Lighting and Branch Wiring

Table D5020-202 General Lighting Loads by Occupancies

Type of Occupancy	Unit Load per S.F. (Watts)
Armories and Auditoriums	1
Banks	5
Barber Shops and Beauty Parlors	3
Churches	1
Clubs	2
Court Rooms	2
*Dwelling Units	3
Garages — Commercial (storage)	½
Hospitals	2
*Hotels and Motels, including apartment houses without provisions for cooking by tenants	2
Industrial Commercial (Loft) Buildings	2
Lodge Rooms	1½
Office Buildings	5
Restaurants	2
Schools	3
Stores	3
Warehouses (storage)	¼
*In any of the above occupancies except one-family dwellings and individual dwelling units of multi-family dwellings:	
Assembly Halls and Auditoriums	1
Halls, Corridors, Closets	½
Storage Spaces	¼

Table D5020-203 Lighting Limit (Connected Load) for Listed Occupancies: New Building Proposed Energy Conservation Guideline

Type of Use	Maximum Watts per S.F.
Interior	
Category A: Classrooms, office areas, automotive mechanical areas, museums, conference rooms, drafting rooms, clerical areas, laboratories, merchandising areas, kitchens, examining rooms, book stacks, athletic facilities.	3.00
Category B: Auditoriums, waiting areas, spectator areas, restrooms, dining areas, transportation terminals, working corridors in prisons and hospitals, book storage areas, active inventory storage, hospital bedrooms, hotel and motel bedrooms, enclosed shopping mall concourse areas, stairways.	1.00
Category C: Corridors, lobbies, elevators, inactive storage areas.	0.50
Category D: Indoor parking.	0.25
Exterior	
Category E: Building perimeter: wall-wash, facade, canopy.	5.00 (per linear foot)
Category F: Outdoor parking.	0.10

Table D5020-204 Procedure for Calculating Footcandles and Watts Per Square Foot

1. Initial footcandles = No. of fixtures × lamps per fixture × lumens per lamp × coefficient of utilization ÷ square feet
2. Maintained footcandles = initial footcandles × maintenance factor
3. Watts per square foot = No. of fixtures × lamps × (lamp watts + ballast watts) ÷ square feet

Example: To find footcandles and watts per S.F. for an office 20′ x 20′ with 11 fluorescent fixtures each having 4–40 watt C.W. lamps.

Based on good reflectance and clean conditions:

Lumens per lamp = 40 watt cool white at 3150 lumens per lamp RD5020-250, Table RD5020-251

Coefficient of utilization = .42 (varies from .62 for light colored areas to .27 for dark)

Maintenance factor = .75 (varies from .80 for clean areas with good maintenance to .50 for poor)

Ballast loss = 8 watts per lamp. (Varies with manufacturer. See manufacturers' catalog.)

1. Initial footcandles:

$$\frac{11 \times 4 \times 3150 \times .42}{400} = \frac{58{,}212}{400} = 145 \text{ footcandles}$$

2. Maintained footcandles:

   145 × .75 = 109 footcandles

3. Watts per S.F.

$$\frac{11 \times 4\ (40 + 8)}{400} = \frac{2{,}112}{400} = 5.3 \text{ watts per S.F.}$$

Table D5020-205 Approximate Watts Per Square Foot for Popular Fixture Types

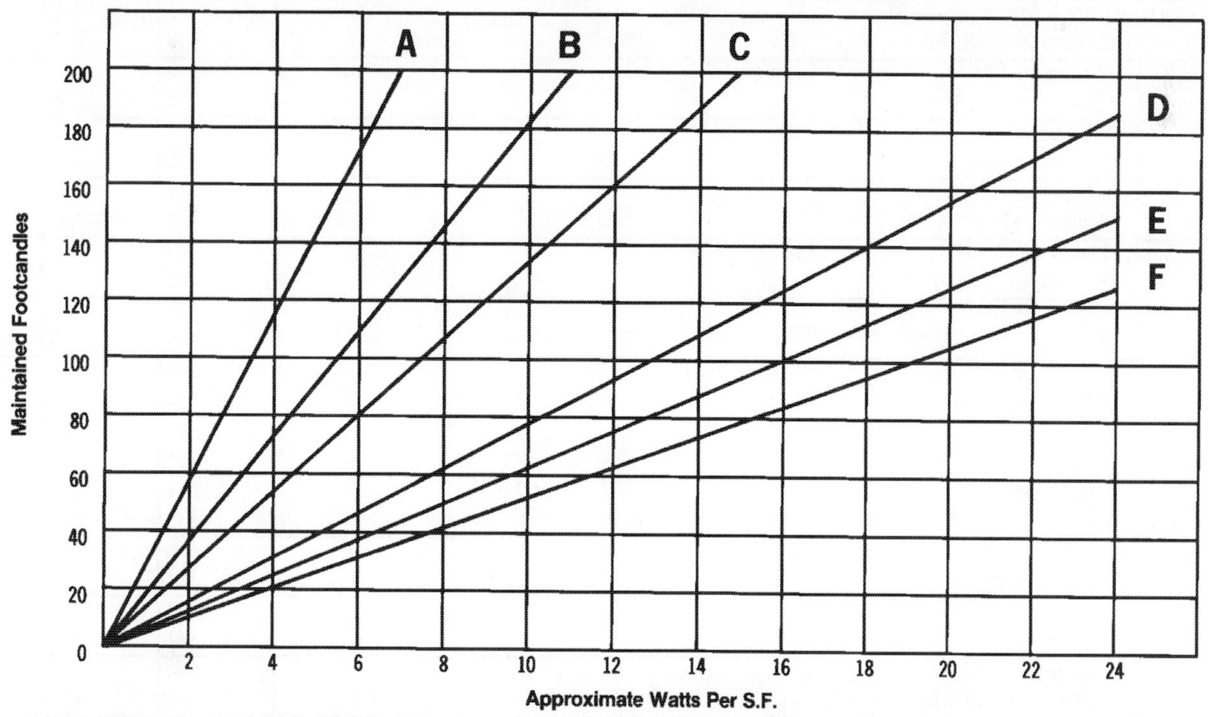

Due to the many variables involved, use for preliminary estimating only:

a. Fluorescent – industrial System D5020 208
b. Fluorescent – lens unit System D5020 208 Fixture types B & C
c. Fluorescent – louvered unit
d. Incandescent – open reflector System D5020 214, Type D
e. Incandescent – lens unit System D5020 214, Type A
f. Incandescent – down light System D5020 214, Type B

D5020 Lighting and Branch Wiring

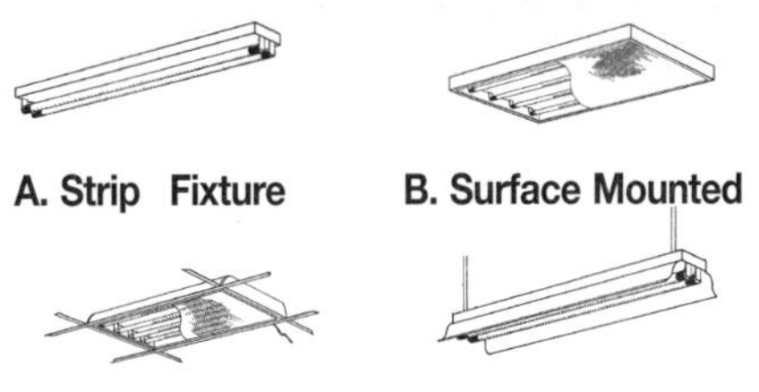

A. Strip Fixture B. Surface Mounted C. Recessed D. Pendent Mounted

Design Assumptions:

1. A 100 footcandle average maintained level of illumination.
2. Ceiling heights range from 9′ to 11′.
3. Average reflectance values are assumed for ceilings, walls and floors.
4. Cool white (CW) fluorescent lamps with 3150 lumens for 40 watt lamps and 6300 lumens for 8′ slimline lamps.
5. Four 40 watt lamps per 4′ fixture and two 8′ lamps per 8′ fixture.
6. Average fixture efficiency values and spacing to mounting height ratios.
7. Installation labor is average U.S. rate as of January 1.

System Components	QUANTITY	UNIT	COST PER S.F.		
			MAT.	INST.	TOTAL
SYSTEM D5020 208 0520					
FLUORESCENT FIXTURES MOUNTED 9′-11″ ABOVE FLOOR, 100 FC					
TYPE A, 8 FIXTURES PER 400 S.F.					
Conduit, steel intermediate, 1/2″ diam.	.185	L.F.	.37	1.16	1.53
Wire, 600V, type THWN-THHN, copper, solid, #12	.004	C.L.F.	.05	.21	.26
Fluorescent strip fixture 8′ long, surface mounted, two 75W SL	.020	Ea.	1.25	2.02	3.27
Steel outlet box 4″ concrete	.020	Ea.	.37	.63	1
Steel outlet box plate with stud, 4″ concrete	.020	Ea.	.25	.16	.41
Fixture hangers, flexible, 1/2″ diameter, 4″ long	.040	Ea.	.52	2.10	2.62
Fixture whip, THHN wire, three #12, 3/8″ greenfield	.040	Ea.	.67	.78	1.45
TOTAL			3.48	7.06	10.54

D5020 208	Fluorescent Fixtures (by Type)	COST PER S.F.		
		MAT.	INST.	TOTAL
0520	Fluorescent fixtures, type A, 8 fixtures per 400 S.F.	3.48	7.05	10.53
0560	11 fixtures per 600 S.F.	3.18	6.45	9.63
0600	17 fixtures per 1000 S.F.	2.95	6	8.95
0640	23 fixtures per 1600 S.F.	2.48	5.05	7.53
0680	28 fixtures per 2000 S.F.	2.44	4.94	7.38
0720	41 fixtures per 3000 S.F.	2.39	4.84	7.23
0800	53 fixtures per 4000 S.F.	2.29	4.68	6.97
0840	64 fixtures per 5000 S.F.	2.26	4.58	6.84
0880	Type B, 11 fixtures per 400 S.F.	5.50	10.05	15.55
0920	15 fixtures per 600 S.F.	4.95	9.05	14
0960	24 fixtures per 1000 S.F.	4.76	8.75	13.51
1000	35 fixtures per 1600 S.F.	4.35	7.95	12.30
1040	42 fixtures per 2000 S.F.	4.15	7.60	11.75
1080	61 fixtures per 3000 S.F.	3.99	7.35	11.34
1160	80 fixtures per 4000 S.F.	2.89	4.66	7.55
1200	98 fixtures per 5000 S.F.	3.91	7.15	11.06
1240	Type C, 11 fixtures per 400 S.F.	4.78	10.45	15.23
1280	14 fixtures per 600 S.F.	3.99	8.75	12.74
1320	23 fixtures per 1000 S.F.	3.96	8.70	12.66
1360	34 fixtures per 1600 S.F.	3.64	8	11.64
1400	43 fixtures per 2000 S.F.	3.74	8.20	11.94
1440	63 fixtures per 3000 S.F.	3.61	7.90	11.51
1520	81 fixtures per 4000 S.F.	3.48	7.65	11.13
1560	101 fixtures per 5000 S.F.	3.45	7.55	11
1600	Type D, 8 fixtures per 400 S.F.	4.35	7.90	12.25
1640	12 fixtures per 600 S.F.	4.35	7.90	12.25
1680	19 fixtures per 1000 S.F.	4.13	7.50	11.63
1720	27 fixtures per 1600 S.F.	3.69	6.70	10.39
1760	34 fixtures per 2000 S.F.	3.69	6.70	10.39
1800	48 fixtures per 3000 S.F.	3.49	6.35	9.84
1880	64 fixtures per 4000 S.F.	3.49	6.35	9.84
1920	79 fixtures per 5000 S.F.	3.48	6.30	9.78

D5020 Lighting and Branch Wiring

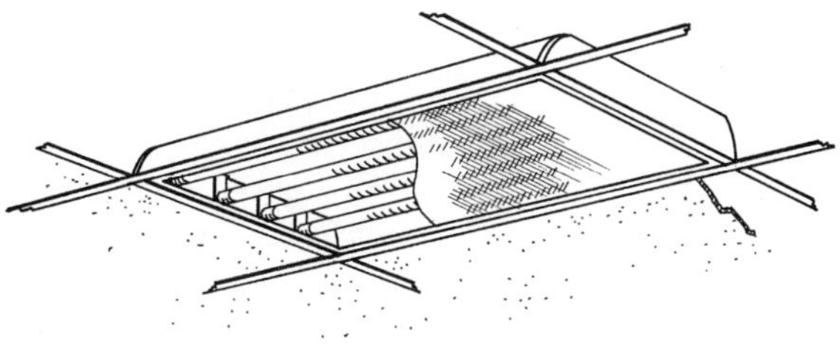

Type C. Recessed, mounted on grid ceiling suspension system, 2′ x 4′, four 40 watt lamps, acrylic prismatic diffusers.

5.3 watts per S.F. for 100 footcandles.

3 watts per S.F. for 57 footcandles.

System Components	QUANTITY	UNIT	COST PER S.F. MAT.	INST.	TOTAL
SYSTEM D5020 210 0200					
FLUORESCENT FIXTURES RECESS MOUNTED IN CEILING					
1 WATT PER S.F., 20 FC, 5 FIXTURES PER 1000 S.F.					
Steel intermediate conduit, (IMC) 1/2″ diam.	.128	L.F.	.26	.80	1.06
Wire, 600 volt, type THW, copper, solid, #12	.003	C.L.F.	.04	.17	.21
Fluorescent fixture, recessed, 2′x 4′, four 40W, w/lens, for grid ceiling	.005	Ea.	.32	.67	.99
Steel outlet box 4″ square	.005	Ea.	.09	.16	.25
Fixture whip, Greenfield w/#12 THHN wire	.005	Ea.	.06	.04	.10
TOTAL			.77	1.84	2.61

D5020 210	Fluorescent Fixtures (by Wattage)	COST PER S.F. MAT.	INST.	TOTAL
0190	Fluorescent fixtures recess mounted in ceiling			
0195	T12, standard 40 watt lamps			
0200	1 watt per S.F., 20 FC, 5 fixtures @40 watts per 1000 S.F.	.77	1.84	2.61
0240	2 watt per S.F., 40 FC, 10 fixtures @40 watt per 1000 S.F.	1.52	3.62	5.14
0280	3 watt per S.F., 60 FC, 15 fixtures @40 watt per 1000 S.F	2.30	5.45	7.75
0320	4 watt per S.F., 80 FC, 20 fixtures @40 watt per 1000 S.F.	3.04	7.20	10.24
0400	5 watt per S.F., 100 FC, 25 fixtures @40 watt per 1000 S.F.	3.82	9.05	12.87
0450	T8, energy saver 32 watt lamps			
0500	0.8 watt per S.F., 20 FC, 5 fixtures @32 watt per 1000 S.F.	.82	1.84	2.66
0520	1.6 watt per S.F., 40 FC, 10 fixtures @32 watt per 1000 S.F.	1.63	3.62	5.25
0540	2.4 watt per S.F., 60 FC, 15 fixtures @ 32 watt per 1000 S.F	2.45	5.45	7.90
0560	3.2 watt per S.F., 80 FC, 20 fixtures @32 watt per 1000 S.F.	3.25	7.20	10.45
0580	4 watt per S.F., 100 FC, 25 fixtures @32 watt per 1000 S.F.	4.08	9.05	13.13

D5020 Lighting and Branch Wiring

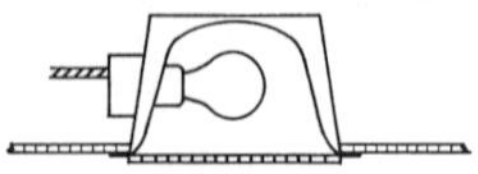

Type A

Type B

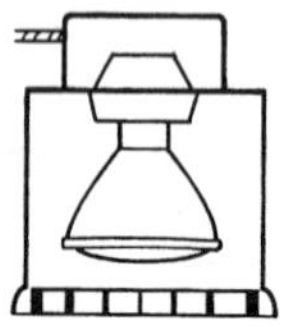

Type C

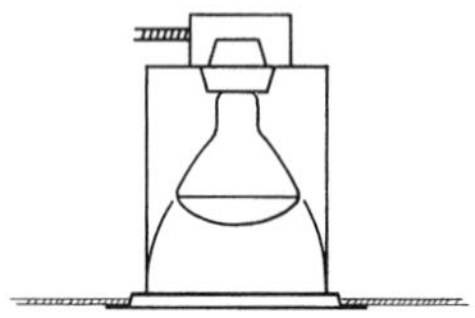

Type D

Type A. Recessed wide distribution reflector with flat glass lens 150 W.

Maximum spacing = 1.2 x mounting height.

13 watts per S.F. for 100 footcandles.

Type B. Recessed reflector down light with baffles 150 W.

Maximum spacing = 0.8 x mounting height.

18 watts per S.F. for 100 footcandles.

Type C. Recessed PAR–38 flood lamp with concentric louver 150 W.

Maximum spacing = 0.5 x mounting height.

19 watts per S.F. for 100 footcandles.

Type D. Recessed R–40 flood lamp with reflector skirt.

Maximum spacing = 0.7 x mounting height.

15 watts per S.F. for 100 footcandles.

System Components	QUANTITY	UNIT	COST PER S.F. MAT.	COST PER S.F. INST.	COST PER S.F. TOTAL
SYSTEM D5020 214 0400					
INCANDESCENT FIXTURE RECESS MOUNTED, 100 FC					
TYPE A, 34 FIXTURES PER 400 S.F.					
Steel intermediate conduit, (IMC) 1/2" diam	1.060	L.F.	2.12	6.63	8.75
Wire, 600V, type THWN-THHN, copper, solid, #12	.033	C.L.F.	.45	1.88	2.33
Steel outlet box 4" square	.085	Ea.	1.59	2.68	4.27
Fixture whip, Greenfield w/#12 THHN wire	.085	Ea.	1.05	.67	1.72
Incandescent fixture, recessed, w/lens, prewired, square trim, 200W	.085	Ea.	8.93	7.95	16.88
TOTAL			14.14	19.81	33.95

D5020 214	Incandescent Fixture (by Type)	COST PER S.F. MAT.	COST PER S.F. INST.	COST PER S.F. TOTAL
0380	Incandescent fixture recess mounted, 100 FC			
0400	Type A, 34 fixtures per 400 S.F.	14.15	19.80	33.95
0440	49 fixtures per 600 S.F.	13.75	19.45	33.20
0480	63 fixtures per 800 S.F.	13.40	19.25	32.65
0520	90 fixtures per 1200 S.F.	12.90	18.85	31.75
0560	116 fixtures per 1600 S.F.	12.65	18.70	31.35
0600	143 fixtures per 2000 S.F.	12.55	18.65	31.20
0640	Type B, 47 fixtures per 400 S.F.	13.55	25	38.55
0680	66 fixtures per 600 S.F.	12.95	24.50	37.45
0720	88 fixtures per 800 S.F.	12.95	24.50	37.45
0760	127 fixtures per 1200 S.F.	12.65	24	36.65
0800	160 fixtures per 1600 S.F.	12.45	24	36.45
0840	206 fixtures per 2000 S.F.	12.45	24	36.45
0880	Type C, 51 fixtures per 400 S.F.	18.15	26	44.15
0920	74 fixtures per 600 S.F.	17.60	25.50	43.10
0960	97 fixtures per 800 S.F.	17.40	25.50	42.90
1000	142 fixtures per 1200 S.F.	17.10	25.50	42.60
1040	186 fixtures per 1600 S.F.	16.85	25	41.85
1080	230 fixtures per 2000 S.F.	16.80	25	41.80
1120	Type D, 39 fixtures per 400 S.F.	16.75	20.50	37.25
1160	57 fixtures per 600 S.F.	16.35	20.50	36.85
1200	75 fixtures per 800 S.F.	16.25	20.50	36.75
1240	109 fixtures per 1200 S.F.	15.85	20	35.85
1280	143 fixtures per 1600 S.F.	15.55	19.85	35.40
1320	176 fixtures per 2000 S.F.	15.45	19.80	35.25

D5020 Lighting and Branch Wiring

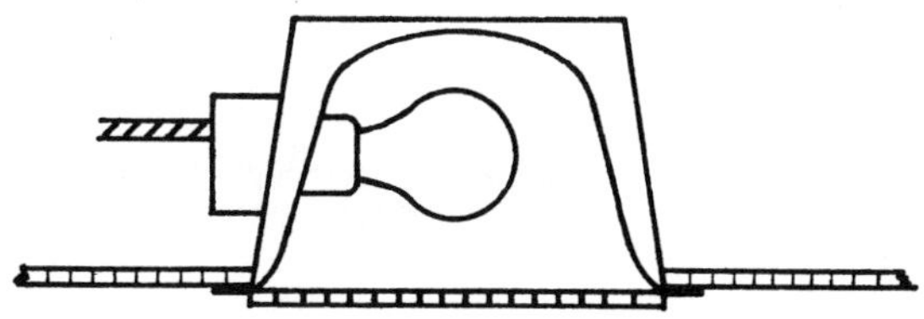

Type A. Recessed, wide distribution reflector with flat glass lens.

150 watt inside frost—2500 lumens per lamp.

PS–25 extended service lamp.

Maximum spacing = 1.2 x mounting height.

13 watts per S.F. for 100 footcandles.

System Components	QUANTITY	UNIT	COST PER S.F. MAT.	COST PER S.F. INST.	COST PER S.F. TOTAL
SYSTEM D5020 216 0200					
INCANDESCENT FIXTURE RECESS MOUNTED, TYPE A					
1 WATT PER S.F., 8 FC, 6 FIXT PER 1000 S.F.					
Steel intermediate conduit, (IMC) 1/2" diam	.091	L.F.	.18	.57	.75
Wire, 600V, type THWN-THHN, copper, solid, #12	.002	C.L.F.	.03	.11	.14
Incandescent fixture, recessed, w/lens, prewired, square trim, 200W	.006	Ea.	.63	.56	1.19
Steel outlet box 4" square	.006	Ea.	.11	.19	.30
Fixture whip, Greenfield w/#12 THHN wire	.006	Ea.	.07	.05	.12
TOTAL			1.02	1.48	2.50

D5020 216	Incandescent Fixture (by Wattage)	COST PER S.F. MAT.	COST PER S.F. INST.	COST PER S.F. TOTAL
0190	Incandescent fixture recess mounted, type A			
0200	1 watt per S.F., 8 FC, 6 fixtures per 1000 S.F.	1.02	1.48	2.50
0240	2 watt per S.F., 16 FC, 12 fixtures per 1000 S.F.	2.04	2.96	5
0280	3 watt per S.F., 24 FC, 18 fixtures, per 1000 S.F.	3.07	4.39	7.46
0320	4 watt per S.F., 32 FC, 24 fixtures per 1000 S.F.	4.10	5.85	9.95
0400	5 watt per S.F., 40 FC, 30 fixtures per 1000 S.F.	5.10	7.35	12.45

D5020 Lighting and Branch Wiring

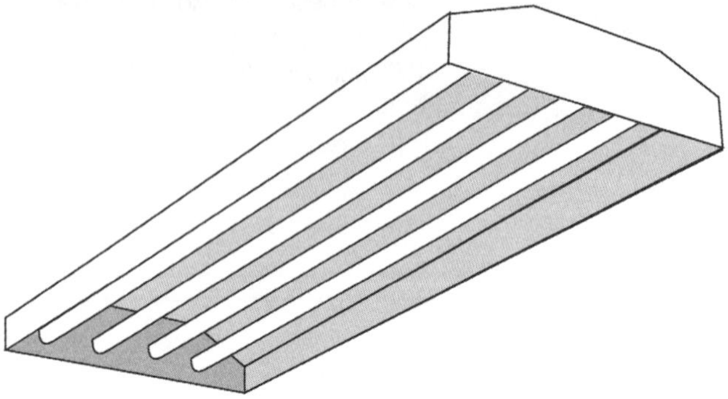

High Bay Fluorescent Fixtures
Four T5 HO/54 watt lamps

System Components	QUANTITY	UNIT	COST PER S.F. MAT.	COST PER S.F. INST.	COST PER S.F. TOTAL
SYSTEM D5020 218 0200					
FLUORESCENT HIGH BAY FIXTURE, 4 LAMP, 8′-10′ ABOVE WORK PLANE					
0.5 WATT/S.F., 29 FC, 2 FIXTURES/1000 S.F.					
Steel intermediate conduit, (IMC) 1/2″ diam	.100	L.F.	.20	.63	.83
Wire, 600V, type THWN-THHN, copper, solid, #10	.002	C.L.F.	.04	.13	.17
Steel outlet box 4″ concrete	.002	Ea.	.04	.06	.10
Steel outlet box plate with stud, 4″ concrete	.002	Ea.	.02	.02	.04
Flourescent , hi bay, 4 lamp, T-5 fixture	.002	Ea.	.49	.28	.77
TOTAL			.79	1.12	1.91

D5020 218	Fluorescent Fixture, High Bay, 8′-10′ (by Wattage)	COST PER S.F. MAT.	COST PER S.F. INST.	COST PER S.F. TOTAL
0190	Fluorescent high bay-4 lamp fixture, 8′-10′ above work plane			
0200	.5 watt/SF, 29 FC, 2 fixtures per 1000 S.F.	.79	1.12	1.91
0400	1 watt/SF, 59 FC, 4 fixtures per 1000 S.F.	1.56	2.13	3.69
0600	1.5 watt/SF, 103 FC, 7 fixtures per 1000 S.F.	2.61	3.33	5.94
0800	2 watt/SF, 133 FC, 9 fixtures per 1000 S.F.	3.38	4.39	7.77
1000	2.5 watt/SF, 162 FC, 11 fixtures per 1000 S.F.	4.18	5.45	9.63

D5020 Lighting and Branch Wiring

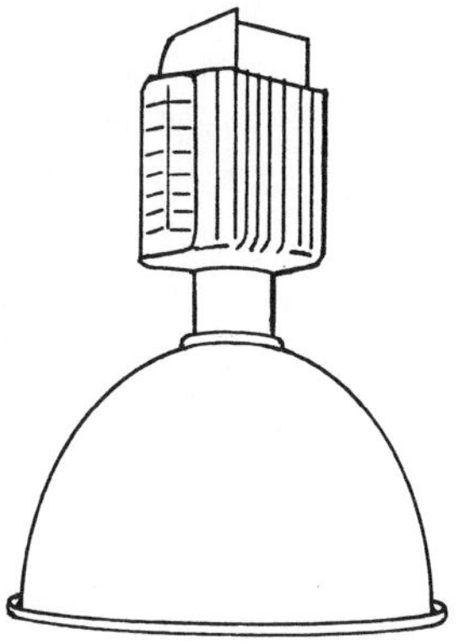

HIGH BAY FIXTURES

B. Metal halide 400 watt
C. High pressure sodium 400 watt
E. Metal halide 1000 watt
F. High pressure sodium 1000 watt
G. Metal halide 1000 watt
125,000 lumen lamp

System Components	QUANTITY	UNIT	COST PER S.F. MAT.	COST PER S.F. INST.	COST PER S.F. TOTAL
SYSTEM D5020 220 0880					
HIGH INTENSITY DISCHARGE FIXTURE, 8′-10′ ABOVE WORK PLANE, 100 FC					
TYPE B, 8 FIXTURES PER 900 S.F.					
Steel intermediate conduit, (IMC) 1/2″ diam	.460	L.F.	.92	2.88	3.80
Wire, 600V, type THWN-THHN, copper, solid, #10	.009	C.L.F.	.19	.56	.75
Steel outlet box 4″ concrete	.009	Ea.	.17	.28	.45
Steel outlet box plate with stud 4″ concrete	.009	Ea.	.11	.07	.18
Metal halide, hi bay, aluminum reflector, 400 W lamp	.009	Ea.	4.01	2.46	6.47
TOTAL			5.40	6.25	11.65

D5020 220	H.I.D. Fixture, High Bay, 8′-10′ (by Type)	COST PER S.F. MAT.	COST PER S.F. INST.	COST PER S.F. TOTAL
0500	High intensity discharge fixture, 8′-10′ above work plane, 100 FC			
0880	Type B, 8 fixtures per 900 S.F.	5.40	6.25	11.65
0920	15 fixtures per 1800 S.F.	4.95	6	10.95
0960	24 fixtures per 3000 S.F.	4.95	6	10.95
1000	31 fixtures per 4000 S.F.	4.94	6	10.94
1040	38 fixtures per 5000 S.F.	4.94	6	10.94
1080	60 fixtures per 8000 S.F.	4.94	6	10.94
1120	72 fixtures per 10000 S.F.	4.43	5.55	9.98
1160	115 fixtures per 16000 S.F.	4.43	5.55	9.98
1200	230 fixtures per 32000 S.F.	4.43	5.55	9.98
1240	Type C, 4 fixtures per 900 S.F.	2.59	3.81	6.40
1280	8 fixtures per 1800 S.F.	2.59	3.81	6.40
1320	13 fixtures per 3000 S.F.	2.59	3.81	6.40
1360	17 fixtures per 4000 S.F.	2.59	3.81	6.40
1400	21 fixtures per 5000 S.F.	2.49	3.51	6
1440	33 fixtures per 8000 S.F.	2.47	3.44	5.91
1480	40 fixtures per 10000 S.F.	2.41	3.25	5.66
1520	63 fixtures per 16000 S.F.	2.41	3.25	5.66
1560	126 fixtures per 32000 S.F.	2.41	3.25	5.66

D5020 Lighting and Branch Wiring

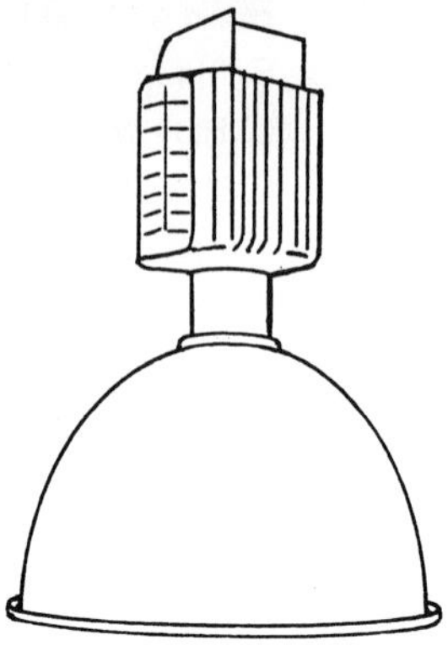

HIGH BAY FIXTURES

B. Metal halide 400 watt
C. High pressure sodium 400 watt
E. Metal halide 1000 watt
F. High pressure sodium 1000 watt
G. Metal halide 1000 watt
125,000 lumen lamp

System Components	QUANTITY	UNIT	COST PER S.F. MAT.	INST.	TOTAL
SYSTEM D5020 222 0240					
HIGH INTENSITY DISCHARGE FIXTURE, 8′-10′ ABOVE WORK PLANE					
1 WATT/S.F., TYPE B, 29 FC, 2 FIXTURES/1000 S.F.					
Steel intermediate conduit, (IMC) 1/2″ diam	.100	L.F.	.20	.63	.83
Wire, 600V, type THWN-THHN, copper, solid, #10	.002	C.L.F.	.04	.13	.17
Steel outlet box 4″ concrete	.002	Ea.	.04	.06	.10
Steel outlet box plate with stud, 4″ concrete	.002	Ea.	.02	.02	.04
Metal halide, hi bay, aluminum reflector, 400 W lamp	.002	Ea.	.89	.55	1.44
TOTAL			1.19	1.39	2.58

D5020 222	H.I.D. Fixture, High Bay, 8′-10′ (by Wattage)	COST PER S.F. MAT.	INST.	TOTAL
0190	High intensity discharge fixture, 8′-10′ above work plane			
0240	1 watt/S.F., type B, 29 FC, 2 fixtures/1000 S.F.	1.19	1.39	2.58
0280	Type C, 54 FC, 2 fixtures/1000 S.F.	1.01	1.05	2.06
0400	2 watt/S.F., type B, 59 FC, 4 fixtures/1000 S.F.	2.36	2.66	5.02
0440	Type C, 108 FC, 4 fixtures/1000 S.F. R265723-05	2.01	2.04	4.05
0560	3 watt/S.F., type B, 103 FC, 7 fixtures/1000 S.F.	4.01	4.25	8.26
0600	Type C, 189 FC, 6 fixtures/1000 S.F.	3.23	2.62	5.85
0720	4 watt/S.F., type B, 133 FC, 9 fixtures/1000 S.F.	5.20	5.60	10.80
0760	Type C, 243 FC, 9 fixtures/1000 S.F.	4.53	4.51	9.04
0880	5 watt/S.F., type B, 162 FC, 11 fixtures/1000 S.F.	6.35	6.90	13.25
0920	Type C, 297 FC, 11 fixtures/1000 S.F.	5.55	5.60	11.15

D5020 Lighting and Branch Wiring

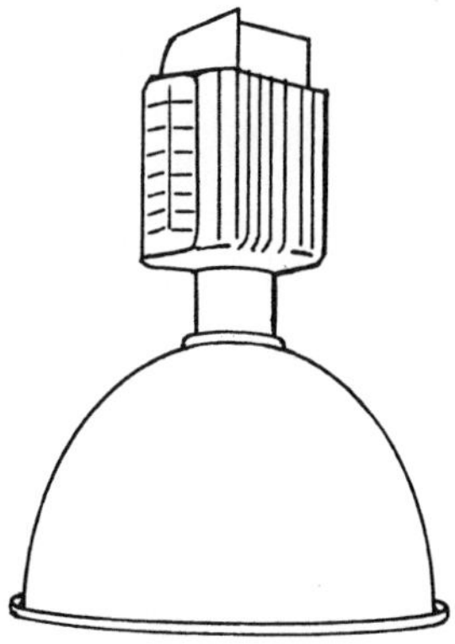

HIGH BAY FIXTURES

B. Metal halide 400 watt
C. High pressure sodium 400 watt
E. Metal halide 1000 watt
F. High pressure sodium 1000 watt
G. Metal halide 1000 watt
125,000 lumen lamp

System Components	QUANTITY	UNIT	COST PER S.F. MAT.	INST.	TOTAL
SYSTEM D5020 224 1240					
HIGH INTENSITY DISCHARGE FIXTURE, 16′ ABOVE WORK PLANE, 100 FC					
TYPE C, 5 FIXTURES PER 900 S.F.					
Steel intermediate conduit, (IMC) 1/2″ diam	.260	L.F.	.52	1.63	2.15
Wire, 600V, type THWN-THHN, copper, solid, #10	.007	C.L.F.	.15	.44	.59
Steel outlet box 4″ concrete	.006	Ea.	.11	.19	.30
Steel outlet box plate with stud, 4″ concrete	.006	Ea.	.07	.05	.12
High pressure sodium, hi bay, aluminum reflector, 400 W lamp	.006	Ea.	2.46	1.64	4.10
TOTAL			3.31	3.95	7.26

D5020 224	H.I.D. Fixture, High Bay, 16′ (by Type)	COST PER S.F. MAT.	INST.	TOTAL
0510	High intensity discharge fixture, 16′ above work plane, 100 FC			
1240	Type C, 5 fixtures per 900 S.F.	3.31	3.95	7.26
1280	9 fixtures per 1800 S.F.	3.03	4.13	7.16
1320	15 fixtures per 3000 S.F.	3.03	4.13	7.16
1360	18 fixtures per 4000 S.F.	2.89	3.70	6.59
1400	22 fixtures per 5000 S.F.	2.89	3.70	6.59
1440	36 fixtures per 8000 S.F.	2.89	3.70	6.59
1480	42 fixtures per 10,000 S.F.	2.59	3.81	6.40
1520	65 fixtures per 16,000 S.F.	2.59	3.81	6.40
1600	Type G, 4 fixtures per 900 S.F.	4.24	6.25	10.49
1640	6 fixtures per 1800 S.F.	3.56	5.80	9.36
1720	9 fixtures per 4000 S.F.	2.94	5.65	8.59
1760	11 fixtures per 5000 S.F.	2.94	5.65	8.59
1840	21 fixtures per 10,000 S.F.	2.76	5.10	7.86
1880	33 fixtures per 16,000 S.F.	2.76	5.10	7.86

D5020 Lighting and Branch Wiring

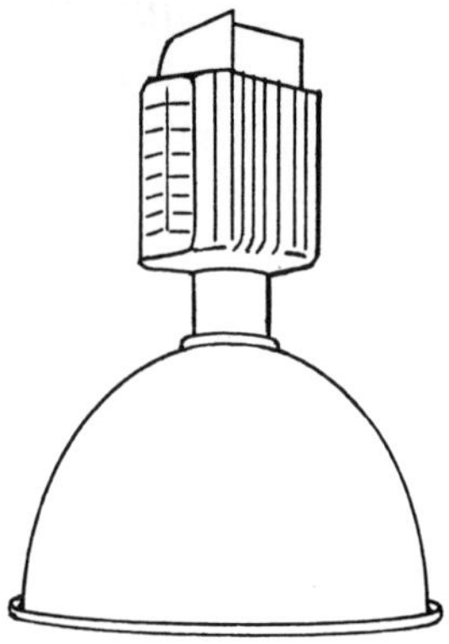

HIGH BAY FIXTURES
B. Metal halide 400 watt
C. High pressure sodium 400 watt
E. Metal halide 1000 watt
F. High pressure sodium 1000 watt
G. Metal halide 1000 watt
125,000 lumen lamp

System Components	QUANTITY	UNIT	COST PER S.F. MAT.	COST PER S.F. INST.	COST PER S.F. TOTAL
SYSTEM D5020 226 0240					
HIGH INTENSITY DISCHARGE FIXTURE, 16′ ABOVE WORK PLANE					
1 WATT/S.F., TYPE E, 42 FC, 1 FIXTURE/1000 S.F.					
Steel intermediate conduit, (IMC) 1/2″ diam	.160	L.F.	.32	1	1.32
Wire, 600V, type THWN-THHN, copper, solid, #10	.003	C.L.F.	.06	.19	.25
Steel outlet box 4″ concrete	.001	Ea.	.02	.03	.05
Steel outlet box plate with stud, 4″ concrete	.001	Ea.	.01	.01	.02
Metal halide, hi bay, aluminum reflector, 1000 W lamp	.001	Ea.	.64	.32	.96
TOTAL			1.05	1.55	2.60

D5020 226	H.I.D. Fixture, High Bay, 16′ (by Wattage)	COST PER S.F. MAT.	COST PER S.F. INST.	COST PER S.F. TOTAL
0190	High intensity discharge fixture, 16′ above work plane			
0240	1 watt/S.F., type E, 42 FC, 1 fixture/1000 S.F.	1.05	1.55	2.60
0280	Type G, 52 FC, 1 fixture/1000 S.F. R265723-05	1.05	1.55	2.60
0320	Type C, 54 FC, 2 fixture/1000 S.F.	1.23	1.73	2.96
0440	2 watt/S.F., type E, 84 FC, 2 fixture/1000 S.F.	2.13	3.15	5.28
0480	Type G, 105 FC, 2 fixture/1000 S.F.	2.13	3.15	5.28
0520	Type C, 108 FC, 4 fixture/1000 S.F.	2.47	3.46	5.93
0640	3 watt/S.F., type E, 126 FC, 3 fixture/1000 S.F.	3.20	4.69	7.89
0680	Type G, 157 FC, 3 fixture/1000 S.F.	3.20	4.69	7.89
0720	Type C, 162 FC, 6 fixture/1000 S.F.	3.71	5.20	8.91
0840	4 watt/S.F., type E, 168 FC, 4 fixture/1000 S.F.	4.26	6.30	10.56
0880	Type G, 210 FC, 4 fixture/1000 S.F.	4.26	6.30	10.56
0920	Type C, 243 FC, 9 fixture/1000 S.F.	5.40	7.20	12.60
1040	5 watt/S.F., type E, 210 FC, 5 fixture/1000 S.F.	5.30	7.85	13.15
1080	Type G, 262 FC, 5 fixture/1000 S.F.	5.30	7.85	13.15
1120	Type C, 297 FC, 11 fixture/1000 S.F.	6.65	8.95	15.60

D5020 Lighting and Branch Wiring

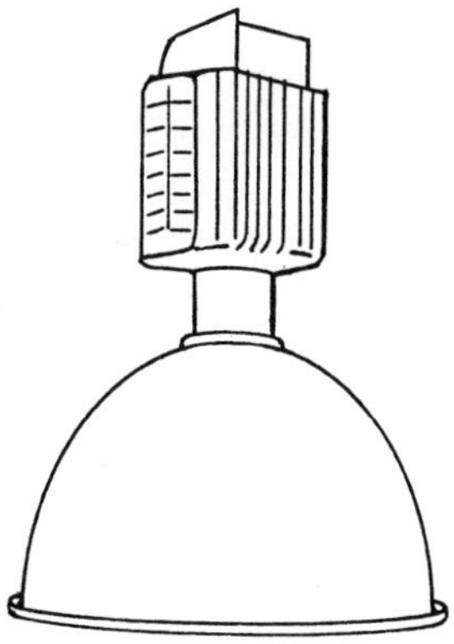

HIGH BAY FIXTURES

B. Metal halide 400 watt
C. High pressure sodium 400 watt
E. Metal halide 1000 watt
F. High pressure sodium 1000 watt
G. Metal halide 1000 watt
125,000 lumen lamp

System Components	QUANTITY	UNIT	COST PER S.F. MAT.	COST PER S.F. INST.	COST PER S.F. TOTAL
SYSTEM D5020 228 1240					
HIGH INTENSITY DISCHARGE FIXTURE, 20′ ABOVE WORK PLANE, 100 FC					
TYPE C, 6 FIXTURES PER 900 S.F.					
Steel intermediate conduit, (IMC) 1/2″ diam	.350	L.F.	.70	2.19	2.89
Wire, 600V, type THWN-THHN, copper, solid, #10.	.011	C.L.F.	.24	.69	.93
Steel outlet box 4″ concrete	.007	Ea.	.13	.22	.35
Steel outlet box plate with stud, 4″ concrete	.007	Ea.	.09	.05	.14
High pressure sodium, hi bay, aluminum reflector, 400 W lamp	.007	Ea.	2.87	1.91	4.78
TOTAL			4.03	5.06	9.09

D5020 228	H.I.D. Fixture, High Bay, 20′ (by Type)	COST PER S.F. MAT.	COST PER S.F. INST.	COST PER S.F. TOTAL
0510	High intensity discharge fixture 20′ above work plane, 100 FC			
1240	Type C, 6 fixtures per 900 S.F.	4.03	5.05	9.08
1280	10 fixtures per 1800 S.F. R265723-05	3.58	4.76	8.34
1320	16 fixtures per 3000 S.F.	3.18	4.57	7.75
1360	20 fixtures per 4000 S.F.	3.18	4.57	7.75
1400	24 fixtures per 5000 S.F.	3.18	4.57	7.75
1440	38 fixtures per 8000 S.F.	3.12	4.38	7.50
1520	68 fixtures per 16000 S.F.	2.92	4.82	7.74
1560	132 fixtures per 32000 S.F.	2.92	4.82	7.74
1600	Type G, 4 fixtures per 900 S.F.	4.21	6.10	10.31
1640	6 fixtures per 1800 S.F.	3.57	5.80	9.37
1680	7 fixtures per 3000 S.F.	3.51	5.65	9.16
1720	10 fixtures per 4000 S.F.	3.49	5.55	9.04
1760	11 fixtures per 5000 S.F.	3.05	5.95	9
1800	18 fixtures per 8000 S.F.	3.05	5.95	9
1840	22 fixtures per 10000 S.F.	3.05	5.95	9
1880	34 fixtures per 16000 S.F.	2.99	5.80	8.79
1920	66 fixtures per 32000 S.F.	2.83	5.30	8.13

D5020 Lighting and Branch Wiring

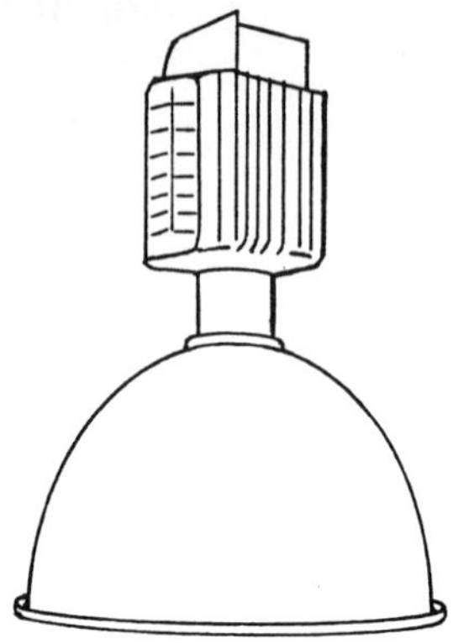

HIGH BAY FIXTURES

B. Metal halide 400 watt
C. High pressure sodium 400 watt
E. Metal halide 1000 watt
F. High pressure sodium 1000 watt
G. Metal halide 1000 watt
125,000 lumen lamp

System Components	QUANTITY	UNIT	COST PER S.F. MAT.	COST PER S.F. INST.	COST PER S.F. TOTAL
SYSTEM D5020 230 0240					
HIGH INTENSITY DISCHARGE FIXTURE, 20′ ABOVE WORK PLANE					
1 WATT/S.F., TYPE E, 40 FC, 1 FIXTURE 1000 S.F.					
Steel intermediate conduit, (IMC) 1/2″ diam	.160	L.F.	.32	1	1.32
Wire, 600V, type THWN-THHN, copper, solid, #10	.005	C.L.F.	.11	.31	.42
Steel outlet box 4″ concrete	.001	Ea.	.02	.03	.05
Steel outlet box plate with stud, 4″ concrete	.001	Ea.	.01	.01	.02
Metal halide, hi bay, aluminum reflector, 1000 W lamp	.001	Ea.	.64	.32	.96
TOTAL			1.10	1.67	2.77

D5020 230	H.I.D. Fixture, High Bay, 20′ (by Wattage)	COST PER S.F. MAT.	COST PER S.F. INST.	COST PER S.F. TOTAL
0190	High intensity discharge fixture, 20′ above work plane			
0240	1 watt/S.F., type E, 40 FC, 1 fixture/1000 S.F.	1.10	1.67	2.77
0280	Type G, 50 FC, 1 fixture/1000 S.F. R265723-05	1.10	1.67	2.77
0320	Type C, 52 FC, 2 fixtures/1000 S.F.	1.31	1.93	3.24
0440	2 watt/S.F., type E, 81 FC, 2 fixtures/1000 S.F.	2.20	3.34	5.54
0480	Type G, 101 FC, 2 fixtures/1000 S.F.	2.20	3.34	5.54
0520	Type C, 104 FC, 4 fixtures/1000 S.F.	2.58	3.79	6.37
0640	3 watt/S.F., type E, 121 FC, 3 fixtures/1000 S.F.	3.30	5	8.30
0680	Type G, 151 FC, 3 fixtures/1000 S.F.	3.30	5	8.30
0720	Type C, 155 FC, 6 fixtures/1000 S.F.	3.89	5.70	9.59
0840	4 watt/S.F., type E, 161 FC, 4 fixtures/1000 S.F.	4.39	6.65	11.04
0880	Type G, 202 FC, 4 fixtures/1000 S.F.	4.39	6.65	11.04
0920	Type C, 233 FC, 9 fixtures/1000 S.F.	5.60	7.90	13.50
1040	5 watt/S.F., type E, 202 FC, 5 fixtures/1000 S.F.	5.50	8.35	13.85
1080	Type G, 252 FC, 5 fixtures/1000 S.F.	5.50	8.35	13.85
1120	Type C, 285 FC, 11 fixtures/1000 S.F.	6.95	9.80	16.75

D5020 Lighting and Branch Wiring

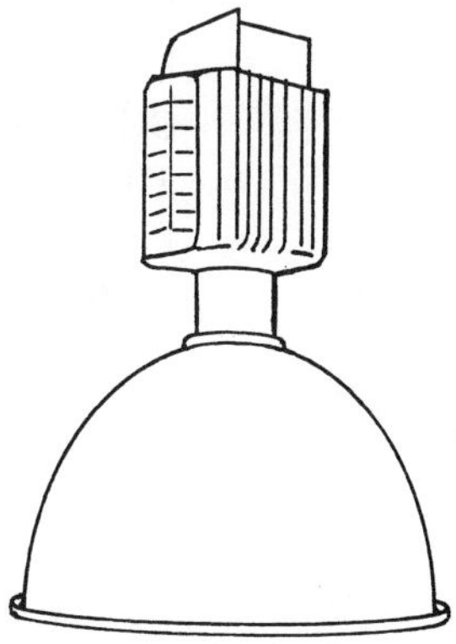

HIGH BAY FIXTURES

B. Metal halide 400 watt
C. High pressure sodium 400 watt
E. Metal halide 1000 watt
F. High pressure sodium 1000 watt
G. Metal halide 1000 watt
125,000 lumen lamp

System Components	QUANTITY	UNIT	COST PER S.F. MAT.	COST PER S.F. INST.	COST PER S.F. TOTAL
SYSTEM D5020 232 1240					
HIGH INTENSITY DISCHARGE FIXTURE, 30′ ABOVE WORK PLANE, 100 FC					
TYPE F, 4 FIXTURES PER 900 S.F.					
Steel intermediate conduit, (IMC) 1/2″ diam	.580	L.F.	1.16	3.63	4.79
Wire, 600V, type THWN-THHN, copper, solid, #10	.018	C.L.F.	.39	1.13	1.52
Steel outlet box 4″ concrete	.004	Ea.	.07	.13	.20
Steel outlet box plate with stud, 4″ concrete	.004	Ea.	.05	.03	.08
High pressure sodium, hi bay, aluminum, 1000 W lamp	.004	Ea.	2.36	1.26	3.62
TOTAL			4.03	6.18	10.21

D5020 232	H.I.D. Fixture, High Bay, 30′ (by Type)	COST PER S.F. MAT.	COST PER S.F. INST.	COST PER S.F. TOTAL
0510	High intensity discharge fixture, 30′ above work plane, 100 FC			
1240	Type F, 4 fixtures per 900 S.F.	4.03	6.20	10.23
1280	6 fixtures per 1800 S.F. R265723-05	3.44	5.90	9.34
1320	8 fixtures per 3000 S.F.	3.44	5.90	9.34
1360	9 fixtures per 4000 S.F.	2.83	5.60	8.43
1400	10 fixtures per 5000 S.F.	2.83	5.60	8.43
1440	17 fixtures per 8000 S.F.	2.83	5.60	8.43
1480	18 fixtures per 10,000 S.F.	2.69	5.15	7.84
1520	27 fixtures per 16,000 S.F.	2.69	5.15	7.84
1560	52 fixtures per 32000 S.F.	2.67	5.10	7.77
1600	Type G, 4 fixtures per 900 S.F.	4.35	6.55	10.90
1640	6 fixtures per 1800 S.F.	3.61	5.95	9.56
1680	9 fixtures per 3000 S.F.	3.55	5.75	9.30
1720	11 fixtures per 4000 S.F.	3.49	5.55	9.04
1760	13 fixtures per 5000 S.F.	3.49	5.55	9.04
1800	21 fixtures per 8000 S.F.	3.49	5.55	9.04
1840	23 fixtures per 10,000 S.F.	3.04	5.90	8.94
1880	36 fixtures per 16,000 S.F.	3.04	5.90	8.94
1920	70 fixtures per 32,000 S.F.	3.04	5.90	8.94

D5020 Lighting and Branch Wiring

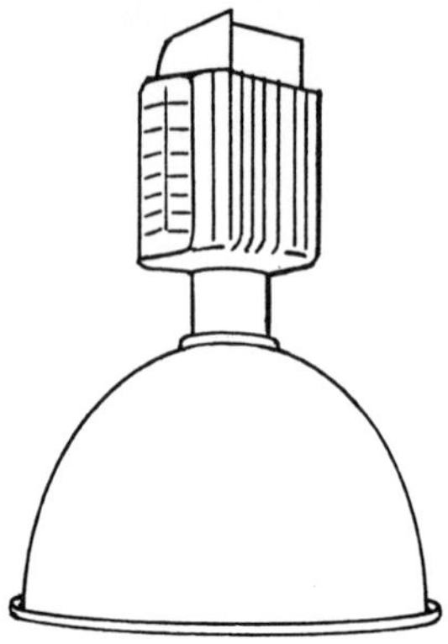

HIGH BAY FIXTURES

B. Metal halide 400 watt
C. High pressure sodium 400 watt
E. Metal halide 1000 watt
F. High pressure sodium 1000 watt
G. Metal halide 1000 watt
125,000 lumen lamp

System Components	QUANTITY	UNIT	COST PER S.F. MAT.	COST PER S.F. INST.	COST PER S.F. TOTAL
SYSTEM D5020 234 0240					
HIGH INTENSITY DISCHARGE FIXTURE, 30′ ABOVE WORK PLANE					
1 WATT/S.F., TYPE E, 37 FC, 1 FIXTURE/1000 S.F.					
Steel intermediate conduit, (IMC) 1/2″ diam	.196	L.F.	.39	1.23	1.62
Wire, 600V type THWN-THHN, copper, solid, #10	.006	C.L.F.	.13	.38	.51
Steel outlet box 4″ concrete	.001	Ea.	.02	.03	.05
Steel outlet box plate with stud, 4″ concrete	.001	Ea.	.01	.01	.02
Metal halide, hi bay, aluminum reflector, 1000 W lamp	.001	Ea.	.64	.32	.96
TOTAL			1.19	1.97	3.16

D5020 234	H.I.D. Fixture, High Bay, 30′ (by Wattage)	COST PER S.F. MAT.	COST PER S.F. INST.	COST PER S.F. TOTAL
0190	High intensity discharge fixture, 30′ above work plane			
0240	1 watt/S.F., type E, 37 FC, 1 fixture/1000 S.F.	1.19	1.97	3.16
0280	Type G, 45 FC., 1 fixture/1000 S.F. R265723 -05	1.19	1.97	3.16
0320	Type F, 50 FC, 1 fixture/1000 S.F.	1.01	1.54	2.55
0440	2 watt/S.F., type E, 74 FC, 2 fixtures/1000 S.F.	2.38	3.91	6.29
0480	Type G, 92 FC, 2 fixtures/1000 S.F.	2.38	3.91	6.29
0520	Type F, 100 FC, 2 fixtures/1000 S.F.	2.02	3.12	5.14
0640	3 watt/S.F., type E, 110 FC, 3 fixtures/1000 S.F.	3.61	5.95	9.56
0680	Type G, 138 FC, 3 fixtures/1000 S.F.	3.61	5.95	9.56
0720	Type F, 150 FC, 3 fixtures/1000 S.F.	3.04	4.65	7.69
0840	4 watt/S.F., type E, 148 FC, 4 fixtures/1000 S.F.	4.79	7.90	12.69
0880	Type G, 185 FC, 4 fixtures/1000 S.F.	4.79	7.90	12.69
0920	Type F, 200 FC, 4 fixtures/1000 S.F.	4.05	6.25	10.30
1040	5 watt/S.F., type E, 185 FC, 5 fixtures/1000 S.F.	6	9.85	15.85
1080	Type G, 230 FC, 5 fixtures/1000 S.F.	6	9.85	15.85
1120	Type F, 250 FC, 5 fixtures/1000 S.F.	5.05	7.80	12.85

D5020 Lighting and Branch Wiring

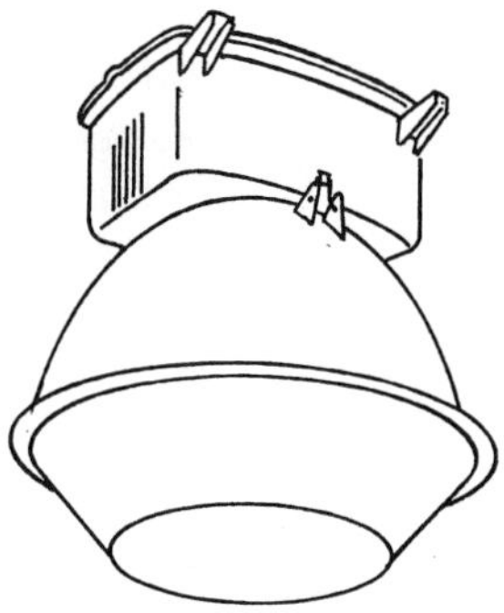

LOW BAY FIXTURES

J. Metal halide 250 watt

K. High pressure sodium 150 watt

System Components	QUANTITY	UNIT	COST PER S.F. MAT.	COST PER S.F. INST.	COST PER S.F. TOTAL
SYSTEM D5020 236 0920					
HIGH INTENSITY DISCHARGE FIXTURE, 8'-10' ABOVE WORK PLANE, 50 FC					
TYPE J, 13 FIXTURES PER 1800 S.F.					
Steel intermediate conduit, (IMC) 1/2" diam	.550	L.F.	1.10	3.44	4.54
Wire, 600V, type THWN-THHN, copper, solid, #10	.012	C.L.F.	.26	.75	1.01
Steel outlet box 4" concrete	.007	Ea.	.13	.22	.35
Steel outlet box plate with stud, 4" concrete	.007	Ea.	.09	.05	.14
Metal halide, lo bay, aluminum reflector, 250 W DX lamp	.007	Ea.	2.73	1.37	4.10
TOTAL			4.31	5.83	10.14

D5020 236	H.I.D. Fixture, Low Bay, 8'-10' (by Type)		COST PER S.F. MAT.	COST PER S.F. INST.	COST PER S.F. TOTAL
0510	High intensity discharge fixture, 8'-10' above work plane, 50 FC				
0880	Type J, 7 fixtures per 900 S.F.		4.67	5.90	10.57
0920	13 fixtures per 1800 S.F.	R265723 -05	4.31	5.85	10.16
0960	21 fixtures per 3000 S.F.		4.31	5.85	10.16
1000	28 fixtures per 4000 S.F.		4.31	5.85	10.16
1040	35 fixtures per 5000 S.F.		4.31	5.85	10.16
1120	62 fixtures per 10,000 S.F.		3.88	5.60	9.48
1160	99 fixtures per 16,000 S.F.		3.88	5.60	9.48
1200	199 fixtures per 32,000 S.F.		3.88	5.60	9.48
1240	Type K, 9 fixtures per 900 S.F.		4.73	5.05	9.78
1280	16 fixtures per 1800 S.F.		4.37	4.86	9.23
1320	26 fixtures per 3000 S.F.		4.35	4.80	9.15
1360	31 fixtures per 4000 S.F.		4	4.69	8.69
1400	39 fixtures per 5000 S.F.		4	4.69	8.69
1440	62 fixtures per 8000 S.F.		4	4.69	8.69
1480	78 fixtures per 10,000 S.F.		4	4.69	8.69
1520	124 fixtures per 16,000 S.F.		3.96	4.57	8.53
1560	248 fixtures per 32,000 S.F.		3.82	5.10	8.92

D5020 Lighting and Branch Wiring

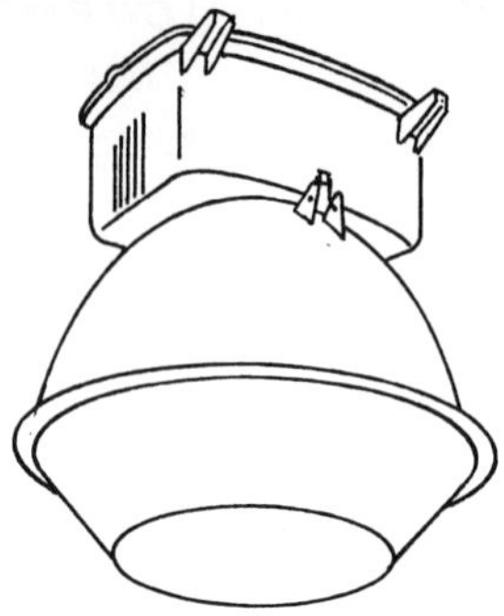

LOW BAY FIXTURES

J. Metal halide 250 watt

K. High pressure sodium 150 watt

System Components	QUANTITY	UNIT	COST PER S.F. MAT.	COST PER S.F. INST.	COST PER S.F. TOTAL
SYSTEM D5020 238 0240					
HIGH INTENSITY DISCHARGE FIXTURE, 8′-10′ ABOVE WORK PLANE					
1 WATT/S.F., TYPE J, 30 FC, 4 FIXTURES/1000 S.F.					
Steel intermediate conduit, (IMC) 1/2″ diam	.280	L.F.	.56	1.75	2.31
Wire, 600V, type THWN-THHN, copper, solid, #10	.008	C.L.F.	.17	.50	.67
Steel outlet box 4″ concrete	.004	Ea.	.07	.13	.20
Steel outlet box plate with stud, 4″ concrete	.004	Ea.	.05	.03	.08
Metal halide, lo bay, aluminum reflector, 250 W DX lamp	.004	Ea.	1.56	.78	2.34
TOTAL			2.41	3.19	5.60

D5020 238	H.I.D. Fixture, Low Bay, 8′-10′ (by Wattage)	COST PER S.F. MAT.	COST PER S.F. INST.	COST PER S.F. TOTAL
0190	High intensity discharge fixture, 8′-10′ above work plane			
0240	1 watt/S.F., type J, 30 FC, 4 fixtures/1000 S.F.	2.41	3.19	5.60
0280	Type K, 29 FC, 5 fixtures/1000 S.F. R265723 -05	2.48	2.87	5.35
0400	2 watt/S.F., type J, 52 FC, 7 fixtures/1000 S.F.	4.31	5.85	10.16
0440	Type K, 63 FC, 11 fixtures/1000 S.F.	5.40	6.05	11.45
0560	3 watt/S.F., type J, 81 FC, 11 fixtures/1000 S.F.	6.65	8.85	15.50
0600	Type K, 92 FC, 16 fixtures/1000 S.F.	7.85	8.90	16.75
0720	4 watt/S.F., type J, 103 FC, 14 fixtures/1000 S.F.	8.60	11.65	20.25
0760	Type K, 127 FC, 22 fixtures/1000 S.F.	10.75	12.05	22.80
0880	5 watt/S.F., type J, 133 FC, 18 fixtures/1000 S.F.	11	14.70	25.70
0920	Type K, 155 FC, 27 fixtures/1000 S.F.	13.20	14.90	28.10

D5020 Lighting and Branch Wiring

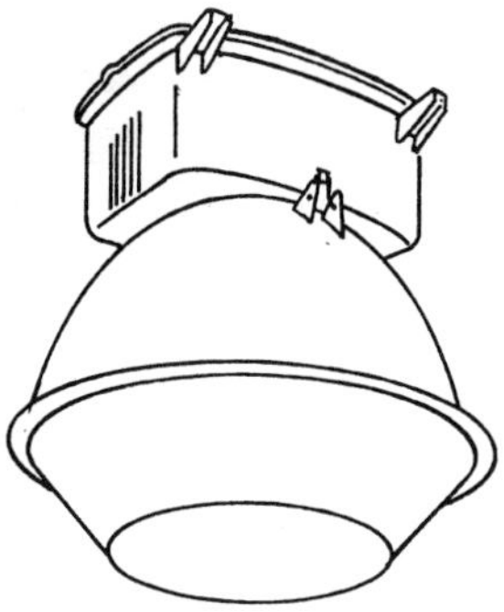

LOW BAY FIXTURES

J. Metal halide 250 watt

K. High pressure sodium 150 watt

System Components	QUANTITY	UNIT	COST PER S.F. MAT.	COST PER S.F. INST.	COST PER S.F. TOTAL
SYSTEM D5020 240 0880					
HIGH INTENSITY DISCHARGE FIXTURE, 16′ ABOVE WORK PLANE, 50 FC					
TYPE J, 9 FIXTURES PER 900 S.F.					
Steel intermediate conduit, (IMC) 1/2″ diam	.630	L.F.	1.26	3.94	5.20
Wire, 600V type, THWN-THHN, copper, solid, #10	.012	C.L.F.	.26	.75	1.01
Steel outlet box 4″ concrete	.010	Ea.	.19	.32	.51
Steel outlet box plate with stud, 4″ concrete	.010	Ea.	.12	.08	.20
Metal halide, lo bay, aluminum reflector, 250 W DX lamp	.010	Ea.	3.90	1.96	5.86
TOTAL			5.73	7.05	12.78

D5020 240	H.I.D. Fixture, Low Bay, 16′ (by Type)	COST PER S.F. MAT.	COST PER S.F. INST.	COST PER S.F. TOTAL
0510	High intensity discharge fixture, 16′ above work plane, 50 FC			
0880	Type J, 9 fixtures per 900 S.F.	5.75	7.05	12.80
0920	14 fixtures per 1800 S.F. R265723-05	4.91	6.65	11.56
0960	24 fixtures per 3000 S.F.	4.95	6.75	11.70
1000	32 fixtures per 4000 S.F.	4.95	6.75	11.70
1040	35 fixtures per 5000 S.F.	4.55	6.60	11.15
1080	56 fixtures per 8000 S.F.	4.55	6.60	11.15
1120	70 fixtures per 10,000 S.F.	4.55	6.60	11.15
1160	111 fixtures per 16,000 S.F.	4.55	6.60	11.15
1200	222 fixtures per 32,000 S.F.	4.55	6.60	11.15
1240	Type K, 11 fixtures per 900 S.F.	5.85	6.60	12.45
1280	20 fixtures per 1800 S.F.	5.40	6.10	11.50
1320	29 fixtures per 3000 S.F.	5.05	6	11.05
1360	39 fixtures per 4000 S.F.	5	5.95	10.95
1400	44 fixtures per 5000 S.F.	4.68	5.80	10.48
1440	62 fixtures per 8000 S.F.	4.68	5.80	10.48
1480	87 fixtures per 10,000 S.F.	4.68	5.80	10.48
1520	138 fixtures per 16,000 S.F.	4.68	5.80	10.48

D5020 Lighting and Branch Wiring

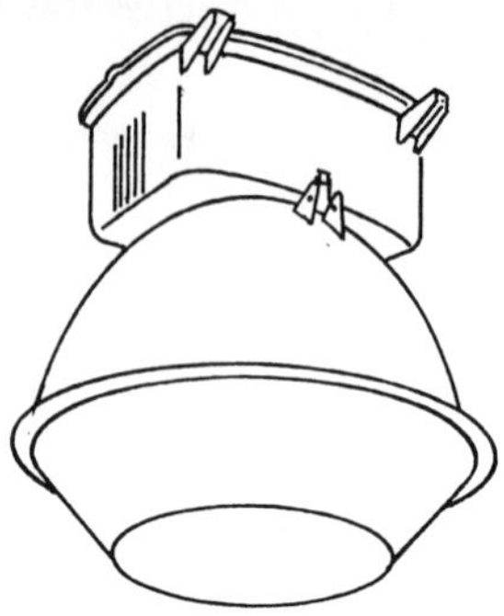

LOW BAY FIXTURES

J. Metal halide 250 watt

K. High pressure sodium 150 watt

System Components	QUANTITY	UNIT	COST PER S.F. MAT.	COST PER S.F. INST.	COST PER S.F. TOTAL
SYSTEM D5020 242 0240					
HIGH INTENSITY DISCHARGE FIXTURE, 16′ ABOVE WORK PLANE					
1 WATT/S.F., TYPE J, 28 FC, 4 FIXTURES/1000 S.F.					
Steel intermediate conduit, (IMC) 1/2″ diam	.328	L.F.	.66	2.05	2.71
Wire, 600V, type THWN-THHN, copper, solid, #10	.010	C.L.F.	.22	.63	.85
Steel outlet box 4″ concrete	.004	Ea.	.07	.13	.20
Steel outlet box plate with stud, 4″ concrete	.004	Ea.	.05	.03	.08
Metal halide, lo bay, aluminum reflector, 250 W DX lamp	.004	Ea.	1.56	.78	2.34
TOTAL			2.56	3.62	6.18

D5020 242	H.I.D. Fixture, Low Bay, 16′ (by Wattage)	COST PER S.F. MAT.	COST PER S.F. INST.	COST PER S.F. TOTAL
0190	High intensity discharge fixture, mounted 16′ above work plane			
0240	1 watt/S.F., type J, 28 FC, 4 fixt./1000 S.F.	2.56	3.62	6.18
0280	Type K, 27 FC, 5 fixt./1000 S.F. R265723-05	2.87	4.06	6.93
0400	2 watt/S.F., type J, 48 FC, 7 fixt/1000 S.F.	4.69	7	11.69
0440	Type K, 58 FC, 11 fixt/1000 S.F.	6.10	8.30	14.40
0560	3 watt/S.F., type J, 75 FC, 11 fixt/1000 S.F.	7.25	10.65	17.90
0600	Type K, 85 FC, 16 fixt/1000 S.F.	8.95	12.35	21.30
0720	4 watt/S.F., type J, 95 FC, 14 fixt/1000 S.F.	9.35	14	23.35
0760	Type K, 117 FC, 22 fixt/1000 S.F.	12.20	16.55	28.75
0880	5 watt/S.F., type J, 122 FC, 18 fixt/1000 S.F.	11.95	17.60	29.55
0920	Type K, 143 FC, 27 fixt/1000 S.F.	15.05	20.50	35.55

D5020 Lighting and Branch Wiring

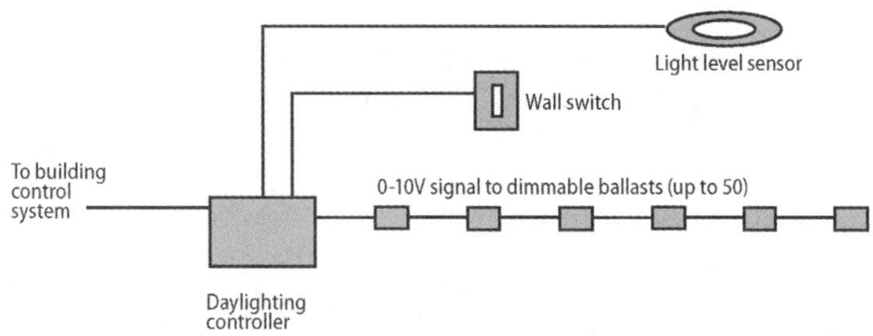

Daylight Dimming System

System Components	QUANTITY	UNIT	COST PER S.F. MAT.	COST PER S.F. INST.	COST PER S.F. TOTAL
SYSTEM D5020 290 1000					
DAYLIGHT DIMMING CONTROL SYSTEM					
10 FIXTURES PER 1000 S.F.					
Tray cable, type TC, copper #16-4 conductor	.600	C.L.F.	.02	.05	.07
Wire, 600 volt, type THWN-THHN, copper, solid, #12	.320	C.L.F.		.02	.02
EMT conduit w/fittings & support	10.000	L.F.	.01	.05	.06
Junction box 12" x12" x 4"	.400	Ea.	.01	.04	.05
Dimmer control	.400	Ea.	.13	.13	.26
Dimmable ballast three lamp	10.000	Ea.	1.33	.91	2.24
Room light sensor	.400	Ea.	.05	.04	.09
Wall switch	.400	Ea.	.03	.01	.04
Remote power pack	.200	Ea.	.01	.01	.02
TOTAL			1.59	1.26	2.85

D5020 290	Daylight Dimming System	COST PER S.F. MAT.	COST PER S.F. INST.	COST PER S.F. TOTAL
0500	Daylight dimming system (no fixtures or fixture power included)			
1000	Daylight dimming control system, 10 fixtures per 1000 S.F.	1.59	1.26	2.85
2000	12 fixtures per 1000 S.F.	1.91	1.49	3.40
3000	15 fixtures per 1000 S.F.	2.38	1.86	4.24
4000	20 fixtures per 1000 S.F.	3.18	2.44	5.62
5000	25 fixtures per 1000 S.F.	3.97	3.06	7.03

D5020 Lighting and Branch Wiring

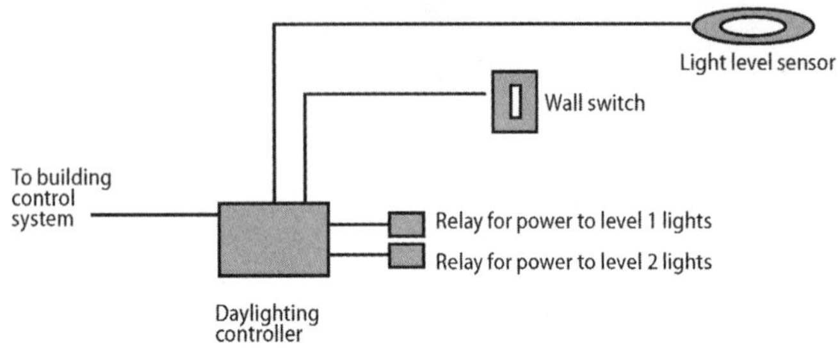

Daylight On/Off Control System

System Components	QUANTITY	UNIT	COST PER S.F. MAT.	COST PER S.F. INST.	COST PER S.F. TOTAL
SYSTEM D5020 295 1000					
DAYLIGHT ON/OFF CONTROL SYSTEM					
10 FIXTURES PER 1000 S.F.					
Tray cable, type TC, copper #16-4 conductor	.600	C.L.F.	.02	.05	.07
Wire, 600 volt, type THWN-THHN, copper, solid, #12	.320	C.L.F.		.02	.02
EMT conduit w/fittings & supports	10.000	L.F.	.01	.05	.06
Enclosure 12″ x12″ x 4″	.400	Ea.	.01	.04	.05
Low voltage relays	.800	Ea.	.04	.04	.08
Time Switch w/reserve power	.400	Ea.	.26	.07	.33
Light sensor	.400	Ea.	.05	.04	.09
Wall Switch	.400	Ea.	.03	.01	.04
Remote power pack	.200	Ea.	.01	.01	.02
TOTAL			.43	.33	.76

D5020 295	Daylight On/Off Control System	COST PER S.F. MAT.	COST PER S.F. INST.	COST PER S.F. TOTAL
0500	Daylight on/off control system (no fixtures or fixture power included)			
1000	Daylight on/off control system, 10 fixtures per 1000 S.F.	.43	.33	.76
2000	12 fixtures per 1000 S.F.	.50	.38	.88
3000	15 fixtures per 1000 S.F.	.61	.46	1.07
4000	20 fixtures per 1000 S.F.	.83	.59	1.42
5000	25 fixtures per 1000 S.F.	1.03	.74	1.77

D5030 Communications and Security

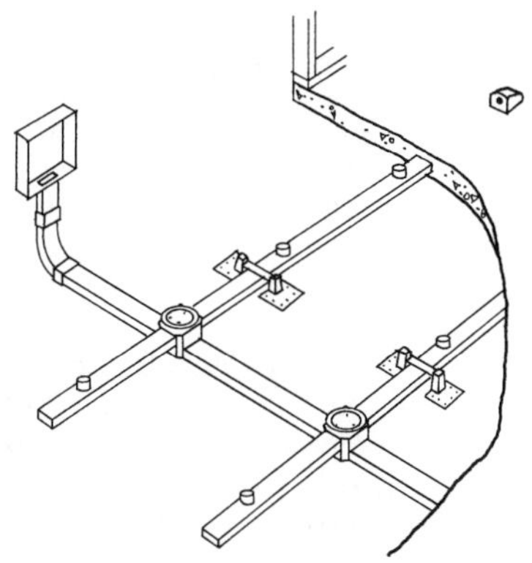

Description: System below includes telephone fitting installed. Does not include cable.

When poke thru fittings and telepoles are used for power, they can also be used for telephones at a negligible additional cost.

System Components	QUANTITY	UNIT	COST PER S.F. MAT.	COST PER S.F. INST.	COST PER S.F. TOTAL
SYSTEM D5030 310 0200					
TELEPHONE SYSTEMS, UNDERFLOOR DUCT, 5′ ON CENTER, LOW DENSITY					
Underfloor duct 7-1/4″ w/insert 2′ O.C. 1-3/8″ x 7-1/4″ super duct	.190	L.F.	6.84	2.38	9.22
Vertical elbow for underfloor superduct, 7-1/4″, included					
Underfloor duct conduit adapter, 2″ x 1-1/4″, included					
Underfloor duct junction box, single duct, 7-1/4″ x 3 1/8″	.003	Ea.	1.61	.47	2.08
Underfloor junction box carpet pan	.003	Ea.	1.14	.02	1.16
Underfloor duct outlet, low tension	.004	Ea.	.43	.31	.74
TOTAL			10.02	3.18	13.20

D5030 310	Telephone Systems	COST PER S.F. MAT.	COST PER S.F. INST.	COST PER S.F. TOTAL
0200	Telephone systems, underfloor duct, 5′ on center, low density	10	3.18	13.18
0240	5′ on center, high density	10.45	3.50	13.95
0280	7′ on center, low density	8	2.65	10.65
0320	7′ on center, high density	8.45	2.97	11.42
0400	Poke thru fittings, low density	1.12	1.13	2.25
0440	High density	2.24	2.25	4.49
0520	Telepoles, low density	1.15	.68	1.83
0560	High density	2.30	1.36	3.66
0640	Conduit system with floor boxes, low density	1.55	1.11	2.66
0680	High density	3.10	2.19	5.29
1020	Telephone wiring for offices & laboratories, 8 jacks/M.S.F.	.44	1.74	2.18

D5030 Communications and Security

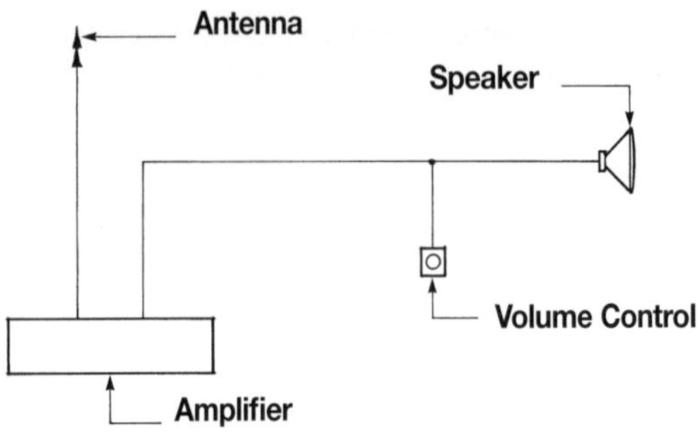

Sound System Includes AM–FM antenna, outlets, rigid conduit, and copper wire.
Fire Detection System Includes pull stations, signals, smoke and heat detectors, rigid conduit, and copper wire.
Intercom System Includes master and remote stations, rigid conduit, and copper wire.
Master Clock System Includes clocks, bells, rigid conduit, and copper wire.
Master TV Antenna Includes antenna, VHF–UHF reception and distribution, rigid conduit, and copper wire.

System Components	QUANTITY	UNIT	COST EACH MAT.	COST EACH INST.	COST EACH TOTAL
SYSTEM D5030 910 0220					
SOUND SYSTEM, INCLUDES OUTLETS, BOXES, CONDUIT & WIRE					
Steel intermediate conduit, (IMC) 1/2" diam	1200.000	L.F.	2,400	7,500	9,900
Wire sound shielded w/drain, #22-2 conductor	15.500	C.L.F.	306.90	1,216.75	1,523.65
Sound system speakers ceiling or wall	12.000	Ea.	1,524	942	2,466
Sound system volume control	12.000	Ea.	1,146	942	2,088
Sound system amplifier, 250 Watts	1.000	Ea.	1,350	625	1,975
Sound system antenna, AM FM	1.000	Ea.	147	157	304
Sound system monitor panel	1.000	Ea.	420	157	577
Sound system cabinet	1.000	Ea.	915	625	1,540
Steel outlet box 4" square	12.000	Ea.	32.52	378	410.52
Steel outlet box 4" plaster rings	12.000	Ea.	43.80	117.60	161.40
TOTAL			8,285.22	12,660.35	20,945.57

D5030 910	Communication & Alarm Systems	COST EACH MAT.	COST EACH INST.	COST EACH TOTAL
0200	Communication & alarm systems, includes outlets, boxes, conduit & wire			
0210	Sound system, 6 outlets	5,750	7,925	13,675
0220	12 outlets	8,275	12,700	20,975
0240	30 outlets	14,800	23,900	38,700
0280	100 outlets	45,200	80,000	125,200
0320	Fire detection systems, non-addressable, 12 detectors	3,300	6,575	9,875
0360	25 detectors	5,775	11,100	16,875
0400	50 detectors	11,400	22,000	33,400
0440	100 detectors	21,200	40,000	61,200
0450	Addressable type, 12 detectors	4,950	6,625	11,575
0452	25 detectors	8,975	11,200	20,175
0454	50 detectors	17,300	22,000	39,300
0456	100 detectors	33,600	40,500	74,100
0458	Fire alarm control panel, 8 zone, excluding wire and conduit	845	1,250	2,095
0459	12 zone	2,575	1,875	4,450
0460	Fire alarm command center, addressable without voice, excl. wire & conduit	4,825	1,100	5,925
0462	Addressable with voice	10,200	1,725	11,925
0480	Intercom systems, 6 stations	3,950	5,400	9,350
0520	12 stations	7,600	10,800	18,400
0560	25 stations	12,700	20,700	33,400
0600	50 stations	24,700	39,000	63,700
0640	100 stations	48,700	76,000	124,700
0680	Master clock systems, 6 rooms	5,125	8,850	13,975
0720	12 rooms	8,050	15,100	23,150
0760	20 rooms	11,200	21,300	32,500
0800	30 rooms	18,300	39,300	57,600

D5030 Communications and Security

D5030 910	Communication & Alarm Systems	COST EACH		
		MAT.	INST.	TOTAL
0840	50 rooms	29,700	66,500	96,200
0880	100 rooms	57,500	131,500	189,000
0920	Master TV antenna systems, 6 outlets	2,625	5,600	8,225
0960	12 outlets	4,900	10,400	15,300
1000	30 outlets	11,400	24,100	35,500
1040	100 outlets	37,900	78,500	116,400

D5030 920	Data Communication	COST PER M.S.F.		
		MAT.	INST.	TOTAL
0100	Data communication system, incl data/voice outlets, boxes, conduit, & cable			
0102	Data and voice system, 2 data/voice outlets per 1000 S.F.	179	465	644
0104	4 data/voice outlets per 1000 S.F.	340	915	1,255
0106	6 data/voice outlets per 1000 S.F.	480	1,350	1,830
0110	8 data/voice outlets per 1000 S.F.	615	1,775	2,390

D5090 Other Electrical Systems

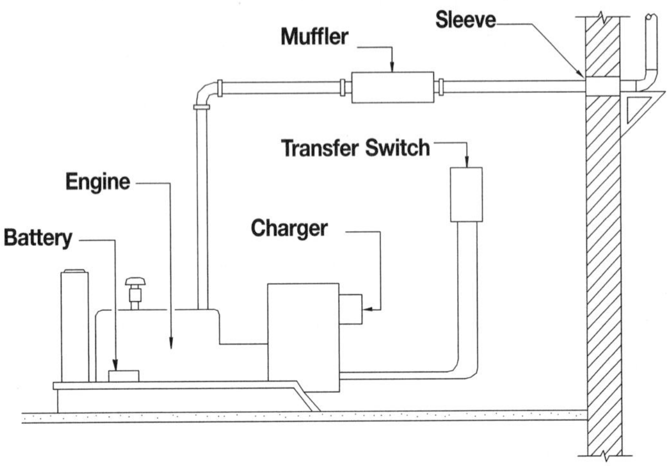

Description: System below tabulates the installed cost for generators by kW. Included in costs are battery, charger, muffler, and transfer switch.

No conduit, wire, or terminations included.

System Components	QUANTITY	UNIT	COST PER kW MAT.	COST PER kW INST.	COST PER kW TOTAL
SYSTEM D5090 210 0200					
GENERATOR SET, INCL. BATTERY, CHARGER, MUFFLER & TRANSFER SWITCH					
GAS/GASOLINE OPER., 3 PHASE, 4 WIRE, 277/480V, 7.5 kW					
Generator set, gas or gasoline operated, 3 ph 4 W, 277/480 V, 7.5 kW	.133	Ea.	1,256.67	274.27	1,530.94
TOTAL			1,256.67	274.27	1,530.94

D5090 210	Generators (by kW)	COST PER kW MAT.	COST PER kW INST.	COST PER kW TOTAL
0190	Generator sets, include battery, charger, muffler & transfer switch			
0200	Gas/gasoline operated, 3 phase, 4 wire, 277/480 volt, 7.5 kW	1,250	275	1,525
0240	11.5 kW	1,150	209	1,359
0280	20 kW	785	136	921
0320	35 kW	535	88.50	623.50
0360	80 kW	385	53.50	438.50
0400	100 kW	335	52	387
0440	125 kW	550	49	599
0480	185 kW	490	37	527
0560	Diesel engine with fuel tank, 30 kW	705	103	808
0600	50 kW	505	81	586
0640	75 kW	435	65.50	500.50
0680	100 kW	365	55.50	420.50
0720	125 kW	305	47	352
0760	150 kW	293	44	337
0800	175 kW	273	39	312
0840	200 kW	247	35.50	282.50
0880	250 kW	232	29.50	261.50
0920	300 kW	208	26	234
0960	350 kW	203	24	227
1000	400 kW	219	22.50	241.50
1040	500 kW	221	19	240
1200	750 kW	242	11.85	253.85
1400	1000 kW	224	12.55	236.55

D5090 Special Construction

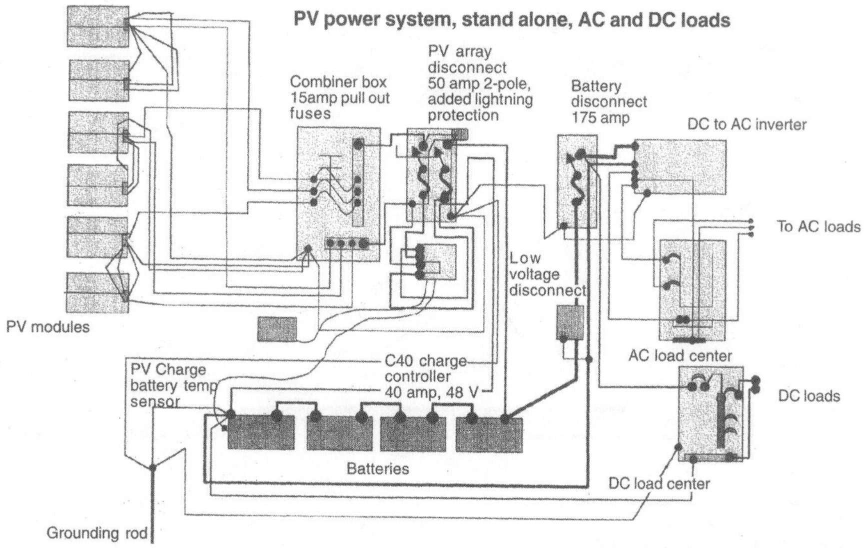

System Components	QUANTITY	UNIT	COST EACH MAT.	COST EACH INST.	COST EACH TOTAL
SYSTEM D5090 420 0100					
PHOTOVOLTAIC POWER SYSTEM, STAND ALONE					
Alternative energy sources, photovoltaic module, 75 watt, 17 volts	12.000	Ea.	5,100	942	6,042
DC to AC inverter for, 48V, 5,500 watt	1.000	Ea.	4,075	315	4,390
PV components, combiner box, 10 lug, NEMA 3R enclosure	1.000	Ea.	206	157	363
Alternative energy sources, PV components, fuse, 15 A for combiner box	10.000	Ea.	156.50	157	313.50
Battery charger controller w/temperature sensor	1.000	Ea.	530	157	687
Digital readout panel, displays hours, volts, amps, etc.	1.000	Ea.	140	157	297
Deep cycle solar battery, 6 V, 180 Ah (C/20)	4.000	Ea.	1,012	314	1,326
Battery interconnection, 15″ AWG #2/0, sealed w/copper ring lugs	3.000	Ea.	41.55	117	158.55
Battery interconn, 24″ AWG #2/0, sealed w/copper ring lugs	2.000	Ea.	39.60	78	117.60
Battery interconn, 60″ AWG #2/0, sealed w/copper ring lugs	2.000	Ea.	94	78	172
Batt temp computer probe, RJ11 jack, 15′ cord	1.000	Ea.	21	39	60
System disconnect, DC 175 amp circuit breaker	1.000	Ea.	151	78.50	229.50
Conduit box for inverter	1.000	Ea.	47.50	78.50	126
Low voltage disconnect	1.000	Ea.	43	78.50	121.50
Vented battery enclosure, wood	1.000	Ea.	188	277	465
Grounding, rod, copper clad, 8′ long, 5/8″ diameter	1.000	Ea.	21	114	135
Grounding, clamp, bronze, 5/8″ dia	1.000	Ea.	6.95	19.60	26.55
Bare copper wire, stranded, #8	1.000	C.L.F.	35.50	57	92.50
Wire, 600 volt, type THW, copper, stranded, #12	3.600	C.L.F.	56.52	205.20	261.72
Wire, 600 volt, type THW, copper, stranded, #8	1.050	C.L.F.	43.05	82.43	125.48
Rigid galvanized steel conduit, 3/4″, including fittings	120.000	L.F.	355.20	942	1,297.20
Conduit,to 15′ H,incl 2 termn,2 elb&11 bm clp per 100′,(EMT), 1″ dia	30.000	L.F.	55.80	163.50	219.30
Lightning surge suppressor	1.000	Ea.	55	19.60	74.60
General duty 240 volt, 2 pole, nonfusible, NEMA 3R, 60 amp	1.000	Ea.	298	273	571
Load centers, 3 wire, 120/240V, 100 amp main lugs, indoor, 8 circuits	2.000	Ea.	360	900	1,260
Plug-in panel or load center, 120/240 volt, to 60 amp, 1 pole	5.000	Ea.	59.75	262.50	322.25
Roof mounting frame, metal, for 6 modules	2.000	Ea.	1,930	258	2,188
Fuses, dual element, time delay, 250 volt, 50 amp	2.000	Ea.	22.20	25.10	47.30
TOTAL			15,144.12	6,345.43	21,489.55

D5090 420	Photovoltaic System, EACH	COST EACH MAT.	COST EACH INST.	COST EACH TOTAL
0100	Photovoltaic power system, stand alone, AC and DC loads	15,100	6,350	21,450

D5090 Special Construction

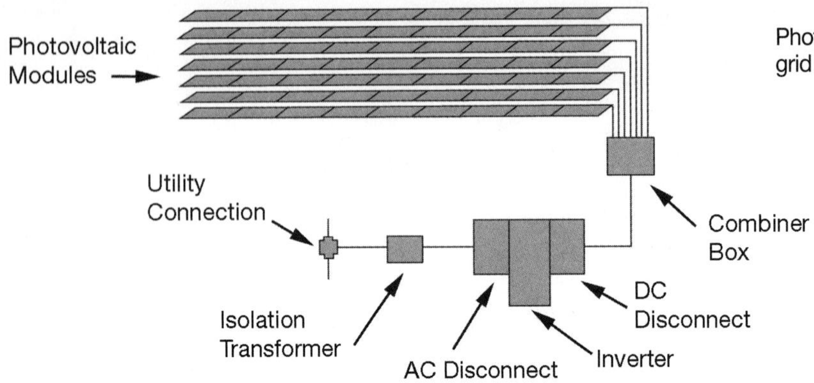

System Components	QUANTITY	UNIT	COST EACH MAT.	COST EACH INST.	COST EACH TOTAL
SYSTEM D5090 430 0100					
PHOTOVOLTAIC POWER SYSTEM, GRID CONNECTED 10 kW					
PV modules, 167 watt, 23.5 V	60.000	Ea.	42,300	4,710	47,010
Mounting frame 6 modules	10.000	Ea.	9,650	1,290	10,940
Steel angle support	400.000	L.F.	192	5,816	6,008
# 12 wire	4.000	C.L.F.	126	228	354
AC Disconnect switch, 60 A	1.000	Ea.	505	285	790
Fuse 75 A	3.000	Ea.	87	47.10	134.10
4/0 wire	2.000	C.L.F.	930	440	1,370
Module connection 4/0	4.000	Ea.	40.20	358	398.20
Combiner box 10 lug, NEMA 3R	1.000	Ea.	206	157	363
Utility connection, 3 pole breaker	1.000	Ea.	243	101	344
15 amp fuses	10.000	Ea.	156.50	157	313.50
Enclosure 24″ x 24″ x 10″, NEMA 4	1.000	Ea.	5,150	1,250	6,400
Inverter 12 kW	1.000	Ea.	10,725	945	11,670
Conduit w/fittings & support	200.000	L.F.	372	1,090	1,462
# 6 ground wire	8.000	C.L.F.	556	772	1,328
Fuse, 60 A	3.000	Ea.	38.40	37.65	76.05
DC disconnect switch, 75 A	1.000	Ea.	940	350	1,290
10 kVA isolation tranformer	1.000	Ea.	1,850	785	2,635
# 6 ground wire	.400	C.L.F.	27.80	38.60	66.40
# 6 ground connection	2.000	Ea.	34.80	105	139.80
TOTAL			74,129.70	18,962.35	93,092.05

D5090 430	Photovoltaic System Grid Connected, AC and DC Loads	COST EACH MAT.	COST EACH INST.	COST EACH TOTAL
0100	Photovoltaic power system, grid connected 10 kW	74,000	19,000	93,000
1100	Photovoltaic power system, grid connected 20 kW	148,500	37,900	186,400

D5090 Other Electrical Systems

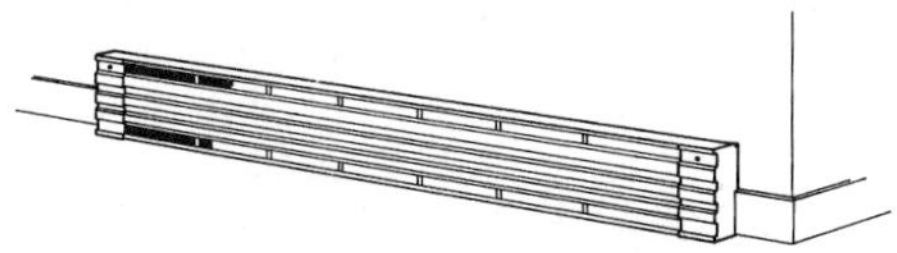

Low Density and Medium Density Baseboard Radiation

The costs shown in Table below are based on the following system considerations:

1. The heat loss per square foot is based on approximately 34 BTU/hr. per S.F. of floor or 10 watts per S.F. of floor.
2. Baseboard radiation is based on the low watt density type rated 187 watts per L.F. and the medium density type rated 250 watts per L.F.
3. Thermostat is not included.
4. Wiring costs include branch circuit wiring.

System Components	QUANTITY	UNIT	COST PER S.F. MAT.	COST PER S.F. INST.	COST PER S.F. TOTAL
SYSTEM D5090 510 1000					
ELECTRIC BASEBOARD RADIATION, LOW DENSITY, 900 S.F., 31 MBH, 9 kW					
Electric baseboard radiator, 5′ long 935 watt	.011	Ea.	.62	1.21	1.83
Steel intermediate conduit, (IMC) 1/2″ diam	.170	L.F.	.34	1.06	1.40
Wire 600 volt, type THW, copper, solid, #12	.005	C.L.F.	.07	.29	.36
TOTAL			1.03	2.56	3.59

D5090 510	Electric Baseboard Radiation (Low Density)	COST PER S.F. MAT.	COST PER S.F. INST.	COST PER S.F. TOTAL
1000	Electric baseboard radiation, low density, 900 S.F., 31 MBH, 9 kW	1.03	2.56	3.59
1200	1500 S.F., 51 MBH, 15 kW	1.01	2.50	3.51
1400	2100 S.F., 72 MBH, 21 kW	.90	2.24	3.14
1600	3000 S.F., 102 MBH, 30 kW	.81	2.04	2.85
2000	Medium density, 900 S.F., 31 MBH, 9 kW	.91	2.34	3.25
2200	1500 S.F., 51 MBH, 15 kW	.89	2.28	3.17
2400	2100 S.F., 72 MBH, 21 kW	.84	2.13	2.97
2600	3000 S.F., 102 MBH, 30 kW	.76	1.93	2.69

D5090 Other Electrical Systems

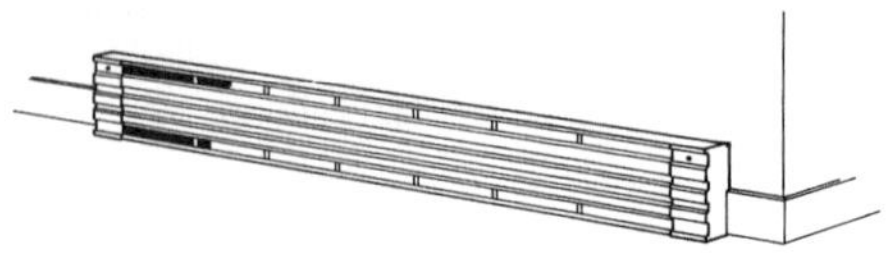

Commercial Duty Baseboard Radiation
The costs shown in Table below are based on the following system considerations:

1. The heat loss per square foot is based on approximately 41 BTU/hr. per S.F. of floor or 12 watts per S.F.
2. The baseboard radiation is of the commercial duty type rated 250 watts per L.F. served by 277 volt, single phase power.
3. Thermostat is not included.
4. Wiring costs include branch circuit wiring.

System Components	QUANTITY	UNIT	COST PER S.F. MAT.	INST.	TOTAL
SYSTEM D5090 520 1000					
ELECTRIC BASEBOARD RADIATION, MEDIUM DENSITY, 1230 S.F., 51 MBH, 15 kW					
Electric baseboard radiator, 5′ long	.013	Ea.	.73	1.43	2.16
Steel intermediate conduit, (IMC) 1/2″ diam	.154	L.F.	.31	.96	1.27
Wire 600 volt, type THW, copper, solid, #12	.004	C.L.F.	.05	.23	.28
TOTAL			1.09	2.62	3.71

D5090 520	Electric Baseboard Radiation (Medium Density)	COST PER S.F. MAT.	INST.	TOTAL
1000	Electric baseboard radiation, medium density, 1230 SF, 51 MBH, 15 kW	1.09	2.62	3.71
1200	2500 S.F. floor area, 106 MBH, 31 kW	1.01	2.46	3.47
1400	3700 S.F. floor area, 157 MBH, 46 kW	.99	2.39	3.38
1600	4800 S.F. floor area, 201 MBH, 59 kW	.92	2.21	3.13
1800	11,300 S.F. floor area, 464 MBH, 136 kW	.87	2.02	2.89
2000	30,000 S.F. floor area, 1229 MBH, 360 kW	.85	1.97	2.82

G Building Sitework

G1030 Site Earthwork

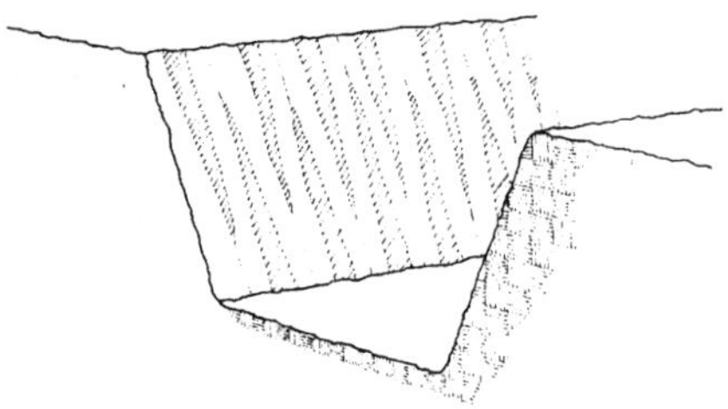

Trenching Systems are shown on a cost per linear foot basis. The systems include: excavation; backfill and removal of spoil; and compaction for various depths and trench bottom widths. The backfill has been reduced to accommodate a pipe of suitable diameter and bedding.

The slope for trench sides varies from none to 1:1.

The Expanded System Listing shows Trenching Systems that range from 2′ to 12′ in width. Depths range from 2′ to 25′.

System Components	QUANTITY	UNIT	COST PER L.F. EQUIP.	COST PER L.F. LABOR	COST PER L.F. TOTAL
SYSTEM G1030 805 1310					
TRENCHING COMMON EARTH, NO SLOPE, 2′ WIDE, 2′ DP, 3/8 C.Y. BUCKET					
Excavation, trench, hyd. backhoe, track mtd., 3/8 C.Y. bucket	.148	B.C.Y.	.40	1	1.40
Backfill and load spoil, from stockpile	.153	L.C.Y.	.12	.30	.42
Compaction by vibrating plate, 6″ lifts, 4 passes	.118	E.C.Y.	.03	.37	.40
Remove excess spoil, 8 C.Y. dump truck, 2 mile roundtrip	.040	L.C.Y.	.15	.15	.30
TOTAL			.70	1.82	2.52

G1030 805	Trenching Common Earth	COST PER L.F. EQUIP.	COST PER L.F. LABOR	COST PER L.F. TOTAL
1310	Trenching, common earth, no slope, 2′ wide, 2′ deep, 3/8 C.Y. bucket	.70	1.82	2.52
1320	3′ deep, 3/8 C.Y. bucket	1	2.75	3.75
1330	4′ deep, 3/8 C.Y. bucket	1.30	3.67	4.97
1340	6′ deep, 3/8 C.Y. bucket	1.68	4.75	6.43
1350	8′ deep, 1/2 C.Y. bucket	2.21	6.30	8.51
1360	10′ deep, 1 C.Y. bucket	3.59	7.50	11.09
1400	4′ wide, 2′ deep, 3/8 C.Y. bucket	1.62	3.58	5.20
1410	3′ deep, 3/8 C.Y. bucket	2.21	5.40	7.61
1420	4′ deep, 1/2 C.Y. bucket	2.57	6.10	8.67
1430	6′ deep, 1/2 C.Y. bucket	4.12	9.80	13.92
1440	8′ deep, 1/2 C.Y. bucket	6.80	12.65	19.45
1450	10′ deep, 1 C.Y. bucket	8.20	15.65	23.85
1460	12′ deep, 1 C.Y. bucket	10.55	20	30.55
1470	15′ deep, 1-1/2 C.Y. bucket	9.30	17.85	27.15
1480	18′ deep, 2-1/2 C.Y. bucket	12.90	25	37.90
1520	6′ wide, 6′ deep, 5/8 C.Y. bucket w/trench box	8.95	14.50	23.45
1530	8′ deep, 3/4 C.Y. bucket	11.90	19.10	31
1540	10′ deep, 1 C.Y. bucket	11.65	19.80	31.45
1550	12′ deep, 1-1/2 C.Y. bucket	12.50	21.50	34
1560	16′ deep, 2-1/2 C.Y. bucket	18	26.50	44.50
1570	20′ deep, 3-1/2 C.Y. bucket	22	31.50	53.50
1580	24′ deep, 3-1/2 C.Y. bucket	26	38	64
1640	8′ wide, 12′ deep, 1-1/2 C.Y. bucket w/trench box	17.50	27	44.50
1650	15′ deep, 1-1/2 C.Y. bucket	23	35.50	58.50
1660	18′ deep, 2-1/2 C.Y. bucket	26	35.50	61.50
1680	24′ deep, 3-1/2 C.Y. bucket	35.50	49.50	85
1730	10′ wide, 20′ deep, 3-1/2 C.Y. bucket w/trench box	28.50	47	75.50
1740	24′ deep, 3-1/2 C.Y. bucket	42.50	56.50	99
1780	12′ wide, 20′ deep, 3-1/2 C.Y. bucket w/trench box	45	59.50	104.50
1790	25′ deep, bucket	56	75.50	131.50
1800	1/2 to 1 slope, 2′ wide, 2′ deep, 3/8 C.Y. bucket	1	2.75	3.75
1810	3′ deep, 3/8 C.Y. bucket	1.68	4.82	6.50

G1030 Site Earthwork

G1030 805	Trenching Common Earth	COST PER L.F.		
		EQUIP.	LABOR	TOTAL
1820	4' deep, 3/8 C.Y. bucket	2.49	7.35	9.84
1840	6' deep, 3/8 C.Y. bucket	4.02	11.90	15.92
1860	8' deep, 1/2 C.Y. bucket	6.40	19.05	25.45
1880	10' deep, 1 C.Y. bucket	12.35	26.50	38.85
2300	4' wide, 2' deep, 3/8 C.Y. bucket	1.92	4.51	6.43
2310	3' deep, 3/8 C.Y. bucket	2.87	7.50	10.37
2320	4' deep, 1/2 C.Y. bucket	3.63	9.30	12.93
2340	6' deep, 1/2 C.Y. bucket	6.90	17.35	24.25
2360	8' deep, 1/2 C.Y. bucket	13.10	25.50	38.60
2380	10' deep, 1 C.Y. bucket	18.15	36	54.15
2400	12' deep, 1 C.Y. bucket	24	48.50	72.50
2430	15' deep, 1-1/2 C.Y. bucket	26.50	52	78.50
2460	18' deep, 2-1/2 C.Y. bucket	49	81.50	130.50
2840	6' wide, 6' deep, 5/8 C.Y. bucket w/trench box	13.25	21.50	34.75
2860	8' deep, 3/4 C.Y. bucket	19.15	32.50	51.65
2880	10' deep, 1 C.Y. bucket	18.75	33	51.75
2900	12' deep, 1-1/2 C.Y. bucket	23.50	43	66.50
2940	16' deep, 2-1/2 C.Y. bucket	41	63.50	104.50
2980	20' deep, 3-1/2 C.Y. bucket	55	85.50	140.50
3020	24' deep, 3-1/2 C.Y. bucket	77.50	117	194.50
3100	8' wide, 12' deep, 1-1/2 C.Y. bucket w/trench box	29.50	49	78.50
3120	15' deep, 1-1/2 C.Y. bucket	42.50	71.50	114
3140	18' deep, 2-1/2 C.Y. bucket	56.50	85	141.50
3180	24' deep, 3-1/2 C.Y. bucket	87	128	215
3270	10' wide, 20' deep, 3-1/2 C.Y. bucket w/trench box	55.50	98.50	154
3280	24' deep, 3-1/2 C.Y. bucket	96	140	236
3370	12' wide, 20' deep, 3-1/2 C.Y. bucket w/trench box	81	114	195
3380	25' deep, 3-1/2 C.Y. bucket	112	163	275
3500	1 to 1 slope, 2' wide, 2' deep, 3/8 C.Y. bucket	1.30	3.67	4.97
3520	3' deep, 3/8 C.Y. bucket	3.72	8.25	11.97
3540	4' deep, 3/8 C.Y. bucket	3.68	11.05	14.73
3560	6' deep, 3/8 C.Y. bucket	4.02	11.90	15.92
3580	8' deep, 1/2 C.Y. bucket	7.95	24	31.95
3600	10' deep, 1 C.Y. bucket	21	45.50	66.50
3800	4' wide, 2' deep, 3/8 C.Y. bucket	2.21	5.45	7.66
3820	3' deep, 3/8 C.Y. bucket	3.55	9.55	13.10
3840	4' deep, 1/2 C.Y. bucket	4.67	12.45	17.12
3860	6' deep, 1/2 C.Y. bucket	9.65	25	34.65
3880	8' deep, 1/2 C.Y. bucket	19.50	38.50	58
3900	10' deep, 1 C.Y. bucket	28	56	84
3920	12' deep, 1 C.Y. bucket	41.50	81.50	123
3940	15' deep, 1-1/2 C.Y. bucket	43.50	86.50	130
3960	18' deep, 2-1/2 C.Y. bucket	67	112	179
4030	6' wide, 6' deep, 5/8 C.Y. bucket w/trench box	17.35	29	46.35
4040	8' deep, 3/4 C.Y. bucket	25	40	65
4050	10' deep, 1 C.Y. bucket	27	48.50	75.50
4060	12' deep, 1-1/2 C.Y. bucket	36	66	102
4070	16' deep, 2-1/2 C.Y. bucket	64.50	100	164.50
4080	20' deep, 3-1/2 C.Y. bucket	88.50	139	227.50
4090	24' deep, 3-1/2 C.Y. bucket	129	196	325
4500	8' wide, 12' deep, 1-1/2 C.Y. bucket w/trench box	41.50	71.50	113
4550	15' deep, 1-1/2 C.Y. bucket	62.50	107	169.50
4600	18' deep, 2-1/2 C.Y. bucket	86	132	218
4650	24' deep, 3-1/2 C.Y. bucket	138	207	345
4800	10' wide, 20' deep, 3-1/2 C.Y. bucket w/trench box	82.50	150	232.50
4850	24' deep, 3-1/2 C.Y. bucket	147	219	366
4950	12' wide, 20' deep, 3-1/2 C.Y. bucket w/trench box	116	169	285
4980	25' deep, 3-1/2 C.Y. bucket	167	247	414

G1030 Site Earthwork

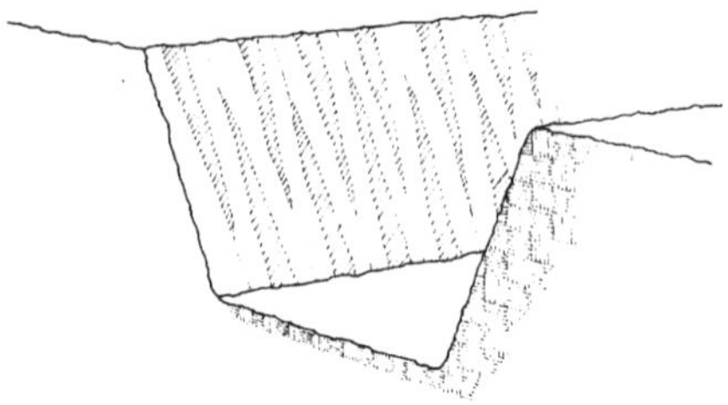

Trenching Systems are shown on a cost per linear foot basis. The systems include: excavation; backfill and removal of spoil; and compaction for various depths and trench bottom widths. The backfill has been reduced to accommodate a pipe of suitable diameter and bedding.

The slope for trench sides varies from none to 1:1.

The Expanded System Listing shows Trenching Systems that range from 2′ to 12′ in width. Depths range from 2′ to 25′.

System Components	QUANTITY	UNIT	COST PER L.F. EQUIP.	COST PER L.F. LABOR	COST PER L.F. TOTAL
SYSTEM G1030 806 1310					
TRENCHING LOAM & SANDY CLAY, NO SLOPE, 2′ WIDE, 2′ DP, 3/8 C.Y. BUCKET					
Excavation, trench, hyd. backhoe, track mtd., 3/8 C.Y. bucket	.148	B.C.Y.	.37	.93	1.30
Backfill and load spoil, from stockpile	.165	L.C.Y.	.13	.33	.46
Compaction by vibrating plate 18″ wide, 6″ lifts, 4 passes	.118	E.C.Y.	.03	.37	.40
Remove excess spoil, 8 C.Y. dump truck, 2 mile roundtrip	.042	L.C.Y.	.16	.16	.32
TOTAL			.70	1.82	2.52

G1030 806	Trenching Loam & Sandy Clay	COST PER L.F. EQUIP.	COST PER L.F. LABOR	COST PER L.F. TOTAL
1310	Trenching, loam & sandy clay, no slope, 2′ wide, 2′ deep, 3/8 C.Y. bucket	.69	1.79	2.48
1320	3′ deep, 3/8 C.Y. bucket	1.08	2.94	4.02
1330	4′ deep, 3/8 C.Y. bucket	1.27	3.58	4.85
1340	6′ deep, 3/8 C.Y. bucket	1.80	4.25	6.05
1350	8′ deep, 1/2 C.Y. bucket	2.38	5.65	8.05
1360	10′ deep, 1 C.Y. bucket	2.68	6.05	8.75
1400	4′ wide, 2′ deep, 3/8 C.Y. bucket	1.61	3.52	5.15
1410	3′ deep, 3/8 C.Y. bucket	2.19	5.30	7.50
1420	4′ deep, 1/2 C.Y. bucket	2.57	6	8.55
1430	6′ deep, 1/2 C.Y. bucket	4.39	8.75	13.15
1440	8′ deep, 1/2 C.Y. bucket	6.65	12.45	19.10
1450	10′ deep, 1 C.Y. bucket	6.40	12.75	19.15
1460	12′ deep, 1 C.Y. bucket	8.05	15.85	24
1470	15′ deep, 1-1/2 C.Y. bucket	9.70	18.45	28
1480	18′ deep, 2-1/2 C.Y. bucket	12.05	20.50	32.50
1520	6′ wide, 6′ deep, 5/8 C.Y. bucket w/trench box	8.65	14.25	23
1530	8′ deep, 3/4 C.Y. bucket	11.45	18.75	30
1540	10′ deep, 1 C.Y. bucket	10.70	19	29.50
1550	12′ deep, 1-1/2 C.Y. bucket	12.20	21.50	33.50
1560	16′ deep, 2-1/2 C.Y. bucket	17.50	27	44.50
1570	20′ deep, 3-1/2 C.Y. bucket	20	32	52
1580	24′ deep, 3-1/2 C.Y. bucket	25.50	39	64.50
1640	8′ wide, 12′ deep, 1-1/4 C.Y. bucket w/trench box	17.25	27	44.50
1650	15′ deep, 1-1/2 C.Y. bucket	21	34.50	55.50
1660	18′ deep, 2-1/2 C.Y. bucket	26.50	38.50	65
1680	24′ deep, 3-1/2 C.Y. bucket	34.50	50	84.50
1730	10′ wide, 20′ deep, 3-1/2 C.Y. bucket w/trench box	35	50	85
1740	24′ deep, 3-1/2 C.Y. bucket	43.50	62	106
1780	12′ wide, 20′ deep, 3-1/2 C.Y. bucket w/trench box	42	59.50	102
1790	25′ deep, bucket	54.50	77	132
1800	1/2:1 slope, 2′ wide, 2′ deep, 3/8 C.Y. bucket	.98	2.68	3.66
1810	3′ deep, 3/8 C.Y. bucket	1.64	4.70	6.35
1820	4′ deep, 3/8 C.Y. bucket	2.43	7.20	9.65
1840	6′ deep, 3/8 C.Y. bucket	4.32	10.65	14.95

G1030 Site Earthwork

G1030 806	Trenching Loam & Sandy Clay	COST PER L.F. EQUIP.	COST PER L.F. LABOR	COST PER L.F. TOTAL
1860	8' deep, 1/2 C.Y. bucket	6.85	17	24
1880	10' deep, 1 C.Y. bucket	9.15	21.50	30.50
2300	4' wide, 2' deep, 3/8 C.Y. bucket	1.90	4.43	6.35
2310	3' deep, 3/8 C.Y. bucket	2.83	7.35	10.20
2320	4' deep, 1/2 C.Y. bucket	3.59	9.10	12.70
2340	6' deep, 1/2 C.Y. bucket	7.35	15.60	23
2360	8' deep, 1/2 C.Y. bucket	12.75	25	38
2380	10' deep, 1 C.Y. bucket	14.05	29.50	43.50
2400	12' deep, 1 C.Y. bucket	23.50	48.50	72
2430	15' deep, 1-1/2 C.Y. bucket	27.50	54	81.50
2460	18' deep, 2-1/2 C.Y. bucket	47.50	82.50	130
2840	6' wide, 6' deep, 5/8 C.Y. bucket w/trench box	12.60	21.50	34
2860	8' deep, 3/4 C.Y. bucket	18.45	32	50.50
2880	10' deep, 1 C.Y. bucket	19.10	36	55
2900	12' deep, 1-1/2 C.Y. bucket	23.50	43.50	67
2940	16' deep, 2-1/2 C.Y. bucket	40	64	104
2980	20' deep, 3-1/2 C.Y. bucket	53	86.50	140
3020	24' deep, 3-1/2 C.Y. bucket	75	119	194
3100	8' wide, 12' deep, 1-1/2 C.Y. bucket w/trench box	29	49.50	78.50
3120	15' deep, 1-1/2 C.Y. bucket	39	69.50	109
3140	18' deep, 2-1/2 C.Y. bucket	55	86	141
3180	24' deep, 3-1/2 C.Y. bucket	84.50	130	215
3270	10' wide, 20' deep, 3-1/2 C.Y. bucket w/trench box	67.50	105	173
3280	24' deep, 3-1/2 C.Y. bucket	93	142	235
3320	12' wide, 20' deep, 3-1/2 C.Y. bucket w/trench box	75	114	189
3380	25' deep, 3-1/2 C.Y. bucket w/trench box	102	153	255
3500	1:1 slope, 2' wide, 2' deep, 3/8 C.Y. bucket	1.27	3.58	4.85
3520	3' deep, 3/8 C.Y. bucket	2.28	6.75	9.05
3540	4' deep, 3/8 C.Y. bucket	3.57	10.80	14.35
3560	6' deep, 3/8 C.Y. bucket	4.32	10.65	14.95
3580	8' deep, 1/2 C.Y. bucket	11.35	28.50	40
3600	10' deep, 1 C.Y. bucket	15.60	36.50	52
3800	4' wide, 2' deep, 3/8 C.Y. bucket	2.19	5.30	7.50
3820	3' deep, 3/8 C.Y. bucket	3.49	9.35	12.85
3840	4' deep, 1/2 C.Y. bucket	4.61	12.25	16.85
3860	6' deep, 1/2 C.Y. bucket	10.30	22.50	33
3880	8' deep, 1/2 C.Y. bucket	18.95	38	57
3900	10' deep, 1 C.Y. bucket	21.50	46	67.50
3920	12' deep, 1 C.Y. bucket	31.50	65	96.50
3940	15' deep, 1-1/2 C.Y. bucket	45	89.50	135
3960	18' deep, 2-1/2 C.Y. bucket	65.50	114	180
4030	6' wide, 6' deep, 5/8 C.Y. bucket w/trench box	16.55	29	45.50
4040	8' deep, 3/4 C.Y. bucket	25.50	45	70.50
4050	10' deep, 1 C.Y. bucket	27.50	52.50	80
4060	12' deep, 1-1/2 C.Y. bucket	35	66	101
4070	16' deep, 2-1/2 C.Y. bucket	62.50	101	164
4080	20' deep, 3-1/2 C.Y. bucket	86	142	228
4090	24' deep, 3-1/2 C.Y. bucket	125	199	325
4500	8' wide, 12' deep, 1-1/4 C.Y. bucket w/trench box	40.50	71.50	112
4550	15' deep, 1-1/2 C.Y. bucket	57	104	161
4600	18' deep, 2-1/2 C.Y. bucket	83.50	133	217
4650	24' deep, 3-1/2 C.Y. bucket	134	210	345
4800	10' wide, 20' deep, 3-1/2 C.Y. bucket w/trench box	100	160	260
4850	24' deep, 3-1/2 C.Y. bucket	143	222	365
4950	12' wide, 20' deep, 3-1/2 C.Y. bucket w/trench box	108	169	277
4980	25' deep, 3-1/2 C.Y. bucket	162	250	410

G1030 Site Earthwork

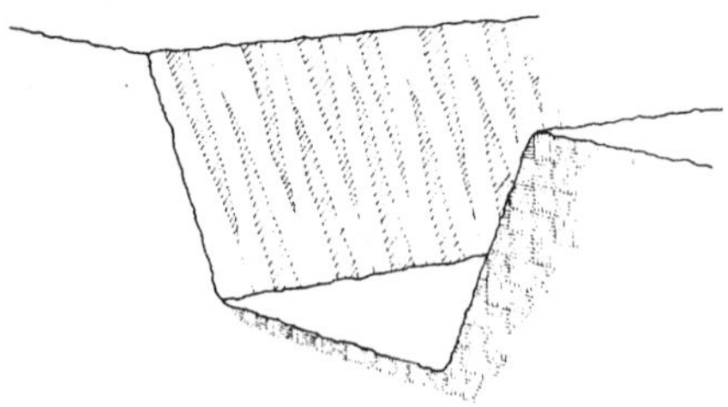

Trenching Systems are shown on a cost per linear foot basis. The systems include: excavation; backfill and removal of spoil; and compaction for various depths and trench bottom widths. The backfill has been reduced to accommodate a pipe of suitable diameter and bedding.

The slope for trench sides varies from none to 1:1.

The Expanded System Listing shows Trenching Systems that range from 2′ to 12′ in width. Depths range from 2′ to 25′.

System Components	QUANTITY	UNIT	COST PER L.F. EQUIP.	COST PER L.F. LABOR	COST PER L.F. TOTAL
SYSTEM G1030 807 1310					
TRENCHING SAND & GRAVEL, NO SLOPE, 2′ WIDE, 2′ DEEP, 3/8 C.Y. BUCKET					
Excavation, trench, hyd. backhoe, track mtd., 3/8 C.Y. bucket	.148	B.C.Y.	.36	.91	1.27
Backfill and load spoil, from stockpile	.140	L.C.Y.	.11	.28	.39
Compaction by vibrating plate 18″ wide, 6″ lifts, 4 passes	.118	E.C.Y.	.03	.37	.40
Remove excess spoil, 8 C.Y. dump truck, 2 mile roundtrip	.035	L.C.Y.	.14	.13	.27
TOTAL			.64	1.69	2.33

G1030 807	Trenching Sand & Gravel	COST PER L.F. EQUIP.	COST PER L.F. LABOR	COST PER L.F. TOTAL
1310	Trenching, sand & gravel, no slope, 2′ wide, 2′ deep, 3/8 C.Y. bucket	.64	1.69	2.33
1320	3′ deep, 3/8 C.Y. bucket	1.03	2.82	3.85
1330	4′ deep, 3/8 C.Y. bucket	1.19	3.40	4.59
1340	6′ deep, 3/8 C.Y. bucket	1.71	4.05	5.75
1350	8′ deep, 1/2 C.Y. bucket	2.25	5.35	7.60
1360	10′ deep, 1 C.Y. bucket	2.50	5.65	8.15
1400	4′ wide, 2′ deep, 3/8 C.Y. bucket	1.47	3.31	4.78
1410	3′ deep, 3/8 C.Y. bucket	2.01	5.05	7.05
1420	4′ deep, 1/2 C.Y. bucket	2.36	5.65	8
1430	6′ deep, 1/2 C.Y. bucket	4.14	8.40	12.55
1440	8′ deep, 1/2 C.Y. bucket	6.25	11.80	18.05
1450	10′ deep, 1 C.Y. bucket	6	12.05	18.05
1460	12′ deep, 1 C.Y. bucket	7.60	15	22.50
1470	15′ deep, 1-1/2 C.Y. bucket	9.15	17.40	26.50
1480	18′ deep, 2-1/2 C.Y. bucket	11.45	19.05	30.50
1520	6′ wide, 6′ deep, 5/8 C.Y. bucket w/trench box	8.20	13.50	21.50
1530	8′ deep, 3/4 C.Y. bucket	10.85	17.75	28.50
1540	10′ deep, 1 C.Y. bucket	10.10	18	28
1550	12′ deep, 1-1/2 C.Y. bucket	11.50	20	31.50
1560	16′ deep, 2 C.Y. bucket	16.75	25.50	42.50
1570	20′ deep, 3-1/2 C.Y. bucket	19.10	30	49
1580	24′ deep, 3-1/2 C.Y. bucket	24	36.50	60.50
1640	8′ wide, 12′ deep, 1-1/2 C.Y. bucket w/trench box	16.20	25.50	41.50
1650	15′ deep, 1-1/2 C.Y. bucket	19.70	32.50	52
1660	18′ deep, 2-1/2 C.Y. bucket	25	36.50	61.50
1680	24′ deep, 3-1/2 C.Y. bucket	32.50	47	79.50
1730	10′ wide, 20′ deep, 3-1/2 C.Y. bucket w/trench box	32.50	47	79.50
1740	24′ deep, 3-1/2 C.Y. bucket	41	58	99
1780	12′ wide, 20′ deep, 3-1/2 C.Y. bucket w/trench box	39.50	55.50	95
1790	25′ deep, 3-1/2 C.Y. bucket	51.50	72	124
1800	1/2:1 slope, 2′ wide, 2′ deep, 3/8 C.Y. bucket	.91	2.55	3.46
1810	3′ deep, 3/8 C.Y. bucket	1.53	4.48	6
1820	4′ deep, 3/8 C.Y. bucket	2.28	6.85	9.15
1840	6′ deep, 3/8 C.Y. bucket	4.10	10.15	14.25

G1030 Site Earthwork

G1030 807	Trenching Sand & Gravel	COST PER L.F.		
		EQUIP.	LABOR	TOTAL
1860	8' deep, 1/2 C.Y. bucket	6.50	16.25	23
1880	10' deep, 1 C.Y. bucket	8.55	20	28.50
2300	4' wide, 2' deep, 3/8 C.Y. bucket	1.74	4.17	5.90
2310	3' deep, 3/8 C.Y. bucket	2.62	6.95	9.55
2320	4' deep, 1/2 C.Y. bucket	3.31	8.65	11.95
2340	6' deep, 1/2 C.Y. bucket	6.95	14.95	22
2360	8' deep, 1/2 C.Y. bucket	12.10	24	36
2380	10' deep, 1 C.Y. bucket	13.25	27.50	41
2400	12' deep, 1 C.Y. bucket	22.50	46	68.50
2430	15' deep, 1-1/2 C.Y. bucket	26	51	77
2460	18' deep, 2-1/2 C.Y. bucket	45	77.50	123
2840	6' wide, 6' deep, 5/8 C.Y. bucket w/trench box	11.95	20.50	32.50
2860	8' deep, 3/4 C.Y. bucket	17.50	30	47.50
2880	10' deep, 1 C.Y. bucket	18.05	34	52
2900	12' deep, 1-1/2 C.Y. bucket	22.50	41	63.50
2940	16' deep, 2 C.Y. bucket	38.50	60.50	99
2980	20' deep, 3-1/2 C.Y. bucket	50.50	81.50	132
3020	24' deep, 3-1/2 C.Y. bucket	71.50	112	184
3100	8' wide, 12' deep, 1-1/4 C.Y. bucket w/trench box	27.50	46.50	74
3120	15' deep, 1-1/2 C.Y. bucket	37	65.50	103
3140	18' deep, 2-1/2 C.Y. bucket	52.50	81	134
3180	24' deep, 3-1/2 C.Y. bucket	80	122	202
3270	10' wide, 20' deep, 3-1/2 C.Y. bucket w/trench box	64	98.50	163
3280	24' deep, 3-1/2 C.Y. bucket	88.50	133	222
3370	12' wide, 20' deep, 3-1/2 C.Y. bucket w/trench box	71.50	109	181
3380	25' deep, 3-1/2 C.Y. bucket	103	154	257
3500	1:1 slope, 2' wide, 2' deep, 3/8 C.Y. bucket	2.04	4.18	6.20
3520	3' deep, 3/8 C.Y. bucket	2.14	6.40	8.55
3540	4' deep, 3/8 C.Y. bucket	3.37	10.30	13.65
3560	6' deep, 3/8 C.Y. bucket	4.10	10.15	14.25
3580	8' deep, 1/2 C.Y. bucket	10.80	27	38
3600	10' deep, 1 C.Y. bucket	14.60	34.50	49
3800	4' wide, 2' deep, 3/8 C.Y. bucket	2.01	5.05	7.05
3820	3' deep, 3/8 C.Y. bucket	3.24	8.90	12.15
3840	4' deep, 1/2 C.Y. bucket	4.26	11.65	15.90
3860	6' deep, 1/2 C.Y. bucket	9.80	21.50	31.50
3880	8' deep, 1/2 C.Y. bucket	17.95	36	54
3900	10' deep, 1 C.Y. bucket	20.50	43.50	64
3920	12' deep, 1 C.Y. bucket	29.50	61.50	91
3940	15' deep, 1-1/2 C.Y. bucket	42.50	84.50	127
3960	18' deep, 2-1/2 C.Y. bucket	62	107	169
4030	6' wide, 6' deep, 5/8 C.Y. bucket w/trench box	15.70	27.50	43
4040	8' deep, 3/4 C.Y. bucket	24	42.50	66.50
4050	10' deep, 1 C.Y. bucket	26	49.50	75.50
4060	12' deep, 1-1/2 C.Y. bucket	33.50	62.50	96
4070	16' deep, 2 C.Y. bucket	60	96	156
4080	20' deep, 3-1/2 C.Y. bucket	81.50	133	215
4090	24' deep, 3-1/2 C.Y. bucket	119	187	305
4500	8' wide, 12' deep, 1-1/2 C.Y. bucket w/trench box	38.50	68	107
4550	15' deep, 1-1/2 C.Y. bucket	54	98.50	153
4600	18' deep, 2-1/2 C.Y. bucket	79.50	125	205
4650	24' deep, 3-1/2 C.Y. bucket	127	197	325
4800	10' wide, 20' deep, 3-1/2 C.Y. bucket w/trench box	95	150	245
4850	24' deep, 3-1/2 C.Y. bucket	136	208	345
4950	12' wide, 20' deep, 3-1/2 C.Y. bucket w/trench box	102	159	261
4980	25' deep, 3-1/2 C.Y. bucket	154	235	390

G4010 Electrical Distribution

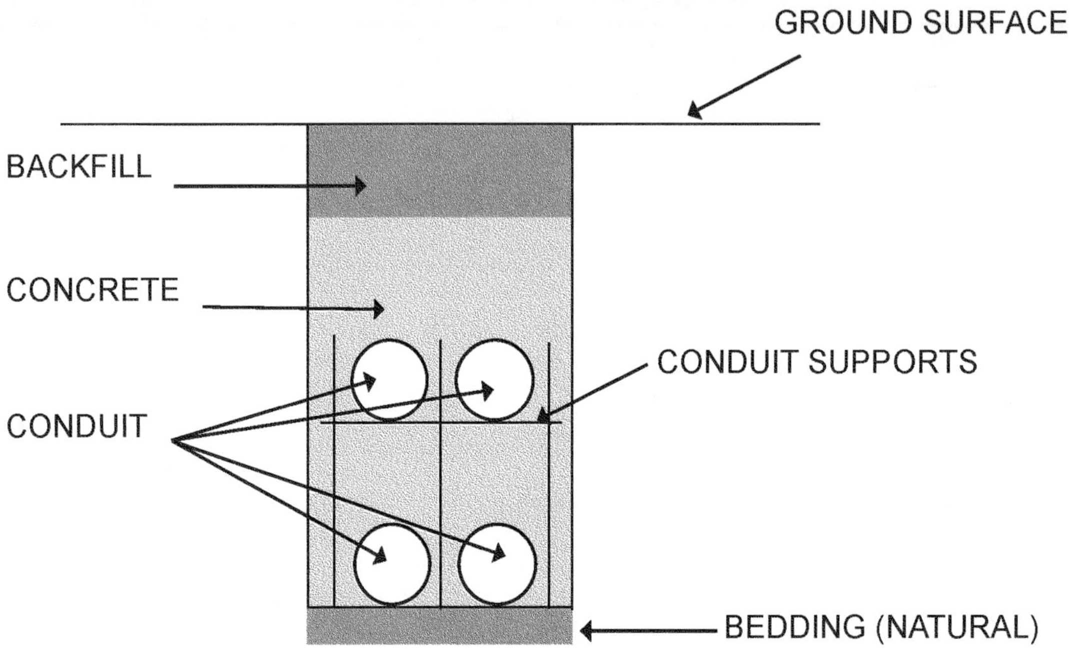

UNDERGROUND ELECTRICAL CONDUIT

System Components	QUANTITY	UNIT	COST EACH MAT.	COST EACH INST.	COST EACH TOTAL
SYSTEM G4010 312 1000					
2000 AMP UNDERGROUND SERVICE INCLUDING COMMON EARTH EXCAVATION, CONCRETE, BACKFILL AND COMPACTION					
Excavate Trench	44.440	B.C.Y.		294.64	294.64
4 inch conduit bank	108.000	L.F.	626.40	2,484	3,110.40
4 inch fitting	8.000	Ea.	212	420	632
4 inch bells	4.000	Ea.	16.08	210	226.08
concrete material	16.580	C.Y.	1,774.06		1,774.06
concrete placement	16.580	C.Y.		383	383
Backfill trench	33.710	L.C.Y.		94.06	94.06
Compact fill material in trench	25.930	E.C.Y.		121.36	121.36
Dispose of excess fill material on-site	24.070	L.C.Y.		268.62	268.62
500 kcmil power cable	18.000	C.L.F.	22,950	7,020	29,970
1/0	6.000	C.L.F.	1,638	1,140	2,778
TOTAL			27,216.54	12,435.68	39,652.22

G4010 312	Underground Power Feed	COST EACH MAT.	COST EACH INST.	COST EACH TOTAL
0950	Underground electrical including common earth excavation,concrete, backfill and			
1000	2000 Amp service, 100′ length, 4′ depth	27,200	12,500	39,700
1100	1600 Amp service	21,400	9,600	31,000
1200	1200 Amp service	16,500	7,975	24,475
1300	1000 Amp service	14,200	7,350	21,550
1400	800 Amp service	12,200	6,900	19,100
1500	600 Amp service	8,075	6,375	14,450

G4010 Electrical Distribution

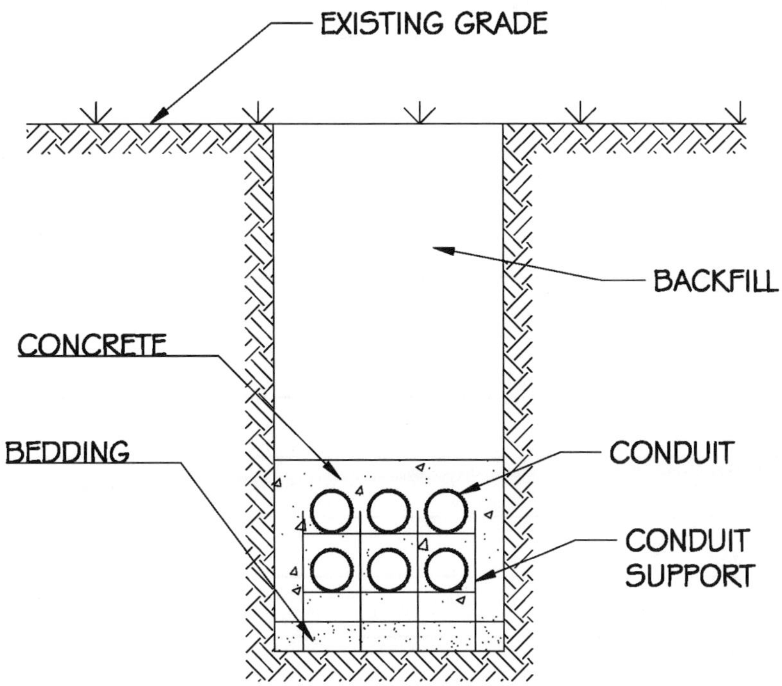

System Components	QUANTITY	UNIT	COST L.F. MAT.	COST L.F. INST.	COST L.F. TOTAL
SYSTEM G4010 320 1012					
UNDERGROUND ELECTRICAL CONDUIT, 2″ DIAMETER, INCLUDING EXCAVATION, CONCRETE, BEDDING, BACKFILL, AND COMPACTION					
Excavate Trench	.150	B.C.Y.		.99	.99
Base spacer	.200	Ea.	.34	2.24	2.58
Conduit	1.000	L.F.	2.26	3.48	5.74
Concrete material	.050	C.Y.	5.35		5.35
Concrete placement	.050	C.Y.		1.16	1.16
Backfill trench	.120	L.C.Y.		.34	.34
Compact fill material in trench	.100	E.C.Y.		.47	.47
Dispose of excess fill material on-site	.120	L.C.Y.		1.34	1.34
Marking tape	.010	C.L.F.	.13	.03	.16
TOTAL			8.08	10.05	18.13

G4010 320	Underground Electrical Conduit	COST L.F. MAT.	COST L.F. INST.	COST L.F. TOTAL
0950	Underground electrical conduit including common earth excavation,			
0952	Concrete, bedding, backfill and compaction			
1012	2″ dia. Schedule 40 PVC, 2′ deep	8.10	10.05	18.15
1014	4′ deep	8.10	14.35	22.45
1016	6′ deep	8.10	18.50	26.60
1022	2@2″ dia. Schedule 40 PVC, 2′ deep	10.65	15.80	26.45
1024	4′ deep	10.65	20	30.65
1026	6′ deep	10.65	24	34.65
1032	3@2″ dia. Schedule 40 PVC, 2′ deep	13.25	21.50	34.75
1034	4′ deep	13.25	26	39.25
1036	6′ deep	13.25	30	43.25
1042	4@2″ dia. Schedule 40 PVC, 2′ deep	17.85	27	44.85
1044	4′ deep	17.85	31	48.85
1046	6′ deep	17.85	35.50	53.35

G4010 Electrical Distribution

G4010 320	Underground Electrical Conduit	COST L.F.		
		MAT.	INST.	TOTAL
1062	6@2" dia. Schedule 40 PVC, 2' deep	23	38	61
1064	4' deep	23	42	65
1066	6' deep	23	46.50	69.50
1082	8@2" dia. Schedule 40 PVC, 2' deep	28	49.50	77.50
1084	4' deep	28	53.50	81.50
1086	6' deep	28	58	86
1092	9@2" dia. Schedule 40 PVC, 2' deep	30.50	55	85.50
1094	4' deep	32.50	59	91.50
1096	6' deep	32.50	63.50	96
1212	3" dia. Schedule 40 PVC, 2' deep	11.20	12.35	23.55
1214	4' deep	11.20	16.70	27.90
1216	6' deep	11.20	21	32.20
1222	2@3" dia. Schedule 40 PVC, 2' deep	14.80	20	34.80
1224	4' deep	14.80	24.50	39.30
1226	6' deep	14.80	28.50	43.30
1232	3@3" dia. Schedule 40 PVC, 2' deep	19.45	28	47.45
1234	4' deep	19.45	32.50	51.95
1236	6' deep	19.45	36.50	55.95
1242	4@3" dia. Schedule 40 PVC, 2' deep	27.50	36	63.50
1244	4' deep	27.50	40	67.50
1246	6' deep	27.50	44.50	72
1262	6@3" dia. Schedule 40 PVC, 2' deep	35.50	52	87.50
1264	4' deep	35.50	56	91.50
1266	6' deep	35.50	60.50	96
1282	8@3" dia. Schedule 40 PVC, 2' deep	45	68	113
1284	4' deep	45	72	117
1286	6' deep	45	77	122
1292	9@3" dia. Schedule 40 PVC, 2' deep	52	77	129
1294	4' deep	53	80	133
1296	6' deep	53	84	137
1312	4" dia. Schedule 40 PVC, 2' deep	12.95	15.10	28.05
1314	4' deep	12.95	19.45	32.40
1316	6' deep	12.95	23.50	36.45
1322	2@4" dia. Schedule 40 PVC, 2' deep	19.35	26	45.35
1324	4' deep	19.35	30.50	49.85
1326	6' deep	19.35	34.50	53.85
1332	3@4" dia. Schedule 40 PVC, 2' deep	24.50	37	61.50
1334	4' deep	24.50	41.50	66
1336	6' deep	24.50	45	69.50
1342	4@4" dia. Schedule 40 PVC, 2' deep	35.50	48	83.50
1344	4' deep	35.50	52	87.50
1346	6' deep	35.50	56.50	92
1362	6@4" dia. Schedule 40 PVC, 2' deep	47	69.50	116.50
1364	4' deep	47	73.50	120.50
1366	6' deep	47	78	125
1382	8@4" dia. Schedule 40 PVC, 2' deep	61	92	153
1384	4' deep	61	97	158
1386	6' deep	61	102	163
1392	9@4" dia. Schedule 40 PVC, 2' deep	68.50	103	171.50
1394	4' deep	68.50	106	174.50
1396	6' deep	68.50	110	178.50
1412	5" dia. Schedule 40 PVC, 2' deep	15.75	18.05	33.80
1414	4' deep	15.75	22	37.75
1416	6' deep	15.75	26.50	42.25
1422	2@5" dia. Schedule 40 PVC, 2' deep	25	32	57
1424	4' deep	25	36	61
1426	6' deep	25	40.50	65.50
1432	3@5" dia. Schedule 40 PVC, 2' deep	34	46	80

G4010 Electrical Distribution

G4010 320	Underground Electrical Conduit	COST L.F.		
		MAT.	INST.	TOTAL
1434	4' deep	34	50.50	84.50
1436	6' deep	34	55	89
1442	4@5" dia. Schedule 40 PVC, 2' deep	46.50	59	105.50
1444	4' deep	46.50	63.50	110
1446	6' deep	46.50	68	114.50
1462	6@5" dia. Schedule 40 PVC, 2' deep	63.50	86.50	150
1464	4' deep	63.50	91	154.50
1466	6' deep	63.50	96	159.50
1482	8@5" dia. Schedule 40 PVC, 2' deep	84	116	200
1484	4' deep	84	122	206
1486	6' deep	84	127	211
1492	9@5" dia. Schedule 40 PVC, 2' deep	94.50	130	224.50
1494	4' deep	94.50	133	227.50
1496	6' deep	94.50	137	231.50
1512	6" dia. Schedule 40 PVC, 2' deep	19.85	22	41.85
1514	4' deep	19.85	26	45.85
1516	6' deep	19.85	30.50	50.35
1522	2@6" dia. Schedule 40 PVC, 2' deep	31	39.50	70.50
1524	4' deep	31	43.50	74.50
1526	6' deep	31	48	79
1532	3@6" dia. Schedule 40 PVC, 2' deep	44.50	57.50	102
1534	4' deep	44.50	62.50	107
1536	6' deep	44.50	68.50	113
1542	4@6" dia. Schedule 40 PVC, 2' deep	58.50	74	132.50
1544	4' deep	58.50	78.50	137
1546	6' deep	58.50	82.50	141
1562	6@6" dia. Schedule 40 PVC, 2' deep	84	110	194
1564	4' deep	84	115	199
1566	6' deep	84	120	204
1582	8@6" dia. Schedule 40 PVC, 2' deep	112	146	258
1584	4' deep	112	153	265
1586	6' deep	112	160	272
1592	9@6" dia. Schedule 40 PVC, 2' deep	125	165	290
1594	4' deep	125	168	293
1596	6' deep	125	173	298

G4020 Site Lighting

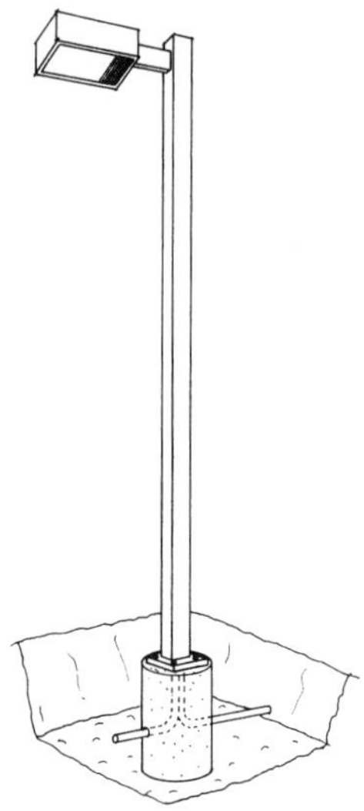

The Site Lighting System includes the complete unit from foundation to electrical fixtures. Each system includes: excavation; concrete base; backfill by hand; compaction with a plate compacter; pole of specified material; all fixtures; and lamps.

The Expanded System Listing shows Site Lighting Systems that use one of two types of lamps: high pressure sodium; and metal halide. Systems are listed for 400-watt and 1000-watt lamps. Pole height varies from 20′ to 40′. There are four types of poles possibly listed: aluminum, fiberglass, steel and wood.

System Components	QUANTITY	UNIT	COST EACH MAT.	COST EACH INST.	COST EACH TOTAL
SYSTEM G4020 110 5820					
SITE LIGHTING,METAL HALIDE, 400 WATT, ALUMINUM POLE, 20′ HIGH					
Excavating, by hand, pits to 6′ deep, heavy soil	2.368	C.Y.		258.11	258.11
Concrete in place incl. forms and reinf. stl. spread footings under 1 CY	.465	C.Y.	82.77	123.83	206.60
Backfill	1.903	C.Y.		75.17	75.17
Compact, vibrating plate	1.903	C.Y.		10.26	10.26
Aluminum light pole, 20′, no concrete base	1.000	Ea.	1,000	587	1,587
Road area luminaire wire	.300	C.L.F.	12.15	23.55	35.70
Roadway area luminaire, metal halide 400 W	1.000	Ea.	600	285	885
TOTAL			1,694.92	1,362.92	3,057.84

G4020 110	Site Lighting	COST EACH MAT.	COST EACH INST.	COST EACH TOTAL
2320	Site lighting, high pressure sodium, 400 watt, aluminum pole, 20′ high	1,725	1,350	3,075
2340	30′ high	2,650	1,650	4,300
2360	40′ high	3,150	2,150	5,300
2520	Fiberglass pole, 20′ high	1,450	1,225	2,675
2540	30′ high	1,650	1,500	3,150
2560	40′ high	2,425	1,900	4,325
2720	Steel pole, 20′ high	1,900	1,425	3,325
2740	30′ high	2,150	1,750	3,900
2760	40′ high	2,675	2,300	4,975
2920	Wood pole, 20′ high	1,025	1,325	2,350
2940	30′ high	1,225	1,650	2,875
2960	40′ high	1,600	2,050	3,650
3120	1000 watt, aluminum pole, 20′ high	1,800	1,400	3,200
3140	30′ high	2,725	1,700	4,425
3160	40′ high	3,225	2,175	5,400
3320	Fiberglass pole, 20′ high	1,525	1,250	2,775
3340	30′ high	1,725	1,525	3,250
3360	40′ high	2,500	1,925	4,425
3420	Steel pole, 20′ high	1,975	1,450	3,425
3440	30′ high	2,225	1,800	4,025

G4020 Site Lighting

G4020 110	Site Lighting	COST EACH		
		MAT.	INST.	TOTAL
3460	40′ high	2,750	2,350	5,100
3520	Wood pole, 20′ high	1,100	1,350	2,450
3540	30′ high	1,300	1,700	3,000
3560	40′ high	1,700	2,075	3,775
5820	Metal halide, 400 watt, aluminum pole, 20′ high	1,700	1,350	3,050
5840	30′ high	2,625	1,650	4,275
5860	40′ high	3,125	2,150	5,275
6120	Fiberglass pole, 20′ high	1,425	1,225	2,650
6140	30′ high	1,625	1,500	3,125
6160	40′ high	2,400	1,900	4,300
6320	Steel pole, 20′ high	1,875	1,425	3,300
6340	30′ high	2,125	1,750	3,875
6360	40′ high	2,650	2,300	4,950
7620	1000 watt, aluminum pole, 20′ high	1,775	1,400	3,175
7640	30′ high	2,700	1,700	4,400
7660	40′ high	3,200	2,175	5,375
7900	Fiberglass pole, 20′ high	1,500	1,250	2,750
7920	30′ high	1,700	1,525	3,225
7940	40′ high	2,475	1,925	4,400
8120	Steel pole, 20′ high	1,950	1,450	3,400
8140	30′ high	2,200	1,800	4,000
8160	40′ high	2,725	2,350	5,075

G4020 Site Lighting

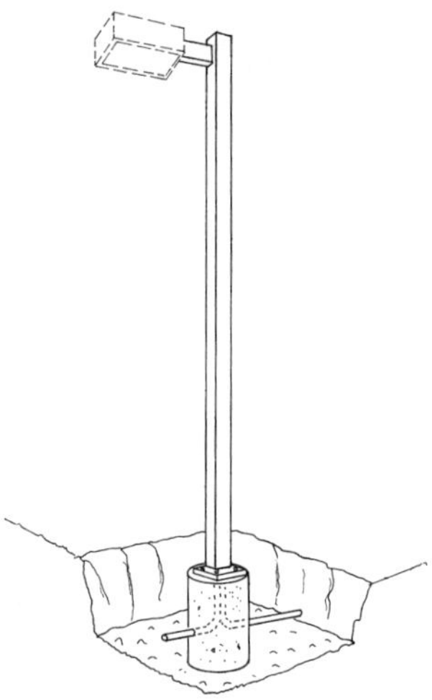

Table G4020 210 Procedure for Calculating Floodlights Required for Various Footcandles
Poles should not be spaced more than 4 times the fixture mounting height for good light distribution.

Estimating Chart
Select Lamp type.

Determine total square feet.

Chart will show quantity of fixtures to provide 1 footcandle initial, at intersection of lines. Multiply fixture quantity by desired footcandle level.

Chart based on use of wide beam luminaires in an area whose dimensions are large compared to mounting height and is approximate only.

To maintain 1 footcandle over a large area use these watts per square foot:

Incandescent	0.15
Metal Halide	0.032
Mercury Vapor	0.05
High Pressure Sodium	0.024

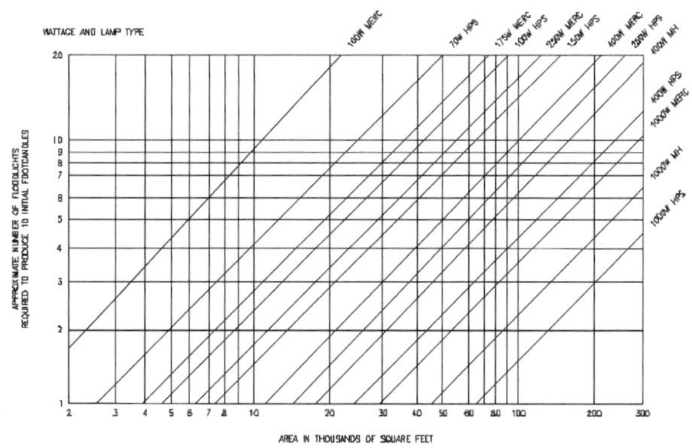

System Components	QUANTITY	UNIT	COST EACH MAT.	INST.	TOTAL
SYSTEM G4020 210 0200					
LIGHT POLES, ALUMINUM, 20′ HIGH, 1 ARM BRACKET					
Aluminum light pole, 20′, no concrete base	1.000	Ea.	1,000	587	1,587
Bracket arm for Aluminum light pole	1.000	Ea.	130	78.50	208.50
Excavation by hand, pits to 6′ deep, heavy soil or clay	2.368	C.Y.		258.11	258.11
Footing, concrete incl forms, reinforcing, spread, under 1 C.Y.	.465	C.Y.	82.77	123.83	206.60
Backfill by hand	1.903	C.Y.		75.17	75.17
Compaction vibrating plate	1.903	C.Y.		10.26	10.26
TOTAL			1,212.77	1,132.87	2,345.64

G4020 210	Light Pole (Installed)	COST EACH MAT.	INST.	TOTAL
0200	Light pole, aluminum, 20′ high, 1 arm bracket	1,225	1,125	2,350
0240	2 arm brackets	1,350	1,125	2,475
0280	3 arm brackets	1,475	1,175	2,650
0320	4 arm brackets	1,600	1,175	2,775
0360	30′ high, 1 arm bracket	2,125	1,425	3,550
0400	2 arm brackets	2,275	1,425	3,700
0440	3 arm brackets	2,400	1,475	3,875
0480	4 arm brackets	2,525	1,475	4,000
0680	40′ high, 1 arm bracket	2,625	1,900	4,525
0720	2 arm brackets	2,750	1,900	4,650
0760	3 arm brackets	2,900	1,950	4,850
0800	4 arm brackets	3,025	1,950	4,975
0840	Steel, 20′ high, 1 arm bracket	1,450	1,200	2,650
0880	2 arm brackets	1,550	1,200	2,750
0920	3 arm brackets	1,575	1,225	2,800
0960	4 arm brackets	1,725	1,225	2,950
1000	30′ high, 1 arm bracket	1,700	1,525	3,225
1040	2 arm brackets	1,800	1,525	3,325
1080	3 arm brackets	1,825	1,575	3,400
1120	4 arm brackets	1,950	1,575	3,525
1320	40′ high, 1 arm bracket	2,225	2,075	4,300
1360	2 arm brackets	2,325	2,075	4,400

G4020 Site Lighting

G4020 210	Light Pole (Installed)	COST EACH MAT.	COST EACH INST.	COST EACH TOTAL
1400	3 arm brackets	2,350	2,100	4,450
1440	4 arm brackets	2,475	2,100	4,575

Reference Section

All the reference information is in one section, making it easy to find what you need to know . . . and easy to use the book on a daily basis. This section is visually identified by a vertical gray bar on the page edges.

In this Reference Section, we've included Equipment Rental Costs, a listing of rental and operating costs; Crew Listings, a full listing of all crews, equipment, and their costs; Historical Cost Indexes for cost comparisons over time; City Cost Indexes and Location Factors for adjusting costs to the region you are in; Reference Tables, where you will find explanations, estimating information and procedures, or technical data; Square Foot Costs that allow you to make a rough estimate for the overall cost of a project; and an explanation of all the Abbreviations in the book.

Table of Contents

Equipment Rental Costs

Estimating Tips

- This section contains the average costs to rent and operate hundreds of pieces of construction equipment. This is useful information when estimating the time and material requirements of any particular operation in order to establish a unit or total cost. Equipment costs include not only rental, but also operating costs for equipment under normal use.

Rental Costs

- Equipment rental rates are obtained from industry sources throughout North America–contractors, suppliers, dealers, manufacturers, and distributors.
- Rental rates vary throughout the country, with larger cities generally having lower rates. Lease plans for new equipment are available for periods in excess of six months, with a percentage of payments applying toward purchase.
- Monthly rental rates vary from 2% to 5% of the purchase price of the equipment depending on the anticipated life of the equipment and its wearing parts.
- Weekly rental rates are about 1/3 the monthly rates, and daily rental rates are about 1/3 the weekly rate.
- Rental rates can also be treated as reimbursement costs for contractor-owned equipment. Owned equipment costs include depreciation, loan payments, interest, taxes, insurance, storage, and major repairs.

Operating Costs

- The operating costs include parts and labor for routine servicing, such as repair and replacement of pumps, filters and worn lines. Normal operating expendables, such as fuel, lubricants, tires and electricity (where applicable), are also included.
- Extraordinary operating expendables with highly variable wear patterns, such as diamond bits and blades, are excluded. These costs can be found as material costs in the Unit Price section.
- The hourly operating costs listed do not include the operator's wages.

Equipment Cost/Day

- Any power equipment required by a crew is shown in the Crew Listings with a daily cost.
- The daily cost of equipment needed by a crew is based on dividing the weekly rental rate by 5 (number of working days in the week), and then adding the hourly operating cost times 8 (the number of hours in a day). This "Equipment Cost/Day" is shown in the far right column of the Equipment Rental pages.
- If equipment is needed for only one or two days, it is best to develop your own cost by including components for daily rent and hourly operating cost. This is important when the listed Crew for a task does not contain the equipment needed, such as a crane for lifting mechanical heating/cooling equipment up onto a roof.
- If the quantity of work is less than the crew's Daily Output shown for a Unit Price line item that includes a bare unit equipment cost, it is recommended to estimate one day's rental cost and operating cost for equipment shown in the Crew Listing for that line item.

Mobilization/ Demobilization

- The cost to move construction equipment from an equipment yard or rental company to the jobsite and back again is not included in equipment rental costs listed in the Reference section, nor in the bare equipment cost of any Unit Price line item, nor in any equipment costs shown in the Crew listings.
- Mobilization (to the site) and demobilization (from the site) costs can be found in the Unit Price section.
- If a piece of equipment is already at the jobsite, it is not appropriate to utilize mobil./demob. costs again in an estimate.

01 54 33 | Equipment Rental

		01 54 33 Equipment Rental	UNIT	HOURLY OPER. COST	RENT PER DAY	RENT PER WEEK	RENT PER MONTH	EQUIPMENT COST/DAY
10	0010	**CONCRETE EQUIPMENT RENTAL** without operators R015433-10						
	0200	Bucket, concrete lightweight, 1/2 C.Y.	Ea.	.80	23.50	70	210	20.40
	0300	1 C.Y.		.85	27.50	83	249	23.40
	0400	1-1/2 C.Y.		1.10	36.50	110	330	30.80
	0500	2 C.Y.		1.20	45	135	405	36.60
	0580	8 C.Y.		5.90	253	760	2,275	199.20
	0600	Cart, concrete, self-propelled, operator walking, 10 C.F.		3.35	56.50	170	510	60.80
	0700	Operator riding, 18 C.F.		5.55	93.50	280	840	100.40
	0800	Conveyer for concrete, portable, gas, 16″ wide, 26′ long		13.30	122	365	1,100	179.40
	0900	46′ long		13.65	147	440	1,325	197.20
	1000	56′ long		13.80	155	465	1,400	203.40
	1100	Core drill, electric, 2-1/2 H.P., 1″ to 8″ bit diameter		1.54	59.50	179	535	48.10
	1150	11 H.P., 8″ to 18″ cores		5.80	110	330	990	112.40
	1200	Finisher, concrete floor, gas, riding trowel, 96″ wide		13.80	143	430	1,300	196.40
	1300	Gas, walk-behind, 3 blade, 36″ trowel		2.30	20	60	180	30.40
	1400	4 blade, 48″ trowel		4.60	27	81	243	53
	1500	Float, hand-operated (Bull float) 48″ wide		.08	14	42	126	9.05
	1570	Curb builder, 14 H.P., gas, single screw		14.45	250	750	2,250	265.60
	1590	Double screw		15.10	287	860	2,575	292.80
	1600	Floor grinder, concrete and terrazzo, electric, 22″ path		2.04	125	374	1,125	91.10
	1700	Edger, concrete, electric, 7″ path		1.01	51.50	155	465	39.10
	1750	Vacuum pick-up system for floor grinders, wet/dry		1.46	81.50	245	735	60.70
	1800	Mixer, powered, mortar and concrete, gas, 6 C.F., 18 H.P.		9.05	117	350	1,050	142.40
	1900	10 C.F., 25 H.P.		11.40	142	425	1,275	176.20
	2000	16 C.F.		11.75	163	490	1,475	192
	2100	Concrete, stationary, tilt drum, 2 C.Y.		6.70	228	685	2,050	190.60
	2120	Pump, concrete, truck mounted 4″ line 80′ boom		24.35	870	2,605	7,825	715.80
	2140	5″ line, 110′ boom		31.50	1,150	3,450	10,400	942
	2160	Mud jack, 50 C.F. per hr.		7.10	123	370	1,100	130.80
	2180	225 C.F. per hr.		9.30	142	425	1,275	159.40
	2190	Shotcrete pump rig, 12 C.Y./hr.		14.70	218	655	1,975	248.60
	2200	35 C.Y./hr.		17.25	235	705	2,125	279
	2600	Saw, concrete, manual, gas, 18 H.P.		7.25	43.50	130	390	84
	2650	Self-propelled, gas, 30 H.P.		14.40	100	300	900	175.20
	2700	Vibrators, concrete, electric, 60 cycle, 2 H.P.		.41	8.65	26	78	8.50
	2800	3 H.P.		.59	11.65	35	105	11.70
	2900	Gas engine, 5 H.P.		2.20	16	48	144	27.20
	3000	8 H.P.		3.00	15	45	135	33
	3050	Vibrating screed, gas engine, 8 H.P.		2.91	71.50	215	645	66.30
	3120	Concrete transit mixer, 6 x 4, 250 H.P., 8 C.Y., rear discharge		62.00	565	1,700	5,100	836
	3200	Front discharge		73.05	695	2,080	6,250	1,000
	3300	6 x 6, 285 H.P., 12 C.Y., rear discharge		72.10	655	1,965	5,900	969.80
	3400	Front discharge	▼	75.40	700	2,105	6,325	1,024
20	0010	**EARTHWORK EQUIPMENT RENTAL** without operators R015433-10						
	0040	Aggregate spreader, push type 8′ to 12′ wide	Ea.	3.20	25.50	76	228	40.80
	0045	Tailgate type, 8′ wide		3.00	32.50	98	294	43.60
	0055	Earth auger, truck-mounted, for fence & sign posts, utility poles		19.65	455	1,370	4,100	431.20
	0060	For borings and monitoring wells		45.00	640	1,920	5,750	744
	0070	Portable, trailer mounted		3.15	31.50	95	285	44.20
	0075	Truck-mounted, for caissons, water wells		96.25	2,825	8,495	25,500	2,469
	0080	Horizontal boring machine, 12″ to 36″ diameter, 45 H.P.		23.80	188	565	1,700	303.40
	0090	12″ to 48″ diameter, 65 H.P.		33.50	330	985	2,950	465
	0095	Auger, for fence posts, gas engine, hand held		.55	6	18	54	8
	0100	Excavator, diesel hydraulic, crawler mounted, 1/2 C.Y. cap.		21.30	385	1,150	3,450	400.40
	0120	5/8 C.Y. capacity		32.75	555	1,670	5,000	596
	0140	3/4 C.Y. capacity		40.10	625	1,870	5,600	694.80
	0150	1 C.Y. capacity	▼	50.20	690	2,070	6,200	815.60

01 54 33 | Equipment Rental

	Line	Description	Unit	Hourly Oper. Cost	Rent per Day	Rent per Week	Rent per Month	Equipment Cost/Day	
20	0200	1-1/2 C.Y. capacity	Ea.	60.35	920	2,765	8,300	1,036	20
	0300	2 C.Y. capacity		81.05	1,200	3,575	10,700	1,363	
	0320	2-1/2 C.Y. capacity		109.55	1,550	4,620	13,900	1,800	
	0325	3-1/2 C.Y. capacity		146.35	2,150	6,415	19,200	2,454	
	0330	4-1/2 C.Y. capacity		175.40	2,750	8,270	24,800	3,057	
	0335	6 C.Y. capacity		223.00	2,925	8,785	26,400	3,541	
	0340	7 C.Y. capacity		225.25	3,025	9,070	27,200	3,616	
	0342	Excavator attachments, bucket thumbs		3.10	240	720	2,150	168.80	
	0345	Grapples		2.65	188	565	1,700	134.20	
	0347	Hydraulic hammer for boom mounting, 5000 ft lb.		14.10	425	1,280	3,850	368.80	
	0349	11,000 ft lb.		22.90	755	2,260	6,775	635.20	
	0350	Gradall type, truck mounted, 3 ton @ 15′ radius, 5/8 C.Y.		43.10	900	2,695	8,075	883.80	
	0370	1 C.Y. capacity		47.25	1,050	3,160	9,475	1,010	
	0400	Backhoe-loader, 40 to 45 H.P., 5/8 C.Y. capacity		14.85	237	710	2,125	260.80	
	0450	45 H.P. to 60 H.P., 3/4 C.Y. capacity		23.80	295	885	2,650	367.40	
	0460	80 H.P., 1-1/4 C.Y. capacity		26.05	320	960	2,875	400.40	
	0470	112 H.P., 1-1/2 C.Y. capacity		41.40	655	1,970	5,900	725.20	
	0482	Backhoe-loader attachment, compactor, 20,000 lb.		5.65	142	425	1,275	130.20	
	0485	Hydraulic hammer, 750 ft lb.		3.20	95	285	855	82.60	
	0486	Hydraulic hammer, 1200 ft lb.		6.20	217	650	1,950	179.60	
	0500	Brush chipper, gas engine, 6″ cutter head, 35 H.P.		11.75	107	320	960	158	
	0550	Diesel engine, 12″ cutter head, 130 H.P.		28.05	287	860	2,575	396.40	
	0600	15″ cutter head, 165 H.P.		33.75	330	990	2,975	468	
	0750	Bucket, clamshell, general purpose, 3/8 C.Y.		1.25	38.50	115	345	33	
	0800	1/2 C.Y.		1.35	45	135	405	37.80	
	0850	3/4 C.Y.		1.50	55	165	495	45	
	0900	1 C.Y.		1.55	58.50	175	525	47.40	
	0950	1-1/2 C.Y.		2.50	80	240	720	68	
	1000	2 C.Y.		2.65	88.50	265	795	74.20	
	1010	Bucket, dragline, medium duty, 1/2 C.Y.		.70	23.50	70	210	19.60	
	1020	3/4 C.Y.		.75	24.50	74	222	20.80	
	1030	1 C.Y.		.75	26	78	234	21.60	
	1040	1-1/2 C.Y.		1.20	40	120	360	33.60	
	1050	2 C.Y.		1.25	45	135	405	37	
	1070	3 C.Y.		1.90	61.50	185	555	52.20	
	1200	Compactor, manually guided 2-drum vibratory smooth roller, 7.5 H.P.		7.00	197	590	1,775	174	
	1250	Rammer/tamper, gas, 8″		2.65	43.50	130	390	47.20	
	1260	15″		2.95	50	150	450	53.60	
	1300	Vibratory plate, gas, 18″ plate, 3000 lb. blow		2.60	23	69	207	34.60	
	1350	21″ plate, 5000 lb. blow		3.25	30.50	92	276	44.40	
	1370	Curb builder/extruder, 14 H.P., gas, single screw		14.45	250	750	2,250	265.60	
	1390	Double screw		15.10	287	860	2,575	292.80	
	1500	Disc harrow attachment, for tractor		.43	71	213	640	46.05	
	1810	Feller buncher, shearing & accumulating trees, 100 H.P.		35.95	550	1,645	4,925	616.60	
	1860	Grader, self-propelled, 25,000 lb.		39.35	605	1,815	5,450	677.80	
	1910	30,000 lb.		43.05	615	1,840	5,525	712.40	
	1920	40,000 lb.		64.35	1,025	3,070	9,200	1,129	
	1930	55,000 lb.		84.30	1,625	4,900	14,700	1,654	
	1950	Hammer, pavement breaker, self-propelled, diesel, 1000 to 1250 lb.		29.95	355	1,060	3,175	451.60	
	2000	1300 to 1500 lb.		44.93	705	2,120	6,350	783.45	
	2050	Pile driving hammer, steam or air, 4150 ft lb. @ 225 bpm		10.25	480	1,435	4,300	369	
	2100	8750 ft lb. @ 145 bpm		12.20	665	2,000	6,000	497.60	
	2150	15,000 ft lb. @ 60 bpm		13.80	805	2,410	7,225	592.40	
	2200	24,450 ft lb. @ 111 bpm		14.80	890	2,675	8,025	653.40	
	2250	Leads, 60′ high for pile driving hammers up to 20,000 ft lb.		3.25	81	243	730	74.60	
	2300	90′ high for hammers over 20,000 ft lb.		4.95	142	426	1,275	124.80	
	2350	Diesel type hammer, 22,400 ft lb.		31.25	615	1,840	5,525	618	
	2400	41,300 ft lb.	↓	40.50	635	1,900	5,700	704	

01 54 33 | Equipment Rental

			UNIT	HOURLY OPER. COST	RENT PER DAY	RENT PER WEEK	RENT PER MONTH	EQUIPMENT COST/DAY	
20	2450	141,000 ft lb.	Ea.	59.15	1,150	3,450	10,400	1,163	20
	2500	Vib. elec. hammer/extractor, 200 kW diesel generator, 34 H.P.		55.90	645	1,935	5,800	834.20	
	2550	80 H.P.		102.85	945	2,830	8,500	1,389	
	2600	150 H.P.		195.65	1,800	5,435	16,300	2,652	
	2800	Log chipper, up to 22″ diameter, 600 H.P.		64.65	640	1,915	5,750	900.20	
	2850	Logger, for skidding & stacking logs, 150 H.P.		53.05	825	2,470	7,400	918.40	
	2860	Mulcher, diesel powered, trailer mounted		24.55	215	645	1,925	325.40	
	2900	Rake, spring tooth, with tractor		18.16	335	1,012	3,025	347.70	
	3000	Roller, vibratory, tandem, smooth drum, 20 H.P.		8.55	143	430	1,300	154.40	
	3050	35 H.P.		10.65	243	730	2,200	231.20	
	3100	Towed type vibratory compactor, smooth drum, 50 H.P.		25.65	335	1,010	3,025	407.20	
	3150	Sheepsfoot, 50 H.P.		27.00	375	1,120	3,350	440	
	3170	Landfill compactor, 220 H.P.		86.90	1,425	4,265	12,800	1,548	
	3200	Pneumatic tire roller, 80 H.P.		16.20	330	985	2,950	326.60	
	3250	120 H.P.		24.15	545	1,630	4,900	519.20	
	3300	Sheepsfoot vibratory roller, 240 H.P.		66.00	1,050	3,130	9,400	1,154	
	3320	340 H.P.		89.30	1,550	4,630	13,900	1,640	
	3350	Smooth drum vibratory roller, 75 H.P.		23.80	565	1,695	5,075	529.40	
	3400	125 H.P.		31.05	680	2,040	6,125	656.40	
	3410	Rotary mower, brush, 60″, with tractor		22.40	305	920	2,750	363.20	
	3420	Rototiller, walk-behind, gas, 5 H.P.		2.30	67.50	203	610	59	
	3422	8 H.P.		3.57	103	310	930	90.55	
	3440	Scrapers, towed type, 7 C.Y. capacity		5.60	110	330	990	110.80	
	3450	10 C.Y. capacity		6.35	150	450	1,350	140.80	
	3500	15 C.Y. capacity		6.75	173	520	1,550	158	
	3525	Self-propelled, single engine, 14 C.Y. capacity		112.95	1,550	4,680	14,000	1,840	
	3550	Dual engine, 21 C.Y. capacity		173.55	2,025	6,110	18,300	2,610	
	3600	31 C.Y. capacity		230.75	2,900	8,690	26,100	3,584	
	3640	44 C.Y. capacity		281.25	3,550	10,635	31,900	4,377	
	3650	Elevating type, single engine, 11 C.Y. capacity		69.90	1,025	3,050	9,150	1,169	
	3700	22 C.Y. capacity		137.35	2,275	6,800	20,400	2,459	
	3710	Screening plant 110 H.P. w/5′ x 10′ screen		37.95	435	1,300	3,900	563.60	
	3720	5′ x 16′ screen		40.20	540	1,615	4,850	644.60	
	3850	Shovel, crawler-mounted, front-loading, 7 C.Y. capacity		238.20	2,650	7,945	23,800	3,495	
	3855	12 C.Y. capacity		321.20	3,500	10,530	31,600	4,676	
	3860	Shovel/backhoe bucket, 1/2 C.Y.		2.40	65	195	585	58.20	
	3870	3/4 C.Y.		2.45	71.50	215	645	62.60	
	3880	1 C.Y.		2.55	81.50	245	735	69.40	
	3890	1-1/2 C.Y.		2.70	95	285	855	78.60	
	3910	3 C.Y.		3.00	128	385	1,150	101	
	3950	Stump chipper, 18″ deep, 30 H.P.		7.47	188	565	1,700	172.75	
	4110	Dozer, crawler, torque converter, diesel 80 H.P.		29.45	410	1,225	3,675	480.60	
	4150	105 H.P.		35.80	510	1,530	4,600	592.40	
	4200	140 H.P.		51.75	790	2,375	7,125	889	
	4260	200 H.P.		76.90	1,200	3,590	10,800	1,333	
	4310	300 H.P.		99.95	1,725	5,145	15,400	1,829	
	4360	410 H.P.		134.95	2,250	6,725	20,200	2,425	
	4370	500 H.P.		173.50	2,925	8,745	26,200	3,137	
	4380	700 H.P.		256.95	4,000	12,015	36,000	4,459	
	4400	Loader, crawler, torque conv., diesel, 1-1/2 C.Y., 80 H.P.		31.20	475	1,420	4,250	533.60	
	4450	1-1/2 to 1-3/4 C.Y., 95 H.P.		34.90	590	1,770	5,300	633.20	
	4510	1-3/4 to 2-1/4 C.Y., 130 H.P.		54.25	870	2,605	7,825	955	
	4530	2-1/2 to 3-1/4 C.Y., 190 H.P.		66.60	1,100	3,270	9,800	1,187	
	4560	3-1/2 to 5 C.Y., 275 H.P.		84.60	1,425	4,265	12,800	1,530	
	4610	Front end loader, 4WD, articulated frame, diesel, 1 to 1-1/4 C.Y., 70 H.P.		19.75	230	690	2,075	296	
	4620	1-1/2 to 1-3/4 C.Y., 95 H.P.		25.15	290	870	2,600	375.20	
	4650	1-3/4 to 2 C.Y., 130 H.P.		20.25	375	1,120	3,350	386	
	4710	2-1/2 to 3-1/2 C.Y., 145 H.P.	▼	33.15	435	1,305	3,925	526.20	

01 54 33 | Equipment Rental

		01 54 33 Equipment Rental	UNIT	HOURLY OPER. COST	RENT PER DAY	RENT PER WEEK	RENT PER MONTH	EQUIPMENT COST/DAY	
20	4730	3 to 4-1/2 C.Y., 185 H.P.	Ea.	42.60	565	1,690	5,075	678.80	20
	4760	5-1/4 to 5-3/4 C.Y., 270 H.P.		67.35	860	2,585	7,750	1,056	
	4810	7 to 9 C.Y., 475 H.P.		115.70	1,800	5,410	16,200	2,008	
	4870	9 - 11 C.Y., 620 H.P.		155.05	2,500	7,470	22,400	2,734	
	4880	Skid steer loader, wheeled, 10 C.F., 30 H.P. gas		10.25	150	450	1,350	172	
	4890	1 C.Y., 78 H.P., diesel		20.20	245	735	2,200	308.60	
	4892	Skid-steer attachment, auger		.55	92.50	277	830	59.80	
	4893	Backhoe		.67	112	336	1,000	72.55	
	4894	Broom		.73	122	367	1,100	79.25	
	4895	Forks		.24	40	120	360	25.90	
	4896	Grapple		.55	91	273	820	59	
	4897	Concrete hammer		1.06	176	528	1,575	114.10	
	4898	Tree spade		.94	156	468	1,400	101.10	
	4899	Trencher		.78	130	391	1,175	84.45	
	4900	Trencher, chain, boom type, gas, operator walking, 12 H.P.		5.30	46.50	140	420	70.40	
	4910	Operator riding, 40 H.P.		20.40	293	880	2,650	339.20	
	5000	Wheel type, diesel, 4′ deep, 12″ wide		86.65	840	2,525	7,575	1,198	
	5100	6′ deep, 20″ wide		92.15	1,925	5,785	17,400	1,894	
	5150	Chain type, diesel, 5′ deep, 8″ wide		38.35	560	1,680	5,050	642.80	
	5200	Diesel, 8′ deep, 16″ wide		153.95	3,550	10,625	31,900	3,357	
	5202	Rock trencher, wheel type, 6″ wide x 18″ deep		21.10	325	975	2,925	363.80	
	5206	Chain type, 18″ wide x 7′ deep		111.50	2,850	8,520	25,600	2,596	
	5210	Tree spade, self-propelled		14.38	267	800	2,400	275.05	
	5250	Truck, dump, 2-axle, 12 ton, 8 C.Y. payload, 220 H.P.		34.95	228	685	2,050	416.60	
	5300	Three axle dump, 16 ton, 12 C.Y. payload, 400 H.P.		62.05	325	980	2,950	692.40	
	5310	Four axle dump, 25 ton, 18 C.Y. payload, 450 H.P.		72.75	475	1,425	4,275	867	
	5350	Dump trailer only, rear dump, 16-1/2 C.Y.		5.35	140	420	1,250	126.80	
	5400	20 C.Y.		5.80	158	475	1,425	141.40	
	5450	Flatbed, single axle, 1-1/2 ton rating		28.10	68.50	205	615	265.80	
	5500	3 ton rating		33.60	98.50	295	885	327.80	
	5550	Off highway rear dump, 25 ton capacity		74.10	1,225	3,640	10,900	1,321	
	5600	35 ton capacity		82.60	1,350	4,080	12,200	1,477	
	5610	50 ton capacity		102.90	1,625	4,870	14,600	1,797	
	5620	65 ton capacity		105.75	1,600	4,805	14,400	1,807	
	5630	100 ton capacity		153.60	2,825	8,455	25,400	2,920	
	6000	Vibratory plow, 25 H.P., walking	↓	9.05	61.50	185	555	109.40	
40	0010	**GENERAL EQUIPMENT RENTAL** without operators R015433-10							40
	0150	Aerial lift, scissor type, to 15′ high, 1000 lb. cap., electric	Ea.	3.05	51.50	155	465	55.40	
	0160	To 25′ high, 2000 lb. capacity		3.50	68.50	205	615	69	
	0170	Telescoping boom to 40′ high, 500 lb. capacity, gas		18.65	305	915	2,750	332.20	
	0180	To 45′ high, 500 lb. capacity		20.80	355	1,060	3,175	378.40	
	0190	To 60′ high, 600 lb. capacity		23.15	460	1,385	4,150	462.20	
	0195	Air compressor, portable, 6.5 CFM, electric		.42	12.65	38	114	10.95	
	0196	Gasoline		.81	19	57	171	17.90	
	0200	Towed type, gas engine, 60 CFM		13.70	46.50	140	420	137.60	
	0300	160 CFM		16.00	50	150	450	158	
	0400	Diesel engine, rotary screw, 250 CFM		16.35	108	325	975	195.80	
	0500	365 CFM		22.10	132	395	1,175	255.80	
	0550	450 CFM		28.05	163	490	1,475	322.40	
	0600	600 CFM		49.35	227	680	2,050	530.80	
	0700	750 CFM	↓	49.55	235	705	2,125	537.40	
	0800	For silenced models, small sizes, add to rent		3%	5%	5%	5%		
	0900	Large sizes, add to rent		5%	7%	7%	7%		
	0930	Air tools, breaker, pavement, 60 lb.	Ea.	.50	9.65	29	87	9.80	
	0940	80 lb.		.50	10.35	31	93	10.20	
	0950	Drills, hand (jackhammer) 65 lb.		.60	16.65	50	150	14.80	
	0960	Track or wagon, swing boom, 4″ drifter		61.00	870	2,605	7,825	1,009	
	0970	5″ drifter	↓	76.60	990	2,970	8,900	1,207	

01 54 33 | Equipment Rental

			UNIT	HOURLY OPER. COST	RENT PER DAY	RENT PER WEEK	RENT PER MONTH	EQUIPMENT COST/DAY	
40	0975	Track mounted quarry drill, 6" diameter drill	Ea.	117.20	1,425	4,275	12,800	1,793	40
	0980	Dust control per drill		.90	21.50	64	192	20	
	0990	Hammer, chipping, 12 lb.		.55	26	78	234	20	
	1000	Hose, air with couplings, 50' long, 3/4" diameter		.03	5	15	45	3.25	
	1100	1" diameter		.04	6.35	19	57	4.10	
	1200	1-1/2" diameter		.05	9	27	81	5.80	
	1300	2" diameter		.07	12	36	108	7.75	
	1400	2-1/2" diameter		.11	19	57	171	12.30	
	1410	3" diameter		.14	23	69	207	14.90	
	1450	Drill, steel, 7/8" x 2'		.05	8.65	26	78	5.60	
	1460	7/8" x 6'		.05	9	27	81	5.80	
	1520	Moil points		.02	4	12	36	2.55	
	1525	Pneumatic nailer w/accessories		.48	31.50	95	285	22.85	
	1530	Sheeting driver for 60 lb. breaker		.04	6	18	54	3.90	
	1540	For 90 lb. breaker		.12	8	24	72	5.75	
	1550	Spade, 25 lb.		.45	6.65	20	60	7.60	
	1560	Tamper, single, 35 lb.		.55	36.50	109	325	26.20	
	1570	Triple, 140 lb.		.82	54.50	164	490	39.35	
	1580	Wrenches, impact, air powered, up to 3/4" bolt		.40	12.65	38	114	10.80	
	1590	Up to 1-1/4" bolt		.50	23.50	70	210	18	
	1600	Barricades, barrels, reflectorized, 1 to 99 barrels		.03	4.60	13.80	41.50	3	
	1610	100 to 200 barrels		.02	3.53	10.60	32	2.30	
	1620	Barrels with flashers, 1 to 99 barrels		.03	5.25	15.80	47.50	3.40	
	1630	100 to 200 barrels		.03	4.20	12.60	38	2.75	
	1640	Barrels with steady burn type C lights		.04	7	21	63	4.50	
	1650	Illuminated board, trailer mounted, with generator		3.50	132	395	1,175	107	
	1670	Portable barricade, stock, with flashers, 1 to 6 units		.03	5.25	15.80	47.50	3.40	
	1680	25 to 50 units		.03	4.90	14.70	44	3.20	
	1685	Butt fusion machine, wheeled, 1.5 HP electric, 2" - 8" diameter pipe		2.62	167	500	1,500	120.95	
	1690	Tracked, 20 HP diesel, 4"-12" diameter pipe		11.20	490	1,465	4,400	382.60	
	1695	83 HP diesel, 8" - 24" diameter pipe		30.71	975	2,930	8,800	831.70	
	1700	Carts, brick, hand powered, 1000 lb. capacity		.44	72.50	218	655	47.10	
	1800	Gas engine, 1500 lb., 7-1/2' lift		4.17	115	345	1,025	102.35	
	1822	Dehumidifier, medium, 6 lb./hr., 150 CFM		.96	60	180	540	43.70	
	1824	Large, 18 lb./hr., 600 CFM		1.95	122	366	1,100	88.80	
	1830	Distributor, asphalt, trailer mounted, 2000 gal., 38 H.P. diesel		9.90	325	980	2,950	275.20	
	1840	3000 gal., 38 H.P. diesel		11.40	355	1,070	3,200	305.20	
	1850	Drill, rotary hammer, electric		.88	26	78	234	22.65	
	1860	Carbide bit, 1-1/2" diameter, add to electric rotary hammer		.02	3.98	11.95	36	2.55	
	1865	Rotary, crawler, 250 H.P.		144.40	2,050	6,120	18,400	2,379	
	1870	Emulsion sprayer, 65 gal., 5 H.P. gas engine		2.91	97	291	875	81.50	
	1880	200 gal., 5 H.P. engine		7.75	162	485	1,450	159	
	1900	Floor auto-scrubbing machine, walk-behind, 28" path		4.02	260	780	2,350	188.15	
	1930	Floodlight, mercury vapor, or quartz, on tripod, 1000 watt		.40	18.65	56	168	14.40	
	1940	2000 watt		.74	36.50	110	330	27.90	
	1950	Floodlights, trailer mounted with generator, 1 - 300 watt light		3.60	71.50	215	645	71.80	
	1960	2 - 1000 watt lights		4.85	98.50	295	885	97.80	
	2000	4 - 300 watt lights		4.55	93.50	280	840	92.40	
	2005	Foam spray rig, incl. box trailer, compressor, generator, proportioner		33.98	485	1,455	4,375	562.85	
	2020	Forklift, straight mast, 12' lift, 5000 lb., 2 wheel drive, gas		25.50	202	605	1,825	325	
	2040	21' lift, 5000 lb., 4 wheel drive, diesel		20.40	245	735	2,200	310.20	
	2050	For rough terrain, 42' lift, 35' reach, 9000 lb., 110 H.P.		28.05	480	1,440	4,325	512.40	
	2060	For plant, 4 ton capacity, 80 H.P., 2 wheel drive, gas		15.60	93.50	280	840	180.80	
	2080	10 ton capacity, 120 H.P., 2 wheel drive, diesel		23.25	163	490	1,475	284	
	2100	Generator, electric, gas engine, 1.5 kW to 3 kW		3.60	11.35	34	102	35.60	
	2200	5 kW		4.65	15	45	135	46.20	
	2300	10 kW		8.80	36.50	110	330	92.40	
	2400	25 kW	↓	10.25	83.50	250	750	132	

01 54 33 | Equipment Rental

			UNIT	HOURLY OPER. COST	RENT PER DAY	RENT PER WEEK	RENT PER MONTH	EQUIPMENT COST/DAY	
40	2500	Diesel engine, 20 kW	Ea.	11.75	68.50	205	615	135	40
	2600	50 kW		22.65	103	310	930	243.20	
	2700	100 kW		41.85	132	395	1,175	413.80	
	2800	250 kW		82.40	235	705	2,125	800.20	
	2850	Hammer, hydraulic, for mounting on boom, to 500 ft lb.		2.50	75	225	675	65	
	2860	1000 ft lb.		4.25	127	380	1,150	110	
	2900	Heaters, space, oil or electric, 50 MBH		1.98	7.65	23	69	20.45	
	3000	100 MBH		3.57	10.65	32	96	34.95	
	3100	300 MBH		11.46	38.50	115	345	114.70	
	3150	500 MBH		23.09	45	135	405	211.70	
	3200	Hose, water, suction with coupling, 20′ long, 2″ diameter		.02	3	9	27	1.95	
	3210	3″ diameter		.03	4.33	13	39	2.85	
	3220	4″ diameter		.03	5	15	45	3.25	
	3230	6″ diameter		.10	17.35	52	156	11.20	
	3240	8″ diameter		.20	33.50	100	300	21.60	
	3250	Discharge hose with coupling, 50′ long, 2″ diameter		.02	2.67	8	24	1.75	
	3260	3″ diameter		.03	4.67	14	42	3.05	
	3270	4″ diameter		.04	7.35	22	66	4.70	
	3280	6″ diameter		.11	18.65	56	168	12.10	
	3290	8″ diameter		.20	33.50	100	300	21.60	
	3295	Insulation blower		.68	6	18	54	9.05	
	3300	Ladders, extension type, 16′ to 36′ long		.14	24	72	216	15.50	
	3400	40′ to 60′ long		.20	32.50	98	294	21.20	
	3405	Lance for cutting concrete		2.73	104	313	940	84.45	
	3407	Lawn mower, rotary, 22″, 5 H.P.		2.21	63.50	190	570	55.70	
	3408	48″ self propelled		2.98	75	225	675	68.85	
	3410	Level, electronic, automatic, with tripod and leveling rod		1.62	108	323	970	77.55	
	3430	Laser type, for pipe and sewer line and grade		.73	48.50	145	435	34.85	
	3440	Rotating beam for interior control		1.16	77	231	695	55.50	
	3460	Builder's optical transit, with tripod and rod		.09	15.65	47	141	10.10	
	3500	Light towers, towable, with diesel generator, 2000 watt		4.55	93.50	280	840	92.40	
	3600	4000 watt		4.85	98.50	295	885	97.80	
	3700	Mixer, powered, plaster and mortar, 6 C.F., 7 H.P.		2.85	19.65	59	177	34.60	
	3800	10 C.F., 9 H.P.		3.00	30.50	92	276	42.40	
	3850	Nailer, pneumatic		.48	31.50	95	285	22.85	
	3900	Paint sprayers complete, 8 CFM		.83	55.50	166	500	39.85	
	4000	17 CFM		1.47	98	294	880	70.55	
	4020	Pavers, bituminous, rubber tires, 8′ wide, 50 H.P., diesel		31.05	490	1,475	4,425	543.40	
	4030	10′ wide, 150 H.P.		104.70	1,800	5,395	16,200	1,917	
	4050	Crawler, 8′ wide, 100 H.P., diesel		88.45	1,825	5,495	16,500	1,807	
	4060	10′ wide, 150 H.P.		111.70	2,200	6,635	19,900	2,221	
	4070	Concrete paver, 12′ to 24′ wide, 250 H.P.		101.10	1,550	4,640	13,900	1,737	
	4080	Placer-spreader-trimmer, 24′ wide, 300 H.P.		146.00	2,500	7,495	22,500	2,667	
	4100	Pump, centrifugal gas pump, 1-1/2″ diam., 65 GPM		3.90	50	150	450	61.20	
	4200	2″ diameter, 130 GPM		5.30	55	165	495	75.40	
	4300	3″ diameter, 250 GPM		5.60	56.50	170	510	78.80	
	4400	6″ diameter, 1500 GPM		29.55	175	525	1,575	341.40	
	4500	Submersible electric pump, 1-1/4″ diameter, 55 GPM		.37	16.35	49	147	12.75	
	4600	1-1/2″ diameter, 83 GPM		.41	18.65	56	168	14.50	
	4700	2″ diameter, 120 GPM		1.35	23.50	70	210	24.80	
	4800	3″ diameter, 300 GPM		2.30	41.50	125	375	43.40	
	4900	4″ diameter, 560 GPM		9.75	158	475	1,425	173	
	5000	6″ diameter, 1590 GPM		14.35	213	640	1,925	242.80	
	5100	Diaphragm pump, gas, single, 1-1/2″ diameter		1.20	50.50	152	455	40	
	5200	2″ diameter		4.20	63.50	190	570	71.60	
	5300	3″ diameter		4.25	63.50	190	570	72	
	5400	Double, 4″ diameter		6.30	107	320	960	114.40	
	5450	Pressure washer 5 GPM, 3000 psi	↓	4.80	51.50	155	465	69.40	

01 54 33 | Equipment Rental			UNIT	HOURLY OPER. COST	RENT PER DAY	RENT PER WEEK	RENT PER MONTH	EQUIPMENT COST/DAY	
40	5460	7 GPM, 3000 psi	Ea.	6.30	60	180	540	86.40	40
	5500	Trash pump, self-priming, gas, 2″ diameter		4.50	21.50	64	192	48.80	
	5600	Diesel, 4″ diameter		8.95	90	270	810	125.60	
	5650	Diesel, 6″ diameter		24.45	150	450	1,350	285.60	
	5655	Grout Pump		26.60	262	785	2,350	369.80	
	5700	Salamanders, L.P. gas fired, 100,000 Btu		3.59	13.65	41	123	36.90	
	5705	50,000 Btu		2.69	10.35	31	93	27.70	
	5720	Sandblaster, portable, open top, 3 C.F. capacity		.55	26	78	234	20	
	5730	6 C.F. capacity		.90	38.50	115	345	30.20	
	5740	Accessories for above		.13	21	63	189	13.65	
	5750	Sander, floor		.76	19.65	59	177	17.90	
	5760	Edger		.62	21.50	64	192	17.75	
	5800	Saw, chain, gas engine, 18″ long		2.25	21	63	189	30.60	
	5900	Hydraulic powered, 36″ long		.75	65	195	585	45	
	5950	60″ long		.75	66.50	200	600	46	
	6000	Masonry, table mounted, 14″ diameter, 5 H.P.		1.31	56.50	170	510	44.50	
	6050	Portable cut-off, 8 H.P.		2.45	32.50	97	291	39	
	6100	Circular, hand held, electric, 7-1/4″ diameter		.18	4.33	13	39	4.05	
	6200	12″ diameter		.25	7.65	23	69	6.60	
	6250	Wall saw, w/hydraulic power, 10 H.P.		10.45	60	180	540	119.60	
	6275	Shot blaster, walk-behind, 20″ wide		4.65	288	865	2,600	210.20	
	6280	Sidewalk broom, walk-behind		2.15	60.50	182	545	53.60	
	6300	Steam cleaner, 100 gallons per hour		3.70	76.50	230	690	75.60	
	6310	200 gallons per hour		5.25	95	285	855	99	
	6340	Tar Kettle/Pot, 400 gallons		5.81	76.50	230	690	92.50	
	6350	Torch, cutting, acetylene-oxygen, 150′ hose, excludes gases		.30	15	45	135	11.40	
	6360	Hourly operating cost includes tips and gas		16.65				133.20	
	6410	Toilet, portable chemical		.12	20.50	61	183	13.15	
	6420	Recycle flush type		.15	24.50	73	219	15.80	
	6430	Toilet, fresh water flush, garden hose,		.17	29	87	261	18.75	
	6440	Hoisted, non-flush, for high rise		.14	24	72	216	15.50	
	6465	Tractor, farm with attachment		21.70	282	845	2,525	342.60	
	6500	Trailers, platform, flush deck, 2 axle, 25 ton capacity		5.40	117	350	1,050	113.20	
	6600	40 ton capacity		6.95	162	485	1,450	152.60	
	6700	3 axle, 50 ton capacity		7.50	180	540	1,625	168	
	6800	75 ton capacity		9.30	235	705	2,125	215.40	
	6810	Trailer mounted cable reel for high voltage line work		5.24	250	749	2,250	191.70	
	6820	Trailer mounted cable tensioning rig		10.43	495	1,490	4,475	381.45	
	6830	Cable pulling rig		70.67	2,775	8,340	25,000	2,233	
	6900	Water tank trailer, engine driven discharge, 5000 gallons		6.90	147	440	1,325	143.20	
	6925	10,000 gallons		9.45	202	605	1,825	196.60	
	6950	Water truck, off highway, 6000 gallons		89.15	790	2,375	7,125	1,188	
	7010	Tram car for high voltage line work, powered, 2 conductor		6.04	136	407	1,225	129.70	
	7020	Transit (builder's level) with tripod		.09	15.65	47	141	10.10	
	7030	Trench box, 3000 lb., 6′ x 8′		.51	84.50	254	760	54.90	
	7040	7200 lb., 6′ x 20′		1.05	175	525	1,575	113.40	
	7050	8000 lb., 8′ x 16′		1.07	178	533	1,600	115.15	
	7060	9500 lb., 8′ x 20′		1.19	199	597	1,800	128.90	
	7065	11,000 lb., 8′ x 24′		1.26	209	628	1,875	135.70	
	7070	12,000 lb., 10′ x 20′		1.36	227	680	2,050	146.90	
	7100	Truck, pickup, 3/4 ton, 2 wheel drive		14.95	58.50	175	525	154.60	
	7200	4 wheel drive		15.20	73.50	220	660	165.60	
	7250	Crew carrier, 9 passenger		21.15	86.50	260	780	221.20	
	7290	Flat bed truck, 20,000 lb. GVW		22.20	127	380	1,150	253.60	
	7300	Tractor, 4 x 2, 220 H.P.		30.85	200	600	1,800	366.80	
	7410	330 H.P.		45.80	275	825	2,475	531.40	
	7500	6 x 4, 380 H.P.		52.55	320	960	2,875	612.40	
	7600	450 H.P.	↓	63.65	385	1,160	3,475	741.20	

01 54 33 | Equipment Rental

			UNIT	HOURLY OPER. COST	RENT PER DAY	RENT PER WEEK	RENT PER MONTH	EQUIPMENT COST/DAY	
40	7610	Tractor, with A frame, boom and winch, 225 H.P.	Ea.	33.65	275	825	2,475	434.20	40
	7620	Vacuum truck, hazardous material, 2500 gallons		11.85	295	885	2,650	271.80	
	7625	5,000 gallons		14.49	415	1,240	3,725	363.90	
	7650	Vacuum, HEPA, 16 gallon, wet/dry		.82	18	54	162	17.35	
	7655	55 gallon, wet/dry		.75	27	81	243	22.20	
	7660	Water tank, portable		.17	28.50	85.50	257	18.45	
	7690	Sewer/catch basin vacuum, 14 C.Y., 1500 gallons		18.83	620	1,860	5,575	522.65	
	7700	Welder, electric, 200 amp		3.48	16.35	49	147	37.65	
	7800	300 amp		5.13	19.65	59	177	52.85	
	7900	Gas engine, 200 amp		14.10	23.50	70	210	126.80	
	8000	300 amp		16.15	24.50	74	222	144	
	8100	Wheelbarrow, any size		.08	12.65	38	114	8.25	
	8200	Wrecking ball, 4000 lb.	▼	2.30	70	210	630	60.40	
50	0010	**HIGHWAY EQUIPMENT RENTAL** without operators R015433-10							50
	0050	Asphalt batch plant, portable drum mixer, 100 ton/hr.	Ea.	71.30	1,425	4,285	12,900	1,427	
	0060	200 ton/hr.		79.90	1,525	4,550	13,700	1,549	
	0070	300 ton/hr.		92.55	1,775	5,350	16,100	1,810	
	0100	Backhoe attachment, long stick, up to 185 H.P., 10.5′ long		.34	22.50	68	204	16.30	
	0140	Up to 250 H.P., 12′ long		.37	24.50	73	219	17.55	
	0180	Over 250 H.P., 15′ long		.51	33.50	101	305	24.30	
	0200	Special dipper arm, up to 100 H.P., 32′ long		1.04	69	207	620	49.70	
	0240	Over 100 H.P., 33′ long		1.29	86	258	775	61.90	
	0280	Catch basin/sewer cleaning truck, 3 ton, 9 C.Y., 1000 gal.		45.45	395	1,190	3,575	601.60	
	0300	Concrete batch plant, portable, electric, 200 C.Y./hr.		29.60	510	1,535	4,600	543.80	
	0520	Grader/dozer attachment, ripper/scarifier, rear mounted, up to 135 H.P.		3.45	65	195	585	66.60	
	0540	Up to 180 H.P.		4.00	83.50	250	750	82	
	0580	Up to 250 H.P.		4.40	95	285	855	92.20	
	0700	Pvmt. removal bucket, for hyd. excavator, up to 90 H.P.		1.85	51.50	155	465	45.80	
	0740	Up to 200 H.P.		2.10	73.50	220	660	60.80	
	0780	Over 200 H.P.		2.20	85	255	765	68.60	
	0900	Aggregate spreader, self-propelled, 187 H.P.		53.20	690	2,065	6,200	838.60	
	1000	Chemical spreader, 3 C.Y.		3.35	43.50	130	390	52.80	
	1900	Hammermill, traveling, 250 H.P.		78.38	2,000	6,000	18,000	1,827	
	2000	Horizontal borer, 3″ diameter, 13 H.P. gas driven		3.20	53.50	160	480	57.60	
	2150	Horizontal directional drill, 20,000 lb. thrust, 78 H.P. diesel		30.30	655	1,965	5,900	635.40	
	2160	30,000 lb. thrust, 115 H.P.		37.80	1,000	3,010	9,025	904.40	
	2170	50,000 lb. thrust, 170 H.P.		54.30	1,275	3,845	11,500	1,203	
	2190	Mud trailer for HDD, 1500 gallons, 175 H.P., gas		34.65	155	465	1,400	370.20	
	2200	Hydromulcher, diesel, 3000 gallon, for truck mounting		23.90	250	750	2,250	341.20	
	2300	Gas, 600 gallon		9.00	98.50	295	885	131	
	2400	Joint & crack cleaner, walk behind, 25 H.P.		4.25	50	150	450	64	
	2500	Filler, trailer mounted, 400 gallons, 20 H.P.		9.55	213	640	1,925	204.40	
	3000	Paint striper, self-propelled, 40 gallon, 22 H.P.		7.50	157	470	1,400	154	
	3100	120 gallon, 120 H.P.		24.45	405	1,210	3,625	437.60	
	3200	Post drivers, 6″ I-Beam frame, for truck mounting		19.55	390	1,175	3,525	391.40	
	3400	Road sweeper, self-propelled, 8′ wide, 90 H.P.		39.85	575	1,720	5,150	662.80	
	3450	Road sweeper, vacuum assisted, 4 C.Y., 220 gallons		76.95	625	1,875	5,625	990.60	
	4000	Road mixer, self-propelled, 130 H.P.		46.55	765	2,295	6,875	831.40	
	4100	310 H.P.		81.85	2,100	6,285	18,900	1,912	
	4220	Cold mix paver, incl. pug mill and bitumen tank, 165 H.P.		96.15	2,275	6,850	20,600	2,139	
	4240	Pavement brush, towed		3.15	93.50	280	840	81.20	
	4250	Paver, asphalt, wheel or crawler, 130 H.P., diesel		95.80	2,275	6,790	20,400	2,124	
	4300	Paver, road widener, gas 1′ to 6′, 67 H.P.		48.00	895	2,685	8,050	921	
	4400	Diesel, 2′ to 14′, 88 H.P.		61.75	1,050	3,135	9,400	1,121	
	4600	Slipform pavers, curb and gutter, 2 track, 75 H.P.		46.65	715	2,140	6,425	801.20	
	4700	4 track, 165 H.P.		41.30	735	2,210	6,625	772.40	
	4800	Median barrier, 215 H.P.	▼	47.30	740	2,225	6,675	823.40	

01 54 33 | Equipment Rental

		01 54 33 | Equipment Rental	UNIT	HOURLY OPER. COST	RENT PER DAY	RENT PER WEEK	RENT PER MONTH	EQUIPMENT COST/DAY
50	4901	Trailer, low bed, 75 ton capacity	Ea.	10.05	233	700	2,100	220.40
	5000	Road planer, walk behind, 10" cutting width, 10 H.P.		3.90	33.50	100	300	51.20
	5100	Self-propelled, 12" cutting width, 64 H.P.		11.25	115	345	1,025	159
	5120	Traffic line remover, metal ball blaster, truck mounted, 115 H.P.		47.40	740	2,220	6,650	823.20
	5140	Grinder, truck mounted, 115 H.P.		53.80	800	2,405	7,225	911.40
	5160	Walk-behind, 11 H.P.		4.45	53.50	160	480	67.60
	5200	Pavement profiler, 4' to 6' wide, 450 H.P.		249.60	3,325	9,960	29,900	3,989
	5300	8' to 10' wide, 750 H.P.		391.40	4,350	13,070	39,200	5,745
	5400	Roadway plate, steel, 1" x 8' x 20'		.08	12.65	38	114	8.25
	5600	Stabilizer, self-propelled, 150 H.P.		48.35	605	1,815	5,450	749.80
	5700	310 H.P.		96.45	1,700	5,075	15,200	1,787
	5800	Striper, truck mounted, 120 gallon paint, 460 H.P.		69.75	490	1,475	4,425	853
	5900	Thermal paint heating kettle, 115 gallons		3.88	25.50	77	231	46.45
	6000	Tar kettle, 330 gallon, trailer mounted		5.46	58.50	175	525	78.70
	7000	Tunnel locomotive, diesel, 8 to 12 ton		31.25	565	1,695	5,075	589
	7005	Electric, 10 ton		25.70	645	1,940	5,825	593.60
	7010	Muck cars, 1/2 C.Y. capacity		2.00	23.50	71	213	30.20
	7020	1 C.Y. capacity		2.20	31.50	95	285	36.60
	7030	2 C.Y. capacity		2.35	36.50	110	330	40.80
	7040	Side dump, 2 C.Y. capacity		2.55	43.50	130	390	46.40
	7050	3 C.Y. capacity		3.45	50	150	450	57.60
	7060	5 C.Y. capacity		4.90	63.50	190	570	77.20
	7100	Ventilating blower for tunnel, 7-1/2 H.P.		1.93	48.50	145	435	44.45
	7110	10 H.P.		2.14	50	150	450	47.10
	7120	20 H.P.		3.23	65	195	585	64.85
	7140	40 H.P.		5.35	93.50	280	840	98.80
	7160	60 H.P.		8.18	145	435	1,300	152.45
	7175	75 H.P.		10.59	193	580	1,750	200.70
	7180	200 H.P.		21.46	290	870	2,600	345.70
	7800	Windrow loader, elevating		46.55	1,300	3,925	11,800	1,157
60	0010	**LIFTING AND HOISTING EQUIPMENT RENTAL** without operators R015433-10						
	0120	Aerial lift truck, 2 person, to 80'	Ea.	29.10	695	2,090	6,275	650.80
	0140	Boom work platform, 40' snorkel R015433-15		18.10	273	820	2,450	308.80
	0150	Crane, flatbed mounted, 3 ton capacity		16.00	192	575	1,725	243
	0200	Crane, climbing, 106' jib, 6000 lb. capacity, 410 fpm R312316-45		36.56	1,575	4,760	14,300	1,244
	0300	101' jib, 10,250 lb. capacity, 270 fpm		42.96	2,025	6,040	18,100	1,552
	0500	Tower, static, 130' high, 106' jib, 6200 lb. capacity at 400 fpm		40.31	1,825	5,510	16,500	1,424
	0600	Crawler mounted, lattice boom, 1/2 C.Y., 15 tons at 12' radius		36.34	630	1,890	5,675	668.70
	0700	3/4 C.Y., 20 tons at 12' radius		48.45	790	2,370	7,100	861.60
	0800	1 C.Y., 25 tons at 12' radius		64.60	1,050	3,155	9,475	1,148
	0900	1-1/2 C.Y., 40 tons at 12' radius		64.55	1,050	3,180	9,550	1,152
	1000	2 C.Y., 50 tons at 12' radius		68.45	1,250	3,720	11,200	1,292
	1100	3 C.Y., 75 tons at 12' radius		73.30	1,450	4,355	13,100	1,457
	1200	100 ton capacity, 60' boom		82.95	1,675	5,010	15,000	1,666
	1300	165 ton capacity, 60' boom		105.60	1,950	5,865	17,600	2,018
	1400	200 ton capacity, 70' boom		127.95	2,450	7,335	22,000	2,491
	1500	350 ton capacity, 80' boom		178.85	3,675	11,040	33,100	3,639
	1600	Truck mounted, lattice boom, 6 x 4, 20 tons at 10' radius		37.54	1,100	3,290	9,875	958.30
	1700	25 tons at 10' radius		40.54	1,200	3,580	10,700	1,040
	1800	8 x 4, 30 tons at 10' radius		44.09	1,275	3,810	11,400	1,115
	1900	40 tons at 12' radius		47.21	1,325	3,980	11,900	1,174
	2000	60 tons at 15' radius		53.66	1,400	4,210	12,600	1,271
	2050	82 tons at 15' radius		60.53	1,500	4,500	13,500	1,384
	2100	90 tons at 15' radius		68.16	1,625	4,900	14,700	1,525
	2200	115 tons at 15' radius		77.06	1,825	5,480	16,400	1,712
	2300	150 tons at 18' radius		88.40	1,925	5,770	17,300	1,861
	2350	165 tons at 18' radius		91.22	2,050	6,120	18,400	1,954
	2400	Truck mounted, hydraulic, 12 ton capacity		41.40	525	1,575	4,725	646.20

01 54 33 | Equipment Rental

			UNIT	HOURLY OPER. COST	RENT PER DAY	RENT PER WEEK	RENT PER MONTH	EQUIPMENT COST/DAY	
60	2500	25 ton capacity	Ea.	43.70	630	1,895	5,675	728.60	60
	2550	33 ton capacity		44.25	650	1,945	5,825	743	
	2560	40 ton capacity		57.30	755	2,270	6,800	912.40	
	2600	55 ton capacity		74.65	865	2,590	7,775	1,115	
	2700	80 ton capacity		97.80	1,375	4,155	12,500	1,613	
	2720	100 ton capacity		91.75	1,425	4,290	12,900	1,592	
	2740	120 ton capacity		106.10	1,525	4,610	13,800	1,771	
	2760	150 ton capacity		124.25	2,025	6,075	18,200	2,209	
	2800	Self-propelled, 4 x 4, with telescoping boom, 5 ton		17.15	228	685	2,050	274.20	
	2900	12-1/2 ton capacity		31.50	365	1,095	3,275	471	
	3000	15 ton capacity		32.15	385	1,150	3,450	487.20	
	3050	20 ton capacity		35.00	445	1,340	4,025	548	
	3100	25 ton capacity		36.55	500	1,505	4,525	593.40	
	3150	40 ton capacity		44.85	565	1,690	5,075	696.80	
	3200	Derricks, guy, 20 ton capacity, 60′ boom, 75′ mast		27.52	390	1,167	3,500	453.55	
	3300	100′ boom, 115′ mast		43.10	670	2,010	6,025	746.80	
	3400	Stiffleg, 20 ton capacity, 70′ boom, 37′ mast		29.99	505	1,520	4,550	543.90	
	3500	100′ boom, 47′ mast		46.04	810	2,430	7,300	854.30	
	3550	Helicopter, small, lift to 1250 lb. maximum, w/pilot		101.04	3,150	9,420	28,300	2,692	
	3600	Hoists, chain type, overhead, manual, 3/4 ton		.10	.33	1	3	1	
	3900	10 ton		.70	6	18	54	9.20	
	4000	Hoist and tower, 5000 lb. cap., portable electric, 40′ high		4.64	224	672	2,025	171.50	
	4100	For each added 10′ section, add		.11	17.65	53	159	11.50	
	4200	Hoist and single tubular tower, 5000 lb. electric, 100′ high		6.29	315	938	2,825	237.90	
	4300	For each added 6′-6″ section, add		.18	30.50	91	273	19.65	
	4400	Hoist and double tubular tower, 5000 lb., 100′ high		6.76	345	1,033	3,100	260.70	
	4500	For each added 6′-6″ section, add		.20	33.50	101	305	21.80	
	4550	Hoist and tower, mast type, 6000 lb., 100′ high		7.27	355	1,072	3,225	272.55	
	4570	For each added 10′ section, add		.13	21	63	189	13.65	
	4600	Hoist and tower, personnel, electric, 2000 lb., 100′ @ 125 fpm		15.53	950	2,850	8,550	694.25	
	4700	3000 lb., 100′ @ 200 fpm		17.75	1,075	3,230	9,700	788	
	4800	3000 lb., 150′ @ 300 fpm		19.70	1,200	3,620	10,900	881.60	
	4900	4000 lb., 100′ @ 300 fpm		20.36	1,225	3,690	11,100	900.90	
	5000	6000 lb., 100′ @ 275 fpm		21.90	1,300	3,870	11,600	949.20	
	5100	For added heights up to 500′, add	L.F.	.01	1.67	5	15	1.10	
	5200	Jacks, hydraulic, 20 ton	Ea.	.05	2	6	18	1.60	
	5500	100 ton		.40	11.65	35	105	10.20	
	6100	Jacks, hydraulic, climbing w/50′ jackrods, control console, 30 ton cap.		1.93	129	386	1,150	92.65	
	6150	For each added 10′ jackrod section, add		.05	3.33	10	30	2.40	
	6300	50 ton capacity		3.11	207	621	1,875	149.10	
	6350	For each added 10′ jackrod section, add		.06	4	12	36	2.90	
	6500	125 ton capacity		8.15	545	1,630	4,900	391.20	
	6550	For each added 10′ jackrod section, add		.56	37	111	335	26.70	
	6600	Cable jack, 10 ton capacity with 200′ cable		1.62	108	323	970	77.55	
	6650	For each added 50′ of cable, add		.19	12.65	38	114	9.10	
70	0010	**WELLPOINT EQUIPMENT RENTAL** without operators R015433-10							70
	0020	Based on 2 months rental							
	0100	Combination jetting & wellpoint pump, 60 H.P. diesel	Ea.	18.31	320	957	2,875	337.90	
	0200	High pressure gas jet pump, 200 H.P., 300 psi	"	44.43	273	818	2,450	519.05	
	0300	Discharge pipe, 8″ diameter	L.F.	.01	.52	1.56	4.68	.40	
	0350	12″ diameter		.01	.76	2.28	6.85	.55	
	0400	Header pipe, flows up to 150 GPM, 4″ diameter		.01	.47	1.41	4.23	.35	
	0500	400 GPM, 6″ diameter		.01	.55	1.66	4.98	.40	
	0600	800 GPM, 8″ diameter		.01	.76	2.28	6.85	.55	
	0700	1500 GPM, 10″ diameter		.01	.80	2.41	7.25	.55	
	0800	2500 GPM, 12″ diameter		.02	1.52	4.56	13.70	1.05	
	0900	4500 GPM, 16″ diameter		.03	1.94	5.83	17.50	1.40	

01 54 33 | Equipment Rental

			UNIT	HOURLY OPER. COST	RENT PER DAY	RENT PER WEEK	RENT PER MONTH	EQUIPMENT COST/DAY	
70	0950	For quick coupling aluminum and plastic pipe, add	L.F.	.03	2.01	6.04	18.10	1.45	70
	1100	Wellpoint, 25′ long, with fittings & riser pipe, 1-1/2″ or 2″ diameter	Ea.	.06	4.02	12.05	36	2.90	
	1200	Wellpoint pump, diesel powered, 4″ suction, 20 H.P.		7.73	184	552	1,650	172.25	
	1300	6″ suction, 30 H.P.		10.59	228	684	2,050	221.50	
	1400	8″ suction, 40 H.P.		14.31	315	938	2,825	302.10	
	1500	10″ suction, 75 H.P.		22.19	365	1,097	3,300	396.90	
	1600	12″ suction, 100 H.P.		31.53	580	1,740	5,225	600.25	
	1700	12″ suction, 175 H.P.	↓	47.37	645	1,930	5,800	764.95	
80	0010	**MARINE EQUIPMENT RENTAL** without operators R015433-10							80
	0200	Barge, 400 Ton, 30′ wide x 90′ long	Ea.	17.10	1,075	3,235	9,700	783.80	
	0240	800 Ton, 45′ wide x 90′ long		20.75	1,300	3,930	11,800	952	
	2000	Tugboat, diesel, 100 H.P.		38.60	218	655	1,975	439.80	
	2040	250 H.P.		81.00	400	1,195	3,575	887	
	2080	380 H.P.		158.50	1,200	3,565	10,700	1,981	
	3000	Small work boat, gas, 16-foot, 50 H.P.		18.80	61.50	185	555	187.40	
	4000	Large, diesel, 48-foot, 200 H.P.	↓	89.95	1,250	3,760	11,300	1,472	

Crew No.	Bare Costs		Incl. Subs O&P		Cost Per Labor-Hour	

Crew A-1	Hr.	Daily	Hr.	Daily	Bare Costs	Incl. O&P
1 Building Laborer	$35.45	$283.60	$54.60	$436.80	$35.45	$54.60
1 Concrete Saw, Gas Manual		84.00		92.40	10.50	11.55
8 L.H., Daily Totals		$367.60		$529.20	$45.95	$66.15

Crew A-1A	Hr.	Daily	Hr.	Daily	Bare Costs	Incl. O&P
1 Skilled Worker	$46.20	$369.60	$71.45	$571.60	$46.20	$71.45
1 Shot Blaster, 20″		210.20		231.22	26.27	28.90
8 L.H., Daily Totals		$579.80		$802.82	$72.47	$100.35

Crew A-1B	Hr.	Daily	Hr.	Daily	Bare Costs	Incl. O&P
1 Building Laborer	$35.45	$283.60	$54.60	$436.80	$35.45	$54.60
1 Concrete Saw		175.20		192.72	21.90	24.09
8 L.H., Daily Totals		$458.80		$629.52	$57.35	$78.69

Crew A-1C	Hr.	Daily	Hr.	Daily	Bare Costs	Incl. O&P
1 Building Laborer	$35.45	$283.60	$54.60	$436.80	$35.45	$54.60
1 Chain Saw, Gas, 18″		30.60		33.66	3.83	4.21
8 L.H., Daily Totals		$314.20		$470.46	$39.27	$58.81

Crew A-1D	Hr.	Daily	Hr.	Daily	Bare Costs	Incl. O&P
1 Building Laborer	$35.45	$283.60	$54.60	$436.80	$35.45	$54.60
1 Vibrating Plate, Gas, 18″		34.60		38.06	4.33	4.76
8 L.H., Daily Totals		$318.20		$474.86	$39.77	$59.36

Crew A-1E	Hr.	Daily	Hr.	Daily	Bare Costs	Incl. O&P
1 Building Laborer	$35.45	$283.60	$54.60	$436.80	$35.45	$54.60
1 Vibrating Plate, Gas, 21″		44.40		48.84	5.55	6.11
8 L.H., Daily Totals		$328.00		$485.64	$41.00	$60.70

Crew A-1F	Hr.	Daily	Hr.	Daily	Bare Costs	Incl. O&P
1 Building Laborer	$35.45	$283.60	$54.60	$436.80	$35.45	$54.60
1 Rammer/Tamper, Gas, 8″		47.20		51.92	5.90	6.49
8 L.H., Daily Totals		$330.80		$488.72	$41.35	$61.09

Crew A-1G	Hr.	Daily	Hr.	Daily	Bare Costs	Incl. O&P
1 Building Laborer	$35.45	$283.60	$54.60	$436.80	$35.45	$54.60
1 Rammer/Tamper, Gas, 15″		53.60		58.96	6.70	7.37
8 L.H., Daily Totals		$337.20		$495.76	$42.15	$61.97

Crew A-1H	Hr.	Daily	Hr.	Daily	Bare Costs	Incl. O&P
1 Building Laborer	$35.45	$283.60	$54.60	$436.80	$35.45	$54.60
1 Exterior Steam Cleaner		75.60		83.16	9.45	10.40
8 L.H., Daily Totals		$359.20		$519.96	$44.90	$65.00

Crew A-1J	Hr.	Daily	Hr.	Daily	Bare Costs	Incl. O&P
1 Building Laborer	$35.45	$283.60	$54.60	$436.80	$35.45	$54.60
1 Cultivator, Walk-Behind, 5 H.P.		59.00		64.90	7.38	8.11
8 L.H., Daily Totals		$342.60		$501.70	$42.83	$62.71

Crew A-1K	Hr.	Daily	Hr.	Daily	Bare Costs	Incl. O&P
1 Building Laborer	$35.45	$283.60	$54.60	$436.80	$35.45	$54.60
1 Cultivator, Walk-Behind, 8 H.P.		90.55		99.61	11.32	12.45
8 L.H., Daily Totals		$374.15		$536.40	$46.77	$67.05

Crew A-1M	Hr.	Daily	Hr.	Daily	Bare Costs	Incl. O&P
1 Building Laborer	$35.45	$283.60	$54.60	$436.80	$35.45	$54.60
1 Snow Blower, Walk-Behind		53.60		58.96	6.70	7.37
8 L.H., Daily Totals		$337.20		$495.76	$42.15	$61.97

Crew A-2	Hr.	Daily	Hr.	Daily	Bare Costs	Incl. O&P
2 Laborers	$35.45	$567.20	$54.60	$873.60	$35.52	$54.60
1 Truck Driver (light)	35.65	285.20	54.60	436.80		
1 Flatbed Truck, Gas, 1.5 Ton		265.80		292.38	11.07	12.18
24 L.H., Daily Totals		$1118.20		$1602.78	$46.59	$66.78

Crew A-2A	Hr.	Daily	Hr.	Daily	Bare Costs	Incl. O&P
2 Laborers	$35.45	$567.20	$54.60	$873.60	$35.52	$54.60
1 Truck Driver (light)	35.65	285.20	54.60	436.80		
1 Flatbed Truck, Gas, 1.5 Ton		265.80		292.38		
1 Concrete Saw		175.20		192.72	18.38	20.21
24 L.H., Daily Totals		$1293.40		$1795.50	$53.89	$74.81

Crew A-2B	Hr.	Daily	Hr.	Daily	Bare Costs	Incl. O&P
1 Truck Driver (light)	$35.65	$285.20	$54.60	$436.80	$35.65	$54.60
1 Flatbed Truck, Gas, 1.5 Ton		265.80		292.38	33.23	36.55
8 L.H., Daily Totals		$551.00		$729.18	$68.88	$91.15

Crew A-3A	Hr.	Daily	Hr.	Daily	Bare Costs	Incl. O&P
1 Truck Driver (light)	$35.65	$285.20	$54.60	$436.80	$35.65	$54.60
1 Pickup Truck, 4 x 4, 3/4 Ton		165.60		182.16	20.70	22.77
8 L.H., Daily Totals		$450.80		$618.96	$56.35	$77.37

Crew A-3B	Hr.	Daily	Hr.	Daily	Bare Costs	Incl. O&P
1 Equip. Oper. (med.)	$47.50	$380.00	$71.75	$574.00	$42.05	$63.90
1 Truck Driver (heav.)	36.60	292.80	56.05	448.40		
1 Dump Truck, 12 C.Y., 400 H.P.		692.40		761.64		
1 F.E. Loader, W.M.,2.5 C.Y.		526.20		578.82	76.16	83.78
16 L.H., Daily Totals		$1891.40		$2362.86	$118.21	$147.68

Crew A-3C	Hr.	Daily	Hr.	Daily	Bare Costs	Incl. O&P
1 Equip. Oper. (light)	$45.80	$366.40	$69.20	$553.60	$45.80	$69.20
1 Loader, Skid Steer, 78 H.P.		308.60		339.46	38.58	42.43
8 L.H., Daily Totals		$675.00		$893.06	$84.38	$111.63

Crew A-3D	Hr.	Daily	Hr.	Daily	Bare Costs	Incl. O&P
1 Truck Driver, Light	$35.65	$285.20	$54.60	$436.80	$35.65	$54.60
1 Pickup Truck, 4 x 4, 3/4 Ton		165.60		182.16		
1 Flatbed Trailer, 25 Ton		113.20		124.52	34.85	38.34
8 L.H., Daily Totals		$564.00		$743.48	$70.50	$92.94

Crew A-3E	Hr.	Daily	Hr.	Daily	Bare Costs	Incl. O&P
1 Equip. Oper. (crane)	$48.80	$390.40	$73.75	$590.00	$42.70	$64.90
1 Truck Driver (heavy)	36.60	292.80	56.05	448.40		
1 Pickup Truck, 4 x 4, 3/4 Ton		165.60		182.16	10.35	11.39
16 L.H., Daily Totals		$848.80		$1220.56	$53.05	$76.28

Crew A-3F	Hr.	Daily	Hr.	Daily	Bare Costs	Incl. O&P
1 Equip. Oper. (crane)	$48.80	$390.40	$73.75	$590.00	$42.70	$64.90
1 Truck Driver (heavy)	36.60	292.80	56.05	448.40		
1 Pickup Truck, 4 x 4, 3/4 Ton		165.60		182.16		
1 Truck Tractor, 6x4, 380 H.P.		612.40		673.64		
1 Lowbed Trailer, 75 Ton		220.40		242.44	62.40	68.64
16 L.H., Daily Totals		$1681.60		$2136.64	$105.10	$133.54

Crew No.	Bare Costs		Incl. Subs O&P		Cost Per Labor-Hour	
Crew A-3G	**Hr.**	**Daily**	**Hr.**	**Daily**	**Bare Costs**	**Incl. O&P**
1 Equip. Oper. (crane)	$48.80	$390.40	$73.75	$590.00	$42.70	$64.90
1 Truck Driver (heavy)	36.60	292.80	56.05	448.40		
1 Pickup Truck, 4 x 4, 3/4 Ton		165.60		182.16		
1 Truck Tractor, 6x4, 450 H.P.		741.20		815.32		
1 Lowbed Trailer, 75 Ton		220.40		242.44	70.45	77.50
16 L.H., Daily Totals		$1810.40		$2278.32	$113.15	$142.40

Crew A-3H	Hr.	Daily	Hr.	Daily	Bare Costs	Incl. O&P
1 Equip. Oper. (crane)	$48.80	$390.40	$73.75	$590.00	$48.80	$73.75
1 Hyd. Crane, 12 Ton (Daily)		856.20		941.82	107.03	117.73
8 L.H., Daily Totals		$1246.60		$1531.82	$155.82	$191.48

Crew A-3I	Hr.	Daily	Hr.	Daily	Bare Costs	Incl. O&P
1 Equip. Oper. (crane)	$48.80	$390.40	$73.75	$590.00	$48.80	$73.75
1 Hyd. Crane, 25 Ton (Daily)		979.60		1077.56	122.45	134.69
8 L.H., Daily Totals		$1370.00		$1667.56	$171.25	$208.44

Crew A-3J	Hr.	Daily	Hr.	Daily	Bare Costs	Incl. O&P
1 Equip. Oper. (crane)	$48.80	$390.40	$73.75	$590.00	$48.80	$73.75
1 Hyd. Crane, 40 Ton (Daily)		1213.00		1334.30	151.63	166.79
8 L.H., Daily Totals		$1603.40		$1924.30	$200.43	$240.54

Crew A-3K	Hr.	Daily	Hr.	Daily	Bare Costs	Incl. O&P
1 Equip. Oper. (crane)	$48.80	$390.40	$73.75	$590.00	$45.52	$68.80
1 Equip. Oper. Oiler	42.25	338.00	63.85	510.80		
1 Hyd. Crane, 55 Ton (Daily)		1462.00		1608.20		
1 P/U Truck, 3/4 Ton (Daily)		179.60		197.56	102.60	112.86
16 L.H., Daily Totals		$2370.00		$2906.56	$148.13	$181.66

Crew A-3L	Hr.	Daily	Hr.	Daily	Bare Costs	Incl. O&P
1 Equip. Oper. (crane)	$48.80	$390.40	$73.75	$590.00	$45.52	$68.80
1 Equip. Oper. Oiler	42.25	338.00	63.85	510.80		
1 Hyd. Crane, 80 Ton (Daily)		2167.00		2383.70		
1 P/U Truck, 3/4 Ton (Daily)		179.60		197.56	146.66	161.33
16 L.H., Daily Totals		$3075.00		$3682.06	$192.19	$230.13

Crew A-3M	Hr.	Daily	Hr.	Daily	Bare Costs	Incl. O&P
1 Equip. Oper. (crane)	$48.80	$390.40	$73.75	$590.00	$45.52	$68.80
1 Equip. Oper. Oiler	42.25	338.00	63.85	510.80		
1 Hyd. Crane, 100 Ton (Daily)		2164.00		2380.40		
1 P/U Truck, 3/4 Ton (Daily)		179.60		197.56	146.47	161.12
16 L.H., Daily Totals		$3072.00		$3678.76	$192.00	$229.92

Crew A-3N	Hr.	Daily	Hr.	Daily	Bare Costs	Incl. O&P
1 Equip. Oper. (crane)	$48.80	$390.40	$73.75	$590.00	$48.80	$73.75
1 Tower Cane (Monthly)		1072.00		1179.20	134.00	147.40
8 L.H., Daily Totals		$1462.40		$1769.20	$182.80	$221.15

Crew A-3P	Hr.	Daily	Hr.	Daily	Bare Costs	Incl. O&P
1 Equip. Oper., Light	$45.80	$366.40	$69.20	$553.60	$45.80	$69.20
1 A.T. Forklift, 42' lift		512.40		563.64	64.05	70.45
8 L.H., Daily Totals		$878.80		$1117.24	$109.85	$139.66

Crew A-4	Hr.	Daily	Hr.	Daily	Bare Costs	Incl. O&P
2 Carpenters	$44.90	$718.40	$69.15	$1106.40	$42.83	$65.45
1 Painter, Ordinary	38.70	309.60	58.05	464.40		
24 L.H., Daily Totals		$1028.00		$1570.80	$42.83	$65.45

Crew No.	Bare Costs		Incl. Subs O&P		Cost Per Labor-Hour	
Crew A-5	**Hr.**	**Daily**	**Hr.**	**Daily**	**Bare Costs**	**Incl. O&P**
2 Laborers	$35.45	$567.20	$54.60	$873.60	$35.47	$54.60
.25 Truck Driver (light)	35.65	71.30	54.60	109.20		
.25 Flatbed Truck, Gas, 1.5 Ton		66.45		73.09	3.69	4.06
18 L.H., Daily Totals		$704.95		$1055.90	$39.16	$58.66

Crew A-6	Hr.	Daily	Hr.	Daily	Bare Costs	Incl. O&P
1 Instrument Man	$46.20	$369.60	$71.45	$571.60	$44.75	$68.50
1 Rodman/Chainman	43.30	346.40	65.55	524.40		
1 Level, Electronic		77.55		85.31	4.85	5.33
16 L.H., Daily Totals		$793.55		$1181.31	$49.60	$73.83

Crew A-7	Hr.	Daily	Hr.	Daily	Bare Costs	Incl. O&P
1 Chief of Party	$57.50	$460.00	$88.50	$708.00	$49.00	$75.17
1 Instrument Man	46.20	369.60	71.45	571.60		
1 Rodman/Chainman	43.30	346.40	65.55	524.40		
1 Level, Electronic		77.55		85.31	3.23	3.55
24 L.H., Daily Totals		$1253.55		$1889.31	$52.23	$78.72

Crew A-8	Hr.	Daily	Hr.	Daily	Bare Costs	Incl. O&P
1 Chief of Party	$57.50	$460.00	$88.50	$708.00	$47.58	$72.76
1 Instrument Man	46.20	369.60	71.45	571.60		
2 Rodmen/Chainmen	43.30	692.80	65.55	1048.80		
1 Level, Electronic		77.55		85.31	2.42	2.67
32 L.H., Daily Totals		$1599.95		$2413.70	$50.00	$75.43

Crew A-9	Hr.	Daily	Hr.	Daily	Bare Costs	Incl. O&P
1 Asbestos Foreman	$50.20	$401.60	$77.95	$623.60	$49.76	$77.29
7 Asbestos Workers	49.70	2783.20	77.20	4323.20		
64 L.H., Daily Totals		$3184.80		$4946.80	$49.76	$77.29

Crew A-10A	Hr.	Daily	Hr.	Daily	Bare Costs	Incl. O&P
1 Asbestos Foreman	$50.20	$401.60	$77.95	$623.60	$49.87	$77.45
2 Asbestos Workers	49.70	795.20	77.20	1235.20		
24 L.H., Daily Totals		$1196.80		$1858.80	$49.87	$77.45

Crew A-10B	Hr.	Daily	Hr.	Daily	Bare Costs	Incl. O&P
1 Asbestos Foreman	$50.20	$401.60	$77.95	$623.60	$49.83	$77.39
3 Asbestos Workers	49.70	1192.80	77.20	1852.80		
32 L.H., Daily Totals		$1594.40		$2476.40	$49.83	$77.39

Crew A-10C	Hr.	Daily	Hr.	Daily	Bare Costs	Incl. O&P
3 Asbestos Workers	$49.70	$1192.80	$77.20	$1852.80	$49.70	$77.20
1 Flatbed Truck, Gas, 1.5 Ton		265.80		292.38	11.07	12.18
24 L.H., Daily Totals		$1458.60		$2145.18	$60.77	$89.38

Crew A-10D	Hr.	Daily	Hr.	Daily	Bare Costs	Incl. O&P
2 Asbestos Workers	$49.70	$795.20	$77.20	$1235.20	$47.61	$73.00
1 Equip. Oper. (crane)	48.80	390.40	73.75	590.00		
1 Equip. Oper. Oiler	42.25	338.00	63.85	510.80		
1 Hydraulic Crane, 33 Ton		743.00		817.30	23.22	25.54
32 L.H., Daily Totals		$2266.60		$3153.30	$70.83	$98.54

Crew A-11	Hr.	Daily	Hr.	Daily	Bare Costs	Incl. O&P
1 Asbestos Foreman	$50.20	$401.60	$77.95	$623.60	$49.76	$77.29
7 Asbestos Workers	49.70	2783.20	77.20	4323.20		
2 Chip. Hammers, 12 Lb., Elec.		40.00		44.00	.63	.69
64 L.H., Daily Totals		$3224.80		$4990.80	$50.39	$77.98

Crew No.	Bare Costs		Incl. Subs O&P		Cost Per Labor-Hour	

Crew A-12	Hr.	Daily	Hr.	Daily	Bare Costs	Incl. O&P
1 Asbestos Foreman	$50.20	$401.60	$77.95	$623.60	$49.76	$77.29
7 Asbestos Workers	49.70	2783.20	77.20	4323.20		
1 Trk-Mtd Vac, 14 CY, 1500 Gal.		522.65		574.91		
1 Flatbed Truck, 20,000 GVW		253.60		278.96	12.13	13.34
64 L.H., Daily Totals		$3961.05		$5800.68	$61.89	$90.64

Crew A-13	Hr.	Daily	Hr.	Daily	Bare Costs	Incl. O&P
1 Equip. Oper. (light)	$45.80	$366.40	$69.20	$553.60	$45.80	$69.20
1 Trk-Mtd Vac, 14 CY, 1500 Gal.		522.65		574.91		
1 Flatbed Truck, 20,000 GVW		253.60		278.96	97.03	106.73
8 L.H., Daily Totals		$1142.65		$1407.47	$142.83	$175.93

Crew B-1	Hr.	Daily	Hr.	Daily	Bare Costs	Incl. O&P
1 Labor Foreman (outside)	$37.45	$299.60	$57.65	$461.20	$36.12	$55.62
2 Laborers	35.45	567.20	54.60	873.60		
24 L.H., Daily Totals		$866.80		$1334.80	$36.12	$55.62

Crew B-1A	Hr.	Daily	Hr.	Daily	Bare Costs	Incl. O&P
1 Labor Foreman (outside)	$37.45	$299.60	$57.65	$461.20	$36.12	$55.62
2 Laborers	35.45	567.20	54.60	873.60		
2 Cutting Torches		22.80		25.08		
2 Sets of Gases		266.40		293.04	12.05	13.26
24 L.H., Daily Totals		$1156.00		$1652.92	$48.17	$68.87

Crew B-1B	Hr.	Daily	Hr.	Daily	Bare Costs	Incl. O&P
1 Labor Foreman (outside)	$37.45	$299.60	$57.65	$461.20	$39.29	$60.15
2 Laborers	35.45	567.20	54.60	873.60		
1 Equip. Oper. (crane)	48.80	390.40	73.75	590.00		
2 Cutting Torches		22.80		25.08		
2 Sets of Gases		266.40		293.04		
1 Hyd. Crane, 12 Ton		646.20		710.82	29.23	32.15
32 L.H., Daily Totals		$2192.60		$2953.74	$68.52	$92.30

Crew B-1C	Hr.	Daily	Hr.	Daily	Bare Costs	Incl. O&P
1 Labor Foreman (outside)	$37.45	$299.60	$57.65	$461.20	$36.12	$55.62
2 Laborers	35.45	567.20	54.60	873.60		
1 Aerial Lift Truck, 60' Boom		462.20		508.42	19.26	21.18
24 L.H., Daily Totals		$1329.00		$1843.22	$55.38	$76.80

Crew B-1D	Hr.	Daily	Hr.	Daily	Bare Costs	Incl. O&P
2 Laborers	$35.45	$567.20	$54.60	$873.60	$35.45	$54.60
1 Small Work Boat, Gas, 50 H.P.		187.40		206.14		
1 Pressure Washer, 7 GPM		86.40		95.04	17.11	18.82
16 L.H., Daily Totals		$841.00		$1174.78	$52.56	$73.42

Crew B-1E	Hr.	Daily	Hr.	Daily	Bare Costs	Incl. O&P
1 Labor Foreman (outside)	$37.45	$299.60	$57.65	$461.20	$35.95	$55.36
3 Laborers	35.45	850.80	54.60	1310.40		
1 Work Boat, Diesel, 200 H.P.		1472.00		1619.20		
2 Pressure Washer, 7 GPM		172.80		190.08	51.40	56.54
32 L.H., Daily Totals		$2795.20		$3580.88	$87.35	$111.90

Crew B-1F	Hr.	Daily	Hr.	Daily	Bare Costs	Incl. O&P
2 Skilled Workers	$46.20	$739.20	$71.45	$1143.20	$42.62	$65.83
1 Laborer	35.45	283.60	54.60	436.80		
1 Small Work Boat, Gas, 50 H.P.		187.40		206.14		
1 Pressure Washer, 7 GPM		86.40		95.04	11.41	12.55
24 L.H., Daily Totals		$1296.60		$1881.18	$54.02	$78.38

Crew B-1G	Hr.	Daily	Hr.	Daily	Bare Costs	Incl. O&P
2 Laborers	$35.45	$567.20	$54.60	$873.60	$35.45	$54.60
1 Small Work Boat, Gas, 50 H.P.		187.40		206.14	11.71	12.88
16 L.H., Daily Totals		$754.60		$1079.74	$47.16	$67.48

Crew B-1H	Hr.	Daily	Hr.	Daily	Bare Costs	Incl. O&P
2 Skilled Workers	$46.20	$739.20	$71.45	$1143.20	$42.62	$65.83
1 Laborer	35.45	283.60	54.60	436.80		
1 Small Work Boat, Gas, 50 H.P.		187.40		206.14	7.81	8.59
24 L.H., Daily Totals		$1210.20		$1786.14	$50.42	$74.42

Crew B-1J	Hr.	Daily	Hr.	Daily	Bare Costs	Incl. O&P
1 Labor Foreman (inside)	$35.95	$287.60	$55.35	$442.80	$35.70	$54.98
1 Laborer	35.45	283.60	54.60	436.80		
16 L.H., Daily Totals		$571.20		$879.60	$35.70	$54.98

Crew B-1K	Hr.	Daily	Hr.	Daily	Bare Costs	Incl. O&P
1 Carpenter Foreman (inside)	$45.40	$363.20	$69.90	$559.20	$45.15	$69.53
1 Carpenter	44.90	359.20	69.15	553.20		
16 L.H., Daily Totals		$722.40		$1112.40	$45.15	$69.53

Crew B-2	Hr.	Daily	Hr.	Daily	Bare Costs	Incl. O&P
1 Labor Foreman (outside)	$37.45	$299.60	$57.65	$461.20	$35.85	$55.21
4 Laborers	35.45	1134.40	54.60	1747.20		
40 L.H., Daily Totals		$1434.00		$2208.40	$35.85	$55.21

Crew B-2A	Hr.	Daily	Hr.	Daily	Bare Costs	Incl. O&P
1 Labor Foreman (outside)	$37.45	$299.60	$57.65	$461.20	$36.12	$55.62
2 Laborers	35.45	567.20	54.60	873.60		
1 Aerial Lift Truck, 60' Boom		462.20		508.42	19.26	21.18
24 L.H., Daily Totals		$1329.00		$1843.22	$55.38	$76.80

Crew B-3	Hr.	Daily	Hr.	Daily	Bare Costs	Incl. O&P
1 Labor Foreman (outside)	$37.45	$299.60	$57.65	$461.20	$38.17	$58.45
2 Laborers	35.45	567.20	54.60	873.60		
1 Equip. Oper. (med.)	47.50	380.00	71.75	574.00		
2 Truck Drivers (heavy)	36.60	585.60	56.05	896.80		
1 Crawler Loader, 3 C.Y.		1187.00		1305.70		
2 Dump Trucks, 12 C.Y., 400 H.P.		1384.80		1523.28	53.58	58.94
48 L.H., Daily Totals		$4404.20		$5634.58	$91.75	$117.39

Crew B-3A	Hr.	Daily	Hr.	Daily	Bare Costs	Incl. O&P
4 Laborers	$35.45	$1134.40	$54.60	$1747.20	$37.86	$58.03
1 Equip. Oper. (med.)	47.50	380.00	71.75	574.00		
1 Hyd. Excavator, 1.5 C.Y.		1036.00		1139.60	25.90	28.49
40 L.H., Daily Totals		$2550.40		$3460.80	$63.76	$86.52

Crew B-3B	Hr.	Daily	Hr.	Daily	Bare Costs	Incl. O&P
2 Laborers	$35.45	$567.20	$54.60	$873.60	$38.75	$59.25
1 Equip. Oper. (med.)	47.50	380.00	71.75	574.00		
1 Truck Driver (heavy)	36.60	292.80	56.05	448.40		
1 Backhoe Loader, 80 H.P.		400.40		440.44		
1 Dump Truck, 12 C.Y., 400 H.P.		692.40		761.64	34.15	37.56
32 L.H., Daily Totals		$2332.80		$3098.08	$72.90	$96.81

Crew B-3C	Hr.	Daily	Hr.	Daily	Bare Costs	Incl. O&P
3 Laborers	$35.45	$850.80	$54.60	$1310.40	$38.46	$58.89
1 Equip. Oper. (med.)	47.50	380.00	71.75	574.00		
1 Crawler Loader, 4 C.Y.		1530.00		1683.00	47.81	52.59
32 L.H., Daily Totals		$2760.80		$3567.40	$86.28	$111.48

Crew No.	Bare Costs		Incl. Subs O&P		Cost Per Labor-Hour	
Crew B-4	**Hr.**	**Daily**	**Hr.**	**Daily**	**Bare Costs**	**Incl. O&P**
1 Labor Foreman (outside)	$37.45	$299.60	$57.65	$461.20	$35.98	$55.35
4 Laborers	35.45	1134.40	54.60	1747.20		
1 Truck Driver (heavy)	36.60	292.80	56.05	448.40		
1 Truck Tractor, 220 H.P.		366.80		403.48		
1 Flatbed Trailer, 40 Ton		152.60		167.86	10.82	11.90
48 L.H., Daily Totals		$2246.20		$3228.14	$46.80	$67.25

Crew B-5	**Hr.**	**Daily**	**Hr.**	**Daily**	**Bare Costs**	**Incl. O&P**
1 Labor Foreman (outside)	$37.45	$299.60	$57.65	$461.20	$39.18	$59.94
4 Laborers	35.45	1134.40	54.60	1747.20		
2 Equip. Oper. (med.)	47.50	760.00	71.75	1148.00		
1 Air Compressor, 250 cfm		195.80		215.38		
2 Breakers, Pavement, 60 lb.		19.60		21.56		
2 -50' Air Hoses, 1.5"		11.60		12.76		
1 Crawler Loader, 3 C.Y.		1187.00		1305.70	25.25	27.77
56 L.H., Daily Totals		$3608.00		$4911.80	$64.43	$87.71

Crew B-5A	**Hr.**	**Daily**	**Hr.**	**Daily**	**Bare Costs**	**Incl. O&P**
1 Labor Foreman (outside)	$37.45	$299.60	$57.65	$461.20	$38.68	$59.17
6 Laborers	35.45	1701.60	54.60	2620.80		
2 Equip. Oper. (med.)	47.50	760.00	71.75	1148.00		
1 Equip. Oper. (light)	45.80	366.40	69.20	553.60		
2 Truck Drivers (heavy)	36.60	585.60	56.05	896.80		
1 Air Compressor, 365 cfm		255.80		281.38		
2 Breakers, Pavement, 60 lb.		19.60		21.56		
8 -50' Air Hoses, 1"		32.80		36.08		
2 Dump Trucks, 8 C.Y., 220 H.P.		833.20		916.52	11.89	13.08
96 L.H., Daily Totals		$4854.60		$6935.94	$50.57	$72.25

Crew B-5B	**Hr.**	**Daily**	**Hr.**	**Daily**	**Bare Costs**	**Incl. O&P**
1 Powderman	$46.20	$369.60	$71.45	$571.60	$41.83	$63.85
2 Equip. Oper. (med.)	47.50	760.00	71.75	1148.00		
3 Truck Drivers (heavy)	36.60	878.40	56.05	1345.20		
1 F.E. Loader, W.M.,2.5 C.Y.		526.20		578.82		
3 Dump Trucks, 12 C.Y., 400 H.P.		2077.20		2284.92		
1 Air Compressor, 365 CFM		255.80		281.38	59.57	65.52
48 L.H., Daily Totals		$4867.20		$6209.92	$101.40	$129.37

Crew B-5C	**Hr.**	**Daily**	**Hr.**	**Daily**	**Bare Costs**	**Incl. O&P**
3 Laborers	$35.45	$850.80	$54.60	$1310.40	$39.76	$60.66
1 Equip. Oper. (med.)	47.50	380.00	71.75	574.00		
2 Truck Drivers (heav.)	36.60	585.60	56.05	896.80		
1 Equip. Oper. (crane)	48.80	390.40	73.75	590.00		
1 Equip. Oper. Oiler	42.25	338.00	63.85	510.80		
2 Dump Trucks, 12 C.Y., 400 H.P.		1384.80		1523.28		
1 Crawler Loader, 4 C.Y.		1530.00		1683.00		
1 S.P. Crane, 4x4, 25 Ton		593.40		652.74	54.82	60.30
64 L.H., Daily Totals		$6053.00		$7741.02	$94.58	$120.95

Crew B-5D	**Hr.**	**Daily**	**Hr.**	**Daily**	**Bare Costs**	**Incl. O&P**
1 Labor Foreman (outside)	$37.45	$299.60	$57.65	$461.20	$38.86	$59.45
4 Laborers	35.45	1134.40	54.60	1747.20		
2 Equip. Oper. (med.)	47.50	760.00	71.75	1148.00		
1 Truck Driver (heavy)	36.60	292.80	56.05	448.40		
1 Air Compressor, 250 cfm		195.80		215.38		
2 Breakers, Pavement, 60 lb.		19.60		21.56		
2 -50' Air Hoses, 1.5"		11.60		12.76		
1 Crawler Loader, 3 C.Y.		1187.00		1305.70		
1 Dump Truck, 12 C.Y., 400 H.P.		692.40		761.64	32.91	36.20
64 L.H., Daily Totals		$4593.20		$6121.84	$71.77	$95.65

Crew No.	Bare Costs		Incl. Subs O&P		Cost Per Labor-Hour	
Crew B-6	**Hr.**	**Daily**	**Hr.**	**Daily**	**Bare Costs**	**Incl. O&P**
2 Laborers	$35.45	$567.20	$54.60	$873.60	$38.90	$59.47
1 Equip. Oper. (light)	45.80	366.40	69.20	553.60		
1 Backhoe Loader, 48 H.P.		367.40		404.14	15.31	16.84
24 L.H., Daily Totals		$1301.00		$1831.34	$54.21	$76.31

Crew B-6A	**Hr.**	**Daily**	**Hr.**	**Daily**	**Bare Costs**	**Incl. O&P**
.5 Labor Foreman (outside)	$37.45	$149.80	$57.65	$230.60	$40.67	$62.07
1 Laborer	35.45	283.60	54.60	436.80		
1 Equip. Oper. (med.)	47.50	380.00	71.75	574.00		
1 Vacuum Truck, 5000 Gal.		363.90		400.29	18.20	20.01
20 L.H., Daily Totals		$1177.30		$1641.69	$58.87	$82.08

Crew B-6B	**Hr.**	**Daily**	**Hr.**	**Daily**	**Bare Costs**	**Incl. O&P**
2 Labor Foremen (outside)	$37.45	$599.20	$57.65	$922.40	$36.12	$55.62
4 Laborers	35.45	1134.40	54.60	1747.20		
1 S.P. Crane, 4x4, 5 Ton		274.20		301.62		
1 Flatbed Truck, Gas, 1.5 Ton		265.80		292.38		
1 Butt Fusion Mach., 4"-12" diam.		382.60		420.86	19.22	21.14
48 L.H., Daily Totals		$2656.20		$3684.46	$55.34	$76.76

Crew B-6C	**Hr.**	**Daily**	**Hr.**	**Daily**	**Bare Costs**	**Incl. O&P**
2 Labor Foremen (outside)	$37.45	$599.20	$57.65	$922.40	$36.12	$55.62
4 Laborers	35.45	1134.40	54.60	1747.20		
1 S.P. Crane, 4x4, 12 Ton		471.00		518.10		
1 Flatbed Truck, Gas, 3 Ton		327.80		360.58		
1 Butt Fusion Mach., 8"-24" diam.		831.70		914.87	33.97	37.37
48 L.H., Daily Totals		$3364.10		$4463.15	$70.09	$92.98

Crew B-7	**Hr.**	**Daily**	**Hr.**	**Daily**	**Bare Costs**	**Incl. O&P**
1 Labor Foreman (outside)	$37.45	$299.60	$57.65	$461.20	$37.79	$57.97
4 Laborers	35.45	1134.40	54.60	1747.20		
1 Equip. Oper. (med.)	47.50	380.00	71.75	574.00		
1 Brush Chipper, 12", 130 H.P.		396.40		436.04		
1 Crawler Loader, 3 C.Y.		1187.00		1305.70		
2 Chain Saws, Gas, 36" Long		90.00		99.00	34.86	38.35
48 L.H., Daily Totals		$3487.40		$4623.14	$72.65	$96.32

Crew B-7A	**Hr.**	**Daily**	**Hr.**	**Daily**	**Bare Costs**	**Incl. O&P**
2 Laborers	$35.45	$567.20	$54.60	$873.60	$38.90	$59.47
1 Equip. Oper. (light)	45.80	366.40	69.20	553.60		
1 Rake w/Tractor		347.70		382.47		
2 Chain Saw, Gas, 18"		61.20		67.32	17.04	18.74
24 L.H., Daily Totals		$1342.50		$1876.99	$55.94	$78.21

Crew B-7B	**Hr.**	**Daily**	**Hr.**	**Daily**	**Bare Costs**	**Incl. O&P**
1 Labor Foreman (outside)	$37.45	$299.60	$57.65	$461.20	$37.62	$57.69
4 Laborers	35.45	1134.40	54.60	1747.20		
1 Equip. Oper. (med.)	47.50	380.00	71.75	574.00		
1 Truck Driver (heavy)	36.60	292.80	56.05	448.40		
1 Brush Chipper, 12", 130 H.P.		396.40		436.04		
1 Crawler Loader, 3 C.Y.		1187.00		1305.70		
2 Chain Saws, Gas, 36" Long		90.00		99.00		
1 Dump Truck, 8 C.Y., 220 H.P.		416.60		458.26	37.32	41.05
56 L.H., Daily Totals		$4196.80		$5529.80	$74.94	$98.75

Crew No.	Bare Costs		Incl. Subs O&P		Cost Per Labor-Hour	
Crew B-7C	**Hr.**	**Daily**	**Hr.**	**Daily**	**Bare Costs**	**Incl. O&P**
1 Labor Foreman (outside)	$37.45	$299.60	$57.65	$461.20	$37.62	$57.69
4 Laborers	35.45	1134.40	54.60	1747.20		
1 Equip. Oper. (med.)	47.50	380.00	71.75	574.00		
1 Truck Driver (heavy)	36.60	292.80	56.05	448.40		
1 Brush Chipper, 12", 130 H.P.		396.40		436.04		
1 Crawler Loader, 3 C.Y.		1187.00		1305.70		
2 Chain Saws, Gas, 36" Long		90.00		99.00		
1 Dump Truck, 12 C.Y., 400 H.P.		692.40		761.64	42.25	46.47
56 L.H., Daily Totals		$4472.60		$5833.18	$79.87	$104.16
Crew B-8	**Hr.**	**Daily**	**Hr.**	**Daily**	**Bare Costs**	**Incl. O&P**
1 Labor Foreman (outside)	$37.45	$299.60	$57.65	$461.20	$39.85	$60.79
2 Laborers	35.45	567.20	54.60	873.60		
2 Equip. Oper. (med.)	47.50	760.00	71.75	1148.00		
1 Equip. Oper. Oiler	42.25	338.00	63.85	510.80		
2 Truck Drivers (heavy)	36.60	585.60	56.05	896.80		
1 Hyd. Crane, 25 Ton		728.60		801.46		
1 Crawler Loader, 3 C.Y.		1187.00		1305.70		
2 Dump Trucks, 12 C.Y., 400 H.P.		1384.80		1523.28	51.57	56.73
64 L.H., Daily Totals		$5850.80		$7520.84	$91.42	$117.51
Crew B-9	**Hr.**	**Daily**	**Hr.**	**Daily**	**Bare Costs**	**Incl. O&P**
1 Labor Foreman (outside)	$37.45	$299.60	$57.65	$461.20	$35.85	$55.21
4 Laborers	35.45	1134.40	54.60	1747.20		
1 Air Compressor, 250 cfm		195.80		215.38		
2 Breakers, Pavement, 60 lb.		19.60		21.56		
2 -50' Air Hoses, 1.5"		11.60		12.76	5.67	6.24
40 L.H., Daily Totals		$1661.00		$2458.10	$41.52	$61.45
Crew B-9A	**Hr.**	**Daily**	**Hr.**	**Daily**	**Bare Costs**	**Incl. O&P**
2 Laborers	$35.45	$567.20	$54.60	$873.60	$35.83	$55.08
1 Truck Driver (heavy)	36.60	292.80	56.05	448.40		
1 Water Tank Trailer, 5000 Gal.		143.20		157.52		
1 Truck Tractor, 220 H.P.		366.80		403.48		
2 -50' Discharge Hoses, 3"		6.10		6.71	21.50	23.65
24 L.H., Daily Totals		$1376.10		$1889.71	$57.34	$78.74
Crew B-9B	**Hr.**	**Daily**	**Hr.**	**Daily**	**Bare Costs**	**Incl. O&P**
2 Laborers	$35.45	$567.20	$54.60	$873.60	$35.83	$55.08
1 Truck Driver (heavy)	36.60	292.80	56.05	448.40		
2 -50' Discharge Hoses, 3"		6.10		6.71		
1 Water Tank Trailer, 5000 Gal.		143.20		157.52		
1 Truck Tractor, 220 H.P.		366.80		403.48		
1 Pressure Washer		69.40		76.34	24.40	26.84
24 L.H., Daily Totals		$1445.50		$1966.05	$60.23	$81.92
Crew B-9D	**Hr.**	**Daily**	**Hr.**	**Daily**	**Bare Costs**	**Incl. O&P**
1 Labor Foreman (Outside)	$37.45	$299.60	$57.65	$461.20	$35.85	$55.21
4 Common Laborers	35.45	1134.40	54.60	1747.20		
1 Air Compressor, 250 cfm		195.80		215.38		
2 -50' Air Hoses, 1.5"		11.60		12.76		
2 Air Powered Tampers		52.40		57.64	6.50	7.14
40 L.H., Daily Totals		$1693.80		$2494.18	$42.34	$62.35
Crew B-10	**Hr.**	**Daily**	**Hr.**	**Daily**	**Bare Costs**	**Incl. O&P**
1 Equip. Oper. (med.)	$47.50	$380.00	$71.75	$574.00	$43.48	$66.03
.5 Laborer	35.45	141.80	54.60	218.40		
12 L.H., Daily Totals		$521.80		$792.40	$43.48	$66.03

Crew No.	Bare Costs		Incl. Subs O&P		Cost Per Labor-Hour	
Crew B-10A	**Hr.**	**Daily**	**Hr.**	**Daily**	**Bare Costs**	**Incl. O&P**
1 Equip. Oper. (med.)	$47.50	$380.00	$71.75	$574.00	$43.48	$66.03
.5 Laborer	35.45	141.80	54.60	218.40		
1 Roller, 2-Drum, W.B., 7.5 H.P.		174.00		191.40	14.50	15.95
12 L.H., Daily Totals		$695.80		$983.80	$57.98	$81.98
Crew B-10B	**Hr.**	**Daily**	**Hr.**	**Daily**	**Bare Costs**	**Incl. O&P**
1 Equip. Oper. (med.)	$47.50	$380.00	$71.75	$574.00	$43.48	$66.03
.5 Laborer	35.45	141.80	54.60	218.40		
1 Dozer, 200 H.P.		1333.00		1466.30	111.08	122.19
12 L.H., Daily Totals		$1854.80		$2258.70	$154.57	$188.22
Crew B-10C	**Hr.**	**Daily**	**Hr.**	**Daily**	**Bare Costs**	**Incl. O&P**
1 Equip. Oper. (med.)	$47.50	$380.00	$71.75	$574.00	$43.48	$66.03
.5 Laborer	35.45	141.80	54.60	218.40		
1 Dozer, 200 H.P.		1333.00		1466.30		
1 Vibratory Roller, Towed, 23 Ton		407.20		447.92	145.02	159.52
12 L.H., Daily Totals		$2262.00		$2706.62	$188.50	$225.55
Crew B-10D	**Hr.**	**Daily**	**Hr.**	**Daily**	**Bare Costs**	**Incl. O&P**
1 Equip. Oper. (med.)	$47.50	$380.00	$71.75	$574.00	$43.48	$66.03
.5 Laborer	35.45	141.80	54.60	218.40		
1 Dozer, 200 H.P.		1333.00		1466.30		
1 Sheepsft. Roller, Towed		440.00		484.00	147.75	162.53
12 L.H., Daily Totals		$2294.80		$2742.70	$191.23	$228.56
Crew B-10E	**Hr.**	**Daily**	**Hr.**	**Daily**	**Bare Costs**	**Incl. O&P**
1 Equip. Oper. (med.)	$47.50	$380.00	$71.75	$574.00	$43.48	$66.03
.5 Laborer	35.45	141.80	54.60	218.40		
1 Tandem Roller, 5 Ton		154.40		169.84	12.87	14.15
12 L.H., Daily Totals		$676.20		$962.24	$56.35	$80.19
Crew B-10F	**Hr.**	**Daily**	**Hr.**	**Daily**	**Bare Costs**	**Incl. O&P**
1 Equip. Oper. (med.)	$47.50	$380.00	$71.75	$574.00	$43.48	$66.03
.5 Laborer	35.45	141.80	54.60	218.40		
1 Tandem Roller, 10 Ton		231.20		254.32	19.27	21.19
12 L.H., Daily Totals		$753.00		$1046.72	$62.75	$87.23
Crew B-10G	**Hr.**	**Daily**	**Hr.**	**Daily**	**Bare Costs**	**Incl. O&P**
1 Equip. Oper. (med.)	$47.50	$380.00	$71.75	$574.00	$43.48	$66.03
.5 Laborer	35.45	141.80	54.60	218.40		
1 Sheepsfoot Roller, 240 H.P.		1154.00		1269.40	96.17	105.78
12 L.H., Daily Totals		$1675.80		$2061.80	$139.65	$171.82
Crew B-10H	**Hr.**	**Daily**	**Hr.**	**Daily**	**Bare Costs**	**Incl. O&P**
1 Equip. Oper. (med.)	$47.50	$380.00	$71.75	$574.00	$43.48	$66.03
.5 Laborer	35.45	141.80	54.60	218.40		
1 Diaphragm Water Pump, 2"		71.60		78.76		
1 -20' Suction Hose, 2"		1.95		2.15		
2 -50' Discharge Hoses, 2"		3.50		3.85	6.42	7.06
12 L.H., Daily Totals		$598.85		$877.15	$49.90	$73.10
Crew B-10I	**Hr.**	**Daily**	**Hr.**	**Daily**	**Bare Costs**	**Incl. O&P**
1 Equip. Oper. (med.)	$47.50	$380.00	$71.75	$574.00	$43.48	$66.03
.5 Laborer	35.45	141.80	54.60	218.40		
1 Diaphragm Water Pump, 4"		114.40		125.84		
1 -20' Suction Hose, 4"		3.25		3.58		
2 -50' Discharge Hoses, 4"		9.40		10.34	10.59	11.65
12 L.H., Daily Totals		$648.85		$932.15	$54.07	$77.68

Crew No.	Bare Costs		Incl. Subs O&P		Cost Per Labor-Hour	
Crew B-10J	**Hr.**	**Daily**	**Hr.**	**Daily**	**Bare Costs**	**Incl. O&P**
1 Equip. Oper. (med.)	$47.50	$380.00	$71.75	$574.00	$43.48	$66.03
.5 Laborer	35.45	141.80	54.60	218.40		
1 Centrifugal Water Pump, 3"		78.80		86.68		
1 -20' Suction Hose, 3"		2.85		3.13		
2 -50' Discharge Hoses, 3"		6.10		6.71	7.31	8.04
12 L.H., Daily Totals		$609.55		$888.92	$50.80	$74.08

Crew B-10K	**Hr.**	**Daily**	**Hr.**	**Daily**	**Bare Costs**	**Incl. O&P**
1 Equip. Oper. (med.)	$47.50	$380.00	$71.75	$574.00	$43.48	$66.03
.5 Laborer	35.45	141.80	54.60	218.40		
1 Centr. Water Pump, 6"		341.40		375.54		
1 -20' Suction Hose, 6"		11.20		12.32		
2 -50' Discharge Hoses, 6"		24.20		26.62	31.40	34.54
12 L.H., Daily Totals		$898.60		$1206.88	$74.88	$100.57

Crew B-10L	**Hr.**	**Daily**	**Hr.**	**Daily**	**Bare Costs**	**Incl. O&P**
1 Equip. Oper. (med.)	$47.50	$380.00	$71.75	$574.00	$43.48	$66.03
.5 Laborer	35.45	141.80	54.60	218.40		
1 Dozer, 80 H.P.		480.60		528.66	40.05	44.06
12 L.H., Daily Totals		$1002.40		$1321.06	$83.53	$110.09

Crew B-10M	**Hr.**	**Daily**	**Hr.**	**Daily**	**Bare Costs**	**Incl. O&P**
1 Equip. Oper. (med.)	$47.50	$380.00	$71.75	$574.00	$43.48	$66.03
.5 Laborer	35.45	141.80	54.60	218.40		
1 Dozer, 300 H.P.		1829.00		2011.90	152.42	167.66
12 L.H., Daily Totals		$2350.80		$2804.30	$195.90	$233.69

Crew B-10N	**Hr.**	**Daily**	**Hr.**	**Daily**	**Bare Costs**	**Incl. O&P**
1 Equip. Oper. (med.)	$47.50	$380.00	$71.75	$574.00	$43.48	$66.03
.5 Laborer	35.45	141.80	54.60	218.40		
1 F.E. Loader, T.M., 1.5 C.Y		533.60		586.96	44.47	48.91
12 L.H., Daily Totals		$1055.40		$1379.36	$87.95	$114.95

Crew B-10O	**Hr.**	**Daily**	**Hr.**	**Daily**	**Bare Costs**	**Incl. O&P**
1 Equip. Oper. (med.)	$47.50	$380.00	$71.75	$574.00	$43.48	$66.03
.5 Laborer	35.45	141.80	54.60	218.40		
1 F.E. Loader, T.M., 2.25 C.Y.		955.00		1050.50	79.58	87.54
12 L.H., Daily Totals		$1476.80		$1842.90	$123.07	$153.57

Crew B-10P	**Hr.**	**Daily**	**Hr.**	**Daily**	**Bare Costs**	**Incl. O&P**
1 Equip. Oper. (med.)	$47.50	$380.00	$71.75	$574.00	$43.48	$66.03
.5 Laborer	35.45	141.80	54.60	218.40		
1 Crawler Loader, 3 C.Y.		1187.00		1305.70	98.92	108.81
12 L.H., Daily Totals		$1708.80		$2098.10	$142.40	$174.84

Crew B-10Q	**Hr.**	**Daily**	**Hr.**	**Daily**	**Bare Costs**	**Incl. O&P**
1 Equip. Oper. (med.)	$47.50	$380.00	$71.75	$574.00	$43.48	$66.03
.5 Laborer	35.45	141.80	54.60	218.40		
1 Crawler Loader, 4 C.Y.		1530.00		1683.00	127.50	140.25
12 L.H., Daily Totals		$2051.80		$2475.40	$170.98	$206.28

Crew B-10R	**Hr.**	**Daily**	**Hr.**	**Daily**	**Bare Costs**	**Incl. O&P**
1 Equip. Oper. (med.)	$47.50	$380.00	$71.75	$574.00	$43.48	$66.03
.5 Laborer	35.45	141.80	54.60	218.40		
1 F.E. Loader, W.M., 1 C.Y.		296.00		325.60	24.67	27.13
12 L.H., Daily Totals		$817.80		$1118.00	$68.15	$93.17

Crew No.	Bare Costs		Incl. Subs O&P		Cost Per Labor-Hour	
Crew B-10S	**Hr.**	**Daily**	**Hr.**	**Daily**	**Bare Costs**	**Incl. O&P**
1 Equip. Oper. (med.)	$47.50	$380.00	$71.75	$574.00	$43.48	$66.03
.5 Laborer	35.45	141.80	54.60	218.40		
1 F.E. Loader, W.M., 1.5 C.Y.		375.20		412.72	31.27	34.39
12 L.H., Daily Totals		$897.00		$1205.12	$74.75	$100.43

Crew B-10T	**Hr.**	**Daily**	**Hr.**	**Daily**	**Bare Costs**	**Incl. O&P**
1 Equip. Oper. (med.)	$47.50	$380.00	$71.75	$574.00	$43.48	$66.03
.5 Laborer	35.45	141.80	54.60	218.40		
1 F.E. Loader, W.M.,2.5 C.Y.		526.20		578.82	43.85	48.23
12 L.H., Daily Totals		$1048.00		$1371.22	$87.33	$114.27

Crew B-10U	**Hr.**	**Daily**	**Hr.**	**Daily**	**Bare Costs**	**Incl. O&P**
1 Equip. Oper. (med.)	$47.50	$380.00	$71.75	$574.00	$43.48	$66.03
.5 Laborer	35.45	141.80	54.60	218.40		
1 F.E. Loader, W.M., 5.5 C.Y.		1056.00		1161.60	88.00	96.80
12 L.H., Daily Totals		$1577.80		$1954.00	$131.48	$162.83

Crew B-10V	**Hr.**	**Daily**	**Hr.**	**Daily**	**Bare Costs**	**Incl. O&P**
1 Equip. Oper. (med.)	$47.50	$380.00	$71.75	$574.00	$43.48	$66.03
.5 Laborer	35.45	141.80	54.60	218.40		
1 Dozer, 700 H.P.		4459.00		4904.90	371.58	408.74
12 L.H., Daily Totals		$4980.80		$5697.30	$415.07	$474.77

Crew B-10W	**Hr.**	**Daily**	**Hr.**	**Daily**	**Bare Costs**	**Incl. O&P**
1 Equip. Oper. (med.)	$47.50	$380.00	$71.75	$574.00	$43.48	$66.03
.5 Laborer	35.45	141.80	54.60	218.40		
1 Dozer, 105 H.P.		592.40		651.64	49.37	54.30
12 L.H., Daily Totals		$1114.20		$1444.04	$92.85	$120.34

Crew B-10X	**Hr.**	**Daily**	**Hr.**	**Daily**	**Bare Costs**	**Incl. O&P**
1 Equip. Oper. (med.)	$47.50	$380.00	$71.75	$574.00	$43.48	$66.03
.5 Laborer	35.45	141.80	54.60	218.40		
1 Dozer, 410 H.P.		2425.00		2667.50	202.08	222.29
12 L.H., Daily Totals		$2946.80		$3459.90	$245.57	$288.32

Crew B-10Y	**Hr.**	**Daily**	**Hr.**	**Daily**	**Bare Costs**	**Incl. O&P**
1 Equip. Oper. (med.)	$47.50	$380.00	$71.75	$574.00	$43.48	$66.03
.5 Laborer	35.45	141.80	54.60	218.40		
1 Vibr. Roller, Towed, 12 Ton		529.40		582.34	44.12	48.53
12 L.H., Daily Totals		$1051.20		$1374.74	$87.60	$114.56

Crew B-11A	**Hr.**	**Daily**	**Hr.**	**Daily**	**Bare Costs**	**Incl. O&P**
1 Equipment Oper. (med.)	$47.50	$380.00	$71.75	$574.00	$41.48	$63.17
1 Laborer	35.45	283.60	54.60	436.80		
1 Dozer, 200 H.P.		1333.00		1466.30	83.31	91.64
16 L.H., Daily Totals		$1996.60		$2477.10	$124.79	$154.82

Crew B-11B	**Hr.**	**Daily**	**Hr.**	**Daily**	**Bare Costs**	**Incl. O&P**
1 Equipment Oper. (light)	$45.80	$366.40	$69.20	$553.60	$40.63	$61.90
1 Laborer	35.45	283.60	54.60	436.80		
1 Air Powered Tamper		26.20		28.82		
1 Air Compressor, 365 cfm		255.80		281.38		
2 -50' Air Hoses, 1.5"		11.60		12.76	18.35	20.18
16 L.H., Daily Totals		$943.60		$1313.36	$58.98	$82.08

Crew B-11C	**Hr.**	**Daily**	**Hr.**	**Daily**	**Bare Costs**	**Incl. O&P**
1 Equipment Oper. (med.)	$47.50	$380.00	$71.75	$574.00	$41.48	$63.17
1 Laborer	35.45	283.60	54.60	436.80		
1 Backhoe Loader, 48 H.P.		367.40		404.14	22.96	25.26
16 L.H., Daily Totals		$1031.00		$1414.94	$64.44	$88.43

Crew No.	Bare Costs		Incl. Subs O&P		Cost Per Labor-Hour	
Crew B-11J	**Hr.**	**Daily**	**Hr.**	**Daily**	**Bare Costs**	**Incl. O&P**
1 Equipment Oper. (med.)	$47.50	$380.00	$71.75	$574.00	$41.48	$63.17
1 Laborer	35.45	283.60	54.60	436.80		
1 Grader, 30,000 Lbs.		712.40		783.64		
1 Ripper, Beam & 1 Shank		82.00		90.20	49.65	54.62
16 L.H., Daily Totals		$1458.00		$1884.64	$91.13	$117.79

Crew B-11K	Hr.	Daily	Hr.	Daily	Bare Costs	Incl. O&P
1 Equipment Oper. (med.)	$47.50	$380.00	$71.75	$574.00	$41.48	$63.17
1 Laborer	35.45	283.60	54.60	436.80		
1 Trencher, Chain Type, 8' D		3357.00		3692.70	209.81	230.79
16 L.H., Daily Totals		$4020.60		$4703.50	$251.29	$293.97

Crew B-11L	Hr.	Daily	Hr.	Daily	Bare Costs	Incl. O&P
1 Equipment Oper. (med.)	$47.50	$380.00	$71.75	$574.00	$41.48	$63.17
1 Laborer	35.45	283.60	54.60	436.80		
1 Grader, 30,000 Lbs.		712.40		783.64	44.52	48.98
16 L.H., Daily Totals		$1376.00		$1794.44	$86.00	$112.15

Crew B-11M	Hr.	Daily	Hr.	Daily	Bare Costs	Incl. O&P
1 Equipment Oper. (med.)	$47.50	$380.00	$71.75	$574.00	$41.48	$63.17
1 Laborer	35.45	283.60	54.60	436.80		
1 Backhoe Loader, 80 H.P.		400.40		440.44	25.02	27.53
16 L.H., Daily Totals		$1064.00		$1451.24	$66.50	$90.70

Crew B-11N	Hr.	Daily	Hr.	Daily	Bare Costs	Incl. O&P
1 Labor Foreman (outside)	$37.45	$299.60	$57.65	$461.20	$39.12	$59.72
2 Equipment Operators (med.)	47.50	760.00	71.75	1148.00		
6 Truck Drivers (hvy.)	36.60	1756.80	56.05	2690.40		
1 F.E. Loader, W.M., 5.5 C.Y.		1056.00		1161.60		
1 Dozer, 410 H.P.		2425.00		2667.50		
6 Dump Trucks, Off Hwy., 50 Ton		10782.00		11860.20	198.10	217.91
72 L.H., Daily Totals		$17079.40		$19988.90	$237.21	$277.62

Crew B-11Q	Hr.	Daily	Hr.	Daily	Bare Costs	Incl. O&P
1 Equipment Operator (med.)	$47.50	$380.00	$71.75	$574.00	$43.48	$66.03
.5 Laborer	35.45	141.80	54.60	218.40		
1 Dozer, 140 H.P.		889.00		977.90	74.08	81.49
12 L.H., Daily Totals		$1410.80		$1770.30	$117.57	$147.53

Crew B-11R	Hr.	Daily	Hr.	Daily	Bare Costs	Incl. O&P
1 Equipment Operator (med.)	$47.50	$380.00	$71.75	$574.00	$43.48	$66.03
.5 Laborer	35.45	141.80	54.60	218.40		
1 Dozer, 200 H.P.		1333.00		1466.30	111.08	122.19
12 L.H., Daily Totals		$1854.80		$2258.70	$154.57	$188.22

Crew B-11S	Hr.	Daily	Hr.	Daily	Bare Costs	Incl. O&P
1 Equipment Operator (med.)	$47.50	$380.00	$71.75	$574.00	$43.48	$66.03
.5 Laborer	35.45	141.80	54.60	218.40		
1 Dozer, 300 H.P.		1829.00		2011.90		
1 Ripper, Beam & 1 Shank		82.00		90.20	159.25	175.18
12 L.H., Daily Totals		$2432.80		$2894.50	$202.73	$241.21

Crew B-11T	Hr.	Daily	Hr.	Daily	Bare Costs	Incl. O&P
1 Equipment Operator (med.)	$47.50	$380.00	$71.75	$574.00	$43.48	$66.03
.5 Laborer	35.45	141.80	54.60	218.40		
1 Dozer, 410 H.P.		2425.00		2667.50		
1 Ripper, Beam & 2 Shanks		92.20		101.42	209.77	230.74
12 L.H., Daily Totals		$3039.00		$3561.32	$253.25	$296.78

Crew B-11U	Hr.	Daily	Hr.	Daily	Bare Costs	Incl. O&P
1 Equipment Operator (med.)	$47.50	$380.00	$71.75	$574.00	$43.48	$66.03
.5 Laborer	35.45	141.80	54.60	218.40		
1 Dozer, 520 H.P.		3137.00		3450.70	261.42	287.56
12 L.H., Daily Totals		$3658.80		$4243.10	$304.90	$353.59

Crew B-11V	Hr.	Daily	Hr.	Daily	Bare Costs	Incl. O&P
3 Laborers	$35.45	$850.80	$54.60	$1310.40	$35.45	$54.60
1 Roller, 2-Drum, W.B., 7.5 H.P.		174.00		191.40	7.25	7.97
24 L.H., Daily Totals		$1024.80		$1501.80	$42.70	$62.58

Crew B-11W	Hr.	Daily	Hr.	Daily	Bare Costs	Incl. O&P
1 Equipment Operator (med.)	$47.50	$380.00	$71.75	$574.00	$37.41	$57.24
1 Common Laborer	35.45	283.60	54.60	436.80		
10 Truck Drivers (hvy.)	36.60	2928.00	56.05	4484.00		
1 Dozer, 200 H.P.		1333.00		1466.30		
1 Vibratory Roller, Towed, 23 Ton		407.20		447.92		
10 Dump Trucks, 8 C.Y., 220 H.P.		4166.00		4582.60	61.52	67.68
96 L.H., Daily Totals		$9497.80		$11991.62	$98.94	$124.91

Crew B-11Y	Hr.	Daily	Hr.	Daily	Bare Costs	Incl. O&P
1 Labor Foreman (Outside)	$37.45	$299.60	$57.65	$461.20	$39.69	$60.66
5 Common Laborers	35.45	1418.00	54.60	2184.00		
3 Equipment Operators (med.)	47.50	1140.00	71.75	1722.00		
1 Dozer, 80 H.P.		480.60		528.66		
2 Roller, 2-Drum, W.B., 7.5 H.P.		348.00		382.80		
4 Vibrating Plate, Gas, 21"		177.60		195.36	13.98	15.37
72 L.H., Daily Totals		$3863.80		$5474.02	$53.66	$76.03

Crew B-12A	Hr.	Daily	Hr.	Daily	Bare Costs	Incl. O&P
1 Equip. Oper. (crane)	$48.80	$390.40	$73.75	$590.00	$42.13	$64.17
1 Laborer	35.45	283.60	54.60	436.80		
1 Hyd. Excavator, 1 C.Y.		815.60		897.16	50.98	56.07
16 L.H., Daily Totals		$1489.60		$1923.96	$93.10	$120.25

Crew B-12B	Hr.	Daily	Hr.	Daily	Bare Costs	Incl. O&P
1 Equip. Oper. (crane)	$48.80	$390.40	$73.75	$590.00	$42.13	$64.17
1 Laborer	35.45	283.60	54.60	436.80		
1 Hyd. Excavator, 1.5 C.Y.		1036.00		1139.60	64.75	71.22
16 L.H., Daily Totals		$1710.00		$2166.40	$106.88	$135.40

Crew B-12C	Hr.	Daily	Hr.	Daily	Bare Costs	Incl. O&P
1 Equip. Oper. (crane)	$48.80	$390.40	$73.75	$590.00	$42.13	$64.17
1 Laborer	35.45	283.60	54.60	436.80		
1 Hyd. Excavator, 2 C.Y.		1363.00		1499.30	85.19	93.71
16 L.H., Daily Totals		$2037.00		$2526.10	$127.31	$157.88

Crew B-12D	Hr.	Daily	Hr.	Daily	Bare Costs	Incl. O&P
1 Equip. Oper. (crane)	$48.80	$390.40	$73.75	$590.00	$42.13	$64.17
1 Laborer	35.45	283.60	54.60	436.80		
1 Hyd. Excavator, 3.5 C.Y.		2454.00		2699.40	153.38	168.71
16 L.H., Daily Totals		$3128.00		$3726.20	$195.50	$232.89

Crew B-12E	Hr.	Daily	Hr.	Daily	Bare Costs	Incl. O&P
1 Equip. Oper. (crane)	$48.80	$390.40	$73.75	$590.00	$42.13	$64.17
1 Laborer	35.45	283.60	54.60	436.80		
1 Hyd. Excavator, .5 C.Y.		400.40		440.44	25.02	27.53
16 L.H., Daily Totals		$1074.40		$1467.24	$67.15	$91.70

Crew No.	Bare Costs		Incl. Subs O&P		Cost Per Labor-Hour	

Crew B-12F	Hr.	Daily	Hr.	Daily	Bare Costs	Incl. O&P
1 Equip. Oper. (crane)	$48.80	$390.40	$73.75	$590.00	$42.13	$64.17
1 Laborer	35.45	283.60	54.60	436.80		
1 Hyd. Excavator, .75 C.Y.		694.80		764.28	43.42	47.77
16 L.H., Daily Totals		$1368.80		$1791.08	$85.55	$111.94

Crew B-12G	Hr.	Daily	Hr.	Daily	Bare Costs	Incl. O&P
1 Equip. Oper. (crane)	$48.80	$390.40	$73.75	$590.00	$42.13	$64.17
1 Laborer	35.45	283.60	54.60	436.80		
1 Crawler Crane, 15 Ton		668.70		735.57		
1 Clamshell Bucket, .5 C.Y.		37.80		41.58	44.16	48.57
16 L.H., Daily Totals		$1380.50		$1803.95	$86.28	$112.75

Crew B-12H	Hr.	Daily	Hr.	Daily	Bare Costs	Incl. O&P
1 Equip. Oper. (crane)	$48.80	$390.40	$73.75	$590.00	$42.13	$64.17
1 Laborer	35.45	283.60	54.60	436.80		
1 Crawler Crane, 25 Ton		1148.00		1262.80		
1 Clamshell Bucket, 1 C.Y.		47.40		52.14	74.71	82.18
16 L.H., Daily Totals		$1869.40		$2341.74	$116.84	$146.36

Crew B-12I	Hr.	Daily	Hr.	Daily	Bare Costs	Incl. O&P
1 Equip. Oper. (crane)	$48.80	$390.40	$73.75	$590.00	$42.13	$64.17
1 Laborer	35.45	283.60	54.60	436.80		
1 Crawler Crane, 20 Ton		861.60		947.76		
1 Dragline Bucket, .75 C.Y.		20.80		22.88	55.15	60.66
16 L.H., Daily Totals		$1556.40		$1997.44	$97.28	$124.84

Crew B-12J	Hr.	Daily	Hr.	Daily	Bare Costs	Incl. O&P
1 Equip. Oper. (crane)	$48.80	$390.40	$73.75	$590.00	$42.13	$64.17
1 Laborer	35.45	283.60	54.60	436.80		
1 Gradall, 5/8 C.Y.		883.80		972.18	55.24	60.76
16 L.H., Daily Totals		$1557.80		$1998.98	$97.36	$124.94

Crew B-12K	Hr.	Daily	Hr.	Daily	Bare Costs	Incl. O&P
1 Equip. Oper. (crane)	$48.80	$390.40	$73.75	$590.00	$42.13	$64.17
1 Laborer	35.45	283.60	54.60	436.80		
1 Gradall, 3 Ton, 1 C.Y.		1010.00		1111.00	63.13	69.44
16 L.H., Daily Totals		$1684.00		$2137.80	$105.25	$133.61

Crew B-12L	Hr.	Daily	Hr.	Daily	Bare Costs	Incl. O&P
1 Equip. Oper. (crane)	$48.80	$390.40	$73.75	$590.00	$42.13	$64.17
1 Laborer	35.45	283.60	54.60	436.80		
1 Crawler Crane, 15 Ton		668.70		735.57		
1 F.E. Attachment, .5 C.Y.		58.20		64.02	45.43	49.97
16 L.H., Daily Totals		$1400.90		$1826.39	$87.56	$114.15

Crew B-12M	Hr.	Daily	Hr.	Daily	Bare Costs	Incl. O&P
1 Equip. Oper. (crane)	$48.80	$390.40	$73.75	$590.00	$42.13	$64.17
1 Laborer	35.45	283.60	54.60	436.80		
1 Crawler Crane, 20 Ton		861.60		947.76		
1 F.E. Attachment, .75 C.Y.		62.60		68.86	57.76	63.54
16 L.H., Daily Totals		$1598.20		$2043.42	$99.89	$127.71

Crew B-12N	Hr.	Daily	Hr.	Daily	Bare Costs	Incl. O&P
1 Equip. Oper. (crane)	$48.80	$390.40	$73.75	$590.00	$42.13	$64.17
1 Laborer	35.45	283.60	54.60	436.80		
1 Crawler Crane, 25 Ton		1148.00		1262.80		
1 F.E. Attachment, 1 C.Y.		69.40		76.34	76.09	83.70
16 L.H., Daily Totals		$1891.40		$2365.94	$118.21	$147.87

Crew B-12O	Hr.	Daily	Hr.	Daily	Bare Costs	Incl. O&P
1 Equip. Oper. (crane)	$48.80	$390.40	$73.75	$590.00	$42.13	$64.17
1 Laborer	35.45	283.60	54.60	436.80		
1 Crawler Crane, 40 Ton		1152.00		1267.20		
1 F.E. Attachment, 1.5 C.Y.		78.60		86.46	76.91	84.60
16 L.H., Daily Totals		$1904.60		$2380.46	$119.04	$148.78

Crew B-12P	Hr.	Daily	Hr.	Daily	Bare Costs	Incl. O&P
1 Equip. Oper. (crane)	$48.80	$390.40	$73.75	$590.00	$42.13	$64.17
1 Laborer	35.45	283.60	54.60	436.80		
1 Crawler Crane, 40 Ton		1152.00		1267.20		
1 Dragline Bucket, 1.5 C.Y.		33.60		36.96	74.10	81.51
16 L.H., Daily Totals		$1859.60		$2330.96	$116.22	$145.69

Crew B-12Q	Hr.	Daily	Hr.	Daily	Bare Costs	Incl. O&P
1 Equip. Oper. (crane)	$48.80	$390.40	$73.75	$590.00	$42.13	$64.17
1 Laborer	35.45	283.60	54.60	436.80		
1 Hyd. Excavator, 5/8 C.Y.		596.00		655.60	37.25	40.98
16 L.H., Daily Totals		$1270.00		$1682.40	$79.38	$105.15

Crew B-12S	Hr.	Daily	Hr.	Daily	Bare Costs	Incl. O&P
1 Equip. Oper. (crane)	$48.80	$390.40	$73.75	$590.00	$42.13	$64.17
1 Laborer	35.45	283.60	54.60	436.80		
1 Hyd. Excavator, 2.5 C.Y.		1800.00		1980.00	112.50	123.75
16 L.H., Daily Totals		$2474.00		$3006.80	$154.63	$187.93

Crew B-12T	Hr.	Daily	Hr.	Daily	Bare Costs	Incl. O&P
1 Equip. Oper. (crane)	$48.80	$390.40	$73.75	$590.00	$42.13	$64.17
1 Laborer	35.45	283.60	54.60	436.80		
1 Crawler Crane, 75 Ton		1457.00		1602.70		
1 F.E. Attachment, 3 C.Y.		101.00		111.10	97.38	107.11
16 L.H., Daily Totals		$2232.00		$2740.60	$139.50	$171.29

Crew B-12V	Hr.	Daily	Hr.	Daily	Bare Costs	Incl. O&P
1 Equip. Oper. (crane)	$48.80	$390.40	$73.75	$590.00	$42.13	$64.17
1 Laborer	35.45	283.60	54.60	436.80		
1 Crawler Crane, 75 Ton		1457.00		1602.70		
1 Dragline Bucket, 3 C.Y.		52.20		57.42	94.33	103.76
16 L.H., Daily Totals		$2183.20		$2686.92	$136.45	$167.93

Crew B-12Y	Hr.	Daily	Hr.	Daily	Bare Costs	Incl. O&P
1 Equip. Oper. (crane)	$48.80	$390.40	$73.75	$590.00	$39.90	$60.98
2 Laborers	35.45	567.20	54.60	873.60		
1 Hyd. Excavator, 3.5 C.Y.		2454.00		2699.40	102.25	112.47
24 L.H., Daily Totals		$3411.60		$4163.00	$142.15	$173.46

Crew B-12Z	Hr.	Daily	Hr.	Daily	Bare Costs	Incl. O&P
1 Equip. Oper. (crane)	$48.80	$390.40	$73.75	$590.00	$39.90	$60.98
2 Laborers	35.45	567.20	54.60	873.60		
1 Hyd. Excavator, 2.5 C.Y.		1800.00		1980.00	75.00	82.50
24 L.H., Daily Totals		$2757.60		$3443.60	$114.90	$143.48

Crew B-13	Hr.	Daily	Hr.	Daily	Bare Costs	Incl. O&P
1 Labor Foreman (outside)	$37.45	$299.60	$57.65	$461.20	$38.61	$59.09
4 Laborers	35.45	1134.40	54.60	1747.20		
1 Equip. Oper. (crane)	48.80	390.40	73.75	590.00		
1 Equip. Oper. Oiler	42.25	338.00	63.85	510.80		
1 Hyd. Crane, 25 Ton		728.60		801.46	13.01	14.31
56 L.H., Daily Totals		$2891.00		$4110.66	$51.63	$73.40

Crew B-13A	Hr.	Daily	Hr.	Daily	Bare Costs	Incl. O&P
1 Labor Foreman (outside)	$37.45	$299.60	$57.65	$461.20	$39.51	$60.35
2 Laborers	35.45	567.20	54.60	873.60		
2 Equipment Operators (med.)	47.50	760.00	71.75	1148.00		
2 Truck Drivers (heavy)	36.60	585.60	56.05	896.80		
1 Crawler Crane, 75 Ton		1457.00		1602.70		
1 Crawler Loader, 4 C.Y.		1530.00		1683.00		
2 Dump Trucks, 8 C.Y., 220 H.P.		833.20		916.52	68.22	75.04
56 L.H., Daily Totals		$6032.60		$7581.82	$107.72	$135.39

Crew B-13B	Hr.	Daily	Hr.	Daily	Bare Costs	Incl. O&P
1 Labor Foreman (outside)	$37.45	$299.60	$57.65	$461.20	$38.61	$59.09
4 Laborers	35.45	1134.40	54.60	1747.20		
1 Equip. Oper. (crane)	48.80	390.40	73.75	590.00		
1 Equip. Oper. Oiler	42.25	338.00	63.85	510.80		
1 Hyd. Crane, 55 Ton		1115.00		1226.50	19.91	21.90
56 L.H., Daily Totals		$3277.40		$4535.70	$58.52	$80.99

Crew B-13C	Hr.	Daily	Hr.	Daily	Bare Costs	Incl. O&P
1 Labor Foreman (outside)	$37.45	$299.60	$57.65	$461.20	$38.61	$59.09
4 Laborers	35.45	1134.40	54.60	1747.20		
1 Equip. Oper. (crane)	48.80	390.40	73.75	590.00		
1 Equip. Oper. Oiler	42.25	338.00	63.85	510.80		
1 Crawler Crane, 100 Ton		1666.00		1832.60	29.75	32.73
56 L.H., Daily Totals		$3828.40		$5141.80	$68.36	$91.82

Crew B-13D	Hr.	Daily	Hr.	Daily	Bare Costs	Incl. O&P
1 Laborer	$35.45	$283.60	$54.60	$436.80	$42.13	$64.17
1 Equip. Oper. (crane)	48.80	390.40	73.75	590.00		
1 Hyd. Excavator, 1 C.Y.		815.60		897.16		
1 Trench Box		113.40		124.74	58.06	63.87
16 L.H., Daily Totals		$1603.00		$2048.70	$100.19	$128.04

Crew B-13E	Hr.	Daily	Hr.	Daily	Bare Costs	Incl. O&P
1 Laborer	$35.45	$283.60	$54.60	$436.80	$42.13	$64.17
1 Equip. Oper. (crane)	48.80	390.40	73.75	590.00		
1 Hyd. Excavator, 1.5 C.Y.		1036.00		1139.60		
1 Trench Box		113.40		124.74	71.84	79.02
16 L.H., Daily Totals		$1823.40		$2291.14	$113.96	$143.20

Crew B-13F	Hr.	Daily	Hr.	Daily	Bare Costs	Incl. O&P
1 Laborer	$35.45	$283.60	$54.60	$436.80	$42.13	$64.17
1 Equip. Oper. (crane)	48.80	390.40	73.75	590.00		
1 Hyd. Excavator, 3.5 C.Y.		2454.00		2699.40		
1 Trench Box		113.40		124.74	160.46	176.51
16 L.H., Daily Totals		$3241.40		$3850.94	$202.59	$240.68

Crew B-13G	Hr.	Daily	Hr.	Daily	Bare Costs	Incl. O&P
1 Laborer	$35.45	$283.60	$54.60	$436.80	$42.13	$64.17
1 Equip. Oper. (crane)	48.80	390.40	73.75	590.00		
1 Hyd. Excavator, .75 C.Y.		694.80		764.28		
1 Trench Box		113.40		124.74	50.51	55.56
16 L.H., Daily Totals		$1482.20		$1915.82	$92.64	$119.74

Crew B-13H	Hr.	Daily	Hr.	Daily	Bare Costs	Incl. O&P
1 Laborer	$35.45	$283.60	$54.60	$436.80	$42.13	$64.17
1 Equip. Oper. (crane)	48.80	390.40	73.75	590.00		
1 Gradall, 5/8 C.Y.		883.80		972.18		
1 Trench Box		113.40		124.74	62.33	68.56
16 L.H., Daily Totals		$1671.20		$2123.72	$104.45	$132.73

Crew B-13I	Hr.	Daily	Hr.	Daily	Bare Costs	Incl. O&P
1 Laborer	$35.45	$283.60	$54.60	$436.80	$42.13	$64.17
1 Equip. Oper. (crane)	48.80	390.40	73.75	590.00		
1 Gradall, 3 Ton, 1 C.Y.		1010.00		1111.00		
1 Trench Box		113.40		124.74	70.21	77.23
16 L.H., Daily Totals		$1797.40		$2262.54	$112.34	$141.41

Crew B-13J	Hr.	Daily	Hr.	Daily	Bare Costs	Incl. O&P
1 Laborer	$35.45	$283.60	$54.60	$436.80	$42.13	$64.17
1 Equip. Oper. (crane)	48.80	390.40	73.75	590.00		
1 Hyd. Excavator, 2.5 C.Y.		1800.00		1980.00		
1 Trench Box		113.40		124.74	119.59	131.55
16 L.H., Daily Totals		$2587.40		$3131.54	$161.71	$195.72

Crew B-14	Hr.	Daily	Hr.	Daily	Bare Costs	Incl. O&P
1 Labor Foreman (outside)	$37.45	$299.60	$57.65	$461.20	$37.51	$57.54
4 Laborers	35.45	1134.40	54.60	1747.20		
1 Equip. Oper. (light)	45.80	366.40	69.20	553.60		
1 Backhoe Loader, 48 H.P.		367.40		404.14	7.65	8.42
48 L.H., Daily Totals		$2167.80		$3166.14	$45.16	$65.96

Crew B-14A	Hr.	Daily	Hr.	Daily	Bare Costs	Incl. O&P
1 Equip. Oper. (crane)	$48.80	$390.40	$73.75	$590.00	$44.35	$67.37
.5 Laborer	35.45	141.80	54.60	218.40		
1 Hyd. Excavator, 4.5 C.Y.		3057.00		3362.70	254.75	280.23
12 L.H., Daily Totals		$3589.20		$4171.10	$299.10	$347.59

Crew B-14B	Hr.	Daily	Hr.	Daily	Bare Costs	Incl. O&P
1 Equip. Oper. (crane)	$48.80	$390.40	$73.75	$590.00	$44.35	$67.37
.5 Laborer	35.45	141.80	54.60	218.40		
1 Hyd. Excavator, 6 C.Y.		3541.00		3895.10	295.08	324.59
12 L.H., Daily Totals		$4073.20		$4703.50	$339.43	$391.96

Crew B-14C	Hr.	Daily	Hr.	Daily	Bare Costs	Incl. O&P
1 Equip. Oper. (crane)	$48.80	$390.40	$73.75	$590.00	$44.35	$67.37
.5 Laborer	35.45	141.80	54.60	218.40		
1 Hyd. Excavator, 7 C.Y.		3616.00		3977.60	301.33	331.47
12 L.H., Daily Totals		$4148.20		$4786.00	$345.68	$398.83

Crew B-14F	Hr.	Daily	Hr.	Daily	Bare Costs	Incl. O&P
1 Equip. Oper. (crane)	$48.80	$390.40	$73.75	$590.00	$44.35	$67.37
.5 Laborer	35.45	141.80	54.60	218.40		
1 Hyd. Shovel, 7 C.Y.		3495.00		3844.50	291.25	320.38
12 L.H., Daily Totals		$4027.20		$4652.90	$335.60	$387.74

Crew B-14G	Hr.	Daily	Hr.	Daily	Bare Costs	Incl. O&P
1 Equip. Oper. (crane)	$48.80	$390.40	$73.75	$590.00	$44.35	$67.37
.5 Laborer	35.45	141.80	54.60	218.40		
1 Hyd. Shovel, 12 C.Y.		4676.00		5143.60	389.67	428.63
12 L.H., Daily Totals		$5208.20		$5952.00	$434.02	$496.00

Crew B-14J	Hr.	Daily	Hr.	Daily	Bare Costs	Incl. O&P
1 Equip. Oper. (med.)	$47.50	$380.00	$71.75	$574.00	$43.48	$66.03
.5 Laborer	35.45	141.80	54.60	218.40		
1 F.E. Loader, 8 C.Y.		2008.00		2208.80	167.33	184.07
12 L.H., Daily Totals		$2529.80		$3001.20	$210.82	$250.10

Crew No.	Bare Costs		Incl. Subs O&P		Cost Per Labor-Hour	
Crew B-14K	**Hr.**	**Daily**	**Hr.**	**Daily**	**Bare Costs**	**Incl. O&P**
1 Equip. Oper. (med.)	$47.50	$380.00	$71.75	$574.00	$43.48	$66.03
.5 Laborer	35.45	141.80	54.60	218.40		
1 F.E. Loader, 10 C.Y.		2734.00		3007.40	227.83	250.62
12 L.H., Daily Totals		$3255.80		$3799.80	$271.32	$316.65

Crew B-15	Hr.	Daily	Hr.	Daily	Bare Costs	Incl. O&P
1 Equipment Oper. (med.)	$47.50	$380.00	$71.75	$574.00	$39.55	$60.33
.5 Laborer	35.45	141.80	54.60	218.40		
2 Truck Drivers (heavy)	36.60	585.60	56.05	896.80		
2 Dump Trucks, 12 C.Y., 400 H.P.		1384.80		1523.28		
1 Dozer, 200 H.P.		1333.00		1466.30	97.06	106.77
28 L.H., Daily Totals		$3825.20		$4678.78	$136.61	$167.10

Crew B-16	Hr.	Daily	Hr.	Daily	Bare Costs	Incl. O&P
1 Labor Foreman (outside)	$37.45	$299.60	$57.65	$461.20	$36.24	$55.73
2 Laborers	35.45	567.20	54.60	873.60		
1 Truck Driver (heavy)	36.60	292.80	56.05	448.40		
1 Dump Truck, 12 C.Y., 400 H.P.		692.40		761.64	21.64	23.80
32 L.H., Daily Totals		$1852.00		$2544.84	$57.88	$79.53

Crew B-17	Hr.	Daily	Hr.	Daily	Bare Costs	Incl. O&P
2 Laborers	$35.45	$567.20	$54.60	$873.60	$38.33	$58.61
1 Equip. Oper. (light)	45.80	366.40	69.20	553.60		
1 Truck Driver (heavy)	36.60	292.80	56.05	448.40		
1 Backhoe Loader, 48 H.P.		367.40		404.14		
1 Dump Truck, 8 C.Y., 220 H.P.		416.60		458.26	24.50	26.95
32 L.H., Daily Totals		$2010.40		$2738.00	$62.83	$85.56

Crew B-17A	Hr.	Daily	Hr.	Daily	Bare Costs	Incl. O&P
2 Labor Foremen (outside)	$37.45	$599.20	$57.65	$922.40	$38.20	$58.88
6 Laborers	35.45	1701.60	54.60	2620.80		
1 Skilled Worker Foreman (out)	48.20	385.60	74.50	596.00		
1 Skilled Worker	46.20	369.60	71.45	571.60		
80 L.H., Daily Totals		$3056.00		$4710.80	$38.20	$58.88

Crew B-17B	Hr.	Daily	Hr.	Daily	Bare Costs	Incl. O&P
2 Laborers	$35.45	$567.20	$54.60	$873.60	$38.33	$58.61
1 Equip. Oper. (light)	45.80	366.40	69.20	553.60		
1 Truck Driver (heavy)	36.60	292.80	56.05	448.40		
1 Backhoe Loader, 48 H.P.		367.40		404.14		
1 Dump Truck, 12 C.Y., 400 H.P.		692.40		761.64	33.12	36.43
32 L.H., Daily Totals		$2286.20		$3041.38	$71.44	$95.04

Crew B-18	Hr.	Daily	Hr.	Daily	Bare Costs	Incl. O&P
1 Labor Foreman (outside)	$37.45	$299.60	$57.65	$461.20	$36.12	$55.62
2 Laborers	35.45	567.20	54.60	873.60		
1 Vibrating Plate, Gas, 21"		44.40		48.84	1.85	2.04
24 L.H., Daily Totals		$911.20		$1383.64	$37.97	$57.65

Crew B-19	Hr.	Daily	Hr.	Daily	Bare Costs	Incl. O&P
1 Pile Driver Foreman (outside)	$45.15	$361.20	$71.90	$575.20	$44.70	$69.76
4 Pile Drivers	43.15	1380.80	68.70	2198.40		
2 Equip. Oper. (crane)	48.80	780.80	73.75	1180.00		
1 Equip. Oper. Oiler	42.25	338.00	63.85	510.80		
1 Crawler Crane, 40 Ton		1152.00		1267.20		
1 Lead, 90' High		124.80		137.28		
1 Hammer, Diesel, 22k ft-lb		618.00		679.80	29.61	32.57
64 L.H., Daily Totals		$4755.60		$6548.68	$74.31	$102.32

Crew B-19A	Hr.	Daily	Hr.	Daily	Bare Costs	Incl. O&P
1 Pile Driver Foreman (outside)	$45.15	$361.20	$71.90	$575.20	$44.70	$69.76
4 Pile Drivers	43.15	1380.80	68.70	2198.40		
2 Equip. Oper. (crane)	48.80	780.80	73.75	1180.00		
1 Equip. Oper. Oiler	42.25	338.00	63.85	510.80		
1 Crawler Crane, 75 Ton		1457.00		1602.70		
1 Lead, 90' high		124.80		137.28		
1 Hammer, Diesel, 41k ft-lb		704.00		774.40	35.72	39.29
64 L.H., Daily Totals		$5146.60		$6978.78	$80.42	$109.04

Crew B-19B	Hr.	Daily	Hr.	Daily	Bare Costs	Incl. O&P
1 Pile Driver Foreman (outside)	$45.15	$361.20	$71.90	$575.20	$44.70	$69.76
4 Pile Drivers	43.15	1380.80	68.70	2198.40		
2 Equip. Oper. (crane)	48.80	780.80	73.75	1180.00		
1 Equip. Oper. Oiler	42.25	338.00	63.85	510.80		
1 Crawler Crane, 40 Ton		1152.00		1267.20		
1 Lead, 90' High		124.80		137.28		
1 Hammer, Diesel, 22k ft-lb		618.00		679.80		
1 Barge, 400 Ton		783.80		862.18	41.85	46.04
64 L.H., Daily Totals		$5539.40		$7410.86	$86.55	$115.79

Crew B-19C	Hr.	Daily	Hr.	Daily	Bare Costs	Incl. O&P
1 Pile Driver Foreman (outside)	$45.15	$361.20	$71.90	$575.20	$44.70	$69.76
4 Pile Drivers	43.15	1380.80	68.70	2198.40		
2 Equip. Oper. (crane)	48.80	780.80	73.75	1180.00		
1 Equip. Oper. Oiler	42.25	338.00	63.85	510.80		
1 Crawler Crane, 75 Ton		1457.00		1602.70		
1 Lead, 90' High		124.80		137.28		
1 Hammer, Diesel, 41k ft-lb		704.00		774.40		
1 Barge, 400 Ton		783.80		862.18	47.96	52.76
64 L.H., Daily Totals		$5930.40		$7840.96	$92.66	$122.52

Crew B-20	Hr.	Daily	Hr.	Daily	Bare Costs	Incl. O&P
1 Labor Foreman (outside)	$37.45	$299.60	$57.65	$461.20	$39.70	$61.23
1 Skilled Worker	46.20	369.60	71.45	571.60		
1 Laborer	35.45	283.60	54.60	436.80		
24 L.H., Daily Totals		$952.80		$1469.60	$39.70	$61.23

Crew B-20A	Hr.	Daily	Hr.	Daily	Bare Costs	Incl. O&P
1 Labor Foreman (outside)	$37.45	$299.60	$57.65	$461.20	$43.34	$65.89
1 Laborer	35.45	283.60	54.60	436.80		
1 Plumber	55.80	446.40	84.05	672.40		
1 Plumber Apprentice	44.65	357.20	67.25	538.00		
32 L.H., Daily Totals		$1386.80		$2108.40	$43.34	$65.89

Crew B-21	Hr.	Daily	Hr.	Daily	Bare Costs	Incl. O&P
1 Labor Foreman (outside)	$37.45	$299.60	$57.65	$461.20	$41.00	$63.02
1 Skilled Worker	46.20	369.60	71.45	571.60		
1 Laborer	35.45	283.60	54.60	436.80		
.5 Equip. Oper. (crane)	48.80	195.20	73.75	295.00		
.5 S.P. Crane, 4x4, 5 Ton		137.10		150.81	4.90	5.39
28 L.H., Daily Totals		$1285.10		$1915.41	$45.90	$68.41

Crew B-21A	Hr.	Daily	Hr.	Daily	Bare Costs	Incl. O&P
1 Labor Foreman (outside)	$37.45	$299.60	$57.65	$461.20	$44.43	$67.46
1 Laborer	35.45	283.60	54.60	436.80		
1 Plumber	55.80	446.40	84.05	672.40		
1 Plumber Apprentice	44.65	357.20	67.25	538.00		
1 Equip. Oper. (crane)	48.80	390.40	73.75	590.00		
1 S.P. Crane, 4x4, 12 Ton		471.00		518.10	11.78	12.95
40 L.H., Daily Totals		$2248.20		$3216.50	$56.20	$80.41

Crew No.	Bare Costs		Incl. Subs O&P		Cost Per Labor-Hour	
Crew B-21B	**Hr.**	**Daily**	**Hr.**	**Daily**	**Bare Costs**	**Incl. O&P**
1 Labor Foreman (outside)	$37.45	$299.60	$57.65	$461.20	$38.52	$59.04
3 Laborers	35.45	850.80	54.60	1310.40		
1 Equip. Oper. (crane)	48.80	390.40	73.75	590.00		
1 Hyd. Crane, 12 Ton		646.20		710.82	16.16	17.77
40 L.H., Daily Totals		$2187.00		$3072.42	$54.67	$76.81

Crew B-21C	Hr.	Daily	Hr.	Daily	Bare Costs	Incl. O&P
1 Labor Foreman (outside)	$37.45	$299.60	$57.65	$461.20	$38.61	$59.09
4 Laborers	35.45	1134.40	54.60	1747.20		
1 Equip. Oper. (crane)	48.80	390.40	73.75	590.00		
1 Equip. Oper. Oiler	42.25	338.00	63.85	510.80		
2 Cutting Torches		22.80		25.08		
2 Sets of Gases		266.40		293.04		
1 Lattice Boom Crane, 90 Ton		1525.00		1677.50	32.40	35.64
56 L.H., Daily Totals		$3976.60		$5304.82	$71.01	$94.73

Crew B-22	Hr.	Daily	Hr.	Daily	Bare Costs	Incl. O&P
1 Labor Foreman (outside)	$37.45	$299.60	$57.65	$461.20	$41.52	$63.74
1 Skilled Worker	46.20	369.60	71.45	571.60		
1 Laborer	35.45	283.60	54.60	436.80		
.75 Equip. Oper. (crane)	48.80	292.80	73.75	442.50		
.75 S.P. Crane, 4x4, 5 Ton		205.65		226.22	6.86	7.54
30 L.H., Daily Totals		$1451.25		$2138.32	$48.38	$71.28

Crew B-22A	Hr.	Daily	Hr.	Daily	Bare Costs	Incl. O&P
1 Labor Foreman (outside)	$37.45	$299.60	$57.65	$461.20	$40.67	$62.41
1 Skilled Worker	46.20	369.60	71.45	571.60		
2 Laborers	35.45	567.20	54.60	873.60		
1 Equipment Oper. (crane)	48.80	390.40	73.75	590.00		
1 S.P. Crane, 4x4, 5 Ton		274.20		301.62		
1 Butt Fusion Mach., 4"-12" diam.		382.60		420.86	16.42	18.06
40 L.H., Daily Totals		$2283.60		$3218.88	$57.09	$80.47

Crew B-22B	Hr.	Daily	Hr.	Daily	Bare Costs	Incl. O&P
1 Labor Foreman (outside)	$37.45	$299.60	$57.65	$461.20	$40.67	$62.41
1 Skilled Worker	46.20	369.60	71.45	571.60		
2 Laborers	35.45	567.20	54.60	873.60		
1 Equip. Oper. (crane)	48.80	390.40	73.75	590.00		
1 S.P. Crane, 4x4, 5 Ton		274.20		301.62		
1 Butt Fusion Mach., 8"-24" diam.		831.70		914.87	27.65	30.41
40 L.H., Daily Totals		$2732.70		$3712.89	$68.32	$92.82

Crew B-22C	Hr.	Daily	Hr.	Daily	Bare Costs	Incl. O&P
1 Skilled Worker	$46.20	$369.60	$71.45	$571.60	$40.83	$63.02
1 Laborer	35.45	283.60	54.60	436.80		
1 Butt Fusion Mach., 2"-8" diam.		120.95		133.04	7.56	8.32
16 L.H., Daily Totals		$774.15		$1141.44	$48.38	$71.34

Crew B-23	Hr.	Daily	Hr.	Daily	Bare Costs	Incl. O&P
1 Labor Foreman (outside)	$37.45	$299.60	$57.65	$461.20	$35.85	$55.21
4 Laborers	35.45	1134.40	54.60	1747.20		
1 Drill Rig, Truck-Mounted		2469.00		2715.90		
1 Flatbed Truck, Gas, 3 Ton		327.80		360.58	69.92	76.91
40 L.H., Daily Totals		$4230.80		$5284.88	$105.77	$132.12

Crew B-23A	Hr.	Daily	Hr.	Daily	Bare Costs	Incl. O&P
1 Labor Foreman (outside)	$37.45	$299.60	$57.65	$461.20	$40.13	$61.33
1 Laborer	35.45	283.60	54.60	436.80		
1 Equip. Operator (med.)	47.50	380.00	71.75	574.00		
1 Drill Rig, Truck-Mounted		2469.00		2715.90		
1 Pickup Truck, 3/4 Ton		154.60		170.06	109.32	120.25
24 L.H., Daily Totals		$3586.80		$4357.96	$149.45	$181.58

Crew B-23B	Hr.	Daily	Hr.	Daily	Bare Costs	Incl. O&P
1 Labor Foreman (outside)	$37.45	$299.60	$57.65	$461.20	$40.13	$61.33
1 Laborer	35.45	283.60	54.60	436.80		
1 Equip. Operator (med.)	47.50	380.00	71.75	574.00		
1 Drill Rig, Truck-Mounted		2469.00		2715.90		
1 Pickup Truck, 3/4 Ton		154.60		170.06		
1 Centr. Water Pump, 6"		341.40		375.54	123.54	135.90
24 L.H., Daily Totals		$3928.20		$4733.50	$163.68	$197.23

Crew B-24	Hr.	Daily	Hr.	Daily	Bare Costs	Incl. O&P
1 Cement Finisher	$43.05	$344.40	$63.40	$507.20	$41.13	$62.38
1 Laborer	35.45	283.60	54.60	436.80		
1 Carpenter	44.90	359.20	69.15	553.20		
24 L.H., Daily Totals		$987.20		$1497.20	$41.13	$62.38

Crew B-25	Hr.	Daily	Hr.	Daily	Bare Costs	Incl. O&P
1 Labor Foreman (outside)	$37.45	$299.60	$57.65	$461.20	$38.92	$59.55
7 Laborers	35.45	1985.20	54.60	3057.60		
3 Equip. Oper. (med.)	47.50	1140.00	71.75	1722.00		
1 Asphalt Paver, 130 H.P.		2124.00		2336.40		
1 Tandem Roller, 10 Ton		231.20		254.32		
1 Roller, Pneum. Whl., 12 Ton		326.60		359.26	30.48	33.52
88 L.H., Daily Totals		$6106.60		$8190.78	$69.39	$93.08

Crew B-25B	Hr.	Daily	Hr.	Daily	Bare Costs	Incl. O&P
1 Labor Foreman (outside)	$37.45	$299.60	$57.65	$461.20	$39.63	$60.57
7 Laborers	35.45	1985.20	54.60	3057.60		
4 Equip. Oper. (med.)	47.50	1520.00	71.75	2296.00		
1 Asphalt Paver, 130 H.P.		2124.00		2336.40		
2 Tandem Rollers, 10 Ton		462.40		508.64		
1 Roller, Pneum. Whl., 12 Ton		326.60		359.26	30.34	33.38
96 L.H., Daily Totals		$6717.80		$9019.10	$69.98	$93.95

Crew B-25C	Hr.	Daily	Hr.	Daily	Bare Costs	Incl. O&P
1 Labor Foreman (outside)	$37.45	$299.60	$57.65	$461.20	$39.80	$60.83
3 Laborers	35.45	850.80	54.60	1310.40		
2 Equip. Oper. (med.)	47.50	760.00	71.75	1148.00		
1 Asphalt Paver, 130 H.P.		2124.00		2336.40		
1 Tandem Roller, 10 Ton		231.20		254.32	49.07	53.97
48 L.H., Daily Totals		$4265.60		$5510.32	$88.87	$114.80

Crew B-25D	Hr.	Daily	Hr.	Daily	Bare Costs	Incl. O&P
1 Labor Foreman (outside)	$37.45	$299.60	$57.65	$461.20	$39.89	$60.95
3 Laborers	35.45	850.80	54.60	1310.40		
2.125 Equip. Oper. (med.)	47.50	807.50	71.75	1219.75		
.125 Truck Driver (hvy.)	36.60	36.60	56.05	56.05		
.125 Truck Tractor, 6x4, 380 H.P.		76.55		84.20		
.125 Dist. Tanker, 3000 Gallon		38.15		41.97		
1 Asphalt Paver, 130 H.P.		2124.00		2336.40		
1 Tandem Roller, 10 Ton		231.20		254.32	49.40	54.34
50 L.H., Daily Totals		$4464.40		$5764.29	$89.29	$115.29

Crew No.	Bare Costs		Incl. Subs O&P		Cost Per Labor-Hour	
Crew B-25E	**Hr.**	**Daily**	**Hr.**	**Daily**	**Bare Costs**	**Incl. O&P**
1 Labor Foreman (outside)	$37.45	$299.60	$57.65	$461.20	$39.97	$61.06
3 Laborers	35.45	850.80	54.60	1310.40		
2.250 Equip. Oper. (med.)	47.50	855.00	71.75	1291.50		
.25 Truck Driver (hvy.)	36.60	73.20	56.05	112.10		
.25 Truck Tractor, 6x4, 380 H.P.		153.10		168.41		
.25 Dist. Tanker, 3000 Gallon		76.30		83.93		
1 Asphalt Paver, 130 H.P.		2124.00		2336.40		
1 Tandem Roller, 10 Ton		231.20		254.32	49.70	54.67
52 L.H., Daily Totals		$4663.20		$6018.26	$89.68	$115.74
Crew B-26	**Hr.**	**Daily**	**Hr.**	**Daily**	**Bare Costs**	**Incl. O&P**
1 Labor Foreman (outside)	$37.45	$299.60	$57.65	$461.20	$39.82	$60.99
6 Laborers	35.45	1701.60	54.60	2620.80		
2 Equip. Oper. (med.)	47.50	760.00	71.75	1148.00		
1 Rodman (reinf.)	49.80	398.40	78.70	629.60		
1 Cement Finisher	43.05	344.40	63.40	507.20		
1 Grader, 30,000 Lbs.		712.40		783.64		
1 Paving Mach. & Equip.		2667.00		2933.70	38.40	42.24
88 L.H., Daily Totals		$6883.40		$9084.14	$78.22	$103.23
Crew B-26A	**Hr.**	**Daily**	**Hr.**	**Daily**	**Bare Costs**	**Incl. O&P**
1 Labor Foreman (outside)	$37.45	$299.60	$57.65	$461.20	$39.82	$60.99
6 Laborers	35.45	1701.60	54.60	2620.80		
2 Equip. Oper. (med.)	47.50	760.00	71.75	1148.00		
1 Rodman (reinf.)	49.80	398.40	78.70	629.60		
1 Cement Finisher	43.05	344.40	63.40	507.20		
1 Grader, 30,000 Lbs.		712.40		783.64		
1 Paving Mach. & Equip.		2667.00		2933.70		
1 Concrete Saw		175.20		192.72	40.39	44.43
88 L.H., Daily Totals		$7058.60		$9276.86	$80.21	$105.42
Crew B-26B	**Hr.**	**Daily**	**Hr.**	**Daily**	**Bare Costs**	**Incl. O&P**
1 Labor Foreman (outside)	$37.45	$299.60	$57.65	$461.20	$40.46	$61.88
6 Laborers	35.45	1701.60	54.60	2620.80		
3 Equip. Oper. (med.)	47.50	1140.00	71.75	1722.00		
1 Rodman (reinf.)	49.80	398.40	78.70	629.60		
1 Cement Finisher	43.05	344.40	63.40	507.20		
1 Grader, 30,000 Lbs.		712.40		783.64		
1 Paving Mach. & Equip.		2667.00		2933.70		
1 Concrete Pump, 110' Boom		942.00		1036.20	45.01	49.52
96 L.H., Daily Totals		$8205.40		$10694.34	$85.47	$111.40
Crew B-26C	**Hr.**	**Daily**	**Hr.**	**Daily**	**Bare Costs**	**Incl. O&P**
1 Labor Foreman (outside)	$37.45	$299.60	$57.65	$461.20	$39.05	$59.91
6 Laborers	35.45	1701.60	54.60	2620.80		
1 Equip. Oper. (med.)	47.50	380.00	71.75	574.00		
1 Rodman (reinf.)	49.80	398.40	78.70	629.60		
1 Cement Finisher	43.05	344.40	63.40	507.20		
1 Paving Mach. & Equip.		2667.00		2933.70		
1 Concrete Saw		175.20		192.72	35.53	39.08
80 L.H., Daily Totals		$5966.20		$7919.22	$74.58	$98.99
Crew B-27	**Hr.**	**Daily**	**Hr.**	**Daily**	**Bare Costs**	**Incl. O&P**
1 Labor Foreman (outside)	$37.45	$299.60	$57.65	$461.20	$35.95	$55.36
3 Laborers	35.45	850.80	54.60	1310.40		
1 Berm Machine		292.80		322.08	9.15	10.07
32 L.H., Daily Totals		$1443.20		$2093.68	$45.10	$65.43

Crew No.	Bare Costs		Incl. Subs O&P		Cost Per Labor-Hour	
Crew B-28	**Hr.**	**Daily**	**Hr.**	**Daily**	**Bare Costs**	**Incl. O&P**
2 Carpenters	$44.90	$718.40	$69.15	$1106.40	$41.75	$64.30
1 Laborer	35.45	283.60	54.60	436.80		
24 L.H., Daily Totals		$1002.00		$1543.20	$41.75	$64.30
Crew B-29	**Hr.**	**Daily**	**Hr.**	**Daily**	**Bare Costs**	**Incl. O&P**
1 Labor Foreman (outside)	$37.45	$299.60	$57.65	$461.20	$38.61	$59.09
4 Laborers	35.45	1134.40	54.60	1747.20		
1 Equip. Oper. (crane)	48.80	390.40	73.75	590.00		
1 Equip. Oper. Oiler	42.25	338.00	63.85	510.80		
1 Gradall, 5/8 C.Y.		883.80		972.18	15.78	17.36
56 L.H., Daily Totals		$3046.20		$4281.38	$54.40	$76.45
Crew B-30	**Hr.**	**Daily**	**Hr.**	**Daily**	**Bare Costs**	**Incl. O&P**
1 Equip. Oper. (med.)	$47.50	$380.00	$71.75	$574.00	$40.23	$61.28
2 Truck Drivers (heavy)	36.60	585.60	56.05	896.80		
1 Hyd. Excavator, 1.5 C.Y.		1036.00		1139.60		
2 Dump Trucks, 12 C.Y., 400 H.P.		1384.80		1523.28	100.87	110.95
24 L.H., Daily Totals		$3386.40		$4133.68	$141.10	$172.24
Crew B-31	**Hr.**	**Daily**	**Hr.**	**Daily**	**Bare Costs**	**Incl. O&P**
1 Labor Foreman (outside)	$37.45	$299.60	$57.65	$461.20	$37.74	$58.12
3 Laborers	35.45	850.80	54.60	1310.40		
1 Carpenter	44.90	359.20	69.15	553.20		
1 Air Compressor, 250 cfm		195.80		215.38		
1 Sheeting Driver		5.75		6.33		
2 -50' Air Hoses, 1.5"		11.60		12.76	5.33	5.86
40 L.H., Daily Totals		$1722.75		$2559.26	$43.07	$63.98
Crew B-32	**Hr.**	**Daily**	**Hr.**	**Daily**	**Bare Costs**	**Incl. O&P**
1 Laborer	$35.45	$283.60	$54.60	$436.80	$44.49	$67.46
3 Equip. Oper. (med.)	47.50	1140.00	71.75	1722.00		
1 Grader, 30,000 Lbs.		712.40		783.64		
1 Tandem Roller, 10 Ton		231.20		254.32		
1 Dozer, 200 H.P.		1333.00		1466.30	71.14	78.26
32 L.H., Daily Totals		$3700.20		$4663.06	$115.63	$145.72
Crew B-32A	**Hr.**	**Daily**	**Hr.**	**Daily**	**Bare Costs**	**Incl. O&P**
1 Laborer	$35.45	$283.60	$54.60	$436.80	$43.48	$66.03
2 Equip. Oper. (med.)	47.50	760.00	71.75	1148.00		
1 Grader, 30,000 Lbs.		712.40		783.64		
1 Roller, Vibratory, 25 Ton		656.40		722.04	57.03	62.74
24 L.H., Daily Totals		$2412.40		$3090.48	$100.52	$128.77
Crew B-32B	**Hr.**	**Daily**	**Hr.**	**Daily**	**Bare Costs**	**Incl. O&P**
1 Laborer	$35.45	$283.60	$54.60	$436.80	$43.48	$66.03
2 Equip. Oper. (med.)	47.50	760.00	71.75	1148.00		
1 Dozer, 200 H.P.		1333.00		1466.30		
1 Roller, Vibratory, 25 Ton		656.40		722.04	82.89	91.18
24 L.H., Daily Totals		$3033.00		$3773.14	$126.38	$157.21
Crew B-32C	**Hr.**	**Daily**	**Hr.**	**Daily**	**Bare Costs**	**Incl. O&P**
1 Labor Foreman (outside)	$37.45	$299.60	$57.65	$461.20	$41.81	$63.68
2 Laborers	35.45	567.20	54.60	873.60		
3 Equip. Oper. (med.)	47.50	1140.00	71.75	1722.00		
1 Grader, 30,000 Lbs.		712.40		783.64		
1 Tandem Roller, 10 Ton		231.20		254.32		
1 Dozer, 200 H.P.		1333.00		1466.30	47.43	52.17
48 L.H., Daily Totals		$4283.40		$5561.06	$89.24	$115.86

Crew No.	Bare Costs		Incl. Subs O&P		Cost Per Labor-Hour	

Crew B-33A	Hr.	Daily	Hr.	Daily	Bare Costs	Incl. O&P
1 Equip. Oper. (med.)	$47.50	$380.00	$71.75	$574.00	$44.06	$66.85
.5 Laborer	35.45	141.80	54.60	218.40		
.25 Equip. Oper. (med.)	47.50	95.00	71.75	143.50		
1 Scraper, Towed, 7 C.Y.		110.80		121.88		
1.25 Dozers, 300 H.P.		2286.25		2514.88	171.22	188.34
14 L.H., Daily Totals		$3013.85		$3572.66	$215.28	$255.19

Crew B-33B	Hr.	Daily	Hr.	Daily	Bare Costs	Incl. O&P
1 Equip. Oper. (med.)	$47.50	$380.00	$71.75	$574.00	$44.06	$66.85
.5 Laborer	35.45	141.80	54.60	218.40		
.25 Equip. Oper. (med.)	47.50	95.00	71.75	143.50		
1 Scraper, Towed, 10 C.Y.		140.80		154.88		
1.25 Dozers, 300 H.P.		2286.25		2514.88	173.36	190.70
14 L.H., Daily Totals		$3043.85		$3605.66	$217.42	$257.55

Crew B-33C	Hr.	Daily	Hr.	Daily	Bare Costs	Incl. O&P
1 Equip. Oper. (med.)	$47.50	$380.00	$71.75	$574.00	$44.06	$66.85
.5 Laborer	35.45	141.80	54.60	218.40		
.25 Equip. Oper. (med.)	47.50	95.00	71.75	143.50		
1 Scraper, Towed, 15 C.Y.		158.00		173.80		
1.25 Dozers, 300 H.P.		2286.25		2514.88	174.59	192.05
14 L.H., Daily Totals		$3061.05		$3624.57	$218.65	$258.90

Crew B-33D	Hr.	Daily	Hr.	Daily	Bare Costs	Incl. O&P
1 Equip. Oper. (med.)	$47.50	$380.00	$71.75	$574.00	$44.06	$66.85
.5 Laborer	35.45	141.80	54.60	218.40		
.25 Equip. Oper. (med.)	47.50	95.00	71.75	143.50		
1 S.P. Scraper, 14 C.Y.		1840.00		2024.00		
.25 Dozer, 300 H.P.		457.25		502.98	164.09	180.50
14 L.H., Daily Totals		$2914.05		$3462.88	$208.15	$247.35

Crew B-33E	Hr.	Daily	Hr.	Daily	Bare Costs	Incl. O&P
1 Equip. Oper. (med.)	$47.50	$380.00	$71.75	$574.00	$44.06	$66.85
.5 Laborer	35.45	141.80	54.60	218.40		
.25 Equip. Oper. (med.)	47.50	95.00	71.75	143.50		
1 S.P. Scraper, 21 C.Y.		2610.00		2871.00		
.25 Dozer, 300 H.P.		457.25		502.98	219.09	241.00
14 L.H., Daily Totals		$3684.05		$4309.88	$263.15	$307.85

Crew B-33F	Hr.	Daily	Hr.	Daily	Bare Costs	Incl. O&P
1 Equip. Oper. (med.)	$47.50	$380.00	$71.75	$574.00	$44.06	$66.85
.5 Laborer	35.45	141.80	54.60	218.40		
.25 Equip. Oper. (med.)	47.50	95.00	71.75	143.50		
1 Elev. Scraper, 11 C.Y.		1169.00		1285.90		
.25 Dozer, 300 H.P.		457.25		502.98	116.16	127.78
14 L.H., Daily Totals		$2243.05		$2724.78	$160.22	$194.63

Crew B-33G	Hr.	Daily	Hr.	Daily	Bare Costs	Incl. O&P
1 Equip. Oper. (med.)	$47.50	$380.00	$71.75	$574.00	$44.06	$66.85
.5 Laborer	35.45	141.80	54.60	218.40		
.25 Equip. Oper. (med.)	47.50	95.00	71.75	143.50		
1 Elev. Scraper, 22 C.Y.		2459.00		2704.90		
.25 Dozer, 300 H.P.		457.25		502.98	208.30	229.13
14 L.H., Daily Totals		$3533.05		$4143.77	$252.36	$295.98

Crew B-33H	Hr.	Daily	Hr.	Daily	Bare Costs	Incl. O&P
.5 Laborer	$35.45	$141.80	$54.60	$218.40	$44.06	$66.85
1 Equipment Operator (med.)	47.50	380.00	71.75	574.00		
.25 Equipment Operator (med.)	47.50	95.00	71.75	143.50		
1 S.P. Scraper, 44 C.Y.		4377.00		4814.70		
.25 Dozer, 410 H.P.		606.25		666.88	355.95	391.54
14 L.H., Daily Totals		$5600.05		$6417.48	$400.00	$458.39

Crew B-33J	Hr.	Daily	Hr.	Daily	Bare Costs	Incl. O&P
1 Equipment Operator (med.)	$47.50	$380.00	$71.75	$574.00	$47.50	$71.75
1 S.P. Scraper, 14 C.Y.		1840.00		2024.00	230.00	253.00
8 L.H., Daily Totals		$2220.00		$2598.00	$277.50	$324.75

Crew B-33K	Hr.	Daily	Hr.	Daily	Bare Costs	Incl. O&P
1 Equipment Operator (med.)	$47.50	$380.00	$71.75	$574.00	$44.06	$66.85
.25 Equipment Operator (med.)	47.50	95.00	71.75	143.50		
.5 Laborer	35.45	141.80	54.60	218.40		
1 S.P. Scraper, 31 C.Y.		3584.00		3942.40		
.25 Dozer, 410 H.P.		606.25		666.88	299.30	329.23
14 L.H., Daily Totals		$4807.05		$5545.18	$343.36	$396.08

Crew B-34A	Hr.	Daily	Hr.	Daily	Bare Costs	Incl. O&P
1 Truck Driver (heavy)	$36.60	$292.80	$56.05	$448.40	$36.60	$56.05
1 Dump Truck, 8 C.Y., 220 H.P.		416.60		458.26	52.08	57.28
8 L.H., Daily Totals		$709.40		$906.66	$88.67	$113.33

Crew B-34B	Hr.	Daily	Hr.	Daily	Bare Costs	Incl. O&P
1 Truck Driver (heavy)	$36.60	$292.80	$56.05	$448.40	$36.60	$56.05
1 Dump Truck, 12 C.Y., 400 H.P.		692.40		761.64	86.55	95.20
8 L.H., Daily Totals		$985.20		$1210.04	$123.15	$151.26

Crew B-34C	Hr.	Daily	Hr.	Daily	Bare Costs	Incl. O&P
1 Truck Driver (heavy)	$36.60	$292.80	$56.05	$448.40	$36.60	$56.05
1 Truck Tractor, 6x4, 380 H.P.		612.40		673.64		
1 Dump Trailer, 16.5 C.Y.		126.80		139.48	92.40	101.64
8 L.H., Daily Totals		$1032.00		$1261.52	$129.00	$157.69

Crew B-34D	Hr.	Daily	Hr.	Daily	Bare Costs	Incl. O&P
1 Truck Driver (heavy)	$36.60	$292.80	$56.05	$448.40	$36.60	$56.05
1 Truck Tractor, 6x4, 380 H.P.		612.40		673.64		
1 Dump Trailer, 20 C.Y.		141.40		155.54	94.22	103.65
8 L.H., Daily Totals		$1046.60		$1277.58	$130.82	$159.70

Crew B-34E	Hr.	Daily	Hr.	Daily	Bare Costs	Incl. O&P
1 Truck Driver (heavy)	$36.60	$292.80	$56.05	$448.40	$36.60	$56.05
1 Dump Truck, Off Hwy., 25 Ton		1321.00		1453.10	165.13	181.64
8 L.H., Daily Totals		$1613.80		$1901.50	$201.72	$237.69

Crew B-34F	Hr.	Daily	Hr.	Daily	Bare Costs	Incl. O&P
1 Truck Driver (heavy)	$36.60	$292.80	$56.05	$448.40	$36.60	$56.05
1 Dump Truck, Off Hwy., 35 Ton		1477.00		1624.70	184.63	203.09
8 L.H., Daily Totals		$1769.80		$2073.10	$221.22	$259.14

Crew B-34G	Hr.	Daily	Hr.	Daily	Bare Costs	Incl. O&P
1 Truck Driver (heavy)	$36.60	$292.80	$56.05	$448.40	$36.60	$56.05
1 Dump Truck, Off Hwy., 50 Ton		1797.00		1976.70	224.63	247.09
8 L.H., Daily Totals		$2089.80		$2425.10	$261.23	$303.14

Crew No.	Bare Costs		Incl. Subs O&P		Cost Per Labor-Hour	

Crew B-34H	Hr.	Daily	Hr.	Daily	Bare Costs	Incl. O&P
1 Truck Driver (heavy)	$36.60	$292.80	$56.05	$448.40	$36.60	$56.05
1 Dump Truck, Off Hwy., 65 Ton		1807.00		1987.70	225.88	248.46
8 L.H., Daily Totals		$2099.80		$2436.10	$262.48	$304.51

Crew B-34I	Hr.	Daily	Hr.	Daily	Bare Costs	Incl. O&P
1 Truck Driver (heavy)	$36.60	$292.80	$56.05	$448.40	$36.60	$56.05
1 Dump Truck, 18 C.Y., 450 H.P.		867.00		953.70	108.38	119.21
8 L.H., Daily Totals		$1159.80		$1402.10	$144.97	$175.26

Crew B-34J	Hr.	Daily	Hr.	Daily	Bare Costs	Incl. O&P
1 Truck Driver (heavy)	$36.60	$292.80	$56.05	$448.40	$36.60	$56.05
1 Dump Truck, Off Hwy., 100 Ton		2920.00		3212.00	365.00	401.50
8 L.H., Daily Totals		$3212.80		$3660.40	$401.60	$457.55

Crew B-34K	Hr.	Daily	Hr.	Daily	Bare Costs	Incl. O&P
1 Truck Driver (heavy)	$36.60	$292.80	$56.05	$448.40	$36.60	$56.05
1 Truck Tractor, 6x4, 450 H.P.		741.20		815.32		
1 Lowbed Trailer, 75 Ton		220.40		242.44	120.20	132.22
8 L.H., Daily Totals		$1254.40		$1506.16	$156.80	$188.27

Crew B-34L	Hr.	Daily	Hr.	Daily	Bare Costs	Incl. O&P
1 Equip. Oper. (light)	$45.80	$366.40	$69.20	$553.60	$45.80	$69.20
1 Flatbed Truck, Gas, 1.5 Ton		265.80		292.38	33.23	36.55
8 L.H., Daily Totals		$632.20		$845.98	$79.03	$105.75

Crew B-34M	Hr.	Daily	Hr.	Daily	Bare Costs	Incl. O&P
1 Equip. Oper. (light)	$45.80	$366.40	$69.20	$553.60	$45.80	$69.20
1 Flatbed Truck, Gas, 3 Ton		327.80		360.58	40.98	45.07
8 L.H., Daily Totals		$694.20		$914.18	$86.78	$114.27

Crew B-34N	Hr.	Daily	Hr.	Daily	Bare Costs	Incl. O&P
1 Truck Driver (heavy)	$36.60	$292.80	$56.05	$448.40	$36.60	$56.05
1 Dump Truck, 8 C.Y., 220 H.P.		416.60		458.26		
1 Flatbed Trailer, 40 Ton		152.60		167.86	71.15	78.27
8 L.H., Daily Totals		$862.00		$1074.52	$107.75	$134.32

Crew B-34P	Hr.	Daily	Hr.	Daily	Bare Costs	Incl. O&P
1 Pipe Fitter	$56.65	$453.20	$85.30	$682.40	$46.60	$70.55
1 Truck Driver (light)	35.65	285.20	54.60	436.80		
1 Equip. Oper. (med.)	47.50	380.00	71.75	574.00		
1 Flatbed Truck, Gas, 3 Ton		327.80		360.58		
1 Backhoe Loader, 48 H.P.		367.40		404.14	28.97	31.86
24 L.H., Daily Totals		$1813.60		$2457.92	$75.57	$102.41

Crew B-34Q	Hr.	Daily	Hr.	Daily	Bare Costs	Incl. O&P
1 Pipe Fitter	$56.65	$453.20	$85.30	$682.40	$47.03	$71.22
1 Truck Driver (light)	35.65	285.20	54.60	436.80		
1 Equip. Oper. (crane)	48.80	390.40	73.75	590.00		
1 Flatbed Trailer, 25 Ton		113.20		124.52		
1 Dump Truck, 8 C.Y., 220 H.P.		416.60		458.26		
1 Hyd. Crane, 25 Ton		728.60		801.46	52.43	57.68
24 L.H., Daily Totals		$2387.20		$3093.44	$99.47	$128.89

Crew B-34R	Hr.	Daily	Hr.	Daily	Bare Costs	Incl. O&P
1 Pipe Fitter	$56.65	$453.20	$85.30	$682.40	$47.03	$71.22
1 Truck Driver (light)	35.65	285.20	54.60	436.80		
1 Equip. Oper. (crane)	48.80	390.40	73.75	590.00		
1 Flatbed Trailer, 25 Ton		113.20		124.52		
1 Dump Truck, 8 C.Y., 220 H.P.		416.60		458.26		
1 Hyd. Crane, 25 Ton		728.60		801.46		
1 Hyd. Excavator, 1 C.Y.		815.60		897.16	86.42	95.06
24 L.H., Daily Totals		$3202.80		$3990.60	$133.45	$166.28

Crew B-34S	Hr.	Daily	Hr.	Daily	Bare Costs	Incl. O&P
2 Pipe Fitters	$56.65	$906.40	$85.30	$1364.80	$49.67	$75.10
1 Truck Driver (heavy)	36.60	292.80	56.05	448.40		
1 Equip. Oper. (crane)	48.80	390.40	73.75	590.00		
1 Flatbed Trailer, 40 Ton		152.60		167.86		
1 Truck Tractor, 6x4, 380 H.P.		612.40		673.64		
1 Hyd. Crane, 80 Ton		1613.00		1774.30		
1 Hyd. Excavator, 2 C.Y.		1363.00		1499.30	116.91	128.60
32 L.H., Daily Totals		$5330.60		$6518.30	$166.58	$203.70

Crew B-34T	Hr.	Daily	Hr.	Daily	Bare Costs	Incl. O&P
2 Pipe Fitters	$56.65	$906.40	$85.30	$1364.80	$49.67	$75.10
1 Truck Driver (heavy)	36.60	292.80	56.05	448.40		
1 Equip. Oper. (crane)	48.80	390.40	73.75	590.00		
1 Flatbed Trailer, 40 Ton		152.60		167.86		
1 Truck Tractor, 6x4, 380 H.P.		612.40		673.64		
1 Hyd. Crane, 80 Ton		1613.00		1774.30	74.31	81.74
32 L.H., Daily Totals		$3967.60		$5019.00	$123.99	$156.84

Crew B-35	Hr.	Daily	Hr.	Daily	Bare Costs	Incl. O&P
1 Labor Foreman (outside)	$37.45	$299.60	$57.65	$461.20	$44.33	$67.56
1 Skilled Worker	46.20	369.60	71.45	571.60		
1 Welder (plumber)	55.80	446.40	84.05	672.40		
1 Laborer	35.45	283.60	54.60	436.80		
1 Equip. Oper. (crane)	48.80	390.40	73.75	590.00		
1 Equip. Oper. Oiler	42.25	338.00	63.85	510.80		
1 Welder, Electric, 300 amp		52.85		58.13		
1 Hyd. Excavator, .75 C.Y.		694.80		764.28	15.58	17.13
48 L.H., Daily Totals		$2875.25		$4065.22	$59.90	$84.69

Crew B-35A	Hr.	Daily	Hr.	Daily	Bare Costs	Incl. O&P
1 Labor Foreman (outside)	$37.45	$299.60	$57.65	$461.20	$43.06	$65.71
2 Laborers	35.45	567.20	54.60	873.60		
1 Skilled Worker	46.20	369.60	71.45	571.60		
1 Welder (plumber)	55.80	446.40	84.05	672.40		
1 Equip. Oper. (crane)	48.80	390.40	73.75	590.00		
1 Equip. Oper. Oiler	42.25	338.00	63.85	510.80		
1 Welder, Gas Engine, 300 amp		144.00		158.40		
1 Crawler Crane, 75 Ton		1457.00		1602.70	28.59	31.45
56 L.H., Daily Totals		$4012.20		$5440.70	$71.65	$97.16

Crew B-36	Hr.	Daily	Hr.	Daily	Bare Costs	Incl. O&P
1 Labor Foreman (outside)	$37.45	$299.60	$57.65	$461.20	$40.67	$62.07
2 Laborers	35.45	567.20	54.60	873.60		
2 Equip. Oper. (med.)	47.50	760.00	71.75	1148.00		
1 Dozer, 200 H.P.		1333.00		1466.30		
1 Aggregate Spreader		40.80		44.88		
1 Tandem Roller, 10 Ton		231.20		254.32	40.13	44.14
40 L.H., Daily Totals		$3231.80		$4248.30	$80.80	$106.21

Crew No.	Bare Costs		Incl. Subs O&P		Cost Per Labor-Hour	
Crew B-36A	**Hr.**	**Daily**	**Hr.**	**Daily**	**Bare Costs**	**Incl. O&P**
1 Labor Foreman (outside)	$37.45	$299.60	$57.65	$461.20	$42.62	$64.84
2 Laborers	35.45	567.20	54.60	873.60		
4 Equip. Oper. (med.)	47.50	1520.00	71.75	2296.00		
1 Dozer, 200 H.P.		1333.00		1466.30		
1 Aggregate Spreader		40.80		44.88		
1 Tandem Roller, 10 Ton		231.20		254.32		
1 Roller, Pneum. Whl., 12 Ton		326.60		359.26	34.49	37.94
56 L.H., Daily Totals		$4318.40		$5755.56	$77.11	$102.78

Crew B-36B	Hr.	Daily	Hr.	Daily	Bare Costs	Incl. O&P
1 Labor Foreman (outside)	$37.45	$299.60	$57.65	$461.20	$41.87	$63.74
2 Laborers	35.45	567.20	54.60	873.60		
4 Equip. Oper. (med.)	47.50	1520.00	71.75	2296.00		
1 Truck Driver, Heavy	36.60	292.80	56.05	448.40		
1 Grader, 30,000 Lbs.		712.40		783.64		
1 F.E. Loader, Crl, 1.5 C.Y.		633.20		696.52		
1 Dozer, 300 H.P.		1829.00		2011.90		
1 Roller, Vibratory, 25 Ton		656.40		722.04		
1 Truck Tractor, 6x4, 450 H.P.		741.20		815.32		
1 Water Tank Trailer, 5000 Gal.		143.20		157.52	73.68	81.05
64 L.H., Daily Totals		$7395.00		$9266.14	$115.55	$144.78

Crew B-36C	Hr.	Daily	Hr.	Daily	Bare Costs	Incl. O&P
1 Labor Foreman (outside)	$37.45	$299.60	$57.65	$461.20	$43.31	$65.79
3 Equip. Oper. (med.)	47.50	1140.00	71.75	1722.00		
1 Truck Driver, Heavy	36.60	292.80	56.05	448.40		
1 Grader, 30,000 Lbs.		712.40		783.64		
1 Dozer, 300 H.P.		1829.00		2011.90		
1 Roller, Vibratory, 25 Ton		656.40		722.04		
1 Truck Tractor, 6x4, 450 H.P.		741.20		815.32		
1 Water Tank Trailer, 5000 Gal.		143.20		157.52	102.06	112.26
40 L.H., Daily Totals		$5814.60		$7122.02	$145.37	$178.05

Crew B-36D	Hr.	Daily	Hr.	Daily	Bare Costs	Incl. O&P
1 Labor Foreman (outside)	$37.45	$299.60	$57.65	$461.20	$44.99	$68.22
3 Equip. Oper. (medium)	47.50	1140.00	71.75	1722.00		
1 Grader, 30,000 Lbs.		712.40		783.64		
1 Dozer, 300 H.P.		1829.00		2011.90		
1 Roller, Vibratory, 25 Ton		656.40		722.04	99.93	109.92
32 L.H., Daily Totals		$4637.40		$5700.78	$144.92	$178.15

Crew B-37	Hr.	Daily	Hr.	Daily	Bare Costs	Incl. O&P
1 Labor Foreman (outside)	$37.45	$299.60	$57.65	$461.20	$37.51	$57.54
4 Laborers	35.45	1134.40	54.60	1747.20		
1 Equip. Oper. (light)	45.80	366.40	69.20	553.60		
1 Tandem Roller, 5 Ton		154.40		169.84	3.22	3.54
48 L.H., Daily Totals		$1954.80		$2931.84	$40.73	$61.08

Crew B-37A	Hr.	Daily	Hr.	Daily	Bare Costs	Incl. O&P
2 Laborers	$35.45	$567.20	$54.60	$873.60	$35.52	$54.60
1 Truck Driver (light)	35.65	285.20	54.60	436.80		
1 Flatbed Truck, Gas, 1.5 Ton		265.80		292.38		
1 Tar Kettle, T.M.		78.70		86.57	14.35	15.79
24 L.H., Daily Totals		$1196.90		$1689.35	$49.87	$70.39

Crew B-37B	Hr.	Daily	Hr.	Daily	Bare Costs	Incl. O&P
3 Laborers	$35.45	$850.80	$54.60	$1310.40	$35.50	$54.60
1 Truck Driver (light)	35.65	285.20	54.60	436.80		
1 Flatbed Truck, Gas, 1.5 Ton		265.80		292.38		
1 Tar Kettle, T.M.		78.70		86.57	10.77	11.84
32 L.H., Daily Totals		$1480.50		$2126.15	$46.27	$66.44

Crew B-37C	Hr.	Daily	Hr.	Daily	Bare Costs	Incl. O&P
2 Laborers	$35.45	$567.20	$54.60	$873.60	$35.55	$54.60
2 Truck Drivers (light)	35.65	570.40	54.60	873.60		
2 Flatbed Trucks, Gas, 1.5 Ton		531.60		584.76		
1 Tar Kettle, T.M.		78.70		86.57	19.07	20.98
32 L.H., Daily Totals		$1747.90		$2418.53	$54.62	$75.58

Crew B-37D	Hr.	Daily	Hr.	Daily	Bare Costs	Incl. O&P
1 Laborer	$35.45	$283.60	$54.60	$436.80	$35.55	$54.60
1 Truck Driver (light)	35.65	285.20	54.60	436.80		
1 Pickup Truck, 3/4 Ton		154.60		170.06	9.66	10.63
16 L.H., Daily Totals		$723.40		$1043.66	$45.21	$65.23

Crew B-37E	Hr.	Daily	Hr.	Daily	Bare Costs	Incl. O&P
3 Laborers	$35.45	$850.80	$54.60	$1310.40	$38.71	$59.14
1 Equip. Oper. (light)	45.80	366.40	69.20	553.60		
1 Equip. Oper. (medium)	47.50	380.00	71.75	574.00		
2 Truck Drivers (light)	35.65	570.40	54.60	873.60		
4 Barrels w/ Flasher		13.60		14.96		
1 Concrete Saw		175.20		192.72		
1 Rotary Hammer Drill		22.65		24.91		
1 Hammer Drill Bit		2.55		2.81		
1 Loader, Skid Steer, 30 H.P.		172.00		189.20		
1 Conc. Hammer Attach.		114.10		125.51		
1 Vibrating Plate, Gas, 18"		34.60		38.06		
2 Flatbed Trucks, Gas, 1.5 Ton		531.60		584.76	19.04	20.95
56 L.H., Daily Totals		$3233.90		$4484.53	$57.75	$80.08

Crew B-37F	Hr.	Daily	Hr.	Daily	Bare Costs	Incl. O&P
3 Laborers	$35.45	$850.80	$54.60	$1310.40	$35.50	$54.60
1 Truck Driver (light)	35.65	285.20	54.60	436.80		
4 Barrels w/ Flasher		13.60		14.96		
1 Concrete Mixer, 10 C.F.		176.20		193.82		
1 Air Compressor, 60 cfm		137.60		151.36		
1 -50' Air Hose, 3/4"		3.25		3.58		
1 Spade (Chipper)		7.60		8.36		
1 Flatbed Truck, Gas, 1.5 Ton		265.80		292.38	18.88	20.76
32 L.H., Daily Totals		$1740.05		$2411.66	$54.38	$75.36

Crew B-37G	Hr.	Daily	Hr.	Daily	Bare Costs	Incl. O&P
1 Labor Foreman (outside)	$37.45	$299.60	$57.65	$461.20	$37.51	$57.54
4 Laborers	35.45	1134.40	54.60	1747.20		
1 Equip. Oper. (light)	45.80	366.40	69.20	553.60		
1 Berm Machine		292.80		322.08		
1 Tandem Roller, 5 Ton		154.40		169.84	9.32	10.25
48 L.H., Daily Totals		$2247.60		$3253.92	$46.83	$67.79

Crew B-37H	Hr.	Daily	Hr.	Daily	Bare Costs	Incl. O&P
1 Labor Foreman (outside)	$37.45	$299.60	$57.65	$461.20	$37.51	$57.54
4 Laborers	35.45	1134.40	54.60	1747.20		
1 Equip. Oper. (light)	45.80	366.40	69.20	553.60		
1 Tandem Roller, 5 Ton		154.40		169.84		
1 Flatbed Trucks, Gas, 1.5 Ton		265.80		292.38		
1 Tar Kettle, T.M.		78.70		86.57	10.39	11.43
48 L.H., Daily Totals		$2299.30		$3310.79	$47.90	$68.97

Crew No.	Bare Costs		Incl. Subs O&P		Cost Per Labor-Hour	

Crew B-37I	Hr.	Daily	Hr.	Daily	Bare Costs	Incl. O&P
3 Laborers	$35.45	$850.80	$54.60	$1310.40	$38.71	$59.14
1 Equip. Oper. (light)	45.80	366.40	69.20	553.60		
1 Equip. Oper. (medium)	47.50	380.00	71.75	574.00		
2 Truck Drivers (light)	35.65	570.40	54.60	873.60		
4 Barrels w/ Flasher		13.60		14.96		
1 Concrete Saw		175.20		192.72		
1 Rotary Hammer Drill		22.65		24.91		
1 Hammer Drill Bit		2.55		2.81		
1 Air Compressor, 60 cfm		137.60		151.36		
1 -50' Air Hose, 3/4"		3.25		3.58		
1 Spade (Chipper)		7.60		8.36		
1 Loader, Skid Steer, 30 H.P.		172.00		189.20		
1 Conc. Hammer Attach.		114.10		125.51		
1 Concrete Mixer, 10 C.F.		176.20		193.82		
1 Vibrating Plate, Gas, 18"		34.60		38.06		
2 Flatbed Trucks, Gas, 1.5 Ton		531.60		584.76	24.84	27.32
56 L.H., Daily Totals		$3558.55		$4841.65	$63.55	$86.46

Crew B-37J	Hr.	Daily	Hr.	Daily	Bare Costs	Incl. O&P
1 Labor Foreman (outside)	$37.45	$299.60	$57.65	$461.20	$37.51	$57.54
4 Laborers	35.45	1134.40	54.60	1747.20		
1 Equip. Oper. (light)	45.80	366.40	69.20	553.60		
1 Air Compressor, 60 cfm		137.60		151.36		
1 -50' Air Hose, 3/4"		3.25		3.58		
2 Concrete Mixer, 10 C.F.		352.40		387.64		
2 Flatbed Trucks, Gas, 1.5 Ton		531.60		584.76		
1 Shot Blaster, 20"		210.20		231.22	25.73	28.30
48 L.H., Daily Totals		$3035.45		$4120.56	$63.24	$85.84

Crew B-37K	Hr.	Daily	Hr.	Daily	Bare Costs	Incl. O&P
1 Labor Foreman (outside)	$37.45	$299.60	$57.65	$461.20	$37.51	$57.54
4 Laborers	35.45	1134.40	54.60	1747.20		
1 Equip. Oper. (light)	45.80	366.40	69.20	553.60		
1 Air Compressor, 60 cfm		137.60		151.36		
1 -50' Air Hose, 3/4"		3.25		3.58		
2 Flatbed Trucks, Gas, 1.5 Ton		531.60		584.76		
1 Shot Blaster, 20"		210.20		231.22	18.39	20.23
48 L.H., Daily Totals		$2683.05		$3732.92	$55.90	$77.77

Crew B-38	Hr.	Daily	Hr.	Daily	Bare Costs	Incl. O&P
1 Labor Foreman (outside)	$37.45	$299.60	$57.65	$461.20	$40.33	$61.56
2 Laborers	35.45	567.20	54.60	873.60		
1 Equip. Oper. (light)	45.80	366.40	69.20	553.60		
1 Equip. Oper. (med.)	47.50	380.00	71.75	574.00		
1 Backhoe Loader, 48 H.P.		367.40		404.14		
1 Hyd. Hammer, (1200 lb.)		179.60		197.56		
1 F.E. Loader, W.M., 4 C.Y.		678.80		746.68		
1 Pvmt. Rem. Bucket		60.80		66.88	32.16	35.38
40 L.H., Daily Totals		$2899.80		$3877.66	$72.50	$96.94

Crew B-39	Hr.	Daily	Hr.	Daily	Bare Costs	Incl. O&P
1 Labor Foreman (outside)	$37.45	$299.60	$57.65	$461.20	$37.51	$57.54
4 Laborers	35.45	1134.40	54.60	1747.20		
1 Equip. Oper. (light)	45.80	366.40	69.20	553.60		
1 Air Compressor, 250 cfm		195.80		215.38		
2 Breakers, Pavement, 60 lb.		19.60		21.56		
2 -50' Air Hoses, 1.5"		11.60		12.76	4.73	5.20
48 L.H., Daily Totals		$2027.40		$3011.70	$42.24	$62.74

Crew B-40	Hr.	Daily	Hr.	Daily	Bare Costs	Incl. O&P
1 Pile Driver Foreman (outside)	$45.15	$361.20	$71.90	$575.20	$44.70	$69.76
4 Pile Drivers	43.15	1380.80	68.70	2198.40		
2 Equip. Oper. (crane)	48.80	780.80	73.75	1180.00		
1 Equip. Oper. Oiler	42.25	338.00	63.85	510.80		
1 Crawler Crane, 40 Ton		1152.00		1267.20		
1 Vibratory Hammer & Gen.		2652.00		2917.20	59.44	65.38
64 L.H., Daily Totals		$6664.80		$8648.80	$104.14	$135.14

Crew B-40B	Hr.	Daily	Hr.	Daily	Bare Costs	Incl. O&P
1 Labor Foreman (outside)	$37.45	$299.60	$57.65	$461.20	$39.14	$59.84
3 Laborers	35.45	850.80	54.60	1310.40		
1 Equip. Oper. (crane)	48.80	390.40	73.75	590.00		
1 Equip. Oper. Oiler	42.25	338.00	63.85	510.80		
1 Lattice Boom Crane, 40 Ton		1174.00		1291.40	24.46	26.90
48 L.H., Daily Totals		$3052.80		$4163.80	$63.60	$86.75

Crew B-41	Hr.	Daily	Hr.	Daily	Bare Costs	Incl. O&P
1 Labor Foreman (outside)	$37.45	$299.60	$57.65	$461.20	$36.73	$56.45
4 Laborers	35.45	1134.40	54.60	1747.20		
.25 Equip. Oper. (crane)	48.80	97.60	73.75	147.50		
.25 Equip. Oper. Oiler	42.25	84.50	63.85	127.70		
.25 Crawler Crane, 40 Ton		288.00		316.80	6.55	7.20
44 L.H., Daily Totals		$1904.10		$2800.40	$43.27	$63.65

Crew B-42	Hr.	Daily	Hr.	Daily	Bare Costs	Incl. O&P
1 Labor Foreman (outside)	$37.45	$299.60	$57.65	$461.20	$40.04	$62.87
4 Laborers	35.45	1134.40	54.60	1747.20		
1 Equip. Oper. (crane)	48.80	390.40	73.75	590.00		
1 Equip. Oper. Oiler	42.25	338.00	63.85	510.80		
1 Welder	50.05	400.40	89.30	714.40		
1 Hyd. Crane, 25 Ton		728.60		801.46		
1 Welder, Gas Engine, 300 amp		144.00		158.40		
1 Horz. Boring Csg. Mch.		465.00		511.50	20.90	22.99
64 L.H., Daily Totals		$3900.40		$5494.96	$60.94	$85.86

Crew B-43	Hr.	Daily	Hr.	Daily	Bare Costs	Incl. O&P
1 Labor Foreman (outside)	$37.45	$299.60	$57.65	$461.20	$39.14	$59.84
3 Laborers	35.45	850.80	54.60	1310.40		
1 Equip. Oper. (crane)	48.80	390.40	73.75	590.00		
1 Equip. Oper. Oiler	42.25	338.00	63.85	510.80		
1 Drill Rig, Truck-Mounted		2469.00		2715.90	51.44	56.58
48 L.H., Daily Totals		$4347.80		$5588.30	$90.58	$116.42

Crew B-44	Hr.	Daily	Hr.	Daily	Bare Costs	Incl. O&P
1 Pile Driver Foreman (outside)	$45.15	$361.20	$71.90	$575.20	$43.85	$68.60
4 Pile Drivers	43.15	1380.80	68.70	2198.40		
2 Equip. Oper. (crane)	48.80	780.80	73.75	1180.00		
1 Laborer	35.45	283.60	54.60	436.80		
1 Crawler Crane, 40 Ton		1152.00		1267.20		
1 Lead, 60' High		74.60		82.06		
1 Hammer, Diesel, 15K ft.-lbs.		592.40		651.64	28.42	31.26
64 L.H., Daily Totals		$4625.40		$6391.30	$72.27	$99.86

Crew B-45	Hr.	Daily	Hr.	Daily	Bare Costs	Incl. O&P
1 Equip. Oper. (med.)	$47.50	$380.00	$71.75	$574.00	$42.05	$63.90
1 Truck Driver (heavy)	36.60	292.80	56.05	448.40		
1 Dist. Tanker, 3000 Gallon		305.20		335.72		
1 Truck Tractor, 6x4, 380 H.P.		612.40		673.64	57.35	63.09
16 L.H., Daily Totals		$1590.40		$2031.76	$99.40	$126.99

Crew B-46	Hr. (Bare Costs)	Daily (Bare Costs)	Hr. (Incl. Subs O&P)	Daily (Incl. Subs O&P)	Bare Costs (Cost Per Labor-Hour)	Incl. O&P (Cost Per Labor-Hour)
1 Pile Driver Foreman (outside)	$45.15	$361.20	$71.90	$575.20	$39.63	$62.18
2 Pile Drivers	43.15	690.40	68.70	1099.20		
3 Laborers	35.45	850.80	54.60	1310.40		
1 Chain Saw, Gas, 36″ Long		45.00		49.50	.94	1.03
48 L.H., Daily Totals		$1947.40		$3034.30	$40.57	$63.21

Crew B-47	Hr.	Daily	Hr.	Daily	Bare Costs	Incl. O&P
1 Blast Foreman (outside)	$37.45	$299.60	$57.65	$461.20	$39.57	$60.48
1 Driller	35.45	283.60	54.60	436.80		
1 Equip. Oper. (light)	45.80	366.40	69.20	553.60		
1 Air Track Drill, 4″		1009.00		1109.90		
1 Air Compressor, 600 cfm		530.80		583.88		
2 -50′ Air Hoses, 3″		29.80		32.78	65.40	71.94
24 L.H., Daily Totals		$2519.20		$3178.16	$104.97	$132.42

Crew B-47A	Hr.	Daily	Hr.	Daily	Bare Costs	Incl. O&P
1 Drilling Foreman (outside)	$37.45	$299.60	$57.65	$461.20	$42.83	$65.08
1 Equip. Oper. (heavy)	48.80	390.40	73.75	590.00		
1 Oiler	42.25	338.00	63.85	510.80		
1 Air Track Drill, 5″		1207.00		1327.70	50.29	55.32
24 L.H., Daily Totals		$2235.00		$2889.70	$93.13	$120.40

Crew B-47C	Hr.	Daily	Hr.	Daily	Bare Costs	Incl. O&P
1 Laborer	$35.45	$283.60	$54.60	$436.80	$40.63	$61.90
1 Equip. Oper. (light)	45.80	366.40	69.20	553.60		
1 Air Compressor, 750 cfm		537.40		591.14		
2 -50′ Air Hoses, 3″		29.80		32.78		
1 Air Track Drill, 4″		1009.00		1109.90	98.51	108.36
16 L.H., Daily Totals		$2226.20		$2724.22	$139.14	$170.26

Crew B-47E	Hr.	Daily	Hr.	Daily	Bare Costs	Incl. O&P
1 Labor Foreman (outside)	$37.45	$299.60	$57.65	$461.20	$35.95	$55.36
3 Laborers	35.45	850.80	54.60	1310.40		
1 Flatbed Truck, Gas, 3 Ton		327.80		360.58	10.24	11.27
32 L.H., Daily Totals		$1478.20		$2132.18	$46.19	$66.63

Crew B-47G	Hr.	Daily	Hr.	Daily	Bare Costs	Incl. O&P
1 Labor Foreman (outside)	$37.45	$299.60	$57.65	$461.20	$38.54	$59.01
2 Laborers	35.45	567.20	54.60	873.60		
1 Equip. Oper. (light)	45.80	366.40	69.20	553.60		
1 Air Track Drill, 4″		1009.00		1109.90		
1 Air Compressor, 600 cfm		530.80		583.88		
2 -50′ Air Hoses, 3″		29.80		32.78		
1 Gunite Pump Rig		369.80		406.78	60.61	66.67
32 L.H., Daily Totals		$3172.60		$4021.74	$99.14	$125.68

Crew B-47H	Hr.	Daily	Hr.	Daily	Bare Costs	Incl. O&P
1 Skilled Worker Foreman (out)	$48.20	$385.60	$74.50	$596.00	$46.70	$72.21
3 Skilled Workers	46.20	1108.80	71.45	1714.80		
1 Flatbed Truck, Gas, 3 Ton		327.80		360.58	10.24	11.27
32 L.H., Daily Totals		$1822.20		$2671.38	$56.94	$83.48

Crew B-48	Hr.	Daily	Hr.	Daily	Bare Costs	Incl. O&P
1 Labor Foreman (outside)	$37.45	$299.60	$57.65	$461.20	$40.09	$61.18
3 Laborers	35.45	850.80	54.60	1310.40		
1 Equip. Oper. (crane)	48.80	390.40	73.75	590.00		
1 Equip. Oper. Oiler	42.25	338.00	63.85	510.80		
1 Equip. Oper. (light)	45.80	366.40	69.20	553.60		
1 Centr. Water Pump, 6″		341.40		375.54		
1 -20′ Suction Hose, 6″		11.20		12.32		
1 -50′ Discharge Hose, 6″		12.10		13.31		
1 Drill Rig, Truck-Mounted		2469.00		2715.90	50.60	55.66
56 L.H., Daily Totals		$5078.90		$6543.07	$90.69	$116.84

Crew B-49	Hr.	Daily	Hr.	Daily	Bare Costs	Incl. O&P
1 Labor Foreman (outside)	$37.45	$299.60	$57.65	$461.20	$41.64	$63.93
3 Laborers	35.45	850.80	54.60	1310.40		
2 Equip. Oper. (crane)	48.80	780.80	73.75	1180.00		
2 Equip. Oper. Oilers	42.25	676.00	63.85	1021.60		
1 Equip. Oper. (light)	45.80	366.40	69.20	553.60		
2 Pile Drivers	43.15	690.40	68.70	1099.20		
1 Hyd. Crane, 25 Ton		728.60		801.46		
1 Centr. Water Pump, 6″		341.40		375.54		
1 -20′ Suction Hose, 6″		11.20		12.32		
1 -50′ Discharge Hose, 6″		12.10		13.31		
1 Drill Rig, Truck-Mounted		2469.00		2715.90	40.48	44.53
88 L.H., Daily Totals		$7226.30		$9544.53	$82.12	$108.46

Crew B-50	Hr.	Daily	Hr.	Daily	Bare Costs	Incl. O&P
2 Pile Driver Foremen (outside)	$45.15	$722.40	$71.90	$1150.40	$42.53	$66.51
6 Pile Drivers	43.15	2071.20	68.70	3297.60		
2 Equip. Oper. (crane)	48.80	780.80	73.75	1180.00		
1 Equip. Oper. Oiler	42.25	338.00	63.85	510.80		
3 Laborers	35.45	850.80	54.60	1310.40		
1 Crawler Crane, 40 Ton		1152.00		1267.20		
1 Lead, 60′ High		74.60		82.06		
1 Hammer, Diesel, 15K ft.-lbs.		592.40		651.64		
1 Air Compressor, 600 cfm		530.80		583.88		
2 -50′ Air Hoses, 3″		29.80		32.78		
1 Chain Saw, Gas, 36″ Long		45.00		49.50	21.65	23.81
112 L.H., Daily Totals		$7187.80		$10116.26	$64.18	$90.32

Crew B-51	Hr.	Daily	Hr.	Daily	Bare Costs	Incl. O&P
1 Labor Foreman (outside)	$37.45	$299.60	$57.65	$461.20	$35.82	$55.11
4 Laborers	35.45	1134.40	54.60	1747.20		
1 Truck Driver (light)	35.65	285.20	54.60	436.80		
1 Flatbed Truck, Gas, 1.5 Ton		265.80		292.38	5.54	6.09
48 L.H., Daily Totals		$1985.00		$2937.58	$41.35	$61.20

Crew B-52	Hr.	Daily	Hr.	Daily	Bare Costs	Incl. O&P
1 Carpenter Foreman (outside)	$46.90	$375.20	$72.25	$578.00	$41.41	$63.40
1 Carpenter	44.90	359.20	69.15	553.20		
3 Laborers	35.45	850.80	54.60	1310.40		
1 Cement Finisher	43.05	344.40	63.40	507.20		
.5 Rodman (reinf.)	49.80	199.20	78.70	314.80		
.5 Equip. Oper. (med.)	47.50	190.00	71.75	287.00		
.5 Crawler Loader, 3 C.Y.		593.50		652.85	10.60	11.66
56 L.H., Daily Totals		$2912.30		$4203.45	$52.01	$75.06

Crew B-53	Hr.	Daily	Hr.	Daily	Bare Costs	Incl. O&P
1 Equip. Oper. (light)	$45.80	$366.40	$69.20	$553.60	$45.80	$69.20
1 Trencher, Chain, 12 H.P.		70.40		77.44	8.80	9.68
8 L.H., Daily Totals		$436.80		$631.04	$54.60	$78.88

Crew B-54	Bare Costs Hr.	Bare Costs Daily	Incl. Subs O&P Hr.	Incl. Subs O&P Daily	Cost Per Labor-Hour Bare Costs	Cost Per Labor-Hour Incl. O&P
1 Equip. Oper. (light)	$45.80	$366.40	$69.20	$553.60	$45.80	$69.20
1 Trencher, Chain, 40 H.P.		339.20		373.12	42.40	46.64
8 L.H., Daily Totals		$705.60		$926.72	$88.20	$115.84

Crew B-54A	Hr.	Daily	Hr.	Daily	Bare Costs	Incl. O&P
.17 Labor Foreman (outside)	$37.45	$50.93	$57.65	$78.40	$46.04	$69.70
1 Equipment Operator (med.)	47.50	380.00	71.75	574.00		
1 Wheel Trencher, 67 H.P.		1198.00		1317.80	127.99	140.79
9.36 L.H., Daily Totals		$1628.93		$1970.20	$174.03	$210.49

Crew B-54B	Hr.	Daily	Hr.	Daily	Bare Costs	Incl. O&P
.25 Labor Foreman (outside)	$37.45	$74.90	$57.65	$115.30	$45.49	$68.93
1 Equipment Operator (med.)	47.50	380.00	71.75	574.00		
1 Wheel Trencher, 150 H.P.		1894.00		2083.40	189.40	208.34
10 L.H., Daily Totals		$2348.90		$2772.70	$234.89	$277.27

Crew B-54C	Hr.	Daily	Hr.	Daily	Bare Costs	Incl. O&P
1 Laborer	$35.45	$283.60	$54.60	$436.80	$41.48	$63.17
1 Equipment Operator (med.)	47.50	380.00	71.75	574.00		
1 Wheel Trencher, 67 H.P.		1198.00		1317.80	74.88	82.36
16 L.H., Daily Totals		$1861.60		$2328.60	$116.35	$145.54

Crew B-54D	Hr.	Daily	Hr.	Daily	Bare Costs	Incl. O&P
1 Laborer	$35.45	$283.60	$54.60	$436.80	$41.48	$63.17
1 Equipment Operator (med.)	47.50	380.00	71.75	574.00		
1 Rock Trencher, 6" Width		363.80		400.18	22.74	25.01
16 L.H., Daily Totals		$1027.40		$1410.98	$64.21	$88.19

Crew B-54E	Hr.	Daily	Hr.	Daily	Bare Costs	Incl. O&P
1 Laborer	$35.45	$283.60	$54.60	$436.80	$41.48	$63.17
1 Equipment Operator (med.)	47.50	380.00	71.75	574.00		
1 Rock Trencher, 18" Width		2596.00		2855.60	162.25	178.47
16 L.H., Daily Totals		$3259.60		$3866.40	$203.72	$241.65

Crew B-55	Hr.	Daily	Hr.	Daily	Bare Costs	Incl. O&P
2 Laborers	$35.45	$567.20	$54.60	$873.60	$35.52	$54.60
1 Truck Driver (light)	35.65	285.20	54.60	436.80		
1 Truck-Mounted Earth Auger		744.00		818.40		
1 Flatbed Truck, Gas, 3 Ton		327.80		360.58	44.66	49.12
24 L.H., Daily Totals		$1924.20		$2489.38	$80.17	$103.72

Crew B-56	Hr.	Daily	Hr.	Daily	Bare Costs	Incl. O&P
1 Laborer	$35.45	$283.60	$54.60	$436.80	$40.63	$61.90
1 Equip. Oper. (light)	45.80	366.40	69.20	553.60		
1 Air Track Drill, 4"		1009.00		1109.90		
1 Air Compressor, 600 cfm		530.80		583.88		
1 -50' Air Hose, 3"		14.90		16.39	97.17	106.89
16 L.H., Daily Totals		$2204.70		$2700.57	$137.79	$168.79

Crew B-57	Hr.	Daily	Hr.	Daily	Bare Costs	Incl. O&P
1 Labor Foreman (outside)	$37.45	$299.60	$57.65	$461.20	$40.87	$62.27
2 Laborers	35.45	567.20	54.60	873.60		
1 Equip. Oper. (crane)	48.80	390.40	73.75	590.00		
1 Equip. Oper. (light)	45.80	366.40	69.20	553.60		
1 Equip. Oper. Oiler	42.25	338.00	63.85	510.80		
1 Crawler Crane, 25 Ton		1148.00		1262.80		
1 Clamshell Bucket, 1 C.Y.		47.40		52.14		
1 Centr. Water Pump, 6"		341.40		375.54		
1 -20' Suction Hose, 6"		11.20		12.32		
20 -50' Discharge Hoses, 6"		242.00		266.20	37.29	41.02
48 L.H., Daily Totals		$3751.60		$4958.20	$78.16	$103.30

Crew B-58	Hr.	Daily	Hr.	Daily	Bare Costs	Incl. O&P
2 Laborers	$35.45	$567.20	$54.60	$873.60	$38.90	$59.47
1 Equip. Oper. (light)	45.80	366.40	69.20	553.60		
1 Backhoe Loader, 48 H.P.		367.40		404.14		
1 Small Helicopter, w/ Pilot		2692.00		2961.20	127.47	140.22
24 L.H., Daily Totals		$3993.00		$4792.54	$166.38	$199.69

Crew B-59	Hr.	Daily	Hr.	Daily	Bare Costs	Incl. O&P
1 Truck Driver (heavy)	$36.60	$292.80	$56.05	$448.40	$36.60	$56.05
1 Truck Tractor, 220 H.P.		366.80		403.48		
1 Water Tank Trailer, 5000 Gal.		143.20		157.52	63.75	70.13
8 L.H., Daily Totals		$802.80		$1009.40	$100.35	$126.18

Crew B-59A	Hr.	Daily	Hr.	Daily	Bare Costs	Incl. O&P
2 Laborers	$35.45	$567.20	$54.60	$873.60	$35.83	$55.08
1 Truck Driver (heavy)	36.60	292.80	56.05	448.40		
1 Water Tank Trailer, 5000 Gal.		143.20		157.52		
1 Truck Tractor, 220 H.P.		366.80		403.48	21.25	23.38
24 L.H., Daily Totals		$1370.00		$1883.00	$57.08	$78.46

Crew B-60	Hr.	Daily	Hr.	Daily	Bare Costs	Incl. O&P
1 Labor Foreman (outside)	$37.45	$299.60	$57.65	$461.20	$41.57	$63.26
2 Laborers	35.45	567.20	54.60	873.60		
1 Equip. Oper. (crane)	48.80	390.40	73.75	590.00		
2 Equip. Oper. (light)	45.80	732.80	69.20	1107.20		
1 Equip. Oper. Oiler	42.25	338.00	63.85	510.80		
1 Crawler Crane, 40 Ton		1152.00		1267.20		
1 Lead, 60' High		74.60		82.06		
1 Hammer, Diesel, 15K ft.-lbs.		592.40		651.64		
1 Backhoe Loader, 48 H.P.		367.40		404.14	39.04	42.95
56 L.H., Daily Totals		$4514.40		$5947.84	$80.61	$106.21

Crew B-61	Hr.	Daily	Hr.	Daily	Bare Costs	Incl. O&P
1 Labor Foreman (outside)	$37.45	$299.60	$57.65	$461.20	$37.92	$58.13
3 Laborers	35.45	850.80	54.60	1310.40		
1 Equip. Oper. (light)	45.80	366.40	69.20	553.60		
1 Cement Mixer, 2 C.Y.		190.60		209.66		
1 Air Compressor, 160 cfm		158.00		173.80	8.71	9.59
40 L.H., Daily Totals		$1865.40		$2708.66	$46.63	$67.72

Crew B-62	Hr.	Daily	Hr.	Daily	Bare Costs	Incl. O&P
2 Laborers	$35.45	$567.20	$54.60	$873.60	$38.90	$59.47
1 Equip. Oper. (light)	45.80	366.40	69.20	553.60		
1 Loader, Skid Steer, 30 H.P.		172.00		189.20	7.17	7.88
24 L.H., Daily Totals		$1105.60		$1616.40	$46.07	$67.35

Crew No.	Bare Costs		Incl. Subs O&P		Cost Per Labor-Hour	
Crew B-63	**Hr.**	**Daily**	**Hr.**	**Daily**	**Bare Costs**	**Incl. O&P**
4 Laborers	$35.45	$1134.40	$54.60	$1747.20	$37.52	$57.52
1 Equip. Oper. (light)	45.80	366.40	69.20	553.60		
1 Loader, Skid Steer, 30 H.P.		172.00		189.20	4.30	4.73
40 L.H., Daily Totals		$1672.80		$2490.00	$41.82	$62.25
Crew B-63B	**Hr.**	**Daily**	**Hr.**	**Daily**	**Bare Costs**	**Incl. O&P**
1 Labor Foreman (inside)	$35.95	$287.60	$55.35	$442.80	$38.16	$58.44
2 Laborers	35.45	567.20	54.60	873.60		
1 Equip. Oper. (light)	45.80	366.40	69.20	553.60		
1 Loader, Skid Steer, 78 H.P.		308.60		339.46	9.64	10.61
32 L.H., Daily Totals		$1529.80		$2209.46	$47.81	$69.05
Crew B-64	**Hr.**	**Daily**	**Hr.**	**Daily**	**Bare Costs**	**Incl. O&P**
1 Laborer	$35.45	$283.60	$54.60	$436.80	$35.55	$54.60
1 Truck Driver (light)	35.65	285.20	54.60	436.80		
1 Power Mulcher (Small)		158.00		173.80		
1 Flatbed Truck, Gas, 1.5 Ton		265.80		292.38	26.49	29.14
16 L.H., Daily Totals		$992.60		$1339.78	$62.04	$83.74
Crew B-65	**Hr.**	**Daily**	**Hr.**	**Daily**	**Bare Costs**	**Incl. O&P**
1 Laborer	$35.45	$283.60	$54.60	$436.80	$35.55	$54.60
1 Truck Driver (light)	35.65	285.20	54.60	436.80		
1 Power Mulcher (Large)		325.40		357.94		
1 Flatbed Truck, Gas, 1.5 Ton		265.80		292.38	36.95	40.65
16 L.H., Daily Totals		$1160.00		$1523.92	$72.50	$95.25
Crew B-66	**Hr.**	**Daily**	**Hr.**	**Daily**	**Bare Costs**	**Incl. O&P**
1 Equip. Oper. (light)	$45.80	$366.40	$69.20	$553.60	$45.80	$69.20
1 Loader-Backhoe, 40 H.P.		260.80		286.88	32.60	35.86
8 L.H., Daily Totals		$627.20		$840.48	$78.40	$105.06
Crew B-67	**Hr.**	**Daily**	**Hr.**	**Daily**	**Bare Costs**	**Incl. O&P**
1 Millwright	$46.50	$372.00	$68.35	$546.80	$46.15	$68.78
1 Equip. Oper. (light)	45.80	366.40	69.20	553.60		
1 Forklift, R/T, 4,000 Lb.		310.20		341.22	19.39	21.33
16 L.H., Daily Totals		$1048.60		$1441.62	$65.54	$90.10
Crew B-67B	**Hr.**	**Daily**	**Hr.**	**Daily**	**Bare Costs**	**Incl. O&P**
1 Millwright Foreman (inside)	$47.00	$376.00	$69.10	$552.80	$46.75	$68.72
1 Millwright	46.50	372.00	68.35	546.80		
16 L.H., Daily Totals		$748.00		$1099.60	$46.75	$68.72
Crew B-68	**Hr.**	**Daily**	**Hr.**	**Daily**	**Bare Costs**	**Incl. O&P**
2 Millwrights	$46.50	$744.00	$68.35	$1093.60	$46.27	$68.63
1 Equip. Oper. (light)	45.80	366.40	69.20	553.60		
1 Forklift, R/T, 4,000 Lb.		310.20		341.22	12.93	14.22
24 L.H., Daily Totals		$1420.60		$1988.42	$59.19	$82.85
Crew B-68A	**Hr.**	**Daily**	**Hr.**	**Daily**	**Bare Costs**	**Incl. O&P**
1 Millwright Foreman (inside)	$47.00	$376.00	$69.10	$552.80	$46.67	$68.60
2 Millwrights	46.50	744.00	68.35	1093.60		
1 Forklift, 8,000 Lb.		180.80		198.88	7.53	8.29
24 L.H., Daily Totals		$1300.80		$1845.28	$54.20	$76.89

Crew No.	Bare Costs		Incl. Subs O&P		Cost Per Labor-Hour	
Crew B-68B	**Hr.**	**Daily**	**Hr.**	**Daily**	**Bare Costs**	**Incl. O&P**
1 Millwright Foreman (inside)	$47.00	$376.00	$69.10	$552.80	$50.91	$75.81
2 Millwrights	46.50	744.00	68.35	1093.60		
2 Electricians	52.40	838.40	78.40	1254.40		
2 Plumbers	55.80	892.80	84.05	1344.80		
1 Forklift, 5,000 Lb.		325.00		357.50	5.80	6.38
56 L.H., Daily Totals		$3176.20		$4603.10	$56.72	$82.20
Crew B-68C	**Hr.**	**Daily**	**Hr.**	**Daily**	**Bare Costs**	**Incl. O&P**
1 Millwright Foreman (inside)	$47.00	$376.00	$69.10	$552.80	$50.42	$74.97
1 Millwright	46.50	372.00	68.35	546.80		
1 Electrician	52.40	419.20	78.40	627.20		
1 Plumber	55.80	446.40	84.05	672.40		
1 Forklift, 5,000 Lb.		325.00		357.50	10.16	11.17
32 L.H., Daily Totals		$1938.60		$2756.70	$60.58	$86.15
Crew B-68D	**Hr.**	**Daily**	**Hr.**	**Daily**	**Bare Costs**	**Incl. O&P**
1 Labor Foreman (inside)	$35.95	$287.60	$55.35	$442.80	$39.07	$59.72
1 Laborer	35.45	283.60	54.60	436.80		
1 Equip. Oper. (light)	45.80	366.40	69.20	553.60		
1 Forklift, 5,000 Lb.		325.00		357.50	13.54	14.90
24 L.H., Daily Totals		$1262.60		$1790.70	$52.61	$74.61
Crew B-68E	**Hr.**	**Daily**	**Hr.**	**Daily**	**Bare Costs**	**Incl. O&P**
1 Struc. Steel Foreman (inside)	$50.55	$404.40	$90.20	$721.60	$50.15	$89.48
3 Struc. Steel Workers	50.05	1201.20	89.30	2143.20		
1 Welder	50.05	400.40	89.30	714.40		
1 Forklift, 8,000 Lb.		180.80		198.88	4.52	4.97
40 L.H., Daily Totals		$2186.80		$3778.08	$54.67	$94.45
Crew B-69	**Hr.**	**Daily**	**Hr.**	**Daily**	**Bare Costs**	**Incl. O&P**
1 Labor Foreman (outside)	$37.45	$299.60	$57.65	$461.20	$39.14	$59.84
3 Laborers	35.45	850.80	54.60	1310.40		
1 Equip Oper. (crane)	48.80	390.40	73.75	590.00		
1 Equip Oper. Oiler	42.25	338.00	63.85	510.80		
1 Hyd. Crane, 80 Ton		1613.00		1774.30	33.60	36.96
48 L.H., Daily Totals		$3491.80		$4646.70	$72.75	$96.81
Crew B-69A	**Hr.**	**Daily**	**Hr.**	**Daily**	**Bare Costs**	**Incl. O&P**
1 Labor Foreman (outside)	$37.45	$299.60	$57.65	$461.20	$39.06	$59.43
3 Laborers	35.45	850.80	54.60	1310.40		
1 Equip. Oper. (med.)	47.50	380.00	71.75	574.00		
1 Concrete Finisher	43.05	344.40	63.40	507.20		
1 Curb/Gutter Paver, 2-Track		801.20		881.32	16.69	18.36
48 L.H., Daily Totals		$2676.00		$3734.12	$55.75	$77.79
Crew B-69B	**Hr.**	**Daily**	**Hr.**	**Daily**	**Bare Costs**	**Incl. O&P**
1 Labor Foreman (outside)	$37.45	$299.60	$57.65	$461.20	$39.06	$59.43
3 Laborers	35.45	850.80	54.60	1310.40		
1 Equip. Oper. (med.)	47.50	380.00	71.75	574.00		
1 Cement Finisher	43.05	344.40	63.40	507.20		
1 Curb/Gutter Paver, 4-Track		772.40		849.64	16.09	17.70
48 L.H., Daily Totals		$2647.20		$3702.44	$55.15	$77.13

Crew No.	Bare Costs		Incl. Subs O&P		Cost Per Labor-Hour	
Crew B-70	**Hr.**	**Daily**	**Hr.**	**Daily**	**Bare Costs**	**Incl. O&P**
1 Labor Foreman (outside)	$37.45	$299.60	$57.65	$461.20	$40.90	$62.39
3 Laborers	35.45	850.80	54.60	1310.40		
3 Equip. Oper. (med.)	47.50	1140.00	71.75	1722.00		
1 Grader, 30,000 Lbs.		712.40		783.64		
1 Ripper, Beam & 1 Shank		82.00		90.20		
1 Road Sweeper, S.P., 8' wide		662.80		729.08		
1 F.E. Loader, W.M., 1.5 C.Y.		375.20		412.72	32.72	35.99
56 L.H., Daily Totals		$4122.80		$5509.24	$73.62	$98.38

Crew B-70A	Hr.	Daily	Hr.	Daily	Bare Costs	Incl. O&P
1 Laborer	$35.45	$283.60	$54.60	$436.80	$45.09	$68.32
4 Equip. Oper. (med.)	47.50	1520.00	71.75	2296.00		
1 Grader, 40,000 Lbs.		1129.00		1241.90		
1 F.E. Loader, W.M.,2.5 C.Y.		526.20		578.82		
1 Dozer, 80 H.P.		480.60		528.66		
1 Roller, Pneum. Whl., 12 Ton		326.60		359.26	61.56	67.72
40 L.H., Daily Totals		$4266.00		$5441.44	$106.65	$136.04

Crew B-71	Hr.	Daily	Hr.	Daily	Bare Costs	Incl. O&P
1 Labor Foreman (outside)	$37.45	$299.60	$57.65	$461.20	$40.90	$62.39
3 Laborers	35.45	850.80	54.60	1310.40		
3 Equip. Oper. (med.)	47.50	1140.00	71.75	1722.00		
1 Pvmt. Profiler, 750 H.P.		5745.00		6319.50		
1 Road Sweeper, S.P., 8' wide		662.80		729.08		
1 F.E. Loader, W.M., 1.5 C.Y.		375.20		412.72	121.13	133.24
56 L.H., Daily Totals		$9073.40		$10954.90	$162.03	$195.62

Crew B-72	Hr.	Daily	Hr.	Daily	Bare Costs	Incl. O&P
1 Labor Foreman (outside)	$37.45	$299.60	$57.65	$461.20	$41.73	$63.56
3 Laborers	35.45	850.80	54.60	1310.40		
4 Equip. Oper. (med.)	47.50	1520.00	71.75	2296.00		
1 Pvmt. Profiler, 750 H.P.		5745.00		6319.50		
1 Hammermill, 250 H.P.		1827.00		2009.70		
1 Windrow Loader		1157.00		1272.70		
1 Mix Paver 165 H.P.		2139.00		2352.90		
1 Roller, Pneum. Whl., 12 Ton		326.60		359.26	174.92	192.41
64 L.H., Daily Totals		$13865.00		$16381.66	$216.64	$255.96

Crew B-73	Hr.	Daily	Hr.	Daily	Bare Costs	Incl. O&P
1 Labor Foreman (outside)	$37.45	$299.60	$57.65	$461.20	$43.23	$65.70
2 Laborers	35.45	567.20	54.60	873.60		
5 Equip. Oper. (med.)	47.50	1900.00	71.75	2870.00		
1 Road Mixer, 310 H.P.		1912.00		2103.20		
1 Tandem Roller, 10 Ton		231.20		254.32		
1 Hammermill, 250 H.P.		1827.00		2009.70		
1 Grader, 30,000 Lbs.		712.40		783.64		
.5 F.E. Loader, W.M., 1.5 C.Y.		187.60		206.36		
.5 Truck Tractor, 220 H.P.		183.40		201.74		
.5 Water Tank Trailer, 5000 Gal.		71.60		78.76	80.08	88.09
64 L.H., Daily Totals		$7892.00		$9842.52	$123.31	$153.79

Crew No.	Bare Costs		Incl. Subs O&P		Cost Per Labor-Hour	
Crew B-74	**Hr.**	**Daily**	**Hr.**	**Daily**	**Bare Costs**	**Incl. O&P**
1 Labor Foreman (outside)	$37.45	$299.60	$57.65	$461.20	$42.01	$63.92
1 Laborer	35.45	283.60	54.60	436.80		
4 Equip. Oper. (med.)	47.50	1520.00	71.75	2296.00		
2 Truck Drivers (heavy)	36.60	585.60	56.05	896.80		
1 Grader, 30,000 Lbs.		712.40		783.64		
1 Ripper, Beam & 1 Shank		82.00		90.20		
2 Stabilizers, 310 H.P.		3574.00		3931.40		
1 Flatbed Truck, Gas, 3 Ton		327.80		360.58		
1 Chem. Spreader, Towed		52.80		58.08		
1 Roller, Vibratory, 25 Ton		656.40		722.04		
1 Water Tank Trailer, 5000 Gal.		143.20		157.52		
1 Truck Tractor, 220 H.P.		366.80		403.48	92.43	101.67
64 L.H., Daily Totals		$8604.20		$10597.74	$134.44	$165.59

Crew B-75	Hr.	Daily	Hr.	Daily	Bare Costs	Incl. O&P
1 Labor Foreman (outside)	$37.45	$299.60	$57.65	$461.20	$42.79	$65.04
1 Laborer	35.45	283.60	54.60	436.80		
4 Equip. Oper. (med.)	47.50	1520.00	71.75	2296.00		
1 Truck Driver (heavy)	36.60	292.80	56.05	448.40		
1 Grader, 30,000 Lbs.		712.40		783.64		
1 Ripper, Beam & 1 Shank		82.00		90.20		
2 Stabilizers, 310 H.P.		3574.00		3931.40		
1 Dist. Tanker, 3000 Gallon		305.20		335.72		
1 Truck Tractor, 6x4, 380 H.P.		612.40		673.64		
1 Roller, Vibratory, 25 Ton		656.40		722.04	106.11	116.73
56 L.H., Daily Totals		$8338.40		$10179.04	$148.90	$181.77

Crew B-76	Hr.	Daily	Hr.	Daily	Bare Costs	Incl. O&P
1 Dock Builder Foreman (outside)	$45.15	$361.20	$71.90	$575.20	$44.53	$69.64
5 Dock Builders	43.15	1726.00	68.70	2748.00		
2 Equip. Oper. (crane)	48.80	780.80	73.75	1180.00		
1 Equip. Oper. Oiler	42.25	338.00	63.85	510.80		
1 Crawler Crane, 50 Ton		1292.00		1421.20		
1 Barge, 400 Ton		783.80		862.18		
1 Hammer, Diesel, 15K ft.-lbs.		592.40		651.64		
1 Lead, 60' High		74.60		82.06		
1 Air Compressor, 600 cfm		530.80		583.88		
2 -50' Air Hoses, 3"		29.80		32.78	45.88	50.47
72 L.H., Daily Totals		$6509.40		$8647.74	$90.41	$120.11

Crew B-76A	Hr.	Daily	Hr.	Daily	Bare Costs	Incl. O&P
1 Labor Foreman (outside)	$37.45	$299.60	$57.65	$461.20	$38.22	$58.53
5 Laborers	35.45	1418.00	54.60	2184.00		
1 Equip. Oper. (crane)	48.80	390.40	73.75	590.00		
1 Equip. Oper. Oiler	42.25	338.00	63.85	510.80		
1 Crawler Crane, 50 Ton		1292.00		1421.20		
1 Barge, 400 Ton		783.80		862.18	32.43	35.68
64 L.H., Daily Totals		$4521.80		$6029.38	$70.65	$94.21

Crew B-77	Hr.	Daily	Hr.	Daily	Bare Costs	Incl. O&P
1 Labor Foreman (outside)	$37.45	$299.60	$57.65	$461.20	$35.89	$55.21
3 Laborers	35.45	850.80	54.60	1310.40		
1 Truck Driver (light)	35.65	285.20	54.60	436.80		
1 Crack Cleaner, 25 H.P.		64.00		70.40		
1 Crack Filler, Trailer Mtd.		204.40		224.84		
1 Flatbed Truck, Gas, 3 Ton		327.80		360.58	14.90	16.40
40 L.H., Daily Totals		$2031.80		$2864.22	$50.80	$71.61

Crew No.	Bare Costs		Incl. Subs O&P		Cost Per Labor-Hour	

Crew B-78	Hr.	Daily	Hr.	Daily	Bare Costs	Incl. O&P
1 Labor Foreman (outside)	$37.45	$299.60	$57.65	$461.20	$35.82	$55.11
4 Laborers	35.45	1134.40	54.60	1747.20		
1 Truck Driver (light)	35.65	285.20	54.60	436.80		
1 Paint Striper, S.P., 40 Gallon		154.00		169.40		
1 Flatbed Truck, Gas, 3 Ton		327.80		360.58		
1 Pickup Truck, 3/4 Ton		154.60		170.06	13.26	14.58
48 L.H., Daily Totals		$2355.60		$3345.24	$49.08	$69.69

Crew B-78A	Hr.	Daily	Hr.	Daily	Bare Costs	Incl. O&P
1 Equip. Oper. (light)	$45.80	$366.40	$69.20	$553.60	$45.80	$69.20
1 Line Rem. (Metal Balls) 115 H.P.		823.20		905.52	102.90	113.19
8 L.H., Daily Totals		$1189.60		$1459.12	$148.70	$182.39

Crew B-78B	Hr.	Daily	Hr.	Daily	Bare Costs	Incl. O&P
2 Laborers	$35.45	$567.20	$54.60	$873.60	$36.60	$56.22
.25 Equip. Oper. (light)	45.80	91.60	69.20	138.40		
1 Pickup Truck, 3/4 Ton		154.60		170.06		
1 Line Rem.,11 H.P.,Walk Behind		67.60		74.36		
.25 Road Sweeper, S.P., 8' wide		165.70		182.27	21.55	23.70
18 L.H., Daily Totals		$1046.70		$1438.69	$58.15	$79.93

Crew B-78C	Hr.	Daily	Hr.	Daily	Bare Costs	Incl. O&P
1 Labor Foreman (outside)	$37.45	$299.60	$57.65	$461.20	$35.82	$55.11
4 Laborers	35.45	1134.40	54.60	1747.20		
1 Truck Driver (light)	35.65	285.20	54.60	436.80		
1 Paint Striper, T.M., 120 Gal.		853.00		938.30		
1 Flatbed Truck, Gas, 3 Ton		327.80		360.58		
1 Pickup Truck, 3/4 Ton		154.60		170.06	27.82	30.60
48 L.H., Daily Totals		$3054.60		$4114.14	$63.64	$85.71

Crew B-78D	Hr.	Daily	Hr.	Daily	Bare Costs	Incl. O&P
2 Labor Foremen (Outside)	$37.45	$599.20	$57.65	$922.40	$35.87	$55.21
7 Laborers	35.45	1985.20	54.60	3057.60		
1 Truck Driver (light)	35.65	285.20	54.60	436.80		
1 Paint Striper, T.M., 120 Gal.		853.00		938.30		
1 Flatbed Truck, Gas, 3 Ton		327.80		360.58		
3 Pickup Trucks, 3/4 Ton		463.80		510.18		
1 Air Compressor, 60 cfm		137.60		151.36		
1 -50' Air Hose, 3/4"		3.25		3.58		
1 Breakers, Pavement, 60 lb.		9.80		10.78	22.44	24.68
80 L.H., Daily Totals		$4664.85		$6391.57	$58.31	$79.89

Crew B-78E	Hr.	Daily	Hr.	Daily	Bare Costs	Incl. O&P
2 Labor Foremen (Outside)	$37.45	$599.20	$57.65	$922.40	$35.80	$55.11
9 Laborers	35.45	2552.40	54.60	3931.20		
1 Truck Driver (light)	35.65	285.20	54.60	436.80		
1 Paint Striper, T.M., 120 Gal.		853.00		938.30		
1 Flatbed Truck, Gas, 3 Ton		327.80		360.58		
4 Pickup Trucks, 3/4 Ton		618.40		680.24		
2 Air Compressor, 60 cfm		275.20		302.72		
2 -50' Air Hose, 3/4"		6.50		7.15		
2 Breakers, Pavement, 60 lb.		19.60		21.56	21.88	24.07
96 L.H., Daily Totals		$5537.30		$7600.95	$57.68	$79.18

Crew B-78F	Hr.	Daily	Hr.	Daily	Bare Costs	Incl. O&P
2 Labor Foremen (Outside)	$37.45	$599.20	$57.65	$922.40	$35.75	$55.04
11 Laborers	35.45	3119.60	54.60	4804.80		
1 Truck Driver (light)	35.65	285.20	54.60	436.80		
1 Paint Striper, T.M., 120 Gal.		853.00		938.30		
1 Flatbed Truck, Gas, 3 Ton		327.80		360.58		
7 Pickup Trucks, 3/4 Ton		1082.20		1190.42		
3 Air Compressor, 60 cfm		412.80		454.08		
3 -50' Air Hose, 3/4"		9.75		10.73		
3 Breakers, Pavement, 60 lb.		29.40		32.34	24.24	26.66
112 L.H., Daily Totals		$6718.95		$9150.44	$59.99	$81.70

Crew B-79	Hr.	Daily	Hr.	Daily	Bare Costs	Incl. O&P
1 Labor Foreman (outside)	$37.45	$299.60	$57.65	$461.20	$35.89	$55.21
3 Laborers	35.45	850.80	54.60	1310.40		
1 Truck Driver (light)	35.65	285.20	54.60	436.80		
1 Paint Striper, T.M., 120 Gal.		853.00		938.30		
1 Heating Kettle, 115 Gallon		46.45		51.09		
1 Flatbed Truck, Gas, 3 Ton		327.80		360.58		
2 Pickup Trucks, 3/4 Ton		309.20		340.12	38.41	42.25
40 L.H., Daily Totals		$2972.05		$3898.49	$74.30	$97.46

Crew B-79A	Hr.	Daily	Hr.	Daily	Bare Costs	Incl. O&P
1.5 Equip. Oper. (light)	$45.80	$549.60	$69.20	$830.40	$45.80	$69.20
.5 Line Remov. (Grinder) 115 H.P.		455.70		501.27		
1 Line Rem. (Metal Balls) 115 H.P.		823.20		905.52	106.58	117.23
12 L.H., Daily Totals		$1828.50		$2237.19	$152.38	$186.43

Crew B-79B	Hr.	Daily	Hr.	Daily	Bare Costs	Incl. O&P
1 Laborer	$35.45	$283.60	$54.60	$436.80	$35.45	$54.60
1 Set of Gases		133.20		146.52	16.65	18.32
8 L.H., Daily Totals		$416.80		$583.32	$52.10	$72.92

Crew B-79C	Hr.	Daily	Hr.	Daily	Bare Costs	Incl. O&P
1 Labor Foreman (outside)	$37.45	$299.60	$57.65	$461.20	$35.76	$55.04
5 Laborers	35.45	1418.00	54.60	2184.00		
1 Truck Driver (light)	35.65	285.20	54.60	436.80		
1 Paint Striper, T.M., 120 Gal.		853.00		938.30		
1 Heating Kettle, 115 Gallon		46.45		51.09		
1 Flatbed Truck, Gas, 3 Ton		327.80		360.58		
3 Pickup Trucks, 3/4 Ton		463.80		510.18		
1 Air Compressor, 60 cfm		137.60		151.36		
1 -50' Air Hose, 3/4"		3.25		3.58		
1 Breakers, Pavement, 60 lb.		9.80		10.78	32.89	36.18
56 L.H., Daily Totals		$3844.50		$5107.87	$68.65	$91.21

Crew B-79D	Hr.	Daily	Hr.	Daily	Bare Costs	Incl. O&P
2 Labor Foremen (Outside)	$37.45	$599.20	$57.65	$922.40	$35.98	$55.36
5 Laborers	35.45	1418.00	54.60	2184.00		
1 Truck Driver (light)	35.65	285.20	54.60	436.80		
1 Paint Striper, T.M., 120 Gal.		853.00		938.30		
1 Heating Kettle, 115 Gallon		46.45		51.09		
1 Flatbed Truck, Gas, 3 Ton		327.80		360.58		
4 Pickup Trucks, 3/4 Ton		618.40		680.24		
1 Air Compressor, 60 cfm		137.60		151.36		
1 -50' Air Hose, 3/4"		3.25		3.58		
1 Breakers, Pavement, 60 lb.		9.80		10.78	31.19	34.31
64 L.H., Daily Totals		$4298.70		$5739.13	$67.17	$89.67

Crew No.	Bare Costs	Incl. Subs O&P	Cost Per Labor-Hour

Crew B-79E	Hr.	Daily	Hr.	Daily	Bare Costs	Incl. O&P
2 Labor Foremen (Outside)	$37.45	$599.20	$57.65	$922.40	$35.87	$55.21
7 Laborers	35.45	1985.20	54.60	3057.60		
1 Truck Driver (light)	35.65	285.20	54.60	436.80		
1 Paint Striper, T.M., 120 Gal.		853.00		938.30		
1 Heating Kettle, 115 Gallon		46.45		51.09		
1 Flatbed Truck, Gas, 3 Ton		327.80		360.58		
5 Pickup Trucks, 3/4 Ton		773.00		850.30		
2 Air Compressors, 60 cfm		275.20		302.72		
2 -50' Air Hoses, 3/4"		6.50		7.15		
2 Breakers, Pavement, 60 lb.		19.60		21.56	28.77	31.65
80 L.H., Daily Totals		$5171.15		$6948.51	$64.64	$86.86

Crew B-80	Hr.	Daily	Hr.	Daily	Bare Costs	Incl. O&P
1 Labor Foreman (outside)	$37.45	$299.60	$57.65	$461.20	$38.59	$59.01
1 Laborer	35.45	283.60	54.60	436.80		
1 Truck Driver (light)	35.65	285.20	54.60	436.80		
1 Equip. Oper. (light)	45.80	366.40	69.20	553.60		
1 Flatbed Truck, Gas, 3 Ton		327.80		360.58		
1 Earth Auger, Truck-Mtd.		431.20		474.32	23.72	26.09
32 L.H., Daily Totals		$1993.80		$2723.30	$62.31	$85.10

Crew B-80A	Hr.	Daily	Hr.	Daily	Bare Costs	Incl. O&P
3 Laborers	$35.45	$850.80	$54.60	$1310.40	$35.45	$54.60
1 Flatbed Truck, Gas, 3 Ton		327.80		360.58	13.66	15.02
24 L.H., Daily Totals		$1178.60		$1670.98	$49.11	$69.62

Crew B-80B	Hr.	Daily	Hr.	Daily	Bare Costs	Incl. O&P
3 Laborers	$35.45	$850.80	$54.60	$1310.40	$38.04	$58.25
1 Equip. Oper. (light)	45.80	366.40	69.20	553.60		
1 Crane, Flatbed Mounted, 3 Ton		243.00		267.30	7.59	8.35
32 L.H., Daily Totals		$1460.20		$2131.30	$45.63	$66.60

Crew B-80C	Hr.	Daily	Hr.	Daily	Bare Costs	Incl. O&P
2 Laborers	$35.45	$567.20	$54.60	$873.60	$35.52	$54.60
1 Truck Driver (light)	35.65	285.20	54.60	436.80		
1 Flatbed Truck, Gas, 1.5 Ton		265.80		292.38		
1 Manual Fence Post Auger, Gas		8.00		8.80	11.41	12.55
24 L.H., Daily Totals		$1126.20		$1611.58	$46.92	$67.15

Crew B-81	Hr.	Daily	Hr.	Daily	Bare Costs	Incl. O&P
1 Laborer	$35.45	$283.60	$54.60	$436.80	$39.85	$60.80
1 Equip. Oper. (med.)	47.50	380.00	71.75	574.00		
1 Truck Driver (heavy)	36.60	292.80	56.05	448.40		
1 Hydromulcher, T.M., 3000 Gal.		341.20		375.32		
1 Truck Tractor, 220 H.P.		366.80		403.48	29.50	32.45
24 L.H., Daily Totals		$1664.40		$2238.00	$69.35	$93.25

Crew B-81A	Hr.	Daily	Hr.	Daily	Bare Costs	Incl. O&P
1 Laborer	$35.45	$283.60	$54.60	$436.80	$35.55	$54.60
1 Truck Driver (light)	35.65	285.20	54.60	436.80		
1 Hydromulcher, T.M., 600 Gal.		131.00		144.10		
1 Flatbed Truck, Gas, 3 Ton		327.80		360.58	28.68	31.54
16 L.H., Daily Totals		$1027.60		$1378.28	$64.22	$86.14

Crew B-82	Hr.	Daily	Hr.	Daily	Bare Costs	Incl. O&P
1 Laborer	$35.45	$283.60	$54.60	$436.80	$40.63	$61.90
1 Equip. Oper. (light)	45.80	366.40	69.20	553.60		
1 Horiz. Borer, 6 H.P.		57.60		63.36	3.60	3.96
16 L.H., Daily Totals		$707.60		$1053.76	$44.23	$65.86

Crew No.	Bare Costs	Incl. Subs O&P	Cost Per Labor-Hour

Crew B-82A	Hr.	Daily	Hr.	Daily	Bare Costs	Incl. O&P
1 Laborer	$35.45	$283.60	$54.60	$436.80	$40.63	$61.90
1 Equip. Oper. (light)	45.80	366.40	69.20	553.60		
1 Flatbed Truck, Gas, 3 Ton		327.80		360.58		
1 Flatbed Trailer, 25 Ton		113.20		124.52		
1 Horiz. Dir. Drill, 20k lb. Thrust		635.40		698.94	67.28	74.00
16 L.H., Daily Totals		$1726.40		$2174.44	$107.90	$135.90

Crew B-82B	Hr.	Daily	Hr.	Daily	Bare Costs	Incl. O&P
2 Laborers	$35.45	$567.20	$54.60	$873.60	$38.90	$59.47
1 Equip. Oper. (light)	45.80	366.40	69.20	553.60		
1 Flatbed Truck, Gas, 3 Ton		327.80		360.58		
1 Flatbed Trailer, 25 Ton		113.20		124.52		
1 Horiz. Dir. Drill, 30k lb. Thrust		904.40		994.84	56.06	61.66
24 L.H., Daily Totals		$2279.00		$2907.14	$94.96	$121.13

Crew B-82C	Hr.	Daily	Hr.	Daily	Bare Costs	Incl. O&P
2 Laborers	$35.45	$567.20	$54.60	$873.60	$38.90	$59.47
1 Equip. Oper. (light)	45.80	366.40	69.20	553.60		
1 Flatbed Truck, Gas, 3 Ton		327.80		360.58		
1 Flatbed Trailer, 25 Ton		113.20		124.52		
1 Horiz. Dir. Drill, 50k lb. Thrust		1203.00		1323.30	68.50	75.35
24 L.H., Daily Totals		$2577.60		$3235.60	$107.40	$134.82

Crew B-82D	Hr.	Daily	Hr.	Daily	Bare Costs	Incl. O&P
1 Equip. Oper. (light)	$45.80	$366.40	$69.20	$553.60	$45.80	$69.20
1 Mud Trailer for HDD, 1500 Gal.		370.20		407.22	46.27	50.90
8 L.H., Daily Totals		$736.60		$960.82	$92.08	$120.10

Crew B-83	Hr.	Daily	Hr.	Daily	Bare Costs	Incl. O&P
1 Tugboat Captain	$47.50	$380.00	$71.75	$574.00	$41.48	$63.17
1 Tugboat Hand	35.45	283.60	54.60	436.80		
1 Tugboat, 250 H.P.		887.00		975.70	55.44	60.98
16 L.H., Daily Totals		$1550.60		$1986.50	$96.91	$124.16

Crew B-84	Hr.	Daily	Hr.	Daily	Bare Costs	Incl. O&P
1 Equip. Oper. (med.)	$47.50	$380.00	$71.75	$574.00	$47.50	$71.75
1 Rotary Mower/Tractor		363.20		399.52	45.40	49.94
8 L.H., Daily Totals		$743.20		$973.52	$92.90	$121.69

Crew B-85	Hr.	Daily	Hr.	Daily	Bare Costs	Incl. O&P
3 Laborers	$35.45	$850.80	$54.60	$1310.40	$38.09	$58.32
1 Equip. Oper. (med.)	47.50	380.00	71.75	574.00		
1 Truck Driver (heavy)	36.60	292.80	56.05	448.40		
1 Aerial Lift Truck, 80'		650.80		715.88		
1 Brush Chipper, 12", 130 H.P.		396.40		436.04		
1 Pruning Saw, Rotary		6.60		7.26	26.34	28.98
40 L.H., Daily Totals		$2577.40		$3491.98	$64.44	$87.30

Crew B-86	Hr.	Daily	Hr.	Daily	Bare Costs	Incl. O&P
1 Equip. Oper. (med.)	$47.50	$380.00	$71.75	$574.00	$47.50	$71.75
1 Stump Chipper, S.P.		172.75		190.03	21.59	23.75
8 L.H., Daily Totals		$552.75		$764.02	$69.09	$95.50

Crew B-86A	Hr.	Daily	Hr.	Daily	Bare Costs	Incl. O&P
1 Equip. Oper. (med.)	$47.50	$380.00	$71.75	$574.00	$47.50	$71.75
1 Grader, 30,000 Lbs.		712.40		783.64	89.05	97.95
8 L.H., Daily Totals		$1092.40		$1357.64	$136.55	$169.71

Crew B-86B	Hr.	Daily	Hr.	Daily	Bare Costs	Incl. O&P
1 Equip. Oper. (med.)	$47.50	$380.00	$71.75	$574.00	$47.50	$71.75
1 Dozer, 200 H.P.		1333.00		1466.30	166.63	183.29
8 L.H., Daily Totals		$1713.00		$2040.30	$214.13	$255.04

Crew B-87	Hr.	Daily	Hr.	Daily	Bare Costs	Incl. O&P
1 Laborer	$35.45	$283.60	$54.60	$436.80	$45.09	$68.32
4 Equip. Oper. (med.)	47.50	1520.00	71.75	2296.00		
2 Feller Bunchers, 100 H.P.		1233.20		1356.52		
1 Log Chipper, 22" Tree		900.20		990.22		
1 Dozer, 105 H.P.		592.40		651.64		
1 Chain Saw, Gas, 36" Long		45.00		49.50	69.27	76.20
40 L.H., Daily Totals		$4574.40		$5780.68	$114.36	$144.52

Crew B-88	Hr.	Daily	Hr.	Daily	Bare Costs	Incl. O&P
1 Laborer	$35.45	$283.60	$54.60	$436.80	$45.78	$69.30
6 Equip. Oper. (med.)	47.50	2280.00	71.75	3444.00		
2 Feller Bunchers, 100 H.P.		1233.20		1356.52		
1 Log Chipper, 22" Tree		900.20		990.22		
2 Log Skidders, 50 H.P.		1836.80		2020.48		
1 Dozer, 105 H.P.		592.40		651.64		
1 Chain Saw, Gas, 36" Long		45.00		49.50	82.28	90.51
56 L.H., Daily Totals		$7171.20		$8949.16	$128.06	$159.81

Crew B-89	Hr.	Daily	Hr.	Daily	Bare Costs	Incl. O&P
1 Equip. Oper. (light)	$45.80	$366.40	$69.20	$553.60	$40.73	$61.90
1 Truck Driver (light)	35.65	285.20	54.60	436.80		
1 Flatbed Truck, Gas, 3 Ton		327.80		360.58		
1 Concrete Saw		175.20		192.72		
1 Water Tank, 65 Gal.		18.45		20.30	32.59	35.85
16 L.H., Daily Totals		$1173.05		$1563.99	$73.32	$97.75

Crew B-89A	Hr.	Daily	Hr.	Daily	Bare Costs	Incl. O&P
1 Skilled Worker	$46.20	$369.60	$71.45	$571.60	$40.83	$63.02
1 Laborer	35.45	283.60	54.60	436.80		
1 Core Drill (Large)		112.40		123.64	7.03	7.73
16 L.H., Daily Totals		$765.60		$1132.04	$47.85	$70.75

Crew B-89B	Hr.	Daily	Hr.	Daily	Bare Costs	Incl. O&P
1 Equip. Oper. (light)	$45.80	$366.40	$69.20	$553.60	$40.73	$61.90
1 Truck Driver, Light	35.65	285.20	54.60	436.80		
1 Wall Saw, Hydraulic, 10 H.P.		119.60		131.56		
1 Generator, Diesel, 100 kW		413.80		455.18		
1 Water Tank, 65 Gal.		18.45		20.30		
1 Flatbed Truck, Gas, 3 Ton		327.80		360.58	54.98	60.48
16 L.H., Daily Totals		$1531.25		$1958.02	$95.70	$122.38

Crew B-90	Hr.	Daily	Hr.	Daily	Bare Costs	Incl. O&P
1 Labor Foreman (outside)	$37.45	$299.60	$57.65	$461.20	$38.58	$58.99
3 Laborers	35.45	850.80	54.60	1310.40		
2 Equip. Oper. (light)	45.80	732.80	69.20	1107.20		
2 Truck Drivers (heavy)	36.60	585.60	56.05	896.80		
1 Road Mixer, 310 H.P.		1912.00		2103.20		
1 Dist. Truck, 2000 Gal.		275.20		302.72	34.17	37.59
64 L.H., Daily Totals		$4656.00		$6181.52	$72.75	$96.59

Crew B-90A	Hr.	Daily	Hr.	Daily	Bare Costs	Incl. O&P
1 Labor Foreman (outside)	$37.45	$299.60	$57.65	$461.20	$42.62	$64.84
2 Laborers	35.45	567.20	54.60	873.60		
4 Equip. Oper. (med.)	47.50	1520.00	71.75	2296.00		
2 Graders, 30,000 Lbs.		1424.80		1567.28		
1 Tandem Roller, 10 Ton		231.20		254.32		
1 Roller, Pneum. Whl., 12 Ton		326.60		359.26	35.40	38.94
56 L.H., Daily Totals		$4369.40		$5811.66	$78.03	$103.78

Crew B-90B	Hr.	Daily	Hr.	Daily	Bare Costs	Incl. O&P
1 Labor Foreman (outside)	$37.45	$299.60	$57.65	$461.20	$41.81	$63.68
2 Laborers	35.45	567.20	54.60	873.60		
3 Equip. Oper. (med.)	47.50	1140.00	71.75	1722.00		
1 Roller, Pneum. Whl., 12 Ton		326.60		359.26		
1 Road Mixer, 310 H.P.		1912.00		2103.20	46.64	51.30
48 L.H., Daily Totals		$4245.40		$5519.26	$88.45	$114.98

Crew B-90C	Hr.	Daily	Hr.	Daily	Bare Costs	Incl. O&P
1 Labor Foreman (outside)	$37.45	$299.60	$57.65	$461.20	$39.23	$59.95
4 Laborers	35.45	1134.40	54.60	1747.20		
3 Equip. Oper. (med.)	47.50	1140.00	71.75	1722.00		
3 Truck Drivers (heavy)	36.60	878.40	56.05	1345.20		
3 Road Mixers, 310 H.P.		5736.00		6309.60	65.18	71.70
88 L.H., Daily Totals		$9188.40		$11585.20	$104.41	$131.65

Crew B-90D	Hr.	Daily	Hr.	Daily	Bare Costs	Incl. O&P
1 Labor Foreman (outside)	$37.45	$299.60	$57.65	$461.20	$38.65	$59.13
6 Laborers	35.45	1701.60	54.60	2620.80		
3 Equip. Oper. (med.)	47.50	1140.00	71.75	1722.00		
3 Truck Drivers (heavy)	36.60	878.40	56.05	1345.20		
3 Road Mixers, 310 H.P.		5736.00		6309.60	55.15	60.67
104 L.H., Daily Totals		$9755.60		$12458.80	$93.80	$119.80

Crew B-90E	Hr.	Daily	Hr.	Daily	Bare Costs	Incl. O&P
1 Labor Foreman (outside)	$37.45	$299.60	$57.65	$461.20	$39.82	$60.82
4 Laborers	35.45	1134.40	54.60	1747.20		
3 Equip. Oper. (med.)	47.50	1140.00	71.75	1722.00		
1 Truck Driver (heavy)	36.60	292.80	56.05	448.40		
1 Road Mixers, 310 H.P.		1912.00		2103.20	26.56	29.21
72 L.H., Daily Totals		$4778.80		$6482.00	$66.37	$90.03

Crew B-91	Hr.	Daily	Hr.	Daily	Bare Costs	Incl. O&P
1 Labor Foreman (outside)	$37.45	$299.60	$57.65	$461.20	$41.87	$63.74
2 Laborers	35.45	567.20	54.60	873.60		
4 Equip. Oper. (med.)	47.50	1520.00	71.75	2296.00		
1 Truck Driver (heavy)	36.60	292.80	56.05	448.40		
1 Dist. Tanker, 3000 Gallon		305.20		335.72		
1 Truck Tractor, 6x4, 380 H.P.		612.40		673.64		
1 Aggreg. Spreader, S.P.		838.60		922.46		
1 Roller, Pneum. Whl., 12 Ton		326.60		359.26		
1 Tandem Roller, 10 Ton		231.20		254.32	36.16	39.77
64 L.H., Daily Totals		$4993.60		$6624.60	$78.03	$103.51

Crew B-91B	Hr.	Daily	Hr.	Daily	Bare Costs	Incl. O&P
1 Laborer	$35.45	$283.60	$54.60	$436.80	$41.48	$63.17
1 Equipment Oper. (med.)	47.50	380.00	71.75	574.00		
1 Road Sweeper, Vac. Assist.		990.60		1089.66	61.91	68.10
16 L.H., Daily Totals		$1654.20		$2100.46	$103.39	$131.28

Crews

Crew B-91C	Hr.	Daily	Hr.	Daily	Bare Costs	Incl. O&P
1 Laborer	$35.45	$283.60	$54.60	$436.80	$35.55	$54.60
1 Truck Driver (light)	35.65	285.20	54.60	436.80		
1 Catch Basin Cleaning Truck		601.60		661.76	37.60	41.36
16 L.H., Daily Totals		$1170.40		$1535.36	$73.15	$95.96

Crew B-91D	Hr.	Daily	Hr.	Daily	Bare Costs	Incl. O&P
1 Labor Foreman (outside)	$37.45	$299.60	$57.65	$461.20	$40.42	$61.65
5 Laborers	35.45	1418.00	54.60	2184.00		
5 Equip. Oper. (med.)	47.50	1900.00	71.75	2870.00		
2 Truck Drivers (heavy)	36.60	585.60	56.05	896.80		
1 Aggreg. Spreader, S.P.		838.60		922.46		
2 Truck Tractor, 6x4, 380 H.P.		1224.80		1347.28		
2 Dist. Tanker, 3000 Gallon		610.40		671.44		
2 Pavement Brush, Towed		162.40		178.64		
2 Roller, Pneum. Whl., 12 Ton		653.20		718.52	33.55	36.91
104 L.H., Daily Totals		$7692.60		$10250.34	$73.97	$98.56

Crew B-92	Hr.	Daily	Hr.	Daily	Bare Costs	Incl. O&P
1 Labor Foreman (outside)	$37.45	$299.60	$57.65	$461.20	$35.95	$55.36
3 Laborers	35.45	850.80	54.60	1310.40		
1 Crack Cleaner, 25 H.P.		64.00		70.40		
1 Air Compressor, 60 cfm		137.60		151.36		
1 Tar Kettle, T.M.		78.70		86.57		
1 Flatbed Truck, Gas, 3 Ton		327.80		360.58	19.00	20.90
32 L.H., Daily Totals		$1758.50		$2440.51	$54.95	$76.27

Crew B-93	Hr.	Daily	Hr.	Daily	Bare Costs	Incl. O&P
1 Equip. Oper. (med.)	$47.50	$380.00	$71.75	$574.00	$47.50	$71.75
1 Feller Buncher, 100 H.P.		616.60		678.26	77.08	84.78
8 L.H., Daily Totals		$996.60		$1252.26	$124.58	$156.53

Crew B-94A	Hr.	Daily	Hr.	Daily	Bare Costs	Incl. O&P
1 Laborer	$35.45	$283.60	$54.60	$436.80	$35.45	$54.60
1 Diaphragm Water Pump, 2"		71.60		78.76		
1 -20' Suction Hose, 2"		1.95		2.15		
2 -50' Discharge Hoses, 2"		3.50		3.85	9.63	10.59
8 L.H., Daily Totals		$360.65		$521.55	$45.08	$65.19

Crew B-94B	Hr.	Daily	Hr.	Daily	Bare Costs	Incl. O&P
1 Laborer	$35.45	$283.60	$54.60	$436.80	$35.45	$54.60
1 Diaphragm Water Pump, 4"		114.40		125.84		
1 -20' Suction Hose, 4"		3.25		3.58		
2 -50' Discharge Hoses, 4"		9.40		10.34	15.88	17.47
8 L.H., Daily Totals		$410.65		$576.55	$51.33	$72.07

Crew B-94C	Hr.	Daily	Hr.	Daily	Bare Costs	Incl. O&P
1 Laborer	$35.45	$283.60	$54.60	$436.80	$35.45	$54.60
1 Centrifugal Water Pump, 3"		78.80		86.68		
1 -20' Suction Hose, 3"		2.85		3.13		
2 -50' Discharge Hoses, 3"		6.10		6.71	10.97	12.07
8 L.H., Daily Totals		$371.35		$533.33	$46.42	$66.67

Crew B-94D	Hr.	Daily	Hr.	Daily	Bare Costs	Incl. O&P
1 Laborer	$35.45	$283.60	$54.60	$436.80	$35.45	$54.60
1 Centr. Water Pump, 6"		341.40		375.54		
1 -20' Suction Hose, 6"		11.20		12.32		
2 -50' Discharge Hoses, 6"		24.20		26.62	47.10	51.81
8 L.H., Daily Totals		$660.40		$851.28	$82.55	$106.41

Crew C-1	Hr.	Daily	Hr.	Daily	Bare Costs	Incl. O&P
3 Carpenters	$44.90	$1077.60	$69.15	$1659.60	$42.54	$65.51
1 Laborer	35.45	283.60	54.60	436.80		
32 L.H., Daily Totals		$1361.20		$2096.40	$42.54	$65.51

Crew C-2	Hr.	Daily	Hr.	Daily	Bare Costs	Incl. O&P
1 Carpenter Foreman (outside)	$46.90	$375.20	$72.25	$578.00	$43.66	$67.24
4 Carpenters	44.90	1436.80	69.15	2212.80		
1 Laborer	35.45	283.60	54.60	436.80		
48 L.H., Daily Totals		$2095.60		$3227.60	$43.66	$67.24

Crew C-2A	Hr.	Daily	Hr.	Daily	Bare Costs	Incl. O&P
1 Carpenter Foreman (outside)	$46.90	$375.20	$72.25	$578.00	$43.35	$66.28
3 Carpenters	44.90	1077.60	69.15	1659.60		
1 Cement Finisher	43.05	344.40	63.40	507.20		
1 Laborer	35.45	283.60	54.60	436.80		
48 L.H., Daily Totals		$2080.80		$3181.60	$43.35	$66.28

Crew C-3	Hr.	Daily	Hr.	Daily	Bare Costs	Incl. O&P
1 Rodman Foreman (outside)	$51.80	$414.40	$81.85	$654.80	$45.96	$71.88
4 Rodmen (reinf.)	49.80	1593.60	78.70	2518.40		
1 Equip. Oper. (light)	45.80	366.40	69.20	553.60		
2 Laborers	35.45	567.20	54.60	873.60		
3 Stressing Equipment		30.60		33.66		
.5 Grouting Equipment		79.70		87.67	1.72	1.90
64 L.H., Daily Totals		$3051.90		$4721.73	$47.69	$73.78

Crew C-4	Hr.	Daily	Hr.	Daily	Bare Costs	Incl. O&P
1 Rodman Foreman (outside)	$51.80	$414.40	$81.85	$654.80	$50.30	$79.49
3 Rodmen (reinf.)	49.80	1195.20	78.70	1888.80		
3 Stressing Equipment		30.60		33.66	.96	1.05
32 L.H., Daily Totals		$1640.20		$2577.26	$51.26	$80.54

Crew C-4A	Hr.	Daily	Hr.	Daily	Bare Costs	Incl. O&P
2 Rodmen (reinf.)	$49.80	$796.80	$78.70	$1259.20	$49.80	$78.70
4 Stressing Equipment		40.80		44.88	2.55	2.81
16 L.H., Daily Totals		$837.60		$1304.08	$52.35	$81.50

Crew C-5	Hr.	Daily	Hr.	Daily	Bare Costs	Incl. O&P
1 Rodman Foreman (outside)	$51.80	$414.40	$81.85	$654.80	$48.86	$76.32
4 Rodmen (reinf.)	49.80	1593.60	78.70	2518.40		
1 Equip. Oper. (crane)	48.80	390.40	73.75	590.00		
1 Equip. Oper. Oiler	42.25	338.00	63.85	510.80		
1 Hyd. Crane, 25 Ton		728.60		801.46	13.01	14.31
56 L.H., Daily Totals		$3465.00		$5075.46	$61.88	$90.63

Crew C-6	Hr.	Daily	Hr.	Daily	Bare Costs	Incl. O&P
1 Labor Foreman (outside)	$37.45	$299.60	$57.65	$461.20	$37.05	$56.58
4 Laborers	35.45	1134.40	54.60	1747.20		
1 Cement Finisher	43.05	344.40	63.40	507.20		
2 Gas Engine Vibrators		66.00		72.60	1.38	1.51
48 L.H., Daily Totals		$1844.40		$2788.20	$38.42	$58.09

Crew No.	Bare Costs		Incl. Subs O&P		Cost Per Labor-Hour	
Crew C-7	**Hr.**	**Daily**	**Hr.**	**Daily**	**Bare Costs**	**Incl. O&P**
1 Labor Foreman (outside)	$37.45	$299.60	$57.65	$461.20	$38.61	$58.85
5 Laborers	35.45	1418.00	54.60	2184.00		
1 Cement Finisher	43.05	344.40	63.40	507.20		
1 Equip. Oper. (med.)	47.50	380.00	71.75	574.00		
1 Equip. Oper. (oiler)	42.25	338.00	63.85	510.80		
2 Gas Engine Vibrators		66.00		72.60		
1 Concrete Bucket, 1 C.Y.		23.40		25.74		
1 Hyd. Crane, 55 Ton		1115.00		1226.50	16.73	18.40
72 L.H., Daily Totals		$3984.40		$5562.04	$55.34	$77.25
Crew C-7A	**Hr.**	**Daily**	**Hr.**	**Daily**	**Bare Costs**	**Incl. O&P**
1 Labor Foreman (outside)	$37.45	$299.60	$57.65	$461.20	$35.99	$55.34
5 Laborers	35.45	1418.00	54.60	2184.00		
2 Truck Drivers (Heavy)	36.60	585.60	56.05	896.80		
2 Conc. Transit Mixers		2048.00		2252.80	32.00	35.20
64 L.H., Daily Totals		$4351.20		$5794.80	$67.99	$90.54
Crew C-7B	**Hr.**	**Daily**	**Hr.**	**Daily**	**Bare Costs**	**Incl. O&P**
1 Labor Foreman (outside)	$37.45	$299.60	$57.65	$461.20	$38.22	$58.53
5 Laborers	35.45	1418.00	54.60	2184.00		
1 Equipment Oper. (crane)	48.80	390.40	73.75	590.00		
1 Equipment Oiler	42.25	338.00	63.85	510.80		
1 Conc. Bucket, 2 C.Y.		36.60		40.26		
1 Lattice Boom Crane, 165 Ton		1954.00		2149.40	31.10	34.21
64 L.H., Daily Totals		$4436.60		$5935.66	$69.32	$92.74
Crew C-7C	**Hr.**	**Daily**	**Hr.**	**Daily**	**Bare Costs**	**Incl. O&P**
1 Labor Foreman (outside)	$37.45	$299.60	$57.65	$461.20	$38.71	$59.27
5 Laborers	35.45	1418.00	54.60	2184.00		
2 Equipment Operators (med.)	47.50	760.00	71.75	1148.00		
2 F.E. Loaders, W.M., 4 C.Y.		1357.60		1493.36	21.21	23.33
64 L.H., Daily Totals		$3835.20		$5286.56	$59.92	$82.60
Crew C-7D	**Hr.**	**Daily**	**Hr.**	**Daily**	**Bare Costs**	**Incl. O&P**
1 Labor Foreman (outside)	$37.45	$299.60	$57.65	$461.20	$37.46	$57.49
5 Laborers	35.45	1418.00	54.60	2184.00		
1 Equip. Oper. (med.)	47.50	380.00	71.75	574.00		
1 Concrete Conveyer		203.40		223.74	3.63	4.00
56 L.H., Daily Totals		$2301.00		$3442.94	$41.09	$61.48
Crew C-8	**Hr.**	**Daily**	**Hr.**	**Daily**	**Bare Costs**	**Incl. O&P**
1 Labor Foreman (outside)	$37.45	$299.60	$57.65	$461.20	$39.63	$60.00
3 Laborers	35.45	850.80	54.60	1310.40		
2 Cement Finishers	43.05	688.80	63.40	1014.40		
1 Equip. Oper. (med.)	47.50	380.00	71.75	574.00		
1 Concrete Pump (Small)		715.80		787.38	12.78	14.06
56 L.H., Daily Totals		$2935.00		$4147.38	$52.41	$74.06
Crew C-8A	**Hr.**	**Daily**	**Hr.**	**Daily**	**Bare Costs**	**Incl. O&P**
1 Labor Foreman (outside)	$37.45	$299.60	$57.65	$461.20	$38.32	$58.04
3 Laborers	35.45	850.80	54.60	1310.40		
2 Cement Finishers	43.05	688.80	63.40	1014.40		
48 L.H., Daily Totals		$1839.20		$2786.00	$38.32	$58.04

Crew No.	Bare Costs		Incl. Subs O&P		Cost Per Labor-Hour	
Crew C-8B	**Hr.**	**Daily**	**Hr.**	**Daily**	**Bare Costs**	**Incl. O&P**
1 Labor Foreman (outside)	$37.45	$299.60	$57.65	$461.20	$38.26	$58.64
3 Laborers	35.45	850.80	54.60	1310.40		
1 Equip. Oper. (med.)	47.50	380.00	71.75	574.00		
1 Vibrating Power Screed		66.30		72.93		
1 Roller, Vibratory, 25 Ton		656.40		722.04		
1 Dozer, 200 H.P.		1333.00		1466.30	51.39	56.53
40 L.H., Daily Totals		$3586.10		$4606.87	$89.65	$115.17
Crew C-8C	**Hr.**	**Daily**	**Hr.**	**Daily**	**Bare Costs**	**Incl. O&P**
1 Labor Foreman (outside)	$37.45	$299.60	$57.65	$461.20	$39.06	$59.43
3 Laborers	35.45	850.80	54.60	1310.40		
1 Cement Finisher	43.05	344.40	63.40	507.20		
1 Equip. Oper. (med.)	47.50	380.00	71.75	574.00		
1 Shotcrete Rig, 12 C.Y./hr		248.60		273.46		
1 Air Compressor, 160 cfm		158.00		173.80		
4 -50' Air Hoses, 1"		16.40		18.04		
4 -50' Air Hoses, 2"		31.00		34.10	9.46	10.40
48 L.H., Daily Totals		$2328.80		$3352.20	$48.52	$69.84
Crew C-8D	**Hr.**	**Daily**	**Hr.**	**Daily**	**Bare Costs**	**Incl. O&P**
1 Labor Foreman (outside)	$37.45	$299.60	$57.65	$461.20	$40.44	$61.21
1 Laborer	35.45	283.60	54.60	436.80		
1 Cement Finisher	43.05	344.40	63.40	507.20		
1 Equipment Oper. (light)	45.80	366.40	69.20	553.60		
1 Air Compressor, 250 cfm		195.80		215.38		
2 -50' Air Hoses, 1"		8.20		9.02	6.38	7.01
32 L.H., Daily Totals		$1498.00		$2183.20	$46.81	$68.22
Crew C-8E	**Hr.**	**Daily**	**Hr.**	**Daily**	**Bare Costs**	**Incl. O&P**
1 Labor Foreman (outside)	$37.45	$299.60	$57.65	$461.20	$38.77	$59.01
3 Laborers	35.45	850.80	54.60	1310.40		
1 Cement Finisher	43.05	344.40	63.40	507.20		
1 Equipment Oper. (light)	45.80	366.40	69.20	553.60		
1 Shotcrete Rig, 35 C.Y./hr		279.00		306.90		
1 Air Compressor, 250 cfm		195.80		215.38		
4 -50' Air Hoses, 1"		16.40		18.04		
4 -50' Air Hoses, 2"		31.00		34.10	10.88	11.97
48 L.H., Daily Totals		$2383.40		$3406.82	$49.65	$70.98
Crew C-10	**Hr.**	**Daily**	**Hr.**	**Daily**	**Bare Costs**	**Incl. O&P**
1 Laborer	$35.45	$283.60	$54.60	$436.80	$40.52	$60.47
2 Cement Finishers	43.05	688.80	63.40	1014.40		
24 L.H., Daily Totals		$972.40		$1451.20	$40.52	$60.47
Crew C-10B	**Hr.**	**Daily**	**Hr.**	**Daily**	**Bare Costs**	**Incl. O&P**
3 Laborers	$35.45	$850.80	$54.60	$1310.40	$38.49	$58.12
2 Cement Finishers	43.05	688.80	63.40	1014.40		
1 Concrete Mixer, 10 C.F.		176.20		193.82		
2 Trowels, 48" Walk-Behind		106.00		116.60	7.05	7.76
40 L.H., Daily Totals		$1821.80		$2635.22	$45.55	$65.88
Crew C-10C	**Hr.**	**Daily**	**Hr.**	**Daily**	**Bare Costs**	**Incl. O&P**
1 Laborer	$35.45	$283.60	$54.60	$436.80	$40.52	$60.47
2 Cement Finishers	43.05	688.80	63.40	1014.40		
1 Trowel, 48" Walk-Behind		53.00		58.30	2.21	2.43
24 L.H., Daily Totals		$1025.40		$1509.50	$42.73	$62.90

Crew No.	Bare Costs		Incl. Subs O&P		Cost Per Labor-Hour	
Crew C-10D	**Hr.**	**Daily**	**Hr.**	**Daily**	**Bare Costs**	**Incl. O&P**
1 Laborer	$35.45	$283.60	$54.60	$436.80	$40.52	$60.47
2 Cement Finishers	43.05	688.80	63.40	1014.40		
1 Vibrating Power Screed		66.30		72.93		
1 Trowel, 48″ Walk-Behind		53.00		58.30	4.97	5.47
24 L.H., Daily Totals		$1091.70		$1582.43	$45.49	$65.93

Crew C-10E	**Hr.**	**Daily**	**Hr.**	**Daily**	**Bare Costs**	**Incl. O&P**
1 Laborer	$35.45	$283.60	$54.60	$436.80	$40.52	$60.47
2 Cement Finishers	43.05	688.80	63.40	1014.40		
1 Vibrating Power Screed		66.30		72.93		
1 Cement Trowel, 96″ Ride-On		196.40		216.04	10.95	12.04
24 L.H., Daily Totals		$1235.10		$1740.17	$51.46	$72.51

Crew C-10F	**Hr.**	**Daily**	**Hr.**	**Daily**	**Bare Costs**	**Incl. O&P**
1 Laborer	$35.45	$283.60	$54.60	$436.80	$40.52	$60.47
2 Cement Finishers	43.05	688.80	63.40	1014.40		
1 Aerial Lift Truck, 60′ Boom		462.20		508.42	19.26	21.18
24 L.H., Daily Totals		$1434.60		$1959.62	$59.77	$81.65

Crew C-11	**Hr.**	**Daily**	**Hr.**	**Daily**	**Bare Costs**	**Incl. O&P**
1 Struc. Steel Foreman (outside)	$52.05	$416.40	$92.85	$742.80	$49.27	$85.14
6 Struc. Steel Workers	50.05	2402.40	89.30	4286.40		
1 Equip. Oper. (crane)	48.80	390.40	73.75	590.00		
1 Equip. Oper. Oiler	42.25	338.00	63.85	510.80		
1 Lattice Boom Crane, 150 Ton		1861.00		2047.10	25.85	28.43
72 L.H., Daily Totals		$5408.20		$8177.10	$75.11	$113.57

Crew C-12	**Hr.**	**Daily**	**Hr.**	**Daily**	**Bare Costs**	**Incl. O&P**
1 Carpenter Foreman (outside)	$46.90	$375.20	$72.25	$578.00	$44.31	$68.01
3 Carpenters	44.90	1077.60	69.15	1659.60		
1 Laborer	35.45	283.60	54.60	436.80		
1 Equip. Oper. (crane)	48.80	390.40	73.75	590.00		
1 Hyd. Crane, 12 Ton		646.20		710.82	13.46	14.81
48 L.H., Daily Totals		$2773.00		$3975.22	$57.77	$82.82

Crew C-13	**Hr.**	**Daily**	**Hr.**	**Daily**	**Bare Costs**	**Incl. O&P**
1 Struc. Steel Worker	$50.05	$400.40	$89.30	$714.40	$48.33	$82.58
1 Welder	50.05	400.40	89.30	714.40		
1 Carpenter	44.90	359.20	69.15	553.20		
1 Welder, Gas Engine, 300 amp		144.00		158.40	6.00	6.60
24 L.H., Daily Totals		$1304.00		$2140.40	$54.33	$89.18

Crew C-14	**Hr.**	**Daily**	**Hr.**	**Daily**	**Bare Costs**	**Incl. O&P**
1 Carpenter Foreman (outside)	$46.90	$375.20	$72.25	$578.00	$43.86	$67.53
5 Carpenters	44.90	1796.00	69.15	2766.00		
4 Laborers	35.45	1134.40	54.60	1747.20		
4 Rodmen (reinf.)	49.80	1593.60	78.70	2518.40		
2 Cement Finishers	43.05	688.80	63.40	1014.40		
1 Equip. Oper. (crane)	48.80	390.40	73.75	590.00		
1 Equip. Oper. Oiler	42.25	338.00	63.85	510.80		
1 Hyd. Crane, 80 Ton		1613.00		1774.30	11.20	12.32
144 L.H., Daily Totals		$7929.40		$11499.10	$55.07	$79.85

Crew No.	Bare Costs		Incl. Subs O&P		Cost Per Labor-Hour	
Crew C-14A	**Hr.**	**Daily**	**Hr.**	**Daily**	**Bare Costs**	**Incl. O&P**
1 Carpenter Foreman (outside)	$46.90	$375.20	$72.25	$578.00	$45.04	$69.51
16 Carpenters	44.90	5747.20	69.15	8851.20		
4 Rodmen (reinf.)	49.80	1593.60	78.70	2518.40		
2 Laborers	35.45	567.20	54.60	873.60		
1 Cement Finisher	43.05	344.40	63.40	507.20		
1 Equip. Oper. (med.)	47.50	380.00	71.75	574.00		
1 Gas Engine Vibrator		33.00		36.30		
1 Concrete Pump (Small)		715.80		787.38	3.74	4.12
200 L.H., Daily Totals		$9756.40		$14726.08	$48.78	$73.63

Crew C-14B	**Hr.**	**Daily**	**Hr.**	**Daily**	**Bare Costs**	**Incl. O&P**
1 Carpenter Foreman (outside)	$46.90	$375.20	$72.25	$578.00	$44.96	$69.28
16 Carpenters	44.90	5747.20	69.15	8851.20		
4 Rodmen (reinf.)	49.80	1593.60	78.70	2518.40		
2 Laborers	35.45	567.20	54.60	873.60		
2 Cement Finishers	43.05	688.80	63.40	1014.40		
1 Equip. Oper. (med.)	47.50	380.00	71.75	574.00		
1 Gas Engine Vibrator		33.00		36.30		
1 Concrete Pump (Small)		715.80		787.38	3.60	3.96
208 L.H., Daily Totals		$10100.80		$15233.28	$48.56	$73.24

Crew C-14C	**Hr.**	**Daily**	**Hr.**	**Daily**	**Bare Costs**	**Incl. O&P**
1 Carpenter Foreman (outside)	$46.90	$375.20	$72.25	$578.00	$42.91	$66.17
6 Carpenters	44.90	2155.20	69.15	3319.20		
2 Rodmen (reinf.)	49.80	796.80	78.70	1259.20		
4 Laborers	35.45	1134.40	54.60	1747.20		
1 Cement Finisher	43.05	344.40	63.40	507.20		
1 Gas Engine Vibrator		33.00		36.30	.29	.32
112 L.H., Daily Totals		$4839.00		$7447.10	$43.21	$66.49

Crew C-14D	**Hr.**	**Daily**	**Hr.**	**Daily**	**Bare Costs**	**Incl. O&P**
1 Carpenter Foreman (outside)	$46.90	$375.20	$72.25	$578.00	$44.65	$68.75
18 Carpenters	44.90	6465.60	69.15	9957.60		
2 Rodmen (reinf.)	49.80	796.80	78.70	1259.20		
2 Laborers	35.45	567.20	54.60	873.60		
1 Cement Finisher	43.05	344.40	63.40	507.20		
1 Equip. Oper. (med.)	47.50	380.00	71.75	574.00		
1 Gas Engine Vibrator		33.00		36.30		
1 Concrete Pump (Small)		715.80		787.38	3.74	4.12
200 L.H., Daily Totals		$9678.00		$14573.28	$48.39	$72.87

Crew C-14E	**Hr.**	**Daily**	**Hr.**	**Daily**	**Bare Costs**	**Incl. O&P**
1 Carpenter Foreman (outside)	$46.90	$375.20	$72.25	$578.00	$44.12	$68.41
2 Carpenters	44.90	718.40	69.15	1106.40		
4 Rodmen (reinf.)	49.80	1593.60	78.70	2518.40		
3 Laborers	35.45	850.80	54.60	1310.40		
1 Cement Finisher	43.05	344.40	63.40	507.20		
1 Gas Engine Vibrator		33.00		36.30	.38	.41
88 L.H., Daily Totals		$3915.40		$6056.70	$44.49	$68.83

Crew C-14F	**Hr.**	**Daily**	**Hr.**	**Daily**	**Bare Costs**	**Incl. O&P**
1 Labor Foreman (outside)	$37.45	$299.60	$57.65	$461.20	$40.74	$60.81
2 Laborers	35.45	567.20	54.60	873.60		
6 Cement Finishers	43.05	2066.40	63.40	3043.20		
1 Gas Engine Vibrator		33.00		36.30	.46	.50
72 L.H., Daily Totals		$2966.20		$4414.30	$41.20	$61.31

Crew C-14G	Hr.	Daily	Hr.	Daily	Bare Costs	Incl. O&P
1 Labor Foreman (outside)	$37.45	$299.60	$57.65	$461.20	$40.08	$60.06
2 Laborers	35.45	567.20	54.60	873.60		
4 Cement Finishers	43.05	1377.60	63.40	2028.80		
1 Gas Engine Vibrator		33.00		36.30	.59	.65
56 L.H., Daily Totals		$2277.40		$3399.90	$40.67	$60.71

Crew C-14H	Hr.	Daily	Hr.	Daily	Bare Costs	Incl. O&P
1 Carpenter Foreman (outside)	$46.90	$375.20	$72.25	$578.00	$44.17	$67.88
2 Carpenters	44.90	718.40	69.15	1106.40		
1 Rodman (reinf.)	49.80	398.40	78.70	629.60		
1 Laborer	35.45	283.60	54.60	436.80		
1 Cement Finisher	43.05	344.40	63.40	507.20		
1 Gas Engine Vibrator		33.00		36.30	.69	.76
48 L.H., Daily Totals		$2153.00		$3294.30	$44.85	$68.63

Crew C-14L	Hr.	Daily	Hr.	Daily	Bare Costs	Incl. O&P
1 Carpenter Foreman (outside)	$46.90	$375.20	$72.25	$578.00	$41.76	$64.08
6 Carpenters	44.90	2155.20	69.15	3319.20		
4 Laborers	35.45	1134.40	54.60	1747.20		
1 Cement Finisher	43.05	344.40	63.40	507.20		
1 Gas Engine Vibrator		33.00		36.30	.34	.38
96 L.H., Daily Totals		$4042.20		$6187.90	$42.11	$64.46

Crew C-14M	Hr.	Daily	Hr.	Daily	Bare Costs	Incl. O&P
1 Carpenter Foreman (outside)	$46.90	$375.20	$72.25	$578.00	$43.49	$66.70
2 Carpenters	44.90	718.40	69.15	1106.40		
1 Rodman (reinf.)	49.80	398.40	78.70	629.60		
2 Laborers	35.45	567.20	54.60	873.60		
1 Cement Finisher	43.05	344.40	63.40	507.20		
1 Equip. Oper. (med.)	47.50	380.00	71.75	574.00		
1 Gas Engine Vibrator		33.00		36.30		
1 Concrete Pump (Small)		715.80		787.38	11.70	12.87
64 L.H., Daily Totals		$3532.40		$5092.48	$55.19	$79.57

Crew C-15	Hr.	Daily	Hr.	Daily	Bare Costs	Incl. O&P
1 Carpenter Foreman (outside)	$46.90	$375.20	$72.25	$578.00	$42.11	$64.43
2 Carpenters	44.90	718.40	69.15	1106.40		
3 Laborers	35.45	850.80	54.60	1310.40		
2 Cement Finishers	43.05	688.80	63.40	1014.40		
1 Rodman (reinf.)	49.80	398.40	78.70	629.60		
72 L.H., Daily Totals		$3031.60		$4638.80	$42.11	$64.43

Crew C-16	Hr.	Daily	Hr.	Daily	Bare Costs	Incl. O&P
1 Labor Foreman (outside)	$37.45	$299.60	$57.65	$461.20	$39.63	$60.00
3 Laborers	35.45	850.80	54.60	1310.40		
2 Cement Finishers	43.05	688.80	63.40	1014.40		
1 Equip. Oper. (med.)	47.50	380.00	71.75	574.00		
1 Gunite Pump Rig		369.80		406.78		
2 -50' Air Hoses, 3/4"		6.50		7.15		
2 -50' Air Hoses, 2"		15.50		17.05	7.00	7.70
56 L.H., Daily Totals		$2611.00		$3790.98	$46.63	$67.70

Crew C-16A	Hr.	Daily	Hr.	Daily	Bare Costs	Incl. O&P
1 Laborer	$35.45	$283.60	$54.60	$436.80	$42.26	$63.29
2 Cement Finishers	43.05	688.80	63.40	1014.40		
1 Equip. Oper. (med.)	47.50	380.00	71.75	574.00		
1 Gunite Pump Rig		369.80		406.78		
2 -50' Air Hoses, 3/4"		6.50		7.15		
2 -50' Air Hoses, 2"		15.50		17.05		
1 Aerial Lift Truck, 60' Boom		462.20		508.42	26.69	29.36
32 L.H., Daily Totals		$2206.40		$2964.60	$68.95	$92.64

Crew C-17	Hr.	Daily	Hr.	Daily	Bare Costs	Incl. O&P
2 Skilled Worker Foremen (out)	$48.20	$771.20	$74.50	$1192.00	$46.60	$72.06
8 Skilled Workers	46.20	2956.80	71.45	4572.80		
80 L.H., Daily Totals		$3728.00		$5764.80	$46.60	$72.06

Crew C-17A	Hr.	Daily	Hr.	Daily	Bare Costs	Incl. O&P
2 Skilled Worker Foremen (out)	$48.20	$771.20	$74.50	$1192.00	$46.63	$72.08
8 Skilled Workers	46.20	2956.80	71.45	4572.80		
.125 Equip. Oper. (crane)	48.80	48.80	73.75	73.75		
.125 Hyd. Crane, 80 Ton		201.63		221.79	2.49	2.74
81 L.H., Daily Totals		$3978.43		$6060.34	$49.12	$74.82

Crew C-17B	Hr.	Daily	Hr.	Daily	Bare Costs	Incl. O&P
2 Skilled Worker Foremen (out)	$48.20	$771.20	$74.50	$1192.00	$46.65	$72.10
8 Skilled Workers	46.20	2956.80	71.45	4572.80		
.25 Equip. Oper. (crane)	48.80	97.60	73.75	147.50		
.25 Hyd. Crane, 80 Ton		403.25		443.57		
.25 Trowel, 48" Walk-Behind		13.25		14.57	5.08	5.59
82 L.H., Daily Totals		$4242.10		$6370.45	$51.73	$77.69

Crew C-17C	Hr.	Daily	Hr.	Daily	Bare Costs	Incl. O&P
2 Skilled Worker Foremen (out)	$48.20	$771.20	$74.50	$1192.00	$46.68	$72.12
8 Skilled Workers	46.20	2956.80	71.45	4572.80		
.375 Equip. Oper. (crane)	48.80	146.40	73.75	221.25		
.375 Hyd. Crane, 80 Ton		604.88		665.36	7.29	8.02
83 L.H., Daily Totals		$4479.27		$6651.41	$53.97	$80.14

Crew C-17D	Hr.	Daily	Hr.	Daily	Bare Costs	Incl. O&P
2 Skilled Worker Foremen (out)	$48.20	$771.20	$74.50	$1192.00	$46.70	$72.14
8 Skilled Workers	46.20	2956.80	71.45	4572.80		
.5 Equip. Oper. (crane)	48.80	195.20	73.75	295.00		
.5 Hyd. Crane, 80 Ton		806.50		887.15	9.60	10.56
84 L.H., Daily Totals		$4729.70		$6946.95	$56.31	$82.70

Crew C-17E	Hr.	Daily	Hr.	Daily	Bare Costs	Incl. O&P
2 Skilled Worker Foremen (out)	$48.20	$771.20	$74.50	$1192.00	$46.60	$72.06
8 Skilled Workers	46.20	2956.80	71.45	4572.80		
1 Hyd. Jack with Rods		92.65		101.92	1.16	1.27
80 L.H., Daily Totals		$3820.65		$5866.72	$47.76	$73.33

Crew C-18	Hr.	Daily	Hr.	Daily	Bare Costs	Incl. O&P
.125 Labor Foreman (outside)	$37.45	$37.45	$57.65	$57.65	$35.67	$54.94
1 Laborer	35.45	283.60	54.60	436.80		
1 Concrete Cart, 10 C.F.		60.80		66.88	6.76	7.43
9 L.H., Daily Totals		$381.85		$561.33	$42.43	$62.37

Crew C-19	Hr.	Daily	Hr.	Daily	Bare Costs	Incl. O&P
.125 Labor Foreman (outside)	$37.45	$37.45	$57.65	$57.65	$35.67	$54.94
1 Laborer	35.45	283.60	54.60	436.80		
1 Concrete Cart, 18 C.F.		100.40		110.44	11.16	12.27
9 L.H., Daily Totals		$421.45		$604.89	$46.83	$67.21

Crew C-20	Hr.	Daily	Hr.	Daily	Bare Costs	Incl. O&P
1 Labor Foreman (outside)	$37.45	$299.60	$57.65	$461.20	$38.16	$58.23
5 Laborers	35.45	1418.00	54.60	2184.00		
1 Cement Finisher	43.05	344.40	63.40	507.20		
1 Equip. Oper. (med.)	47.50	380.00	71.75	574.00		
2 Gas Engine Vibrators		66.00		72.60		
1 Concrete Pump (Small)		715.80		787.38	12.22	13.44
64 L.H., Daily Totals		$3223.80		$4586.38	$50.37	$71.66

Crew No.	Bare Costs		Incl. Subs O&P		Cost Per Labor-Hour	
Crew C-21	**Hr.**	**Daily**	**Hr.**	**Daily**	**Bare Costs**	**Incl. O&P**
1 Labor Foreman (outside)	$37.45	$299.60	$57.65	$461.20	$38.16	$58.23
5 Laborers	35.45	1418.00	54.60	2184.00		
1 Cement Finisher	43.05	344.40	63.40	507.20		
1 Equip. Oper. (med.)	47.50	380.00	71.75	574.00		
2 Gas Engine Vibrators		66.00		72.60		
1 Concrete Conveyer		203.40		223.74	4.21	4.63
64 L.H., Daily Totals		$2711.40		$4022.74	$42.37	$62.86
Crew C-22	**Hr.**	**Daily**	**Hr.**	**Daily**	**Bare Costs**	**Incl. O&P**
1 Rodman Foreman (outside)	$51.80	$414.40	$81.85	$654.80	$49.98	$78.83
4 Rodmen (reinf.)	49.80	1593.60	78.70	2518.40		
.125 Equip. Oper. (crane)	48.80	48.80	73.75	73.75		
.125 Equip. Oper. Oiler	42.25	42.25	63.85	63.85		
.125 Hyd. Crane, 25 Ton		91.08		100.18	2.17	2.39
42 L.H., Daily Totals		$2190.13		$3410.98	$52.15	$81.21
Crew C-23	**Hr.**	**Daily**	**Hr.**	**Daily**	**Bare Costs**	**Incl. O&P**
2 Skilled Worker Foremen (out)	$48.20	$771.20	$74.50	$1192.00	$46.47	$71.53
6 Skilled Workers	46.20	2217.60	71.45	3429.60		
1 Equip. Oper. (crane)	48.80	390.40	73.75	590.00		
1 Equip. Oper. Oiler	42.25	338.00	63.85	510.80		
1 Lattice Boom Crane, 90 Ton		1525.00		1677.50	19.06	20.97
80 L.H., Daily Totals		$5242.20		$7399.90	$65.53	$92.50
Crew C-23A	**Hr.**	**Daily**	**Hr.**	**Daily**	**Bare Costs**	**Incl. O&P**
1 Labor Foreman (outside)	$37.45	$299.60	$57.65	$461.20	$39.88	$60.89
2 Laborers	35.45	567.20	54.60	873.60		
1 Equip. Oper. (crane)	48.80	390.40	73.75	590.00		
1 Equip. Oper. Oiler	42.25	338.00	63.85	510.80		
1 Crawler Crane, 100 Ton		1666.00		1832.60		
3 Conc. Buckets, 8 C.Y.		597.60		657.36	56.59	62.25
40 L.H., Daily Totals		$3858.80		$4925.56	$96.47	$123.14
Crew C-24	**Hr.**	**Daily**	**Hr.**	**Daily**	**Bare Costs**	**Incl. O&P**
2 Skilled Worker Foremen (out)	$48.20	$771.20	$74.50	$1192.00	$46.47	$71.53
6 Skilled Workers	46.20	2217.60	71.45	3429.60		
1 Equip. Oper. (crane)	48.80	390.40	73.75	590.00		
1 Equip. Oper. Oiler	42.25	338.00	63.85	510.80		
1 Lattice Boom Crane, 150 Ton		1861.00		2047.10	23.26	25.59
80 L.H., Daily Totals		$5578.20		$7769.50	$69.73	$97.12
Crew C-25	**Hr.**	**Daily**	**Hr.**	**Daily**	**Bare Costs**	**Incl. O&P**
2 Rodmen (reinf.)	$49.80	$796.80	$78.70	$1259.20	$39.10	$63.27
2 Rodmen Helpers	28.40	454.40	47.85	765.60		
32 L.H., Daily Totals		$1251.20		$2024.80	$39.10	$63.27
Crew C-27	**Hr.**	**Daily**	**Hr.**	**Daily**	**Bare Costs**	**Incl. O&P**
2 Cement Finishers	$43.05	$688.80	$63.40	$1014.40	$43.05	$63.40
1 Concrete Saw		175.20		192.72	10.95	12.05
16 L.H., Daily Totals		$864.00		$1207.12	$54.00	$75.44
Crew C-28	**Hr.**	**Daily**	**Hr.**	**Daily**	**Bare Costs**	**Incl. O&P**
1 Cement Finisher	$43.05	$344.40	$63.40	$507.20	$43.05	$63.40
1 Portable Air Compressor, Gas		17.90		19.69	2.24	2.46
8 L.H., Daily Totals		$362.30		$526.89	$45.29	$65.86
Crew C-29	**Hr.**	**Daily**	**Hr.**	**Daily**	**Bare Costs**	**Incl. O&P**
1 Laborer	$35.45	$283.60	$54.60	$436.80	$35.45	$54.60
1 Pressure Washer		69.40		76.34	8.68	9.54
8 L.H., Daily Totals		$353.00		$513.14	$44.13	$64.14

Crew No.	Bare Costs		Incl. Subs O&P		Cost Per Labor-Hour	
Crew C-30	**Hr.**	**Daily**	**Hr.**	**Daily**	**Bare Costs**	**Incl. O&P**
1 Laborer	$35.45	$283.60	$54.60	$436.80	$35.45	$54.60
1 Concrete Mixer, 10 C.F.		176.20		193.82	22.02	24.23
8 L.H., Daily Totals		$459.80		$630.62	$57.48	$78.83
Crew C-31	**Hr.**	**Daily**	**Hr.**	**Daily**	**Bare Costs**	**Incl. O&P**
1 Cement Finisher	$43.05	$344.40	$63.40	$507.20	$43.05	$63.40
1 Grout Pump		369.80		406.78	46.23	50.85
8 L.H., Daily Totals		$714.20		$913.98	$89.28	$114.25
Crew D-1	**Hr.**	**Daily**	**Hr.**	**Daily**	**Bare Costs**	**Incl. O&P**
1 Bricklayer	$44.80	$358.40	$67.85	$542.80	$40.38	$61.15
1 Bricklayer Helper	35.95	287.60	54.45	435.60		
16 L.H., Daily Totals		$646.00		$978.40	$40.38	$61.15
Crew D-2	**Hr.**	**Daily**	**Hr.**	**Daily**	**Bare Costs**	**Incl. O&P**
3 Bricklayers	$44.80	$1075.20	$67.85	$1628.40	$41.59	$63.10
2 Bricklayer Helpers	35.95	575.20	54.45	871.20		
.5 Carpenter	44.90	179.60	69.15	276.60		
44 L.H., Daily Totals		$1830.00		$2776.20	$41.59	$63.10
Crew D-3	**Hr.**	**Daily**	**Hr.**	**Daily**	**Bare Costs**	**Incl. O&P**
3 Bricklayers	$44.80	$1075.20	$67.85	$1628.40	$41.43	$62.81
2 Bricklayer Helpers	35.95	575.20	54.45	871.20		
.25 Carpenter	44.90	89.80	69.15	138.30		
42 L.H., Daily Totals		$1740.20		$2637.90	$41.43	$62.81
Crew D-4	**Hr.**	**Daily**	**Hr.**	**Daily**	**Bare Costs**	**Incl. O&P**
1 Bricklayer	$44.80	$358.40	$67.85	$542.80	$40.63	$61.49
2 Bricklayer Helpers	35.95	575.20	54.45	871.20		
1 Equip. Oper. (light)	45.80	366.40	69.20	553.60		
1 Grout Pump, 50 C.F./hr.		130.80		143.88	4.09	4.50
32 L.H., Daily Totals		$1430.80		$2111.48	$44.71	$65.98
Crew D-5	**Hr.**	**Daily**	**Hr.**	**Daily**	**Bare Costs**	**Incl. O&P**
1 Bricklayer	$44.80	$358.40	$67.85	$542.80	$44.80	$67.85
8 L.H., Daily Totals		$358.40		$542.80	$44.80	$67.85
Crew D-6	**Hr.**	**Daily**	**Hr.**	**Daily**	**Bare Costs**	**Incl. O&P**
3 Bricklayers	$44.80	$1075.20	$67.85	$1628.40	$40.56	$61.47
3 Bricklayer Helpers	35.95	862.80	54.45	1306.80		
.25 Carpenter	44.90	89.80	69.15	138.30		
50 L.H., Daily Totals		$2027.80		$3073.50	$40.56	$61.47
Crew D-7	**Hr.**	**Daily**	**Hr.**	**Daily**	**Bare Costs**	**Incl. O&P**
1 Tile Layer	$41.25	$330.00	$60.60	$484.80	$36.98	$54.33
1 Tile Layer Helper	32.70	261.60	48.05	384.40		
16 L.H., Daily Totals		$591.60		$869.20	$36.98	$54.33
Crew D-8	**Hr.**	**Daily**	**Hr.**	**Daily**	**Bare Costs**	**Incl. O&P**
3 Bricklayers	$44.80	$1075.20	$67.85	$1628.40	$41.26	$62.49
2 Bricklayer Helpers	35.95	575.20	54.45	871.20		
40 L.H., Daily Totals		$1650.40		$2499.60	$41.26	$62.49
Crew D-9	**Hr.**	**Daily**	**Hr.**	**Daily**	**Bare Costs**	**Incl. O&P**
3 Bricklayers	$44.80	$1075.20	$67.85	$1628.40	$40.38	$61.15
3 Bricklayer Helpers	35.95	862.80	54.45	1306.80		
48 L.H., Daily Totals		$1938.00		$2935.20	$40.38	$61.15

Crew D-10	Bare Costs Hr.	Bare Costs Daily	Incl. Subs O&P Hr.	Incl. Subs O&P Daily	Cost Per Labor-Hour Bare Costs	Cost Per Labor-Hour Incl. O&P
1 Bricklayer Foreman (outside)	$46.80	$374.40	$70.85	$566.80	$44.09	$66.72
1 Bricklayer	44.80	358.40	67.85	542.80		
1 Bricklayer Helper	35.95	287.60	54.45	435.60		
1 Equip. Oper. (crane)	48.80	390.40	73.75	590.00		
1 S.P. Crane, 4x4, 12 Ton		471.00		518.10	14.72	16.19
32 L.H., Daily Totals		$1881.80		$2653.30	$58.81	$82.92

Crew D-11	Hr.	Daily	Hr.	Daily	Bare Costs	Incl. O&P
1 Bricklayer Foreman (outside)	$46.80	$374.40	$70.85	$566.80	$42.52	$64.38
1 Bricklayer	44.80	358.40	67.85	542.80		
1 Bricklayer Helper	35.95	287.60	54.45	435.60		
24 L.H., Daily Totals		$1020.40		$1545.20	$42.52	$64.38

Crew D-12	Hr.	Daily	Hr.	Daily	Bare Costs	Incl. O&P
1 Bricklayer Foreman (outside)	$46.80	$374.40	$70.85	$566.80	$40.88	$61.90
1 Bricklayer	44.80	358.40	67.85	542.80		
2 Bricklayer Helpers	35.95	575.20	54.45	871.20		
32 L.H., Daily Totals		$1308.00		$1980.80	$40.88	$61.90

Crew D-13	Hr.	Daily	Hr.	Daily	Bare Costs	Incl. O&P
1 Bricklayer Foreman (outside)	$46.80	$374.40	$70.85	$566.80	$42.87	$65.08
1 Bricklayer	44.80	358.40	67.85	542.80		
2 Bricklayer Helpers	35.95	575.20	54.45	871.20		
1 Carpenter	44.90	359.20	69.15	553.20		
1 Equip. Oper. (crane)	48.80	390.40	73.75	590.00		
1 S.P. Crane, 4x4, 12 Ton		471.00		518.10	9.81	10.79
48 L.H., Daily Totals		$2528.60		$3642.10	$52.68	$75.88

Crew E-1	Hr.	Daily	Hr.	Daily	Bare Costs	Incl. O&P
1 Welder Foreman (outside)	$52.05	$416.40	$92.85	$742.80	$49.30	$83.78
1 Welder	50.05	400.40	89.30	714.40		
1 Equip. Oper. (light)	45.80	366.40	69.20	553.60		
1 Welder, Gas Engine, 300 amp		144.00		158.40	6.00	6.60
24 L.H., Daily Totals		$1327.20		$2169.20	$55.30	$90.38

Crew E-2	Hr.	Daily	Hr.	Daily	Bare Costs	Incl. O&P
1 Struc. Steel Foreman (outside)	$52.05	$416.40	$92.85	$742.80	$49.04	$83.95
4 Struc. Steel Workers	50.05	1601.60	89.30	2857.60		
1 Equip. Oper. (crane)	48.80	390.40	73.75	590.00		
1 Equip. Oper. Oiler	42.25	338.00	63.85	510.80		
1 Lattice Boom Crane, 90 Ton		1525.00		1677.50	27.23	29.96
56 L.H., Daily Totals		$4271.40		$6378.70	$76.28	$113.91

Crew E-3	Hr.	Daily	Hr.	Daily	Bare Costs	Incl. O&P
1 Struc. Steel Foreman (outside)	$52.05	$416.40	$92.85	$742.80	$50.72	$90.48
1 Struc. Steel Worker	50.05	400.40	89.30	714.40		
1 Welder	50.05	400.40	89.30	714.40		
1 Welder, Gas Engine, 300 amp		144.00		158.40	6.00	6.60
24 L.H., Daily Totals		$1361.20		$2330.00	$56.72	$97.08

Crew E-3A	Hr.	Daily	Hr.	Daily	Bare Costs	Incl. O&P
1 Struc. Steel Foreman (outside)	$52.05	$416.40	$92.85	$742.80	$50.72	$90.48
1 Struc. Steel Worker	50.05	400.40	89.30	714.40		
1 Welder	50.05	400.40	89.30	714.40		
1 Welder, Gas Engine, 300 amp		144.00		158.40		
1 Aerial Lift Truck, 40' Boom		332.20		365.42	19.84	21.83
24 L.H., Daily Totals		$1693.40		$2695.42	$70.56	$112.31

Crew E-4	Hr.	Daily	Hr.	Daily	Bare Costs	Incl. O&P
1 Struc. Steel Foreman (outside)	$52.05	$416.40	$92.85	$742.80	$50.55	$90.19
3 Struc. Steel Workers	50.05	1201.20	89.30	2143.20		
1 Welder, Gas Engine, 300 amp		144.00		158.40	4.50	4.95
32 L.H., Daily Totals		$1761.60		$3044.40	$55.05	$95.14

Crew E-5	Hr.	Daily	Hr.	Daily	Bare Costs	Incl. O&P
2 Struc. Steel Foremen (outside)	$52.05	$832.80	$92.85	$1485.60	$49.55	$85.91
5 Struc. Steel Workers	50.05	2002.00	89.30	3572.00		
1 Equip. Oper. (crane)	48.80	390.40	73.75	590.00		
1 Welder	50.05	400.40	89.30	714.40		
1 Equip. Oper. Oiler	42.25	338.00	63.85	510.80		
1 Lattice Boom Crane, 90 Ton		1525.00		1677.50		
1 Welder, Gas Engine, 300 amp		144.00		158.40	20.86	22.95
80 L.H., Daily Totals		$5632.60		$8708.70	$70.41	$108.86

Crew E-6	Hr.	Daily	Hr.	Daily	Bare Costs	Incl. O&P
3 Struc. Steel Foremen (outside)	$52.05	$1249.20	$92.85	$2228.40	$49.59	$86.15
9 Struc. Steel Workers	50.05	3603.60	89.30	6429.60		
1 Equip. Oper. (crane)	48.80	390.40	73.75	590.00		
1 Welder	50.05	400.40	89.30	714.40		
1 Equip. Oper. Oiler	42.25	338.00	63.85	510.80		
1 Equip. Oper. (light)	45.80	366.40	69.20	553.60		
1 Lattice Boom Crane, 90 Ton		1525.00		1677.50		
1 Welder, Gas Engine, 300 amp		144.00		158.40		
1 Air Compressor, 160 cfm		158.00		173.80		
2 Impact Wrenches		36.00		39.60	14.55	16.01
128 L.H., Daily Totals		$8211.00		$13076.10	$64.15	$102.16

Crew E-7	Hr.	Daily	Hr.	Daily	Bare Costs	Incl. O&P
1 Struc. Steel Foreman (outside)	$52.05	$416.40	$92.85	$742.80	$49.55	$85.91
4 Struc. Steel Workers	50.05	1601.60	89.30	2857.60		
1 Equip. Oper. (crane)	48.80	390.40	73.75	590.00		
1 Equip. Oper. Oiler	42.25	338.00	63.85	510.80		
1 Welder Foreman (outside)	52.05	416.40	92.85	742.80		
2 Welders	50.05	800.80	89.30	1428.80		
1 Lattice Boom Crane, 90 Ton		1525.00		1677.50		
2 Welder, Gas Engine, 300 amp		288.00		316.80	22.66	24.93
80 L.H., Daily Totals		$5776.60		$8867.10	$72.21	$110.84

Crew E-8	Hr.	Daily	Hr.	Daily	Bare Costs	Incl. O&P
1 Struc. Steel Foreman (outside)	$52.05	$416.40	$92.85	$742.80	$49.33	$85.15
4 Struc. Steel Workers	50.05	1601.60	89.30	2857.60		
1 Welder Foreman (outside)	52.05	416.40	92.85	742.80		
4 Welders	50.05	1601.60	89.30	2857.60		
1 Equip. Oper. (crane)	48.80	390.40	73.75	590.00		
1 Equip. Oper. Oiler	42.25	338.00	63.85	510.80		
1 Equip. Oper. (light)	45.80	366.40	69.20	553.60		
1 Lattice Boom Crane, 90 Ton		1525.00		1677.50		
4 Welder, Gas Engine, 300 amp		576.00		633.60	20.20	22.22
104 L.H., Daily Totals		$7231.80		$11166.30	$69.54	$107.37

Crew E-9	Hr.	Daily	Hr.	Daily	Bare Costs	Incl. O&P
2 Struc. Steel Foremen (outside)	$52.05	$832.80	$92.85	$1485.60	$49.59	$86.15
5 Struc. Steel Workers	50.05	2002.00	89.30	3572.00		
1 Welder Foreman (outside)	52.05	416.40	92.85	742.80		
5 Welders	50.05	2002.00	89.30	3572.00		
1 Equip. Oper. (crane)	48.80	390.40	73.75	590.00		
1 Equip. Oper. Oiler	42.25	338.00	63.85	510.80		
1 Equip. Oper. (light)	45.80	366.40	69.20	553.60		
1 Lattice Boom Crane, 90 Ton		1525.00		1677.50		
5 Welder, Gas Engine, 300 amp		720.00		792.00	17.54	19.29
128 L.H., Daily Totals		$8593.00		$13496.30	$67.13	$105.44

Crew No.	Bare Costs		Incl. Subs O&P		Cost Per Labor-Hour	

Crew E-10	Hr.	Daily	Hr.	Daily	Bare Costs	Incl. O&P
1 Welder Foreman (outside)	$52.05	$416.40	$92.85	$742.80	$51.05	$91.08
1 Welder	50.05	400.40	89.30	714.40		
1 Welder, Gas Engine, 300 amp		144.00		158.40		
1 Flatbed Truck, Gas, 3 Ton		327.80		360.58	29.49	32.44
16 L.H., Daily Totals		$1288.60		$1976.18	$80.54	$123.51

Crew E-11	Hr.	Daily	Hr.	Daily	Bare Costs	Incl. O&P
2 Painters, Struc. Steel	$39.70	$635.20	$71.90	$1150.40	$40.16	$66.90
1 Building Laborer	35.45	283.60	54.60	436.80		
1 Equip. Oper. (light)	45.80	366.40	69.20	553.60		
1 Air Compressor, 250 cfm		195.80		215.38		
1 Sandblaster, Portable, 3 C.F.		20.00		22.00		
1 Set Sand Blasting Accessories		13.65		15.02	7.17	7.89
32 L.H., Daily Totals		$1514.65		$2393.20	$47.33	$74.79

Crew E-11A	Hr.	Daily	Hr.	Daily	Bare Costs	Incl. O&P
2 Painters, Struc. Steel	$39.70	$635.20	$71.90	$1150.40	$40.16	$66.90
1 Building Laborer	35.45	283.60	54.60	436.80		
1 Equip. Oper. (light)	45.80	366.40	69.20	553.60		
1 Air Compressor, 250 cfm		195.80		215.38		
1 Sandblaster, Portable, 3 C.F.		20.00		22.00		
1 Set Sand Blasting Accessories		13.65		15.02		
1 Aerial Lift Truck, 60' Boom		462.20		508.42	21.61	23.78
32 L.H., Daily Totals		$1976.85		$2901.61	$61.78	$90.68

Crew E-11B	Hr.	Daily	Hr.	Daily	Bare Costs	Incl. O&P
2 Painters, Struc. Steel	$39.70	$635.20	$71.90	$1150.40	$38.28	$66.13
1 Building Laborer	35.45	283.60	54.60	436.80		
2 Paint Sprayer, 8 C.F.M.		79.70		87.67		
1 Aerial Lift Truck, 60' Boom		462.20		508.42	22.58	24.84
24 L.H., Daily Totals		$1460.70		$2183.29	$60.86	$90.97

Crew E-12	Hr.	Daily	Hr.	Daily	Bare Costs	Incl. O&P
1 Welder Foreman (outside)	$52.05	$416.40	$92.85	$742.80	$48.92	$81.03
1 Equip. Oper. (light)	45.80	366.40	69.20	553.60		
1 Welder, Gas Engine, 300 amp		144.00		158.40	9.00	9.90
16 L.H., Daily Totals		$926.80		$1454.80	$57.92	$90.92

Crew E-13	Hr.	Daily	Hr.	Daily	Bare Costs	Incl. O&P
1 Welder Foreman (outside)	$52.05	$416.40	$92.85	$742.80	$49.97	$84.97
.5 Equip. Oper. (light)	45.80	183.20	69.20	276.80		
1 Welder, Gas Engine, 300 amp		144.00		158.40	12.00	13.20
12 L.H., Daily Totals		$743.60		$1178.00	$61.97	$98.17

Crew E-14	Hr.	Daily	Hr.	Daily	Bare Costs	Incl. O&P
1 Welder Foreman (outside)	$52.05	$416.40	$92.85	$742.80	$52.05	$92.85
1 Welder, Gas Engine, 300 amp		144.00		158.40	18.00	19.80
8 L.H., Daily Totals		$560.40		$901.20	$70.05	$112.65

Crew E-16	Hr.	Daily	Hr.	Daily	Bare Costs	Incl. O&P
1 Welder Foreman (outside)	$52.05	$416.40	$92.85	$742.80	$51.05	$91.08
1 Welder	50.05	400.40	89.30	714.40		
1 Welder, Gas Engine, 300 amp		144.00		158.40	9.00	9.90
16 L.H., Daily Totals		$960.80		$1615.60	$60.05	$100.97

Crew E-17	Hr.	Daily	Hr.	Daily	Bare Costs	Incl. O&P
1 Struc. Steel Foreman (outside)	$52.05	$416.40	$92.85	$742.80	$51.05	$91.08
1 Structural Steel Worker	50.05	400.40	89.30	714.40		
16 L.H., Daily Totals		$816.80		$1457.20	$51.05	$91.08

Crew E-18	Hr.	Daily	Hr.	Daily	Bare Costs	Incl. O&P
1 Struc. Steel Foreman (outside)	$52.05	$416.40	$92.85	$742.80	$49.94	$86.50
3 Structural Steel Workers	50.05	1201.20	89.30	2143.20		
1 Equipment Operator (med.)	47.50	380.00	71.75	574.00		
1 Lattice Boom Crane, 20 Ton		958.30		1054.13	23.96	26.35
40 L.H., Daily Totals		$2955.90		$4514.13	$73.90	$112.85

Crew E-19	Hr.	Daily	Hr.	Daily	Bare Costs	Incl. O&P
1 Structural Steel Worker	$50.05	$400.40	$89.30	$714.40	$49.30	$83.78
1 Struc. Steel Foreman (outside)	52.05	416.40	92.85	742.80		
1 Equip. Oper. (light)	45.80	366.40	69.20	553.60		
1 Lattice Boom Crane, 20 Ton		958.30		1054.13	39.93	43.92
24 L.H., Daily Totals		$2141.50		$3064.93	$89.23	$127.71

Crew E-20	Hr.	Daily	Hr.	Daily	Bare Costs	Incl. O&P
1 Struc. Steel Foreman (outside)	$52.05	$416.40	$92.85	$742.80	$49.17	$84.62
5 Structural Steel Workers	50.05	2002.00	89.30	3572.00		
1 Equip. Oper. (crane)	48.80	390.40	73.75	590.00		
1 Oiler	42.25	338.00	63.85	510.80		
1 Lattice Boom Crane, 40 Ton		1174.00		1291.40	18.34	20.18
64 L.H., Daily Totals		$4320.80		$6707.00	$67.51	$104.80

Crew E-22	Hr.	Daily	Hr.	Daily	Bare Costs	Incl. O&P
1 Skilled Worker Foreman (out)	$48.20	$385.60	$74.50	$596.00	$46.87	$72.47
2 Skilled Workers	46.20	739.20	71.45	1143.20		
24 L.H., Daily Totals		$1124.80		$1739.20	$46.87	$72.47

Crew E-24	Hr.	Daily	Hr.	Daily	Bare Costs	Incl. O&P
3 Structural Steel Workers	$50.05	$1201.20	$89.30	$2143.20	$49.41	$84.91
1 Equipment Operator (med.)	47.50	380.00	71.75	574.00		
1 Hyd. Crane, 25 Ton		728.60		801.46	22.77	25.05
32 L.H., Daily Totals		$2309.80		$3518.66	$72.18	$109.96

Crew E-25	Hr.	Daily	Hr.	Daily	Bare Costs	Incl. O&P
1 Welder Foreman (outside)	$52.05	$416.40	$92.85	$742.80	$52.05	$92.85
1 Cutting Torch		11.40		12.54		
1 Set of Gases		133.20		146.52	18.07	19.88
8 L.H., Daily Totals		$561.00		$901.86	$70.13	$112.73

Crew F-3	Hr.	Daily	Hr.	Daily	Bare Costs	Incl. O&P
4 Carpenters	$44.90	$1436.80	$69.15	$2212.80	$45.68	$70.07
1 Equip. Oper. (crane)	48.80	390.40	73.75	590.00		
1 Hyd. Crane, 12 Ton		646.20		710.82	16.16	17.77
40 L.H., Daily Totals		$2473.40		$3513.62	$61.84	$87.84

Crew F-4	Hr.	Daily	Hr.	Daily	Bare Costs	Incl. O&P
4 Carpenters	$44.90	$1436.80	$69.15	$2212.80	$45.11	$69.03
1 Equip. Oper. (crane)	48.80	390.40	73.75	590.00		
1 Equip. Oper. Oiler	42.25	338.00	63.85	510.80		
1 Hyd. Crane, 55 Ton		1115.00		1226.50	23.23	25.55
48 L.H., Daily Totals		$3280.20		$4540.10	$68.34	$94.59

Crew F-5	Hr.	Daily	Hr.	Daily	Bare Costs	Incl. O&P
1 Carpenter Foreman (outside)	$46.90	$375.20	$72.25	$578.00	$45.40	$69.92
3 Carpenters	44.90	1077.60	69.15	1659.60		
32 L.H., Daily Totals		$1452.80		$2237.60	$45.40	$69.92

Crew No.	Bare Costs		Incl. Subs O&P		Cost Per Labor-Hour	
Crew F-6	**Hr.**	**Daily**	**Hr.**	**Daily**	**Bare Costs**	**Incl. O&P**
2 Carpenters	$44.90	$718.40	$69.15	$1106.40	$41.90	$64.25
2 Building Laborers	35.45	567.20	54.60	873.60		
1 Equip. Oper. (crane)	48.80	390.40	73.75	590.00		
1 Hyd. Crane, 12 Ton		646.20		710.82	16.16	17.77
40 L.H., Daily Totals		$2322.20		$3280.82	$58.06	$82.02

Crew F-7	Hr.	Daily	Hr.	Daily	Bare Costs	Incl. O&P
2 Carpenters	$44.90	$718.40	$69.15	$1106.40	$40.17	$61.88
2 Building Laborers	35.45	567.20	54.60	873.60		
32 L.H., Daily Totals		$1285.60		$1980.00	$40.17	$61.88

Crew G-1	Hr.	Daily	Hr.	Daily	Bare Costs	Incl. O&P
1 Roofer Foreman (outside)	$40.20	$321.60	$67.75	$542.00	$35.69	$60.12
4 Roofers, Composition	38.20	1222.40	64.35	2059.20		
2 Roofer Helpers	28.40	454.40	47.85	765.60		
1 Application Equipment		197.20		216.92		
1 Tar Kettle/Pot		92.50		101.75		
1 Crew Truck		221.20		243.32	9.12	10.04
56 L.H., Daily Totals		$2509.30		$3928.79	$44.81	$70.16

Crew G-2	Hr.	Daily	Hr.	Daily	Bare Costs	Incl. O&P
1 Plasterer	$41.45	$331.60	$62.10	$496.80	$37.67	$56.93
1 Plasterer Helper	36.10	288.80	54.10	432.80		
1 Building Laborer	35.45	283.60	54.60	436.80		
1 Grout Pump, 50 C.F./hr.		130.80		143.88	5.45	6.00
24 L.H., Daily Totals		$1034.80		$1510.28	$43.12	$62.93

Crew G-2A	Hr.	Daily	Hr.	Daily	Bare Costs	Incl. O&P
1 Roofer, composition	$38.20	$305.60	$64.35	$514.80	$34.02	$55.60
1 Roofer Helper	28.40	227.20	47.85	382.80		
1 Building Laborer	35.45	283.60	54.60	436.80		
1 Foam Spray Rig, Trailer-Mtd.		562.85		619.13		
1 Pickup Truck, 3/4 Ton		154.60		170.06	29.89	32.88
24 L.H., Daily Totals		$1533.85		$2123.59	$63.91	$88.48

Crew G-3	Hr.	Daily	Hr.	Daily	Bare Costs	Incl. O&P
2 Sheet Metal Workers	$53.30	$852.80	$81.30	$1300.80	$44.38	$67.95
2 Building Laborers	35.45	567.20	54.60	873.60		
32 L.H., Daily Totals		$1420.00		$2174.40	$44.38	$67.95

Crew G-4	Hr.	Daily	Hr.	Daily	Bare Costs	Incl. O&P
1 Labor Foreman (outside)	$37.45	$299.60	$57.65	$461.20	$36.12	$55.62
2 Building Laborers	35.45	567.20	54.60	873.60		
1 Flatbed Truck, Gas, 1.5 Ton		265.80		292.38		
1 Air Compressor, 160 cfm		158.00		173.80	17.66	19.42
24 L.H., Daily Totals		$1290.60		$1800.98	$53.77	$75.04

Crew G-5	Hr.	Daily	Hr.	Daily	Bare Costs	Incl. O&P
1 Roofer Foreman (outside)	$40.20	$321.60	$67.75	$542.00	$34.68	$58.43
2 Roofers, Composition	38.20	611.20	64.35	1029.60		
2 Roofer Helpers	28.40	454.40	47.85	765.60		
1 Application Equipment		197.20		216.92	4.93	5.42
40 L.H., Daily Totals		$1584.40		$2554.12	$39.61	$63.85

Crew G-6A	Hr.	Daily	Hr.	Daily	Bare Costs	Incl. O&P
2 Roofers Composition	$38.20	$611.20	$64.35	$1029.60	$38.20	$64.35
1 Small Compressor, Electric		10.95		12.05		
2 Pneumatic Nailers		45.70		50.27	3.54	3.89
16 L.H., Daily Totals		$667.85		$1091.92	$41.74	$68.24

Crew G-7	Hr.	Daily	Hr.	Daily	Bare Costs	Incl. O&P
1 Carpenter	$44.90	$359.20	$69.15	$553.20	$44.90	$69.15
1 Small Compressor, Electric		10.95		12.05		
1 Pneumatic Nailer		22.85		25.14	4.22	4.65
8 L.H., Daily Totals		$393.00		$590.38	$49.13	$73.80

Crew H-1	Hr.	Daily	Hr.	Daily	Bare Costs	Incl. O&P
2 Glaziers	$43.30	$692.80	$65.55	$1048.80	$46.67	$77.42
2 Struc. Steel Workers	50.05	800.80	89.30	1428.80		
32 L.H., Daily Totals		$1493.60		$2477.60	$46.67	$77.42

Crew H-2	Hr.	Daily	Hr.	Daily	Bare Costs	Incl. O&P
2 Glaziers	$43.30	$692.80	$65.55	$1048.80	$40.68	$61.90
1 Building Laborer	35.45	283.60	54.60	436.80		
24 L.H., Daily Totals		$976.40		$1485.60	$40.68	$61.90

Crew H-3	Hr.	Daily	Hr.	Daily	Bare Costs	Incl. O&P
1 Glazier	$43.30	$346.40	$65.55	$524.40	$38.52	$58.77
1 Helper	33.75	270.00	52.00	416.00		
16 L.H., Daily Totals		$616.40		$940.40	$38.52	$58.77

Crew H-4	Hr.	Daily	Hr.	Daily	Bare Costs	Incl. O&P
1 Carpenter	$44.90	$359.20	$69.15	$553.20	$41.94	$64.14
1 Carpenter Helper	33.75	270.00	52.00	416.00		
.5 Electrician	52.40	209.60	78.40	313.60		
20 L.H., Daily Totals		$838.80		$1282.80	$41.94	$64.14

Crew J-1	Hr.	Daily	Hr.	Daily	Bare Costs	Incl. O&P
3 Plasterers	$41.45	$994.80	$62.10	$1490.40	$39.31	$58.90
2 Plasterer Helpers	36.10	577.60	54.10	865.60		
1 Mixing Machine, 6 C.F.		142.40		156.64	3.56	3.92
40 L.H., Daily Totals		$1714.80		$2512.64	$42.87	$62.82

Crew J-2	Hr.	Daily	Hr.	Daily	Bare Costs	Incl. O&P
3 Plasterers	$41.45	$994.80	$62.10	$1490.40	$39.38	$58.80
2 Plasterer Helpers	36.10	577.60	54.10	865.60		
1 Lather	39.70	317.60	58.30	466.40		
1 Mixing Machine, 6 C.F.		142.40		156.64	2.97	3.26
48 L.H., Daily Totals		$2032.40		$2979.04	$42.34	$62.06

Crew J-3	Hr.	Daily	Hr.	Daily	Bare Costs	Incl. O&P
1 Terrazzo Worker	$41.00	$328.00	$60.25	$482.00	$37.73	$55.42
1 Terrazzo Helper	34.45	275.60	50.60	404.80		
1 Floor Grinder, 22" Path		91.10		100.21		
1 Terrazzo Mixer		192.00		211.20	17.69	19.46
16 L.H., Daily Totals		$886.70		$1198.21	$55.42	$74.89

Crew J-4	Hr.	Daily	Hr.	Daily	Bare Costs	Incl. O&P
2 Cement Finishers	$43.05	$688.80	$63.40	$1014.40	$40.52	$60.47
1 Laborer	35.45	283.60	54.60	436.80		
1 Floor Grinder, 22" Path		91.10		100.21		
1 Floor Edger, 7" Path		39.10		43.01		
1 Vacuum Pick-Up System		60.70		66.77	7.95	8.75
24 L.H., Daily Totals		$1163.30		$1661.19	$48.47	$69.22

Crew J-4A	Bare Costs Hr.	Bare Costs Daily	Incl. Subs O&P Hr.	Incl. Subs O&P Daily	Cost Per Labor-Hour Bare Costs	Cost Per Labor-Hour Incl. O&P
2 Cement Finishers	$43.05	$688.80	$63.40	$1014.40	$39.25	$59.00
2 Laborers	35.45	567.20	54.60	873.60		
1 Floor Grinder, 22" Path		91.10		100.21		
1 Floor Edger, 7" Path		39.10		43.01		
1 Vacuum Pick-Up System		60.70		66.77		
1 Floor Auto Scrubber		188.15		206.97	11.85	13.03
32 L.H., Daily Totals		$1635.05		$2304.95	$51.10	$72.03

Crew J-4B	Hr.	Daily	Hr.	Daily	Bare Costs	Incl. O&P
1 Laborer	$35.45	$283.60	$54.60	$436.80	$35.45	$54.60
1 Floor Auto Scrubber		188.15		206.97	23.52	25.87
8 L.H., Daily Totals		$471.75		$643.76	$58.97	$80.47

Crew J-6	Hr.	Daily	Hr.	Daily	Bare Costs	Incl. O&P
2 Painters	$38.70	$619.20	$58.05	$928.80	$39.66	$59.98
1 Building Laborer	35.45	283.60	54.60	436.80		
1 Equip. Oper. (light)	45.80	366.40	69.20	553.60		
1 Air Compressor, 250 cfm		195.80		215.38		
1 Sandblaster, Portable, 3 C.F.		20.00		22.00		
1 Set Sand Blasting Accessories		13.65		15.02	7.17	7.89
32 L.H., Daily Totals		$1498.65		$2171.59	$46.83	$67.86

Crew K-1	Hr.	Daily	Hr.	Daily	Bare Costs	Incl. O&P
1 Carpenter	$44.90	$359.20	$69.15	$553.20	$40.27	$61.88
1 Truck Driver (light)	35.65	285.20	54.60	436.80		
1 Flatbed Truck, Gas, 3 Ton		327.80		360.58	20.49	22.54
16 L.H., Daily Totals		$972.20		$1350.58	$60.76	$84.41

Crew K-2	Hr.	Daily	Hr.	Daily	Bare Costs	Incl. O&P
1 Struc. Steel Foreman (outside)	$52.05	$416.40	$92.85	$742.80	$45.92	$78.92
1 Struc. Steel Worker	50.05	400.40	89.30	714.40		
1 Truck Driver (light)	35.65	285.20	54.60	436.80		
1 Flatbed Truck, Gas, 3 Ton		327.80		360.58	13.66	15.02
24 L.H., Daily Totals		$1429.80		$2254.58	$59.58	$93.94

Crew L-1	Hr.	Daily	Hr.	Daily	Bare Costs	Incl. O&P
1 Electrician	$52.40	$419.20	$78.40	$627.20	$54.10	$81.22
1 Plumber	55.80	446.40	84.05	672.40		
16 L.H., Daily Totals		$865.60		$1299.60	$54.10	$81.22

Crew L-2	Hr.	Daily	Hr.	Daily	Bare Costs	Incl. O&P
1 Carpenter	$44.90	$359.20	$69.15	$553.20	$39.33	$60.58
1 Carpenter Helper	33.75	270.00	52.00	416.00		
16 L.H., Daily Totals		$629.20		$969.20	$39.33	$60.58

Crew L-3	Hr.	Daily	Hr.	Daily	Bare Costs	Incl. O&P
1 Carpenter	$44.90	$359.20	$69.15	$553.20	$48.88	$74.50
.5 Electrician	52.40	209.60	78.40	313.60		
.5 Sheet Metal Worker	53.30	213.20	81.30	325.20		
16 L.H., Daily Totals		$782.00		$1192.00	$48.88	$74.50

Crew L-3A	Hr.	Daily	Hr.	Daily	Bare Costs	Incl. O&P
1 Carpenter Foreman (outside)	$46.90	$375.20	$72.25	$578.00	$49.03	$75.27
.5 Sheet Metal Worker	53.30	213.20	81.30	325.20		
12 L.H., Daily Totals		$588.40		$903.20	$49.03	$75.27

Crew L-4	Hr.	Daily	Hr.	Daily	Bare Costs	Incl. O&P
2 Skilled Workers	$46.20	$739.20	$71.45	$1143.20	$42.05	$64.97
1 Helper	33.75	270.00	52.00	416.00		
24 L.H., Daily Totals		$1009.20		$1559.20	$42.05	$64.97

Crew L-5	Hr.	Daily	Hr.	Daily	Bare Costs	Incl. O&P
1 Struc. Steel Foreman (outside)	$52.05	$416.40	$92.85	$742.80	$50.16	$87.59
5 Struc. Steel Workers	50.05	2002.00	89.30	3572.00		
1 Equip. Oper. (crane)	48.80	390.40	73.75	590.00		
1 Hyd. Crane, 25 Ton		728.60		801.46	13.01	14.31
56 L.H., Daily Totals		$3537.40		$5706.26	$63.17	$101.90

Crew L-5A	Hr.	Daily	Hr.	Daily	Bare Costs	Incl. O&P
1 Struc. Steel Foreman (outside)	$52.05	$416.40	$92.85	$742.80	$50.24	$86.30
2 Structural Steel Workers	50.05	800.80	89.30	1428.80		
1 Equip. Oper. (crane)	48.80	390.40	73.75	590.00		
1 S.P. Crane, 4x4, 25 Ton		593.40		652.74	18.54	20.40
32 L.H., Daily Totals		$2201.00		$3414.34	$68.78	$106.70

Crew L-5B	Hr.	Daily	Hr.	Daily	Bare Costs	Incl. O&P
1 Struc. Steel Foreman (outside)	$52.05	$416.40	$92.85	$742.80	$51.26	$81.83
2 Structural Steel Workers	50.05	800.80	89.30	1428.80		
2 Electricians	52.40	838.40	78.40	1254.40		
2 Steamfitters/Pipefitters	56.65	906.40	85.30	1364.80		
1 Equip. Oper. (crane)	48.80	390.40	73.75	590.00		
1 Equip. Oper. Oiler	42.25	338.00	63.85	510.80		
1 Hyd. Crane, 80 Ton		1613.00		1774.30	22.40	24.64
72 L.H., Daily Totals		$5303.40		$7665.90	$73.66	$106.47

Crew L-6	Hr.	Daily	Hr.	Daily	Bare Costs	Incl. O&P
1 Plumber	$55.80	$446.40	$84.05	$672.40	$54.67	$82.17
.5 Electrician	52.40	209.60	78.40	313.60		
12 L.H., Daily Totals		$656.00		$986.00	$54.67	$82.17

Crew L-7	Hr.	Daily	Hr.	Daily	Bare Costs	Incl. O&P
2 Carpenters	$44.90	$718.40	$69.15	$1106.40	$43.27	$66.31
1 Building Laborer	35.45	283.60	54.60	436.80		
.5 Electrician	52.40	209.60	78.40	313.60		
28 L.H., Daily Totals		$1211.60		$1856.80	$43.27	$66.31

Crew L-8	Hr.	Daily	Hr.	Daily	Bare Costs	Incl. O&P
2 Carpenters	$44.90	$718.40	$69.15	$1106.40	$47.08	$72.13
.5 Plumber	55.80	223.20	84.05	336.20		
20 L.H., Daily Totals		$941.60		$1442.60	$47.08	$72.13

Crew L-9	Hr.	Daily	Hr.	Daily	Bare Costs	Incl. O&P
1 Labor Foreman (inside)	$35.95	$287.60	$55.35	$442.80	$40.69	$65.12
2 Building Laborers	35.45	567.20	54.60	873.60		
1 Struc. Steel Worker	50.05	400.40	89.30	714.40		
.5 Electrician	52.40	209.60	78.40	313.60		
36 L.H., Daily Totals		$1464.80		$2344.40	$40.69	$65.12

Crew L-10	Hr.	Daily	Hr.	Daily	Bare Costs	Incl. O&P
1 Struc. Steel Foreman (outside)	$52.05	$416.40	$92.85	$742.80	$50.30	$85.30
1 Structural Steel Worker	50.05	400.40	89.30	714.40		
1 Equip. Oper. (crane)	48.80	390.40	73.75	590.00		
1 Hyd. Crane, 12 Ton		646.20		710.82	26.93	29.62
24 L.H., Daily Totals		$1853.40		$2758.02	$77.22	$114.92

Crew L-11	Hr.	Daily	Hr.	Daily	Bare Costs	Incl. O&P
2 Wreckers	$35.45	$567.20	$60.35	$965.60	$41.38	$65.91
1 Equip. Oper. (crane)	48.80	390.40	73.75	590.00		
1 Equip. Oper. (light)	45.80	366.40	69.20	553.60		
1 Hyd. Excavator, 2.5 C.Y.		1800.00		1980.00		
1 Loader, Skid Steer, 78 H.P.		308.60		339.46	65.89	72.48
32 L.H., Daily Totals		$3432.60		$4428.66	$107.27	$138.40

Crew M-1	Bare Costs Hr.	Bare Costs Daily	Incl. Subs O&P Hr.	Incl. Subs O&P Daily	Cost Per Labor-Hour Bare Costs	Cost Per Labor-Hour Incl. O&P
3 Elevator Constructors	$71.10	$1706.40	$106.30	$2551.20	$67.55	$100.99
1 Elevator Apprentice	56.90	455.20	85.05	680.40		
5 Hand Tools		46.00		50.60	1.44	1.58
32 L.H., Daily Totals		$2207.60		$3282.20	$68.99	$102.57

Crew M-3	Hr.	Daily	Hr.	Daily	Bare Costs	Incl. O&P
1 Electrician Foreman (outside)	$54.40	$435.20	$81.40	$651.20	$54.05	$81.24
1 Common Laborer	35.45	283.60	54.60	436.80		
.25 Equipment Operator, Med.	47.50	95.00	71.75	143.50		
1 Elevator Constructor	71.10	568.80	106.30	850.40		
1 Elevator Apprentice	56.90	455.20	85.05	680.40		
.25 S.P. Crane, 4x4, 20 Ton		137.00		150.70	4.03	4.43
34 L.H., Daily Totals		$1974.80		$2913.00	$58.08	$85.68

Crew M-4	Hr.	Daily	Hr.	Daily	Bare Costs	Incl. O&P
1 Electrician Foreman (outside)	$54.40	$435.20	$81.40	$651.20	$53.47	$80.39
1 Common Laborer	35.45	283.60	54.60	436.80		
.25 Equipment Operator, Crane	48.80	97.60	73.75	147.50		
.25 Equipment Operator, Oiler	42.25	84.50	63.85	127.70		
1 Elevator Constructor	71.10	568.80	106.30	850.40		
1 Elevator Apprentice	56.90	455.20	85.05	680.40		
.25 S.P. Crane, 4x4, 40 Ton		174.20		191.62	4.84	5.32
36 L.H., Daily Totals		$2099.10		$3085.62	$58.31	$85.71

Crew Q-1	Hr.	Daily	Hr.	Daily	Bare Costs	Incl. O&P
1 Plumber	$55.80	$446.40	$84.05	$672.40	$50.23	$75.65
1 Plumber Apprentice	44.65	357.20	67.25	538.00		
16 L.H., Daily Totals		$803.60		$1210.40	$50.23	$75.65

Crew Q-1A	Hr.	Daily	Hr.	Daily	Bare Costs	Incl. O&P
.25 Plumber Foreman (outside)	$57.80	$115.60	$87.05	$174.10	$56.20	$84.65
1 Plumber	55.80	446.40	84.05	672.40		
10 L.H., Daily Totals		$562.00		$846.50	$56.20	$84.65

Crew Q-1C	Hr.	Daily	Hr.	Daily	Bare Costs	Incl. O&P
1 Plumber	$55.80	$446.40	$84.05	$672.40	$49.32	$74.35
1 Plumber Apprentice	44.65	357.20	67.25	538.00		
1 Equip. Oper. (medium)	47.50	380.00	71.75	574.00		
1 Trencher, Chain Type, 8' D		3357.00		3692.70	139.88	153.86
24 L.H., Daily Totals		$4540.60		$5477.10	$189.19	$228.21

Crew Q-2	Hr.	Daily	Hr.	Daily	Bare Costs	Incl. O&P
2 Plumbers	$55.80	$892.80	$84.05	$1344.80	$52.08	$78.45
1 Plumber Apprentice	44.65	357.20	67.25	538.00		
24 L.H., Daily Totals		$1250.00		$1882.80	$52.08	$78.45

Crew Q-3	Hr.	Daily	Hr.	Daily	Bare Costs	Incl. O&P
1 Plumber Foreman (inside)	$56.30	$450.40	$84.80	$678.40	$53.14	$80.04
2 Plumbers	55.80	892.80	84.05	1344.80		
1 Plumber Apprentice	44.65	357.20	67.25	538.00		
32 L.H., Daily Totals		$1700.40		$2561.20	$53.14	$80.04

Crew Q-4	Hr.	Daily	Hr.	Daily	Bare Costs	Incl. O&P
1 Plumber Foreman (inside)	$56.30	$450.40	$84.80	$678.40	$53.14	$80.04
1 Plumber	55.80	446.40	84.05	672.40		
1 Welder (plumber)	55.80	446.40	84.05	672.40		
1 Plumber Apprentice	44.65	357.20	67.25	538.00		
1 Welder, Electric, 300 amp		52.85		58.13	1.65	1.82
32 L.H., Daily Totals		$1753.25		$2619.34	$54.79	$81.85

Crew Q-5	Hr.	Daily	Hr.	Daily	Bare Costs	Incl. O&P
1 Steamfitter	$56.65	$453.20	$85.30	$682.40	$50.98	$76.75
1 Steamfitter Apprentice	45.30	362.40	68.20	545.60		
16 L.H., Daily Totals		$815.60		$1228.00	$50.98	$76.75

Crew Q-6	Hr.	Daily	Hr.	Daily	Bare Costs	Incl. O&P
2 Steamfitters	$56.65	$906.40	$85.30	$1364.80	$52.87	$79.60
1 Steamfitter Apprentice	45.30	362.40	68.20	545.60		
24 L.H., Daily Totals		$1268.80		$1910.40	$52.87	$79.60

Crew Q-7	Hr.	Daily	Hr.	Daily	Bare Costs	Incl. O&P
1 Steamfitter Foreman (inside)	$57.15	$457.20	$86.05	$688.40	$53.94	$81.21
2 Steamfitters	56.65	906.40	85.30	1364.80		
1 Steamfitter Apprentice	45.30	362.40	68.20	545.60		
32 L.H., Daily Totals		$1726.00		$2598.80	$53.94	$81.21

Crew Q-8	Hr.	Daily	Hr.	Daily	Bare Costs	Incl. O&P
1 Steamfitter Foreman (inside)	$57.15	$457.20	$86.05	$688.40	$53.94	$81.21
1 Steamfitter	56.65	453.20	85.30	682.40		
1 Welder (steamfitter)	56.65	453.20	85.30	682.40		
1 Steamfitter Apprentice	45.30	362.40	68.20	545.60		
1 Welder, Electric, 300 amp		52.85		58.13	1.65	1.82
32 L.H., Daily Totals		$1778.85		$2656.93	$55.59	$83.03

Crew Q-9	Hr.	Daily	Hr.	Daily	Bare Costs	Incl. O&P
1 Sheet Metal Worker	$53.30	$426.40	$81.30	$650.40	$47.98	$73.17
1 Sheet Metal Apprentice	42.65	341.20	65.05	520.40		
16 L.H., Daily Totals		$767.60		$1170.80	$47.98	$73.17

Crew Q-10	Hr.	Daily	Hr.	Daily	Bare Costs	Incl. O&P
2 Sheet Metal Workers	$53.30	$852.80	$81.30	$1300.80	$49.75	$75.88
1 Sheet Metal Apprentice	42.65	341.20	65.05	520.40		
24 L.H., Daily Totals		$1194.00		$1821.20	$49.75	$75.88

Crew Q-11	Hr.	Daily	Hr.	Daily	Bare Costs	Incl. O&P
1 Sheet Metal Foreman (inside)	$53.80	$430.40	$82.05	$656.40	$50.76	$77.42
2 Sheet Metal Workers	53.30	852.80	81.30	1300.80		
1 Sheet Metal Apprentice	42.65	341.20	65.05	520.40		
32 L.H., Daily Totals		$1624.40		$2477.60	$50.76	$77.42

Crew Q-12	Hr.	Daily	Hr.	Daily	Bare Costs	Incl. O&P
1 Sprinkler Installer	$54.65	$437.20	$82.35	$658.80	$49.17	$74.10
1 Sprinkler Apprentice	43.70	349.60	65.85	526.80		
16 L.H., Daily Totals		$786.80		$1185.60	$49.17	$74.10

Crew Q-13	Hr.	Daily	Hr.	Daily	Bare Costs	Incl. O&P
1 Sprinkler Foreman (inside)	$55.15	$441.20	$83.10	$664.80	$52.04	$78.41
2 Sprinkler Installers	54.65	874.40	82.35	1317.60		
1 Sprinkler Apprentice	43.70	349.60	65.85	526.80		
32 L.H., Daily Totals		$1665.20		$2509.20	$52.04	$78.41

Crew Q-14	Hr.	Daily	Hr.	Daily	Bare Costs	Incl. O&P
1 Asbestos Worker	$49.70	$397.60	$77.20	$617.60	$44.73	$69.47
1 Asbestos Apprentice	39.75	318.00	61.75	494.00		
16 L.H., Daily Totals		$715.60		$1111.60	$44.73	$69.47

Crew No.	Bare Costs		Incl. Subs O&P		Cost Per Labor-Hour	

Crew Q-15	Hr.	Daily	Hr.	Daily	Bare Costs	Incl. O&P
1 Plumber	$55.80	$446.40	$84.05	$672.40	$50.23	$75.65
1 Plumber Apprentice	44.65	357.20	67.25	538.00		
1 Welder, Electric, 300 amp		52.85		58.13	3.30	3.63
16 L.H., Daily Totals		$856.45		$1268.54	$53.53	$79.28

Crew Q-16	Hr.	Daily	Hr.	Daily	Bare Costs	Incl. O&P
2 Plumbers	$55.80	$892.80	$84.05	$1344.80	$52.08	$78.45
1 Plumber Apprentice	44.65	357.20	67.25	538.00		
1 Welder, Electric, 300 amp		52.85		58.13	2.20	2.42
24 L.H., Daily Totals		$1302.85		$1940.93	$54.29	$80.87

Crew Q-17	Hr.	Daily	Hr.	Daily	Bare Costs	Incl. O&P
1 Steamfitter	$56.65	$453.20	$85.30	$682.40	$50.98	$76.75
1 Steamfitter Apprentice	45.30	362.40	68.20	545.60		
1 Welder, Electric, 300 amp		52.85		58.13	3.30	3.63
16 L.H., Daily Totals		$868.45		$1286.14	$54.28	$80.38

Crew Q-17A	Hr.	Daily	Hr.	Daily	Bare Costs	Incl. O&P
1 Steamfitter	$56.65	$453.20	$85.30	$682.40	$50.25	$75.75
1 Steamfitter Apprentice	45.30	362.40	68.20	545.60		
1 Equip. Oper. (crane)	48.80	390.40	73.75	590.00		
1 Hyd. Crane, 12 Ton		646.20		710.82		
1 Welder, Electric, 300 amp		52.85		58.13	29.13	32.04
24 L.H., Daily Totals		$1905.05		$2586.95	$79.38	$107.79

Crew Q-18	Hr.	Daily	Hr.	Daily	Bare Costs	Incl. O&P
2 Steamfitters	$56.65	$906.40	$85.30	$1364.80	$52.87	$79.60
1 Steamfitter Apprentice	45.30	362.40	68.20	545.60		
1 Welder, Electric, 300 amp		52.85		58.13	2.20	2.42
24 L.H., Daily Totals		$1321.65		$1968.54	$55.07	$82.02

Crew Q-19	Hr.	Daily	Hr.	Daily	Bare Costs	Incl. O&P
1 Steamfitter	$56.65	$453.20	$85.30	$682.40	$51.45	$77.30
1 Steamfitter Apprentice	45.30	362.40	68.20	545.60		
1 Electrician	52.40	419.20	78.40	627.20		
24 L.H., Daily Totals		$1234.80		$1855.20	$51.45	$77.30

Crew Q-20	Hr.	Daily	Hr.	Daily	Bare Costs	Incl. O&P
1 Sheet Metal Worker	$53.30	$426.40	$81.30	$650.40	$48.86	$74.22
1 Sheet Metal Apprentice	42.65	341.20	65.05	520.40		
.5 Electrician	52.40	209.60	78.40	313.60		
20 L.H., Daily Totals		$977.20		$1484.40	$48.86	$74.22

Crew Q-21	Hr.	Daily	Hr.	Daily	Bare Costs	Incl. O&P
2 Steamfitters	$56.65	$906.40	$85.30	$1364.80	$52.75	$79.30
1 Steamfitter Apprentice	45.30	362.40	68.20	545.60		
1 Electrician	52.40	419.20	78.40	627.20		
32 L.H., Daily Totals		$1688.00		$2537.60	$52.75	$79.30

Crew Q-22	Hr.	Daily	Hr.	Daily	Bare Costs	Incl. O&P
1 Plumber	$55.80	$446.40	$84.05	$672.40	$50.23	$75.65
1 Plumber Apprentice	44.65	357.20	67.25	538.00		
1 Hyd. Crane, 12 Ton		646.20		710.82	40.39	44.43
16 L.H., Daily Totals		$1449.80		$1921.22	$90.61	$120.08

Crew No.	Bare Costs		Incl. Subs O&P		Cost Per Labor-Hour	

Crew Q-22A	Hr.	Daily	Hr.	Daily	Bare Costs	Incl. O&P
1 Plumber	$55.80	$446.40	$84.05	$672.40	$46.17	$69.91
1 Plumber Apprentice	44.65	357.20	67.25	538.00		
1 Laborer	35.45	283.60	54.60	436.80		
1 Equip. Oper. (crane)	48.80	390.40	73.75	590.00		
1 Hyd. Crane, 12 Ton		646.20		710.82	20.19	22.21
32 L.H., Daily Totals		$2123.80		$2948.02	$66.37	$92.13

Crew Q-23	Hr.	Daily	Hr.	Daily	Bare Costs	Incl. O&P
1 Plumber Foreman (outside)	$57.80	$462.40	$87.05	$696.40	$53.70	$80.95
1 Plumber	55.80	446.40	84.05	672.40		
1 Equip. Oper. (med.)	47.50	380.00	71.75	574.00		
1 Lattice Boom Crane, 20 Ton		958.30		1054.13	39.93	43.92
24 L.H., Daily Totals		$2247.10		$2996.93	$93.63	$124.87

Crew R-1	Hr.	Daily	Hr.	Daily	Bare Costs	Incl. O&P
1 Electrician Foreman	$52.90	$423.20	$79.15	$633.20	$46.27	$69.72
3 Electricians	52.40	1257.60	78.40	1881.60		
2 Helpers	33.75	540.00	52.00	832.00		
48 L.H., Daily Totals		$2220.80		$3346.80	$46.27	$69.72

Crew R-1A	Hr.	Daily	Hr.	Daily	Bare Costs	Incl. O&P
1 Electrician	$52.40	$419.20	$78.40	$627.20	$43.08	$65.20
1 Helper	33.75	270.00	52.00	416.00		
16 L.H., Daily Totals		$689.20		$1043.20	$43.08	$65.20

Crew R-2	Hr.	Daily	Hr.	Daily	Bare Costs	Incl. O&P
1 Electrician Foreman	$52.90	$423.20	$79.15	$633.20	$46.63	$70.30
3 Electricians	52.40	1257.60	78.40	1881.60		
2 Helpers	33.75	540.00	52.00	832.00		
1 Equip. Oper. (crane)	48.80	390.40	73.75	590.00		
1 S.P. Crane, 4x4, 5 Ton		274.20		301.62	4.90	5.39
56 L.H., Daily Totals		$2885.40		$4238.42	$51.52	$75.69

Crew R-3	Hr.	Daily	Hr.	Daily	Bare Costs	Incl. O&P
1 Electrician Foreman	$52.90	$423.20	$79.15	$633.20	$51.88	$77.77
1 Electrician	52.40	419.20	78.40	627.20		
.5 Equip. Oper. (crane)	48.80	195.20	73.75	295.00		
.5 S.P. Crane, 4x4, 5 Ton		137.10		150.81	6.86	7.54
20 L.H., Daily Totals		$1174.70		$1706.21	$58.73	$85.31

Crew R-4	Hr.	Daily	Hr.	Daily	Bare Costs	Incl. O&P
1 Struc. Steel Foreman (outside)	$52.05	$416.40	$92.85	$742.80	$50.92	$87.83
3 Struc. Steel Workers	50.05	1201.20	89.30	2143.20		
1 Electrician	52.40	419.20	78.40	627.20		
1 Welder, Gas Engine, 300 amp		144.00		158.40	3.60	3.96
40 L.H., Daily Totals		$2180.80		$3671.60	$54.52	$91.79

Crew R-5	Hr.	Daily	Hr.	Daily	Bare Costs	Incl. O&P
1 Electrician Foreman	$52.90	$423.20	$79.15	$633.20	$45.66	$68.87
4 Electrician Linemen	52.40	1676.80	78.40	2508.80		
2 Electrician Operators	52.40	838.40	78.40	1254.40		
4 Electrician Groundmen	33.75	1080.00	52.00	1664.00		
1 Crew Truck		221.20		243.32		
1 Flatbed Truck, 20,000 GVW		253.60		278.96		
1 Pickup Truck, 3/4 Ton		154.60		170.06		
.2 Hyd. Crane, 55 Ton		223.00		245.30		
.2 Hyd. Crane, 12 Ton		129.24		142.16		
.2 Earth Auger, Truck-Mtd.		86.24		94.86		
1 Tractor w/Winch		434.20		477.62	17.07	18.78
88 L.H., Daily Totals		$5520.48		$7712.69	$62.73	$87.64

Crew R-6	Hr.	Daily	Hr.	Daily	Bare Costs	Incl. O&P
1 Electrician Foreman	$52.90	$423.20	$79.15	$633.20	$45.66	$68.87
4 Electrician Linemen	52.40	1676.80	78.40	2508.80		
2 Electrician Operators	52.40	838.40	78.40	1254.40		
4 Electrician Groundmen	33.75	1080.00	52.00	1664.00		
1 Crew Truck		221.20		243.32		
1 Flatbed Truck, 20,000 GVW		253.60		278.96		
1 Pickup Truck, 3/4 Ton		154.60		170.06		
.2 Hyd. Crane, 55 Ton		223.00		245.30		
.2 Hyd. Crane, 12 Ton		129.24		142.16		
.2 Earth Auger, Truck-Mtd.		86.24		94.86		
1 Tractor w/Winch		434.20		477.62		
3 Cable Trailers		575.10		632.61		
.5 Tensioning Rig		190.72		209.80		
.5 Cable Pulling Rig		1116.50		1228.15	38.46	42.31
88 L.H., Daily Totals		$7402.81		$9783.25	$84.12	$111.17

Crew R-7	Hr.	Daily	Hr.	Daily	Bare Costs	Incl. O&P
1 Electrician Foreman	$52.90	$423.20	$79.15	$633.20	$36.94	$56.52
5 Electrician Groundmen	33.75	1350.00	52.00	2080.00		
1 Crew Truck		221.20		243.32	4.61	5.07
48 L.H., Daily Totals		$1994.40		$2956.52	$41.55	$61.59

Crew R-8	Hr.	Daily	Hr.	Daily	Bare Costs	Incl. O&P
1 Electrician Foreman	$52.90	$423.20	$79.15	$633.20	$46.27	$69.72
3 Electrician Linemen	52.40	1257.60	78.40	1881.60		
2 Electrician Groundmen	33.75	540.00	52.00	832.00		
1 Pickup Truck, 3/4 Ton		154.60		170.06		
1 Crew Truck		221.20		243.32	7.83	8.61
48 L.H., Daily Totals		$2596.60		$3760.18	$54.10	$78.34

Crew R-9	Hr.	Daily	Hr.	Daily	Bare Costs	Incl. O&P
1 Electrician Foreman	$52.90	$423.20	$79.15	$633.20	$43.14	$65.29
1 Electrician Lineman	52.40	419.20	78.40	627.20		
2 Electrician Operators	52.40	838.40	78.40	1254.40		
4 Electrician Groundmen	33.75	1080.00	52.00	1664.00		
1 Pickup Truck, 3/4 Ton		154.60		170.06		
1 Crew Truck		221.20		243.32	5.87	6.46
64 L.H., Daily Totals		$3136.60		$4592.18	$49.01	$71.75

Crew R-10	Hr.	Daily	Hr.	Daily	Bare Costs	Incl. O&P
1 Electrician Foreman	$52.90	$423.20	$79.15	$633.20	$49.38	$74.13
4 Electrician Linemen	52.40	1676.80	78.40	2508.80		
1 Electrician Groundman	33.75	270.00	52.00	416.00		
1 Crew Truck		221.20		243.32		
3 Tram Cars		389.10		428.01	12.71	13.99
48 L.H., Daily Totals		$2980.30		$4229.33	$62.09	$88.11

Crew R-11	Hr.	Daily	Hr.	Daily	Bare Costs	Incl. O&P
1 Electrician Foreman	$52.90	$423.20	$79.15	$633.20	$49.54	$74.44
4 Electricians	52.40	1676.80	78.40	2508.80		
1 Equip. Oper. (crane)	48.80	390.40	73.75	590.00		
1 Common Laborer	35.45	283.60	54.60	436.80		
1 Crew Truck		221.20		243.32		
1 Hyd. Crane, 12 Ton		646.20		710.82	15.49	17.04
56 L.H., Daily Totals		$3641.40		$5122.94	$65.03	$91.48

Crew R-12	Hr.	Daily	Hr.	Daily	Bare Costs	Incl. O&P
1 Carpenter Foreman (inside)	$45.40	$363.20	$69.90	$559.20	$42.21	$66.00
4 Carpenters	44.90	1436.80	69.15	2212.80		
4 Common Laborers	35.45	1134.40	54.60	1747.20		
1 Equip. Oper. (med.)	47.50	380.00	71.75	574.00		
1 Steel Worker	50.05	400.40	89.30	714.40		
1 Dozer, 200 H.P.		1333.00		1466.30		
1 Pickup Truck, 3/4 Ton		154.60		170.06	16.90	18.59
88 L.H., Daily Totals		$5202.40		$7443.96	$59.12	$84.59

Crew R-13	Hr.	Daily	Hr.	Daily	Bare Costs	Incl. O&P
1 Electrician Foreman	$52.90	$423.20	$79.15	$633.20	$50.39	$75.55
3 Electricians	52.40	1257.60	78.40	1881.60		
.25 Equip. Oper. (crane)	48.80	97.60	73.75	147.50		
1 Equipment Oiler	42.25	338.00	63.85	510.80		
.25 Hydraulic Crane, 33 Ton		185.75		204.32	4.42	4.86
42 L.H., Daily Totals		$2302.15		$3377.43	$54.81	$80.41

Crew R-15	Hr.	Daily	Hr.	Daily	Bare Costs	Incl. O&P
1 Electrician Foreman	$52.90	$423.20	$79.15	$633.20	$51.38	$76.99
4 Electricians	52.40	1676.80	78.40	2508.80		
1 Equipment Oper. (light)	45.80	366.40	69.20	553.60		
1 Aerial Lift Truck, 40' Boom		332.20		365.42	6.92	7.61
48 L.H., Daily Totals		$2798.60		$4061.02	$58.30	$84.60

Crew R-18	Hr.	Daily	Hr.	Daily	Bare Costs	Incl. O&P
.25 Electrician Foreman	$52.90	$105.80	$79.15	$158.30	$40.96	$62.21
1 Electrician	52.40	419.20	78.40	627.20		
2 Helpers	33.75	540.00	52.00	832.00		
26 L.H., Daily Totals		$1065.00		$1617.50	$40.96	$62.21

Crew R-19	Hr.	Daily	Hr.	Daily	Bare Costs	Incl. O&P
.5 Electrician Foreman	$52.90	$211.60	$79.15	$316.60	$52.50	$78.55
2 Electricians	52.40	838.40	78.40	1254.40		
20 L.H., Daily Totals		$1050.00		$1571.00	$52.50	$78.55

Crew R-21	Hr.	Daily	Hr.	Daily	Bare Costs	Incl. O&P
1 Electrician Foreman	$52.90	$423.20	$79.15	$633.20	$52.40	$78.42
3 Electricians	52.40	1257.60	78.40	1881.60		
.1 Equip. Oper. (med.)	47.50	38.00	71.75	57.40		
.1 S.P. Crane, 4x4, 25 Ton		59.34		65.27	1.81	1.99
32.8 L.H., Daily Totals		$1778.14		$2637.47	$54.21	$80.41

Crew R-22	Hr.	Daily	Hr.	Daily	Bare Costs	Incl. O&P
.66 Electrician Foreman	$52.90	$279.31	$79.15	$417.91	$44.47	$67.18
2 Helpers	33.75	540.00	52.00	832.00		
2 Electricians	52.40	838.40	78.40	1254.40		
37.28 L.H., Daily Totals		$1657.71		$2504.31	$44.47	$67.18

Crew R-30	Hr.	Daily	Hr.	Daily	Bare Costs	Incl. O&P
.25 Electrician Foreman (outside)	$54.40	$108.80	$81.40	$162.80	$42.12	$63.98
1 Electrician	52.40	419.20	78.40	627.20		
2 Laborers, (Semi-Skilled)	35.45	567.20	54.60	873.60		
26 L.H., Daily Totals		$1095.20		$1663.60	$42.12	$63.98

Crew R-31	Hr.	Daily	Hr.	Daily	Bare Costs	Incl. O&P
1 Electrician	$52.40	$419.20	$78.40	$627.20	$52.40	$78.40
1 Core Drill, Electric, 2.5 H.P.		48.10		52.91	6.01	6.61
8 L.H., Daily Totals		$467.30		$680.11	$58.41	$85.01

(Column groups: Bare Costs — Hr., Daily; Incl. Subs O&P — Hr., Daily; Cost Per Labor-Hour — Bare Costs, Incl. O&P)

Crew No.	Bare Costs		Incl. Subs O & P		Cost Per Labor-Hour	
Crew W-41E	**Hr.**	**Daily**	**Hr.**	**Daily**	**Bare Costs**	**Incl. O&P**
.5 Plumber Foreman (outside)	$57.80	$231.20	$87.05	$348.20	$48.06	$72.87
1 Plumber	55.80	446.40	84.05	672.40		
1 Laborer	35.45	283.60	54.60	436.80		
20 L.H., Daily Totals		$961.20		$1457.40	$48.06	$72.87

Historical Cost Indexes

The table below lists both the RSMeans® historical cost index based on Jan. 1, 1993 = 100 as well as the computed value of an index based on Jan. 1, 2013 costs. Since the Jan. 1, 2013 figure is estimated, space is left to write in the actual index figures as they become available through either the quarterly *RSMeans Construction Cost Indexes* or as printed in the *Engineering News-Record.* To compute the actual index based on Jan. 1, 2013 = 100, divide the historical cost index for a particular year by the actual Jan. 1, 2013 construction cost index. Space has been left to advance the index figures as the year progresses.

Year	Historical Cost Index Jan. 1, 1993 = 100		Current Index Based on Jan. 1, 2013 = 100	
	Est.	Actual	Est.	Actual
Oct 2013*				
July 2013*				
April 2013*				
Jan 2013*	197.6		100.0	100.0
July 2012		194.6	98.5	
2011		191.2	96.8	
2010		183.5	92.9	
2009		180.1	91.1	
2008		180.4	91.3	
2007		169.4	85.7	
2006		162.0	82.0	
2005		151.6	76.7	
2004		143.7	72.7	
2003		132.0	66.8	
2002		128.7	65.1	
2001		125.1	63.3	
2000		120.9	61.2	
1999		117.6	59.5	

Year	Historical Cost Index Jan. 1, 1993 = 100	Current Index Based on Jan. 1, 2013 = 100	
	Actual	Est.	Actual
July 1998	115.1	59.7	
1997	112.8	57.1	
1996	110.2	55.8	
1995	107.6	54.5	
1994	104.4	52.8	
1993	101.7	51.5	
1992	99.4	50.3	
1991	96.8	49.0	
1990	94.3	47.7	
1989	92.1	46.6	
1988	89.9	45.5	
1987	87.7	44.4	
1986	84.2	42.6	
1985	82.6	41.8	
1984	82.0	41.5	
1983	80.2	40.6	
1982	76.1	38.5	
1981	70.0	35.4	

Year	Historical Cost Index Jan. 1, 1993 = 100	Current Index Based on Jan. 1, 2013 = 100	
	Actual	Est.	Actual
July 1980	62.9	31.8	
1979	57.8	29.3	
1978	53.5	27.1	
1977	49.5	25.1	
1976	46.9	23.7	
1975	44.8	22.7	
1974	41.4	21.0	
1973	37.7	19.1	
1972	34.8	17.6	
1971	32.1	16.2	
1970	28.7	14.5	
1969	26.9	13.6	
1968	24.9	12.6	
1967	23.5	11.9	
1966	22.7	11.5	
1965	21.7	11.0	
1964	21.2	10.7	
1963	20.7	10.5	

Adjustments to Costs

The "Historical Cost Index" can be used to convert national average building costs at a particular time to the approximate building costs for some other time.

Time Adjustment Using the Historical Cost Indexes:

$$\frac{\text{Index for Year A}}{\text{Index for Year B}} \times \text{Cost in Year B} = \text{Cost in Year A}$$

Example:

Estimate and compare construction costs for different years in the same city.

To estimate the national average construction cost of a building in 1970, knowing that it cost $900,000 in 2013:

INDEX in 1970 = 28.7

INDEX in 2013 = 197.6

$$\frac{\text{INDEX 1970}}{\text{INDEX 2013}} \times \text{Cost 2013} = \text{Cost 1970}$$

$$\frac{28.7}{197.6} \times \$900{,}000 = .145 \times \$900{,}000 = \$130{,}500$$

The construction cost of the building in 1970 is $130,500.

Note: The city cost indexes for Canada can be used to convert U.S. national averages to local costs in Canadian dollars.

*Historical Cost Index updates and other resources are provided on the following website.
http://www.reedconstructiondata.com/rsmeans/chgnotice/878796

How to Use the City Cost Indexes

What you should know before you begin

RSMeans City Cost Indexes (CCI) are an extremely useful tool to use when you want to compare costs from city to city and region to region.

This publication contains average construction cost indexes for 731 U.S. and Canadian cities covering over 930 three-digit zip code locations, as listed directly under each city.

Keep in mind that a City Cost Index number is a percentage ratio of a specific city's cost to the national average cost of the same item at a stated time period.

In other words, these index figures represent relative construction factors (or, if you prefer, multipliers) for Material and Installation costs, as well as the weighted average for Total In Place costs for each CSI MasterFormat division. Installation costs include both labor and equipment rental costs. When estimating equipment rental rates only, for a specific location, use 01 54 33 EQUIPMENT RENTAL COSTS in the Reference Section at the back of the book.

The 30 City Average Index is the average of 30 major U.S. cities and serves as a National Average.

Index figures for both material and installation are based on the 30 major city average of 100 and represent the cost relationship as of July 1, 2012. The index for each division is computed from representative material and labor quantities for that division. The weighted average for each city is a weighted total of the components listed above it, but does not include relative productivity between trades or cities.

As changes occur in local material prices, labor rates, and equipment rental rates, (including fuel costs) the impact of these changes should be accurately measured by the change in the City Cost Index for each particular city (as compared to the 30 City Average).

Therefore, if you know (or have estimated) building costs in one city today, you can easily convert those costs to expected building costs in another city.

In addition, by using the Historical Cost Index, you can easily convert National Average building costs at a particular time to the approximate building costs for some other time. The City Cost Indexes can then be applied to calculate the costs for a particular city.

Quick Calculations

Location Adjustment Using the City Cost Indexes:

$$\frac{\text{Index for City A}}{\text{Index for City B}} \times \text{Cost in City B} = \text{Cost in City A}$$

Time Adjustment for the National Average Using the Historical Cost Index:

$$\frac{\text{Index for Year A}}{\text{Index for Year B}} \times \text{Cost in Year B} = \text{Cost in Year A}$$

Adjustment from the National Average:

$$\frac{\text{Index for City A}}{100} \times \text{National Average Cost} = \text{Cost in City A}$$

Since each of the other RSMeans publications contains many different items, any *one* item multiplied by the particular city index may give incorrect results. However, the larger the number of items compiled, the closer the results should be to actual costs for that particular city.

The City Cost Indexes for Canadian cities are calculated using Canadian material and equipment prices and labor rates, in Canadian dollars. Therefore, indexes for Canadian cities can be used to convert U.S. National Average prices to local costs in Canadian dollars.

How to use this section

1. Compare costs from city to city.

In using the RSMeans Indexes, remember that an index number is not a fixed number but a ratio: It's a percentage ratio of a building component's cost at any stated time to the National Average cost of that same component at the same time period. Put in the form of an equation:

$$\frac{\text{Specific City Cost}}{\text{National Average Cost}} \times 100 = \text{City Index Number}$$

Therefore, when making cost comparisons between cities, do not subtract one city's index number from the index number of another city and read the result as a percentage difference. Instead, divide one city's index number by that of the other city. The resulting number may then be used as a multiplier to calculate cost differences from city to city.

The formula used to find cost differences between cities for the purpose of comparison is as follows:

$$\frac{\text{City A Index}}{\text{City B Index}} \times \text{City B Cost (Known)} = \text{City A Cost (Unknown)}$$

In addition, you can use RSMeans CCI to calculate and compare costs division by division between cities using the same basic formula. (Just be sure that you're comparing similar divisions.)

2. Compare a specific city's construction costs with the National Average.

When you're studying construction location feasibility, it's advisable to compare a prospective project's cost index with an index of the National Average cost.

For example, divide the weighted average index of construction costs of a specific city by that of the 30 City Average, which = 100.

$$\frac{\text{City Index}}{100} = \text{\% of National Average}$$

As a result, you get a ratio that indicates the relative cost of construction in that city in comparison with the National Average.

3. Convert U.S. National Average to actual costs in Canadian City.

$$\frac{\text{Index for Canadian City}}{100} \times \text{National Average Cost} = \text{Cost in Canadian City in \$ CAN}$$

4. **Adjust construction cost data based on a National Average.**

When you use a source of construction cost data which is based on a National Average (such as RSMeans cost data publications), it is necessary to adjust those costs to a specific location.

$$\frac{\text{City Index}}{100} \times \frac{\text{"Book" Cost Based on}}{\text{National Average Costs}} = \frac{\text{City Cost}}{\text{(Unknown)}}$$

5. **When applying the City Cost Indexes to demolition projects, use the appropriate division installation index.** For example, for removal of existing doors and windows, use Division 8 (Openings) index.

What you might like to know about how we developed the Indexes

The information presented in the CCI is organized according to the Construction Specifications Institute (CSI) MasterFormat 2010 classification system.

To create a reliable index, RSMeans researched the building type most often constructed in the United States and Canada. Because it was concluded that no one type of building completely represented the building construction industry, nine different types of buildings were combined to create a composite model.

The exact material, labor, and equipment quantities are based on detailed analyses of these nine building types, and then each quantity is weighted in proportion to expected usage. These various material items, labor hours, and equipment rental rates are thus combined to form a composite building representing as closely as possible the actual usage of materials, labor and equipment used in the North American building construction industry.

The following structures were chosen to make up that composite model:

1. Factory, 1 story
2. Office, 2–4 story
3. Store, Retail
4. Town Hall, 2–3 story
5. High School, 2–3 story
6. Hospital, 4–8 story
7. Garage, Parking
8. Apartment, 1–3 story
9. Hotel/Motel, 2–3 story

For the purposes of ensuring the timeliness of the data, the components of the index for the composite model have been streamlined. They currently consist of:

- specific quantities of 66 commonly used construction materials;
- specific labor-hours for 21 building construction trades; and
- specific days of equipment rental for 6 types of construction equipment (normally used to install the 66 material items by the 21 trades.) Fuel costs and routine maintenance costs are included in the equipment cost.

A sophisticated computer program handles the updating of all costs for each city on a quarterly basis. Material and equipment price quotations are gathered quarterly from 731 cities in the United States and Canada. These prices and the latest negotiated labor wage rates for 21 different building trades are used to compile the quarterly update of the City Cost Index.

The 30 major U.S. cities used to calculate the National Average are:

Atlanta, GA	Memphis, TN
Baltimore, MD	Milwaukee, WI
Boston, MA	Minneapolis, MN
Buffalo, NY	Nashville, TN
Chicago, IL	New Orleans, LA
Cincinnati, OH	New York, NY
Cleveland, OH	Philadelphia, PA
Columbus, OH	Phoenix, AZ
Dallas, TX	Pittsburgh, PA
Denver, CO	St. Louis, MO
Detroit, MI	San Antonio, TX
Houston, TX	San Diego, CA
Indianapolis, IN	San Francisco, CA
Kansas City, MO	Seattle, WA
Los Angeles, CA	Washington, DC

What the CCI does not indicate

The weighted average for each city is a total of the divisional components weighted to reflect typical usage, but it does not include the productivity variations between trades or cities.

In addition, the CCI does not take into consideration factors such as the following:

- managerial efficiency
- competitive conditions
- automation
- restrictive union practices
- unique local requirements
- regional variations due to specific building codes

	DIVISION	UNITED STATES			ALABAMA															
		30 CITY			ANNISTON			BIRMINGHAM			BUTLER			DECATUR			DOTHAN			
		AVERAGE			362			350 - 352			369			356			363			
		MAT.	INST.	TOTAL	MAT.	INST.	TOTAL	MAT.	INST.	TOTAL	MAT.	INST.	TOTAL	MAT.	INST.	TOTAL	MAT.	INST.	TOTAL	
015433	CONTRACTOR EQUIPMENT		100.0	100.0		101.7	101.7		101.8	101.8		98.9	98.9		101.7	101.7		98.9	98.9	
0241, 31 - 34	SITE & INFRASTRUCTURE, DEMOLITION	100.0	100.0	100.0	88.1	94.4	92.5	86.3	95.0	92.4	102.3	88.4	92.5	81.6	93.1	89.7	99.9	88.4	91.8	
0310	Concrete Forming & Accessories	100.0	100.0	100.0	92.8	49.1	55.0	93.5	71.3	74.3	89.3	44.5	50.5	97.0	47.3	54.0	97.9	43.9	51.2	
0320	Concrete Reinforcing	100.0	100.0	100.0	92.3	82.2	87.3	92.3	82.6	87.5	97.6	48.2	73.0	92.3	77.1	84.8	97.6	47.6	72.7	
0330	Cast-in-Place Concrete	100.0	100.0	100.0	94.9	51.3	77.0	101.3	65.9	86.7	92.6	56.7	77.8	94.4	64.2	82.0	92.6	47.7	74.1	
03	CONCRETE	100.0	100.0	100.0	95.3	57.7	76.7	94.2	72.6	83.5	96.1	51.3	74.0	91.1	60.4	75.9	95.6	47.8	72.0	
04	MASONRY	100.0	100.0	100.0	91.9	69.5	78.2	94.9	77.9	84.5	96.8	50.5	68.6	93.1	53.4	68.9	98.0	41.3	63.4	
05	METALS	100.0	100.0	100.0	93.6	90.8	92.7	94.5	92.6	93.9	92.7	78.6	88.2	95.8	87.4	93.2	92.7	77.3	87.9	
06	WOOD, PLASTICS & COMPOSITES	100.0	100.0	100.0	92.5	44.6	65.2	94.4	69.8	80.3	87.4	44.3	62.8	97.6	44.9	67.5	99.1	44.3	67.8	
07	THERMAL & MOISTURE PROTECTION	100.0	100.0	100.0	95.2	54.9	79.0	95.5	81.0	89.7	95.2	58.8	80.6	95.0	60.6	81.2	95.2	50.9	77.4	
08	OPENINGS	100.0	100.0	100.0	93.6	50.3	83.1	97.1	72.2	91.0	93.7	46.1	82.1	97.4	51.8	86.3	93.7	46.1	82.1	
0920	Plaster & Gypsum Board	100.0	100.0	100.0	93.6	43.3	58.4	96.8	69.3	77.5	91.1	43.0	57.4	98.4	43.6	60.0	101.5	43.0	60.6	
0950, 0980	Ceilings & Acoustic Treatment	100.0	100.0	100.0	88.9	43.3	58.5	96.4	69.3	78.3	88.9	43.0	58.3	92.6	43.6	59.9	88.9	43.0	58.3	
0960	Flooring	100.0	100.0	100.0	102.4	41.7	84.3	104.4	62.5	91.9	107.8	58.0	93.0	104.8	52.4	89.2	114.2	28.5	88.7	
0970, 0990	Wall Finishes & Painting/Coating	100.0	100.0	100.0	93.2	32.9	57.1	93.2	58.1	72.1	93.2	55.7	70.7	93.2	59.3	72.9	93.2	55.7	70.7	
09	FINISHES	100.0	100.0	100.0	94.6	44.9	66.7	97.2	67.8	80.7	98.1	47.7	69.9	96.1	48.7	69.5	101.0	41.9	67.9	
COVERS	DIVS. 10 - 14, 25, 28, 41, 43, 44, 46	100.0	100.0	100.0	100.0	71.3	94.1	100.0	88.8	97.7	100.0	43.1	88.4	100.0	45.0	88.8	100.0	43.2	88.4	
21, 22, 23	FIRE SUPPRESSION, PLUMBING & HVAC	100.0	100.0	100.0	99.3	58.3	82.5	99.9	64.7	85.5	96.7	35.1	71.5	99.9	42.9	76.6	96.7	35.0	71.4	
26, 27, 3370	ELECTRICAL, COMMUNICATIONS & UTIL.	100.0	100.0	100.0	95.8	59.7	76.8	103.3	61.3	81.2	97.9	41.9	68.4	97.6	64.4	80.1	96.7	59.8	77.3	
MF2010	WEIGHTED AVERAGE	100.0	100.0	100.0	95.9	63.8	81.7	97.4	73.8	87.0	96.3	51.4	76.5	96.5	59.4	80.1	96.4	51.4	76.6	

	DIVISION	ALABAMA																	
		EVERGREEN			GADSDEN			HUNTSVILLE			JASPER			MOBILE			MONTGOMERY		
		364			359			357 - 358			355			365 - 366			360 - 361		
		MAT.	INST.	TOTAL	MAT.	INST.	TOTAL	MAT.	INST.	TOTAL	MAT.	INST.	TOTAL	MAT.	INST.	TOTAL	MAT.	INST.	TOTAL
015433	CONTRACTOR EQUIPMENT		98.9	98.9		101.7	101.7		101.7	101.7		101.7	101.7		98.9	98.9		98.9	98.9
0241, 31 - 34	SITE & INFRASTRUCTURE, DEMOLITION	102.7	88.7	92.9	86.9	94.0	91.8	81.4	94.1	90.3	86.4	93.7	91.6	94.9	89.4	91.0	93.6	89.3	90.6
0310	Concrete Forming & Accessories	85.8	46.2	51.5	89.4	43.2	49.4	97.0	63.3	67.9	94.5	35.9	43.8	97.1	56.1	61.6	96.6	45.2	52.1
0320	Concrete Reinforcing	97.7	48.2	73.0	97.9	81.4	89.7	92.3	77.6	85.0	92.3	81.3	86.9	95.3	80.7	88.0	95.3	81.3	88.3
0330	Cast-in-Place Concrete	92.6	59.3	78.9	94.4	50.4	76.3	91.9	62.6	79.9	104.6	45.4	80.2	97.1	68.6	85.3	99.1	48.1	78.1
03	CONCRETE	96.3	52.9	74.9	95.1	54.7	75.1	90.0	67.0	78.6	98.5	49.6	74.4	92.9	66.5	79.8	93.8	54.7	74.5
04	MASONRY	96.8	55.3	71.5	91.4	44.7	62.9	94.5	59.4	73.1	88.4	42.9	60.6	95.2	54.9	70.6	93.4	40.9	61.4
05	METALS	92.7	78.5	88.2	93.7	90.0	92.6	95.8	89.0	93.7	93.7	89.3	92.3	94.7	91.1	93.5	94.5	89.7	93.0
06	WOOD, PLASTICS & COMPOSITES	83.8	44.3	61.2	88.0	40.4	60.9	97.6	64.1	78.5	94.4	33.0	59.3	97.8	55.2	73.5	94.0	44.3	65.6
07	THERMAL & MOISTURE PROTECTION	95.2	59.1	80.7	95.1	65.3	83.1	94.9	73.7	86.4	95.1	47.1	75.8	95.0	72.4	85.9	94.8	64.1	82.5
08	OPENINGS	93.7	46.1	82.1	93.6	49.2	82.8	97.4	62.2	88.8	93.6	48.1	82.5	97.4	59.8	88.2	97.4	53.5	86.7
0920	Plaster & Gypsum Board	90.3	43.0	57.2	90.7	39.0	54.6	98.4	63.4	73.9	94.0	31.3	50.2	98.4	54.2	67.5	97.3	43.0	59.3
0950, 0980	Ceilings & Acoustic Treatment	88.9	43.0	58.3	88.9	39.0	55.7	94.6	63.4	73.8	88.9	31.3	50.6	94.6	54.2	67.7	98.2	43.0	61.4
0960	Flooring	105.4	58.0	91.3	100.7	46.6	84.6	104.8	56.0	90.3	103.1	41.7	84.8	113.6	58.0	97.1	111.4	30.7	87.4
0970, 0990	Wall Finishes & Painting/Coating	93.2	55.7	70.7	93.2	63.4	75.4	93.2	61.1	74.0	93.2	40.3	61.5	96.9	56.7	72.8	93.2	55.7	70.7
09	FINISHES	97.4	48.9	70.2	93.6	44.3	66.0	96.6	61.3	76.8	94.6	36.0	61.7	101.2	55.4	75.5	100.7	42.2	67.9
COVERS	DIVS. 10 - 14, 25, 28, 41, 43, 44, 46	100.0	44.7	88.7	100.0	80.8	96.1	100.0	85.4	97.0	100.0	41.4	88.0	100.0	82.0	96.3	100.0	80.7	96.1
21, 22, 23	FIRE SUPPRESSION, PLUMBING & HVAC	96.7	37.5	72.5	101.9	37.9	75.7	99.9	54.5	81.3	101.9	55.4	82.9	99.9	63.5	85.0	100.0	35.8	73.7
26, 27, 3370	ELECTRICAL, COMMUNICATIONS & UTIL.	95.1	41.9	67.1	97.6	61.3	78.5	98.8	64.4	80.7	97.1	61.3	78.3	98.8	59.6	78.2	99.4	63.4	80.5
MF2010	WEIGHTED AVERAGE	95.9	52.8	77.0	96.5	57.1	79.1	96.6	67.4	83.7	96.8	56.9	79.2	97.5	66.8	84.0	97.5	56.1	79.2

	DIVISION	ALABAMA									ALASKA								
		PHENIX CITY			SELMA			TUSCALOOSA			ANCHORAGE			FAIRBANKS			JUNEAU		
		368			367			354			995 - 996			997			998		
		MAT.	INST.	TOTAL	MAT.	INST.	TOTAL	MAT.	INST.	TOTAL	MAT.	INST.	TOTAL	MAT.	INST.	TOTAL	MAT.	INST.	TOTAL
015433	CONTRACTOR EQUIPMENT		98.9	98.9		98.9	98.9		101.7	101.7		114.3	114.3		114.3	114.3		114.3	114.3
0241, 31 - 34	SITE & INFRASTRUCTURE, DEMOLITION	106.5	89.3	94.5	99.7	89.2	92.4	81.9	94.1	90.5	123.0	129.2	127.4	115.6	129.2	125.2	122.5	129.2	127.2
0310	Concrete Forming & Accessories	92.8	40.1	47.2	90.5	44.4	50.6	96.9	48.8	55.3	140.2	118.7	121.6	140.3	118.4	121.4	139.4	118.6	121.4
0320	Concrete Reinforcing	97.6	64.3	81.0	97.6	81.0	89.4	92.3	81.4	86.9	147.5	111.2	129.5	142.8	111.3	127.1	110.8	111.2	111.0
0330	Cast-in-Place Concrete	92.6	48.7	74.5	92.6	47.9	74.2	95.7	50.4	77.1	143.0	117.5	132.5	135.2	117.9	128.1	142.9	117.5	132.5
03	CONCRETE	98.8	49.5	74.5	95.1	54.2	74.9	91.8	57.1	74.7	131.1	116.1	123.7	118.8	116.2	117.5	121.3	116.1	118.8
04	MASONRY	96.8	42.7	63.8	100.1	40.2	63.6	93.3	47.8	65.5	188.3	123.9	149.0	190.1	123.9	149.7	173.8	123.9	143.4
05	METALS	92.6	83.4	89.7	92.6	88.6	91.3	95.0	90.0	93.4	121.6	102.8	115.7	118.4	103.0	113.6	118.4	102.8	113.5
06	WOOD, PLASTICS & COMPOSITES	92.2	37.0	60.7	89.2	44.3	63.5	97.6	46.4	68.4	116.8	117.4	117.1	121.5	116.8	118.9	116.8	117.4	117.1
07	THERMAL & MOISTURE PROTECTION	95.5	64.0	82.8	95.1	54.1	78.6	95.1	67.4	83.9	185.7	116.4	157.8	179.2	117.6	154.4	176.5	116.4	152.3
08	OPENINGS	93.6	45.5	81.9	93.6	53.5	83.9	97.4	59.5	88.1	117.2	114.3	116.5	114.1	114.0	114.0	113.3	114.3	113.5
0920	Plaster & Gypsum Board	94.9	35.5	53.3	92.8	43.0	57.9	98.4	45.1	61.1	137.7	117.7	123.7	149.1	117.2	126.8	124.0	117.7	119.6
0950, 0980	Ceilings & Acoustic Treatment	88.9	35.5	53.3	88.9	43.0	58.3	94.6	45.1	61.7	138.3	117.7	124.6	123.7	117.2	119.4	120.1	117.7	118.5
0960	Flooring	110.1	30.7	86.5	108.5	29.5	84.9	104.8	38.9	85.2	159.5	130.6	150.9	150.4	130.6	144.5	150.6	130.6	144.7
0970, 0990	Wall Finishes & Painting/Coating	93.2	55.7	70.7	93.2	55.7	70.7	93.2	48.0	66.1	154.3	115.8	131.2	153.8	124.6	136.3	154.3	115.8	131.2
09	FINISHES	99.7	38.2	65.2	98.2	42.0	66.7	96.5	45.7	68.0	149.2	121.0	133.4	142.7	121.7	130.9	139.2	121.0	129.0
COVERS	DIVS. 10 - 14, 25, 28, 41, 43, 44, 46	100.0	80.3	96.0	100.0	43.1	88.4	100.0	82.6	96.4	100.0	105.8	101.2	100.0	105.7	101.2	100.0	105.8	101.2
21, 22, 23	FIRE SUPPRESSION, PLUMBING & HVAC	96.7	36.2	71.9	96.7	35.3	71.6	99.9	35.6	73.6	100.3	104.3	101.9	100.3	107.6	103.3	100.3	103.2	101.5
26, 27, 3370	ELECTRICAL, COMMUNICATIONS & UTIL.	97.3	72.5	84.3	96.4	41.9	67.7	98.3	61.3	78.8	133.9	116.5	124.8	145.4	116.5	130.2	130.4	116.5	123.1
MF2010	WEIGHTED AVERAGE	96.8	55.4	78.5	96.1	51.3	76.4	96.5	58.1	79.6	123.8	114.6	119.8	122.0	115.4	119.1	119.7	114.3	117.3

DIVISION		ALASKA			ARIZONA														
		KETCHIKAN			CHAMBERS			FLAGSTAFF			GLOBE			KINGMAN			MESA/TEMPE		
		999			865			860			855			864			852		
		MAT.	INST.	TOTAL	MAT.	INST.	TOTAL	MAT.	INST.	TOTAL	MAT.	INST.	TOTAL	MAT.	INST.	TOTAL	MAT.	INST.	TOTAL
015433	CONTRACTOR EQUIPMENT		114.3	114.3		92.2	92.2		92.2	92.2		93.6	93.6		92.2	92.2		93.6	93.6
0241, 31 - 34	SITE & INFRASTRUCTURE, DEMOLITION	171.1	129.2	141.7	66.2	94.0	85.7	84.1	95.9	92.4	99.6	95.7	96.8	66.2	96.0	87.1	90.1	97.3	95.1
0310	Concrete Forming & Accessories	132.1	118.6	120.5	97.8	44.8	51.9	103.0	67.0	71.9	97.7	44.6	51.8	96.1	66.4	70.4	100.5	66.2	70.8
0320	Concrete Reinforcing	111.4	111.2	111.3	102.1	77.0	89.6	101.9	84.1	93.1	105.5	76.9	91.3	102.2	84.1	93.2	106.2	84.1	95.2
0330	Cast-in-Place Concrete	273.3	117.5	209.2	91.5	53.2	75.8	91.6	74.3	84.5	91.7	53.2	75.9	91.3	70.8	82.9	92.5	70.8	83.6
03	CONCRETE	199.2	116.1	158.1	96.9	54.1	75.8	115.6	72.8	94.5	109.9	54.1	82.3	96.5	71.3	84.0	101.5	71.2	86.5
04	MASONRY	198.9	123.9	153.1	92.6	40.6	60.8	92.8	61.9	73.9	106.0	40.5	66.0	92.6	60.4	73.0	106.2	60.4	78.2
05	METALS	118.6	102.8	113.6	94.7	67.6	86.2	95.2	75.8	89.1	91.7	67.9	84.2	95.3	75.7	89.1	92.0	75.9	87.0
06	WOOD, PLASTICS & COMPOSITES	113.0	117.4	115.5	96.4	42.6	65.7	102.0	67.0	82.0	94.0	42.7	64.7	92.2	67.0	77.8	97.4	67.1	80.1
07	THERMAL & MOISTURE PROTECTION	182.3	116.4	155.8	94.3	50.0	76.5	96.1	67.0	84.4	101.0	47.6	79.5	94.3	63.3	81.8	100.3	63.9	85.6
08	OPENINGS	116.0	114.3	115.6	106.0	50.0	92.4	106.2	72.1	97.9	102.3	50.1	89.6	106.2	70.9	97.6	102.4	70.9	94.8
0920	Plaster & Gypsum Board	137.8	117.7	123.8	86.5	40.9	54.6	89.7	66.1	73.2	92.2	40.9	56.3	80.1	66.1	70.3	94.7	66.1	74.7
0950, 0980	Ceilings & Acoustic Treatment	117.6	117.7	117.7	110.4	40.9	64.1	111.3	66.1	81.2	95.8	40.9	59.2	111.3	66.1	81.2	95.8	66.1	76.0
0960	Flooring	150.4	130.6	144.5	92.5	41.0	77.1	94.7	41.2	78.7	99.1	41.0	81.8	91.5	55.6	80.8	100.4	51.3	85.8
0970, 0990	Wall Finishes & Painting/Coating	153.8	115.8	131.1	93.2	38.6	60.5	93.2	57.2	71.6	101.5	38.6	63.8	93.2	57.2	71.6	101.5	57.2	74.9
09	FINISHES	144.8	121.0	131.5	96.1	42.2	65.9	99.4	60.5	77.6	95.9	42.3	65.8	95.2	62.6	76.9	95.4	61.8	76.6
COVERS	DIVS. 10 - 14, 25, 28, 41, 43, 44, 46	100.0	105.8	101.2	100.0	79.0	95.7	100.0	83.9	96.7	100.0	79.2	95.7	100.0	83.4	96.6	100.0	83.6	96.7
21, 22, 23	FIRE SUPPRESSION, PLUMBING & HVAC	98.4	103.2	100.4	96.3	72.7	86.7	100.2	81.6	92.6	94.2	72.4	85.3	96.3	80.6	89.9	100.2	75.5	90.1
26, 27, 3370	ELECTRICAL, COMMUNICATIONS & UTIL.	145.3	116.5	130.2	101.4	67.2	83.4	100.3	61.1	79.6	96.0	62.5	78.4	101.4	61.1	80.2	92.7	61.1	76.0
MF2010	WEIGHTED AVERAGE	132.8	114.3	124.7	96.8	62.0	81.5	100.7	72.7	88.4	98.2	61.4	82.0	96.8	72.3	86.0	98.1	71.3	86.3

DIVISION		ARIZONA												ARKANSAS					
		PHOENIX			PRESCOTT			SHOW LOW			TUCSON			BATESVILLE			CAMDEN		
		850,853			863			859			856 - 857			725			717		
		MAT.	INST.	TOTAL	MAT.	INST.	TOTAL	MAT.	INST.	TOTAL	MAT.	INST.	TOTAL	MAT.	INST.	TOTAL	MAT.	INST.	TOTAL
015433	CONTRACTOR EQUIPMENT		94.2	94.2		92.2	92.2		93.6	93.6		93.6	93.6		88.7	88.7		88.7	88.7
0241, 31 - 34	SITE & INFRASTRUCTURE, DEMOLITION	90.6	97.8	95.6	72.5	94.0	87.6	101.7	95.7	97.5	85.9	97.5	94.1	71.8	84.9	81.0	72.4	84.4	80.8
0310	Concrete Forming & Accessories	101.4	69.4	73.7	99.2	44.5	51.9	104.3	44.9	52.9	101.0	68.9	73.2	83.9	46.8	51.8	83.7	31.8	38.8
0320	Concrete Reinforcing	104.5	84.4	94.5	101.9	77.2	89.6	106.2	77.1	91.7	87.7	84.1	85.9	91.8	63.7	77.8	93.1	63.5	78.4
0330	Cast-in-Place Concrete	92.5	74.5	85.1	91.6	53.1	75.8	91.8	53.3	75.9	95.2	74.3	86.6	81.1	46.8	67.0	83.2	39.5	65.2
03	CONCRETE	101.1	73.9	87.7	101.9	54.0	78.3	112.1	54.3	83.5	99.6	73.6	86.7	87.0	50.8	69.1	88.7	41.6	65.4
04	MASONRY	94.1	64.9	76.3	92.7	45.7	64.0	106.0	42.0	67.0	92.1	61.9	73.7	105.3	42.5	66.9	113.5	32.7	64.2
05	METALS	93.5	77.3	88.4	95.2	67.7	86.6	91.5	68.2	84.2	92.8	76.0	87.5	97.3	66.2	87.5	97.3	65.7	87.4
06	WOOD, PLASTICS & COMPOSITES	98.4	69.9	82.1	97.8	42.6	66.3	101.9	42.7	68.1	97.7	69.9	81.8	88.3	47.5	65.0	88.0	31.0	55.5
07	THERMAL & MOISTURE PROTECTION	99.9	67.6	86.9	94.8	49.4	76.6	101.2	50.0	80.6	101.3	63.4	86.1	97.3	45.6	76.5	97.1	37.0	72.9
08	OPENINGS	104.0	73.6	96.6	106.2	50.0	92.5	101.4	50.1	88.9	98.4	73.6	92.4	95.1	46.2	83.2	91.5	40.7	79.1
0920	Plaster & Gypsum Board	97.1	69.0	77.4	86.7	40.9	54.7	97.2	40.9	57.8	100.8	69.0	78.5	77.2	46.4	55.6	77.2	29.3	43.7
0950, 0980	Ceilings & Acoustic Treatment	103.1	69.0	80.4	109.5	40.9	63.8	95.8	40.9	59.2	96.8	69.0	78.2	88.2	46.4	60.3	88.2	29.3	49.0
0960	Flooring	100.6	53.6	86.6	93.3	41.0	77.7	101.9	46.7	85.5	91.1	41.2	76.3	102.2	61.2	90.0	102.1	41.0	83.9
0970, 0990	Wall Finishes & Painting/Coating	101.5	63.4	78.6	93.2	38.6	60.5	101.5	38.6	63.8	102.5	57.2	75.4	100.7	43.0	66.1	100.7	49.4	70.0
09	FINISHES	97.4	65.1	79.3	96.8	42.2	66.2	97.6	43.2	67.1	93.6	62.2	76.0	90.2	48.8	67.0	90.3	34.6	59.1
COVERS	DIVS. 10 - 14, 25, 28, 41, 43, 44, 46	100.0	84.5	96.8	100.0	78.9	95.7	100.0	79.2	95.7	100.0	84.4	96.8	100.0	41.1	88.0	100.0	37.0	87.1
21, 22, 23	FIRE SUPPRESSION, PLUMBING & HVAC	100.1	81.6	92.5	100.2	69.9	87.8	94.2	72.7	85.4	100.2	72.2	88.7	94.2	52.7	77.2	94.2	50.0	76.1
26, 27, 3370	ELECTRICAL, COMMUNICATIONS & UTIL.	99.8	69.5	83.9	99.9	61.0	79.4	93.1	62.4	76.9	94.9	58.7	75.8	97.8	62.1	79.0	94.0	59.2	75.7
MF2010	WEIGHTED AVERAGE	98.7	75.4	88.4	98.5	61.1	82.0	98.3	61.8	82.2	96.9	70.9	85.5	94.6	55.3	77.3	94.4	49.5	74.6

DIVISION		ARKANSAS																	
		FAYETTEVILLE			FORT SMITH			HARRISON			HOT SPRINGS			JONESBORO			LITTLE ROCK		
		727			729			726			719			724			720 - 722		
		MAT.	INST.	TOTAL	MAT.	INST.	TOTAL	MAT.	INST.	TOTAL	MAT.	INST.	TOTAL	MAT.	INST.	TOTAL	MAT.	INST.	TOTAL
015433	CONTRACTOR EQUIPMENT		88.7	88.7		88.7	88.7		88.7	88.7		88.7	88.7		107.3	107.3		88.7	88.7
0241, 31 - 34	SITE & INFRASTRUCTURE, DEMOLITION	71.3	86.3	81.8	76.0	86.2	83.2	76.2	84.9	82.3	75.5	85.8	82.7	94.2	100.4	98.5	82.3	86.3	85.1
0310	Concrete Forming & Accessories	79.6	40.2	45.5	98.0	57.3	62.8	88.3	46.8	52.4	81.2	37.1	43.1	87.4	50.2	55.2	93.2	61.6	65.9
0320	Concrete Reinforcing	91.8	68.5	80.2	92.9	69.2	81.1	91.4	63.6	77.5	91.3	63.6	77.5	88.8	65.9	77.4	93.1	64.4	78.8
0330	Cast-in-Place Concrete	81.1	51.1	68.8	92.7	65.7	81.6	90.0	44.8	71.4	85.1	40.0	66.5	88.3	56.8	75.4	93.2	65.8	81.9
03	CONCRETE	86.7	50.2	68.6	94.9	62.9	79.1	94.4	50.0	72.5	92.5	44.1	68.6	91.6	56.9	74.5	94.8	64.0	79.6
04	MASONRY	95.0	43.1	63.3	100.5	51.2	70.4	105.7	40.7	66.0	85.3	34.9	54.5	96.4	43.7	64.2	97.4	51.2	69.2
05	METALS	97.3	68.3	88.2	99.6	70.0	90.3	98.4	65.9	88.2	97.3	66.2	87.5	94.0	79.5	89.4	95.6	68.5	87.1
06	WOOD, PLASTICS & COMPOSITES	84.0	37.1	57.2	104.7	59.6	79.0	93.7	47.5	67.3	85.1	37.3	57.8	91.7	50.4	68.2	99.2	65.3	79.8
07	THERMAL & MOISTURE PROTECTION	98.1	47.1	77.6	98.4	55.1	81.0	97.6	44.3	76.1	97.3	39.3	74.0	102.1	51.0	81.6	96.0	55.7	79.8
08	OPENINGS	95.1	45.7	83.1	96.0	56.1	86.3	96.0	46.8	84.0	91.4	42.5	79.5	98.2	54.6	87.6	96.0	59.3	87.1
0920	Plaster & Gypsum Board	74.7	35.6	47.4	82.6	58.8	66.0	81.4	46.4	56.9	75.5	35.8	47.8	92.6	49.1	62.2	88.7	64.7	71.9
0950, 0980	Ceilings & Acoustic Treatment	88.2	35.6	53.2	90.0	58.8	69.2	90.0	46.4	60.9	88.2	35.8	53.3	92.4	49.1	63.5	95.5	64.7	75.0
0960	Flooring	99.4	61.2	88.0	109.0	63.0	95.3	104.6	61.2	91.7	100.8	61.2	89.0	73.0	55.4	67.7	109.6	63.0	95.7
0970, 0990	Wall Finishes & Painting/Coating	100.7	30.7	58.7	100.7	56.2	74.0	100.7	43.0	66.1	100.7	49.4	70.0	90.2	50.5	66.4	100.7	57.7	74.9
09	FINISHES	89.2	41.8	62.6	93.5	58.4	73.9	92.2	48.8	67.9	90.0	41.9	63.1	87.4	50.7	66.8	95.9	62.0	76.9
COVERS	DIVS. 10 - 14, 25, 28, 41, 43, 44, 46	100.0	50.7	89.9	100.0	77.8	95.5	100.0	51.2	90.0	100.0	38.0	87.3	100.0	47.0	89.2	100.0	78.4	95.6
21, 22, 23	FIRE SUPPRESSION, PLUMBING & HVAC	94.3	51.2	76.7	100.3	50.7	80.0	94.2	50.4	76.3	94.2	49.5	75.9	100.5	53.6	81.3	100.1	58.2	82.9
26, 27, 3370	ELECTRICAL, COMMUNICATIONS & UTIL.	91.2	53.4	71.3	94.9	67.3	80.3	96.2	38.6	65.9	96.3	65.9	80.3	102.3	62.2	81.2	102.6	68.4	84.6
MF2010	WEIGHTED AVERAGE	93.3	53.4	75.7	97.4	61.8	81.7	95.8	51.6	76.4	93.7	52.2	75.4	97.0	60.0	80.7	97.5	64.2	82.8

	DIVISION	ARKANSAS												CALIFORNIA					
		PINE BLUFF			RUSSELLVILLE			TEXARKANA			WEST MEMPHIS			ALHAMBRA			ANAHEIM		
		716			728			718			723			917 - 918			928		
		MAT.	INST.	TOTAL	MAT.	INST.	TOTAL	MAT.	INST.	TOTAL	MAT.	INST.	TOTAL	MAT.	INST.	TOTAL	MAT.	INST.	TOTAL
015433	CONTRACTOR EQUIPMENT		88.7	88.7		88.7	88.7		89.4	89.4		107.3	107.3		98.2	98.2		100.5	100.5
0241, 31 - 34	SITE & INFRASTRUCTURE, DEMOLITION	77.5	86.2	83.6	72.9	84.8	81.2	89.1	86.9	87.6	100.2	100.4	100.3	96.7	109.2	105.5	96.1	107.9	104.4
0310	Concrete Forming & Accessories	80.9	61.5	64.1	84.6	55.6	59.5	87.2	41.9	48.0	92.6	50.4	56.1	116.6	118.0	117.8	106.5	121.7	119.6
0320	Concrete Reinforcing	93.0	64.4	78.7	92.5	63.5	78.0	92.6	63.7	78.2	88.8	65.9	77.4	109.9	116.7	113.3	95.4	116.7	106.0
0330	Cast-in-Place Concrete	85.1	65.8	77.1	85.0	46.8	69.3	92.7	44.7	73.0	92.6	56.9	77.9	93.8	119.8	104.5	92.0	122.7	104.6
03	CONCRETE	93.3	63.9	78.8	90.2	54.6	72.6	91.1	47.8	69.7	99.1	57.0	78.3	104.4	117.4	110.8	100.6	120.1	110.2
04	MASONRY	122.0	51.2	78.8	99.8	38.9	62.6	100.5	34.5	60.2	81.8	43.7	58.5	117.4	116.5	116.9	79.3	115.8	101.5
05	METALS	98.1	68.4	88.8	97.3	65.8	87.4	91.0	66.3	83.3	93.0	79.7	88.8	83.2	100.3	88.6	101.8	101.6	101.7
06	WOOD, PLASTICS & COMPOSITES	84.6	65.3	73.6	89.6	59.6	72.5	93.1	43.8	65.0	97.7	50.4	70.7	97.3	115.5	107.7	97.0	120.1	110.2
07	THERMAL & MOISTURE PROTECTION	97.4	55.7	80.6	98.3	45.4	77.0	98.1	45.9	77.1	102.5	51.0	81.8	96.6	116.4	104.6	100.0	121.3	108.6
08	OPENINGS	91.4	59.3	83.6	95.1	55.4	85.5	96.5	46.9	84.4	98.2	54.6	87.6	88.2	117.8	95.4	100.9	120.5	105.6
0920	Plaster & Gypsum Board	75.1	64.7	67.8	77.2	58.8	64.3	79.4	42.6	53.7	95.6	49.1	63.0	98.6	116.2	110.9	100.2	120.7	114.5
0950, 0980	Ceilings & Acoustic Treatment	88.2	64.7	72.6	88.2	58.8	68.6	91.9	42.6	59.0	90.4	49.1	62.8	96.4	116.2	109.6	107.6	120.7	116.3
0960	Flooring	100.6	63.0	89.4	101.8	61.2	89.7	102.8	52.8	87.9	75.3	55.4	69.3	97.5	103.9	99.4	108.7	112.6	109.8
0970, 0990	Wall Finishes & Painting/Coating	100.7	57.7	74.9	100.7	36.5	62.3	100.7	31.7	59.3	90.2	50.5	66.4	110.6	108.8	109.5	104.9	108.6	107.1
09	FINISHES	90.0	62.0	74.3	90.4	55.2	70.7	92.5	42.1	64.3	88.7	50.7	67.4	97.5	114.3	106.9	103.3	118.5	111.8
COVERS	DIVS. 10 - 14, 25, 28, 41, 43, 44, 46	100.0	78.4	95.6	100.0	42.4	88.2	100.0	33.9	86.5	100.0	47.0	89.2	100.0	110.2	102.1	100.0	111.2	102.3
21, 22, 23	FIRE SUPPRESSION, PLUMBING & HVAC	100.3	53.0	80.9	94.3	52.5	77.2	100.3	56.3	82.3	94.5	65.1	82.5	94.0	109.1	100.2	100.0	116.6	106.8
26, 27, 3370	ELECTRICAL, COMMUNICATIONS & UTIL.	94.2	68.4	80.6	94.9	46.6	69.4	96.2	37.5	65.3	104.1	59.1	80.4	117.5	113.6	115.5	92.4	105.2	99.1
MF2010	WEIGHTED AVERAGE	97.1	63.1	82.1	94.5	54.6	76.9	96.0	50.6	76.0	96.0	62.0	81.0	97.3	112.1	103.8	98.8	113.7	105.4

	DIVISION	CALIFORNIA																	
		BAKERSFIELD			BERKELEY			EUREKA			FRESNO			INGLEWOOD			LONG BEACH		
		932 - 933			947			955			936 - 938			903 - 905			906 - 908		
		MAT.	INST.	TOTAL	MAT.	INST.	TOTAL	MAT.	INST.	TOTAL	MAT.	INST.	TOTAL	MAT.	INST.	TOTAL	MAT.	INST.	TOTAL
015433	CONTRACTOR EQUIPMENT		98.4	98.4		98.1	98.1		98.1	98.1		98.4	98.4		98.4	98.4		98.4	98.4
0241, 31 - 34	SITE & INFRASTRUCTURE, DEMOLITION	97.8	105.5	103.2	116.2	105.0	108.3	107.9	102.0	103.8	100.6	104.9	103.6	91.9	103.9	100.3	99.0	103.9	102.4
0310	Concrete Forming & Accessories	106.4	121.1	119.1	119.8	147.8	144.0	116.1	130.0	128.2	102.4	131.1	127.3	111.4	118.2	117.3	106.6	118.3	116.7
0320	Concrete Reinforcing	107.3	116.6	111.9	92.2	118.1	105.1	104.1	117.3	110.7	91.1	117.3	104.2	108.2	116.7	112.4	107.3	116.7	112.0
0330	Cast-in-Place Concrete	93.8	121.6	105.2	124.0	120.4	122.5	99.8	116.4	106.6	97.0	117.2	105.3	86.6	121.6	101.0	98.5	121.6	108.0
03	CONCRETE	101.0	119.4	110.1	110.2	130.8	120.4	111.3	121.6	116.4	102.0	122.4	112.1	97.7	118.3	107.9	107.4	118.3	112.8
04	MASONRY	98.2	115.2	108.6	124.7	134.9	130.9	101.0	125.4	115.9	101.0	123.1	114.5	75.3	116.6	100.5	83.6	116.6	103.7
05	METALS	97.1	100.7	98.2	105.9	100.7	104.3	101.6	99.2	100.8	101.8	101.1	101.6	91.5	103.0	95.1	91.4	103.0	95.0
06	WOOD, PLASTICS & COMPOSITES	92.0	120.2	108.1	113.3	153.1	136.0	110.5	134.8	124.4	103.6	134.8	121.4	103.1	115.8	110.3	96.8	115.8	107.6
07	THERMAL & MOISTURE PROTECTION	101.2	110.9	105.1	109.7	135.9	120.3	104.1	113.7	108.0	93.9	114.8	102.3	102.8	117.9	108.9	103.1	118.2	109.2
08	OPENINGS	98.6	116.1	102.9	97.8	140.8	108.3	100.1	112.8	103.2	100.3	124.0	106.1	92.6	117.9	98.7	92.6	117.9	98.7
0920	Plaster & Gypsum Board	97.0	120.7	113.5	106.1	154.3	139.8	105.4	135.7	126.6	95.4	135.7	123.6	102.7	116.2	112.1	98.2	116.2	110.8
0950, 0980	Ceilings & Acoustic Treatment	116.4	120.7	119.3	109.6	154.3	139.4	113.1	135.7	128.2	110.6	135.7	127.3	102.1	116.2	111.5	102.1	116.2	111.5
0960	Flooring	109.1	110.2	109.5	113.6	126.8	117.5	113.1	120.8	115.4	115.9	138.3	122.6	99.7	103.9	100.9	97.1	103.9	99.1
0970, 0990	Wall Finishes & Painting/Coating	107.2	97.6	101.5	105.5	140.3	126.3	106.5	48.3	71.6	122.1	110.0	114.8	102.9	108.8	106.4	102.9	108.8	106.4
09	FINISHES	106.5	117.0	112.4	107.9	145.3	128.8	109.3	122.3	116.6	108.7	132.1	121.8	100.7	114.4	108.4	99.9	114.4	108.0
COVERS	DIVS. 10 - 14, 25, 28, 41, 43, 44, 46	100.0	103.8	100.8	100.0	126.2	105.4	100.0	122.3	104.6	100.0	122.3	104.6	100.0	110.7	102.2	100.0	110.7	102.2
21, 22, 23	FIRE SUPPRESSION, PLUMBING & HVAC	100.0	107.7	103.2	94.1	145.6	115.1	94.0	101.7	97.1	100.1	111.3	104.7	93.6	109.2	100.0	93.6	109.2	100.0
26, 27, 3370	ELECTRICAL, COMMUNICATIONS & UTIL.	102.5	100.1	101.3	107.0	135.8	122.2	100.2	114.6	107.8	91.8	101.5	96.9	101.6	113.6	108.0	101.3	113.6	107.8
MF2010	WEIGHTED AVERAGE	100.1	109.8	104.4	103.8	132.5	116.5	101.3	112.7	106.3	100.3	115.0	106.8	95.0	112.1	102.5	96.5	112.2	103.4

	DIVISION	CALIFORNIA																	
		LOS ANGELES			MARYSVILLE			MODESTO			MOJAVE			OAKLAND			OXNARD		
		900 - 902			959			953			935			946			930		
		MAT.	INST.	TOTAL	MAT.	INST.	TOTAL	MAT.	INST.	TOTAL	MAT.	INST.	TOTAL	MAT.	INST.	TOTAL	MAT.	INST.	TOTAL
015433	CONTRACTOR EQUIPMENT		102.1	102.1		98.1	98.1		98.1	98.1		98.4	98.4		98.1	98.1		97.2	97.2
0241, 31 - 34	SITE & INFRASTRUCTURE, DEMOLITION	99.4	107.6	105.2	104.2	104.5	104.4	99.0	104.8	103.1	94.3	105.5	102.2	122.4	105.0	110.2	100.9	103.6	102.8
0310	Concrete Forming & Accessories	108.4	121.9	120.1	106.5	131.1	127.8	103.1	131.4	127.6	115.5	114.4	114.5	109.0	147.8	142.6	107.1	121.7	119.7
0320	Concrete Reinforcing	109.0	116.9	113.0	104.1	117.0	110.5	107.9	117.4	112.6	109.3	116.4	112.8	94.3	118.1	106.2	107.3	116.5	111.9
0330	Cast-in-Place Concrete	93.5	122.4	105.4	111.5	117.7	114.0	99.8	117.8	107.2	83.9	121.4	99.4	117.6	122.2	119.5	97.7	122.0	107.6
03	CONCRETE	103.2	120.3	111.7	112.7	122.6	117.6	103.8	122.8	113.2	94.7	116.4	105.4	109.9	131.5	120.6	102.9	119.8	111.3
04	MASONRY	89.9	119.7	108.1	101.9	117.5	111.4	101.9	124.3	115.5	99.5	113.5	108.0	132.8	134.9	134.1	102.2	112.9	108.7
05	METALS	98.3	104.8	100.3	101.1	102.1	101.4	98.3	102.9	99.8	98.6	100.2	99.1	100.4	100.7	100.5	96.9	101.1	98.2
06	WOOD, PLASTICS & COMPOSITES	102.9	120.2	112.8	98.3	134.8	119.1	94.3	134.8	117.4	102.9	111.6	107.9	101.4	153.1	130.9	98.0	120.2	110.7
07	THERMAL & MOISTURE PROTECTION	102.2	121.2	109.9	103.6	116.1	108.6	103.2	116.9	108.7	100.0	108.7	103.5	107.7	135.9	119.0	103.1	118.0	109.1
08	OPENINGS	99.6	120.7	104.7	99.3	124.0	105.3	98.0	124.6	104.5	95.0	111.4	99.0	97.9	140.8	108.3	97.8	120.6	103.3
0920	Plaster & Gypsum Board	102.3	120.7	115.1	98.1	135.7	124.4	100.6	135.7	125.2	107.3	111.8	110.4	100.2	154.3	138.0	100.2	120.7	114.5
0950, 0980	Ceilings & Acoustic Treatment	112.3	120.7	117.9	112.1	135.7	127.8	107.6	135.7	126.3	112.1	111.8	111.9	112.5	154.3	140.4	113.7	120.7	118.3
0960	Flooring	97.4	112.6	101.9	108.7	109.4	108.9	109.1	115.5	111.0	115.2	96.4	109.6	108.7	126.8	114.1	107.6	112.6	109.1
0970, 0990	Wall Finishes & Painting/Coating	101.9	108.8	106.0	106.5	114.2	111.1	106.5	117.3	113.0	106.5	92.9	98.3	105.5	140.3	126.3	106.5	103.5	104.7
09	FINISHES	102.3	118.6	111.4	106.4	127.7	118.3	105.3	129.0	118.6	108.6	109.0	108.8	106.9	145.3	128.4	106.0	118.0	112.7
COVERS	DIVS. 10 - 14, 25, 28, 41, 43, 44, 46	100.0	111.5	102.3	100.0	122.3	104.6	100.0	122.3	104.6	100.0	102.8	100.6	100.0	126.3	105.4	100.0	111.4	102.3
21, 22, 23	FIRE SUPPRESSION, PLUMBING & HVAC	100.1	116.7	106.9	94.0	107.0	99.3	100.0	111.3	104.6	94.0	99.2	96.1	100.1	145.6	118.7	100.0	116.7	106.8
26, 27, 3370	ELECTRICAL, COMMUNICATIONS & UTIL.	99.8	120.8	110.8	96.6	101.8	99.3	99.3	108.4	104.1	90.4	96.4	93.6	106.1	140.8	124.4	96.4	105.4	101.2
MF2010	WEIGHTED AVERAGE	99.8	116.6	107.2	100.5	113.1	106.1	100.4	116.0	107.3	96.8	105.4	100.6	104.6	133.3	117.3	99.9	112.9	105.6

	DIVISION	CALIFORNIA																	
		PALM SPRINGS			PALO ALTO			PASADENA			REDDING			RICHMOND			RIVERSIDE		
		922			943			910 - 912			960			948			925		
		MAT.	INST.	TOTAL	MAT.	INST.	TOTAL	MAT.	INST.	TOTAL	MAT.	INST.	TOTAL	MAT.	INST.	TOTAL	MAT.	INST.	TOTAL
015433	CONTRACTOR EQUIPMENT		99.3	99.3		98.1	98.1		98.2	98.2		98.1	98.1		98.1	98.1		99.3	99.3
0241, 31 - 34	SITE & INFRASTRUCTURE, DEMOLITION	87.7	105.9	100.5	112.4	104.9	107.1	93.8	109.2	104.6	104.4	104.5	104.5	121.7	104.9	109.9	94.5	106.0	102.5
0310	Concrete Forming & Accessories	103.1	118.0	116.0	107.1	138.3	134.1	106.9	118.1	116.6	106.3	131.1	127.7	122.3	147.4	144.0	106.8	121.7	119.7
0320	Concrete Reinforcing	109.5	116.5	113.0	92.2	117.6	104.9	110.9	116.7	113.8	104.1	117.0	110.5	92.2	117.7	104.9	106.4	116.7	111.5
0330	Cast-in-Place Concrete	87.7	122.5	102.0	104.9	121.9	111.9	89.0	119.8	101.7	110.5	117.6	113.5	120.6	122.0	121.2	95.4	122.7	106.6
03	CONCRETE	95.7	118.4	106.9	99.2	127.1	113.0	99.9	117.4	108.6	111.6	122.6	117.0	111.9	131.1	121.4	101.6	120.1	110.8
04	MASONRY	77.0	113.7	99.4	106.2	128.1	119.6	103.0	116.5	111.3	102.0	117.5	111.4	124.4	128.1	126.7	78.0	115.4	100.8
05	METALS	102.4	101.0	102.0	97.8	99.6	98.4	83.2	100.4	88.6	101.4	102.1	101.6	97.9	99.8	98.5	102.0	101.6	101.9
06	WOOD, PLASTICS & COMPOSITES	91.9	115.8	105.5	98.9	140.9	122.9	84.8	115.5	102.3	98.0	134.8	119.0	117.1	153.1	137.6	97.0	120.1	110.2
07	THERMAL & MOISTURE PROTECTION	99.7	117.4	106.8	107.3	127.2	115.3	96.4	116.9	104.6	103.6	115.7	108.4	107.9	133.1	118.0	100.2	121.2	108.6
08	OPENINGS	96.7	116.4	101.5	97.9	132.5	106.3	88.3	117.8	95.4	100.2	124.0	106.0	97.9	139.2	107.9	99.5	120.5	104.6
0920	Plaster & Gypsum Board	94.9	116.2	109.8	98.1	141.7	128.6	91.1	116.2	108.7	98.4	135.7	124.5	108.5	154.3	140.5	99.5	120.7	114.3
0950, 0980	Ceilings & Acoustic Treatment	104.9	116.2	112.4	110.5	141.7	131.3	96.4	116.2	109.6	119.2	135.7	130.2	110.5	154.3	139.7	111.0	120.7	117.4
0960	Flooring	111.1	103.9	109.0	107.8	119.5	111.3	92.7	103.9	96.1	108.7	109.4	108.9	115.5	119.5	116.7	112.7	112.6	112.7
0970, 0990	Wall Finishes & Painting/Coating	103.2	116.2	111.0	105.5	137.9	124.9	110.6	108.8	109.5	106.5	114.2	111.1	105.5	137.9	124.9	103.2	108.6	106.4
09	FINISHES	101.9	115.3	109.4	105.2	136.6	122.8	95.0	114.3	105.8	107.8	127.7	118.9	109.5	143.8	128.7	104.7	118.5	112.4
COVERS	DIVS. 10 - 14, 25, 28, 41, 43, 44, 46	100.0	110.7	102.2	100.0	124.9	105.1	100.0	110.2	102.1	100.0	122.3	104.6	100.0	126.2	105.4	100.0	111.2	102.3
21, 22, 23	FIRE SUPPRESSION, PLUMBING & HVAC	94.0	105.4	98.7	94.1	139.4	112.6	94.0	109.2	100.2	100.0	107.0	102.9	94.1	134.7	110.7	100.0	116.6	106.8
26, 27, 3370	ELECTRICAL, COMMUNICATIONS & UTIL.	96.2	102.4	99.5	105.9	139.3	123.5	114.4	113.6	114.0	99.7	101.8	100.8	106.5	132.4	120.2	92.5	106.5	99.9
MF2010	WEIGHTED AVERAGE	96.3	109.5	102.2	99.8	128.5	112.5	95.4	112.1	102.8	102.4	113.1	107.1	102.9	128.7	114.3	98.8	113.7	105.4

	DIVISION	CALIFORNIA																	
		SACRAMENTO			SALINAS			SAN BERNARDINO			SAN DIEGO			SAN FRANCISCO			SAN JOSE		
		942,956 - 958			939			923 - 924			919 - 921			940 - 941			951		
		MAT.	INST.	TOTAL	MAT.	INST.	TOTAL	MAT.	INST.	TOTAL	MAT.	INST.	TOTAL	MAT.	INST.	TOTAL	MAT.	INST.	TOTAL
015433	CONTRACTOR EQUIPMENT		97.7	97.7		98.4	98.4		99.3	99.3		98.2	98.2		107.9	107.9		98.8	98.8
0241, 31 - 34	SITE & INFRASTRUCTURE, DEMOLITION	98.7	111.0	107.3	113.9	105.4	108.0	74.7	105.9	96.6	100.3	102.8	102.0	124.7	111.3	115.3	130.9	100.0	109.2
0310	Concrete Forming & Accessories	107.2	134.2	130.6	109.8	134.6	131.2	110.2	117.9	116.9	104.8	109.1	108.5	108.6	148.8	143.4	109.0	148.0	142.7
0320	Concrete Reinforcing	87.3	117.5	102.4	107.9	117.8	112.8	106.4	116.5	111.4	107.4	116.6	112.0	107.4	118.4	112.9	95.0	118.1	106.5
0330	Cast-in-Place Concrete	99.5	118.3	107.2	96.5	118.1	105.4	66.0	122.5	89.2	97.5	105.5	100.8	120.6	123.9	121.9	115.9	122.4	118.6
03	CONCRETE	100.0	123.9	111.8	112.8	124.4	118.6	76.5	118.3	97.1	105.0	108.6	106.8	113.4	133.1	123.2	109.8	132.1	120.8
04	MASONRY	108.2	124.3	118.0	99.0	129.5	117.6	84.8	113.7	102.5	93.4	112.9	105.3	133.3	138.2	136.3	129.7	135.1	133.0
05	METALS	96.0	97.1	96.4	101.2	104.0	102.1	102.0	100.8	101.6	98.5	101.9	99.6	106.6	110.6	107.9	96.5	108.6	100.3
06	WOOD, PLASTICS & COMPOSITES	94.8	138.1	119.5	102.3	137.9	122.6	100.7	115.8	109.3	98.4	105.2	102.3	101.4	153.3	131.0	106.6	153.0	133.0
07	THERMAL & MOISTURE PROTECTION	116.3	122.5	118.8	100.7	127.8	111.6	98.8	117.4	106.3	101.2	104.8	102.7	109.5	138.1	121.0	99.6	138.9	115.4
08	OPENINGS	110.8	126.4	114.6	99.0	132.6	107.2	96.8	116.4	101.6	95.9	110.1	99.3	102.4	140.9	111.7	90.4	140.7	102.6
0920	Plaster & Gypsum Board	94.7	139.0	125.6	101.5	139.0	127.7	101.5	116.2	111.8	98.9	105.3	103.4	102.9	154.3	138.8	95.4	154.3	136.6
0950, 0980	Ceilings & Acoustic Treatment	110.5	139.0	129.5	112.1	139.0	130.0	107.6	116.2	113.3	103.7	105.3	104.8	120.5	154.3	143.0	103.7	154.3	137.4
0960	Flooring	107.7	121.8	111.9	109.9	125.5	114.5	114.7	103.9	111.5	100.5	112.6	104.1	108.7	126.8	114.1	101.8	126.8	109.2
0970, 0990	Wall Finishes & Painting/Coating	103.1	117.3	111.6	107.6	140.3	127.2	103.2	104.0	103.6	107.1	108.8	108.1	105.5	152.7	133.8	106.8	140.3	126.9
09	FINISHES	103.6	132.1	119.5	108.5	135.1	123.4	103.1	113.9	109.2	100.5	109.0	105.3	109.2	146.8	130.2	103.6	145.2	126.9
COVERS	DIVS. 10 - 14, 25, 28, 41, 43, 44, 46	100.0	123.0	104.7	100.0	122.7	104.6	100.0	103.1	100.6	100.0	108.7	101.8	100.0	126.8	105.5	100.0	126.0	105.3
21, 22, 23	FIRE SUPPRESSION, PLUMBING & HVAC	100.0	119.8	108.1	94.0	117.9	103.8	94.0	105.5	98.7	100.0	115.2	106.2	100.1	167.3	127.6	100.0	144.8	118.3
26, 27, 3370	ELECTRICAL, COMMUNICATIONS & UTIL.	101.2	113.3	107.6	91.8	123.6	108.5	96.2	101.6	99.0	99.2	99.1	99.1	106.1	157.1	133.0	102.5	153.2	129.2
MF2010	WEIGHTED AVERAGE	101.7	119.2	109.5	100.3	121.8	109.8	94.3	109.0	100.8	99.6	108.0	103.3	106.8	142.4	122.5	102.5	135.3	116.9

	DIVISION	CALIFORNIA																	
		SAN LUIS OBISPO			SAN MATEO			SAN RAFAEL			SANTA ANA			SANTA BARBARA			SANTA CRUZ		
		934			944			949			926 - 927			931			950		
		MAT.	INST.	TOTAL	MAT.	INST.	TOTAL	MAT.	INST.	TOTAL	MAT.	INST.	TOTAL	MAT.	INST.	TOTAL	MAT.	INST.	TOTAL
015433	CONTRACTOR EQUIPMENT		98.4	98.4		98.1	98.1		98.2	98.2		99.3	99.3		98.4	98.4		98.8	98.8
0241, 31 - 34	SITE & INFRASTRUCTURE, DEMOLITION	106.0	105.5	105.7	119.2	104.9	109.2	111.4	110.9	111.0	86.3	105.9	100.1	100.8	105.6	104.2	130.4	99.7	108.9
0310	Concrete Forming & Accessories	117.1	117.8	117.7	112.9	147.4	142.7	116.2	147.2	143.1	110.4	118.0	117.0	108.0	121.7	119.8	109.0	134.7	131.2
0320	Concrete Reinforcing	109.3	116.4	112.9	92.2	117.8	105.0	92.8	117.8	105.3	110.0	116.5	113.3	107.3	116.7	112.0	117.6	117.9	117.7
0330	Cast-in-Place Concrete	103.6	121.5	111.0	116.8	122.0	118.9	135.9	121.4	129.9	84.2	122.5	99.9	97.3	121.8	107.4	115.1	120.0	117.1
03	CONCRETE	110.8	117.9	114.3	108.3	131.2	119.6	129.7	130.8	130.2	93.4	118.4	105.7	102.8	119.8	111.2	113.1	125.4	119.2
04	MASONRY	100.9	113.5	108.6	124.1	131.5	128.6	101.8	131.6	120.0	74.4	114.0	98.6	99.6	116.1	109.7	133.5	129.6	131.1
05	METALS	99.2	100.4	99.6	97.7	100.0	98.4	99.2	98.3	98.9	102.0	101.0	101.7	97.3	101.4	98.6	103.7	107.4	104.8
06	WOOD, PLASTICS & COMPOSITES	105.1	115.9	111.3	106.5	153.1	133.1	102.3	152.9	131.2	102.3	115.8	110.0	98.0	120.2	110.7	106.6	138.1	124.5
07	THERMAL & MOISTURE PROTECTION	100.9	114.1	106.2	107.7	134.1	118.3	111.7	130.3	119.2	100.0	117.5	107.0	100.2	117.0	107.0	99.5	129.5	111.6
08	OPENINGS	97.1	113.7	101.1	97.8	139.2	107.9	108.2	139.0	115.7	96.1	116.4	101.0	98.6	120.6	104.0	91.7	132.6	101.7
0920	Plaster & Gypsum Board	108.9	116.2	114.0	103.9	154.3	139.2	105.5	154.3	139.6	102.3	116.2	112.0	100.2	120.7	114.5	101.4	139.0	127.7
0950, 0980	Ceilings & Acoustic Treatment	112.1	116.2	114.8	110.5	154.3	139.7	118.7	154.3	142.4	107.6	116.2	113.3	113.7	120.7	118.3	109.4	139.0	129.1
0960	Flooring	116.2	96.4	110.3	110.6	119.5	113.2	120.6	119.5	120.3	115.3	103.9	111.9	109.1	110.2	109.5	106.1	125.5	111.9
0970, 0990	Wall Finishes & Painting/Coating	106.5	98.8	101.9	105.5	137.9	124.9	101.8	137.9	123.4	103.2	104.0	103.6	106.5	103.5	104.7	107.0	140.3	126.9
09	FINISHES	110.2	111.3	110.8	107.3	143.3	127.5	109.9	143.6	128.8	104.7	113.9	109.8	106.7	117.6	112.8	106.9	135.2	122.8
COVERS	DIVS. 10 - 14, 25, 28, 41, 43, 44, 46	100.0	113.2	102.7	100.0	126.3	105.4	100.0	125.6	105.2	100.0	110.7	102.2	100.0	111.4	102.3	100.0	123.0	104.7
21, 22, 23	FIRE SUPPRESSION, PLUMBING & HVAC	94.0	105.5	98.7	94.1	136.7	111.5	94.0	144.2	114.6	94.0	105.5	98.7	100.0	116.7	106.8	100.0	117.9	107.3
26, 27, 3370	ELECTRICAL, COMMUNICATIONS & UTIL.	90.4	96.2	93.5	105.9	138.4	123.0	102.7	114.0	108.7	96.2	103.0	99.8	89.4	109.7	100.1	101.4	123.6	113.1
MF2010	WEIGHTED AVERAGE	99.4	107.9	103.2	102.1	130.3	114.5	104.4	128.7	115.1	96.1	109.5	102.0	99.2	113.9	105.7	104.4	121.9	112.1

	DIVISION	CALIFORNIA															COLORADO		
		SANTA ROSA			STOCKTON			SUSANVILLE			VALLEJO			VAN NUYS			ALAMOSA		
		954			952			961			945			913 - 916			811		
		MAT.	INST.	TOTAL	MAT.	INST.	TOTAL	MAT.	INST.	TOTAL	MAT.	INST.	TOTAL	MAT.	INST.	TOTAL	MAT.	INST.	TOTAL
015433	**CONTRACTOR EQUIPMENT**		98.6	98.6		98.1	98.1		98.1	98.1		98.2	98.2		98.2	98.2		93.3	93.3
0241, 31 - 34	**SITE & INFRASTRUCTURE, DEMOLITION**	99.7	104.8	103.3	98.8	104.8	103.0	110.3	104.5	106.2	98.5	111.0	107.2	110.2	109.2	109.5	131.8	88.9	101.7
0310	Concrete Forming & Accessories	104.1	146.5	140.8	106.8	131.5	128.1	107.5	131.1	127.9	106.8	146.1	140.8	112.7	118.0	117.3	105.5	70.4	75.2
0320	Concrete Reinforcing	105.0	118.3	111.6	107.9	117.4	112.6	104.1	117.0	110.5	93.9	118.2	106.0	110.9	116.7	113.8	110.3	79.7	95.1
0330	Cast-in-Place Concrete	109.5	119.4	113.6	97.3	117.8	105.7	100.6	117.6	107.6	108.3	119.5	112.9	93.9	119.8	104.5	102.1	80.8	93.3
03	**CONCRETE**	112.3	130.3	121.2	102.8	122.9	112.7	113.2	122.6	117.8	105.2	129.7	117.3	114.1	117.4	115.8	115.6	76.1	96.1
04	**MASONRY**	100.3	134.7	121.3	101.9	124.3	115.5	100.0	117.5	110.6	78.1	134.6	112.6	117.4	116.5	116.9	126.4	74.4	94.7
05	**METALS**	102.3	105.7	103.4	98.5	103.0	99.9	100.5	102.1	101.0	99.2	99.0	99.1	82.4	100.3	88.0	96.8	81.8	92.1
06	**WOOD, PLASTICS & COMPOSITES**	93.4	152.7	127.3	99.6	134.8	119.7	99.9	134.8	119.8	91.9	152.9	126.7	92.6	115.5	105.7	94.7	70.5	80.9
07	**THERMAL & MOISTURE PROTECTION**	100.4	133.5	113.7	103.3	120.1	110.1	104.1	115.7	108.7	109.6	133.2	119.1	97.3	116.4	105.0	104.6	79.6	94.6
08	**OPENINGS**	97.7	140.6	108.1	98.0	124.6	104.5	100.1	124.0	105.9	110.3	140.7	117.7	88.1	117.8	95.3	96.5	76.8	91.7
0920	Plaster & Gypsum Board	96.9	154.3	137.0	100.6	135.7	125.2	99.4	135.7	124.8	100.1	154.3	138.0	96.2	116.2	110.2	76.8	69.4	71.6
0950, 0980	Ceilings & Acoustic Treatment	107.6	154.3	138.7	113.7	135.7	128.4	112.1	135.7	127.8	120.7	154.3	143.1	93.7	116.2	108.7	106.7	69.4	81.9
0960	Flooring	111.6	119.1	113.9	109.1	115.5	111.0	109.1	109.4	109.2	116.0	126.8	119.2	95.4	103.9	97.9	112.2	56.8	95.7
0970, 0990	Wall Finishes & Painting/Coating	103.2	140.3	125.4	106.5	117.3	113.0	106.5	114.2	111.1	102.7	140.3	125.2	110.6	108.8	109.5	107.6	24.7	57.9
09	**FINISHES**	104.0	142.8	125.7	106.6	129.0	119.2	107.5	127.7	118.8	106.8	144.2	127.8	97.2	114.3	106.8	105.7	62.5	81.5
COVERS	**DIVS. 10 - 14, 25, 28, 41, 43, 44, 46**	100.0	124.2	104.9	100.0	122.3	104.6	100.0	122.3	104.6	100.0	124.5	105.0	100.0	110.2	102.1	100.0	89.8	97.9
21, 22, 23	**FIRE SUPPRESSION, PLUMBING & HVAC**	94.0	165.3	123.2	100.0	111.3	104.6	94.0	107.0	99.3	100.1	132.3	113.3	94.0	109.1	100.2	94.0	73.5	85.6
26, 27, 3370	**ELECTRICAL, COMMUNICATIONS & UTIL.**	96.5	118.7	108.2	99.3	112.1	106.0	100.1	101.8	101.0	98.3	127.4	113.6	114.4	113.6	114.0	95.8	75.1	84.9
MF2010	**WEIGHTED AVERAGE**	100.0	134.2	115.1	100.5	116.6	107.6	101.1	113.1	106.4	101.0	128.5	113.1	98.3	112.1	104.4	101.8	75.7	90.3

	DIVISION	COLORADO																	
		BOULDER			COLORADO SPRINGS			DENVER			DURANGO			FORT COLLINS			FORT MORGAN		
		803			808 - 809			800 - 802			813			805			807		
		MAT.	INST.	TOTAL	MAT.	INST.	TOTAL	MAT.	INST.	TOTAL	MAT.	INST.	TOTAL	MAT.	INST.	TOTAL	MAT.	INST.	TOTAL
015433	**CONTRACTOR EQUIPMENT**		94.6	94.6		93.1	93.1		97.5	97.5		93.3	93.3		94.6	94.6		94.6	94.6
0241, 31 - 34	**SITE & INFRASTRUCTURE, DEMOLITION**	92.4	95.1	94.3	94.2	93.1	93.4	93.3	101.3	98.9	125.1	88.9	99.7	104.1	94.7	97.5	95.0	94.6	94.8
0310	Concrete Forming & Accessories	106.6	83.5	86.6	97.1	79.1	81.6	102.8	79.4	82.5	111.6	70.6	76.1	103.8	78.1	81.5	106.9	78.3	82.1
0320	Concrete Reinforcing	104.5	79.8	92.2	103.6	84.1	93.9	103.6	84.1	93.9	110.3	79.7	95.1	104.6	79.9	92.3	104.7	79.7	92.2
0330	Cast-in-Place Concrete	102.3	81.4	93.7	105.1	84.6	96.6	99.4	84.2	93.2	117.3	80.9	102.4	115.5	80.1	100.9	100.4	80.2	92.1
03	**CONCRETE**	102.4	82.2	92.4	105.5	82.2	94.0	100.1	82.2	91.3	118.3	76.2	97.5	112.6	79.3	96.2	101.0	79.3	90.3
04	**MASONRY**	92.5	75.0	81.8	93.3	75.0	82.1	95.3	75.0	82.9	113.5	74.4	89.7	110.1	78.7	90.9	107.0	74.6	87.2
05	**METALS**	99.7	85.3	95.2	102.9	87.5	98.0	105.4	87.6	99.8	96.8	82.0	92.1	101.0	82.1	95.0	99.5	81.9	94.0
06	**WOOD, PLASTICS & COMPOSITES**	95.6	86.7	90.5	86.4	80.8	83.2	93.5	80.7	86.2	102.8	70.5	84.4	93.3	80.6	86.1	95.6	80.6	87.1
07	**THERMAL & MOISTURE PROTECTION**	102.7	81.8	94.3	103.5	81.8	94.8	101.9	75.8	91.4	104.5	79.6	94.5	103.1	73.2	91.1	102.6	81.2	94.0
08	**OPENINGS**	98.9	85.6	95.7	103.4	83.7	98.6	103.8	83.6	98.9	103.8	76.8	97.3	98.9	82.4	94.9	98.9	82.4	94.9
0920	Plaster & Gypsum Board	104.9	86.5	92.0	86.6	80.3	82.2	100.8	80.3	86.5	88.4	69.4	75.1	99.1	80.3	85.9	104.9	80.3	87.7
0950, 0980	Ceilings & Acoustic Treatment	97.3	86.5	90.1	104.6	80.3	88.4	107.8	80.3	89.5	106.7	69.4	81.9	97.3	80.3	86.0	97.3	80.3	86.0
0960	Flooring	108.7	86.9	102.2	100.4	70.7	91.5	105.5	87.9	100.2	116.8	56.8	98.9	105.3	56.8	90.9	109.1	56.8	93.5
0970, 0990	Wall Finishes & Painting/Coating	99.1	69.6	81.4	98.8	42.3	65.0	99.1	79.0	87.1	107.6	24.7	57.9	99.1	42.0	64.9	99.1	55.9	73.2
09	**FINISHES**	104.2	82.9	92.3	101.1	73.6	85.7	104.9	80.7	91.3	107.6	62.5	82.3	103.1	70.4	84.8	104.2	71.9	86.1
COVERS	**DIVS. 10 - 14, 25, 28, 41, 43, 44, 46**	100.0	90.8	98.1	100.0	90.5	98.1	100.0	90.2	98.0	100.0	89.8	97.9	100.0	90.1	98.0	100.0	90.1	98.0
21, 22, 23	**FIRE SUPPRESSION, PLUMBING & HVAC**	94.0	79.4	88.0	100.2	74.2	89.5	99.9	79.5	91.6	94.0	87.5	91.3	100.0	79.3	91.5	94.0	79.3	88.0
26, 27, 3370	**ELECTRICAL, COMMUNICATIONS & UTIL.**	99.3	85.7	92.1	103.2	83.6	92.8	104.9	85.8	94.8	95.1	72.3	83.1	99.3	85.1	91.8	99.6	85.7	92.3
MF2010	**WEIGHTED AVERAGE**	98.4	83.3	91.7	101.4	80.6	92.2	101.7	83.4	93.6	102.1	78.3	91.6	102.3	80.8	92.8	99.0	80.9	91.0

	DIVISION	COLORADO																	
		GLENWOOD SPRINGS			GOLDEN			GRAND JUNCTION			GREELEY			MONTROSE			PUEBLO		
		816			804			815			806			814			810		
		MAT.	INST.	TOTAL	MAT.	INST.	TOTAL	MAT.	INST.	TOTAL	MAT.	INST.	TOTAL	MAT.	INST.	TOTAL	MAT.	INST.	TOTAL
015433	**CONTRACTOR EQUIPMENT**		96.1	96.1		94.6	94.6		96.1	96.1		94.6	94.6		94.7	94.7		93.3	93.3
0241, 31 - 34	**SITE & INFRASTRUCTURE, DEMOLITION**	140.7	95.8	109.2	104.9	94.9	97.9	124.7	95.5	104.2	91.4	94.1	93.3	134.3	92.0	104.7	116.7	91.4	98.9
0310	Concrete Forming & Accessories	102.7	78.4	81.7	99.5	78.2	81.1	110.3	77.8	82.2	101.9	82.0	84.6	102.3	78.1	81.4	107.5	79.5	83.2
0320	Concrete Reinforcing	109.2	79.7	94.5	104.7	79.7	92.2	109.5	79.6	94.6	104.5	78.3	91.4	109.0	79.7	94.4	105.5	84.1	94.9
0330	Cast-in-Place Concrete	102.0	80.3	93.1	100.4	80.2	92.1	112.9	79.6	99.2	96.5	59.5	81.3	102.0	80.2	93.1	101.3	85.3	94.7
03	**CONCRETE**	120.5	79.5	100.2	110.3	79.3	95.0	114.7	78.9	97.0	97.8	73.6	85.9	111.8	79.3	95.7	104.7	82.6	93.8
04	**MASONRY**	99.2	74.6	84.2	109.7	74.7	88.3	132.8	74.4	97.2	104.2	50.8	71.6	106.6	74.4	87.0	95.6	75.0	83.0
05	**METALS**	96.5	82.5	92.1	99.7	82.0	94.1	98.1	81.2	92.8	100.9	79.3	94.1	95.7	81.6	91.2	99.6	88.3	96.0
06	**WOOD, PLASTICS & COMPOSITES**	90.0	80.7	84.7	88.5	80.6	84.0	100.7	80.7	89.3	90.8	86.7	88.4	91.2	80.9	85.3	97.1	81.0	87.9
07	**THERMAL & MOISTURE PROTECTION**	104.5	80.7	94.9	103.5	75.3	92.2	103.5	69.8	89.9	102.5	65.9	87.8	104.6	80.7	95.0	103.0	81.3	94.3
08	**OPENINGS**	102.9	82.4	97.9	98.9	82.4	94.9	103.7	82.4	98.5	98.9	85.6	95.7	104.0	82.5	98.8	98.4	83.8	94.9
0920	Plaster & Gypsum Board	108.5	80.3	88.8	96.6	80.3	85.2	120.1	80.3	92.3	97.4	86.5	89.8	76.0	80.3	79.0	81.3	80.3	80.6
0950, 0980	Ceilings & Acoustic Treatment	105.8	80.3	88.8	97.3	80.3	86.0	105.8	80.3	88.8	97.3	86.5	90.1	106.7	80.3	89.1	115.9	80.3	92.2
0960	Flooring	111.5	52.3	93.8	103.2	56.8	89.4	116.2	56.8	98.5	104.3	56.8	90.2	114.4	47.0	94.3	113.2	87.9	105.6
0970, 0990	Wall Finishes & Painting/Coating	107.6	55.9	76.7	99.1	69.6	81.4	107.6	69.6	84.8	99.1	26.2	55.4	107.6	24.7	57.9	107.6	39.2	66.7
09	**FINISHES**	109.9	71.2	88.2	103.0	74.2	86.8	110.6	73.5	89.8	101.7	72.2	85.2	106.3	66.7	84.1	106.5	76.5	89.7
COVERS	**DIVS. 10 - 14, 25, 28, 41, 43, 44, 46**	100.0	90.3	98.0	100.0	90.1	98.0	100.0	90.2	98.0	100.0	90.8	98.1	100.0	90.6	98.1	100.0	91.0	98.2
21, 22, 23	**FIRE SUPPRESSION, PLUMBING & HVAC**	94.0	87.5	91.4	94.0	78.7	87.8	99.9	87.0	94.6	100.0	79.2	91.5	94.0	87.5	91.3	99.9	74.2	89.4
26, 27, 3370	**ELECTRICAL, COMMUNICATIONS & UTIL.**	92.4	72.4	81.9	99.6	85.7	92.3	94.8	56.1	74.4	99.3	85.7	92.1	94.8	58.8	75.8	95.8	75.1	84.9
MF2010	**WEIGHTED AVERAGE**	101.8	80.9	92.6	100.4	80.9	91.8	104.4	78.2	92.9	99.8	77.3	89.9	101.0	78.0	90.8	100.7	79.8	91.5

	DIVISION	COLORADO			CONNECTICUT														
		SALIDA			BRIDGEPORT			BRISTOL			HARTFORD			MERIDEN			NEW BRITAIN		
		812			066			060			061			064			060		
		MAT.	INST.	TOTAL	MAT.	INST.	TOTAL	MAT.	INST.	TOTAL	MAT.	INST.	TOTAL	MAT.	INST.	TOTAL	MAT.	INST.	TOTAL
015433	CONTRACTOR EQUIPMENT		94.7	94.7		100.5	100.5		100.5	100.5		100.5	100.5		100.9	100.9		100.5	100.5
0241, 31 - 34	SITE & INFRASTRUCTURE, DEMOLITION	124.8	92.3	102.0	116.1	103.9	107.5	115.4	103.9	107.3	108.0	103.9	105.1	113.4	104.6	107.2	115.5	103.9	107.4
0310	Concrete Forming & Accessories	110.2	78.0	82.4	95.7	123.7	119.9	95.7	123.6	119.8	96.1	123.6	119.9	95.4	123.6	119.8	96.1	123.5	119.8
0320	Concrete Reinforcing	108.8	79.6	94.3	110.7	127.4	119.0	110.7	127.3	119.0	110.7	127.3	119.0	110.7	127.3	119.0	110.7	127.3	119.0
0330	Cast-in-Place Concrete	116.9	80.2	101.8	106.5	126.9	114.9	99.8	126.8	110.9	105.8	126.8	114.4	96.0	126.8	108.7	101.4	126.8	111.9
03	CONCRETE	113.5	79.2	96.5	107.0	125.2	116.0	103.8	125.1	114.3	106.7	125.1	115.8	102.0	125.1	113.4	104.6	125.1	114.7
04	MASONRY	134.9	74.4	98.0	102.6	132.5	120.8	95.3	132.5	118.0	95.7	132.5	118.2	94.9	132.5	117.8	96.3	132.5	118.4
05	METALS	95.4	81.5	91.0	98.8	124.1	106.8	98.8	124.0	106.7	103.6	124.0	110.0	96.4	124.0	105.1	95.6	124.0	104.5
06	WOOD, PLASTICS & COMPOSITES	98.2	80.9	88.3	95.5	122.5	110.9	95.5	122.5	110.9	94.1	122.5	110.3	95.5	122.5	110.9	95.5	122.5	110.9
07	THERMAL & MOISTURE PROTECTION	103.5	80.7	94.4	97.0	129.3	110.0	97.1	125.9	108.7	102.9	125.9	112.2	97.1	125.9	108.7	97.1	125.0	108.3
08	OPENINGS	96.6	82.5	93.2	103.4	131.2	110.2	103.4	131.2	110.2	104.0	131.2	110.7	106.0	131.2	112.1	103.4	131.2	110.2
0920	Plaster & Gypsum Board	76.4	80.3	79.1	98.4	122.6	115.3	98.4	122.6	115.3	92.9	122.6	113.6	100.1	122.6	115.8	98.4	122.6	115.3
0950, 0980	Ceilings & Acoustic Treatment	106.7	80.3	89.1	97.7	122.6	114.3	97.7	122.6	114.3	102.2	122.6	115.8	102.0	122.6	115.7	97.7	122.6	114.3
0960	Flooring	119.2	47.0	97.7	98.3	133.0	108.6	98.3	133.0	108.6	98.1	133.0	108.5	98.3	133.0	108.6	98.3	133.0	108.6
0970, 0990	Wall Finishes & Painting/Coating	107.6	24.7	57.9	96.1	116.4	108.3	96.1	116.4	108.3	96.1	116.4	108.3	96.1	116.4	108.3	96.1	116.4	108.3
09	FINISHES	106.6	66.7	84.2	101.4	124.3	114.2	101.5	124.3	114.3	100.5	124.3	113.9	102.6	124.3	114.8	101.5	124.3	114.3
COVERS	DIVS. 10 - 14, 25, 28, 41, 43, 44, 46	100.0	90.6	98.1	100.0	108.5	101.7	100.0	108.5	101.7	100.0	108.6	101.7	100.0	108.5	101.7	100.0	108.5	101.7
21, 22, 23	FIRE SUPPRESSION, PLUMBING & HVAC	94.0	74.0	85.8	100.0	117.6	107.2	100.0	117.6	107.2	100.0	117.6	107.2	94.0	117.6	103.6	100.0	117.6	107.2
26, 27, 3370	ELECTRICAL, COMMUNICATIONS & UTIL.	95.1	75.1	84.5	101.9	107.8	105.0	101.9	110.4	106.4	101.3	111.4	106.6	101.8	110.4	106.4	102.0	110.4	106.4
MF2010	WEIGHTED AVERAGE	101.5	77.4	90.9	101.7	119.7	109.6	100.9	120.0	109.3	102.0	120.1	109.9	99.2	120.0	108.4	100.6	119.9	109.1

	DIVISION	CONNECTICUT																	
		NEW HAVEN			NEW LONDON			NORWALK			STAMFORD			WATERBURY			WILLIMANTIC		
		065			063			068			069			067			062		
		MAT.	INST.	TOTAL	MAT.	INST.	TOTAL	MAT.	INST.	TOTAL	MAT.	INST.	TOTAL	MAT.	INST.	TOTAL	MAT.	INST.	TOTAL
015433	CONTRACTOR EQUIPMENT		100.9	100.9		100.9	100.9		100.5	100.5		100.5	100.5		100.5	100.5		100.5	100.5
0241, 31 - 34	SITE & INFRASTRUCTURE, DEMOLITION	115.3	104.6	107.8	107.0	104.6	105.3	115.9	103.9	107.5	116.7	103.9	107.7	115.7	103.9	107.4	116.2	103.9	107.5
0310	Concrete Forming & Accessories	95.4	123.6	119.8	95.4	123.6	119.8	95.7	124.1	120.3	95.6	124.1	120.3	95.7	123.5	119.8	95.6	123.5	119.8
0320	Concrete Reinforcing	110.7	127.3	119.0	86.8	127.3	107.0	110.7	127.5	119.1	110.7	127.5	119.1	110.7	127.3	119.0	110.7	127.3	119.0
0330	Cast-in-Place Concrete	103.2	126.8	112.9	88.0	126.8	103.9	104.8	128.4	114.5	106.5	128.4	115.5	106.5	126.8	114.9	99.5	126.8	110.7
03	CONCRETE	118.7	125.1	121.9	91.5	125.1	108.1	106.2	125.9	115.9	107.0	125.9	116.3	107.0	125.1	115.9	103.7	125.1	114.2
04	MASONRY	95.5	132.5	118.1	93.8	132.5	117.4	95.0	134.0	118.8	95.8	134.0	119.1	95.8	132.5	118.2	95.1	132.5	117.9
05	METALS	95.8	124.0	104.6	95.5	124.0	104.4	98.8	124.6	106.9	98.8	124.6	106.9	98.8	124.0	106.7	98.6	124.0	106.6
06	WOOD, PLASTICS & COMPOSITES	95.5	122.5	110.9	95.5	122.5	110.9	95.5	122.5	110.9	95.5	122.5	110.9	95.5	122.5	110.9	95.5	122.5	110.9
07	THERMAL & MOISTURE PROTECTION	97.2	125.9	108.7	97.1	125.9	108.7	97.2	129.9	110.4	97.1	129.9	110.3	97.1	125.9	108.7	97.3	125.1	108.5
08	OPENINGS	103.4	131.2	110.2	106.4	131.2	112.5	103.4	131.2	110.2	103.4	131.2	110.2	103.4	131.2	110.2	106.4	131.2	112.5
0920	Plaster & Gypsum Board	98.4	122.6	115.3	98.4	122.6	115.3	98.4	122.6	115.3	98.4	122.6	115.3	98.4	122.6	115.3	98.4	122.6	115.3
0950, 0980	Ceilings & Acoustic Treatment	97.7	122.6	114.3	95.6	122.6	113.6	97.7	122.6	114.3	97.7	122.6	114.3	97.7	122.6	114.3	95.6	122.6	113.6
0960	Flooring	98.3	133.0	108.6	98.3	133.0	108.6	98.3	133.0	108.6	98.3	133.0	108.6	98.3	133.0	108.6	98.3	138.1	110.1
0970, 0990	Wall Finishes & Painting/Coating	96.1	116.4	108.3	96.1	116.4	108.3	96.1	116.4	108.3	96.1	116.4	108.3	96.1	116.4	108.3	96.1	116.4	108.3
09	FINISHES	101.5	124.3	114.3	100.4	124.3	113.8	101.5	124.3	114.3	101.6	124.3	114.3	101.3	124.3	114.2	101.2	125.2	114.6
COVERS	DIVS. 10 - 14, 25, 28, 41, 43, 44, 46	100.0	108.5	101.7	100.0	108.5	101.7	100.0	108.8	101.8	100.0	108.8	101.8	100.0	108.5	101.7	100.0	108.6	101.7
21, 22, 23	FIRE SUPPRESSION, PLUMBING & HVAC	100.0	117.6	107.2	94.0	117.6	103.6	100.0	117.7	107.2	100.0	117.7	107.2	100.0	117.6	107.2	100.0	117.6	107.2
26, 27, 3370	ELECTRICAL, COMMUNICATIONS & UTIL.	101.8	110.4	106.4	98.2	110.4	104.7	101.9	159.2	132.1	101.9	159.2	132.1	101.3	107.8	104.7	101.9	111.4	106.9
MF2010	WEIGHTED AVERAGE	102.1	120.0	110.0	97.2	120.0	107.2	101.2	127.1	112.6	101.4	127.1	112.7	101.3	119.6	109.3	101.2	120.2	109.6

	DIVISION	D.C.			DELAWARE									FLORIDA					
		WASHINGTON			DOVER			NEWARK			WILMINGTON			DAYTONA BEACH			FORT LAUDERDALE		
		200 - 205			199			197			198			321			333		
		MAT.	INST.	TOTAL	MAT.	INST.	TOTAL	MAT.	INST.	TOTAL	MAT.	INST.	TOTAL	MAT.	INST.	TOTAL	MAT.	INST.	TOTAL
015433	CONTRACTOR EQUIPMENT		103.7	103.7		118.8	118.8		118.8	118.8		119.0	119.0		98.9	98.9		91.6	91.6
0241, 31 - 34	SITE & INFRASTRUCTURE, DEMOLITION	108.0	93.7	98.0	101.3	113.6	109.9	102.0	113.6	110.1	93.5	113.9	107.8	114.2	90.6	97.6	98.8	78.4	84.5
0310	Concrete Forming & Accessories	99.6	80.2	82.8	95.4	103.1	102.0	96.2	103.1	102.1	97.1	103.1	102.3	97.8	71.0	74.6	96.4	71.6	75.0
0320	Concrete Reinforcing	105.7	89.4	97.6	98.6	100.9	99.7	99.4	100.9	100.2	99.4	100.9	100.2	95.3	76.0	85.7	95.3	75.7	85.5
0330	Cast-in-Place Concrete	125.6	89.1	110.6	91.9	100.9	95.6	82.6	100.9	90.1	89.1	100.9	94.0	94.2	73.3	85.6	98.8	80.1	91.1
03	CONCRETE	112.6	86.3	99.6	94.7	102.8	98.7	90.4	102.8	96.6	93.6	102.8	98.2	91.7	73.9	82.9	93.8	76.5	85.3
04	MASONRY	93.9	80.9	86.0	108.4	96.2	101.0	106.8	96.2	100.3	113.1	96.2	102.8	93.7	68.4	78.3	93.9	71.1	80.0
05	METALS	98.0	106.8	100.8	99.9	116.8	105.2	99.9	116.8	105.2	97.0	116.8	103.2	96.2	92.0	94.9	96.1	92.7	95.0
06	WOOD, PLASTICS & COMPOSITES	96.8	78.4	86.3	95.2	103.3	99.8	97.9	103.3	101.0	100.0	103.3	101.9	98.6	72.3	83.6	94.4	70.2	80.6
07	THERMAL & MOISTURE PROTECTION	100.3	85.8	94.5	99.2	111.4	104.1	100.9	111.4	105.1	99.2	111.4	104.1	97.2	76.5	88.9	97.2	84.1	91.9
08	OPENINGS	98.2	87.7	95.7	94.2	111.2	98.4	94.2	111.2	98.4	93.9	111.2	98.1	99.6	69.4	92.3	97.4	68.5	90.3
0920	Plaster & Gypsum Board	107.2	77.9	86.7	107.3	103.2	104.4	106.9	103.2	104.3	109.8	103.2	105.2	98.4	71.9	79.8	97.5	69.7	78.0
0950, 0980	Ceilings & Acoustic Treatment	106.6	77.9	87.4	107.7	103.2	104.7	101.3	103.2	102.6	105.8	103.2	104.1	94.6	71.9	79.5	94.6	69.7	78.0
0960	Flooring	110.1	97.0	106.2	97.5	108.2	100.7	97.9	108.2	101.0	97.3	108.2	100.5	117.7	76.3	105.4	117.7	71.9	104.1
0970, 0990	Wall Finishes & Painting/Coating	109.8	85.9	95.5	98.4	105.3	102.6	98.4	105.3	102.6	98.4	105.3	102.6	106.8	77.4	89.2	101.6	73.4	84.7
09	FINISHES	104.5	83.0	92.5	102.7	103.9	103.4	101.7	103.9	102.9	102.8	103.9	103.4	105.5	72.4	86.9	102.5	71.1	84.9
COVERS	DIVS. 10 - 14, 25, 28, 41, 43, 44, 46	100.0	99.3	99.9	100.0	90.7	98.1	100.0	90.7	98.1	100.0	90.7	98.1	100.0	75.5	95.0	100.0	84.8	96.9
21, 22, 23	FIRE SUPPRESSION, PLUMBING & HVAC	100.1	94.6	97.9	99.9	116.9	106.9	100.1	116.9	107.0	100.1	116.9	107.0	99.9	75.6	90.0	99.9	70.2	87.8
26, 27, 3370	ELECTRICAL, COMMUNICATIONS & UTIL.	99.5	103.9	101.8	98.5	111.0	105.1	98.8	111.0	105.2	97.3	111.0	104.5	97.4	57.3	76.3	97.4	72.9	84.5
MF2010	WEIGHTED AVERAGE	101.1	92.4	97.3	99.3	108.8	103.5	98.8	108.8	103.2	98.6	108.9	103.1	98.6	74.3	87.9	97.9	75.2	87.9

DIVISION		FLORIDA																	
		FORT MYERS			GAINESVILLE			JACKSONVILLE			LAKELAND			MELBOURNE			MIAMI		
		339,341			326,344			320,322			338			329			330 - 332,340		
		MAT.	INST.	TOTAL	MAT.	INST.	TOTAL	MAT.	INST.	TOTAL	MAT.	INST.	TOTAL	MAT.	INST.	TOTAL	MAT.	INST.	TOTAL
015433	CONTRACTOR EQUIPMENT		98.9	98.9		98.9	98.9		98.9	98.9		98.9	98.9		98.9	98.9		91.6	91.6
0241, 31 - 34	SITE & INFRASTRUCTURE, DEMOLITION	111.2	89.8	96.2	123.8	89.6	99.9	114.2	90.0	97.2	113.3	90.1	97.0	121.9	89.9	99.5	100.8	78.2	84.9
0310	Concrete Forming & Accessories	91.9	78.3	80.1	92.8	56.6	61.5	97.5	57.0	62.4	88.1	78.9	80.1	94.0	72.7	75.6	102.9	72.1	76.2
0320	Concrete Reinforcing	96.4	97.3	96.9	101.2	66.7	84.0	95.3	66.8	81.1	98.8	98.2	98.5	96.4	76.0	86.2	95.3	75.8	85.6
0330	Cast-in-Place Concrete	103.1	70.8	89.8	108.2	64.8	90.3	95.1	70.8	85.1	105.4	72.3	91.8	113.5	75.7	98.0	100.7	81.1	92.7
03	CONCRETE	94.7	80.3	87.5	102.6	62.9	83.0	92.1	65.1	78.8	96.5	81.1	88.9	103.1	75.5	89.5	95.1	77.0	86.2
04	MASONRY	87.9	64.5	73.6	108.7	63.7	81.3	93.7	63.8	75.4	105.5	79.0	89.3	92.6	72.7	80.5	94.7	74.9	82.6
05	METALS	98.2	99.9	98.8	95.1	86.9	92.5	94.8	87.3	92.4	98.1	100.9	99.0	104.2	92.1	100.4	100.9	91.4	97.9
06	WOOD, PLASTICS & COMPOSITES	91.2	80.9	85.3	92.4	54.1	70.5	98.6	54.1	73.2	86.7	80.9	83.4	94.0	72.3	81.6	100.3	70.2	83.1
07	THERMAL & MOISTURE PROTECTION	97.0	83.1	91.4	97.5	63.7	83.9	97.4	64.4	84.1	96.9	87.9	93.3	97.6	79.0	90.1	102.7	78.2	92.9
08	OPENINGS	98.3	77.8	93.3	97.7	54.8	87.3	99.6	58.1	89.5	98.2	78.5	93.4	98.8	72.8	92.4	97.4	68.4	90.3
0920	Plaster & Gypsum Board	92.8	80.7	84.4	93.6	53.1	65.3	98.4	53.1	66.7	90.3	80.7	83.6	93.6	71.9	78.4	94.5	69.7	77.1
0950, 0980	Ceilings & Acoustic Treatment	88.9	80.7	83.5	88.9	53.1	65.0	94.6	53.1	66.9	88.9	80.7	83.5	91.0	71.9	78.2	98.2	69.7	79.2
0960	Flooring	114.3	57.1	97.3	114.9	44.7	94.0	117.7	66.9	102.6	111.8	58.6	96.0	115.1	76.3	103.5	125.2	75.6	110.5
0970, 0990	Wall Finishes & Painting/Coating	106.8	63.5	80.8	106.8	63.5	80.8	106.8	64.7	81.6	106.8	63.5	80.8	106.8	96.0	100.3	101.6	73.4	84.7
09	FINISHES	102.3	73.0	85.9	103.9	54.0	76.0	105.5	58.6	79.2	101.4	73.3	85.7	104.0	75.5	88.1	105.3	72.3	86.8
COVERS	DIVS. 10 - 14, 25, 28, 41, 43, 44, 46	100.0	75.8	95.1	100.0	73.9	94.7	100.0	72.0	94.3	100.0	75.8	95.1	100.0	77.0	95.3	100.0	85.4	97.0
21, 22, 23	FIRE SUPPRESSION, PLUMBING & HVAC	96.7	65.2	83.8	98.8	67.4	85.9	99.9	67.4	86.6	96.7	85.2	92.0	99.9	77.8	90.9	99.9	70.1	87.7
26, 27, 3370	ELECTRICAL, COMMUNICATIONS & UTIL.	99.9	65.5	81.8	97.8	74.8	85.7	97.1	64.9	80.1	97.8	64.2	80.1	98.8	66.7	81.9	102.0	76.9	88.8
MF2010	WEIGHTED AVERAGE	97.9	75.1	87.8	100.0	68.9	86.3	98.4	68.6	85.2	98.7	81.0	90.9	101.2	77.3	90.7	99.8	76.0	89.3

DIVISION		FLORIDA																	
		ORLANDO			PANAMA CITY			PENSACOLA			SARASOTA			ST. PETERSBURG			TALLAHASSEE		
		327 - 328,347			324			325			342			337			323		
		MAT.	INST.	TOTAL	MAT.	INST.	TOTAL	MAT.	INST.	TOTAL	MAT.	INST.	TOTAL	MAT.	INST.	TOTAL	MAT.	INST.	TOTAL
015433	CONTRACTOR EQUIPMENT		98.9	98.9		98.9	98.9		98.9	98.9		98.9	98.9		98.9	98.9		98.9	98.9
0241, 31 - 34	SITE & INFRASTRUCTURE, DEMOLITION	114.6	89.8	97.2	128.0	88.9	100.5	127.4	89.3	100.7	115.6	89.9	97.5	114.9	89.5	97.1	110.6	89.1	95.5
0310	Concrete Forming & Accessories	98.6	75.2	78.4	96.8	45.9	52.8	94.5	54.1	59.6	97.8	78.4	81.0	95.7	53.3	59.0	100.1	46.1	53.4
0320	Concrete Reinforcing	95.3	75.3	85.3	99.5	76.1	87.9	102.1	76.6	89.4	95.3	98.1	96.7	98.8	85.1	92.0	95.3	66.6	81.0
0330	Cast-in-Place Concrete	118.0	73.1	99.6	100.0	58.8	83.1	123.5	68.3	100.8	107.7	72.1	93.1	106.6	66.8	90.2	102.9	58.4	84.6
03	CONCRETE	104.4	75.6	90.2	100.2	57.8	79.3	110.9	64.8	88.1	98.2	80.9	89.6	98.1	65.5	82.0	96.0	56.0	76.2
04	MASONRY	99.4	68.4	80.5	98.6	49.2	68.5	116.3	56.7	79.9	94.5	79.0	85.0	147.1	51.0	88.4	96.9	56.1	72.0
05	METALS	101.7	91.4	98.4	95.9	89.4	93.9	97.0	90.6	95.0	98.9	100.5	99.4	99.1	93.7	97.4	89.8	86.7	88.8
06	WOOD, PLASTICS & COMPOSITES	98.2	78.6	87.0	97.4	43.6	66.7	95.4	53.8	71.7	98.6	80.9	88.5	95.8	52.3	71.0	96.0	43.2	65.8
07	THERMAL & MOISTURE PROTECTION	97.1	77.2	89.1	97.6	58.7	82.0	97.6	64.7	84.4	97.2	87.9	93.5	97.2	60.0	82.2	102.8	74.3	91.3
08	OPENINGS	99.6	71.8	92.9	97.3	48.6	85.4	97.2	59.5	88.0	99.6	77.5	94.2	98.2	62.3	89.5	100.1	49.1	87.7
0920	Plaster & Gypsum Board	102.5	78.3	85.6	96.5	42.3	58.6	96.2	52.8	65.9	98.4	80.7	86.0	95.3	51.3	64.5	107.2	41.8	61.5
0950, 0980	Ceilings & Acoustic Treatment	104.6	78.3	87.1	91.0	42.3	58.5	91.0	52.8	65.6	92.6	80.7	84.7	91.0	51.3	64.5	104.6	41.8	62.8
0960	Flooring	117.5	76.3	105.2	117.2	44.6	95.6	112.5	65.2	98.4	117.7	60.0	100.5	116.4	57.2	98.8	117.5	64.3	101.7
0970, 0990	Wall Finishes & Painting/Coating	106.8	73.4	86.7	106.8	67.8	83.4	106.8	67.8	83.4	106.8	63.5	80.8	106.8	63.5	80.8	106.8	67.8	83.4
09	FINISHES	107.9	75.7	89.8	105.9	46.8	72.8	104.0	56.9	77.6	105.0	73.6	87.4	103.9	54.0	76.0	108.1	50.4	75.8
COVERS	DIVS. 10 - 14, 25, 28, 41, 43, 44, 46	100.0	76.2	95.1	100.0	48.1	89.4	100.0	48.2	89.4	100.0	75.8	95.1	100.0	58.2	91.5	100.0	68.7	93.6
21, 22, 23	FIRE SUPPRESSION, PLUMBING & HVAC	99.9	59.6	83.4	99.9	54.8	81.5	99.9	55.2	81.6	99.8	66.1	86.0	99.9	61.5	84.2	100.0	41.1	75.9
26, 27, 3370	ELECTRICAL, COMMUNICATIONS & UTIL.	101.9	62.5	81.2	96.0	61.8	78.0	100.3	56.1	77.0	97.4	64.2	79.9	97.8	64.2	80.1	103.1	61.8	81.4
MF2010	WEIGHTED AVERAGE	101.8	72.3	88.8	99.8	60.2	82.4	102.2	63.4	85.1	99.7	76.9	89.7	102.2	65.6	86.1	99.1	59.1	81.5

DIVISION		FLORIDA						GEORGIA											
		TAMPA			WEST PALM BEACH			ALBANY			ATHENS			ATLANTA			AUGUSTA		
		335 - 336,346			334,349			317,398			306			300 - 303,399			308 - 309		
		MAT.	INST.	TOTAL	MAT.	INST.	TOTAL	MAT.	INST.	TOTAL	MAT.	INST.	TOTAL	MAT.	INST.	TOTAL	MAT.	INST.	TOTAL
015433	CONTRACTOR EQUIPMENT		98.9	98.9		91.6	91.6		92.5	92.5		93.6	93.6		94.1	94.1		93.6	93.6
0241, 31 - 34	SITE & INFRASTRUCTURE, DEMOLITION	115.4	90.0	97.6	95.6	78.4	83.6	99.0	79.8	85.6	98.6	92.8	94.6	95.2	94.7	94.8	92.0	92.6	92.4
0310	Concrete Forming & Accessories	98.6	78.9	81.6	100.0	71.3	75.1	97.0	45.1	52.1	95.0	42.8	49.8	98.4	74.6	77.8	95.9	67.9	71.6
0320	Concrete Reinforcing	95.3	98.2	96.7	98.0	75.4	86.8	95.0	82.4	88.7	98.6	81.7	90.1	97.9	83.8	90.9	98.9	75.1	87.1
0330	Cast-in-Place Concrete	104.2	72.3	91.1	93.9	77.1	87.0	100.6	47.4	78.7	114.4	52.4	88.9	114.4	71.5	96.7	108.0	49.4	83.9
03	CONCRETE	96.5	81.2	88.9	90.9	75.2	83.2	94.6	54.5	74.8	105.6	54.2	80.2	103.3	75.3	89.5	98.7	63.1	81.1
04	MASONRY	95.0	79.0	85.2	93.4	69.3	78.7	96.1	44.0	64.3	81.0	56.0	65.8	94.9	67.9	78.4	95.2	40.7	61.9
05	METALS	98.1	100.9	99.0	95.2	92.1	94.2	95.4	86.5	92.6	91.1	74.5	85.9	92.0	78.7	87.8	90.8	72.3	85.0
06	WOOD, PLASTICS & COMPOSITES	99.9	80.9	89.1	99.3	70.2	82.7	97.6	39.7	64.5	90.6	36.7	59.8	94.7	76.6	84.4	92.0	74.1	81.8
07	THERMAL & MOISTURE PROTECTION	97.4	87.9	93.6	96.9	72.9	87.3	94.9	60.8	81.2	93.3	53.7	77.4	93.2	73.9	85.4	93.0	58.1	79.0
08	OPENINGS	99.6	80.9	95.1	96.5	68.5	89.7	97.4	45.9	84.8	90.8	44.5	79.5	96.5	73.2	90.8	90.8	65.0	84.5
0920	Plaster & Gypsum Board	98.4	80.7	86.0	101.5	69.7	79.3	98.1	38.2	56.2	96.1	35.1	53.4	98.6	76.3	83.0	97.3	73.6	80.8
0950, 0980	Ceilings & Acoustic Treatment	94.6	80.7	85.4	88.9	69.7	76.1	93.7	38.2	56.7	96.5	35.1	55.6	96.5	76.3	83.0	97.6	73.6	81.6
0960	Flooring	117.7	58.6	100.1	119.9	68.9	104.7	117.7	37.1	93.7	89.7	55.8	79.6	91.0	67.6	84.0	90.0	40.7	75.3
0970, 0990	Wall Finishes & Painting/Coating	106.8	63.5	80.8	101.6	73.4	84.7	101.6	55.2	73.8	91.9	36.6	58.7	91.9	83.8	87.0	91.9	38.2	59.7
09	FINISHES	105.5	73.3	87.5	102.1	70.5	84.4	102.4	42.7	68.9	92.5	42.9	64.7	92.8	74.3	82.4	92.3	60.6	74.6
COVERS	DIVS. 10 - 14, 25, 28, 41, 43, 44, 46	100.0	76.3	95.2	100.0	84.8	96.9	100.0	81.4	96.2	100.0	40.7	87.9	100.0	86.3	97.2	100.0	52.0	90.2
21, 22, 23	FIRE SUPPRESSION, PLUMBING & HVAC	99.9	85.2	93.9	96.7	66.1	84.2	99.9	67.7	86.7	94.0	71.6	84.8	99.9	72.7	88.8	100.0	64.5	85.5
26, 27, 3370	ELECTRICAL, COMMUNICATIONS & UTIL.	97.5	64.2	80.0	98.8	72.9	85.2	92.8	59.5	75.3	101.0	72.5	86.0	100.2	74.1	86.4	101.6	64.4	82.0
MF2010	WEIGHTED AVERAGE	99.5	81.1	91.4	96.6	73.5	86.4	97.4	61.1	81.4	95.0	63.3	81.0	97.5	75.9	88.0	96.3	64.2	82.1

DIVISION		GEORGIA																	
		COLUMBUS			DALTON			GAINESVILLE			MACON			SAVANNAH			STATESBORO		
		318 - 319			307			305			310 - 312			313 - 314			304		
		MAT.	INST.	TOTAL	MAT.	INST.	TOTAL	MAT.	INST.	TOTAL	MAT.	INST.	TOTAL	MAT.	INST.	TOTAL	MAT.	INST.	TOTAL
015433	**CONTRACTOR EQUIPMENT**		92.5	92.5		105.3	105.3		93.6	93.6		103.0	103.0		93.4	93.4		92.8	92.8
0241, 31 - 34	**SITE & INFRASTRUCTURE, DEMOLITION**	99.0	80.0	85.6	99.0	97.0	97.6	98.4	92.8	94.5	99.9	95.0	96.5	99.9	80.8	86.5	99.7	77.2	84.0
0310	Concrete Forming & Accessories	96.9	56.3	61.8	88.1	48.2	53.6	98.1	44.6	51.8	96.3	53.5	59.3	98.1	50.4	56.8	82.8	41.8	47.3
0320	Concrete Reinforcing	95.3	83.2	89.2	98.0	77.3	87.7	98.4	81.6	90.0	96.6	82.5	89.5	96.2	75.3	85.8	97.6	43.4	70.6
0330	Cast-in-Place Concrete	100.2	55.7	81.9	111.1	46.5	84.5	120.2	52.3	92.3	98.9	49.7	78.6	110.2	49.4	85.2	114.2	61.0	92.3
03	**CONCRETE**	94.4	62.4	78.6	104.5	54.6	79.9	107.7	55.0	81.7	93.9	59.0	76.7	99.4	56.2	78.1	104.9	50.5	78.0
04	**MASONRY**	96.0	57.5	72.5	82.4	33.4	52.5	89.8	56.4	69.4	108.4	38.9	65.9	95.1	52.0	68.8	83.9	41.0	57.7
05	**METALS**	95.0	88.2	92.8	91.9	84.3	89.5	90.3	74.2	85.3	91.1	87.3	89.9	91.5	84.1	89.1	95.4	71.9	88.0
06	**WOOD, PLASTICS & COMPOSITES**	97.6	54.4	73.0	74.4	49.5	60.2	94.5	39.5	63.1	106.5	53.5	76.2	112.5	46.7	74.9	67.8	41.1	52.6
07	**THERMAL & MOISTURE PROTECTION**	94.9	64.7	82.7	95.3	51.9	77.8	93.3	56.6	78.5	93.4	62.5	80.9	96.0	58.2	80.8	94.0	50.2	76.4
08	**OPENINGS**	97.4	55.8	87.3	92.1	46.4	81.0	90.8	41.7	78.8	95.9	54.4	85.8	98.6	47.9	86.2	92.9	32.9	78.3
0920	Plaster & Gypsum Board	98.1	53.5	66.9	81.7	48.3	58.4	98.6	38.0	56.2	105.4	52.4	68.3	95.5	45.5	60.5	82.3	39.7	52.5
0950, 0980	Ceilings & Acoustic Treatment	93.7	53.5	66.9	109.7	48.3	68.8	96.5	38.0	57.5	88.5	52.4	64.5	101.9	45.5	64.3	104.8	39.7	61.4
0960	Flooring	117.7	52.2	98.2	90.2	37.1	74.4	91.0	37.1	74.9	92.3	37.6	76.0	118.2	48.1	97.3	106.5	46.5	88.6
0970, 0990	Wall Finishes & Painting/Coating	101.6	69.4	82.3	83.1	55.8	66.8	91.9	36.6	58.7	104.0	55.2	74.7	103.0	60.9	77.8	89.8	34.3	56.6
09	**FINISHES**	102.3	55.9	76.3	100.9	46.3	70.3	93.0	40.3	63.5	90.4	49.6	67.5	104.1	50.1	73.9	103.5	41.1	68.5
COVERS	**DIVS. 10 - 14, 25, 28, 41, 43, 44, 46**	100.0	83.1	96.5	100.0	23.4	84.3	100.0	41.0	87.9	100.0	81.2	96.2	100.0	51.1	90.0	100.0	37.8	87.3
21, 22, 23	**FIRE SUPPRESSION, PLUMBING & HVAC**	99.9	49.1	79.1	94.0	58.1	79.3	94.0	71.5	84.8	99.9	65.6	85.9	100.0	63.2	85.0	94.5	55.7	78.6
26, 27, 3370	**ELECTRICAL, COMMUNICATIONS & UTIL.**	92.8	72.5	82.1	111.7	70.6	90.1	100.9	72.5	86.0	91.9	62.1	76.2	97.7	59.8	77.7	101.1	56.3	77.5
MF2010	**WEIGHTED AVERAGE**	97.4	63.9	82.6	96.9	59.3	80.3	95.6	63.1	81.3	96.2	63.9	82.0	98.2	61.1	81.9	96.9	53.3	77.7

DIVISION		GEORGIA						HAWAII									IDAHO		
		VALDOSTA			WAYCROSS			HILO			HONOLULU			STATES & POSS., GUAM			BOISE		
		316			315			967			968			969			836 - 837		
		MAT.	INST.	TOTAL	MAT.	INST.	TOTAL	MAT.	INST.	TOTAL	MAT.	INST.	TOTAL	MAT.	INST.	TOTAL	MAT.	INST.	TOTAL
015433	**CONTRACTOR EQUIPMENT**		92.5	92.5		92.5	92.5		99.0	99.0		99.0	99.0		161.6	161.6		97.5	97.5
0241, 31 - 34	**SITE & INFRASTRUCTURE, DEMOLITION**	108.3	79.9	88.4	105.2	78.6	86.5	121.6	106.7	111.2	124.4	106.7	112.0	161.6	103.8	121.0	81.1	95.6	91.2
0310	Concrete Forming & Accessories	86.5	44.6	50.3	88.5	67.5	70.3	109.2	136.5	132.8	118.0	136.5	134.0	111.5	66.0	72.1	100.8	76.1	79.4
0320	Concrete Reinforcing	97.3	70.5	84.0	97.3	76.2	86.8	108.4	123.5	115.9	107.3	123.5	115.3	192.9	31.6	112.6	104.1	80.1	92.1
0330	Cast-in-Place Concrete	98.4	51.7	79.2	111.2	49.8	86.0	185.7	127.2	161.7	156.7	127.2	144.6	161.0	108.1	139.3	92.4	73.4	84.6
03	**CONCRETE**	98.4	53.6	76.2	102.4	63.9	83.4	145.4	129.6	137.6	131.9	129.6	130.8	147.6	74.6	111.5	101.0	75.9	88.6
04	**MASONRY**	102.1	51.5	71.2	102.8	40.9	65.0	118.3	129.6	125.2	111.8	129.6	122.6	170.1	45.4	94.0	121.5	61.6	84.9
05	**METALS**	94.6	82.7	90.8	93.6	79.5	89.2	108.7	109.1	108.9	130.4	109.1	123.7	140.3	77.2	120.5	100.7	76.4	93.1
06	**WOOD, PLASTICS & COMPOSITES**	85.0	38.7	58.5	86.9	75.7	80.5	101.4	138.7	122.7	111.7	138.7	127.1	112.2	69.9	88.0	89.5	79.3	83.7
07	**THERMAL & MOISTURE PROTECTION**	95.0	64.3	82.7	94.8	52.3	77.7	103.0	126.1	112.3	115.4	126.1	119.7	118.7	68.8	98.6	94.5	66.0	83.0
08	**OPENINGS**	93.0	42.1	80.6	93.0	59.8	84.9	97.7	131.9	106.0	102.7	131.9	109.8	102.3	56.2	91.0	97.3	73.6	91.5
0920	Plaster & Gypsum Board	90.3	37.2	53.2	90.7	75.3	80.0	100.7	139.7	128.0	131.4	139.7	137.2	197.1	58.1	99.9	81.6	78.6	79.5
0950, 0980	Ceilings & Acoustic Treatment	91.0	37.2	55.2	88.9	75.3	79.9	109.4	139.7	129.6	122.8	139.7	134.1	215.4	58.1	110.6	109.7	78.6	89.0
0960	Flooring	111.0	48.2	92.3	112.2	31.0	88.0	130.8	135.8	132.3	155.3	135.8	149.5	154.3	48.1	122.7	96.2	65.9	87.2
0970, 0990	Wall Finishes & Painting/Coating	101.6	54.5	73.4	101.6	41.7	65.7	100.9	145.4	127.5	105.5	145.4	129.4	106.5	37.5	65.2	98.3	41.2	64.1
09	**FINISHES**	99.7	44.7	68.9	99.3	59.3	76.9	113.4	138.5	127.4	128.3	138.5	134.0	185.6	61.1	115.8	97.9	71.7	83.2
COVERS	**DIVS. 10 - 14, 25, 28, 41, 43, 44, 46**	100.0	50.3	89.8	100.0	41.5	88.1	100.0	115.5	103.2	100.0	115.5	103.2	100.0	77.1	95.3	100.0	60.4	91.9
21, 22, 23	**FIRE SUPPRESSION, PLUMBING & HVAC**	99.9	67.5	86.7	96.0	57.1	80.1	100.0	112.9	105.3	100.1	112.9	105.3	102.0	39.3	76.3	100.0	68.3	87.0
26, 27, 3370	**ELECTRICAL, COMMUNICATIONS & UTIL.**	90.8	58.6	73.9	95.3	59.3	76.4	107.5	124.7	116.6	108.8	124.7	117.2	156.9	42.9	96.9	95.9	75.9	85.4
MF2010	**WEIGHTED AVERAGE**	97.4	60.4	81.1	97.1	60.6	81.0	109.5	122.3	115.2	113.6	122.3	117.4	130.4	60.0	99.4	99.6	73.3	88.0

DIVISION		IDAHO															ILLINOIS		
		COEUR D'ALENE			IDAHO FALLS			LEWISTON			POCATELLO			TWIN FALLS			BLOOMINGTON		
		838			834			835			832			833			617		
		MAT.	INST.	TOTAL	MAT.	INST.	TOTAL	MAT.	INST.	TOTAL	MAT.	INST.	TOTAL	MAT.	INST.	TOTAL	MAT.	INST.	TOTAL
015433	**CONTRACTOR EQUIPMENT**		92.0	92.0		97.5	97.5		92.0	92.0		97.5	97.5		97.5	97.5		100.6	100.6
0241, 31 - 34	**SITE & INFRASTRUCTURE, DEMOLITION**	78.4	91.6	87.6	79.0	94.7	90.0	84.7	92.1	89.9	81.9	95.6	91.5	88.4	93.9	92.3	97.5	96.8	97.0
0310	Concrete Forming & Accessories	115.3	57.8	65.5	95.2	40.7	48.0	120.2	58.8	67.0	101.0	75.7	79.1	101.8	29.2	39.0	86.8	118.1	113.9
0320	Concrete Reinforcing	111.9	96.9	104.4	106.1	79.1	92.6	111.9	97.1	104.5	104.5	80.2	92.4	106.4	79.3	92.9	97.8	106.2	102.0
0330	Cast-in-Place Concrete	99.9	86.8	94.5	88.0	51.1	72.8	103.8	87.2	97.0	95.0	73.2	86.0	97.5	39.1	73.5	96.7	112.5	103.2
03	**CONCRETE**	108.0	75.6	92.0	93.4	52.4	73.1	111.6	76.2	94.1	100.3	75.7	88.2	107.7	43.2	75.9	96.7	113.9	105.2
04	**MASONRY**	123.1	83.2	98.7	116.5	34.0	66.2	123.5	83.8	99.3	119.2	63.4	85.2	121.7	32.1	67.0	116.3	117.5	117.0
05	**METALS**	94.7	84.8	91.6	108.5	73.7	97.6	94.2	85.9	91.6	108.6	76.2	98.4	108.6	73.2	97.5	95.6	110.1	100.1
06	**WOOD, PLASTICS & COMPOSITES**	94.4	49.6	68.8	83.8	39.1	58.2	99.3	49.6	70.9	89.5	79.3	83.7	90.3	27.7	54.6	84.2	117.7	103.3
07	**THERMAL & MOISTURE PROTECTION**	146.7	74.6	117.7	94.1	43.9	73.9	146.9	74.5	117.8	94.6	60.7	81.0	95.3	38.1	72.3	99.2	112.1	104.4
08	**OPENINGS**	116.0	56.2	101.5	101.2	46.7	87.9	116.0	58.9	102.1	98.0	68.5	90.8	101.2	38.4	85.9	90.0	105.3	93.7
0920	Plaster & Gypsum Board	148.9	48.1	78.4	76.2	37.1	48.9	149.7	48.1	78.6	78.8	78.6	78.7	79.5	25.4	41.7	96.3	118.2	111.6
0950, 0980	Ceilings & Acoustic Treatment	143.3	48.1	79.9	107.7	37.1	60.7	143.3	48.1	79.9	115.9	78.6	91.0	110.4	25.4	53.8	94.6	118.2	110.3
0960	Flooring	136.7	47.5	110.1	96.2	44.7	80.9	139.6	89.7	124.7	99.4	65.9	89.4	100.5	44.7	83.9	98.3	111.1	102.1
0970, 0990	Wall Finishes & Painting/Coating	115.9	67.8	87.1	98.4	33.9	59.7	115.9	67.8	87.1	98.3	42.9	65.1	98.4	25.3	54.6	95.3	123.1	111.9
09	**FINISHES**	165.3	55.2	103.6	96.2	40.0	64.7	166.4	63.6	108.8	99.8	71.9	84.2	99.8	30.5	60.9	97.8	118.0	109.2
COVERS	**DIVS. 10 - 14, 25, 28, 41, 43, 44, 46**	100.0	51.1	90.0	100.0	37.9	87.3	100.0	62.9	92.4	100.0	60.3	91.9	100.0	34.1	86.5	100.0	104.2	100.9
21, 22, 23	**FIRE SUPPRESSION, PLUMBING & HVAC**	99.5	86.4	94.1	100.9	62.3	85.1	100.7	89.4	96.1	99.9	68.3	87.0	99.9	58.4	83.0	93.9	104.8	98.4
26, 27, 3370	**ELECTRICAL, COMMUNICATIONS & UTIL.**	87.4	80.8	83.9	87.4	72.1	79.3	85.3	80.8	83.0	93.3	72.1	82.1	88.9	62.6	75.0	95.0	94.9	94.9
MF2010	**WEIGHTED AVERAGE**	107.4	77.1	94.0	99.4	58.3	81.3	108.1	79.5	95.5	100.7	72.5	88.3	101.8	52.7	80.1	96.4	107.8	101.4

DIVISION		ILLINOIS																	
		CARBONDALE			CENTRALIA			CHAMPAIGN			CHICAGO			DECATUR			EAST ST. LOUIS		
		629			628			618 - 619			606 - 608			625			620 - 622		
		MAT.	INST.	TOTAL	MAT.	INST.	TOTAL	MAT.	INST.	TOTAL	MAT.	INST.	TOTAL	MAT.	INST.	TOTAL	MAT.	INST.	TOTAL
015433	**CONTRACTOR EQUIPMENT**		106.9	106.9		106.9	106.9		101.4	101.4		92.7	92.7		101.4	101.4		106.9	106.9
0241, 31 - 34	**SITE & INFRASTRUCTURE, DEMOLITION**	102.5	97.4	98.9	102.8	98.7	99.9	106.7	97.7	100.4	106.6	94.2	97.9	92.7	97.9	96.3	104.9	98.7	100.5
0310	Concrete Forming & Accessories	95.3	102.4	101.4	97.0	110.4	108.6	92.8	116.6	113.4	98.0	157.1	149.2	97.2	116.9	114.3	92.8	110.8	108.4
0320	Concrete Reinforcing	96.6	93.4	95.0	96.6	103.1	99.9	97.8	101.3	99.6	96.2	159.9	127.9	94.7	101.8	98.2	96.6	107.1	101.8
0330	Cast-in-Place Concrete	90.5	97.6	93.4	91.0	114.4	100.6	112.0	108.3	110.5	103.4	147.2	121.4	98.8	112.5	104.4	92.5	114.4	101.5
03	**CONCRETE**	87.5	99.8	93.6	88.0	111.1	99.4	108.9	110.8	109.8	100.0	152.9	126.1	97.9	112.5	105.1	88.9	112.0	100.3
04	**MASONRY**	78.2	101.1	92.2	78.2	114.9	100.6	139.9	117.0	125.9	100.7	151.0	131.4	75.0	118.2	101.4	78.5	114.9	100.7
05	**METALS**	94.8	110.0	99.6	94.8	116.8	101.7	95.6	106.3	98.9	95.4	133.1	107.2	98.2	107.0	101.0	95.9	118.6	103.1
06	**WOOD, PLASTICS & COMPOSITES**	98.1	101.6	100.1	100.7	107.9	104.8	90.9	116.0	105.2	99.5	157.6	132.6	98.3	116.0	108.4	95.4	107.9	102.5
07	**THERMAL & MOISTURE PROTECTION**	94.2	97.3	95.5	94.3	111.0	101.0	99.8	112.1	104.8	103.3	143.1	119.3	99.9	109.0	103.6	94.3	110.6	100.8
08	**OPENINGS**	85.7	106.8	90.8	85.7	112.2	92.1	90.6	111.2	95.6	99.4	158.4	113.7	96.3	111.4	99.9	85.7	113.9	92.6
0920	Plaster & Gypsum Board	100.1	101.6	101.1	101.3	108.1	106.0	98.8	116.5	111.2	97.1	159.3	140.6	103.0	116.5	112.4	98.8	108.1	105.3
0950, 0980	Ceilings & Acoustic Treatment	94.6	101.6	99.2	94.6	108.1	103.6	94.6	116.5	109.2	106.2	159.3	141.6	101.0	116.5	111.3	94.6	108.1	103.6
0960	Flooring	116.7	124.5	119.0	117.7	117.3	117.6	101.1	124.5	108.1	95.4	147.0	110.8	104.5	120.9	109.4	115.6	117.3	116.1
0970, 0990	Wall Finishes & Painting/Coating	103.4	100.1	101.4	103.4	107.3	105.8	95.3	111.3	104.9	91.9	159.6	132.5	95.3	109.6	103.8	103.4	107.3	105.8
09	**FINISHES**	101.2	105.4	103.6	101.7	111.3	107.1	99.8	118.0	110.0	99.7	156.7	131.7	101.9	117.9	110.8	100.9	111.5	106.8
COVERS	**DIVS. 10 - 14, 25, 28, 41, 43, 44, 46**	100.0	96.7	99.3	100.0	100.0	100.0	100.0	103.9	100.8	100.0	124.1	104.9	100.0	104.2	100.9	100.0	100.0	100.0
21, 22, 23	**FIRE SUPPRESSION, PLUMBING & HVAC**	93.8	95.0	94.3	93.8	91.2	92.7	93.9	103.6	97.8	99.9	133.3	113.6	99.9	100.8	100.3	99.9	95.3	98.0
26, 27, 3370	**ELECTRICAL, COMMUNICATIONS & UTIL.**	93.7	99.6	96.8	95.2	103.3	99.5	98.5	94.0	96.1	97.8	130.8	115.2	98.0	88.5	93.0	94.7	103.3	99.2
MF2010	**WEIGHTED AVERAGE**	93.1	100.5	96.4	93.4	105.4	98.7	99.8	106.8	102.9	99.3	138.3	116.5	97.6	105.8	101.2	95.0	106.6	100.1

DIVISION		ILLINOIS																	
		EFFINGHAM			GALESBURG			JOLIET			KANKAKEE			LA SALLE			NORTH SUBURBAN		
		624			614			604			609			613			600 - 603		
		MAT.	INST.	TOTAL	MAT.	INST.	TOTAL	MAT.	INST.	TOTAL	MAT.	INST.	TOTAL	MAT.	INST.	TOTAL	MAT.	INST.	TOTAL
015433	**CONTRACTOR EQUIPMENT**		101.4	101.4		100.6	100.6		90.8	90.8		90.8	90.8		100.6	100.6		90.8	90.8
0241, 31 - 34	**SITE & INFRASTRUCTURE, DEMOLITION**	98.7	97.2	97.7	100.0	96.7	97.7	106.4	93.1	97.1	100.0	92.4	94.7	99.3	97.6	98.1	105.6	92.3	96.3
0310	Concrete Forming & Accessories	101.8	109.6	108.6	92.7	113.7	110.9	100.0	152.1	145.1	93.9	133.1	127.8	105.9	119.1	117.3	99.4	146.3	140.0
0320	Concrete Reinforcing	97.8	98.6	98.2	97.3	105.8	101.5	96.2	144.3	120.2	97.1	138.6	117.8	97.4	131.7	114.5	96.2	153.2	124.6
0330	Cast-in-Place Concrete	98.4	103.7	100.6	99.5	103.6	101.2	103.4	140.2	118.5	96.4	125.1	108.2	99.4	112.8	104.9	103.4	137.7	117.5
03	**CONCRETE**	98.8	105.6	102.1	99.5	108.8	104.1	100.1	145.2	122.4	94.4	130.5	112.2	100.3	119.3	109.7	100.0	143.4	121.5
04	**MASONRY**	82.2	102.9	94.8	116.5	112.0	113.8	103.6	144.1	128.3	100.2	130.6	118.8	116.5	115.3	115.7	100.7	140.3	124.8
05	**METALS**	95.5	104.4	98.3	95.6	108.8	99.7	93.3	124.6	103.1	93.3	119.8	101.6	95.6	124.3	104.6	94.4	127.1	104.7
06	**WOOD, PLASTICS & COMPOSITES**	100.8	111.3	106.8	90.8	113.0	103.4	101.0	153.3	130.8	94.2	132.0	115.8	105.7	118.2	112.8	99.5	147.9	127.1
07	**THERMAL & MOISTURE PROTECTION**	99.5	104.3	101.4	99.4	104.9	101.6	103.1	139.0	117.6	102.4	129.8	113.4	99.5	113.1	105.0	103.6	136.3	116.7
08	**OPENINGS**	90.4	108.0	94.7	90.0	107.2	94.2	97.4	151.9	110.6	89.3	137.2	100.9	90.0	120.8	97.5	97.4	149.7	110.1
0920	Plaster & Gypsum Board	103.0	111.6	109.0	98.8	113.3	109.0	93.8	155.0	136.6	91.7	133.0	120.6	105.9	118.7	114.9	97.1	149.4	133.7
0950, 0980	Ceilings & Acoustic Treatment	94.6	111.6	105.9	94.6	113.3	107.1	106.2	155.0	138.7	106.2	133.0	124.1	94.6	118.7	110.7	106.2	149.4	135.0
0960	Flooring	105.6	124.5	111.2	101.0	111.1	104.0	94.9	136.7	107.4	92.1	136.7	105.4	107.4	115.1	109.7	95.4	136.7	107.7
0970, 0990	Wall Finishes & Painting/Coating	95.3	105.0	101.1	95.3	93.1	93.9	90.1	149.8	125.9	90.1	123.1	109.9	95.3	123.1	111.9	91.9	149.8	126.6
09	**FINISHES**	100.9	113.0	107.7	99.1	111.9	106.3	99.1	150.4	127.8	97.6	131.1	116.3	101.8	118.5	111.2	99.7	146.2	125.8
COVERS	**DIVS. 10 - 14, 25, 28, 41, 43, 44, 46**	100.0	69.7	93.8	100.0	97.1	99.4	100.0	122.2	104.5	100.0	110.4	102.1	100.0	97.2	99.4	100.0	112.6	102.6
21, 22, 23	**FIRE SUPPRESSION, PLUMBING & HVAC**	93.9	99.3	96.1	93.9	100.8	96.7	100.0	130.9	112.6	94.0	125.0	106.6	93.9	122.5	105.6	99.9	121.0	108.5
26, 27, 3370	**ELECTRICAL, COMMUNICATIONS & UTIL.**	95.8	99.6	97.8	96.0	86.0	90.8	97.0	135.3	117.2	91.5	134.4	114.1	93.0	132.2	113.6	96.9	130.8	114.7
MF2010	**WEIGHTED AVERAGE**	95.4	102.3	98.4	97.0	103.2	99.7	98.8	134.5	114.5	94.8	125.4	108.3	97.1	119.0	106.8	98.8	130.3	112.7

DIVISION		ILLINOIS																	
		PEORIA			QUINCY			ROCK ISLAND			ROCKFORD			SOUTH SUBURBAN			SPRINGFIELD		
		615 - 616			623			612			610 - 611			605			626 - 627		
		MAT.	INST.	TOTAL	MAT.	INST.	TOTAL	MAT.	INST.	TOTAL	MAT.	INST.	TOTAL	MAT.	INST.	TOTAL	MAT.	INST.	TOTAL
015433	**CONTRACTOR EQUIPMENT**		100.6	100.6		101.4	101.4		100.6	100.6		100.6	100.6		90.8	90.8		101.4	101.4
0241, 31 - 34	**SITE & INFRASTRUCTURE, DEMOLITION**	100.3	96.8	97.8	97.5	97.3	97.4	98.1	96.2	96.7	99.6	98.0	98.5	105.6	92.3	96.3	98.9	97.9	98.2
0310	Concrete Forming & Accessories	95.6	114.9	112.3	99.6	112.6	110.9	94.3	98.5	97.9	99.8	130.1	126.0	99.4	146.3	140.0	98.0	117.3	114.7
0320	Concrete Reinforcing	94.7	105.9	100.3	97.4	98.5	98.0	97.3	98.5	97.9	89.6	138.2	113.8	96.2	153.2	124.6	94.7	101.7	98.2
0330	Cast-in-Place Concrete	96.5	113.6	103.5	98.6	96.2	97.6	97.4	94.6	96.2	98.8	124.1	109.2	103.4	137.7	117.5	88.8	107.6	96.6
03	**CONCRETE**	96.7	112.8	104.6	98.4	104.3	101.3	97.5	97.6	97.6	97.2	129.2	113.0	100.0	143.4	121.5	93.2	110.9	101.9
04	**MASONRY**	116.9	112.7	114.3	102.1	99.6	100.6	116.3	94.2	102.8	91.4	135.3	118.2	100.7	140.3	124.8	79.6	119.3	103.8
05	**METALS**	98.2	109.8	101.9	95.5	104.1	98.2	95.6	105.1	98.6	98.2	127.3	107.4	94.4	127.1	104.7	100.7	106.9	102.6
06	**WOOD, PLASTICS & COMPOSITES**	98.4	113.0	106.7	98.2	116.0	108.4	92.4	97.5	95.3	98.3	128.7	115.7	99.5	147.9	127.1	99.2	116.0	108.8
07	**THERMAL & MOISTURE PROTECTION**	100.0	110.3	104.2	99.5	101.8	100.4	99.3	95.7	97.9	102.4	127.6	112.5	103.6	136.5	116.8	100.7	111.9	105.2
08	**OPENINGS**	96.4	111.1	100.0	91.3	110.7	96.0	90.0	96.9	91.7	96.4	129.9	104.6	97.4	149.7	110.1	97.3	111.4	100.7
0920	Plaster & Gypsum Board	103.0	113.3	110.2	101.3	116.5	111.9	98.8	97.4	97.8	103.0	129.5	121.6	97.1	149.4	133.7	101.2	116.5	111.9
0950, 0980	Ceilings & Acoustic Treatment	101.0	113.3	109.2	94.6	116.5	109.2	94.6	97.4	96.4	101.0	129.5	120.0	106.2	149.4	135.0	107.4	116.5	113.4
0960	Flooring	104.5	111.1	106.4	104.5	102.9	104.0	102.2	99.8	101.5	104.5	115.1	107.6	95.4	136.7	107.7	104.5	107.9	105.5
0970, 0990	Wall Finishes & Painting/Coating	95.3	123.1	111.9	95.3	107.2	102.4	95.3	96.1	95.8	95.3	130.2	116.2	91.9	149.8	126.6	95.3	107.2	102.4
09	**FINISHES**	102.0	115.4	109.5	100.3	111.9	106.8	99.3	98.4	98.8	102.0	127.8	116.4	99.7	146.2	125.8	102.8	115.3	109.8
COVERS	**DIVS. 10 - 14, 25, 28, 41, 43, 44, 46**	100.0	103.9	100.8	100.0	72.6	94.4	100.0	97.5	99.5	100.0	90.8	98.1	100.0	112.6	102.6	100.0	104.6	100.9
21, 22, 23	**FIRE SUPPRESSION, PLUMBING & HVAC**	99.9	103.8	101.5	93.9	96.6	95.0	93.9	100.6	96.6	99.9	113.6	105.5	99.9	121.0	108.5	99.8	103.9	101.5
26, 27, 3370	**ELECTRICAL, COMMUNICATIONS & UTIL.**	97.1	95.3	96.2	93.0	80.1	86.2	87.7	96.6	92.3	97.5	129.3	114.2	96.9	130.8	114.7	101.1	90.7	95.6
MF2010	**WEIGHTED AVERAGE**	99.7	106.7	102.8	96.1	98.6	97.2	96.0	98.3	97.0	98.6	122.3	109.0	98.8	130.3	112.7	98.3	106.5	101.9

	DIVISION	INDIANA																	
		ANDERSON			BLOOMINGTON			COLUMBUS			EVANSVILLE			FORT WAYNE			GARY		
		460			474			472			476 - 477			467 - 468			463 - 464		
		MAT.	INST.	TOTAL	MAT.	INST.	TOTAL	MAT.	INST.	TOTAL	MAT.	INST.	TOTAL	MAT.	INST.	TOTAL	MAT.	INST.	TOTAL
015433	**CONTRACTOR EQUIPMENT**		97.2	97.2		87.5	87.5		87.5	87.5		116.7	116.7		97.2	97.2		97.2	97.2
0241, 31 - 34	**SITE & INFRASTRUCTURE, DEMOLITION**	93.4	95.4	94.8	82.4	94.7	91.0	79.1	94.5	89.9	87.6	124.3	113.3	94.0	95.1	94.8	94.0	98.7	97.3
0310	Concrete Forming & Accessories	94.5	80.2	82.1	100.2	81.0	83.6	94.3	78.0	80.2	93.9	79.1	81.1	92.6	73.5	76.1	94.6	108.4	106.6
0320	Concrete Reinforcing	100.6	81.7	91.2	90.6	81.4	86.1	91.1	81.4	86.2	99.3	75.7	87.5	100.6	76.8	88.8	100.6	101.9	101.3
0330	Cast-in-Place Concrete	106.3	78.7	95.0	101.3	79.6	92.3	100.8	73.4	89.5	96.8	86.3	92.4	113.0	74.4	97.1	111.1	106.6	109.3
03	**CONCRETE**	98.4	80.5	89.5	101.7	80.4	91.2	101.1	76.9	89.1	102.1	81.2	91.7	101.4	75.1	88.4	100.7	106.5	103.5
04	**MASONRY**	85.0	80.5	82.2	90.0	78.0	82.6	89.8	76.8	81.8	85.8	81.8	83.3	88.6	78.0	82.1	86.2	111.5	101.7
05	**METALS**	90.8	88.7	90.1	95.5	77.6	89.8	95.5	76.9	89.7	88.8	83.7	87.2	90.8	86.1	89.3	90.8	103.8	94.9
06	**WOOD, PLASTICS & COMPOSITES**	111.6	80.3	93.7	120.0	80.7	97.5	114.8	77.4	93.4	98.7	77.6	86.7	111.3	72.7	89.3	109.2	106.4	107.6
07	**THERMAL & MOISTURE PROTECTION**	99.8	78.9	91.4	94.1	81.4	89.0	93.7	80.6	88.5	98.1	83.0	92.0	99.6	81.2	92.2	98.5	107.8	102.2
08	**OPENINGS**	100.5	80.8	95.7	107.1	81.0	100.7	102.3	79.2	96.7	99.2	77.7	93.9	100.5	73.1	93.8	100.5	109.9	102.8
0920	Plaster & Gypsum Board	94.8	80.1	84.5	101.0	80.7	86.8	97.2	77.3	83.3	95.8	76.4	82.2	94.0	72.3	78.8	88.2	107.0	101.3
0950, 0980	Ceilings & Acoustic Treatment	89.0	80.1	83.1	83.1	80.7	81.5	83.1	77.3	79.3	89.1	76.4	80.6	89.0	72.3	77.9	89.0	107.0	101.0
0960	Flooring	96.8	80.4	91.9	104.7	77.7	96.6	100.1	77.7	93.4	99.2	83.6	94.5	96.8	77.7	91.1	96.8	120.2	103.8
0970, 0990	Wall Finishes & Painting/Coating	93.1	69.4	78.9	88.8	82.6	85.1	88.8	82.6	85.1	93.9	85.0	88.6	93.1	74.5	81.9	93.1	121.5	110.1
09	**FINISHES**	93.8	79.2	85.6	97.3	80.6	87.9	95.5	78.4	85.9	96.4	80.4	87.5	93.5	74.4	82.8	92.9	112.0	103.6
COVERS	**DIVS. 10 - 14, 25, 28, 41, 43, 44, 46**	100.0	92.2	98.4	100.0	84.5	96.8	100.0	83.7	96.7	100.0	88.6	97.7	100.0	84.3	96.8	100.0	99.7	99.9
21, 22, 23	**FIRE SUPPRESSION, PLUMBING & HVAC**	99.9	76.0	90.2	99.8	79.4	91.4	93.7	77.3	87.0	99.9	79.7	91.6	99.9	73.2	89.0	99.9	103.9	101.6
26, 27, 3370	**ELECTRICAL, COMMUNICATIONS & UTIL.**	86.6	84.7	85.6	99.4	84.0	91.3	98.5	84.6	91.2	95.4	84.0	89.4	87.4	72.7	79.7	99.1	112.1	106.0
MF2010	**WEIGHTED AVERAGE**	95.7	82.3	89.8	98.8	81.6	91.2	96.4	80.1	89.3	96.5	85.2	91.5	96.3	77.7	88.1	97.2	107.0	101.5

	DIVISION	INDIANA																	
		INDIANAPOLIS			KOKOMO			LAFAYETTE			LAWRENCEBURG			MUNCIE			NEW ALBANY		
		461 - 462			469			479			470			473			471		
		MAT.	INST.	TOTAL	MAT.	INST.	TOTAL	MAT.	INST.	TOTAL	MAT.	INST.	TOTAL	MAT.	INST.	TOTAL	MAT.	INST.	TOTAL
015433	**CONTRACTOR EQUIPMENT**		92.9	92.9		97.2	97.2		87.5	87.5		105.5	105.5		95.9	95.9		95.4	95.4
0241, 31 - 34	**SITE & INFRASTRUCTURE, DEMOLITION**	93.7	98.7	97.2	90.0	95.3	93.8	80.0	94.5	90.2	77.7	111.2	101.2	82.2	94.8	91.1	74.9	97.4	90.7
0310	Concrete Forming & Accessories	95.1	84.1	85.5	98.0	75.9	78.8	92.1	80.7	82.2	90.9	74.6	76.8	92.2	80.0	81.6	88.7	72.5	74.6
0320	Concrete Reinforcing	100.0	81.7	90.9	91.1	81.5	86.3	90.6	81.6	86.2	89.8	77.8	83.8	100.3	81.6	91.0	91.3	75.5	83.5
0330	Cast-in-Place Concrete	102.3	85.6	95.4	105.3	82.9	96.1	101.4	81.9	93.4	94.8	72.4	85.6	106.4	78.7	95.0	97.9	71.4	87.0
03	**CONCRETE**	100.1	83.9	92.1	95.1	79.9	87.6	101.3	81.0	91.3	94.5	75.1	84.9	100.9	80.4	90.8	99.6	73.1	86.5
04	**MASONRY**	91.4	82.0	85.7	84.6	78.4	80.8	94.9	80.3	86.0	74.9	76.6	76.0	91.9	80.4	84.9	81.9	69.9	74.6
05	**METALS**	92.8	79.9	88.8	87.4	88.4	87.7	93.9	77.8	88.8	90.6	85.5	89.0	97.2	88.9	94.6	92.5	80.8	88.9
06	**WOOD, PLASTICS & COMPOSITES**	108.1	84.0	94.3	114.7	74.2	91.6	112.1	80.4	94.0	97.3	73.8	83.9	114.4	80.1	94.8	99.0	72.9	84.1
07	**THERMAL & MOISTURE PROTECTION**	97.7	83.4	92.0	99.4	78.6	91.0	93.7	81.9	89.0	98.1	78.3	90.1	96.8	79.2	89.7	85.7	69.2	79.1
08	**OPENINGS**	109.4	83.0	103.0	94.5	77.5	90.3	100.6	80.8	95.8	101.0	75.3	94.8	101.2	80.7	96.2	98.7	74.9	92.9
0920	Plaster & Gypsum Board	92.9	83.7	86.5	100.2	73.8	81.7	94.3	80.4	84.6	73.3	73.5	73.5	95.8	80.1	84.8	93.7	72.4	78.8
0950, 0980	Ceilings & Acoustic Treatment	93.6	83.7	87.0	89.0	73.8	78.9	78.6	80.4	79.8	93.7	73.5	80.2	84.0	80.1	81.4	89.1	72.4	78.0
0960	Flooring	96.6	86.6	93.6	100.7	87.9	96.9	99.1	88.1	95.8	73.6	86.6	77.5	99.2	80.4	93.6	97.1	65.1	87.6
0970, 0990	Wall Finishes & Painting/Coating	93.1	83.7	87.5	93.1	74.9	82.2	88.8	97.2	93.9	89.7	73.7	80.1	88.8	69.4	77.2	93.9	86.4	89.4
09	**FINISHES**	94.8	84.6	89.1	95.4	77.7	85.5	93.8	83.8	88.2	86.9	77.1	81.4	94.9	79.0	86.0	95.8	72.9	82.9
COVERS	**DIVS. 10 - 14, 25, 28, 41, 43, 44, 46**	100.0	93.6	98.7	100.0	84.0	96.7	100.0	91.7	98.3	100.0	44.6	88.7	100.0	91.6	98.3	100.0	44.0	88.6
21, 22, 23	**FIRE SUPPRESSION, PLUMBING & HVAC**	99.9	82.0	92.6	93.9	79.3	87.9	93.7	78.2	87.4	94.5	74.3	86.2	99.8	76.0	90.0	93.9	75.6	86.4
26, 27, 3370	**ELECTRICAL, COMMUNICATIONS & UTIL.**	101.0	84.6	92.4	91.4	77.7	84.2	98.0	81.8	89.5	93.2	74.2	83.2	90.9	78.8	84.5	93.8	74.9	83.8
MF2010	**WEIGHTED AVERAGE**	98.9	84.7	92.6	93.3	81.1	87.9	96.1	82.0	89.9	93.0	78.4	86.6	97.5	81.4	90.4	94.1	75.4	85.9

	DIVISION	INDIANA									IOWA								
		SOUTH BEND			TERRE HAUTE			WASHINGTON			BURLINGTON			CARROLL			CEDAR RAPIDS		
		465 - 466			478			475			526			514			522 - 524		
		MAT.	INST.	TOTAL	MAT.	INST.	TOTAL	MAT.	INST.	TOTAL	MAT.	INST.	TOTAL	MAT.	INST.	TOTAL	MAT.	INST.	TOTAL
015433	**CONTRACTOR EQUIPMENT**		105.5	105.5		116.7	116.7		116.7	116.7		100.1	100.1		100.1	100.1		96.7	96.7
0241, 31 - 34	**SITE & INFRASTRUCTURE, DEMOLITION**	96.2	95.6	95.8	89.4	124.6	114.1	89.1	122.4	112.5	100.3	96.6	97.7	88.1	96.7	94.2	102.1	96.1	97.9
0310	Concrete Forming & Accessories	100.7	77.5	80.6	94.7	79.2	81.3	95.5	77.9	80.3	98.7	75.9	79.0	87.3	51.9	56.7	104.3	80.3	83.5
0320	Concrete Reinforcing	100.6	76.9	88.8	99.3	81.8	90.6	91.8	51.1	71.5	98.9	81.3	90.2	99.7	80.7	90.3	99.7	83.7	91.8
0330	Cast-in-Place Concrete	103.5	78.5	93.2	93.7	87.7	91.2	101.9	86.9	95.8	108.0	56.4	86.8	108.0	61.2	88.8	108.3	79.5	96.5
03	**CONCRETE**	93.4	79.1	86.3	104.6	82.9	93.9	110.0	76.1	93.3	102.5	71.0	87.0	101.4	61.9	81.9	102.7	81.3	92.2
04	**MASONRY**	84.7	78.8	81.1	93.1	80.9	85.7	85.8	81.9	83.4	98.9	67.3	79.6	101.0	73.8	84.4	105.0	79.1	89.2
05	**METALS**	90.8	98.6	93.3	89.5	87.1	88.7	84.2	69.9	79.8	88.8	90.9	89.5	88.9	90.5	89.4	91.1	92.8	91.6
06	**WOOD, PLASTICS & COMPOSITES**	108.9	77.2	90.8	100.6	78.0	87.7	101.0	77.5	87.6	95.6	76.5	84.7	83.3	46.9	62.5	101.7	79.3	88.9
07	**THERMAL & MOISTURE PROTECTION**	91.9	82.5	88.1	98.2	81.4	91.4	98.3	83.1	92.2	104.5	76.8	93.3	104.8	70.4	90.9	105.3	79.1	94.8
08	**OPENINGS**	93.0	76.9	89.1	99.8	79.5	94.8	96.1	68.0	89.2	91.2	70.1	86.0	95.7	55.3	85.9	96.7	80.5	92.7
0920	Plaster & Gypsum Board	91.1	76.9	81.1	95.8	76.7	82.4	95.3	76.3	82.0	103.9	75.8	84.3	99.3	45.3	61.5	108.6	78.9	87.8
0950, 0980	Ceilings & Acoustic Treatment	89.9	76.9	81.2	89.1	76.7	80.8	82.7	76.3	78.4	116.3	75.8	89.3	116.3	45.3	69.0	119.9	78.9	92.6
0960	Flooring	94.9	82.5	91.2	99.2	86.6	95.4	100.1	79.2	93.9	114.7	39.2	92.2	108.5	34.4	86.4	132.4	81.1	117.1
0970, 0990	Wall Finishes & Painting/Coating	92.6	86.4	88.9	93.9	84.4	88.2	93.9	85.0	88.6	97.3	82.0	88.1	97.3	77.6	85.5	99.2	71.7	82.7
09	**FINISHES**	92.9	79.4	85.3	96.4	80.9	87.7	95.5	79.6	86.6	110.8	69.8	87.9	106.5	49.6	74.6	117.3	79.5	96.1
COVERS	**DIVS. 10 - 14, 25, 28, 41, 43, 44, 46**	100.0	68.5	93.6	100.0	94.1	98.8	100.0	88.5	97.7	100.0	84.2	96.8	100.0	67.4	93.3	100.0	85.2	97.0
21, 22, 23	**FIRE SUPPRESSION, PLUMBING & HVAC**	99.9	76.0	90.1	99.9	78.5	91.2	93.9	78.9	87.8	94.3	76.6	87.0	94.3	69.8	84.3	100.3	80.5	92.2
26, 27, 3370	**ELECTRICAL, COMMUNICATIONS & UTIL.**	100.6	82.6	91.1	93.5	84.0	88.5	94.1	83.9	88.7	101.4	70.5	85.1	102.1	80.9	90.9	98.8	79.4	88.6
MF2010	**WEIGHTED AVERAGE**	95.5	81.8	89.5	97.2	85.7	92.1	94.7	82.4	89.3	97.2	76.2	88.0	97.0	71.4	85.7	100.3	82.8	92.6

	DIVISION	IOWA																	
		COUNCIL BLUFFS			CRESTON			DAVENPORT			DECORAH			DES MOINES			DUBUQUE		
		515			508			527 - 528			521			500 - 503,509			520		
		MAT.	INST.	TOTAL	MAT.	INST.	TOTAL	MAT.	INST.	TOTAL	MAT.	INST.	TOTAL	MAT.	INST.	TOTAL	MAT.	INST.	TOTAL
015433	CONTRACTOR EQUIPMENT		96.2	96.2		100.1	100.1		100.1	100.1		100.1	100.1		101.7	101.7		95.6	95.6
0241, 31 - 34	SITE & INFRASTRUCTURE, DEMOLITION	108.0	92.3	97.0	88.1	96.3	93.9	100.4	99.0	99.5	98.9	96.2	97.0	89.6	99.8	96.8	99.9	93.3	95.3
0310	Concrete Forming & Accessories	86.8	71.0	73.1	87.9	67.6	70.4	103.6	86.7	88.9	96.4	45.7	52.6	106.8	77.1	81.1	88.0	69.7	72.2
0320	Concrete Reinforcing	101.7	79.1	90.4	99.0	80.8	89.9	99.7	94.5	97.1	98.9	80.3	89.6	99.7	81.0	90.4	98.3	83.6	91.0
0330	Cast-in-Place Concrete	112.6	75.9	97.5	107.7	65.2	90.2	104.3	86.0	96.7	105.1	57.9	85.7	90.8	89.5	90.3	106.0	86.7	98.1
03	CONCRETE	105.1	75.1	90.2	101.2	70.2	85.9	100.7	88.5	94.7	100.5	57.9	79.5	93.4	82.9	88.2	99.5	79.1	89.4
04	MASONRY	106.3	74.7	87.0	101.0	80.8	88.7	101.9	84.5	91.3	120.0	70.1	89.5	95.7	81.9	87.2	105.9	75.5	87.3
05	METALS	96.5	90.4	94.6	88.8	90.7	89.4	91.1	100.8	94.2	89.0	88.7	88.9	91.3	93.0	91.8	89.8	92.3	90.6
06	WOOD, PLASTICS & COMPOSITES	82.2	70.2	75.4	83.7	64.4	72.7	101.8	85.6	92.5	92.9	38.7	61.9	101.6	74.8	86.3	83.7	66.1	73.6
07	THERMAL & MOISTURE PROTECTION	104.8	68.7	90.3	105.5	77.7	94.3	104.9	84.6	96.7	104.7	54.0	84.3	104.9	78.8	94.4	105.1	71.0	91.4
08	OPENINGS	95.7	76.1	90.9	106.8	65.2	96.7	96.7	87.7	94.5	94.2	50.8	83.7	96.7	80.9	92.9	95.7	75.0	90.7
0920	Plaster & Gypsum Board	99.3	69.5	78.5	99.3	63.3	74.1	108.6	85.1	92.2	102.6	36.8	56.6	96.4	74.1	80.8	99.3	65.2	75.5
0950, 0980	Ceilings & Acoustic Treatment	116.3	69.5	85.1	116.3	63.3	81.0	119.9	85.1	96.7	116.3	36.8	63.3	119.9	74.1	89.4	116.3	65.2	82.3
0960	Flooring	107.3	83.6	100.3	108.8	34.4	86.6	117.2	91.6	109.6	114.4	48.9	94.9	116.4	83.7	106.6	122.5	78.9	109.5
0970, 0990	Wall Finishes & Painting/Coating	94.1	66.9	77.8	97.3	77.6	85.5	97.3	87.8	91.6	97.3	34.4	59.6	97.3	81.7	88.0	98.2	66.6	79.3
09	FINISHES	107.7	72.5	88.0	106.6	61.7	81.5	112.9	87.5	98.7	110.5	43.4	72.9	109.6	78.4	92.1	112.2	70.3	88.7
COVERS	DIVS. 10 - 14, 25, 28, 41, 43, 44, 46	100.0	85.1	97.0	100.0	71.6	94.2	100.0	87.1	97.4	100.0	79.0	95.7	100.0	85.1	97.0	100.0	83.3	96.6
21, 22, 23	FIRE SUPPRESSION, PLUMBING & HVAC	100.3	76.9	90.7	94.3	79.3	88.1	100.3	88.0	95.3	94.3	72.0	85.2	100.1	79.9	91.8	100.3	75.1	90.0
26, 27, 3370	ELECTRICAL, COMMUNICATIONS & UTIL.	104.5	82.1	92.7	96.5	80.9	88.3	95.9	85.0	90.2	98.8	46.2	71.1	109.5	80.9	94.4	102.9	77.6	89.6
MF2010	WEIGHTED AVERAGE	101.2	79.1	91.5	97.6	77.7	88.8	99.3	89.2	94.8	98.1	64.8	83.4	98.9	83.5	92.1	99.5	78.6	90.3

	DIVISION	IOWA																	
		FORT DODGE			MASON CITY			OTTUMWA			SHENANDOAH			SIBLEY			SIOUX CITY		
		505			504			525			516			512			510 - 511		
		MAT.	INST.	TOTAL	MAT.	INST.	TOTAL	MAT.	INST.	TOTAL	MAT.	INST.	TOTAL	MAT.	INST.	TOTAL	MAT.	INST.	TOTAL
015433	CONTRACTOR EQUIPMENT		100.1	100.1		100.1	100.1		95.6	95.6		96.2	96.2		100.1	100.1		100.1	100.1
0241, 31 - 34	SITE & INFRASTRUCTURE, DEMOLITION	96.0	95.2	95.4	96.0	96.2	96.2	100.4	91.5	94.1	106.1	91.4	95.8	111.3	95.0	99.9	112.8	95.8	100.9
0310	Concrete Forming & Accessories	88.4	46.8	52.4	92.9	46.5	52.7	94.6	58.4	63.3	88.3	58.4	62.4	88.7	39.1	45.7	104.3	64.7	70.1
0320	Concrete Reinforcing	99.0	63.6	81.4	98.9	80.2	89.6	98.9	81.5	90.3	101.7	67.8	84.8	101.7	63.5	82.7	99.7	63.9	81.9
0330	Cast-in-Place Concrete	101.0	45.6	78.3	101.0	58.5	83.6	108.7	55.7	86.9	108.8	61.5	89.4	106.7	46.1	81.8	107.3	56.7	86.5
03	CONCRETE	96.8	51.0	74.2	97.0	58.4	78.0	102.1	63.0	82.8	102.5	62.4	82.7	101.5	47.7	74.9	102.1	62.9	82.7
04	MASONRY	99.6	39.4	62.9	113.2	68.9	86.1	102.3	58.5	75.6	105.9	74.7	86.9	124.6	39.7	72.8	98.9	55.5	72.4
05	METALS	88.9	80.5	86.2	88.9	89.0	88.9	88.8	90.4	89.3	95.5	84.1	92.0	89.0	79.3	86.0	91.1	82.0	88.3
06	WOOD, PLASTICS & COMPOSITES	84.2	47.2	63.0	88.3	38.7	60.0	90.4	59.8	72.9	83.7	55.1	67.4	84.5	37.4	57.6	101.7	64.8	80.6
07	THERMAL & MOISTURE PROTECTION	104.9	58.7	86.3	104.4	64.0	88.2	105.2	63.5	88.4	104.2	67.0	89.2	104.5	49.8	82.5	104.9	64.2	88.5
08	OPENINGS	100.3	49.7	88.0	92.0	50.7	82.0	95.7	65.3	88.3	86.7	56.1	79.2	92.3	44.4	80.6	96.7	61.1	88.0
0920	Plaster & Gypsum Board	99.3	45.6	61.7	99.3	36.8	55.6	100.5	58.8	71.3	99.3	54.0	67.6	99.3	35.5	54.7	108.6	63.7	77.2
0950, 0980	Ceilings & Acoustic Treatment	116.3	45.6	69.2	116.3	36.8	63.3	116.3	58.8	78.0	116.3	54.0	74.8	116.3	35.5	62.5	119.9	63.7	82.5
0960	Flooring	110.5	48.9	92.2	112.8	48.9	93.8	125.8	52.0	103.8	108.1	35.5	86.5	109.4	35.2	87.3	117.4	55.6	99.0
0970, 0990	Wall Finishes & Painting/Coating	97.3	61.4	75.8	97.3	31.4	57.8	98.2	80.5	87.6	94.1	66.9	77.8	97.3	61.4	75.8	97.8	61.4	76.0
09	FINISHES	108.7	47.8	74.6	109.4	43.3	72.3	113.5	58.7	82.8	107.8	54.0	77.7	110.5	39.4	70.7	114.9	62.6	85.6
COVERS	DIVS. 10 - 14, 25, 28, 41, 43, 44, 46	100.0	75.1	94.9	100.0	79.4	95.8	100.0	77.3	95.4	100.0	68.5	93.6	100.0	74.1	94.7	100.0	81.3	96.2
21, 22, 23	FIRE SUPPRESSION, PLUMBING & HVAC	94.3	65.6	82.5	94.3	73.0	85.6	94.3	71.3	84.9	94.3	70.8	84.7	94.3	61.6	80.9	100.3	68.1	87.1
26, 27, 3370	ELECTRICAL, COMMUNICATIONS & UTIL.	104.0	43.6	72.2	102.9	58.0	79.3	101.2	66.6	83.0	98.8	80.4	89.1	98.8	43.2	69.5	98.8	72.9	85.2
MF2010	WEIGHTED AVERAGE	97.5	58.9	80.5	97.3	66.9	83.9	98.0	69.7	85.5	97.8	71.3	86.1	98.5	55.8	79.7	100.0	69.7	86.7

	DIVISION	IOWA						KANSAS											
		SPENCER			WATERLOO			BELLEVILLE			COLBY			DODGE CITY			EMPORIA		
		513			506 - 507			669			677			678			668		
		MAT.	INST.	TOTAL	MAT.	INST.	TOTAL	MAT.	INST.	TOTAL	MAT.	INST.	TOTAL	MAT.	INST.	TOTAL	MAT.	INST.	TOTAL
015433	CONTRACTOR EQUIPMENT		100.1	100.1		100.1	100.1		102.3	102.3		102.3	102.3		102.3	102.3		100.6	100.6
0241, 31 - 34	SITE & INFRASTRUCTURE, DEMOLITION	111.4	95.0	99.9	100.9	96.2	97.6	114.4	94.0	100.1	111.8	94.0	99.3	114.3	93.6	99.8	105.7	91.7	95.9
0310	Concrete Forming & Accessories	94.5	39.1	46.5	104.7	48.9	56.4	95.4	55.6	61.0	101.6	60.0	65.6	95.0	60.0	64.7	87.3	67.3	70.0
0320	Concrete Reinforcing	101.7	63.5	82.7	99.7	83.4	91.6	102.5	54.8	78.7	105.1	54.8	80.0	102.5	54.9	78.8	101.1	55.4	78.3
0330	Cast-in-Place Concrete	106.7	46.1	81.8	108.3	55.9	86.7	117.0	57.0	92.3	112.5	56.9	89.7	114.5	56.6	90.7	113.1	47.7	86.2
03	CONCRETE	101.9	47.7	75.1	102.7	59.3	81.3	116.7	57.3	87.3	114.0	59.2	86.9	115.2	59.0	87.4	109.2	59.4	84.6
04	MASONRY	124.6	39.7	72.8	100.3	74.7	84.7	100.1	57.6	74.2	101.5	57.6	74.7	109.0	60.0	79.1	105.9	67.7	82.6
05	METALS	89.0	79.3	86.0	91.1	91.8	91.3	93.8	77.6	88.7	94.1	77.6	89.0	95.6	76.6	89.7	93.5	78.9	88.9
06	WOOD, PLASTICS & COMPOSITES	90.3	37.4	60.1	102.3	38.7	66.0	92.5	54.4	70.7	99.8	60.4	77.3	91.8	60.4	73.9	84.4	69.6	75.9
07	THERMAL & MOISTURE PROTECTION	105.3	49.8	83.0	104.6	71.0	91.1	100.7	58.3	83.6	100.7	58.9	83.9	100.7	59.9	84.2	99.3	73.1	88.8
08	OPENINGS	103.9	44.4	89.4	93.0	53.3	83.3	96.0	49.8	84.8	96.1	53.1	85.6	96.0	53.1	85.6	93.7	58.1	85.1
0920	Plaster & Gypsum Board	100.5	35.5	55.1	108.6	36.8	58.4	100.9	52.9	67.4	105.9	59.1	73.2	100.1	59.1	71.4	97.6	68.6	77.3
0950, 0980	Ceilings & Acoustic Treatment	116.3	35.5	62.5	119.9	36.8	64.6	94.6	52.9	66.8	94.6	59.1	71.0	94.6	59.1	71.0	94.6	68.6	77.3
0960	Flooring	112.3	35.2	89.3	118.9	78.2	106.8	104.0	40.6	85.1	107.4	40.6	87.5	103.8	40.6	85.0	99.4	37.7	81.0
0970, 0990	Wall Finishes & Painting/Coating	97.3	61.4	75.8	97.8	81.2	87.9	95.3	40.7	62.5	95.3	40.7	62.5	95.3	40.7	62.5	95.3	40.7	62.5
09	FINISHES	111.5	39.4	71.1	113.4	55.6	81.0	101.6	51.1	73.3	103.0	54.6	75.9	101.4	54.6	75.2	98.8	59.5	76.8
COVERS	DIVS. 10 - 14, 25, 28, 41, 43, 44, 46	100.0	74.1	94.7	100.0	80.6	96.0	100.0	41.9	88.1	100.0	42.5	88.3	100.0	42.6	88.3	100.0	41.6	88.1
21, 22, 23	FIRE SUPPRESSION, PLUMBING & HVAC	94.3	61.6	80.9	100.3	79.8	91.9	93.9	69.5	83.9	93.9	66.2	82.6	99.9	66.2	86.1	93.9	69.7	84.0
26, 27, 3370	ELECTRICAL, COMMUNICATIONS & UTIL.	100.6	43.2	70.4	98.8	58.0	77.3	100.9	66.0	82.5	100.9	66.0	82.5	97.4	72.3	84.2	97.8	72.2	84.3
MF2010	WEIGHTED AVERAGE	100.1	55.8	80.5	99.4	71.2	86.9	99.5	64.6	84.1	99.5	64.8	84.2	101.2	65.8	85.6	97.9	68.7	85.0

DIVISION		KANSAS																	
		FORT SCOTT			HAYS			HUTCHINSON			INDEPENDENCE			KANSAS CITY			LIBERAL		
		667			676			675			673			660 - 662			679		
		MAT.	INST.	TOTAL	MAT.	INST.	TOTAL	MAT.	INST.	TOTAL	MAT.	INST.	TOTAL	MAT.	INST.	TOTAL	MAT.	INST.	TOTAL
015433	CONTRACTOR EQUIPMENT		101.6	101.6		102.3	102.3		102.3	102.3		102.3	102.3		99.0	99.0		102.3	102.3
0241, 31 - 34	SITE & INFRASTRUCTURE, DEMOLITION	102.2	92.2	95.2	118.4	94.0	101.3	95.5	94.1	94.5	115.6	94.2	100.6	95.5	91.8	92.9	118.1	94.1	101.3
0310	Concrete Forming & Accessories	102.3	79.5	82.6	99.2	60.0	65.3	89.8	59.9	63.9	109.9	68.3	73.9	99.2	95.5	96.0	95.4	59.9	64.7
0320	Concrete Reinforcing	100.4	87.9	94.2	102.5	54.8	78.7	102.5	54.9	78.8	101.9	59.7	80.9	97.4	93.1	95.3	104.0	54.9	79.5
0330	Cast-in-Place Concrete	104.9	55.8	84.7	88.8	56.9	75.7	82.3	47.1	67.8	115.0	47.3	87.2	90.0	94.6	91.9	88.8	47.1	71.7
03	CONCRETE	104.3	73.5	89.1	105.6	59.2	82.7	88.6	55.7	72.4	116.4	60.4	88.8	96.4	95.0	95.7	107.7	55.8	82.1
04	MASONRY	106.6	58.5	77.2	109.6	57.6	77.9	101.4	59.7	75.9	98.6	62.3	76.4	109.5	96.0	101.3	107.0	59.7	78.1
05	METALS	93.5	90.2	92.5	93.7	77.6	88.7	93.6	76.5	88.2	93.5	79.0	88.9	100.6	97.6	99.7	94.0	76.5	88.5
06	WOOD, PLASTICS & COMPOSITES	101.9	87.2	93.5	96.9	60.4	76.1	86.8	60.4	71.7	109.7	71.0	87.6	97.9	95.9	96.8	92.5	60.4	74.2
07	THERMAL & MOISTURE PROTECTION	100.0	72.4	88.9	100.9	58.9	84.1	99.7	59.8	83.6	100.7	71.8	89.1	99.6	95.7	98.1	101.1	59.8	84.5
08	OPENINGS	93.7	79.3	90.2	96.0	53.1	85.6	96.0	53.1	85.5	93.7	60.0	85.5	95.1	89.4	93.7	96.1	53.1	85.6
0920	Plaster & Gypsum Board	103.8	86.8	91.9	103.8	59.1	72.6	98.8	59.1	71.1	112.1	70.1	82.7	95.9	95.7	95.8	100.9	59.1	71.7
0950, 0980	Ceilings & Acoustic Treatment	94.6	86.8	89.4	94.6	59.1	71.0	94.6	59.1	71.0	94.6	70.1	78.3	94.6	95.7	95.4	94.6	59.1	71.0
0960	Flooring	115.1	40.9	93.0	106.3	40.6	86.7	101.0	40.6	83.0	111.7	40.6	90.5	91.8	97.3	93.4	104.0	40.6	85.1
0970, 0990	Wall Finishes & Painting/Coating	97.1	40.7	63.3	95.3	40.7	62.5	95.3	40.7	62.5	95.3	40.7	62.5	103.4	70.7	83.8	95.3	40.7	62.5
09	FINISHES	104.3	69.7	84.9	103.1	54.6	75.9	98.6	54.6	74.0	105.2	60.9	80.4	98.5	92.4	95.1	102.4	54.6	75.6
COVERS	DIVS. 10 - 14, 25, 28, 41, 43, 44, 46	100.0	47.8	89.3	100.0	42.5	88.3	100.0	42.5	88.3	100.0	43.7	88.5	100.0	63.8	92.6	100.0	42.5	88.3
21, 22, 23	FIRE SUPPRESSION, PLUMBING & HVAC	93.9	67.8	83.2	93.9	66.2	82.6	93.9	66.2	82.6	93.9	68.1	83.3	99.8	97.3	98.8	93.9	66.2	82.6
26, 27, 3370	ELECTRICAL, COMMUNICATIONS & UTIL.	97.0	72.2	83.9	99.7	66.0	81.9	94.3	66.0	79.4	96.6	72.1	83.7	103.3	94.1	98.5	97.4	68.3	82.1
MF2010	WEIGHTED AVERAGE	97.8	72.9	86.8	99.0	64.8	83.9	95.0	64.5	81.5	99.2	68.4	85.6	99.7	93.9	97.2	98.8	64.8	83.8

DIVISION		KANSAS									KENTUCKY								
		SALINA			TOPEKA			WICHITA			ASHLAND			BOWLING GREEN			CAMPTON		
		674			664 - 666			670 - 672			411 - 412			421 - 422			413 - 414		
		MAT.	INST.	TOTAL	MAT.	INST.	TOTAL	MAT.	INST.	TOTAL	MAT.	INST.	TOTAL	MAT.	INST.	TOTAL	MAT.	INST.	TOTAL
015433	CONTRACTOR EQUIPMENT		102.3	102.3		100.6	100.6		102.3	102.3		97.9	97.9		95.4	95.4		101.7	101.7
0241, 31 - 34	SITE & INFRASTRUCTURE, DEMOLITION	104.3	94.0	97.1	97.0	90.8	92.6	96.9	92.8	94.0	109.1	86.9	93.5	75.2	98.1	91.3	83.6	99.3	94.6
0310	Concrete Forming & Accessories	91.7	56.0	60.8	98.7	42.1	49.7	98.1	45.6	52.7	88.0	104.1	102.0	85.0	78.8	79.6	89.1	86.2	86.6
0320	Concrete Reinforcing	101.9	63.5	82.8	94.7	98.7	96.7	94.7	74.8	84.8	92.9	105.3	99.1	90.0	88.3	89.1	90.9	104.2	97.5
0330	Cast-in-Place Concrete	99.4	50.6	79.3	89.5	46.8	72.0	87.5	49.7	72.0	89.1	99.8	93.5	88.4	98.4	92.5	98.6	62.8	83.9
03	CONCRETE	102.3	56.9	79.9	93.6	56.0	75.0	92.5	53.9	73.5	95.5	103.3	99.3	93.7	87.5	90.6	97.8	81.5	89.7
04	MASONRY	124.3	57.6	83.6	103.3	55.9	74.4	97.0	52.4	69.7	91.9	103.3	98.8	93.7	81.5	86.2	91.4	71.9	79.5
05	METALS	95.5	81.2	91.0	100.7	96.3	99.3	100.7	83.1	95.2	91.7	109.2	97.2	93.3	86.2	91.1	92.5	93.1	92.7
06	WOOD, PLASTICS & COMPOSITES	88.3	54.4	68.9	93.8	38.8	62.4	95.3	44.3	66.2	80.2	104.3	93.9	94.0	77.3	84.5	92.6	93.0	92.8
07	THERMAL & MOISTURE PROTECTION	100.2	58.3	83.4	100.4	62.2	85.1	97.0	53.7	79.6	90.2	98.0	93.4	85.6	82.3	84.3	98.4	71.1	87.4
08	OPENINGS	96.0	50.7	85.0	96.4	56.4	86.7	96.4	52.2	85.6	97.9	98.8	98.2	98.7	75.8	93.1	100.3	91.0	98.0
0920	Plaster & Gypsum Board	98.8	52.9	66.7	100.0	36.9	55.9	97.0	42.6	58.9	61.0	104.5	91.4	89.2	76.9	80.6	89.2	92.2	91.3
0950, 0980	Ceilings & Acoustic Treatment	94.6	52.9	66.8	102.8	36.9	58.9	101.9	42.6	62.4	83.0	104.5	97.3	89.1	76.9	81.0	89.1	92.2	91.2
0960	Flooring	102.3	64.2	91.0	105.0	76.4	96.5	104.3	65.3	92.6	78.5	103.9	86.0	95.1	89.2	93.3	96.8	42.7	80.7
0970, 0990	Wall Finishes & Painting/Coating	95.3	40.7	62.5	95.3	54.9	71.1	95.3	50.1	68.2	95.2	94.7	94.9	93.9	69.2	79.1	93.9	68.3	78.6
09	FINISHES	99.9	55.6	75.1	101.9	47.9	71.7	101.3	49.1	72.1	82.9	103.6	94.5	94.4	79.6	86.1	95.2	77.3	85.1
COVERS	DIVS. 10 - 14, 25, 28, 41, 43, 44, 46	100.0	52.0	90.2	100.0	49.0	89.6	100.0	48.3	89.4	100.0	96.4	99.3	100.0	59.1	91.6	100.0	58.1	91.4
21, 22, 23	FIRE SUPPRESSION, PLUMBING & HVAC	99.9	66.3	86.2	99.9	56.4	82.1	99.7	62.4	84.5	93.8	89.7	92.1	99.9	82.4	92.8	94.0	79.4	88.0
26, 27, 3370	ELECTRICAL, COMMUNICATIONS & UTIL.	97.2	72.3	84.1	101.9	72.2	86.3	102.1	72.3	86.4	91.5	95.4	93.6	94.3	82.1	87.9	91.5	68.9	79.6
MF2010	WEIGHTED AVERAGE	100.1	65.9	85.0	99.4	63.8	83.7	98.8	63.2	83.1	93.7	98.0	95.6	95.6	83.3	90.1	94.9	80.0	88.3

DIVISION		KENTUCKY																	
		CORBIN			COVINGTON			ELIZABETHTOWN			FRANKFORT			HAZARD			HENDERSON		
		407 - 409			410			427			406			417 - 418			424		
		MAT.	INST.	TOTAL	MAT.	INST.	TOTAL	MAT.	INST.	TOTAL	MAT.	INST.	TOTAL	MAT.	INST.	TOTAL	MAT.	INST.	TOTAL
015433	CONTRACTOR EQUIPMENT		101.7	101.7		105.5	105.5		95.4	95.4		101.7	101.7		101.7	101.7		116.7	116.7
0241, 31 - 34	SITE & INFRASTRUCTURE, DEMOLITION	81.0	99.1	93.7	79.1	113.2	103.0	70.1	97.9	89.6	86.2	100.0	95.9	81.5	100.8	95.0	78.2	124.9	111.0
0310	Concrete Forming & Accessories	86.1	68.0	70.4	85.0	94.9	93.5	80.5	77.8	78.1	98.3	74.0	77.3	86.1	87.2	87.0	92.1	81.7	83.1
0320	Concrete Reinforcing	90.4	67.9	79.2	89.4	94.3	91.9	90.4	91.3	90.9	91.7	91.3	91.5	91.3	104.2	97.7	90.1	88.8	89.5
0330	Cast-in-Place Concrete	93.3	60.3	79.7	94.3	92.6	93.6	79.9	74.9	77.9	90.2	70.7	82.2	94.8	86.1	91.2	78.1	88.1	82.2
03	CONCRETE	92.9	65.8	79.5	96.0	94.0	95.0	86.1	79.5	82.9	92.5	76.4	84.6	94.6	90.0	92.3	90.8	85.4	88.1
04	MASONRY	91.8	64.5	75.1	105.0	105.8	105.5	78.1	78.5	78.3	88.6	80.6	83.7	89.0	74.6	80.2	97.5	91.1	93.6
05	METALS	92.5	77.8	87.9	90.6	92.6	91.2	92.4	87.5	90.9	99.7	88.4	96.1	92.5	93.5	92.9	84.0	88.6	85.4
06	WOOD, PLASTICS & COMPOSITES	89.8	71.8	79.5	90.9	87.8	89.1	89.5	77.3	82.5	101.6	71.8	84.6	89.8	93.0	91.6	96.5	79.2	86.6
07	THERMAL & MOISTURE PROTECTION	98.2	66.4	85.4	98.3	96.4	97.6	85.3	77.8	82.3	99.4	77.0	90.4	98.3	73.1	88.2	97.8	91.5	95.3
08	OPENINGS	99.6	63.1	90.7	102.1	87.0	98.4	98.7	81.5	94.5	98.8	78.2	93.8	100.7	81.3	96.0	96.6	81.7	93.0
0920	Plaster & Gypsum Board	88.3	70.3	75.7	70.2	88.0	82.6	88.3	76.9	80.4	98.7	70.3	78.8	88.3	92.2	91.0	92.0	77.9	82.2
0950, 0980	Ceilings & Acoustic Treatment	89.1	70.3	76.6	92.8	88.0	89.6	89.1	76.9	81.0	97.3	70.3	79.3	89.1	92.2	91.2	82.7	77.9	79.5
0960	Flooring	95.4	42.7	79.7	71.3	87.2	76.1	92.7	89.2	91.6	99.5	56.4	86.6	95.4	43.5	79.9	98.5	89.2	95.7
0970, 0990	Wall Finishes & Painting/Coating	93.9	51.6	68.6	89.7	84.7	86.7	93.9	77.6	84.1	93.9	83.4	87.6	93.9	68.3	78.6	93.9	92.9	93.3
09	FINISHES	94.4	61.9	76.2	85.9	92.2	89.4	93.2	79.2	85.3	99.0	71.2	83.4	94.4	78.1	85.3	93.8	84.2	88.4
COVERS	DIVS. 10 - 14, 25, 28, 41, 43, 44, 46	100.0	49.5	89.7	100.0	103.9	100.8	100.0	87.0	97.4	100.0	67.5	93.4	100.0	58.9	91.6	100.0	67.0	93.3
21, 22, 23	FIRE SUPPRESSION, PLUMBING & HVAC	93.9	68.3	83.4	94.6	92.4	93.7	94.3	80.9	88.8	100.0	82.1	92.7	94.0	80.4	88.4	94.3	76.8	87.2
26, 27, 3370	ELECTRICAL, COMMUNICATIONS & UTIL.	91.7	82.1	86.6	95.6	79.6	87.2	91.3	82.1	86.5	100.9	90.8	95.6	91.5	60.8	75.3	93.6	82.3	87.6
MF2010	WEIGHTED AVERAGE	94.2	71.4	84.1	95.0	94.1	94.6	91.9	82.6	87.8	98.0	82.3	91.1	94.4	80.4	88.2	92.8	86.6	90.1

DIVISION		KENTUCKY																	
		LEXINGTON			LOUISVILLE			OWENSBORO			PADUCAH			PIKEVILLE			SOMERSET		
		403 - 405			400 - 402			423			420			415 - 416			425 - 426		
		MAT.	INST.	TOTAL	MAT.	INST.	TOTAL	MAT.	INST.	TOTAL	MAT.	INST.	TOTAL	MAT.	INST.	TOTAL	MAT.	INST.	TOTAL
015433	CONTRACTOR EQUIPMENT		101.7	101.7		95.4	95.4		116.7	116.7		116.7	116.7		97.9	97.9		101.7	101.7
0241, 31 - 34	SITE & INFRASTRUCTURE, DEMOLITION	82.9	101.0	95.6	73.1	97.9	90.5	87.6	124.8	113.7	80.8	124.4	111.4	119.8	86.2	96.2	74.9	99.7	92.3
0310	Concrete Forming & Accessories	97.4	74.8	77.9	96.3	77.8	80.3	90.6	81.4	82.7	88.6	77.9	79.4	96.3	92.7	93.2	86.8	72.8	74.7
0320	Concrete Reinforcing	99.3	91.0	95.2	99.3	91.3	95.3	90.1	88.5	89.3	90.6	85.0	87.8	93.4	105.1	99.2	90.4	91.3	90.9
0330	Cast-in-Place Concrete	95.4	85.5	91.4	97.5	74.8	88.1	91.1	96.6	93.4	83.2	86.3	84.5	97.9	97.2	97.6	78.1	101.4	87.7
03	CONCRETE	96.2	81.9	89.1	97.1	79.5	88.4	102.4	88.1	95.4	95.1	82.3	88.8	108.5	97.3	103.0	81.6	86.4	84.0
04	MASONRY	90.0	73.4	79.9	91.1	78.2	83.3	90.1	82.5	85.4	93.0	90.2	91.3	89.3	99.5	95.5	84.3	78.3	80.6
05	METALS	94.8	88.5	92.8	101.9	87.6	97.4	85.5	88.4	86.4	82.5	86.5	83.8	91.6	107.9	96.7	92.5	88.0	91.0
06	WOOD, PLASTICS & COMPOSITES	104.3	71.8	85.7	101.6	77.3	87.8	94.7	79.2	85.8	92.4	76.3	83.2	89.0	92.3	90.9	90.3	71.8	79.7
07	THERMAL & MOISTURE PROTECTION	98.4	86.1	93.4	95.8	77.7	88.5	98.1	90.0	94.8	97.8	80.5	90.8	90.8	82.6	87.5	97.8	69.7	86.5
08	OPENINGS	99.8	75.8	93.9	99.8	81.5	95.3	96.6	82.5	93.2	95.7	76.2	90.9	98.7	89.5	96.5	99.6	76.1	93.9
0920	Plaster & Gypsum Board	98.3	70.3	78.7	93.2	76.9	81.8	90.8	77.9	81.8	89.5	75.0	79.4	64.3	92.2	83.8	88.3	70.3	75.7
0950, 0980	Ceilings & Acoustic Treatment	93.7	70.3	78.1	97.3	76.9	83.7	82.7	77.9	79.5	82.7	75.0	77.6	83.0	92.2	89.1	89.1	70.3	76.6
0960	Flooring	99.9	72.8	91.8	98.2	89.2	95.5	97.8	89.2	95.2	96.7	61.5	86.2	82.4	103.9	88.8	95.6	42.7	79.9
0970, 0990	Wall Finishes & Painting/Coating	93.9	83.4	87.6	93.9	77.6	84.1	93.9	96.1	95.3	93.9	80.3	85.7	95.2	84.8	89.0	93.9	77.6	84.1
09	FINISHES	98.0	74.6	84.9	97.2	79.7	87.4	94.0	84.0	88.4	93.2	74.9	83.0	85.4	94.5	90.5	93.8	67.3	79.0
COVERS	DIVS. 10 - 14, 25, 28, 41, 43, 44, 46	100.0	89.2	97.8	100.0	86.9	97.3	100.0	95.0	99.0	100.0	63.0	92.4	100.0	58.7	91.6	100.0	59.5	91.7
21, 22, 23	FIRE SUPPRESSION, PLUMBING & HVAC	99.9	75.4	89.9	100.0	80.8	92.1	99.9	75.8	90.1	94.3	81.1	88.9	93.8	87.2	91.1	94.3	73.8	85.9
26, 27, 3370	ELECTRICAL, COMMUNICATIONS & UTIL.	94.8	82.1	88.1	101.7	79.2	89.9	93.7	82.3	87.7	96.2	79.0	87.1	94.7	76.5	85.1	91.9	82.1	86.7
MF2010	WEIGHTED AVERAGE	97.0	81.0	90.0	98.6	82.2	91.3	95.6	86.7	91.7	93.0	84.5	89.2	96.0	90.2	93.4	92.5	79.3	86.7

DIVISION		LOUISIANA																	
		ALEXANDRIA			BATON ROUGE			HAMMOND			LAFAYETTE			LAKE CHARLES			MONROE		
		713 - 714			707 - 708			704			705			706			712		
		MAT.	INST.	TOTAL	MAT.	INST.	TOTAL	MAT.	INST.	TOTAL	MAT.	INST.	TOTAL	MAT.	INST.	TOTAL	MAT.	INST.	TOTAL
015433	CONTRACTOR EQUIPMENT		89.4	89.4		89.3	89.3		90.0	90.0		90.0	90.0		89.3	89.3		89.4	89.4
0241, 31 - 34	SITE & INFRASTRUCTURE, DEMOLITION	95.2	86.3	88.9	100.3	86.0	90.3	101.2	87.1	91.3	102.1	87.0	91.5	102.8	85.8	90.8	95.2	85.6	88.5
0310	Concrete Forming & Accessories	82.8	35.8	42.1	94.4	61.1	65.6	77.8	47.9	51.9	94.1	52.9	58.4	95.2	56.5	61.7	82.3	36.1	42.3
0320	Concrete Reinforcing	94.3	56.8	75.6	97.9	59.9	78.9	96.5	60.3	78.5	97.9	59.9	78.9	97.9	57.5	77.8	93.3	56.5	75.0
0330	Cast-in-Place Concrete	96.9	39.0	73.1	96.2	57.2	80.2	99.0	44.6	76.7	98.5	45.3	76.6	103.7	65.9	88.2	96.9	55.1	79.7
03	CONCRETE	96.8	42.2	69.8	95.8	60.1	78.2	95.9	50.0	73.2	96.9	52.3	74.9	99.4	60.6	80.3	96.6	47.7	72.5
04	MASONRY	118.4	48.6	75.8	107.9	50.3	72.8	107.3	52.0	73.6	107.3	49.5	72.0	106.3	55.9	75.5	113.4	45.1	71.7
05	METALS	89.7	71.1	83.9	101.8	70.6	92.0	97.2	70.2	88.7	96.4	70.4	88.2	96.4	70.0	88.1	89.7	68.8	83.1
06	WOOD, PLASTICS & COMPOSITES	88.0	34.5	57.5	99.9	66.7	80.9	85.2	49.1	64.6	106.2	56.0	77.6	104.5	60.0	79.1	87.3	34.9	57.3
07	THERMAL & MOISTURE PROTECTION	98.5	50.1	79.0	98.5	57.0	81.8	96.8	57.9	81.1	97.3	55.5	80.5	96.4	58.0	80.9	98.5	49.3	78.7
08	OPENINGS	98.1	40.9	84.2	100.7	59.1	90.6	96.7	53.5	86.2	100.7	53.3	89.2	100.7	56.0	89.8	98.1	44.8	85.1
0920	Plaster & Gypsum Board	75.7	33.0	45.8	103.5	66.0	77.3	95.0	47.8	62.0	106.1	55.0	70.4	106.1	59.1	73.3	75.3	33.3	46.0
0950, 0980	Ceilings & Acoustic Treatment	89.1	33.0	51.7	106.7	66.0	79.6	99.3	47.8	65.0	97.4	55.0	69.2	98.5	59.1	72.3	89.1	33.3	52.0
0960	Flooring	100.8	66.2	90.5	104.0	66.2	92.8	95.7	66.2	86.9	104.0	66.2	92.8	104.0	75.6	95.5	100.4	59.2	88.1
0970, 0990	Wall Finishes & Painting/Coating	100.7	33.3	60.3	105.1	38.1	65.0	105.1	50.5	72.4	105.1	38.1	65.0	105.1	35.1	63.2	100.7	39.7	64.2
09	FINISHES	91.4	40.1	62.7	102.8	60.2	79.0	97.7	51.6	71.9	101.1	53.7	74.6	101.3	57.9	77.0	91.3	39.8	62.4
COVERS	DIVS. 10 - 14, 25, 28, 41, 43, 44, 46	100.0	49.1	89.6	100.0	80.1	95.9	100.0	47.3	89.2	100.0	78.7	95.7	100.0	79.5	95.8	100.0	45.1	88.8
21, 22, 23	FIRE SUPPRESSION, PLUMBING & HVAC	100.3	50.3	79.8	100.0	58.1	82.8	94.0	45.1	74.0	100.1	58.4	83.0	100.1	58.7	83.2	100.3	49.2	79.4
26, 27, 3370	ELECTRICAL, COMMUNICATIONS & UTIL.	93.0	55.4	73.2	110.6	61.1	84.6	101.3	55.5	77.2	102.6	65.7	83.2	102.1	65.7	82.9	95.1	56.6	74.8
MF2010	WEIGHTED AVERAGE	97.2	52.8	77.6	101.5	62.6	84.4	97.5	55.4	78.9	99.9	61.0	82.7	100.1	63.4	83.9	97.1	52.9	77.6

DIVISION		LOUISIANA									MAINE								
		NEW ORLEANS			SHREVEPORT			THIBODAUX			AUGUSTA			BANGOR			BATH		
		700 - 701			710 - 711			703			043			044			045		
		MAT.	INST.	TOTAL	MAT.	INST.	TOTAL	MAT.	INST.	TOTAL	MAT.	INST.	TOTAL	MAT.	INST.	TOTAL	MAT.	INST.	TOTAL
015433	CONTRACTOR EQUIPMENT		90.2	90.2		89.4	89.4		90.0	90.0		100.5	100.5		100.5	100.5		100.5	100.5
0241, 31 - 34	SITE & INFRASTRUCTURE, DEMOLITION	105.2	88.6	93.5	97.4	85.5	89.1	103.6	87.0	91.9	91.7	100.8	98.1	94.7	101.0	99.1	92.7	100.8	98.4
0310	Concrete Forming & Accessories	96.5	67.9	71.8	99.1	38.1	46.3	89.6	60.7	64.6	97.0	99.1	98.9	90.0	101.0	99.5	86.3	99.2	97.5
0320	Concrete Reinforcing	97.9	61.0	79.5	92.9	56.9	75.0	96.5	60.2	78.4	90.4	112.0	101.2	90.1	113.3	101.6	89.1	113.0	101.0
0330	Cast-in-Place Concrete	102.7	69.2	88.9	96.4	44.4	75.0	106.3	47.5	82.2	94.2	61.2	80.7	82.6	115.3	96.1	82.6	61.3	73.8
03	CONCRETE	99.0	67.6	83.5	96.7	45.0	71.2	101.2	56.6	79.2	102.1	87.8	95.0	95.6	107.4	101.4	95.6	88.0	91.9
04	MASONRY	107.6	61.1	79.2	103.9	46.7	69.0	135.0	47.8	81.7	93.8	66.3	77.0	107.0	104.5	105.5	113.5	86.0	96.7
05	METALS	109.5	73.8	98.3	86.3	68.9	80.8	97.2	69.9	88.7	86.4	88.9	87.2	86.0	91.1	87.6	84.5	90.3	86.3
06	WOOD, PLASTICS & COMPOSITES	100.8	69.3	82.8	104.0	37.8	66.2	93.3	65.9	77.7	94.1	110.0	103.2	89.2	100.8	95.8	84.7	110.0	99.2
07	THERMAL & MOISTURE PROTECTION	96.7	69.9	85.9	100.0	48.3	79.2	96.7	53.6	79.4	102.1	67.8	88.3	97.3	87.4	93.3	97.3	73.8	87.8
08	OPENINGS	101.8	67.8	93.5	96.0	42.9	83.1	101.7	62.7	92.2	100.4	91.9	98.3	100.3	86.8	97.0	100.3	91.8	98.2
0920	Plaster & Gypsum Board	104.1	68.7	79.3	86.8	36.4	51.6	98.7	65.2	75.3	94.5	109.7	105.1	93.5	100.2	98.2	91.0	109.7	104.1
0950, 0980	Ceilings & Acoustic Treatment	97.4	68.7	78.3	95.5	36.4	56.1	99.3	65.2	76.6	94.9	109.7	104.8	90.4	100.2	96.9	88.3	109.7	102.6
0960	Flooring	104.4	66.2	93.0	108.0	63.5	94.7	101.5	45.7	84.9	98.1	57.3	85.9	96.3	111.6	100.9	94.7	57.3	83.6
0970, 0990	Wall Finishes & Painting/Coating	106.5	67.0	82.8	100.7	33.3	60.3	106.5	50.5	72.9	96.1	66.2	78.2	96.1	46.1	66.2	96.1	36.1	60.1
09	FINISHES	101.5	67.6	82.5	96.5	41.5	65.7	100.2	57.5	76.2	98.0	89.6	93.3	96.8	97.7	97.3	95.5	86.3	90.3
COVERS	DIVS. 10 - 14, 25, 28, 41, 43, 44, 46	100.0	84.0	96.7	100.0	62.5	92.3	100.0	80.6	96.0	100.0	102.0	100.4	100.0	111.8	102.4	100.0	102.0	100.4
21, 22, 23	FIRE SUPPRESSION, PLUMBING & HVAC	100.0	67.1	86.5	100.1	54.1	81.3	94.0	58.6	79.5	99.8	66.7	86.3	100.0	77.2	90.7	94.0	66.7	82.8
26, 27, 3370	ELECTRICAL, COMMUNICATIONS & UTIL.	102.7	72.9	87.0	100.8	65.3	82.1	99.7	68.0	83.0	100.9	78.6	89.2	99.7	78.6	88.6	97.5	78.6	87.5
MF2010	WEIGHTED AVERAGE	102.4	70.5	88.4	97.0	55.6	78.8	100.1	62.7	83.6	97.5	81.7	90.5	97.1	92.1	94.9	95.3	83.6	90.1

	DIVISION	MAINE																	
		HOULTON			KITTERY			LEWISTON			MACHIAS			PORTLAND			ROCKLAND		
		047			039			042			046			040 - 041			048		
		MAT.	INST.	TOTAL	MAT.	INST.	TOTAL	MAT.	INST.	TOTAL	MAT.	INST.	TOTAL	MAT.	INST.	TOTAL	MAT.	INST.	TOTAL
015433	**CONTRACTOR EQUIPMENT**		100.5	100.5		100.5	100.5		100.5	100.5		100.5	100.5		100.5	100.5		100.5	100.5
0241, 31 - 34	**SITE & INFRASTRUCTURE, DEMOLITION**	94.5	101.8	99.6	88.3	100.9	97.1	91.5	101.0	98.2	93.8	101.8	99.4	88.9	101.0	97.4	89.7	101.8	98.2
0310	Concrete Forming & Accessories	93.1	105.8	104.1	86.9	99.6	97.9	94.9	101.0	100.2	90.7	105.8	103.7	96.2	101.0	100.3	91.6	105.8	103.9
0320	Concrete Reinforcing	90.1	111.7	100.9	88.6	113.0	100.7	110.7	113.3	112.0	90.1	111.7	100.9	110.7	113.3	112.0	90.1	111.7	100.9
0330	Cast-in-Place Concrete	82.6	73.9	79.0	82.6	62.1	74.2	84.3	115.3	97.0	82.6	73.3	78.8	104.5	115.3	108.9	84.3	73.3	79.8
03	**CONCRETE**	96.5	95.1	95.8	91.4	88.5	89.9	96.3	107.4	101.8	96.0	94.9	95.4	106.1	107.4	106.7	93.4	94.9	94.1
04	**MASONRY**	92.1	78.1	83.6	103.7	87.0	93.5	92.8	104.5	99.9	92.1	78.1	83.6	93.1	104.5	100.0	87.1	78.1	81.6
05	**METALS**	84.7	88.8	86.0	84.4	90.3	86.3	88.9	91.1	89.6	84.7	88.7	86.0	90.2	91.1	90.5	84.6	88.7	85.9
06	**WOOD, PLASTICS & COMPOSITES**	92.9	110.0	102.7	85.2	110.0	99.4	94.8	100.8	98.2	90.1	110.0	101.5	94.0	100.8	97.9	90.9	110.0	101.8
07	**THERMAL & MOISTURE PROTECTION**	97.4	75.1	88.4	96.9	73.1	87.3	97.1	87.4	93.2	97.3	74.1	88.0	104.5	87.4	97.6	97.1	74.1	87.8
08	**OPENINGS**	100.4	90.0	97.9	100.3	91.8	98.2	103.4	86.8	99.3	100.4	90.0	97.9	103.4	86.8	99.3	100.3	90.0	97.8
0920	Plaster & Gypsum Board	96.0	109.7	105.6	91.0	109.7	104.1	99.2	100.2	99.9	94.3	109.7	105.1	98.7	100.2	99.7	94.3	109.7	105.1
0950, 0980	Ceilings & Acoustic Treatment	88.3	109.7	102.6	88.3	109.7	102.6	97.7	100.2	99.3	88.3	109.7	102.6	100.4	100.2	100.2	88.3	109.7	102.6
0960	Flooring	97.5	53.6	84.4	94.8	59.7	84.4	98.9	111.6	102.7	96.6	53.6	83.8	98.1	111.6	102.1	97.0	53.6	84.0
0970, 0990	Wall Finishes & Painting/Coating	96.1	140.6	122.8	96.1	40.2	62.6	96.1	46.1	66.2	96.1	140.6	122.8	96.1	46.1	66.2	96.1	140.6	122.8
09	**FINISHES**	97.0	101.5	99.6	95.1	87.4	90.8	99.4	97.7	98.4	96.5	101.5	99.3	99.1	97.7	98.3	96.1	101.5	99.2
COVERS	**DIVS. 10 - 14, 25, 28, 41, 43, 44, 46**	100.0	107.7	101.6	100.0	102.3	100.5	100.0	111.8	102.4	100.0	107.7	101.6	100.0	111.8	102.4	100.0	107.7	101.6
21, 22, 23	**FIRE SUPPRESSION, PLUMBING & HVAC**	94.0	75.5	86.4	94.0	67.3	83.0	100.0	77.2	90.7	94.0	75.5	86.4	99.8	77.2	90.6	94.0	75.5	86.4
26, 27, 3370	**ELECTRICAL, COMMUNICATIONS & UTIL.**	102.0	78.6	89.7	97.5	78.6	87.5	102.0	78.6	89.7	102.0	78.6	89.7	104.5	78.6	90.9	101.9	78.6	89.6
MF2010	**WEIGHTED AVERAGE**	95.0	87.6	91.7	94.2	84.0	89.7	97.6	92.1	95.2	94.9	87.5	91.6	99.2	92.1	96.1	94.2	87.5	91.2

	DIVISION	MAINE			MARYLAND														
		WATERVILLE			ANNAPOLIS			BALTIMORE			COLLEGE PARK			CUMBERLAND			EASTON		
		049			214			210 - 212			207 - 208			215			216		
		MAT.	INST.	TOTAL	MAT.	INST.	TOTAL	MAT.	INST.	TOTAL	MAT.	INST.	TOTAL	MAT.	INST.	TOTAL	MAT.	INST.	TOTAL
015433	**CONTRACTOR EQUIPMENT**		100.5	100.5		99.4	99.4		102.9	102.9		103.6	103.6		99.4	99.4		99.4	99.4
0241, 31 - 34	**SITE & INFRASTRUCTURE, DEMOLITION**	94.7	100.8	99.0	99.5	91.1	93.6	99.9	95.1	96.6	105.0	93.7	97.1	91.2	90.9	91.0	98.1	88.3	91.2
0310	Concrete Forming & Accessories	85.8	99.1	97.3	96.5	71.2	74.6	97.0	74.2	77.3	87.0	72.0	74.0	88.2	80.5	81.5	86.3	71.3	73.3
0320	Concrete Reinforcing	90.1	112.0	101.0	102.9	83.4	93.2	102.9	83.5	93.2	105.7	79.8	92.8	90.9	75.6	83.3	90.1	79.7	84.9
0330	Cast-in-Place Concrete	82.6	61.2	73.8	134.3	78.7	111.5	119.6	79.5	103.1	126.7	79.3	107.2	106.8	83.7	97.3	118.6	50.1	90.4
03	**CONCRETE**	97.1	87.8	92.5	116.8	77.1	97.2	109.8	78.8	94.5	112.5	77.2	95.1	98.0	81.6	89.9	106.4	66.6	86.7
04	**MASONRY**	102.0	66.3	80.2	98.9	74.7	84.1	100.1	74.7	84.6	106.5	69.5	83.9	98.8	82.8	89.0	112.2	44.7	71.0
05	**METALS**	84.6	88.8	86.0	97.3	95.0	96.6	98.4	95.6	97.5	85.1	96.2	88.6	94.8	92.0	93.9	95.0	88.3	92.9
06	**WOOD, PLASTICS & COMPOSITES**	84.2	110.0	99.0	98.0	71.3	82.7	100.8	75.3	86.2	81.9	71.5	76.0	90.9	79.5	84.4	88.8	78.1	82.7
07	**THERMAL & MOISTURE PROTECTION**	97.4	67.8	85.5	99.2	79.7	91.3	97.4	80.4	90.5	100.6	78.2	91.6	96.8	78.5	89.4	96.9	61.4	82.6
08	**OPENINGS**	100.4	91.9	98.3	92.5	78.7	89.1	94.8	81.5	91.6	92.7	74.7	88.3	93.5	79.2	90.0	91.8	72.3	87.1
0920	Plaster & Gypsum Board	91.0	109.7	104.1	96.7	70.7	78.5	99.7	74.7	82.2	98.2	70.7	79.0	93.9	79.2	83.6	93.4	77.8	82.5
0950, 0980	Ceilings & Acoustic Treatment	88.3	109.7	102.6	103.1	70.7	81.5	99.5	74.7	82.9	96.5	70.7	79.3	99.5	79.2	86.0	99.5	77.8	85.0
0960	Flooring	94.4	57.3	83.3	91.4	80.0	88.0	91.4	80.0	88.0	103.0	81.2	96.5	87.9	89.7	88.4	87.1	53.6	77.1
0970, 0990	Wall Finishes & Painting/Coating	96.1	66.2	78.2	98.0	82.8	88.8	98.0	82.8	88.8	109.8	78.0	90.7	98.0	76.6	85.2	98.0	78.0	86.0
09	**FINISHES**	95.7	89.6	92.3	96.8	73.1	83.5	96.6	75.5	84.8	99.0	73.6	84.8	94.2	81.7	87.2	94.4	70.3	80.9
COVERS	**DIVS. 10 - 14, 25, 28, 41, 43, 44, 46**	100.0	102.0	100.4	100.0	86.7	97.3	100.0	87.5	97.4	100.0	82.2	96.4	100.0	90.9	98.1	100.0	78.2	95.5
21, 22, 23	**FIRE SUPPRESSION, PLUMBING & HVAC**	94.0	66.7	82.8	100.0	82.3	92.8	100.0	82.4	92.8	94.2	87.0	91.2	93.9	72.8	85.3	93.9	69.3	83.8
26, 27, 3370	**ELECTRICAL, COMMUNICATIONS & UTIL.**	102.0	78.6	89.7	101.5	91.4	96.2	101.1	93.3	97.0	99.0	101.1	100.1	98.7	83.1	90.5	98.1	64.9	80.6
MF2010	**WEIGHTED AVERAGE**	95.4	81.7	89.4	100.4	82.7	92.6	100.1	84.1	93.0	97.0	84.5	91.5	95.8	81.9	89.7	97.3	69.7	85.1

	DIVISION	MARYLAND															MASSACHUSETTS		
		ELKTON			HAGERSTOWN			SALISBURY			SILVER SPRING			WALDORF			BOSTON		
		219			217			218			209			206			020 - 022, 024		
		MAT.	INST.	TOTAL	MAT.	INST.	TOTAL	MAT.	INST.	TOTAL	MAT.	INST.	TOTAL	MAT.	INST.	TOTAL	MAT.	INST.	TOTAL
015433	**CONTRACTOR EQUIPMENT**		99.4	99.4		99.4	99.4		99.4	99.4		96.4	96.4		96.4	96.4		106.1	106.1
0241, 31 - 34	**SITE & INFRASTRUCTURE, DEMOLITION**	85.6	89.0	88.0	89.5	91.8	91.1	98.0	88.3	91.2	92.0	86.7	88.3	98.8	86.7	90.3	99.6	108.2	105.6
0310	Concrete Forming & Accessories	92.0	84.4	85.4	87.3	77.4	78.7	99.5	54.1	60.2	95.7	70.2	73.6	102.9	70.3	74.7	98.8	143.9	137.8
0320	Concrete Reinforcing	90.1	103.3	96.7	90.9	75.7	83.3	90.1	66.7	78.5	104.4	75.1	89.8	105.1	75.1	90.2	109.5	153.9	131.6
0330	Cast-in-Place Concrete	96.1	78.1	88.7	101.8	83.8	94.4	118.6	48.7	89.9	129.7	78.2	108.5	145.4	78.2	117.8	106.5	155.0	126.5
03	**CONCRETE**	90.2	86.5	88.4	94.1	80.2	87.3	107.3	56.1	82.0	111.1	75.1	93.3	122.3	75.2	99.0	109.7	148.5	128.9
04	**MASONRY**	98.1	61.4	75.7	104.0	82.8	91.0	112.0	49.3	73.7	105.4	67.6	82.3	90.8	67.6	76.7	109.8	165.9	144.1
05	**METALS**	95.1	100.6	96.8	94.9	92.1	94.0	95.1	83.5	91.4	89.3	92.6	90.3	89.3	92.7	90.4	100.8	131.2	110.3
06	**WOOD, PLASTICS & COMPOSITES**	95.3	90.7	92.7	89.9	75.1	81.5	104.8	57.5	77.8	88.2	70.7	78.2	95.7	70.7	81.4	98.1	143.2	123.9
07	**THERMAL & MOISTURE PROTECTION**	96.5	74.6	87.7	96.5	76.8	88.6	97.0	66.5	84.8	103.2	81.7	94.6	103.7	81.7	94.9	97.4	153.0	119.8
08	**OPENINGS**	91.8	83.1	89.7	91.8	75.6	87.8	92.0	63.7	85.1	85.7	74.3	82.9	86.4	74.3	83.5	98.9	145.0	110.1
0920	Plaster & Gypsum Board	96.3	90.8	92.5	93.4	74.7	80.3	102.6	56.5	70.3	103.9	70.7	80.7	107.6	70.7	81.8	101.9	144.0	131.3
0950, 0980	Ceilings & Acoustic Treatment	99.5	90.8	93.7	100.5	74.7	83.3	99.5	56.5	70.8	106.6	70.7	82.7	106.6	70.7	82.7	96.6	144.0	128.2
0960	Flooring	89.4	60.6	80.8	87.6	89.7	88.2	92.6	67.1	85.0	109.3	81.2	100.9	113.0	81.2	103.5	98.7	184.0	124.1
0970, 0990	Wall Finishes & Painting/Coating	98.0	78.0	86.0	98.0	78.0	86.0	98.0	78.0	86.0	116.8	78.0	93.5	116.8	78.0	93.5	98.5	152.4	130.8
09	**FINISHES**	94.5	80.0	86.4	94.1	79.3	85.8	97.1	58.4	75.4	99.4	71.8	83.9	101.3	72.4	85.1	99.2	152.7	129.2
COVERS	**DIVS. 10 - 14, 25, 28, 41, 43, 44, 46**	100.0	62.4	92.3	100.0	90.4	98.0	100.0	40.3	87.8	100.0	79.8	95.9	100.0	79.8	95.9	100.0	119.6	104.0
21, 22, 23	**FIRE SUPPRESSION, PLUMBING & HVAC**	93.9	81.9	89.0	99.9	83.4	93.2	93.9	63.2	81.3	94.2	85.7	90.7	94.2	85.6	90.7	100.0	128.2	111.6
26, 27, 3370	**ELECTRICAL, COMMUNICATIONS & UTIL.**	100.1	91.4	95.5	98.5	83.1	90.4	96.8	70.0	82.7	96.1	101.1	98.7	93.5	101.1	97.5	101.1	139.7	121.4
MF2010	**WEIGHTED AVERAGE**	94.8	83.2	89.7	96.8	83.5	90.9	97.7	64.6	83.1	96.3	82.6	90.3	97.0	82.7	90.7	101.5	139.2	118.1

DIVISION		MASSACHUSETTS																	
		BROCKTON			BUZZARDS BAY			FALL RIVER			FITCHBURG			FRAMINGHAM			GREENFIELD		
		023			025			027			014			017			013		
		MAT.	INST.	TOTAL	MAT.	INST.	TOTAL	MAT.	INST.	TOTAL	MAT.	INST.	TOTAL	MAT.	INST.	TOTAL	MAT.	INST.	TOTAL
015433	CONTRACTOR EQUIPMENT		102.2	102.2		102.2	102.2		103.2	103.2		100.5	100.5		101.6	101.6		100.5	100.5
0241, 31 - 34	SITE & INFRASTRUCTURE, DEMOLITION	96.9	104.7	102.4	86.9	104.6	99.3	96.0	104.8	102.2	88.8	104.6	99.9	85.1	104.3	98.6	92.7	103.1	100.0
0310	Concrete Forming & Accessories	99.1	133.4	128.8	96.9	133.2	128.3	99.1	133.5	128.9	91.1	128.4	123.4	97.9	133.1	128.4	89.5	110.8	108.0
0320	Concrete Reinforcing	110.7	153.8	132.1	88.7	167.9	128.2	110.7	167.9	139.2	88.7	150.2	119.3	88.7	153.1	120.8	92.5	127.6	110.0
0330	Cast-in-Place Concrete	101.4	155.7	123.7	84.3	155.6	113.6	98.1	156.1	122.0	88.4	145.3	111.8	88.4	141.4	110.2	90.9	125.0	104.9
03	CONCRETE	107.0	143.9	125.2	90.3	146.4	118.0	105.4	146.7	125.8	88.8	137.3	112.8	91.6	138.8	114.9	92.6	118.2	105.3
04	MASONRY	105.5	163.4	140.9	97.5	163.4	137.7	106.4	163.3	141.1	94.1	152.1	129.5	100.0	152.2	131.9	98.2	128.7	116.8
05	METALS	97.8	128.9	107.6	92.9	134.3	105.9	97.8	134.7	109.4	92.8	123.8	102.5	92.9	127.6	103.8	95.0	109.7	99.6
06	WOOD, PLASTICS & COMPOSITES	97.0	131.1	116.4	94.2	131.1	115.2	97.0	131.3	116.6	90.7	125.3	110.5	96.6	130.8	116.1	88.8	107.5	99.5
07	THERMAL & MOISTURE PROTECTION	97.4	149.4	118.3	96.8	144.1	115.8	97.4	143.3	115.8	96.9	138.7	113.7	96.9	144.3	116.0	96.9	120.0	106.2
08	OPENINGS	96.8	137.9	106.8	92.5	136.0	103.1	96.8	136.1	106.3	102.1	134.0	109.9	92.9	137.8	103.8	102.2	112.9	104.8
0920	Plaster & Gypsum Board	96.4	131.4	120.9	92.4	131.4	119.7	96.4	131.4	120.9	93.5	125.5	115.9	96.4	131.4	120.9	94.2	107.1	103.2
0950, 0980	Ceilings & Acoustic Treatment	98.7	131.4	120.5	89.3	131.4	117.4	98.7	131.4	120.5	88.3	125.5	113.1	88.3	131.4	117.0	95.6	107.1	103.3
0960	Flooring	99.3	184.0	124.5	98.2	184.0	123.8	99.1	184.0	124.4	96.3	184.0	122.5	97.7	184.0	123.4	95.6	139.2	108.6
0970, 0990	Wall Finishes & Painting/Coating	98.1	137.9	121.9	98.1	137.9	121.9	98.1	137.9	121.9	96.1	137.9	121.1	97.1	137.9	121.5	96.1	108.4	103.5
09	FINISHES	99.2	143.3	123.9	95.6	143.3	122.3	99.1	143.5	124.0	94.9	139.9	120.1	95.4	143.1	122.1	96.8	115.7	107.3
COVERS	DIVS. 10 - 14, 25, 28, 41, 43, 44, 46	100.0	117.4	103.6	100.0	117.4	103.6	100.0	118.0	103.7	100.0	104.1	100.8	100.0	117.0	103.5	100.0	106.2	101.3
21, 22, 23	FIRE SUPPRESSION, PLUMBING & HVAC	100.0	112.3	105.0	94.0	112.4	101.5	100.0	112.5	105.1	94.9	115.9	103.5	94.9	124.2	106.8	94.9	103.7	98.5
26, 27, 3370	ELECTRICAL, COMMUNICATIONS & UTIL.	100.2	99.1	99.6	97.1	99.1	98.1	100.2	99.1	99.6	102.5	99.9	101.1	98.6	126.0	113.0	102.5	94.5	98.3
MF2010	WEIGHTED AVERAGE	100.1	127.1	112.0	94.3	127.7	109.0	99.9	127.9	112.3	95.7	124.1	108.2	94.9	131.2	110.9	96.9	109.9	102.6

DIVISION		MASSACHUSETTS																	
		HYANNIS			LAWRENCE			LOWELL			NEW BEDFORD			PITTSFIELD			SPRINGFIELD		
		026			019			018			027			012			010 - 011		
		MAT.	INST.	TOTAL	MAT.	INST.	TOTAL	MAT.	INST.	TOTAL	MAT.	INST.	TOTAL	MAT.	INST.	TOTAL	MAT.	INST.	TOTAL
015433	CONTRACTOR EQUIPMENT		102.2	102.2		102.2	102.2		100.5	100.5		103.2	103.2		100.5	100.5		100.5	100.5
0241, 31 - 34	SITE & INFRASTRUCTURE, DEMOLITION	93.2	104.6	101.2	97.6	104.7	102.6	96.9	104.7	102.3	94.5	104.8	101.7	97.8	102.8	101.3	97.3	103.2	101.4
0310	Concrete Forming & Accessories	91.4	133.2	127.6	98.9	133.7	129.0	95.9	133.5	128.4	99.1	133.5	128.9	95.9	108.1	106.5	96.1	110.4	108.5
0320	Concrete Reinforcing	88.7	167.9	128.2	109.7	144.0	126.8	110.7	143.4	127.0	110.7	167.9	139.2	91.8	127.6	109.7	110.7	127.6	119.1
0330	Cast-in-Place Concrete	92.6	155.6	118.5	102.3	152.6	123.0	93.0	152.5	117.5	86.5	156.1	115.1	101.4	121.0	109.5	96.8	124.3	108.1
03	CONCRETE	96.9	146.4	121.3	107.2	141.1	124.0	98.4	140.8	119.4	99.8	146.7	123.0	99.4	115.6	107.4	100.3	117.8	108.9
04	MASONRY	104.5	163.4	140.4	104.7	163.4	140.5	93.4	152.1	129.2	104.5	163.3	140.4	94.1	121.4	110.8	93.7	127.4	114.3
05	METALS	94.4	134.3	106.9	95.4	125.2	104.8	95.4	121.9	103.7	97.8	134.7	109.4	95.2	109.7	99.7	97.7	109.7	101.5
06	WOOD, PLASTICS & COMPOSITES	88.1	131.1	112.6	97.0	131.1	116.4	96.4	131.1	116.2	97.0	131.3	116.6	96.4	107.5	102.7	96.4	107.5	102.7
07	THERMAL & MOISTURE PROTECTION	97.0	144.1	115.9	97.3	149.0	118.1	97.2	145.0	116.4	97.3	143.3	115.8	97.2	116.8	105.1	97.2	119.4	106.1
08	OPENINGS	93.1	136.0	103.5	96.8	135.3	106.1	103.4	135.3	111.2	96.8	136.1	106.3	103.4	112.9	105.7	103.4	112.9	105.7
0920	Plaster & Gypsum Board	88.7	131.4	118.6	99.2	131.4	121.7	99.2	131.4	121.7	96.4	131.4	120.9	99.2	107.1	104.7	99.2	107.1	104.7
0950, 0980	Ceilings & Acoustic Treatment	91.4	131.4	118.1	97.7	131.4	120.1	97.7	131.4	120.1	98.7	131.4	120.5	97.7	107.1	104.0	97.7	107.1	104.0
0960	Flooring	96.2	184.0	122.3	98.3	184.0	123.8	98.3	184.0	123.8	99.1	184.0	124.4	98.5	139.2	110.6	98.0	139.2	110.3
0970, 0990	Wall Finishes & Painting/Coating	98.1	137.9	121.9	96.2	137.9	121.2	96.1	137.9	121.1	98.1	137.9	121.9	96.1	108.4	103.5	97.5	108.4	104.0
09	FINISHES	95.4	143.3	122.2	98.9	143.3	123.8	98.8	143.3	123.8	99.0	143.5	123.9	98.9	113.8	107.3	98.9	115.3	108.1
COVERS	DIVS. 10 - 14, 25, 28, 41, 43, 44, 46	100.0	117.4	103.6	100.0	117.5	103.6	100.0	117.5	103.6	100.0	118.0	103.7	100.0	103.8	100.8	100.0	105.8	101.2
21, 22, 23	FIRE SUPPRESSION, PLUMBING & HVAC	100.0	112.4	105.1	100.0	124.2	109.9	100.0	124.1	109.9	100.0	112.5	105.1	100.0	98.0	99.2	100.0	103.1	101.3
26, 27, 3370	ELECTRICAL, COMMUNICATIONS & UTIL.	97.6	99.1	98.4	101.4	126.0	114.4	102.0	126.0	114.6	101.1	99.1	100.0	102.0	94.5	98.0	102.0	94.5	98.0
MF2010	WEIGHTED AVERAGE	97.3	127.7	110.7	99.8	132.5	114.2	99.0	130.9	113.1	99.3	127.9	111.9	99.1	107.1	102.7	99.6	109.5	104.0

DIVISION		MASSACHUSETTS			MICHIGAN														
		WORCESTER			ANN ARBOR			BATTLE CREEK			BAY CITY			DEARBORN			DETROIT		
		015 - 016			481			490			487			481			482		
		MAT.	INST.	TOTAL	MAT.	INST.	TOTAL	MAT.	INST.	TOTAL	MAT.	INST.	TOTAL	MAT.	INST.	TOTAL	MAT.	INST.	TOTAL
015433	CONTRACTOR EQUIPMENT		100.5	100.5		110.4	110.4		102.4	102.4		110.4	110.4		110.4	110.4		98.5	98.5
0241, 31 - 34	SITE & INFRASTRUCTURE, DEMOLITION	97.2	104.6	102.4	79.8	97.2	92.0	87.9	87.6	87.7	71.4	96.0	88.7	79.6	97.4	92.1	94.8	99.0	97.8
0310	Concrete Forming & Accessories	96.4	128.4	124.1	99.4	108.9	107.6	97.7	88.3	89.6	99.5	87.6	89.2	99.2	113.8	111.9	99.9	113.9	112.0
0320	Concrete Reinforcing	110.7	152.4	131.4	101.9	123.3	112.5	99.3	89.9	94.6	101.9	122.4	112.1	101.9	123.4	112.6	101.2	123.4	112.3
0330	Cast-in-Place Concrete	96.2	145.3	116.4	87.0	110.7	96.7	95.3	101.3	97.8	83.2	91.7	86.7	85.0	111.5	95.9	93.3	111.5	100.8
03	CONCRETE	100.0	137.7	118.6	92.3	112.5	102.3	96.1	92.5	94.3	90.5	96.3	93.4	91.4	115.0	103.0	95.3	113.8	104.4
04	MASONRY	93.2	152.1	129.2	99.1	106.9	103.9	97.1	85.5	90.0	98.6	87.2	91.6	98.9	113.2	107.6	98.1	113.2	107.3
05	METALS	97.8	124.6	106.2	93.6	118.6	101.5	93.7	86.1	91.3	94.2	115.2	100.8	93.6	119.1	101.6	97.6	101.6	98.9
06	WOOD, PLASTICS & COMPOSITES	96.8	125.3	113.1	99.5	109.1	105.0	100.6	88.1	93.5	99.5	86.8	92.3	99.5	113.6	107.6	99.9	113.6	107.7
07	THERMAL & MOISTURE PROTECTION	97.2	138.7	113.9	101.0	108.4	104.0	94.2	86.0	90.9	98.7	94.4	97.0	99.5	114.3	105.5	98.0	114.3	104.6
08	OPENINGS	103.4	134.6	111.0	98.4	108.4	100.9	94.1	82.5	91.3	98.4	94.4	97.4	98.4	110.8	101.4	100.2	110.9	102.8
0920	Plaster & Gypsum Board	99.2	125.5	117.6	106.1	108.4	107.7	97.0	83.9	87.8	106.1	85.4	91.6	106.1	113.1	111.0	106.1	113.1	111.0
0950, 0980	Ceilings & Acoustic Treatment	97.7	125.5	116.2	94.7	108.4	103.8	93.7	83.9	87.1	95.7	85.4	88.9	94.7	113.1	106.9	95.7	113.1	107.3
0960	Flooring	98.3	171.6	120.1	95.7	115.9	101.7	98.8	95.2	97.7	95.5	79.7	90.8	95.3	116.6	101.6	95.4	116.6	101.7
0970, 0990	Wall Finishes & Painting/Coating	96.1	137.9	121.1	88.6	101.1	96.1	93.9	81.9	86.7	88.6	82.6	85.0	88.6	101.2	96.1	90.0	101.2	96.7
09	FINISHES	98.8	137.4	120.5	96.5	109.5	103.8	97.4	88.9	92.6	96.2	84.9	89.9	96.4	113.3	105.9	98.0	113.3	106.6
COVERS	DIVS. 10 - 14, 25, 28, 41, 43, 44, 46	100.0	110.3	102.1	100.0	106.3	101.3	100.0	99.3	99.9	100.0	93.3	98.6	100.0	108.1	101.6	100.0	108.1	101.6
21, 22, 23	FIRE SUPPRESSION, PLUMBING & HVAC	100.0	115.9	106.5	100.0	98.4	99.4	99.9	89.2	95.6	100.0	86.4	94.4	100.0	110.5	104.3	100.0	111.7	104.8
26, 27, 3370	ELECTRICAL, COMMUNICATIONS & UTIL.	102.0	100.6	101.3	96.3	108.3	102.6	94.3	81.8	87.7	95.1	89.5	92.1	96.3	107.1	102.0	97.3	107.0	102.4
MF2010	WEIGHTED AVERAGE	99.6	124.3	110.4	96.8	106.7	101.2	96.5	87.7	92.6	96.2	92.3	94.5	96.6	111.0	102.9	98.4	109.6	103.4

	DIVISION	MICHIGAN																	
		FLINT			GAYLORD			GRAND RAPIDS			IRON MOUNTAIN			JACKSON			KALAMAZOO		
		484 - 485			497			493,495			498 - 499			492			491		
		MAT.	INST.	TOTAL	MAT.	INST.	TOTAL	MAT.	INST.	TOTAL	MAT.	INST.	TOTAL	MAT.	INST.	TOTAL	MAT.	INST.	TOTAL
015433	CONTRACTOR EQUIPMENT		110.4	110.4		105.0	105.0		102.4	102.4		94.4	94.4		105.0	105.0		102.4	102.4
0241, 31 - 34	SITE & INFRASTRUCTURE, DEMOLITION	69.0	96.4	88.2	82.4	85.9	84.9	89.4	87.5	88.1	90.6	94.3	93.2	104.3	87.6	92.6	88.3	87.6	87.8
0310	Concrete Forming & Accessories	102.2	89.9	91.6	95.4	55.1	60.5	100.0	79.7	82.4	87.4	84.6	85.0	92.6	86.5	87.3	97.7	88.0	89.3
0320	Concrete Reinforcing	101.9	122.7	112.3	92.5	91.5	92.0	99.3	89.6	94.5	92.4	93.1	92.8	89.9	122.5	106.2	99.3	89.9	94.6
0330	Cast-in-Place Concrete	87.6	93.8	90.1	95.1	70.4	85.0	101.5	94.4	98.6	112.5	88.0	102.5	95.0	93.2	94.2	97.2	101.1	98.8
03	CONCRETE	92.8	98.1	95.4	92.6	68.7	80.8	99.2	86.3	92.8	102.0	87.5	94.8	87.6	96.2	91.9	99.4	92.3	95.9
04	MASONRY	99.1	94.8	96.5	107.4	81.7	91.7	93.1	80.7	85.5	93.6	87.5	89.9	87.9	91.9	90.3	95.8	85.5	89.5
05	METALS	93.6	115.8	100.6	95.0	99.8	96.5	95.0	85.2	91.9	94.4	89.6	92.9	95.2	113.3	100.9	93.7	85.8	91.2
06	WOOD, PLASTICS & COMPOSITES	102.7	88.1	94.3	93.1	48.0	67.3	97.7	77.8	86.3	87.6	84.5	85.8	91.7	84.8	87.7	100.6	88.1	93.5
07	THERMAL & MOISTURE PROTECTION	98.8	97.7	98.3	92.7	69.7	83.4	97.3	75.8	88.6	96.1	82.7	90.7	92.1	95.0	93.3	94.2	86.0	90.9
08	OPENINGS	98.4	95.1	97.6	95.6	53.1	85.2	97.9	81.0	93.8	101.9	72.2	94.7	94.6	86.2	92.6	94.1	82.5	91.3
0920	Plaster & Gypsum Board	107.7	86.8	93.1	96.8	45.1	60.7	98.7	73.2	80.9	59.3	84.5	77.0	94.7	83.1	86.6	97.0	83.9	87.8
0950, 0980	Ceilings & Acoustic Treatment	94.7	86.8	89.4	92.8	45.1	61.0	97.3	73.2	81.3	91.6	84.5	86.9	92.8	83.1	86.3	93.7	83.9	87.1
0960	Flooring	95.5	82.3	91.6	91.1	49.9	78.8	99.0	69.8	90.3	113.2	95.8	108.0	89.8	68.7	83.5	98.8	75.0	91.7
0970, 0990	Wall Finishes & Painting/Coating	88.6	89.2	89.0	89.7	45.2	63.0	93.9	80.6	85.9	106.7	58.3	77.7	89.7	75.0	80.9	93.9	81.9	86.7
09	FINISHES	95.7	87.7	91.2	97.5	51.1	71.5	98.6	77.3	86.7	99.6	84.0	90.9	98.5	81.2	88.8	97.4	85.0	90.5
COVERS	DIVS. 10 - 14, 25, 28, 41, 43, 44, 46	100.0	94.4	98.9	100.0	88.4	97.6	100.0	98.0	99.6	100.0	90.1	98.0	100.0	94.4	98.9	100.0	99.3	99.9
21, 22, 23	FIRE SUPPRESSION, PLUMBING & HVAC	100.0	92.5	96.9	94.3	81.4	89.0	99.9	81.7	92.5	94.2	85.2	90.6	94.3	90.3	92.7	99.9	81.6	92.4
26, 27, 3370	ELECTRICAL, COMMUNICATIONS & UTIL.	96.3	93.2	94.6	92.2	48.0	69.0	99.7	73.6	86.0	98.8	85.9	92.0	96.3	108.2	102.6	94.2	85.2	89.5
MF2010	WEIGHTED AVERAGE	96.4	95.7	96.1	95.2	71.7	84.8	98.0	81.6	90.8	97.2	86.4	92.4	94.7	94.5	94.6	96.8	86.0	92.1

	DIVISION	MICHIGAN															MINNESOTA		
		LANSING			MUSKEGON			ROYAL OAK			SAGINAW			TRAVERSE CITY			BEMIDJI		
		488 - 489			494			480,483			486			496			566		
		MAT.	INST.	TOTAL	MAT.	INST.	TOTAL	MAT.	INST.	TOTAL	MAT.	INST.	TOTAL	MAT.	INST.	TOTAL	MAT.	INST.	TOTAL
015433	CONTRACTOR EQUIPMENT		110.4	110.4		102.4	102.4		95.9	95.9		110.4	110.4		94.4	94.4		98.9	98.9
0241, 31 - 34	SITE & INFRASTRUCTURE, DEMOLITION	88.5	96.1	93.8	86.0	87.5	87.1	83.9	96.3	92.6	72.4	96.0	89.0	77.2	94.0	89.0	97.4	98.3	98.0
0310	Concrete Forming & Accessories	102.9	86.8	89.0	98.0	86.6	88.2	95.5	109.8	107.9	99.4	87.4	89.1	87.4	69.8	72.2	90.6	92.6	92.4
0320	Concrete Reinforcing	101.9	122.5	112.2	100.0	89.3	94.6	92.3	119.0	105.6	101.9	122.4	112.1	93.8	89.0	91.4	101.8	108.0	104.9
0330	Cast-in-Place Concrete	92.4	92.1	92.3	95.0	95.3	95.1	76.2	108.6	89.5	85.9	91.6	88.3	87.8	62.7	77.5	102.5	105.4	103.7
03	CONCRETE	95.2	96.1	95.6	94.6	89.6	92.1	79.9	110.1	94.8	91.8	96.2	94.0	84.4	71.6	78.1	97.8	101.0	99.4
04	MASONRY	92.6	92.4	92.5	94.3	75.1	82.6	94.1	110.3	104.0	100.6	87.2	92.4	91.8	81.5	85.5	99.2	105.5	103.0
05	METALS	92.2	115.2	99.4	91.5	84.8	89.4	96.8	98.4	97.3	93.7	115.0	100.4	94.3	87.8	92.3	92.2	121.5	101.4
06	WOOD, PLASTICS & COMPOSITES	102.1	84.9	92.3	96.6	86.9	91.1	94.8	110.1	103.5	95.3	86.8	90.4	87.6	68.2	76.5	71.8	89.4	81.8
07	THERMAL & MOISTURE PROTECTION	98.5	90.5	95.3	93.3	76.9	86.7	97.5	112.2	103.4	99.6	94.4	97.5	95.3	70.2	85.2	106.2	98.2	103.0
08	OPENINGS	98.4	93.3	97.2	93.4	86.4	91.7	98.5	108.4	100.9	96.1	94.4	95.7	101.9	63.9	92.7	96.4	110.2	99.8
0920	Plaster & Gypsum Board	100.8	83.5	88.7	76.1	82.7	80.7	102.7	109.5	107.4	106.1	85.4	91.6	59.3	67.7	65.2	104.5	89.3	93.9
0950, 0980	Ceilings & Acoustic Treatment	100.2	83.5	89.1	94.6	82.7	86.7	93.9	109.5	104.3	94.7	85.4	88.5	91.6	67.7	75.7	144.5	89.3	107.7
0960	Flooring	105.5	82.3	98.6	97.6	64.7	87.8	93.5	115.2	100.0	95.7	79.7	90.9	113.2	49.9	94.4	111.2	125.2	115.4
0970, 0990	Wall Finishes & Painting/Coating	99.3	83.4	89.7	92.3	80.6	85.3	90.0	101.1	96.6	88.6	82.6	85.0	106.7	45.2	69.8	94.1	100.1	97.7
09	FINISHES	101.5	85.0	92.3	93.8	81.8	87.1	95.6	110.2	103.8	96.1	84.9	89.9	98.5	62.8	78.5	115.4	98.8	106.1
COVERS	DIVS. 10 - 14, 25, 28, 41, 43, 44, 46	100.0	93.8	98.7	100.0	99.1	99.8	100.0	98.8	99.8	100.0	93.3	98.6	100.0	86.9	97.3	100.0	92.3	98.4
21, 22, 23	FIRE SUPPRESSION, PLUMBING & HVAC	99.9	90.5	96.0	99.8	83.7	93.2	94.4	107.6	99.8	100.0	86.0	94.2	94.2	80.9	88.8	94.4	86.5	91.2
26, 27, 3370	ELECTRICAL, COMMUNICATIONS & UTIL.	99.9	91.2	95.3	94.7	78.6	86.2	98.5	92.2	95.2	93.5	89.5	91.4	93.9	75.2	84.1	105.3	102.0	103.6
MF2010	WEIGHTED AVERAGE	97.4	93.8	95.8	95.4	83.6	90.2	94.4	104.5	98.9	96.0	92.2	94.3	94.2	77.4	86.8	98.4	99.8	99.0

	DIVISION	MINNESOTA																	
		BRAINERD			DETROIT LAKES			DULUTH			MANKATO			MINNEAPOLIS			ROCHESTER		
		564			565			556 - 558			560			553 - 555			559		
		MAT.	INST.	TOTAL	MAT.	INST.	TOTAL	MAT.	INST.	TOTAL	MAT.	INST.	TOTAL	MAT.	INST.	TOTAL	MAT.	INST.	TOTAL
015433	CONTRACTOR EQUIPMENT		101.6	101.6		98.9	98.9		100.5	100.5		101.6	101.6		104.0	104.0		100.5	100.5
0241, 31 - 34	SITE & INFRASTRUCTURE, DEMOLITION	97.3	103.1	101.4	95.5	98.7	97.8	100.1	100.6	100.5	93.8	103.8	100.8	102.4	105.8	104.8	101.6	100.3	100.6
0310	Concrete Forming & Accessories	93.0	95.4	95.1	87.8	95.4	94.3	101.3	114.5	112.8	100.3	103.0	102.6	102.0	129.9	126.1	102.6	108.9	108.0
0320	Concrete Reinforcing	100.6	108.2	104.4	101.8	108.0	104.9	98.6	108.5	103.5	100.4	120.1	110.2	98.8	120.5	109.6	98.6	120.2	109.4
0330	Cast-in-Place Concrete	111.4	110.3	110.9	99.5	109.6	103.7	103.2	108.5	105.4	102.7	109.4	105.4	115.5	119.3	117.1	113.2	101.8	108.5
03	CONCRETE	101.8	103.9	102.9	95.4	103.6	99.5	99.0	111.9	105.4	97.4	109.3	103.3	106.6	124.8	115.6	103.9	109.4	106.6
04	MASONRY	122.2	115.8	118.3	122.3	113.1	116.7	108.1	119.5	115.0	110.7	112.8	111.9	107.4	127.2	119.5	106.7	115.2	111.9
05	METALS	93.4	121.5	102.3	92.2	121.4	101.4	97.6	124.1	105.9	93.3	128.2	104.3	100.3	132.0	110.2	97.5	130.2	107.7
06	WOOD, PLASTICS & COMPOSITES	91.5	89.2	90.2	69.0	89.4	80.6	110.2	114.3	112.5	100.1	100.8	100.5	113.4	128.7	122.1	113.4	107.5	110.0
07	THERMAL & MOISTURE PROTECTION	105.1	110.5	107.2	106.0	110.7	107.9	105.7	118.7	110.9	105.4	101.6	103.9	104.8	130.8	115.3	105.1	102.9	104.2
08	OPENINGS	84.3	110.1	90.6	96.4	110.2	99.8	92.2	119.3	98.8	88.7	120.1	96.4	95.9	136.0	105.6	92.2	124.5	100.0
0920	Plaster & Gypsum Board	86.4	89.3	88.5	103.3	89.3	93.5	92.9	115.1	108.4	91.0	101.3	98.2	101.6	129.8	121.3	101.1	108.1	106.0
0950, 0980	Ceilings & Acoustic Treatment	70.4	89.3	83.0	144.5	89.3	107.7	102.2	115.1	110.8	70.4	101.3	91.0	99.5	129.8	119.7	97.7	108.1	104.6
0960	Flooring	110.1	125.2	114.6	110.0	125.2	114.5	104.3	126.5	110.9	111.8	125.2	115.8	102.3	125.2	109.1	104.7	125.2	110.8
0970, 0990	Wall Finishes & Painting/Coating	88.5	100.1	95.4	94.1	100.1	97.7	87.4	110.6	101.3	99.6	104.2	102.3	96.1	130.9	117.0	90.7	104.2	98.8
09	FINISHES	96.2	100.5	98.6	114.7	100.7	106.8	101.5	116.7	110.1	97.8	107.1	103.0	102.2	129.7	117.6	102.0	111.5	107.3
COVERS	DIVS. 10 - 14, 25, 28, 41, 43, 44, 46	100.0	94.4	98.9	100.0	94.7	98.9	100.0	101.7	100.3	100.0	94.6	98.9	100.0	109.3	101.9	100.0	103.0	100.6
21, 22, 23	FIRE SUPPRESSION, PLUMBING & HVAC	93.6	90.4	92.3	94.4	90.1	92.7	99.9	99.2	99.6	93.6	90.1	92.2	99.9	120.1	108.2	100.0	97.7	99.1
26, 27, 3370	ELECTRICAL, COMMUNICATIONS & UTIL.	102.7	106.8	104.8	105.0	69.2	86.1	102.1	106.8	104.6	109.8	87.3	98.0	101.7	111.0	106.6	100.8	87.3	93.7
MF2010	WEIGHTED AVERAGE	97.1	103.7	100.0	99.1	97.9	98.6	99.7	110.3	104.3	97.3	103.2	99.9	101.4	122.0	110.5	100.1	106.1	102.7

	DIVISION	MINNESOTA															MISSISSIPPI		
		SAINT PAUL			ST. CLOUD			THIEF RIVER FALLS			WILLMAR			WINDOM			BILOXI		
		550 - 551			563			567			562			561			395		
		MAT.	INST.	TOTAL	MAT.	INST.	TOTAL	MAT.	INST.	TOTAL	MAT.	INST.	TOTAL	MAT.	INST.	TOTAL	MAT.	INST.	TOTAL
015433	CONTRACTOR EQUIPMENT		100.5	100.5		101.6	101.6		98.9	98.9		101.6	101.6		101.6	101.6		99.4	99.4
0241, 31 - 34	SITE & INFRASTRUCTURE, DEMOLITION	102.1	101.2	101.5	92.3	105.2	101.3	96.3	98.2	97.6	91.8	102.9	99.6	85.5	101.6	96.8	101.0	89.8	93.2
0310	Concrete Forming & Accessories	99.1	127.2	123.5	90.3	122.9	118.5	91.2	92.0	91.9	90.1	94.8	94.2	94.3	86.9	87.9	95.7	55.2	60.7
0320	Concrete Reinforcing	95.1	120.5	107.8	100.6	120.3	110.4	102.1	107.9	105.0	100.2	120.0	110.1	100.2	119.1	109.6	95.3	59.3	77.4
0330	Cast-in-Place Concrete	115.5	118.6	116.8	98.4	118.1	106.5	101.6	90.3	97.0	99.9	88.3	95.1	86.4	66.4	78.2	109.4	54.6	86.9
03	CONCRETE	108.2	123.4	115.7	93.3	121.2	107.0	96.5	95.5	96.0	93.2	98.4	95.8	84.2	87.2	85.7	98.7	57.4	78.3
04	MASONRY	116.5	127.2	123.0	107.5	124.6	117.9	99.1	105.5	103.0	111.8	115.8	114.3	121.9	88.8	101.7	95.6	45.0	64.7
05	METALS	97.0	131.6	107.9	94.1	129.6	105.2	92.4	120.8	101.3	93.2	127.2	103.9	93.1	125.0	103.1	92.5	83.3	89.6
06	WOOD, PLASTICS & COMPOSITES	106.3	125.4	117.2	89.0	120.1	106.8	72.7	89.4	82.2	88.8	89.4	89.1	92.7	86.6	89.2	97.9	58.6	75.5
07	THERMAL & MOISTURE PROTECTION	106.3	130.3	116.0	105.2	120.9	111.5	107.1	99.9	104.2	105.1	110.2	107.1	105.0	90.1	99.0	95.0	56.1	79.3
08	OPENINGS	90.4	134.2	101.1	88.7	131.4	99.1	96.4	110.2	99.8	86.4	95.4	88.6	90.0	93.9	91.0	97.4	59.8	88.2
0920	Plaster & Gypsum Board	89.1	126.6	115.3	86.4	121.2	110.8	104.1	89.3	93.8	86.4	89.6	88.6	86.4	86.6	86.6	102.5	57.8	71.2
0950, 0980	Ceilings & Acoustic Treatment	97.7	126.6	116.9	70.4	121.2	104.2	144.5	89.3	107.7	70.4	89.6	83.2	70.4	86.6	81.2	94.6	57.8	70.1
0960	Flooring	98.9	125.2	106.7	106.4	125.2	112.0	110.9	125.2	115.2	108.3	125.2	113.3	110.7	125.2	115.1	117.7	54.8	99.0
0970, 0990	Wall Finishes & Painting/Coating	96.1	126.5	114.3	99.6	130.9	118.4	94.1	100.1	97.7	94.1	100.1	97.7	94.1	104.2	100.1	101.6	45.4	67.9
09	FINISHES	99.2	127.2	114.9	95.5	124.4	111.7	115.2	98.8	106.0	95.6	98.2	97.1	95.7	93.7	94.5	103.1	54.3	75.7
COVERS	DIVS. 10 - 14, 25, 28, 41, 43, 44, 46	100.0	108.6	101.8	100.0	100.8	100.2	100.0	92.2	98.4	100.0	94.3	98.8	100.0	89.6	97.9	100.0	57.8	91.4
21, 22, 23	FIRE SUPPRESSION, PLUMBING & HVAC	99.9	115.1	106.1	99.6	119.7	107.9	94.4	86.2	91.1	93.6	108.1	99.5	93.6	84.7	90.0	99.9	49.5	79.3
26, 27, 3370	ELECTRICAL, COMMUNICATIONS & UTIL.	99.2	111.0	105.4	102.7	111.0	107.0	102.3	69.2	84.8	102.7	88.8	95.4	109.8	87.3	98.0	98.2	57.9	77.0
MF2010	WEIGHTED AVERAGE	100.4	119.9	109.0	97.2	119.4	107.0	98.0	94.5	96.4	95.6	103.9	99.2	96.0	92.8	94.6	98.1	59.5	81.0

	DIVISION	MISSISSIPPI																	
		CLARKSDALE			COLUMBUS			GREENVILLE			GREENWOOD			JACKSON			LAUREL		
		386			397			387			389			390 - 392			394		
		MAT.	INST.	TOTAL	MAT.	INST.	TOTAL	MAT.	INST.	TOTAL	MAT.	INST.	TOTAL	MAT.	INST.	TOTAL	MAT.	INST.	TOTAL
015433	CONTRACTOR EQUIPMENT		99.4	99.4		99.4	99.4		99.4	99.4		99.4	99.4		99.4	99.4		99.4	99.4
0241, 31 - 34	SITE & INFRASTRUCTURE, DEMOLITION	98.6	89.6	92.3	99.5	89.9	92.8	104.4	89.8	94.2	101.4	89.3	92.9	96.6	89.8	91.9	105.4	89.7	94.4
0310	Concrete Forming & Accessories	86.7	51.5	56.3	84.6	53.6	57.8	83.3	66.6	68.9	95.6	51.5	57.4	95.4	60.1	64.8	84.7	55.4	59.3
0320	Concrete Reinforcing	102.7	43.6	73.3	102.1	44.6	73.5	103.3	63.2	83.3	102.7	55.1	79.0	95.3	58.3	76.9	102.7	41.0	72.0
0330	Cast-in-Place Concrete	106.7	55.5	85.6	111.6	60.7	90.7	109.9	57.7	88.4	114.7	55.1	90.2	93.1	58.4	78.9	109.1	57.2	87.8
03	CONCRETE	96.9	53.1	75.3	100.4	56.0	78.5	101.9	64.2	83.3	103.2	55.0	79.4	90.9	60.7	76.0	102.5	55.0	79.0
04	MASONRY	95.4	46.9	65.8	122.0	53.8	80.3	141.2	50.4	85.8	95.9	46.7	65.9	96.6	50.4	68.4	117.8	50.2	76.5
05	METALS	89.6	75.8	85.3	89.5	76.7	85.5	90.7	85.1	88.9	89.6	77.3	85.7	92.4	83.1	89.5	89.6	75.7	85.2
06	WOOD, PLASTICS & COMPOSITES	85.3	53.5	67.2	83.3	54.5	66.9	82.0	71.2	75.8	98.0	53.5	72.6	96.5	62.3	77.0	84.4	57.1	68.8
07	THERMAL & MOISTURE PROTECTION	94.8	49.0	76.4	94.9	53.4	78.2	95.1	62.0	81.8	95.1	48.9	76.5	96.8	61.0	82.4	95.0	55.1	79.0
08	OPENINGS	96.7	51.6	85.7	96.6	51.7	85.7	96.7	64.2	88.8	96.7	53.0	86.0	97.7	58.0	88.1	93.5	52.5	83.5
0920	Plaster & Gypsum Board	90.7	52.5	64.0	90.3	53.6	64.6	90.3	70.7	76.6	101.5	52.5	67.2	95.5	61.5	71.7	90.3	56.2	66.4
0950, 0980	Ceilings & Acoustic Treatment	88.9	52.5	64.7	88.9	53.6	65.4	91.0	70.7	77.5	88.9	52.5	64.7	98.2	61.5	73.8	88.9	56.2	67.1
0960	Flooring	111.9	54.8	94.9	110.8	61.4	96.1	110.0	54.8	93.6	117.7	54.8	99.0	117.9	54.8	99.1	109.2	54.8	93.0
0970, 0990	Wall Finishes & Painting/Coating	101.6	48.2	69.6	101.6	55.3	73.9	101.6	68.8	81.9	101.6	48.2	69.6	101.6	68.8	81.9	101.6	63.6	78.8
09	FINISHES	98.4	52.1	72.5	98.3	55.6	74.4	99.0	65.7	80.3	101.9	52.1	74.0	103.2	60.4	79.2	98.4	56.7	75.1
COVERS	DIVS. 10 - 14, 25, 28, 41, 43, 44, 46	100.0	58.1	91.4	100.0	59.3	91.7	100.0	61.0	92.0	100.0	58.1	91.4	100.0	60.0	91.8	100.0	39.0	87.5
21, 22, 23	FIRE SUPPRESSION, PLUMBING & HVAC	97.4	43.6	75.4	96.7	39.6	73.3	99.9	50.2	79.6	97.4	41.5	74.5	100.0	55.5	81.8	96.7	48.3	76.9
26, 27, 3370	ELECTRICAL, COMMUNICATIONS & UTIL.	97.4	48.3	71.6	95.5	51.2	72.1	97.4	65.3	80.5	97.4	45.3	70.0	100.0	65.3	81.7	97.1	58.4	76.7
MF2010	WEIGHTED AVERAGE	96.1	54.9	77.9	97.4	56.2	79.3	99.9	64.3	84.2	97.3	54.4	78.4	97.4	63.6	82.5	97.5	58.1	80.1

	DIVISION	MISSISSIPPI									MISSOURI								
		MCCOMB			MERIDIAN			TUPELO			BOWLING GREEN			CAPE GIRARDEAU			CHILLICOTHE		
		396			393			388			633			637			646		
		MAT.	INST.	TOTAL	MAT.	INST.	TOTAL	MAT.	INST.	TOTAL	MAT.	INST.	TOTAL	MAT.	INST.	TOTAL	MAT.	INST.	TOTAL
015433	CONTRACTOR EQUIPMENT		99.4	99.4		99.4	99.4		99.4	99.4		105.5	105.5		105.5	105.5		102.2	102.2
0241, 31 - 34	SITE & INFRASTRUCTURE, DEMOLITION	92.5	89.5	90.4	96.5	90.1	92.0	96.0	89.5	91.5	93.4	94.4	94.1	94.9	94.2	94.4	102.1	94.5	96.7
0310	Concrete Forming & Accessories	84.7	53.3	57.6	82.1	66.1	68.3	83.8	53.7	57.7	96.0	90.1	90.9	89.4	75.5	77.4	89.9	96.6	95.7
0320	Concrete Reinforcing	103.3	43.0	73.3	102.1	58.4	80.3	100.5	55.1	77.9	100.4	99.8	100.1	101.6	85.4	93.5	98.2	103.8	101.0
0330	Cast-in-Place Concrete	96.9	55.0	79.6	103.6	60.9	86.1	106.7	56.9	86.2	92.5	86.1	89.9	91.6	84.4	88.6	95.8	88.7	92.9
03	CONCRETE	90.6	53.5	72.3	94.7	64.2	79.6	96.4	56.5	76.7	96.5	91.6	94.1	95.7	81.9	88.9	102.3	95.7	99.0
04	MASONRY	123.4	46.2	76.3	95.3	54.5	70.4	130.9	49.9	81.4	124.6	97.6	108.2	121.5	77.7	94.7	105.5	95.6	99.4
05	METALS	89.7	74.0	84.8	90.6	83.3	88.3	89.5	77.5	85.8	94.8	112.9	100.4	95.8	105.2	98.8	94.9	107.0	98.7
06	WOOD, PLASTICS & COMPOSITES	83.3	56.4	68.0	80.8	68.1	73.6	82.6	54.5	66.6	90.4	89.8	90.1	83.3	72.9	77.4	90.0	97.5	94.3
07	THERMAL & MOISTURE PROTECTION	94.5	48.1	75.9	94.7	63.4	82.1	94.8	54.9	78.8	101.7	99.1	100.7	101.6	81.8	93.6	97.3	96.1	96.8
08	OPENINGS	96.7	53.3	86.1	96.6	66.9	89.4	96.6	52.9	86.0	97.8	99.8	98.3	97.8	75.8	92.4	91.2	98.1	92.9
0920	Plaster & Gypsum Board	90.3	55.5	66.0	90.3	67.5	74.4	90.3	53.6	64.6	103.3	89.5	93.7	101.2	72.2	80.9	101.6	97.2	98.5
0950, 0980	Ceilings & Acoustic Treatment	88.9	55.5	66.7	91.0	67.5	75.3	88.9	53.6	65.4	93.4	89.5	90.8	93.4	72.2	79.2	89.0	97.2	94.5
0960	Flooring	110.8	54.8	94.1	109.2	54.8	92.9	110.3	54.8	93.7	103.0	98.2	101.6	99.8	87.1	96.0	94.9	106.2	98.3
0970, 0990	Wall Finishes & Painting/Coating	101.6	44.2	67.2	101.6	79.5	88.4	101.6	61.3	77.5	104.6	105.1	104.9	104.6	76.5	87.8	93.5	91.6	92.4
09	FINISHES	97.7	53.2	72.8	97.9	66.1	80.1	97.9	54.9	73.8	101.3	92.5	96.4	100.0	76.4	86.8	96.3	98.0	97.3
COVERS	DIVS. 10 - 14, 25, 28, 41, 43, 44, 46	100.0	61.6	92.2	100.0	62.1	92.2	100.0	59.3	91.7	100.0	91.2	98.2	100.0	88.5	97.7	100.0	92.0	98.4
21, 22, 23	FIRE SUPPRESSION, PLUMBING & HVAC	96.7	36.4	72.0	99.9	57.7	82.7	97.7	45.6	76.4	93.9	94.9	94.3	99.9	93.2	97.2	94.1	96.5	95.1
26, 27, 3370	ELECTRICAL, COMMUNICATIONS & UTIL.	93.9	51.5	71.6	97.1	58.1	76.6	97.2	51.3	73.0	99.9	82.3	90.6	99.9	105.8	103.0	95.6	81.6	88.2
MF2010	WEIGHTED AVERAGE	96.1	54.0	77.5	96.5	65.3	82.7	97.7	57.3	79.9	98.2	94.5	96.6	99.5	89.5	95.1	96.6	95.2	96.0

DIVISION		MISSOURI																	
		COLUMBIA			FLAT RIVER			HANNIBAL			HARRISONVILLE			JEFFERSON CITY			JOPLIN		
		652			636			634			647			650 - 651			648		
		MAT.	INST.	TOTAL	MAT.	INST.	TOTAL	MAT.	INST.	TOTAL	MAT.	INST.	TOTAL	MAT.	INST.	TOTAL	MAT.	INST.	TOTAL
015433	**CONTRACTOR EQUIPMENT**		106.9	106.9		105.5	105.5		105.5	105.5		102.2	102.2		106.9	106.9		105.9	105.9
0241, 31 - 34	**SITE & INFRASTRUCTURE, DEMOLITION**	103.3	96.3	98.4	96.2	94.3	94.9	91.2	94.4	93.4	93.4	95.3	94.8	100.2	96.2	97.4	102.4	99.6	100.4
0310	Concrete Forming & Accessories	88.0	79.7	80.8	101.7	86.6	88.6	94.3	80.4	82.2	87.2	102.4	100.4	97.9	79.6	82.0	101.3	75.0	78.6
0320	Concrete Reinforcing	98.3	118.2	108.2	101.6	108.0	104.8	99.8	95.9	97.9	97.7	110.4	104.1	97.1	110.2	103.6	101.3	103.4	102.4
0330	Cast-in-Place Concrete	88.5	86.1	87.5	95.5	94.0	94.9	87.6	85.9	86.9	98.1	105.4	101.1	89.0	86.1	87.8	103.7	78.9	93.5
03	**CONCRETE**	86.1	90.5	88.3	99.5	94.3	96.9	92.8	86.5	89.7	98.5	105.3	101.9	87.7	89.0	88.3	101.8	82.6	92.3
04	**MASONRY**	136.6	89.8	108.1	121.9	79.1	95.8	115.7	97.6	104.7	99.9	106.3	103.8	99.1	89.8	93.4	98.5	88.2	92.2
05	**METALS**	93.2	120.8	101.9	94.7	115.3	101.1	94.8	111.0	99.8	95.3	111.4	100.4	92.6	117.5	100.4	98.0	103.4	99.7
06	**WOOD, PLASTICS & COMPOSITES**	90.7	75.8	82.2	97.9	85.8	91.0	88.3	77.0	81.9	86.5	101.9	95.3	101.5	75.8	86.9	101.9	73.3	85.6
07	**THERMAL & MOISTURE PROTECTION**	94.4	88.7	92.1	101.9	93.6	98.5	101.5	97.8	100.0	96.6	106.0	100.4	97.2	88.7	93.8	96.8	79.8	89.9
08	**OPENINGS**	90.3	94.1	91.3	97.8	100.1	98.3	97.8	84.2	94.5	91.4	104.4	94.6	87.7	92.0	88.8	92.4	79.8	89.3
0920	Plaster & Gypsum Board	98.8	75.1	82.2	109.1	85.4	92.5	101.6	76.4	84.0	96.1	101.8	100.1	102.8	75.1	83.4	109.3	72.3	83.4
0950, 0980	Ceilings & Acoustic Treatment	94.6	75.1	81.6	93.4	85.4	88.1	93.4	76.4	82.0	89.0	101.8	97.5	101.0	75.1	83.7	89.9	72.3	78.2
0960	Flooring	113.1	99.2	108.9	106.1	87.1	100.4	102.4	98.2	101.1	90.6	106.2	95.2	120.4	99.2	114.1	119.7	77.7	107.2
0970, 0990	Wall Finishes & Painting/Coating	103.4	78.8	88.7	104.6	81.9	91.0	104.6	105.1	104.9	98.2	106.0	102.9	103.4	78.8	88.7	93.1	78.2	84.2
09	**FINISHES**	99.9	81.6	89.7	103.1	85.0	92.9	100.6	85.0	91.9	93.9	103.4	99.2	104.2	81.6	91.6	103.0	75.9	87.8
COVERS	**DIVS. 10 - 14, 25, 28, 41, 43, 44, 46**	100.0	95.8	99.1	100.0	90.4	98.0	100.0	89.8	97.9	100.0	94.1	98.8	100.0	95.8	99.1	100.0	90.8	98.1
21, 22, 23	**FIRE SUPPRESSION, PLUMBING & HVAC**	99.9	95.1	97.9	93.9	94.4	94.1	93.9	94.9	94.3	94.0	101.4	97.1	99.9	90.6	96.1	100.2	73.5	89.3
26, 27, 3370	**ELECTRICAL, COMMUNICATIONS & UTIL.**	96.4	87.3	91.6	104.6	105.8	105.2	98.6	82.3	90.0	102.8	99.7	101.2	102.3	87.3	94.4	93.2	67.8	79.8
MF2010	**WEIGHTED AVERAGE**	97.7	93.2	95.7	99.1	95.2	97.4	97.1	91.8	94.7	96.2	102.9	99.1	96.6	91.7	94.4	98.6	81.7	91.2

DIVISION		MISSOURI																	
		KANSAS CITY			KIRKSVILLE			POPLAR BLUFF			ROLLA			SEDALIA			SIKESTON		
		640 - 641			635			639			654 - 655			653			638		
		MAT.	INST.	TOTAL	MAT.	INST.	TOTAL	MAT.	INST.	TOTAL	MAT.	INST.	TOTAL	MAT.	INST.	TOTAL	MAT.	INST.	TOTAL
015433	**CONTRACTOR EQUIPMENT**		103.5	103.5		97.8	97.8		100.2	100.2		106.9	106.9		99.0	99.0		100.2	100.2
0241, 31 - 34	**SITE & INFRASTRUCTURE, DEMOLITION**	95.4	97.9	97.2	94.7	90.4	91.7	81.4	94.2	90.4	102.5	96.1	98.0	96.9	91.9	93.4	85.0	94.3	91.5
0310	Concrete Forming & Accessories	100.2	109.3	108.0	87.1	78.9	80.0	87.2	75.5	77.1	95.2	95.4	95.4	92.4	78.8	80.7	88.3	75.5	77.2
0320	Concrete Reinforcing	96.2	119.0	107.5	100.7	103.5	102.1	103.7	99.1	101.4	98.8	99.4	99.1	97.4	103.7	100.5	103.1	103.0	103.1
0330	Cast-in-Place Concrete	96.5	109.4	101.8	95.4	85.8	91.5	73.5	84.9	78.2	90.5	86.7	88.9	94.5	85.1	90.6	78.6	84.9	81.2
03	**CONCRETE**	97.9	111.3	104.5	111.0	86.8	99.0	86.5	84.1	85.3	87.9	94.1	91.0	100.0	86.5	93.3	90.1	84.8	87.5
04	**MASONRY**	101.9	109.3	106.4	128.2	89.7	104.7	120.0	77.7	94.2	112.7	90.5	99.1	118.5	87.2	99.4	119.7	77.7	94.1
05	**METALS**	105.6	115.8	108.8	94.4	105.3	97.8	94.9	102.8	97.4	92.6	112.3	98.8	91.2	106.5	96.0	95.2	104.3	98.1
06	**WOOD, PLASTICS & COMPOSITES**	101.2	109.1	105.7	76.6	75.8	76.1	75.8	73.0	74.2	98.0	96.9	97.4	91.0	75.9	82.4	77.4	73.0	74.9
07	**THERMAL & MOISTURE PROTECTION**	96.7	109.3	101.8	108.0	89.7	100.6	106.6	83.6	97.3	94.5	96.8	95.4	100.2	94.4	97.8	106.8	82.1	96.8
08	**OPENINGS**	98.6	111.3	101.7	102.5	86.2	98.6	103.6	79.4	97.7	90.3	92.3	90.8	94.0	85.9	92.0	103.6	80.5	98.0
0920	Plaster & Gypsum Board	105.6	109.2	108.1	96.2	75.1	81.4	96.7	72.2	79.5	100.1	96.8	97.8	93.4	75.1	80.6	98.7	72.2	80.2
0950, 0980	Ceilings & Acoustic Treatment	96.3	109.2	104.9	91.5	75.1	80.6	93.4	72.2	79.2	94.6	96.8	96.1	94.6	75.1	81.6	93.4	72.2	79.2
0960	Flooring	95.6	109.6	99.7	79.5	94.9	84.1	94.4	83.1	91.0	116.6	98.1	111.1	88.8	97.5	91.4	94.8	83.1	91.4
0970, 0990	Wall Finishes & Painting/Coating	98.2	115.2	108.4	99.7	78.8	87.2	99.2	76.5	85.6	103.4	105.1	104.4	103.4	104.7	104.2	99.2	76.5	85.6
09	**FINISHES**	98.1	109.9	104.7	99.5	80.8	89.0	99.7	75.8	86.3	101.2	97.1	98.9	97.5	83.6	89.7	100.5	75.8	86.6
COVERS	**DIVS. 10 - 14, 25, 28, 41, 43, 44, 46**	100.0	101.8	100.4	100.0	89.7	97.9	100.0	88.5	97.7	100.0	92.5	98.5	100.0	90.8	98.1	100.0	88.5	97.7
21, 22, 23	**FIRE SUPPRESSION, PLUMBING & HVAC**	100.0	109.6	103.9	93.9	94.5	94.2	93.9	92.8	93.5	93.8	97.1	95.2	93.8	95.5	94.5	93.9	92.8	93.5
26, 27, 3370	**ELECTRICAL, COMMUNICATIONS & UTIL.**	104.5	101.9	103.1	98.8	82.2	90.1	99.1	105.8	102.6	94.7	87.3	90.8	96.3	99.7	98.1	98.1	105.8	102.1
MF2010	**WEIGHTED AVERAGE**	100.7	108.2	104.0	100.3	89.4	95.5	97.0	89.7	93.8	95.1	95.7	95.3	96.7	92.4	94.8	97.5	89.9	94.2

DIVISION		MISSOURI									MONTANA								
		SPRINGFIELD			ST. JOSEPH			ST. LOUIS			BILLINGS			BUTTE			GREAT FALLS		
		656 - 658			644 - 645			630 - 631			590 - 591			597			594		
		MAT.	INST.	TOTAL	MAT.	INST.	TOTAL	MAT.	INST.	TOTAL	MAT.	INST.	TOTAL	MAT.	INST.	TOTAL	MAT.	INST.	TOTAL
015433	**CONTRACTOR EQUIPMENT**		101.6	101.6		102.2	102.2		106.5	106.5		99.2	99.2		98.9	98.9		98.9	98.9
0241, 31 - 34	**SITE & INFRASTRUCTURE, DEMOLITION**	99.2	94.0	95.6	96.9	93.6	94.6	95.1	97.7	96.9	98.7	97.1	97.6	107.2	96.5	99.7	111.0	97.0	101.2
0310	Concrete Forming & Accessories	100.4	75.8	79.1	100.1	91.0	92.2	101.8	104.6	104.2	102.0	70.1	74.4	89.6	74.2	76.3	101.7	70.1	74.4
0320	Concrete Reinforcing	94.7	117.9	106.3	95.1	110.1	102.5	92.9	110.9	101.9	99.6	83.2	91.4	108.1	83.4	95.8	99.7	83.3	91.5
0330	Cast-in-Place Concrete	96.0	78.0	88.6	96.4	105.0	99.9	91.6	105.4	97.3	111.8	69.3	94.3	123.0	70.8	101.5	130.0	65.2	103.3
03	**CONCRETE**	96.8	85.3	91.1	97.6	100.0	98.8	95.1	106.8	100.9	104.5	73.2	89.0	108.0	75.5	92.0	112.9	71.8	92.6
04	**MASONRY**	92.3	88.1	89.7	101.7	96.5	98.5	102.0	110.9	107.4	124.0	75.2	94.2	119.2	78.1	94.1	123.4	76.9	95.1
05	**METALS**	97.1	108.4	100.6	101.7	110.3	104.4	100.5	119.3	106.4	109.7	91.5	104.0	102.5	91.7	99.1	106.4	91.6	101.8
06	**WOOD, PLASTICS & COMPOSITES**	98.6	74.3	84.7	101.9	89.5	94.8	97.8	102.9	100.7	96.5	68.5	80.5	85.0	74.0	78.7	97.6	68.5	81.0
07	**THERMAL & MOISTURE PROTECTION**	98.7	81.0	91.6	97.2	96.5	96.9	101.7	107.0	103.8	105.0	71.0	91.3	104.9	72.6	91.9	105.4	69.5	91.0
08	**OPENINGS**	96.4	82.9	93.1	97.0	98.4	97.3	97.6	110.6	100.8	95.5	68.3	88.8	93.7	72.5	88.5	96.7	67.6	89.6
0920	Plaster & Gypsum Board	101.3	73.5	81.8	110.8	89.0	95.5	110.2	103.0	105.2	97.7	67.8	76.8	99.8	73.5	81.4	108.6	67.8	80.1
0950, 0980	Ceilings & Acoustic Treatment	94.6	73.5	80.5	95.4	89.0	91.1	98.8	103.0	101.6	111.8	67.8	82.5	118.1	73.5	88.4	119.9	67.8	85.2
0960	Flooring	114.0	77.7	103.2	98.7	106.2	100.9	105.8	98.7	103.7	110.4	72.8	99.2	108.1	76.9	98.8	115.7	76.9	104.1
0970, 0990	Wall Finishes & Painting/Coating	97.1	78.2	85.8	93.5	90.6	91.7	104.6	105.7	105.3	95.9	54.0	70.8	94.1	53.5	69.8	94.1	54.0	70.0
09	**FINISHES**	103.3	76.4	88.2	99.2	93.2	95.8	104.3	103.0	103.6	107.4	68.4	85.5	108.1	72.4	88.1	112.1	69.1	88.0
COVERS	**DIVS. 10 - 14, 25, 28, 41, 43, 44, 46**	100.0	93.3	98.6	100.0	97.2	99.4	100.0	102.2	100.4	100.0	94.7	98.9	100.0	95.3	99.0	100.0	94.7	98.9
21, 22, 23	**FIRE SUPPRESSION, PLUMBING & HVAC**	99.9	73.5	89.1	100.2	90.4	96.2	99.9	103.3	101.3	100.2	74.9	89.9	100.3	70.5	88.1	100.3	73.6	89.4
26, 27, 3370	**ELECTRICAL, COMMUNICATIONS & UTIL.**	100.7	71.1	85.1	102.8	81.6	91.7	102.5	106.2	104.5	97.6	75.2	85.8	105.7	72.2	88.1	97.7	72.2	84.3
MF2010	**WEIGHTED AVERAGE**	98.6	82.8	91.7	99.9	94.3	97.4	99.9	106.3	102.7	103.3	77.5	91.9	103.1	77.5	91.8	104.5	76.8	92.3

DIVISION		MONTANA																	
		HAVRE			HELENA			KALISPELL			MILES CITY			MISSOULA			WOLF POINT		
		595			596			599			593			598			592		
		MAT.	INST.	TOTAL	MAT.	INST.	TOTAL	MAT.	INST.	TOTAL	MAT.	INST.	TOTAL	MAT.	INST.	TOTAL	MAT.	INST.	TOTAL
015433	CONTRACTOR EQUIPMENT		98.9	98.9		98.9	98.9		98.9	98.9		98.9	98.9		98.9	98.9		98.9	98.9
0241, 31 - 34	SITE & INFRASTRUCTURE, DEMOLITION	115.7	96.8	102.4	100.5	96.1	97.4	96.7	96.4	96.5	103.7	96.4	98.6	88.5	96.6	94.2	123.6	96.6	104.6
0310	Concrete Forming & Accessories	83.2	68.3	70.3	103.6	67.2	72.1	92.6	71.8	74.6	100.5	66.8	71.3	92.6	72.1	74.8	93.3	67.1	70.6
0320	Concrete Reinforcing	109.0	83.3	96.2	103.1	83.2	93.2	111.0	83.3	97.2	108.5	83.3	96.0	109.9	83.4	96.7	110.1	83.3	96.7
0330	Cast-in-Place Concrete	132.5	62.9	103.8	100.9	64.3	85.8	106.8	64.0	89.2	116.9	62.7	94.6	90.7	66.6	80.8	131.0	63.1	103.1
03	CONCRETE	116.2	70.2	93.5	99.7	70.1	85.1	97.9	72.1	85.2	105.0	69.5	87.4	86.4	73.1	79.9	120.1	69.7	95.2
04	MASONRY	120.0	72.8	91.2	115.3	76.9	91.8	118.2	78.2	93.8	125.6	68.1	90.5	143.4	78.1	103.6	126.7	68.8	91.3
05	METALS	99.0	91.5	96.6	105.6	90.4	100.8	98.9	91.6	96.6	98.2	91.7	96.2	99.4	91.8	97.0	98.3	91.7	96.2
06	WOOD, PLASTICS & COMPOSITES	77.5	68.5	72.4	97.7	65.1	79.1	88.0	71.0	78.3	95.0	68.5	79.9	88.0	71.0	78.3	87.3	68.5	76.6
07	THERMAL & MOISTURE PROTECTION	105.2	67.3	90.0	104.6	72.2	91.6	104.5	72.4	91.6	104.8	65.5	89.0	104.2	80.8	94.8	105.6	65.3	89.4
08	OPENINGS	93.7	67.6	87.4	96.1	64.7	88.5	93.7	69.0	87.7	93.2	67.6	86.9	93.7	68.7	87.6	93.2	67.4	86.9
0920	Plaster & Gypsum Board	95.2	67.8	76.1	96.7	64.3	74.0	99.8	70.4	79.2	107.6	67.8	79.8	99.8	70.4	79.2	102.2	67.8	78.2
0950, 0980	Ceilings & Acoustic Treatment	118.1	67.8	84.6	120.9	64.3	83.2	118.1	70.4	86.3	116.3	67.8	84.0	118.1	70.4	86.3	116.3	67.8	84.0
0960	Flooring	105.2	76.9	96.8	114.9	61.2	98.9	110.1	76.9	100.2	115.5	72.8	102.8	110.1	76.9	100.2	111.6	72.8	100.0
0970, 0990	Wall Finishes & Painting/Coating	94.1	50.8	68.1	94.1	45.2	64.8	94.1	39.1	61.1	94.1	46.9	65.8	94.1	64.7	76.5	94.1	50.8	68.1
09	FINISHES	107.4	67.7	85.1	109.9	63.4	83.9	108.0	69.0	86.1	110.6	65.4	85.2	107.3	71.9	87.4	110.3	65.9	85.4
COVERS	DIVS. 10 - 14, 25, 28, 41, 43, 44, 46	100.0	73.4	94.6	100.0	94.3	98.8	100.0	95.0	99.0	100.0	91.8	98.3	100.0	95.0	99.0	100.0	92.0	98.4
21, 22, 23	FIRE SUPPRESSION, PLUMBING & HVAC	94.3	71.4	84.9	100.2	70.5	88.0	94.3	70.8	84.7	94.3	70.4	84.5	100.3	70.8	88.2	94.3	70.7	84.6
26, 27, 3370	ELECTRICAL, COMMUNICATIONS & UTIL.	97.7	73.2	84.8	104.9	72.4	87.8	102.3	70.6	85.6	97.7	78.7	87.7	103.4	70.6	86.1	97.7	78.7	87.7
MF2010	WEIGHTED AVERAGE	101.4	75.0	89.7	102.7	75.0	90.5	99.3	76.3	89.1	100.3	75.1	89.2	100.6	77.0	90.2	102.5	75.4	90.6

DIVISION		NEBRASKA																	
		ALLIANCE			COLUMBUS			GRAND ISLAND			HASTINGS			LINCOLN			MCCOOK		
		693			686			688			689			683 - 685			690		
		MAT.	INST.	TOTAL	MAT.	INST.	TOTAL	MAT.	INST.	TOTAL	MAT.	INST.	TOTAL	MAT.	INST.	TOTAL	MAT.	INST.	TOTAL
015433	CONTRACTOR EQUIPMENT		96.9	96.9		100.6	100.6		100.6	100.6		100.6	100.6		100.6	100.6		100.6	100.6
0241, 31 - 34	SITE & INFRASTRUCTURE, DEMOLITION	100.0	97.7	98.4	98.9	91.1	93.4	103.6	92.2	95.6	102.5	91.4	94.7	92.4	92.2	92.3	103.4	91.2	94.9
0310	Concrete Forming & Accessories	89.3	55.3	59.9	97.7	58.1	63.4	97.3	74.9	77.9	100.4	73.0	76.7	101.7	59.4	65.1	95.0	55.6	60.9
0320	Concrete Reinforcing	113.2	47.0	80.2	103.7	45.8	74.9	103.1	76.6	89.9	103.1	46.5	74.9	94.7	76.1	85.4	105.1	74.1	89.7
0330	Cast-in-Place Concrete	107.2	57.5	86.8	108.8	54.5	86.5	115.2	68.8	96.1	115.2	63.0	93.7	86.1	71.3	80.0	115.2	57.2	91.4
03	CONCRETE	119.5	55.1	87.7	104.2	55.7	80.3	108.7	73.8	91.5	108.9	65.4	87.4	92.1	67.7	80.1	108.9	60.8	85.1
04	MASONRY	113.1	65.8	84.2	118.7	70.6	89.3	111.4	77.9	91.0	120.4	87.3	100.2	99.9	68.6	80.8	108.5	65.9	82.5
05	METALS	99.4	64.2	88.3	91.6	73.8	86.0	93.3	86.3	91.1	94.1	75.3	88.2	98.3	85.4	94.2	94.3	83.2	90.8
06	WOOD, PLASTICS & COMPOSITES	84.6	54.5	67.4	95.9	59.2	75.0	95.1	75.4	83.9	98.9	75.4	85.5	98.1	54.8	73.4	93.0	54.5	71.1
07	THERMAL & MOISTURE PROTECTION	105.6	66.7	90.0	99.6	63.7	85.2	99.7	75.1	89.8	99.7	85.3	93.9	99.0	69.5	87.1	99.7	63.9	85.3
08	OPENINGS	90.0	50.9	80.5	90.2	55.3	81.7	90.2	72.0	85.8	90.2	64.2	83.9	95.5	56.9	86.1	90.3	57.8	82.4
0920	Plaster & Gypsum Board	88.7	53.1	63.8	99.7	57.9	70.5	98.8	74.6	81.9	100.9	74.6	82.5	102.3	53.4	68.1	98.8	53.1	66.8
0950, 0980	Ceilings & Acoustic Treatment	98.9	53.1	68.4	94.6	57.9	70.2	94.6	74.6	81.3	94.6	74.6	81.3	102.8	53.4	69.9	94.6	53.1	66.9
0960	Flooring	104.9	85.1	99.0	103.2	104.0	103.4	103.0	111.1	105.4	104.3	104.0	104.2	104.5	86.6	99.1	102.1	85.1	97.0
0970, 0990	Wall Finishes & Painting/Coating	175.9	57.2	104.8	95.3	66.6	78.1	95.3	70.6	80.5	95.3	66.6	78.1	95.3	81.4	86.9	95.3	49.9	68.1
09	FINISHES	104.1	60.2	79.5	100.3	66.4	81.3	100.4	80.2	89.1	101.1	77.4	87.8	102.5	65.7	81.9	100.2	59.3	77.3
COVERS	DIVS. 10 - 14, 25, 28, 41, 43, 44, 46	100.0	65.4	92.9	100.0	74.3	94.8	100.0	85.7	97.1	100.0	78.1	95.5	100.0	83.4	96.6	100.0	64.7	92.8
21, 22, 23	FIRE SUPPRESSION, PLUMBING & HVAC	94.1	58.4	79.5	93.9	71.0	84.5	99.9	76.9	90.5	93.9	74.1	85.8	99.8	76.9	90.4	93.9	71.5	84.7
26, 27, 3370	ELECTRICAL, COMMUNICATIONS & UTIL.	93.1	72.3	82.2	92.8	80.3	86.2	91.4	66.6	78.4	90.8	87.1	88.8	103.1	66.6	83.9	95.4	72.3	83.2
MF2010	WEIGHTED AVERAGE	100.0	64.8	84.5	96.8	70.6	85.3	98.7	77.8	89.5	97.8	78.1	89.1	98.6	73.0	87.3	97.6	69.6	85.3

DIVISION		NEBRASKA												NEVADA					
		NORFOLK			NORTH PLATTE			OMAHA			VALENTINE			CARSON CITY			ELKO		
		687			691			680 - 681			692			897			898		
		MAT.	INST.	TOTAL	MAT.	INST.	TOTAL	MAT.	INST.	TOTAL	MAT.	INST.	TOTAL	MAT.	INST.	TOTAL	MAT.	INST.	TOTAL
015433	CONTRACTOR EQUIPMENT		92.0	92.0		100.6	100.6		92.0	92.0		95.0	95.0		97.5	97.5		97.5	97.5
0241, 31 - 34	SITE & INFRASTRUCTURE, DEMOLITION	81.5	90.6	87.9	104.6	91.4	95.4	82.1	91.3	88.6	87.2	95.2	92.8	72.3	98.5	90.7	58.8	98.3	86.5
0310	Concrete Forming & Accessories	84.2	72.2	73.8	97.2	72.8	76.1	94.2	72.7	75.6	85.5	54.9	59.0	100.9	96.0	96.6	108.9	82.6	86.2
0320	Concrete Reinforcing	103.9	48.4	76.3	104.5	75.1	89.9	99.2	76.8	88.0	105.2	45.5	75.5	109.5	116.0	112.7	111.2	90.9	101.1
0330	Cast-in-Place Concrete	110.3	62.2	90.5	115.2	64.4	94.3	92.8	77.5	86.5	102.2	54.9	82.8	97.0	86.5	92.7	97.8	82.2	91.4
03	CONCRETE	103.1	64.6	84.1	108.9	71.0	90.2	94.7	75.4	85.1	106.3	54.0	80.5	102.1	96.3	99.3	100.0	84.3	92.2
04	MASONRY	125.1	68.5	90.6	95.6	87.3	90.5	105.5	80.5	90.3	108.4	66.0	82.5	117.1	75.4	91.7	119.4	74.8	92.2
05	METALS	94.9	67.0	86.1	93.5	85.2	90.9	97.6	78.8	91.7	105.4	65.3	92.8	94.8	101.9	97.0	93.9	90.0	92.7
06	WOOD, PLASTICS & COMPOSITES	79.6	74.9	76.9	95.0	75.4	83.8	89.9	72.2	79.8	79.3	54.0	64.9	89.3	98.6	94.6	98.7	82.9	89.7
07	THERMAL & MOISTURE PROTECTION	99.5	72.5	88.7	99.7	79.3	91.5	95.4	80.8	89.5	100.4	66.5	86.8	99.7	85.6	94.0	98.9	76.3	89.8
08	OPENINGS	91.8	64.7	85.2	89.5	70.9	85.0	98.3	72.3	92.0	92.0	52.0	82.2	97.9	110.9	101.0	100.5	80.9	95.8
0920	Plaster & Gypsum Board	99.6	74.6	82.1	98.8	74.6	81.9	105.8	71.8	82.0	101.3	53.1	67.6	89.3	98.5	95.7	89.4	82.3	84.4
0950, 0980	Ceilings & Acoustic Treatment	108.6	74.6	86.0	94.6	74.6	81.3	117.7	71.8	87.1	111.9	53.1	72.7	114.0	98.5	103.7	110.4	82.3	91.7
0960	Flooring	131.9	104.0	123.6	102.9	104.0	103.2	135.9	84.9	120.7	134.2	82.7	118.8	101.4	81.1	95.3	106.0	53.0	90.2
0970, 0990	Wall Finishes & Painting/Coating	173.1	66.6	109.3	95.3	66.6	78.1	173.1	68.3	110.3	173.1	69.0	110.7	99.3	82.5	89.2	98.4	82.5	88.9
09	FINISHES	122.1	77.4	97.1	100.4	77.4	87.5	126.1	74.2	97.0	124.2	60.5	88.5	101.6	92.0	96.2	100.5	77.1	87.4
COVERS	DIVS. 10 - 14, 25, 28, 41, 43, 44, 46	100.0	76.9	95.3	100.0	68.1	93.5	100.0	84.2	96.8	100.0	62.7	92.4	100.0	93.4	98.7	100.0	65.2	92.9
21, 22, 23	FIRE SUPPRESSION, PLUMBING & HVAC	93.6	74.0	85.6	99.9	72.3	88.6	99.7	79.0	91.2	93.6	56.5	78.4	99.9	82.7	92.9	97.1	82.3	91.1
26, 27, 3370	ELECTRICAL, COMMUNICATIONS & UTIL.	91.6	81.0	86.0	93.5	80.3	86.5	100.5	82.9	91.2	90.3	90.2	90.3	96.2	96.6	96.4	93.7	96.6	95.2
MF2010	WEIGHTED AVERAGE	98.6	74.0	87.8	98.1	78.3	89.4	100.4	79.5	91.2	100.0	66.6	85.3	98.9	91.8	95.8	97.6	84.5	91.8

DIVISION		NEVADA									NEW HAMPSHIRE								
		ELY			LAS VEGAS			RENO			CHARLESTON			CLAREMONT			CONCORD		
		893			889 - 891			894 - 895			036			037			032 - 033		
		MAT.	INST.	TOTAL	MAT.	INST.	TOTAL	MAT.	INST.	TOTAL	MAT.	INST.	TOTAL	MAT.	INST.	TOTAL	MAT.	INST.	TOTAL
015433	**CONTRACTOR EQUIPMENT**		97.5	97.5		97.5	97.5		97.5	97.5		100.5	100.5		100.5	100.5		100.5	100.5
0241, 31 - 34	**SITE & INFRASTRUCTURE, DEMOLITION**	63.6	98.4	88.0	66.0	101.0	90.5	63.6	98.5	88.1	89.8	99.3	96.4	83.3	99.3	94.5	96.1	101.4	99.8
0310	Concrete Forming & Accessories	102.3	103.4	103.2	103.5	112.8	111.6	98.8	95.9	96.3	84.7	81.3	81.7	90.0	81.3	82.4	94.0	92.1	92.4
0320	Concrete Reinforcing	110.0	85.3	97.7	101.4	123.7	112.5	103.9	122.5	113.2	88.6	95.4	92.0	88.6	95.4	92.0	88.6	95.9	92.2
0330	Cast-in-Place Concrete	105.0	101.7	103.6	101.7	109.2	104.8	111.0	86.4	100.9	98.8	73.4	88.4	90.8	73.4	83.7	107.5	94.0	102.0
03	**CONCRETE**	108.1	99.2	103.7	103.2	113.2	108.1	107.8	97.5	102.7	101.0	81.3	91.3	92.9	81.3	87.2	103.8	93.4	98.6
04	**MASONRY**	123.5	103.2	111.1	111.6	105.2	107.7	118.4	75.4	92.2	89.3	84.2	86.2	89.0	84.2	86.1	91.7	104.6	99.6
05	**METALS**	93.8	91.2	93.0	100.8	109.1	103.4	95.2	104.5	98.1	92.2	90.9	91.8	92.2	90.9	91.8	92.9	94.3	93.4
06	**WOOD, PLASTICS & COMPOSITES**	90.1	104.9	98.6	89.3	112.8	102.7	85.6	98.6	93.0	83.6	89.6	87.0	89.3	89.6	89.5	93.0	89.6	91.1
07	**THERMAL & MOISTURE PROTECTION**	99.5	93.2	97.0	112.5	100.2	107.6	99.0	85.6	93.6	96.6	73.0	87.1	96.5	73.0	87.0	102.3	92.8	98.5
08	**OPENINGS**	100.5	91.1	98.2	99.5	120.3	104.6	98.3	106.4	100.3	100.2	76.7	94.5	101.4	76.7	95.4	101.4	86.7	97.8
0920	Plaster & Gypsum Board	86.1	105.0	99.3	85.3	113.1	104.8	78.8	98.5	92.6	91.0	88.6	89.3	92.2	88.6	89.7	93.2	88.6	90.0
0950, 0980	Ceilings & Acoustic Treatment	110.4	105.0	106.8	119.5	113.1	115.3	115.9	98.5	104.3	88.3	88.6	88.5	88.3	88.6	88.5	92.0	88.6	89.7
0960	Flooring	103.6	59.3	90.4	94.8	111.3	99.7	100.6	81.1	94.8	93.7	33.3	75.7	95.6	33.3	77.1	93.3	114.2	99.5
0970, 0990	Wall Finishes & Painting/Coating	98.4	117.2	109.6	101.2	121.0	113.1	98.4	82.5	88.9	96.1	47.8	67.1	96.1	48.0	67.3	96.1	99.7	98.2
09	**FINISHES**	100.0	97.5	98.6	99.4	113.6	107.4	99.2	92.0	95.1	93.7	70.1	80.4	93.8	70.1	80.5	95.6	96.7	96.2
COVERS	**DIVS. 10 - 14, 25, 28, 41, 43, 44, 46**	100.0	64.4	92.7	100.0	89.7	97.9	100.0	93.4	98.7	100.0	90.9	98.1	100.0	90.9	98.1	100.0	103.0	100.6
21, 22, 23	**FIRE SUPPRESSION, PLUMBING & HVAC**	97.1	102.9	99.5	100.1	106.6	102.7	99.9	82.7	92.9	94.0	41.7	72.6	94.0	41.8	72.6	99.8	88.8	95.3
26, 27, 3370	**ELECTRICAL, COMMUNICATIONS & UTIL.**	94.0	109.9	102.4	98.3	114.3	106.7	94.3	96.6	95.5	99.1	55.1	75.9	99.1	55.1	75.9	98.5	79.9	88.7
MF2010	**WEIGHTED AVERAGE**	98.7	99.3	99.0	100.2	109.0	104.1	99.0	92.0	95.9	95.8	70.6	84.7	94.9	70.7	84.2	98.4	92.8	95.9

DIVISION		NEW HAMPSHIRE															NEW JERSEY		
		KEENE			LITTLETON			MANCHESTER			NASHUA			PORTSMOUTH			ATLANTIC CITY		
		034			035			031			030			038			082,084		
		MAT.	INST.	TOTAL	MAT.	INST.	TOTAL	MAT.	INST.	TOTAL	MAT.	INST.	TOTAL	MAT.	INST.	TOTAL	MAT.	INST.	TOTAL
015433	**CONTRACTOR EQUIPMENT**		100.5	100.5		100.5	100.5		100.5	100.5		100.5	100.5		100.5	100.5		98.7	98.7
0241, 31 - 34	**SITE & INFRASTRUCTURE, DEMOLITION**	98.3	99.5	99.2	83.2	99.5	94.7	94.0	101.4	99.2	99.4	101.4	100.8	92.9	101.5	98.9	103.1	103.4	103.3
0310	Concrete Forming & Accessories	88.8	83.3	84.0	99.2	83.3	85.4	96.5	92.6	93.1	96.1	92.6	93.0	85.9	92.9	91.9	103.8	126.3	123.2
0320	Concrete Reinforcing	88.6	95.5	92.0	89.3	95.5	92.4	110.7	96.0	103.4	110.7	96.0	103.4	88.6	96.1	92.3	84.4	120.4	102.3
0330	Cast-in-Place Concrete	99.3	75.7	89.6	89.1	75.7	83.6	110.6	115.6	112.6	94.0	115.6	102.9	89.2	115.7	100.1	84.3	131.3	103.6
03	**CONCRETE**	100.7	83.0	92.0	92.4	83.0	87.8	109.0	101.0	105.1	101.0	101.0	101.0	92.3	101.2	96.7	93.7	125.8	109.5
04	**MASONRY**	93.0	88.0	90.0	99.1	88.0	92.3	94.1	104.6	100.5	93.3	104.6	100.2	89.4	104.6	98.6	96.1	127.6	115.3
05	**METALS**	92.9	91.6	92.4	92.9	91.6	92.5	97.7	94.8	96.8	97.7	94.8	96.8	94.4	95.5	94.7	92.6	106.6	97.0
06	**WOOD, PLASTICS & COMPOSITES**	87.8	89.6	88.8	97.7	89.6	93.1	93.2	89.6	91.1	96.4	89.6	92.5	84.7	89.6	87.5	106.4	126.6	117.9
07	**THERMAL & MOISTURE PROTECTION**	97.0	76.1	88.6	96.6	74.5	87.7	103.4	96.0	100.5	97.3	96.0	96.8	97.0	115.8	104.6	97.0	124.9	108.2
08	**OPENINGS**	98.8	81.7	94.6	102.4	77.4	96.3	103.4	86.7	99.3	103.4	86.7	99.3	104.0	83.4	99.0	101.0	124.1	106.6
0920	Plaster & Gypsum Board	91.4	88.6	89.5	103.9	88.6	93.2	98.2	88.6	91.5	99.2	88.6	91.8	91.0	88.6	89.3	103.4	126.8	119.8
0950, 0980	Ceilings & Acoustic Treatment	88.3	88.6	88.5	88.3	88.6	88.5	98.6	88.6	91.9	97.7	88.6	91.6	90.4	88.6	89.2	88.3	126.8	114.0
0960	Flooring	95.3	54.5	83.2	103.6	33.3	82.6	99.0	114.2	103.5	98.3	114.2	103.0	93.8	114.2	99.9	101.6	146.6	115.0
0970, 0990	Wall Finishes & Painting/Coating	96.1	47.8	67.1	96.1	62.6	76.0	96.1	99.7	98.2	96.1	99.7	98.2	96.1	99.7	98.2	96.1	120.0	110.4
09	**FINISHES**	95.7	75.0	84.1	97.6	72.6	83.6	98.8	96.7	97.6	99.5	96.7	97.9	95.0	96.7	96.0	99.6	130.5	116.9
COVERS	**DIVS. 10 - 14, 25, 28, 41, 43, 44, 46**	100.0	92.2	98.4	100.0	97.4	99.5	100.0	103.0	100.6	100.0	103.0	100.6	100.0	103.0	100.6	100.0	110.9	102.2
21, 22, 23	**FIRE SUPPRESSION, PLUMBING & HVAC**	94.0	45.6	74.2	94.0	67.3	83.0	99.8	88.8	95.3	100.0	88.8	95.4	100.0	88.9	95.4	99.5	123.4	109.3
26, 27, 3370	**ELECTRICAL, COMMUNICATIONS & UTIL.**	99.1	65.9	81.6	100.3	58.3	78.2	102.0	79.9	90.4	101.9	79.9	90.3	99.6	79.9	89.3	96.6	136.9	117.8
MF2010	**WEIGHTED AVERAGE**	96.4	74.6	86.8	96.0	77.7	88.0	100.6	94.0	97.7	99.8	94.0	97.2	97.3	94.5	96.1	97.6	123.3	108.9

DIVISION		NEW JERSEY																	
		CAMDEN			DOVER			ELIZABETH			HACKENSACK			JERSEY CITY			LONG BRANCH		
		081			078			072			076			073			077		
		MAT.	INST.	TOTAL	MAT.	INST.	TOTAL	MAT.	INST.	TOTAL	MAT.	INST.	TOTAL	MAT.	INST.	TOTAL	MAT.	INST.	TOTAL
015433	**CONTRACTOR EQUIPMENT**		98.7	98.7		100.5	100.5		100.5	100.5		100.5	100.5		98.7	98.7		98.3	98.3
0241, 31 - 34	**SITE & INFRASTRUCTURE, DEMOLITION**	103.9	104.0	103.9	115.6	104.8	108.0	120.5	104.8	109.5	115.9	105.0	108.2	103.9	104.9	104.6	109.3	104.8	106.1
0310	Concrete Forming & Accessories	96.1	126.2	122.1	92.4	126.8	122.1	103.1	126.7	123.6	92.4	126.8	122.1	96.1	126.8	122.6	96.6	122.9	119.3
0320	Concrete Reinforcing	110.7	115.1	112.9	85.3	122.1	103.6	85.3	122.1	103.6	85.3	122.1	103.6	110.7	122.1	116.4	85.3	122.1	103.6
0330	Cast-in-Place Concrete	81.8	129.9	101.6	104.8	129.1	114.8	90.0	129.1	106.1	102.4	129.1	113.4	81.8	129.1	101.2	90.9	131.0	107.4
03	**CONCRETE**	95.2	124.3	109.6	101.8	125.8	113.7	97.0	125.8	111.2	100.0	125.8	112.7	95.2	125.6	110.2	98.9	124.6	111.6
04	**MASONRY**	87.8	127.6	112.1	91.9	128.1	114.0	106.1	128.1	119.5	95.5	128.1	115.4	84.8	128.1	111.2	99.3	127.6	116.6
05	**METALS**	97.6	103.9	99.6	92.6	111.0	98.4	94.1	111.0	99.4	92.7	111.0	98.4	97.7	108.7	101.1	92.7	108.5	97.7
06	**WOOD, PLASTICS & COMPOSITES**	96.4	126.6	113.7	95.8	126.6	113.4	109.4	126.6	119.2	95.8	126.6	113.4	96.4	126.6	113.7	97.4	121.7	111.3
07	**THERMAL & MOISTURE PROTECTION**	96.9	123.8	107.7	97.4	126.2	109.0	97.5	126.2	109.1	97.1	125.1	108.4	96.9	126.2	108.7	97.0	124.6	108.1
08	**OPENINGS**	103.4	122.4	108.0	107.3	124.4	111.5	105.4	124.4	110.0	104.6	124.4	109.4	103.4	124.4	108.5	99.6	121.7	105.0
0920	Plaster & Gypsum Board	99.2	126.8	118.5	96.0	126.8	117.6	103.4	126.8	119.8	96.0	126.8	117.6	99.2	126.8	118.5	98.0	121.8	114.6
0950, 0980	Ceilings & Acoustic Treatment	97.7	126.8	117.1	88.3	126.8	114.0	90.4	126.8	114.7	88.3	126.8	114.0	97.7	126.8	117.1	88.3	121.8	110.6
0960	Flooring	98.3	146.6	112.7	97.4	169.0	118.7	102.1	169.0	122.0	97.4	169.0	118.7	98.3	169.0	119.3	98.5	169.0	119.5
0970, 0990	Wall Finishes & Painting/Coating	96.1	120.0	110.4	96.0	122.0	111.6	96.0	122.0	111.6	96.0	122.0	111.6	96.1	122.0	111.6	96.1	120.0	110.4
09	**FINISHES**	100.0	130.5	117.1	97.6	133.6	117.8	101.0	133.6	119.3	97.4	133.6	117.7	100.0	133.6	118.8	98.6	131.3	116.9
COVERS	**DIVS. 10 - 14, 25, 28, 41, 43, 44, 46**	100.0	110.9	102.2	100.0	114.7	103.0	100.0	114.7	103.0	100.0	114.7	103.0	100.0	114.7	103.0	100.0	110.2	102.1
21, 22, 23	**FIRE SUPPRESSION, PLUMBING & HVAC**	100.0	123.4	109.6	99.6	127.2	110.9	100.0	126.6	110.9	99.6	127.2	110.9	100.0	127.2	111.1	99.6	122.5	109.0
26, 27, 3370	**ELECTRICAL, COMMUNICATIONS & UTIL.**	102.3	136.9	120.5	98.1	134.4	117.2	98.8	134.4	117.6	98.1	134.3	117.1	103.5	134.3	119.7	97.7	130.1	114.8
MF2010	**WEIGHTED AVERAGE**	99.0	122.8	109.5	99.2	124.9	110.5	100.1	124.8	111.0	98.9	124.9	110.4	99.0	124.7	110.3	98.4	122.3	108.9

DIVISION		NEW JERSEY																	
		NEW BRUNSWICK			NEWARK			PATERSON			POINT PLEASANT			SUMMIT			TRENTON		
		088 - 089			070 - 071			074 - 075			087			079			085 - 086		
		MAT.	INST.	TOTAL	MAT.	INST.	TOTAL	MAT.	INST.	TOTAL	MAT.	INST.	TOTAL	MAT.	INST.	TOTAL	MAT.	INST.	TOTAL
015433	**CONTRACTOR EQUIPMENT**		98.3	98.3		100.5	100.5		100.5	100.5		98.3	98.3		100.5	100.5		98.3	98.3
0241, 31 - 34	**SITE & INFRASTRUCTURE, DEMOLITION**	118.1	104.8	108.8	119.8	104.8	109.3	117.8	105.0	108.8	119.7	104.8	109.2	117.7	104.8	108.6	100.5	104.8	103.5
0310	Concrete Forming & Accessories	98.7	126.6	122.9	94.6	126.7	122.4	94.3	126.8	122.4	93.9	122.7	118.8	95.0	126.7	122.5	96.7	122.8	119.3
0320	Concrete Reinforcing	85.3	122.1	103.6	110.7	122.1	116.4	110.7	122.1	116.4	85.3	122.1	103.6	85.3	122.1	103.6	110.7	117.6	114.1
0330	Cast-in-Place Concrete	104.1	131.3	115.3	108.6	129.1	117.0	104.1	129.1	114.4	104.1	131.0	115.2	87.1	129.1	104.3	100.4	131.0	113.0
03	**CONCRETE**	110.7	126.3	118.4	107.9	125.8	116.8	105.8	125.8	115.7	110.4	124.5	117.3	94.0	125.8	109.7	104.1	123.7	113.8
04	**MASONRY**	94.0	128.1	114.8	97.0	128.1	116.0	92.5	128.1	114.2	85.0	127.6	111.0	94.0	128.1	114.8	84.2	127.6	110.7
05	**METALS**	92.7	108.5	97.6	97.6	111.0	101.8	92.7	111.0	98.5	92.7	108.3	97.6	92.6	111.0	98.4	92.7	106.3	97.0
06	**WOOD, PLASTICS & COMPOSITES**	100.5	126.6	115.4	95.1	126.6	113.1	98.1	126.6	114.4	94.2	121.7	109.9	99.2	126.6	114.9	93.6	121.7	109.6
07	**THERMAL & MOISTURE PROTECTION**	97.2	125.3	108.5	102.9	126.2	112.3	97.4	125.1	108.6	97.2	124.6	108.2	97.7	126.2	109.2	101.1	124.2	110.4
08	**OPENINGS**	95.8	124.4	102.8	110.0	124.4	113.5	110.0	124.4	113.5	97.8	122.6	103.8	112.2	124.4	115.2	103.9	120.2	107.9
0920	Plaster & Gypsum Board	100.1	126.8	118.8	95.4	126.8	117.4	99.2	126.8	118.5	96.0	121.8	114.0	98.0	126.8	118.2	95.9	121.8	114.0
0950, 0980	Ceilings & Acoustic Treatment	88.3	126.8	114.0	98.6	126.8	117.4	97.7	126.8	117.1	88.3	121.8	110.6	88.3	126.8	114.0	100.4	121.8	114.6
0960	Flooring	99.5	169.0	120.2	98.3	169.0	119.3	98.3	169.0	119.3	97.4	146.6	112.1	98.6	169.0	119.6	98.5	169.0	119.5
0970, 0990	Wall Finishes & Painting/Coating	96.1	122.0	111.6	96.0	122.0	111.6	96.0	122.0	111.6	96.1	120.0	110.4	96.0	122.0	111.6	96.1	120.0	110.4
09	**FINISHES**	100.2	133.5	118.9	99.6	133.6	118.6	100.2	133.6	118.9	99.0	127.5	115.0	98.6	133.6	118.2	99.6	131.3	117.4
COVERS	**DIVS. 10 - 14, 25, 28, 41, 43, 44, 46**	100.0	114.6	103.0	100.0	114.7	103.0	100.0	114.7	103.0	100.0	105.9	101.2	100.0	114.7	103.0	100.0	110.2	102.1
21, 22, 23	**FIRE SUPPRESSION, PLUMBING & HVAC**	99.5	122.8	109.0	100.0	126.6	110.9	100.0	127.2	111.1	99.5	122.4	108.9	99.6	126.6	110.6	100.0	122.5	109.2
26, 27, 3370	**ELECTRICAL, COMMUNICATIONS & UTIL.**	97.4	135.4	117.4	107.7	134.3	121.7	103.5	134.4	119.8	96.6	130.1	114.3	98.8	134.4	117.6	106.1	134.9	121.3
MF2010	**WEIGHTED AVERAGE**	99.4	123.9	110.2	102.7	124.8	112.4	100.9	124.9	111.5	98.9	121.6	108.9	99.2	124.8	110.5	99.5	122.5	109.6

DIVISION		NEW JERSEY			NEW MEXICO														
		VINELAND			ALBUQUERQUE			CARRIZOZO			CLOVIS			FARMINGTON			GALLUP		
		080,083			870 - 872			883			881			874			873		
		MAT.	INST.	TOTAL	MAT.	INST.	TOTAL	MAT.	INST.	TOTAL	MAT.	INST.	TOTAL	MAT.	INST.	TOTAL	MAT.	INST.	TOTAL
015433	**CONTRACTOR EQUIPMENT**		98.7	98.7		109.2	109.2		109.2	109.2		109.2	109.2		109.2	109.2		109.2	109.2
0241, 31 - 34	**SITE & INFRASTRUCTURE, DEMOLITION**	108.2	104.0	105.2	81.3	104.1	97.3	102.3	104.1	103.5	90.2	104.1	99.9	87.6	104.1	99.1	95.4	104.1	101.5
0310	Concrete Forming & Accessories	91.6	126.3	121.6	99.3	65.6	70.1	99.3	65.6	70.1	99.3	65.4	70.0	99.3	65.6	70.1	99.3	65.6	70.1
0320	Concrete Reinforcing	84.4	120.4	102.3	104.7	71.5	88.2	112.3	71.5	92.0	113.5	71.5	92.6	114.4	71.5	93.0	109.5	71.5	90.6
0330	Cast-in-Place Concrete	90.9	131.3	107.5	103.9	72.4	90.9	96.7	72.4	86.7	96.6	72.3	86.6	104.8	72.4	91.5	98.7	72.4	87.9
03	**CONCRETE**	98.5	125.8	112.0	104.5	70.1	87.5	118.9	70.1	94.8	107.9	70.0	89.2	108.4	70.1	89.5	114.1	70.1	92.4
04	**MASONRY**	85.1	127.6	111.1	107.6	62.0	79.8	103.5	62.0	78.2	103.5	62.0	78.2	114.9	62.0	82.6	103.6	62.0	78.2
05	**METALS**	92.6	106.6	97.0	101.3	88.0	97.1	98.1	88.0	94.9	97.7	87.9	94.6	99.0	88.0	95.6	98.1	88.0	94.9
06	**WOOD, PLASTICS & COMPOSITES**	91.4	126.6	111.5	89.4	66.0	76.1	89.5	66.0	76.1	89.5	66.0	76.1	89.5	66.0	76.1	89.5	66.0	76.1
07	**THERMAL & MOISTURE PROTECTION**	96.8	124.9	108.1	98.7	73.1	88.5	100.2	73.1	89.3	99.0	73.1	88.6	98.9	73.1	88.5	99.9	73.1	89.1
08	**OPENINGS**	97.3	124.1	103.8	98.1	69.2	91.1	96.8	69.2	90.1	96.9	69.2	90.2	100.7	69.2	93.1	100.8	69.2	93.1
0920	Plaster & Gypsum Board	94.3	126.8	117.1	83.8	64.7	70.4	76.4	64.7	68.2	76.4	64.7	68.2	76.4	64.7	68.2	76.4	64.7	68.2
0950, 0980	Ceilings & Acoustic Treatment	88.3	126.8	114.0	110.0	64.7	79.8	106.7	64.7	78.7	106.7	64.7	78.7	106.7	64.7	78.7	106.7	64.7	78.7
0960	Flooring	96.6	146.6	111.5	98.8	68.4	89.8	100.6	68.4	91.0	100.6	68.4	91.0	100.6	68.4	91.0	100.6	68.4	91.0
0970, 0990	Wall Finishes & Painting/Coating	96.1	120.0	110.4	104.1	51.2	72.4	98.4	51.2	70.1	98.4	51.2	70.1	98.4	51.2	70.1	98.4	51.2	70.1
09	**FINISHES**	97.6	130.5	116.0	99.2	64.3	79.7	99.9	64.3	80.0	98.4	64.3	79.3	97.9	64.3	79.1	99.2	64.3	79.7
COVERS	**DIVS. 10 - 14, 25, 28, 41, 43, 44, 46**	100.0	110.9	102.2	100.0	77.3	95.4	100.0	77.3	95.4	100.0	77.3	95.4	100.0	77.3	95.4	100.0	77.3	95.4
21, 22, 23	**FIRE SUPPRESSION, PLUMBING & HVAC**	99.5	123.5	109.3	100.1	69.1	87.4	96.5	69.1	85.3	96.5	68.7	85.1	99.9	69.1	87.3	96.3	69.1	85.1
26, 27, 3370	**ELECTRICAL, COMMUNICATIONS & UTIL.**	96.6	136.9	117.8	89.4	72.9	80.7	89.4	72.9	80.7	86.9	72.9	79.5	87.6	72.9	79.8	86.7	72.9	79.4
MF2010	**WEIGHTED AVERAGE**	97.1	123.4	108.7	99.1	73.5	87.9	99.7	73.5	88.2	97.7	73.4	87.0	99.7	73.5	88.2	99.0	73.5	87.8

DIVISION		NEW MEXICO																	
		LAS CRUCES			LAS VEGAS			ROSWELL			SANTA FE			SOCORRO			TRUTH/CONSEQUENCES		
		880			877			882			875			878			879		
		MAT.	INST.	TOTAL	MAT.	INST.	TOTAL	MAT.	INST.	TOTAL	MAT.	INST.	TOTAL	MAT.	INST.	TOTAL	MAT.	INST.	TOTAL
015433	**CONTRACTOR EQUIPMENT**		85.5	85.5		109.2	109.2		109.2	109.2		109.2	109.2		109.2	109.2		85.5	85.5
0241, 31 - 34	**SITE & INFRASTRUCTURE, DEMOLITION**	91.4	83.6	85.9	87.0	104.1	99.0	92.3	104.1	100.6	85.6	104.1	98.5	83.4	104.1	97.9	102.2	83.6	89.1
0310	Concrete Forming & Accessories	96.5	64.4	68.7	99.3	65.6	70.1	99.3	65.6	70.1	97.8	65.6	69.9	99.3	65.6	70.1	96.5	64.4	68.7
0320	Concrete Reinforcing	108.1	71.4	89.8	111.3	71.5	91.5	113.5	71.5	92.6	112.3	71.5	92.0	113.5	71.5	92.6	106.8	71.4	89.2
0330	Cast-in-Place Concrete	91.2	64.8	80.3	102.0	72.4	89.8	96.6	72.4	86.6	106.1	72.4	92.2	99.9	72.4	88.6	109.5	64.8	91.1
03	**CONCRETE**	86.8	66.6	76.9	105.6	70.1	88.1	108.6	70.1	89.6	106.7	70.1	88.7	104.5	70.1	87.5	98.5	66.7	82.8
04	**MASONRY**	99.5	61.7	76.4	103.9	62.0	78.3	114.6	62.0	82.5	107.9	62.0	79.9	103.8	62.0	78.3	101.1	61.7	77.0
05	**METALS**	98.9	81.4	93.4	97.8	88.0	94.8	99.0	88.0	95.5	99.0	88.0	95.5	98.1	88.0	94.9	97.7	81.5	92.6
06	**WOOD, PLASTICS & COMPOSITES**	80.9	64.9	71.8	89.5	66.0	76.1	89.5	66.0	76.1	89.0	66.0	75.9	89.5	66.0	76.1	80.9	64.9	71.8
07	**THERMAL & MOISTURE PROTECTION**	85.8	68.5	78.8	98.5	73.1	88.3	99.1	73.1	88.7	98.7	73.1	88.4	98.5	73.1	88.3	86.3	68.5	79.1
08	**OPENINGS**	90.7	68.6	85.3	96.9	69.2	90.2	96.7	69.2	90.0	96.9	69.2	90.2	96.8	69.2	90.1	90.6	68.6	85.2
0920	Plaster & Gypsum Board	78.4	64.7	68.8	76.4	64.7	68.2	76.4	64.7	68.2	87.5	64.7	71.5	76.4	64.7	68.2	78.4	64.7	68.8
0950, 0980	Ceilings & Acoustic Treatment	94.6	64.7	74.7	106.7	64.7	78.7	106.7	64.7	78.7	111.3	64.7	80.2	106.7	64.7	78.7	94.6	64.7	74.7
0960	Flooring	132.0	68.4	113.0	100.6	68.4	91.0	100.6	68.4	91.0	101.0	68.4	91.3	100.6	68.4	91.0	132.0	68.4	113.0
0970, 0990	Wall Finishes & Painting/Coating	89.3	51.2	66.5	98.4	51.2	70.1	98.4	51.2	70.1	98.4	51.2	70.1	98.4	51.2	70.1	89.3	51.2	66.5
09	**FINISHES**	110.3	63.5	84.0	97.8	64.3	79.0	98.5	64.3	79.4	100.6	64.3	80.3	97.7	64.3	79.0	111.0	63.5	84.4
COVERS	**DIVS. 10 - 14, 25, 28, 41, 43, 44, 46**	100.0	74.6	94.8	100.0	77.3	95.4	100.0	77.3	95.4	100.0	77.3	95.4	100.0	77.3	95.4	100.0	74.6	94.8
21, 22, 23	**FIRE SUPPRESSION, PLUMBING & HVAC**	100.3	68.8	87.4	96.3	69.1	85.1	99.9	69.1	87.3	99.9	69.1	87.3	96.3	69.1	85.1	96.2	68.8	85.0
26, 27, 3370	**ELECTRICAL, COMMUNICATIONS & UTIL.**	88.1	66.6	76.8	89.4	72.9	80.7	88.3	72.9	80.2	102.3	72.9	86.8	87.3	72.9	79.7	91.1	72.9	81.5
MF2010	**WEIGHTED AVERAGE**	96.2	69.4	84.4	97.6	73.5	87.0	99.6	73.5	88.1	100.4	73.5	88.5	97.1	73.5	86.7	97.1	70.2	85.2

DIVISION		NEW MEXICO			NEW YORK															
		TUCUMCARI			ALBANY			BINGHAMTON			BRONX			BROOKLYN			BUFFALO			
		884			120 - 122			137 - 139			104			112			140 - 142			
		MAT.	INST.	TOTAL	MAT.	INST.	TOTAL	MAT.	INST.	TOTAL	MAT.	INST.	TOTAL	MAT.	INST.	TOTAL	MAT.	INST.	TOTAL	
015433	CONTRACTOR EQUIPMENT		109.2	109.2		114.1	114.1		114.7	114.7		113.0	113.0		115.0	115.0		97.4	97.4	
0241, 31 - 34	SITE & INFRASTRUCTURE, DEMOLITION	89.9	104.1	99.8	70.7	107.2	96.3	92.6	92.9	92.8	110.9	123.3	119.6	117.6	129.1	125.7	97.4	98.3	98.0	
0310	Concrete Forming & Accessories	99.3	65.4	70.0	97.6	95.3	95.6	100.7	83.9	86.1	103.2	175.2	165.5	106.9	174.9	165.7	97.7	114.2	112.0	
0320	Concrete Reinforcing	111.3	71.5	91.5	97.1	108.1	102.6	96.1	96.1	96.1	94.0	177.6	135.6	97.6	185.2	141.2	100.9	100.5	100.7	
0330	Cast-in-Place Concrete	96.6	72.3	86.6	75.6	102.1	86.5	96.9	130.5	110.7	89.4	177.6	125.7	99.0	176.0	130.7	102.6	116.7	108.4	
03	CONCRETE	107.0	70.0	88.7	93.2	100.7	96.9	95.1	103.8	99.4	94.3	174.8	134.0	107.1	175.2	140.7	106.1	111.8	108.9	
04	MASONRY	114.3	62.0	82.4	93.6	101.0	98.1	108.9	129.6	121.5	93.8	175.3	143.6	118.6	175.3	153.2	105.6	115.1	111.4	
05	METALS	97.7	87.9	94.6	98.7	110.8	102.5	95.3	118.2	102.5	103.6	145.2	116.7	103.6	141.2	115.4	100.5	94.8	98.7	
06	WOOD, PLASTICS & COMPOSITES	89.5	66.0	76.1	94.5	92.7	93.5	104.7	79.8	90.5	100.6	175.2	143.2	106.2	174.8	145.4	96.0	114.5	106.5	
07	THERMAL & MOISTURE PROTECTION	99.0	73.1	88.6	92.3	97.0	94.2	104.4	98.3	102.0	103.5	164.4	128.0	106.5	163.4	129.4	100.9	108.6	104.0	
08	OPENINGS	96.7	69.2	90.0	95.2	91.7	94.3	90.2	81.8	88.1	86.9	172.1	107.7	88.7	169.5	108.3	90.1	102.8	93.2	
0920	Plaster & Gypsum Board	76.4	64.7	68.2	102.9	92.2	95.4	112.9	78.7	89.0	104.4	177.1	155.3	110.4	177.1	157.0	105.2	114.6	111.8	
0950, 0980	Ceilings & Acoustic Treatment	106.7	64.7	78.7	98.7	92.2	94.4	97.8	78.7	85.1	83.8	177.1	146.0	93.4	177.1	149.2	103.4	114.6	110.9	
0960	Flooring	100.6	68.4	91.0	92.2	107.0	96.6	105.1	94.7	102.0	98.6	183.9	124.0	112.3	183.9	133.6	104.5	115.5	107.7	
0970, 0990	Wall Finishes & Painting/Coating	98.4	51.2	70.1	90.9	88.6	89.5	92.4	91.8	92.0	108.1	151.3	134.0	123.9	151.3	140.3	104.1	111.7	108.7	
09	FINISHES	98.3	64.3	79.3	96.6	96.4	96.5	98.4	85.4	91.1	98.9	174.7	141.4	114.1	174.4	147.9	102.8	115.0	109.7	
COVERS	DIVS. 10 - 14, 25, 28, 41, 43, 44, 46	100.0	77.3	95.4	100.0	93.8	98.7	100.0	94.9	99.0	100.0	131.7	106.5	100.0	131.0	106.3	100.0	105.8	101.2	
21, 22, 23	FIRE SUPPRESSION, PLUMBING & HVAC	96.5	68.7	85.1	100.0	97.8	99.1	100.4	86.9	94.9	100.2	167.4	127.7	99.8	167.2	127.4	100.0	96.8	98.7	
26, 27, 3370	ELECTRICAL, COMMUNICATIONS & UTIL.	89.4	72.9	80.7	101.5	97.3	99.3	100.7	109.8	105.5	96.4	178.1	139.4	100.6	178.1	141.5	98.6	97.8	98.2	
MF2010	WEIGHTED AVERAGE	98.4	73.4	87.4	97.0	99.8	98.2	98.2	100.1	99.0	98.3	164.8	127.6	103.0	164.8	130.2	100.0	104.1	101.8	

DIVISION		NEW YORK																	
		ELMIRA			FAR ROCKAWAY			FLUSHING			GLENS FALLS			HICKSVILLE			JAMAICA		
		148 - 149			116			113			128			115,117,118			114		
		MAT.	INST.	TOTAL	MAT.	INST.	TOTAL	MAT.	INST.	TOTAL	MAT.	INST.	TOTAL	MAT.	INST.	TOTAL	MAT.	INST.	TOTAL
015433	CONTRACTOR EQUIPMENT		116.2	116.2		115.0	115.0		115.0	115.0		114.1	114.1		115.0	115.0		115.0	115.0
0241, 31 - 34	SITE & INFRASTRUCTURE, DEMOLITION	92.9	94.3	93.9	120.9	129.1	126.7	120.9	129.1	126.7	62.8	106.5	93.4	110.6	127.9	122.7	115.2	129.1	125.0
0310	Concrete Forming & Accessories	83.0	84.4	84.2	94.0	174.9	164.0	97.6	174.9	164.5	85.3	87.9	87.6	90.7	145.0	137.7	97.6	174.9	164.5
0320	Concrete Reinforcing	102.2	90.2	96.2	97.6	185.2	141.2	99.3	185.2	142.1	96.7	93.7	95.2	97.6	177.3	137.3	97.6	185.2	141.2
0330	Cast-in-Place Concrete	96.0	96.7	96.3	107.1	176.0	135.4	107.1	176.0	135.4	73.8	98.9	84.1	91.0	167.1	122.3	99.0	176.0	130.7
03	CONCRETE	99.9	91.5	95.8	113.0	175.2	143.7	113.5	175.2	144.0	86.0	93.7	89.8	99.0	157.3	127.8	106.4	175.2	140.4
04	MASONRY	102.2	93.1	96.7	122.5	175.3	154.7	116.5	175.3	152.4	95.0	95.4	95.3	113.2	160.5	142.1	120.7	175.3	154.0
05	METALS	95.9	115.3	102.0	103.6	141.2	115.4	103.6	141.2	115.4	94.9	105.6	98.2	105.1	136.4	114.9	103.6	141.2	115.4
06	WOOD, PLASTICS & COMPOSITES	79.9	83.0	81.7	90.5	174.8	138.7	95.0	174.8	140.6	82.2	86.0	84.4	87.2	143.3	119.2	95.0	174.8	140.6
07	THERMAL & MOISTURE PROTECTION	101.7	84.3	94.7	106.4	163.4	129.4	106.5	163.4	129.4	88.1	93.9	90.4	106.1	152.4	124.7	106.3	163.4	129.3
08	OPENINGS	90.6	82.7	88.7	87.2	169.5	107.2	87.2	169.5	107.2	89.1	84.8	88.1	87.2	150.6	102.6	87.2	169.5	107.2
0920	Plaster & Gypsum Board	99.2	82.2	87.3	99.1	177.1	153.7	101.6	177.1	154.4	93.6	85.3	87.8	98.7	144.6	130.8	101.6	177.1	154.4
0950, 0980	Ceilings & Acoustic Treatment	98.8	82.2	87.7	82.4	177.1	145.5	82.4	177.1	145.5	91.5	85.3	87.4	81.4	144.6	123.5	82.4	177.1	145.5
0960	Flooring	94.4	94.7	94.5	107.4	183.9	130.2	108.8	183.9	131.2	84.3	107.0	91.1	106.5	147.8	118.8	108.8	183.9	131.2
0970, 0990	Wall Finishes & Painting/Coating	102.7	88.1	93.9	123.9	151.3	140.3	123.9	151.3	140.3	90.9	88.6	89.5	123.9	151.3	140.3	123.9	151.3	140.3
09	FINISHES	97.8	86.3	91.4	109.4	174.4	145.8	110.1	174.4	146.2	90.9	91.1	91.0	107.8	145.1	128.7	109.6	174.4	145.9
COVERS	DIVS. 10 - 14, 25, 28, 41, 43, 44, 46	100.0	90.7	98.1	100.0	131.0	106.3	100.0	131.0	106.3	100.0	91.2	98.2	100.0	122.4	104.6	100.0	131.0	106.3
21, 22, 23	FIRE SUPPRESSION, PLUMBING & HVAC	94.2	87.8	91.6	93.7	167.2	123.8	93.7	167.2	123.8	94.2	89.9	92.4	99.8	147.8	119.4	93.7	167.2	123.8
26, 27, 3370	ELECTRICAL, COMMUNICATIONS & UTIL.	97.1	97.0	97.0	108.4	178.1	145.1	108.4	178.1	145.1	95.9	97.3	96.7	99.8	138.3	120.1	98.7	178.1	140.5
MF2010	WEIGHTED AVERAGE	96.3	92.7	94.8	102.6	164.8	130.0	102.4	164.8	129.9	92.2	94.9	93.4	101.0	145.4	120.5	100.7	164.8	129.0

DIVISION		NEW YORK																	
		JAMESTOWN			KINGSTON			LONG ISLAND CITY			MONTICELLO			MOUNT VERNON			NEW ROCHELLE		
		147			124			111			127			105			108		
		MAT.	INST.	TOTAL	MAT.	INST.	TOTAL	MAT.	INST.	TOTAL	MAT.	INST.	TOTAL	MAT.	INST.	TOTAL	MAT.	INST.	TOTAL
015433	CONTRACTOR EQUIPMENT		94.3	94.3		115.0	115.0		115.0	115.0		115.0	115.0		113.0	113.0		113.0	113.0
0241, 31 - 34	SITE & INFRASTRUCTURE, DEMOLITION	94.1	94.0	94.0	116.8	124.8	122.4	118.7	129.1	126.0	112.4	124.8	121.1	118.0	117.1	117.4	117.1	117.1	117.1
0310	Concrete Forming & Accessories	83.0	80.1	80.5	86.9	107.2	104.5	101.7	174.9	165.0	93.6	107.1	105.3	93.7	127.6	123.0	108.4	127.6	125.0
0320	Concrete Reinforcing	102.3	98.4	100.3	97.0	133.1	115.0	97.6	185.2	141.2	96.3	132.9	114.6	92.8	176.3	134.4	92.9	176.3	134.4
0330	Cast-in-Place Concrete	99.5	93.0	96.8	99.8	132.0	113.0	102.2	176.0	132.5	93.5	129.7	108.4	99.9	130.6	112.5	99.9	130.6	112.5
03	CONCRETE	103.1	88.0	95.7	107.2	120.3	113.7	109.4	175.2	141.9	102.2	119.4	110.7	103.4	136.8	119.9	103.0	136.8	119.7
04	MASONRY	110.2	93.1	99.7	110.0	136.7	126.3	115.4	175.3	151.9	103.3	134.1	122.1	99.6	128.0	117.0	99.6	128.0	117.0
05	METALS	93.2	91.0	92.5	103.6	114.7	107.1	103.6	141.2	115.4	103.5	114.4	107.0	103.3	132.6	112.5	103.6	132.6	112.7
06	WOOD, PLASTICS & COMPOSITES	78.5	76.7	77.4	83.4	100.8	93.3	100.6	174.8	143.0	91.0	100.8	96.6	90.7	130.1	113.2	107.6	130.1	120.4
07	THERMAL & MOISTURE PROTECTION	101.2	85.4	94.8	108.0	136.9	119.6	106.4	163.4	129.3	107.6	135.9	119.0	104.4	136.7	117.4	104.4	136.7	117.4
08	OPENINGS	90.4	80.6	88.0	92.1	117.7	98.3	87.2	169.5	107.2	87.2	117.7	94.7	86.9	143.4	100.7	87.0	143.4	100.7
0920	Plaster & Gypsum Board	94.1	75.7	81.2	95.1	100.7	99.1	105.8	177.1	155.7	97.6	100.7	99.8	99.4	130.6	121.2	111.4	130.6	124.9
0950, 0980	Ceilings & Acoustic Treatment	96.1	75.7	82.5	76.9	100.7	92.8	82.4	177.1	145.5	76.9	100.7	92.8	81.9	130.6	114.4	81.9	130.6	114.4
0960	Flooring	97.5	91.3	95.7	99.8	75.1	92.5	110.4	183.9	132.3	102.2	75.1	94.1	91.1	177.2	116.8	97.2	177.2	121.1
0970, 0990	Wall Finishes & Painting/Coating	104.1	77.5	88.2	121.1	110.9	115.0	123.9	151.3	140.3	121.1	110.9	115.0	106.3	151.3	133.3	106.3	151.3	133.3
09	FINISHES	97.7	80.9	88.3	104.9	99.3	101.8	110.8	174.4	146.5	105.5	99.3	102.1	96.5	139.0	120.3	99.5	139.0	121.6
COVERS	DIVS. 10 - 14, 25, 28, 41, 43, 44, 46	100.0	93.0	98.6	100.0	107.9	101.6	100.0	131.0	106.3	100.0	107.9	101.6	100.0	116.0	103.3	100.0	116.0	103.3
21, 22, 23	FIRE SUPPRESSION, PLUMBING & HVAC	94.0	84.7	90.2	94.1	122.6	105.8	99.8	167.2	127.4	94.1	120.3	104.8	94.2	122.3	105.7	94.2	122.3	105.7
26, 27, 3370	ELECTRICAL, COMMUNICATIONS & UTIL.	95.8	92.0	93.8	97.6	113.5	106.0	99.2	178.1	140.8	97.6	113.5	106.0	94.6	154.5	126.2	94.6	154.5	126.2
MF2010	WEIGHTED AVERAGE	96.5	87.9	92.7	100.4	118.5	108.4	102.5	164.8	130.0	99.0	117.6	107.2	98.0	133.0	113.4	98.3	133.0	113.6

DIVISION		NEW YORK																	
		NEW YORK			NIAGARA FALLS			PLATTSBURGH			POUGHKEEPSIE			QUEENS			RIVERHEAD		
		100 - 102			143			129			125 - 126			110			119		
		MAT.	INST.	TOTAL	MAT.	INST.	TOTAL	MAT.	INST.	TOTAL	MAT.	INST.	TOTAL	MAT.	INST.	TOTAL	MAT.	INST.	TOTAL
015433	CONTRACTOR EQUIPMENT		113.4	113.4		94.3	94.3		99.8	99.8		115.0	115.0		115.0	115.0		115.0	115.0
0241, 31 - 34	SITE & INFRASTRUCTURE, DEMOLITION	119.4	123.9	122.5	96.2	95.5	95.7	85.9	103.2	98.0	113.0	124.6	121.1	113.9	129.1	124.6	111.7	127.9	123.0
0310	Concrete Forming & Accessories	107.5	182.4	172.3	83.0	115.6	111.2	90.0	85.4	86.0	86.9	151.3	142.6	90.9	174.9	163.6	95.0	145.0	138.2
0320	Concrete Reinforcing	99.5	184.8	142.0	100.9	101.0	100.9	101.3	93.3	97.3	97.0	133.4	115.1	99.3	185.2	142.1	99.4	174.4	136.8
0330	Cast-in-Place Concrete	100.6	182.2	134.2	103.0	121.2	110.5	90.5	94.6	92.2	96.6	129.2	110.0	94.1	176.0	127.8	92.5	167.1	123.2
03	CONCRETE	104.2	180.7	142.0	105.4	113.9	109.6	99.3	89.8	94.6	104.5	138.9	121.5	102.3	175.2	138.3	99.8	156.8	127.9
04	MASONRY	103.7	175.6	147.6	118.2	121.6	120.3	90.5	89.4	89.9	103.2	130.1	119.6	109.7	175.3	149.7	118.5	160.5	144.1
05	METALS	116.8	145.4	125.8	96.0	92.9	95.0	99.4	85.8	95.1	103.6	116.9	107.8	103.6	141.2	115.4	105.5	135.3	114.9
06	WOOD, PLASTICS & COMPOSITES	104.9	184.5	150.3	78.4	113.3	98.3	89.4	83.6	86.1	83.4	161.7	128.1	87.3	174.8	137.3	92.2	143.3	121.4
07	THERMAL & MOISTURE PROTECTION	103.5	165.7	128.5	101.2	110.4	104.9	102.1	90.9	97.6	107.9	140.7	121.1	106.1	163.4	129.2	106.7	152.4	125.1
08	OPENINGS	92.6	177.2	113.2	90.4	102.3	93.3	97.2	81.4	93.4	92.1	149.5	106.1	87.2	169.5	107.2	87.2	149.9	102.5
0920	Plaster & Gypsum Board	112.6	186.7	164.4	94.1	113.5	107.6	117.3	82.4	92.9	95.1	163.5	143.0	98.7	177.1	153.5	100.2	144.6	131.2
0950, 0980	Ceilings & Acoustic Treatment	104.7	186.7	159.4	96.1	113.5	107.7	105.8	82.4	90.2	76.9	163.5	134.6	82.4	177.1	145.5	82.3	144.6	123.8
0960	Flooring	99.6	183.9	124.7	97.5	115.8	103.0	108.1	107.0	107.8	99.8	148.7	114.4	106.5	183.9	129.5	107.4	147.8	119.5
0970, 0990	Wall Finishes & Painting/Coating	108.1	154.9	136.1	104.1	112.1	108.9	119.1	78.1	94.6	121.1	110.9	115.0	123.9	151.3	140.3	123.9	151.3	140.3
09	FINISHES	105.5	180.7	147.6	97.8	116.1	108.1	102.2	88.3	94.4	104.7	148.5	129.3	108.2	174.4	145.3	108.4	145.1	129.0
COVERS	DIVS. 10 - 14, 25, 28, 41, 43, 44, 46	100.0	136.1	107.4	100.0	103.5	100.7	100.0	49.6	89.7	100.0	112.3	102.5	100.0	131.0	106.3	100.0	122.4	104.6
21, 22, 23	FIRE SUPPRESSION, PLUMBING & HVAC	100.1	167.5	127.7	94.0	100.1	96.5	94.1	89.9	92.4	94.1	120.6	105.0	99.8	167.2	127.4	99.9	147.7	119.5
26, 27, 3370	ELECTRICAL, COMMUNICATIONS & UTIL.	104.8	178.2	143.4	94.3	98.7	96.6	93.8	88.7	91.1	97.6	113.5	106.0	99.8	178.1	141.1	101.6	138.3	120.9
MF2010	WEIGHTED AVERAGE	104.2	167.0	131.9	97.5	105.5	101.0	96.8	88.6	93.2	99.6	128.6	112.4	101.1	164.8	129.1	101.7	145.1	120.9

DIVISION		NEW YORK																	
		ROCHESTER			SCHENECTADY			STATEN ISLAND			SUFFERN			SYRACUSE			UTICA		
		144 - 146			123			103			109			130 - 132			133 - 135		
		MAT.	INST.	TOTAL	MAT.	INST.	TOTAL	MAT.	INST.	TOTAL	MAT.	INST.	TOTAL	MAT.	INST.	TOTAL	MAT.	INST.	TOTAL
015433	CONTRACTOR EQUIPMENT		116.9	116.9		114.1	114.1		113.0	113.0		113.0	113.0		114.1	114.1		114.1	114.1
0241, 31 - 34	SITE & INFRASTRUCTURE, DEMOLITION	74.7	109.2	98.9	72.1	107.1	96.6	122.7	123.3	123.2	113.9	118.2	116.9	91.6	106.7	102.2	70.5	104.9	94.6
0310	Concrete Forming & Accessories	101.5	92.0	93.3	100.5	94.3	95.1	93.2	175.2	164.2	102.0	124.5	121.5	99.6	89.2	90.6	101.0	83.5	85.9
0320	Concrete Reinforcing	101.8	90.3	96.1	95.7	108.1	101.9	94.0	177.6	135.6	92.9	133.4	113.1	97.1	94.5	95.8	97.1	91.5	94.3
0330	Cast-in-Place Concrete	96.4	100.2	97.9	89.1	100.6	93.8	99.9	177.6	131.8	96.9	127.4	109.4	90.1	104.9	96.2	82.4	92.3	86.5
03	CONCRETE	113.2	95.4	104.4	99.6	99.8	99.7	105.1	174.8	139.5	100.0	126.5	113.1	98.5	96.4	97.5	96.7	89.1	92.9
04	MASONRY	102.5	99.8	100.8	93.0	98.4	96.3	106.5	175.3	148.5	98.9	123.7	114.1	100.2	103.9	102.5	91.7	100.0	96.8
05	METALS	97.1	104.9	99.5	98.8	110.8	102.6	101.4	145.2	115.2	101.4	117.6	106.5	98.7	105.0	100.7	96.8	103.5	98.9
06	WOOD, PLASTICS & COMPOSITES	93.1	90.1	91.4	101.0	92.7	96.3	89.2	175.2	138.3	100.3	130.3	117.4	101.0	86.8	92.9	101.0	81.0	89.6
07	THERMAL & MOISTURE PROTECTION	99.9	98.0	99.1	89.3	95.9	91.9	103.8	164.4	128.2	104.3	134.9	116.6	100.2	96.9	98.9	89.0	94.6	91.2
08	OPENINGS	93.3	86.6	91.7	95.2	91.7	94.3	86.9	172.1	107.7	87.0	134.8	98.6	92.8	80.2	89.7	95.2	79.2	91.3
0920	Plaster & Gypsum Board	111.7	89.7	96.4	104.7	92.2	96.0	99.4	177.1	153.8	103.5	130.9	122.6	104.7	86.1	91.7	104.7	80.2	87.5
0950, 0980	Ceilings & Acoustic Treatment	109.8	89.7	96.5	97.8	92.2	94.1	83.8	177.1	146.0	81.9	130.9	114.5	97.8	86.1	90.0	97.8	80.2	86.0
0960	Flooring	90.8	106.8	95.6	91.0	107.0	95.8	95.0	183.9	121.5	94.0	147.8	110.0	93.6	96.6	94.5	91.0	96.7	92.7
0970, 0990	Wall Finishes & Painting/Coating	104.0	96.8	99.7	90.9	88.6	89.5	108.1	151.3	134.0	106.3	120.3	114.7	98.4	97.6	97.9	90.9	97.6	94.9
09	FINISHES	104.9	95.3	99.5	96.3	95.8	96.0	98.5	174.7	141.2	97.4	125.9	113.4	98.4	91.1	94.3	96.3	87.0	91.1
COVERS	DIVS. 10 - 14, 25, 28, 41, 43, 44, 46	100.0	95.1	99.0	100.0	92.9	98.6	100.0	131.7	106.5	100.0	114.4	102.9	100.0	96.0	99.2	100.0	90.5	98.1
21, 22, 23	FIRE SUPPRESSION, PLUMBING & HVAC	99.9	89.1	95.5	100.2	96.4	98.6	100.2	167.4	127.7	94.2	113.7	102.2	100.2	90.1	96.1	100.2	88.5	95.4
26, 27, 3370	ELECTRICAL, COMMUNICATIONS & UTIL.	101.5	94.0	97.5	100.9	97.3	99.0	96.4	178.1	139.4	102.9	121.4	112.7	100.9	97.3	99.0	99.0	97.3	98.1
MF2010	WEIGHTED AVERAGE	100.1	96.0	98.3	97.6	99.0	98.2	100.0	164.8	128.6	98.1	121.5	108.4	98.7	96.2	97.6	96.7	93.3	95.2

DIVISION		NEW YORK									NORTH CAROLINA								
		WATERTOWN			WHITE PLAINS			YONKERS			ASHEVILLE			CHARLOTTE			DURHAM		
		136			106			107			287 - 288			281 - 282			277		
		MAT.	INST.	TOTAL	MAT.	INST.	TOTAL	MAT.	INST.	TOTAL	MAT.	INST.	TOTAL	MAT.	INST.	TOTAL	MAT.	INST.	TOTAL
015433	CONTRACTOR EQUIPMENT		114.1	114.1		113.0	113.0		113.0	113.0		95.0	95.0		95.0	95.0		100.5	100.5
0241, 31 - 34	SITE & INFRASTRUCTURE, DEMOLITION	77.9	106.5	97.9	110.6	117.1	115.2	118.9	117.0	117.6	107.0	76.2	85.4	108.0	76.3	85.8	107.3	84.9	91.6
0310	Concrete Forming & Accessories	86.9	89.0	88.7	106.8	127.6	124.8	107.1	127.6	124.8	93.2	38.6	45.9	98.7	40.8	48.6	95.7	44.8	51.6
0320	Concrete Reinforcing	97.7	94.5	96.1	92.9	176.3	134.4	96.6	176.3	136.3	98.4	62.0	80.3	98.9	57.4	78.2	100.1	56.7	78.5
0330	Cast-in-Place Concrete	95.9	93.6	94.9	88.7	130.6	105.9	99.2	130.6	112.1	113.7	48.8	87.0	119.9	49.9	91.1	115.5	46.4	87.1
03	CONCRETE	109.0	92.4	100.8	93.9	136.8	115.1	103.2	136.8	119.8	105.9	48.4	77.5	108.7	48.9	79.2	106.6	49.4	78.3
04	MASONRY	92.7	102.6	98.7	98.5	128.0	116.5	103.3	128.0	118.4	83.0	44.1	59.3	88.8	53.3	67.1	88.2	36.3	56.5
05	METALS	96.8	104.8	99.4	102.9	132.6	112.3	112.6	132.6	118.9	89.0	82.3	86.9	93.1	80.8	89.2	102.5	79.8	95.4
06	WOOD, PLASTICS & COMPOSITES	83.8	87.8	86.1	105.9	130.1	119.7	105.8	130.1	119.7	95.6	36.8	62.1	101.5	38.9	65.7	98.5	46.6	68.9
07	THERMAL & MOISTURE PROTECTION	89.3	96.2	92.0	104.2	136.7	117.2	104.4	136.7	117.4	100.2	43.4	77.3	100.6	46.9	79.0	100.9	45.0	78.4
08	OPENINGS	95.2	84.4	92.6	87.0	143.4	100.7	90.3	143.4	103.2	90.5	42.1	78.7	94.2	43.0	81.7	94.2	47.3	82.8
0920	Plaster & Gypsum Board	95.6	87.2	89.7	106.8	130.6	123.5	112.1	130.6	125.1	102.5	34.7	55.1	101.6	36.8	56.3	107.8	44.8	63.7
0950, 0980	Ceilings & Acoustic Treatment	97.8	87.2	90.7	81.9	130.6	114.4	102.9	130.6	121.4	93.4	34.7	54.3	98.8	36.8	57.5	97.0	44.8	62.2
0960	Flooring	84.7	96.7	88.2	95.8	177.2	120.1	95.4	177.2	119.8	100.6	44.2	83.8	102.0	44.4	84.9	103.8	44.2	86.1
0970, 0990	Wall Finishes & Painting/Coating	90.9	91.4	91.2	106.3	151.3	133.3	106.3	151.3	133.3	109.6	42.1	69.1	109.6	42.1	69.1	109.6	35.6	65.2
09	FINISHES	94.1	90.3	92.0	97.8	139.0	120.9	103.7	139.0	123.4	99.6	39.0	65.6	100.6	40.7	67.1	102.0	43.5	69.2
COVERS	DIVS. 10 - 14, 25, 28, 41, 43, 44, 46	100.0	95.3	99.0	100.0	116.0	103.3	100.0	119.2	103.9	100.0	78.4	95.6	100.0	79.3	95.8	100.0	73.9	94.7
21, 22, 23	FIRE SUPPRESSION, PLUMBING & HVAC	100.2	83.4	93.3	100.3	122.3	109.3	100.3	122.3	109.3	100.3	53.7	81.2	99.9	54.6	81.4	100.4	52.3	80.7
26, 27, 3370	ELECTRICAL, COMMUNICATIONS & UTIL.	100.9	90.0	95.1	94.6	154.5	126.2	103.0	154.5	130.1	97.9	58.1	76.9	96.8	57.6	76.2	98.7	46.4	71.2
MF2010	WEIGHTED AVERAGE	98.2	93.1	95.9	98.3	133.0	113.6	102.9	133.1	116.2	97.1	55.1	78.6	98.7	56.5	80.1	100.3	53.9	79.9

	DIVISION	NORTH CAROLINA																	
		ELIZABETH CITY			FAYETTEVILLE			GASTONIA			GREENSBORO			HICKORY			KINSTON		
		279			283			280			270,272 - 274			286			285		
		MAT.	INST.	TOTAL	MAT.	INST.	TOTAL	MAT.	INST.	TOTAL	MAT.	INST.	TOTAL	MAT.	INST.	TOTAL	MAT.	INST.	TOTAL
015433	CONTRACTOR EQUIPMENT		104.6	104.6		100.5	100.5		95.0	95.0		100.5	100.5		100.5	100.5		100.5	100.5
0241, 31 - 34	SITE & INFRASTRUCTURE, DEMOLITION	112.3	86.9	94.5	106.0	85.0	91.3	106.7	76.3	85.4	107.2	85.0	91.7	106.0	85.1	91.3	104.9	85.0	90.9
0310	Concrete Forming & Accessories	82.3	40.3	46.0	92.8	61.4	65.6	99.4	40.2	48.2	95.4	45.8	52.5	89.5	38.6	45.5	85.9	43.0	48.8
0320	Concrete Reinforcing	97.9	48.1	73.1	102.3	56.8	79.7	98.9	57.4	78.2	98.9	56.7	77.9	98.4	57.0	77.8	97.9	48.0	73.1
0330	Cast-in-Place Concrete	115.7	47.9	87.8	119.2	48.8	90.2	111.1	54.1	87.7	114.7	48.1	87.3	113.6	49.4	87.2	109.7	45.8	83.4
03	CONCRETE	106.7	46.3	76.8	108.2	57.6	83.2	104.5	50.1	77.7	106.0	50.4	78.5	105.7	47.7	77.1	102.5	46.8	75.0
04	MASONRY	100.0	49.9	69.4	86.3	40.1	58.1	87.3	52.9	66.3	86.2	38.6	57.1	73.5	45.5	56.4	78.8	50.1	61.3
05	METALS	90.9	77.2	86.6	106.7	79.9	98.3	89.8	80.9	87.0	96.3	79.7	91.1	89.0	79.9	86.2	88.0	76.3	84.3
06	WOOD, PLASTICS & COMPOSITES	82.5	40.2	58.4	94.7	68.0	79.5	104.2	38.2	66.5	98.2	46.8	68.9	90.2	36.7	59.7	87.1	43.6	62.3
07	THERMAL & MOISTURE PROTECTION	100.2	44.4	77.8	100.0	45.7	78.2	100.4	46.9	78.9	100.6	43.1	77.5	100.4	43.3	77.5	100.3	44.4	77.8
08	OPENINGS	91.4	37.3	78.3	90.6	59.0	82.9	94.2	42.3	81.6	94.2	47.4	82.8	90.5	37.9	77.7	90.6	44.4	79.4
0920	Plaster & Gypsum Board	99.7	37.6	56.2	107.1	66.8	78.9	109.7	36.1	58.2	109.7	45.0	64.4	102.5	34.6	55.0	102.6	41.7	60.0
0950, 0980	Ceilings & Acoustic Treatment	97.0	37.6	57.4	94.3	66.8	76.0	97.0	36.1	56.4	97.0	45.0	62.4	93.4	34.6	54.2	97.0	41.7	60.2
0960	Flooring	95.5	24.0	74.2	100.7	44.2	83.9	103.8	44.2	86.1	103.8	40.9	85.1	100.5	34.0	80.7	97.9	23.4	75.7
0970, 0990	Wall Finishes & Painting/Coating	109.6	43.9	70.2	109.6	33.9	64.3	109.6	42.1	69.1	109.6	33.0	63.7	109.6	42.1	69.1	109.6	41.0	68.5
09	FINISHES	98.9	37.2	64.3	100.3	56.5	75.8	102.1	40.2	67.4	102.3	43.3	69.2	99.7	36.9	64.5	99.6	38.9	65.6
COVERS	DIVS. 10 - 14, 25, 28, 41, 43, 44, 46	100.0	72.7	94.4	100.0	77.0	95.3	100.0	79.1	95.7	100.0	78.6	95.6	100.0	78.8	95.7	100.0	74.4	94.8
21, 22, 23	FIRE SUPPRESSION, PLUMBING & HVAC	94.3	53.0	77.4	100.2	53.5	81.1	100.3	54.4	81.5	100.3	53.5	81.2	94.3	54.2	77.9	94.3	53.2	77.4
26, 27, 3370	ELECTRICAL, COMMUNICATIONS & UTIL.	98.4	36.0	65.5	96.6	52.8	73.6	97.2	57.6	76.4	97.6	58.1	76.8	95.2	57.6	75.4	95.0	48.0	70.3
MF2010	WEIGHTED AVERAGE	97.0	52.2	77.2	100.2	59.1	82.1	97.8	56.5	79.6	99.0	56.2	80.2	94.8	55.3	77.4	94.5	54.3	76.8

	DIVISION	NORTH CAROLINA															NORTH DAKOTA		
		MURPHY			RALEIGH			ROCKY MOUNT			WILMINGTON			WINSTON-SALEM			BISMARCK		
		289			275 - 276			278			284			271			585		
		MAT.	INST.	TOTAL	MAT.	INST.	TOTAL	MAT.	INST.	TOTAL	MAT.	INST.	TOTAL	MAT.	INST.	TOTAL	MAT.	INST.	TOTAL
015433	CONTRACTOR EQUIPMENT		95.0	95.0		100.5	100.5		100.5	100.5		95.0	95.0		100.5	100.5		98.9	98.9
0241, 31 - 34	SITE & INFRASTRUCTURE, DEMOLITION	108.4	76.2	85.8	107.5	85.1	91.8	110.2	85.2	92.6	108.2	76.4	85.9	107.5	85.1	91.8	99.3	97.4	98.0
0310	Concrete Forming & Accessories	100.1	41.0	48.9	97.0	48.1	54.7	88.3	46.1	51.8	94.5	50.1	56.0	97.1	49.8	56.1	102.3	42.0	50.2
0320	Concrete Reinforcing	98.0	47.6	72.9	98.9	56.8	77.9	97.9	56.7	77.4	99.2	56.7	78.0	98.9	56.7	77.9	108.1	84.6	96.4
0330	Cast-in-Place Concrete	117.5	45.5	87.9	122.9	52.1	93.8	113.2	48.9	86.7	113.2	49.6	87.1	117.5	48.8	89.3	98.2	49.0	77.9
03	CONCRETE	109.0	45.6	77.7	110.0	52.8	81.8	107.3	50.8	79.4	105.8	52.8	79.6	107.4	52.4	80.3	99.1	53.8	76.7
04	MASONRY	75.9	42.9	55.8	86.9	41.0	58.9	79.2	41.1	55.9	73.8	41.3	54.0	86.4	39.9	58.0	103.2	58.8	76.1
05	METALS	87.0	75.6	83.5	93.9	79.9	89.5	90.1	79.3	86.7	88.7	79.8	85.9	93.9	79.8	89.5	96.1	86.0	92.9
06	WOOD, PLASTICS & COMPOSITES	104.9	41.3	68.6	98.5	49.6	70.6	89.5	46.6	65.0	97.5	51.1	71.0	98.2	51.5	71.5	81.7	36.3	55.8
07	THERMAL & MOISTURE PROTECTION	100.3	42.3	77.0	100.8	45.8	78.7	100.6	42.4	77.2	100.2	47.9	79.1	100.6	44.2	77.9	110.0	51.8	86.6
08	OPENINGS	90.5	40.6	78.3	91.4	48.7	81.0	90.6	43.3	79.1	90.6	50.5	80.9	94.2	50.0	83.4	96.8	48.5	85.0
0920	Plaster & Gypsum Board	108.7	39.4	60.2	101.6	47.9	64.0	101.9	44.8	62.0	104.4	49.4	65.9	109.7	49.8	67.8	97.4	34.5	53.4
0950, 0980	Ceilings & Acoustic Treatment	93.4	39.4	57.4	98.8	47.9	64.9	94.3	44.8	61.3	94.3	49.4	64.4	97.0	49.8	65.6	157.3	34.5	75.5
0960	Flooring	104.2	25.0	80.6	102.6	44.2	85.2	99.2	22.6	76.4	101.3	46.1	84.9	103.8	44.2	86.1	115.6	74.7	103.4
0970, 0990	Wall Finishes & Painting/Coating	109.6	41.2	68.6	109.6	37.4	66.3	109.6	40.2	68.0	109.6	40.4	68.1	109.6	34.6	64.7	94.1	30.7	56.1
09	FINISHES	101.6	37.6	65.7	101.0	46.2	70.3	99.8	41.0	66.9	100.2	48.3	71.1	102.3	47.2	71.4	118.7	44.7	77.2
COVERS	DIVS. 10 - 14, 25, 28, 41, 43, 44, 46	100.0	78.7	95.6	100.0	75.3	95.0	100.0	75.5	95.0	100.0	76.1	95.1	100.0	80.4	96.0	100.0	84.2	96.8
21, 22, 23	FIRE SUPPRESSION, PLUMBING & HVAC	94.3	52.9	77.3	100.0	54.0	81.2	94.3	54.7	78.1	100.3	54.9	81.7	100.3	54.2	81.4	100.2	61.2	84.3
26, 27, 3370	ELECTRICAL, COMMUNICATIONS & UTIL.	98.9	30.6	62.9	101.1	41.9	70.0	100.6	41.4	69.4	98.1	52.8	74.2	97.6	58.1	76.8	102.5	72.4	86.7
MF2010	WEIGHTED AVERAGE	95.7	49.9	75.5	99.0	55.1	79.7	96.1	53.9	77.5	96.7	56.5	79.0	98.8	57.5	80.6	101.0	64.4	84.9

	DIVISION	NORTH DAKOTA																	
		DEVILS LAKE			DICKINSON			FARGO			GRAND FORKS			JAMESTOWN			MINOT		
		583			586			580 - 581			582			584			587		
		MAT.	INST.	TOTAL	MAT.	INST.	TOTAL	MAT.	INST.	TOTAL	MAT.	INST.	TOTAL	MAT.	INST.	TOTAL	MAT.	INST.	TOTAL
015433	CONTRACTOR EQUIPMENT		98.9	98.9		98.9	98.9		98.9	98.9		98.9	98.9		98.9	98.9		98.9	98.9
0241, 31 - 34	SITE & INFRASTRUCTURE, DEMOLITION	107.9	95.2	99.0	117.4	93.9	100.9	99.9	97.4	98.2	112.5	93.9	99.4	106.9	93.9	97.7	109.6	97.4	101.1
0310	Concrete Forming & Accessories	104.0	36.5	45.6	93.3	35.6	43.4	103.0	42.8	50.9	97.0	35.5	43.8	94.7	35.3	43.3	93.1	66.9	70.4
0320	Concrete Reinforcing	108.1	84.9	96.6	109.1	41.6	75.5	99.7	84.6	92.2	106.4	84.7	95.6	108.7	49.2	79.1	110.1	85.0	97.6
0330	Cast-in-Place Concrete	119.2	47.5	89.7	107.9	45.9	82.4	97.2	50.8	78.2	107.8	45.9	82.3	117.7	45.8	88.1	107.8	48.1	83.2
03	CONCRETE	112.1	50.8	81.8	111.7	41.7	77.2	107.1	54.8	81.2	108.7	49.6	79.5	110.6	42.9	77.2	107.3	64.6	86.2
04	MASONRY	111.4	65.9	83.6	113.5	60.0	80.8	105.0	58.8	76.8	106.1	65.4	81.2	123.6	34.1	69.0	105.4	65.1	80.8
05	METALS	96.1	85.0	92.6	96.1	64.6	86.2	98.4	86.3	94.6	96.1	80.1	91.1	96.1	65.6	86.5	96.4	86.8	93.4
06	WOOD, PLASTICS & COMPOSITES	85.2	33.4	55.6	74.4	33.4	51.0	81.6	36.8	56.0	78.2	33.4	52.6	76.1	33.4	51.7	74.1	69.5	71.5
07	THERMAL & MOISTURE PROTECTION	106.0	50.4	83.6	106.5	48.1	83.0	106.9	52.9	85.2	106.2	50.4	83.7	105.8	42.3	80.3	106.0	56.9	86.2
08	OPENINGS	96.8	41.7	83.4	96.7	31.5	80.9	96.7	48.8	85.0	96.8	41.7	83.3	96.8	33.6	81.4	96.9	66.6	89.5
0920	Plaster & Gypsum Board	118.6	31.6	57.8	109.1	31.6	54.9	96.9	35.1	53.7	110.7	31.6	55.4	110.3	31.6	55.3	109.1	68.8	80.9
0950, 0980	Ceilings & Acoustic Treatment	153.6	31.6	72.3	153.6	31.6	72.3	155.5	35.1	75.3	153.6	31.6	72.3	153.6	31.6	72.3	153.6	68.8	97.1
0960	Flooring	119.9	37.2	95.3	112.5	37.2	90.1	115.6	74.7	103.4	114.7	37.2	91.6	113.4	37.2	90.7	112.3	86.9	104.7
0970, 0990	Wall Finishes & Painting/Coating	94.1	23.5	51.8	94.1	23.5	51.8	94.1	73.8	81.9	94.1	23.5	51.8	94.1	23.5	51.8	94.1	25.7	53.1
09	FINISHES	122.6	33.6	72.7	120.3	33.6	71.7	118.3	49.8	79.9	120.5	33.6	71.8	119.5	33.6	71.4	119.3	66.2	89.6
COVERS	DIVS. 10 - 14, 25, 28, 41, 43, 44, 46	100.0	33.7	86.5	100.0	33.9	86.5	100.0	84.3	96.8	100.0	33.8	86.5	100.0	82.2	96.4	100.0	87.9	97.5
21, 22, 23	FIRE SUPPRESSION, PLUMBING & HVAC	94.4	63.8	81.9	94.4	69.2	84.1	100.3	66.9	86.6	100.5	35.0	73.7	94.4	36.9	70.9	100.5	61.3	84.4
26, 27, 3370	ELECTRICAL, COMMUNICATIONS & UTIL.	96.7	37.3	65.4	106.2	73.4	88.9	99.7	67.8	82.9	100.5	55.8	77.0	96.7	37.2	65.4	104.0	78.9	90.8
MF2010	WEIGHTED AVERAGE	101.3	56.9	81.7	102.3	58.7	83.1	101.9	65.8	86.0	102.4	52.6	80.5	101.4	46.0	77.0	102.4	71.6	88.9

City Cost Indexes

	DIVISION	NORTH DAKOTA			OHIO															
		WILLISTON			AKRON			ATHENS			CANTON			CHILLICOTHE			CINCINNATI			
		588			442 - 443			457			446 - 447			456			451 - 452			
		MAT.	INST.	TOTAL	MAT.	INST.	TOTAL	MAT.	INST.	TOTAL	MAT.	INST.	TOTAL	MAT.	INST.	TOTAL	MAT.	INST.	TOTAL	
015433	CONTRACTOR EQUIPMENT		98.9	98.9		96.3	96.3		90.5	90.5		96.3	96.3		100.3	100.3		100.1	100.1	
0241, 31 - 34	SITE & INFRASTRUCTURE, DEMOLITION	110.4	93.9	98.8	94.4	104.7	101.6	111.7	92.6	98.3	94.5	104.2	101.3	96.5	104.1	101.8	93.5	103.6	100.6	
0310	Concrete Forming & Accessories	98.8	35.7	44.2	99.7	91.1	92.2	92.7	87.6	88.3	99.7	81.2	83.7	94.6	93.8	93.9	96.4	81.8	83.8	
0320	Concrete Reinforcing	111.2	41.6	76.6	96.2	93.1	94.6	89.8	88.0	88.9	96.2	76.3	86.3	86.9	80.0	83.5	92.1	81.9	87.0	
0330	Cast-in-Place Concrete	107.8	45.9	82.4	91.2	96.7	93.5	111.7	94.0	104.4	92.1	94.3	93.0	101.4	94.6	98.6	93.4	85.7	90.2	
03	CONCRETE	108.8	41.7	75.7	91.1	92.7	91.9	104.6	89.3	97.0	91.5	84.4	88.0	98.9	91.3	95.2	93.6	83.3	88.5	
04	MASONRY	100.0	60.0	75.6	89.7	94.2	92.4	79.9	91.7	87.1	90.3	85.2	87.2	87.2	99.0	94.4	86.9	86.3	86.5	
05	METALS	96.2	64.7	86.3	95.7	81.7	91.3	101.7	79.2	94.6	95.7	74.6	89.1	93.7	84.4	90.8	96.0	84.3	92.3	
06	WOOD, PLASTICS & COMPOSITES	79.4	33.4	53.2	97.2	89.7	92.9	87.3	88.7	88.1	97.6	79.4	87.2	100.6	92.5	96.0	103.0	79.7	89.7	
07	THERMAL & MOISTURE PROTECTION	106.1	48.1	82.8	107.9	95.9	103.1	100.6	92.7	97.4	109.0	91.8	102.1	102.1	95.1	99.3	100.4	90.5	96.4	
08	OPENINGS	96.8	31.5	80.9	112.3	91.6	107.2	100.0	85.3	96.4	105.9	76.8	98.8	92.0	84.8	90.2	100.4	79.3	95.2	
0920	Plaster & Gypsum Board	110.7	31.6	55.4	99.3	89.1	92.2	89.5	88.0	88.5	100.5	78.5	85.1	91.2	92.5	92.1	93.1	79.3	83.5	
0950, 0980	Ceilings & Acoustic Treatment	153.6	31.6	72.3	96.4	89.1	91.6	103.6	88.0	93.2	96.4	78.5	84.5	97.7	92.5	94.2	98.6	79.3	85.7	
0960	Flooring	115.6	37.2	92.2	98.4	91.1	96.3	128.8	102.5	121.0	98.6	81.3	93.5	104.7	102.5	104.1	105.7	91.6	101.5	
0970, 0990	Wall Finishes & Painting/Coating	94.1	32.9	57.4	99.3	101.3	100.5	102.7	95.2	98.2	99.3	82.6	89.3	99.9	91.7	95.0	99.9	86.7	92.0	
09	FINISHES	120.6	34.7	72.5	100.1	91.9	95.5	105.4	92.0	97.9	100.4	80.8	89.4	102.7	95.7	98.7	103.1	83.7	92.2	
COVERS	DIVS. 10 - 14, 25, 28, 41, 43, 44, 46	100.0	33.9	86.5	100.0	90.3	98.0	100.0	54.1	90.6	100.0	87.9	97.5	100.0	88.7	97.7	100.0	92.3	98.4	
21, 22, 23	FIRE SUPPRESSION, PLUMBING & HVAC	94.4	69.2	84.1	100.0	96.9	98.7	94.1	54.0	77.7	100.0	82.4	92.8	94.5	91.7	93.4	100.0	82.7	92.9	
26, 27, 3370	ELECTRICAL, COMMUNICATIONS & UTIL.	101.0	73.4	86.4	98.9	92.7	95.6	99.1	100.6	99.9	98.1	89.1	93.4	98.6	83.6	90.7	97.4	79.7	88.1	
MF2010	WEIGHTED AVERAGE	100.7	58.8	82.3	99.1	93.6	96.6	98.9	82.2	91.6	98.5	85.0	92.5	96.2	91.9	94.3	97.9	85.1	92.3	

	DIVISION	OHIO																	
		CLEVELAND			COLUMBUS			DAYTON			HAMILTON			LIMA			LORAIN		
		441			430 - 432			453 - 454			450			458			440		
		MAT.	INST.	TOTAL	MAT.	INST.	TOTAL	MAT.	INST.	TOTAL	MAT.	INST.	TOTAL	MAT.	INST.	TOTAL	MAT.	INST.	TOTAL
015433	CONTRACTOR EQUIPMENT		96.6	96.6		95.0	95.0		95.4	95.4		100.3	100.3		93.0	93.0		96.3	96.3
0241, 31 - 34	SITE & INFRASTRUCTURE, DEMOLITION	94.4	105.0	101.9	94.1	99.2	97.7	92.3	103.4	100.1	92.1	103.6	100.2	105.3	92.3	96.2	93.9	103.6	100.7
0310	Concrete Forming & Accessories	99.8	99.4	99.5	98.1	87.6	89.0	96.4	80.3	82.4	96.4	81.2	83.2	92.6	85.9	86.8	99.8	86.8	88.5
0320	Concrete Reinforcing	96.7	93.5	95.1	105.7	82.0	93.9	92.1	80.2	86.2	92.1	81.9	87.0	89.8	80.4	85.1	96.2	93.4	94.8
0330	Cast-in-Place Concrete	89.5	107.0	96.7	91.5	94.4	92.7	86.9	86.4	86.7	93.1	84.9	89.7	102.6	96.1	100.0	86.9	104.1	93.9
03	CONCRETE	90.3	100.1	95.2	95.0	88.6	91.8	90.4	82.2	86.3	93.4	82.8	88.2	97.7	88.2	93.0	89.0	93.4	91.2
04	MASONRY	94.1	105.3	100.9	96.5	95.4	95.8	86.4	82.5	84.0	86.7	84.8	85.5	109.5	83.6	93.7	86.5	99.4	94.4
05	METALS	97.3	84.6	93.3	97.9	80.7	92.5	95.2	77.1	89.5	95.3	84.3	91.9	101.7	80.3	95.0	96.3	82.6	92.0
06	WOOD, PLASTICS & COMPOSITES	96.3	96.7	96.5	95.8	85.5	89.9	104.4	78.9	89.9	103.0	79.7	89.7	87.2	85.8	86.4	97.2	83.7	89.5
07	THERMAL & MOISTURE PROTECTION	106.7	109.1	107.6	100.9	96.2	99.0	105.7	87.6	98.4	102.3	89.8	97.2	100.1	94.5	97.8	109.0	101.8	106.1
08	OPENINGS	101.8	95.5	100.3	104.2	81.5	98.7	100.2	78.7	95.0	97.7	79.3	93.2	100.0	79.9	95.1	105.9	88.4	101.6
0920	Plaster & Gypsum Board	98.4	96.4	97.0	97.0	85.0	88.6	93.1	78.5	82.9	93.1	79.3	83.5	89.5	85.0	86.4	99.3	83.0	87.9
0950, 0980	Ceilings & Acoustic Treatment	94.6	96.4	95.8	96.2	85.0	88.8	99.6	78.5	85.6	98.6	79.3	85.7	102.5	85.0	90.9	96.4	83.0	87.5
0960	Flooring	98.2	105.3	100.3	95.0	89.5	93.3	108.5	82.2	100.7	105.7	91.6	101.5	128.0	89.9	116.6	98.6	105.3	100.6
0970, 0990	Wall Finishes & Painting/Coating	99.3	107.6	104.3	96.1	95.4	95.7	99.9	85.2	91.1	99.9	85.3	91.2	102.7	84.2	91.6	99.3	107.6	104.3
09	FINISHES	99.7	101.2	100.5	96.4	88.5	92.0	104.1	80.8	91.0	103.0	83.2	91.9	104.4	86.4	94.3	100.2	92.0	95.6
COVERS	DIVS. 10 - 14, 25, 28, 41, 43, 44, 46	100.0	95.2	99.0	100.0	94.5	98.9	100.0	91.6	98.3	100.0	91.8	98.3	100.0	89.8	97.9	100.0	91.7	98.3
21, 22, 23	FIRE SUPPRESSION, PLUMBING & HVAC	100.0	100.2	100.1	100.0	92.4	96.9	100.9	83.3	93.7	100.6	81.9	92.9	94.1	84.7	90.3	100.0	83.7	93.3
26, 27, 3370	ELECTRICAL, COMMUNICATIONS & UTIL.	98.4	108.1	103.5	96.6	86.4	91.2	95.9	82.2	88.7	96.3	80.3	87.9	99.5	79.2	88.8	98.3	88.2	93.0
MF2010	WEIGHTED AVERAGE	98.2	100.7	99.3	98.6	90.0	94.8	97.7	83.9	91.6	97.5	84.7	91.9	99.4	85.0	93.1	98.1	90.8	94.9

	DIVISION	OHIO																	
		MANSFIELD			MARION			SPRINGFIELD			STEUBENVILLE			TOLEDO			YOUNGSTOWN		
		448 - 449			433			455			439			434 - 436			444 - 445		
		MAT.	INST.	TOTAL	MAT.	INST.	TOTAL	MAT.	INST.	TOTAL	MAT.	INST.	TOTAL	MAT.	INST.	TOTAL	MAT.	INST.	TOTAL
015433	CONTRACTOR EQUIPMENT		96.3	96.3		94.6	94.6		95.4	95.4		99.4	99.4		97.4	97.4		96.3	96.3
0241, 31 - 34	SITE & INFRASTRUCTURE, DEMOLITION	90.4	104.5	100.3	90.3	98.0	95.7	92.6	102.1	99.3	133.8	107.5	115.4	93.4	99.9	97.9	94.3	105.2	101.9
0310	Concrete Forming & Accessories	90.1	85.1	85.8	94.8	83.0	84.6	96.4	84.9	86.4	96.7	85.4	86.9	98.0	94.4	94.9	99.7	84.8	86.8
0320	Concrete Reinforcing	87.5	76.9	82.2	97.4	82.0	89.7	92.1	80.2	86.2	94.7	84.3	89.5	105.7	88.6	97.2	96.2	87.5	91.9
0330	Cast-in-Place Concrete	84.5	92.9	87.9	83.5	91.5	86.8	89.2	86.5	88.1	90.7	91.9	91.2	91.5	102.2	95.9	90.4	98.5	93.7
03	CONCRETE	83.9	85.7	84.8	87.2	85.4	86.3	91.5	84.2	87.9	90.9	86.9	88.9	95.0	95.7	95.3	90.7	89.5	90.1
04	MASONRY	88.8	95.9	93.2	98.1	94.6	96.0	86.6	82.8	84.3	84.0	89.8	87.5	104.9	98.6	101.1	90.0	92.0	91.2
05	METALS	96.5	75.5	89.9	96.9	78.4	91.1	95.2	76.9	89.5	93.3	78.8	88.8	97.7	86.9	94.3	95.8	79.4	90.6
06	WOOD, PLASTICS & COMPOSITES	85.6	83.7	84.5	91.9	83.5	87.1	106.0	85.2	94.1	88.0	83.6	85.5	95.8	93.4	94.4	97.2	82.3	88.7
07	THERMAL & MOISTURE PROTECTION	108.4	95.1	103.0	100.5	83.7	93.8	105.5	88.4	98.6	112.5	92.8	104.6	102.5	102.0	102.3	109.2	93.7	102.9
08	OPENINGS	105.6	78.3	99.0	98.1	79.0	93.5	98.1	80.2	93.7	98.9	81.6	94.7	101.7	90.0	98.9	105.9	83.7	100.5
0920	Plaster & Gypsum Board	92.9	83.0	85.9	94.5	82.9	86.4	93.1	85.0	87.5	95.0	82.6	86.3	97.0	93.2	94.3	99.3	81.5	86.8
0950, 0980	Ceilings & Acoustic Treatment	97.3	83.0	87.8	96.2	82.9	87.4	99.6	85.0	89.9	93.0	82.6	86.1	96.2	93.2	94.2	96.4	81.5	86.5
0960	Flooring	94.2	103.5	96.9	93.7	72.3	87.3	108.5	82.2	100.7	122.8	94.2	114.3	94.2	103.3	96.9	98.6	92.7	96.9
0970, 0990	Wall Finishes & Painting/Coating	99.3	90.3	93.9	96.1	49.6	68.3	99.9	85.2	91.1	108.6	94.6	100.2	96.1	101.9	99.6	99.3	94.7	96.6
09	FINISHES	98.0	89.0	92.9	95.3	77.7	85.4	104.1	84.6	93.2	113.3	87.4	98.8	96.1	96.6	96.4	100.3	86.6	92.6
COVERS	DIVS. 10 - 14, 25, 28, 41, 43, 44, 46	100.0	88.8	97.7	100.0	55.8	91.0	100.0	92.4	98.4	100.0	92.8	98.5	100.0	97.9	99.6	100.0	88.9	97.7
21, 22, 23	FIRE SUPPRESSION, PLUMBING & HVAC	94.0	81.9	89.1	94.0	87.4	91.3	100.9	83.5	93.8	94.3	89.7	92.4	100.1	100.3	100.2	100.0	90.8	96.3
26, 27, 3370	ELECTRICAL, COMMUNICATIONS & UTIL.	95.7	79.1	87.0	90.4	79.1	84.4	95.9	82.4	88.8	85.5	109.4	98.1	96.7	102.1	99.5	98.3	84.4	91.0
MF2010	WEIGHTED AVERAGE	95.6	86.1	91.4	94.8	84.1	90.1	97.6	84.8	91.9	96.6	92.0	94.6	98.8	97.5	98.2	98.4	89.1	94.3

	DIVISION	OHIO			OKLAHOMA															
		ZANESVILLE			ARDMORE			CLINTON			DURANT			ENID			GUYMON		
		437 - 438			734			736			747			737			739		
		MAT.	INST.	TOTAL	MAT.	INST.	TOTAL	MAT.	INST.	TOTAL	MAT.	INST.	TOTAL	MAT.	INST.	TOTAL	MAT.	INST.	TOTAL
015433	CONTRACTOR EQUIPMENT		94.6	94.6		82.3	82.3		81.4	81.4		81.4	81.4		81.4	81.4		81.4	81.4
0241, 31 - 34	SITE & INFRASTRUCTURE, DEMOLITION	93.0	99.3	97.4	97.4	91.1	93.0	99.0	89.6	92.4	92.7	89.4	90.4	100.5	89.6	92.9	103.2	88.3	92.7
0310	Concrete Forming & Accessories	92.0	83.2	84.3	92.3	39.0	46.1	91.1	33.7	41.4	84.2	38.9	45.0	94.5	33.1	41.4	98.0	21.6	31.9
0320	Concrete Reinforcing	96.8	83.8	90.4	92.8	76.7	84.8	93.3	76.7	85.0	93.6	60.8	77.3	92.7	76.7	84.7	93.3	26.9	60.2
0330	Cast-in-Place Concrete	88.0	90.5	89.0	98.9	42.6	75.7	95.6	42.6	73.8	92.5	42.5	71.9	95.6	44.9	74.7	95.6	30.8	69.0
03	CONCRETE	90.6	85.5	88.1	98.6	47.8	73.5	98.2	45.5	72.1	93.2	44.8	69.3	98.7	46.0	72.7	101.7	26.5	64.6
04	MASONRY	95.4	81.0	86.6	97.5	55.0	71.5	125.3	55.0	82.4	96.1	62.7	75.7	105.9	55.0	74.8	100.5	21.9	52.5
05	METALS	98.4	79.6	92.5	91.4	68.6	84.2	91.5	68.6	84.3	91.4	62.2	82.2	92.9	68.6	85.3	91.9	40.0	75.6
06	WOOD, PLASTICS & COMPOSITES	88.3	83.5	85.5	99.2	37.6	64.0	98.2	30.5	59.5	89.7	37.6	60.0	101.9	29.7	60.7	105.7	19.7	56.6
07	THERMAL & MOISTURE PROTECTION	100.7	87.9	95.5	98.5	59.3	82.7	98.6	58.5	82.5	98.2	56.6	81.5	98.6	58.4	82.5	98.9	27.7	70.3
08	OPENINGS	98.1	80.9	93.9	94.3	47.9	83.0	94.4	44.1	82.1	94.4	44.0	82.1	94.3	43.5	82.0	94.5	21.1	76.6
0920	Plaster & Gypsum Board	92.0	82.9	85.7	80.1	36.3	49.4	79.7	29.0	44.2	75.1	36.3	47.9	80.9	28.1	44.0	81.7	17.9	37.0
0950, 0980	Ceilings & Acoustic Treatment	96.2	82.9	87.4	83.7	36.3	52.1	83.7	29.0	47.2	83.7	36.3	52.1	83.7	28.1	46.7	86.4	17.9	40.7
0960	Flooring	92.2	86.5	90.5	106.3	44.7	87.9	105.0	42.6	86.4	101.5	64.2	90.4	106.9	42.6	87.8	108.7	25.2	83.9
0970, 0990	Wall Finishes & Painting/Coating	96.1	77.7	85.1	100.7	53.6	72.5	100.7	53.6	72.5	100.7	53.6	72.5	100.7	53.6	72.5	100.7	19.1	51.8
09	FINISHES	94.8	83.0	88.2	92.9	39.8	63.1	92.7	35.2	60.5	90.6	43.8	64.3	93.5	34.7	60.5	95.2	21.6	53.9
COVERS	DIVS. 10 - 14, 25, 28, 41, 43, 44, 46	100.0	84.8	96.9	100.0	70.0	93.9	100.0	69.2	93.7	100.0	70.1	93.9	100.0	69.1	93.7	100.0	67.2	93.3
21, 22, 23	FIRE SUPPRESSION, PLUMBING & HVAC	94.0	87.4	91.3	94.3	60.2	80.3	94.3	60.2	80.3	94.3	60.3	80.3	100.3	60.2	83.9	94.3	22.7	65.0
26, 27, 3370	ELECTRICAL, COMMUNICATIONS & UTIL.	90.8	79.6	84.9	93.5	67.2	79.7	94.6	67.2	80.2	96.3	50.7	72.3	94.6	67.2	80.2	96.3	16.0	54.1
MF2010	WEIGHTED AVERAGE	95.3	84.8	90.7	95.1	59.3	79.3	96.6	58.0	79.6	94.3	56.9	77.8	97.5	58.0	80.1	96.4	30.7	67.4

	DIVISION	OKLAHOMA																	
		LAWTON			MCALESTER			MIAMI			MUSKOGEE			OKLAHOMA CITY			PONCA CITY		
		735			745			743			744			730 - 731			746		
		MAT.	INST.	TOTAL	MAT.	INST.	TOTAL	MAT.	INST.	TOTAL	MAT.	INST.	TOTAL	MAT.	INST.	TOTAL	MAT.	INST.	TOTAL
015433	CONTRACTOR EQUIPMENT		82.3	82.3		81.4	81.4		89.4	89.4		89.4	89.4		82.5	82.5		81.4	81.4
0241, 31 - 34	SITE & INFRASTRUCTURE, DEMOLITION	96.6	91.1	92.7	86.3	89.7	88.7	87.6	86.9	87.1	87.8	85.6	86.3	95.2	91.6	92.7	93.1	89.6	90.7
0310	Concrete Forming & Accessories	97.9	44.5	51.7	82.3	40.2	45.9	94.7	68.1	71.6	99.0	29.6	39.0	98.1	38.7	46.7	90.4	39.3	46.2
0320	Concrete Reinforcing	92.9	76.7	84.8	93.3	40.4	66.9	91.8	76.8	84.3	92.7	30.7	61.8	92.9	76.7	84.8	92.7	76.7	84.7
0330	Cast-in-Place Concrete	92.5	44.9	72.9	81.1	44.9	66.2	85.0	44.3	68.3	86.0	35.5	65.2	93.6	47.5	74.6	94.9	37.9	71.5
03	CONCRETE	94.8	51.1	73.2	83.7	42.4	63.3	88.4	62.1	75.4	90.2	33.2	62.0	95.3	49.3	72.6	95.2	46.3	71.1
04	MASONRY	100.5	55.0	72.7	114.9	55.6	78.7	98.8	56.1	72.8	117.4	44.6	73.0	100.1	56.7	73.6	91.2	55.5	69.4
05	METALS	96.5	68.7	87.7	91.4	53.0	79.3	91.3	80.8	88.0	92.8	55.7	81.2	98.3	68.6	89.0	91.3	68.3	84.1
06	WOOD, PLASTICS & COMPOSITES	104.7	45.0	70.7	87.3	40.7	60.7	101.9	76.2	87.2	106.1	29.4	62.3	104.0	36.3	65.3	97.5	38.1	63.6
07	THERMAL & MOISTURE PROTECTION	98.4	60.0	83.0	97.9	55.5	80.8	98.2	60.0	82.9	98.3	40.8	75.2	93.5	59.9	80.0	98.4	58.5	82.3
08	OPENINGS	96.0	51.8	85.3	94.3	39.6	81.0	94.3	69.4	88.3	94.3	28.5	78.3	96.0	47.1	84.1	94.3	48.2	83.1
0920	Plaster & Gypsum Board	82.6	44.0	55.6	73.9	39.5	49.8	80.9	75.9	77.4	82.6	27.7	44.2	88.7	34.9	51.1	79.7	36.8	49.7
0950, 0980	Ceilings & Acoustic Treatment	90.0	44.0	59.3	83.7	39.5	54.2	83.7	75.9	78.5	90.0	27.7	48.5	95.5	34.9	55.2	83.7	36.8	52.5
0960	Flooring	109.0	42.6	89.2	100.4	42.6	83.2	107.6	64.2	94.7	110.1	41.2	89.6	108.6	42.6	88.9	104.6	42.6	86.2
0970, 0990	Wall Finishes & Painting/Coating	100.7	53.6	72.5	100.7	39.5	64.0	100.7	80.9	88.8	100.7	28.7	57.5	100.7	53.6	72.5	100.7	53.6	72.5
09	FINISHES	95.2	43.8	66.4	89.5	39.8	61.7	92.4	69.8	79.7	94.8	31.6	59.4	96.8	39.0	64.4	92.2	39.9	62.9
COVERS	DIVS. 10 - 14, 25, 28, 41, 43, 44, 46	100.0	70.8	94.0	100.0	70.5	94.0	100.0	75.0	94.9	100.0	68.8	93.6	100.0	70.4	94.0	100.0	70.2	93.9
21, 22, 23	FIRE SUPPRESSION, PLUMBING & HVAC	100.3	60.2	83.9	94.3	29.6	67.8	94.3	59.2	79.9	100.3	23.0	68.7	100.0	61.0	84.1	94.3	58.8	79.8
26, 27, 3370	ELECTRICAL, COMMUNICATIONS & UTIL.	96.4	67.2	81.0	94.7	65.1	79.1	96.2	65.2	79.9	94.2	27.5	59.0	102.1	67.2	83.8	94.2	61.4	76.9
MF2010	WEIGHTED AVERAGE	97.7	60.6	81.3	93.7	49.9	74.4	94.0	66.9	82.1	96.8	38.8	71.3	98.5	59.8	81.4	94.3	57.9	78.3

	DIVISION	OKLAHOMA												OREGON					
		POTEAU			SHAWNEE			TULSA			WOODWARD			BEND			EUGENE		
		749			748			740 - 741			738			977			974		
		MAT.	INST.	TOTAL	MAT.	INST.	TOTAL	MAT.	INST.	TOTAL	MAT.	INST.	TOTAL	MAT.	INST.	TOTAL	MAT.	INST.	TOTAL
015433	CONTRACTOR EQUIPMENT		88.7	88.7		81.4	81.4		89.4	89.4		81.4	81.4		98.4	98.4		98.4	98.4
0241, 31 - 34	SITE & INFRASTRUCTURE, DEMOLITION	73.9	85.5	82.1	96.2	89.6	91.5	93.9	86.9	89.0	99.3	89.6	92.5	110.4	103.0	105.2	100.0	103.0	102.1
0310	Concrete Forming & Accessories	88.4	41.5	47.8	84.1	38.0	44.2	99.1	38.5	46.6	91.2	33.7	41.4	110.5	101.2	102.5	107.0	101.1	101.9
0320	Concrete Reinforcing	93.7	76.7	85.3	92.7	44.4	68.6	92.9	76.6	84.8	92.7	76.7	84.7	100.2	103.2	101.7	104.6	103.2	103.9
0330	Cast-in-Place Concrete	85.0	43.9	68.1	98.0	42.0	75.0	93.7	43.9	73.2	95.6	42.6	73.8	98.4	104.9	101.1	95.3	104.8	99.2
03	CONCRETE	90.7	50.1	70.7	96.9	41.2	69.4	95.4	48.8	72.4	98.5	45.5	72.3	110.3	102.5	106.4	101.3	102.4	101.9
04	MASONRY	99.8	55.6	72.8	116.1	54.4	78.4	99.7	56.7	73.5	93.9	55.0	70.2	104.9	107.0	106.2	102.0	107.0	105.1
05	METALS	91.4	80.4	87.9	91.3	54.8	79.8	96.0	79.6	90.8	91.5	68.6	84.3	87.3	97.5	90.5	88.0	97.4	91.0
06	WOOD, PLASTICS & COMPOSITES	94.3	40.8	63.8	89.6	37.3	59.8	105.4	37.4	66.5	98.3	30.4	59.5	102.0	100.7	101.3	97.9	100.7	99.5
07	THERMAL & MOISTURE PROTECTION	98.3	55.8	81.2	98.4	58.5	82.3	98.3	54.3	80.6	98.6	58.5	82.5	103.7	95.5	100.4	102.9	92.6	98.8
08	OPENINGS	94.3	50.2	83.6	94.3	36.8	80.3	96.0	47.3	84.1	94.4	44.0	82.1	96.0	104.6	98.1	96.4	104.6	98.4
0920	Plaster & Gypsum Board	78.0	39.5	51.1	75.1	36.0	47.8	82.6	35.9	50.0	80.4	28.9	44.4	99.8	100.6	100.3	97.3	100.6	99.6
0950, 0980	Ceilings & Acoustic Treatment	83.7	39.5	54.2	83.7	36.0	51.9	90.0	35.9	54.0	86.4	28.9	48.1	96.6	100.6	99.3	97.7	100.6	99.6
0960	Flooring	104.0	64.2	92.1	101.5	33.9	81.3	108.8	44.4	89.6	105.0	44.7	87.0	109.2	107.1	108.6	107.5	107.1	107.4
0970, 0990	Wall Finishes & Painting/Coating	100.7	53.6	72.5	100.7	35.6	61.7	100.7	42.3	65.7	100.7	53.6	72.5	110.6	72.8	88.0	110.6	72.8	88.0
09	FINISHES	90.1	45.7	65.2	90.8	35.6	59.9	94.6	38.8	63.3	93.5	35.5	61.0	104.3	99.1	101.4	102.5	99.1	100.6
COVERS	DIVS. 10 - 14, 25, 28, 41, 43, 44, 46	100.0	70.8	94.0	100.0	69.7	93.8	100.0	70.8	94.0	100.0	69.2	93.7	100.0	103.4	100.7	100.0	103.4	100.7
21, 22, 23	FIRE SUPPRESSION, PLUMBING & HVAC	94.3	58.9	79.8	94.3	59.5	80.0	100.3	46.5	78.3	94.3	60.2	80.3	94.0	99.2	96.1	100.0	99.2	99.7
26, 27, 3370	ELECTRICAL, COMMUNICATIONS & UTIL.	94.3	65.2	79.0	96.4	67.2	81.1	96.4	35.8	64.5	96.2	67.2	80.9	100.9	95.9	98.3	99.5	95.9	97.6
MF2010	WEIGHTED AVERAGE	93.5	60.6	79.0	95.8	55.8	78.2	97.5	52.8	77.8	95.3	58.0	78.9	98.3	100.4	99.3	98.2	100.3	99.1

City Cost Indexes

	DIVISION	OREGON																	
		KLAMATH FALLS			MEDFORD			PENDLETON			PORTLAND			SALEM			VALE		
		976			975			978			970 - 972			973			979		
		MAT.	INST.	TOTAL	MAT.	INST.	TOTAL	MAT.	INST.	TOTAL	MAT.	INST.	TOTAL	MAT.	INST.	TOTAL	MAT.	INST.	TOTAL
015433	CONTRACTOR EQUIPMENT		98.4	98.4		98.4	98.4		95.8	95.8		98.4	98.4		98.4	98.4		95.8	95.8
0241, 31 - 34	SITE & INFRASTRUCTURE, DEMOLITION	114.6	103.0	106.5	107.9	103.0	104.5	106.8	96.6	99.6	102.6	103.0	102.9	95.5	103.0	100.8	94.4	96.5	95.9
0310	Concrete Forming & Accessories	103.8	101.0	101.4	103.0	101.0	101.3	103.6	101.5	101.7	108.1	101.3	102.2	107.8	101.2	102.1	109.7	100.4	101.6
0320	Concrete Reinforcing	100.2	103.1	101.7	101.9	103.1	102.5	99.3	103.2	101.3	105.4	103.2	104.3	105.6	103.2	104.4	96.9	103.0	99.9
0330	Cast-in-Place Concrete	98.4	104.8	101.0	98.4	104.8	101.0	99.2	106.2	102.1	97.9	104.9	100.8	87.4	104.9	94.6	78.2	105.8	89.6
03	CONCRETE	113.2	102.4	107.9	107.6	102.4	105.0	93.0	103.1	98.0	102.8	102.5	102.7	97.8	102.5	100.1	78.4	102.4	90.2
04	MASONRY	118.0	107.0	111.3	99.0	107.0	103.9	108.2	107.1	107.5	102.6	107.0	105.3	106.9	107.0	107.0	106.6	107.1	106.9
05	METALS	87.3	97.3	90.5	87.6	97.3	90.7	93.4	98.1	94.9	88.9	97.6	91.6	88.4	97.5	91.2	93.3	97.0	94.4
06	WOOD, PLASTICS & COMPOSITES	93.4	100.7	97.6	92.3	100.7	97.1	94.9	100.9	98.3	99.0	100.7	100.0	95.2	100.7	98.4	103.0	100.9	101.8
07	THERMAL & MOISTURE PROTECTION	103.9	94.0	99.9	103.5	94.0	99.7	96.6	95.9	96.3	102.9	98.1	101.0	102.1	95.5	99.4	96.0	91.6	94.2
08	OPENINGS	96.1	104.6	98.1	98.7	104.6	100.2	91.2	104.7	94.5	94.2	104.6	96.8	96.2	104.6	98.2	91.2	96.8	92.5
0920	Plaster & Gypsum Board	94.9	100.6	98.9	94.1	100.6	98.6	80.1	100.6	94.4	97.9	100.6	99.8	93.9	100.6	98.6	86.3	100.6	96.3
0950, 0980	Ceilings & Acoustic Treatment	104.9	100.6	102.0	111.0	100.6	104.1	67.3	100.6	89.5	99.7	100.6	100.3	107.3	100.6	102.8	67.3	100.6	89.5
0960	Flooring	106.2	107.1	106.4	105.6	107.1	106.1	73.9	107.1	83.8	105.1	107.1	105.7	108.3	107.1	108.0	75.8	107.1	85.1
0970, 0990	Wall Finishes & Painting/Coating	110.6	60.7	80.7	110.6	60.7	80.7	100.7	68.4	81.3	110.4	68.4	85.2	110.6	72.8	88.0	100.7	72.8	84.0
09	FINISHES	105.1	97.8	101.0	105.3	97.8	101.1	72.5	98.7	87.2	102.4	98.6	100.3	103.9	99.1	101.2	72.7	99.2	87.6
COVERS	DIVS. 10 - 14, 25, 28, 41, 43, 44, 46	100.0	103.3	100.7	100.0	103.3	100.7	100.0	93.5	98.7	100.0	103.4	100.7	100.0	103.4	100.7	100.0	103.6	100.7
21, 22, 23	FIRE SUPPRESSION, PLUMBING & HVAC	94.0	99.2	96.1	100.0	99.2	99.7	95.8	114.3	103.4	100.0	99.2	99.7	100.0	99.2	99.7	95.8	73.7	86.8
26, 27, 3370	ELECTRICAL, COMMUNICATIONS & UTIL.	99.6	82.8	90.7	103.2	82.8	92.4	91.4	96.0	93.8	99.7	101.3	100.5	108.2	95.9	101.7	91.4	75.3	82.9
MF2010	WEIGHTED AVERAGE	99.3	98.3	98.9	99.7	98.3	99.1	93.8	102.9	97.8	98.4	101.2	99.6	98.9	100.4	99.5	91.8	91.1	91.5

	DIVISION	PENNSYLVANIA																	
		ALLENTOWN			ALTOONA			BEDFORD			BRADFORD			BUTLER			CHAMBERSBURG		
		181			166			155			167			160			172		
		MAT.	INST.	TOTAL	MAT.	INST.	TOTAL	MAT.	INST.	TOTAL	MAT.	INST.	TOTAL	MAT.	INST.	TOTAL	MAT.	INST.	TOTAL
015433	CONTRACTOR EQUIPMENT		114.1	114.1		114.1	114.1		110.9	110.9		114.1	114.1		114.1	114.1		113.3	113.3
0241, 31 - 34	SITE & INFRASTRUCTURE, DEMOLITION	90.6	106.0	101.4	94.1	105.9	102.4	103.9	102.6	103.0	89.6	105.3	100.6	85.6	107.3	100.8	83.9	103.2	97.5
0310	Concrete Forming & Accessories	99.0	116.5	114.2	85.1	80.5	81.1	87.2	82.9	83.5	87.1	81.1	81.9	86.5	95.1	94.0	85.3	82.1	82.5
0320	Concrete Reinforcing	97.1	110.3	103.7	94.1	107.7	100.9	92.4	82.8	87.6	96.1	108.1	102.1	94.8	108.8	101.8	94.5	105.2	99.8
0330	Cast-in-Place Concrete	81.7	103.7	90.7	91.0	89.3	90.3	105.4	72.6	91.9	86.9	92.0	89.0	80.4	98.5	87.9	88.3	72.5	81.8
03	CONCRETE	93.3	111.6	102.4	88.6	90.2	89.4	100.3	80.6	90.6	95.1	91.4	93.3	80.9	100.1	90.4	100.7	84.6	92.8
04	MASONRY	96.9	104.0	101.2	98.9	67.2	79.5	108.9	88.8	96.6	96.1	86.6	90.3	101.1	98.5	99.5	97.1	90.4	93.0
05	METALS	99.1	123.4	106.8	93.0	119.9	101.4	97.1	105.6	99.7	97.0	119.8	104.2	92.7	122.1	101.9	97.0	116.5	103.1
06	WOOD, PLASTICS & COMPOSITES	100.4	119.8	111.5	79.9	82.5	81.4	84.9	82.4	83.4	86.3	78.9	82.1	81.2	93.5	88.2	89.5	82.5	85.5
07	THERMAL & MOISTURE PROTECTION	100.2	116.2	106.6	99.7	89.8	95.7	105.5	89.3	99.0	100.1	91.8	96.8	99.4	100.6	99.9	99.4	75.2	89.7
08	OPENINGS	92.8	116.4	98.6	87.0	90.9	87.9	96.4	83.6	93.3	93.0	92.6	92.9	87.0	104.5	91.2	89.7	84.6	88.4
0920	Plaster & Gypsum Board	102.3	120.2	114.8	94.3	81.7	85.5	98.0	81.7	86.6	95.3	78.0	83.2	94.3	93.0	93.4	98.6	81.7	86.8
0950, 0980	Ceilings & Acoustic Treatment	88.7	120.2	109.7	93.3	81.7	85.5	95.1	81.7	86.1	91.5	78.0	82.5	94.3	93.0	93.5	91.5	81.7	85.0
0960	Flooring	93.6	99.4	95.3	87.5	101.0	91.5	90.2	91.4	90.5	88.1	104.9	93.1	88.4	83.9	87.0	91.6	47.8	78.6
0970, 0990	Wall Finishes & Painting/Coating	98.4	73.7	83.6	94.7	109.2	103.4	97.3	85.2	90.1	98.4	92.0	94.6	94.7	109.2	103.4	98.4	85.2	90.5
09	FINISHES	96.1	109.4	103.6	94.7	86.4	90.1	96.5	83.3	89.1	94.3	84.4	88.8	94.6	93.4	93.9	94.7	75.9	84.1
COVERS	DIVS. 10 - 14, 25, 28, 41, 43, 44, 46	100.0	106.3	101.3	100.0	95.9	99.2	100.0	99.2	99.8	100.0	99.1	99.8	100.0	102.8	100.6	100.0	96.6	99.3
21, 22, 23	FIRE SUPPRESSION, PLUMBING & HVAC	100.2	110.5	104.4	99.8	85.6	94.0	93.9	90.6	92.5	94.2	91.0	92.9	93.8	95.1	94.3	94.2	91.8	93.2
26, 27, 3370	ELECTRICAL, COMMUNICATIONS & UTIL.	99.9	98.8	99.3	89.4	108.4	99.4	94.0	108.4	101.6	93.0	108.4	101.1	90.0	104.8	97.8	91.6	86.9	89.1
MF2010	WEIGHTED AVERAGE	97.7	109.5	102.9	94.4	93.2	93.9	97.5	92.8	95.5	95.2	96.3	95.7	92.0	101.5	96.2	95.2	90.6	93.2

	DIVISION	PENNSYLVANIA																	
		DOYLESTOWN			DUBOIS			ERIE			GREENSBURG			HARRISBURG			HAZLETON		
		189			158			164 - 165			156			170 - 171			182		
		MAT.	INST.	TOTAL	MAT.	INST.	TOTAL	MAT.	INST.	TOTAL	MAT.	INST.	TOTAL	MAT.	INST.	TOTAL	MAT.	INST.	TOTAL
015433	CONTRACTOR EQUIPMENT		95.0	95.0		110.9	110.9		114.1	114.1		110.9	110.9		113.3	113.3		114.1	114.1
0241, 31 - 34	SITE & INFRASTRUCTURE, DEMOLITION	103.6	91.6	95.2	109.0	102.9	104.7	90.9	106.3	101.7	100.0	104.9	103.4	81.0	105.0	97.9	84.0	106.3	99.6
0310	Concrete Forming & Accessories	84.1	129.3	123.2	86.6	83.5	84.0	98.6	87.4	88.9	93.1	95.1	94.8	94.1	89.8	90.4	81.8	90.5	89.3
0320	Concrete Reinforcing	93.9	137.2	115.5	91.8	108.3	100.1	96.1	108.2	102.1	91.8	108.8	100.3	97.1	105.6	101.3	94.2	110.4	102.3
0330	Cast-in-Place Concrete	77.2	88.1	81.7	101.6	93.9	98.5	89.4	83.3	86.9	97.8	97.9	97.9	86.0	97.8	90.9	77.2	93.0	83.7
03	CONCRETE	89.2	116.4	102.6	101.5	93.1	97.4	87.6	91.2	89.4	95.5	99.7	97.6	93.4	96.9	95.1	86.2	96.4	91.3
04	MASONRY	100.1	128.2	117.2	109.0	89.7	97.2	88.8	90.8	90.0	119.5	98.4	106.6	96.7	92.1	93.9	109.3	97.3	102.0
05	METALS	96.6	127.3	106.3	97.1	119.0	104.0	93.2	118.7	101.2	97.0	120.6	104.4	101.0	120.6	107.2	98.8	122.3	106.2
06	WOOD, PLASTICS & COMPOSITES	81.9	132.2	110.6	83.8	81.3	82.4	96.6	85.7	90.4	91.3	93.4	92.5	94.3	88.9	91.2	80.6	88.9	85.3
07	THERMAL & MOISTURE PROTECTION	98.5	130.7	111.5	105.8	95.7	101.7	100.1	92.9	97.2	105.4	100.3	103.4	103.5	107.3	105.0	99.7	105.9	102.2
08	OPENINGS	94.9	140.1	105.9	96.4	93.9	95.8	87.1	92.6	88.5	96.3	104.4	98.3	92.8	94.4	93.2	93.5	95.0	93.8
0920	Plaster & Gypsum Board	92.6	132.9	120.8	96.7	80.6	85.5	102.3	85.0	90.2	99.4	93.0	94.9	102.6	88.3	92.6	93.1	88.3	89.7
0950, 0980	Ceilings & Acoustic Treatment	87.9	132.9	117.9	95.1	80.6	85.4	88.7	85.0	86.3	94.2	93.0	93.4	97.8	88.3	91.5	89.7	88.3	88.8
0960	Flooring	78.3	139.2	96.5	90.0	104.9	94.4	92.3	95.1	93.1	93.3	70.5	86.5	94.2	95.0	94.4	85.7	98.0	89.4
0970, 0990	Wall Finishes & Painting/Coating	98.0	72.1	82.4	97.3	105.9	102.5	102.6	92.0	96.2	97.3	109.2	104.4	98.4	88.2	92.3	98.4	99.6	99.1
09	FINISHES	87.2	124.4	108.1	96.8	88.5	92.2	96.5	88.7	92.2	97.0	91.7	94.0	96.9	90.2	93.1	92.4	90.9	91.5
COVERS	DIVS. 10 - 14, 25, 28, 41, 43, 44, 46	100.0	72.0	94.3	100.0	99.1	99.8	100.0	100.9	100.2	100.0	102.6	100.5	100.0	97.9	99.6	100.0	97.3	99.4
21, 22, 23	FIRE SUPPRESSION, PLUMBING & HVAC	93.8	126.3	107.1	93.9	91.0	92.7	99.8	93.3	97.1	93.9	94.3	94.1	100.0	93.6	97.4	94.2	101.5	97.2
26, 27, 3370	ELECTRICAL, COMMUNICATIONS & UTIL.	92.4	119.5	106.6	94.6	108.4	101.9	91.4	89.0	90.1	94.6	108.4	101.9	99.4	87.0	92.9	94.0	84.1	88.8
MF2010	WEIGHTED AVERAGE	94.5	120.2	105.8	97.9	97.2	97.6	94.2	95.2	94.6	97.5	101.2	99.2	97.8	96.5	97.2	94.9	98.6	96.5

DIVISION		PENNSYLVANIA																	
		INDIANA			JOHNSTOWN			KITTANNING			LANCASTER			LEHIGH VALLEY			MONTROSE		
		157			159			162			175 - 176			180			188		
		MAT.	INST.	TOTAL	MAT.	INST.	TOTAL	MAT.	INST.	TOTAL	MAT.	INST.	TOTAL	MAT.	INST.	TOTAL	MAT.	INST.	TOTAL
015433	**CONTRACTOR EQUIPMENT**		110.9	110.9		110.9	110.9		114.1	114.1		113.3	113.3		114.1	114.1		114.1	114.1
0241, 31 - 34	**SITE & INFRASTRUCTURE, DEMOLITION**	98.1	103.6	102.0	104.4	104.2	104.2	88.1	107.3	101.6	76.6	105.0	96.5	87.8	105.7	100.3	86.7	104.0	98.8
0310	Concrete Forming & Accessories	87.7	85.8	86.1	86.6	84.4	84.7	86.5	93.6	92.7	87.2	89.1	88.9	93.1	115.7	112.7	82.7	87.7	87.1
0320	Concrete Reinforcing	91.1	108.9	99.9	92.4	108.6	100.5	94.8	108.9	101.8	94.2	105.5	99.8	94.2	110.3	102.2	98.7	110.2	104.4
0330	Cast-in-Place Concrete	95.9	97.7	96.7	106.3	94.8	101.6	83.5	98.5	89.7	75.2	97.3	84.3	83.5	102.7	91.4	81.9	89.8	85.2
03	**CONCRETE**	93.2	95.6	94.3	101.1	93.9	97.5	83.4	99.4	91.3	88.8	96.4	92.5	92.3	110.9	101.5	91.1	93.8	92.5
04	**MASONRY**	104.8	100.2	102.0	105.6	89.2	95.6	104.8	100.3	102.0	102.4	91.6	95.8	96.8	102.7	100.4	96.8	96.8	96.8
05	**METALS**	97.1	120.2	104.4	97.1	119.2	104.0	92.8	122.3	102.0	97.0	120.1	104.2	98.8	122.2	106.1	97.1	117.0	103.3
06	**WOOD, PLASTICS & COMPOSITES**	85.6	81.3	83.2	83.8	82.4	83.0	81.2	91.3	87.0	92.0	88.9	90.2	92.7	119.8	108.2	81.4	87.0	84.6
07	**THERMAL & MOISTURE PROTECTION**	105.3	98.1	102.4	105.5	94.3	101.0	99.5	100.9	100.0	99.0	96.1	97.8	100.0	104.7	101.9	99.7	93.6	97.3
08	**OPENINGS**	96.4	94.0	95.8	96.4	91.0	95.0	87.0	99.5	90.0	89.7	98.2	91.7	93.5	116.4	99.0	89.9	94.0	90.9
0920	Plaster & Gypsum Board	98.4	80.6	86.0	96.5	81.7	86.1	94.3	90.8	91.9	100.7	88.3	92.0	96.4	120.2	113.0	93.6	86.3	88.5
0950, 0980	Ceilings & Acoustic Treatment	95.1	80.6	85.4	94.2	81.7	85.8	94.3	90.8	92.0	91.5	88.3	89.4	89.7	120.2	110.0	91.5	86.3	88.1
0960	Flooring	90.7	104.9	94.9	90.0	101.8	93.5	88.4	104.9	93.3	92.5	95.0	93.2	90.7	99.4	93.3	86.3	60.0	78.5
0970, 0990	Wall Finishes & Painting/Coating	97.3	109.2	104.4	97.3	109.2	104.4	94.7	109.2	103.4	98.4	59.7	75.2	98.4	69.4	81.0	98.4	99.6	99.1
09	**FINISHES**	96.2	89.8	92.6	96.1	89.2	92.2	94.8	95.7	95.3	94.7	86.9	90.3	94.6	107.6	101.9	93.3	83.6	87.9
COVERS	**DIVS. 10 - 14, 25, 28, 41, 43, 44, 46**	100.0	101.3	100.3	100.0	99.4	99.9	100.0	102.6	100.5	100.0	97.5	99.5	100.0	102.5	100.5	100.0	97.1	99.4
21, 22, 23	**FIRE SUPPRESSION, PLUMBING & HVAC**	93.9	91.3	92.8	93.9	91.2	92.8	93.8	98.4	95.7	94.2	93.2	93.8	94.2	109.8	100.6	94.2	100.8	96.9
26, 27, 3370	**ELECTRICAL, COMMUNICATIONS & UTIL.**	94.6	108.4	101.9	94.6	108.4	101.9	89.4	108.4	99.4	93.1	45.9	68.2	94.0	135.9	116.0	93.0	93.2	93.1
MF2010	**WEIGHTED AVERAGE**	96.4	99.1	97.6	97.5	97.4	97.4	92.5	102.9	97.1	94.1	90.0	92.3	95.3	113.4	103.3	94.2	97.3	95.6

DIVISION		PENNSYLVANIA																	
		NEW CASTLE			NORRISTOWN			OIL CITY			PHILADELPHIA			PITTSBURGH			POTTSVILLE		
		161			194			163			190 - 191			150 - 152			179		
		MAT.	INST.	TOTAL	MAT.	INST.	TOTAL	MAT.	INST.	TOTAL	MAT.	INST.	TOTAL	MAT.	INST.	TOTAL	MAT.	INST.	TOTAL
015433	**CONTRACTOR EQUIPMENT**		114.1	114.1		99.6	99.6		114.1	114.1		99.1	99.1		112.1	112.1		113.3	113.3
0241, 31 - 34	**SITE & INFRASTRUCTURE, DEMOLITION**	86.0	107.3	100.9	95.1	100.4	98.8	84.6	105.5	99.3	101.8	99.8	100.4	103.5	107.0	105.9	79.2	105.1	97.4
0310	Concrete Forming & Accessories	86.5	97.2	95.8	83.3	130.0	123.7	86.5	82.2	82.7	98.3	136.3	131.2	100.2	98.2	98.5	79.4	90.4	88.9
0320	Concrete Reinforcing	93.6	94.7	94.1	98.1	144.4	121.1	94.8	94.5	94.6	101.2	144.5	122.8	92.9	109.2	101.0	93.5	100.7	97.1
0330	Cast-in-Place Concrete	81.1	98.9	88.5	78.0	127.2	98.2	78.9	96.6	86.1	95.6	128.7	109.2	101.6	98.1	100.2	80.0	98.6	87.7
03	**CONCRETE**	81.2	98.5	89.7	85.4	131.4	108.2	79.8	90.9	85.3	97.1	134.8	115.7	98.7	101.3	100.0	92.3	96.5	94.4
04	**MASONRY**	100.5	98.5	99.2	114.3	127.1	122.1	100.9	93.1	96.1	100.7	128.5	117.6	98.4	104.5	102.2	96.5	93.8	94.8
05	**METALS**	92.8	115.1	99.8	95.8	132.1	107.2	92.8	113.6	99.3	99.6	133.1	110.1	98.4	121.5	105.7	97.2	118.4	103.9
06	**WOOD, PLASTICS & COMPOSITES**	81.2	97.2	90.4	83.0	130.0	109.8	81.2	78.9	79.9	100.2	137.9	121.7	100.6	97.1	98.6	83.0	88.9	86.3
07	**THERMAL & MOISTURE PROTECTION**	99.4	97.7	98.7	100.6	130.5	112.6	99.3	94.1	97.2	101.4	131.9	113.6	105.6	102.4	104.3	99.1	105.6	101.7
08	**OPENINGS**	87.0	98.2	89.7	88.8	141.0	101.5	87.0	80.7	85.4	96.9	145.3	108.7	99.7	106.4	101.3	89.7	92.1	90.3
0920	Plaster & Gypsum Board	94.3	96.9	96.1	91.0	130.6	118.7	94.3	78.0	82.9	98.9	138.8	126.8	106.3	96.9	99.7	95.7	88.3	90.5
0950, 0980	Ceilings & Acoustic Treatment	94.3	96.9	96.0	96.8	130.6	119.3	94.3	78.0	83.4	96.8	138.8	124.8	95.1	96.9	96.3	91.5	88.3	89.4
0960	Flooring	88.4	59.5	79.8	93.5	137.9	106.7	88.4	104.9	93.3	100.5	141.7	112.8	96.4	106.2	99.4	89.0	95.0	90.8
0970, 0990	Wall Finishes & Painting/Coating	94.7	109.2	103.4	99.8	145.4	127.1	94.7	109.2	103.4	99.8	146.4	127.7	97.3	113.0	106.7	98.4	99.6	99.1
09	**FINISHES**	94.6	91.2	92.7	96.9	133.1	117.2	94.5	87.1	90.4	100.3	138.5	121.7	99.2	100.7	100.0	93.2	91.8	92.4
COVERS	**DIVS. 10 - 14, 25, 28, 41, 43, 44, 46**	100.0	103.2	100.6	100.0	111.0	102.2	100.0	100.2	100.0	100.0	119.9	104.1	100.0	103.0	100.6	100.0	94.9	99.0
21, 22, 23	**FIRE SUPPRESSION, PLUMBING & HVAC**	93.8	94.8	94.2	94.0	126.1	107.1	93.8	92.8	93.4	100.0	129.0	111.9	99.9	99.3	99.7	94.2	100.9	97.0
26, 27, 3370	**ELECTRICAL, COMMUNICATIONS & UTIL.**	90.0	90.3	90.1	92.8	140.1	117.7	92.1	104.8	98.8	97.5	149.1	124.7	97.1	108.4	103.1	91.1	92.8	92.0
MF2010	**WEIGHTED AVERAGE**	92.0	98.0	94.7	94.7	128.4	109.5	92.1	96.1	93.8	99.2	132.2	113.7	99.4	104.7	101.8	94.0	98.8	96.1

DIVISION		PENNSYLVANIA																	
		READING			SCRANTON			STATE COLLEGE			STROUDSBURG			SUNBURY			UNIONTOWN		
		195 - 196			184 - 185			168			183			178			154		
		MAT.	INST.	TOTAL	MAT.	INST.	TOTAL	MAT.	INST.	TOTAL	MAT.	INST.	TOTAL	MAT.	INST.	TOTAL	MAT.	INST.	TOTAL
015433	**CONTRACTOR EQUIPMENT**		118.8	118.8		114.1	114.1		113.3	113.3		114.1	114.1		114.1	114.1		110.9	110.9
0241, 31 - 34	**SITE & INFRASTRUCTURE, DEMOLITION**	99.8	112.8	108.9	91.0	106.4	101.8	81.3	104.7	97.7	85.6	104.1	98.6	89.1	106.2	101.1	98.7	104.9	103.0
0310	Concrete Forming & Accessories	97.8	90.7	91.6	99.2	89.4	90.7	84.4	80.3	80.8	87.9	88.9	88.7	91.0	89.2	89.5	81.3	95.1	93.3
0320	Concrete Reinforcing	99.4	100.8	100.1	97.1	110.5	103.8	95.4	108.4	101.9	97.4	117.1	107.2	96.1	105.5	100.8	91.8	108.8	100.3
0330	Cast-in-Place Concrete	70.0	96.4	80.8	85.3	93.3	88.6	82.3	67.9	76.3	80.4	74.6	78.0	87.5	97.1	91.4	95.9	98.0	96.8
03	**CONCRETE**	84.5	95.9	90.1	95.1	96.1	95.6	95.3	82.9	89.2	90.0	90.4	90.2	95.7	96.4	96.0	92.8	99.8	96.3
04	**MASONRY**	102.6	94.8	97.9	97.2	97.4	97.3	102.5	80.4	89.0	93.7	102.3	98.9	96.1	91.2	93.1	122.1	98.4	107.7
05	**METALS**	96.1	119.1	103.3	101.1	123.0	108.0	96.8	120.7	104.3	98.9	119.6	105.4	96.9	120.4	104.3	96.8	120.6	104.3
06	**WOOD, PLASTICS & COMPOSITES**	100.3	88.9	93.8	100.4	87.0	92.7	88.3	82.5	85.0	87.2	87.0	87.1	90.4	88.9	89.5	77.9	93.4	86.8
07	**THERMAL & MOISTURE PROTECTION**	101.0	108.4	103.9	100.1	97.5	99.0	99.2	90.3	95.6	99.9	86.9	94.6	100.0	104.5	101.8	105.2	100.3	103.2
08	**OPENINGS**	93.9	96.7	94.6	92.8	94.0	93.1	89.6	90.9	90.0	93.5	89.2	92.4	89.8	93.6	90.7	96.3	104.4	98.3
0920	Plaster & Gypsum Board	103.2	88.3	92.8	104.7	86.3	91.8	96.6	81.7	86.2	94.7	86.3	88.8	95.4	88.3	90.4	94.4	93.0	93.5
0950, 0980	Ceilings & Acoustic Treatment	86.6	88.3	87.7	97.8	86.3	90.1	88.8	81.7	84.0	87.9	86.3	86.8	88.8	88.3	88.4	94.2	93.0	93.4
0960	Flooring	97.9	95.0	97.0	93.6	109.6	98.4	91.2	101.0	94.1	88.7	53.2	78.1	89.6	90.9	90.0	87.5	104.9	92.7
0970, 0990	Wall Finishes & Painting/Coating	98.4	100.0	99.4	98.4	99.6	99.1	98.4	109.2	104.9	98.4	67.5	79.9	98.4	99.6	99.1	97.3	109.2	104.4
09	**FINISHES**	97.9	91.5	94.3	98.3	93.5	95.6	93.5	86.4	89.5	93.2	79.9	85.7	94.1	90.5	92.1	94.6	96.7	95.8
COVERS	**DIVS. 10 - 14, 25, 28, 41, 43, 44, 46**	100.0	100.4	100.1	100.0	97.1	99.4	100.0	93.1	98.6	100.0	63.9	92.6	100.0	94.0	98.8	100.0	102.6	100.5
21, 22, 23	**FIRE SUPPRESSION, PLUMBING & HVAC**	100.1	106.6	102.8	100.2	101.5	100.7	94.2	87.3	91.4	94.2	103.3	97.9	94.2	92.8	93.6	93.9	94.3	94.1
26, 27, 3370	**ELECTRICAL, COMMUNICATIONS & UTIL.**	99.8	92.8	96.1	99.9	93.3	96.4	92.2	108.4	100.7	94.0	139.9	118.2	91.5	85.0	88.1	91.4	108.4	100.3
MF2010	**WEIGHTED AVERAGE**	97.1	101.1	98.8	98.4	99.9	99.1	94.7	93.7	94.3	94.7	102.7	98.2	94.7	95.8	95.2	96.7	101.9	99.0

DIVISION		PENNSYLVANIA																	
		WASHINGTON			WELLSBORO			WESTCHESTER			WILKES-BARRE			WILLIAMSPORT			YORK		
		153			169			193			186 - 187			177			173 - 174		
		MAT.	INST.	TOTAL	MAT.	INST.	TOTAL	MAT.	INST.	TOTAL	MAT.	INST.	TOTAL	MAT.	INST.	TOTAL	MAT.	INST.	TOTAL
015433	CONTRACTOR EQUIPMENT		110.9	110.9		114.1	114.1		99.6	99.6		114.1	114.1		114.1	114.1		113.3	113.3
0241, 31 - 34	SITE & INFRASTRUCTURE, DEMOLITION	98.8	104.9	103.0	93.3	103.9	100.7	101.0	98.2	99.0	83.7	106.3	99.6	81.1	104.6	97.6	79.9	105.0	97.5
0310	Concrete Forming & Accessories	87.8	95.3	94.3	86.5	87.5	87.4	89.5	130.4	124.9	90.3	88.2	88.4	87.9	61.0	64.6	82.6	89.6	88.7
0320	Concrete Reinforcing	91.8	108.9	100.4	95.4	110.1	102.7	97.1	111.5	104.3	96.1	110.5	103.3	95.4	67.7	81.6	96.1	105.6	100.8
0330	Cast-in-Place Concrete	95.9	98.0	96.8	86.2	87.8	86.8	86.3	127.9	103.4	77.2	93.2	83.8	74.1	73.3	73.7	80.4	97.5	87.4
03	CONCRETE	93.3	99.9	96.5	97.5	93.0	95.3	92.5	125.6	108.9	87.1	95.5	91.2	84.0	68.5	76.3	93.6	96.7	95.1
04	MASONRY	104.6	100.2	101.9	102.4	91.2	95.6	108.2	128.2	120.4	109.9	97.3	102.2	88.2	91.0	89.9	97.3	91.6	93.8
05	METALS	96.8	120.9	104.4	96.9	117.1	103.3	95.8	116.5	102.3	97.1	122.7	105.1	97.0	101.6	98.4	98.6	120.6	105.5
06	WOOD, PLASTICS & COMPOSITES	85.7	93.4	90.1	85.7	88.9	87.5	89.8	132.1	114.0	89.5	85.2	87.0	87.2	53.3	67.8	86.4	88.9	87.8
07	THERMAL & MOISTURE PROTECTION	105.3	100.8	103.5	100.3	91.6	96.8	100.9	130.6	112.8	99.7	105.5	102.0	99.6	100.2	99.8	99.2	107.0	102.3
08	OPENINGS	96.3	104.4	98.3	93.0	95.0	93.5	88.8	127.1	98.2	89.9	97.6	91.7	89.8	55.9	81.6	89.7	94.4	90.9
0920	Plaster & Gypsum Board	98.1	93.0	94.6	94.5	88.3	90.2	93.1	132.9	120.9	95.7	84.5	87.9	95.7	51.6	64.8	96.9	88.3	90.9
0950, 0980	Ceilings & Acoustic Treatment	94.2	93.0	93.4	88.8	88.3	88.4	96.8	132.9	120.8	91.5	84.5	86.8	91.5	51.6	64.9	90.5	88.3	89.0
0960	Flooring	90.7	104.9	94.9	87.9	52.6	77.4	96.6	141.7	110.0	89.5	98.0	92.1	88.7	51.5	77.6	90.2	95.0	91.6
0970, 0990	Wall Finishes & Painting/Coating	97.3	109.2	104.4	98.4	99.6	99.1	99.8	145.4	127.1	98.4	102.3	100.7	98.4	102.3	100.7	98.4	88.2	92.3
09	FINISHES	96.0	97.0	96.6	93.9	82.3	87.4	98.6	132.9	117.8	94.2	90.4	92.1	93.8	62.6	76.3	93.5	90.0	91.6
COVERS	DIVS. 10 - 14, 25, 28, 41, 43, 44, 46	100.0	102.6	100.5	100.0	94.0	98.8	100.0	114.3	102.9	100.0	96.9	99.4	100.0	93.1	98.6	100.0	97.7	99.5
21, 22, 23	FIRE SUPPRESSION, PLUMBING & HVAC	93.9	98.4	95.8	94.2	92.1	93.3	94.0	126.4	107.2	94.2	101.6	97.2	94.2	91.8	93.2	100.2	93.3	97.3
26, 27, 3370	ELECTRICAL, COMMUNICATIONS & UTIL.	93.9	108.4	101.6	93.0	84.8	88.7	92.7	118.5	106.2	94.0	84.1	88.8	92.0	48.6	69.1	93.1	87.0	89.9
MF2010	WEIGHTED AVERAGE	96.3	103.0	99.2	95.8	93.4	94.7	95.5	122.7	107.5	94.6	98.5	96.3	92.8	79.3	86.9	96.0	96.4	96.2

DIVISION		PUERTO RICO			RHODE ISLAND						SOUTH CAROLINA								
		SAN JUAN			NEWPORT			PROVIDENCE			AIKEN			BEAUFORT			CHARLESTON		
		009			028			029			298			299			294		
		MAT.	INST.	TOTAL	MAT.	INST.	TOTAL	MAT.	INST.	TOTAL	MAT.	INST.	TOTAL	MAT.	INST.	TOTAL	MAT.	INST.	TOTAL
015433	CONTRACTOR EQUIPMENT		90.1	90.1		102.0	102.0		102.0	102.0		100.1	100.1		100.1	100.1		100.1	100.1
0241, 31 - 34	SITE & INFRASTRUCTURE, DEMOLITION	135.3	90.6	104.0	92.0	103.8	100.3	93.0	103.8	100.6	125.1	87.0	98.4	119.7	85.2	95.5	102.1	84.8	89.9
0310	Concrete Forming & Accessories	90.6	17.0	26.9	99.1	122.0	118.9	101.1	122.0	119.2	97.2	70.5	74.1	96.2	38.7	46.4	95.2	40.5	47.8
0320	Concrete Reinforcing	199.0	12.1	105.9	110.7	156.2	133.4	110.7	156.2	133.4	100.0	68.4	84.3	99.1	27.0	63.2	98.9	58.1	78.6
0330	Cast-in-Place Concrete	106.9	30.3	75.4	82.6	118.7	97.4	101.4	118.7	108.5	89.7	72.2	82.5	89.7	48.0	72.6	105.2	48.6	82.0
03	CONCRETE	113.7	21.7	68.3	97.9	126.7	112.1	107.1	126.7	116.8	108.2	71.7	90.2	105.5	41.8	74.0	101.4	48.3	75.2
04	MASONRY	84.4	16.0	42.7	99.2	128.7	117.2	102.6	128.7	118.5	78.0	64.0	69.4	92.7	34.2	57.0	94.0	38.3	60.0
05	METALS	111.2	34.4	87.0	97.7	127.6	107.1	97.7	127.6	107.1	87.7	86.0	87.2	87.8	69.1	81.9	89.6	76.6	85.5
06	WOOD, PLASTICS & COMPOSITES	89.9	16.8	48.2	96.9	121.3	110.8	96.1	121.3	110.5	101.1	71.6	84.3	99.5	39.4	65.2	98.2	39.4	64.6
07	THERMAL & MOISTURE PROTECTION	125.2	21.0	83.3	97.3	119.3	106.1	102.0	119.3	109.0	101.4	69.1	88.4	101.1	37.2	75.4	100.3	41.6	76.7
08	OPENINGS	147.8	14.3	115.3	96.8	129.2	104.7	97.4	129.2	105.1	90.5	67.7	85.0	90.5	38.3	77.8	94.2	43.1	81.8
0920	Plaster & Gypsum Board	114.4	14.2	44.3	96.0	121.4	113.7	92.5	121.4	112.7	106.6	70.6	81.4	110.5	37.4	59.4	112.4	37.4	59.9
0950, 0980	Ceilings & Acoustic Treatment	228.4	14.2	85.7	94.8	121.4	112.5	98.7	121.4	113.8	93.4	70.6	78.2	97.0	37.4	57.3	97.0	37.4	57.3
0960	Flooring	213.7	13.6	154.1	99.1	139.1	111.0	98.5	139.1	110.6	102.8	70.9	93.3	104.3	53.1	89.0	103.8	60.3	90.9
0970, 0990	Wall Finishes & Painting/Coating	205.5	12.5	89.9	98.1	127.0	115.4	98.1	127.0	115.4	109.6	73.5	87.9	109.6	36.2	65.6	109.6	40.9	68.4
09	FINISHES	206.2	16.1	99.7	98.3	126.1	113.9	98.3	126.1	113.9	103.9	70.9	85.4	105.1	41.4	69.4	103.0	43.7	69.8
COVERS	DIVS. 10 - 14, 25, 28, 41, 43, 44, 46	100.0	16.2	82.9	100.0	104.4	100.9	100.0	104.4	100.9	100.0	74.8	94.8	100.0	45.0	88.8	100.0	67.8	93.4
21, 22, 23	FIRE SUPPRESSION, PLUMBING & HVAC	102.8	13.1	66.1	100.0	111.1	104.5	99.8	111.1	104.4	94.3	66.5	82.9	94.3	24.9	65.9	100.3	39.2	75.3
26, 27, 3370	ELECTRICAL, COMMUNICATIONS & UTIL.	131.0	12.4	68.5	101.1	97.1	99.0	101.2	97.1	99.1	95.3	68.7	81.3	99.0	31.6	63.5	97.4	92.5	94.8
MF2010	WEIGHTED AVERAGE	121.0	23.8	78.2	98.7	116.7	106.6	100.1	116.7	107.4	96.1	71.8	85.4	96.8	42.1	72.7	97.7	56.7	79.6

DIVISION		SOUTH CAROLINA															SOUTH DAKOTA		
		COLUMBIA			FLORENCE			GREENVILLE			ROCK HILL			SPARTANBURG			ABERDEEN		
		290 - 292			295			296			297			293			574		
		MAT.	INST.	TOTAL	MAT.	INST.	TOTAL	MAT.	INST.	TOTAL	MAT.	INST.	TOTAL	MAT.	INST.	TOTAL	MAT.	INST.	TOTAL
015433	CONTRACTOR EQUIPMENT		100.1	100.1		100.1	100.1		100.1	100.1		100.1	100.1		100.1	100.1		98.9	98.9
0241, 31 - 34	SITE & INFRASTRUCTURE, DEMOLITION	99.7	84.7	89.2	113.0	84.7	93.2	107.6	84.4	91.3	105.1	84.3	90.5	107.4	84.4	91.3	97.8	94.6	95.6
0310	Concrete Forming & Accessories	98.1	39.5	47.4	83.3	39.7	45.5	94.7	39.5	47.0	93.2	37.2	44.8	97.8	39.5	47.4	100.2	38.3	46.7
0320	Concrete Reinforcing	98.9	57.9	78.5	98.5	58.1	78.4	98.4	46.3	72.5	99.3	46.1	72.8	98.4	46.3	72.5	106.6	39.9	73.4
0330	Cast-in-Place Concrete	108.0	49.2	83.8	89.7	48.4	72.7	89.7	48.4	72.7	89.7	43.5	70.7	89.7	48.4	72.7	102.8	45.3	79.1
03	CONCRETE	103.0	48.0	75.8	100.1	47.9	74.3	98.8	45.6	72.5	96.9	42.9	70.2	99.0	45.6	72.6	101.6	42.5	72.4
04	MASONRY	88.5	36.2	56.6	78.1	38.3	53.8	76.0	38.3	53.0	99.7	33.6	59.3	78.1	38.3	53.8	107.1	57.1	76.6
05	METALS	89.6	75.8	85.3	88.6	76.1	84.6	88.6	71.9	83.3	87.8	71.4	82.6	88.6	71.9	83.3	102.6	65.8	91.0
06	WOOD, PLASTICS & COMPOSITES	100.7	39.0	65.5	83.7	39.0	58.2	97.8	39.0	64.2	96.1	37.1	62.4	102.2	39.0	66.1	95.5	37.6	62.5
07	THERMAL & MOISTURE PROTECTION	98.7	40.3	75.2	100.6	41.5	76.8	100.5	41.5	76.8	100.4	39.2	75.8	100.6	41.5	76.8	104.8	49.8	82.7
08	OPENINGS	94.2	42.9	81.7	90.6	42.9	79.0	90.5	40.1	78.2	90.5	39.2	78.0	90.5	40.1	78.3	92.9	38.4	79.6
0920	Plaster & Gypsum Board	104.5	36.9	57.3	99.0	36.9	55.6	105.0	36.9	57.4	104.1	35.0	55.8	108.3	36.9	58.4	110.8	35.9	58.4
0950, 0980	Ceilings & Acoustic Treatment	99.7	36.9	57.9	94.3	36.9	56.1	93.4	36.9	55.8	93.4	35.0	54.5	93.4	36.9	55.8	118.5	35.9	63.5
0960	Flooring	99.8	45.2	83.5	96.1	45.2	80.9	101.6	59.2	89.0	100.8	46.1	84.5	103.1	59.2	90.0	116.5	52.1	97.3
0970, 0990	Wall Finishes & Painting/Coating	107.2	40.9	67.5	109.6	40.9	68.4	109.6	40.9	68.4	109.6	40.9	68.4	109.6	40.9	68.4	94.1	38.8	60.9
09	FINISHES	100.7	40.5	66.9	99.8	40.9	66.8	101.5	43.3	68.9	100.9	38.9	66.2	102.4	43.3	69.3	111.8	40.5	71.8
COVERS	DIVS. 10 - 14, 25, 28, 41, 43, 44, 46	100.0	67.5	93.4	100.0	67.5	93.4	100.0	67.5	93.4	100.0	66.7	93.2	100.0	67.5	93.4	100.0	42.3	88.2
21, 22, 23	FIRE SUPPRESSION, PLUMBING & HVAC	99.9	34.7	73.2	100.3	34.7	73.5	100.3	33.3	72.9	94.3	32.3	68.9	100.3	33.3	72.9	100.2	41.9	76.4
26, 27, 3370	ELECTRICAL, COMMUNICATIONS & UTIL.	97.8	47.3	71.2	95.2	45.0	68.8	97.4	48.7	71.7	97.4	48.7	71.7	97.4	48.7	71.7	101.1	53.3	75.9
MF2010	WEIGHTED AVERAGE	97.3	48.8	75.9	96.0	48.7	75.2	96.0	48.4	75.0	95.3	46.6	73.8	96.3	48.4	75.2	101.3	51.7	79.4

	DIVISION	SOUTH DAKOTA																	
		MITCHELL			MOBRIDGE			PIERRE			RAPID CITY			SIOUX FALLS			WATERTOWN		
		573			576			575			577			570 - 571			572		
		MAT.	INST.	TOTAL	MAT.	INST.	TOTAL	MAT.	INST.	TOTAL	MAT.	INST.	TOTAL	MAT.	INST.	TOTAL	MAT.	INST.	TOTAL
015433	CONTRACTOR EQUIPMENT		98.9	98.9		98.9	98.9		98.9	98.9		98.9	98.9		99.9	99.9		98.9	98.9
0241, 31 - 34	SITE & INFRASTRUCTURE, DEMOLITION	94.6	94.6	94.6	94.5	94.6	94.6	93.2	94.7	94.2	96.1	94.8	95.1	94.0	96.3	95.6	94.5	94.6	94.6
0310	Concrete Forming & Accessories	99.4	38.7	46.8	90.6	38.5	45.5	100.5	39.5	47.7	107.7	37.6	47.0	101.2	42.0	50.0	87.4	38.3	44.9
0320	Concrete Reinforcing	106.0	48.6	77.4	108.7	39.9	74.5	106.2	58.4	82.4	99.7	58.5	79.2	99.7	58.5	79.2	103.1	40.0	71.6
0330	Cast-in-Place Concrete	99.9	44.0	76.9	99.9	45.3	77.4	86.2	43.7	68.7	99.1	44.3	76.6	91.7	44.6	72.3	99.9	45.3	77.4
03	CONCRETE	99.4	43.7	71.9	99.3	42.5	71.2	92.9	45.9	69.7	98.5	45.3	72.3	93.5	47.4	70.7	98.1	42.5	70.6
04	MASONRY	95.7	55.5	71.2	104.5	57.1	75.6	102.8	55.8	74.1	104.6	57.1	75.6	100.3	55.8	73.1	129.5	58.1	85.9
05	METALS	101.6	66.0	90.4	101.7	65.8	90.4	102.6	74.4	93.7	104.3	74.9	95.1	104.5	75.0	95.2	101.6	66.2	90.5
06	WOOD, PLASTICS & COMPOSITES	94.6	38.1	62.3	84.5	37.8	57.9	93.5	38.4	62.0	98.9	34.8	62.3	94.3	41.0	63.9	81.3	37.6	56.3
07	THERMAL & MOISTURE PROTECTION	104.6	47.7	81.7	104.7	49.8	82.6	106.9	48.0	83.2	105.1	49.6	82.8	111.4	50.4	86.8	104.5	50.0	82.6
08	OPENINGS	91.6	39.0	78.8	94.1	38.0	80.5	95.7	44.3	83.2	96.7	42.3	83.5	96.9	45.7	84.5	91.6	38.1	78.5
0920	Plaster & Gypsum Board	109.0	36.5	58.3	102.5	36.2	56.1	99.6	36.7	55.6	110.3	33.0	56.3	105.3	39.5	59.3	100.3	35.9	55.3
0950, 0980	Ceilings & Acoustic Treatment	114.8	36.5	62.6	118.5	36.2	63.6	116.7	36.7	63.4	121.2	33.0	62.5	117.6	39.5	65.5	114.8	35.9	62.3
0960	Flooring	116.0	52.1	97.0	111.1	52.1	93.5	115.5	37.2	92.2	115.7	78.5	104.6	115.3	77.4	104.0	109.6	52.1	92.5
0970, 0990	Wall Finishes & Painting/Coating	94.1	42.4	63.1	94.1	43.5	63.8	94.1	46.6	65.6	94.1	46.6	65.6	94.1	46.6	65.6	94.1	38.8	60.9
09	FINISHES	110.4	41.2	71.7	109.0	41.2	71.0	109.3	39.0	69.9	112.0	45.2	74.6	110.3	48.5	75.6	107.5	40.5	69.9
COVERS	DIVS. 10 - 14, 25, 28, 41, 43, 44, 46	100.0	39.2	87.6	100.0	42.3	88.2	100.0	77.3	95.4	100.0	77.2	95.3	100.0	77.7	95.4	100.0	42.2	88.2
21, 22, 23	FIRE SUPPRESSION, PLUMBING & HVAC	94.2	41.6	72.7	94.2	42.1	72.8	100.0	41.7	76.2	100.2	42.2	76.5	100.1	40.5	75.7	94.2	42.1	72.9
26, 27, 3370	ELECTRICAL, COMMUNICATIONS & UTIL.	99.3	45.8	71.2	101.1	46.7	72.5	104.2	52.1	76.8	97.4	52.1	73.5	102.0	66.3	83.2	98.4	46.7	71.2
MF2010	WEIGHTED AVERAGE	98.4	50.5	77.3	99.1	50.9	77.8	100.5	53.7	79.8	101.2	54.6	80.6	100.9	57.1	81.6	99.5	50.9	78.1

	DIVISION	TENNESSEE																	
		CHATTANOOGA			COLUMBIA			COOKEVILLE			JACKSON			JOHNSON CITY			KNOXVILLE		
		373 - 374			384			385			383			376			377 - 379		
		MAT.	INST.	TOTAL	MAT.	INST.	TOTAL	MAT.	INST.	TOTAL	MAT.	INST.	TOTAL	MAT.	INST.	TOTAL	MAT.	INST.	TOTAL
015433	CONTRACTOR EQUIPMENT		105.3	105.3		98.9	98.9		98.9	98.9		105.2	105.2		99.0	99.0		99.0	99.0
0241, 31 - 34	SITE & INFRASTRUCTURE, DEMOLITION	103.0	100.0	100.9	89.0	88.2	88.5	94.5	87.6	89.7	98.0	98.3	98.2	110.0	88.7	95.0	89.0	89.1	89.1
0310	Concrete Forming & Accessories	97.5	56.2	61.8	83.6	60.3	63.4	83.8	36.8	43.1	90.4	46.9	52.8	84.6	40.1	46.1	96.2	43.2	50.3
0320	Concrete Reinforcing	89.0	61.8	75.5	90.0	61.8	75.9	90.0	61.8	75.9	90.0	62.1	76.1	89.6	60.3	75.0	89.0	60.3	74.7
0330	Cast-in-Place Concrete	99.9	61.5	84.1	95.1	52.7	77.7	107.8	44.7	81.9	105.3	51.0	83.0	80.4	62.1	72.9	93.9	49.6	75.7
03	CONCRETE	92.2	60.8	76.7	93.9	59.6	76.9	103.5	46.4	75.3	95.6	53.1	74.6	97.0	53.5	75.5	89.5	50.7	70.3
04	MASONRY	104.9	50.9	71.9	120.9	55.7	81.1	115.2	45.5	72.7	121.1	49.1	77.2	120.1	46.6	75.2	82.7	54.2	65.3
05	METALS	96.2	87.0	93.3	90.6	87.5	89.6	90.7	86.4	89.3	92.9	87.3	91.1	93.5	85.7	91.0	96.8	86.0	93.4
06	WOOD, PLASTICS & COMPOSITES	105.5	57.8	78.3	72.6	61.8	66.4	72.8	35.1	51.3	88.7	47.4	65.1	78.1	38.6	55.6	92.0	38.6	61.5
07	THERMAL & MOISTURE PROTECTION	99.4	58.6	83.0	92.6	60.1	79.5	93.0	47.4	74.7	94.8	53.2	78.1	94.7	54.6	78.6	92.9	59.4	79.4
08	OPENINGS	101.4	56.4	90.5	91.9	56.8	83.4	91.9	43.3	80.1	100.5	53.1	89.0	97.3	46.2	84.9	94.4	46.3	82.7
0920	Plaster & Gypsum Board	80.4	57.0	64.0	87.8	61.0	69.1	87.8	33.6	49.9	89.7	46.2	59.3	101.3	37.1	56.4	109.0	37.1	58.8
0950, 0980	Ceilings & Acoustic Treatment	96.5	57.0	70.2	82.5	61.0	68.2	82.5	33.6	49.9	91.5	46.2	61.3	92.5	37.1	55.6	93.4	37.1	55.9
0960	Flooring	98.5	58.7	86.6	94.2	21.3	72.5	94.3	60.7	84.3	92.3	42.3	77.4	96.8	42.1	80.5	101.0	57.8	88.1
0970, 0990	Wall Finishes & Painting/Coating	102.1	55.2	74.0	91.5	34.9	57.6	91.5	38.1	59.5	93.1	51.8	68.3	99.3	44.0	66.2	99.3	50.7	70.2
09	FINISHES	94.7	56.6	73.3	93.8	50.8	69.7	94.4	40.1	64.0	93.8	45.7	66.8	99.1	39.8	65.9	92.3	45.5	66.1
COVERS	DIVS. 10 - 14, 25, 28, 41, 43, 44, 46	100.0	42.9	88.3	100.0	48.9	89.6	100.0	42.7	88.3	100.0	44.8	88.7	100.0	45.0	88.8	100.0	47.6	89.3
21, 22, 23	FIRE SUPPRESSION, PLUMBING & HVAC	100.2	62.2	84.6	96.9	76.0	88.3	96.9	70.8	86.2	100.0	70.3	87.9	99.9	55.3	81.7	99.9	65.5	85.9
26, 27, 3370	ELECTRICAL, COMMUNICATIONS & UTIL.	102.3	67.9	84.2	94.3	59.9	76.1	96.1	60.7	77.5	101.6	61.0	80.2	91.9	48.3	68.9	97.9	56.9	76.3
MF2010	WEIGHTED AVERAGE	98.9	65.5	84.2	95.5	66.2	82.6	96.7	59.5	80.3	98.9	63.2	83.2	98.4	56.0	79.8	95.5	60.8	80.2

	DIVISION	TENNESSEE									TEXAS								
		MCKENZIE			MEMPHIS			NASHVILLE			ABILENE			AMARILLO			AUSTIN		
		382			375,380 - 381			370 - 372			795 - 796			790 - 791			786 - 787		
		MAT.	INST.	TOTAL	MAT.	INST.	TOTAL	MAT.	INST.	TOTAL	MAT.	INST.	TOTAL	MAT.	INST.	TOTAL	MAT.	INST.	TOTAL
015433	CONTRACTOR EQUIPMENT		98.9	98.9		103.0	103.0		106.2	106.2		89.4	89.4		89.4	89.4		88.8	88.8
0241, 31 - 34	SITE & INFRASTRUCTURE, DEMOLITION	94.2	87.9	89.8	91.8	94.6	93.8	98.2	101.0	100.2	94.2	86.6	88.9	94.4	87.7	89.7	100.1	86.9	90.8
0310	Concrete Forming & Accessories	91.5	39.8	46.7	95.0	61.8	66.2	99.1	64.0	68.7	96.4	39.6	47.2	97.4	53.2	59.2	95.7	52.6	58.4
0320	Concrete Reinforcing	90.1	62.1	76.2	95.5	65.6	80.6	95.5	64.7	80.1	90.2	51.1	70.7	90.2	49.4	69.9	84.4	48.2	66.4
0330	Cast-in-Place Concrete	105.5	59.4	86.5	91.4	62.4	79.5	90.4	65.9	80.3	97.2	48.1	77.0	96.3	54.7	79.2	98.4	48.6	77.9
03	CONCRETE	102.2	52.9	77.8	89.5	64.3	77.0	89.3	66.3	77.9	96.5	45.9	71.5	96.1	53.9	75.3	91.6	51.2	71.6
04	MASONRY	119.4	48.1	75.9	95.6	63.7	76.1	89.7	58.3	70.5	103.3	56.7	74.8	103.3	57.4	75.3	101.9	46.9	68.3
05	METALS	90.7	87.5	89.7	96.6	90.1	94.6	98.9	90.1	96.1	99.6	68.2	89.7	99.6	67.9	89.6	94.8	64.3	85.2
06	WOOD, PLASTICS & COMPOSITES	81.6	37.9	56.7	97.1	62.8	77.5	102.6	65.1	81.2	101.7	38.0	65.3	100.2	55.6	74.7	98.8	55.6	74.1
07	THERMAL & MOISTURE PROTECTION	93.0	50.6	75.9	96.0	64.4	83.3	95.6	61.3	81.8	98.4	50.2	79.0	98.0	53.3	80.0	91.8	50.2	75.1
08	OPENINGS	91.9	45.2	80.6	99.1	63.1	90.3	101.9	64.5	92.8	92.7	41.5	80.2	92.7	52.9	83.0	97.9	53.4	87.1
0920	Plaster & Gypsum Board	91.1	36.4	52.8	93.4	62.0	71.4	97.5	64.3	74.3	82.6	36.6	50.4	88.3	54.7	64.8	85.4	54.8	64.0
0950, 0980	Ceilings & Acoustic Treatment	82.5	36.4	51.8	98.6	62.0	74.2	89.5	64.3	72.7	90.0	36.6	54.4	95.5	54.7	68.3	96.5	54.8	68.7
0960	Flooring	97.2	40.8	80.4	94.1	37.4	77.2	101.9	66.6	91.4	109.0	76.5	99.3	108.8	76.5	99.2	94.6	41.0	78.6
0970, 0990	Wall Finishes & Painting/Coating	91.5	51.8	67.7	96.4	54.8	71.4	99.9	68.3	81.0	99.1	51.3	70.5	99.1	50.6	70.0	102.3	41.7	66.0
09	FINISHES	95.5	39.9	64.4	95.0	56.1	73.2	96.9	64.7	78.9	94.6	46.8	67.8	96.6	57.3	74.6	93.1	49.2	68.5
COVERS	DIVS. 10 - 14, 25, 28, 41, 43, 44, 46	100.0	28.5	85.4	100.0	81.8	96.3	100.0	82.2	96.4	100.0	76.5	95.2	100.0	68.7	93.6	100.0	78.3	95.6
21, 22, 23	FIRE SUPPRESSION, PLUMBING & HVAC	96.9	70.6	86.1	100.0	69.5	87.5	100.0	78.8	91.3	100.3	44.4	77.4	100.0	50.2	79.7	100.0	54.3	81.3
26, 27, 3370	ELECTRICAL, COMMUNICATIONS & UTIL.	95.8	65.5	79.9	102.1	66.0	83.1	96.9	64.3	79.7	96.5	49.5	71.8	102.0	65.8	82.9	97.1	63.2	79.3
MF2010	WEIGHTED AVERAGE	96.8	61.1	81.1	97.4	70.0	85.3	97.6	73.1	86.8	98.1	53.6	78.5	98.6	60.1	81.7	97.1	58.1	79.9

		TEXAS																	
	DIVISION	BEAUMONT			BROWNWOOD			BRYAN			CHILDRESS			CORPUS CHRISTI			DALLAS		
		776 - 777			768			778			792			783 - 784			752 - 753		
		MAT.	INST.	TOTAL	MAT.	INST.	TOTAL	MAT.	INST.	TOTAL	MAT.	INST.	TOTAL	MAT.	INST.	TOTAL	MAT.	INST.	TOTAL
015433	**CONTRACTOR EQUIPMENT**		90.1	90.1		89.4	89.4		90.1	90.1		89.4	89.4		96.1	96.1		99.0	99.0
0241, 31 - 34	**SITE & INFRASTRUCTURE, DEMOLITION**	86.7	87.1	87.0	99.6	86.0	90.1	78.4	88.6	85.6	104.1	86.5	91.7	142.0	82.7	100.4	105.3	88.3	93.4
0310	Concrete Forming & Accessories	103.5	49.3	56.6	94.9	33.5	41.8	82.5	41.0	46.6	94.9	60.5	65.1	98.6	37.5	45.7	96.8	64.4	68.8
0320	Concrete Reinforcing	87.3	42.1	64.8	90.5	48.4	69.5	89.5	48.3	68.9	90.3	48.7	69.6	83.8	46.7	65.3	96.2	51.7	74.0
0330	Cast-in-Place Concrete	83.8	52.4	70.9	104.8	43.3	79.5	67.2	65.2	66.4	99.7	49.5	79.1	109.2	45.6	83.0	96.5	55.7	79.7
03	**CONCRETE**	94.2	50.0	72.4	104.1	40.9	72.9	79.0	51.9	65.7	105.6	55.1	80.7	99.6	43.9	72.1	99.2	60.4	80.0
04	**MASONRY**	101.9	57.7	74.9	131.9	49.5	81.6	140.6	56.5	89.2	107.9	50.7	73.0	90.9	48.5	65.0	102.3	57.8	75.1
05	**METALS**	96.9	64.5	86.7	97.3	61.9	86.2	96.5	67.4	87.4	97.0	63.8	86.6	94.4	75.1	88.3	102.0	79.6	94.9
06	**WOOD, PLASTICS & COMPOSITES**	115.3	49.4	77.7	100.3	29.7	60.0	80.4	34.3	54.1	101.1	65.4	80.7	118.8	36.7	71.9	100.1	67.7	81.6
07	**THERMAL & MOISTURE PROTECTION**	102.9	55.9	84.0	98.5	44.9	77.0	95.0	56.0	79.3	99.0	49.4	79.0	98.9	45.3	77.3	94.1	63.2	81.7
08	**OPENINGS**	98.7	46.1	85.9	88.8	36.9	76.2	101.2	42.4	86.9	89.8	55.0	81.3	104.5	38.0	88.3	100.3	58.3	90.1
0920	Plaster & Gypsum Board	93.0	48.4	61.8	82.1	28.1	44.3	81.2	32.8	47.4	82.1	64.9	70.1	86.0	35.1	50.4	93.2	67.0	74.9
0950, 0980	Ceilings & Acoustic Treatment	94.9	48.4	63.9	88.2	28.1	48.1	88.2	32.8	51.3	88.2	64.9	72.7	92.9	35.1	54.4	95.3	67.0	76.5
0960	Flooring	105.0	73.9	95.7	108.1	46.3	89.7	79.9	54.6	72.4	107.1	42.4	87.8	107.6	65.4	95.0	98.3	49.5	83.8
0970, 0990	Wall Finishes & Painting/Coating	92.4	49.7	66.8	99.1	29.9	57.6	88.8	61.9	72.7	99.1	39.5	63.4	115.5	59.0	81.6	105.5	52.7	73.9
09	**FINISHES**	89.4	53.5	69.3	94.4	34.7	60.9	78.5	43.9	59.1	94.9	56.0	73.1	100.4	44.3	69.0	97.6	60.9	77.0
COVERS	**DIVS. 10 - 14, 25, 28, 41, 43, 44, 46**	100.0	79.7	95.9	100.0	37.8	87.3	100.0	79.0	95.7	100.0	70.7	94.0	100.0	77.3	95.4	100.0	82.4	96.4
21, 22, 23	**FIRE SUPPRESSION, PLUMBING & HVAC**	100.2	58.2	83.0	94.3	47.5	75.1	94.1	66.2	82.7	94.3	51.8	76.9	100.1	39.4	75.3	99.9	63.3	84.9
26, 27, 3370	**ELECTRICAL, COMMUNICATIONS & UTIL.**	94.3	71.2	82.2	92.3	44.5	67.1	92.0	68.6	79.7	96.5	44.2	69.0	92.6	53.7	72.1	93.5	66.3	79.2
MF2010	**WEIGHTED AVERAGE**	97.3	61.3	81.4	97.8	48.4	76.0	94.5	61.6	80.0	97.5	56.5	79.4	99.7	51.7	78.6	99.5	66.5	85.0

		TEXAS																	
	DIVISION	DEL RIO			DENTON			EASTLAND			EL PASO			FORT WORTH			GALVESTON		
		788			762			764			798 - 799,885			760 - 761			775		
		MAT.	INST.	TOTAL	MAT.	INST.	TOTAL	MAT.	INST.	TOTAL	MAT.	INST.	TOTAL	MAT.	INST.	TOTAL	MAT.	INST.	TOTAL
015433	**CONTRACTOR EQUIPMENT**		88.8	88.8		95.2	95.2		89.4	89.4		89.4	89.4		89.4	89.4		99.0	99.0
0241, 31 - 34	**SITE & INFRASTRUCTURE, DEMOLITION**	121.1	85.2	95.9	99.9	79.2	85.4	102.2	86.0	90.8	91.7	86.0	87.7	96.1	87.0	89.7	105.8	86.4	92.2
0310	Concrete Forming & Accessories	93.3	37.1	44.6	100.8	33.0	42.1	95.5	36.5	44.4	96.4	44.3	51.3	93.1	63.8	67.8	91.6	64.4	68.1
0320	Concrete Reinforcing	84.4	43.8	64.2	91.8	49.7	70.8	90.7	48.4	69.6	90.2	46.6	68.5	90.2	51.6	71.0	89.0	63.3	76.2
0330	Cast-in-Place Concrete	117.5	43.1	86.9	81.2	45.1	66.3	110.7	43.9	83.3	90.8	41.3	70.5	100.0	51.5	80.1	89.1	66.3	79.7
03	**CONCRETE**	118.1	41.5	80.3	82.1	42.3	62.4	108.9	42.4	76.1	93.4	44.7	69.4	97.6	57.9	78.0	95.2	66.3	80.9
04	**MASONRY**	105.8	51.1	72.4	140.4	48.2	84.1	101.7	49.3	69.7	101.1	47.2	68.2	97.6	57.8	73.3	97.6	56.5	72.5
05	**METALS**	93.2	59.1	82.5	96.9	75.2	90.1	97.1	62.3	86.2	99.4	62.8	87.9	96.5	68.1	87.6	98.1	86.5	94.5
06	**WOOD, PLASTICS & COMPOSITES**	98.1	35.3	62.2	110.4	32.5	66.0	107.4	33.4	65.1	101.3	46.7	70.1	102.7	67.5	82.6	98.1	65.4	79.5
07	**THERMAL & MOISTURE PROTECTION**	95.5	44.4	75.0	96.7	44.9	75.9	98.9	44.1	76.9	95.5	52.9	78.4	95.4	53.5	78.6	94.1	65.3	82.5
08	**OPENINGS**	98.7	33.6	82.9	103.3	37.7	87.3	75.1	38.9	66.3	92.7	43.0	80.6	84.5	58.3	78.1	105.8	63.3	95.4
0920	Plaster & Gypsum Board	80.7	33.8	47.9	85.4	30.8	47.2	82.1	31.8	47.0	91.1	45.5	59.2	86.7	67.0	73.0	87.8	64.8	71.7
0950, 0980	Ceilings & Acoustic Treatment	88.3	33.8	52.0	91.0	30.8	50.9	88.2	31.8	50.7	95.5	45.5	62.2	98.2	67.0	77.4	91.0	64.8	73.6
0960	Flooring	91.5	35.3	74.8	101.0	44.3	84.1	140.1	46.3	112.2	108.6	64.7	95.5	140.5	42.2	111.2	93.1	54.6	81.7
0970, 0990	Wall Finishes & Painting/Coating	102.3	26.3	56.8	109.4	30.3	62.0	100.7	31.2	59.1	99.1	35.7	61.1	100.7	52.2	71.6	99.3	48.1	68.6
09	**FINISHES**	92.0	35.3	60.3	90.7	34.1	59.0	104.2	37.0	66.5	96.4	47.3	68.9	106.4	59.3	80.0	87.8	60.8	72.7
COVERS	**DIVS. 10 - 14, 25, 28, 41, 43, 44, 46**	100.0	37.6	87.2	100.0	36.5	87.0	100.0	38.9	87.5	100.0	78.2	95.5	100.0	82.1	96.3	100.0	84.2	96.8
21, 22, 23	**FIRE SUPPRESSION, PLUMBING & HVAC**	94.0	45.4	74.1	94.3	39.1	71.7	94.3	47.9	75.3	100.0	34.8	73.3	100.1	58.8	83.2	94.1	65.0	82.2
26, 27, 3370	**ELECTRICAL, COMMUNICATIONS & UTIL.**	94.2	30.6	60.7	94.8	44.6	68.4	92.1	44.5	67.1	93.1	52.3	71.6	93.7	61.2	76.6	93.8	66.0	79.1
MF2010	**WEIGHTED AVERAGE**	98.9	45.9	75.6	97.3	47.3	75.2	96.3	49.1	75.5	97.2	50.6	76.7	97.1	62.9	82.0	96.6	68.3	84.1

		TEXAS																	
	DIVISION	GIDDINGS			GREENVILLE			HOUSTON			HUNTSVILLE			LAREDO			LONGVIEW		
		789			754			770 - 772			773			780			756		
		MAT.	INST.	TOTAL	MAT.	INST.	TOTAL	MAT.	INST.	TOTAL	MAT.	INST.	TOTAL	MAT.	INST.	TOTAL	MAT.	INST.	TOTAL
015433	**CONTRACTOR EQUIPMENT**		88.8	88.8		96.6	96.6		98.8	98.8		90.1	90.1		88.8	88.8		91.5	91.5
0241, 31 - 34	**SITE & INFRASTRUCTURE, DEMOLITION**	106.5	85.5	91.8	98.7	83.2	87.8	104.4	86.2	91.6	93.3	86.8	88.7	100.2	86.7	90.7	95.1	90.5	91.8
0310	Concrete Forming & Accessories	91.1	56.6	61.3	88.9	27.7	36.0	93.5	64.6	68.5	89.3	56.3	60.7	93.3	37.8	45.3	85.9	28.6	36.3
0320	Concrete Reinforcing	84.9	45.6	65.3	96.7	25.3	61.1	88.8	63.8	76.4	89.7	48.1	69.0	84.4	47.1	65.9	95.6	23.1	59.5
0330	Cast-in-Place Concrete	99.6	43.5	76.5	87.9	37.8	67.3	86.4	66.4	78.2	92.3	47.9	74.0	84.1	58.4	73.6	102.2	35.5	74.7
03	**CONCRETE**	95.5	50.7	73.4	91.8	32.9	62.7	92.9	66.5	79.9	101.6	52.6	77.4	90.2	47.8	69.2	108.0	31.3	70.1
04	**MASONRY**	113.9	51.2	75.6	157.9	47.0	90.2	97.4	66.4	78.5	138.5	52.6	86.0	99.5	56.3	73.2	153.6	43.0	86.1
05	**METALS**	92.7	61.4	82.9	99.3	63.4	88.0	101.0	87.4	96.7	96.5	63.8	86.2	95.8	64.1	85.8	92.5	50.8	79.4
06	**WOOD, PLASTICS & COMPOSITES**	97.1	60.7	76.3	91.0	26.8	54.4	100.4	65.4	80.5	88.8	59.3	72.0	98.1	36.6	63.0	85.2	29.0	53.1
07	**THERMAL & MOISTURE PROTECTION**	96.2	48.6	77.1	94.2	40.2	72.5	93.7	67.5	83.1	95.9	49.1	77.1	94.6	51.1	77.1	95.6	35.6	71.5
08	**OPENINGS**	97.9	49.4	86.1	98.8	26.5	81.2	108.3	63.3	97.4	101.2	48.6	88.4	98.6	38.6	84.0	89.3	26.5	74.0
0920	Plaster & Gypsum Board	79.4	60.0	65.8	87.4	25.0	43.8	90.5	64.8	72.5	84.9	58.6	66.6	81.9	35.1	49.2	85.8	27.4	44.9
0950, 0980	Ceilings & Acoustic Treatment	88.3	60.0	69.4	90.7	25.0	46.9	96.4	64.8	75.4	88.2	58.6	68.5	92.9	35.1	54.4	88.0	27.4	47.6
0960	Flooring	91.7	35.3	74.9	94.5	48.6	80.8	94.1	54.6	82.4	83.2	35.3	68.9	91.4	65.4	83.6	98.9	35.7	80.0
0970, 0990	Wall Finishes & Painting/Coating	102.3	31.2	59.7	105.5	25.7	57.7	99.3	61.9	76.9	88.8	32.0	54.8	102.3	55.6	74.3	96.1	30.0	56.5
09	**FINISHES**	90.7	51.2	68.6	94.2	31.0	58.8	95.4	62.3	76.9	80.9	50.9	64.1	91.3	43.8	64.7	99.2	30.0	60.4
COVERS	**DIVS. 10 - 14, 25, 28, 41, 43, 44, 46**	100.0	43.5	88.5	100.0	34.9	86.7	100.0	84.2	96.8	100.0	43.8	88.5	100.0	75.3	94.9	100.0	22.1	84.1
21, 22, 23	**FIRE SUPPRESSION, PLUMBING & HVAC**	94.0	59.1	79.7	94.0	44.9	73.9	100.0	66.4	86.3	94.1	59.9	80.1	100.0	38.1	74.7	93.9	28.4	67.1
26, 27, 3370	**ELECTRICAL, COMMUNICATIONS & UTIL.**	90.9	44.2	66.3	90.1	27.6	57.2	96.0	68.7	81.6	92.0	44.2	66.8	94.4	61.0	76.8	90.5	44.7	66.4
MF2010	**WEIGHTED AVERAGE**	95.8	55.4	78.0	98.6	42.9	74.0	99.3	70.3	86.5	97.6	56.3	79.4	96.7	53.2	77.5	98.4	40.0	72.7

DIVISION		TEXAS																	
		LUBBOCK			LUFKIN			MCALLEN			MCKINNEY			MIDLAND			ODESSA		
		793 - 794			759			785			750			797			797		
		MAT.	INST.	TOTAL	MAT.	INST.	TOTAL	MAT.	INST.	TOTAL	MAT.	INST.	TOTAL	MAT.	INST.	TOTAL	MAT.	INST.	TOTAL
015433	CONTRACTOR EQUIPMENT		97.8	97.8		91.5	91.5		96.1	96.1		96.6	96.6		97.8	97.8		89.4	89.4
0241, 31 - 34	SITE & INFRASTRUCTURE, DEMOLITION	117.8	84.9	94.7	90.4	91.5	91.2	146.9	82.6	101.8	95.2	83.7	87.2	120.7	84.9	95.6	94.5	86.9	89.1
0310	Concrete Forming & Accessories	94.9	51.8	57.6	88.7	57.8	62.0	97.7	35.2	43.6	88.0	37.0	43.9	98.6	39.1	47.1	96.3	38.9	46.7
0320	Concrete Reinforcing	91.3	51.1	71.3	97.3	60.6	79.0	84.0	46.6	65.4	96.7	49.7	73.3	92.3	50.4	71.5	90.2	50.4	70.4
0330	Cast-in-Place Concrete	97.3	48.4	77.2	91.4	44.8	72.3	118.6	44.0	87.9	82.5	45.9	67.4	103.4	45.8	79.7	97.2	44.7	75.6
03	CONCRETE	95.1	52.2	73.9	100.2	54.5	77.6	105.7	42.3	74.4	87.2	44.3	66.0	99.7	45.6	73.0	96.5	44.3	70.7
04	MASONRY	102.6	55.0	73.6	118.7	53.4	78.8	105.4	48.5	70.7	169.6	49.5	96.3	120.8	43.0	73.3	103.3	42.9	66.5
05	METALS	103.0	80.0	95.8	99.4	67.7	89.4	93.2	74.6	87.4	99.2	74.3	91.4	101.3	78.8	94.2	98.9	66.8	88.8
06	WOOD, PLASTICS & COMPOSITES	101.0	55.7	75.1	93.2	60.7	74.6	117.4	34.2	69.9	89.8	37.1	59.7	105.6	38.6	67.4	101.7	38.5	65.6
07	THERMAL & MOISTURE PROTECTION	88.3	52.4	73.8	95.4	49.0	76.7	98.8	44.2	76.8	94.0	46.8	75.0	88.5	44.1	70.7	98.4	43.3	76.2
08	OPENINGS	102.7	49.9	89.9	69.9	53.8	66.0	102.9	37.2	86.9	98.8	40.0	84.5	101.6	40.3	86.7	92.7	40.2	79.9
0920	Plaster & Gypsum Board	83.1	54.7	63.2	84.7	60.0	67.4	86.0	32.6	48.6	86.6	35.6	50.9	84.9	37.1	51.4	82.6	37.1	50.8
0950, 0980	Ceilings & Acoustic Treatment	91.9	54.7	67.1	80.7	60.0	66.9	92.9	32.6	52.7	90.7	35.6	54.0	89.1	37.1	54.4	90.0	37.1	54.7
0960	Flooring	102.1	40.0	83.6	131.1	38.8	103.6	107.3	65.4	94.8	94.1	36.2	76.9	103.5	49.4	87.4	109.0	49.4	91.2
0970, 0990	Wall Finishes & Painting/Coating	110.3	50.6	74.6	96.1	38.6	61.7	115.5	29.5	64.0	105.5	42.1	67.5	110.3	50.6	74.6	99.1	50.6	70.0
09	FINISHES	97.2	49.6	70.5	106.4	53.1	76.5	100.9	39.5	66.5	93.7	37.4	62.2	97.6	41.4	66.1	94.6	41.3	64.7
COVERS	DIVS. 10 - 14, 25, 28, 41, 43, 44, 46	100.0	78.2	95.5	100.0	71.9	94.3	100.0	75.2	94.9	100.0	40.3	87.8	100.0	77.9	95.5	100.0	77.6	95.4
21, 22, 23	FIRE SUPPRESSION, PLUMBING & HVAC	99.8	43.5	76.8	93.9	60.0	80.0	94.0	29.9	67.8	94.0	42.7	73.0	93.7	45.7	74.1	100.3	45.7	77.9
26, 27, 3370	ELECTRICAL, COMMUNICATIONS & UTIL.	95.1	50.5	71.6	91.9	44.2	66.8	91.7	34.8	61.8	90.2	46.0	66.9	95.1	41.1	66.6	96.7	41.0	67.4
MF2010	WEIGHTED AVERAGE	99.7	56.2	80.6	95.5	58.8	79.3	99.4	46.1	75.9	98.5	49.7	77.0	99.5	51.3	78.2	98.0	50.1	76.9

DIVISION		TEXAS																	
		PALESTINE			SAN ANGELO			SAN ANTONIO			TEMPLE			TEXARKANA			TYLER		
		758			769			781 - 782			765			755			757		
		MAT.	INST.	TOTAL	MAT.	INST.	TOTAL	MAT.	INST.	TOTAL	MAT.	INST.	TOTAL	MAT.	INST.	TOTAL	MAT.	INST.	TOTAL
015433	CONTRACTOR EQUIPMENT		91.5	91.5		89.4	89.4		91.3	91.3		89.4	89.4		91.5	91.5		91.5	91.5
0241, 31 - 34	SITE & INFRASTRUCTURE, DEMOLITION	96.4	91.3	92.8	96.0	87.3	89.9	99.9	91.5	94.0	85.0	85.8	85.5	85.0	91.2	89.4	94.5	91.3	92.2
0310	Concrete Forming & Accessories	80.8	56.9	60.1	95.2	36.8	44.7	93.4	56.0	61.0	98.1	34.6	43.1	95.3	34.6	42.8	90.3	33.6	41.2
0320	Concrete Reinforcing	94.9	48.5	71.8	90.4	50.4	70.5	90.5	48.0	69.3	90.5	48.1	69.4	94.8	51.3	73.1	95.6	51.5	73.7
0330	Cast-in-Place Concrete	83.6	43.3	67.0	98.9	43.0	75.9	82.6	66.7	76.0	81.1	48.3	67.6	84.2	44.1	67.7	100.3	43.0	76.8
03	CONCRETE	102.2	51.2	77.0	99.4	42.7	71.4	90.4	59.0	74.9	85.1	43.1	64.3	93.1	42.3	68.0	107.5	41.6	74.9
04	MASONRY	113.8	49.6	74.7	128.5	42.3	75.9	99.4	68.1	80.3	139.8	48.1	83.8	173.8	45.7	95.6	163.5	48.5	93.3
05	METALS	99.1	61.9	87.4	97.5	64.0	87.0	96.5	68.6	87.7	97.1	62.9	86.4	92.4	66.7	84.3	99.0	67.2	89.0
06	WOOD, PLASTICS & COMPOSITES	83.9	60.7	70.7	100.5	36.8	64.2	98.1	53.9	72.9	109.7	33.4	66.2	96.0	32.9	60.0	94.8	31.1	58.4
07	THERMAL & MOISTURE PROTECTION	95.9	47.4	76.4	98.3	42.9	76.1	94.3	65.4	82.6	97.9	44.5	76.5	95.2	48.0	76.3	95.7	52.3	78.3
08	OPENINGS	69.8	50.9	65.2	88.8	39.2	76.7	100.6	52.2	88.8	73.1	35.7	64.0	89.3	38.2	76.8	69.8	37.1	61.8
0920	Plaster & Gypsum Board	80.6	60.0	66.2	82.1	35.3	49.4	81.9	52.9	61.6	82.1	31.8	47.0	89.6	31.4	48.9	84.7	29.5	46.1
0950, 0980	Ceilings & Acoustic Treatment	80.7	60.0	66.9	88.2	35.3	53.0	92.9	52.9	66.3	88.2	31.8	50.7	88.0	31.4	50.3	80.7	29.5	46.6
0960	Flooring	123.4	60.0	104.5	108.1	35.3	86.4	91.4	65.5	83.7	142.4	42.4	112.6	106.0	60.4	92.4	132.8	37.7	104.5
0970, 0990	Wall Finishes & Painting/Coating	96.1	31.2	57.2	99.1	50.6	70.0	102.3	55.6	74.3	100.7	34.4	61.0	96.1	40.3	62.6	96.1	40.3	62.6
09	FINISHES	104.4	55.6	77.1	94.2	37.2	62.2	91.3	56.6	71.8	103.5	35.5	65.4	101.0	38.9	66.2	107.3	33.3	65.8
COVERS	DIVS. 10 - 14, 25, 28, 41, 43, 44, 46	100.0	41.3	88.0	100.0	75.2	94.9	100.0	81.1	96.1	100.0	36.0	86.9	100.0	75.3	95.0	100.0	75.0	94.9
21, 22, 23	FIRE SUPPRESSION, PLUMBING & HVAC	93.9	56.7	78.7	94.3	42.9	73.2	99.9	66.2	86.1	94.3	34.3	69.7	93.9	33.8	69.3	93.9	58.8	79.5
26, 27, 3370	ELECTRICAL, COMMUNICATIONS & UTIL.	87.9	44.2	64.9	96.5	46.6	70.2	94.8	61.1	77.0	93.4	66.1	79.0	91.7	62.2	76.2	90.5	52.9	70.7
MF2010	WEIGHTED AVERAGE	95.0	55.9	77.8	97.5	49.1	76.2	97.0	65.6	83.2	95.0	48.8	74.6	97.8	50.5	77.0	98.6	54.1	79.0

DIVISION		TEXAS															UTAH		
		VICTORIA			WACO			WAXAHACKIE			WHARTON			WICHITA FALLS			LOGAN		
		779			766 - 767			751			774			763			843		
		MAT.	INST.	TOTAL	MAT.	INST.	TOTAL	MAT.	INST.	TOTAL	MAT.	INST.	TOTAL	MAT.	INST.	TOTAL	MAT.	INST.	TOTAL
015433	CONTRACTOR EQUIPMENT		97.8	97.8		89.4	89.4		96.6	96.6		99.0	99.0		89.4	89.4		96.8	96.8
0241, 31 - 34	SITE & INFRASTRUCTURE, DEMOLITION	110.5	81.7	90.3	93.2	86.9	88.8	96.8	84.2	88.0	116.1	84.6	94.0	93.8	86.6	88.7	92.5	95.9	94.9
0310	Concrete Forming & Accessories	91.5	35.8	43.3	96.8	39.9	47.6	88.1	58.2	62.2	86.4	37.3	43.9	96.8	37.8	45.7	105.1	59.2	65.4
0320	Concrete Reinforcing	85.5	30.8	58.2	90.2	48.2	69.3	96.7	49.7	73.3	88.9	52.4	70.7	90.2	50.4	70.4	104.5	74.9	89.8
0330	Cast-in-Place Concrete	100.0	41.5	76.0	87.7	52.6	73.3	87.0	42.0	68.5	102.8	51.4	81.6	93.7	45.2	73.8	88.1	74.0	82.3
03	CONCRETE	102.3	39.0	71.0	92.0	47.1	69.8	90.8	52.5	71.9	106.5	47.0	77.1	94.8	44.0	69.7	107.8	67.8	88.1
04	MASONRY	114.1	34.9	65.8	100.1	55.0	72.6	158.5	49.6	92.0	98.6	45.5	66.2	100.6	56.7	73.8	106.0	61.1	78.6
05	METALS	96.6	70.4	88.4	99.5	66.1	89.0	99.3	76.3	92.0	98.1	77.3	91.6	99.4	68.2	89.6	100.6	76.3	93.0
06	WOOD, PLASTICS & COMPOSITES	102.0	36.6	64.7	108.2	35.8	66.9	89.8	65.5	75.9	91.9	33.2	58.4	108.2	35.7	66.8	81.6	57.4	67.7
07	THERMAL & MOISTURE PROTECTION	97.9	40.2	74.7	98.9	47.9	78.4	94.1	46.8	75.1	94.4	50.3	76.6	98.9	51.1	79.7	97.4	62.7	83.4
08	OPENINGS	105.6	35.9	88.6	84.5	36.9	72.9	98.8	55.5	88.2	105.8	37.7	89.2	84.5	40.7	73.8	92.6	56.7	83.8
0920	Plaster & Gypsum Board	85.5	35.1	50.2	82.6	34.3	48.8	87.1	64.9	71.5	83.6	31.6	47.2	82.6	34.2	48.7	76.6	56.0	62.2
0950, 0980	Ceilings & Acoustic Treatment	91.9	35.1	54.0	90.0	34.3	52.9	92.6	64.9	74.1	91.0	31.6	51.4	90.0	34.2	52.8	107.7	56.0	73.2
0960	Flooring	92.5	38.8	76.5	141.3	45.9	112.9	94.1	38.8	77.6	90.8	46.3	77.5	142.2	76.5	122.6	100.6	49.5	85.3
0970, 0990	Wall Finishes & Painting/Coating	99.5	38.6	63.0	100.7	34.4	61.0	105.5	38.6	65.4	99.3	32.0	59.0	103.0	50.6	71.6	98.4	63.0	77.2
09	FINISHES	85.9	36.7	58.4	104.1	39.2	67.8	94.4	54.0	71.7	87.4	37.4	59.4	104.6	45.3	71.4	99.3	57.2	75.7
COVERS	DIVS. 10 - 14, 25, 28, 41, 43, 44, 46	100.0	37.0	87.1	100.0	78.6	95.6	100.0	40.8	87.9	100.0	39.2	87.6	100.0	66.2	93.1	100.0	54.8	90.8
21, 22, 23	FIRE SUPPRESSION, PLUMBING & HVAC	94.1	32.2	68.8	100.3	50.4	79.9	94.0	41.7	72.6	94.1	60.0	80.1	100.3	45.8	78.0	99.9	70.7	88.0
26, 27, 3370	ELECTRICAL, COMMUNICATIONS & UTIL.	99.0	40.5	68.2	96.7	66.3	80.7	90.1	66.3	77.6	97.8	44.6	69.8	98.4	56.8	76.5	94.3	72.3	82.7
MF2010	WEIGHTED AVERAGE	98.7	43.4	74.3	97.3	55.8	79.0	98.5	56.7	80.1	98.5	53.4	78.6	97.9	54.1	78.6	99.4	69.1	86.1

	DIVISION	UTAH												VERMONT					
		OGDEN			PRICE			PROVO			SALT LAKE CITY			BELLOWS FALLS			BENNINGTON		
		842,844			845			846 - 847			840 - 841			051			052		
		MAT.	INST.	TOTAL	MAT.	INST.	TOTAL	MAT.	INST.	TOTAL	MAT.	INST.	TOTAL	MAT.	INST.	TOTAL	MAT.	INST.	TOTAL
015433	CONTRACTOR EQUIPMENT		96.8	96.8		95.8	95.8		95.8	95.8		96.7	96.7		100.5	100.5		100.5	100.5
0241, 31 - 34	SITE & INFRASTRUCTURE, DEMOLITION	81.1	95.9	91.4	90.1	94.2	93.0	88.9	94.2	92.6	80.6	95.8	91.3	86.1	98.6	94.9	85.5	98.1	94.4
0310	Concrete Forming & Accessories	105.1	59.2	65.4	107.2	50.4	58.0	106.4	59.1	65.5	107.6	59.1	65.6	95.9	84.6	86.1	93.6	84.6	85.8
0320	Concrete Reinforcing	104.1	74.9	89.6	112.1	74.8	93.5	113.0	74.9	94.0	106.5	74.9	90.7	88.6	85.0	86.8	88.6	86.8	87.7
0330	Cast-in-Place Concrete	89.5	74.0	83.1	88.2	60.9	77.0	88.2	74.0	82.4	97.8	74.0	88.0	95.8	70.6	85.5	95.8	67.3	84.1
03	CONCRETE	97.9	67.8	83.1	109.3	59.4	84.6	107.8	67.7	88.0	116.7	67.7	92.5	97.4	79.7	88.7	97.3	78.8	88.1
04	MASONRY	99.9	61.1	76.2	111.6	61.1	80.8	111.7	61.1	80.8	113.3	61.1	81.5	92.8	59.8	72.6	100.5	94.4	96.8
05	METALS	101.1	76.3	93.3	97.9	75.8	91.0	98.9	76.2	91.8	105.5	76.2	96.3	92.4	78.7	88.1	92.4	80.0	88.5
06	WOOD, PLASTICS & COMPOSITES	81.6	57.4	67.7	84.3	46.2	62.6	82.9	57.4	68.3	83.3	57.4	68.5	96.4	86.3	90.6	93.6	86.3	89.4
07	THERMAL & MOISTURE PROTECTION	96.3	62.7	82.8	99.1	60.6	83.6	99.1	62.7	84.5	103.3	62.7	86.9	96.6	68.6	85.3	96.5	77.7	88.9
08	OPENINGS	92.6	56.7	83.8	96.7	49.5	85.2	96.8	56.7	87.0	94.5	56.7	85.3	100.3	72.4	93.5	100.3	71.4	93.2
0920	Plaster & Gypsum Board	76.6	56.0	62.2	78.7	44.5	54.8	77.1	56.0	62.3	79.4	56.0	63.0	97.2	85.2	88.8	95.1	85.2	88.2
0950, 0980	Ceilings & Acoustic Treatment	107.7	56.0	73.2	107.7	44.5	65.6	107.7	56.0	73.2	101.5	56.0	71.2	88.3	85.2	86.3	88.3	85.2	86.3
0960	Flooring	98.3	49.5	83.7	101.6	36.8	82.3	101.3	49.5	85.8	102.2	49.5	86.5	98.3	31.6	78.4	97.4	31.6	77.8
0970, 0990	Wall Finishes & Painting/Coating	98.4	63.0	77.2	98.4	39.2	62.9	98.4	50.1	69.5	101.4	50.1	70.7	96.1	50.0	68.5	96.1	50.0	68.5
09	FINISHES	97.2	57.2	74.8	100.4	45.4	69.6	99.9	55.7	75.2	98.1	55.7	74.4	95.4	72.6	82.7	94.9	72.6	82.4
COVERS	DIVS. 10 - 14, 25, 28, 41, 43, 44, 46	100.0	54.8	90.8	100.0	47.6	89.3	100.0	54.8	90.8	100.0	54.8	90.8	100.0	49.4	89.7	100.0	49.3	89.6
21, 22, 23	FIRE SUPPRESSION, PLUMBING & HVAC	99.9	70.7	88.0	96.9	70.3	86.0	99.9	70.7	88.0	100.1	70.7	88.1	94.0	94.3	94.1	94.0	88.2	91.6
26, 27, 3370	ELECTRICAL, COMMUNICATIONS & UTIL.	94.7	72.3	82.9	100.2	72.3	85.5	95.2	62.3	77.9	97.7	62.3	79.1	101.7	93.0	97.1	101.7	63.6	81.6
MF2010	WEIGHTED AVERAGE	97.6	69.1	85.1	99.8	65.5	84.7	100.0	67.4	85.6	101.9	67.5	86.7	96.0	81.8	89.7	96.3	80.1	89.1

	DIVISION	VERMONT																	
		BRATTLEBORO			BURLINGTON			GUILDHALL			MONTPELIER			RUTLAND			ST. JOHNSBURY		
		053			054			059			056			057			058		
		MAT.	INST.	TOTAL	MAT.	INST.	TOTAL	MAT.	INST.	TOTAL	MAT.	INST.	TOTAL	MAT.	INST.	TOTAL	MAT.	INST.	TOTAL
015433	CONTRACTOR EQUIPMENT		100.5	100.5		100.5	100.5		100.5	100.5		100.5	100.5		100.5	100.5		100.5	100.5
0241, 31 - 34	SITE & INFRASTRUCTURE, DEMOLITION	87.0	98.6	95.2	88.3	98.6	95.5	85.1	97.2	93.6	88.9	98.4	95.5	89.3	98.4	95.7	85.1	97.2	93.6
0310	Concrete Forming & Accessories	96.1	85.1	86.6	95.7	85.0	86.4	94.5	77.3	79.6	101.0	85.0	87.1	96.3	85.0	86.5	93.6	77.3	79.5
0320	Concrete Reinforcing	87.7	86.9	87.3	110.7	85.8	98.3	89.3	85.0	87.2	88.1	85.8	87.0	110.7	85.8	98.3	87.7	85.0	86.3
0330	Cast-in-Place Concrete	98.8	68.0	86.2	104.5	108.3	106.1	92.9	57.7	78.4	110.6	108.3	109.7	93.9	108.3	99.8	92.9	57.7	78.4
03	CONCRETE	99.6	79.3	89.6	106.0	92.9	99.6	95.0	72.0	83.6	105.6	92.9	99.4	101.0	92.9	97.0	94.6	72.0	83.5
04	MASONRY	99.2	94.6	96.4	103.3	71.1	83.7	99.4	43.6	65.3	88.5	71.1	77.9	84.4	71.1	76.3	122.8	43.6	74.5
05	METALS	92.4	81.7	89.0	99.3	80.9	93.5	92.4	78.0	87.9	92.4	80.9	88.8	97.7	80.9	92.4	92.4	78.0	87.9
06	WOOD, PLASTICS & COMPOSITES	96.7	86.3	90.7	91.7	86.3	88.6	93.2	86.3	89.2	95.7	86.3	90.3	96.8	86.3	90.8	89.3	86.3	87.5
07	THERMAL & MOISTURE PROTECTION	96.6	77.7	89.0	103.6	75.7	92.4	96.4	56.9	80.5	103.3	76.2	92.4	96.7	76.2	88.4	96.4	56.9	80.5
08	OPENINGS	100.3	72.4	93.5	103.4	72.5	95.9	100.3	72.4	93.5	100.3	72.5	93.5	103.4	72.5	95.9	100.3	72.4	93.5
0920	Plaster & Gypsum Board	97.2	85.2	88.8	97.2	85.2	88.8	103.0	85.2	90.6	104.4	85.2	91.0	98.0	85.2	89.0	105.5	85.2	91.3
0950, 0980	Ceilings & Acoustic Treatment	88.3	85.2	86.3	94.9	85.2	88.5	88.3	85.2	86.3	92.0	85.2	87.5	93.1	85.2	87.9	88.3	85.2	86.3
0960	Flooring	98.4	31.6	78.5	98.3	63.4	87.9	101.2	31.6	80.5	104.2	63.4	92.0	98.3	63.4	87.9	104.4	31.6	82.7
0970, 0990	Wall Finishes & Painting/Coating	96.1	50.0	68.5	96.1	50.0	68.5	96.1	50.0	68.5	96.1	50.0	68.5	96.1	50.0	68.5	96.1	50.0	68.5
09	FINISHES	95.6	72.6	82.7	97.0	78.1	86.4	96.9	68.2	80.8	99.0	78.1	87.3	96.7	78.1	86.3	98.2	68.2	81.4
COVERS	DIVS. 10 - 14, 25, 28, 41, 43, 44, 46	100.0	49.4	89.7	100.0	78.3	95.6	100.0	43.3	88.4	100.0	78.3	95.6	100.0	78.3	95.6	100.0	43.3	88.4
21, 22, 23	FIRE SUPPRESSION, PLUMBING & HVAC	94.0	96.4	95.0	99.8	71.5	88.2	94.0	62.3	81.0	93.8	71.5	84.6	100.0	71.5	88.3	94.0	62.3	81.0
26, 27, 3370	ELECTRICAL, COMMUNICATIONS & UTIL.	101.7	93.0	97.1	100.2	63.7	81.0	101.7	63.7	81.7	102.0	63.7	81.8	101.7	63.7	81.7	101.7	63.7	81.7
MF2010	WEIGHTED AVERAGE	96.6	86.2	92.0	100.5	77.8	90.5	96.1	67.1	83.3	97.2	77.8	88.7	98.8	77.8	89.6	97.3	67.1	84.0

	DIVISION	VERMONT			VIRGINIA														
		WHITE RIVER JCT.			ALEXANDRIA			ARLINGTON			BRISTOL			CHARLOTTESVILLE			CULPEPER		
		050			223			222			242			229			227		
		MAT.	INST.	TOTAL	MAT.	INST.	TOTAL	MAT.	INST.	TOTAL	MAT.	INST.	TOTAL	MAT.	INST.	TOTAL	MAT.	INST.	TOTAL
015433	CONTRACTOR EQUIPMENT		100.5	100.5		101.8	101.8		100.5	100.5		100.5	100.5		104.7	104.7		100.5	100.5
0241, 31 - 34	SITE & INFRASTRUCTURE, DEMOLITION	89.5	97.3	95.0	117.3	89.2	97.6	128.2	87.1	99.4	111.8	85.7	93.5	116.8	87.6	96.3	115.2	87.0	95.4
0310	Concrete Forming & Accessories	91.1	78.0	79.7	90.7	74.6	76.7	89.9	74.2	76.3	86.2	44.3	50.0	84.7	50.6	55.2	81.8	73.5	74.6
0320	Concrete Reinforcing	88.6	85.0	86.8	87.9	84.8	86.3	99.2	84.7	92.0	99.2	68.4	83.8	98.5	74.1	86.4	99.2	84.6	92.0
0330	Cast-in-Place Concrete	98.8	58.7	82.3	117.6	83.3	103.5	114.5	82.8	101.5	114.0	49.9	87.6	118.5	56.9	93.2	117.2	69.9	97.8
03	CONCRETE	101.4	72.6	87.2	108.2	80.5	94.5	112.7	80.2	96.6	109.0	52.6	81.2	110.0	58.9	84.7	107.1	75.4	91.4
04	MASONRY	110.9	44.7	70.5	89.3	72.1	78.8	102.7	71.4	83.6	91.7	53.6	68.4	116.1	55.4	79.1	104.6	71.3	84.3
05	METALS	92.4	78.1	87.9	97.7	96.5	97.4	96.4	96.4	96.4	95.3	86.7	92.6	95.5	91.1	94.2	95.6	94.6	95.3
06	WOOD, PLASTICS & COMPOSITES	90.7	86.3	88.2	97.3	73.1	83.5	94.0	73.1	82.1	86.7	41.7	61.0	85.2	47.5	63.7	83.9	73.1	77.8
07	THERMAL & MOISTURE PROTECTION	96.7	57.5	80.9	99.2	81.7	92.2	101.1	80.4	92.8	100.6	62.3	85.2	100.3	70.1	88.1	100.5	77.8	91.4
08	OPENINGS	100.3	74.6	94.0	94.2	75.8	89.7	92.3	75.7	88.2	95.2	47.7	83.7	93.5	53.7	83.8	93.8	75.7	89.4
0920	Plaster & Gypsum Board	93.5	85.2	87.7	109.7	72.1	83.4	106.0	72.1	82.3	101.6	39.7	58.3	101.6	45.1	62.1	101.9	72.1	81.1
0950, 0980	Ceilings & Acoustic Treatment	88.3	85.2	86.3	97.0	72.1	80.4	94.3	72.1	79.5	93.4	39.7	57.6	93.4	45.1	61.2	94.3	72.1	79.5
0960	Flooring	96.3	31.6	77.0	103.8	86.0	98.5	102.3	83.5	96.7	98.9	68.9	90.0	97.6	68.9	89.1	97.6	83.5	93.4
0970, 0990	Wall Finishes & Painting/Coating	96.1	50.7	68.9	124.6	80.4	98.1	124.6	80.4	98.1	109.6	56.7	77.9	109.6	79.5	91.5	124.6	79.5	97.6
09	FINISHES	94.8	68.7	80.2	103.3	76.6	88.3	103.2	75.8	87.9	99.6	49.2	71.4	99.3	55.8	74.9	100.1	75.6	86.4
COVERS	DIVS. 10 - 14, 25, 28, 41, 43, 44, 46	100.0	43.9	88.5	100.0	89.3	97.8	100.0	84.4	96.8	100.0	70.5	94.0	100.0	76.3	95.2	100.0	84.4	96.8
21, 22, 23	FIRE SUPPRESSION, PLUMBING & HVAC	94.0	63.2	81.4	100.3	89.3	95.8	100.3	88.3	95.4	94.3	53.2	77.5	94.3	70.9	84.7	94.3	75.5	86.6
26, 27, 3370	ELECTRICAL, COMMUNICATIONS & UTIL.	101.7	63.7	81.7	97.5	100.6	99.1	95.0	100.6	97.9	97.2	36.9	65.4	97.1	72.1	83.9	99.8	100.5	100.2
MF2010	WEIGHTED AVERAGE	97.3	67.7	84.3	99.9	86.0	93.8	100.8	85.3	93.9	97.9	56.7	79.8	99.2	68.4	85.6	98.6	81.6	91.1

	DIVISION	VIRGINIA																	
		FAIRFAX			FARMVILLE			FREDERICKSBURG			GRUNDY			HARRISONBURG			LYNCHBURG		
		220 - 221			239			224 - 225			246			228			245		
		MAT.	INST.	TOTAL	MAT.	INST.	TOTAL	MAT.	INST.	TOTAL	MAT.	INST.	TOTAL	MAT.	INST.	TOTAL	MAT.	INST.	TOTAL
015433	CONTRACTOR EQUIPMENT		100.5	100.5		104.7	104.7		100.5	100.5		100.5	100.5		100.5	100.5		100.5	100.5
0241, 31 - 34	SITE & INFRASTRUCTURE, DEMOLITION	127.0	87.2	99.1	112.7	87.2	94.8	114.6	87.2	95.4	109.5	84.4	91.9	123.7	85.6	97.0	110.4	85.9	93.2
0310	Concrete Forming & Accessories	84.6	73.9	75.3	96.1	45.6	52.4	84.6	72.1	73.8	89.2	38.0	44.9	80.9	44.2	49.1	86.2	63.0	66.1
0320	Concrete Reinforcing	99.2	84.7	92.0	99.2	52.4	75.9	100.0	84.7	92.4	97.8	52.9	75.4	99.2	68.0	83.7	98.5	68.5	83.6
0330	Cast-in-Place Concrete	114.5	82.7	101.4	118.5	55.0	92.4	116.2	65.8	95.5	114.0	48.8	87.2	114.5	59.5	91.9	114.0	59.4	91.5
03	CONCRETE	112.3	80.0	96.4	109.2	51.8	80.8	106.7	73.4	90.3	107.8	46.1	77.3	110.2	55.6	83.2	107.8	64.2	86.2
04	MASONRY	102.5	71.4	83.5	99.7	46.7	67.4	103.6	71.4	83.9	94.4	49.3	66.9	100.9	52.3	71.2	108.6	55.4	76.1
05	METALS	95.7	96.1	95.8	95.5	78.9	90.3	95.6	95.7	95.6	95.3	71.6	87.8	95.5	84.8	92.2	95.4	87.2	92.9
06	WOOD, PLASTICS & COMPOSITES	86.7	73.1	78.9	99.3	45.9	68.8	86.7	71.1	77.8	89.5	36.5	59.3	83.0	40.6	58.8	86.7	65.3	74.5
07	THERMAL & MOISTURE PROTECTION	101.0	70.9	88.9	100.3	51.2	80.6	100.5	78.3	91.6	100.6	47.0	79.0	100.8	64.6	86.2	100.4	66.2	86.7
08	OPENINGS	92.3	75.7	88.2	93.8	44.9	81.9	93.5	74.6	88.9	95.2	35.9	80.8	93.8	51.6	83.5	93.7	60.6	85.7
0920	Plaster & Gypsum Board	101.9	72.1	81.1	108.7	43.4	63.0	101.9	70.0	79.6	101.6	34.4	54.6	101.6	38.7	57.6	101.6	64.1	75.4
0950, 0980	Ceilings & Acoustic Treatment	94.3	72.1	79.5	93.4	43.4	60.1	94.3	70.0	78.1	93.4	34.4	54.1	93.4	38.7	56.9	93.4	64.1	73.9
0960	Flooring	99.2	83.5	94.5	103.8	62.5	91.5	99.2	83.5	94.5	100.2	34.9	80.8	97.4	76.7	91.2	98.9	68.9	90.0
0970, 0990	Wall Finishes & Painting/Coating	124.6	80.4	98.1	109.6	40.5	68.2	124.6	79.5	97.6	109.6	36.7	65.9	124.6	56.7	83.9	109.6	56.7	77.9
09	FINISHES	101.8	75.8	87.2	101.6	48.0	71.6	100.5	74.4	85.9	99.8	36.6	64.4	100.6	49.9	72.2	99.4	62.8	78.9
COVERS	DIVS. 10 - 14, 25, 28, 41, 43, 44, 46	100.0	84.4	96.8	100.0	47.4	89.2	100.0	80.4	96.0	100.0	45.4	88.8	100.0	70.3	93.9	100.0	73.8	94.7
21, 22, 23	FIRE SUPPRESSION, PLUMBING & HVAC	94.3	88.3	91.8	94.3	45.7	74.4	94.3	88.0	91.7	94.3	65.5	82.5	94.3	69.3	84.0	94.3	70.6	84.6
26, 27, 3370	ELECTRICAL, COMMUNICATIONS & UTIL.	98.4	100.6	99.5	94.2	50.3	71.1	95.2	100.5	98.0	97.2	46.5	70.5	97.4	100.5	99.0	98.3	54.9	75.5
MF2010	WEIGHTED AVERAGE	99.3	84.9	93.0	98.2	54.4	78.9	98.1	83.8	91.8	97.9	54.5	78.8	98.8	69.3	85.8	98.5	67.4	84.8

	DIVISION	VIRGINIA																	
		NEWPORT NEWS			NORFOLK			PETERSBURG			PORTSMOUTH			PULASKI			RICHMOND		
		236			233 - 235			238			237			243			230 - 232		
		MAT.	INST.	TOTAL	MAT.	INST.	TOTAL	MAT.	INST.	TOTAL	MAT.	INST.	TOTAL	MAT.	INST.	TOTAL	MAT.	INST.	TOTAL
015433	CONTRACTOR EQUIPMENT		104.7	104.7		105.3	105.3		104.7	104.7		104.6	104.6		100.5	100.5		104.6	104.6
0241, 31 - 34	SITE & INFRASTRUCTURE, DEMOLITION	111.3	88.3	95.2	109.2	89.3	95.2	115.3	88.5	96.5	109.8	88.0	94.5	108.8	84.7	91.9	108.5	88.5	94.4
0310	Concrete Forming & Accessories	95.4	66.7	70.6	99.8	66.9	71.3	89.2	59.5	63.5	85.4	55.2	59.3	89.2	40.0	46.6	98.3	59.5	64.8
0320	Concrete Reinforcing	98.9	73.9	86.4	98.9	73.9	86.5	98.5	74.3	86.5	98.5	73.9	86.3	97.8	67.8	82.9	98.9	74.3	86.7
0330	Cast-in-Place Concrete	115.2	66.6	95.2	125.8	66.7	101.5	122.3	54.9	94.6	114.0	66.6	94.5	114.0	49.1	87.3	111.0	54.9	87.9
03	CONCRETE	106.2	69.3	88.0	111.6	69.4	90.8	111.8	62.2	87.3	104.9	64.1	84.8	107.8	50.1	79.3	104.4	62.2	83.6
04	MASONRY	94.8	56.2	71.2	100.6	56.2	73.5	107.3	54.8	75.2	100.1	56.2	73.3	88.5	50.1	65.1	92.7	54.8	69.6
05	METALS	97.8	90.9	95.6	96.8	91.0	95.0	95.6	91.7	94.4	96.7	90.2	94.7	95.3	83.1	91.5	99.9	91.7	97.3
06	WOOD, PLASTICS & COMPOSITES	98.2	69.1	81.6	100.3	69.1	82.5	89.9	61.1	73.5	86.5	53.6	67.7	89.5	38.3	60.3	97.3	61.1	76.6
07	THERMAL & MOISTURE PROTECTION	100.3	69.7	88.0	101.7	69.7	88.8	100.3	70.7	88.4	100.3	68.1	87.3	100.6	52.0	81.1	103.4	70.7	90.3
08	OPENINGS	94.2	64.5	87.0	94.2	64.5	87.0	93.5	61.1	85.6	94.3	56.1	85.0	95.3	44.0	82.8	94.2	61.1	86.1
0920	Plaster & Gypsum Board	109.7	67.3	80.0	104.5	67.3	78.5	101.9	59.0	71.9	102.2	51.4	66.6	101.6	36.3	55.9	111.0	59.0	74.6
0950, 0980	Ceilings & Acoustic Treatment	97.0	67.3	77.2	99.7	67.3	78.1	94.3	59.0	70.8	97.0	51.4	66.6	93.4	36.3	55.3	103.4	59.0	73.8
0960	Flooring	103.8	68.9	93.4	102.4	68.9	92.4	99.8	73.8	92.0	96.9	68.9	88.5	100.2	68.9	90.9	102.4	73.8	93.9
0970, 0990	Wall Finishes & Painting/Coating	109.6	43.9	70.2	109.6	79.5	91.5	109.6	79.5	91.5	109.6	79.5	91.5	109.6	56.7	77.9	109.6	79.5	91.5
09	FINISHES	102.3	64.9	81.3	101.4	68.8	83.1	100.0	63.8	79.7	99.4	59.2	76.8	99.8	45.6	69.4	102.9	63.8	81.0
COVERS	DIVS. 10 - 14, 25, 28, 41, 43, 44, 46	100.0	76.5	95.2	100.0	76.5	95.2	100.0	76.6	95.2	100.0	74.8	94.9	100.0	45.6	88.9	100.0	76.6	95.2
21, 22, 23	FIRE SUPPRESSION, PLUMBING & HVAC	100.3	66.0	86.3	100.0	65.5	85.9	94.3	69.0	83.9	100.3	65.4	86.0	94.3	56.3	78.7	100.0	69.0	87.3
26, 27, 3370	ELECTRICAL, COMMUNICATIONS & UTIL.	97.2	61.4	78.4	101.9	58.7	79.2	97.4	72.1	84.1	95.4	58.7	76.1	97.2	56.7	75.9	100.2	72.1	85.4
MF2010	WEIGHTED AVERAGE	99.8	69.3	86.3	100.8	69.4	87.0	99.0	70.1	86.3	99.2	66.7	84.9	97.6	57.3	79.8	100.1	70.1	86.9

	DIVISION	VIRGINIA									WASHINGTON								
		ROANOKE			STAUNTON			WINCHESTER			CLARKSTON			EVERETT			OLYMPIA		
		240 - 241			244			226			994			982			985		
		MAT.	INST.	TOTAL	MAT.	INST.	TOTAL	MAT.	INST.	TOTAL	MAT.	INST.	TOTAL	MAT.	INST.	TOTAL	MAT.	INST.	TOTAL
015433	CONTRACTOR EQUIPMENT		100.5	100.5		104.7	104.7		100.5	100.5		90.6	90.6		102.5	102.5		102.5	102.5
0241, 31 - 34	SITE & INFRASTRUCTURE, DEMOLITION	108.9	85.9	92.7	113.2	87.3	95.0	122.4	87.0	97.6	95.3	89.9	91.5	90.8	110.6	104.7	92.1	110.6	105.1
0310	Concrete Forming & Accessories	95.2	63.1	67.4	88.8	52.2	57.1	83.1	72.2	73.6	122.5	69.9	77.0	114.1	98.1	100.3	105.9	98.2	99.3
0320	Concrete Reinforcing	98.9	68.5	83.8	98.5	51.3	75.0	98.6	84.1	91.4	106.0	87.6	96.8	116.1	101.8	109.0	116.7	101.8	109.3
0330	Cast-in-Place Concrete	129.4	59.5	100.7	118.5	52.9	91.5	114.5	62.8	93.2	101.5	83.2	94.0	98.8	104.4	101.1	89.9	104.5	95.9
03	CONCRETE	113.0	64.2	88.9	109.3	54.0	82.0	109.6	72.3	91.2	99.6	77.9	88.9	99.4	100.5	100.0	96.0	100.6	98.3
04	MASONRY	95.7	55.4	71.1	104.5	53.5	73.4	98.5	61.4	75.8	101.0	79.9	88.1	118.2	99.2	106.6	109.1	99.2	103.1
05	METALS	97.5	87.3	94.3	95.5	82.5	91.4	95.6	94.0	95.1	87.7	81.7	85.8	103.2	92.3	99.8	98.0	92.4	96.3
06	WOOD, PLASTICS & COMPOSITES	98.2	65.3	79.4	89.5	52.0	68.1	85.2	71.1	77.2	101.0	66.6	81.4	108.0	97.2	101.9	94.8	97.2	96.2
07	THERMAL & MOISTURE PROTECTION	100.2	68.3	87.4	100.2	54.0	81.7	100.9	75.7	90.8	152.0	77.5	122.0	102.5	100.8	101.8	102.4	98.4	100.8
08	OPENINGS	94.2	60.6	86.0	93.8	50.2	83.2	95.3	75.4	90.5	104.4	70.9	96.2	101.0	98.7	100.4	100.9	98.7	100.4
0920	Plaster & Gypsum Board	109.7	64.1	77.8	101.6	49.7	65.3	101.9	70.0	79.6	130.4	65.4	85.0	109.4	97.1	100.8	100.5	97.1	98.1
0950, 0980	Ceilings & Acoustic Treatment	97.0	64.1	75.1	93.4	49.7	64.3	94.3	70.0	78.1	99.4	65.4	76.8	104.9	97.1	99.7	103.8	97.1	99.3
0960	Flooring	103.8	68.9	93.4	99.7	40.9	82.2	98.6	83.5	94.1	103.9	48.8	87.5	117.1	94.9	110.5	110.9	95.1	106.2
0970, 0990	Wall Finishes & Painting/Coating	109.6	56.7	77.9	109.6	36.1	65.5	124.6	91.1	104.5	112.5	61.1	81.7	106.9	79.3	90.4	106.9	83.4	92.8
09	FINISHES	102.1	63.7	80.6	99.6	48.5	71.0	101.0	75.7	86.8	116.0	63.8	86.7	109.0	95.2	101.3	105.9	95.7	100.2
COVERS	DIVS. 10 - 14, 25, 28, 41, 43, 44, 46	100.0	73.8	94.7	100.0	73.7	94.6	100.0	84.2	96.8	100.0	78.3	95.6	100.0	93.0	98.6	100.0	100.1	100.0
21, 22, 23	FIRE SUPPRESSION, PLUMBING & HVAC	100.3	62.7	84.9	94.3	62.0	81.1	94.3	88.3	91.8	94.3	86.9	91.3	100.1	102.6	101.1	99.9	102.7	101.0
26, 27, 3370	ELECTRICAL, COMMUNICATIONS & UTIL.	97.2	56.7	75.9	96.0	75.9	85.4	95.7	100.5	98.2	100.3	93.9	96.9	103.6	99.0	101.2	101.1	101.4	101.3
MF2010	WEIGHTED AVERAGE	100.4	66.2	85.3	98.4	63.9	83.2	98.6	82.7	91.6	100.0	81.4	91.8	102.4	99.7	101.2	100.1	100.3	100.2

	DIVISION	WASHINGTON																	
		RICHLAND			SEATTLE			SPOKANE			TACOMA			VANCOUVER			WENATCHEE		
		993			980 - 981,987			990 - 992			983 - 984			986			988		
		MAT.	INST.	TOTAL	MAT.	INST.	TOTAL	MAT.	INST.	TOTAL	MAT.	INST.	TOTAL	MAT.	INST.	TOTAL	MAT.	INST.	TOTAL
015433	CONTRACTOR EQUIPMENT		90.6	90.6		102.4	102.4		90.6	90.6		102.5	102.5		97.3	97.3		102.5	102.5
0241, 31 - 34	SITE & INFRASTRUCTURE, DEMOLITION	97.7	90.7	92.8	94.6	108.8	104.6	96.9	90.7	92.5	93.8	110.6	105.6	103.9	98.2	99.9	103.3	109.3	107.5
0310	Concrete Forming & Accessories	122.6	79.3	85.1	105.2	98.7	99.6	127.0	78.9	85.4	105.2	98.2	99.2	106.2	92.8	94.6	106.9	77.2	81.2
0320	Concrete Reinforcing	101.7	87.7	94.7	114.8	101.9	108.3	102.4	87.6	95.1	114.8	101.8	108.3	115.7	101.6	108.7	115.6	90.6	103.2
0330	Cast-in-Place Concrete	101.8	85.3	95.0	103.6	104.7	104.1	105.7	85.1	97.3	101.5	104.5	102.8	113.3	100.2	107.9	103.7	78.3	93.2
03	CONCRETE	99.1	83.0	91.1	102.2	101.0	101.6	101.4	82.7	92.2	101.2	100.6	100.9	111.2	96.8	104.1	109.0	80.2	94.8
04	MASONRY	103.1	82.4	90.5	113.2	99.2	104.7	103.7	82.4	90.7	113.0	99.2	104.6	113.7	98.4	104.3	115.7	90.3	100.2
05	METALS	88.1	84.1	86.8	104.9	94.2	101.5	90.4	83.6	88.3	104.9	92.4	101.0	102.3	93.0	99.4	102.4	86.8	97.5
06	WOOD, PLASTICS & COMPOSITES	101.1	77.3	87.5	99.3	97.2	98.1	109.6	77.3	91.1	97.9	97.2	97.5	89.7	92.3	91.2	99.1	75.3	85.5
07	THERMAL & MOISTURE PROTECTION	153.2	79.9	123.8	102.3	100.8	101.7	149.5	79.6	121.4	102.2	98.7	100.8	102.5	93.6	98.9	102.8	82.7	94.7
08	OPENINGS	106.8	72.4	98.4	103.5	98.7	102.4	107.4	72.7	98.9	101.7	98.7	101.0	98.4	94.2	97.4	101.2	71.8	94.0
0920	Plaster & Gypsum Board	130.4	76.5	92.7	103.1	97.1	98.9	125.3	76.5	91.1	105.8	97.1	99.7	103.7	92.3	95.7	107.6	74.5	84.5
0950, 0980	Ceilings & Acoustic Treatment	105.5	76.5	86.1	107.4	97.1	100.5	101.4	76.5	84.8	107.7	97.1	100.6	102.3	92.3	95.6	100.1	74.5	83.1
0960	Flooring	104.1	83.0	97.8	110.4	102.8	108.1	103.8	88.4	99.2	110.9	95.1	106.2	116.9	105.9	113.6	113.5	64.9	99.1
0970, 0990	Wall Finishes & Painting/Coating	112.5	67.7	85.6	106.9	83.4	92.8	112.3	61.1	81.6	106.9	83.4	92.8	109.3	68.9	85.1	106.9	67.7	83.4
09	FINISHES	117.6	78.2	95.5	107.1	97.2	101.5	116.0	78.5	95.0	107.5	95.7	100.9	105.2	92.8	98.2	107.9	73.0	88.3
COVERS	DIVS. 10 - 14, 25, 28, 41, 43, 44, 46	100.0	64.8	92.8	100.0	100.1	100.0	100.0	64.7	92.8	100.0	100.1	100.0	100.0	62.8	92.4	100.0	79.2	95.8
21, 22, 23	FIRE SUPPRESSION, PLUMBING & HVAC	100.4	108.1	103.6	100.0	120.0	108.2	100.3	88.1	95.3	100.1	102.7	101.2	100.2	91.4	96.6	94.1	88.9	92.0
26, 27, 3370	ELECTRICAL, COMMUNICATIONS & UTIL.	97.5	93.9	95.6	103.4	107.2	105.4	95.2	79.1	86.7	103.4	101.4	102.4	108.5	99.1	103.5	104.3	99.0	101.5
MF2010	WEIGHTED AVERAGE	101.8	88.7	96.0	102.8	105.1	103.8	102.0	82.4	93.4	102.5	100.3	101.6	103.5	94.1	99.4	102.1	87.5	95.6

	DIVISION	WASHINGTON			WEST VIRGINIA														
		YAKIMA			BECKLEY			BLUEFIELD			BUCKHANNON			CHARLESTON			CLARKSBURG		
		989			258 - 259			247 - 248			262			250 - 253			263 - 264		
		MAT.	INST.	TOTAL	MAT.	INST.	TOTAL	MAT.	INST.	TOTAL	MAT.	INST.	TOTAL	MAT.	INST.	TOTAL	MAT.	INST.	TOTAL
015433	CONTRACTOR EQUIPMENT		102.5	102.5		100.5	100.5		100.5	100.5		100.5	100.5		100.5	100.5		100.5	100.5
0241, 31 - 34	SITE & INFRASTRUCTURE, DEMOLITION	96.2	110.1	105.9	103.4	87.8	92.4	103.2	87.7	92.3	109.7	87.5	94.1	101.4	89.2	92.9	110.3	87.5	94.3
0310	Concrete Forming & Accessories	105.6	93.7	95.3	84.8	85.7	85.6	86.1	85.4	85.5	85.5	83.3	83.6	101.7	87.7	89.6	83.2	83.2	83.2
0320	Concrete Reinforcing	115.3	90.2	102.8	97.5	83.2	90.4	97.5	65.6	81.6	98.1	84.6	91.4	98.9	83.4	91.2	98.1	82.8	90.4
0330	Cast-in-Place Concrete	108.5	84.2	98.5	111.5	102.3	107.7	111.5	102.1	107.6	111.1	100.8	106.9	103.1	100.5	102.0	121.8	98.9	112.4
03	CONCRETE	105.9	89.5	97.8	103.5	91.6	97.6	103.6	88.2	96.0	106.5	90.3	98.5	100.9	92.0	96.5	111.0	89.3	100.3
04	MASONRY	106.4	80.7	90.7	95.2	91.5	92.9	90.1	91.5	90.9	100.9	87.4	92.6	91.7	92.7	92.3	103.4	87.4	93.6
05	METALS	103.1	86.3	97.8	95.5	95.0	95.3	95.5	88.5	93.3	95.7	96.4	95.9	97.8	96.2	97.3	95.7	95.7	95.7
06	WOOD, PLASTICS & COMPOSITES	98.4	97.2	97.7	86.9	86.3	86.5	88.6	86.3	87.3	87.9	81.9	84.5	104.5	86.3	94.1	84.5	81.9	83.0
07	THERMAL & MOISTURE PROTECTION	102.3	81.4	93.9	100.3	87.4	95.1	100.4	87.4	95.2	100.6	86.3	94.9	100.0	88.3	95.3	100.6	86.3	94.8
08	OPENINGS	101.1	80.4	96.1	94.8	80.1	91.2	95.6	76.1	90.8	95.6	78.0	91.4	95.4	80.1	91.7	95.6	79.2	91.6
0920	Plaster & Gypsum Board	105.6	97.1	99.6	99.9	85.7	90.0	100.7	85.7	90.2	101.2	81.2	87.2	104.8	85.7	91.5	98.7	81.2	86.4
0950, 0980	Ceilings & Acoustic Treatment	102.2	97.1	98.8	91.5	85.7	87.7	91.5	85.7	87.7	93.4	81.2	85.2	100.7	85.7	90.7	93.4	81.2	85.2
0960	Flooring	111.8	60.1	96.4	96.3	112.9	101.2	96.9	112.9	101.6	96.6	106.4	99.5	102.6	112.9	105.7	95.5	106.4	98.7
0970, 0990	Wall Finishes & Painting/Coating	106.9	61.1	79.5	109.6	36.6	65.8	109.6	36.6	65.8	109.6	88.9	97.2	109.6	89.2	97.3	109.6	88.9	97.2
09	FINISHES	106.7	84.3	94.1	97.5	86.2	91.2	97.8	86.2	91.3	98.8	87.9	92.7	101.4	92.5	96.4	98.0	87.9	92.3
COVERS	DIVS. 10 - 14, 25, 28, 41, 43, 44, 46	100.0	97.4	99.5	100.0	55.6	90.9	100.0	55.5	90.9	100.0	94.4	98.9	100.0	95.8	99.1	100.0	94.4	98.9
21, 22, 23	FIRE SUPPRESSION, PLUMBING & HVAC	100.1	107.4	103.1	94.3	81.9	89.2	94.3	81.7	89.1	94.3	89.9	92.5	100.0	83.9	93.4	94.3	89.7	92.4
26, 27, 3370	ELECTRICAL, COMMUNICATIONS & UTIL.	106.5	93.9	99.9	93.3	92.6	92.9	96.0	72.8	83.8	97.6	96.9	97.2	98.7	92.6	95.5	97.6	96.9	97.2
MF2010	WEIGHTED AVERAGE	102.7	93.6	98.7	96.7	87.2	92.5	96.8	83.2	90.8	98.1	90.3	94.7	98.9	90.1	95.0	98.7	90.1	94.9

	DIVISION	WEST VIRGINIA																	
		GASSAWAY			HUNTINGTON			LEWISBURG			MARTINSBURG			MORGANTOWN			PARKERSBURG		
		266			255 - 257			249			254			265			261		
		MAT.	INST.	TOTAL	MAT.	INST.	TOTAL	MAT.	INST.	TOTAL	MAT.	INST.	TOTAL	MAT.	INST.	TOTAL	MAT.	INST.	TOTAL
015433	CONTRACTOR EQUIPMENT		100.5	100.5		100.5	100.5		100.5	100.5		100.5	100.5		100.5	100.5		100.5	100.5
0241, 31 - 34	SITE & INFRASTRUCTURE, DEMOLITION	106.9	87.9	93.6	108.0	90.1	95.4	119.7	87.7	97.3	107.2	85.6	92.0	104.2	87.7	92.6	112.6	89.2	96.2
0310	Concrete Forming & Accessories	85.0	83.5	83.7	96.3	94.9	95.1	83.5	85.4	85.1	84.9	78.9	79.7	83.5	83.0	83.0	87.5	85.3	85.6
0320	Concrete Reinforcing	98.1	83.4	90.8	98.9	89.4	94.2	98.1	65.6	81.9	97.5	74.9	86.2	98.1	82.7	90.4	97.5	84.6	91.1
0330	Cast-in-Place Concrete	116.3	99.0	109.2	121.5	104.7	114.6	111.5	102.1	107.7	116.7	88.7	105.2	111.1	100.6	106.8	113.6	94.1	105.6
03	CONCRETE	107.2	89.6	98.5	109.3	97.7	103.6	112.8	88.2	100.6	107.3	82.1	94.8	103.3	89.8	96.6	108.5	88.9	98.8
04	MASONRY	104.5	90.0	95.7	95.2	91.9	93.1	93.8	91.5	92.4	96.3	80.9	86.9	122.0	87.4	100.9	80.7	88.8	85.7
05	METALS	95.6	96.3	95.8	97.8	98.2	97.9	95.6	88.5	93.4	95.9	85.7	92.7	95.7	95.4	95.6	96.3	96.4	96.4
06	WOOD, PLASTICS & COMPOSITES	87.0	81.9	84.1	98.2	94.5	96.1	84.8	86.3	85.7	86.9	81.0	83.5	84.8	81.9	83.1	88.0	83.4	85.4
07	THERMAL & MOISTURE PROTECTION	100.4	87.5	95.2	100.4	88.8	95.8	101.2	87.4	95.6	100.6	73.3	89.6	100.4	86.3	94.7	100.3	86.7	94.8
08	OPENINGS	93.8	77.7	89.9	94.2	86.0	92.2	95.6	76.1	90.9	96.8	68.5	89.9	96.9	79.2	92.6	94.6	78.9	90.8
0920	Plaster & Gypsum Board	100.4	81.2	86.9	108.7	94.1	98.5	98.7	85.7	89.6	100.4	80.2	86.3	98.7	81.2	86.4	101.6	82.7	88.4
0950, 0980	Ceilings & Acoustic Treatment	93.4	81.2	85.2	93.4	94.1	93.9	93.4	85.7	88.3	93.4	80.2	84.6	93.4	81.2	85.2	93.4	82.7	86.3
0960	Flooring	96.4	112.9	101.3	103.6	104.0	103.7	95.6	112.9	100.8	96.3	55.6	84.2	95.6	106.4	98.8	99.3	103.8	100.7
0970, 0990	Wall Finishes & Painting/Coating	109.6	89.2	97.3	109.6	82.4	93.3	109.6	36.6	65.8	109.6	43.2	69.8	109.6	88.9	97.2	109.6	87.6	96.4
09	FINISHES	98.3	89.2	93.2	101.1	95.6	98.0	99.3	86.2	91.9	98.2	71.6	83.3	97.6	87.9	92.1	99.7	88.6	93.5
COVERS	DIVS. 10 - 14, 25, 28, 41, 43, 44, 46	100.0	94.4	98.9	100.0	97.7	99.5	100.0	55.5	90.9	100.0	53.8	90.6	100.0	64.2	92.7	100.0	95.3	99.0
21, 22, 23	FIRE SUPPRESSION, PLUMBING & HVAC	94.3	90.8	92.8	100.3	86.0	94.4	94.3	81.7	89.1	94.3	84.8	90.4	94.3	89.7	92.4	100.2	88.1	95.2
26, 27, 3370	ELECTRICAL, COMMUNICATIONS & UTIL.	97.6	92.6	94.9	97.2	93.7	95.4	93.3	72.8	82.5	99.4	83.0	90.8	97.8	96.9	97.3	97.7	92.6	95.0
MF2010	WEIGHTED AVERAGE	98.0	90.2	94.6	99.9	92.5	96.6	98.3	83.2	91.7	98.2	80.3	90.3	98.7	89.2	94.5	98.9	89.6	94.8

	DIVISION	WEST VIRGINIA									WISCONSIN								
		PETERSBURG			ROMNEY			WHEELING			BELOIT			EAU CLAIRE			GREEN BAY		
		268			267			260			535			547			541 - 543		
		MAT.	INST.	TOTAL	MAT.	INST.	TOTAL	MAT.	INST.	TOTAL	MAT.	INST.	TOTAL	MAT.	INST.	TOTAL	MAT.	INST.	TOTAL
015433	CONTRACTOR EQUIPMENT		100.5	100.5		100.5	100.5		100.5	100.5		102.4	102.4		101.1	101.1		98.9	98.9
0241, 31 - 34	SITE & INFRASTRUCTURE, DEMOLITION	103.4	88.1	92.6	106.3	88.1	93.5	113.4	88.3	95.8	96.8	107.8	104.5	97.9	103.2	101.6	101.7	99.4	100.1
0310	Concrete Forming & Accessories	86.5	81.7	82.3	82.8	81.8	81.9	89.2	83.4	84.2	103.6	93.1	94.5	101.4	97.2	97.8	108.0	96.9	98.4
0320	Concrete Reinforcing	97.5	75.3	86.4	98.1	72.5	85.3	96.8	82.8	89.8	98.8	116.9	107.8	98.2	100.4	99.3	96.4	93.6	95.0
0330	Cast-in-Place Concrete	111.1	93.0	103.6	116.3	93.0	106.7	113.6	100.8	108.3	101.7	91.8	97.6	100.6	92.5	97.3	104.0	99.5	102.2
03	CONCRETE	103.4	85.2	94.4	107.0	84.7	96.0	108.5	90.0	99.4	100.2	97.4	98.8	99.7	96.4	98.1	102.7	97.4	100.1
04	MASONRY	96.4	87.4	90.9	94.2	82.5	87.1	103.7	87.4	93.8	102.8	92.5	96.5	91.6	96.5	94.6	123.0	98.4	108.0
05	METALS	95.8	92.2	94.7	95.8	91.1	94.3	96.5	95.8	96.3	97.1	107.8	100.5	95.3	102.3	97.5	97.5	99.5	98.2
06	WOOD, PLASTICS & COMPOSITES	88.8	81.0	84.3	84.0	81.0	82.3	89.5	81.9	85.1	98.0	95.3	96.5	106.3	98.7	102.0	108.6	98.7	102.9
07	THERMAL & MOISTURE PROTECTION	100.5	74.8	90.1	100.5	73.5	89.7	100.7	86.4	94.9	104.8	89.9	98.8	103.9	83.9	95.9	105.6	85.8	97.6
08	OPENINGS	96.9	75.4	91.7	96.8	74.8	91.5	95.4	79.2	91.5	101.2	103.7	101.8	99.3	96.2	98.6	96.5	96.7	96.6
0920	Plaster & Gypsum Board	101.2	80.2	86.5	98.3	80.2	85.7	101.6	81.2	87.3	101.0	95.5	97.2	104.6	98.9	100.6	97.6	98.9	98.5
0950, 0980	Ceilings & Acoustic Treatment	93.4	80.2	84.6	93.4	80.2	84.6	93.4	81.2	85.2	91.8	95.5	94.3	107.2	98.9	101.6	98.0	98.9	98.6
0960	Flooring	97.4	106.4	100.0	95.4	96.8	95.8	100.2	106.4	102.1	107.3	116.3	110.0	97.2	112.1	101.6	117.0	117.0	117.0
0970, 0990	Wall Finishes & Painting/Coating	109.6	88.9	97.2	109.6	88.9	97.2	109.6	88.9	97.2	89.7	96.5	93.8	86.6	90.3	88.8	96.4	82.8	88.2
09	FINISHES	98.4	87.3	92.2	97.7	85.5	90.9	99.9	87.9	93.2	101.7	97.9	99.6	102.4	99.8	101.0	106.6	99.8	102.8
COVERS	DIVS. 10 - 14, 25, 28, 41, 43, 44, 46	100.0	64.1	92.7	100.0	94.3	98.8	100.0	96.1	99.2	100.0	86.2	97.2	100.0	88.9	97.7	100.0	92.3	98.4
21, 22, 23	FIRE SUPPRESSION, PLUMBING & HVAC	94.3	90.8	92.8	94.3	90.6	92.8	100.3	89.7	96.0	99.9	87.5	94.8	100.2	83.3	93.3	100.5	84.3	93.8
26, 27, 3370	ELECTRICAL, COMMUNICATIONS & UTIL.	101.1	83.1	91.6	100.3	83.1	91.2	94.7	96.9	95.8	102.1	85.9	93.5	104.7	88.5	96.2	98.8	81.2	89.5
MF2010	WEIGHTED AVERAGE	97.8	86.1	92.6	98.0	86.0	92.7	100.0	90.3	95.7	100.2	94.8	97.8	99.5	93.5	96.9	101.5	92.6	97.6

	DIVISION	WISCONSIN																	
		KENOSHA			LA CROSSE			LANCASTER			MADISON			MILWAUKEE			NEW RICHMOND		
		531			546			538			537			530,532			540		
		MAT.	INST.	TOTAL	MAT.	INST.	TOTAL	MAT.	INST.	TOTAL	MAT.	INST.	TOTAL	MAT.	INST.	TOTAL	MAT.	INST.	TOTAL
015433	CONTRACTOR EQUIPMENT		100.3	100.3		101.1	101.1		102.4	102.4		102.4	102.4		90.5	90.5		101.6	101.6
0241, 31 - 34	SITE & INFRASTRUCTURE, DEMOLITION	103.2	104.9	104.4	91.4	103.3	99.8	96.1	108.1	104.5	92.5	108.2	103.5	97.6	98.5	98.2	94.6	103.4	100.8
0310	Concrete Forming & Accessories	108.6	101.7	102.6	89.2	97.3	96.2	102.9	93.9	95.1	105.3	95.7	97.0	105.7	122.5	120.2	97.3	101.4	100.9
0320	Concrete Reinforcing	98.6	99.8	99.2	97.9	89.4	93.7	100.1	89.2	94.7	98.8	89.6	94.2	98.8	100.4	99.6	95.4	100.3	97.9
0330	Cast-in-Place Concrete	110.6	96.4	104.8	90.4	90.0	90.3	101.1	95.4	98.8	93.0	96.1	94.3	99.3	112.1	104.6	104.7	80.2	94.6
03	CONCRETE	104.7	99.6	102.2	91.0	93.6	92.3	99.9	93.7	96.9	96.1	94.9	95.5	99.3	113.8	106.5	97.5	94.1	95.8
04	MASONRY	100.1	105.4	103.4	90.9	98.4	95.5	102.9	98.1	100.0	100.4	100.6	100.5	102.8	121.1	114.0	117.8	93.0	102.6
05	METALS	97.9	101.7	99.1	95.2	97.2	95.8	94.7	94.1	94.5	99.6	96.1	98.5	100.3	95.8	98.9	95.4	101.4	97.3
06	WOOD, PLASTICS & COMPOSITES	100.5	100.8	100.7	92.6	98.7	96.1	97.4	95.3	96.2	95.2	95.3	95.3	100.0	123.9	113.7	95.7	107.1	102.2
07	THERMAL & MOISTURE PROTECTION	104.5	94.8	100.6	103.5	83.5	95.4	104.6	78.5	94.1	107.7	89.4	100.4	102.7	114.3	107.4	105.2	90.6	99.4
08	OPENINGS	95.9	104.9	98.1	99.3	84.4	95.7	96.6	83.4	93.4	101.2	96.0	99.9	103.7	117.5	107.0	86.3	100.8	89.9
0920	Plaster & Gypsum Board	90.3	101.2	97.9	99.1	98.9	98.9	99.5	95.5	96.7	104.7	95.5	98.3	102.2	124.8	118.0	87.4	107.8	101.7
0950, 0980	Ceilings & Acoustic Treatment	91.8	101.2	98.0	105.4	98.9	101.0	86.3	95.5	92.4	102.0	95.5	97.7	96.3	124.8	115.3	69.5	107.8	95.0
0960	Flooring	127.1	111.0	122.3	90.8	112.1	97.1	106.8	105.1	106.3	100.9	105.1	102.1	110.5	118.1	112.8	109.8	112.1	110.5
0970, 0990	Wall Finishes & Painting/Coating	98.0	102.7	100.8	86.6	71.8	77.7	89.7	69.4	77.5	89.7	96.5	93.8	91.5	126.9	112.7	99.6	90.3	94.0
09	FINISHES	106.9	103.9	105.2	98.9	98.3	98.6	100.1	93.1	96.2	102.0	97.9	99.7	103.9	123.3	114.8	96.7	102.2	99.8
COVERS	DIVS. 10 - 14, 25, 28, 41, 43, 44, 46	100.0	98.7	99.7	100.0	89.5	97.9	100.0	49.3	89.6	100.0	89.0	97.7	100.0	104.2	100.8	100.0	86.3	97.2
21, 22, 23	FIRE SUPPRESSION, PLUMBING & HVAC	100.1	93.2	97.3	100.2	84.1	93.6	93.9	83.8	89.8	99.9	93.0	97.0	99.9	102.7	101.0	93.6	80.8	88.4
26, 27, 3370	ELECTRICAL, COMMUNICATIONS & UTIL.	102.6	97.5	99.9	105.0	88.7	96.4	101.8	88.7	94.9	101.3	92.3	96.6	103.6	100.9	102.2	102.8	88.7	95.4
MF2010	WEIGHTED AVERAGE	100.8	99.8	100.4	98.0	92.4	95.5	97.7	90.3	94.4	99.9	96.0	98.2	101.1	108.6	104.4	97.0	93.0	95.2

	DIVISION	WISCONSIN																	
		OSHKOSH			PORTAGE			RACINE			RHINELANDER			SUPERIOR			WAUSAU		
		549			539			534			545			548			544		
		MAT.	INST.	TOTAL	MAT.	INST.	TOTAL	MAT.	INST.	TOTAL	MAT.	INST.	TOTAL	MAT.	INST.	TOTAL	MAT.	INST.	TOTAL
015433	CONTRACTOR EQUIPMENT		98.9	98.9		102.4	102.4		102.4	102.4		98.9	98.9		101.6	101.6		98.9	98.9
0241, 31 - 34	SITE & INFRASTRUCTURE, DEMOLITION	93.1	98.1	96.6	86.5	108.2	101.7	96.7	108.8	105.2	106.2	99.3	101.3	91.2	103.7	100.0	88.8	99.3	96.2
0310	Concrete Forming & Accessories	91.9	83.2	84.4	95.9	94.9	95.1	103.9	101.7	102.0	89.6	94.0	93.4	95.7	91.2	91.8	91.3	96.6	95.9
0320	Concrete Reinforcing	96.5	76.7	86.6	100.2	89.4	94.8	98.8	99.8	99.3	96.6	91.5	94.1	95.4	92.5	94.0	96.6	89.4	93.1
0330	Cast-in-Place Concrete	96.5	87.3	92.7	86.8	97.7	91.3	99.7	96.1	98.2	109.4	97.7	104.6	98.6	95.8	97.5	89.9	97.0	92.8
03	CONCRETE	92.5	84.0	88.3	88.0	95.0	91.5	99.3	99.5	99.4	104.5	95.1	99.9	92.1	93.4	92.7	87.6	95.6	91.6
04	MASONRY	104.5	86.6	93.6	101.9	98.1	99.6	102.8	105.4	104.4	121.2	97.6	106.8	117.0	102.4	108.1	104.1	95.8	99.0
05	METALS	95.6	89.2	93.6	95.4	95.2	95.3	98.7	101.7	99.7	95.5	97.4	96.1	96.3	99.2	97.2	95.4	97.4	96.0
06	WOOD, PLASTICS & COMPOSITES	90.2	84.7	87.1	88.8	95.3	92.5	98.3	100.8	99.7	88.0	95.4	92.2	94.1	89.7	91.6	89.8	98.7	94.9
07	THERMAL & MOISTURE PROTECTION	104.8	77.1	93.6	104.0	90.6	98.6	104.8	94.4	100.6	105.5	82.2	96.1	104.9	93.6	100.4	104.6	83.2	96.0
08	OPENINGS	92.3	82.0	89.8	96.8	90.2	95.2	101.2	104.9	102.1	92.3	85.5	90.7	86.0	90.9	87.2	92.5	95.6	93.3
0920	Plaster & Gypsum Board	85.9	84.5	84.9	93.6	95.5	94.9	101.0	101.2	101.1	85.9	95.5	92.6	87.3	89.8	89.1	85.9	98.9	95.0
0950, 0980	Ceilings & Acoustic Treatment	98.0	84.5	89.0	89.0	95.5	93.3	91.8	101.2	98.0	98.0	95.5	96.3	70.4	89.8	83.3	98.0	98.9	98.6
0960	Flooring	108.3	98.7	105.4	103.2	111.6	105.7	107.3	111.0	108.4	107.5	112.1	108.9	111.3	121.2	114.3	108.1	112.1	109.3
0970, 0990	Wall Finishes & Painting/Coating	94.1	67.7	78.3	89.7	69.4	77.5	89.6	107.7	100.4	94.1	63.0	75.5	88.5	99.5	95.0	94.1	79.5	85.4
09	FINISHES	101.8	84.0	91.8	98.2	95.0	96.4	101.7	104.4	103.2	102.8	94.2	97.9	96.1	97.5	96.9	101.4	98.5	99.7
COVERS	DIVS. 10 - 14, 25, 28, 41, 43, 44, 46	100.0	61.8	92.2	100.0	63.6	92.6	100.0	98.7	99.7	100.0	54.9	90.8	100.0	86.0	97.1	100.0	92.3	98.4
21, 22, 23	FIRE SUPPRESSION, PLUMBING & HVAC	94.4	73.3	85.8	93.9	88.6	91.7	99.9	93.2	97.2	94.4	83.3	89.9	93.6	90.7	92.4	94.4	82.1	89.4
26, 27, 3370	ELECTRICAL, COMMUNICATIONS & UTIL.	103.4	74.7	88.3	105.9	91.9	98.5	101.9	95.5	98.5	102.7	79.8	90.6	108.1	95.0	101.2	104.6	79.8	91.6
MF2010	WEIGHTED AVERAGE	96.9	81.4	90.1	96.3	93.4	95.0	100.3	99.9	100.1	99.5	89.2	94.9	96.8	95.5	96.2	96.3	91.0	94.0

	DIVISION	WYOMING																	
		CASPER			CHEYENNE			NEWCASTLE			RAWLINS			RIVERTON			ROCK SPRINGS		
		826			820			827			823			825			829 - 831		
		MAT.	INST.	TOTAL	MAT.	INST.	TOTAL	MAT.	INST.	TOTAL	MAT.	INST.	TOTAL	MAT.	INST.	TOTAL	MAT.	INST.	TOTAL
015433	**CONTRACTOR EQUIPMENT**		97.5	97.5		97.5	97.5		97.5	97.5		97.5	97.5		97.5	97.5		97.5	97.5
0241, 31 - 34	**SITE & INFRASTRUCTURE, DEMOLITION**	97.6	95.1	95.8	90.8	95.1	93.8	81.9	94.7	90.9	96.4	94.8	95.2	89.7	94.7	93.2	85.5	94.6	91.8
0310	Concrete Forming & Accessories	102.7	44.4	52.2	105.0	55.8	62.5	95.9	63.3	67.7	100.0	63.7	68.5	94.9	63.2	67.5	102.0	62.5	67.8
0320	Concrete Reinforcing	111.4	70.3	90.9	105.0	71.1	88.1	112.9	71.5	92.3	112.5	71.6	92.1	113.6	71.2	92.5	113.6	71.2	92.5
0330	Cast-in-Place Concrete	95.2	73.9	86.4	93.6	74.1	85.5	94.5	63.0	81.5	94.6	63.1	81.7	94.6	53.4	77.6	94.5	52.2	77.1
03	**CONCRETE**	101.7	60.2	81.2	100.5	65.5	83.2	101.1	65.0	83.3	114.0	65.3	89.9	108.9	61.7	85.6	101.7	60.9	81.5
04	**MASONRY**	101.3	35.3	61.0	98.0	48.7	67.9	94.6	57.0	71.7	94.6	57.0	71.7	94.6	51.0	68.0	151.4	48.3	88.4
05	**METALS**	98.1	71.0	89.6	99.3	72.4	90.9	94.6	72.1	87.5	94.6	72.5	87.6	94.7	71.6	87.4	95.4	71.9	88.0
06	**WOOD, PLASTICS & COMPOSITES**	93.5	41.8	64.0	95.3	56.7	73.3	85.6	68.3	75.7	89.4	68.3	77.4	84.5	68.3	75.3	93.0	68.3	78.9
07	**THERMAL & MOISTURE PROTECTION**	100.5	50.3	80.3	97.4	55.4	80.5	98.9	56.8	82.0	100.5	56.8	82.9	99.9	60.7	84.1	99.0	59.6	83.1
08	**OPENINGS**	97.2	49.6	85.6	99.4	57.7	89.2	103.2	60.9	92.9	102.9	60.9	92.7	103.1	60.9	92.8	103.6	60.9	93.2
0920	Plaster & Gypsum Board	93.0	39.9	55.9	83.5	55.4	63.8	78.7	67.3	70.7	79.1	67.3	70.8	78.7	67.3	70.7	87.8	67.3	73.5
0950, 0980	Ceilings & Acoustic Treatment	117.7	39.9	65.9	104.7	55.4	71.8	106.7	67.3	80.4	106.7	67.3	80.4	106.7	67.3	80.4	106.7	67.3	80.4
0960	Flooring	99.9	34.2	80.3	103.8	54.9	89.2	97.5	53.6	84.4	100.3	53.6	86.4	97.1	53.6	84.1	102.5	53.6	87.9
0970, 0990	Wall Finishes & Painting/Coating	98.5	45.0	66.4	101.9	45.0	67.8	97.7	57.8	73.8	97.7	57.8	73.8	97.7	57.8	73.8	97.7	57.8	73.8
09	**FINISHES**	104.9	41.0	69.1	102.6	53.9	75.3	97.2	61.0	76.9	99.8	61.0	78.1	98.2	61.0	77.3	99.8	60.3	77.7
COVERS	**DIVS. 10 - 14, 25, 28, 41, 43, 44, 46**	100.0	88.3	97.6	100.0	89.9	97.9	100.0	58.6	91.6	100.0	90.7	98.1	100.0	58.6	91.5	100.0	81.5	96.2
21, 22, 23	**FIRE SUPPRESSION, PLUMBING & HVAC**	100.0	66.7	86.4	100.0	66.7	86.4	96.9	66.0	84.2	96.9	66.0	84.3	96.9	66.0	84.2	99.9	64.6	85.5
26, 27, 3370	**ELECTRICAL, COMMUNICATIONS & UTIL.**	94.5	66.8	79.9	97.5	71.5	83.8	95.0	71.9	82.9	95.0	71.9	82.9	95.0	65.6	79.5	93.3	65.6	78.7
MF2010	**WEIGHTED AVERAGE**	99.4	61.5	82.7	99.4	66.7	85.0	97.2	67.5	84.1	99.3	68.6	85.8	98.4	65.6	84.0	101.2	65.6	85.5

	DIVISION	WYOMING												CANADA					
		SHERIDAN			WHEATLAND			WORLAND			YELLOWSTONE NAT'L PA			BARRIE, ONTARIO			BATHURST, NEW BRUNSWICK		
		828			822			824			821								
		MAT.	INST.	TOTAL	MAT.	INST.	TOTAL	MAT.	INST.	TOTAL	MAT.	INST.	TOTAL	MAT.	INST.	TOTAL	MAT.	INST.	TOTAL
015433	**CONTRACTOR EQUIPMENT**		97.5	97.5		97.5	97.5		97.5	97.5		97.5	97.5		101.6	101.6		101.7	101.7
0241, 31 - 34	**SITE & INFRASTRUCTURE, DEMOLITION**	90.3	95.1	93.7	86.2	94.7	92.2	83.6	94.7	91.4	83.7	94.9	91.5	116.0	101.1	105.6	96.7	97.4	97.2
0310	Concrete Forming & Accessories	102.9	52.7	59.4	97.9	52.1	58.3	97.9	62.9	67.6	98.0	64.2	68.7	124.1	92.0	96.4	102.8	66.6	71.4
0320	Concrete Reinforcing	113.6	71.6	92.7	112.9	71.6	92.3	113.6	70.9	92.3	115.5	70.9	93.3	180.3	86.5	133.6	148.9	59.3	104.3
0330	Cast-in-Place Concrete	97.8	74.0	88.0	98.7	62.8	83.9	94.5	53.0	77.4	94.5	55.0	78.3	167.6	91.0	136.1	138.2	63.9	107.7
03	**CONCRETE**	109.0	64.2	86.9	105.8	60.0	83.2	101.4	61.3	81.6	101.7	62.6	82.4	152.9	90.8	122.2	135.7	64.9	100.8
04	**MASONRY**	94.9	58.6	72.7	94.9	56.6	71.5	94.6	46.0	64.9	94.7	49.6	67.2	166.4	100.5	126.2	157.8	67.7	102.8
05	**METALS**	98.1	73.0	90.2	94.5	72.3	87.5	94.7	71.2	87.3	95.3	71.1	87.7	107.2	91.7	102.4	104.8	73.3	94.9
06	**WOOD, PLASTICS & COMPOSITES**	96.5	52.6	71.5	87.3	53.1	67.8	87.4	68.3	76.5	87.4	68.3	76.5	117.9	90.6	102.3	94.6	66.8	78.7
07	**THERMAL & MOISTURE PROTECTION**	100.0	57.6	82.9	99.2	56.0	81.8	98.9	56.9	82.0	98.5	58.5	82.4	111.7	91.6	103.6	104.9	64.9	88.8
08	**OPENINGS**	103.9	53.2	91.5	101.9	55.5	90.6	103.4	60.9	93.1	96.7	60.9	88.0	92.1	88.9	91.3	87.5	58.4	80.4
0920	Plaster & Gypsum Board	106.6	51.1	67.8	78.7	51.6	59.8	78.7	67.3	70.7	79.0	67.3	70.8	154.1	90.3	109.5	164.9	65.6	95.5
0950, 0980	Ceilings & Acoustic Treatment	109.8	51.1	70.7	106.7	51.6	70.0	106.7	67.3	80.4	107.7	67.3	80.7	95.6	90.3	92.0	96.5	65.6	76.0
0960	Flooring	100.9	53.6	86.8	99.0	52.6	85.2	99.0	53.6	85.5	99.0	53.6	85.5	130.8	100.7	121.9	115.7	48.9	95.8
0970, 0990	Wall Finishes & Painting/Coating	100.1	45.0	67.1	97.7	36.0	60.7	97.7	57.8	73.8	97.7	57.8	73.8	113.0	92.3	100.6	112.2	53.9	77.3
09	**FINISHES**	105.6	50.7	74.8	98.0	49.3	70.7	97.6	60.8	77.0	97.9	61.7	77.6	118.3	94.0	104.7	113.5	62.2	84.8
COVERS	**DIVS. 10 - 14, 25, 28, 41, 43, 44, 46**	100.0	89.5	97.8	100.0	81.4	96.2	100.0	58.4	91.5	100.0	59.7	91.8	139.2	76.3	126.4	131.1	67.9	118.2
21, 22, 23	**FIRE SUPPRESSION, PLUMBING & HVAC**	96.9	66.7	84.5	96.9	65.8	84.2	96.9	65.7	84.1	96.9	67.3	84.8	102.7	103.0	102.8	102.9	73.6	90.9
26, 27, 3370	**ELECTRICAL, COMMUNICATIONS & UTIL.**	97.8	65.6	80.8	95.0	71.5	82.6	95.0	65.6	79.5	94.0	65.6	79.0	119.5	92.7	105.4	120.1	64.5	90.8
MF2010	**WEIGHTED AVERAGE**	100.0	66.1	85.1	97.8	65.5	83.6	97.4	64.9	83.0	96.7	65.9	83.2	118.1	95.4	108.1	112.9	70.0	94.0

	DIVISION	CANADA																	
		BRANDON, MANITOBA			BRANTFORD, ONTARIO			BRIDGEWATER, NOVA SCOTIA			CALGARY, ALBERTA			CAP-DE-LA-MADELEINE, QUEBEC			CHARLESBOURG, QUEBEC		
		MAT.	INST.	TOTAL	MAT.	INST.	TOTAL	MAT.	INST.	TOTAL	MAT.	INST.	TOTAL	MAT.	INST.	TOTAL	MAT.	INST.	TOTAL
015433	**CONTRACTOR EQUIPMENT**		103.8	103.8		101.6	101.6		101.4	101.4		106.4	106.4		102.3	102.3		102.3	102.3
0241, 31 - 34	**SITE & INFRASTRUCTURE, DEMOLITION**	112.4	100.1	103.8	114.0	101.4	105.2	99.1	98.9	98.9	126.5	106.0	112.1	95.1	100.5	98.9	95.1	100.5	98.9
0310	Concrete Forming & Accessories	123.4	75.7	82.2	124.0	99.7	103.0	96.9	73.7	76.8	129.3	99.0	103.1	129.1	90.5	95.7	129.1	90.5	95.7
0320	Concrete Reinforcing	162.0	55.8	109.1	175.2	85.2	130.3	148.9	48.2	98.8	144.1	75.3	109.8	148.9	81.3	115.3	148.9	81.3	115.3
0330	Cast-in-Place Concrete	151.1	79.3	121.6	163.5	112.9	142.7	168.1	73.8	129.3	201.8	108.6	163.4	132.0	100.1	118.9	132.0	100.1	118.9
03	**CONCRETE**	135.4	73.9	105.0	150.1	101.5	126.1	147.5	69.7	109.0	165.8	97.9	132.3	131.0	92.2	111.9	131.0	92.2	111.9
04	**MASONRY**	161.3	70.5	105.9	162.8	105.1	127.6	158.4	73.7	106.7	211.8	92.9	139.2	158.7	89.2	116.3	158.7	89.2	116.3
05	**METALS**	107.4	78.6	98.4	106.6	92.4	102.2	105.8	75.4	96.2	140.7	90.3	124.8	105.0	89.0	100.0	105.0	89.0	100.0
06	**WOOD, PLASTICS & COMPOSITES**	113.6	76.7	92.5	115.9	98.3	105.9	87.0	73.0	79.0	114.1	98.6	105.2	127.2	90.4	106.2	127.2	90.4	106.2
07	**THERMAL & MOISTURE PROTECTION**	105.2	74.9	93.0	111.6	97.3	105.8	106.6	71.9	92.6	123.1	96.2	112.3	105.3	92.4	100.1	105.3	92.4	100.1
08	**OPENINGS**	93.0	67.7	86.9	92.0	95.2	92.8	84.3	66.6	80.0	93.0	87.9	91.8	93.0	83.3	90.7	93.0	83.3	90.7
0920	Plaster & Gypsum Board	131.3	75.5	92.3	154.1	98.3	115.1	158.4	72.0	98.0	179.7	98.1	122.6	189.1	89.9	119.7	189.1	89.9	119.7
0950, 0980	Ceilings & Acoustic Treatment	95.6	75.5	82.2	95.6	98.3	97.4	95.6	72.0	79.9	148.8	98.1	115.0	95.6	89.9	91.8	95.6	89.9	91.8
0960	Flooring	130.8	72.8	113.5	130.8	100.7	121.9	112.3	68.5	99.3	132.8	96.0	121.8	130.8	101.4	122.1	130.8	101.4	122.1
0970, 0990	Wall Finishes & Painting/Coating	112.3	61.3	81.7	112.2	101.3	105.7	112.2	65.9	84.4	112.1	115.0	113.8	112.2	94.2	101.4	112.2	94.2	101.4
09	**FINISHES**	114.3	74.1	91.8	116.9	100.6	107.7	111.2	72.1	89.3	134.1	100.6	115.3	120.4	93.3	105.2	120.4	93.3	105.2
COVERS	**DIVS. 10 - 14, 25, 28, 41, 43, 44, 46**	131.1	70.2	118.7	131.1	78.5	120.4	131.1	68.2	118.3	131.1	100.0	124.8	131.1	87.3	122.2	131.1	87.3	122.2
21, 22, 23	**FIRE SUPPRESSION, PLUMBING & HVAC**	102.9	88.7	97.1	102.9	106.1	104.2	102.9	87.6	96.6	100.5	95.2	98.3	103.4	94.4	99.7	103.4	94.4	99.7
26, 27, 3370	**ELECTRICAL, COMMUNICATIONS & UTIL.**	122.1	72.6	96.0	117.4	92.2	104.1	123.6	66.0	93.3	120.9	102.7	111.3	117.0	74.4	94.6	117.0	74.4	94.6
MF2010	**WEIGHTED AVERAGE**	114.9	78.9	99.0	116.4	99.4	108.9	114.3	76.6	97.7	127.9	97.4	114.5	113.7	89.9	103.2	113.7	89.9	103.2

DIVISION		CANADA																	
		CHARLOTTETOWN, PRINCE EDWARD ISLAND			CHICOUTIMI, QUEBEC			CORNER BROOK, NEWFOUNDLAND			CORNWALL, ONTARIO			DALHOUSIE, NEW BRUNSWICK			DARTMOUTH, NOVA SCOTIA		
		MAT.	INST.	TOTAL	MAT.	INST.	TOTAL	MAT.	INST.	TOTAL	MAT.	INST.	TOTAL	MAT.	INST.	TOTAL	MAT.	INST.	TOTAL
015433	**CONTRACTOR EQUIPMENT**		101.4	101.4		102.3	102.3		102.9	102.9		101.6	101.6		101.7	101.7		101.4	101.4
0241, 31 - 34	**SITE & INFRASTRUCTURE, DEMOLITION**	109.9	96.4	100.5	94.6	99.8	98.2	114.7	97.9	102.9	112.1	100.9	104.3	96.5	97.4	97.1	105.9	98.9	101.0
0310	Concrete Forming & Accessories	100.9	59.9	65.4	129.1	94.9	99.5	104.4	63.4	68.9	121.9	92.2	96.2	102.8	66.8	71.6	97.0	73.7	76.8
0320	Concrete Reinforcing	157.5	48.3	103.1	112.3	98.0	105.2	148.9	50.1	99.7	175.2	84.9	130.2	148.9	59.4	104.3	155.4	48.2	102.0
0330	Cast-in-Place Concrete	160.5	62.4	120.2	130.2	101.3	118.3	172.2	72.1	131.1	147.1	103.0	128.9	138.2	64.0	107.7	169.8	73.8	130.3
03	**CONCRETE**	154.1	59.4	107.3	124.1	97.5	111.0	161.7	64.9	113.9	142.1	94.6	118.6	135.7	65.1	100.8	149.4	69.7	110.0
04	**MASONRY**	163.9	64.2	103.0	159.2	96.3	120.8	158.7	65.8	102.0	161.6	96.4	121.8	155.4	67.7	101.9	170.0	73.7	111.2
05	**METALS**	118.2	68.6	102.7	104.6	93.3	101.0	107.8	74.1	97.3	106.6	91.0	101.7	104.8	73.5	94.9	108.0	75.4	97.7
06	**WOOD, PLASTICS & COMPOSITES**	87.2	59.3	71.3	127.2	95.2	108.9	95.6	62.4	76.7	114.7	91.4	101.4	94.6	66.8	78.7	87.0	73.0	79.0
07	**THERMAL & MOISTURE PROTECTION**	114.5	61.4	93.2	105.3	98.0	102.4	108.6	63.6	90.5	111.4	91.7	103.5	109.1	64.9	91.4	106.6	71.9	92.6
08	**OPENINGS**	95.2	51.3	84.5	92.6	82.0	90.0	98.9	58.7	89.1	93.0	88.6	92.0	87.5	58.4	80.4	84.3	66.6	80.0
0920	Plaster & Gypsum Board	161.2	58.0	89.0	189.1	94.9	123.2	164.9	61.1	92.3	226.5	91.1	131.8	164.9	65.6	95.5	160.8	72.0	98.7
0950, 0980	Ceilings & Acoustic Treatment	114.8	58.0	76.9	95.6	94.9	95.1	96.5	61.1	72.9	98.4	91.1	93.6	96.5	65.6	76.0	104.7	72.0	83.0
0960	Flooring	113.0	64.1	98.4	130.8	101.4	122.1	116.4	58.3	99.1	130.8	99.3	121.4	115.7	74.7	103.5	112.3	68.5	99.3
0970, 0990	Wall Finishes & Painting/Coating	112.2	44.1	71.4	112.2	108.1	109.7	112.2	63.4	83.0	112.2	94.4	101.5	112.2	53.9	77.3	112.2	65.9	84.4
09	**FINISHES**	117.4	59.4	84.9	120.4	98.3	108.0	114.8	62.5	85.5	126.6	94.0	108.4	113.5	67.3	87.6	113.6	72.1	90.3
COVERS	**DIVS. 10 - 14, 25, 28, 41, 43, 44, 46**	131.1	66.4	117.9	131.1	88.5	122.4	131.1	67.2	118.1	131.1	75.8	119.8	131.1	67.9	118.2	131.1	68.2	118.3
21, 22, 23	**FIRE SUPPRESSION, PLUMBING & HVAC**	103.1	66.2	88.0	102.9	95.5	99.9	102.9	74.5	91.3	103.4	103.9	103.6	102.9	73.6	90.9	102.9	87.6	96.6
26, 27, 3370	**ELECTRICAL, COMMUNICATIONS & UTIL.**	118.6	53.6	84.4	116.6	88.1	101.6	119.2	59.7	87.9	118.4	93.2	105.1	120.2	60.9	89.0	124.1	66.0	93.5
MF2010	**WEIGHTED AVERAGE**	119.0	64.5	95.0	112.6	94.6	104.7	118.2	69.4	96.7	116.5	95.7	107.3	113.0	70.2	94.1	115.9	76.6	98.6

DIVISION		CANADA																	
		EDMONTON, ALBERTA			FORT MCMURRAY, ALBERTA			FREDERICTON, NEW BRUNSWICK			GATINEAU, QUEBEC			GRANBY, QUEBEC			HALIFAX, NOVA SCOTIA		
		MAT.	INST.	TOTAL	MAT.	INST.	TOTAL	MAT.	INST.	TOTAL	MAT.	INST.	TOTAL	MAT.	INST.	TOTAL	MAT.	INST.	TOTAL
015433	**CONTRACTOR EQUIPMENT**		106.4	106.4		103.7	103.7		101.7	101.7		102.3	102.3		102.3	102.3		101.4	101.4
0241, 31 - 34	**SITE & INFRASTRUCTURE, DEMOLITION**	150.4	106.0	119.3	119.8	102.1	107.4	98.2	97.4	97.7	94.9	100.5	98.8	95.4	100.5	99.0	101.7	98.9	99.8
0310	Concrete Forming & Accessories	129.5	99.0	103.1	122.1	96.5	99.9	124.9	67.1	74.9	129.1	90.4	95.6	129.1	90.3	95.6	99.7	80.0	82.7
0320	Concrete Reinforcing	153.6	75.3	114.6	162.0	75.3	118.8	141.8	59.5	100.8	157.5	81.3	119.6	157.5	81.3	119.6	158.4	63.9	111.3
0330	Cast-in-Place Concrete	210.3	108.6	168.5	217.1	107.1	171.9	132.6	64.1	104.4	130.2	100.0	117.8	134.5	100.0	120.3	172.4	82.6	135.5
03	**CONCRETE**	171.5	97.9	135.1	173.5	96.3	135.3	133.5	65.3	99.8	131.6	92.2	112.1	133.6	92.1	113.1	151.3	78.4	115.3
04	**MASONRY**	199.8	92.9	134.6	201.4	91.8	134.5	175.2	69.3	110.6	158.6	89.2	116.3	158.9	89.2	116.4	170.6	85.4	118.6
05	**METALS**	141.8	90.3	125.6	132.2	90.1	119.0	121.3	74.1	106.5	105.0	88.9	100.0	105.0	88.8	99.9	130.8	82.4	115.6
06	**WOOD, PLASTICS & COMPOSITES**	110.4	98.6	103.7	110.3	95.8	102.0	116.6	66.8	88.2	127.2	90.4	106.2	127.2	90.4	106.2	87.1	79.0	82.5
07	**THERMAL & MOISTURE PROTECTION**	126.8	96.2	114.5	118.9	94.6	109.1	112.9	66.0	94.0	105.3	92.4	100.1	105.3	90.8	99.4	115.3	79.6	100.9
08	**OPENINGS**	93.0	87.9	91.8	93.0	86.4	91.4	87.4	57.2	80.0	93.0	78.4	89.5	93.0	78.4	89.5	84.3	73.3	81.6
0920	Plaster & Gypsum Board	163.1	98.1	117.6	153.1	95.3	112.7	181.1	65.6	100.3	148.1	89.9	107.4	150.9	89.9	108.2	159.5	78.3	102.7
0950, 0980	Ceilings & Acoustic Treatment	154.6	98.1	116.9	104.7	95.3	98.5	108.4	65.6	79.9	95.6	89.9	91.8	95.6	89.9	91.8	108.4	78.3	88.3
0960	Flooring	135.6	96.0	123.8	130.8	96.0	120.4	127.3	78.0	112.6	130.8	101.4	122.1	130.8	101.4	122.1	112.9	90.0	106.1
0970, 0990	Wall Finishes & Painting/Coating	112.3	115.0	113.9	112.3	99.7	104.7	112.2	68.7	86.1	112.2	94.2	101.4	112.2	94.2	101.4	112.2	83.8	95.2
09	**FINISHES**	136.0	100.6	116.1	120.2	97.0	107.2	121.9	69.5	92.6	115.2	93.3	102.9	115.6	93.3	103.1	114.6	82.7	96.7
COVERS	**DIVS. 10 - 14, 25, 28, 41, 43, 44, 46**	131.1	100.0	124.8	131.1	99.1	124.6	131.1	67.9	118.2	131.1	87.3	122.2	131.1	87.3	122.2	131.1	69.9	118.6
21, 22, 23	**FIRE SUPPRESSION, PLUMBING & HVAC**	100.6	95.2	98.4	103.3	104.3	103.7	103.0	83.7	95.1	103.4	94.4	99.7	102.9	94.4	99.4	99.8	80.2	91.8
26, 27, 3370	**ELECTRICAL, COMMUNICATIONS & UTIL.**	115.3	102.7	108.7	111.0	89.1	99.5	123.0	80.3	100.5	117.0	74.4	94.6	117.7	74.4	94.9	126.0	81.9	102.8
MF2010	**WEIGHTED AVERAGE**	128.5	97.4	114.8	125.2	96.2	112.4	117.6	75.5	99.1	113.3	89.7	102.9	113.5	89.6	103.0	119.5	82.2	103.0

DIVISION		CANADA																	
		HAMILTON, ONTARIO			HULL, QUEBEC			JOLIETTE, QUEBEC			KAMLOOPS, BRITISH COLUMBIA			KINGSTON, ONTARIO			KITCHENER, ONTARIO		
		MAT.	INST.	TOTAL	MAT.	INST.	TOTAL	MAT.	INST.	TOTAL	MAT.	INST.	TOTAL	MAT.	INST.	TOTAL	MAT.	INST.	TOTAL
015433	**CONTRACTOR EQUIPMENT**		108.4	108.4		102.3	102.3		102.3	102.3		105.5	105.5		103.9	103.9		103.8	103.8
0241, 31 - 34	**SITE & INFRASTRUCTURE, DEMOLITION**	117.5	113.0	114.4	94.9	100.5	98.8	95.5	100.5	99.0	118.5	104.1	108.4	112.1	104.7	106.9	101.8	105.3	104.3
0310	Concrete Forming & Accessories	130.7	96.1	100.7	129.1	90.4	95.6	129.1	90.5	95.7	123.2	95.0	98.8	122.1	92.3	96.3	118.3	88.5	92.5
0320	Concrete Reinforcing	161.9	94.7	128.4	157.5	81.3	119.6	148.9	81.3	115.3	116.8	80.6	98.8	175.2	84.9	130.2	110.6	94.6	102.6
0330	Cast-in-Place Concrete	153.0	102.0	132.0	130.2	100.0	117.8	135.6	100.1	121.0	117.4	105.3	112.4	147.1	102.9	128.9	140.8	92.0	120.7
03	**CONCRETE**	143.3	97.7	120.8	131.6	92.2	112.1	132.7	92.2	112.7	138.6	95.8	117.5	142.1	94.7	118.7	122.0	90.9	106.6
04	**MASONRY**	178.5	101.5	131.5	158.6	89.2	116.3	159.0	89.2	116.4	165.2	99.0	124.8	168.1	96.4	124.4	159.4	97.6	121.7
05	**METALS**	124.6	93.7	114.9	105.0	88.9	100.0	105.0	89.0	100.0	107.2	89.7	101.7	108.2	91.0	102.8	115.4	93.4	108.5
06	**WOOD, PLASTICS & COMPOSITES**	115.2	95.7	104.1	127.2	90.4	106.2	127.2	90.4	106.2	98.4	93.3	95.5	114.7	91.6	101.5	109.1	87.2	96.6
07	**THERMAL & MOISTURE PROTECTION**	124.1	96.7	113.1	105.3	92.4	100.1	105.3	92.4	100.1	120.1	90.6	108.2	111.4	92.9	104.0	111.0	93.5	104.0
08	**OPENINGS**	93.0	93.7	93.2	93.0	78.4	89.5	93.0	83.3	90.7	89.1	89.5	89.2	93.0	88.3	91.9	84.8	87.3	85.4
0920	Plaster & Gypsum Board	193.0	95.5	124.8	148.1	89.9	107.4	189.1	89.9	119.7	135.6	92.6	105.5	230.2	91.3	133.0	151.6	86.7	106.2
0950, 0980	Ceilings & Acoustic Treatment	128.2	95.5	106.4	95.6	89.9	91.8	95.6	89.9	91.8	95.6	92.6	93.6	112.0	91.3	98.2	104.7	86.7	92.7
0960	Flooring	132.8	104.2	124.3	130.8	101.4	122.1	130.8	101.4	122.1	129.5	57.6	108.1	130.8	99.3	121.4	127.2	104.2	120.4
0970, 0990	Wall Finishes & Painting/Coating	112.2	104.8	107.8	112.2	94.2	101.4	112.2	94.2	101.4	112.2	87.7	97.5	112.2	87.2	97.2	112.2	94.6	101.6
09	**FINISHES**	130.6	98.7	112.7	115.2	93.3	102.9	120.4	93.3	105.2	116.4	87.9	100.4	130.1	93.3	109.5	116.6	91.7	102.7
COVERS	**DIVS. 10 - 14, 25, 28, 41, 43, 44, 46**	131.1	100.7	124.9	131.1	87.3	122.2	131.1	87.3	122.2	131.1	99.5	124.7	131.1	75.9	119.8	131.1	98.7	124.5
21, 22, 23	**FIRE SUPPRESSION, PLUMBING & HVAC**	100.8	88.0	95.5	102.9	94.4	99.4	102.9	94.4	99.4	102.9	99.6	101.5	103.4	104.0	103.6	99.7	89.2	95.4
26, 27, 3370	**ELECTRICAL, COMMUNICATIONS & UTIL.**	117.0	97.8	106.9	119.1	74.4	95.6	117.7	74.4	94.9	121.5	85.4	102.5	118.4	91.8	104.4	117.7	95.3	105.9
MF2010	**WEIGHTED AVERAGE**	120.4	97.0	110.1	113.4	89.7	102.9	113.8	89.9	103.3	115.7	94.3	106.3	117.3	95.8	107.8	112.6	93.5	104.2

DIVISION		CANADA																	
		LAVAL, QUEBEC			LETHBRIDGE, ALBERTA			LLOYDMINSTER, ALBERTA			LONDON, ONTARIO			MEDICINE HAT, ALBERTA			MONCTON, NEW BRUNSWICK		
		MAT.	INST.	TOTAL	MAT.	INST.	TOTAL	MAT.	INST.	TOTAL	MAT.	INST.	TOTAL	MAT.	INST.	TOTAL	MAT.	INST.	TOTAL
015433	CONTRACTOR EQUIPMENT		102.3	102.3		103.7	103.7		103.7	103.7		104.0	104.0		103.7	103.7		101.7	101.7
0241, 31 - 34	SITE & INFRASTRUCTURE, DEMOLITION	95.4	100.5	99.0	112.4	102.7	105.6	112.4	102.1	105.2	116.3	105.5	108.7	111.1	102.2	104.8	96.1	97.6	97.2
0310	Concrete Forming & Accessories	129.3	90.4	95.6	123.3	96.5	100.1	121.6	86.3	91.0	130.9	89.7	95.3	123.2	86.2	91.2	102.8	67.3	72.1
0320	Concrete Reinforcing	157.5	81.3	119.6	162.0	75.3	118.8	162.0	75.3	118.8	141.2	94.6	118.0	162.0	75.3	118.8	148.9	62.5	105.9
0330	Cast-in-Place Concrete	134.5	100.0	120.3	162.9	107.1	139.9	151.1	103.2	131.4	164.1	99.8	137.7	151.1	103.2	131.4	133.1	81.5	111.9
03	CONCRETE	133.6	92.2	113.2	147.6	96.3	122.3	141.8	90.4	116.4	145.2	94.1	120.0	141.9	90.4	116.5	133.3	72.2	103.1
04	MASONRY	158.8	89.2	116.4	175.8	91.8	124.6	158.3	84.9	113.5	182.5	98.7	131.4	158.3	84.9	113.5	157.4	67.7	102.7
05	METALS	105.1	88.9	100.0	126.1	90.1	114.8	107.4	90.0	101.9	125.2	93.0	115.1	107.4	89.9	101.9	104.8	82.4	97.7
06	WOOD, PLASTICS & COMPOSITES	127.4	90.4	106.2	113.7	95.8	103.5	110.3	85.6	96.2	115.2	88.2	99.8	113.7	85.6	97.6	94.6	66.8	78.7
07	THERMAL & MOISTURE PROTECTION	105.8	92.4	100.4	116.4	94.6	107.6	113.4	90.0	104.0	124.9	94.4	112.7	119.8	90.0	107.8	109.2	67.2	92.3
08	OPENINGS	93.0	78.4	89.5	93.0	86.4	91.4	93.0	80.8	90.1	94.1	88.5	92.7	93.0	80.8	90.1	87.5	63.7	81.7
0920	Plaster & Gypsum Board	151.3	89.9	108.3	139.4	95.3	108.6	134.3	84.8	99.7	194.7	87.9	120.0	137.2	84.8	100.5	164.9	65.6	95.5
0950, 0980	Ceilings & Acoustic Treatment	95.6	89.9	91.8	103.8	95.3	98.2	95.6	84.8	88.4	134.6	87.9	103.5	95.6	84.8	88.4	96.5	65.6	76.0
0960	Flooring	130.8	101.4	122.1	130.8	96.0	120.4	130.8	96.0	120.4	133.0	104.2	124.4	130.8	96.0	120.4	115.7	74.7	103.5
0970, 0990	Wall Finishes & Painting/Coating	112.2	94.2	101.4	112.1	108.9	110.2	112.3	84.8	95.8	112.2	101.5	105.8	112.1	84.8	95.8	112.2	53.9	77.3
09	FINISHES	115.6	93.3	103.1	117.3	98.0	106.5	115.0	87.6	99.6	131.8	93.4	110.3	115.2	87.6	99.7	113.5	67.3	87.6
COVERS	DIVS. 10 - 14, 25, 28, 41, 43, 44, 46	131.1	87.3	122.2	131.1	99.1	124.6	131.1	95.6	123.9	131.1	99.2	124.6	131.1	95.6	123.9	131.1	67.9	118.2
21, 22, 23	FIRE SUPPRESSION, PLUMBING & HVAC	99.7	94.4	97.5	103.1	100.8	102.2	103.4	100.8	102.3	100.8	85.2	94.4	102.9	97.3	100.6	102.9	73.9	91.0
26, 27, 3370	ELECTRICAL, COMMUNICATIONS & UTIL.	118.9	74.4	95.5	112.5	89.1	100.2	109.9	89.1	99.0	113.5	95.3	103.9	109.9	89.1	98.9	125.1	64.5	93.2
MF2010	WEIGHTED AVERAGE	112.9	89.7	102.7	119.7	95.6	109.1	114.7	92.2	104.8	120.8	93.5	108.8	114.8	91.4	104.5	113.3	72.9	95.5

DIVISION		CANADA																	
		MONTREAL, QUEBEC			MOOSE JAW, SASKATCHEWAN			NEW GLASGOW, NOVA SCOTIA			NEWCASTLE, NEW BRUNSWICK			NORTH BAY, ONTARIO			OSHAWA, ONTARIO		
		MAT.	INST.	TOTAL	MAT.	INST.	TOTAL	MAT.	INST.	TOTAL	MAT.	INST.	TOTAL	MAT.	INST.	TOTAL	MAT.	INST.	TOTAL
015433	CONTRACTOR EQUIPMENT		103.9	103.9		100.1	100.1		101.4	101.4		101.7	101.7		101.6	101.6		103.8	103.8
0241, 31 - 34	SITE & INFRASTRUCTURE, DEMOLITION	104.7	100.0	101.4	112.0	96.6	101.2	99.3	98.9	99.0	96.7	97.3	97.2	114.0	100.5	104.5	113.9	104.6	107.4
0310	Concrete Forming & Accessories	135.7	95.2	100.6	106.9	62.3	68.3	96.9	73.7	76.8	102.8	66.7	71.6	124.0	89.6	94.2	124.1	91.7	96.1
0320	Concrete Reinforcing	151.0	98.0	124.6	114.3	63.8	89.2	148.9	48.2	98.8	148.9	58.4	103.8	175.2	84.4	130.0	175.2	87.1	131.3
0330	Cast-in-Place Concrete	182.2	102.6	149.5	146.6	72.7	116.2	169.8	73.8	130.3	138.2	64.0	107.7	163.5	89.0	132.9	162.8	89.9	132.8
03	CONCRETE	155.8	98.1	127.3	124.6	66.8	96.0	148.3	69.7	109.5	135.7	64.8	100.7	150.1	88.6	119.7	149.7	90.4	120.4
04	MASONRY	165.7	96.4	123.4	156.9	66.0	101.4	158.5	73.7	106.8	157.8	67.7	102.8	162.8	92.2	119.7	162.4	96.7	122.3
05	METALS	131.1	93.7	119.3	103.9	75.1	94.9	105.8	75.4	96.2	104.8	73.0	94.8	106.6	90.7	101.6	106.7	92.9	102.4
06	WOOD, PLASTICS & COMPOSITES	127.4	95.4	109.1	96.1	60.7	75.9	87.0	73.0	79.0	94.6	66.8	78.7	115.9	90.0	101.1	116.2	90.6	101.6
07	THERMAL & MOISTURE PROTECTION	121.5	98.6	112.3	104.7	66.0	89.1	106.6	71.9	92.6	109.2	64.5	91.2	111.6	88.0	102.1	112.0	89.8	103.0
08	OPENINGS	93.0	83.9	90.8	88.2	58.3	80.9	84.3	66.6	80.0	87.5	58.0	80.3	92.0	86.0	90.5	92.0	90.3	91.6
0920	Plaster & Gypsum Board	172.6	94.9	118.2	131.8	59.5	81.2	158.4	72.0	98.0	164.9	65.6	95.5	154.1	89.7	109.0	156.2	90.3	110.1
0950, 0980	Ceilings & Acoustic Treatment	112.8	94.9	100.9	95.6	59.5	71.5	95.6	72.0	79.9	96.5	65.6	76.0	95.6	89.7	91.6	100.2	90.3	93.6
0960	Flooring	133.2	101.4	123.7	120.5	64.5	103.8	112.3	68.5	99.3	115.7	73.5	103.2	130.8	99.3	121.4	130.8	106.9	123.7
0970, 0990	Wall Finishes & Painting/Coating	112.2	108.1	109.7	112.2	68.6	86.1	112.2	65.9	84.4	112.2	53.0	76.7	112.2	93.7	101.1	112.2	109.0	110.3
09	FINISHES	123.4	98.5	109.4	111.5	63.1	84.3	111.2	72.1	89.3	113.5	67.0	87.4	116.9	92.2	103.1	118.1	96.3	105.9
COVERS	DIVS. 10 - 14, 25, 28, 41, 43, 44, 46	131.1	89.0	122.5	131.1	66.1	117.9	131.1	68.2	118.3	131.1	67.1	118.1	131.1	74.5	119.6	131.1	99.7	124.7
21, 22, 23	FIRE SUPPRESSION, PLUMBING & HVAC	100.7	95.6	98.6	103.4	78.9	93.4	102.9	87.6	96.6	102.9	72.6	90.5	102.9	101.8	102.4	99.7	104.5	101.7
26, 27, 3370	ELECTRICAL, COMMUNICATIONS & UTIL.	122.8	88.1	104.5	120.1	64.3	90.8	119.6	66.0	91.3	119.3	63.4	89.9	119.7	93.1	105.7	119.0	92.7	105.2
MF2010	WEIGHTED AVERAGE	121.8	94.9	110.0	112.0	71.3	94.1	114.0	76.6	97.6	113.0	70.2	94.1	116.6	93.5	106.4	115.8	96.8	107.4

DIVISION		CANADA																	
		OTTAWA, ONTARIO			OWEN SOUND, ONTARIO			PETERBOROUGH, ONTARIO			PORTAGE LA PRAIRIE, MANITOBA			PRINCE ALBERT, SASKATCHEWAN			PRINCE GEORGE, BRITISH COLUMBIA		
		MAT.	INST.	TOTAL	MAT.	INST.	TOTAL	MAT.	INST.	TOTAL	MAT.	INST.	TOTAL	MAT.	INST.	TOTAL	MAT.	INST.	TOTAL
015433	CONTRACTOR EQUIPMENT		103.8	103.8		101.6	101.6		101.6	101.6		103.8	103.8		100.1	100.1		105.5	105.5
0241, 31 - 34	SITE & INFRASTRUCTURE, DEMOLITION	113.6	105.2	107.7	116.0	101.0	105.5	114.0	100.9	104.8	112.9	100.1	103.9	106.8	96.7	99.8	121.9	104.1	109.4
0310	Concrete Forming & Accessories	130.7	92.7	97.8	124.1	88.0	92.9	124.0	90.6	95.1	123.4	75.3	81.7	106.9	62.1	68.1	112.4	89.4	92.5
0320	Concrete Reinforcing	156.9	94.6	125.9	180.3	86.5	133.5	175.2	84.9	130.2	162.0	55.8	109.1	119.5	63.8	91.7	116.8	80.6	98.8
0330	Cast-in-Place Concrete	168.5	100.8	140.7	167.6	85.0	133.6	163.5	90.7	133.6	151.1	78.7	121.3	132.9	72.6	108.1	147.1	105.3	129.9
03	CONCRETE	149.9	95.8	123.2	152.9	86.9	120.3	150.1	89.8	120.3	135.4	73.4	104.8	118.8	66.7	93.1	152.1	93.4	123.1
04	MASONRY	179.8	98.7	130.3	166.4	98.0	124.6	162.8	99.1	123.9	161.3	69.4	105.2	156.0	66.0	101.0	167.4	99.0	125.6
05	METALS	125.7	92.9	115.4	107.2	91.5	102.3	106.6	91.2	101.8	107.4	78.6	98.4	104.0	74.9	94.9	107.2	89.8	101.8
06	WOOD, PLASTICS & COMPOSITES	115.0	92.5	102.1	117.9	86.6	100.0	115.9	88.6	100.3	113.6	76.7	92.5	96.1	60.7	75.9	98.4	85.7	91.1
07	THERMAL & MOISTURE PROTECTION	127.2	94.4	114.0	111.7	88.8	102.5	111.6	93.7	104.4	105.2	74.4	92.8	104.6	64.8	88.6	114.4	89.8	104.5
08	OPENINGS	93.0	90.7	92.5	92.1	85.3	90.5	92.0	87.8	91.0	93.0	67.7	86.9	86.8	58.3	79.9	89.1	85.4	88.2
0920	Plaster & Gypsum Board	228.2	92.2	133.1	154.1	86.2	106.6	154.1	88.2	108.0	131.3	75.5	92.3	131.8	59.5	81.2	135.6	84.8	100.1
0950, 0980	Ceilings & Acoustic Treatment	138.2	92.2	107.6	95.6	86.2	89.3	95.6	88.2	90.7	95.6	75.5	82.2	95.6	59.5	71.5	95.6	84.8	88.4
0960	Flooring	132.8	99.3	122.8	130.8	100.7	121.9	130.8	99.3	121.4	130.8	72.8	113.5	120.5	64.5	103.8	126.2	79.0	112.1
0970, 0990	Wall Finishes & Painting/Coating	112.2	96.2	102.6	113.0	92.3	100.6	112.2	96.0	102.5	112.3	61.3	81.7	112.2	58.5	80.0	112.2	87.7	97.5
09	FINISHES	136.2	94.4	112.8	118.3	91.0	103.0	116.9	92.9	103.5	114.4	73.8	91.6	111.5	61.9	83.7	115.4	87.0	99.5
COVERS	DIVS. 10 - 14, 25, 28, 41, 43, 44, 46	131.1	97.4	124.2	139.2	75.0	126.1	131.1	76.1	119.9	131.1	69.8	118.6	131.1	66.1	117.9	131.1	98.6	124.5
21, 22, 23	FIRE SUPPRESSION, PLUMBING & HVAC	100.8	85.6	94.6	102.7	101.7	102.3	102.9	105.5	104.0	102.9	88.1	96.8	103.4	71.1	90.2	102.9	99.6	101.5
26, 27, 3370	ELECTRICAL, COMMUNICATIONS & UTIL.	109.7	96.1	102.6	120.9	91.8	105.6	117.4	92.7	104.4	119.8	62.6	89.7	120.1	64.3	90.8	117.6	85.4	100.7
MF2010	WEIGHTED AVERAGE	121.1	94.1	109.2	118.2	93.6	107.4	116.4	95.5	107.2	114.7	77.2	98.2	111.1	69.5	92.7	116.8	93.6	106.5

CANADA

	DIVISION	QUEBEC, QUEBEC			RED DEER, ALBERTA			REGINA, SASKATCHEWAN			RIMOUSKI, QUEBEC			ROUYN-NORANDA, QUEBEC			SAINT HYACINTHE, QUEBEC		
		MAT.	INST.	TOTAL	MAT.	INST.	TOTAL	MAT.	INST.	TOTAL	MAT.	INST.	TOTAL	MAT.	INST.	TOTAL	MAT.	INST.	TOTAL
015433	CONTRACTOR EQUIPMENT		104.2	104.2		103.7	103.7		100.1	100.1		102.3	102.3		102.3	102.3		102.3	102.3
0241, 31 - 34	SITE & INFRASTRUCTURE, DEMOLITION	102.9	100.1	100.9	111.1	102.2	104.8	116.9	98.6	104.0	95.4	99.8	98.5	94.9	100.5	98.8	95.4	100.5	99.0
0310	Concrete Forming & Accessories	136.7	95.4	101.0	138.8	86.2	93.3	110.7	92.4	94.9	129.1	94.9	99.5	129.1	90.4	95.6	129.1	90.4	95.6
0320	Concrete Reinforcing	144.4	98.1	121.3	162.0	75.3	118.8	139.4	84.8	112.2	112.3	98.0	105.2	157.5	81.3	119.6	157.5	81.3	119.6
0330	Cast-in-Place Concrete	149.3	103.0	130.3	151.1	103.2	131.4	170.7	96.9	140.4	137.0	101.3	122.3	130.2	100.0	117.8	134.5	100.0	120.3
03	CONCRETE	139.1	98.4	119.0	143.0	90.4	117.0	140.5	92.5	116.8	127.4	97.5	112.6	131.6	92.2	112.1	133.6	92.2	113.1
04	MASONRY	173.6	96.4	126.5	158.3	84.9	113.5	167.2	96.7	124.2	158.4	96.3	120.5	158.6	89.2	116.3	158.8	89.2	116.4
05	METALS	130.0	94.0	118.7	107.4	89.9	101.9	122.3	86.6	111.1	104.6	93.3	101.0	105.0	88.9	100.0	105.0	88.9	100.0
06	WOOD, PLASTICS & COMPOSITES	128.0	95.5	109.4	113.7	85.6	97.6	96.4	93.3	94.6	127.2	95.2	108.9	127.2	90.4	106.2	127.2	90.4	106.2
07	THERMAL & MOISTURE PROTECTION	120.5	98.8	111.8	129.4	90.0	113.6	114.2	87.3	103.4	105.3	98.0	102.4	105.3	92.4	100.1	105.6	92.4	100.3
08	OPENINGS	93.0	91.8	92.7	93.0	80.8	90.1	88.2	81.7	86.6	92.6	82.0	90.0	93.0	78.4	89.5	93.0	78.4	89.5
0920	Plaster & Gypsum Board	193.4	94.9	124.5	137.2	84.8	100.5	170.8	93.0	116.4	188.9	94.9	123.1	147.9	89.9	107.3	150.6	89.9	108.1
0950, 0980	Ceilings & Acoustic Treatment	117.9	94.9	102.6	95.6	84.8	88.4	125.7	93.0	103.9	94.7	94.9	94.8	94.7	89.9	91.5	94.7	89.9	91.5
0960	Flooring	132.8	101.4	123.5	133.2	96.0	122.1	121.5	66.7	105.2	130.8	101.4	122.1	130.8	101.4	122.1	130.8	101.4	122.1
0970, 0990	Wall Finishes & Painting/Coating	112.7	108.1	109.9	112.1	84.8	95.8	112.2	92.9	100.6	112.2	108.1	109.7	112.2	94.2	101.4	112.2	94.2	101.4
09	FINISHES	127.4	98.5	111.2	115.8	87.6	100.0	123.9	88.4	104.0	120.2	98.3	107.9	115.0	93.3	102.8	115.3	93.3	103.0
COVERS	DIVS. 10 - 14, 25, 28, 41, 43, 44, 46	131.1	89.1	122.6	131.1	95.6	123.9	131.1	73.8	119.4	131.1	88.5	122.4	131.1	87.3	122.2	131.1	87.3	122.2
21, 22, 23	FIRE SUPPRESSION, PLUMBING & HVAC	100.5	95.6	98.5	102.9	97.3	100.6	99.9	86.9	94.6	102.9	95.5	99.9	102.9	94.4	99.4	98.6	94.4	96.9
26, 27, 3370	ELECTRICAL, COMMUNICATIONS & UTIL.	113.7	88.1	100.2	109.9	89.1	98.9	118.0	93.1	104.9	117.7	88.1	102.1	117.7	74.4	94.9	118.5	74.4	95.3
MF2010	WEIGHTED AVERAGE	119.5	95.3	108.9	115.3	91.4	104.8	117.6	90.1	105.5	113.1	94.6	104.9	113.2	89.7	102.9	112.6	89.7	102.5

CANADA

	DIVISION	SAINT JOHN, NEW BRUNSWICK			SARNIA, ONTARIO			SASKATOON, SASKATCHEWAN			SAULT STE MARIE, ONTARIO			SHERBROOKE, QUEBEC			SOREL, QUEBEC		
		MAT.	INST.	TOTAL	MAT.	INST.	TOTAL	MAT.	INST.	TOTAL	MAT.	INST.	TOTAL	MAT.	INST.	TOTAL	MAT.	INST.	TOTAL
015433	CONTRACTOR EQUIPMENT		101.7	101.7		101.6	101.6		100.1	100.1		101.6	101.6		102.3	102.3		102.3	102.3
0241, 31 - 34	SITE & INFRASTRUCTURE, DEMOLITION	96.8	98.6	98.1	112.6	101.0	104.4	107.8	98.6	101.3	102.7	100.5	101.2	95.4	100.5	99.0	95.5	100.5	99.0
0310	Concrete Forming & Accessories	121.3	64.4	72.0	122.8	98.0	101.4	107.0	92.4	94.4	112.3	88.7	91.9	129.1	90.4	95.6	129.1	90.5	95.7
0320	Concrete Reinforcing	148.9	62.9	106.1	124.3	86.3	105.4	119.5	84.8	102.2	112.5	84.7	98.6	157.5	81.3	119.6	148.9	81.3	115.3
0330	Cast-in-Place Concrete	136.2	82.3	114.1	150.9	104.5	131.8	142.0	96.9	123.5	135.6	89.2	116.5	134.5	100.0	120.3	135.6	100.1	121.0
03	CONCRETE	136.2	71.3	104.1	135.6	98.0	117.0	123.2	92.5	108.0	119.4	88.3	104.0	133.6	92.2	113.1	132.7	92.2	112.7
04	MASONRY	175.9	73.9	113.7	173.5	101.9	129.8	163.8	96.7	122.8	159.3	96.2	120.8	158.9	89.2	116.4	159.0	89.2	116.4
05	METALS	104.7	83.5	98.0	106.6	91.6	101.9	105.0	86.6	99.2	105.8	91.7	101.4	105.0	88.9	100.0	105.0	89.0	100.0
06	WOOD, PLASTICS & COMPOSITES	116.6	61.3	85.0	115.1	97.2	104.9	94.8	93.3	93.9	102.9	88.3	94.6	127.2	90.4	106.2	127.2	90.4	106.2
07	THERMAL & MOISTURE PROTECTION	109.4	70.6	93.8	111.6	97.3	105.9	105.3	87.3	98.1	110.5	90.5	102.4	105.3	92.4	100.1	105.3	92.4	100.1
08	OPENINGS	87.4	59.4	80.6	94.1	91.8	93.5	87.3	81.7	85.9	85.1	86.7	85.5	93.0	78.4	89.5	93.0	83.3	90.7
0920	Plaster & Gypsum Board	182.4	60.0	96.8	181.2	97.1	122.4	144.0	93.0	108.4	143.1	87.9	104.5	150.6	89.9	108.1	188.9	89.9	119.6
0950, 0980	Ceilings & Acoustic Treatment	102.9	60.0	74.3	100.2	97.1	98.2	117.5	93.0	101.2	95.6	87.9	90.5	94.7	89.9	91.5	94.7	89.9	91.5
0960	Flooring	126.7	74.7	111.2	130.8	108.6	124.2	120.5	66.7	104.5	123.5	102.7	117.3	130.8	101.4	122.1	130.8	101.4	122.1
0970, 0990	Wall Finishes & Painting/Coating	112.2	80.6	93.3	112.2	108.6	110.0	112.2	92.9	100.6	112.2	100.6	105.2	112.2	94.2	101.4	112.2	94.2	101.4
09	FINISHES	120.4	67.3	90.6	121.3	101.6	110.3	117.8	88.4	101.3	112.5	92.6	101.3	115.3	93.3	103.0	120.2	93.3	105.1
COVERS	DIVS. 10 - 14, 25, 28, 41, 43, 44, 46	131.1	67.0	118.0	131.1	77.5	120.2	131.1	74.6	119.6	131.1	98.6	124.5	131.1	87.3	122.2	131.1	87.3	122.2
21, 22, 23	FIRE SUPPRESSION, PLUMBING & HVAC	102.9	83.6	95.0	102.9	111.8	106.5	99.8	86.9	94.5	102.9	98.3	101.0	103.4	94.4	99.7	102.9	94.4	99.4
26, 27, 3370	ELECTRICAL, COMMUNICATIONS & UTIL.	128.6	91.9	109.3	120.5	95.0	107.1	122.0	93.1	106.8	119.2	93.1	105.4	117.7	74.4	94.9	117.7	74.4	94.9
MF2010	WEIGHTED AVERAGE	115.5	79.2	99.5	116.2	100.1	109.1	112.0	90.1	102.4	111.4	94.1	103.8	113.6	89.7	103.1	113.8	89.9	103.3

CANADA

	DIVISION	ST CATHARINES, ONTARIO			ST JEROME, QUEBEC			ST JOHNS, NEWFOUNDLAND			SUDBURY, ONTARIO			SUMMERSIDE, PRINCE EDWARD ISLAND			SYDNEY, NOVA SCOTIA		
		MAT.	INST.	TOTAL	MAT.	INST.	TOTAL	MAT.	INST.	TOTAL	MAT.	INST.	TOTAL	MAT.	INST.	TOTAL	MAT.	INST.	TOTAL
015433	CONTRACTOR EQUIPMENT		101.6	101.6		102.3	102.3		102.9	102.9		101.6	101.6		101.4	101.4		101.4	101.4
0241, 31 - 34	SITE & INFRASTRUCTURE, DEMOLITION	102.7	101.9	102.2	94.9	100.5	98.8	115.1	99.9	104.5	102.7	101.5	101.9	106.9	96.4	99.6	94.9	98.9	97.7
0310	Concrete Forming & Accessories	116.2	95.7	98.5	129.1	90.4	95.6	107.9	84.2	87.4	112.3	89.1	92.2	97.2	60.0	65.0	96.9	73.7	76.8
0320	Concrete Reinforcing	111.6	94.6	103.1	157.5	81.3	119.6	167.5	75.6	121.7	112.5	94.5	103.5	146.7	48.3	97.7	148.9	48.2	98.8
0330	Cast-in-Place Concrete	134.5	101.8	121.0	130.2	100.0	117.8	173.5	88.6	138.6	135.6	97.9	120.1	161.2	62.5	120.6	131.0	73.8	107.4
03	CONCRETE	118.9	97.5	108.4	131.6	92.2	112.1	165.6	84.4	125.5	119.4	93.2	106.4	152.4	59.4	106.5	129.7	69.7	100.1
04	MASONRY	158.9	101.5	123.8	158.6	89.2	116.3	163.9	91.9	120.0	159.0	96.7	120.9	157.6	64.2	100.6	155.7	73.7	105.7
05	METALS	105.8	93.4	101.9	105.0	88.9	100.0	124.2	84.1	111.6	105.8	92.9	101.7	105.8	68.7	94.2	105.8	75.4	96.2
06	WOOD, PLASTICS & COMPOSITES	106.8	95.6	100.4	127.2	90.4	106.2	95.9	82.6	88.3	102.9	88.3	94.6	87.2	59.3	71.3	87.0	73.0	79.0
07	THERMAL & MOISTURE PROTECTION	111.0	97.7	105.7	105.3	92.4	100.1	118.3	86.8	105.7	110.5	91.9	103.0	105.8	62.9	88.6	106.6	71.9	92.6
08	OPENINGS	84.2	92.2	86.2	93.0	78.4	89.5	98.9	74.7	93.0	85.1	86.7	85.5	95.2	51.3	84.5	84.3	66.6	80.0
0920	Plaster & Gypsum Board	137.8	95.5	108.2	147.9	89.9	107.3	175.3	81.9	110.0	143.1	87.9	104.5	158.8	58.0	88.3	158.4	72.0	98.0
0950, 0980	Ceilings & Acoustic Treatment	100.2	95.5	97.0	94.7	89.9	91.5	112.0	81.9	91.9	95.6	87.9	90.5	95.6	58.0	70.6	95.6	72.0	79.9
0960	Flooring	125.7	100.7	118.3	130.8	101.4	122.1	116.9	60.4	100.1	123.5	102.7	117.3	112.4	64.1	98.0	112.3	68.5	99.3
0970, 0990	Wall Finishes & Painting/Coating	112.2	104.8	107.8	112.2	94.2	101.4	112.2	77.5	91.4	112.2	95.6	102.3	112.2	44.1	71.4	112.2	65.9	84.4
09	FINISHES	113.4	98.0	104.8	115.0	93.3	102.8	119.7	79.5	97.1	112.5	92.1	101.1	112.3	59.4	82.6	111.2	72.1	89.3
COVERS	DIVS. 10 - 14, 25, 28, 41, 43, 44, 46	131.1	76.5	120.0	131.1	87.3	122.2	131.1	73.6	119.4	131.1	98.7	124.5	131.1	66.4	117.9	131.1	68.2	118.3
21, 22, 23	FIRE SUPPRESSION, PLUMBING & HVAC	99.7	86.8	94.4	102.9	94.4	99.4	100.1	76.2	90.3	102.9	87.8	96.7	102.9	66.2	87.9	102.9	87.6	96.6
26, 27, 3370	ELECTRICAL, COMMUNICATIONS & UTIL.	119.4	96.2	107.2	118.4	74.4	95.2	120.8	73.7	96.0	116.4	96.5	105.9	118.4	53.6	84.3	119.6	66.0	91.3
MF2010	WEIGHTED AVERAGE	110.6	94.6	103.6	113.3	89.7	102.9	121.7	81.9	104.2	111.1	93.2	103.2	115.7	64.5	93.2	111.7	76.6	96.3

	DIVISION	CANADA																	
		THUNDER BAY, ONTARIO			TIMMINS, ONTARIO			TORONTO, ONTARIO			TROIS RIVIERES, QUEBEC			TRURO, NOVA SCOTIA			VANCOUVER, BRITISH COLUMBIA		
		MAT.	INST.	TOTAL	MAT.	INST.	TOTAL	MAT.	INST.	TOTAL	MAT.	INST.	TOTAL	MAT.	INST.	TOTAL	MAT.	INST.	TOTAL
015433	**CONTRACTOR EQUIPMENT**		101.6	101.6		101.6	101.6		104.0	104.0		102.3	102.3		101.4	101.4		112.0	112.0
0241, 31 - 34	**SITE & INFRASTRUCTURE, DEMOLITION**	107.7	101.8	103.5	114.0	100.5	104.5	138.5	106.0	115.7	95.5	100.5	99.0	99.3	98.9	99.0	121.7	108.4	112.4
0310	Concrete Forming & Accessories	124.1	92.6	96.9	124.0	89.6	94.2	131.5	101.4	105.5	129.1	90.5	95.7	96.9	73.7	76.8	126.2	93.7	98.1
0320	Concrete Reinforcing	99.7	93.4	96.6	175.2	84.4	130.0	161.9	94.8	128.5	148.9	81.3	115.3	148.9	48.2	98.8	151.3	83.1	117.3
0330	Cast-in-Place Concrete	148.0	100.2	128.4	163.5	89.0	132.9	154.4	110.1	136.2	135.6	100.1	121.0	169.8	73.8	130.3	164.1	101.3	138.3
03	**CONCRETE**	127.4	95.4	111.6	150.1	88.6	119.7	144.0	102.9	123.7	132.7	92.2	112.7	148.3	69.7	109.5	152.5	94.6	123.9
04	**MASONRY**	159.6	100.7	123.7	162.8	92.2	119.7	192.7	105.9	139.7	159.0	89.2	116.4	158.5	73.7	106.8	177.1	95.2	127.1
05	**METALS**	105.6	92.3	101.4	106.6	90.7	101.6	125.6	94.0	115.7	105.0	89.0	100.0	105.8	75.5	96.3	152.8	92.5	133.9
06	**WOOD, PLASTICS & COMPOSITES**	116.2	91.4	102.0	115.9	90.0	101.1	116.2	100.8	107.4	127.2	90.4	106.2	87.0	73.0	79.0	114.9	93.6	102.8
07	**THERMAL & MOISTURE PROTECTION**	111.2	94.8	104.6	111.6	88.0	102.1	122.8	101.4	114.2	105.3	92.4	100.1	106.6	71.9	92.6	132.4	90.4	115.5
08	**OPENINGS**	83.3	89.3	84.7	92.0	86.0	90.5	92.0	98.5	93.5	93.0	83.3	90.7	84.3	66.6	80.0	94.8	89.7	93.5
0920	Plaster & Gypsum Board	174.1	91.1	116.1	154.1	89.7	109.0	178.7	100.8	124.2	188.9	89.9	119.6	158.4	72.0	98.0	162.9	92.9	113.9
0950, 0980	Ceilings & Acoustic Treatment	95.6	91.1	92.6	95.6	89.7	91.6	130.1	100.8	110.6	94.7	89.9	91.5	95.6	72.0	79.9	156.0	92.9	113.9
0960	Flooring	130.8	107.9	124.0	130.8	99.3	121.4	135.6	110.5	128.1	130.8	101.4	122.1	112.3	68.5	99.3	133.2	99.4	123.1
0970, 0990	Wall Finishes & Painting/Coating	112.2	94.1	101.4	112.2	93.7	101.1	114.3	109.0	111.1	112.2	94.2	101.4	112.2	65.9	84.4	112.1	99.9	104.8
09	**FINISHES**	119.0	95.7	105.9	116.9	92.2	103.1	131.7	104.2	116.3	120.2	93.3	105.1	111.2	72.1	89.3	133.4	95.6	112.2
COVERS	**DIVS. 10 - 14, 25, 28, 41, 43, 44, 46**	131.1	76.2	119.9	131.1	74.5	119.6	131.1	102.3	125.3	131.1	87.3	122.2	131.1	68.2	118.3	131.1	98.5	124.5
21, 22, 23	**FIRE SUPPRESSION, PLUMBING & HVAC**	99.7	86.7	94.4	102.9	101.8	102.4	100.6	96.0	98.8	102.9	94.4	99.4	102.9	87.6	96.6	100.6	84.9	94.2
26, 27, 3370	**ELECTRICAL, COMMUNICATIONS & UTIL.**	117.7	94.6	105.6	119.2	93.1	105.4	113.7	98.3	105.6	119.9	74.4	96.0	117.4	66.0	90.3	119.2	82.8	100.0
MF2010	**WEIGHTED AVERAGE**	111.9	93.3	103.7	116.6	93.5	106.4	121.6	100.5	112.3	114.0	89.9	103.4	113.8	76.6	97.4	126.8	91.9	111.4

	DIVISION	CANADA																	
		VICTORIA, BRITISH COLUMBIA			WHITEHORSE, YUKON			WINDSOR, ONTARIO			WINNIPEG, MANITOBA			YARMOUTH, NOVA SCOTIA			YELLOWKNIFE, NWT		
		MAT.	INST.	TOTAL	MAT.	INST.	TOTAL	MAT.	INST.	TOTAL	MAT.	INST.	TOTAL	MAT.	INST.	TOTAL	MAT.	INST.	TOTAL
015433	**CONTRACTOR EQUIPMENT**		108.7	108.7		101.6	101.6		101.6	101.6		105.8	105.8		101.4	101.4		101.5	101.5
0241, 31 - 34	**SITE & INFRASTRUCTURE, DEMOLITION**	121.8	107.9	112.1	110.0	97.4	101.2	97.5	101.8	100.5	122.7	101.5	107.8	99.1	98.9	98.9	128.4	101.8	109.8
0310	Concrete Forming & Accessories	112.4	93.1	95.7	110.5	61.3	67.9	124.1	91.8	96.1	133.2	69.7	78.2	96.9	73.7	76.8	110.4	83.9	87.5
0320	Concrete Reinforcing	116.8	83.0	100.0	133.7	61.9	97.9	109.4	94.5	102.0	146.5	58.4	102.6	148.9	48.2	98.8	132.3	65.8	99.2
0330	Cast-in-Place Concrete	147.1	100.6	128.0	162.2	71.8	125.0	137.7	102.1	123.1	192.3	75.3	144.1	168.1	73.8	129.3	244.8	91.3	181.7
03	**CONCRETE**	152.1	93.8	123.3	135.4	65.8	101.0	120.7	95.8	108.4	161.9	70.1	116.5	147.5	69.7	109.0	174.8	83.3	129.6
04	**MASONRY**	166.5	95.2	123.0	170.0	64.3	105.5	159.1	99.5	122.7	187.4	71.0	116.3	158.4	73.7	106.7	176.8	78.0	116.5
05	**METALS**	107.2	88.4	101.3	121.9	74.8	107.1	105.7	92.8	101.6	152.3	75.3	128.1	105.8	75.4	96.2	117.6	80.7	106.0
06	**WOOD, PLASTICS & COMPOSITES**	98.4	93.5	95.6	96.1	59.8	75.4	116.2	91.0	101.8	114.0	70.4	89.2	87.0	73.0	79.0	96.1	85.5	90.0
07	**THERMAL & MOISTURE PROTECTION**	114.4	90.2	104.7	113.4	63.8	93.5	111.0	93.9	104.2	118.3	72.2	99.7	106.6	71.9	92.6	113.4	81.8	100.7
08	**OPENINGS**	89.1	85.8	88.3	86.8	57.0	79.6	83.0	89.6	84.6	93.0	62.8	85.7	84.3	66.6	80.0	86.8	72.0	83.2
0920	Plaster & Gypsum Board	135.6	92.9	105.7	147.5	58.4	85.2	157.7	90.7	110.9	159.3	69.0	96.2	158.4	72.0	98.0	184.1	84.9	114.7
0950, 0980	Ceilings & Acoustic Treatment	95.6	92.9	93.8	102.9	58.4	73.3	95.6	90.7	92.4	140.2	69.0	92.8	95.6	72.0	79.9	127.5	84.9	99.1
0960	Flooring	126.2	79.0	112.1	121.5	62.5	103.9	130.8	105.0	123.1	132.4	75.3	115.4	112.3	68.5	99.3	121.5	95.9	113.9
0970, 0990	Wall Finishes & Painting/Coating	112.2	99.9	104.8	112.2	57.5	79.4	112.2	97.5	103.4	112.3	56.7	79.0	112.2	65.9	84.4	112.2	83.9	95.2
09	**FINISHES**	115.4	92.1	102.3	115.5	60.8	84.9	116.5	95.0	104.4	128.9	69.7	95.7	111.2	72.1	89.3	125.5	85.5	103.1
COVERS	**DIVS. 10 - 14, 25, 28, 41, 43, 44, 46**	131.1	74.0	119.5	131.1	65.4	117.7	131.1	75.6	119.8	131.1	68.9	118.4	131.1	68.2	118.3	131.1	69.8	118.6
21, 22, 23	**FIRE SUPPRESSION, PLUMBING & HVAC**	102.9	84.8	95.5	103.1	77.6	92.6	99.7	86.3	94.2	100.5	67.8	87.1	102.9	87.6	96.6	103.3	99.1	101.6
26, 27, 3370	**ELECTRICAL, COMMUNICATIONS & UTIL.**	118.6	82.8	99.7	134.1	63.8	97.1	123.0	96.2	108.9	119.7	68.1	92.6	119.6	66.0	91.3	135.2	88.8	110.8
MF2010	**WEIGHTED AVERAGE**	116.8	90.0	105.0	118.4	70.3	97.2	111.2	93.4	103.3	127.3	72.3	103.1	113.9	76.6	97.5	123.9	87.6	107.9

Location Factors

Costs shown in RSMeans cost data publications are based on national averages for materials and installation. To adjust these costs to a specific location, simply multiply the base cost by the factor and divide by 100 for that city. The data is arranged alphabetically by state and postal zip code numbers. For a city not listed, use the factor for a nearby city with similar economic characteristics.

STATE/ZIP	CITY	MAT.	INST.	TOTAL
ALABAMA				
350-352	Birmingham	97.4	73.8	87.0
354	Tuscaloosa	96.5	58.1	79.6
355	Jasper	96.8	56.9	79.2
356	Decatur	96.5	59.4	80.1
357-358	Huntsville	96.6	67.4	83.7
359	Gadsden	96.5	57.1	79.1
360-361	Montgomery	97.5	56.1	79.2
362	Anniston	95.9	63.8	81.7
363	Dothan	96.4	51.4	76.6
364	Evergreen	95.9	52.8	77.0
365-366	Mobile	97.5	66.8	84.0
367	Selma	96.1	51.3	76.4
368	Phenix City	96.8	55.4	78.5
369	Butler	96.3	51.4	76.5
ALASKA				
995-996	Anchorage	123.8	114.6	119.8
997	Fairbanks	122.0	115.4	119.1
998	Juneau	119.7	114.3	117.3
999	Ketchikan	132.8	114.3	124.7
ARIZONA				
850,853	Phoenix	98.7	75.4	88.4
851,852	Mesa/Tempe	98.1	71.3	86.3
855	Globe	98.2	61.4	82.0
856-857	Tucson	96.9	70.9	85.5
859	Show Low	98.3	61.8	82.2
860	Flagstaff	100.7	72.7	88.4
863	Prescott	98.5	61.1	82.0
864	Kingman	96.8	72.3	86.0
865	Chambers	96.8	62.0	81.5
ARKANSAS				
716	Pine Bluff	97.1	63.1	82.1
717	Camden	94.4	49.5	74.6
718	Texarkana	96.0	50.6	76.0
719	Hot Springs	93.7	52.2	75.4
720-722	Little Rock	97.5	64.2	82.8
723	West Memphis	96.0	62.0	81.0
724	Jonesboro	97.0	60.0	80.7
725	Batesville	94.6	55.3	77.3
726	Harrison	95.8	51.6	76.4
727	Fayetteville	93.3	53.4	75.7
728	Russellville	94.5	54.6	76.9
729	Fort Smith	97.4	61.8	81.7
CALIFORNIA				
900-902	Los Angeles	99.8	116.6	107.2
903-905	Inglewood	95.0	112.1	102.5
906-908	Long Beach	96.5	112.2	103.4
910-912	Pasadena	95.4	112.1	102.8
913-916	Van Nuys	98.3	112.1	104.4
917-918	Alhambra	97.3	112.1	103.8
919-921	San Diego	99.6	108.0	103.3
922	Palm Springs	96.3	109.5	102.2
923-924	San Bernardino	94.3	109.0	100.8
925	Riverside	98.8	113.7	105.4
926-927	Santa Ana	96.1	109.5	102.0
928	Anaheim	98.8	113.7	105.4
930	Oxnard	99.9	112.9	105.6
931	Santa Barbara	99.2	113.9	105.7
932-933	Bakersfield	100.1	109.8	104.4
934	San Luis Obispo	99.4	107.9	103.2
935	Mojave	96.8	105.4	100.6
936-938	Fresno	100.3	115.0	106.8
939	Salinas	100.3	121.8	109.8
940-941	San Francisco	106.8	142.4	122.5
942,956-958	Sacramento	101.7	119.2	109.5
943	Palo Alto	99.8	128.5	112.5
944	San Mateo	102.1	130.3	114.5
945	Vallejo	101.0	128.5	113.1
946	Oakland	104.6	133.3	117.3
947	Berkeley	103.8	132.5	116.5
948	Richmond	102.9	128.7	114.3
949	San Rafael	104.4	128.7	115.1
950	Santa Cruz	104.4	121.9	112.1

STATE/ZIP	CITY	MAT.	INST.	TOTAL
CALIFORNIA (CONT'D)				
951	San Jose	102.5	135.3	116.9
952	Stockton	100.5	116.6	107.6
953	Modesto	100.4	116.0	107.3
954	Santa Rosa	100.0	134.2	115.1
955	Eureka	101.3	112.7	106.3
959	Marysville	100.5	113.1	106.1
960	Redding	102.4	113.1	107.1
961	Susanville	101.1	113.1	106.4
COLORADO				
800-802	Denver	101.7	83.4	93.6
803	Boulder	98.4	83.3	91.7
804	Golden	100.4	80.9	91.8
805	Fort Collins	102.3	80.8	92.8
806	Greeley	99.8	77.3	89.9
807	Fort Morgan	99.0	80.9	91.0
808-809	Colorado Springs	101.4	80.6	92.2
810	Pueblo	100.7	79.8	91.5
811	Alamosa	101.8	75.7	90.3
812	Salida	101.5	77.4	90.9
813	Durango	102.1	78.3	91.6
814	Montrose	101.0	78.0	90.8
815	Grand Junction	104.4	78.2	92.9
816	Glenwood Springs	101.8	80.9	92.6
CONNECTICUT				
060	New Britain	100.6	119.9	109.1
061	Hartford	102.0	120.1	109.9
062	Willimantic	101.2	120.2	109.6
063	New London	97.2	120.0	107.2
064	Meriden	99.2	120.0	108.4
065	New Haven	102.1	120.0	110.0
066	Bridgeport	101.7	119.7	109.6
067	Waterbury	101.3	119.6	109.3
068	Norwalk	101.2	127.1	112.6
069	Stamford	101.4	127.1	112.7
D.C.				
200-205	Washington	101.1	92.4	97.3
DELAWARE				
197	Newark	98.8	108.8	103.2
198	Wilmington	98.6	108.9	103.1
199	Dover	99.3	108.8	103.5
FLORIDA				
320,322	Jacksonville	98.4	68.6	85.2
321	Daytona Beach	98.6	74.3	87.9
323	Tallahassee	99.1	59.1	81.5
324	Panama City	99.8	60.2	82.4
325	Pensacola	102.2	63.4	85.1
326,344	Gainesville	100.0	68.9	86.3
327-328,347	Orlando	101.8	72.3	88.8
329	Melbourne	101.2	77.3	90.7
330-332,340	Miami	99.8	76.0	89.3
333	Fort Lauderdale	97.9	75.2	87.9
334,349	West Palm Beach	96.6	73.5	86.4
335-336,346	Tampa	99.5	81.1	91.4
337	St. Petersburg	102.2	65.6	86.1
338	Lakeland	98.7	81.0	90.9
339,341	Fort Myers	97.9	75.1	87.8
342	Sarasota	99.7	76.9	89.7
GEORGIA				
300-303,399	Atlanta	97.5	75.9	88.0
304	Statesboro	96.9	53.3	77.7
305	Gainesville	95.6	63.1	81.3
306	Athens	95.0	63.3	81.0
307	Dalton	96.9	59.3	80.3
308-309	Augusta	96.3	64.2	82.1
310-312	Macon	96.2	63.9	82.0
313-314	Savannah	98.2	61.1	81.9
315	Waycross	97.1	60.6	81.0
316	Valdosta	97.4	60.4	81.1
317,398	Albany	97.4	61.1	81.4
318-319	Columbus	97.4	63.9	82.6

Location Factors

STATE/ZIP	CITY	MAT.	INST.	TOTAL
HAWAII				
967	Hilo	109.5	122.3	115.2
968	Honolulu	113.6	122.3	117.4
STATES & POSS.				
969	Guam	130.4	60.0	99.4
IDAHO				
832	Pocatello	100.7	72.5	88.3
833	Twin Falls	101.8	52.7	80.1
834	Idaho Falls	99.4	58.3	81.3
835	Lewiston	108.1	79.5	95.5
836-837	Boise	99.6	73.3	88.0
838	Coeur d'Alene	107.4	77.1	94.0
ILLINOIS				
600-603	North Suburban	98.8	130.3	112.7
604	Joliet	98.8	134.5	114.5
605	South Suburban	98.8	130.3	112.7
606-608	Chicago	99.3	138.3	116.5
609	Kankakee	94.8	125.4	108.3
610-611	Rockford	98.6	122.3	109.0
612	Rock Island	96.0	98.3	97.0
613	La Salle	97.1	119.0	106.8
614	Galesburg	97.0	103.2	99.7
615-616	Peoria	99.7	106.7	102.8
617	Bloomington	96.4	107.8	101.4
618-619	Champaign	99.8	106.8	102.9
620-622	East St. Louis	95.0	106.6	100.1
623	Quincy	96.1	98.6	97.2
624	Effingham	95.4	102.3	98.4
625	Decatur	97.6	105.8	101.2
626-627	Springfield	98.3	106.5	101.9
628	Centralia	93.4	105.4	98.7
629	Carbondale	93.1	100.5	96.4
INDIANA				
460	Anderson	95.7	82.3	89.8
461-462	Indianapolis	98.9	84.7	92.6
463-464	Gary	97.2	107.0	101.5
465-466	South Bend	95.5	81.8	89.5
467-468	Fort Wayne	96.3	77.7	88.1
469	Kokomo	93.3	81.1	87.9
470	Lawrenceburg	93.0	78.4	86.6
471	New Albany	94.1	75.4	85.9
472	Columbus	96.4	80.1	89.3
473	Muncie	97.5	81.4	90.4
474	Bloomington	98.8	81.6	91.2
475	Washington	94.7	82.4	89.3
476-477	Evansville	96.5	85.2	91.5
478	Terre Haute	97.2	85.7	92.1
479	Lafayette	96.1	82.0	89.9
IOWA				
500-503,509	Des Moines	98.9	83.5	92.1
504	Mason City	97.3	66.9	83.9
505	Fort Dodge	97.5	58.9	80.5
506-507	Waterloo	99.4	71.2	86.9
508	Creston	97.6	77.7	88.8
510-511	Sioux City	100.0	69.7	86.7
512	Sibley	98.5	55.8	79.7
513	Spencer	100.1	55.8	80.5
514	Carroll	97.0	71.4	85.7
515	Council Bluffs	101.2	79.1	91.5
516	Shenandoah	97.8	71.3	86.1
520	Dubuque	99.5	78.6	90.3
521	Decorah	98.1	64.8	83.4
522-524	Cedar Rapids	100.3	82.8	92.6
525	Ottumwa	98.0	69.7	85.5
526	Burlington	97.2	76.2	88.0
527-528	Davenport	99.3	89.2	94.8
KANSAS				
660-662	Kansas City	99.7	93.9	97.2
664-666	Topeka	99.4	63.8	83.7
667	Fort Scott	97.8	72.9	86.8
668	Emporia	97.9	68.7	85.0
669	Belleville	99.5	64.6	84.1
670-672	Wichita	98.8	63.2	83.1
673	Independence	99.2	68.4	85.6
674	Salina	100.1	65.9	85.0
675	Hutchinson	95.0	64.5	81.5
676	Hays	99.0	64.8	83.9
677	Colby	99.5	64.8	84.2

STATE/ZIP	CITY	MAT.	INST.	TOTAL
KANSAS (CONT'D)				
678	Dodge City	101.2	65.8	85.6
679	Liberal	98.8	64.8	83.8
KENTUCKY				
400-402	Louisville	98.6	82.2	91.3
403-405	Lexington	97.0	81.0	90.0
406	Frankfort	98.0	82.3	91.1
407-409	Corbin	94.2	71.4	84.1
410	Covington	95.0	94.1	94.6
411-412	Ashland	93.7	98.0	95.6
413-414	Campton	94.9	80.0	88.3
415-416	Pikeville	96.0	90.2	93.4
417-418	Hazard	94.4	80.4	88.2
420	Paducah	93.0	84.5	89.2
421-422	Bowling Green	95.6	83.3	90.1
423	Owensboro	95.6	86.7	91.7
424	Henderson	92.8	86.6	90.1
425-426	Somerset	92.5	79.3	86.7
427	Elizabethtown	91.9	82.6	87.8
LOUISIANA				
700-701	New Orleans	102.4	70.5	88.4
703	Thibodaux	100.1	62.7	83.6
704	Hammond	97.5	55.4	78.9
705	Lafayette	99.9	61.0	82.7
706	Lake Charles	100.1	63.4	83.9
707-708	Baton Rouge	101.5	62.6	84.4
710-711	Shreveport	97.0	55.6	78.8
712	Monroe	97.1	52.9	77.6
713-714	Alexandria	97.2	52.8	77.6
MAINE				
039	Kittery	94.2	84.0	89.7
040-041	Portland	99.2	92.1	96.1
042	Lewiston	97.6	92.1	95.2
043	Augusta	97.5	81.7	90.5
044	Bangor	97.1	92.1	94.9
045	Bath	95.3	83.6	90.1
046	Machias	94.9	87.5	91.6
047	Houlton	95.0	87.6	91.7
048	Rockland	94.2	87.5	91.2
049	Waterville	95.4	81.7	89.4
MARYLAND				
206	Waldorf	97.0	82.7	90.7
207-208	College Park	97.0	84.5	91.5
209	Silver Spring	96.3	82.6	90.3
210-212	Baltimore	100.1	84.1	93.0
214	Annapolis	100.4	82.7	92.6
215	Cumberland	95.8	81.9	89.7
216	Easton	97.3	69.7	85.1
217	Hagerstown	96.8	83.5	90.9
218	Salisbury	97.7	64.6	83.1
219	Elkton	94.8	83.2	89.7
MASSACHUSETTS				
010-011	Springfield	99.6	109.5	104.0
012	Pittsfield	99.1	107.1	102.7
013	Greenfield	96.9	109.9	102.6
014	Fitchburg	95.7	124.1	108.2
015-016	Worcester	99.6	124.3	110.4
017	Framingham	94.9	131.2	110.9
018	Lowell	99.0	130.9	113.1
019	Lawrence	99.8	132.5	114.2
020-022, 024	Boston	101.5	139.2	118.1
023	Brockton	100.1	127.1	112.0
025	Buzzards Bay	94.3	127.7	109.0
026	Hyannis	97.3	127.7	110.7
027	New Bedford	99.3	127.9	111.9
MICHIGAN				
480,483	Royal Oak	94.4	104.5	98.9
481	Ann Arbor	96.8	106.7	101.2
482	Detroit	98.4	109.6	103.4
484-485	Flint	96.4	95.7	96.1
486	Saginaw	96.0	92.2	94.3
487	Bay City	96.2	92.3	94.5
488-489	Lansing	97.4	93.8	95.8
490	Battle Creek	96.5	87.7	92.6
491	Kalamazoo	96.8	86.0	92.1
492	Jackson	94.7	94.5	94.6
493,495	Grand Rapids	98.0	81.6	90.8
494	Muskegon	95.4	83.6	90.2

Location Factors

STATE/ZIP	CITY	MAT.	INST.	TOTAL
MICHIGAN (CONT'D)				
496	Traverse City	94.2	77.4	86.8
497	Gaylord	95.2	71.7	84.8
498-499	Iron mountain	97.2	86.4	92.4
MINNESOTA				
550-551	Saint Paul	100.4	119.9	109.0
553-555	Minneapolis	101.4	122.0	110.5
556-558	Duluth	99.7	110.3	104.3
559	Rochester	100.1	106.1	102.7
560	Mankato	97.3	103.2	99.9
561	Windom	96.0	92.8	94.6
562	Willmar	95.6	103.9	99.2
563	St. Cloud	97.2	119.4	107.0
564	Brainerd	97.1	103.7	100.0
565	Detroit Lakes	99.1	97.9	98.6
566	Bemidji	98.4	99.8	99.0
567	Thief River Falls	98.0	94.5	96.4
MISSISSIPPI				
386	Clarksdale	96.1	54.9	77.9
387	Greenville	99.9	64.3	84.2
388	Tupelo	97.7	57.3	79.9
389	Greenwood	97.3	54.4	78.4
390-392	Jackson	97.4	63.6	82.5
393	Meridian	96.5	65.3	82.7
394	Laurel	97.5	58.1	80.1
395	Biloxi	98.1	59.5	81.0
396	Mccomb	96.1	54.0	77.5
397	Columbus	97.4	56.2	79.3
MISSOURI				
630-631	St. Louis	99.9	106.3	102.7
633	Bowling Green	98.2	94.5	96.6
634	Hannibal	97.1	91.8	94.7
635	Kirksville	100.3	89.4	95.5
636	Flat River	99.1	95.2	97.4
637	Cape Girardeau	99.5	89.5	95.1
638	Sikeston	97.5	89.9	94.2
639	Poplar Bluff	97.0	89.7	93.8
640-641	Kansas City	100.7	108.2	104.0
644-645	St. Joseph	99.9	94.3	97.4
646	Chillicothe	96.6	95.2	96.0
647	Harrisonville	96.2	102.9	99.1
648	Joplin	98.6	81.7	91.2
650-651	Jefferson City	96.6	91.7	94.4
652	Columbia	97.7	93.2	95.7
653	Sedalia	96.7	92.4	94.8
654-655	Rolla	95.1	95.7	95.3
656-658	Springfield	98.6	82.8	91.7
MONTANA				
590-591	Billings	103.3	77.5	91.9
592	Wolf Point	102.5	75.4	90.6
593	Miles City	100.3	75.1	89.2
594	Great Falls	104.5	76.8	92.3
595	Havre	101.4	75.0	89.7
596	Helena	102.7	75.0	90.5
597	Butte	103.1	77.5	91.8
598	Missoula	100.6	77.0	90.2
599	Kalispell	99.3	76.3	89.1
NEBRASKA				
680-681	Omaha	100.4	79.5	91.2
683-685	Lincoln	98.6	73.0	87.3
686	Columbus	96.8	70.6	85.3
687	Norfolk	98.6	74.0	87.8
688	Grand Island	98.7	77.8	89.5
689	Hastings	97.8	78.1	89.1
690	Mccook	97.6	69.6	85.3
691	North Platte	98.1	78.3	89.4
692	Valentine	100.0	66.6	85.3
693	Alliance	100.0	64.8	84.5
NEVADA				
889-891	Las Vegas	100.2	109.0	104.1
893	Ely	98.7	99.3	99.0
894-895	Reno	99.0	92.0	95.9
897	Carson City	98.9	91.8	95.8
898	Elko	97.6	84.5	91.8
NEW HAMPSHIRE				
030	Nashua	99.8	94.0	97.2
031	Manchester	100.6	94.0	97.7

STATE/ZIP	CITY	MAT.	INST.	TOTAL
NEW HAMPSHIRE (CONT'D)				
032-033	Concord	98.4	92.8	95.9
034	Keene	96.4	74.6	86.8
035	Littleton	96.0	77.7	88.0
036	Charleston	95.8	70.6	84.7
037	Claremont	94.9	70.7	84.2
038	Portsmouth	97.3	94.5	96.1
NEW JERSEY				
070-071	Newark	102.7	124.8	112.4
072	Elizabeth	100.1	124.8	111.0
073	Jersey City	99.0	124.7	110.3
074-075	Paterson	100.9	124.9	111.5
076	Hackensack	98.9	124.9	110.4
077	Long Branch	98.4	122.3	108.9
078	Dover	99.2	124.9	110.5
079	Summit	99.2	124.8	110.5
080,083	Vineland	97.1	123.4	108.7
081	Camden	99.0	122.8	109.5
082,084	Atlantic City	97.6	123.3	108.9
085-086	Trenton	99.5	122.5	109.6
087	Point Pleasant	98.9	121.6	108.9
088-089	New Brunswick	99.4	123.9	110.2
NEW MEXICO				
870-872	Albuquerque	99.1	73.5	87.9
873	Gallup	99.0	73.5	87.8
874	Farmington	99.7	73.5	88.2
875	Santa Fe	100.4	73.5	88.5
877	Las Vegas	97.6	73.5	87.0
878	Socorro	97.1	73.5	86.7
879	Truth/Consequences	97.1	70.2	85.2
880	Las Cruces	96.2	69.4	84.4
881	Clovis	97.7	73.4	87.0
882	Roswell	99.6	73.5	88.1
883	Carrizozo	99.7	73.5	88.2
884	Tucumcari	98.4	73.4	87.4
NEW YORK				
100-102	New York	104.2	167.0	131.9
103	Staten Island	100.0	164.8	128.6
104	Bronx	98.3	164.8	127.6
105	Mount Vernon	98.0	133.0	113.4
106	White Plains	98.3	133.0	113.6
107	Yonkers	102.9	133.1	116.2
108	New Rochelle	98.3	133.0	113.6
109	Suffern	98.1	121.5	108.4
110	Queens	101.1	164.8	129.1
111	Long Island City	102.5	164.8	130.0
112	Brooklyn	103.0	164.8	130.2
113	Flushing	102.4	164.8	129.9
114	Jamaica	100.7	164.8	129.0
115,117,118	Hicksville	101.0	145.4	120.5
116	Far Rockaway	102.6	164.8	130.0
119	Riverhead	101.7	145.1	120.9
120-122	Albany	97.0	99.8	98.2
123	Schenectady	97.6	99.0	98.2
124	Kingston	100.4	118.5	108.4
125-126	Poughkeepsie	99.6	128.6	112.4
127	Monticello	99.0	117.6	107.2
128	Glens Falls	92.2	94.9	93.4
129	Plattsburgh	96.8	88.6	93.2
130-132	Syracuse	98.7	96.2	97.6
133-135	Utica	96.7	93.3	95.2
136	Watertown	98.2	93.1	95.9
137-139	Binghamton	98.2	100.1	99.0
140-142	Buffalo	100.0	104.1	101.8
143	Niagara Falls	97.5	105.5	101.0
144-146	Rochester	100.1	96.0	98.3
147	Jamestown	96.5	87.9	92.7
148-149	Elmira	96.3	92.7	94.8
NORTH CAROLINA				
270,272-274	Greensboro	99.0	56.2	80.2
271	Winston-Salem	98.8	57.5	80.6
275-276	Raleigh	99.0	55.1	79.7
277	Durham	100.3	53.9	79.9
278	Rocky Mount	96.1	53.9	77.5
279	Elizabeth City	97.0	52.2	77.2
280	Gastonia	97.8	56.5	79.6
281-282	Charlotte	98.7	56.5	80.1
283	Fayetteville	100.2	59.1	82.1
284	Wilmington	96.7	56.5	79.0
285	Kinston	94.5	54.3	76.8

STATE/ZIP	CITY	MAT.	INST.	TOTAL
NORTH CAROLINA (CONT'D)				
286	Hickory	94.8	55.3	77.4
287-288	Asheville	97.1	55.1	78.6
289	Murphy	95.7	49.9	75.5
NORTH DAKOTA				
580-581	Fargo	101.9	65.8	86.0
582	Grand Forks	102.4	52.6	80.5
583	Devils Lake	101.3	56.9	81.7
584	Jamestown	101.4	46.0	77.0
585	Bismarck	101.0	64.4	84.9
586	Dickinson	102.3	58.7	83.1
587	Minot	102.4	71.6	88.9
588	Williston	100.7	58.8	82.3
OHIO				
430-432	Columbus	98.6	90.0	94.8
433	Marion	94.8	84.1	90.1
434-436	Toledo	98.8	97.5	98.2
437-438	Zanesville	95.3	84.8	90.7
439	Steubenville	96.6	92.0	94.6
440	Lorain	98.1	90.8	94.9
441	Cleveland	98.2	100.7	99.3
442-443	Akron	99.1	93.6	96.6
444-445	Youngstown	98.4	89.1	94.3
446-447	Canton	98.5	85.0	92.5
448-449	Mansfield	95.6	86.1	91.4
450	Hamilton	97.5	84.7	91.9
451-452	Cincinnati	97.9	85.1	92.3
453-454	Dayton	97.7	83.9	91.6
455	Springfield	97.6	84.8	91.9
456	Chillicothe	96.2	91.9	94.3
457	Athens	98.9	82.2	91.6
458	Lima	99.4	85.0	93.1
OKLAHOMA				
730-731	Oklahoma City	98.5	59.8	81.4
734	Ardmore	95.1	59.3	79.3
735	Lawton	97.7	60.6	81.3
736	Clinton	96.6	58.0	79.6
737	Enid	97.5	58.0	80.1
738	Woodward	95.3	58.0	78.9
739	Guymon	96.4	30.7	67.4
740-741	Tulsa	97.5	52.8	77.8
743	Miami	94.0	66.9	82.1
744	Muskogee	96.8	38.8	71.3
745	Mcalester	93.7	49.9	74.4
746	Ponca City	94.3	57.9	78.3
747	Durant	94.3	56.9	77.8
748	Shawnee	95.8	55.8	78.2
749	Poteau	93.5	60.6	79.0
OREGON				
970-972	Portland	98.4	101.2	99.6
973	Salem	98.9	100.4	99.5
974	Eugene	98.2	100.3	99.1
975	Medford	99.7	98.3	99.1
976	Klamath Falls	99.3	98.3	98.9
977	Bend	98.3	100.4	99.3
978	Pendleton	93.8	102.9	97.8
979	Vale	91.8	91.1	91.5
PENNSYLVANIA				
150-152	Pittsburgh	99.4	104.7	101.8
153	Washington	96.3	103.0	99.2
154	Uniontown	96.7	101.9	99.0
155	Bedford	97.5	92.8	95.5
156	Greensburg	97.5	101.2	99.2
157	Indiana	96.4	99.1	97.6
158	Dubois	97.9	97.2	97.6
159	Johnstown	97.5	97.4	97.4
160	Butler	92.0	101.5	96.2
161	New Castle	92.0	98.0	94.7
162	Kittanning	92.5	102.9	97.1
163	Oil City	92.1	96.1	93.8
164-165	Erie	94.2	95.2	94.6
166	Altoona	94.4	93.2	93.9
167	Bradford	95.2	96.3	95.7
168	State College	94.7	93.7	94.3
169	Wellsboro	95.8	93.4	94.7
170-171	Harrisburg	97.8	96.5	97.2
172	Chambersburg	95.2	90.6	93.2
173-174	York	96.0	96.4	96.2
175-176	Lancaster	94.1	90.0	92.3

STATE/ZIP	CITY	MAT.	INST.	TOTAL
PENNSYLVANIA (CONT'D)				
177	Williamsport	92.8	79.3	86.9
178	Sunbury	94.7	95.8	95.2
179	Pottsville	94.0	98.8	96.1
180	Lehigh Valley	95.3	113.4	103.3
181	Allentown	97.7	109.5	102.9
182	Hazleton	94.9	98.6	96.5
183	Stroudsburg	94.7	102.7	98.2
184-185	Scranton	98.4	99.9	99.1
186-187	Wilkes-Barre	94.6	98.5	96.3
188	Montrose	94.2	97.3	95.6
189	Doylestown	94.5	120.2	105.8
190-191	Philadelphia	99.2	132.2	113.7
193	Westchester	95.5	122.7	107.5
194	Norristown	94.7	128.4	109.5
195-196	Reading	97.1	101.1	98.8
PUERTO RICO				
009	San Juan	121.0	23.8	78.2
RHODE ISLAND				
028	Newport	98.7	116.7	106.6
029	Providence	100.1	116.7	107.4
SOUTH CAROLINA				
290-292	Columbia	97.3	48.8	75.9
293	Spartanburg	96.3	48.4	75.2
294	Charleston	97.7	56.7	79.6
295	Florence	96.0	48.7	75.2
296	Greenville	96.0	48.4	75.0
297	Rock Hill	95.3	46.6	73.8
298	Aiken	96.1	71.8	85.4
299	Beaufort	96.8	42.1	72.7
SOUTH DAKOTA				
570-571	Sioux Falls	100.9	57.1	81.6
572	Watertown	99.5	50.9	78.1
573	Mitchell	98.4	50.5	77.3
574	Aberdeen	101.3	51.7	79.4
575	Pierre	100.5	53.7	79.8
576	Mobridge	99.1	50.9	77.8
577	Rapid City	101.2	54.6	80.6
TENNESSEE				
370-372	Nashville	97.6	73.1	86.8
373-374	Chattanooga	98.9	65.5	84.2
375,380-381	Memphis	97.4	70.0	85.3
376	Johnson City	98.4	56.0	79.8
377-379	Knoxville	95.5	60.8	80.2
382	Mckenzie	96.8	61.1	81.1
383	Jackson	98.9	63.2	83.2
384	Columbia	95.5	66.2	82.6
385	Cookeville	96.7	59.5	80.3
TEXAS				
750	Mckinney	98.5	49.7	77.0
751	Waxahackie	98.5	56.7	80.1
752-753	Dallas	99.5	66.5	85.0
754	Greenville	98.6	42.9	74.0
755	Texarkana	97.8	50.5	77.0
756	Longview	98.4	40.0	72.7
757	Tyler	98.6	54.1	79.0
758	Palestine	95.0	55.9	77.8
759	Lufkin	95.5	58.8	79.3
760-761	Fort Worth	97.1	62.9	82.0
762	Denton	97.3	47.3	75.2
763	Wichita Falls	97.9	54.1	78.6
764	Eastland	96.3	49.1	75.5
765	Temple	95.0	48.8	74.6
766-767	Waco	97.3	55.8	79.0
768	Brownwood	97.8	48.4	76.0
769	San Angelo	97.5	49.1	76.2
770-772	Houston	99.3	70.3	86.5
773	Huntsville	97.6	56.3	79.4
774	Wharton	98.5	53.4	78.6
775	Galveston	96.6	68.3	84.1
776-777	Beaumont	97.3	61.3	81.4
778	Bryan	94.5	61.6	80.0
779	Victoria	98.7	43.4	74.3
780	Laredo	96.7	53.2	77.5
781-782	San Antonio	97.0	65.6	83.2
783-784	Corpus Christi	99.7	51.7	78.6
785	Mc Allen	99.4	46.1	75.9
786-787	Austin	97.1	58.1	79.9

Location Factors

STATE/ZIP	CITY	MAT.	INST.	TOTAL
TEXAS (CONT'D)				
788	Del Rio	98.9	45.9	75.6
789	Giddings	95.8	55.4	78.0
790-791	Amarillo	98.6	60.1	81.7
792	Childress	97.5	56.5	79.4
793-794	Lubbock	99.7	56.2	80.6
795-796	Abilene	98.1	53.6	78.5
797	Midland	99.5	51.3	78.2
798-799,885	El Paso	97.2	50.6	76.7
UTAH				
840-841	Salt Lake City	101.9	67.5	86.7
842,844	Ogden	97.6	69.1	85.1
843	Logan	99.4	69.1	86.1
845	Price	99.8	65.5	84.7
846-847	Provo	100.0	67.4	85.6
VERMONT				
050	White River Jct.	97.3	67.7	84.3
051	Bellows Falls	96.0	81.8	89.7
052	Bennington	96.3	80.1	89.1
053	Brattleboro	96.6	86.2	92.0
054	Burlington	100.5	77.8	90.5
056	Montpelier	97.2	77.8	88.7
057	Rutland	98.8	77.8	89.6
058	St. Johnsbury	97.3	67.1	84.0
059	Guildhall	96.1	67.1	83.3
VIRGINIA				
220-221	Fairfax	99.3	84.9	93.0
222	Arlington	100.8	85.3	93.9
223	Alexandria	99.9	86.0	93.8
224-225	Fredericksburg	98.1	83.8	91.8
226	Winchester	98.6	82.7	91.6
227	Culpeper	98.6	81.6	91.1
228	Harrisonburg	98.8	69.3	85.8
229	Charlottesville	99.2	68.4	85.6
230-232	Richmond	100.1	70.1	86.9
233-235	Norfolk	100.8	69.4	87.0
236	Newport News	99.8	69.3	86.3
237	Portsmouth	99.2	66.7	84.9
238	Petersburg	99.0	70.1	86.3
239	Farmville	98.2	54.4	78.9
240-241	Roanoke	100.4	66.2	85.3
242	Bristol	97.9	56.7	79.8
243	Pulaski	97.6	57.3	79.8
244	Staunton	98.4	63.9	83.2
245	Lynchburg	98.5	67.4	84.8
246	Grundy	97.9	54.5	78.8
WASHINGTON				
980-981,987	Seattle	102.8	105.1	103.8
982	Everett	102.4	99.7	101.2
983-984	Tacoma	102.5	100.3	101.6
985	Olympia	100.1	100.3	100.2
986	Vancouver	103.5	94.1	99.4
988	Wenatchee	102.1	87.5	95.6
989	Yakima	102.7	93.6	98.7
990-992	Spokane	102.0	82.4	93.4
993	Richland	101.8	88.7	96.0
994	Clarkston	100.0	81.4	91.8
WEST VIRGINIA				
247-248	Bluefield	96.8	83.2	90.8
249	Lewisburg	98.3	83.2	91.7
250-253	Charleston	98.9	90.1	95.0
254	Martinsburg	98.2	80.3	90.3
255-257	Huntington	99.9	92.5	96.6
258-259	Beckley	96.7	87.2	92.5
260	Wheeling	100.0	90.3	95.7
261	Parkersburg	98.9	89.6	94.8
262	Buckhannon	98.1	90.3	94.7
263-264	Clarksburg	98.7	90.1	94.9
265	Morgantown	98.7	89.2	94.5
266	Gassaway	98.0	90.2	94.6
267	Romney	98.0	86.0	92.7
268	Petersburg	97.8	86.1	92.6
WISCONSIN				
530,532	Milwaukee	101.1	108.6	104.4
531	Kenosha	100.8	99.8	100.4
534	Racine	100.3	99.9	100.1
535	Beloit	100.2	94.8	97.8
537	Madison	99.9	96.0	98.2

STATE/ZIP	CITY	MAT.	INST.	TOTAL
WISCONSIN (CONT'D)				
538	Lancaster	97.7	90.3	94.4
539	Portage	96.3	93.4	95.0
540	New Richmond	97.0	93.0	95.2
541-543	Green Bay	101.5	92.6	97.6
544	Wausau	96.3	91.0	94.0
545	Rhinelander	99.5	89.2	94.9
546	La Crosse	98.0	92.4	95.5
547	Eau Claire	99.5	93.5	96.9
548	Superior	96.8	95.5	96.2
549	Oshkosh	96.9	81.4	90.1
WYOMING				
820	Cheyenne	99.4	66.7	85.0
821	Yellowstone Nat'l Park	96.7	65.9	83.2
822	Wheatland	97.8	65.5	83.6
823	Rawlins	99.3	68.6	85.8
824	Worland	97.4	64.9	83.0
825	Riverton	98.4	65.6	84.0
826	Casper	99.4	61.5	82.7
827	Newcastle	97.2	67.5	84.1
828	Sheridan	100.0	66.1	85.1
829-831	Rock Springs	101.2	65.6	85.5
CANADIAN FACTORS (reflect Canadian currency)				
ALBERTA				
	Calgary	127.9	97.4	114.5
	Edmonton	128.5	97.4	114.8
	Fort McMurray	125.2	96.2	112.4
	Lethbridge	119.7	95.6	109.1
	Lloydminster	114.7	92.2	104.8
	Medicine Hat	114.8	91.4	104.5
	Red Deer	115.3	91.4	104.8
BRITISH COLUMBIA				
	Kamloops	115.7	94.3	106.3
	Prince George	116.8	93.6	106.5
	Vancouver	126.8	91.9	111.4
	Victoria	116.8	90.0	105.0
MANITOBA				
	Brandon	114.9	78.9	99.0
	Portage la Prairie	114.7	77.2	98.2
	Winnipeg	127.3	72.3	103.1
NEW BRUNSWICK				
	Bathurst	112.9	70.0	94.0
	Dalhousie	113.0	70.2	94.1
	Fredericton	117.6	75.5	99.1
	Moncton	113.3	72.9	95.5
	Newcastle	113.0	70.2	94.1
	St. John	115.5	79.2	99.5
NEWFOUNDLAND				
	Corner Brook	118.2	69.4	96.7
	St Johns	121.7	81.9	104.2
NORTHWEST TERRITORIES				
	Yellowknife	123.9	87.6	107.9
NOVA SCOTIA				
	Bridgewater	114.3	76.6	97.7
	Dartmouth	115.9	76.6	98.6
	Halifax	119.5	82.2	103.0
	New Glasgow	114.0	76.6	97.6
	Sydney	111.7	76.6	96.3
	Truro	113.8	76.6	97.4
	Yarmouth	113.9	76.6	97.5
ONTARIO				
	Barrie	118.1	95.4	108.1
	Brantford	116.4	99.4	108.9
	Cornwall	116.5	95.7	107.3
	Hamilton	120.4	97.0	110.1
	Kingston	117.3	95.8	107.8
	Kitchener	112.6	93.5	104.2
	London	120.8	93.5	108.8
	North Bay	116.6	93.5	106.4
	Oshawa	115.8	96.8	107.4
	Ottawa	121.1	94.1	109.2
	Owen Sound	118.2	93.6	107.4
	Peterborough	116.4	95.5	107.2
	Sarnia	116.2	100.1	109.1

STATE/ZIP	CITY	MAT.	INST.	TOTAL
ONTARIO (CONT'D)				
	Sault Ste Marie	111.4	94.1	103.8
	St. Catharines	110.6	94.6	103.6
	Sudbury	111.1	93.2	103.2
	Thunder Bay	111.9	93.3	103.7
	Timmins	116.6	93.5	106.4
	Toronto	121.6	100.5	112.3
	Windsor	111.2	93.4	103.3
PRINCE EDWARD ISLAND				
	Charlottetown	119.0	64.5	95.0
	Summerside	115.7	64.5	93.2
QUEBEC				
	Cap-de-la-Madeleine	113.7	89.9	103.2
	Charlesbourg	113.7	89.9	103.2
	Chicoutimi	112.6	94.6	104.7
	Gatineau	113.3	89.7	102.9
	Granby	113.5	89.6	103.0
	Hull	113.4	89.7	102.9
	Joliette	113.8	89.9	103.3
	Laval	112.9	89.7	102.7
	Montreal	121.8	94.9	110.0
	Quebec	119.5	95.3	108.9
	Rimouski	113.1	94.6	104.9
	Rouyn-Noranda	113.2	89.7	102.9
	Saint Hyacinthe	112.6	89.7	102.5
	Sherbrooke	113.6	89.7	103.1
	Sorel	113.8	89.9	103.3
	St Jerome	113.3	89.7	102.9
	Trois Rivieres	114.0	89.9	103.4
SASKATCHEWAN				
	Moose Jaw	112.0	71.3	94.1
	Prince Albert	111.1	69.5	92.7
	Regina	117.6	90.1	105.5
	Saskatoon	112.0	90.1	102.4
YUKON				
	Whitehorse	118.4	70.3	97.2

R011105-05 Tips for Accurate Estimating

1. Use pre-printed or columnar forms for orderly sequence of dimensions and locations and for recording telephone quotations.
2. Use only the front side of each paper or form except for certain pre-printed summary forms.
3. Be consistent in listing dimensions: For example, length x width x height. This helps in rechecking to ensure that the total length of partitions is appropriate for the building area.
4. Use printed (rather than measured) dimensions where given.
5. Add up multiple printed dimensions for a single entry where possible.
6. Measure all other dimensions carefully.
7. Use each set of dimensions to calculate multiple related quantities.
8. Convert foot and inch measurements to decimal feet when listing. Memorize decimal equivalents to .01 parts of a foot (1/8″ equals approximately .01′).
9. Do not "round off" quantities until the final summary.
10. Mark drawings with different colors as items are taken off.
11. Keep similar items together, different items separate.
12. Identify location and drawing numbers to aid in future checking for completeness.
13. Measure or list everything on the drawings or mentioned in the specifications.
14. It may be necessary to list items not called for to make the job complete.
15. Be alert for: Notes on plans such as N.T.S. (not to scale); changes in scale throughout the drawings; reduced size drawings; discrepancies between the specifications and the drawings.
16. Develop a consistent pattern of performing an estimate. For example:
    a. Start the quantity takeoff at the lower floor and move to the next higher floor.
    b. Proceed from the main section of the building to the wings.
    c. Proceed from south to north or vice versa, clockwise or counterclockwise.
    d. Take off floor plan quantities first, elevations next, then detail drawings.
17. List all gross dimensions that can be either used again for different quantities, or used as a rough check of other quantities for verification (exterior perimeter, gross floor area, individual floor areas, etc.).
18. Utilize design symmetry or repetition (repetitive floors, repetitive wings, symmetrical design around a center line, similar room layouts, etc.). Note: Extreme caution is needed here so as not to omit or duplicate an area.
19. Do not convert units until the final total is obtained. For instance, when estimating concrete work, keep all units to the nearest cubic foot, then summarize and convert to cubic yards.
20. When figuring alternatives, it is best to total all items involved in the basic system, then total all items involved in the alternates. Therefore, you work with positive numbers in all cases. When adds and deducts are used, it is often confusing whether to add or subtract a portion of an item; especially on a complicated or involved alternate.

R011105-10 Unit Gross Area Requirements

The figures in the table below indicate typical ranges in square feet as a function of the "occupant" unit. This table is best used in the preliminary design stages to help determine the probable size requirement for the total project.

		Gross Area in S.F.		
Building Type	Unit	1/4	Median	3/4
Apartments	Unit	660	860	1,100
Auditorium & Play Theaters	Seat	18	25	38
Bowling Alleys	Lane		940	
Churches & Synagogues	Seat	20	28	39
Dormitories	Bed	200	230	275
Fraternity & Sorority Houses	Bed	220	315	370
Garages, Parking	Car	325	355	385
Hospitals	Bed	685	850	1,075
Hotels	Rental Unit	475	600	710
Housing for the elderly	Unit	515	635	755
Housing, Public	Unit	700	875	1,030
Ice Skating Rinks	Total	27,000	30,000	36,000
Motels	Rental Unit	360	465	620
Nursing Homes	Bed	290	350	450
Restaurants	Seat	23	29	39
Schools, Elementary	Pupil	65	77	90
Junior High & Middle	↓	85	110	129
Senior High	↓	102	130	145
Vocational	↓	110	135	195
Shooting Ranges	Point		450	
Theaters & Movies	Seat		15	

General Requirements — R0111 Summary of Work

R011110-30 Engineering Fees

Typical **Structural Engineering Fees** based on type of construction and total project size. These fees are included in Architectural Fees.

Type of Construction	Total Project Size (in thousands of dollars)			
	$500	$500-$1,000	$1,000-$5,000	Over $5000
Industrial buildings, factories & warehouses	Technical payroll times 2.0 to 2.5	1.60%	1.25%	1.00%
Hotels, apartments, offices, dormitories, hospitals, public buildings, food stores	↓	2.00%	1.70%	1.20%
Museums, banks, churches and cathedrals	↓	2.00%	1.75%	1.25%
Thin shells, prestressed concrete, earthquake resistive	↓	2.00%	1.75%	1.50%
Parking ramps, auditoriums, stadiums, convention halls, hangars & boiler houses	↓	2.50%	2.00%	1.75%
Special buildings, major alterations, underpinning & future expansion	↓	Add to above 0.5%	Add to above 0.5%	Add to above 0.5%

For complex reinforced concrete or unusually complicated structures, add 20% to 50%.

Typical **Mechanical and Electrical Engineering Fees** are based on the size of the subcontract. The fee structure for both are shown below. These fees are included in Architectural Fees.

Type of Construction	Subcontract Size							
	$25,000	$50,000	$100,000	$225,000	$350,000	$500,000	$750,000	$1,000,000
Simple structures	6.4%	5.7%	4.8%	4.5%	4.4%	4.3%	4.2%	4.1%
Intermediate structures	8.0	7.3	6.5	5.6	5.1	5.0	4.9	4.8
Complex structures	10.1	9.0	9.0	8.0	7.5	7.5	7.0	7.0

For renovations, add 15% to 25% to applicable fee.

General Requirements — R0121 Allowances

R012153-10 Repair and Remodeling

Cost figures are based on new construction utilizing the most cost-effective combination of labor, equipment and material with the work scheduled in proper sequence to allow the various trades to accomplish their work in an efficient manner.

The costs for repair and remodeling work must be modified due to the following factors that may be present in any given repair and remodeling project.

1. Equipment usage curtailment due to the physical limitations of the project, with only hand-operated equipment being used.
2. Increased requirement for shoring and bracing to hold up the building while structural changes are being made and to allow for temporary storage of construction materials on above-grade floors.
3. Material handling becomes more costly due to having to move within the confines of an enclosed building. For multi-story construction, low capacity elevators and stairwells may be the only access to the upper floors.
4. Large amounts of cutting and patching and attempting to match the existing construction is required. It is often more economical to remove entire walls rather than create many new door and window openings. This sort of trade-off has to be carefully analyzed.
5. Cost of protection of completed work is increased since the usual sequence of construction usually cannot be accomplished.
6. Economies of scale usually associated with new construction may not be present. If small quantities of components must be custom fabricated due to job requirements, unit costs will naturally increase. Also, if only small work areas are available at a given time, job scheduling between trades becomes difficult and subcontractor quotations may reflect the excessive start-up and shut-down phases of the job.
7. Work may have to be done on other than normal shifts and may have to be done around an existing production facility which has to stay in production during the course of the repair and remodeling.
8. Dust and noise protection of adjoining non-construction areas can involve substantial special protection and alter usual construction methods.
9. Job may be delayed due to unexpected conditions discovered during demolition or removal. These delays ultimately increase construction costs.
10. Piping and ductwork runs may not be as simple as for new construction. Wiring may have to be snaked through walls and floors.
11. Matching "existing construction" may be impossible because materials may no longer be manufactured. Substitutions may be expensive.
12. Weather protection of existing structure requires additional temporary structures to protect building at openings.
13. On small projects, because of local conditions, it may be necessary to pay a tradesman for a minimum of four hours for a task that is completed in one hour.

All of the above areas can contribute to increased costs for a repair and remodeling project. Each of the above factors should be considered in the planning, bidding and construction stage in order to minimize the increased costs associated with repair and remodeling jobs.

R012909-80 Sales Tax by State

State sales tax on materials is tabulated below (5 states have no sales tax). Many states allow local jurisdictions, such as a county or city, to levy additional sales tax.

Some projects may be sales tax exempt, particularly those constructed with public funds.

State	Tax (%)	State	Tax (%)	State	Tax (%)	State	Tax (%)
Alabama	4	Illinois	6.25	Montana	0	Rhode Island	7
Alaska	0	Indiana	7	Nebraska	5.5	South Carolina	6
Arizona	6.6	Iowa	6	Nevada	6.85	South Dakota	4
Arkansas	6	Kansas	5.3	New Hampshire	0	Tennessee	7
California	7.25	Kentucky	6	New Jersey	7	Texas	6.25
Colorado	2.9	Louisiana	4	New Mexico	5.13	Utah	4.75
Connecticut	6	Maine	5	New York	4	Vermont	6
Delaware	0	Maryland	6	North Carolina	5.75	Virginia	4
District of Columbia	6	Massachusetts	6.25	North Dakota	5	Washington	6.5
Florida	6	Michigan	6	Ohio	5.5	West Virginia	6
Georgia	4	Minnesota	6.88	Oklahoma	4.5	Wisconsin	5
Hawaii	4	Mississippi	7	Oregon	0	Wyoming	4
Idaho	6	Missouri	4.23	Pennsylvania	6	Average	5.03 %

Sales Tax by Province (Canada)

GST - a value-added tax, which the government imposes on most goods and services provided in or imported into Canada. PST - a retail sales tax, which three of the provinces impose on the price of most goods and some services. QST - a value-added tax, similar to the federal GST, which Quebec imposes. HST - Five provinces have combined their retail sales tax with the federal GST into one harmonized tax.

Province	PST (%)	QST (%)	GST(%)	HST(%)
Alberta	0	0	5	0
British Columbia	0	0	0	12
Manitoba	7	0	5	0
New Brunswick	0	0	0	13
Newfoundland	0	0	0	13
Northwest Territories	0	0	5	0
Nova Scotia	0	0	0	15
Ontario	0	0	0	13
Prince Edward Island	10	0	5	0
Quebec	0	9.5	5	0
Saskatchewan	5	0	5	0
Yukon	0	0	5	0

R012909-85 Unemployment Taxes and Social Security Taxes

State unemployment tax rates vary not only from state to state, but also with the experience rating of the contractor. The federal unemployment tax rate is 6.2% of the first $7,000 of wages. This is reduced by a credit of up to 5.4% for timely payment to the state. The minimum federal unemployment tax is 0.8% after all credits.

Social security (FICA) for 2013 is estimated at time of publication to be 7.65% of wages up to $114,000.

R012909-90 Overtime

One way to improve the completion date of a project or eliminate negative float from a schedule is to compress activity duration times. This can be achieved by increasing the crew size or working overtime with the proposed crew.

To determine the costs of working overtime to compress activity duration times, consider the following examples. Below is an overtime efficiency and cost chart based on a five, six, or seven day week with an eight through twelve hour day. Payroll percentage increases for time and one half and double time are shown for the various working days.

Days per Week	Hours per Day	Production Efficiency					Payroll Cost Factors	
		1st Week	2nd Week	3rd Week	4th Week	Average 4 Weeks	@ 1-1/2 Times	@ 2 Times
5	8	100%	100%	100%	100%	100 %	100 %	100 %
	9	100	100	95	90	96.25	105.6	111.1
	10	100	95	90	85	91.25	110.0	120.0
	11	95	90	75	65	81.25	113.6	127.3
	12	90	85	70	60	76.25	116.7	133.3
6	8	100	100	95	90	96.25	108.3	116.7
	9	100	95	90	85	92.50	113.0	125.9
	10	95	90	85	80	87.50	116.7	133.3
	11	95	85	70	65	78.75	119.7	139.4
	12	90	80	65	60	73.75	122.2	144.4
7	8	100	95	85	75	88.75	114.3	128.6
	9	95	90	80	70	83.75	118.3	136.5
	10	90	85	75	65	78.75	121.4	142.9
	11	85	80	65	60	72.50	124.0	148.1
	12	85	75	60	55	68.75	126.2	152.4

General Requirements R0131 Project Management & Coordination

R013113-40 Builder's Risk Insurance

Builder's Risk Insurance is insurance on a building during construction. Premiums are paid by the owner or the contractor. Blasting, collapse and underground insurance would raise total insurance costs above those listed. Floater policy for materials delivered to the job runs $.75 to $1.25 per $100 value. Contractor equipment insurance runs $.50 to $1.50 per $100 value. Insurance for miscellaneous tools to $1,500 value runs from $3.00 to $7.50 per $100 value.

Tabulated below are New England Builder's Risk insurance rates in dollars per $100 value for $1,000 deductible. For $25,000 deductible, rates can be reduced 13% to 34%. On contracts over $1,000,000, rates may be lower than those tabulated. Policies are written annually for the total completed value in place. For "all risk" insurance (excluding flood, earthquake and certain other perils) add $.025 to total rates below.

Coverage	Frame Construction (Class 1)				Brick Construction (Class 4)				Fire Resistive (Class 6)			
	Range			Average	Range			Average	Range			Average
Fire Insurance	$.350	to	$ .850	$.600	$.158	to	$.189	$.174	$.052	to	$.080	$.070
Extended Coverage	.115	to	.200	.158	.080	to	.105	.101	.081	to	.105	.100
Vandalism	.012	to	.016	.014	.008	to	.011	.011	.008	to	.011	.010
Total Annual Rate	$.477	to	$1.066	$.772	$.246	to	$.305	$.286	$.141	to	$.196	$.180

R013113-60 Workers' Compensation Insurance Rates by Trade

The table below tabulates the national averages for workers' compensation insurance rates by trade and type of building. The average "Insurance Rate" is multiplied by the "% of Building Cost" for each trade. This produces the "Workers' Compensation" cost by % of total labor cost, to be added for each trade by building type to determine the weighted average workers' compensation rate for the building types analyzed.

Trade	Insurance Rate (% Labor Cost)		% of Building Cost			Workers' Compensation		
	Range	Average	Office Bldgs.	Schools & Apts.	Mfg.	Office Bldgs.	Schools & Apts.	Mfg.
Excavation, Grading, etc.	3.9 % to 18.0%	9.2%	4.8%	4.9%	4.5%	0.44%	0.45%	0.41%
Piles & Foundations	5.5 to 50.2	15.3	7.1	5.2	8.7	1.09	0.80	1.33
Concrete	4.7 to 28.1	12.6	5.0	14.8	3.7	0.63	1.86	0.47
Masonry	5.1 to 33.0	12.5	6.9	7.5	1.9	0.86	0.94	0.24
Structural Steel	5.5 to 125.8	36.5	10.7	3.9	17.6	3.91	1.42	6.42
Miscellaneous & Ornamental Metals	4.5 to 22.6	10.6	2.8	4.0	3.6	0.30	0.42	0.38
Carpentry & Millwork	5.5 to 36.6	15.1	3.7	4.0	0.5	0.56	0.60	0.08
Metal or Composition Siding	4.3 to 61.9	16.4	2.3	0.3	4.3	0.38	0.05	0.71
Roofing	5.5 to 90.6	29.6	2.3	2.6	3.1	0.68	0.77	0.92
Doors & Hardware	3.5 to 36.6	10.8	0.9	1.4	0.4	0.10	0.15	0.04
Sash & Glazing	5.4 to 35.7	12.5	3.5	4.0	1.0	0.44	0.50	0.13
Lath & Plaster	3.3 to 28.8	10.9	3.3	6.9	0.8	0.36	0.75	0.09
Tile, Marble & Floors	3.0 to 17.6	8.0	2.6	3.0	0.5	0.21	0.24	0.04
Acoustical Ceilings	3.3 to 36.7	7.9	2.4	0.2	0.3	0.19	0.02	0.02
Painting	4.6 to 31.5	11.1	1.5	1.6	1.6	0.17	0.18	0.18
Interior Partitions	5.5 to 36.6	15.1	3.9	4.3	4.4	0.59	0.65	0.66
Miscellaneous Items	2.5 to 125.0	14.4	5.2	3.7	9.7	0.75	0.53	1.40
Elevators	1.8 to 13.1	5.6	2.1	1.1	2.2	0.12	0.06	0.12
Sprinklers	2.3 to 17.2	6.8	0.5	—	2.0	0.03	—	0.14
Plumbing	2.6 to 12.9	6.7	4.9	7.2	5.2	0.33	0.48	0.35
Heat., Vent., Air Conditioning	3.7 to 20.5	8.6	13.5	11.0	12.9	1.16	0.95	1.11
Electrical	2.8 to 13.1	5.7	10.1	8.4	11.1	0.58	0.48	0.63
Total	1.8 % to 125.8%	—	100.0%	100.0%	100.0%	13.88%	12.30%	15.87%
	Overall Weighted Average	14.02%						

Workers' Compensation Insurance Rates by States

The table below lists the weighted average workers' compensation base rate for each state with a factor comparing this with the national average of 13.7%.

State	Weighted Average	Factor	State	Weighted Average	Factor	State	Weighted Average	Factor
Alabama	19.5%	142	Kentucky	16.0%	117	North Dakota	9.4%	69
Alaska	16.0	117	Louisiana	16.9	123	Ohio	8.8	64
Arizona	9.5	69	Maine	13.2	96	Oklahoma	14.5	106
Arkansas	9.6	70	Maryland	15.6	114	Oregon	12.2	89
California	26.5	193	Massachusetts	12.3	90	Pennsylvania	17.0	124
Colorado	7.4	54	Michigan	18.9	138	Rhode Island	11.9	87
Connecticut	24.0	175	Minnesota	24.7	180	South Carolina	20.1	147
Delaware	10.4	76	Mississippi	13.8	101	South Dakota	16.1	118
District of Columbia	11.6	85	Missouri	16.9	123	Tennessee	14.6	107
Florida	10.7	78	Montana	11.5	84	Texas	10.5	77
Georgia	30.3	221	Nebraska	18.9	138	Utah	8.4	61
Hawaii	8.8	64	Nevada	9.0	66	Vermont	14.3	104
Idaho	10.7	78	New Hampshire	22.9	167	Virginia	9.4	69
Illinois	22.5	164	New Jersey	14.1	103	Washington	11.2	82
Indiana	5.7	42	New Mexico	16.1	118	West Virginia	10.7	78
Iowa	10.7	78	New York	6.2	45	Wisconsin	14.1	103
Kansas	9.1	66	North Carolina	17.5	128	Wyoming	6.1	45
			Weighted Average for U.S. is	14.1% of payroll = 100%				

Rates in the following table are the base or manual costs per $100 of payroll for workers' compensation in each state. Rates are usually applied to straight time wages only and not to premium time wages and bonuses.

The weighted average skilled worker rate for 35 trades is 13.7%. For bidding purposes, apply the full value of workers' compensation directly to total labor costs, or if labor is 38%, materials 42% and overhead and profit 20% of total cost, carry 38/80 × 13.7% =6.5% of cost (before overhead and profit) into overhead. Rates vary not only from state to state but also with the experience rating of the contractor.

Rates are the most current available at the time of publication.

R013113-60 Workers' Compensation Insurance Rates by Trade and State (cont.)

State	Carpentry — 3 stories or less 5651	Carpentry — interior cab. work 5437	Carpentry — general 5403	Concrete Work — NOC 5213	Concrete Work — flat (flr., sdwk.) 5221	Electrical Wiring — inside 5190	Excavation — earth NOC 6217	Excavation — rock 6217	Glaziers 5462	Insulation Work 5479	Lathing 5443	Masonry 5022	Painting & Decorating 5474	Pile Driving 6003	Plastering 5480	Plumbing 5183	Roofing 5551	Sheet Metal Work (HVAC) 5538	Steel Erection — door & sash 5102	Steel Erection — inter., ornam. 5102	Steel Erection — structure 5040	Steel Erection — NOC 5057	Tile Work — (interior ceramic) 5348	Waterproofing 9014	Wrecking 5701
AL	33.16	14.38	24.57	12.56	9.62	9.05	13.51	13.51	17.48	16.31	8.34	22.18	17.10	17.88	15.08	7.53	46.99	9.68	11.01	11.01	42.67	23.81	12.67	9.00	42.67
AK	15.60	10.19	12.62	10.76	10.50	6.72	9.35	9.35	31.18	16.18	10.41	14.44	12.20	30.90	11.89	6.88	28.31	7.92	7.95	7.95	29.64	33.41	5.25	5.38	29.64
AZ	13.84	5.84	16.22	7.78	4.67	4.54	5.48	5.48	6.75	10.41	3.94	8.00	7.92	8.19	5.86	4.82	18.65	6.85	11.02	11.02	20.46	9.79	3.05	3.72	20.46
AR	12.64	5.52	10.90	6.45	5.52	4.56	6.98	6.98	7.80	8.42	4.23	8.42	7.44	9.52	13.32	4.58	18.55	5.18	8.04	8.04	19.86	14.63	6.30	2.90	19.86
CA	36.56	36.56	36.56	18.18	18.18	13.06	15.93	15.93	27.92	15.97	17.81	25.50	23.38	22.56	28.82	6.44	67.92	20.51	16.42	16.42	32.53	20.51	12.92	23.38	20.51
CO	9.14	4.76	5.76	6.29	4.46	2.90	5.10	5.10	5.36	7.05	3.69	7.41	5.85	6.81	5.42	3.94	14.82	3.74	4.51	4.51	26.55	9.93	3.98	2.88	19.25
CT	24.20	17.13	32.36	23.82	14.23	7.81	14.76	14.76	20.15	20.20	10.64	29.90	19.14	20.98	18.42	11.02	51.86	13.90	18.99	18.99	65.52	27.25	14.41	6.82	65.52
DE	10.15	10.15	7.92	8.38	7.06	3.56	6.26	6.26	8.82	7.92	8.82	9.44	10.62	12.26	8.82	5.43	21.11	6.82	10.50	10.50	19.57	10.50	4.61	9.44	19.57
DC	9.43	7.29	8.15	13.63	7.29	4.11	9.26	9.26	12.04	6.67	8.28	10.03	6.01	15.97	8.79	7.69	16.04	6.43	19.51	19.51	22.71	9.28	7.60	4.71	22.71
FL	10.88	8.00	11.48	11.88	5.72	5.65	6.59	6.59	9.56	10.12	4.66	9.77	8.86	17.52	10.48	5.31	17.10	7.37	9.25	9.25	22.79	11.00	5.25	4.78	22.79
GA	61.87	20.95	29.49	21.35	16.06	11.54	17.98	17.98	19.86	18.38	12.52	33.03	31.51	24.06	20.88	12.92	90.59	17.73	22.61	22.61	73.34	31.15	13.64	11.47	73.34
HI	8.48	6.62	14.65	6.90	3.88	4.10	5.15	5.15	8.04	5.53	7.12	7.86	6.88	7.68	7.24	4.84	14.83	4.22	6.73	6.73	27.79	7.53	5.28	5.39	27.79
ID	13.87	7.07	11.31	12.31	5.33	3.88	6.92	6.92	9.01	8.49	4.97	9.39	9.72	8.39	10.73	5.22	22.39	7.91	9.62	9.62	28.90	8.36	6.12	4.95	28.90
IL	23.38	15.06	20.46	28.05	11.77	7.55	10.37	10.37	17.20	16.69	10.75	19.99	12.36	17.62	22.12	9.56	33.65	11.09	20.83	20.83	62.17	19.02	17.57	5.48	62.17
IN	7.76	4.17	6.49	4.81	3.67	2.95	3.93	3.93	5.43	6.15	3.51	5.11	4.63	6.30	3.60	2.64	10.53	4.15	5.10	5.10	11.78	4.77	2.98	2.81	11.78
IA	10.34	8.42	12.90	12.72	8.24	4.47	6.88	6.88	9.20	5.82	4.64	8.40	6.76	8.73	10.98	6.40	19.92	4.80	5.70	5.70	36.26	14.65	6.70	4.60	24.10
KS	14.38	6.41	9.59	8.14	6.41	4.94	4.91	4.91	8.62	7.32	4.14	7.47	8.33	8.31	6.21	5.27	16.38	5.64	6.57	6.57	21.39	14.64	5.42	4.41	14.64
KY	18.75	12.00	24.00	11.00	7.25	5.25	12.50	12.50	19.84	11.27	7.23	7.50	10.00	18.59	12.49	5.50	34.00	12.00	10.67	10.67	48.00	13.78	15.43	4.00	48.00
LA	23.72	13.00	21.51	13.00	10.19	7.26	12.45	12.45	15.57	10.97	6.96	14.17	15.60	20.40	13.16	6.05	35.41	13.54	15.88	15.88	43.26	9.29	8.36	6.48	43.26
ME	12.86	9.34	18.66	15.82	8.40	4.84	10.34	10.34	14.07	13.62	5.16	10.73	13.40	9.74	10.18	6.24	23.06	6.89	7.53	7.53	42.08	13.38	5.77	4.87	42.08
MD	12.84	14.69	13.94	19.92	5.91	6.13	7.34	7.34	13.74	13.95	6.39	10.39	7.28	18.75	9.37	6.93	30.19	11.62	12.36	12.36	62.69	20.35	6.49	5.91	34.01
MA	8.68	5.23	9.61	18.85	6.24	2.84	4.35	4.35	9.58	7.78	5.27	10.55	5.09	12.92	4.68	3.50	30.99	5.72	6.89	6.89	54.08	33.00	5.81	2.48	23.75
MI	18.76	12.20	18.76	17.13	11.09	6.05	14.34	14.34	12.16	12.23	12.20	17.36	15.21	50.15	10.05	7.06	41.77	10.97	12.36	12.36	50.15	11.72	12.17	6.73	50.15
MN	19.08	19.63	32.94	10.69	13.05	5.59	10.82	10.82	35.70	23.79	9.93	14.01	15.96	34.92	9.93	7.70	68.07	18.46	9.02	9.02	125.84	6.99	13.08	7.38	6.99
MS	18.84	9.11	13.10	9.88	6.18	5.44	10.83	10.83	13.36	9.02	5.50	11.97	10.74	14.86	10.19	8.67	34.07	7.00	13.91	13.91	38.39	8.30	8.30	5.37	38.39
MO	20.18	12.02	13.09	14.94	10.39	6.92	11.53	11.53	10.35	12.21	8.01	15.54	11.65	16.04	13.35	9.62	38.70	8.65	12.57	12.57	44.61	30.73	11.93	6.78	44.61
MT	10.19	10.19	16.22	8.28	7.10	4.79	10.08	10.08	10.07	12.78	5.36	9.49	8.64	11.25	8.99	5.69	34.41	5.97	8.60	8.60	20.45	7.77	7.08	6.02	7.77
NE	21.40	15.45	20.10	19.40	10.83	8.43	17.60	17.60	11.63	21.53	8.53	19.73	12.58	15.20	14.93	9.58	36.13	12.95	13.88	13.88	52.18	18.15	8.40	7.08	67.55
NV	9.92	6.70	12.24	9.79	5.97	4.28	8.50	8.50	6.28	7.41	3.32	6.54	5.68	8.18	5.81	5.97	17.37	7.36	7.25	7.25	19.00	11.91	4.01	4.36	11.91
NH	22.39	11.73	18.76	25.48	13.80	5.76	12.76	12.76	13.34	13.60	7.44	18.16	25.71	16.95	11.36	10.77	39.32	7.96	11.58	11.58	125.75	36.22	13.21	7.76	125.75
NJ	16.94	12.03	16.94	13.84	11.18	4.86	9.96	9.96	9.40	11.59	14.39	17.16	11.07	13.95	14.39	6.66	36.02	6.81	11.88	11.88	14.78	12.26	9.40	6.16	16.81
NM	24.37	9.11	19.35	12.09	10.28	7.25	7.11	7.11	11.76	10.64	6.21	14.95	11.19	13.88	14.09	6.83	37.02	9.49	11.71	11.71	53.67	22.25	7.23	6.93	53.67
NY	4.33	3.54	6.47	7.73	5.97	2.75	3.94	3.94	6.58	3.77	4.72	6.95	4.90	7.10	3.33	3.81	11.62	4.70	6.32	6.32	11.22	6.07	3.67	3.28	6.06
NC	15.50	11.55	18.63	17.12	8.14	9.96	13.11	13.11	11.87	12.57	9.34	11.05	11.85	16.82	13.63	8.92	33.88	11.87	11.97	11.97	70.40	20.61	8.12	5.41	70.40
ND	9.54	9.54	9.54	5.43	5.43	3.75	4.13	4.13	9.54	9.54	8.78	6.85	5.64	9.33	8.78	5.76	16.48	5.76	9.33	9.33	9.33	9.33	9.54	16.48	9.04
OH	7.73	4.31	5.52	4.67	4.92	3.43	4.79	4.79	8.83	11.38	36.65	6.93	8.15	6.99	7.80	3.52	14.89	4.00	6.58	6.58	11.33	5.84	4.85	4.58	5.84
OK	18.56	9.92	13.23	13.71	8.14	7.02	11.13	11.13	12.71	12.47	5.73	15.58	11.38	12.79	17.44	7.23	24.19	10.52	11.32	11.32	33.69	18.85	7.23	6.79	33.69
OR	22.03	8.09	13.21	9.91	9.68	5.04	9.25	9.25	11.50	11.30	6.46	13.57	12.29	11.17	10.00	6.02	25.38	6.95	7.37	7.37	21.11	12.91	7.22	6.11	21.11
PA	17.84	17.84	13.93	17.58	12.59	6.86	10.29	10.29	13.07	13.93	13.07	14.91	15.32	17.84	13.07	8.68	34.96	8.55	17.39	17.39	27.42	17.39	10.29	14.94	27.42
RI	9.69	10.13	12.42	13.18	10.51	4.12	8.56	8.56	10.26	13.56	4.93	8.95	9.48	20.95	8.02	5.77	19.37	6.98	6.06	6.06	28.84	20.64	4.60	4.61	24.05
SC	24.05	18.13	22.49	16.89	9.59	11.45	11.78	11.78	16.99	18.58	11.06	13.93	15.50	20.61	15.75	12.36	50.08	13.06	13.91	13.91	41.22	32.53	10.50	6.33	41.22
SD	20.93	15.47	19.14	18.15	9.09	5.14	9.75	9.75	12.75	14.72	6.60	13.21	10.64	14.24	10.47	11.11	35.39	8.50	13.06	13.06	39.43	23.87	5.91	5.78	39.43
TN	23.85	11.59	12.99	12.86	6.71	6.45	12.53	12.53	12.51	11.17	7.83	13.19	9.84	18.04	12.47	6.13	34.24	8.48	12.29	12.29	21.99	22.19	8.37	4.68	21.99
TX	9.68	7.76	9.68	8.61	6.25	5.65	7.33	7.33	7.82	10.32	4.91	8.32	8.04	12.94	8.04	5.47	17.96	13.94	7.16	7.16	29.32	11.95	5.11	5.97	8.11
UT	12.88	5.69	8.30	6.49	5.64	3.19	7.02	7.02	6.89	5.87	5.20	7.33	6.66	6.78	5.02	3.49	20.72	5.58	5.07	5.07	22.57	10.74	4.22	4.58	18.77
VT	15.08	11.19	16.10	12.21	10.23	4.29	10.77	10.77	14.25	15.36	5.94	12.50	8.94	11.44	9.33	7.98	27.37	7.17	10.63	10.63	40.22	21.76	6.92	7.60	40.22
VA	10.37	7.72	8.56	9.91	5.11	4.15	7.23	7.23	7.71	6.23	6.03	7.08	8.23	7.98	6.24	4.64	21.77	5.23	6.70	6.70	28.82	11.93	4.63	2.80	28.82
WA	9.36	9.36	9.36	8.89	8.84	3.25	6.31	6.31	13.96	7.06	9.36	12.14	14.60	21.48	11.32	4.26	20.35	4.48	8.15	8.15	8.15	8.15	7.73	20.35	8.15
WV	13.06	8.27	9.67	9.38	5.25	5.27	7.90	7.90	9.46	7.92	5.44	8.80	8.73	12.01	8.28	5.23	24.07	6.86	7.49	7.49	28.94	13.43	6.48	3.54	29.69
WI	11.59	12.19	16.32	11.25	8.29	5.40	7.88	7.88	12.04	12.28	4.23	15.07	12.86	27.64	10.08	6.29	26.71	9.04	11.42	11.42	20.66	22.20	15.04	5.37	20.66
WY	5.50	5.50	5.50	5.50	5.50	5.50	5.50	5.50	5.50	5.50	5.50	5.50	5.50	5.50	5.50	5.50	5.50	5.50	5.50	5.50	5.50	5.50	5.50	5.50	5.50
AVG.	16.40	10.76	15.13	12.62	8.36	5.68	9.20	9.20	12.46	11.44	7.89	12.48	11.12	15.32	10.91	6.66	29.63	8.56	10.55	10.55	36.45	16.12	7.97	6.57	31.38

R013113-60 Workers' Compensation (cont.) (Canada in Canadian dollars)

Province		Alberta	British Columbia	Manitoba	Ontario	New Brunswick	Newfndld. & Labrador	Northwest Territories	Nova Scotia	Prince Edward Island	Quebec	Saskat-chewan	Yukon
Carpentry—3 stories or less	Rate	5.89	3.35	4.04	4.44	3.80	8.38	4.16	5.71	6.55	12.17	3.54	6.63
	Code	42143	721028	40102	723	238130	4226	4-41	4226	403	80110	B1226	202
Carpentry—interior cab. work	Rate	2.21	2.75	4.04	4.44	3.80	4.10	4.16	5.39	3.89	12.17	2.03	6.63
	Code	42133	721021	40102	723	238350	4279	4-41	4274	402	80110	B1127	202
CARPENTRY—general	Rate	5.89	3.35	4.04	4.44	3.80	4.10	4.16	5.71	6.55	12.17	3.54	6.63
	Code	42143	721028	40102	723	238130	4299	4-41	4226	403	80110	B1202	202
CONCRETE WORK—NOC	Rate	3.86	3.24	6.35	17.86	3.80	8.38	4.16	5.04	3.89	12.32	3.38	6.63
	Code	42104	721010	40110	748	238110	4224	4-41	4224	402	80100	B1314	203
CONCRETE WORK—flat (flr. sidewalk)	Rate	3.86	3.24	6.35	17.86	3.80	8.38	4.16	5.04	3.89	12.32	3.38	6.63
	Code	42104	721010	40110	748	238110	4224	4-41	4224	402	80100	B1314	203
ELECTRICAL Wiring—inside	Rate	1.66	1.81	1.70	3.60	1.73	2.42	2.45	2.14	3.89	4.94	2.03	3.89
	Code	42124	721019	40203	704	238210	4261	4-46	4261	402	80170	B1105	206
EXCAVATION—earth NOC	Rate	2.13	3.55	4.09	5.16	3.80	3.22	3.68	3.38	3.89	7.71	2.17	3.89
	Code	40604	721031	40706	711	238190	4214	4-43	4214	402	80030	R1106	207
EXCAVATION—rock	Rate	2.13	3.55	4.09	5.16	3.80	3.22	3.68	3.38	3.89	7.71	2.17	3.89
	Code	40604	721031	40706	711	238190	4214	4-43	4214	402	80030	R1106	207
GLAZIERS	Rate	2.38	3.15	4.04	10.00	3.80	5.14	4.16	5.71	3.89	13.74	3.38	3.89
	Code	42121	715020	40109	751	238150	4233	4-41	4233	402	80150	B1304	212
INSULATION WORK	Rate	1.81	8.22	4.04	10.00	3.80	5.14	4.16	5.71	3.89	12.17	2.54	6.63
	Code	42184	721029	40102	751	238310	4234	4-41	4234	402	80110	B1207	202
LATHING	Rate	4.16	6.65	4.04	4.44	3.80	4.10	4.16	5.39	3.89	12.17	3.38	6.63
	Code	42135	721042	40102	723	238390	4279	4-41	4271	402	80110	B1316	202
MASONRY	Rate	3.86	2.90	4.04	12.39	3.80	5.14	4.16	5.71	3.89	12.32	3.38	6.63
	Code	42102	721037	40102	741	238140	4231	4-41	4231	402	80100	B1318	202
PAINTING & DECORATING	Rate	3.51	3.30	3.39	7.33	3.80	4.10	4.16	5.39	3.89	12.17	3.54	6.63
	Code	42111	721041	40105	719	238320	4275	4-41	4275	402	80110	B1201	202
PILE DRIVING	Rate	3.82	3.47	4.09	6.86	3.80	3.22	3.68	5.04	6.55	7.71	3.38	6.63
	Code	42159	722004	40706	732	238190	4129	4-43	4221	403	80030	B1310	202
PLASTERING	Rate	4.16	6.65	4.64	7.33	3.80	4.10	4.16	5.39	3.89	12.17	3.54	6.63
	Code	42135	721042	40108	719	238390	4271	4-41	4271	402	80110	B1221	202
PLUMBING	Rate	1.66	2.39	2.93	4.06	1.73	2.50	2.45	3.38	2.20	6.38	2.03	3.89
	Code	42122	721043	40204	707	238220	4241	4-46	4241	401	80160	B1101	214
ROOFING	Rate	5.53	5.19	6.85	14.44	3.80	8.38	4.16	7.06	3.89	19.12	3.38	6.63
	Code	42118	721036	40403	728	238160	4236	4-41	4236	402	80130	B1320	202
SHEET METAL WORK (HVAC)	Rate	1.66	2.39	6.85	4.06	1.73	2.50	2.45	2.78	2.20	6.38	2.03	3.89
	Code	42117	721043	40402	707	238220	4244	4-46	4244	401	80160	B1107	208
STEEL ERECTION—door & sash	Rate	1.81	11.51	7.42	17.86	3.80	8.38	4.16	5.71	3.89	18.60	3.38	6.63
	Code	42106	722005	40502	748	238120	4227	4-41	4227	402	80080	B1322	202
STEEL ERECTION—inter., ornam.	Rate	1.81	11.51	7.42	17.86	3.80	8.38	4.16	5.71	3.89	18.68	3.38	6.63
	Code	42106	722005	40502	748	238120	4227	4-41	4227	402	80080	B1322	202
STEEL ERECTION—structure	Rate	1.81	11.51	7.42	17.86	3.80	8.38	4.16	5.71	3.89	18.68	3.38	6.63
	Code	42106	722005	40502	748	238120	4227	4-41	4227	402	80080	B1322	202
STEEL ERECTION—NOC	Rate	1.81	11.51	7.42	17.86	3.80	8.38	4.16	5.71	3.89	18.68	3.38	6.63
	Code	42106	722005	40502	748	238120	4227	4-41	4227	402	80080	B1322	202
TILE WORK—inter. (ceramic)	Rate	2.84	3.44	1.87	7.33	3.80	4.10	4.16	4.83	3.89	12.17	3.38	6.63
	Code	42113	721054	40103	719	238340	4276	4-41	4276	402	80110	B1301	202
WATERPROOFING	Rate	3.51	3.29	4.04	4.44	3.80	4.10	4.16	5.71	3.89	19.12	3.54	6.63
	Code	42139	721016	40102	723	238190	4299	4-41	4239	402	80130	B1217	202
WRECKING	Rate	2.13	5.32	6.86	17.86	3.80	3.22	3.68	3.38	3.89	12.17	2.17	6.63
	Code	40604	721005	40106	748	238190	4211	4-43	4211	402	80110	R1125	202

General Requirements — R0131 Project Management & Coordination

R013113-80 Performance Bond

This table shows the cost of a Performance Bond for a construction job scheduled to be completed in 12 months. Add 1% of the premium cost per month for jobs requiring more than 12 months to complete. The rates are "standard" rates offered to contractors that the bonding company considers financially sound and capable of doing the work. Preferred rates are offered by some bonding companies based upon financial strength of the contractor. Actual rates vary from contractor to contractor and from bonding company to bonding company. Contractors should prequalify through a bonding agency before submitting a bid on a contract that requires a bond.

Contract Amount	Building Construction Class B Projects	Highways & Bridges: Class A New Construction	Highways & Bridges: Class A-1 Highway Resurfacing
First $ 100,000 bid	$25.00 per M	$15.00 per M	$9.40 per M
Next 400,000 bid	$ 2,500 plus $15.00 per M	$ 1,500 plus $10.00 per M	$ 940 plus $7.20 per M
Next 2,000,000 bid	8,500 plus 10.00 per M	5,500 plus 7.00 per M	3,820 plus 5.00 per M
Next 2,500,000 bid	28,500 plus 7.50 per M	19,500 plus 5.50 per M	15,820 plus 4.50 per M
Next 2,500,000 bid	47,250 plus 7.00 per M	33,250 plus 5.00 per M	28,320 plus 4.50 per M
Over 7,500,000 bid	64,750 plus 6.00 per M	45,750 plus 4.50 per M	39,570 plus 4.00 per M

General Requirements — R0151 Temporary Utilities

R015113-65 Temporary Power Equipment

Cost data for the temporary equipment was developed utilizing the following information.

1) Re-usable material-services, transformers, equipment and cords are based on new purchase and prorated to three projects.
2) PVC feeder includes trench and backfill.
3) Connections include disconnects and fuses.
4) Labor units include an allowance for removal.
5) No utility company charges or fees are included.
6) Concrete pads or vaults are not included.
7) Utility company conduits not included.

R015423-10 Steel Tubular Scaffolding

On new construction, tubular scaffolding is efficient up to 60′ high or five stories. Above this it is usually better to use a hung scaffolding if construction permits. Swing scaffolding operations may interfere with tenants. In this case, the tubular is more practical at all heights.

In repairing or cleaning the front of an existing building the cost of tubular scaffolding per S.F. of building front increases as the height increases above the first tier. The first tier cost is relatively high due to leveling and alignment.

The minimum efficient crew for erecting and dismantling is three workers. They can set up and remove 18 frame sections per day up to 5 stories high. For 6 to 12 stories high, a crew of four is most efficient. Use two or more on top and two on the bottom for handing up or hoisting. They can also set up and remove 18 frame sections per day. At 7' horizontal spacing, this will run about 800 S.F. per day of erecting and dismantling. Time for placing and removing planks must be added to the above. A crew of three can place and remove 72 planks per day up to 5 stories. For over 5 stories, a crew of four can place and remove 80 planks per day.

The table below shows the number of pieces required to erect tubular steel scaffolding for 1000 S.F. of building frontage. This area is made up of a scaffolding system that is 12 frames (11 bays) long by 2 frames high.

For jobs under twenty-five frames, add 50% to rental cost. Rental rates will be lower for jobs over three months duration. Large quantities for long periods can reduce rental rates by 20%.

Description of Component	Number of Pieces for 1000 S.F. of Building Front	Unit
5′ Wide Standard Frame, 6′-4″ High	24	Ea.
Leveling Jack & Plate	24	↓
Cross Brace	44	↓
Side Arm Bracket, 21″	12	↓
Guardrail Post	12	↓
Guardrail, 7′ section	22	↓
Stairway Section	2	↓
Stairway Starter Bar	1	↓
Stairway Inside Handrail	2	↓
Stairway Outside Handrail	2	↓
Walk-Thru Frame Guardrail	2	↓

Scaffolding is often used as falsework over 15′ high during construction of cast-in-place concrete beams and slabs. Two foot wide scaffolding is generally used for heavy beam construction. The span between frames depends upon the load to be carried with a maximum span of 5′.

Heavy duty shoring frames with a capacity of 10,000#/leg can be spaced up to 10′ O.C. depending upon form support design and loading.

Scaffolding used as horizontal shoring requires less than half the material required with conventional shoring.

On new construction, erection is done by carpenters.

Rolling towers supporting horizontal shores can reduce labor and speed the job. For maintenance work, catwalks with spans up to 70′ can be supported by the rolling towers.

R015433-10 Contractor Equipment

Rental Rates shown elsewhere in the book pertain to late model high quality machines in excellent working condition, rented from equipment dealers. Rental rates from contractors may be substantially lower than the rental rates from equipment dealers depending upon economic conditions; for older, less productive machines, reduce rates by a maximum of 15%. Any overtime must be added to the base rates. For shift work, rates are lower. Usual rule of thumb is 150% of one shift rate for two shifts; 200% for three shifts.

For periods of less than one week, operated equipment is usually more economical to rent than renting bare equipment and hiring an operator.

Costs to move equipment to a job site (mobilization) or from a job site (demobilization) are not included in rental rates, nor in any Equipment costs on any Unit Price line items or crew listings. These costs can be found elsewhere. If a piece of equipment is already at a job site, it is not appropriate to utilize mob/demob costs in an estimate again.

Rental rates vary throughout the country with larger cities generally having lower rates. Lease plans for new equipment are available for periods in excess of six months with a percentage of payments applying toward purchase.

Rental rates can also be treated as reimbursement costs for contractor-owned equipment. Owned equipment costs include depreciation, loan payments, interest, taxes, insurance, storage, and major repairs.

Monthly rental rates vary from 2% to 5% of the cost of the equipment depending on the anticipated life of the equipment and its wearing parts. Weekly rates are about 1/3 the monthly rates and daily rental rates about 1/3 the weekly rate.

The hourly operating costs for each piece of equipment include costs to the user such as fuel, oil, lubrication, normal expendables for the equipment, and a percentage of mechanic's wages chargeable to maintenance. The hourly operating costs listed do not include the operator's wages.

The daily cost for equipment used in the standard crews is figured by dividing the weekly rate by five, then adding eight times the hourly operating cost to give the total daily equipment cost, not including the operator. This figure is in the right hand column of the Equipment listings under Equipment Cost/Day.

Pile Driving rates shown for pile hammer and extractor do not include leads, crane, boiler or compressor. Vibratory pile driving requires an added field specialist during set-up and pile driving operation for the electric model. The hydraulic model requires a field specialist for set-up only. Up to 125 reuses of sheet piling are possible using vibratory drivers. For normal conditions, crane capacity for hammer type and size are as follows.

Crane Capacity	Hammer Type and Size		
	Air or Steam	Diesel	Vibratory
25 ton	to 8,750 ft.-lb.		70 H.P.
40 ton	15,000 ft.-lb.	to 32,000 ft.-lb.	170 H.P.
60 ton	25,000 ft.-lb.		300 H.P.
100 ton		112,000 ft.-lb.	

Cranes should be specified for the job by size, building and site characteristics, availability, performance characteristics, and duration of time required.

Backhoes & Shovels rent for about the same as equivalent size cranes but maintenance and operating expense is higher. Crane operators rate must be adjusted for high boom heights. Average adjustments: for 150′ boom add 2% per hour; over 185′, add 4% per hour; over 210′, add 6% per hour; over 250′, add 8% per hour and over 295′, add 12% per hour.

Tower Cranes of the climbing or static type have jibs from 50′ to 200′ and capacities at maximum reach range from 4,000 to 14,000 pounds. Lifting capacities increase up to maximum load as the hook radius decreases.

Typical rental rates, based on purchase price are about 2% to 3% per month.

Erection and dismantling runs between 500 and 2000 labor hours. Climbing operation takes 10 labor hours per 20′ climb. Crane dead time is about 5 hours per 40′ climb. If crane is bolted to side of the building add cost of ties and extra mast sections. Climbing cranes have from 80′ to 180′ of mast while static cranes have 80′ to 800′ of mast.

Truck Cranes can be converted to tower cranes by using tower attachments. Mast heights over 400′ have been used.

A single 100′ high material **Hoist and Tower** can be erected and dismantled in about 400 labor hours; a double 100′ high hoist and tower in about 600 labor hours. Erection times for additional heights are 3 and 4 labor hours per vertical foot respectively up to 150′, and 4 to 5 labor hours per vertical foot over 150′ high. A 40′ high portable Buck hoist takes about 160 labor hours to erect and dismantle. Additional heights take 2 labor hours per vertical foot to 80′ and 3 labor hours per vertical foot for the next 100′. Most material hoists do not meet local code requirements for carrying personnel.

A 150′ high **Personnel Hoist** requires about 500 to 800 labor hours to erect and dismantle. Budget erection time at 5 labor hours per vertical foot for all trades. Local code requirements or labor scarcity requiring overtime can add up to 50% to any of the above erection costs.

Earthmoving Equipment: The selection of earthmoving equipment depends upon the type and quantity of material, moisture content, haul distance, haul road, time available, and equipment available. Short haul cut and fill operations may require dozers only, while another operation may require excavators, a fleet of trucks, and spreading and compaction equipment. Stockpiled material and granular material are easily excavated with front end loaders. Scrapers are most economically used with hauls between 300′ and 1-1/2 miles if adequate haul roads can be maintained. Shovels are often used for blasted rock and any material where a vertical face of 8′ or more can be excavated. Special conditions may dictate the use of draglines, clamshells, or backhoes. Spreading and compaction equipment must be matched to the soil characteristics, the compaction required and the rate the fill is being supplied.

R024119-10 Demolition Defined

Whole Building Demolition - Demolition of the whole building with no concern for any particular building element, component, or material type being demolished. This type of demolition is accomplished with large pieces of construction equipment that break up the structure, load it into trucks and haul it to a disposal site, but disposal or dump fees are not included. Demolition of below-grade foundation elements, such as footings, foundation walls, grade beams, slabs on grade, etc., is not included. Certain mechanical equipment containing flammable liquids or ozone-depleting refrigerants, electric lighting elements, communication equipment components, and other building elements may contain hazardous waste, and must be removed, either selectively or carefully, as hazardous waste before the building can be demolished.

Foundation Demolition - Demolition of below-grade foundation footings, foundation walls, grade beams, and slabs on grade. This type of demolition is accomplished by hand or pneumatic hand tools, and does not include saw cutting, or handling, loading, hauling, or disposal of the debris.

Gutting - Removal of building interior finishes and electrical/mechanical systems down to the load-bearing and sub-floor elements of the rough building frame, with no concern for any particular building element, component, or material type being demolished. This type of demolition is accomplished by hand or pneumatic hand tools, and includes loading into trucks, but not hauling, disposal or dump fees, scaffolding, or shoring. Certain mechanical equipment containing flammable liquids or ozone-depleting refrigerants, electric lighting elements, communication equipment components, and other building elements may contain hazardous waste, and must be removed, either selectively or carefully, as hazardous waste, before the building is gutted.

Selective Demolition - Demolition of a selected building element, component, or finish, with some concern for surrounding or adjacent elements, components, or finishes (see the first Subdivision (s) at the beginning of appropriate Divisions). This type of demolition is accomplished by hand or pneumatic hand tools, and does not include handling, loading, storing, hauling, or disposal of the debris, scaffolding, or shoring. "Gutting" methods may be used in order to save time, but damage that is caused to surrounding or adjacent elements, components, or finishes may have to be repaired at a later time.

Careful Removal - Removal of a piece of service equipment, building element or component, or material type, with great concern for both the removed item and surrounding or adjacent elements, components or finishes. The purpose of careful removal may be to protect the removed item for later re-use, preserve a higher salvage value of the removed item, or replace an item while taking care to protect surrounding or adjacent elements, components, connections, or finishes from cosmetic and/or structural damage. An approximation of the time required to perform this type of removal is 1/3 to 1/2 the time it would take to install a new item of like kind (see Reference Number R260105-30). This type of removal is accomplished by hand or pneumatic hand tools, and does not include loading, hauling, or storing the removed item, scaffolding, shoring, or lifting equipment.

Cutout Demolition - Demolition of a small quantity of floor, wall, roof, or other assembly, with concern for the appearance and structural integrity of the surrounding materials. This type of demolition is accomplished by hand or pneumatic hand tools, and does not include saw cutting, handling, loading, hauling, or disposal of debris, scaffolding, or shoring.

Rubbish Handling - Work activities that involve handling, loading or hauling of debris. Generally, the cost of rubbish handling must be added to the cost of all types of demolition, with the exception of whole building demolition.

Minor Site Demolition - Demolition of site elements outside the footprint of a building. This type of demolition is accomplished by hand or pneumatic hand tools, or with larger pieces of construction equipment, and may include loading a removed item onto a truck (check the Crew for equipment used). It does not include saw cutting, hauling or disposal of debris, and, sometimes, handling or loading.

R024119-20 Dumpsters

Dumpster rental costs on construction sites are presented in two ways.

The cost per week rental includes the delivery of the dumpster; its pulling or emptying once per week, and its final removal. The assumption is made that the dumpster contractor could choose to empty a dumpster by simply bringing in an empty unit and removing the full one. These costs also include the disposal of the materials in the dumpster.

The Alternate Pricing can be used when actual planned conditions are not approximated by the weekly numbers. For example, these lines can be used when a dumpster is needed for 4 weeks and will need to be emptied 2 or 3 times per week. Conversely the Alternate Pricing lines can be used when a dumpster will be rented for several weeks or months but needs to be emptied only a few times over this period.

Metals R0531 Steel Decking

R053100-10 Decking Descriptions

General - All Deck Products

Steel deck is made by cold forming structural grade sheet steel into a repeating pattern of parallel ribs. The strength and stiffness of the panels are the result of the ribs and the material properties of the steel. Deck lengths can be varied to suit job conditions, but because of shipping considerations, are usually less than 40 feet. Standard deck width varies with the product used but full sheets are usually 12″, 18″, 24″, 30″, or 36″. Deck is typically furnished in a standard width with the ends cut square. Any cutting for width, such as at openings or for angular fit, is done at the job site.

Deck is typically attached to the building frame with arc puddle welds, self-drilling screws, or powder or pneumatically driven pins. Sheet to sheet fastening is done with screws, button punching (crimping), or welds.

Composite Floor Deck

After installation and adequate fastening, floor deck serves several purposes. It (a) acts as a working platform, (b) stabilizes the frame, (c) serves as a concrete form for the slab, and (d) reinforces the slab to carry the design loads applied during the life of the building. Composite decks are distinguished by the presence of shear connector devices as part of the deck. These devices are designed to mechanically lock the concrete and deck together so that the concrete and the deck work together to carry subsequent floor loads. These shear connector devices can be rolled-in embossments, lugs, holes, or wires welded to the panels. The deck profile can also be used to interlock concrete and steel.

Composite deck finishes are either galvanized (zinc coated) or phosphatized/painted. Galvanized deck has a zinc coating on both the top and bottom surfaces. The phosphatized/painted deck has a bare (phosphatized) top surface that will come into contact with the concrete. This bare top surface can be expected to develop rust before the concrete is placed. The bottom side of the deck has a primer coat of paint.

Composite floor deck is normally installed so the panel ends do not overlap on the supporting beams. Shear lugs or panel profile shape often prevent a tight metal to metal fit if the panel ends overlap; the air gap caused by overlapping will prevent proper fusion with the structural steel supports when the panel end laps are shear stud welded.

Adequate end bearing of the deck must be obtained as shown on the drawings. If bearing is actually less in the field than shown on the drawings, further investigation is required.

Roof Deck

Roof deck is not designed to act compositely with other materials. Roof deck acts alone in transferring horizontal and vertical loads into the building frame. Roof deck rib openings are usually narrower than floor deck rib openings. This provides adequate support of rigid thermal insulation board.

Roof deck is typically installed to endlap approximately 2″ over supports. However, it can be butted (or lapped more than 2″) to solve field fit problems. Since designers frequently use the installed deck system as part of the horizontal bracing system (the deck as a diaphragm), any fastening substitution or change should be approved by the designer. Continuous perimeter support of the deck is necessary to limit edge deflection in the finished roof and may be required for diaphragm shear transfer.

Standard roof deck finishes are galvanized or primer painted. The standard factory applied paint for roof deck is a primer paint and is not intended to weather for extended periods of time. Field painting or touching up of abrasions and deterioration of the primer coat or other protective finishes is the responsibility of the contractor.

Cellular Deck

Cellular deck is made by attaching a bottom steel sheet to a roof deck or composite floor deck panel. Cellular deck can be used in the same manner as floor deck. Electrical, telephone, and data wires are easily run through the chase created between the deck panel and the bottom sheet.

When used as part of the electrical distribution system, the cellular deck must be installed so that the ribs line up and create a smooth cell transition at abutting ends. The joint that occurs at butting cell ends must be taped or otherwise sealed to prevent wet concrete from seeping into the cell. Cell interiors must be free of welding burrs, or other sharp intrusions, to prevent damage to wires.

When used as a roof deck, the bottom flat plate is usually left exposed to view. Care must be maintained during erection to keep good alignment and prevent damage.

Cellular deck is sometimes used with the flat plate on the top side to provide a flat working surface. Installation of the deck for this purpose requires special methods for attachment to the frame because the flat plate, now on the top, can prevent direct access to the deck material that is bearing on the structural steel. It may be advisable to treat the flat top surface to prevent slipping.

Cellular deck is always furnished galvanized or painted over galvanized.

Form Deck

Form deck can be any floor or roof deck product used as a concrete form. Connections to the frame are by the same methods used to anchor floor and roof deck. Welding washers are recommended when welding deck that is less than 20 gauge thickness.

Form deck is furnished galvanized, prime painted, or uncoated. Galvanized deck must be used for those roof deck systems where form deck is used to carry a lightweight insulating concrete fill.

Thermal & Moist. Protec. R0784 Firestopping

R078413-30 Firestopping

Firestopping is the sealing of structural, mechanical, electrical and other penetrations through fire-rated assemblies. The basic components of firestop systems are safing insulation and firestop sealant on both sides of wall penetrations and the top side of floor penetrations.

Pipe penetrations are assumed to be through concrete, grout, or joint compound and can be sleeved or unsleeved. Costs for the penetrations and sleeves are not included. An annular space of 1″ is assumed. Escutcheons are not included.

Metallic pipe is assumed to be copper, aluminum, cast iron or similar metallic material. Insulated metallic pipe is assumed to be covered with a thermal insulating jacket of varying thickness and materials.

Non-metallic pipe is assumed to be PVC, CPVC, FR Polypropylene or similar plastic piping material. Intumescent firestop sealant or wrap strips are included. Collars on both sides of wall penetrations and a sheet metal plate on the underside of floor penetrations are included.

Ductwork is assumed to be sheet metal, stainless steel or similar metallic material. Duct penetrations are assumed to be through concrete, grout or joint compound. Costs for penetrations and sleeves are not included. An annular space of 1/2″ is assumed.

Multi-trade openings include costs for sheet metal forms, firestop mortar, wrap strips, collars and sealants as necessary.

Structural penetrations joints are assumed to be 1/2″ or less. CMU walls are assumed to be within 1-1/2″ of metal deck. Drywall walls are assumed to be tight to the underside of metal decking.

Metal panel, glass or curtain wall systems include a spandrel area of 5′ filled with mineral wool foil-faced insulation. Fasteners and stiffeners are included.

R220102-20 Labor Adjustment Factors

Labor Adjustment Factors are provided for Divisions 21, 22, and 23 to assist the mechanical estimator account for the various complexities and special conditions of any particular project. While a single percentage has been entered on each line of Division 22 01 02.20, it should be understood that these are just suggested midpoints of ranges of values commonly used by mechanical estimators. They may be increased or decreased depending on the severity of the special conditions.

The group for "existing occupied buildings", has been the subject of requests for explanation. Actually there are two stages to this group: buildings that are existing and "finished" but unoccupied, and those that also are occupied. Buildings that are "finished" may result in higher labor costs due to the workers having to be more careful not to damage finished walls, ceilings, floors, etc. and may necessitate special protective coverings and barriers. Also corridor bends and doorways may not accommodate long pieces of pipe or larger pieces of equipment. Work above an already hung ceiling can be very time consuming. The addition of occupants may force the work to be done on premium time (nights and/or weekends), eliminate the possible use of some preferred tools such as pneumatic drivers, powder charged drivers etc. The estimator should evaluate the access to the work area and just how the work is going to be accomplished to arrive at an increase in labor costs over "normal" new construction productivity.

R224000-10 Hot Water Consumption Rates

Type of Building	Size Factor	Maximum Hourly Demand	Average Day Demand
Apartment Dwellings	No. of Apartments:		
	Up to 20	12.0 Gal. per apt.	42.0 Gal. per apt.
	21 to 50	10.0 Gal. per apt.	40.0 Gal. per apt.
	51 to 75	8.5 Gal. per apt.	38.0 Gal. per apt.
	76 to 100	7.0 Gal. per apt.	37.0 Gal. per apt.
	101 to 200	6.0 Gal. per apt.	36.0 Gal. per apt.
	201 up	5.0 Gal. per apt.	35.0 Gal. per apt.
Dormitories	Men	3.8 Gal. per man	13.1 Gal. per man
	Women	5.0 Gal. per woman	12.3 Gal. per woman
Hospitals	Per bed	23.0 Gal. per patient	90.0 Gal. per patient
Hotels	Single room with bath	17.0 Gal. per unit	50.0 Gal. per unit
	Double room with bath	27.0 Gal. per unit	80.0 Gal. per unit
Motels	No. of units:		
	Up to 20	6.0 Gal. per unit	20.0 Gal. per unit
	21 to 100	5.0 Gal. per unit	14.0 Gal. per unit
	101 Up	4.0 Gal. per unit	10.0 Gal. per unit
Nursing Homes		4.5 Gal. per bed	18.4 Gal. per bed
Office buildings		0.4 Gal. per person	1.0 Gal. per person
Restaurants	Full meal type	1.5 Gal./max. meals/hr.	2.4 Gal. per meal
	Drive-in snack type	0.7 Gal./max. meals/hr.	0.7 Gal. per meal
Schools	Elementary	0.6 Gal. per student	0.6 Gal. per student
	Secondary & High	1.0 Gal. per student	1.8 Gal. per student

For evaluation purposes, recovery rate and storage capacity are inversely proportional. Water heaters should be sized so that the maximum hourly demand anticipated can be met in addition to allowance for the heat loss from the pipes and storage tank.

R224000-20 Fixture Demands in Gallons Per Fixture Per Hour

Table below is based on 140° F final temperature except for dishwashers in public places (*) where 180° F water is mandatory.

Fixture	Apartment House	Club	Gym	Hospital	Hotel	Indust. Plant	Office	Private Home	School
Bathtubs	20	20	30	20	20			20	
Dishwashers, automatic	15	50-150*		50-150*	50-200*	20-100*		15	20-100*
Kitchen sink	10	20		20	30	20	20	10	20
Laundry, stationary tubs	20	28		28	28			20	
Laundry, automatic wash	75	75		100	150			75	
Private lavatory	2	2	2	2	2	2	2	2	2
Public lavatory	4	6	8	6	8	12	6		15
Showers	30	150	225	75	75	225	30	30	225
Service sink	20	20		20	30	20	20	15	20
Demand factor	0.30	0.30	0.40	0.25	0.25	0.40	0.30	0.30	0.40
Storage capacity factor	1.25	0.90	1.00	0.60	0.80	1.00	2.00	0.70	1.00

To obtain the probable maximum demand multiply the total demands for the fixtures (gal./fixture/hour) by the demand factor. The heater should have a heating capacity in gallons per hour equal to this maximum. The storage tank should have a capacity in gallons equal to the probable maximum demand multiplied by the storage capacity factor.

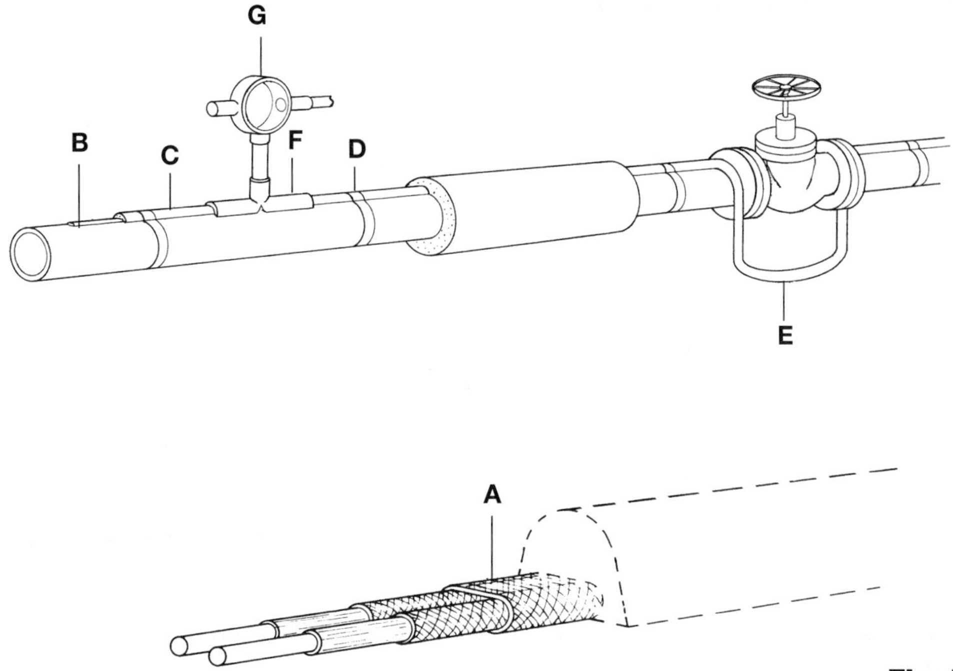

Fig. R238313-11

R238313-10 Heat Trace Systems

Before you can determine the cost of a HEAT TRACE installation, the method of attachment must be established. There are (4) common methods:

1. Cable is simply attached to the pipe with polyester tape every 12′.
2. Cable is attached with a continuous cover of 2″ wide aluminum tape.
3. Cable is attached with factory extruded heat transfer cement and covered with metallic raceway with clips every 10′.
4. Cable is attached between layers of pipe insulation using either clips or polyester tape.

Example: Components for method 3 must include:

A. Heat trace cable by voltage and watts per linear foot.
B. Heat transfer cement, 1 gallon per 60 linear feet of cover.
C. Metallic raceway by size and type.
D. Raceway clips by size of pipe.

When taking off linear foot lengths of cable add the following for each valve in the system. (E)

In all of the above methods each component of the system must be priced individually.

SCREWED OR WELDED VALVE:

Size		Length
1/2″	=	6″
3/4″	=	9″
1″	=	1′ -0″
1-1/2″	=	1′ -6″
2″	=	2′
2-1/2″	=	2′ -6″
3″	=	2′ -6″
4″	=	4′ -0″
6″	=	7′ -0″
8″	=	9′ -6″
10″	=	12′ -6″
12″	=	15′ -0″
14″	=	18′ -0″
16″	=	21′ -6″
18″	=	25′ -6″
20″	=	28′ -6″
24″	=	34′ -0″
30″	=	40′ -0″

FLANGED VALVE:

Size		Length
1/2″	=	1′ -0″
3/4″	=	1′ -6″
1″	=	2′ -0″
1-1/2″	=	2′ -6″
2″	=	2′ -6″
2-1/2″	=	3′ -0″
3″	=	3′ -6″
4″	=	4′ -0″
6″	=	8′ -0″
8″	=	11′ -0″
10″	=	14′ -0″
12″	=	16′ -6″
14″	=	19′ -6″
16″	=	23′ -0″
18″	=	27′ -0″
20″	=	30′ -0″
24″	=	36′ -0″
30″	=	42′ -0″

BUTTERFLY VALVES:

Size		Length
1/2″	=	0′
3/4″	=	0′
1″	=	1′ -0″
1-1/2″	=	1′ -6″
2″	=	2′ -0″
2-1/2″	=	2′ -6″
3″	=	2′ -6″
4″	=	3′ -0″
6″	=	3′ -6″
8″	=	4′ -0″
10″	=	4′ -0″
12″	=	5′ -0″
14″	=	5′ -6″
16″	=	6′ -0″
18″	=	6′ -6″
20″	=	7′ -0″
24″	=	8′ -0″
30″	=	10′ -0″

R238313-10 Heat Trace Systems (cont.)

Add the following quantities of heat transfer cement to linear foot totals for each valve:

Nominal Valve Size	Gallons of Cement per Valve
1/2″	0.14
3/4″	0.21
1″	0.29
1-1/2″	0.36
2″	0.43
2-1/2″	0.70
3″	0.71
4″	1.00
6″	1.43
8″	1.48
10″	1.50
12″	1.60
14″	1.75
16″	2.00
18″	2.25
20″	2.50
24″	3.00
30″	3.75

The following must be added to the list of components to accurately price HEAT TRACE systems:

1. Expediter fitting and clamp fasteners (F)
2. Junction box and nipple connected to expediter fitting (G)
3. Field installed terminal blocks within junction box
4. Ground lugs
5. Piping from power source to expediter fitting
6. Controls
7. Thermostats
8. Branch wiring
9. Cable splices
10. End of cable terminations
11. Branch piping fittings and boxes

Deduct the following percentages from labor if cable lengths in the same area exceed:

150′ to 250′	10%	351′ to 500′	20%
251′ to 350′	15%	Over 500′	25%

Add the following percentages to labor for elevated installations:

15′ to 20′ high	10%	31′ to 35′ high	40%
21′ to 25′ high	20%	36′ to 40′ high	50%
26′ to 30′ high	30%	Over 40′ high	60%

R238313-20 Spiral-Wrapped Heat Trace Cable (Pitch Table)

In order to increase the amount of heat, occasionally heat trace cable is wrapped in a spiral fashion around a pipe; increasing the number of feet of heater cable per linear foot of pipe.

Engineers first determine the heat loss per foot of pipe (based on the insulating material, its thickness, and the temperature differential across it). A ratio is then calculated by the formula:

$$\text{Feet of Heat Trace per Foot of Pipe} = \frac{\text{Watts/Foot of Heat Loss}}{\text{Watts/Foot of the Cable}}$$

The linear distance between wraps (pitch) is then taken from a chart or table. Generally, the pitch is listed on a drawing leaving the estimator to calculate the total length of heat tape required. An approximation may be taken from this table.

	Feet of Heat Trace Per Foot of Pipe															
	Nominal Pipe Size in Inches															
Pitch In Inches	1	1¼	1½	2	2½	3	4	6	8	10	12	14	16	18	20	24
3.5	1.80															
4	1.65															
5	1.46	1.60	1.80													
6	1.34	1.45	1.55	1.75												
7	1.25	1.35	1.43	1.57	1.75											
8	1.20	1.28	1.34	1.45	1.60	1.80										
9	1.16	1.23	1.28	1.37	1.51	1.68										
10	1.13	1.19	1.24	1.32	1.44	1.57	1.82									
15	1.06	1.08	1.10	1.15	1.21	1.29	1.42	1.78								
20	1.04	1.05	1.06	1.08	1.13	1.17	1.25	1.49	1.73							
25		1.04	1.04	1.06	1.08	1.11	1.17	1.33	1.51	1.72						
30				1.04	1.05	1.07	1.12	1.24	1.37	1.54	1.70	1.80				
35					1.06	1.06	1.09	1.17	1.28	1.42	1.54	1.64	1.78			
40						1.05	1.07	1.14	1.22	1.33	1.44	1.52	1.64	1.75		
50							1.05	1.09	1.15	1.22	1.29	1.35	1.44	1.53	1.64	1.83
60								1.06	1.11	1.16	1.21	1.25	1.31	1.39	1.46	1.62
70								1.05	1.08	1.12	1.17	1.19	1.24	1.30	1.35	1.47
80									1.06	1.09	1.13	1.15	1.19	1.24	1.28	1.38
90									1.04	1.06	1.10	1.13	1.16	1.19	1.23	1.32
100										1.05	1.08	1.10	1.13	1.15	1.19	1.23

Note: Common practice would normally limit the lower end of the table to 5% of additional heat and above 80% an engineer would likely opt for two (2) parallel cables.

R260105-30 Electrical Demolition (Removal for Replacement)

The purpose of this reference number is to provide a guide to users for electrical "removal for replacement" by applying the rule of thumb: 1/3 of new installation time (typical range from 20% to 50%) for removal. Remember to use reasonable judgment when applying the suggested percentage factor. For example:

Contractors have been requested to remove an existing fluorescent lighting fixture and replace with a new fixture utilizing energy saver lamps and electronic ballast:

In order to fully understand the extent of the project, contractors should visit the job site and estimate the time to perform the renovation work in accordance with applicable national, state and local regulation and codes.

The contractor may need to add extra labor hours to his estimate if he discovers unknown concealed conditions such as: contaminated asbestos ceiling, broken acoustical ceiling tile and need to repair, patch and touch-up paint all the damaged or disturbed areas, tasks normally assigned to general contractors. In addition, the owner could request that the contractors salvage the materials removed and turn over the materials to the owner or dispose of the materials to a reclamation station. The normal removal item is 0.5 labor hour for a lighting fixture and 1.5 labor-hours for new installation time. Revise the estimate times from 2 labor-hours work up to a minimum 4 labor-hours work for just fluorescent lighting fixture.

For removal of large concentrations of lighting fixtures in the same area, apply an "economy of scale" to reduce estimating labor hours.

R260519-20 Armored Cable

Armored Cable – Quantities are taken off in the same manner as wire.

Bx Type Cable – Productivities are based on an average run of 50′ before terminating at a box fixture, etc. Each 50′ section includes field preparation of (2) ends with hacksaw, identification and tagging of wire. Set up is open coil type without reels attaching cable to snake and pulling across a suspended ceiling or open face wood or steel studding, price does not include drilling of studs.

Cable in Tray – Productivities are based on an average run of 100 L.F. with set up of pulling equipment for (2) 90° bends, attaching cable to pull-in means, identification and tagging of wires, set up of reels. Wire termination to breakers equipment, etc., are not included.

Job Conditions – Productivities are based on new construction to a height of 15′ using rolling staging in an unobstructed area. Material staging is assumed to be within 100′ of work being performed.

R260519-80 Undercarpet Systems

Takeoff Procedure for Power Systems: List components for each fitting type, tap, splice, and bend on your quantity takeoff sheet. Each component must be priced separately. Start at the power supply transition fittings and survey each circuit for the components needed. List the quantities of each component under a specific circuit number. Use the floor plan layout scale to get cable footage.

Reading across the list, combine the totals of each component in each circuit and list the total quantity in the last column. Calculate approximately 5% for scrap for items such as cable, top shield, tape, and spray adhesive. Also provide for final variations that may occur on-site.

Suggested guidelines are:
1. Equal amounts of cable and top shield should be priced.
2. For each roll of cable, price a set of cable splices.
3. For every 1 ft. of cable, price 2-1/2 ft. of hold-down tape.
4. For every 3 rolls of hold-down tape, price 1 can of spray adhesive.

Adjust final figures wherever possible to accommodate standard packaging of the product. This information is available from the distributor.

Each transition fitting requires:
1. 1 base
2. 1 cover
3. 1 transition block

Each floor fitting requires:
1. 1 frame/base kit
2. 1 transition block
3. 2 covers (duplex/blank)

Each tap requires:
1. 1 tap connector for each conductor
2. 1 pair insulating patches
3. 2 top shield connectors

Each splice requires:
1. 1 splice connector for each conductor
2. 1 pair insulating patches
3. 3 top shield connectors

Each cable bend requires:
1. 2 top shield connectors

Each cable dead end (outside of transition block) requires:
1. 1 pair insulating patches

Labor does not include:
1. Patching or leveling uneven floors.
2. Filling in holes or removing projections from concrete slabs.
3. Sealing porous floors.
4. Sweeping and vacuuming floors.
5. Removal of existing carpeting.
6. Carpet square cut-outs.
7. Installation of carpet squares.

Takeoff Procedures for Telephone Systems: After reviewing floor plans identify each transition. Number or letter each cable run from that fitting.

Start at the transition fitting and survey each circuit for the components needed. List the cable type, terminations, cable length, and floor fitting type under the specific circuit number. Use the floor plan layout scale to get the cable footage. Add some extra length (next higher increment of 5 feet) to preconnectorized cable.

Transition fittings require:
1. 1 base plate
2. 1 cover
3. 1 transition block

Floor fittings require:
1. 1 frame/base kit
2. 2 covers
3. Modular jacks

Reading across the list, combine the list of components in each circuit and list the total quantity in the last column. Calculate the necessary scrap factors for such items as tape, bottom shield and spray adhesive. Also provide for final variations that may occur on-site.

Adjust final figures whenever possible to accommodate standard packaging. Check that items such as transition fittings, floor boxes, and floor fittings that are to utilize both power and telephone have been priced as combination fittings, so as to avoid duplication.

Make sure to include marking of floors and drilling of fasteners if fittings specified are not the adhesive type.

Labor does not include:
1. Conduit or raceways before transition of floor boxes
2. Telephone cable before transition boxes
3. Terminations before transition boxes
4. Floor preparation as described in power section

Be sure to include all cable folds when pricing labor.

Takeoff Procedure for Data Systems: Start at the transition fittings and take off quantities in the same manner as the telephone system, keeping in mind that data cable does not require top or bottom shields.

The data cable is simply cross-taped on the cable run to the floor fitting.

Data cable can be purchased in either bulk form in which case coaxial connector material and labor must be priced, or in preconnectorized cut lengths.

Data cable cannot be folded and must be notched at 1 inch intervals. A count of all turns must be added to the labor portion of the estimate. (Note: Some manufacturers have prenotched cable.)

Notching required:
1. 90 degree turn requires 8 notches per side.
2. 180 degree turn requires 16 notches per side.

Floor boxes, transition boxes, and fittings are the same as described in the power and telephone procedures.

Since undercarpet systems require special hand tools, be sure to include this cost in proportion to number of crews involved in the installation.

Job Conditions: Productivity is based on new construction in an unobstructed area. Staging area is assumed to be within 200′ of work being performed.

R260519-90 Wire

Wire quantities are taken off by either measuring each cable run or by extending the conduit and raceway quantities times the number of conductors in the raceway. Ten percent should be added for waste and tie-ins. Keep in mind that the unit of measure of wire is C.L.F. not L.F. as in raceways so the formula would read:

$$\frac{(\text{L.F. Raceway x No. of Conductors}) \times 1.10}{100} = \text{C.L.F.}$$

Price per C.L.F. of wire includes:
1. Set up wire coils or spools on racks
2. Attaching wire to pull in means
3. Measuring and cutting wire
4. Pulling wire into a raceway
5. Identifying and tagging

Price does not include:
1. Connections to breakers, panelboards or equipment
2. Splices

Job Conditions: Productivity is based on new construction to a height of 15′ using rolling staging in an unobstructed area. Material staging is assumed to be within 100′ of work being performed.

Economy of Scale: If more than three wires at a time are being pulled, deduct the following percentages from the labor of that grouping:

4-5 wires	25%
6-10 wires	30%
11-15 wires	35%
over 15	40%

If a wire pull is less than 100′ in length and is interrupted several times by boxes, lighting outlets, etc., it may be necessary to add the following lengths to each wire being pulled:

Junction box to junction box	2 L.F.
Lighting panel to junction box	6 L.F.
Distribution panel to sub panel	8 L.F.
Switchboard to distribution panel	12 L.F.
Switchboard to motor control center	20 L.F.
Switchboard to cable tray	40 L.F.

Measure of Drops and Riser: It is important when taking off wire quantities to include the wire for drops to electrical equipment. If heights of electrical equipment are not clearly stated, use the following guide:

	Bottom A.F.F.	Top A.F.F.	Inside Cabinet
Safety switch to 100A	5′	6′	2′
Safety switch 400 to 600A	4′	6′	3′
100A panel 12 to 30 circuit	4′	6′	3′
42 circuit panel	3′	6′	4′
Switch box	3′	3′6″	1′
Switchgear	0′	8′	8′
Motor control centers	0′	8′	8′
Transformers - wall mount	4′	8′	2′
Transformers - floor mount	0′	12′	4′

R260519-91 Maximum Circuit Length (approximate) for Various Power Requirements Assuming THW, Copper Wire @ 75°C, Based Upon a 4% Voltage Drop

Maximum Circuit Length: Table R260519-91 indicates typical maximum installed length a circuit can have and still maintain an adequate voltage level at the point of use. The circuit length is similar to the conduit length.

If the circuit length for an ampere load and a copper wire size exceeds the length obtained from Table R260519-91, use the next largest wire size to compensate voltage drop.

Amperes	Wire Size	Maximum Circuit Length in Feet				
		2 Wire, 1 Phase		3 Wire, 3 Phase		
		120V	240V	240V	480V	600V
15	14*	50	105	120	240	300
	14	50	100	120	235	295
20	12*	60	125	145	290	360
	12	60	120	140	280	350
30	10*	65	130	155	305	380
	10	65	130	150	300	375
50	8	60	125	145	285	355
65	6	75	150	175	345	435
85	4	90	185	210	425	530
115	2	110	215	250	500	620
130	1	120	240	275	555	690
150	1/0	130	260	305	605	760
175	2/0	140	285	330	655	820
200	3/0	155	315	360	725	904
230	4/0	170	345	395	795	990
255	250	185	365	420	845	1055
285	300	195	395	455	910	1140
310	350	210	420	485	975	1220
380	500	245	490	565	1130	1415

*Solid Conductor

Note: The circuit length is the one-way distance between the origin and the load.

Example: A 130 ampere load at 480 volts, 3 phase, 3 wire with No. 1 wire can be run a maximum of 555 L.F. and provide satisfactory operation. If the same load is to be wired at the end of a 625 L.F. circuit, then a larger wire must be used.

R260519-92 Minimum Copper and Aluminum Wire Size Allowed for Various Types of Insulation

Minimum Wire Sizes									
	Copper		Aluminum			Copper		Aluminum	
Amperes	THW THWN or XHHW	THHN XHHW *	THW XHHW	THHN XHHW *	Amperes	THW THWN or XHHW	THHN XHHW *	THW XHHW	THHN XHHW *
15A	#14	#14	#12	#12	195	3/0	2/0	250kcmil	4/0
20	#12	#12	#10	#10	200	3/0	3/0	250kcmil	4/0
25	#10	#10	#10	#10	205	4/0	3/0	250kcmil	4/0
30	#10	#10	# 8	# 8	225	4/0	3/0	300kcmil	250kcmil
40	# 8	# 8	# 8	# 8	230	4/0	4/0	300kcmil	250kcmil
45	# 8	# 8	# 6	# 8	250	250kcmil	4/0	350kcmil	300kcmil
50	# 8	# 8	# 6	# 6	255	250kcmil	4/0	400kcmil	300kcmil
55	# 6	# 8	# 4	# 6	260	300kcmil	4/0	400kcmil	350kcmil
60	# 6	# 6	# 4	# 6	270	300kcmil	250kcmil	400kcmil	350kcmil
65	# 6	# 6	# 4	# 4	280	300kcmil	250kcmil	500kcmil	350kcmil
75	# 4	# 6	# 3	# 4	285	300kcmil	250kcmil	500kcmil	400kcmil
85	# 4	# 4	# 2	# 3	290	350kcmil	250kcmil	500kcmil	400kcmil
90	# 3	# 4	# 2	# 2	305	350kcmil	300kcmil	500kcmil	400kcmil
95	# 3	# 4	# 1	# 2	310	350kcmil	300kcmil	500kcmil	500kcmil
100	# 3	# 3	# 1	# 2	320	400kcmil	300kcmil	600kcmil	500kcmil
110	# 2	# 3	1/0	# 1	335	400kcmil	350kcmil	600kcmil	500kcmil
115	# 2	# 2	1/0	# 1	340	500kcmil	350kcmil	600kcmil	500kcmil
120	# 1	# 2	1/0	1/0	350	500kcmil	350kcmil	700kcmil	500kcmil
130	# 1	# 2	2/0	1/0	375	500kcmil	400kcmil	700kcmil	600kcmil
135	1/0	# 1	2/0	1/0	380	500kcmil	400kcmil	750kcmil	600kcmil
150	1/0	# 1	3/0	2/0	385	600kcmil	500kcmil	750kcmil	600kcmil
155	2/0	1/0	3/0	3/0	420	600kcmil	500kcmil		700kcmil
170	2/0	1/0	4/0	3/0	430		500kcmil		750kcmil
175	2/0	2/0	4/0	3/0	435		600kcmil		750kcmil
180	3/0	2/0	4/0	4/0	475		600kcmil		

*Dry Locations Only

Notes:

1. Size #14 to 4/0 is in AWG units (American Wire Gauge).
2. Size 250 to 750 is in kcmil units (Thousand Circular Mils).
3. Use next higher ampere value if exact value is not listed in table.
4. For loads that operate continuously increase ampere value by 25% to obtain proper wire size.
5. Refer to Table R260519-91 for the maximum circuit length for the various size wires.
6. Table R260519-92 has been written for estimating purpose only, based on ambient temperature of 30°C (86° F); for ambient temperature other than 30°C (86° F), ampacity correction factors will be applied.

R260519-93 Metric Equivalent, Wire

U.S. vs. European Wire – Approximate Equivalents			
United States		European	
Size AWG or kcmil	Area Cir. Mils.(cmil) mm²	Size mm²	Area Cir. Mils.
18	1620/.82	.75	1480
16	2580/1.30	1.0	1974
14	4110/2.08	1.5	2961
12	6530/3.30	2.5	4935
10	10,380/5.25	4	7896
8	16,510/8.36	6	11,844
6	26,240/13.29	10	19,740
4	41,740/21.14	16	31,584
3	52,620/26.65	25	49,350
2	66,360/33.61	-	-
1	83,690/42.39	35	69,090
1/0	105,600/53.49	50	98,700
2/0	133,100/67.42	-	-
3/0	167,800/85.00	70	138,180
4/0	211,600/107.19	95	187,530
250	250,000/126.64	120	236,880
300	300,000/151.97	150	296,100
350	350,000/177.30	-	-
400	400,000/202.63	185	365,190
500	500,000/253.29	240	473,760
600	600,000/303.95	300	592,200
700	700,000/354.60	-	-
750	750,000/379.93	-	-

R260519-94 Size Required and Weight (Lbs./1000 L.F.) of Aluminum and Copper THW Wire by Ampere Load

Amperes	Copper Size	Aluminum Size	Copper Weight	Aluminum Weight
15	14	12	24	11
20	12	10	33	17
30	10	8	48	39
45	8	6	77	52
65	6	4	112	72
85	4	2	167	101
100	3	1	205	136
115	2	1/0	252	162
130	1	2/0	324	194
150	1/0	3/0	397	233
175	2/0	4/0	491	282
200	3/0	250	608	347
230	4/0	300	753	403
255	250	400	899	512
285	300	500	1068	620
310	350	500	1233	620
335	400	600	1396	772
380	500	750	1732	951

R260526-80 Grounding

Grounding When taking off grounding systems, identify separately the type and size of wire.

Example:

Bare copper & size

Bare aluminum & size

Insulated copper & size

Insulated aluminum & size

Count the number of ground rods and their size.

Example:

1. 8′ grounding rod – 5/8″ dia. 20 Ea.
2. 10′ grounding rod – 5/8″ dia. 12 Ea.
3. 15′ grounding rod – 3/4″ dia. 4 Ea.

Count the number of connections; the size of the largest wire will determine the productivity.

Example:

Braze a #2 wire to a #4/0 cable
The 4/0 cable will determine the L.H. and cost to be used.

Include individual connections to:
1. Ground rods
2. Building steel
3. Equipment
4. Raceways

Price does not include:
1. Excavation
2. Backfill
3. Sleeves or raceways used to protect grounding wires
4. Wall penetrations
5. Floor cutting
6. Core drilling

Job Conditions: Productivity is based on a ground floor area, using cable reels in an unobstructed area. Material staging area assumed to be within 100′ of work being performed.

R260533-20 Conduit To 15′ High

List conduit by quantity, size, and type. Do not deduct for lengths occupied by fittings, since this will be allowance for scrap. Example:

A. Aluminum — size
B. Rigid galvanized — size
C. Steel intermediate (IMC) — size
D. Rigid steel, plastic-coated 20 Mil. — size
E. Rigid steel, plastic-coated 40 Mil. — size
F. Electric metallic tubing (EMT) — size
G. PVC Schedule 40 — size

Types (A) thru (E) listed above contain the following per 100 L.F.:
1. (11) Threaded couplings
2. (11) Beam-type hangers
3. (2) Factory sweeps
4. (2) Fiber bushings
5. (4) Locknuts
6. (2) Field threaded pipe terminations
7. (2) Removal of concentric knockouts

Type (F) contains per 100 L.F.:
1. (11) Set screw couplings
2. (11) Beam clamps
3. (2) Field bends on 1/2″ and 3/4″ diameter
4. (2) Factory sweeps for 1″ and above
5. (2) Set screw steel connectors
6. (2) Removal of concentric knockouts

Type (G) contains per 100 L.F.:
1. (11) Field cemented couplings
2. (34) Beam clamps
3. (2) Factory sweeps
4. (2) Adapters
5. (2) Locknuts
6. (2) Removal of concentric knockouts

Labor-hours for all conduit to 15′ include:
1. Unloading by hand
2. Hauling by hand to an area up to 200′ from loading dock
3. Setup of rolling staging
4. Installation of conduit and fittings as described in Conduit models (A) thru (G)

Not included in the material and labor are:
1. Staging rental or purchase
2. Structural modifications
3. Wire
4. Junction boxes
5. Fittings in excess of those described in conduit models (A) thru (G)
6. Painting of conduit

Fittings

Only those fittings listed above are included in the linear foot totals, although they should be listed separately from conduit lengths, without prices, to ensure proper quantities for material procurement.

If the fittings required exceed the quantities included in the model conduit runs, then material and labor costs must be added to the difference. If actual needs per 100 L.F. of conduit are: (2) sweeps, (4) LBs and (1) field bend, then, (4) LBs and (1) field bend must be priced additionally.

R260533-21 Hangers

It is sometimes desirable to substitute an alternate style of hanger if the support being used is not the type described in the conduit models.

One approach is the substitution method:

1. Find the cost of the type hanger described in the conduit model.
2. Calculate the cost of the desired type hanger (it may be necessary to calculate individual components such as drilling, expansion shields, etc.).
3. Calculate the cost difference (delta) between the two types of hangers.
4. Multiply the cost delta by the number of hangers in the model.
5. Divide the total delta cost for hangers in the model by the length of the model to find the delta cost per L.F. for that model.
6. Modify the given unit costs per L.F. for the model by the delta cost per L.F.

Another approach to hanger configurations would be to start with the conduit only and add all the supports and any other items as separate lines. This procedure is most useful if the project involves racking many runs of conduit on a single hanger, for instance, a trapeze type hanger.

Example: Five (5) 2″ RGS conduits, 50 L.F. each, are to be run on trapeze hangers from one pull box to another. The run includes one 90° bend.

1. List the hangers' components to create an assembly cost for each 2′-wide trapeze.
2. List the components for the 50′ conduit run, noting that 6 trapeze supports will be required.

Job Conditions: Productivities are based on new construction to 15′ high, using scaffolding in an unobstructed area. Material storage is assumed to be within 100′ of work being performed.

Add to labor for elevated installations:

15′ to 20′ High–10%	30′ to 35 High–30%
20′ to 25′ High–20%	35′ to 40′ High–35%
25′ to 30′ High–25%	Over 40′ High–40%

Add these percentages to the L.F. labor cost, but not to fittings. Add these percentages only to quantities exceeding the different height levels, rather than the total conduit quantities.

Linear foot price for labor does not include penetrations in walls or floors and must be added to the estimate

R260533-22 Conductors in Conduit

Table below lists maximum number of conductors for various sized conduit using THW, TW or THWN insulations.

Copper Wire Size	Conduit Size																										
	½″			¾″			1″			1-¼″			1-½″			2″			2-½″			3″		3-½″		4″	
	TW	THW	THWN	TW	THW	THWN	TW	THW	THWN	TW	THW	THWN	TW	THW	THWN	TW	THW	THWN	TW	THW	THWN	THW	THWN	THW	THWN	THW	THWN
#14	9	6	13	15	10	24	25	16	39	44	29	69	60	40	94	99	65	154	142	93		143		192			
#12	7	4	10	12	8	18	19	13	29	35	24	51	47	32	70	78	53	114	111	76	164	117		157			
#10	5	4	6	9	6	11	15	11	18	26	19	32	36	26	44	60	43	73	85	61	104	95	160	127		163	
#8	2	1	3	4	3	5	7	5	9	12	10	16	17	13	22	28	22	36	40	32	51	49	79	66	106	85	136
#6		1	1		2	4		4	6		7	11		10	15		16	26		23	37	36	57	48	76	62	98
#4		1	1		1	2		3	4		5	7		7	9		12	16		17	22	27	35	36	47	47	60
#3		1	1		1	1		2	3		4	6		6	8		10	13		15	19	23	29	31	39	40	51
#2		1	1		1	1		2	3		4	5		5	7		9	11		13	16	20	25	27	33	34	43
#1					1	1		1	1		3	3		4	5		6	8		9	12	14	18	19	25	25	32
1/0					1	1		1	1		2	3		3	4		5	7		8	10	12	15	16	21	21	27
2/0					1	1		1	1		1	2		3	3		5	6		7	8	10	13	14	17	18	22
3/0					1	1		1	1		1	1		2	3		4	5		6	7	9	11	12	14	15	18
4/0						1		1	1		1	1		1	2		3	4		5	6	7	9	10	12	13	15
250 kcmil								1	1		1	1		1	1		2	3		4	4	6	7	8	10	10	12
300								1	1		1	1		1	1		2	3		3	4	5	6	7	8	9	11
350									1		1	1		1	1		1	2		3	3	4	5	6	7	8	9
400											1	1		1	1		1	1		2	3	4	5	5	6	7	8
500											1	1		1	1		1	1		1	2	3	4	4	5	6	7
600												1		1	1		1	1		1	1	3	3	4	4	5	5
700														1	1		1	1		1	1	2	3	3	4	4	5
750 ▼														1	1		1	1		1	1	2	2	3	3	4	4

R260533-23 Metric Equivalent, Conduit

U.S. vs. European Conduit – Approximate Equivalents			
United States		European	
Trade Size	Inside Diameter Inch/mm	Trade Size	Inside Diameter mm
½	.622/15.8	11	16.4
¾	.824/20.9	16	19.9
1	1.049/26.6	21	25.5
1¼	1.380/35.0	29	34.2
1½	1.610/40.9	36	44.0
2	2.067/52.5	42	51.0
2½	2.469/62.7		
3	3.068/77.9		
3½	3.548/90.12		
4	4.026/102.3		
5	5.047/128.2		
6	6.065/154.1		

R260533-24 Conduit Weight Comparisons (Lbs. per 100 ft.) Empty

Type	1/2″	3/4″	1″	1-1/4″	1-1/2″	2″	2-1/2″	3″	3-1/2″	4″	5″	6″
Rigid Aluminum	28	37	55	72	89	119	188	246	296	350	479	630
Rigid Steel	79	105	153	201	249	332	527	683	831	972	1314	1745
Intermediate Steel (IMC)	60	82	116	150	182	242	401	493	573	638		
Electrical Metallic Tubing (EMT)	29	45	65	96	111	141	215	260	365	390		
Polyvinyl Chloride, Schedule 40	16	22	32	43	52	69	109	142	170	202	271	350
Polyvinyl Chloride Encased Burial						38		67	88	105	149	202
Fibre Duct Encased Burial						127		164	180	206	400	511
Fibre Duct Direct Burial						150		251	300	354		
Transite Encased Burial						160		240	290	330	450	550
Transite Direct Burial						220		310		400	540	640

R260533-25 Conduit Weight Comparisons (Lbs. per 100 ft.) with Maximum Cable Fill*

Type	1/2″	3/4″	1″	1-1/4″	1-1/2″	2″	2-1/2″	3″	3-1/2″	4″	5″	6″
Rigid Galvanized Steel (RGS)	104	140	235	358	455	721	1022	1451	1749	2148	3083	4343
Intermediate Steel (IMC)	84	113	186	293	379	611	883	1263	1501	1830		
Electrical Metallic Tubing (EMT)	54	116	183	296	368	445	641	930	1215	1540		

*Conduit & Heaviest Conductor Combination

R260533-30 Labor for Couplings and Fittings

The labor included in the unit price lines for couplings is for their installation separate from a fitting. However, the labor included in the unit price lines for fittings covers their complete installation, which may include the labor installation of the coupling(s). If coupling(s) are required to complete a fitting installation, only the material for the required number of couplings needs to be added to the appropriate unit price lines.

R260533-60 Wireway

When "taking off" Wireway, list by size and type.

Example:
1. Screw cover, unflanged + size
2. Screw cover, flanged + size
3. Hinged cover, flanged + size
4. Hinged cover, unflanged + size

Each 10′ length on Wireway contains:
1. 10′ of cover either screw or hinged type
2. (1) Coupling or flange gasket
3. (1) Wall type mount

All fittings must be priced separately.

Substitution of hanger types is done the same as described in R260533-21, "HANGERS," keeping in mind that the wireway model is based on 10′ sections instead of a 100′ conduit run.

Labor-hours for wireway include:
1. Unloading by hand
2. Hauling by hand up to 100′ from loading dock
3. Measuring and marking
4. Mounting wall bracket using (2) anchor type lead fasteners
5. Installing wireway on brackets, to 15′ high (For higher elevations use factors in R260533-20)

Job Conditions: Productivity is based on new construction, to a height of 15′ using rolling staging in an unobstructed area.

Material staging area is assumed to be within 100′ of work being performed.

R260533-65 Outlet Boxes

Outlet boxes should be included on the same takeoff sheet as branch piping or devices to better explain what is included in each circuit.

Each unit price in this section is a stand alone item and contains no other component unless specified. For example, to estimate a duplex outlet, components that must be added are:
1. 4″ square box
2. 4″ plaster ring
3. Duplex receptacle
4. Device cover

The method of mounting outlet boxes is (2) plastic shield fasteners.

Outlet boxes plastic, labor-hours include:
1. Marking box location on wood studding
2. Mounting box

Economy of Scale – For large concentrations of plastic boxes in the same area deduct the following percentages from labor-hour totals:

1	to	10	0%
11	to	25	20%
26	to	50	25%
51	to	100	30%
	over	100	35%

Note: It is important to understand that these percentages are not used on the total job quantities, but only areas where concentrations exceed the levels specified.

R260533-70 Pull Boxes and Cabinets

List cabinets and pull boxes by NEMA type and size.

Example:	TYPE	SIZE
	NEMA 1	6″W x 6″H x 4″D
	NEMA 3R	6″W x 6″H x 4″D

Labor-hours for wall mount (indoor or outdoor) installations include:
1. Unloading and uncrating
2. Handling of enclosures up to 200′ from loading dock using a dolly or pipe rollers
3. Measuring and marking
4. Drilling (4) anchor type lead fasteners using a hammer drill
5. Mounting and leveling boxes

Note: A plywood backboard is not included.

Labor-hours for ceiling mounting include:
1. Unloading and uncrating
2. Handling boxes up to 100′ from loading dock
3. Measuring and marking
4. Drilling (4) anchor type lead fasteners using a hammer drill
5. Installing and leveling boxes to a height of 15′ using rolling staging

Labor-hours for free standing cabinets include:
1. Unloading and uncrating
2. Handling of cabinets up to 200′ from loading dock using a dolly or pipe rollers
3. Marking of floor
4. Drilling (4) anchor type lead fasteners using a hammer drill
5. Leveling and shimming

Labor-hours for telephone cabinets include:
1. Unloading and uncrating
2. Handling cabinets up to 200′ using a dolly or pipe rollers
3. Measuring and marking
4. Mounting and leveling, using (4) lead anchor type fasteners

R260533-75 Weight Comparisons of Common Size Cast Boxes in Lbs.

Size NEMA 4 or 9	Cast Iron	Cast Aluminum	Size NEMA 7	Cast Iron	Cast Aluminum
6″ x 6″ x 6″	17	7	6″ x 6″ x 6″	40	15
8″ x 6″ x 6″	21	8	8″ x 6″ x 6″	50	19
10″ x 6″ x 6″	23	9	10″ x 6″ x 6″	55	21
12″ x 12″ x 6″	52	20	12″ x 6″ x 6″	100	37
16″ x 16″ x 6″	97	36	16″ x 16″ x 6″	140	52
20″ x 20″ x 6″	133	50	20″ x 20″ x 6″	180	67
24″ x 18″ x 8″	149	56	24″ x 18″ x 8″	250	93
24″ x 24″ x 10″	238	88	24″ x 24″ x 10″	358	133
30″ x 24″ x 12″	324	120	30″ x 24″ x 10″	475	176
36″ x 36″ x 12″	500	185	30″ x 24″ x 12″	510	189

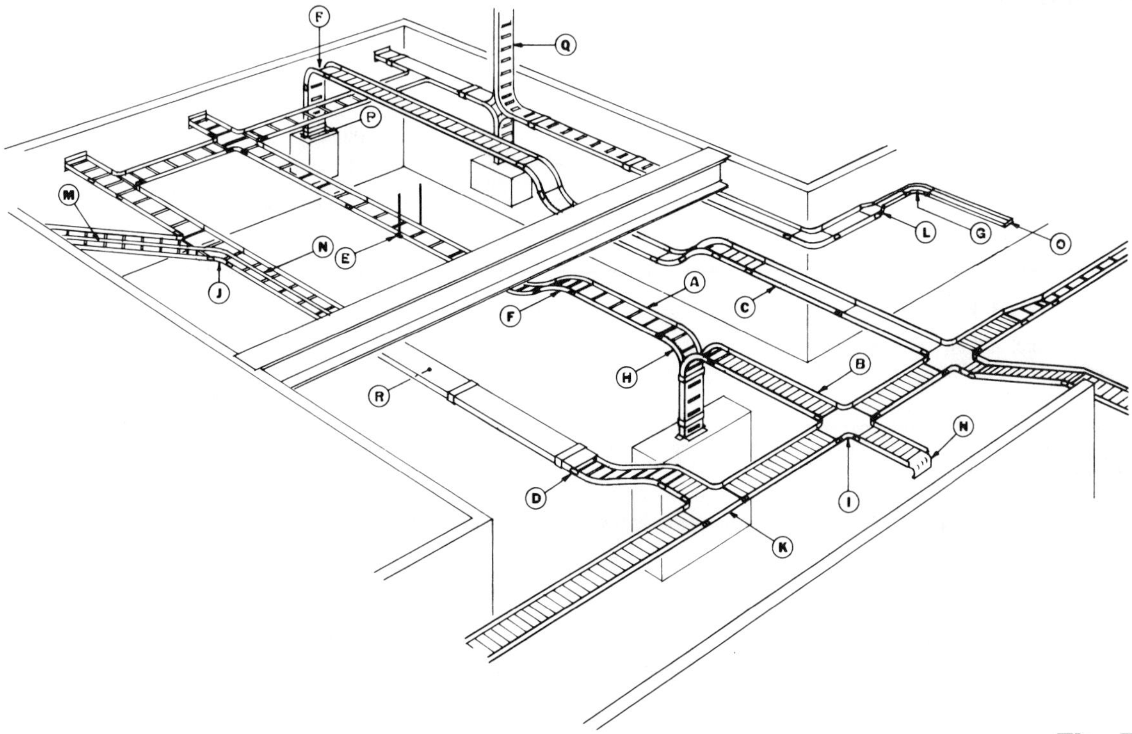

Fig. R260536-11

R260536-10 Cable Tray

Cable Tray - When taking off cable tray it is important to identify separately the different types and sizes involved in the system being estimated. (Fig. R260536-11)

A. – Ladder Type, galvanized or aluminum
B. – Trough Type, galvanized or aluminum
C. – Solid Bottom, galvanized or aluminum

The unit of measure is calculated in linear feet; do not deduct from this footage any length occupied by fittings, this will be the only allowance for scrap. Be sure to include all vertical drops to panels, switch gear, etc.

Hangers – Included in the linear footage of cable tray is

D. – 1 – Pair of connector plates per 12 L.F.

E. – 1 – Pair clamp type hangers and 4′ of 3/8″ threaded rod per 12 L.F.

Not included are structural supports, which must be priced in addition to the hangers.

Fittings – Identify separately the different types of fittings
1.) Ladder Type, galvanized or aluminum
2.) Trough Type, galvanized or aluminum
3.) Solid Bottom Type, galvanized or aluminum

The configuration, radius and rung spacing must also be listed. The unit of measure is "Ea."

F. – Elbow, vertical Ea.
G. – Elbow, horizontal Ea.
H. – Tee, vertical Ea.
I. – Cross, horizontal Ea.
J. – Wye, horizontal Ea.
K. – Tee, horizontal Ea.
L. – Reducing fitting Ea.

Depending on the use of the system other examples of units which must be included are:

M. – Divider strip L.F.
N. – Drop-outs Ea.
O. – End caps Ea.
P. – Panel connectors Ea.

Wire and cable are not included and should be taken off separately, see Unit Price sections.

Job Conditions – Unit prices are based on a new installation to a work plane of 15′ using rolling staging.

Add to labor for elevated installations:

15′ to 20′ High	10%
20′ to 25′ High	20%
25′ to 30′ High	25%
30′ to 35′ High	30%
35′ to 40′ High	35%
Over 40′ High	40%

Add these percentages for L.F. totals but not to fittings. Add percentages to only those quantities that fall in the different elevations, in other words, if the total quantity of cable tray is 200′ but only 75′ is above 15′ then the 10% is added to the 75′ only.

Linear foot costs do not include penetrations through walls and floors which must be added to the estimate.

Cable Tray Covers

Covers – Cable tray covers are taken off in the same manner as the tray itself, making distinctions as to the type of cover. (Fig. R260536-11)

Q. – Vented, galvanized or aluminum
R. – Solid, galvanized or aluminum

Cover configurations are taken off separately noting type, specific radius and widths.

Note: Care should be taken to identify from plans and specifications exactly what is being covered. In many systems only vertical fittings are covered to retain wire and cable.

R260539-30 Conduit In Concrete Slab

List conduit by quantity, size and type.

Example:

A. Rigid galvanized steel + size
B. P.V.C. + size

Rigid galvanized steel (A) contains per 100 L.F.:

1. (20) Ties to slab reinforcing
2. (11) Threaded steel couplings
3. (2) Factory sweeps
4. (2) Field threaded conduit terminations
5. (2) Fiber bushings + locknuts
6. (2) Removal of concentric knockouts

P.V.C. (B) contains per 100 L.F.:

1. (20) Ties to slab reinforcing
2. (11) Field cemented couplings
3. (2) Factory sweeps
4. (2) Adapters
5. (2) Removal of concentric knockouts

R260539-40 Conduit In Trench

Conduit in trench is galvanized steel and contains per 100 L.F.:

1. (11) Threaded couplings
2. (2) Factory sweeps
3. (2) Fiber bushings + (4) locknuts
4. (2) Field threaded conduit terminations
5. (2) Removal of concentric knockouts

Note:

Conduit in Unit Price sections do not include:

1. Floor cutting
2. Excavation or backfill
3. Grouting or patching

Conduit fittings in excess of those listed in the above Conduit model must be added. (Refer to R260533-20 for Procedure example.)

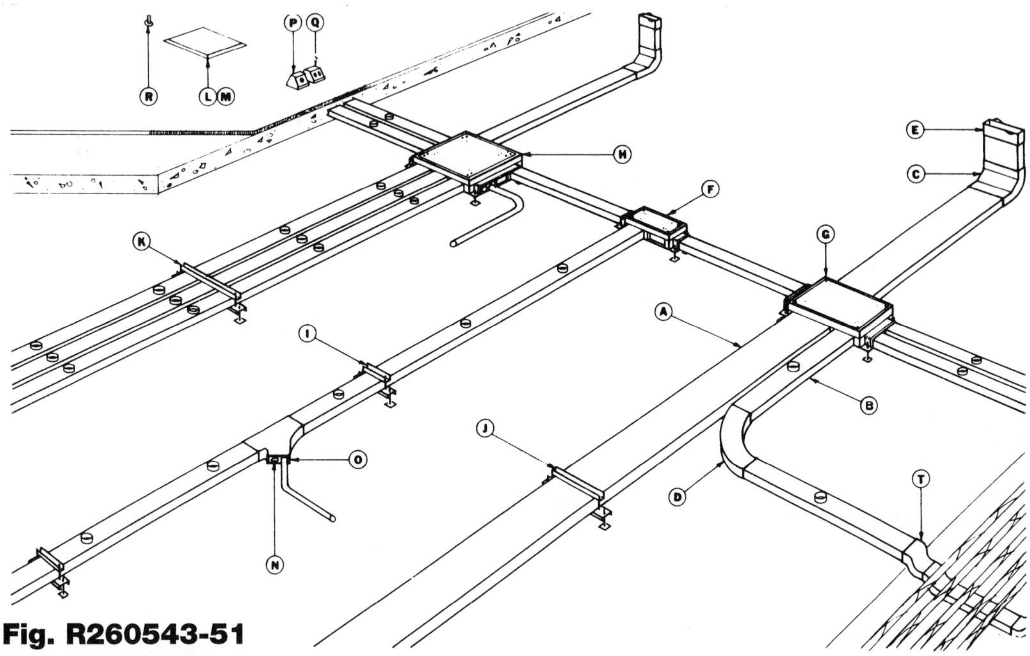

Fig. R260543-51

R260543-50 Underfloor Duct

When pricing Underfloor Duct it is important to identify and list each component, since costs vary significantly from one type of fitting to another. Do not deduct boxes or fittings from linear foot totals; this will be your allowance for scrap.

The first step is to identify the system as either:

FIG. R260543-51 Single Level

FIG. R260543-52 Dual Level

Single Level System

Include on your "takeoff sheet" the following unit price items, making sure to distinquish between Standard and Super duct:

A. Feeder duct (blank) in L.F.
B. Distribution duct (Inserts 2′ on center) in L.F.
C. Elbows (Vertical) Ea.
D. Elbows (Horizontal) Ea.
E. Cabinet connector Ea.
F. Single duct junction box Ea.
G. Double duct junction box Ea.
H. Triple duct junction box Ea.
I. Support, single cell Ea.
J. Support, double cell Ea.
K. Support, triple cell Ea.
L. Carpet pan Ea.
M. Terrazzo pan Ea.
N. Insert to conduit adapter Ea.
O. Conduit adapter Ea.
P. Low tension outlet Ea.
Q. High tension outlet Ea.
R. Galvanized nipple Ea.
S. Wire per C.L.F.
T. Offset (Duct type) Ea.

**Dual Level System + Labor
see next page**

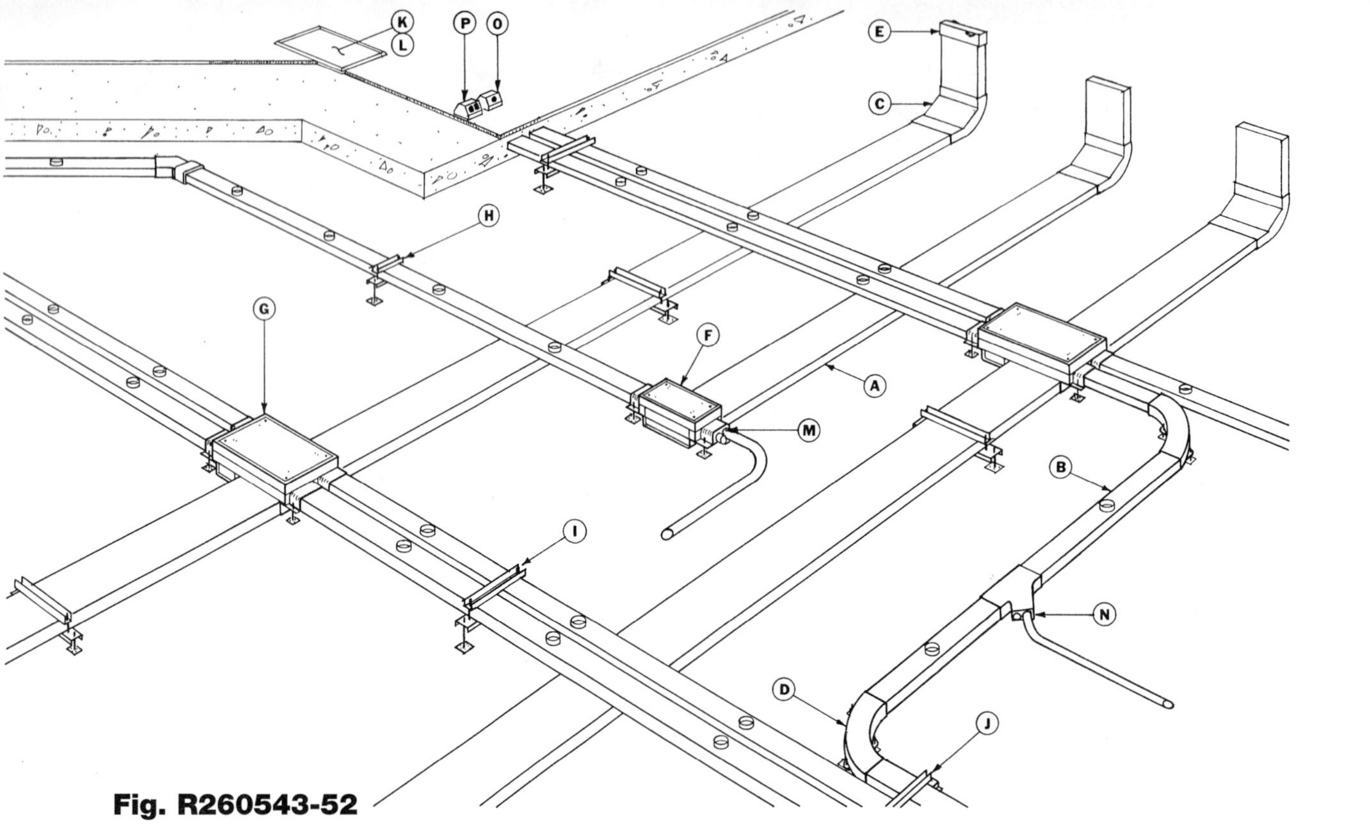

Fig. R260543-52

R260543-50 Underfloor Duct (cont.)

Dual Level

Include the following when "taking off" Dual Level systems:

Distinguish between Standard and Super duct.

A. Feeder duct (blank) in L.F.
B. Distribution duct (Inserts 2′ on center) in L.F.
C. Elbows (Vertical) Ea.
D. Elbows (Horizontal) Ea.
E. Cabinet connector Ea.
F. Single duct, 2 level, junction box Ea.
G. Double duct, 2 level, junction box Ea.
H. Support, single cell Ea.
I. Support, double cell Ea.
J. Support, triple cell Ea.
K. Carpet pan Ea.
L. Terrazzo pan Ea.
M. Insert to conduit adapter Ea.
N. Conduit adapter Ea.
O. Low tension outlet Ea.
P. High tension outlet Ea.
Q. Wire per C.L.F.

Note: Make sure to include risers in linear foot totals. High tension outlets include box, receptacle, covers and related mounting hardware.

Labor-hours for both Single and Dual Level systems include:

1. Unloading and uncrating
2. Hauling up to 200′ from loading dock
3. Measuring and marking
4. Setting raceway and fittings in slab or on grade
5. Leveling raceway and fittings

Labor-hours do not include:

1. Floor cutting
2. Excavation or backfill
3. Concrete pour
4. Grouting or patching
5. Wire or wire pulls
6. Additional outlets after concrete is poured
7. Piping to or from Underfloor Duct

Note: Installation is based on installing up to 150′ of duct. If quantities exceed this, deduct the following percentages:

1. 150′ to 250′ - 10%
2. 250′ to 350′ - 15%
3. 350′ to 500′ - 20%
4. over 500′ - 25%

Deduct these percentages from labor only.

Deduct these percentages from straight sections only.

Do not deduct from fittings or junction boxes.

Job Conditions: Productivity is based on new construction.

Underfloor duct to be installed on first three floors.

Material staging area within 100′ of work being performed.

Area unobstructed and duct not subject to physical damage.

Electrical — R2605 Common Work Results for Electrical

R260580-75 Motor Connections

Motor connections should be listed by size and type of motor. Included in the material and labor cost is:

1. (2) Flex connectors
2. 18″ of flexible metallic wireway
3. Wire identification and termination
4. Test for rotation
5. (2) or (3) Conductors

Price does not include:

1. Mounting of motor
2. Disconnect Switch
3. Motor Starter
4. Controls
5. Conduit or wire ahead of flex

Note: When "Taking off" Motor connections, it is advisable to list connections on the same quantity sheet as Motors, Motor Starters and controls.

Electrical — R2609 Instrumen. & Cont. of Elect. Systems

R260913-80 Switchboard Instruments

Switchboard instruments are added to the price of switchboards according to job specifications. This equipment is usually included when ordering "Gear" from the manufacturer and will arrive factory installed.

Included in the labor cost is:

1. Internal wiring connections
2. Wire identification
3. Wire tagging

Transition sections include:

1. Uncrating
2. Hauling sections up to 100′ from loading dock
3. Positioning sections
4. Leveling sections
5. Bolting enclosures
6. Bolting vertical bus bars

Price does not include:

1. Equipment pads
2. Steel channels embedded or grouted in concrete
3. Special knockouts
4. Rigging

Job Conditions: Productivity is based on new construction, equipment to be installed on the first floor within 200′ of the loading dock.

R262213-10 Electric Circuit Voltages

General: The following method provides the user with a simple non-technical means of obtaining comparative costs of wiring circuits. The circuits considered serve the electrical loads of motors, electric heating, lighting and transformers, for example, that require low voltage 60 Hertz alternating current.

The method used here is suitable only for obtaining estimated costs. It is **not** intended to be used as a substitute for electrical engineering design applications.

Conduit and wire circuits can represent from twenty to thirty percent of the total building electrical cost. By following the described steps and using the tables the user can translate the various types of electric circuits into estimated costs.

Wire Size: Wire size is a function of the electric load which is usually listed in one of the following units:

1. Amperes (A)
2. Watts (W)
3. Kilowatts (kW)
4. Volt amperes (VA)
5. Kilovolt amperes (kVA)
6. Horsepower (HP)

These units of electric load must be converted to amperes in order to obtain the size of wire necessary to carry the load. To convert electric load units to amperes one must have an understanding of the voltage classification of the power source and the voltage characteristics of the electrical equipment or load to be energized. The seven A.C. circuits commonly used are illustrated in Figures R262213-11 thru R262213-17 showing the tranformer load voltage and the point of use voltage at the point on the circuit where the load is connected. The difference between the source and point of use voltages is attributed to the circuit voltage drop and is considered to be approximately 4%.

Motor Voltages: Motor voltages are listed by their point of use voltage and not the power source voltage.

For example: 460 volts instead of 480 volts
200 instead of 208 volts
115 volts instead of 120 volts

Lighting and Heating Voltages: Lighting and heating equipment voltages are listed by the power source voltage and not the point of wire voltage.

For example: 480, 277, 120 volt lighting
480 volt heating or air conditioning unit
208 volt heating unit

Transformer Voltages: Transformer primary (input) and secondary (output) voltages are listed by the power source voltage.

For example: Single phase 10 kVA
Primary 240/480 volts
Secondary 120/240 volts

In this case, the primary voltage may be 240 volts with a 120 volts secondary or may be 480 volts with either a 120V or a 240V secondary.

For example: Three phase 10 kVA
Primary 480 volts
Secondary 208Y/120 volts

In this case the transformer is suitable for connection to a circuit with a 3 phase 3 wire or 3 phase 4 wire circuit with a 480 voltage. This application will provide a secondary circuit of 3 phase 4 wire with 208 volts between phase wires and 120 volts between any phase wire and the neutral (white) wire.

R262213-11

3 Wire, 1 Phase, 120/240 Volt System

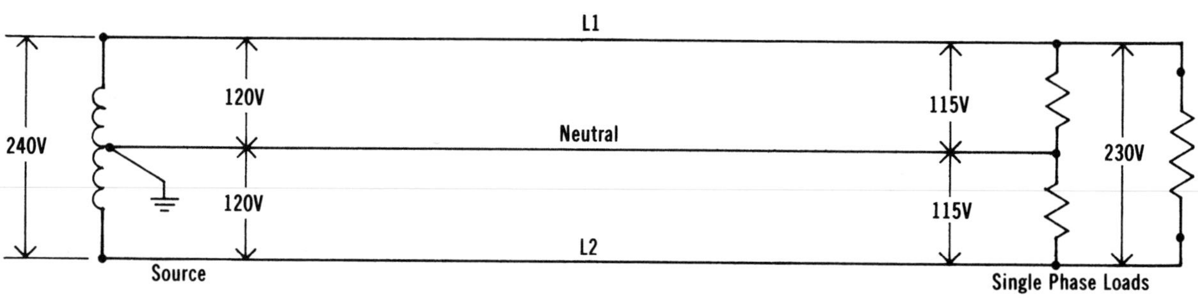

R262213-12

4 wire, 3 Phase, 208Y/120 Volt System

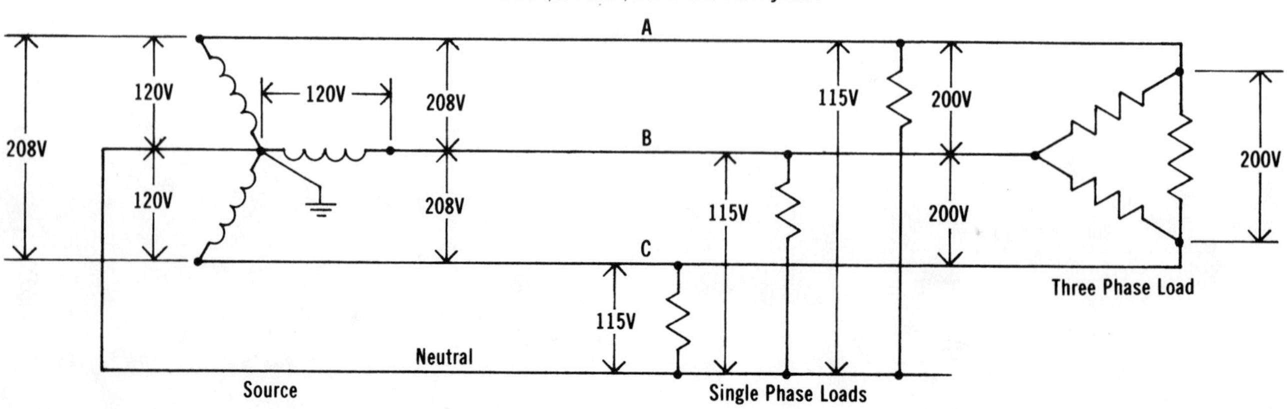

R262213-13 Electric Circuit Voltages (cont.)

3 Wire, 3 Phase 240 Volt System

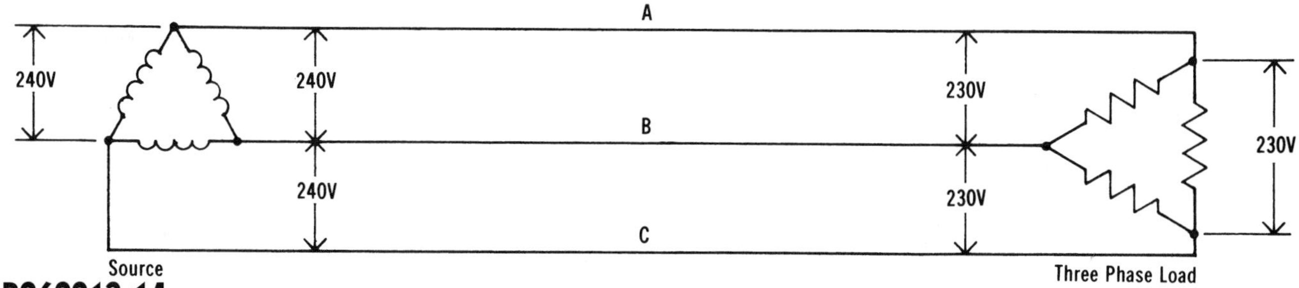

R262213-14

4 Wire, 3 Phase, 240/120 Volt System

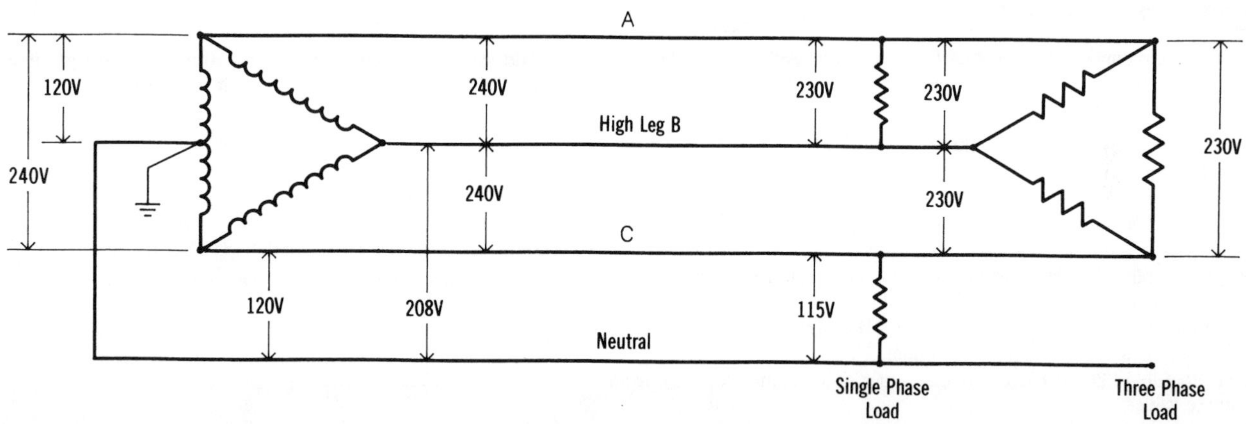

R262213-15

3 Wire, 3 Phase 480 Volt System

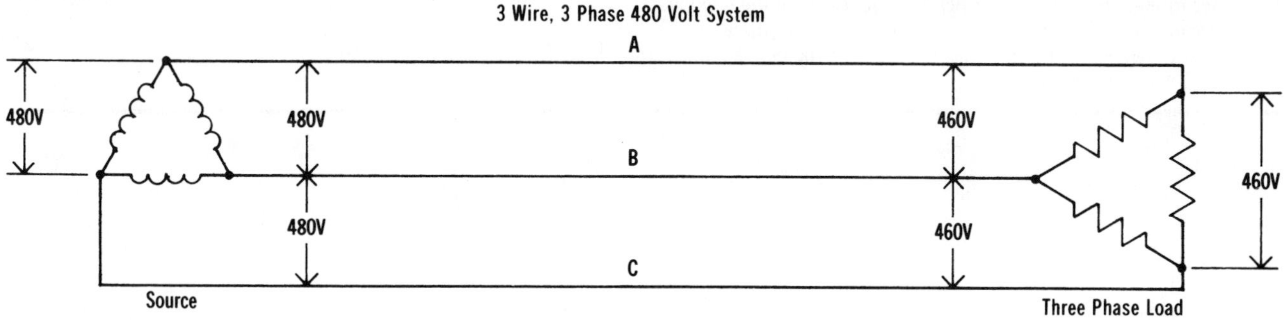

R262213-16

4 Wire, 3 Phase, 480Y/277 Volt System

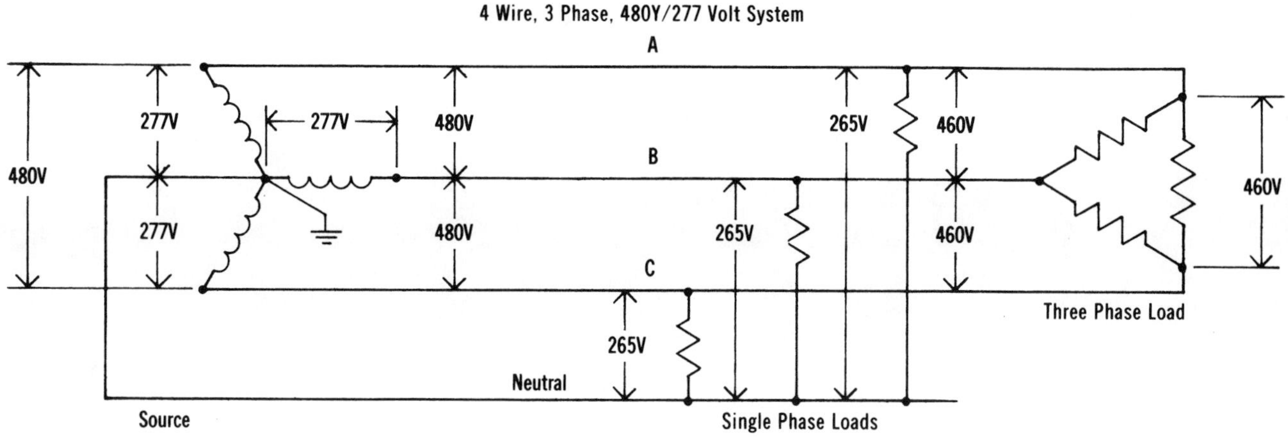

R262213-17 Electric Circuit Voltages (cont.)

3 Wire, 3 Phase, 600 Volt System

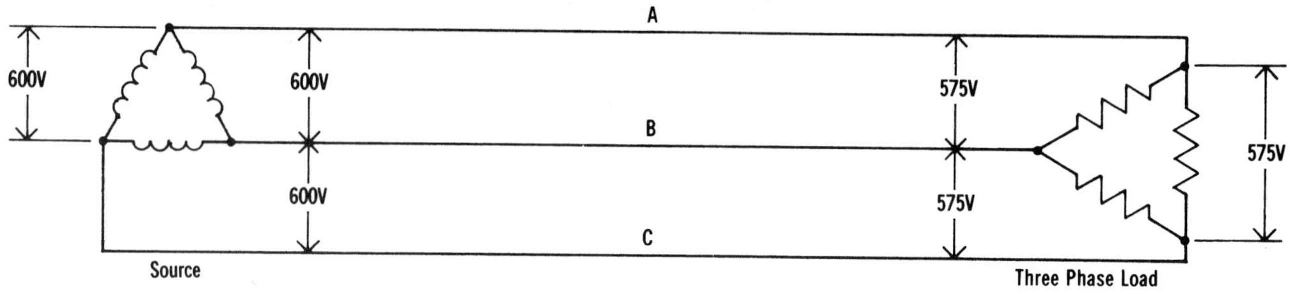

R262213-20 kW Value/Cost Determination

General: Lighting and electric heating loads are expressed in watts and kilowatts.

Cost Determination:

The proper ampere values can be obtained as follows:

1. Convert watts to kilowatts (watts 1000 ÷ kilowatts)
2. Determine voltage rating of equipment.
3. Determine whether equipment is single phase or three phase.
4. Refer to Table R262213-21 to find ampere value from kW, Ton and Btu/hr. values.
5. Determine type of wire insulation – TW, THW, THWN.
6. Determine if wire is copper or aluminum.
7. Refer to Table R260519-92 to obtain copper or aluminum wire size from ampere values.
8. Next refer to Table R260533-22 for the proper conduit size to accommodate the number and size of wires in each particular case.
9. Next refer to unit cost data for the per linear foot cost of the conduit.
10. Next refer to unit cost data for the per linear foot cost of the wire. Multiply cost of wire per L.F. x number of wires in the circuits to obtain total wire cost per L.F.
11. Add values obtained in Step 9 and 10 for total cost per linear foot for conduit and wire x length of circuit = Total Cost.

Notes:

1. 1 Phase refers to single phase, 2 wire circuits.
2. 3 Phase refers to three phase, 3 wire circuits.
3. For circuits which operate continuously for 3 hours or more, multiply the ampere values by 1.25 for a given kw requirement.
4. For kW ratings not listed, add ampere values.

   For example: Find the ampere value of 9 kW at 208 volt, single phase.

   4 kW = 19.2A
   5 kW = 24.0A
   9 kW = 43.2A
5. "Length of Circuit" refers to the one way distance of the run, not to the total sum of wire lengths.

R262213-21 Ampere Values as Determined by kW Requirements, BTU/HR or Ton, Voltage and Phase Values

			Ampere Values						
			120V	208V		240V		277V	480V
kW	Ton	BTU/HR	1 Phase	1 Phase	3 Phase	1 Phase	3 Phase	1 Phase	3 Phase
0.5	.1422	1,707	4.2A	2.4A	1.4A	2.1A	1.2A	1.8A	0.6A
0.75	.2133	2,560	6.2	3.6	2.1	3.1	1.9	2.7	.9
1.0	.2844	3,413	8.3	4.9	2.8	4.2	2.4	3.6	1.2
1.25	.3555	4,266	10.4	6.0	3.5	5.2	3.0	4.5	1.5
1.5	.4266	5,120	12.5	7.2	4.2	6.3	3.1	5.4	1.8
2.0	.5688	6,826	16.6	9.7	5.6	8.3	4.8	7.2	2.4
2.5	.7110	8,533	20.8	12.0	7.0	10.4	6.1	9.1	3.1
3.0	.8532	10,239	25.0	14.4	8.4	12.5	7.2	10.8	3.6
4.0	1.1376	13,652	33.4	19.2	11.1	16.7	9.6	14.4	4.8
5.0	1.4220	17,065	41.6	24.0	13.9	20.8	12.1	18.1	6.1
7.5	2.1331	25,598	62.4	36.0	20.8	31.2	18.8	27.0	9.0
10.0	2.8441	34,130	83.2	48.0	27.7	41.6	24.0	36.5	12.0
12.5	3.5552	42,663	104.2	60.1	35.0	52.1	30.0	45.1	15.0
15.0	4.2662	51,195	124.8	72.0	41.6	62.4	37.6	54.0	18.0
20.0	5.6883	68,260	166.4	96.0	55.4	83.2	48.0	73.0	24.0
25.0	7.1104	85,325	208.4	120.2	70.0	104.2	60.0	90.2	30.0
30.0	8.5325	102,390		144.0	83.2	124.8	75.2	108.0	36.0
35.0	9.9545	119,455		168.0	97.1	145.6	87.3	126.0	42.1
40.0	11.3766	136,520		192.0	110.8	166.4	96.0	146.0	48.0
45.0	12.7987	153,585			124.8	187.5	112.8	162.0	54.0
50.0	14.2208	170,650			140.0	208.4	120.0	180.4	60.0
60.0	17.0650	204,780			166.4		150.4	216.0	72.0
70.0	19.9091	238,910			194.2		174.6		84.2
80.0	22.7533	273,040			221.6		192.0		96.0
90.0	25.5975	307,170					225.6		108.0
100.0	28.4416	341,300							120.0

R262213-25 kVA Value/Cost Determination

General: Control transformers are listed in VA. Step-down and power transformers are listed in kVA.

Cost Determination:

1. Convert VA to kVA. Volt amperes (VA) ÷ 1000 = Kilovolt amperes (kVA).
2. Determine voltage rating of equipment.
3. Determine whether equipment is single phase or three phase.
4. Refer to Table R262213-26 to find ampere value from kVA value.
5. Determine type of wire insulation – TW, THW, THWN.
6. Determine if wire is copper or aluminum.
7. Refer to Table R260519-92 to obtain copper or aluminum wire size from ampere values.

Example: A transformer rated 10 kVA 480 volts primary, 240 volts secondary, 3 phase has the capacity to furnish the following:

1. Primary amperes = 10 kVA x 1.20 = 12 amperes (from Table R262213-26)
2. Secondary amperes = 10 kVA x 2.40 = 24 amperes (from Table R262213-26)

Note: Transformers can deliver generally 125% of their rated kVA. For instance, a 10 kVA rated transformer can safely deliver 12.5 kVA.

8. Next refer to Table R260533-22 for the proper conduit size to accommodate the number and size of wires in each particular case.
9. Next refer to unit price data for the per linear foot cost of the conduit.
10. Next refer to unit price data for the per linear foot cost of the wire. Multiply cost of wire per L.F. x number of wires in the circuits to obtain total wire cost.
11. Add values obtained in Step 9 and 10 for total cost per linear foot for conduit and wire x length of circuit = Total Cost.

R262213-26 Multiplier Values for kVA to Amperes Determined by Voltage and Phase Values

	Multiplier for Circuits	
Volts	2 Wire, 1 Phase	3 Wire, 3 Phase
115	8.70	
120	8.30	
230	4.30	2.51
240	4.16	2.40
200	5.00	2.89
208	4.80	2.77
265	3.77	2.18
277	3.60	2.08
460	2.17	1.26
480	2.08	1.20
575	1.74	1.00
600	1.66	0.96

R262213-27 Central Air Conditioning Watts per S.F., BTUs per Hour per S.F. of Floor Area and S.F. per Ton of Air Conditioning

Type Building	Watts per S.F.	BTUH per S.F.	S.F. per Ton	Type Building	Watts per S.F.	BTUH per S.F.	S.F. per Ton	Type Building	Watts per S.F.	BTUH per S.F.	S.F. per Ton
Apartments, Individual	3	26	450	Dormitory, Rooms	4.5	40	300	Libraries	5.7	50	240
Corridors	2.5	22	550	Corridors	3.4	30	400	Low Rise Office, Ext.	4.3	38	320
Auditoriums & Theaters	3.3	40	300/18*	Dress Shops	4.9	43	280	Interior	3.8	33	360
Banks	5.7	50	240	Drug Stores	9	80	150	Medical Centers	3.2	28	425
Barber Shops	5.5	48	250	Factories	4.5	40	300	Motels	3.2	28	425
Bars & Taverns	15	133	90	High Rise Off.-Ext. Rms.	5.2	46	263	Office (small suite)	4.9	43	280
Beauty Parlors	7.6	66	180	Interior Rooms	4.2	37	325	Post Office, Int. Office	4.9	42	285
Bowling Alleys	7.8	68	175	Hospitals, Core	4.9	43	280	Central Area	5.3	46	260
Churches	3.3	36	330/20*	Perimeter	5.3	46	260	Residences	2.3	20	600
Cocktail Lounges	7.8	68	175	Hotels, Guest Rooms	5	44	275	Restaurants	6.8	60	200
Computer Rooms	16	141	85	Public Spaces	6.2	55	220	Schools & Colleges	5.3	46	260
Dental Offices	6	52	230	Corridors	3.4	30	400	Shoe Stores	6.2	55	220
Dept. Stores, Basement	4	34	350	Industrial Plants, Offices	4.3	38	320	Shop'g. Ctrs., Sup. Mkts.	4	34	350
Main Floor	4.5	40	300	General Offices	4	34	350	Retail Stores	5.5	48	250
Upper Floor	3.4	30	400	Plant Areas	4.5	40	300	Specialty Shops	6.8	60	200

*Persons per ton

12,000 BTUH = 1 ton of air conditioning

R262213-60 Oil Filled Transformers

Transformers in this section include:
1. Rigging (as required)
2. Rental of crane and operator
3. Setting of oil filled transformer
4. (4) Anchor bolts, nuts and washers in concrete pad

Price does not include:
1. Primary and secondary terminations
2. Transformer pad
3. Equipment grounding
4. Cable
5. Conduit locknuts or bushings

Transformers in Unit Price sections for dry type, back-boost and isolating transformers include:
1. Unloading and uncrating
2. Hauling transformer to within 200′ of loading dock
3. Setting in place
4. Wall mounting hardware
5. Testing

Price does not include:
1. Structural supports
2. Suspension systems
3. Welding or fabrication
4. Primary & secondary terminations

Add the following percentages to the labor for ceiling mounted transformers:

10' to 15′	= + 15%
15′ to 25′	= + 30%
Over 25′	= + 35%

Job Conditions: Productivities are based on new construction. Installation is assumed to be on the first floor, in an obstructed area to a height of 10′. Material staging area is within 100′ of final transformer location.

Electrical — R2622 Low Voltage Transformers

R262213-65 Transformer Weight (Lbs.) by kVA

Oil Filled 3 Phase 5/15 KV To 480/277			
kVA	Lbs.	kVA	Lbs.
150	1800	1000	6200
300	2900	1500	8400
500	4700	2000	9700
750	5300	3000	15000
Dry 240/480 To 120/240 Volt			
1 Phase		3 Phase	
kVA	Lbs.	kVA	Lbs.
1	23	3	90
2	36	6	135
3	59	9	170
5	73	15	220
7.5	131	30	310
10	149	45	400
15	205	75	600
25	255	112.5	950
37.5	295	150	1140
50	340	225	1575
75	550	300	1870
100	670	500	2850
167	900	750	4300

Electrical — R2624 Switchboards & Panelboards

R262416-50 Load Centers and Panelboards

When pricing Load Centers list panels by size and type. List Breakers in a separate column of the "Quantity Sheet," and define by phase and ampere rating.

Material and Labor prices include breakers; for example: a 100A, 3-Wire, 102/240V, 18 circuit panel w/ main breaker, as described in the unit cost section, contains 18 single pole 20A breakers.

If you do not choose to include a full panel of single pole breakers, use the following method to adjust material and labor costs.

Example: In an 18 circuit panel only 16 single pole breakers are desired, requiring that the cost of 2 breakers be subtracted from the panel cost.

1. Go to the appropriate unit cost section of the book to find the unit prices of the given circuit breaker type.
2. Modify those costs as follows: Bare material price x 0.50 Bare labor cost x 0.60.
3. Multiply those modified bare costs by 2 (breakers in this example).
4. Subtract the modified costs for the 2 breakers from the given cost of the panel.

Labor-hours for Load Center installation includes:

1. Unloading, uncrating, and handling enclosures 200′ from unloading area
2. Measuring and marking
3. Drilling (4) lead anchor type fasteners using a hammer drill
4. Mounting and leveling panel to a height of 6′
5. Preparation and termination of feeder cable to lugs or main breaker
6. Branch circuit identification
7. Lacing using tie wraps
8. Testing and load balancing
9. Marking panel directory

Not included in the material and labor are:

1. Modifications to enclosure
2. Structural supports
3. Additional lugs
4. Plywood backboards
5. Painting or lettering

Note: Knockouts are included in the price of terminating pipe runs and need not be added to the Load Center costs.

Job Conditions: Productivity is based on new construction to a height of 6′, in an unobstructed area. Material staging area is assumed to be within 100′ of work being performed.

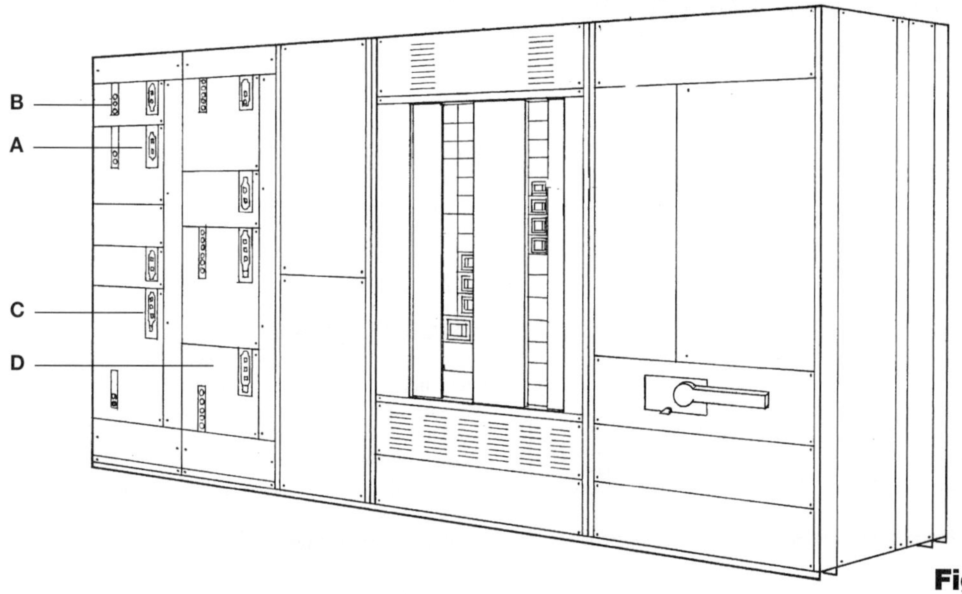

Fig. R262419-61

R262419-60 Motor Control Centers

When taking off Motor Control Centers, list the size, type and height of structures.

Example:

1. 600A, 22,000 RMS, 72″ high
2. 600A, back to back, 72″ high

Next take off individual starters; the number of structures can also be determined by adding the height in inches of starters divided by the height of the structure, and list on the same quantity sheet as the structures. Identify starters by Type, Horsepower rating, Size and Height in inches.

Example:

A. Class I, Type B, FVNR starter, 25 H.P., 18″ high.
Add to the list with Starters, factory installed controls.

Example:

B. Pilot lights

C. Push buttons

D. Auxiliary contacts

Identify starters and structures as either copper or aluminum and by the NEMA type of enclosure.

When pricing starters and structures, be sure to add or deduct adjustments, using lines in the unit price section.

Included in the cost of Motor Control Structures are:

1. Uncrating
2. Hauling to location within 100′ of loading dock
3. Setting structures
4. Leveling
5. Aligning
6. Bolting together structure frames
7. Bolting horizontal bus bars

Labor-hours do not include:

1. Equipment pad
2. Steel channels embedded or grouted in concrete
3. Pull boxes
4. Special knockouts
5. Main switchboard section
6. Transition section
7. Instrumentation
8. External control wiring
9. Conduit or wire

Material for Starters includes:

1. Circuit breaker or fused disconnect
2. Magnetic motor starter
3. Control transformer
4. Control fuse and fuse block

Labor-hours for Starters include:

1. Handling
2. Installing starter within structure
3. Internal wiring connections
4. Lacing within enclosure
5. Testing
6. Phasing

Job Conditions: Productivity is based on new construction. Motor control location assumed to be on first floor in an unobstructed area. Material staging area within 100′ of final location.

Note: Additional labor-hours must be added if M.C.C. is to be installed on other than the first floor, or if rigging is required.

R262419-65 Motor Starters and Controls

Motor starters should be listed on the same "Quantity Sheet" as Motors and Motor Connections. Identify each starter by:

1. Size
2. Voltage
3. Type

Example:

A. FVNR, 480V, 2HP, Size 00
B. FVNR, 480V, 5HP, Size 0, Combination type
C. FVR, 480V, Size 2

The NEMA type of enclosure should also be identified.

Included in the labor-hours are:

1. Unloading, uncrating and handling of starters up to 200′ from loading area
2. Measuring and marking
3. Drilling (4) anchor type lead fasteners, using a hammer drill
4. Mounting and leveling starter
5. Connecting wire or cable to line and load sides of starter (when already lugged)
6. Installation of (3) thermal type heaters
7. Testing

The following is not included unless specified in the unit price description.

1. Control transformer
2. Controls, either factory or field installed
3. Conduit and wire to or from starter
4. Plywood backboard
5. Cable terminations

The following material and labor has been included for Combination type starters:

1. Unloading, uncrating and handling of starter up to 200′ of loading dock
2. Measuring and marking
3. Drilling (4) anchor type lead fasteners using a hammer drill
4. Mounting and leveling
5. Connecting prepared cable conductors
6. Installation of (3) dual element cartridge type fuses
7. Installation of (3) thermal type heaters
8. Test for rotation

MOTOR CONTROLS

When pricing motor controls make sure you consider the type of control system being utilized. If the controls are factory installed and located in the enclosure itself, then you would add to your starter price the items 5200 thru 5800 in the Unit Price section. If control voltage is different from line voltage, add the material and labor cost of a control transformer.

For external control of starters include the following items:

1. Raceways
2. Wire & terminations
3. Control enclosures
4. Fittings
5. Push button stations
6. Indicators

Job Conditions: Productivity is based on new construction to a height of 10′.

Material staging area is assumed to be within 100′ of work being performed.

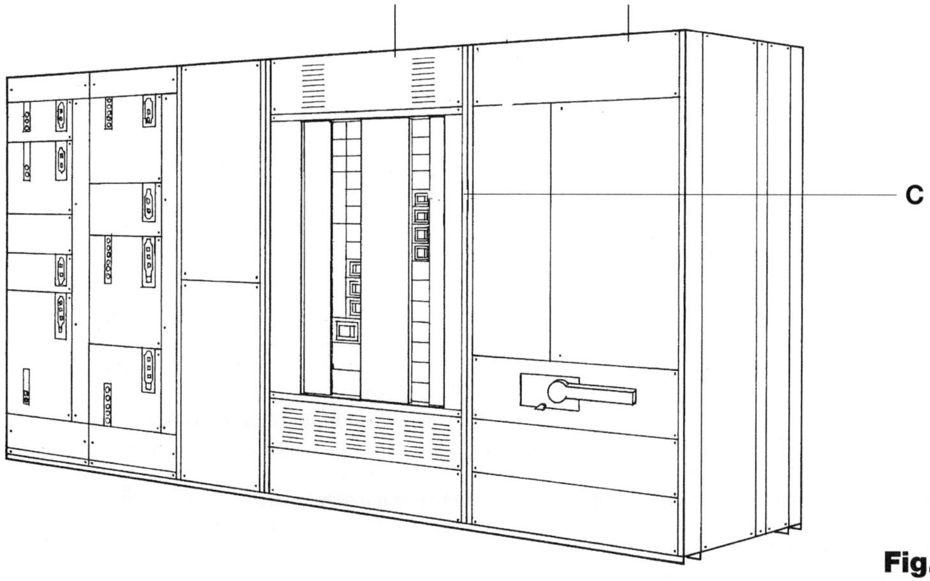

Fig. R262419-81

R262419-80 Distribution Section

After "Taking off" the Switchboard section, include on the same "Quantity sheet" the Distribution section; identify by:

1. Voltage
2. Ampere rating
3. Type

Example:

(Fig. R262419-81) (B) Distribution Section

Included in the labor costs of the Distribution section is:

1. Uncrating
2. Hauling to 200′ of loading dock
3. Setting of distribution panel
4. Leveling & shimming
5. Anchoring of equipment to pad or floor
6. Bolting of horizontal bus bars between
7. Testing of equipment

Not included in the Distribution section is:

1. Breakers (C)
2. Equipment pads
3. Steel channels embedded or grouted in concrete
4. Pull boxes
5. Special knockouts
6. Transition section
7. Conduit or wire

R262419-82 Feeder Section

List quantities on the same sheet as the Distribution section. Identify breakers by: (Fig. R262419-81)

1. Frame type
2. Number of poles
3. Ampere rating

Installation includes:

1. Handling
2. Placing breakers in distribution panel
3. Preparing wire or cable
4. Lacing wire
5. Marking each phase with colored tape
6. Marking panel legend
7. Testing
8. Balancing

R262419-84 Switchgear

It is recommended that "Switchgear" or those items contained in the following sections be quoted from equipment manufacturers as a package price.

Switchboard instruments

Switchboard distribution sections

Switchboard feeder sections

Switchboard Service Disc.

Switchboard In-plant Dist.

Included in these sections are the most common types and sizes of factory assembled equipment.

The recommended procedure for low voltage switchgear would be to price (Fig. R262419-81) (A) Main Switchboard

Identify by:

1. Voltage
2. Amperage
3. Type

Example:

1. 120/208V, 4-wire, 600A, nonfused
2. 277/480V, 4-wire, 600A, nonfused
3. 120/208V, 4-wire, 400A w/fused switch & CT compartment
4. 277/480V, 4-wire, 400A, w/fused switch & CT compartment
5. 120/208V, 4-wire, 800A, w/pressure switch & CT compartment
6. 277/480V, 4-wire, 800A, w/molded CB & CT compartment

Included in the labor costs for Switchboards are:

1. Uncrating
2. Hauling to 200′ of loading dock
3. Setting equipment
4. Leveling and shimming
5. Anchoring
6. Cable identification
7. Testing of equipment

Not included in the Switchboard price is:

1. Rigging
2. Equipment pads
3. Steel channels embedded or grouted in concrete
4. Special knockouts
5. Transition or Auxiliary sections
6. Instrumentation
7. External control
8. Conduit and wire
9. Conductor terminations

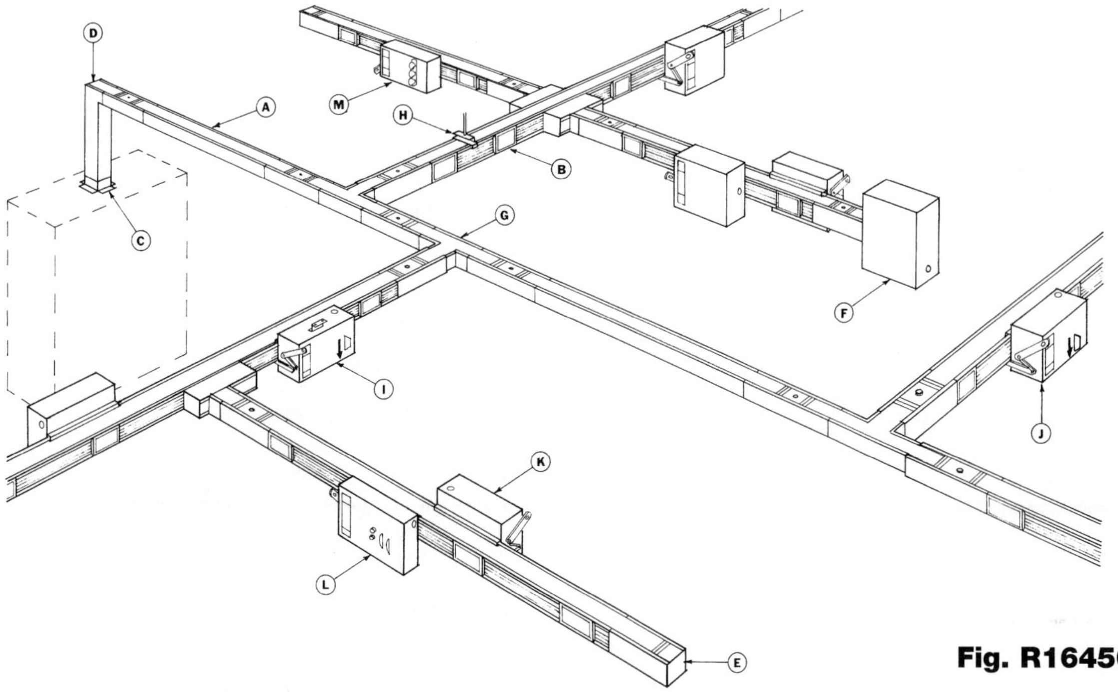

Fig. R16450-101

R262513-10 Aluminum Bus Duct

When taking off bus duct identify the system as either:

1. Aluminum
2. Copper

List straight lengths by type and size

A. Feeder — 800 A
B. Plug in — 800 A

Do not measure thru fittings as you would on conduit, since there is no allowance for scrap in bus duct systems.

If upon taking off linear foot quantities of bus duct you find your quantities are not divisible evenly by 10 ft., then the remainder must be priced as a special item and quoted from the manufactuer. Do not use the bare material cost per L.F. for these special items. You can, however, safely use the bare labor cost per L.F. for the entire length.

Identify fittings by type and ampere rating.

Example:

C. Switchboard stub 800 A

D. Elbows 800 A

E. End box 800 A

F. Cable tap box 800 A

G. Tee Fittings 800 A

H. Hangers

Plug-in Units – List separately plug-in units and identify by type and ampere rating

I. Plug-in switches 600 Volt 3 phase 60 A

J. Plug-in molded case C.B. 60 A

K. Combination starter FVNR NEMA 1

L. Combination contactor & fused switch NEMA 1

M. Combination fusible switch & lighting control 60 A

Labor-hours for feeder and plug-in sections include:

1. Unloading and uncrating
2. Hauling up to 200 ft. from loading dock
3. Measuring and marking
4. Setup of rolling staging
5. Installing hangers
6. Hanging and bolting sections
7. Aligning and leveling
8. Testing

Labor-hours do not include:

1. Modifications to existing structure for hanger supports
2. Threaded rod in excess of 2 ft.
3. Welding
4. Penetrations thru walls
5. Staging rental

Deduct the following percentages from labor only:

150 ft. to 250 ft. — 10%
251 ft. to 350 ft. — 15%
351 ft. to 500 ft. — 20%
Over 500 ft. — 25%

Deduct percentage only if runs are contained in the same area.

Example: If the job entails running 100 ft. in 5 different locations do not deduct 20%, but if the duct is being run in 1 area and the quantity is 500 ft. then you would deduct 20%.

Deduct only from straight lengths, not fittings or plug-in units.

R262513-10 Aluminum Bus Duct (cont.)

Add to labor for elevated installations:

15 ft. to 20 ft. high	10%
21 ft. to 25 ft. high	20%
26 ft. to 30 ft. high	30%
31 ft. to 35 ft. high	40%
36 ft. to 40 ft. high	50%
Over 40 ft. high	60%

Bus Duct Fittings:

Labor-hours for fittings include:
1. Unloading and uncrating
2. Hauling up to 200 ft. from loading dock
3. Installing, fitting, and bolting all ends to in-place sections

Plug-in units include:
1. Unloading and uncrating
2. Hauling up to 200 ft. from loading dock
3. Installing plug-in into in-place duct
4. Setup of rolling staging
5. Connecting load wire to lugs
6. Marking wire
7. Checking phase rotation

Labor-hours for plug-ins do not include:
1. Conduit runs from plug-in
2. Wire from plug-in
3. Conduit termination

Economy of Scale – For large concentrations of plug-in units in the same area deduct the following percentages:

11	to	25	15%
26	to	50	20%
51	to	75	25%
76	to	100	30%
100	and	over	35%

Job Conditions: Productivities are based on new construction in an unobstructed first floor area to a height of 15 ft. using rolling staging.

Material staging area is within 100 ft. of work being performed.

Add to the duct fittings and hangers:
- Plug-in switches (fused)
- Plug-in breakers
- Combination starters
- Combination contactors
- Combination fusible switch and lighting control

Labor-hours for duct and fittings include:
1. Unloading and uncrating
2. Hauling up to 200 ft. from loading dock
3. Measuring and marking
4. Installing duct runs
5. Leveling
6. Sound testing

Labor-hours for plug-ins include:
1. Unloading and uncrating
2. Hauling up to 200 ft. from loading dock
3. Installing plug-ins
4. Preparing wire
5. Wire connections and marking

R262513-15 Weight (Lbs./L.F.) of 4 Pole Aluminum and Copper Bus Duct by Ampere Load

Amperes	Aluminum Feeder	Copper Feeder	Aluminum Plug-In	Copper Plug-In
225			7	7
400			8	13
600	10	10	11	14
800	10	19	13	18
1000	11	19	16	22
1350	14	24	20	30
1600	17	26	25	39
2000	19	30	29	46
2500	27	43	36	56
3000	30	48	42	73
4000	39	67		
5000		78		

R262716-40 Standard Electrical Enclosure Types

NEMA Enclosures

Electrical enclosures serve two basic purposes; they protect people from accidental contact with enclosed electrical devices and connections, and they protect the enclosed devices and connections from specified external conditions. The National Electrical Manufacturers Association (NEMA) has established the following standards. Because these descriptions are not intended to be complete representations of NEMA listings, consultation of NEMA literature is advised for detailed information.

The following definitions and descriptions pertain to NONHAZARDOUS locations.

NEMA Type 1: General purpose enclosures intended for use indoors, primarily to prevent accidental contact of personnel with the enclosed equipment in areas that do not involve unusual conditions.

NEMA Type 2: Dripproof indoor enclosures intended to protect the enclosed equipment against dripping noncorrosive liquids and falling dirt.

NEMA Type 3: Dustproof, raintight and sleet-resistant (ice-resistant) enclosures intended for use outdoors to protect the enclosed equipment against wind-blown dust, rain, sleet, and external ice formation.

NEMA Type 3R: Rainproof and sleet-resistant (ice-resistant) enclosures which are intended for use outdoors to protect the enclosed equipment against rain. These enclosures are constructed so that the accumulation and melting of sleet (ice) will not damage the enclosure and its internal mechanisms.

NEMA Type 3S: Enclosures intended for outdoor use to provide limited protection against wind-blown dust, rain, and sleet (ice) and to allow operation of external mechanisms when ice-laden.

NEMA Type 4: Watertight and dust-tight enclosures intended for use indoors and out – to protect the enclosed equipment against splashing water, see page of water, falling or hose-directed water, and severe external condensation.

NEMA Type 4X: Watertight, dust-tight, and corrosion-resistant indoor and outdoor enclosures featuring the same provisions as Type 4 enclosures, plus corrosion resistance.

NEMA Type 5: Indoor enclosures intended primarily to provide limited protection against dust and falling dirt.

NEMA Type 6: Enclosures intended for indoor and outdoor use – primarily to provide limited protection against the entry of water during occasional temporary submersion at a limited depth.

NEMA Type 6R: Enclosures intended for indoor and outdoor use – primarily to provide limited protection against the entry of water during prolonged submersion at a limited depth.

NEMA Type 11: Enclosures intended for indoor use – primarily to provide, by means of oil immersion, limited protection to enclosed equipment against the corrosive effects of liquids and gases.

NEMA Type 12: Dust-tight and driptight indoor enclosures intended for use indoors in industrial locations to protect the enclosed equipment against fibers, flyings, lint, dust, and dirt, as well as light splashing, see page, dripping, and external condensation of noncorrosive liquids.

NEMA Type 13: Oil-tight and dust-tight indoor enclosures intended primarily to house pilot devices, such as limit switches, foot switches, push buttons, selector switches, and pilot lights, and to protect these devices against lint and dust, see page, external condensation, and sprayed water, oil, and noncorrosive coolant.

The following definitions and descriptions pertain to HAZARDOUS, or CLASSIFIED, locations:

NEMA Type 7: Enclosures intended to use in indoor locations classified as Class 1, Groups A, B, C, or D, as defined in the National Electrical Code.

NEMA Type 9: Enclosures intended for use in indoor locations classified as Class 2, Groups E, F, or G, as defined in the National Electrical Code.

R262726-90 Wiring Devices

Wiring devices should be priced on a separate takeoff form which includes boxes, covers, conduit and wire.

Labor-hours for devices include:

1. Stripping of wire
2. Attaching wire to device using terminators on the device itself, lugs, set screws etc.
3. Mounting of device in box

Labor-hours do not include:

1. Conduit
2. Wire
3. Boxes
4. Plates

Economy of Scale – for large concentrations of devices in the same area deduct the following percentages from labor-hours:

1	to	10	0%
11	to	25	20%
26	to	50	25%
51	to	100	30%
	over	100	35%

R262726-90 Wiring Devices (cont.)

<table>
<tr><th>NEMA No.</th><th>15 R</th><th>20 R</th><th>30 R</th><th>50 R</th><th>60 R</th></tr>
<tr><td>1
125V
2 Pole, 2 Wire</td><td colspan="5">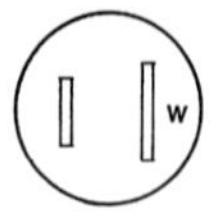</td></tr>
<tr><td>2
250V
2 Pole, 2 Wire</td><td colspan="5">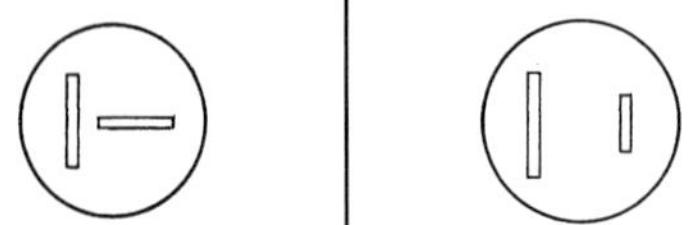</td></tr>
<tr><td>5
125V
2 Pole, 3 Wire</td><td colspan="5">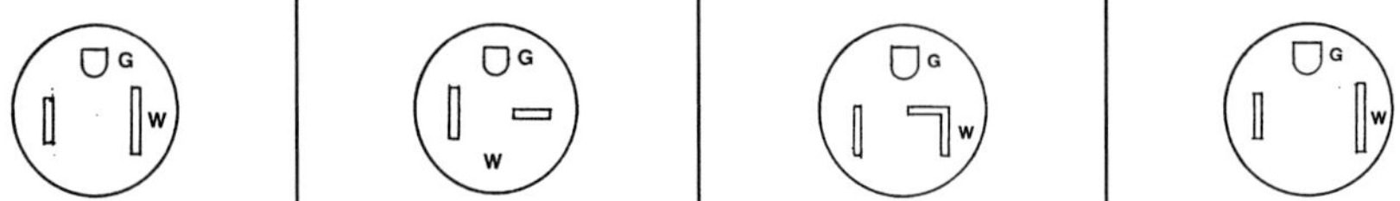</td></tr>
<tr><td>6
250V
2 Pole, 3 Wire</td><td colspan="5">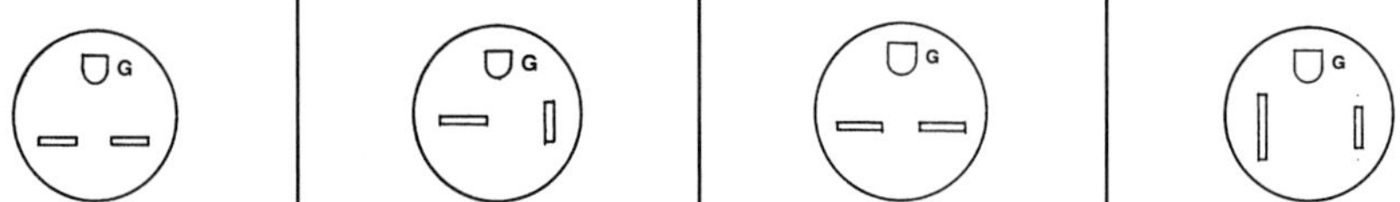</td></tr>
<tr><td>7
277V, AC
2 Pole, 3 Wire</td><td colspan="5">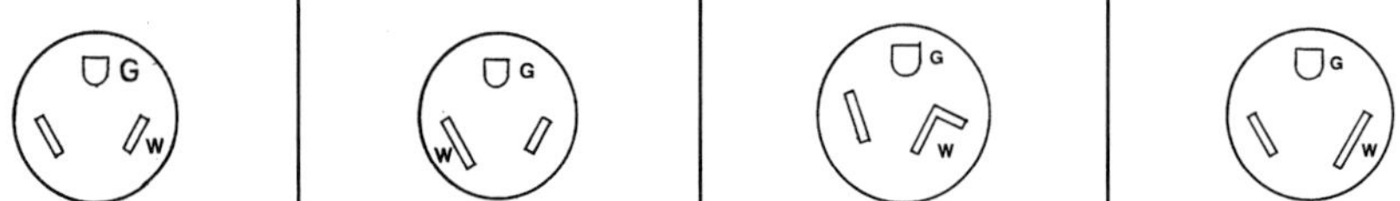</td></tr>
<tr><td>10
125/250V
3 Pole, 3 Wire</td><td colspan="5">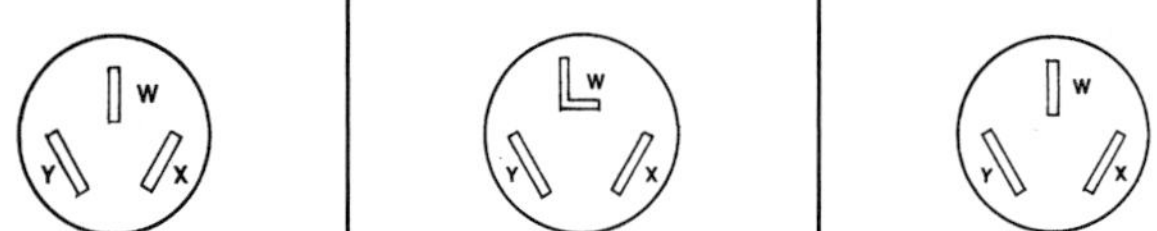</td></tr>
<tr><td>11
3 Phase
250V
3 Pole, 3 Wire</td><td colspan="5">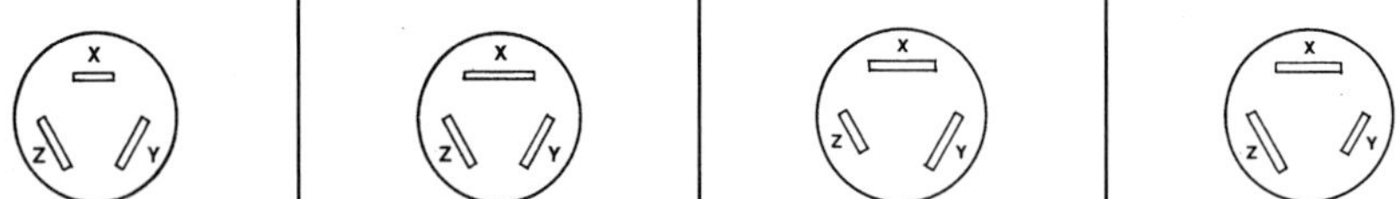</td></tr>
<tr><td>14
125/250V
3 Pole, 4 Wire</td><td colspan="5">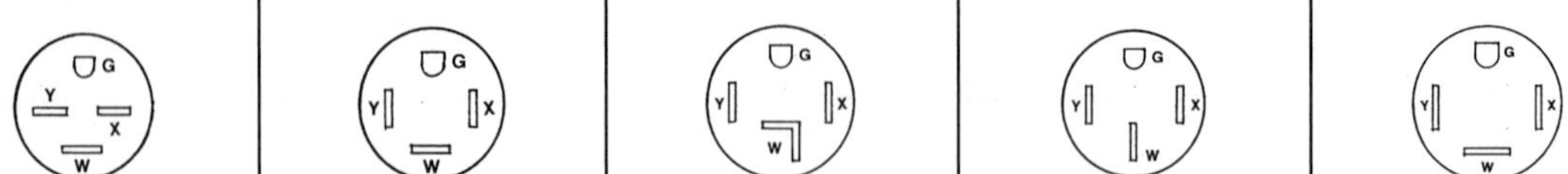</td></tr>
<tr><td>15
3 Phase
250V
3 Pole, 4 Wire</td><td colspan="5">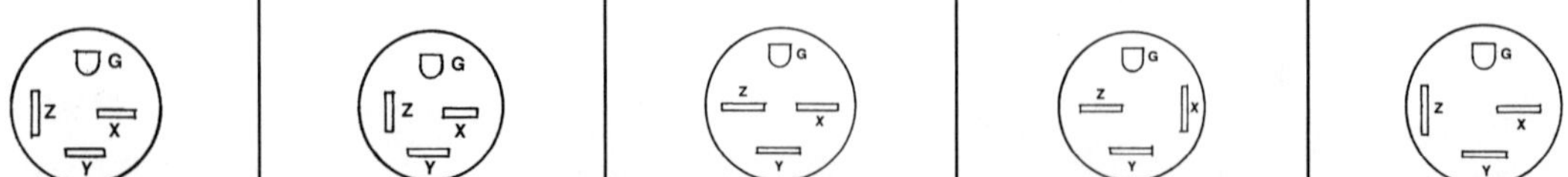</td></tr>
<tr><td>18
3 Phase
208Y/120V
4 Pole, 4 Wire</td><td colspan="5">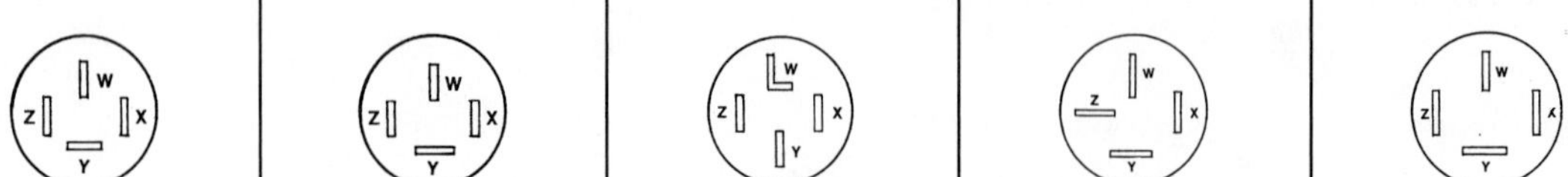</td></tr>
</table>

R262726-90 Wiring Devices (cont.)

NEMA No.	15 R	20 R	30 R
L 1 125V 2 Pole, 2 Wire	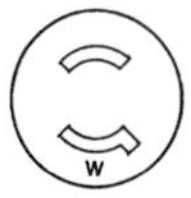		
L 2 250V 2 Pole, 2 Wire			
L 5 125 V 2 Pole, 3 Wire	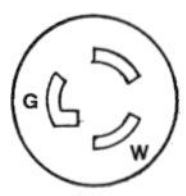		
L 6 250 V 2 Pole, 3 Wire			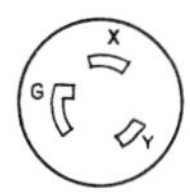
L 7 227 V, AC 2 Pole, 3 Wire		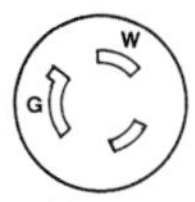	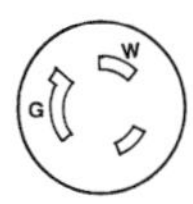
L 8 480 V 2 Pole, 3 Wire			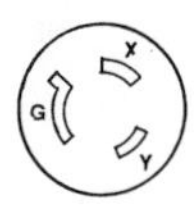
L 9 600 V 2 Pole, 3 Wire		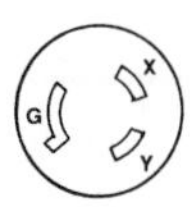	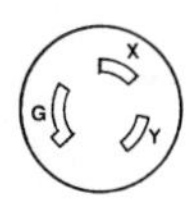
L 10 125 /250V 3 Pole, 3 Wire		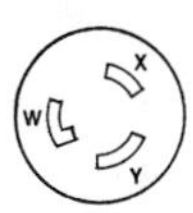	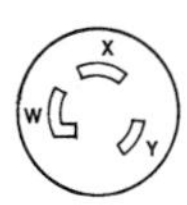
L 11 3 Phase 250 V 3 Pole, 3 Wire		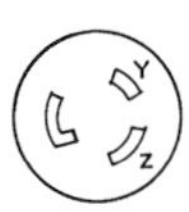	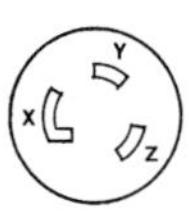
L 12 3 Phase 480 V 3 Pole, 3 Wire		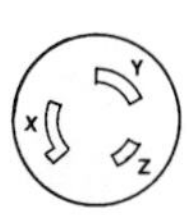	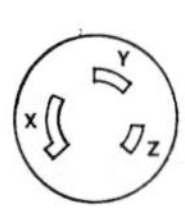
L 13 3 Phase 600V 3 Pole, 3 Wire			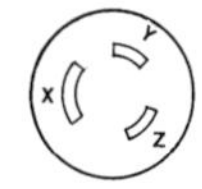
L 14 125/250V 3 Pole, 4 Wire		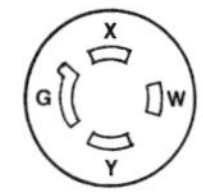	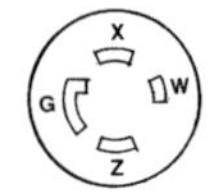
L 15 3 Phase 250 V 3 Pole, 4 Wire		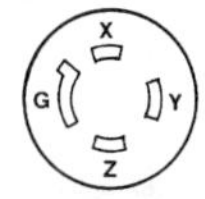	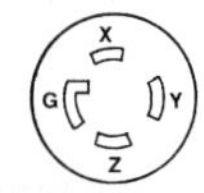
L 16 3 Phase 480V 3 Pole, 4 Wire		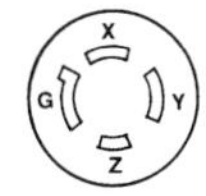	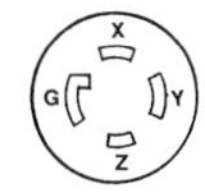
L 17 3 Phase 600 V 3 Pole, 4 Wire			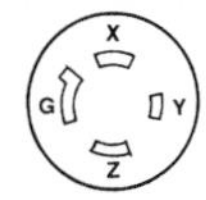
L 18 3 Phase 208Y/120V 4 Pole, 4 Wire		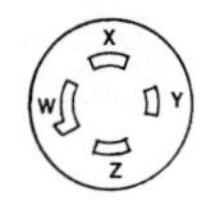	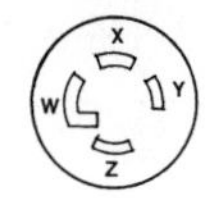
L 19 3 Phase 480Y/277V 4 Pole, 4 Wire		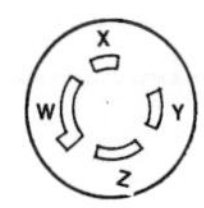	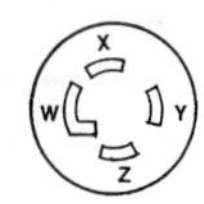
L 20 3 Phase 600Y/347V 4 Pole, 4 Wire		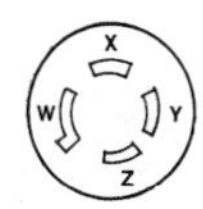	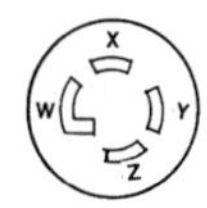
L 21 3 Phase 208Y/120V 4 Pole, 5 Wire		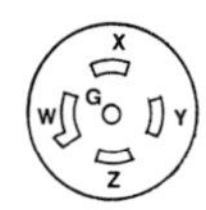	
L 22 3 Phase 480Y/277V 4 Pole, 5 Wire		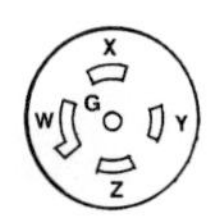	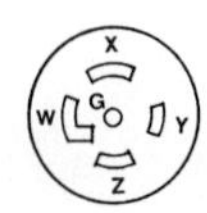
L 23 3 Phase 600Y/347V 4 Pole, 5 Wire		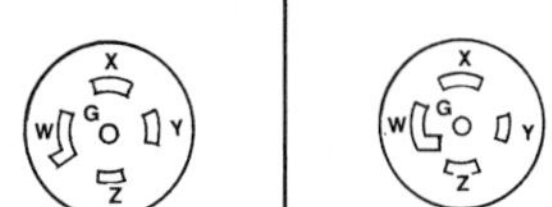	

Electrical — R2628 Low-Voltage Circuit Protective Devices

R262816-80 Safety Switches

List each Safety Switch by type, ampere rating, voltage, single or three phase, fused or nonfused.

Example:

A. General duty, 240V, 3-pole, fused
B. Heavy Duty, 600V, 3-pole, nonfused
C. Heavy Duty, 240V, 2-pole, fused
D. Heavy Duty, 600V, 3-pole, fused

Also include NEMA enclosure type and identify as:

1. Indoor
2. Weatherproof
3. Explosionproof

Installation of Safety Switches includes:

1. Unloading and uncrating
2. Handling disconnects up to 200′ from loading dock
3. Measuring and marking location
4. Drilling (4) anchor type lead fasteners, using a hammer drill
5. Mounting and leveling Safety Switch
6. Installing (3) fuses
7. Phasing and tagging line and load wires

Price does not include:

1. Modifications to enclosure
2. Plywood backboard
3. Conduit or wire
4. Fuses
5. Termination of wires

Job Conditions: Productivities are based on new construction to an installed height of 6′ above finished floor.

Material staging area is assumed to be within 100′ of work in progress.

Electrical — R2632 Packaged Generator Assemblies

R263213-45 Generator Weight (Lbs.) by kW

3 Phase 4 Wire /480 Volt			
Gas		Diesel	
kW	Lbs.	kW	Lbs.
7.5	600	30	1800
10	630	50	2230
15	960	75	2250
30	1500	100	3840
65	2350	125	4030
85	2570	150	5500
115	4310	175	5650
170	6530	200	5930
		250	6320
		300	7840
		350	8220
		400	10750
		500	11900

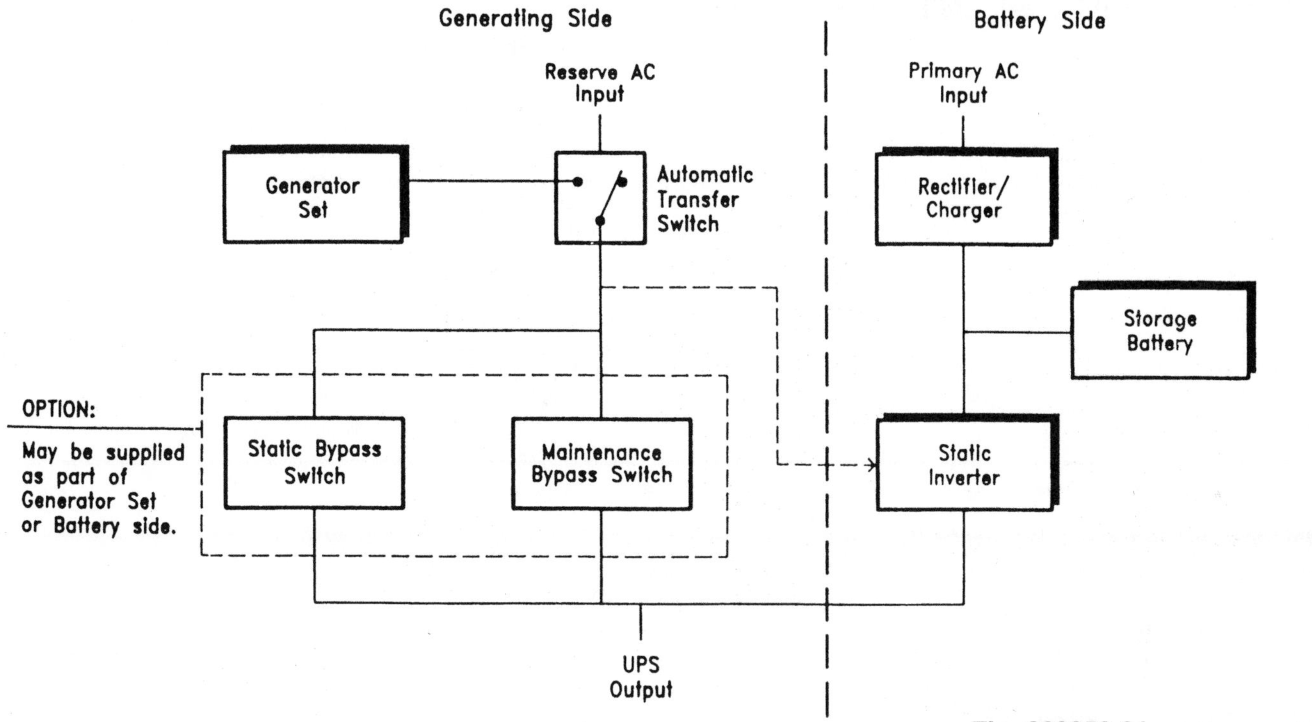

Fig. 263353-81

R263353-80 Uninterruptible Power Supply Systems

General: Uninterruptible Power Supply (UPS) Systems are used to provide power for legally required standby systems. They are also designed to protect computers and provide optional additional coverage for any or all loads in a facility. Figure R263353-81 shows a typical configuration for a UPS System with generator set.

Cost Determination:

It is recommended that UPS System material costs be obtained from equipment manufacturers as a package. Installation costs should be obtained from a vendor or contractor.

The recommended procedure for pricing UPS Systems would be to identify:

1. Frequency – by Hertz (Hz.)
   For example: 50, 60, and 400/415 Hz.
2. Apparent power and real power
   For example: 200 kVA/170 kW
3. Input/Output voltage For example: 120V, 208V or 240V
4. Phase – either single or three-phase
5. Options and accessories – For extended run time more batteries would be required. Other accessories include battery cabinets, battery racks, remote control panel, power distribution unit (PDU), and warranty enhancement plans.

Note:

1. Larger systems can be configured by paralleling two or more standard small size single modules.
   For example: two 15 kVA modules, combined together and configured to 30 kVA.
2. Maximum input current during battery recharge is typically 15% higher than normal current.
3. UPS Systems weights vary depending on options purchased and power rating in kVA.

R263413-30 HP Value/Cost Determination

General: Motors can be powered by any of the seven systems shown in Figure R262213-11 thru Figure R262213-17 provided the motor voltage characteristics are compatible with the power system characteristics.

Cost Determination:

Motor Amperes for the various size H.P. and voltage are listed in Table R263413-31. To find the amperes, locate the required H.P. rating and locate the amperes under the appropriate circuit characteristics.

For example:

A. 100 H.P., 3 phase, 460 volt motor = 124 amperes (Table R263413-31)

B. 10 H.P., 3 phase, 200 volt motor = 32.2 amperes (Table R263413-31)

Motor Wire Size: After the amperes are found in Table R263413-31 the amperes must be increased 25% to compensate for power losses. Next refer to Table R260519-92. Find the appropriate insulation column for copper or aluminum wire to determine the proper wire size.

For example:

A. 100 H.P., 3 phase, 460 volt motor has an ampere value of 124 amperes from Table R263413-31

B. 124A x 1.25 = 155 amperes

C. Refer to Table R260519-92 for THW or THWN wire insulations to find the proper wire size. For a 155 ampere load using copper wire a size 2/0 wire is needed.

D. For the 3 phase motor three wires of 2/0 size are required.

Conduit Size: To obtain the proper conduit size for the wires and type of insulation used, refer to Table R260533-22.

For example: For the 100 H.P., 460V, 3 phase motor, it was determined that three 2/0 wires are required. Assuming THWN insulated copper wire, use Table R260533-22 to determine that three 2/0 wires require 1-1/2″ conduit.

Material Cost of the conduit and wire system depends on:

1. Wire size required
2. Copper or aluminum wire
3. Wire insulation type selected
4. Steel or plastic conduit
5. Type of conduit raceway selected.

Labor Cost of the conduit and wire system depends on:

1. Type and size of conduit
2. Type and size of wires installed
3. Location and height of installation in building or depth of trench
4. Support system for conduit.

R263413-31 Ampere Values Determined by Horsepower, Voltage and Phase Values

H.P.	Amperes							
	Single Phase			Three Phase				
	115V	208V	230V	200V	208V	230V	460V	575V
1/6	4.4A	2.4A	2.2A					
1/4	5.8	3.2	2.9					
1/3	7.2	4.0	3.6					
1/2	9.8	5.4	4.9	2.5A	2.4A	2.2A	1.1A	0.9A
3/4	13.8	7.6	6.9	3.7	3.5	3.2	1.6	1.3
1	16	8.8	8	4.8	4.6	4.2	2.1	1.7
1-1/2	20	11	10	6.9	6.6	6.0	3.0	2.4
2	24	13.2	12	7.8	7.5	6.8	3.4	2.7
3	34	18.7	17	11.0	10.6	9.6	4.8	3.9
5	56	30.8	28	17.5	16.7	15.2	7.6	6.1
7-1/2	80	44	40	25.3	24.2	22	11	9
10	100	55	50	32.2	30.8	28	14	11
15				48.3	46.2	42	21	17
20				62.1	59.4	54	27	22
25				78.2	74.8	68	34	27
30				92	88	80	40	32
40				120	114	104	52	41
50				150	143	130	65	52
60				177	169	154	77	62
75				221	211	192	96	77
100				285	273	248	124	99
125				359	343	312	156	125
150				414	396	360	180	144
200				552	528	480	240	192
250							302	242
300							361	289
350							414	336
400							477	382

R263413-30 Cost Determination (cont.)

Magnetic starters, switches, and motor connection:

To complete the cost picture from H.P. to Costs additional items must be added to the cost of the conduit and wire system to arrive at a total cost.

1. Assembly Table D5020-160 Magnetic Starters Installed Cost lists the various size starters for single phase and three phase motors.
2. Assembly Table D5020-165 Heavy Duty Safety Switches Installed Cost lists safety switches required at the beginning of a motor circuit and also one required in the vicinity of the motor location.
3. Assembly Table D5020-170 Motor Connection lists the various costs for single and three phase motors.

Worksheet to obtain total motor wiring costs:

It is assumed that the motors or motor driven equipment are furnished and installed under other sections for this estimate and the following work is done under this section:

1. Conduit
2. Wire (add 10% for additional wire beyond conduit ends for connections to switches, boxes, starters, etc.)
3. Starters
4. Safety switches
5. Motor connections

Figure R263413-32

Worksheet for Motor Circuits					
				Cost	
Item	Type	Size	Quantity	Unit	Total
Wire					
Conduit					
Switch					
Starter					
Switch					
Motor Connection					
Other					
Total Cost					

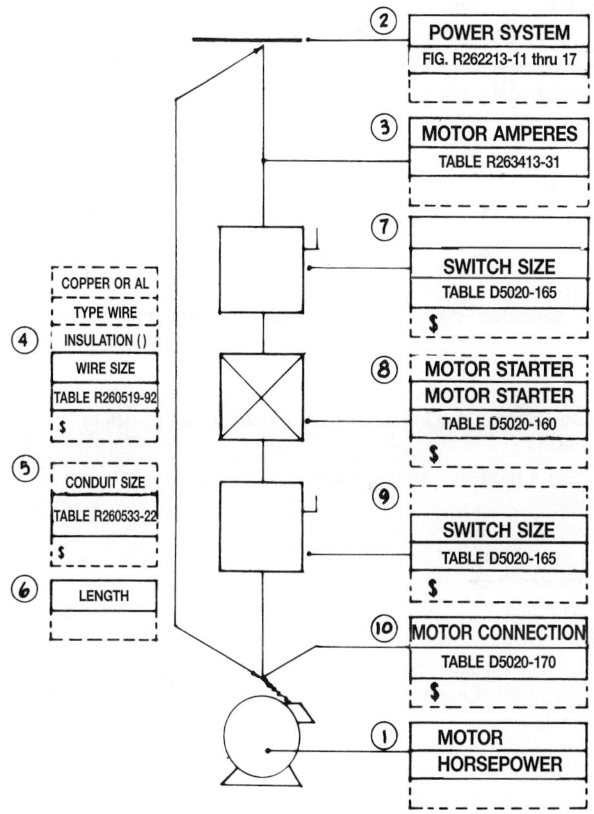

R263413-33 Maximum Horsepower for Starter Size by Voltage

Starter	Maximum HP (3ϕ)			
Size	208V	240V	480V	600V
00	1½	1½	2	2
0	3	3	5	5
1	7½	7½	10	10
2	10	15	25	25
3	25	30	50	50
4	40	50	100	100
5		100	200	200
6		200	300	300
7		300	600	600
8		450	900	900
8L		700	1500	1500

Electrical R2636 Transfer Switches

R263623-60 Automatic Transfer Switches

When taking off Automatic transfer switches identify by voltage, amperage and number of poles.

Example: Automatic transfer switch, 480V, 3 phase, 30A, NEMA 1 enclosure

Labor-hours for transfer switches include:

1. Unloading, uncrating and handling switches up to 200′ from loading dock
2. Measuring and marking
3. Drilling (4) lead type anchors, using a hammer drill
4. Mounting and leveling
5. Circuit identification
6. Testing and load balancing

Labor-hours do not include:

1. Modifications in enclosure
2. Structural supports
3. Additional lugs
4. Plywood backboard
5. Painting or lettering
6. Conduit runs to or from transfer switch
7. Wire
8. Termination of wires

Electrical R2651 Interior Lighting

R265113-40 Interior Lighting Fixtures

When taking off interior lighting fixtures, it is advisable to set up your quantity worksheet to conform to the lighting schedule as it appears on the print. Include the alpha-numeric code plus the symbol on your work sheet.

Take off a particular section or floor of the building and count each type of fixture before going on to another type. It would also be advantageous to include on the same worksheet the pipe, wire, fittings and circuit number associated with each type of lighting fixture. This will help you identify the costs associated with any particular lighting system and in turn make material purchases more specific as to when and how much to order under the classification of lighting.

By taking off lighting first you can get a complete "WALK THRU" of the job. This will become helpful when doing other phases of the project.

Materials for a recessed fixture include:

1. Fixture
2. Lamps
3. 6′ of jack chain
4. (2) S hooks
5. (2) Wire nuts

Labor for interior recessed fixtures include:

1. Unloading by hand
2. Hauling by hand to an area up to 200′ from loading dock
3. Uncrating
4. Layout
5. Installing fixture
6. Attaching jack chain & S hooks
7. Connecting circuit power
8. Reassembling fixture
9. Installing lamps
10. Testing

Material for surface mounted fixtures includes:

1. Fixture
2. Lamps
3. Either (4) lead type anchors, (4) toggle bolts, or (4) ceiling grid clips
4. (2) Wire nuts

Material for pendent mounted fixtures includes:

1. Fixture
2. Lamps
3. (2) Wire nuts
4. Rigid pendents as required by type of fixtures
5. Canopies as required by type of fixture

Labor hours include the following for both surface and pendent fixtures:

1. Unloading by hand
2. Hauling by hand to an area up to 200′ from loading dock
3. Uncrating
4. Layout and marking
5. Drilling (4) holes for either lead anchors or toggle bolts using a hammer drill
6. Installing fixture
7. Leveling fixture
8. Connecting circuit power
9. Installing lamps
10. Testing

Labor for surface or pendent fixtures does not include:

1. Conduit
2. Boxes or covers
3. Connectors
4. Fixture whips
5. Special support
6. Switching
7. Wire

Economy of Scale: For large concentrations of lighting fixtures in the same area deduct the following percentages from labor:

25	to	50	fixtures	15%
51	to	75	fixtures	20%
76	to	100	fixtures	25%
101 and over				30%

Job Conditions: Productivity is based on new construction in a unobstructed first floor location, using rolling staging to 15′ high.

Material staging is assumed to be within 100′ of work being performed.

Add the following percentages to labor for elevated installations:

15′	to	20′	high	10%
21′	to	25′	high	20%
26′	to	30′	high	30%
31′	to	35′	high	40%
36′	to	40′	high	50%
41′ and over				60%

R265723-05 Comparison - Operation of High Intensity Discharge Lamps

Lamp Type	Wattage	Life (Hours)	Circuit Wattage [1]	Average Initial Lumens	L.L.D. [2]	Mean Lumens [3]
M.V.	100 DX	24,000	125	4,000	61%	2,440
L.P.S.	SOX-35	18,000	65	4,800	100%	4,800
H.P.S.	LU-70	12,000	84	5,800	90%	5,220
M.H.	No Equivalent					
M.V.	175 DX	24,000	210	8,500	66%	5,676
M.V.	250 DX	24,000	295	13,000	66%	7,986
L.P.S.	SOX-55	18,000	82	8,000	100%	8,000
H.P.S.	LU-100	12,000	120	9,500	90%	8,550
M.H.	No Equivalent					
M.V.	400 DX	24,000	465	24,000	64%	14,400
L.P.S.	SOX-90	18,000	141	13,500	100%	13,500
H.P.S.	LU-150	16,000	188	16,000	90%	14,400
M.H.	MH-175	7,500	210	14,000	73%	10,200
M.V.	No Equivalent					
L.P.S.	SOX-135	18,000	147	22,500	100%	22,500
H.P.S.	LU-250	20,000	310	25,500	92%	23,205
M.H.	MH-250	7,500	295	20,500	78%	16,000
M.V.	1000 DX	24,000	1,085	63,000	61%	37,820
L.P.S.	SOX-180	18,000	248	33,000	100%	33,000
H.P.S.	LU-400	20,000	480	50,000	90%	45,000
M.H.	MH-400	15,000	465	34,000	72%	24,600
H.P.S.	LU-1000	15,000	1,100	140,000	91%	127,400

1. Includes ballast losses and average lamp watts
2. Lamp lumen depreciation (% of initial light output at 70% rated life)
3. Lamp lumen output at 70% rated life (L.L.D. x initial)
4. Cost of yearly operation = Cicuit Wattage/1000 x annual operating hours x average cost per KWH

M.V. = Mercury Vapor
L.P.S. = Low pressure sodium
H.P.S. = High pressure sodium
L.H. = Metal halide

R265723-10 For Other than Regular Cool White (CW) Lamps

Multiply Material Costs as Follows:					
Regular Lamps	Cool white deluxe (CWX)	x 1.35	Energy Saving Lamps	Cool white (CW/ES)	x 1.35
	Warm white deluxe (WWX)	x 1.35		Cool white deluxe (CWX/ES)	x 1.65
	Warm white (WW)	x 1.30		Warm white (WW/ES)	x 1.55
	Natural (N)	x 2.05		Warm white deluxe (WWX/ES)	x 1.65

R265723-20 Lamp Comparison Chart with Enclosed Floodlight, Ballast, & Lamp for Pole Mounting

Type	Watts	Initial Lumens	Lumens per Watt	Lumens @ 40% Life	Life (Hours)
Incandescent	150	2,880	19	85	750
	300	6,360	21	84	750
	500	10,850	22	80	1,000
	1,000	23,740	24	80	1,000
	1,500	34,400	23	80	1,000
Tungsten Halogen	500	10,950	22	97	2,000
	1,500	35,800	24	97	2,000
Fluorescent Cool White	40	3,150	79	88	20,000
	110	9,200	84	87	12,000
	215	16,000	74	81	12,000
Deluxe Mercury	250	12,100	48	86	24,000
	400	22,500	56	85	24,000
	1,000	63,000	63	75	24,000
Metal Halide	175	14,000	80	77	7,500
	400	34,000	85	75	15,000
	1,000	100,000	100	83	10,000
	1,500	155,000	103	92	1,500
High Pressure Sodium	70	5,800	83	90	20,000
	100	9,500	95	90	20,000
	150	16,000	107	90	24,000
	400	50,000	125	90	24,000
	1,000	140,000	140	90	24,000
Low Pressure Sodium	55	4,600	131	98	18,000
	90	12,750	142	98	18,000
	180	33,000	183	98	18,000

Color: High Pressure Sodium — Slightly Yellow
Low Pressure Sodium — Yellow
Mercury Vapor — Green-Blue
Metal Halide — Blue White

Note: Pole not included.

R265723-25 Energy Efficiency Rating for Luminaires

The energy efficiency program for luminaires recommends the use of a metric called the Luminaire Efficacy Rating (LER). The LER value expresses the total luminaires generated by a lamp, to the watts consumed by the lamp.

Lamp Type	Lumens per Watt
Incandescent	17
Tungsten Halogen	14 - 20
Fluorescent	50 - 104
Metal Halide	64 - 96
High Pressure Sodium	76 - 116

Communications — R2713 Communications Backbone Cabling

R271323-40 Fiber Optics

Fiber optic systems use optical fiber such as plastic, glass, or fused silica, a transparent material, to transmit radiant power (i.e., light) for control, communication, and signaling applications. The types of fiber optic cables can be nonconductive, conductive, or composite. The composite cables contain fiber optics and current-carrying electrical conductors. The configuration for one of the fiber optic systems is as follows:

The transceiver module acts as transmitting and receiving equipment in a common house, which converts electrical energy to light energy or vice versa.

Pricing the fiber optic system is not an easy task. The performance of the whole system will affect the cost significantly. New specialized tools and techniques decrease the installing cost tremendously. In the fiber optic section of Means Electrical Cost Data, a benchmark for labor-hours and material costs is set up so that users can adjust their costs according to unique project conditions.

Units for Measure: Fiber optic cable is measured in hundred linear feet (C.L.F.) or industry units of measure - meter (m) or kilometer (km). The connectors are counted as units (EA.)

Material Units: Generally, the material costs include only the cable. All the accessories shall be priced separately.

Labor Units: The following procedures are generally included for the installation of fiber optic cables:

- Receiving
- Material handling
- Setting up pulling equipment
- Measuring and cutting cable
- Pulling cable

These additional items are listed and extended: Terminations

Takeoff Procedure: Cable should be taken off by type, size, number of fibers, and number of terminations required. List the lengths of each type of cable on the takeoff sheets. Total and add 10% for waste. Transfer the figures to a cost analysis sheet and extend.

Communications — R2715 Communications Horizontal Cabling

R271513-75 High Performance Cable

There are several categories used to describe high performance cable. The following information includes a description of categories CAT 3, 5, 5e, 6, and 7, and details classifications of frequency and specific standards. The category standards have evolved under the sponsorship of organizations such as the Telecommunication Industry Association (TIA), the Electronic Industries Alliance (EIA), the American National Standards Institute (ANSI), the International Organization for Standardization (ISO), and the International Electrotechnical Commission (IEC), all of which have catered to the increasing complexities of modern network technology. For network cabling, users must comply with national or international standards. A breakdown of these categories is as follows:

Category 3: Designed to handle frequencies up to 16 MHz.

Category 5: (TIA/EIA 568A) Designed to handle frequencies up to 100 MHz.

Category 5e: Additional transmission performance to exceed Category 5.

Category 6: Development by TIA and other international groups to handle frequencies of 250 MHz.

Category 7: Development to handle a frequency range up to 600 MHz.

R337116-60 Average Transmission Line Material Requirements (Per Mile)

Terrain:		Flat				Rolling				Mountain			
Item		69kV	161kV	161kV	500kV	69kV	161kV	161kV	500kV	69kV	161kV	161kV	500kV
Pole Type	Unit	Wood	Wood	Steel	Steel	Wood	Wood	Steel	Steel	Wood	Wood	Steel	Steel
Conductor: 397,500 – Cir. Mil., 26/7-ACSR													
Structures	Ea.	12[1]				9[2]				7[2]			
Poles	Ea.	12				18				14			
Crossarms	Ea.	24[3]				9[4]				7[4]			
Conductor	Ft.	15,990				15,990				15,990			
Insulators[5]	Ea.	180				135				105			
Ground Wire	Ft.	5,330				10,660				10,660			
Conductor: 636,000 – Cir. Mil., 26/7-ACSR													
Structures	Ea.	13[1]	11[6]			10[2]	9[2]	6[9]		8[2]	8[2]	6[9]	
Excavation	C.Y.	–	–			–	–	120		–	–	120	
Concrete	C.Y.	–	–			–	–	10		–	–	10	
Steel Towers	Tons	–	–			–	–	32		–	–	32	
Poles	Ea.	13	11			20	18	–		16	16	–	
Crossarms	Ea.	26[3]	33[7]			10[4]	9[8]	–		8[4]	8[8]	–	
Conductor	Ft.	15,990	15,990			15,990	15,990	15,990		15,990	15,990	15,990	
Insulators[5]	Ea.	195	165			150	297	297		120	264	297	
Ground Wire	Ft.	5330	5330			10,660	10,660	10,660		10,660	10,660	10,660	
Conductor: 954,000 – Cir. Mil., 45/7-ACSR													
Structures	Ea.	14[1]	12[6]		4[11]	10[2]	9[2]	6[9]	4[11]	8[2]	8[2]	6[9]	4[13]
Excavation	C.Y.	–	–		200	–	–	125	214	–	–	125	233
Concrete	C.Y.	–	–		20	–	–	10	21	–	–	10	21
Steel Towers	Tons	–	–		57	–	–	33	57	–	–	33	63
Poles	Ea.	14	12		–	20	18	–	–	16	16	–	–
Crossarms	Ea.	28[10]	36[7]		–	10[4]	9[8]	–	–	8[4]	8[8]	–	–
Conductor	Ft.	15,990	15,990		47,970[12]	15,990	15,990	15,990	47,970[12]	15,990	15,990	15,990	47,970[12]
Insulators[5]	Ea.	210	180		288	150	297	297	288	120	264	297	576
Ground Wire	Ft.	5330	5330		10,660	10,660	10,660	10,660	10,660	10,660	10,660	10,660	10,660
Conductor: 1,351,500 – Cir. Mil., 45/7-ACSR													
Structures	Ea.	8[14]				8[14]							
Excavation	C.Y.	220				220							
Concrete	C.Y.	28				28							
Steel Towers	Ton	46				46							
Conductor	Ft.	31,680[15]				31,680[15]							
Insulators	Ea.	528				528							
Ground Wire	Ft.	10,660				10,660							

1. Single pole two-arm suspension type construction
2. Two-pole wood H-frame construction
3. 4¾″ x 5¾″ x 8′ and 4¾″ x 5¾″ x 10′ wood crossarm
4. 6″ x 8″ x 26′-0″ wood crossarm
5. 5¾″ x 10″ disc insulator
6. Single pole construction with 3 fiberglass crossarms (5 fog- type insulators per phase)
7. 7′-0″ fiberglass crossarms
8. 6″ x 10″ x 35′-0″ wood crossarm
9. Laced steel tower, single circuit construction
10. 5″ x 7″ x 8′ and 5″ x 7″ x 10′ wood crossarm
11. Laced steel tower, single circuit 500-kV construction
12. Bundled conductor (3 sub-conductors per phase)
13. Laced steel tower, single circuit restrained phases (500-kV)
14. Laced steel tower, double circuit construction
15. Both sides of double circuit strung

Note: To allow for sagging, a mile (5280 Ft.) of transmission line uses 5330 Ft. of conductor per wire (called a wire mile).

R337119-30 Concrete for Conduit Encasement

Table below lists C.Y. of concrete for 100 L.F. of trench. Conduits separation center to center should meet 7.5″ (N.E.C.).

Number of Conduits	1	2	3	4	6	8	9	Number of Conduits
Trench Dimension	11.5″ x 11.5″	11.5″ x 19″	11.5″ x 27″	19″ x 19″	19″ x 27″	19″ x 38″	27″ x 27″	Trench Dimension
Conduit Diameter 2.0″	3.29	5.39	7.64	8.83	12.51	17.66	17.72	Conduit Diameter 2.0″
2.5″	3.23	5.29	7.49	8.62	12.19	17.23	17.25	2.5″
3.0″	3.15	5.13	7.24	8.29	11.71	16.59	16.52	3.0″
3.5″	3.08	4.97	7.02	7.99	11.26	15.98	15.84	3.5″
4.0″	2.99	4.80	6.76	7.65	10.74	15.30	15.07	4.0″
5.0″	2.78	4.37	6.11	6.78	9.44	13.57	13.12	5.0″
6.0″	2.52	3.84	5.33	5.74	7.87	11.48	10.77	6.0″
					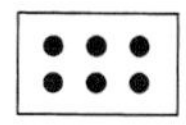		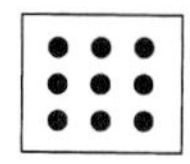	

Estimating Tips

- The cost figures in this Square Foot Cost section were derived from approximately 11,000 projects contained in the RSMeans database of completed construction projects. They include the contractor's overhead and profit, but do not generally include architectural fees or land costs. The figures have been adjusted to January of the current year. New projects are added to our files each year, and outdated projects are discarded. For this reason, certain costs may not show a uniform annual progression. In no case are all subdivisions of a project listed.
- These projects were located throughout the U.S. and reflect a tremendous variation in square foot (S.F.) and cubic foot (C.F.) costs. This is due to differences, not only in labor and material costs, but also in individual owners' requirements. For instance, a bank in a large city would have different features than one in a rural area. This is true of all the different types of buildings analyzed. Therefore, caution should be exercised when using these square foot costs. For example, for courthouses, costs in the database are local courthouse costs and will not apply to the larger, more elaborate federal courthouses. As a general rule, the projects in the 1/4 column do not include any site work or equipment, while the projects in the 3/4 column may include both equipment and site work. The median figures do not generally include site work.
- None of the figures "go with" any others. All individual cost items were computed and tabulated separately. Thus, the sum of the median figures for plumbing, HVAC and electrical will not normally total up to the total mechanical and electrical costs arrived at by separate analysis and tabulation of the projects.
- Each building was analyzed as to total and component costs and percentages. The figures were arranged in ascending order with the results tabulated as shown. The 1/4 column shows that 25% of the projects had lower costs and 75% had higher. The 3/4 column shows that 75% of the projects had lower costs and 25% had higher. The median column shows that 50% of the projects had lower costs and 50% had higher.
- There are two times when square foot costs are useful. The first is in the conceptual stage when no details are available. Then square foot costs make a useful starting point. The second is after the bids are in and the costs can be worked back into their appropriate categories for information purposes. As soon as details become available in the project design, the square foot approach should be discontinued and the project priced as to its particular components. When more precision is required, or for estimating the replacement cost of specific buildings, the current edition of *RSMeans Square Foot Costs* should be used.
- In using the figures in this section, it is recommended that the median column be used for preliminary figures if no additional information is available. The median figures, when multiplied by the total city construction cost index figures (see city cost indexes) and then multiplied by the project size modifier at the end of this section, should present a fairly accurate base figure, which would then have to be adjusted in view of the estimator's experience, local economic conditions, code requirements, and the owner's particular requirements. There is no need to factor the percentage figures, as these should remain constant from city to city. All tabulations mentioning air conditioning had at least partial air conditioning.
- The editors of this book would greatly appreciate receiving cost figures on one or more of your recent projects, which would then be included in the averages for next year. All cost figures received will be kept confidential, except that they will be averaged with other similar projects to arrive at square foot cost figures for next year's book. See the last page of the book for details and the discount available for submitting one or more of your projects.

50 17 00 | S.F. Costs

		50 17 00 | S.F. Costs	UNIT	UNIT COSTS 1/4	UNIT COSTS MEDIAN	UNIT COSTS 3/4	% OF TOTAL 1/4	% OF TOTAL MEDIAN	% OF TOTAL 3/4	
01	0010	**APARTMENTS Low Rise (1 to 3 story)**	S.F.	71.50	90	119				01
	0020	Total project cost	C.F.	6.40	8.50	10.45				
	0100	Site work	S.F.	6.10	8.35	14.65	6.05%	10.55%	14.05%	
	0500	Masonry		1.40	3.26	5.65	1.54%	3.67%	6.35%	
	1500	Finishes		7.55	10.40	12.85	9.05%	10.75%	12.85%	
	1800	Equipment		2.34	3.54	5.25	2.73%	4.03%	5.95%	
	2720	Plumbing		5.55	7.15	9.10	6.65%	8.95%	10.05%	
	2770	Heating, ventilating, air conditioning		3.54	4.36	6.40	4.20%	5.60%	7.60%	
	2900	Electrical		4.14	5.50	7.45	5.20%	6.65%	8.40%	
	3100	Total: Mechanical & Electrical		14.30	18.20	22.50	15.90%	18.05%	23%	
	9000	Per apartment unit, total cost	Apt.	66,500	101,500	149,500				
	9500	Total: Mechanical & Electrical	"	12,600	19,800	25,800				
02	0010	**APARTMENTS Mid Rise (4 to 7 story)**	S.F.	94.50	114	141				02
	0020	Total project costs	C.F.	7.40	10.20	13.90				
	0100	Site work	S.F.	3.78	7.65	14.80	5.25%	6.70%	9.20%	
	0500	Masonry		6.30	8.65	11.85	5.10%	7.25%	10.50%	
	1500	Finishes		12.35	16.60	19.55	10.70%	13.50%	17.70%	
	1800	Equipment		2.75	4.11	5.40	2.54%	3.47%	4.31%	
	2500	Conveying equipment		2.14	2.64	3.18	2.05%	2.27%	2.69%	
	2720	Plumbing		5.55	8.90	9.40	5.70%	7.20%	8.95%	
	2900	Electrical		6.25	8.45	10.30	6.35%	7.20%	8.95%	
	3100	Total: Mechanical & Electrical		19.95	25	30.50	18.25%	21%	23%	
	9000	Per apartment unit, total cost	Apt.	107,000	126,000	209,000				
	9500	Total: Mechanical & Electrical	"	20,200	23,400	24,400				
03	0010	**APARTMENTS High Rise (8 to 24 story)**	S.F.	107	123	148				03
	0020	Total project costs	C.F.	10.40	12.10	15.50				
	0100	Site work	S.F.	3.89	6.30	8.80	2.58%	4.84%	6.15%	
	0500	Masonry		6.20	11.30	14.05	4.74%	9.65%	11.05%	
	1500	Finishes		11.90	14.85	17.55	9.75%	11.80%	13.70%	
	1800	Equipment		3.45	4.24	5.60	2.78%	3.49%	4.35%	
	2500	Conveying equipment		2.44	3.70	5.05	2.23%	2.78%	3.37%	
	2720	Plumbing		7.90	9.30	11.40	6.80%	7.20%	10.45%	
	2900	Electrical		7.35	9.30	12.60	6.45%	7.65%	8.80%	
	3100	Total: Mechanical & Electrical		22	28	34	17.95%	22.50%	24.50%	
	9000	Per apartment unit, total cost	Apt.	111,500	122,500	170,000				
	9500	Total: Mechanical & Electrical	"	24,100	27,500	29,100				
04	0010	**AUDITORIUMS**	S.F.	111	152	219				04
	0020	Total project costs	C.F.	6.95	9.70	13.90				
	2720	Plumbing	S.F.	6.60	9.75	11.65	5.85%	7.20%	8.70%	
	2900	Electrical		9	12.70	21.50	6.85%	9.45%	11.35%	
	3100	Total: Mechanical & Electrical		59.50	79.50	96.50	24.50%	28%	31%	
05	0010	**AUTOMOTIVE SALES**	S.F.	82.50	112	139				05
	0020	Total project costs	C.F.	5.45	6.50	8.45				
	2720	Plumbing	S.F.	3.76	6.55	7.10	2.89%	6.05%	6.50%	
	2770	Heating, ventilating, air conditioning		5.80	8.85	9.60	4.61%	10%	10.35%	
	2900	Electrical		6.50	10.20	14.15	7.25%	8.80%	12.15%	
	3100	Total: Mechanical & Electrical		21	29.50	35.50	18.90%	20.50%	22%	
06	0010	**BANKS**	S.F.	162	202	256				06
	0020	Total project costs	C.F.	11.55	15.70	20.50				
	0100	Site work	S.F.	18.55	28	41	7.90%	12.95%	16.95%	
	0500	Masonry		8.40	16.10	28.50	3.46%	7.65%	10.10%	
	1500	Finishes		15	22	27.50	5.95%	8.65%	11.70%	
	1800	Equipment		6.10	13.40	27	1.34%	5.55%	10.50%	
	2720	Plumbing		5.05	7.25	10.60	2.82%	3.90%	4.93%	
	2770	Heating, ventilating, air conditioning		9.65	12.85	17.15	4.86%	7.15%	8.50%	
	2900	Electrical		15.30	20.50	26.50	8.20%	10.20%	12.20%	
	3100	Total: Mechanical & Electrical		36.50	49.50	59	16.25%	19.45%	23%	
	3500	See also Divisions 11 16 00 & 11 17 00								

	50 17 00 | S.F. Costs		UNIT	UNIT COSTS 1/4	UNIT COSTS MEDIAN	UNIT COSTS 3/4	% OF TOTAL 1/4	% OF TOTAL MEDIAN	% OF TOTAL 3/4	
13	0010	**CHURCHES**	S.F.	109	138	180				13
	0020	Total project costs	C.F.	6.75	8.50	11.20				
	1800	Equipment	S.F.	1.30	3.11	6.60	.95%	2.11%	4.31%	
	2720	Plumbing		4.24	5.95	8.75	3.51%	4.96%	6.25%	
	2770	Heating, ventilating, air conditioning		9.90	12.90	18.30	7.50%	10%	12%	
	2900	Electrical		9.15	12.60	17.10	7.30%	8.80%	10.95%	
	3100	Total: Mechanical & Electrical	↓	28.50	38	50.50	18.30%	22%	25%	
	3500	See also division 11040 (MF2004 11 91 00)								
15	0010	**CLUBS, COUNTRY**	S.F.	117	141	177				15
	0020	Total project costs	C.F.	9.40	11.45	15.85				
	2720	Plumbing	S.F.	7.05	10.45	24	5.60%	7.90%	10%	
	2900	Electrical		9.20	12.60	16.40	7%	8.95%	11%	
	3100	Total: Mechanical & Electrical	↓	49	61	64.50	19%	26.50%	29.50%	
17	0010	**CLUBS, SOCIAL Fraternal**	S.F.	98.50	135	179				17
	0020	Total project costs	C.F.	5.80	8.85	10.55				
	2720	Plumbing	S.F.	5.85	7.30	11.05	5.60%	6.90%	8.55%	
	2770	Heating, ventilating, air conditioning		8.45	10.20	13.15	8.20%	9.25%	14.40%	
	2900	Electrical		7	11.50	13.20	5.95%	9.30%	10.55%	
	3100	Total: Mechanical & Electrical	↓	36	39	50	21%	23%	23.50%	
18	0010	**CLUBS, Y.M.C.A.**	S.F.	122	158	241				18
	0020	Total project costs	C.F.	5.40	9	13.45				
	2720	Plumbing	S.F.	7.40	14.70	16.50	5.65%	7.60%	10.85%	
	2900	Electrical		9.60	12.10	19.40	6.25%	8%	9.20%	
	3100	Total: Mechanical & Electrical	↓	39	44.50	72.50	20.50%	21.50%	28.50%	
19	0010	**COLLEGES Classrooms & Administration**	S.F.	117	160	208				19
	0020	Total project costs	C.F.	8.30	12.30	19.35				
	0500	Masonry	S.F.	11.65	16.40	20	5.65%	8.05%	10.50%	
	2720	Plumbing		6	12.30	23	5.10%	6.60%	8.95%	
	2900	Electrical		9.60	14.95	20.50	7.70%	9.75%	12%	
	3100	Total: Mechanical & Electrical	↓	40.50	55	74	23.50%	28%	32%	
21	0010	**COLLEGES Science, Engineering, Laboratories**	S.F.	220	257	298				21
	0020	Total project costs	C.F.	12.60	18.40	21				
	1800	Equipment	S.F.	12.25	27.50	30.50	2%	6.45%	12.65%	
	2900	Electrical		18.15	26	39.50	7.10%	9.40%	12.10%	
	3100	Total: Mechanical & Electrical	↓	67.50	80	124	28.50%	31.50%	41%	
	3500	See also division 11600 (MF2004 11 53 00)								
23	0010	**COLLEGES Student Unions**	S.F.	140	190	230				23
	0020	Total project costs	C.F.	7.80	10.25	12.70				
	3100	Total: Mechanical & Electrical	S.F.	52.50	57	67.50	23.50%	26%	29%	
25	0010	**COMMUNITY CENTERS**	S.F.	116	144	194				25
	0020	Total project costs	C.F.	7.55	10.80	14				
	1800	Equipment	S.F.	2.80	4.59	7.55	1.47%	3.01%	5.30%	
	2720	Plumbing		5.50	9.60	13.10	4.85%	7%	8.95%	
	2770	Heating, ventilating, air conditioning		8.80	12.80	18.30	6.80%	10.35%	12.90%	
	2900	Electrical		9.80	12.95	19	7.20%	8.90%	10.40%	
	3100	Total: Mechanical & Electrical	↓	32.50	41	59	19.05%	23.50%	31%	
28	0010	**COURT HOUSES**	S.F.	167	197	271				28
	0020	Total project costs	C.F.	12.80	15.30	19.30				
	2720	Plumbing	S.F.	7.95	11.10	12.55	5.95%	7.45%	8.20%	
	2900	Electrical		17.75	19.70	29	8.90%	10.45%	11.85%	
	3100	Total: Mechanical & Electrical	↓	49.50	65	71	22.50%	27.50%	30%	
30	0010	**DEPARTMENT STORES**	S.F.	61.50	83.50	105				30
	0020	Total project costs	C.F.	3.31	4.29	5.80				
	2720	Plumbing	S.F.	1.92	2.43	3.68	1.82%	4.21%	5.90%	
	2770	Heating, ventilating, air conditioning		5.60	8.65	13.05	8.20%	9.10%	14.80%	

	50 17 00 | S.F. Costs		UNIT	UNIT COSTS 1/4	UNIT COSTS MEDIAN	UNIT COSTS 3/4	% OF TOTAL 1/4	% OF TOTAL MEDIAN	% OF TOTAL 3/4	
30	2900	Electrical	S.F.	7.10	9.70	11.45	9.05%	12.15%	14.95%	30
	3100	Total: Mechanical & Electrical		12.45	15.95	28	13.20%	21.50%	50%	
31	0010	**DORMITORIES Low Rise (1 to 3 story)**	S.F.	117	162	203				31
	0020	Total project costs	C.F.	6.60	10.70	16				
	2720	Plumbing	S.F.	7.05	9.40	11.90	8.05%	9%	9.65%	
	2770	Heating, ventilating, air conditioning		7.45	8.90	11.90	4.61%	8.05%	10%	
	2900	Electrical		7.75	11.80	16.15	6.40%	8.65%	9.50%	
	3100	Total: Mechanical & Electrical		40.50	43	67.50	22%	25%	27%	
	9000	Per bed, total cost	Bed	49,600	55,000	118,000				
32	0010	**DORMITORIES Mid Rise (4 to 8 story)**	S.F.	144	187	231				32
	0020	Total project costs	C.F.	15.85	17.40	21				
	2900	Electrical	S.F.	15.25	17.35	23.50	8.20%	10.20%	11.95%	
	3100	Total: Mechanical & Electrical	"	42.50	85	86.50	25%	30.50%	35.50%	
	9000	Per bed, total cost	Bed	20,500	46,700	270,000				
34	0010	**FACTORIES**	S.F.	54.50	81	125				34
	0020	Total project costs	C.F.	3.49	5.20	8.65				
	0100	Site work	S.F.	6.20	11.35	17.90	6.95%	11.45%	17.95%	
	2720	Plumbing		2.93	5.45	9.05	3.73%	6.05%	8.10%	
	2770	Heating, ventilating, air conditioning		5.70	8.20	11.05	5.25%	8.45%	11.35%	
	2900	Electrical		6.75	10.70	16.30	8.10%	10.50%	14.20%	
	3100	Total: Mechanical & Electrical		19.30	29.50	39	21%	28.50%	35.50%	
36	0010	**FIRE STATIONS**	S.F.	108	149	199				36
	0020	Total project costs	C.F.	6.30	8.65	11.55				
	0500	Masonry	S.F.	15.85	28.50	37	8.10%	11.30%	15.70%	
	1140	Roofing		3.52	9.55	10.85	1.90%	4.94%	5.05%	
	1580	Painting		2.73	4.08	4.17	1.37%	1.57%	2.07%	
	1800	Equipment		1.34	2.57	4.75	.62%	1.63%	3.42%	
	2720	Plumbing		6.05	9.60	13.60	5.85%	7.35%	9.45%	
	2770	Heating, ventilating, air conditioning		6	9.70	14.95	5.15%	7.40%	9.40%	
	2900	Electrical		7.70	13.45	18.35	6.85%	8.60%	10.60%	
	3100	Total: Mechanical & Electrical		39.50	50.50	57	18.40%	23%	26%	
37	0010	**FRATERNITY HOUSES & Sorority Houses**	S.F.	108	139	190				37
	0020	Total project costs	C.F.	10.75	11.20	13.50				
	2720	Plumbing	S.F.	8.15	9.35	17.10	6.80%	8%	10.85%	
	2900	Electrical		7.10	15.40	18.85	6.60%	9.90%	10.65%	
	3100	Total: Mechanical & Electrical		8.70	27	32.50		15.10%	15.90%	
38	0010	**FUNERAL HOMES**	S.F.	114	155	282				38
	0020	Total project costs	C.F.	11.60	12.95	25				
	2900	Electrical	S.F.	5.05	9.20	10.10	3.58%	4.44%	5.95%	
	3100	Total: Mechanical & Electrical		17.80	26	36.50	12.90%	12.90%	12.90%	
39	0010	**GARAGES, COMMERCIAL (Service)**	S.F.	64.50	99.50	138				39
	0020	Total project costs	C.F.	4.24	6.25	9.10				
	1800	Equipment	S.F.	3.63	8.15	12.70	2.21%	4.62%	6.80%	
	2720	Plumbing		4.46	6.90	12.55	5.45%	7.85%	10.65%	
	2730	Heating & ventilating		5.85	7.60	9.70	5.25%	6.85%	8.20%	
	2900	Electrical		6.10	9.35	13.45	7.15%	9.25%	10.85%	
	3100	Total: Mechanical & Electrical		13.50	26	38	12.35%	17.40%	26%	
40	0010	**GARAGES, MUNICIPAL (Repair)**	S.F.	100	126	178				40
	0020	Total project costs	C.F.	5.90	7.50	12.90				
	0500	Masonry	S.F.	8.90	17.40	27	4.03%	9.15%	12.50%	
	2720	Plumbing		4.25	8.15	15.35	3.59%	6.70%	7.95%	
	2730	Heating & ventilating		7.25	10.55	20.50	6.15%	7.45%	13.50%	
	2900	Electrical		7	11	16	6.65%	8.15%	11.15%	
	3100	Total: Mechanical & Electrical		32.50	45	65	21.50%	23%	28.50%	

50 17 00 | S.F. Costs

		50 17 00 | S.F. Costs	UNIT	UNIT COSTS			% OF TOTAL			
				1/4	MEDIAN	3/4	1/4	MEDIAN	3/4	
41	0010	**GARAGES, PARKING**	S.F.	36.50	53.50	92.50				41
	0020	Total project costs	C.F.	3.46	4.70	6.85				
	2720	Plumbing	S.F.	1.04	1.62	2.50	1.72%	2.70%	3.85%	
	2900	Electrical		2.02	2.48	3.90	4.33%	5.20%	6.30%	
	3100	Total: Mechanical & Electrical		4.14	5.75	7.20	7%	8.80%	9.70%	
	3200									
	9000	Per car, total cost	Car	15,500	19,500	24,900				
43	0010	**GYMNASIUMS**	S.F.	103	137	185				43
	0020	Total project costs	C.F.	5.10	6.95	8.50				
	1800	Equipment	S.F.	2.44	4.57	8.75	1.81%	3.30%	6.70%	
	2720	Plumbing		6.50	7.55	9.95	4.65%	6.40%	7.75%	
	2770	Heating, ventilating, air conditioning		7	10.60	21.50	5.15%	9.05%	11.10%	
	2900	Electrical		8.15	10.75	14.30	6.75%	8.50%	10.35%	
	3100	Total: Mechanical & Electrical		29	38	46.50	19.60%	23.50%	29%	
	3500	See also division 11480 (MF2004 11 67 00)								
46	0010	**HOSPITALS**	S.F.	196	246	335				46
	0020	Total project costs	C.F.	14.90	18.55	26.50				
	1800	Equipment	S.F.	4.99	9.60	16.55	.80%	2.53%	4.80%	
	2720	Plumbing		16.95	24	30.50	7.60%	9.10%	10.85%	
	2770	Heating, ventilating, air conditioning		25	32	44.50	7.80%	12.95%	16.65%	
	2900	Electrical		21.50	29	41.50	9.90%	11.75%	14.10%	
	3100	Total: Mechanical & Electrical		63.50	88.50	131	28%	33.50%	37%	
	9000	Per bed or person, total cost	Bed	228,000	314,000	362,000				
	9900	See also division 11700 (MF2004 11 71 00)								
48	0010	**HOUSING For the Elderly**	S.F.	97	123	150				48
	0020	Total project costs	C.F.	6.90	9.60	12.25				
	0100	Site work	S.F.	6.75	10.50	15.35	5.05%	7.90%	12.10%	
	0500	Masonry		2.95	11	16.10	1.30%	6.05%	11%	
	1800	Equipment		2.34	3.22	5.15	1.88%	3.23%	4.43%	
	2510	Conveying systems		2.35	3.16	4.29	1.78%	2.20%	2.81%	
	2720	Plumbing		7.20	9.15	11.55	8.15%	9.55%	10.50%	
	2730	Heating, ventilating, air conditioning		3.69	5.20	7.80	3.30%	5.60%	7.25%	
	2900	Electrical		7.20	9.80	12.55	7.30%	8.50%	10.25%	
	3100	Total: Mechanical & Electrical		26.50	31	39.50	18.10%	22.50%	29%	
	9000	Per rental unit, total cost	Unit	90,000	105,500	117,000				
	9500	Total: Mechanical & Electrical	"	20,000	23,100	26,900				
50	0010	**HOUSING Public (Low Rise)**	S.F.	81.50	113	147				50
	0020	Total project costs	C.F.	7.25	9.05	11.25				
	0100	Site work	S.F.	10.35	14.90	24	8.35%	11.75%	16.50%	
	1800	Equipment		2.21	3.61	5.50	2.26%	3.03%	4.24%	
	2720	Plumbing		5.85	7.75	9.80	7.15%	9.05%	11.60%	
	2730	Heating, ventilating, air conditioning		2.95	5.70	6.25	4.26%	6.05%	6.45%	
	2900	Electrical		4.92	7.35	10.15	5.10%	6.55%	8.25%	
	3100	Total: Mechanical & Electrical		23.50	30	33.50	14.50%	17.55%	26.50%	
	9000	Per apartment, total cost	Apt.	89,500	101,500	127,500				
	9500	Total: Mechanical & Electrical	"	19,100	23,500	26,000				
51	0010	**ICE SKATING RINKS**	S.F.	69.50	163	178				51
	0020	Total project costs	C.F.	5.10	5.25	6.05				
	2720	Plumbing	S.F.	2.60	4.87	4.98	3.12%	3.23%	5.65%	
	2900	Electrical		7.45	11.40	12.10	6.30%	10.15%	15.05%	
	3100	Total: Mechanical & Electrical		31	31	31	18.95%	18.95%	18.95%	
52	0010	**JAILS**	S.F.	211	273	355				52
	0020	Total project costs	C.F.	19.10	26.50	31				
	1800	Equipment	S.F.	8.25	24.50	41.50	2.80%	6.05%	9.80%	
	2720	Plumbing		20.50	27.50	36	7%	8.90%	13.35%	
	2770	Heating, ventilating, air conditioning		19.10	25.50	49.50	7.50%	9.45%	17.75%	
	2900	Electrical		22	30	37.50	8.80%	11.65%	15%	

50 17 00 | S.F. Costs

		50 17 00 | S.F. Costs	UNIT	UNIT COSTS 1/4	UNIT COSTS MEDIAN	UNIT COSTS 3/4	% OF TOTAL 1/4	% OF TOTAL MEDIAN	% OF TOTAL 3/4	
52	3100	Total: Mechanical & Electrical	S.F.	59.50	105	125	28%	30%	34%	52
53	0010	**LIBRARIES**	S.F.	136	178	232				53
	0020	Total project costs	C.F.	9.15	11.45	14.65				
	0500	Masonry	S.F.	10.70	18.50	31	5.65%	7.45%	11.60%	
	1800	Equipment		1.83	4.92	7.45	.28%	1.39%	4.07%	
	2720	Plumbing		4.94	7	9.75	3.38%	4.60%	5.70%	
	2770	Heating, ventilating, air conditioning		10.95	18.50	22.50	7.80%	10.95%	12.80%	
	2900	Electrical		13.75	18.15	24	8.40%	10.50%	12%	
	3100	Total: Mechanical & Electrical		41.50	52.50	64	20.50%	23%	26.50%	
54	0010	**LIVING, ASSISTED**	S.F.	124	146	172				54
	0020	Total project costs	C.F.	10.40	12.15	13.80				
	0500	Masonry	S.F.	3.59	4.33	5.30	2.36%	3.16%	3.90%	
	1800	Equipment		2.74	3.21	4.19	2.06%	2.45%	2.87%	
	2720	Plumbing		10.35	13.90	14.35	6.05%	8.15%	10.60%	
	2770	Heating, ventilating, air conditioning		12.30	12.85	14.10	7.95%	9.35%	9.70%	
	2900	Electrical		12.10	13.40	15.35	8.95%	9.95%	10.65%	
	3100	Total: Mechanical & Electrical		34	39.50	45.50	25%	28.50%	31.50%	
55	0010	**MEDICAL CLINICS**	S.F.	126	155	197				55
	0020	Total project costs	C.F.	9.25	11.95	15.85				
	1800	Equipment	S.F.	3.40	7.15	11.15	1.05%	2.94%	6.35%	
	2720	Plumbing		8.35	11.80	15.75	6.15%	8.40%	10.10%	
	2770	Heating, ventilating, air conditioning		10	13.10	19.20	6.65%	8.85%	11.35%	
	2900	Electrical		10.80	15.35	20	8.10%	10%	12.25%	
	3100	Total: Mechanical & Electrical		34.50	47	64.50	22.50%	27%	33.50%	
	3500	See also division 11700 (MF2004 11 71 00)								
57	0010	**MEDICAL OFFICES**	S.F.	118	147	180				57
	0020	Total project costs	C.F.	8.85	11.95	16.20				
	1800	Equipment	S.F.	3.91	7.75	11	.68%	4.97%	7.05%	
	2720	Plumbing		6.55	10.10	13.60	5.60%	6.80%	8.50%	
	2770	Heating, ventilating, air conditioning		7.90	11.40	15.10	6.10%	8%	9.70%	
	2900	Electrical		9.50	13.80	19.25	7.60%	9.80%	11.70%	
	3100	Total: Mechanical & Electrical		25.50	37	53.50	19.30%	23%	27.50%	
59	0010	**MOTELS**	S.F.	74.50	108	141				59
	0020	Total project costs	C.F.	6.65	8.90	14.55				
	2720	Plumbing	S.F.	7.60	9.65	11.50	9.45%	10.60%	12.55%	
	2770	Heating, ventilating, air conditioning		4.62	6.90	12.35	5.60%	5.60%	10%	
	2900	Electrical		7.05	8.95	11.15	7.45%	9.05%	10.45%	
	3100	Total: Mechanical & Electrical		22.50	30	51.50	18.50%	24%	25.50%	
	5000									
	9000	Per rental unit, total cost	Unit	38,100	72,500	78,000				
	9500	Total: Mechanical & Electrical	"	7,400	11,200	13,000				
60	0010	**NURSING HOMES**	S.F.	117	151	188				60
	0020	Total project costs	C.F.	9.20	11.50	15.70				
	1800	Equipment	S.F.	3.69	4.89	8.15	2%	3.62%	4.99%	
	2720	Plumbing		10.05	15.20	18.35	8.75%	10.10%	12.70%	
	2770	Heating, ventilating, air conditioning		10.60	16.05	21.50	9.70%	11.45%	11.80%	
	2900	Electrical		11.60	14.55	19.75	9.40%	10.60%	12.50%	
	3100	Total: Mechanical & Electrical		27.50	38.50	65	26%	28%	30%	
	9000	Per bed or person, total cost	Bed	52,000	65,000	84,000				
61	0010	**OFFICES Low Rise (1 to 4 story)**	S.F.	99	129	167				61
	0020	Total project costs	C.F.	7.05	9.70	12.80				
	0100	Site work	S.F.	7.95	13.85	20	6.20%	9.70%	13.55%	
	0500	Masonry		3.93	7.70	14.05	2.62%	5.45%	8.45%	
	1800	Equipment		1.05	2.06	5.60	.60%	1.50%	3.50%	
	2720	Plumbing		3.53	5.45	8	3.66%	4.50%	6.10%	
	2770	Heating, ventilating, air conditioning		7.80	10.90	15.90	7.20%	10.30%	11.70%	
	2900	Electrical		8.05	11.50	16.35	7.45%	9.65%	11.40%	

		50 17 00 | S.F. Costs	UNIT	UNIT COSTS 1/4	UNIT COSTS MEDIAN	UNIT COSTS 3/4	% OF TOTAL 1/4	% OF TOTAL MEDIAN	% OF TOTAL 3/4	
61	3100	Total: Mechanical & Electrical	S.F.	22	30.50	46	18.15%	22%	27%	61
62	0010	**OFFICES Mid Rise (5 to 10 story)**	S.F.	105	127	172				62
	0020	Total project costs	C.F.	7.40	9.45	13.40				
	2720	Plumbing	S.F.	3.16	4.90	7.05	2.83%	3.74%	4.50%	
	2770	Heating, ventilating, air conditioning		7.95	11.35	18.10	7.65%	9.40%	11%	
	2900	Electrical		7.75	9.95	15.05	6.35%	7.80%	10%	
	3100	Total: Mechanical & Electrical	↓	20	26	49	18.95%	21%	27.50%	
63	0010	**OFFICES High Rise (11 to 20 story)**	S.F.	128	162	199				63
	0020	Total project costs	C.F.	9	11.20	16.10				
	2900	Electrical	S.F.	7.80	9.50	14.15	5.80%	7.85%	10.50%	
	3100	Total: Mechanical & Electrical	↓	25	33.50	57	16.90%	23.50%	34%	
64	0010	**POLICE STATIONS**	S.F.	154	202	255				64
	0020	Total project costs	C.F.	12.25	15.05	20.50				
	0500	Masonry	S.F.	14.45	25.50	32	6.70%	9.10%	11.35%	
	1800	Equipment		2.45	10.75	17.05	.98%	3.35%	6.70%	
	2720	Plumbing		8.60	17.15	21.50	5.65%	6.90%	10.75%	
	2770	Heating, ventilating, air conditioning		13.45	17.85	25	5.85%	10.55%	11.70%	
	2900	Electrical		16.85	24	32	9.80%	11.85%	14.80%	
	3100	Total: Mechanical & Electrical	↓	55	66.50	90	25%	31.50%	32.50%	
65	0010	**POST OFFICES**	S.F.	122	150	191				65
	0020	Total project costs	C.F.	7.30	9.25	10.55				
	2720	Plumbing	S.F.	5.45	6.80	8.55	4.24%	5.30%	5.60%	
	2770	Heating, ventilating, air conditioning		8.55	10.55	11.75	6.65%	7.15%	9.35%	
	2900	Electrical		10.05	14.10	16.75	7.25%	9%	11%	
	3100	Total: Mechanical & Electrical	↓	29	38	43	16.25%	18.80%	22%	
66	0010	**POWER PLANTS**	S.F.	845	1,125	2,050				66
	0020	Total project costs	C.F.	23	50.50	108				
	2900	Electrical	S.F.	59.50	126	189	9.30%	12.75%	21.50%	
	8100	Total: Mechanical & Electrical	↓	148	485	1,075	32.50%	32.50%	52.50%	
67	0010	**RELIGIOUS EDUCATION**	S.F.	98	128	157				67
	0020	Total project costs	C.F.	5.45	7.80	9.75				
	2720	Plumbing	S.F.	4.09	5.80	8.15	4.40%	5.30%	7.10%	
	2770	Heating, ventilating, air conditioning		10.35	11.70	16.55	10.05%	11.45%	12.35%	
	2900	Electrical		7.75	10.95	14.50	7.60%	9.05%	10.35%	
	3100	Total: Mechanical & Electrical	↓	31.50	41.50	50.50	22%	23%	27%	
69	0010	**RESEARCH Laboratories & Facilities**	S.F.	150	212	310				69
	0020	Total project costs	C.F.	10.90	21.50	26				
	1800	Equipment	S.F.	6.50	12.80	31	.94%	4.58%	9.10%	
	2720	Plumbing		14.70	18.80	30	6.15%	8.30%	10.80%	
	2770	Heating, ventilating, air conditioning		13.20	44.50	52.50	7.25%	16.50%	17.50%	
	2900	Electrical		17.65	28.50	46.50	9.55%	11.50%	15.40%	
	3100	Total: Mechanical & Electrical	↓	56.50	97.50	140	29.50%	37%	41%	
70	0010	**RESTAURANTS**	S.F.	142	183	238				70
	0020	Total project costs	C.F.	11.95	15.70	20.50				
	1800	Equipment	S.F.	9.20	22.50	34	6.10%	13%	15.65%	
	2720	Plumbing		11.25	13.65	17.95	6.10%	8.15%	9%	
	2770	Heating, ventilating, air conditioning		14.30	19.75	24	9.20%	12%	12.40%	
	2900	Electrical		15	18.55	24	8.35%	10.55%	11.55%	
	3100	Total: Mechanical & Electrical	↓	46.50	50	64.50	21%	25%	29.50%	
	9000	Per seat unit, total cost	Seat	5,225	6,950	8,225				
	9500	Total: Mechanical & Electrical	"	1,300	1,725	2,050				
72	0010	**RETAIL STORES**	S.F.	66.50	89	118				72
	0020	Total project costs	C.F.	4.50	6.40	8.90				
	2720	Plumbing	S.F.	2.41	4.02	6.85	3.26%	4.60%	6.80%	
	2770	Heating, ventilating, air conditioning		5.20	7.10	10.70	6.75%	8.75%	10.15%	
	2900	Electrical		6	8.15	11.80	7.25%	9.90%	11.60%	
	3100	Total: Mechanical & Electrical	↓	15.90	20.50	28	17.05%	21%	23.50%	

	50 17 00 | S.F. Costs		UNIT	UNIT COSTS 1/4	UNIT COSTS MEDIAN	UNIT COSTS 3/4	% OF TOTAL 1/4	% OF TOTAL MEDIAN	% OF TOTAL 3/4	
74	0010	**SCHOOLS Elementary**	S.F.	109	135	165				74
	0020	Total project costs	C.F.	7.10	9.10	11.70				
	0500	Masonry	S.F.	9.40	16.55	24.50	5.25%	10.65%	14.05%	
	1800	Equipment		2.96	4.91	9.25	1.80%	3.13%	4.61%	
	2720	Plumbing		6.25	8.80	11.75	5.70%	7.15%	9.35%	
	2730	Heating, ventilating, air conditioning		9.35	14.85	21	8.15%	10.80%	14.90%	
	2900	Electrical		10.25	13.65	17.40	8.50%	10.05%	11.80%	
	3100	Total: Mechanical & Electrical	↓	37	46	57.50	25%	27.50%	30%	
	9000	Per pupil, total cost	Ea.	12,400	18,500	41,400				
	9500	Total: Mechanical & Electrical	"	3,500	4,450	11,200				
76	0010	**SCHOOLS Junior High & Middle**	S.F.	112	138	169				76
	0020	Total project costs	C.F.	7.10	9.15	10.30				
	0500	Masonry	S.F.	13.90	17.95	22	7.55%	11.10%	14.30%	
	1800	Equipment		3.57	5.75	8.45	1.79%	3.03%	4.80%	
	2720	Plumbing		6.50	8	9.95	5.30%	6.80%	7.25%	
	2770	Heating, ventilating, air conditioning		13	15.80	28	8.90%	11.55%	14.20%	
	2900	Electrical		11.10	13.90	17.75	7.90%	9.40%	10.60%	
	3100	Total: Mechanical & Electrical	↓	35.50	45.50	57	23%	25.50%	29.50%	
	9000	Per pupil, total cost	Ea.	14,200	18,600	25,000				
78	0010	**SCHOOLS Senior High**	S.F.	117	145	181				78
	0020	Total project costs	C.F.	7	10	16.65				
	1800	Equipment	S.F.	3.10	7.20	10	1.88%	2.67%	4.30%	
	2720	Plumbing		6.50	9.80	17.90	5.60%	6.90%	8.30%	
	2770	Heating, ventilating, air conditioning		13.30	15.25	23	8.95%	11.60%	15%	
	2900	Electrical		11.75	15.65	23	8.70%	10.35%	12.50%	
	3100	Total: Mechanical & Electrical	↓	39.50	46	75	24%	26.50%	28.50%	
	9000	Per pupil, total cost	Ea.	11,000	22,300	27,900				
80	0010	**SCHOOLS Vocational**	S.F.	94	137	169				80
	0020	Total project costs	C.F.	5.85	8.40	11.60				
	0500	Masonry	S.F.	5.50	13.65	21	3.20%	4.61%	10.95%	
	1800	Equipment		2.94	7.30	10.10	1.24%	3.10%	4.26%	
	2720	Plumbing		6	8.95	13.20	5.40%	6.90%	8.55%	
	2770	Heating, ventilating, air conditioning		8.40	15.65	26	8.60%	11.90%	14.65%	
	2900	Electrical		9.80	13.40	18.30	8.45%	11%	13.20%	
	3100	Total: Mechanical & Electrical	↓	37	37.50	64.50	27.50%	29.50%	31%	
	9000	Per pupil, total cost	Ea.	13,100	35,100	52,500				
83	0010	**SPORTS ARENAS**	S.F.	82	110	169				83
	0020	Total project costs	C.F.	4.46	8	10.30				
	2720	Plumbing	S.F.	4.77	7.25	15.25	4.35%	6.35%	9.40%	
	2770	Heating, ventilating, air conditioning		10.25	12.15	16.85	8.80%	10.20%	13.55%	
	2900	Electrical		8.55	11.65	15	7.65%	9.75%	11.90%	
	3100	Total: Mechanical & Electrical	↓	21.50	38	51	18.05%	21.50%	25%	
85	0010	**SUPERMARKETS**	S.F.	76	88	103				85
	0020	Total project costs	C.F.	4.23	5.10	7.75				
	2720	Plumbing	S.F.	4.24	5.35	6.20	5.40%	6%	7.45%	
	2770	Heating, ventilating, air conditioning		6.25	8.30	10.10	8.60%	8.65%	9.60%	
	2900	Electrical		9.50	10.85	12.90	10.40%	12.45%	13.60%	
	3100	Total: Mechanical & Electrical	↓	24.50	26	34.50	23.50%	27.50%	28.50%	
86	0010	**SWIMMING POOLS**	S.F.	123	206	440				86
	0020	Total project costs	C.F.	9.85	12.30	13.40				
	2720	Plumbing	S.F.	11.35	13	17.45	4.80%	9.70%	20.50%	
	2900	Electrical		9.25	15	22	5.75%	6.95%	7.60%	
	3100	Total: Mechanical & Electrical	↓	22.50	57.50	77	11.15%	14.10%	23.50%	
87	0010	**TELEPHONE EXCHANGES**	S.F.	163	232	305				87
	0020	Total project costs	C.F.	10.15	16.30	22.50				
	2720	Plumbing	S.F.	6.90	10.65	15.60	4.52%	5.80%	6.90%	
	2770	Heating, ventilating, air conditioning		16	32	40	11.80%	16.05%	18.40%	

	50 17 00 | S.F. Costs		UNIT	UNIT COSTS 1/4	UNIT COSTS MEDIAN	UNIT COSTS 3/4	% OF TOTAL 1/4	% OF TOTAL MEDIAN	% OF TOTAL 3/4	
87	2900	Electrical	S.F.	16.65	26.50	46.50	10.90%	14%	17.85%	87
	3100	Total: Mechanical & Electrical	↓	49	93	132	29.50%	33.50%	44.50%	
91	0010	**THEATERS**	S.F.	103	132	194				91
	0020	Total project costs	C.F.	4.74	7	10.30				
	2720	Plumbing	S.F.	3.42	3.71	15.15	2.92%	4.70%	6.80%	
	2770	Heating, ventilating, air conditioning		9.95	12.10	14.95	8%	12.25%	13.40%	
	2900	Electrical		9	12.10	24.50	8.05%	9.35%	12.25%	
	3100	Total: Mechanical & Electrical	↓	23	31	36.50	21.50%	25.50%	27.50%	
94	0010	**TOWN HALLS City Halls & Municipal Buildings**	S.F.	115	146	190				94
	0020	Total project costs	C.F.	10.45	12.05	17.65				
	2720	Plumbing	S.F.	4.79	8.95	15.70	4.31%	5.95%	7.95%	
	2770	Heating, ventilating, air conditioning		8.65	17.20	25	7.05%	9.05%	13.45%	
	2900	Electrical		10.90	15	20.50	8.05%	9.45%	11.65%	
	3100	Total: Mechanical & Electrical	↓	38	48	73	22%	26.50%	31%	
97	0010	**WAREHOUSES & Storage Buildings**	S.F.	42.50	61.50	91				97
	0020	Total project costs	C.F.	2.24	3.51	5.80				
	0100	Site work	S.F.	4.42	8.75	13.20	6.05%	12.95%	19.55%	
	0500	Masonry		2.53	6.10	13.10	3.60%	7.35%	12%	
	1800	Equipment		.69	1.48	8.30	.72%	1.69%	5.55%	
	2720	Plumbing		1.43	2.56	4.78	2.90%	4.80%	6.55%	
	2730	Heating, ventilating, air conditioning		1.63	4.59	6.15	2.41%	5%	8.90%	
	2900	Electrical		2.53	4.77	7.80	5.15%	7.20%	10.05%	
	3100	Total: Mechanical & Electrical	↓	7.05	10.85	21.50	13.30%	18.90%	26%	
99	0010	**WAREHOUSE & OFFICES Combination**	S.F.	52.50	70	96.50				99
	0020	Total project costs	C.F.	2.69	3.91	5.75				
	1800	Equipment	S.F.	.91	1.76	2.62	.52%	1.19%	2.40%	
	2720	Plumbing		2.03	3.53	5.25	3.74%	4.76%	6.30%	
	2770	Heating, ventilating, air conditioning		3.21	5	7	5%	5.65%	10.05%	
	2900	Electrical		3.59	5.25	8.30	5.85%	8%	10%	
	3100	Total: Mechanical & Electrical	↓	10	15.15	24	14.55%	19.95%	24.50%	

Square Foot Project Size Modifier

One factor that affects the S.F. cost of a particular building is the size. In general, for buildings built to the same specifications in the same locality, the larger building will have the lower S.F. cost. This is due mainly to the decreasing contribution of the exterior walls, plus the economy of scale usually achievable in larger buildings. The area conversion scale shown below will give a factor to convert costs for the typical size building to an adjusted cost for the particular project.

The square foot base size table lists the median costs, most typical project size in our accumulated data, and the range in size of the projects.

The size factor for your project is determined by dividing your project area in S.F. by the typical project size for the particular building type. With this factor, enter the area conversion scale at the appropriate size factor and determine the appropriate cost multiplier for your building size.

Example: Determine the cost per S.F. for a 100,000 S.F. Mid-rise apartment building.

$$\frac{\text{Proposed building area} = 100{,}000 \text{ S.F.}}{\text{Typical size from below} = 50{,}000 \text{ S.F.}} = 2.00$$

Enter area conversion scale at 2.0, intersect curve, read horizontally the appropriate cost multiplier of .94. Size adjusted cost becomes .94 × \$114.00 = \$107.00 based on national average costs.

Note: For size factors less than .50, the cost multiplier is 1.1
For size factors greater than 3.5, the cost multiplier is .90

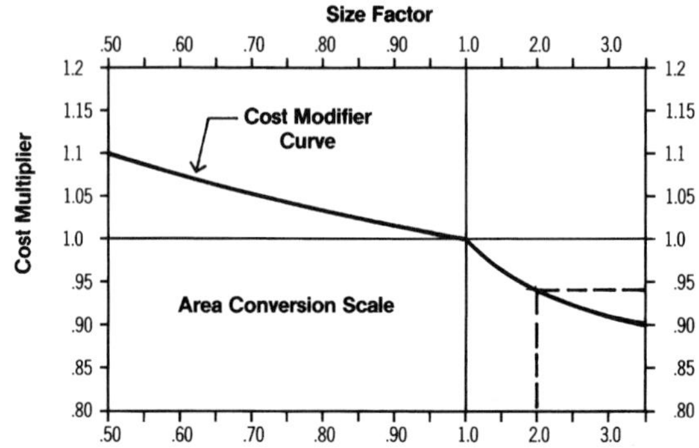

Square Foot Base Size							
Building Type	**Median Cost per S.F.**	**Typical Size Gross S.F.**	**Typical Range Gross S.F.**	**Building Type**	**Median Cost per S.F.**	**Typical Size Gross S.F.**	**Typical Range Gross S.F.**
Apartments, Low Rise	\$ 90.00	21,000	9,700 - 37,200	Jails	\$ 273.00	40,000	5,500 - 145,000
Apartments, Mid Rise	114.00	50,000	32,000 - 100,000	Libraries	178.00	12,000	7,000 - 31,000
Apartments, High Rise	123.00	145,000	95,000 - 600,000	Living, Assisted	146.00	32,300	23,500 - 50,300
Auditoriums	152.00	25,000	7,600 - 39,000	Medical Clinics	155.00	7,200	4,200 - 15,700
Auto Sales	112.00	20,000	10,800 - 28,600	Medical Offices	147.00	6,000	4,000 - 15,000
Banks	202.00	4,200	2,500 - 7,500	Motels	108.00	40,000	15,800 - 120,000
Churches	138.00	17,000	2,000 - 42,000	Nursing Homes	151.00	23,000	15,000 - 37,000
Clubs, Country	141.00	6,500	4,500 - 15,000	Offices, Low Rise	129.00	20,000	5,000 - 80,000
Clubs, Social	135.00	10,000	6,000 - 13,500	Offices, Mid Rise	127.00	120,000	20,000 - 300,000
Clubs, YMCA	158.00	28,300	12,800 - 39,400	Offices, High Rise	162.00	260,000	120,000 - 800,000
Colleges (Class)	160.00	50,000	15,000 - 150,000	Police Stations	202.00	10,500	4,000 - 19,000
Colleges (Science Lab)	257.00	45,600	16,600 - 80,000	Post Offices	150.00	12,400	6,800 - 30,000
College (Student Union)	190.00	33,400	16,000 - 85,000	Power Plants	1125.00	7,500	1,000 - 20,000
Community Center	144.00	9,400	5,300 - 16,700	Religious Education	128.00	9,000	6,000 - 12,000
Court Houses	197.00	32,400	17,800 - 106,000	Research	212.00	19,000	6,300 - 45,000
Dept. Stores	83.50	90,000	44,000 - 122,000	Restaurants	183.00	4,400	2,800 - 6,000
Dormitories, Low Rise	162.00	25,000	10,000 - 95,000	Retail Stores	89.00	7,200	4,000 - 17,600
Dormitories, Mid Rise	187.00	85,000	20,000 - 200,000	Schools, Elementary	135.00	41,000	24,500 - 55,000
Factories	81.00	26,400	12,900 - 50,000	Schools, Jr. High	138.00	92,000	52,000 - 119,000
Fire Stations	149.00	5,800	4,000 - 8,700	Schools, Sr. High	145.00	101,000	50,500 - 175,000
Fraternity Houses	139.00	12,500	8,200 - 14,800	Schools, Vocational	137.00	37,000	20,500 - 82,000
Funeral Homes	155.00	10,000	4,000 - 20,000	Sports Arenas	110.00	15,000	5,000 - 40,000
Garages, Commercial	99.50	9,300	5,000 - 13,600	Supermarkets	88.00	44,000	12,000 - 60,000
Garages, Municipal	126.00	8,300	4,500 - 12,600	Swimming Pools	206.00	20,000	10,000 - 32,000
Garages, Parking	53.50	163,000	76,400 - 225,300	Telephone Exchange	232.00	4,500	1,200 - 10,600
Gymnasiums	137.00	19,200	11,600 - 41,000	Theaters	132.00	10,500	8,800 - 17,500
Hospitals	246.00	55,000	27,200 - 125,000	Town Halls	146.00	10,800	4,800 - 23,400
House (Elderly)	123.00	37,000	21,000 - 66,000	Warehouses	61.50	25,000	8,000 - 72,000
Housing (Public)	113.00	36,000	14,400 - 74,400	Warehouse & Office	70.00	25,000	8,000 - 72,000
Ice Rinks	163.00	29,000	27,200 - 33,600				

Abbreviations

A	Area Square Feet; Ampere
AAFES	Army and Air Force Exchange Service
ABS	Acrylonitrile Butadiene Stryrene; Asbestos Bonded Steel
A.C., AC	Alternating Current; Air-Conditioning; Asbestos Cement; Plywood Grade A & C
ACI	American Concrete Institute
ACR	Air Conditioning Refrigeration
ADA	Americans with Disabilities Act
AD	Plywood, Grade A & D
Addit.	Additional
Adj.	Adjustable
af	Audio-frequency
AFUE	Annual Fuel Utilization Efficiency
AGA	American Gas Association
Agg.	Aggregate
A.H., Ah	Ampere Hours
A hr.	Ampere-hour
A.H.U., AHU	Air Handling Unit
A.I.A.	American Institute of Architects
AIC	Ampere Interrupting Capacity
Allow.	Allowance
alt., alt	Alternate
Alum.	Aluminum
a.m.	Ante Meridiem
Amp.	Ampere
Anod.	Anodized
ANSI	American National Standards Institute
APA	American Plywood Association
Approx.	Approximate
Apt.	Apartment
Asb.	Asbestos
A.S.B.C.	American Standard Building Code
Asbe.	Asbestos Worker
ASCE.	American Society of Civil Engineers
A.S.H.R.A.E.	American Society of Heating, Refrig. & AC Engineers
ASME	American Society of Mechanical Engineers
ASTM	American Society for Testing and Materials
Attchmt.	Attachment
Avg., Ave.	Average
AWG	American Wire Gauge
AWWA	American Water Works Assoc.
Bbl.	Barrel
B&B, BB	Grade B and Better; Balled & Burlapped
B&S	Bell and Spigot
B.&W.	Black and White
b.c.c.	Body-centered Cubic
B.C.Y.	Bank Cubic Yards
BE	Bevel End
B.F.	Board Feet
Bg. cem.	Bag of Cement
BHP	Boiler Horsepower; Brake Horsepower
B.I.	Black Iron
bidir.	bidirectional
Bit., Bitum.	Bituminous
Bit., Conc.	Bituminous Concrete
Bk.	Backed
Bkrs.	Breakers
Bldg., bldg	Building
Blk.	Block
Bm.	Beam
Boil.	Boilermaker
bpm	Blows per Minute
BR	Bedroom
Brg.	Bearing
Brhe.	Bricklayer Helper
Bric.	Bricklayer
Brk., brk	Brick
brkt	Bracket
Brng.	Bearing
Brs.	Brass
Brz.	Bronze
Bsn.	Basin
Btr.	Better
Btu	British Thermal Unit
BTUH	BTU per Hour
Bu.	bushels
BUR	Built-up Roofing
BX	Interlocked Armored Cable
°C	degree centegrade
c	Conductivity, Copper Sweat
C	Hundred; Centigrade
C/C	Center to Center, Cedar on Cedar
C-C	Center to Center
Cab	Cabinet
Cair.	Air Tool Laborer
Cal.	caliper
Calc	Calculated
Cap.	Capacity
Carp.	Carpenter
C.B.	Circuit Breaker
C.C.A.	Chromate Copper Arsenate
C.C.F.	Hundred Cubic Feet
cd	Candela
cd/sf	Candela per Square Foot
CD	Grade of Plywood Face & Back
CDX	Plywood, Grade C & D, exterior glue
Cefi.	Cement Finisher
Cem.	Cement
CF	Hundred Feet
C.F.	Cubic Feet
CFM	Cubic Feet per Minute
c.g.	Center of Gravity
CHW	Chilled Water; Commercial Hot Water
C.I., CI	Cast Iron
C.I.P., CIP	Cast in Place
Circ.	Circuit
C.L.	Carload Lot
CL	Chain Link
Clab.	Common Laborer
Clam	Common maintenance laborer
C.L.F.	Hundred Linear Feet
CLF	Current Limiting Fuse
CLP	Cross Linked Polyethylene
cm	Centimeter
CMP	Corr. Metal Pipe
CMU	Concrete Masonry Unit
CN	Change Notice
Col.	Column
CO_2	Carbon Dioxide
Comb.	Combination
comm.	Commercial, Communication
Compr.	Compressor
Conc.	Concrete
Cont., cont	Continuous; Continued, Container
Corr.	Corrugated
Cos	Cosine
Cot	Cotangent
Cov.	Cover
C/P	Cedar on Paneling
CPA	Control Point Adjustment
Cplg.	Coupling
CPM	Critical Path Method
CPVC	Chlorinated Polyvinyl Chloride
C.Pr.	Hundred Pair
CRC	Cold Rolled Channel
Creos.	Creosote
Crpt.	Carpet & Linoleum Layer
CRT	Cathode-ray Tube
CS	Carbon Steel, Constant Shear Bar Joist
Csc	Cosecant
C.S.F.	Hundred Square Feet
CSI	Construction Specifications Institute
CT	Current Transformer
CTS	Copper Tube Size
Cu	Copper, Cubic
Cu. Ft.	Cubic Foot
cw	Continuous Wave
C.W.	Cool White; Cold Water
Cwt.	100 Pounds
C.W.X.	Cool White Deluxe
C.Y.	Cubic Yard (27 cubic feet)
C.Y./Hr.	Cubic Yard per Hour
Cyl.	Cylinder
d	Penny (nail size)
D	Deep; Depth; Discharge
Dis., Disch.	Discharge
Db	Decibel
Dbl.	Double
DC	Direct Current
DDC	Direct Digital Control
Demob.	Demobilization
d.f.t.	Dry Film Thickness
d.f.u.	Drainage Fixture Units
D.H.	Double Hung
DHW	Domestic Hot Water
DI	Ductile Iron
Diag.	Diagonal
Diam., Dia	Diameter
Distrib.	Distribution
Div.	Division
Dk.	Deck
D.L.	Dead Load; Diesel
DLH	Deep Long Span Bar Joist
dlx	Deluxe
Do.	Ditto
DOP	Dioctyl Phthalate Penetration Test (Air Filters)
Dp., dp	Depth
D.P.S.T.	Double Pole, Single Throw
Dr.	Drive
DR	Dimension Ratio
Drink.	Drinking
D.S.	Double Strength
D.S.A.	Double Strength A Grade
D.S.B.	Double Strength B Grade
Dty.	Duty
DWV	Drain Waste Vent
DX	Deluxe White, Direct Expansion
dyn	Dyne
e	Eccentricity
E	Equipment Only; East; emissivity
Ea.	Each
EB	Encased Burial
Econ.	Economy
E.C.Y	Embankment Cubic Yards
EDP	Electronic Data Processing
EIFS	Exterior Insulation Finish System
E.D.R.	Equiv. Direct Radiation
Eq.	Equation
EL	elevation
Elec.	Electrician; Electrical
Elev.	Elevator; Elevating
EMT	Electrical Metallic Conduit; Thin Wall Conduit
Eng.	Engine, Engineered
EPDM	Ethylene Propylene Diene Monomer
EPS	Expanded Polystyrene
Eqhv.	Equip. Oper., Heavy
Eqlt.	Equip. Oper., Light
Eqmd.	Equip. Oper., Medium
Eqmm.	Equip. Oper., Master Mechanic
Eqol.	Equip. Oper., Oilers
Equip.	Equipment
ERW	Electric Resistance Welded

Abbreviations

E.S.	Energy Saver
Est.	Estimated
esu	Electrostatic Units
E.W.	Each Way
EWT	Entering Water Temperature
Excav.	Excavation
excl	Excluding
Exp., exp	Expansion, Exposure
Ext., ext	Exterior; Extension
Extru.	Extrusion
f.	Fiber stress
F	Fahrenheit; Female; Fill
Fab., fab	Fabricated; fabric
FBGS	Fiberglass
F.C.	Footcandles
f.c.c.	Face-centered Cubic
f'c.	Compressive Stress in Concrete; Extreme Compressive Stress
F.E.	Front End
FEP	Fluorinated Ethylene Propylene (Teflon)
F.G.	Flat Grain
F.H.A.	Federal Housing Administration
Fig.	Figure
Fin.	Finished
FIPS	Female Iron Pipe Size
Fixt.	Fixture
FJP	Finger jointed and primed
Fl. Oz.	Fluid Ounces
Flr.	Floor
FM	Frequency Modulation; Factory Mutual
Fmg.	Framing
FM/UL	Factory Mutual/Underwriters Labs
Fdn.	Foundation
FNPT	Female National Pipe Thread
Fori.	Foreman, Inside
Foro.	Foreman, Outside
Fount.	Fountain
fpm	Feet per Minute
FPT	Female Pipe Thread
Fr	Frame
F.R.	Fire Rating
FRK	Foil Reinforced Kraft
FSK	Foil/scrim/kraft
FRP	Fiberglass Reinforced Plastic
FS	Forged Steel
FSC	Cast Body; Cast Switch Box
Ft., ft	Foot; Feet
Ftng.	Fitting
Ftg.	Footing
Ft lb.	Foot Pound
Furn.	Furniture
FVNR	Full Voltage Non-Reversing
FVR	Full Voltage Reversing
FXM	Female by Male
Fy.	Minimum Yield Stress of Steel
g	Gram
G	Gauss
Ga.	Gauge
Gal., gal.	Gallon
gpm, GPM	Gallon per Minute
Galv., galv	Galvanized
GC/MS	Gas Chromatograph/Mass Spectrometer
Gen.	General
GFI	Ground Fault Interrupter
GFRC	Glass Fiber Reinforced Concrete
Glaz.	Glazier
GPD	Gallons per Day
gpf	Gallon per flush
GPH	Gallons per Hour
GPM	Gallons per Minute
GR	Grade
Gran.	Granular
Grnd.	Ground
GVW	Gross Vehicle Weight
GWB	Gypsum wall board
H	High Henry
HC	High Capacity
H.D., HD	Heavy Duty; High Density
H.D.O.	High Density Overlaid
HDPE	High density polyethelene plastic
Hdr.	Header
Hdwe.	Hardware
H.I.D., HID	High Intensity Discharge
Help.	Helper Average
HEPA	High Efficiency Particulate Air Filter
Hg	Mercury
HIC	High Interrupting Capacity
HM	Hollow Metal
HMWPE	high molecular weight polyethylene
HO	High Output
Horiz.	Horizontal
H.P., HP	Horsepower; High Pressure
H.P.F.	High Power Factor
Hr.	Hour
Hrs./Day	Hours per Day
HSC	High Short Circuit
Ht.	Height
Htg.	Heating
Htrs.	Heaters
HVAC	Heating, Ventilation & Air-Conditioning
Hvy.	Heavy
HW	Hot Water
Hyd.; Hydr.	Hydraulic
Hz	Hertz (cycles)
I.	Moment of Inertia
IBC	International Building Code
I.C.	Interrupting Capacity
ID	Inside Diameter
I.D.	Inside Dimension; Identification
I.F.	Inside Frosted
I.M.C.	Intermediate Metal Conduit
In.	Inch
Incan.	Incandescent
Incl.	Included; Including
Int.	Interior
Inst.	Installation
Insul., insul	Insulation/Insulated
I.P.	Iron Pipe
I.P.S., IPS	Iron Pipe Size
IPT	Iron Pipe Threaded
I.W.	Indirect Waste
J	Joule
J.I.C.	Joint Industrial Council
K	Thousand; Thousand Pounds; Heavy Wall Copper Tubing, Kelvin
K.A.H.	Thousand Amp. Hours
kcmil	Thousand Circular Mils
KD	Knock Down
K.D.A.T.	Kiln Dried After Treatment
kg	Kilogram
kG	Kilogauss
kgf	Kilogram Force
kHz	Kilohertz
Kip	1000 Pounds
KJ	Kiljoule
K.L.	Effective Length Factor
K.L.F.	Kips per Linear Foot
Km	Kilometer
KO	Knock Out
K.S.F.	Kips per Square Foot
K.S.I.	Kips per Square Inch
kV	Kilovolt
kVA	Kilovolt Ampere
kVAR	Kilovar (Reactance)
KW	Kilowatt
KWh	Kilowatt-hour
L	Labor Only; Length; Long; Medium Wall Copper Tubing
Lab.	Labor
lat	Latitude
Lath.	Lather
Lav.	Lavatory
lb.; #	Pound
L.B., LB	Load Bearing; L Conduit Body
L. & E.	Labor & Equipment
lb./hr.	Pounds per Hour
lb./L.F.	Pounds per Linear Foot
lbf/sq.in.	Pound-force per Square Inch
L.C.L.	Less than Carload Lot
L.C.Y.	Loose Cubic Yard
Ld.	Load
LE	Lead Equivalent
LED	Light Emitting Diode
L.F.	Linear Foot
L.F. Nose	Linear Foot of Stair Nosing
L.F. Rsr	Linear Foot of Stair Riser
Lg.	Long; Length; Large
L & H	Light and Heat
LH	Long Span Bar Joist
L.H.	Labor Hours
L.L., LL	Live Load
L.L.D.	Lamp Lumen Depreciation
lm	Lumen
lm/sf	Lumen per Square Foot
lm/W	Lumen per Watt
LOA	Length Over All
log	Logarithm
L-O-L	Lateralolet
long.	longitude
L.P., LP	Liquefied Petroleum; Low Pressure
L.P.F.	Low Power Factor
LR	Long Radius
L.S.	Lump Sum
Lt.	Light
Lt. Ga.	Light Gauge
L.T.L.	Less than Truckload Lot
Lt. Wt.	Lightweight
L.V.	Low Voltage
M	Thousand; Material; Male; Light Wall Copper Tubing
M^2CA	Meters Squared Contact Area
m/hr.; M.H.	Man-hour
mA	Milliampere
Mach.	Machine
Mag. Str.	Magnetic Starter
Maint.	Maintenance
Marb.	Marble Setter
Mat; Mat'l.	Material
Max.	Maximum
MBF	Thousand Board Feet
MBH	Thousand BTU's per hr.
MC	Metal Clad Cable
MCC	Motor Control Center
M.C.F.	Thousand Cubic Feet
MCFM	Thousand Cubic Feet per Minute
M.C.M.	Thousand Circular Mils
MCP	Motor Circuit Protector
MD	Medium Duty
MDF	Medium-density fibreboard
M.D.O.	Medium Density Overlaid
Med.	Medium
MF	Thousand Feet
M.F.B.M.	Thousand Feet Board Measure
Mfg.	Manufacturing
Mfrs.	Manufacturers
mg	Milligram
MGD	Million Gallons per Day
MGPH	Thousand Gallons per Hour
MH, M.H.	Manhole; Metal Halide; Man-Hour
MHz	Megahertz
Mi.	Mile
MI	Malleable Iron; Mineral Insulated
MIPS	Male Iron Pipe Size
mj	Mechanical Joint
m	Meter
mm	Millimeter
Mill.	Millwright
Min., min.	Minimum, minute

Abbreviations

Misc.	Miscellaneous
ml	Milliliter, Mainline
M.L.F.	Thousand Linear Feet
Mo.	Month
Mobil.	Mobilization
Mog.	Mogul Base
MPH	Miles per Hour
MPT	Male Pipe Thread
MRGWB	Moisture Resistant Gypsum Wallboard
MRT	Mile Round Trip
ms	Millisecond
M.S.F.	Thousand Square Feet
Mstz.	Mosaic & Terrazzo Worker
M.S.Y.	Thousand Square Yards
Mtd., mtd., mtd	Mounted
Mthe.	Mosaic & Terrazzo Helper
Mtng.	Mounting
Mult.	Multi; Multiply
M.V.A.	Million Volt Amperes
M.V.A.R.	Million Volt Amperes Reactance
MV	Megavolt
MW	Megawatt
MXM	Male by Male
MYD	Thousand Yards
N	Natural; North
nA	Nanoampere
NA	Not Available; Not Applicable
N.B.C.	National Building Code
NC	Normally Closed
NEMA	National Electrical Manufacturers Assoc.
NEHB	Bolted Circuit Breaker to 600V.
NFPA	National Fire Protection Association
NLB	Non-Load-Bearing
NM	Non-Metallic Cable
nm	Nanometer
No.	Number
NO	Normally Open
N.O.C.	Not Otherwise Classified
Nose.	Nosing
NPT	National Pipe Thread
NQOD	Combination Plug-on/Bolt on Circuit Breaker to 240V.
N.R.C., NRC	Noise Reduction Coefficient/ Nuclear Regulator Commission
N.R.S.	Non Rising Stem
ns	Nanosecond
nW	Nanowatt
OB	Opposing Blade
OC	On Center
OD	Outside Diameter
O.D.	Outside Dimension
ODS	Overhead Distribution System
O.G.	Ogee
O.H.	Overhead
O&P	Overhead and Profit
Oper.	Operator
Opng.	Opening
Orna.	Ornamental
OSB	Oriented Strand Board
OS&Y	Outside Screw and Yoke
OSHA	Occupational Safety and Health Act
Ovhd.	Overhead
OWG	Oil, Water or Gas
Oz.	Ounce
P.	Pole; Applied Load; Projection
p.	Page
Pape.	Paperhanger
P.A.P.R.	Powered Air Purifying Respirator
PAR	Parabolic Reflector
P.B., PB	Push Button
Pc., Pcs.	Piece, Pieces
P.C.	Portland Cement; Power Connector
P.C.F.	Pounds per Cubic Foot
PCM	Phase Contrast Microscopy
PDCA	Painting and Decorating Contractors of America
P.E., PE	Professional Engineer; Porcelain Enamel; Polyethylene; Plain End
Perf.	Perforated
PEX	Cross linked polyethylene
Ph.	Phase
P.I.	Pressure Injected
Pile.	Pile Driver
Pkg.	Package
Pl.	Plate
Plah.	Plasterer Helper
Plas.	Plasterer
plf	Pounds Per Linear Foot
Pluh.	Plumbers Helper
Plum.	Plumber
Ply.	Plywood
p.m.	Post Meridiem
Pntd.	Painted
Pord.	Painter, Ordinary
pp	Pages
PP, PPL	Polypropylene
P.P.M.	Parts per Million
Pr.	Pair
P.E.S.B.	Pre-engineered Steel Building
Prefab.	Prefabricated
Prefin.	Prefinished
Prop.	Propelled
PSF, psf	Pounds per Square Foot
PSI, psi	Pounds per Square Inch
PSIG	Pounds per Square Inch Gauge
PSP	Plastic Sewer Pipe
Pspr.	Painter, Spray
Psst.	Painter, Structural Steel
P.T.	Potential Transformer
P. & T.	Pressure & Temperature
Ptd.	Painted
Ptns.	Partitions
Pu	Ultimate Load
PVC	Polyvinyl Chloride
Pvmt.	Pavement
PRV	Pressure Relief Valve
Pwr.	Power
Q	Quantity Heat Flow
Qt.	Quart
Quan., Qty.	Quantity
Q.C.	Quick Coupling
r	Radius of Gyration
R	Resistance
R.C.P.	Reinforced Concrete Pipe
Rect.	Rectangle
recpt.	receptacle
Reg.	Regular
Reinf.	Reinforced
Req'd.	Required
Res.	Resistant
Resi.	Residential
RF	Radio Frequency
RFID	Radio-frequency identification
Rgh.	Rough
RGS	Rigid Galvanized Steel
RHW	Rubber, Heat & Water Resistant; Residential Hot Water
rms	Root Mean Square
Rnd.	Round
Rodm.	Rodman
Rofc.	Roofer, Composition
Rofp.	Roofer, Precast
Rohe.	Roofer Helpers (Composition)
Rots.	Roofer, Tile & Slate
R.O.W.	Right of Way
RPM	Revolutions per Minute
R.S.	Rapid Start
Rsr	Riser
RT	Round Trip
S.	Suction; Single Entrance; South
SC	Screw Cover
SCFM	Standard Cubic Feet per Minute
Scaf.	Scaffold
Sch., Sched.	Schedule
S.C.R.	Modular Brick
S.D.	Sound Deadening
SDR	Standard Dimension Ratio
S.E.	Surfaced Edge
Sel.	Select
SER, SEU	Service Entrance Cable
S.F.	Square Foot
S.F.C.A.	Square Foot Contact Area
S.F. Flr.	Square Foot of Floor
S.F.G.	Square Foot of Ground
S.F. Hor.	Square Foot Horizontal
SFR	Square Feet of Radiation
S.F. Shlf.	Square Foot of Shelf
S4S	Surface 4 Sides
Shee.	Sheet Metal Worker
Sin.	Sine
Skwk.	Skilled Worker
SL	Saran Lined
S.L.	Slimline
Sldr.	Solder
SLH	Super Long Span Bar Joist
S.N.	Solid Neutral
SO	Stranded with oil resistant inside insulation
S-O-L	Socketolet
sp	Standpipe
S.P.	Static Pressure; Single Pole; Self-Propelled
Spri.	Sprinkler Installer
spwg	Static Pressure Water Gauge
S.P.D.T.	Single Pole, Double Throw
SPF	Spruce Pine Fir
S.P.S.T.	Single Pole, Single Throw
SPT	Standard Pipe Thread
Sq.	Square; 100 Square Feet
Sq. Hd.	Square Head
Sq. In.	Square Inch
S.S.	Single Strength; Stainless Steel
S.S.B.	Single Strength B Grade
sst, ss	Stainless Steel
Sswk.	Structural Steel Worker
Sswl.	Structural Steel Welder
St.; Stl.	Steel
STC	Sound Transmission Coefficient
Std.	Standard
Stg.	Staging
STK	Select Tight Knot
STP	Standard Temperature & Pressure
Stpi.	Steamfitter, Pipefitter
Str.	Strength; Starter; Straight
Strd.	Stranded
Struct.	Structural
Sty.	Story
Subj.	Subject
Subs.	Subcontractors
Surf.	Surface
Sw.	Switch
Swbd.	Switchboard
S.Y.	Square Yard
Syn.	Synthetic
S.Y.P.	Southern Yellow Pine
Sys.	System
t.	Thickness
T	Temperature; Ton
Tan	Tangent
T.C.	Terra Cotta
T & C	Threaded and Coupled
T.D.	Temperature Difference
Tdd	Telecommunications Device for the Deaf
T.E.M.	Transmission Electron Microscopy
temp	Temperature, Tempered, Temporary
TFFN	Nylon Jacketed Wire

Abbreviations

TFE	Tetrafluoroethylene (Teflon)
T. & G.	Tongue & Groove; Tar & Gravel
Th., Thk.	Thick
Thn.	Thin
Thrded	Threaded
Tilf.	Tile Layer, Floor
Tilh.	Tile Layer, Helper
THHN	Nylon Jacketed Wire
THW.	Insulated Strand Wire
THWN	Nylon Jacketed Wire
T.L., TL	Truckload
T.M.	Track Mounted
Tot.	Total
T-O-L	Threadolet
tmpd	Tempered
T.S.	Trigger Start
Tr.	Trade
Transf.	Transformer
Trhv.	Truck Driver, Heavy
Trlr	Trailer
Trlt.	Truck Driver, Light
TTY	Teletypewriter
TV	Television
T.W.	Thermoplastic Water Resistant Wire
UCI	Uniform Construction Index
UF	Underground Feeder
UGND	Underground Feeder
UHF	Ultra High Frequency
U.I.	United Inch
U.L., UL	Underwriters Laboratory
Uld.	unloading
Unfin.	Unfinished
UPS	Uninterruptible Power Supply
URD	Underground Residential Distribution
US	United States
USGBC	U.S. Green Building Council
USP	United States Primed
UTMCD	Uniform Traffic Manual For Control Devices
UTP	Unshielded Twisted Pair
V	Volt
VA	Volt Amperes
V.C.T.	Vinyl Composition Tile
VAV	Variable Air Volume
VC	Veneer Core
VDC	Volts Direct Current
Vent.	Ventilation
Vert.	Vertical
V.F.	Vinyl Faced
V.G.	Vertical Grain
VHF	Very High Frequency
VHO	Very High Output
Vib.	Vibrating
VLF	Vertical Linear Foot
VOC	Volitile Organic Compound
Vol.	Volume
VRP	Vinyl Reinforced Polyester
W	Wire; Watt; Wide; West
w/	With
W.C., WC	Water Column; Water Closet
W.F.	Wide Flange
W.G.	Water Gauge
Wldg.	Welding
W. Mile	Wire Mile
W-O-L	Weldolet
W.R.	Water Resistant
Wrck.	Wrecker
W.S.P.	Water, Steam, Petroleum
WT., Wt.	Weight
WWF	Welded Wire Fabric
XFER	Transfer
XFMR	Transformer
XHD	Extra Heavy Duty
XHHW, XLPE	Cross-Linked Polyethylene Wire Insulation
XLP	Cross-linked Polyethylene
Xport	Transport
Y	Wye
yd	Yard
yr	Year
Δ	Delta
%	Percent
~	Approximately
Ø	Phase; diameter
@	At
#	Pound; Number
<	Less Than
>	Greater Than
Z	zone

Index

A

B

D

E

M

N

O

S

T

U

V

W

X

Z

Notes

Notes

Notes

Notes

Notes

Reed Construction Data/RSMeans— a tradition of excellence in construction cost information and services since 1942

Table of Contents

Annual Cost Guides
Seminars
New Titles
Reference Books
Order Form

For more information visit the RSMeans website at **www.rsmeans.com**

Unit prices according to the latest MasterFormat!

Book Selection Guide

The following table provides definitive information on the content of each cost data publication. The number of lines of data provided in each unit price or assemblies division, as well as the number of reference tables and crews, is listed for each book. The presence of other elements such as equipment rental costs, historical cost indexes, city cost indexes, square foot models, or cross-referenced indexes is also indicated. You can use the table to help select the RSMeans book that has the quantity and type of information you most need in your work.

Unit Cost Divisions	Building Construction	Mechanical	Electrical	Commercial Renovation	Square Foot	Site Work Landsc.	Green Building	Interior	Concrete Masonry	Open Shop	Heavy Construction	Light Commercial	Facilities Construction	Plumbing	Residential
1	573	395	410	499		517	124	319	463	572	507	233	1038	411	176
2	755	281	87	706		1005	180	366	223	751	744	447	1197	291	258
3	1678	346	223	1053		1474	1004	323	2037	1678	1680	469	1774	315	376
4	900	22	0	698		701	183	573	1135	876	591	490	1147	0	405
5	1844	159	156	1040		809	1777	1047	712	1835	1031	924	1845	204	720
6	2294	67	69	1881		411	534	1421	300	2273	383	1968	1908	22	2480
7	1467	215	125	1501		563	776	572	497	1464	26	1178	1529	224	942
8	2122	98	48	2351		299	1157	1768	657	2066	0	1790	2581	0	1410
9	1927	72	26	1720		313	449	2012	389	1868	15	1639	2171	54	1416
10	987	17	10	610		215	27	807	156	987	29	497	1073	230	221
11	1022	208	165	486		126	54	878	28	1007	0	231	1042	169	110
12	563	0	2	325		248	150	1698	12	552	19	371	1748	23	324
13	727	117	112	246		346	128	251	66	726	253	82	754	71	80
14	277	36	0	225		28	0	259	0	276	0	12	296	18	6
21	78	0	16	30		0	0	248	0	78	0	60	429	435	224
22	1140	7318	154	1105		1536	1100	605	20	1102	1659	830	7152	9029	659
23	1211	7086	599	846		158	940	695	38	1132	111	752	5176	1860	416
26	1344	499	10087	1019		816	644	1132	55	1329	562	1218	10,011	444	618
27	72	0	266	34		13	0	60	0	72	39	52	268	0	4
28	96	58	124	71		0	21	78	0	99	0	40	139	44	25
31	1482	735	610	803		3207	290	7	1208	1426	3284	600	1545	660	611
32	770	54	0	836		4341	341	362	274	741	1796	381	1647	166	428
33	509	1014	461	207		2027	41	0	231	508	1937	119	1557	1205	117
34	109	0	20	4		192	0	0	31	64	170	0	129	0	0
35	18	0	0	0		327	0	0	0	18	442	0	83	0	0
41	52	0	0	24		7	0	22	0	52	30	0	59	14	0
44	75	83	0	0		0	0	0	0	0	0	0	75	75	0
46	23	16	0	0		274	181	0		23	264	0	33	33	0
48	13	0	28	0		4	28	0		13	4	28	13	0	26
Totals	24128	18896	13798	18320		19957	10,129	15503	8532	23588	15576	14411	48419	15997	12052

Assem Div	Building Construction	Mechanical	Elec-trical	Commercial Renovation	Square Foot	Site Work Landscape	Assemblies	Green Building	Interior	Concrete Masonry	Heavy Construction	Light Commercial	Facilities Construction	Plumbing	Asm Div	Residential
A		15	0	188	150	577	598	0	0	536	571	154	24	0	1	368
B		0	0	848	2483	0	5643	56	329	1966	368	2074	174	0	2	211
C		0	0	647	863	0	1233	0	1558	146	0	763	245	0	3	588
D		1067	886	712	1864	72	2464	265	826	0	0	1348	1084	1083	4	839
E		0	0	86	260	0	298	0	5	0	0	257	5	0	5	392
F		0	0	0	115	0	115	0	0	0	0	115	3	0	6	357
G		527	447	318	202	3456	668	0	0	535	1431	200	293	707	7	296
															8	760
															9	80
															10	0
															11	0
															12	0
Totals		1609	1333	2799	5937	4105	11019	321	2718	3183	2370	4911	1828	1790		3891

Reference Section	Building Construction Costs	Mechanical	Electrical	Commercial Renovation	Square Foot	Site Work Landscape	Assem.	Green Building	Interior	Concrete Masonry	Open Shop	Heavy Construction	Light Commercial	Facilities Construction	Plumbing	Resi.
Reference Tables	yes	yes	yes	yes	no	yes	yes	yes	yes	yes	yes	yes	yes	yes	yes	yes
Models					109								48			28
Crews	558	558	558	536		558		558	558	558	534	558	534	536	558	534
Equipment Rental Costs	yes	yes	yes	yes		yes		yes	yes	yes	yes	yes	yes	yes	yes	yes
Historical Cost Indexes	yes	yes	yes	yes	yes	yes	yes	yes	yes	yes	yes	yes	yes	yes	yes	no
City Cost Indexes	yes	yes	yes	yes	yes	yes	yes	yes	yes	yes	yes	yes	yes	yes	yes	yes

Annual Cost Guides

Cost Guides also available in CD and at rsmeansonline.com

For more information visit the RSMeans website at **www.rsmeans.com**

Unit prices according to the latest MasterFormat!

RSMeans Building Construction Cost Data 2013

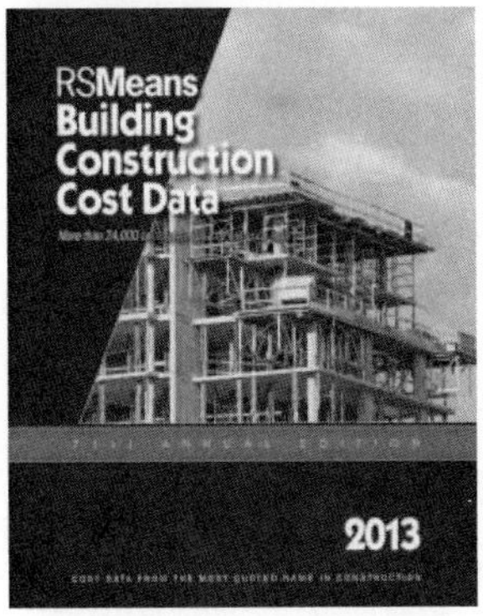

Offers you unchallenged unit price reliability in an easy-to-use format. Whether used for verifying complete, finished estimates or for periodic checks, it supplies more cost facts better and faster than any comparable source. Over 23,000 unit prices have been updated for 2013. The City Cost Indexes and Location Factors cover over 930 areas, for indexing to any project location in North America. Order and get *RSMeans Quarterly Update Service* FREE. You'll have year-long access to the RSMeans Estimating **HOTLINE** FREE with your subscription. Expert assistance when using RSMeans data is just a phone call away.

$186.95 per copy | Available Sept. 2012 | Catalog no. 60013

RSMeans Green Building Cost Data 2013

Estimate, plan, and budget the costs of green building for both new commercial construction and renovation work with this first edition of RSMeans Green Building Cost Data. More than 9,000 unit costs for a wide array of green building products plus assemblies costs. Easily identified cross references to LEED and Green Globes building rating systems criteria.

$155.95 per copy | Available Nov. 2012 | Catalog no. 60553

RSMeans Mechanical Cost Data 2013

Total unit and systems price guidance for mechanical construction. . . materials, parts, fittings, and complete labor cost information. Includes prices for piping, heating, air conditioning, ventilation, and all related construction.

Plus new 2013 unit costs for:

- Thousands of installed HVAC/ controls assemblies components
- "On-site" Location Factors for over 930 cities and towns in the U.S. and Canada
- Crews, labor, and equipment

$183.95 per copy | Available Oct. 2012 | Catalog no. 60023

RSMeans Facilities Construction Cost Data 2013

For the maintenance and construction of commercial, industrial, municipal, and institutional properties. Costs are shown for new and remodeling construction and are broken down into materials, labor, equipment, and overhead and profit. Special emphasis is given to sections on mechanical, electrical, furnishings, site work, building maintenance, finish work, and demolition.

More than 47,000 unit costs, plus assemblies costs and a comprehensive Reference Section are included.

$476.95 per copy | Available Nov. 2012 | Catalog no. 60203

RSMeans Square Foot Costs 2013

Accurate and Easy To Use

- **Updated price information** based on nationwide figures from suppliers, estimators, labor experts, and contractors
- "How-to-Use" sections, with **clear examples** of commercial, residential, industrial, and institutional structures
- Realistic graphics, offering true-to-life illustrations of building projects
- Extensive information on using square foot cost data, including sample estimates and alternate pricing methods

$197.95 per copy | Available Oct. 2012 | Catalog no. 60053

RSMeans Commercial Renovation Cost Data 2013

Commercial/Multifamily Residential

Use this valuable tool to estimate commercial and multifamily residential renovation and remodeling.

Includes: New costs for hundreds of unique methods, materials, and conditions that only come up in repair and remodeling, PLUS:

- Unit costs for more than 17,000 construction components
- Installed costs for more than 2,000 assemblies
- More than 930 "on-site" localization factors for the U.S. and Canada

$152.95 per copy | Available Oct. 2012 | Catalog no. 60043

Annual Cost Guides

Cost Guides also available in CD and at rsmeansonline.com

For more information visit the RSMeans website at **www.rsmeans.com**

Unit prices according to the latest MasterFormat!

RSMeans Electrical Cost Data 2013

Pricing information for every part of electrical cost planning. More than 13,000 unit and systems costs with design tables; clear specifications and drawings; engineering guides; illustrated estimating procedures; complete labor-hour and materials costs for better scheduling and procurement; and the latest electrical products and construction methods.

- A variety of special electrical systems including cathodic protection
- Costs for maintenance, demolition, HVAC/mechanical, specialties, equipment, and more

$188.95 per copy | Available Oct. 2012 | Catalog no. 60033

RSMeans Electrical Change Order Cost Data 2013

RSMeans Electrical Change Order Cost Data provides you with electrical unit prices exclusively for pricing change orders—based on the recent, direct experience of contractors and suppliers. Analyze and check your own change order estimates against the experience others have had doing the same work. It also covers productivity analysis and change order cost justifications. With useful information for calculating the effects of change orders and dealing with their administration.

$183.95 per copy | Available Dec. 2012 | Catalog no. 60233

RSMeans Assemblies Cost Data 2013

RSMeans Assemblies Cost Data takes the guesswork out of preliminary or conceptual estimates. Now you don't have to try to calculate the assembled cost by working up individual component costs. We've done all the work for you.

Presents detailed illustrations, descriptions, specifications, and costs for every conceivable building assembly—over 350 types in all—arranged in the easy-to-use UNIFORMAT II system. Each illustrated "assembled" cost includes a complete grouping of materials and associated installation costs, including the installing contractor's overhead and profit.

$305.95 per copy | Available Sept. 2012 | Catalog no. 60063

RSMeans Open Shop Building Construction Cost Data 2013

The latest costs for accurate budgeting and estimating of new commercial and residential construction. . . renovation work. . . change orders. . . cost engineering.

RSMeans Open Shop "BCCD" will assist you to:

- Develop benchmark prices for change orders
- Plug gaps in preliminary estimates and budgets
- Estimate complex projects
- Substantiate invoices on contracts
- Price ADA-related renovations

$160.95 per copy | Available Dec. 2012 | Catalog no. 60153

RSMeans Residential Cost Data 2013

Contains square foot costs for 28 basic home models with the look of today, plus hundreds of custom additions and modifications you can quote right off the page. Includes costs for more than 700 residential systems. Complete with blank estimating forms, sample estimates, and step-by-step instructions.

Now contains line items for cultured stone and brick, PVC trim lumber, and TPO roofing.

$134.95 per copy | Available Oct. 2012 | Catalog no. 60173

RSMeans Site Work & Landscape Cost Data 2013

Includes unit and assemblies costs for earthwork, sewerage, piped utilities, site improvements, drainage, paving, trees and shrubs, street openings/repairs, underground tanks, and more. Contains 78 types of assemblies costs for accurate conceptual estimates.

Includes:

- Estimating for infrastructure improvements
- Environmentally-oriented construction
- ADA-mandated handicapped access
- Hazardous waste line items

$181.95 per copy | Available Dec. 2012 | Catalog no. 60283

Annual Cost Guides

Cost Guides also available in CD and at rsmeansonline.com

For more information visit the RSMeans website at **www.rsmeans.com**

Unit prices according to the latest MasterFormat!

RSMeans Facilities Maintenance & Repair Cost Data 2013

RSMeans Facilities Maintenance & Repair Cost Data gives you a complete system to manage and plan your facility repair and maintenance costs and budget efficiently. Guidelines for auditing a facility and developing an annual maintenance plan. Budgeting is included, along with reference tables on cost and management, and information on frequency and productivity of maintenance operations.

The only nationally recognized source of maintenance and repair costs. Developed in cooperation with the Civil Engineering Research Laboratory (CERL) of the Army Corps of Engineers.

$417.95 per copy | Available Nov. 2012 | Catalog no. 60303

RSMeans Light Commercial Cost Data 2013

Specifically addresses the light commercial market, which is a specialized niche in the construction industry. Aids you, the owner/designer/contractor, in preparing all types of estimates—from budgets to detailed bids. Includes new advances in methods and materials.

Assemblies Section allows you to evaluate alternatives in the early stages of design/planning.

Over 13,000 unit costs ensure that you have the prices you need. . . when you need them.

$139.95 per copy | Available Nov. 2012 | Catalog no. 60183

RSMeans Concrete & Masonry Cost Data 2013

Provides you with cost facts for virtually all concrete/masonry estimating needs, from complicated formwork to various sizes and face finishes of brick and block—all in great detail. The comprehensive Unit Price Section contains more than 8,000 selected entries. Also contains an Assemblies [Cost] Section, and a detailed Reference Section that supplements the cost data.

$171.95 per copy | Available Dec. 2012 | Catalog no. 60113

RSMeans Labor Rates for the Construction Industry 2013

Complete information for estimating labor costs, making comparisons, and negotiating wage rates by trade for more than 300 U.S. and Canadian cities. With 46 construction trades listed by local union number in each city, and historical wage rates included for comparison. Each city chart lists the county and is alphabetically arranged with handy visual flip tabs for quick reference.

$415.95 per copy | Available Dec. 2012 | Catalog no. 60123

RSMeans Construction Cost Indexes 2013

What materials and labor costs will change unexpectedly this year? By how much?

- Breakdowns for 318 major cities
- National averages for 30 key cities
- Expanded five major city indexes
- Historical construction cost indexes

$362.00 per year (subscription) | Catalog no. 50143
$90.50 individual quarters | Catalog no. 60143 A,B,C,D

RSMeans Interior Cost Data 2013

Provides you with prices and guidance needed to make accurate interior work estimates. Contains costs on materials, equipment, hardware, custom installations, furnishings, and labor costs. . . for new and remodel commercial and industrial interior construction, including updated information on office furnishings, and reference information.

$192.95 per copy | Available Nov. 2012 | Catalog no. 60093

Annual Cost Guides

Cost Guides also available in CD and at rsmeansonline.com

For more information visit the RSMeans website at **www.rsmeans.com**

Unit prices according to the latest MasterFormat!

RSMeans Heavy Construction Cost Data 2013

A comprehensive guide to heavy construction costs. Includes costs for highly specialized projects such as tunnels, dams, highways, airports, and waterways. Information on labor rates, equipment, and materials costs is included. Features unit price costs, systems costs, and numerous reference tables for costs and design.

$186.95 per copy | Available Dec. 2012 | Catalog no. 60163

RSMeans Plumbing Cost Data 2013

Comprehensive unit prices and assemblies for plumbing, irrigation systems, commercial and residential fire protection, point-of-use water heaters, and the latest approved materials. This publication and its companion, *RSMeans Mechanical Cost Data*, provide full-range cost estimating coverage for all the mechanical trades.

Now contains updated costs for potable water and radiant heat systems and high efficiency fixtures.

$188.95 per copy | Available Oct. 2012 | Catalog no. 60213

2013 RSMeans Seminar Schedule

Note: call for exact dates and details.

Location	Dates
Las Vegas, NV	March
Washington, DC	April
San Francisco, CA	June
Washington, DC	September
Dallas, TX	September
Las Vegas, NV	October
Orlando, FL	November
San Diego, CA	December

1-800-334-3509, Press 1

2013 Regional Seminar Schedule

Note: call for exact dates and details.

Location	Dates
Seattle, WA	January and August
Dallas/Ft. Worth, TX	January
Austin, TX	February
Oklahoma City, OK	February
Anchorage, AK	March and September
Phoenix, AZ	March
San Bernardino, CA	April
Kansas City, MO	April
Toronto	May
Denver, CO	May
Bethesda, MD	June
Columbus, GA	June
El Segundo, CA	August
Jacksonville FL	September
St. Louis, MO	October
Houston, TX	October
Vancouver, BC	November
Baltimore, MD	November
San Antonio, TX	December
Raleigh, NC	December

1-800-334-3509, Press 1

For more information visit the RSMeans website at **www.rsmeans.com**

Registration Information

Register early... Save up to $100! Register 30 days before the start date of a seminar and save $100 off your total fee. *Note: This discount can be applied only once per order. It cannot be applied to team discount registrations or any other special offer.*

How to register Register by phone today! The RSMeans toll-free number for making reservations is **1-800-334-3509, Press 1.**

Individual seminar registration fee - $935. *Means CostWorks®* training registration fee - $375. To register by mail, complete the registration form and return, with your full fee, to: RSMeans Seminars, 700 Longwater Drive, Norwell, MA 02061.

Government pricing All federal government employees save off the regular seminar price. Other promotional discounts cannot be combined with the government discount.

Team discount program for two to four seminar registrations. Call for pricing: 1-800-334-3509, Press 1

Multiple course discounts When signing up for two or more courses, call for pricing.

Refund policy Cancellations will be accepted up to ten business days prior to the seminar start. There are no refunds for cancellations received later than ten working days prior to the first day of the seminar. A $150 processing fee will be applied for all cancellations. Written notice of cancellation is required. Substitutions can be made at any time before the session starts. **No-shows are subject to the full seminar fee.**

AACE approved courses Many seminars described and offered here have been approved for 14 hours (1.4 recertification credits) of credit by the AACE International Certification Board toward meeting the continuing education requirements for recertification as a Certified Cost Engineer/Certified Cost Consultant.

AIA Continuing Education We are registered with the AIA Continuing Education System (AIA/CES) and are committed to developing quality learning activities in accordance with the CES criteria. Many seminars meet the AIA/CES criteria for Quality Level 2. AIA members may receive (14) learning units (LUs) for each two-day RSMeans course.

Daily course schedule The first day of each seminar session begins at 8:30 a.m. and ends at 4:30 p.m. The second day begins at 8:00 a.m. and ends at 4:00 p.m. Participants are urged to bring a hand-held calculator, since many actual problems will be worked out in each session.

Continental breakfast Your registration includes the cost of a continental breakfast, and a morning and afternoon refreshment break. These informal segments allow you to discuss topics of mutual interest with other seminar attendees. (You are free to make your own lunch and dinner arrangements.)

Hotel/transportation arrangements RSMeans arranges to hold a block of rooms at most host hotels. To take advantage of special group rates when making your reservation, be sure to mention that you are attending the RSMeans seminar. You are, of course, free to stay at the lodging place of your choice. **(Hotel reservations and transportation arrangements should be made directly by seminar attendees.)**

Important Class sizes are limited, so please register as soon as possible.

Note: *Pricing subject to change.*

Registration Form

ADDS-1000

Call 1-800-334-3509, Press 1 to register or FAX this form 1-800-632-6732. Visit our website: www.rsmeans.com

Please register the following people for the RSMeans construction seminars as shown here. We understand that we must make our own hotel reservations if overnight stays are necessary.

☐ Full payment of $________________ enclosed.

☐ Bill me.

Please print name of registrant(s).

(To appear on certificate of completion)

____

____

____

____

P.O. #: ________________

GOVERNMENT AGENCIES MUST SUPPLY PURCHASE ORDER NUMBER OR TRAINING FORM.

Firm name________________

Address________________

City/State/Zip________________

Telephone no.__________ Fax no.__________

E-mail address________________

Charge registration(s) to: ☐ MasterCard ☐ VISA ☐ American Express

Account no.__________ Exp. date__________

Cardholder's signature________________

Seminar name________________

Seminar City ________________

Please mail check to: RSMeans Seminars, 700 Longwater Drive, Norwell, MA 02061 USA

Professional Development

For more information visit the RSMeans website at **www.rsmeans.com**

RSMeans Online™ Training

Construction estimating is vital to the decision-making process at each state of every project. Means CostWorks.com works the way you do. It's systematic, flexible and intuitive. In this one day class you will see how you can estimate any phase of any project faster and better.

Some of what you'll learn:
- Customizing Means CostWorks.com
- Making the most of RSMeans "Circle Reference" numbers
- How to integrate your cost data
- Generate reports, exporting estimates to MS Excel, sharing, collaborating and more

Also available: RSMeans Online™ training webinar

Maintenance & Repair Estimating for Facilities

This *two-day* course teaches attendees how to plan, budget, and estimate the cost of ongoing and preventive maintenance and repair for existing buildings and grounds.

Some of what you'll learn:
- The most financially favorable maintenance, repair, and replacement scheduling and estimating
- Auditing and value engineering facilities
- Preventive planning and facilities upgrading
- Determining both in-house and contract-out service costs
- Annual, asset-protecting M&R plan

Who should attend: facility managers, maintenance supervisors, buildings and grounds superintendents, plant managers, planners, estimators, and others involved in facilities planning and budgeting.

Facilities Construction Estimating

In this *two-day* course, professionals working in facilities management can get help with their daily challenges to establish budgets for all phases of a project.

Some of what you'll learn:
- Determining the full scope of a project
- Identifying the scope of risks and opportunities
- Creative solutions to estimating issues
- Organizing estimates for presentation and discussion
- Special techniques for repair/remodel and maintenance projects
- Negotiating project change orders

Who should attend: facility managers, engineers, contractors, facility tradespeople, planners, and project managers.

Practical Project Management for Construction Professionals

In this *two-day* course, acquire the essential knowledge and develop the skills to effectively and efficiently execute the day-to-day responsibilities of the construction project manager.

Covers:
- General conditions of the construction contract
- Contract modifications: change orders and construction change directives
- Negotiations with subcontractors and vendors
- Effective writing: notification and communications
- Dispute resolution: claims and liens

Who should attend: architects, engineers, owner's representatives, project managers.

Scheduling with Microsoft Project

Two days of hands-on training gives you a look at the basics to putting it all together with MS Project to create a complete project schedule. Learn to work better and faster, manage changes and updates, enhance tracking, generate actionable reports, and boost control of your time and budget.

Some of what you'll learn:
- Defining task/activity relationships: link tasks, establish predecessor/successor relationships and create logs
- Baseline scheduling: the essential what/when
- Using MS Project to determine the critical path
- Integrate RSMeans data with MS Project
- The dollar-loaded schedule: enhance the reliability of this "must-know" technique for scheduling public projects.

Who should attend: contractors' project-management teams, architects and engineers, project owners and their representatives, and those interested in improving their project planning and scheduling and management skills.

Mechanical & Electrical Estimating

This *two-day* course teaches attendees how to prepare more accurate and complete mechanical/electrical estimates, avoiding the pitfalls of omission and double-counting, while understanding the composition and rationale within the RSMeans mechanical/electrical database.

Some of what you'll learn:
- The unique way mechanical and electrical systems are interrelated
- M&E estimates—conceptual, planning, budgeting, and bidding stages
- Order of magnitude, square foot, assemblies, and unit price estimating
- Comparative cost analysis of equipment and design alternatives

Who should attend: architects, engineers, facilities managers, mechanical and electrical contractors, and others who need a highly reliable method for developing, understanding, and evaluating mechanical and electrical contracts.

Professional Development

For more information visit the RSMeans website at **www.rsmeans.com**

Unit Price Estimating

This interactive *two-day* seminar teaches attendees how to interpret project information and process it into final, detailed estimates with the greatest accuracy level.

The most important credential an estimator can take to the job is the ability to visualize construction and estimate accurately.

Some of what you'll learn:
- Interpreting the design in terms of cost
- The most detailed, time-tested methodology for accurate pricing
- Key cost drivers—material, labor, equipment, staging, and subcontracts
- Understanding direct and indirect costs for accurate job cost accounting and change order management

Who should attend: corporate and government estimators and purchasers, architects, engineers, and others who need to produce accurate project estimates.

Conceptual Estimating Using RSMeans CostWorks® CD

This *two day* class uses the leading industry data and a powerful software package to develop highly accurate conceptual estimates for your construction projects. All attendees must bring a laptop computer loaded with the current year *Square Foot Model Costs* and the *Assemblies Cost Data* CostWorks titles.

Some of what you'll learn:
- Introduction to conceptual estimating
- Types of conceptual estimates
- Helpful hints
- Order of magnitude estimating
- Square foot estimating
- Assemblies estimating

Who should attend: architects, engineers, contractors, construction estimators, owner's representatives, and anyone looking for an electronic method for performing square foot estimating.

RSMeans CostWorks® CD Training

This one-day course helps users become more familiar with the functionality of *RSMeans CostWork*s program. Each menu, icon, screen, and function found in the program is explained in depth. Time is devoted to hands-on estimating exercises.

Some of what you'll learn:
- Searching the database using all navigation methods
- Exporting RSMeans data to your preferred spreadsheet format
- Viewing crews, assembly components, and much more
- Automatically regionalizing the database

This training session requires you to bring a laptop computer to class.

When you register for this course you will receive an outline for your laptop requirements.

Also offering web training for CostWorks CD!

Assessing Scope of Work for Facility Construction Estimating

This *two-day* practical training program addresses the vital importance of understanding the SCOPE of projects in order to produce accurate cost estimates for facility repair and remodeling.

Some of what you'll learn:
- Discussions of site visits, plans/specs, record drawings of facilities, and site-specific lists
- Review of CSI divisions, including means, methods, materials, and the challenges of scoping each topic
- Exercises in SCOPE identification and SCOPE writing for accurate estimating of projects
- Hands-on exercises that require SCOPE, take-off, and pricing

Who should attend: corporate and government estimators, planners, facility managers, and others who need to produce accurate project estimates.

Facilities Est. Using RSMeans CostWorks® CD

Combines hands-on skill building with best estimating practices and real-life problems. Brings you up-to-date with key concepts, and provides tips, pointers, and guidelines to save time and avoid cost oversights and errors.

Some of what you'll learn:
- Estimating process concepts
- Customizing and adapting RSMeans cost data
- Establishing scope of work to account for all known variables
- Budget estimating: when, why, and how
- Site visits: what to look for—what you can't afford to overlook
- How to estimate repair and remodeling variables

This training session requires you to bring a laptop computer to class.

Who should attend: facility managers, architects, engineers, contractors, facility tradespeople, planners, project managers and anyone involved with JOC, SABRE, or IDIQ.

Unit Price Est. Using RSMeans CostWorks® CD

Step-by-step instruction and practice problems to identify and track key cost drivers—material, labor, equipment, staging, and subcontractors—for each specific task. Learn the most detailed, time-tested methodology for accurately "pricing" these variables, their impact on each other and on total cost.

Some of what you'll learn:
- Unit price cost estimating
- Order of magnitude, square foot, and assemblies estimating
- Quantity takeoff
- Direct and indirect construction costs
- Development of contractor's bill rates
- How to use *RSMeans Building Construction Cost Data*

This training session requires you to bring a laptop computer to class.

Who should attend: architects, engineers, corporate and government estimators, facility managers, and government procurement staff.

New Titles - Reference Books

For more information visit the RSMeans website at **www.rsmeans.com**

Complete Book of Framing, 2nd Edition

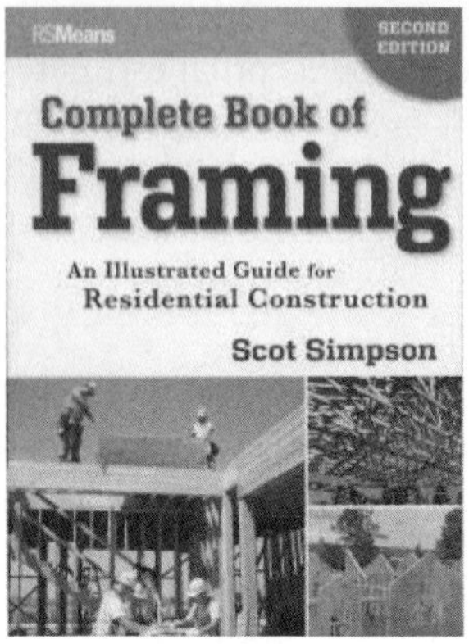

by Scot Simpson

This updated, easy-to-learn guide to rough carpentry and framing is written by an expert with more than thirty years of framing experience. Starting with the basics, this book begins with types of lumber, nails, and what tools are needed, followed by detailed, fully illustrated steps for framing each building element. Framer-Friendly Tips throughout the book show how to get a task done right—and more easily.

$29.95 per copy | 352 pages, softcover | Catalog no. 67353A

How to Estimate with Means Data & CostWorks, 4th Edition

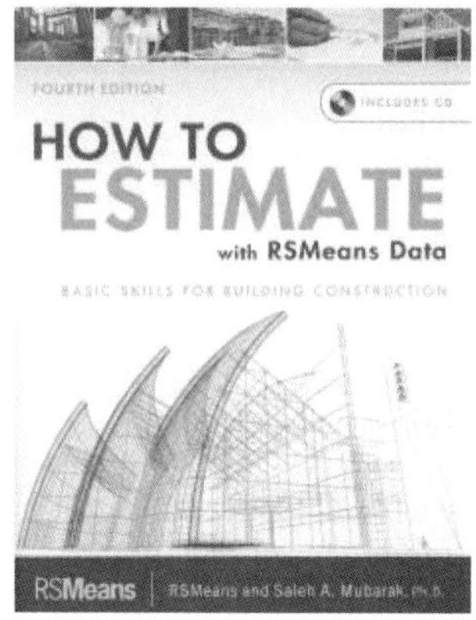

by RSMeans and Saleh A. Mubarak, Ph.D.

This step-by-step guide takes you through all the major construction items with extensive coverage of site work, concrete and masonry, wood and metal framing, doors and windows, and other divisions. The only construction cost estimating handbook that uses the most popular source of construction data, RSMeans, this indispensible guide features access to the instructional version of CostWorks in electronic form, enabling you to practice techniques to solve real-world estimating problems.

$70.00 per copy | 292 pages, softcover | Includes CostWorks CD
Catalog no. 67324C

Estimating Building Costs, 2nd Edition

For the Residential & Light Commercial Construction Professional

by Wayne J. DelPico

This book guides readers through the entire estimating process, explaining in detail how to put together a reliable estimate that can be used not only for budgeting, but also for developing a schedule, managing a project, dealing with contingencies, and ultimately making a profit.

Completely revised and updated to reflect the new CSI MasterFormat 2010 system, this practical guide describes estimating techniques for each building system and how to apply them according to the latest industry standards.

$65.00 per copy | 398 pages, softcover | Catalog no. 67343A

How Your House Works, 2nd Edition

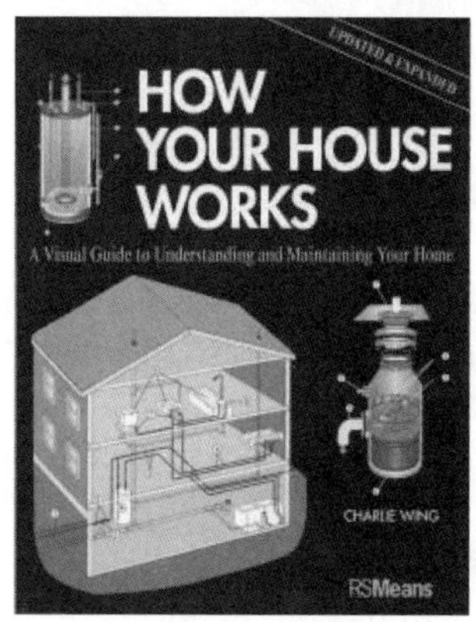

by Charlie Wing

Knowledge of your home's systems helps you control repair and construction costs and makes sure the correct elements are being installed or replaced. This book uncovers the mysteries behind just about every major appliance and building element in your house. See-through, cross-section drawings in full color show you exactly how these things should be put together and how they function, including what to check if they don't work. It just might save you having to call in a professional.

$22.95 per copy | 192 pages, softcover | Catalog no. 67351A

Risk Management for Design and Construction

This book introduces risk as a central pillar of project management and shows how a project manager can be prepared for dealing with uncertainty. Written by experts in the field, *Risk Management for Design and Construction* uses clear, straightforward terminology to demystify the concepts of project uncertainty and risk.

Highlights include:

- Integrated cost and schedule risk analysis
- An introduction to a ready-to-use system of analyzing a project's risks and tools to proactively manage risks
- A methodology that was developed and used by the Washington State Department of Transportation
- Case studies and examples on the proper application of principles
- Information about combining value analysis with risk analysis

$125.00 per copy | Over 250 pages, softcover | Catalog no. 67359

RSMeans Cost Data, Student Edition

This book provides a thorough introduction to cost estimating in a self-contained print and online package. With clear explanations and a hands-on, example-driven approach, it is the ideal reference for students and new professionals.

Features include:

- Commercial and residential construction cost data in print and online formats
- Complete how-to guidance on the essentials of cost estimating
- A supplemental website with plans, problem sets, and a full sample estimate

$99.00 per copy | 512 pages, softcover | Catalog no. 67363

Value Engineering: Practical Applications

For Design, Construction, Maintenance & Operations
by Alphonse Dell'Isola, PE

A tool for immediate application—for engineers, architects, facility managers, owners, and contractors. Includes making the case for VE—the management briefing; integrating VE into planning, budgeting, and design; conducting life cycle costing; using VE methodology in design review and consultant selection; case studies; VE workbook; and a life cycle costing program on disk.

$79.95 per copy | Over 450 pages, illustrated, softcover | Catalog no. 67319A

Facilities Operations & Engineering Reference

by the Association for Facilities Engineering and RSMeans

An all-in-one technical reference for planning and managing facility projects and solving day-to-day operations problems. Selected as the official certified plant engineer reference, this handbook covers financial analysis, maintenance, HVAC and energy efficiency, and more.

$109.95 per copy | Over 700 pages, illustrated, hardcover | Catalog no. 67318

The Building Professional's Guide to Contract Documents

3rd Edition
by Waller S. Poage, AIA, CSI, CVS

A comprehensive reference for owners, design professionals, contractors, and students

- Structure your documents for maximum efficiency.
- Effectively communicate construction requirements.
- Understand the roles and responsibilities of construction professionals.
- Improve methods of project delivery.

$70.00 per copy | 400 pages Diagrams and construction forms, hardcover Catalog no. 67261A

Green Home Improvement

by Daniel D. Chiras, PhD

With energy costs rising and environmental awareness increasing, people are looking to make their homes greener. This book, with 65 projects and actual costs and projected savings help homeowners prioritize their green improvements.

Projects range from simple water savers that cost only a few dollars, to bigger-ticket items such as HVAC systems. With color photos and cost estimates, each project compares options and describes the work involved, the benefits, and the savings.

$34.95 | 320 pages, illustrated, softcover | Catalog no. 67355

Cost Planning & Estimating for Facilities Maintenance

In this unique book, a team of facilities management authorities shares their expertise on:

- Evaluating and budgeting maintenance operations
- Maintaining and repairing key building components
- Applying *RSMeans Facilities Maintenance & Repair Cost Data* to your estimating

Covers special maintenance requirements of the ten major building types

$89.95 per copy | Over 475 pages, hardcover | Catalog no. 67314

Life Cycle Costing for Facilities

by Alphonse Dell'Isola and Dr. Steven Kirk

Guidance for achieving higher quality design and construction projects at lower costs! Cost-cutting efforts often sacrifice quality to yield the cheapest product. Life cycle costing enables building designers and owners to achieve both. The authors of this book show how LCC can work for a variety of projects — from roads to HVAC upgrades to different types of buildings.

$99.95 per copy | 396 pages, hardcover | Catalog no. 67341

Interior Home Improvement Costs

9th Edition

Updated estimates for the most popular remodeling and repair projects—from small, do-it-yourself jobs to major renovations and new construction. Includes: kitchens & baths; new living space from your attic, basement, or garage; new floors, paint, and wallpaper; tearing out or building new walls; closets, stairs, and fireplaces; new energy-saving improvements, home theaters, and more!

$24.95 per copy | 250 pages, illustrated, softcover | Catalog no. 67308E

Exterior Home Improvement Costs

9th Edition

Updated estimates for the most popular remodeling and repair projects—from small, do-it-yourself jobs, to major renovations and new construction. Includes: curb appeal projects—landscaping, patios, porches, driveways, and walkways; new windows and doors; decks, greenhouses, and sunrooms; room additions and garages; roofing, siding, and painting; "green" improvements to save energy & water.

$24.95 per copy | Over 275 pages, illustrated, softcover | Catalog no. 67309E

Builder's Essentials: Plan Reading & Material Takeoff

For Residential and Light Commercial Construction

by Wayne J. DelPico

A valuable tool for understanding plans and specs, and accurately calculating material quantities. Step-by-step instructions and takeoff procedures based on a full set of working drawings.

$35.95 per copy | Over 420 pages, softcover | Catalog no. 67307

Means Unit Price Estimating Methods

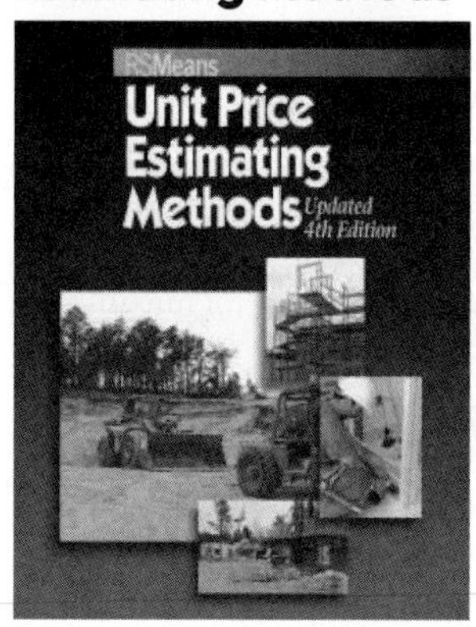

New 4th Edition

This new edition includes up-to-date cost data and estimating examples, updated to reflect changes to the CSI numbering system and new features of RSMeans cost data. It describes the most productive, universally accepted ways to estimate, and uses checklists and forms to illustrate shortcuts and timesavers. A model estimate demonstrates procedures. A new chapter explores computer estimating alternatives.

$65.00 per copy | Over 350 pages, illustrated, softcover | Catalog no. 67303B

Concrete Repair and Maintenance Illustrated

by Peter Emmons

Hundreds of illustrations show users how to analyze, repair, clean, and maintain concrete structures for optimal performance and cost effectiveness. From parking garages to roads and bridges to structural concrete, this comprehensive book describes the causes, effects, and remedies for concrete wear and failure. Invaluable for planning jobs, selecting materials, and training employees, this book is a must-have for concrete specialists, general contractors, facility managers, civil and structural engineers, and architects.

$69.95 per copy | 300 pages, illustrated, softcover | Catalog no. 67146

Total Productive Facilities Management

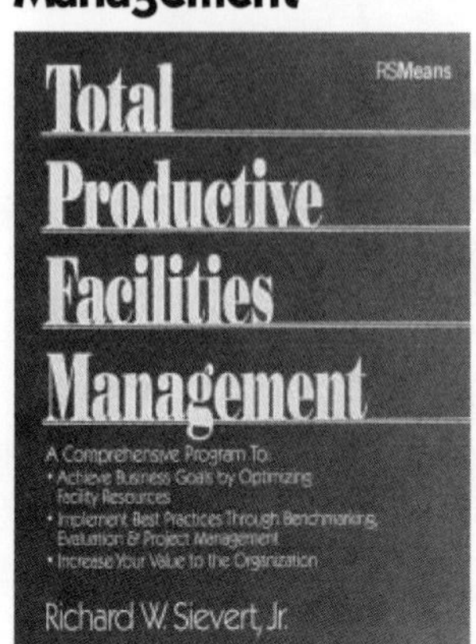

by Richard W. Sievert, Jr.

Today, facilities are viewed as strategic resources. . . elevating the facility manager to the role of asset manager supporting the organization's overall business goals. Now, Richard Sievert Jr., in this well-articulated guidebook, sets forth a new operational standard for the facility manager's emerging role. . . a comprehensive program for managing facilities as a true profit center.

$79.95 per copy | 275 pages, softcover | Catalog no. 67321

Reference Books

For more information visit the RSMeans website at **www.rsmeans.com**

Unit prices according to the latest MasterFormat!

Means Illustrated Construction Dictionary

Condensed, 2nd Edition

The best portable dictionary for office or field use—an essential tool for contractors, architects, insurance and real estate personnel, facility managers, homeowners, and anyone who needs quick, clear definitions for construction terms. The second edition has been further enhanced with updates and hundreds of new terms and illustrations . . . in keeping with the most recent developments in the construction industry.

Now with a quick-reference Spanish section. Includes tools and equipment, materials, tasks, and more!

$59.95 per copy | Over 500 pages, softcover | Catalog no. 67282A

Square Foot & UNIFORMAT Assemblies Estimating Methods

3rd Edition

Develop realistic square foot and assemblies costs for budgeting and construction funding. The new edition features updated guidance on square foot and assemblies estimating using UNIFORMAT II. An essential reference for anyone who performs conceptual estimates.

$69.95 per copy | Over 300 pages, illustrated, softcover | Catalog no. 67145B

Facilities Planning & Relocation

by David D. Owen

A complete system for planning space needs and managing relocations. Includes step-by-step manual, over 50 forms, and extensive reference section on materials and furnishings.

New lower price and user-friendly format.

$89.95 per copy | Over 450 pages, softcover | Catalog no. 67301

Mechanical Estimating Methods

4th Edition

Completely updated, this guide assists you in making a review of plans, specs, and bid packages, with suggestions for takeoff procedures, listings, substitutions, and pre-bid scheduling for all components of HVAC. Includes suggestions for budgeting labor and equipment usage. Compares materials and construction methods to allow you to select the best option.

$64.95 per copy | Over 350 pages, illustrated, softcover | Catalog no. 67294B

Electrical Estimating Methods

3rd Edition

Expanded edition includes sample estimates and cost information in keeping with the latest version of the CSI MasterFormat and UNIFORMAT II. Complete coverage of fiber optic and uninterruptible power supply electrical systems, broken down by components, and explained in detail. Includes a new chapter on computerized estimating methods. A practical companion to *RSMeans Electrical Cost Data.*

$64.95 per copy | Over 325 pages, hardcover | Catalog no. 67230B

Residential & Light Commercial Construction Standards

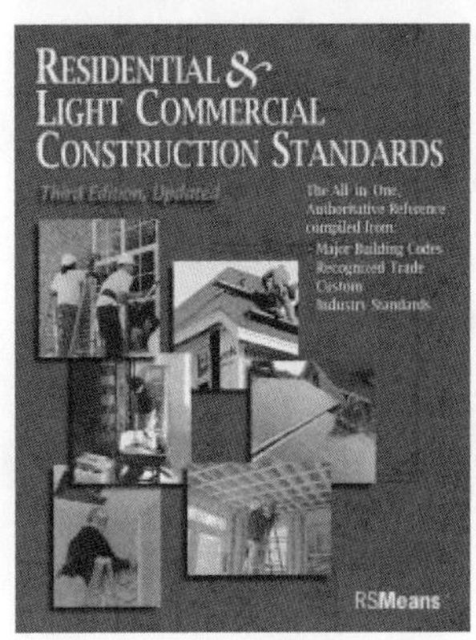

3rd Edition, Updated by RSMeans and contributing authors

This book provides authoritative requirements and recommendations compiled from leading professional associations, industry publications, and building code organizations. This all-in-one reference helps establish a standard for workmanship, quickly resolve disputes, and avoid defect claims. Updated third edition includes new coverage of green building, seismic, hurricane, and mold-resistant construction.

$59.95 | Over 550 pages, illustrated, softcover | Catalog no. 67322B

For more information visit the RSMeans website at **www.rsmeans.com**

Unit prices according to the latest MasterFormat!

Construction Business Management

by Nick Ganaway

Only 43% of construction firms stay in business after four years. Make sure your company thrives with valuable guidance from a pro with 25 years of success as a commercial contractor. Find out what it takes to build all aspects of a business that is profitable, enjoyable, and enduring. With a bonus chapter on retail construction.

$49.95 per copy | 200 pages, softcover | Catalog no. 67352

Project Scheduling & Management for Construction

3rd Edition

by David R. Pierce, Jr.

A comprehensive yet easy-to-follow guide to construction project scheduling and control—from vital project management principles through the latest scheduling, tracking, and controlling techniques. The author is a leading authority on scheduling, with years of field and teaching experience at leading academic institutions. Spend a few hours with this book and come away with a solid understanding of this essential management topic.

$64.95 per copy | Over 300 pages, illustrated, hardcover | Catalog no. 67247B

The Practice of Cost Segregation Analysis

by Bruce A. Desrosiers and Wayne J. DelPico

This expert guide walks you through the practice of cost segregation analysis, which enables property owners to defer taxes and benefit from "accelerated cost recovery" through depreciation deductions on assets that are properly identified and classified.

With a glossary of terms, sample cost segregation estimates for various building types, key information resources, and updates via a dedicated website, this book is a critical resource for anyone involved in cost segregation analysis.

$99.95 per copy | Over 225 pages | Catalog no. 67345

Preventive Maintenance for Multi-Family Housing

by John C. Maciha

Prepared by one of the nation's leading experts on multi-family housing.

This complete PM system for apartment and condominium communities features expert guidance, checklists for buildings and grounds maintenance tasks and their frequencies, a reusable wall chart to track maintenance, and a dedicated website featuring customizable electronic forms. A must-have for anyone involved with multi-family housing maintenance and upkeep.

$89.95 per copy | 225 pages | Catalog no. 67346

Green Building: Project Planning & Cost Estimating

3rd Edition

Since the widely read first edition of this book, green building has gone from a growing trend to a major force in design and construction.

This new edition has been updated with the latest in green building technologies, design concepts, standards, and costs. Full-color with all new case studies—plus a new chapter on commercial real estate.

$99.95 per copy | Over 450 pages, softcover | Catalog no. 67338B

Job Order Contracting Expediting Construction Project Delivery

by Allen Henderson

Expert guidance to help you implement JOC—fast becoming the preferred project delivery method for repair and renovation, minor new construction, and maintenance projects in the public sector and in many states and municipalities. The author, a leading JOC expert and practitioner, shows how to:

- Establish a JOC program
- Evaluate proposals and award contracts
- Handle general requirements and estimating
- Partner for maximum benefits

$89.95 per copy | 192 pages, illustrated, hardcover | Catalog no. 67348

Reference Books

For more information visit the RSMeans website at **www.rsmeans.com**

Unit prices according to the latest MasterFormat!

Construction Supervision

This brand new title inspires supervisory excellence with proven tactics and techniques applied by thousands of construction supervisors over the past decade. Recognizing the unique and critical role the supervisor plays in project success, the book's leadership guidelines carve out a practical blueprint for motivating work performance and increasing productivity through effective communication.

Features:

- A unique focus on field supervision and crew management
- Coverage of supervision from the foreman to the superintendent level
- An overview of technical skills whose mastery will build confidence and success for the supervisor
- A detailed view of "soft" management and communication skills

$90.00 per copy | Over 450 pages, softcover | Catalog no. 67358

Building & Renovating Schools

This all-inclusive guide covers every step of the school construction process—from initial planning, needs assessment, and design, right through moving into the new facility. A must-have resource for anyone concerned with new school construction or renovation. With square foot cost models for elementary, middle, and high school facilities, and real-life case studies of recently completed school projects.

The contributors to this book—architects, construction project managers, contractors, and estimators who specialize in school construction—provide start-to-finish, expert guidance on the process.

$99.95 per copy | Over 425 pages, hardcover | Catalog no. 67342

The Homeowner's Guide to Mold

By Michael Pugliese

Expert guidance to protect your health and your home.

Mold, whether caused by leaks, humidity or flooding, is a real health and financial issue—for homeowners and contractors.

This full-color book explains:

- Construction and maintenance practices to prevent mold
- How to inspect for and remove mold
- Mold remediation procedures and costs
- What to do after a flood
- How to deal with insurance companies

$21.95 per copy | 144 pages, softcover | Catalog no. 67344

Universal Design Ideas for Style, Comfort & Safety

by RSMeans and Lexicon Consulting, Inc.

Incorporating universal design when building or remodeling helps people of any age and physical ability more fully and safely enjoy their living spaces. This book shows how universal design can be artfully blended into the most attractive homes. It discusses specialized products like adjustable countertops and chair lifts, as well as simple ways to enhance a home's safety and comfort. With color photos and expert guidance, every area of the home is covered. Includes budget estimates that give an idea how much projects will cost.

$21.95 | 160 pages, illustrated, softcover | Catalog no. 67354

Landscape Estimating Methods

5th Edition

Answers questions about preparing competitive landscape construction estimates, with up-to-date cost estimates and the new MasterFormat classification system. Expanded and revised to address the latest materials and methods, including new coverage on approaches to green building. Includes:

- Step-by-step explanation of the estimating process
- Sample forms and worksheets that save time and prevent errors

$69.95 per copy | Over 350 pages, softcover | Catalog no. 67295C

Plumbing Estimating Methods

3rd Edition
by Joseph Galeno and Sheldon Greene

Updated and revised! This practical guide walks you through a plumbing estimate, from basic materials and installation methods through change order analysis. *Plumbing Estimating Methods* covers residential, commercial, industrial, and medical systems, and features sample takeoff and estimate forms and detailed illustrations of systems and components.

$59.95 per copy | 330+ pages, softcover | Catalog no. 67283B

Understanding & Negotiating Construction Contracts

by Kit Werremeyer

Take advantage of the author's 30 years' experience in small-to-large (including international) construction projects. Learn how to identify, understand, and evaluate high risk terms and conditions typically found in all construction contracts—then negotiate to lower or eliminate the risk, improve terms of payment, and reduce exposure to claims and disputes. The author avoids "legalese" and gives real-life examples from actual projects.

$74.95 per copy | 300 pages, softcover | Catalog no. 67350

Spanish/English Construction Dictionary

2nd Edition
by RSMeans and the International Code Council

This expanded edition features thousands of the most common words and useful phrases in the construction industry with easy-to-follow pronunciations in both Spanish and English. Over 800 new terms, phrases, and illustrations have been added. It also features a new stand-alone "Safety & Emergencies" section, with colored pages for quick access.

Unique to this dictionary are the systems illustrations showing the relationship of components in the most common building systems for all major trades.

$23.95 per copy | Over 400 pages | Catalog no. 67327A

Means Illustrated Construction Dictionary

Unabridged 4th Edition, with CD-ROM

Long regarded as the industry's finest, *Means Illustrated Construction Dictionary* is now even better. With nearly 20,000 terms and more than 1,400 illustrations and photos, it is the clear choice for the most comprehensive and current information. The companion CD-ROM that comes with this new edition adds extra features such as: larger graphics and expanded definitions.

$99.95 per copy | Over 790 pages, illust., hardcover | Catalog no. 67292B

The Gypsum Construction Handbook

by USG

An invaluable reference of construction procedures for gypsum drywall, cement board, veneer plaster, and conventional plaster. This new edition includes the newest product developments, installation methods, fire- and sound-rated construction information, illustrated framing-to-finish application instructions, estimating and planning information, and more. Great for architects and engineers; contractors, builders, and dealers; apprentices and training programs; building inspectors and code officials; and anyone interested in gypsum construction. Features information on tools and safety practices, a glossary of construction terms, and a list of important agencies and associations. Also available in Spanish.

$29.95 | Over 575 pages, illustrated, softcover | Catalog no. 67357A

RSMeans Estimating Handbook

3rd Edition

Widely used in the industry for tasks ranging from routine estimates to special cost analysis projects, this handbook has been completely updated and reorganized with new and expanded technical information.

RSMeans Estimating Handbook will help construction professionals:

- Evaluate architectural plans and specifications
- Prepare accurate quantity takeoffs
- Compare design alternatives and costs
- Perform value engineering
- Double-check estimates and quotes
- Estimate change orders

$99.95 per copy | Over 900 pages, hardcover | Catalog No. 67276B

2013 Order Form

ORDER TOLL FREE 1-800-334-3509
OR FAX 1-800-632-6732

For more information visit the RSMeans website at **www.rsmeans.com**

Qty.	Book no.	COST ESTIMATING BOOKS	Unit Price	Total
	60063	Assemblies Cost Data 2013	$305.95	
	60013	Building Construction Cost Data 2013	186.95	
	60113	Concrete & Masonry Cost Data 2013	171.95	
	50143	Construction Cost Indexes 2013 (subscription)	362.00	
	60143A	Construction Cost Index–January 2013	90.50	
	60143B	Construction Cost Index–April 2013	90.50	
	60143C	Construction Cost Index–July 2013	90.50	
	60143D	Construction Cost Index–October 2013	90.50	
	60343	Contr. Pricing Guide: Resid. R & R Costs 2013	39.95	
	60233	Electrical Change Order Cost Data 2013	183.95	
	60033	Electrical Cost Data 2013	188.95	
	60203	Facilities Construction Cost Data 2013	476.95	
	60303	Facilities Maintenance & Repair Cost Data 2013	417.95	
	60553	Green Building Cost Data 2013	155.95	
	60163	Heavy Construction Cost Data 2013	186.95	
	60093	Interior Cost Data 2013	192.95	
	60123	Labor Rates for the Const. Industry 2013	415.95	
	60183	Light Commercial Cost Data 2013	139.95	
	60023	Mechanical Cost Data 2013	183.95	
	60153	Open Shop Building Const. Cost Data 2013	160.95	
	60213	Plumbing Cost Data 2013	188.95	
	60043	Commercial Renovation Cost Data 2013	152.95	
	60173	Residential Cost Data 2013	134.95	
	60283	Site Work & Landscape Cost Data 2013	181.95	
	60053	Square Foot Costs 2013	197.95	
	62012	Yardsticks for Costing (2012)	176.95	
	62013	Yardsticks for Costing (2013)	185.95	
		REFERENCE BOOKS		
	67329	Bldrs Essentials: Best Bus. Practices for Bldrs	29.95	
	67298A	Bldrs Essentials: Framing/Carpentry 2nd Ed.	24.95	
	67298AS	Bldrs Essentials: Framing/Carpentry Spanish	24.95	
	67307	Bldrs Essentials: Plan Reading & Takeoff	35.95	
	67342	Building & Renovating Schools	99.95	
	67261A	The Building Prof. Guide to Contract Documents	70.00	
	67339	Building Security: Strategies & Costs	89.95	
	67353A	Complete Book of Framing, 2nd Ed.	29.95	
	67146	Concrete Repair & Maintenance Illustrated	69.95	
	67352	Construction Business Management	49.95	
	67358	Construction Supervision	90.00	
	67363	Cost Data Student Edition	99.00	
	67314	Cost Planning & Est. for Facil. Maint.	89.95	
	67328A	Designing & Building with the IBC, 2nd Ed.	99.95	
	67230B	Electrical Estimating Methods, 3rd Ed.	64.95	
	67343A	Est. Bldg. Costs for Resi. & Lt. Comm., 2nd Ed.	65.00	
	67276B	Estimating Handbook, 3rd Ed.	99.95	
	67318	Facilities Operations & Engineering Reference	109.95	
	67301	Facilities Planning & Relocation	89.95	
	67338B	Green Building: Proj. Planning & Cost Est., 3rd Ed.	99.95	
	67355	Green Home Improvement	34.95	
	67357A	The Gypsum Construction Handbook	29.95	
	67357S	The Gypsum Construction Handbook (Spanish)	29.95	

Qty.	Book no.	REFERENCE BOOKS (Cont.)	Unit Price	Total
	67148	Heavy Construction Handbook	99.95	
	67308E	Home Improvement Costs–Int. Projects, 9th Ed.	24.95	
	67309E	Home Improvement Costs–Ext. Projects, 9th Ed.	24.95	
	67344	Homeowner's Guide to Mold	21.95	
	67324C	How to Est.w/Means Data & CostWorks, 4th Ed.	70.00	
	67351A	How Your House Works, 2nd Ed.	22.95	
	67282A	Illustrated Const. Dictionary, Condensed, 2nd Ed.	59.95	
	67292B	Illustrated Const. Dictionary, w/CD-ROM, 4th Ed.	99.95	
	67348	Job Order Contracting	89.95	
	67295C	Landscape Estimating Methods, 5th Ed.	69.95	
	67341	Life Cycle Costing for Facilities	99.95	
	67294B	Mechanical Estimating Methods, 4th Ed.	64.95	
	67283B	Plumbing Estimating Methods, 3rd Ed.	59.95	
	67345	Practice of Cost Segregation Analysis	99.95	
	67346	Preventive Maint. for Multi-Family Housing	89.95	
	67326	Preventive Maint. Guidelines for School Facil.	149.95	
	67247B	Project Scheduling & Manag. for Constr., 3rd Ed.	64.95	
	67322B	Resi. & Light Commercial Const. Stds., 3rd Ed.	59.95	
	67359	Risk Management for Design & Construction	125.00	
	67327A	Spanish/English Construction Dictionary, 2nd Ed.	23.95	
	67145B	Sq. Ft. & Assem. Estimating Methods, 3rd Ed.	69.95	
	67321	Total Productive Facilities Management	79.95	
	67350	Understanding and Negotiating Const. Contracts	74.95	
	67303B	Unit Price Estimating Methods, 4th Ed.	65.00	
	67354	Universal Design	21.95	
	67319A	Value Engineering: Practical Applications	79.95	

MA residents add 6.25% state sales tax	
Shipping & Handling**	
Total (U.S. Funds)*	

Prices are subject to change and are for U.S. delivery only. *Canadian customers may call for current prices. **Shipping & handling charges: Add 7% of total order for check and credit card payments. Add 9% of total order for invoiced orders.

Send order to: ADDV-1000

Name (please print) ______________________

Company ______________________

☐ Company

☐ Home Address ______________________

____

City/State/Zip ______________________

Phone # ______________ P.O. # ______________

(Must accompany all orders being billed)

ail to: RSMeans, 700 Longwater Drive, Norwell, MA 02061**

RSMeans Project Cost Report

By filling out this report, your project data will contribute to the database that supports the RSMeans® Project Cost Square Foot Data. When you fill out this form, RSMeans will provide a $30 discount off one of the RSMeans products advertised in the preceding pages. Please complete the form including all items where you have cost data, and all the items marked (☑).

$30.00 Discount per product for each report you submit

Project Description (NEW construction only, please)

☑ Building Use (Office School...) ______________________

☑ Address (City, State) ______________________

____

☑ Total Building Area (SF) ______________________

☑ Ground Floor (SF) ______________________

☑ Frame (Wood, Steel...) ______________________

☑ Exterior Wall (Brick, Tilt-up...) ______________________

☑ Basement: (checkone) ☐ Full ☐ Partial ☐ None

☑ Number of Stories ______________________

☑ Floor-to-Floor Height ______________________

☑ Volume (C.F.) ______________________

% Air Conditioned ____________ Tons ____________

Total Project Cost $ ____________

Owner ______________________

Architect ______________________

General Contractor ______________________

☑ Bid Date ______________________

☑ Typical Bay Size ______________________

☑ Occupant Capacity ______________________

☑ Labor Force: _____ % Union _____ % Non-Union

☑ Project Description (Circle one number in each line.)

1. Economy 2. Average 3. Custom 4. Luxury

1. Square 2. Rectangular 3. Irregular 4. Very Irregular

Comments ______________________

____

A	☑	General Conditions	$
B	☑	Site Work	$
C	☑	Concrete	$
D	☑	Masonry	$
E	☑	Metals	$
F	☑	Wood & Plastics	$
G	☑	Thermal &Moisture Protection	$
GR		Roofing & Flashing	$
H	☑	Doors and Windows	$
J	☑	Finishes	$
JP		Painting & Wall Covering	$

K	☑	Specialties	$
L	☑	Equipment	$
M	☑	Furnishings	$
N	☑	SpecialConstruction	$
P	☑	ConveyingSystems	$
Q	☑	Mechanical	$
QP		Plumbing	$
QB		HVAC	$
R	☑	Electrical	$
S	☑	Mech./Elec. Combined	$

Please specify the RSMeans product you wish to receive.
Complete the address information.

Product Name ______________________

Product Number ______________________

Your Name ______________________

Title ______________________

Company ______________________

☐ Company

☐ Home Street Address ______________________

City, State, Zip ______________________

Email Address ______________________

Return by mail or fax 888-492-6770.

Method of Payment:

Credit Card # ______________________

Expiration Date ______________________

Check ______________________

Purchase Order ______________________

Phone Number ______________________
(if you would prefer a sales call)

RSMeans
a division of Reed Construction Data
Square Foot Costs Department
700 Longwater Drive
Norwell, MA 02061